W9-DIR-635

FLUID MECHANICS, HEAT TRANSFER, AND MASS TRANSFER

FLUID MECHANICS, HEAT TRANSFER, AND MASS TRANSFER

Chemical Engineering Practice

K. S. N. RAJU

A JOHN WILEY & SONS, INC., PUBLICATION

Published by John Wiley & Sons, Inc., Hoboken, New Jersey
Published simultaneously in Canada

Library of Congress Cataloging-in-Publication Data:

Raju, K. S. N.
Fluid mechanics, heat transfer, and mass transfer: chemical engineering practice/K. S. N. Raju.
p. cm.
Includes index.
ISBN 978-0-470-63774-6 (cloth)
1. Chemical engineering—Miscellanea. 2. Fluid mechanics—Miscellanea. 3. Thermodynamics—Miscellanea. 4. Mass transfer—Miscellanea. I. Title.
TP155.R28 2010
660'.29–dc22

2010019512

Printed in Singapore

oBook ISBN: 978-0-470-90997-3
ePDF ISBN: 978-0-470-90996-6
ePub ISBN: 978-0-470-92292-7

10 9 8 7 6 5 4 3 2 1

Practice of Science is Engineering

CONTENTS

FOREWORD ix

PREFACE xi

ACKNOWLEDGMENTS xiii

ABOUT THE AUTHOR xv

LIST OF FIGURES xvii

LIST OF TABLES xxv

SECTION I FLUID MECHANICS

1 Fluid Mechanics Basics 3

2 Fluid Flow 21

3 Piping, Seals, and Valves 35

4 Flow Measurement 59

5 Pumps, Ejectors, Blowers, and Compressors 101

6 Mixing 163

7 Two-Phase Flow Systems 195

SECTION II HEAT TRANSFER

8 Dimensionless Numbers, Temperature Measurement, and Conduction Heat Transfer 225

9 **Convective Heat Transfer Basics** 245

10 **Shell and Tube Heat Exchangers** 271

11 **Heat Transfer Equipment Involving Phase Transfer** 331

12 **Refrigeration, Heat Pumps, Heat Tracing, Coiled and Jacketed Vessels, Steam Traps, and Immersion Heaters** 371

13 **Compact Heat Exchangers, Regenerators, and Recuperators** 395

14 **Radiant Heat Transfer and Fired Heaters** 425

SECTION III MASS TRANSFER

15 **Mass Transfer Basics** 455

16 **Mass Transfer Equipment** 475

17 **Absorption, Distillation, and Extraction** 527

18 **Crystallization, Air–Water Operations, Drying, Adsorption, Membrane Separations, and Other Mass Transfer Processes** 613

INDEX 717

FOREWORD

Professor K. S. N. Raju has presented the technical community with an interesting, valuable, and unique book on the practice of chemical engineering in the broad areas of fluid mechanics, heat transfer, and mass transfer. Based upon his five decades of experience as an educator, researcher, and consultant, Professor Raju has chosen to adopt the question–answer format.

Consider, for example, an engineer faced with the analysis and design of a fired heater. This book on chemical engineering practice immediately answers design questions such as how the tubes are arranged in the furnace and how many rows are usually provided and also the rationale for the optimum design choice. In addition, the book introduces the theoretical background of radiant heat transfer by explaining concepts such as emissivity and absorptivity and key design relationships like the Stefan–Boltzmann equation and Kirchoff's law. Finally, this thorough book presents and resolves operational issues, for example, hot spots, high-temperature creep, corrosion, and tube life. Professor Raju's book equips the practicing engineer with the tools to design a fired heater as well as to diagnose and resolve operational problems.

Radiant Heat Transfer is one of the eight chapters in the section on Heat Transfer, which cover the theory and application of heat transfer in the process industries. In addition to Heat Transfer, the book has two other sections, Fluid Mechanics and Mass Transfer. Each section introduces the theoretical background, describes the applications and equipment, and anticipates and resolves operational issues. The Mass Transfer section introduces underlying concepts (phase equilibria, mass transfer coefficients, correlations involving dimensionless numbers, polymorphic structures), describes applications (absorption, distillation, crystallization, adsorption), and equipment (tray and packed columns, crystallizers, dryers, and membrane modules), and anticipates and resolves operational issues (column flooding, liquid inclusion in crystals). It will be difficult to find an area in the chemical process industries not covered in this comprehensive book!

While this book is wide in scope, it is also quite detailed. As an example, an engineer who drills down into the chapter on crystallization will learn about the factors that restrict the productivity and purity of crystals (agglomeration, liquid impurities inside and outside the crystals, cavitation).

Professor Raju has structured his book in the question–answer format, which he feels stimulates interest in the subject matter and focuses attention on specific topics. I completely agree. As I read the book, I found that it precisely explained concepts and applications in areas where I have some expertise, and also sparked my interest and gave me new understanding in subjects outside my specialization.

The style, structure, preciseness, and clarity of Professor Raju's book are a reflection of his five decades of experience as an educator, researcher, and consultant. As an educator, he has taught graduate and undergraduate students, created and delivered on-site courses for industry, and developed and nurtured new chemical engineering departments. He has published over 90 papers in international journals. His consultancy has covered the chemical, petroleum, petrochemical, and fertilizer industries and government organizations. Professor Raju's students report that his teaching style was always practical, focusing on solving real-world problems rather than just teaching a concept; he invariably used examples from his extensive experience to help students understand chemical engineering problems. That practical teaching style is clearly evident in this book.

Fluid Mechanics, Heat Transfer, and Mass Transfer: Chemical Engineering Practice is intended as a text book to undergraduate and graduate students and a reference book for practicing engineers in the chemical process industries. I plan to keep a copy handy as a reference to understand and resolve new technical issues I am confronted with.

Paul. M. Mathias

Fluor Corporation
Irvine, California

PREFACE

The book is intended for use by students at undergraduate and graduate levels, faculty in Chemical Engineering Departments across the world, working and consulting engineers in areas such as petroleum refineries, petrochemical, gas processing and fertilizer plants, design organizations, food and pharmaceutical processing, environmental engineering, and the like. The book is also useful to mechanical engineering students and faculty.

The book is written with emphasis on *practice* with brief theoretical concepts in the form of Questions and Answers, bridging the two areas of theory and practice with respect to the core areas of chemical engineering.

The author considers that the question–answer approach adopted stimulates interest in the subject matter and focuses attention to specific topics in a better and concise manner than running matter given in normal text and reference books.

The approach was used by the author in the classroom for several years, spanning a period of over four decades. Feedback from faculty, students, alumni, and practicing engineers in several institutions/organizations appreciated this approach when the author used this approach during continuing education and training courses, besides classroom instruction. This prompted the author to embark upon writing this book.

The book is an attempt to bridge the gap between theory and practice in a *balanced manner*, so that it will be easy for students and academics to get a grasp of practice and industry personnel to understand theoretical concepts necessary to appreciate the genesis involved in practice.

At the teaching level, the book is suitable for different courses involving fluid mechanics, heat transfer, mass transfer, and membrane technology as well as design courses at both undergraduate and graduate levels. The author considers it to be useful in design project work by students and others. It can be used as a textbook and/or a reference book.

In the market, there is no such book in these areas linking theory with broad practical aspects and to this effect it is original in nature with almost no competitors on the subject. It avoids referring to several text and reference books, to get information on specific topics/content.

The vast literature available in the form of articles in chemical engineering magazines, monographs, and manufacturers literature has been used in its preparation along with the active interaction of the author with practicing world over a period spanning nearly five decades.

To summarize, features include emphasis on practical aspects in the learning process for students and faculty and ease and convenience in the use of the theoretical and practical aspects of the subject in the practice of professional engineers.

Question–answer approach focuses on individual topics in a more effective way than the normally used running matter in textbooks. Adds to learning process, creating better interest among students, faculty, and others for whom time factor is important.

Involves the core areas of chemical engineering and fulfills the need for a single source, avoiding the requirement of using several books, articles, and other sources for the topics covered in the book.

The core areas of fluid mechanics, heat transfer, mass transfer, and membrane processes are covered in the book in a balanced way, avoiding making the book bulky and unwieldy in its use.

Most parts of the book are *easily understandable* by those who are *not experts* in the field. For example, it covers types on pumps, valves, process equipment, membranes, and areas of their use, their merits, and drawbacks and selection in a simple way.

No author of any book, other than those written as purely research publications or biographies, can claim the contents of their books as *original* contributions by them. To this extent, the author admits that the book is a collection and compilation of the material available in literature with the interpretations and comments by the author and acknowledges the individuals and groups of pioneers who contributed to the evolution of the subject matter over centuries. The author wishes to dedicate the book to all such pioneers who contributed to the evolution of chemical engineering as a profession.

K. S. N. Raju

ACKNOWLEDGMENTS

The author wishes to acknowledge the help, assistance, and encouragement received from many organizations, which include, specifically, American Institute of Chemical Engineers and John Wiley for accepting the book for publication. Equipment manufacturers, who include Swenson Technology, Spirax Sarco, Alfa Laval, Sulzer, Koch-Knight, Koch-Glitsch, Pace Engineering, and other industrial organizations. Acknowledgments are due to professional bodies such as Gas Processors Suppliers Association, Hydraulic Institute, Cheresources, Society of Automotive Engineers, American Conference of Governmental Industrial Hygienists, Tubular Exchanger Manufacturers Association, and American Membrane Technology Association whose literature is of help in making the book toward realizing the goal of making it practice-oriented.

Encouragement and support received from Sri K. V. Vishnu Raju, Chairman, Sri Vishnu Educational Society, Faculty, Associates, Alumni, and Students from Panjab University, B. V. R. Institute of Technology, B. S. University of Technology, Libya, and several other institutions provided the necessary motivation to write the book, which the author wishes to acknowledge.

Comments from several individuals, who include Dr. J. M. Alford, Chairman of AIChE New Books Committee, Peer Reviewers, Academic and Industry Professionals, and scores of others, have been valuable and the author duly acknowledges their inputs.

Involvement and review of the content of the book by Dr. Paul M. Mathias of Fluor Corporation, who happened to be one of the authors to *Perry's Chemical Engineers' Handbook*, has been particularly valuable. He has been very considerate in offering valuable suggestions and comments during the process of preparation of the book. He readily agreed to write Foreword for the book unhesitatingly. The author is very much indebted to him.

Ms. Haeja Han of AIChE, Bob Esposito, Michael Leventhal, and Rosalyn Farkas of John Wiley, Sanchari Sil and Joseph Varghese of Thomson Digital and others at AIChE and John Wiley have been very helpful in processing the contents of the book and making it more presentable. Special appreciation is due to all of them. The author duly acknowledges the help received from D. V. S. S. Prasad of Shanna Technologies (India) Private Limited, Srinivas Vadla and their Associates in Graphics Work.

Patience and tolerance by my wife Bangaramma, computer and other inputs by my son Prasad, daughter Anuradha, granddaughter Suhasa, and daughter-in-law Usha have made me to accomplish my work and they deserve my appreciation.

K. S. N. Raju

ABOUT THE AUTHOR

The author is a Retired Professor of Chemical Engineering with involvement in Chemical Engineering Education and Research covering graduate and undergraduate students for *50 years*. Published over 90 papers and articles in International magazines and journals. Supervised Graduate Research at Doctoral and Post Doctoral levels. Acted as Reviewer for publications in International Journals and Magazines in Engineering, including the prestigious assignment by Applied Mechanics Reviews of American Institute of Mechanical Engineers for 3 years.

Delivered invited lectures on Plate Heat Exchangers at NATO Advanced Institute along with highly distinguished speakers at International level. This material appeared in the book form by Hemisphere Publishing Corporation, Washington D.C.

Gave *Onsite Courses* in industry, covering refineries, gas processing, petrochemical, and fertilizer plants with practice-oriented approach. Delivered lectures on several areas of chemical engineering to trainees and practicing engineers in petroleum and petrochemical industry and at International forums.

Involved in active interaction with industry taking up projects and executing them with the induction of graduate students on-site in selected industries.

Has been responsible for developing Chemical Engineering Departments at Panjab University, Chandigarh, India, B. S. University of Technology, Libya, and B. V. R. Institute of Technology, Andhra Pradesh, India.

Received B.Sc. (Honors) Degree in Chemical Engineering from Andhra University in 1958, M.Tech from I.I.T., Kharagpur in 1959, and Ph.D. from Panjab University in 1971.

Contact Details:
Professor K. S. N. Raju
Jubilee Hills, Hyderabad 500 033, India
Email: k_snraju@hotmail.com

LIST OF FIGURES

Fluid Mechanics

1.1 Shear rate versus shear stress diagrams for Newtonian and non-Newtonian fluids. 6
1.2 Classification of viscometers. 8
1.3 Rotating disk and parallel plate viscometers. 8
1.4 Cup and bob coaxial viscometers. 9
1.5 Cambridge moving piston viscometer. 9
1.6 Cone and plate viscometers. 9
1.7 Tube viscometers. 9
1.8 Mercury barometer. 11
1.9 Aneroid barograph. 12
1.10 Different types of manometers. 12
1.11 Two-liquid manometer. 13
1.12 Bourdon tube pressure gauge. 13
1.13 Helical Bourdon tube. 14
1.14 Basic metallic bellows. 14
1.15 Capsule device for measurement of differential pressure. 14
1.16 Float-type level measurement. 16
1.17 Liquid level measurement in an open tank. 16
1.18 Ultrasonic level measurement. 17
1.19 Liquid level measurement. 19
1.20 Inaccuracies in level measurements for foaming liquids. 19

2.1 Changeover of laminar flow into turbulent eddies. 22
2.2 Velocity profiles for laminar and turbulent flows. 23
2.3 Reynolds number demonstration experiment. 24
2.4 Equivalent diameter for annulus. 25
2.5 Development of boundary layer. 25
2.6 Drag coefficients for different shaped objects. *Note*: All objects have the same projected (frontal) area. 27
2.7 Drag coefficients for spheres, disks, and cylinders. 28
2.8 Vortex street phenomenon. 29
2.9 Oil damper to cushion valve closure. 31
2.10 Air gap cushions shock waves when valve is suddenly closed. 32

3.1 D'Arcy friction factors as function of N_{Re}. 37
3.2 Tee entry arrangements. 38
3.3 Losses on fluid entry into a pipe for different entry configurations. 38
3.4 Strainer. 39
3.5 Important types of valves. 47
3.6 Gate valve. 47
3.7 Globe valve. 48
3.8 Ball valve. (*Courtesy*: Vaishnavi Engineering.) 49
3.9 Diaphragm valve. 50
3.10 Butterfly valve. 50
3.11 Swing check valve. 51
3.12 Ball check valves. 52
3.13 Lift check valve. 52
3.14 Symbols for some common types of valves. 54
3.15 Relief valve. 55
3.16 Different types of valves for solids flow. 56
3.17 Collapse of a bubble. (*Source:* Samson AG, Frankfurt.) 56

4.1 Turndown ratios. 60
4.2 Meter installations for gas and liquid flows. 60
4.3 Vena contracta. 61
4.4 Types of orifice plates. (a) Sharp-edged, (b) thick plate, and (c) thick plate with curved radius. 62
4.5 Sharp-edged orifice meter showing flow pattern with flange taps. 62

4.6 Pressure tap location alternatives. 63
4.7 Flow straighteners. 63
4.8 Different types of orifice plates. 64
4.9 Orifices for viscous flows. 65
4.10 Venturi meter. 67
4.11 Effect of Reynolds numbers on different flow meters. 67
4.12 Flow nozzle. 68
4.13 Pressure losses for different head flow meters as function of β-ratio. 69
4.14 Elbow flow meter. 70
4.15 Segmental-wedge V-element flow meter. 71
4.16 V-cone flow meter. 71
4.17 Pitot tube arrangements for flow measurement. 72
4.18 Annubar. 73
4.19 Rotameter showing different types of float designs. 74
4.20 Purge flow meter design. 76
4.21 Rotameter installed in a bypass line around an orifice plate in the main line. 77
4.22 Paddle wheel flow meter. 80
4.23 Propeller flow meter. 81
4.24 Transit time flow meter. 81
4.25 Doppler flow meter. 82
4.26 Elements of an electromagnetic flow meter. 84
4.27 Magnetic flow meter and its components. 84
4.28 Single straight tube Coriolis mass flow meter. 86
4.29 Single U-tube Coriolis mass flow meter. 86
4.30 Double U-tube Coriolis mass flow meter designs. 87
4.31 Nutating disk positive displacement flow meter. 89
4.32 Rotating vane positive displacement flow meter. 90
4.33 Oscillating piston positive displacement flow meter. 91
4.34 Single piston reciprocating positive displacement flow meter. 91
4.35 Piston and diaphragm positive displacement metering pumps. 92
4.36 Oval gear lobe flow meter. 93
4.37 Target meter. 94
4.38 Major components in a vortex shedding flow meter. 94
4.39 Vortex shedding flow meters. *Note:* Different shapes of shedders are used in the designs for getting the desired flow rate measurements. 95
4.40 Rectangular and triangular types of weirs. 99

5.1 Pump classification. 102
5.2 Classification of kinetic pumps. (*Source*: Hydraulic Institute.) 102
5.3 Classification of positive displacement pumps. (*Source*: Hydraulic Institute.) 103
5.4 Centrifugal pump showing important parameters. (*Source*: www.cheresources.com.) 104
5.5 Open, semiopen, and enclosed impellers. 105
5.6 Double suction to a centrifugal pump. 105
5.7 Suction head and suction lift for a centrifugal pump. 106
5.8 Bubble formation, collapse, and metal damage. (*Source*: www.cheresources.com.) 113
5.9 Potential cavitation damage areas in the eye of pump impeller. 113
5.10 Typical blade damage due to cavitation of a mixed flow pump. 115
5.11 Heat exchanger tube damage at the entrance area into the tube due to cavitation. 116
5.12 Illustrative diagram for cavitation conditions. 117
5.13 Pump suction side problems. 118
5.14 Pump suction through a sump. 119
5.15 Pump performance curve and system curve illustrated. 121
5.16 Pump performance curves. 121
5.17 Effect of impeller size on capacity versus head developed for a centrifugal pump. 122
5.18 Two pumps draining liquid from a tank. 123
5.19 Performance curves for centrifugal pumps operating in parallel. (*Courtesy*: GPSA Engineering Data Book, 12th ed.) 124
5.20 Performance curves for centrifugal pumps operating in series. (*Courtesy*: GPSA Engineering Data Book, 12th ed.) 124
5.21 Minimum flow bypass. 124
5.22 Valve action for a double-acting reciprocating piston pump. 128
5.23 Discharge curves for different reciprocating flow configurations. 128
5.24 Sliding vane rotary pump. 130
5.25 Gear pumps. 131
5.26 Rotary screw pump. 133
5.27 Rotary lobe pumps. 135
5.28 Air-operated double diaphragm pump. 136
5.29 Airlift pump. 139
5.30 Peristaltic pump. 140
5.31 Liquid ring vacuum pump. 142
5.32 Rotary claw pump. 144
5.33 Steam jet ejector. 148
5.34 Two-stage ejector system. 148
5.35 Condensate drain leg layouts. 149
5.36 Types of fans. 154
5.37 Types of compressors. (*Courtesy*: GPSA Engineering Data Book, 12th ed.) 155
5.38 Centrifugal compressor operating curve. 159
5.39 Illustration of operable range for the compressor with surge-free conditions. 160

6.1 Power number as a function of Reynolds number for different turbine impellers. 166

6.2 A basic stirred tank design showing a lower radial impeller and an upper axial impeller housed in a draft tube (not to scale). Four equally spaced baffles are standard. 171
6.3 Multistage agitator with baffles and sparger for gas–liquid reactors. 172
6.4 Turbine and propeller mixers. 173
6.5 Marine, saw-toothed, and perforated propellers. 174
6.6 Flat plate impeller with sawtooth edges. 175
6.7 Different types of turbine impellers. 175
6.8 Gate and leaf impellers. 176
6.9 Draft tube agitator. 177
6.10 Incorporating floating solids into liquids. 179
6.11 Side-entry propeller mixers used for blending liquids. 180
6.12 Double arm kneader. 182
6.13 Helical ribbon impeller. 182
6.14 Helical coil and anchor mixers. 183
6.15 Impeller selection. 183
6.16 Tee and injection mixers. 184
6.17 An in-line static mixer. 184
6.18 Ribbon blender and a double cone mixer. 187
6.19 Solids mixer. 188
6.20 Static mixer. 192
6.21 Viscosity ranges for different types of mixers. 193

7.1 Flow patterns in gas–liquid flow in vertical pipes. 196
7.2 Annular droplet or mist flows. 196
7.3 Two-phase flow patterns in a vertical evaporator tube. 197
7.4 Flow pattern as a function of fraction of air in air–water flow in a vertical pipe. 197
7.5 Flow patterns in gas/vapor–liquid flows in horizontal pipes. 197
7.6 Stratified and wavy flows. 198
7.7 Flow patterns for two-phase flow in horizontal pipes. 199
7.8 Six regimes of fluidization identified with increasing gas superficial velocities: (a) fixed bed; (b) particulate fluidization; (c) bubbling fluidization; (d) slugging fluidization; (e) turbulent regime. 201
7.9 Particle diameter versus gas velocity, showing minimum fluidization velocity for good fluidization and total carryover bands. 202
7.10 Recirculating fluidized bed concept. The draft tube operates as a dilute phase pneumatic transport tube. 204
7.11 Pressure drop in a fluid–solid bed as a function of fluid superficial velocity. 205
7.12 Typical fluid bed catalytic cracking unit. 206
7.13 Fluid catalytic cracking unit with a two-stage regenerator (UOP). 207
7.14 Pneumatic conveyance: negative system. 213
7.15 Pneumatic conveyor: positive system. 213
7.16 Blind tee and blind bend with arrows showing impact points. 216
7.17 The regimes of flow for settling slurries in horizontal pipelines. 219
7.18 Schematic representation of the boundaries between the flow regimes for settling slurries in horizontal pipelines. 219

Heat Transfer

8.1 Illustration for triple point. 226
8.2 Internal construction of a typical thermocouple. 227
8.3 Simple thermocouple circuit. 227
8.4 Resistance—temperature curve of a thermistor. 229
8.5 (a) Effect of temperature change on a bimetallic strip. (b) Bimetallic strip thermometer. 230
8.6 Vapor pressure thermometer. 231
8.7 Vapor pressure curve for methyl chloride. 231
8.8 Approximate ranges of thermal conductivities of materials at normal temperatures and pressures. 234
8.9 Multilayer slab. 235
8.10 Multilayer hollow cylinder. 236
8.11 Multilayer hollow sphere. 236
8.12 Heat transfer shape factors. 236

9.1 Prandtl numbers of light hydrocarbon gases at 1 bar. 246
9.2 Hydrodynamic and thermal boundary layers. 247
9.3 J_H factors as function of N_{Re}. 248
9.4 In-line and staggered tube arrangements. 250
9.5 Comparison of heat flux for film-type and dropwise condensation. 251
9.6 Condensation on horizontal tube banks. 252
9.7 Film condensation on a vertical plate. 253
9.8 Nucleate boiling illustrated. 254
9.9 Linde porous boiling surface. 254
9.10 Pool boiling curve for water at atmospheric pressure. 255
9.11 Stages in pool boiling curve. 256
9.12 Bubble agitation. 257
9.13 Vapor–liquid exchange. 257
9.14 Vaporization. 257
9.15 Heat transfer mechanisms in convective boiling in a vertical tube. 259
9.16 Temperature gradients in forced convection. 260
9.17 Overall coefficients. Join process side duty to service side and read U from center scale (see dotted line for illustration). 262
9.18 Dowtherm and Syltherm heat transfer fluids. 264
9.19 Dowtherm, Dowfrost, and Dowcal heat transfer fluids. 264
9.20 Condenser vacuum deaerator for power plant boiler. 268

10.1 Double pipe (hairpin) heat exchanger with annuli connected in series and inner pipes connected in parallel. 273
10.2 TEMA designations for shell and tube heat exchangers. 274
10.3 1–1 Fixed tube sheet shell and tube heat exchanger with baffles. 275
10.4 Pull-through 1–2 floating head heat exchanger with baffles (TEMA S). 275
10.5 2–4 Floating head heat exchanger with baffles. 276
10.6 U-bundle heat exchanger with baffles. 276
10.7 Pull-through floating head heat exchanger, suitable for kettle reboilers (TEMA T). 276
10.8 Flow arrangement for two heat exchangers in series. 277
10.9 Grooves are made in the tube sheet for increased jointing between tube and tube sheet. 280
10.10 Different tube-to-tube joints. 281
10.11 Different types of expansion joints. 281
10.12 Illustration showing shell expansion joint. 281
10.13 Double tube sheet design. 281
10.14 Typical pass partitions for two to eight tube passes. 282
10.15 Types of tube pitch. 285
10.16 Shell side flow patterns in triangular pitch. 285
10.17 Vapor bubbles rising through boiling liquid inside a heat exchanger with square tube pitch. 286
10.18 Photograph of a cutaway of a baffled shell and tube heat exchanger. 287
10.19 Different arrangements for segmental baffles. 288
10.20 Segmental baffles. 288
10.21 Disk and doughnut baffle. 288
10.22 Orifice baffle. 288
10.23 Rod baffles. 289
10.24 Baffle cut. 289
10.25 Effect of small and large baffle cuts. 290
10.26 Large clearance between baffle and tube. 290
10.27 Baffle cut orientations. 291
10.28 Leaking paths for flow bypassing the tube matrix. Both through baffle clearances between the tube matrix and the shell. 291
10.29 Seal strips reduce bypassing around tube bundle. 292
10.30 Helixchanger heat exchanger. 293
10.31 Use of impingement baffle. 294
10.32 Rotating helical coil tube insert. 295
10.33 Twisted tape tube insert. 295
10.34 Wire mesh insert. 295
10.35 Idealized fouling curve. 300
10.36 Temperature profiles for countercurrent flow. 314
10.37 Temperature profiles for cocurrent flow. 314
10.38 Comparison of E and J shells for flow directions. 315
10.39 Temperature profiles for a 1–2 heat exchanger. 315
10.40 LMTD correction factors, *F*, for a 1–2 heat exchanger. 316
10.41 LMTD correction factors, *F*, for a 2–4 heat exchanger. 317
10.42 Heat exchanger effectiveness for countercurrent flow. 323
10.43 Heat exchanger effectiveness for cocurrent flow. 324

11.1 Illustration showing condensate backup. 332
11.2 Condensate removal system for a reboiler. 333
11.3 Annular and stratified flows inside horizontal condensers. 334
11.4 (a) Horizontal once-through reboiler with shell side boiling. (b) Horizontal recirculating reboiler with shell side boiling. 337
11.5 Horizontal reboilers. 338
11.6 (a) Vertical single pass, once-through with tube side boiling. (b) Vertical, recirculating with shell side boiling. 338
11.7 Recirculating baffled bottoms reboiler system. 338
11.8 Column internal reboiler. 340
11.9 Kettle reboiler. 340
11.10 Vertical thermosiphon reboiler. 341
11.11 Two reboilers in parallel. 344
11.12 Two reboilers in series. 344
11.13 New reboiler installed at an upper section of the column. 345
11.14 Quick selection guide for reboilers. 347
11.15 Evaporator selection guide. 349
11.16 Energy-efficient evaporation systems. 350
11.17 Short tube vertical calandria-type evaporator. 350
11.18 Swenson rising film evaporator unit. (*Courtesy*: Swenson Technology, Inc.) 353
11.19 Swenson falling film evaporation unit. (*Courtesy*: Swenson Technology, Inc.) 354
11.20 Tube showing falling liquid film. 355
11.21 Vertical forced circulation evaporator. 356
11.22 Mechanical vapor recompression evaporation system. 359
11.23 Double effect evaporator with forward feed operation. 362
11.24 Backward feed operation for a double effect evaporator. 363
11.25 Duhring plot for sodium chloride solutions. 366
11.26 Different types of entrainment separators. 369
11.27 Barometric condenser. 369

12.1 Typical vapor compression refrigeration cycle. 372
12.2 Cascade refrigeration cycle. 372
12.3 Comparison of a steam turbine and a heat pump. 375
12.4 Heat pump as applied to evaporation. 376
12.5 Distillation column with a separate refrigerant circuit. 376
12.6 Distillation column using process fluid as a refrigerant. 376
12.7 Heat tracer over a pipe carrying a fluid. 378
12.8 Heating coils. 379
12.9 Jackets with nozzles to admit heat transfer fluids. 380
12.10 Dimple jacket vessel. (*Source*: www.reimec.co.za.) 380
12.11 Dimple jacket cross section. (*Courtesy*: Santosh Singh (process.santosh@googlemail.com).) 380
12.12 Half-pipe jacket angles. (*Courtesy*: Santosh Singh (process.santosh@googlemail.com).) 381
12.13 Jacketed vessel with a half-pipe jacket. (*Source*: www.reimec.co.za.) 381
12.14 Half-pipe coil dimensions. (*Courtesy*: Santosh Singh (process.santosh@googlemail.com).) 382
12.15 Conventional jacket with baffles. (*Courtesy*: Santosh Singh (process.santosh@googlemail.com).) 383
12.16 Constant flux heat transfer jacket. 384
12.17 An inverted bucket steam trap. (*Courtesy*: Spirax Sarco.) 386
12.18 Ball float trap with (a) air cock and (b) thermostatic air vent. (*Courtesy*: Spirax Sarco.) 387
12.19 Liquid expansion steam trap. (*Courtesy*: Spirax Sarco.) 387
12.20 Balanced pressure steam trap. (*Courtesy*: Spirax Sarco.) 388
12.21 Bimetallic element made out of two laminated dissimilar metal strips. (*Courtesy*: Spirax Sarco.) 388
12.22 Operation of a bimetallic steam trap with a two-leaf element. (*Courtesy*: Spirax Sarco.) 389
12.23 Multicross elements as used in the Spirax Sarco SM range of bimetallic steam traps. (*Courtesy*: Spirax Sarco.) 389
12.24 Operation of a thermodynamic steam trap. (*Courtesy*: Spirax Sarco.) 390

13.1 L-footed tension wound aluminum fin. 398
13.2 Embedded fin. 399
13.3 Extruded fin. 399
13.4 Double L-footed fin. 399
13.5 Extended axial finned tube. 400
13.6 Continuous circular fins on a tube. 400
13.7 Serrated fins. 400
13.8 Air-cooled heat exchanger. 402
13.9 Construction of a typical plate heat exchanger. (*Courtesy*: Alfa Laval.) 403
13.10 Flow patterns in a plate heat exchanger. 404
13.11 Gaskets for a plate exchanger. 405
13.12 Ring and field gaskets to prevent intermixing of the fluids. 405
13.13 Conventional heat transfer plates and channel combinations. 406
13.14 Asymmetric heat transfer plates and channel combinations. 407
13.15 Flow patterns. 408
13.16 Single PHE handling three process streams. 408
13.17 Spiral heat exchanger. (*Courtesy*: Alfa Laval.) 415
13.18 Scraped surface heat exchanger. 416
13.19 Heat pipes. 417
13.20 Flat plate type heat pipe. 418
13.21 Micro-heat pipe operation. 418
13.22 Variable conductance heat pipe. 418
13.23 Capillary pumped looped heat pipe. 419
13.24 Heat pipe performance curves. 419
13.25 Typical heat pipe wick configurations and structures. 420
13.26 Heat pipe heat sink for power transistors. 420

14.1 Emissivity ranges of different materials. 426
14.2 Tube arrangements in small cylindrical fired heaters. 428
14.3 Fired heater with vertical radiant tubes and side view of top section. 430
14.4 A cabin heater with horizontal tubes and a rectangular firebox. 431
14.5 Large box-type cabin heater showing three separate radiant sections. 431
14.6 Absorption efficiency of the tube banks. 435
14.7 Partial pressure of $CO_2 + H_2O$ in flue gases. 436
14.8 Gas emissivity as a function of gas temperature. (Lobo WE, Evans JE. Heat transfer in radiant section of petroleum heaters. *Transactions of the American Institute of Chemical Engineers* 1939;35:743.) 436
14.9 Overall radiant exchange factor *F*. (Lobo WE, Evans JE. Heat transfer in radiant section of petroleum heaters. *Transactions of the American Institute of Chemical Engineers* 1939;35:743.) 437
14.10 Convective heater with flue gas recirculation. 442
14.11 Dimpled tube. 443
14.12 Premix gas burner. 444
14.13 Regenerative two-bed oxidizer. 451
14.14 Sankey diagram. 452

Mass Transfer

15.1 One-dimensional diffusion. 459
15.2 Concentration driving force. 460
15.3 Laminar hydrodynamic and concentration boundary layers for a flat plate. 464
15.4 Simplified diagram illustrating two-film theory. 465
15.5 Schematic representation of the situation at the interface. 466
15.6 Rising gas bubble in liquid. 466
15.7 Gas–liquid contacting. 466
15.8 Shapes of bubbles. 471
15.9 Bubble collapse and droplet formation phenomena. 472

16.1 Schematic diagram for a spray column. 476
16.2 Details of Venturi scrubbers. 476
16.3 Typical bubble cap design. 477
16.4 Types of valves used on valve trays. 478
16.5 Sieve tray indicating different parameters. 478
16.6 Types of flows on distillation trays. 480
16.7 Picket weir. 481
16.8 Generalized performance diagram for cross-flow trays. 482
16.9 Tray performance versus throughput. 482
16.10 O'Connell correlation for the estimation of overall column efficiency. 483
16.11 Murphree tray efficiencies illustrated. 484
16.12 Weir height. 489
16.13 Downcomer and active areas illustrated. 492
16.14 Flooding correlation for cross-flow trays (sieve, valve, and bubble cap trays). 492
16.15 Packed column. 493
16.16 Angle of wettability. 495
16.17 Some common types of random packings. (*Courtesy*: Koch Knight LLC for permission to use FLEXISADDLE™.) 497
16.18 Different structured packings (Mellapak). (*Courtesy:* Copyright © Sulzer Chemtech Ltd.) 499
16.19 Structured packing assembled to fit into a given column diameter. (*Courtesy*: Copyright © Sulzer Chemtech Ltd.) 500
16.20 Honeycomb packings. 500
16.21 Ceramic structured packing. (*Courtesy*: Koch Knight LLC for permission to use FLEXERAMIC® TYPE 28 Packing.) 500
16.22 Typical arrangement of horizontal expanded metal sheets with opposing angles. 500
16.23 Stacked packing to support dumped packing. 503
16.24 Vapor injection support plate. (*Source*: Saint Gobain Norpro.) 503
16.25 Schematic of vapor injection grid. 503
16.26 Trough and weir type distributor. (*Courtesy*: Kotch-Glitsch, LP.) 504
16.27 Orifice plate liquid distributor. 505
16.28 Perforated tube type distributor. 505
16.29 Spray nozzle type liquid distributor. (*Courtesy*: Copyright © Sulzer Chemtech Ltd.) 506
16.30 Multipan liquid distributor. 506
16.31 Wall wiper liquid redistributor. (*Source*: Norton.) 508
16.32 Generalized pressure drop and flooding correlation for packed columns. 511
16.33 Practices of location of bottom feed or reboiler return lines. 515
16.34 Forces acting on a liquid droplet suspended in a gas stream. 516
16.35 Typical droplet size distribution from entrainment. 516
16.36 Important dimensions of vertical and horizontal knockout drums. 518
16.37 Typical gas/vapor–liquid separators. (*Courtesy*: Pace Engineering.) 519
16.38 Typical gas/vapor–liquid separators. (*Courtesy*: Pace Engineering.) 520
16.39 Vapor–liquid separator for different cases. 520
16.40 Cross section of vane element mist extractor showing corrugated plates with liquid drainage traps. 521
16.41 Gas–Liquid separators. (*Courtesy*: Copyright © Sulzer Chemtech Ltd.) 522
16.42 Coalescer plate pack orientations. 524
16.43 Typical coalescer designs. 524
16.44 Three-phase horizontal coalescer. 524

17.1 Absorption equilibrium diagrams for SO_2–water system at different temperatures. 534
17.2 Equilibrium curve and operating line for absorption systems without heat effects. 536
17.3 Equilibrium curve and operating line for absorption with heat effects. 536
17.4 Limiting operating line for systems involving heat effects. 536
17.5 Graphical determination of number of trays for absorbers. 536
17.6 Tray column design for strippers. 537
17.7 Colburn diagram for estimation of N_{OG}. 540
17.8 T–x–y diagram for benzene–toluene system at 1 atm. 548
17.9 Equilibrium diagram for benzene–toluene system at 1 atm. 548
17.10 Effect of relative volatility on x–y diagrams. 548
17.11 Relative positions of EFV, ASTM, and TBP curves on a plot of percent distilled versus temperature. 550

17.12 Inverted batch distillation. 551
17.13 Equilibrium flash vaporization. 552
17.14 Operation of a simple distillation unit showing different parts. 552
17.15 Optimum column pressure. 553
17.16 Optimum reflux ratio. 554
17.17 McCabe–Thiele construction for number of theoretical trays. 555
17.18 Flow chart for multistage separations. 556
17.19 Enthalpy–concentration method for number of trays for binary distillation at a given reflux ratio. 557
17.20 Batch fractionation at constant reflux ratio (for four theoretical trays). 557
17.21 Batch fractionation at constant distillate composition (for four theoretical trays). 558
17.22 Instrumentation for constant vaporization rate and constant overhead composition in batch distillation. 558
17.23 Batch, continuous, and semicontinuous processes for separation of *n*-hexane, *n*-heptane, and *n*-octane mixtures. 559
17.24 Gilliland correlation on log–log coordinates. 561
17.25 Gilliland correlation as a function of reflux ratio. 562
17.26 Erbar–Maddox correlation for number of theoretical trays. 562
17.27 *T*–*x*–*y* and *x*–*y* diagrams for isopropyl ether–isopropyl alcohol system at 101.3 kPa pressure. 564
17.28 *T*–*x*–*y* and *x*–*y* diagrams for carbon disulfide–water system at 101.3 kPa pressure. 564
17.29 *T*–*x*–*y* and *x*–*y* diagrams for acetone–chloroform system at 101.3 kPa pressure. 565
17.30 *T*–*x*–*y* and *x*–*y* diagrams for water–1-butanol system at 101.3 kPa. *Note*: The diagrams are based on NRTL equation. 565
17.31 Pressure swing distillation for a minimum boiling binary azeotrope that is sensitive to changes to pressure. 566
17.32 A common heteroazeotropic distillation scheme with distillate decanter. 566
17.33 Two-column system for extractive distillation. 568
17.34 Extractive distillation process for separation of C_4 hydrocarbons. 569
17.35 Centrifugal type molecular distillation still. 574
17.36 Details of a divided wall column (Montz). 575
17.37 Different arrangements for separation of a three-component mixture. 576
17.38 Remixing in the conventional direct distillation sequence. 576
17.39 Feed preheater duty versus bottoms composition. 583
17.40 Level control for reflux drum. 584
17.41 Bypassing and flooding the condenser illustrated. 585
17.42 Equilateral triangular diagram. 588
17.43 Binodal and spinodal curves illustrated. Solid curve represents binodal curve. Dotted curve represents the spinal curve. Five tie lines (dashed lines) for the system, water–methanol–benzene (estimated) and plait point are also shown. 588
17.44 Triangular diagrams for Type I and Type II liquid–liquid equilibria. 589
17.45 Ternary equilibria for Type I system. 1–Hexene (A)–tetramethylene sulfone (B)–benzene (C) at 50°C. (a) Equilateral triangular plot, (b) right angular triangular plot, and (c) rectangular coordinate plot (Janecke Diagram, solvent-free coordinates). 589
17.46 Ternary equilibria for Type II system. Hexane (A)–aniline (B)–methylcyclopentane (C) at 34.5°C. (a) Equilateral triangular plot, (b) right-angled triangular plot, and (c) rectangular coordinate plot (Janecke and solvent-free coordinates). 590
17.47 Single-stage extraction 591
17.48 Multistage extraction with cross-flow. 592
17.49 Multistage extraction with counterflow. 592
17.50 Pressure–temperature diagram for a pure compound. 595
17.51 Simplified flow diagram for a supercritical solvent extraction process. 598
17.52 Settling chambers of different designs. 600
17.53 Combination of mixer–settler unit. 600
17.54 (a) Two-compartment mixing system and (b) drop in weir box system. 601
17.55 Pulse column. 603
17.56 Karr extraction column. 603
17.57 Scheibel column. 604
17.58 Rotating disk column. 604
17.59 Batch solid–liquid extractor. 609
17.60 Batch solid–liquid extractor (horizontal). 609
17.61 Thickener type countercurrent leaching equipment. 609
17.62 Hildebrand extractor. 610
17.63 Bollman extractor. 610
17.64 Rotocell extractor. 610
17.65 Pachuca tank. 611
17.66 Percolator type extractor for extraction of oils from oil seeds. 611
17.67 Diffuser for leaching process. 611

18.1 Depiction of supersaturation, metastable, and unsaturation zones in a crystallization process. 615

18.2 Crystal growth rate versus solution mixing velocity. 618
18.3 Cooling crystallizer. 621
18.4 Forced circulation crystallizer. (*Courtesy*: Swenson Technology, Inc.) 623
18.5 Direct contact refrigeration DTB crystallizer. 624
18.6 Surface-cooled baffled crystallizer using external heat exchanger surface to generate supersaturation by cooling. 625
18.7 Oslo type crystallizer. 626
18.8 Generalized melt crystallization process with a wash column. 627
18.9 Adiabatic process. 632
18.10 Illustration for obtaining wet bulb temperature. 633
18.11 Humidity chart. 633
18.12 Hygrometer using metal–wood laminate. 634
18.13 Sling hygrometer. 635
18.14 Vapor-compression cooling-based dehumidification process. 635
18.15 Spray humidification. 636
18.16 Packed bed humidifier. 636
18.17 Tubular humidifier. 636
18.18 Atmospheric and natural draft cooling towers. 637
18.19 Induced draft cooling tower. 638
18.20 Cooling tower performance curves. 643
18.21 Types of moisture content. 645
18.22 Equilibrium moisture content curves for different types of solids. 646
18.23 Water activity versus moisture content for different types of foods. 647
18.24 Movement of moisture during drying of porous materials. 648
18.25 Typical drying rate curve for constant drying conditions. 649
18.26 Examples of normalized drying rate curves for some typical materials. 650
18.27 Tray dryer. 654
18.28 Three-stage conveyor dryer. 655
18.29 Simplified diagram of a direct heat rotary dryer. 656
18.30 Principle of operation of a spouted bed dryer. 658
18.31 Pneumatic dryer. 659
18.32 Spray dryer. 662
18.33 Phase diagram of water. 667
18.34 Main components of a batch freeze dryer. 667
18.35 Types of pores on adsorbents. 673
18.36 A and X type zeolites. 674
18.37 Pressure swing adsorption for the dehydration of air. 676
18.38 Temperature swing adsorption. 676
18.39 Chromatographic unit. 679
18.40 Classification of analytical chromatographic systems. 679
18.41 IUPAC classification of gas adsorption isotherms. 681
18.42 Adsorption column mass transfer zone and idealized breakthrough zone. 683
18.43 Ion exchange process in solid ion exchange resin. 684
18.44 Symmetrical and asymmetrical membranes. 686
18.45 Liquid membrane. 687
18.46 Classification of membranes. 687
18.47 Asymmetric composite membrane. 688
18.48 Hollow fiber module (vertical). 690
18.49 Hollow fiber module (horizontal). 690
18.50 Hollow fiber membrane module. 691
18.51 Schematic of a spiral wound membrane module. 691
18.52 Plate and frame module. 692
18.53 Schematic diagram of flows inside a multichannel membrane element operating in cross-flow mode. 692
18.54 Cut section view of a typical membrane module. 692
18.55 Flow disruption around spacer netting to promote turbulence. 695
18.56 Range of pore diameters used in reverse osmosis, ultrafiltration, microfiltration, and conventional filtration. 695
18.57 Osmosis and reverse osmosis. 696
18.58 RO cascade to produce high-quality permeate. 699
18.59 Pervaporation and vapor permeation processes. 700
18.60 Polymer-enhanced ultrafiltration flow diagram. 703
18.61 Dead-end filtration and cross-flow filtration. 704
18.62 General transport mechanisms for gas permeation through porous and dense gas separation membranes. 707
18.63 Electrodialysis. 708
18.64 Illustration of the three operating modes for foam separating columns. 714
18.65 Froth flotation cell. 714

LIST OF TABLES

Fluid Mechanics

1.1 Rheological Characteristics of Non-Newtonian Fluids 8
1.2 Viscosity Conversions 11
1.3 Modern Level Measuring Methods and Their Applications 19
1.4 Problems and Possible Solutions in Pressure and Level Measurements 20

3.1 Values of Roughness for Different Materials 37
3.2 Loss Coefficients for Some Pipe Fittings 38
3.3 Design Velocities for Different Applications 46
3.4 Maximum Hydrocarbon Flow Velocities to Avoid Static Electricity Problems 46
3.5 Recommended Equivalent Lengths of Valves and Fittings 57
3.6 Causes and Consequences of Operational Deviations of Piping Systems 57

4.1 Comparison of Head Flow Meters 70
4.2 Summary of Plus and Minus Points of Different Types of Flow Meters 98
4.3 Summary of Flow Measurement Problems 98

5.1 Capacities, Heads Developed, and Efficiencies of Different Types of Pumps 110
5.2 Additional Requirements for Different Pump Options 126
5.3 Symptoms and Possible Causes for Centrifugal Pump Problems 127
5.4 Comparison of Capacities, Heads, and Efficiencies for Positive Displacement Pumps 130
5.5 Comparison between Centrifugal and Positive Displacement Pumps 137
5.6 Comparison of Magnetic Drive and Canned Motor Pumps 138
5.7 Pressures Obtainable with Different Vacuum Producing Equipment 142
5.8 Troubleshooting Guidelines for Steam Jet Ejectors 153
5.9 Pressures and Capacities Obtainable for Vacuum Equipment 153
5.10 Summary of Characteristics of Different Types of Compressors 157
5.11 Compression Ratio Versus Efficiency 159
5.12 Summary of Centrifugal Compressor Problems 160
5.13 Summary of Reciprocating Compressor Problems 161

6.1 Power Numbers for Different Impeller Designs 166
6.2 Examples of Mixing Processes Based on Mixing Intensity Scales 167
6.3 Mixing Intensity Versus Impeller Tip Speed 167
6.4 Agitation Achievable for Different Fluid Velocities 168
6.5 Power Requirements for Baffled Vessels 169
6.6 Power Input and Impeller Tip Speeds for Baffled Tanks 172
6.7 Advantages and Disadvantages of Selected Liquid Mixers 177
6.8 Impellers Used for Different Liquid Viscosity Ranges 183
6.9 Solids Mixer Selection for Different Applications 190

6.10 Selection of Solids Mixers Based on Operational Requirements 190
6.11 Types of Mixers and Their Mixing Action and Applications 193

7.1 Examples of Porosities of Different Materials 200
7.2 Materials Transported by Pneumatic Conveyance 209
7.3 Typical Conveying Velocities for Different Materials 214
7.4 Minimum Recommended Conveying Velocities for Good Ventilation 217

Heat Transfer

8.1 Reference Temperatures 226
8.2 Characteristics of Thermocouples 228
8.3 Commonly Observed Temperature Measurement Problems 232
8.4 Characteristics, Advantages, and Disadvantages of Temperature Measuring Devices 232

9.1 Magnitude of Heat transfer Coefficients in Increasing Order are Illustrated 260
9.2 Heat Transfer Coefficients for Different Applications 260
9.3 Typical Ranges of Overall Heat Transfer Coefficients 261
9.4 Recommended U for Broad Categories of Design Applications 262
9.5 Operating Temperature Ranges of Different Heat Transfer Fluids 267
9.6 Heat Transfer Coefficients for Preliminary Design of Heat Exchangers in Refinery Service 269

10.1 Selection of Shell and Tube Heat Exchangers 279
10.2 Tube Side Pressure Drops in Shell and Tube Heat Exchangers 284
10.3 Maximum Unsupported Span for the Tubes 287
10.4 Fractions of Different Shell Side Flow Streams 292
10.5 Selection of Materials of Construction for Heat Exchangers 294
10.6 Influence of Flow Velocity on Fouling Mechanisms 299
10.7 Influence of Temperature on Fouling Mechanisms 299
10.8 Some Considerations for the Choice of Heat Exchanger Design 302
10.9 Fouling Factors for Different Streams (10^4 m^2 K/W) 307
10.10 Specification Sheet for a Shell and Tube Heat Exchanger 312
10.11 TEMA Fouling Factors 328

11.1 Condenser Troubleshooting 336
11.2 Summary of Reboiler Types and their Characteristic Features 345
11.3 Advantages and Disadvantages of Different Types of Reboilers 346
11.4 Reboiler Selection Guide 347
11.5 Advantages and Limitations of Different Arrangements in Multiple Effect Evaporators 364
11.6 Boiler Feed Water Purity Requirements as Function of Boiler Pressure 370

12.1 Different Types of Refrigeration Systems and Refrigerants Used 372
12.2 Advantages and Disadvantages of Different Types of Jackets 382
12.3 Values of a in Equation 12.26 383
12.4 Values of Overall Heat Transfer Coefficients for Jacketed Vessels 384
12.5 Steam Trap Selection Criteria 392

13.1 Classification of Passive and Active Heat Transfer Enhancement Techniques 396
13.2 Types, Features, Applications, and Limitations of Heat Exchangers 416
13.3 Some Working Fluids for Heat Pipes 420

14.1 Values of Emissivities of Some Materials as a Function of Temperature 427
14.2 Heating Values and Stoichiometric Combustion Air Requirements for Different Fuels 433
14.3 Fuel Savings as a Function of Preheated Air Temperature 433
14.4 Mean Beam Length for Different Furnace Shapes/Dimensions 436
14.5 Average Heat Fluxes and Temperatures Employed in Process Heaters 438
14.6 Maximum Permissible Radiant Heat Fluxes for Personal Exposure 438

Mass Transfer

15.1 Definitions of Mass Transfer Coefficients 461
15.2 Examples of Gas or Liquid Film Controlled Processes 462
15.3 Relationships Among Drop Size, Drop Surface Area, and Drop Count 470

16.1 Recommended Tray Spacings 479
16.2 Recommended Minimum Residence Times for Liquid in a Downcomer 480
16.3 Recommended Design Pressure Drops for Packed Columns 495
16.4 Advantages and Disadvantages of Different Types of Liquid Distributors 506

16.5 Drum Pressure Versus K-Values 521
16.6 Comparison of Different Media for Droplet Capture 525
16.7 Recommended Residence Times for Gas–Liquid and Liquid–Liquid Separators 526

17.1 HETP Values for Preliminary Packed Column Design 541
17.2 Comparison of Control Systems for Emergency Releases 542
17.3 Deviations from Raoult's Law as Related to Molecular Interactions 545
17.4 Classification of Molecules Based on Their Hydrogen Bonding Nature 545
17.5 Comparison of Batch, Continuous, and Semicontinuous Distillation Processes 559
17.6 Number of Trays Used Between Side-Draw Products 560
17.7 Ewell, Harrison, and Berg Classification for Entrainers 563
17.8 Examples of Solvents Used in Different Extractive Distillation Processes 568
17.9 Salt Effect on Vapor–Liquid Equilibria for Ethanol–Water System 569
17.10 Examples of Reactive Distillation Processes 571
17.11 Comparison Between Extraction and Distillation 585
17.12 Critical Properties of Different Solvents used for Supercritical Extractions 597
17.13 General Features of Liquid–Liquid Extraction Equipment 607

18.1 Choice of Operating Mechanism for a Crystallizer 620
18.2 Comparison of Layer and Suspension Crystallization Processes 628
18.3 Differences Between Melt and Solution Crystallization 628
18.4 Troubleshooting of Crystallizer Operation 631
18.5 Temperature Approach Versus Relative Tower Size 637
18.6 Critical Moisture Content for Different Materials 648
18.7 Criteria for Classification of Dryers 653
18.8 Some Materials Dried in Flash Dryers 660
18.9 Droplet and Product Sizes (μm) Obtainable for Different Atomizers 664
18.10 Effect of Different Variables on the Operation of Spray Dryer 665
18.11 Characteristics and Features of Kneader Dryers 666
18.12 Applications of Different Types of Dryers 670
18.13 Differences Between Physical Adsorption and Chemisorption 671
18.14 Structures and Applications of Different Molecular Sieves 675
18.15 Applications of Type X Molecular Sieves 675
18.16 Applications of Some Commercially Used Adsorbents 675
18.17 Types of Ion Exchange Processes 683
18.18 Comparison of Membrane Element Configurations 693
18.19 Different Types of Membrane Modules and their Characteristics 693
18.20 Examples of Membrane Processes in Separation of Gas Mixtures 694
18.21 Foulants and Pretreatment Techniques 698
18.22 Applications of Reverse Osmosis 698
18.23 Applications of Ultrafiltration 702
18.24 Advantages and Disadvantages of Different Types of Membranes for Applications to Reverse Osmosis and'Ultrafiltration 704
18.25 Summary of Size Ranges of Materials Separated by Different Membrane Processes 709
18.26 Classification and Principles of Major Bubble Separation Techniques 712
18.27 Some Applications of Floatation 715

SECTION I

FLUID MECHANICS

1

FLUID MECHANICS BASICS

1.1 Dimensional Analysis 3
1.2 Fluid Properties 4
1.3 Newtonian and Non-Newtonian Fluids 5
1.4 Viscosity Measurement 8
1.5 Fluid Statics 10
1.5.1 Liquid Level 15

1.1 DIMENSIONAL ANALYSIS

- Differentiate between units and dimensions by means of examples.
  - Examples of dimensions include weight, time, length, and so on.
  - Examples of units include seconds, days, years, inches, centimeters, kilometers, grams, pounds, and so on.
- What are the methods used to carry out dimensional analysis?
  - Rayleigh's method.
  - Buckingham π-theorem.
- Under what circumstances dimensional analysis becomes a tool for obtaining solutions to problems?
  - When two variables are to be correlated, a simple plot of one variable versus the other will describe the problem.
  - When three variables are to be correlated, for *each* value of the third variable, a plot of the other two as in the above case will describe the problem; that is, a number of plots, each for one value of the third variable, will be required.
  - When more than three variables are involved in a correlation, the correlation becomes complex and requires a set of curves for each of the fourth variable.
  - In such situations involving four or more variables, dimensional analysis becomes a necessity as correlations progressively become very complex.
- Name the dimensionless numbers of significance in fluid mechanics. Give their physical significance.
  - *Reynolds number*, $N_{Re} = DV\rho/\mu$ = inertial forces/viscous forces.
  - *Weber number*, $N_{We} = LV^2\rho/\sigma$ = inertial forces/surface tension forces. L is the characteristic length and σ is surface tension. It can be considered as a measure of the relative importance of the inertia of the fluid compared to its surface tension. It is useful in analyzing thin film flows and the formation of droplets and bubbles.
  - *Froude number*, $N_{Fr} = V^2/Lg$ = inertial forces/gravity forces. In the study of stirred tanks, the Froude number governs the formation of surface vortices. Since the impeller tip velocity is proportional to ND, where N is the impeller speed (rev/s) and D is the impeller diameter, the Froude number then takes the form $Fr = N^2D/g$.
  - *Euler number*, $N_{Eu} = (-\Delta P)/\rho V^2$ = frictional pressure loss/(2 × velocity head).
  - *Critical cavitation number*, $\sigma = (P - P^0)/(\rho V^2/2)$ = excess pressure above vapor pressure/velocity head. Cavitation number is useful for analyzing fluid flow dynamics problems where cavitation may occur.
  - *Cauchy number*, $C = \rho v^2/\beta$ = inertial force/compressibility force. v is the local fluid velocity (m/s) and β is the bulk modulus of elasticity (Pa). Cauchy number is defined as the ratio between inertial force and the compressibility force (elastic force) in a flow. It is used in the study of compressible flows. Cauchy number is related to Mach number. It is equal to square of the Mach number for isentropic flow of a perfect gas.
  - *Capillary number*, $Ca = \mu V/\rho$ = viscous force/surface tension force. Capillary number represents the relative effect of viscous forces versus surface

Fluid Mechanics, Heat Transfer, and Mass Transfer: Chemical Engineering Practice, By K. S. N. Raju

tension acting across an interface between a liquid and a gas, or between two immiscible liquids.

- *Mach number*, $M = V/c$ = fluid velocity/sonic velocity. Mach number is commonly used both with objects traveling at high speed in a fluid and with high-speed fluid flows inside channels such as nozzles, diffusers, or wind tunnels.

  The speed represented by Mach 1 is not a constant. For example, it depends on temperature and atmospheric composition. In the stratosphere, it remains constant irrespective of altitude even though the air pressure varies with altitude.

  Since the speed of sound increases as the temperature increases, the actual speed of an object traveling at Mach 1 will depend on the fluid temperature around it. Mach number is useful because the fluid behaves in a similar way at the same Mach number.
- *Knudsen number*, $Kn = \lambda/L = k_B T/\sqrt{(2\pi\sigma^2 PL)}$ = molecular mean free path length/representative physical length scale. Example of length scale could be the radius of a body in a fluid. L is the representative physical length scale (m), k_B is the Boltzmann constant (1.38×10^{-23} J/K), T is the temperature (K), σ is the particle diameter (m), and P is the total pressure (Pa).
- *Fanning friction factor*, $f = D\Delta P/2\rho V^2 L \rightarrow 2\tau_w/\rho V^2$ = wall shear stress/velocity head.
- *Drag coefficient*, $C_D = F_D/(A\rho V^2/2)$ = drag force/(projected area × velocity head).
- *Flow number*, $N_Q = Q'/(ND^3)$, represents actual flow during mixing in a vessel. Q' is the flow rate/pumping capacity, N is the speed of rotation of the impeller (rev/s), and D is the impeller diameter.
- *Power number*, $P_o = P/N^3 D^5 \rho$ = ratio of energy causing local turbulence to that providing bulk flow.
- *Deborah number*, $D_e = t_c/t_p$ = ratio of the relaxation time, characterizing the intrinsic fluidity of a material, and the characteristic timescale of an experiment (or a computer simulation) probing the response of the material. t_c refers to the relaxation timescale and t_p refers to the timescale of observation. The smaller the Deborah number, the more fluid the material appears.
- *Archimedes number*, $A_r = gL^3 \rho_L(\rho - \rho_L)/\mu^2$, is used to determine the motion of fluids due to density differences. It is useful in applications involving gravitational settling of particles in a fluid. $\rho - \rho_L$ is the density difference between the body and the fluid and μ is the viscosity of the fluid.

1.2 FLUID PROPERTIES

- Define API gravity? Where is it used? What is its use?
  - Degree API = (141.5/sp.gr.)−131.5. (1.1)
  - It is used in characterizing densities of petroleum oils and other liquids.
  - It provides a stretched scale for expressing densities and is useful in easily distinguishing different oils.
  - Sp. gr. in the above equation corresponds to 15.6°C (60°F), referenced to that of water at the same temperature.
- What is viscosity? What are its units?
  - Viscosity is the measure of the internal friction of a fluid. This friction becomes apparent when a layer of fluid is made to move in relation to another layer. The greater the friction, the greater the amount of force required to cause this movement, which is called *shear*. Shearing occurs whenever the fluid is physically moved or distributed, as in pouring, spreading, spraying, mixing, and so on. Highly viscous fluids, therefore, require more force to move than less viscous materials.
  - Consider two parallel planes of fluid of equal area, separated by unit distance and moving in the same direction at different velocities. The force required to maintain this difference in velocity is proportional to the difference in velocity through the liquid, called *velocity gradient*.
  - Velocity gradient is a measure of the change in velocity at which the intermediate layers move with respect to each other. It describes the shearing the liquid experiences and is thus called *shear rate*. Its unit of measure is called the *reciprocal second* (s^{-1}). The force per unit area required to produce the shearing action is called *shear stress*.
  - The defining equation for viscosity is called Newton's equation, which is

$$\tau = -\mu(du/dy), \tag{1.2}$$

where τ is the *shear stress*, μ is the viscosity, and du/dy is the *shear rate*, γ.

  - The equation for viscosity becomes

$$\mu = \tau/\gamma, \tag{1.3}$$

which is the ratio of shear stress to shear rate. μ is the absolute viscosity that is sometimes called shear viscosity, τ is the shear stress, and γ is the shear rate.

  - Absolute (dynamic) viscosity is a measure of how resistive the flow of a fluid is between two layers of fluid in motion.
  - Viscosity affects the magnitude of energy loss in a flowing fluid.
  - High-viscosity fluids require greater shearing forces than low-viscosity fluids at a given shear rate.

    Units of Viscosity: 1 poise is equal to 100 centipoise (cP). SI units of viscosity are kg/(m s) or Pa s.

- What is kinematic viscosity? Give its units.
  - Kinematic viscosity, ν, is the ratio of viscosity to density, μ/ρ. SI units of kinematic viscosity are m^2/s. The cgs unit is Stoke, which is 1 cm^2/s.
  - Kinematic viscosity is a measure of how resistive the flow of a fluid is under the influence of gravity.
  - Kinematic viscometers usually use the force of gravity to cause the flow through a calibrated orifice, while timing the flow.
- How does viscosity (μ) of a gas change with temperature and pressure?
  - Viscosity of gases increases as temperature increases and is approximately proportional to the square root of temperature. This is due to the increase in the frequency of intermolecular collisions at higher temperatures. Since most of the time the molecules in a gas are flying freely through the void, anything that increases the number of times one molecule is in contact with another will decrease the ability of the molecules as a whole to engage in the coordinated movement. The more these molecules collide with one another, the more disorganized their motion becomes.
- What is SAE classification?
  - Society of Automotive Engineers (SAE) has developed a numbering system based on viscosities for application to lubricants. The SAE numbering scheme describes the behavior of motor oils under low- and high-temperature conditions—conditions that correspond to starting and operating temperatures. The first number, which is always followed by the letter W, describes the low-temperature behavior of the oil at start-up while the second number describes the high-temperature behavior of the oil after the engine has been running for some time. Lower SAE numbers describe oils that are meant to be used at lower temperatures. Oils with low SAE numbers are generally less viscous than oils with high SAE numbers, which tend to be thicker.
  - SAE 0W, 5W, 10W, 15W, 20W, and 25W are grades of motor lubricating oils for low-temperature applications. SAE numbers 20, 30, 40, 50, and 60 are indicative of high-temperature applications.
  - For example, 10W-40 oil would have a viscosity not more than 7000 mPa s in a cold engine crankcase even if its temperature should drop to −25°C on a cold winter night and a viscosity not less than 2.9 mPa s in the high-pressure parts of an engine very near the point of overheating (150°C).
- Define compressibility factor and state its significance.
  - Compressibility factor quantifies the departure from ideal conditions for a gas.
  - An ideal gas obeys the equation

$$PV = nRT. \tag{1.4}$$

  - For a nonideal gas,

$$PV = ZnRT \quad \text{or} \quad Z = PV/nRT. \tag{1.5}$$

  - For example, a gas for which $Z = 0.90$ will occupy only 90% of the volume occupied by an ideal gas at the same temperature and pressure.
  - The values of Z range from about 0.2 to a little over 1.0 for pressures and temperatures of up to 10 times the critical values.
  - Z is a complex function of reduced temperature (T_r), reduced pressure (P_r), critical compressibility factor (Z_c), acentric factor (ω), or other parameters.

1.3 NEWTONIAN AND NON-NEWTONIAN FLUIDS

- What is rheology?
  - Rheology is defined as the study of the change in form and the flow of matter, embracing elasticity, viscosity, and plasticity.
- What are the differences between Newtonian and non-Newtonian fluids?
  - *Newtonian Fluids*: In these fluids, viscosity is constant regardless of the shear forces applied to the layers of fluid. At constant temperature, the viscosity is constant with changes in shear rate or agitation.
  - These are unaffected by magnitude and kind of fluid motion.
  - Examples of Newtonian fluids include water, milk, alcohol, aqueous solutions, hydrocarbons, and so on.
  - Fluids for which a plot of shear stress versus shear rate at a given temperature is a straight line with a constant slope, independent of the shear rate, are called *Newtonian fluids*. This slope is called absolute viscosity of the fluid, represented by the equation $\mu = \tau/\gamma$, as given earlier.
    Absolute viscosity is sometimes called shear viscosity. When written as $\tau = -\mu\, du/dy$, the equation represents Newton's law and the fluids that follow Newton's law are *Newtonian fluids*.
  - All fluids for which viscosity varies with shear rate are called *non-Newtonian fluids*. For such fluids, viscosity, defined as the ratio of shear stress to shear rate, is called *apparent viscosity* to emphasize the distinction from Newtonian behavior. For such fluids, apparent viscosity is given by

$$\tau = \eta\gamma, \tag{1.6}$$

    where τ is the shear stress, η is the apparent viscosity, and γ is the shear rate.

- Any fluid that does not obey the Newtonian relationship between the shear stress and shear rate is called non-Newtonian. The subject of *rheology* is devoted to the study of the behavior of such fluids. High molecular weight liquids that include polymer melts and solutions of polymers, as well as liquids in which fine particles are suspended (slurries and pastes), are usually non-Newtonian. Fluids such as water, ethanol, and benzene and air and all gases are Newtonian. Also, low molecular weight liquids and solutions of low molecular weight substances in liquids are usually Newtonian. Some examples are aqueous solutions of sugar or salt.

- What are broad classes of non-Newtonian fluids?
  - Non-Newtonian fluids can be divided into two broad categories on the basis of their shear stress/shear rate behavior: those whose shear stress is independent of time or duration of shear (*time-independent*) and those whose shear stress depends on time or duration of shear (*time-dependent*).
- How is a power law fluid characterized?
  - A power law fluid exhibits the following nonlinear relationship between shear stress and shear rate:

$$\tau = K'\gamma^{n'}. \tag{1.7}$$

  - When $n' = 1$, the fluid is Newtonian.
  - Apparent viscosity for a power law fluid under shear conditions is

$$\mu_a = \tau/\gamma = K'\gamma^{n'-1}. \tag{1.8}$$

  - When $n' < 1$, μ_a decreases with increasing γ, and the fluid is called shear thinning or pseudoplastic.
  - When $n' > 1$, μ_a increases with increasing γ, and the fluid is called shear thickening or dilatant.
- What are shear thinning and shear thickening fluids?
  - The slope of the shear stress versus shear rate curve will not be constant as we change the shear rate. When the viscosity decreases with increasing shear rate, the fluid is called *shear thinning*. However, the rate of increase is less than linear as shown in Figure 1.1.
  - Where the viscosity increases as the fluid is subjected to a higher shear rate, the fluid is called *shear thickening*. For a shear thickening fluid, τ increases faster than linearly with increasing γ. Shear thinning behavior is more common than shear thickening behavior.
- What are *pseudoplastic* fluids?
  - Shear thinning fluids are called pseudoplastic fluids.
  - Apparent viscosity for these fluids decreases with increasing shear rate, but initial viscosity may be high enough to prevent start of flow.
  - Most shear thinning fluids behave in a Newtonian manner (i.e., their viscosity is independent of γ) at very high and very low shear rates.
  - Pseudoplastics do not have a yield value.
  - Examples of shear thinning fluids are polymer melts such as molten polystyrene, polymer solutions such as polyethylene oxide in water, emulsions, molten sulfur, molasses, greases, starch suspensions, paper pulp, biological fluids, soaps, detergent slurries, toothpaste, cosmetics and some pharmaceuticals, and inks and paints.
- Give examples of shear thickening fluids.
  - Some examples of shear thickening fluids are cornstarch, pastes, slurries of clay and titanium dioxide, wet beach sand and sand-filled emulsions, sugar, paper coatings, pigment–vehicle suspensions such

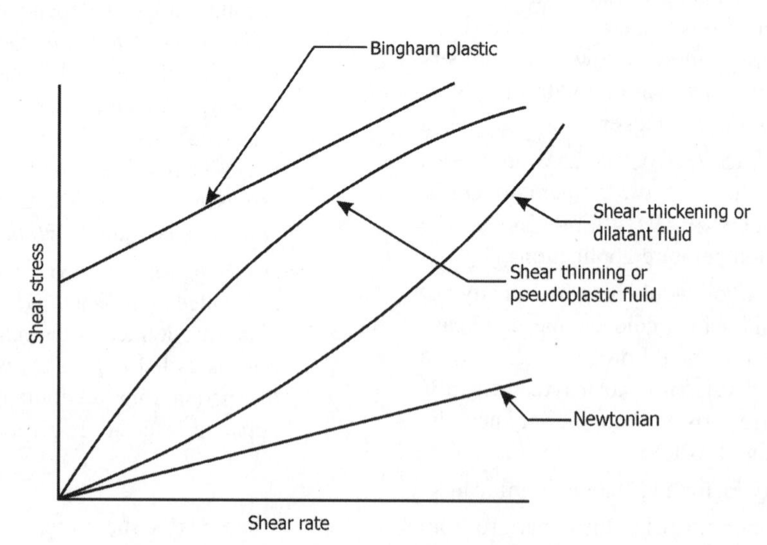

FIGURE 1.1 Shear rate versus shear stress diagrams for Newtonian and non-Newtonian fluids.

as paints and printing inks of high concentrations, gum arabic in water, and solutions of certain surfactants.

- Most shear thickening fluids tend to show shear thinning behavior at very low shear rates. However, at high shear rates, the suspension expands (dilates), such that the amount of liquid present cannot overcome the frictional forces between the particles and the consequent increase in apparent viscosity, μ_a.

• What are power law fluids?
  - Power law fluids are fluids with non-Newtonian behavior, exhibiting a nonlinear relationship between the shear stress (τ) and shear rate (λ).

• What is a Bingham plastic?
  - Non-Newtonian fluids for which a finite stress is required before continuous deformation takes place are called *yield stress materials*, which are also called *viscoplastic materials*. These fluids will not flow when only a small shear stress is applied. The shear stress must exceed a critical value, known as the yield stress τ_0, for the fluid to flow. For example, when a tube of toothpaste is opened, certain amount of force is needed to be applied before the toothpaste will start flowing. In other words, viscoplastic fluids behave like solids when the applied shear stress is less than the yield stress. Once it exceeds the yield stress, the viscoplastic fluid will flow just like a fluid.
  - *Bingham plastics* are a special class of viscoplastic fluids that exhibit a linear behavior of shear stress against shear rate.
  - Bingham plastics have definite yield value that must be exceeded before flow starts. After flow starts, viscosity decreases with increase in agitation.
  - Examples of viscoplastic fluids are drilling mud, toothpaste, paper pulp, greases, soap, nuclear fuel slurries, mayonnaise, margarine, chocolate mixtures, blood, suspensions of chalk, grain, and thoria, and sewage sludge.
  - *Bingham plastic materials* are the simplest of the yield stress materials. Highly concentrated suspensions of fine solid particles frequently exhibit Bingham plastic behavior. Slope of the line shown in Figure 1.1 for a Bingham plastic is called *infinite shear viscosity*.
  - The equation for a Bingham plastic material is given as

$$\tau = \tau_y + \mu_\infty \gamma. \tag{1.9}$$

  - Figure 1.1 illustrates shear rate–shear stress behavior of Newtonian and non-Newtonian fluids.

• What are thixotropic fluids?
  - Thixotropic fluids are *time-dependent* fluids for which structural rearrangements occur during deformation at a rate too slow to maintain equilibrium configurations. As a result, shear stress changes with duration of shear. These show decreasing shear stress with time at constant shear rate. In other words, apparent viscosity decreases with time under shear conditions. After shear ceases, apparent viscosity returns to its original value, with the time for recovery varying for different fluids. Increased agitation normally decreases apparent viscosity, but this depends on duration of agitation, viscosity of the fluid, and rate of motion before agitation.
  - Examples of thixotropic fluids include mayonnaise, margarine, honey, shaving cream, clay suspensions used as drilling muds, ketchup, gelatin solutions, some polymer solutions and food materials, and some paints and inks.

• What are *rheopectic* fluids?
  - For *rheopectic* fluids, shear stress increases with time at constant shear rate, their behavior being opposite to thixotropy.
  - Rheopectic behavior has been observed in bentonite clay suspensions, gypsum suspensions, certain sols such as vanadium pentoxide, and some polyester solutions.

• What are *rheomalectic* fluids?
  - In rheomalectic fluids, viscosity decreases with time under shear conditions but does not recover, with the fluid structure being irreversibly destroyed.

• What are *viscoelastic* fluids?
  - Viscoelastic fluids exhibit both viscous and elastic behavior. They behave like elastic rubber-like solids and as viscous liquids.
  - They exhibit elastic recovery from deformation when stress is removed.
  - Viscoelastic fluids exhibit tendencies such as climbing up a rotating shaft, swelling when extruded out of a dye, and the like. This phenomenon is called *Weissenberg effect*. Polymeric liquids, flour dough, egg white, and bitumen are examples of this class.
  - *Relaxation time*, the time required for elastic effects to decay, is a property of these fluids.
  - Viscoelastic effects may be important with sudden changes in rates of deformation, as during flow start-up or stoppage, rapidly oscillating flows, and during expansions and contractions involving accelerations.
  - In many fully developed flows, viscoelastic fluids behave as purely viscous fluids.

TABLE 1.1 Rheological Characteristics of Non-Newtonian Fluids

Fluid Type	Effect of Increased Shear Rate	Time-Dependent	Examples
Pseudoplastic	Thinning	No	Polymer solutions, starch suspensions, paints, greases, emulsions, biofluids, detergent slurries, and so on
Thixotropic	Thinning	Yes	Clay suspensions used in drilling muds, honey, ketchup, gelatin solutions, shaving cream, and some food materials
Dilatant	Thickening	No	Cornstarch pastes, slurries of clay and TiO_2, wet beach sand, and sand-filled emulsions
Rheopectic	Thickening	Yes	Bentonite clay suspensions, gypsum suspensions, and some polyester solutions

- "Apparent viscosity of a pseudoplastic fluid flowing in a pipe decreases as the flow rate increases." *True/False*?
  - *False*.
- Summarize rheological characteristics of non-Newtonian fluids.
  - Table 1.1 summarizes rheological characteristics of non-Newtonian fluids.

1.4 VISCOSITY MEASUREMENT

- How are viscometers classified?
  - Viscometers are broadly divided into two categories, namely, rotational and tube type.
  - Figure 1.2 gives subgroups of viscometers under the above classification.
    - ➢ Glass capillary viscometers (e.g., Cannon–Fenske viscometers) are suitable for Newtonian fluids because the shear rate varies during discharge.
    - ➢ Cone and plate viscometers are limited to moderate shear rates.
    - ➢ Pipe and mixer viscometers can handle much larger particles than cone and plate or parallel plate instruments.
    - ➢ High-pressure capillaries operate at high shear rates.
- What are the various types of viscometers and describe their principles of operation?
  - Flow viscometers work on the principle of time taken for a fixed volume of liquid flowing through an orifice of finite diameter. These include Oswald glass viscometer, capillary viscometer, Redwood No. 1 and No. 2 viscometers, and Saybolt Universal and Furol viscometers.
  - Rising bubble or falling ball viscometers.
    - ➢ Kits of bubble viscometers are available for different viscosity ranges in which a bubble of a standard size is enclosed in sealed glass tubes having different, but known viscosities. The liquid under test is enclosed in a tube of same geometry and the time of bubble rise in vertical position in the tube with liquid under test is compared with bubble rise in the sealed tubes with liquids of known viscosities. Viscosity of the test liquid is taken as the viscosity in a particular sealed tube having the same rate of bubble rise, that is, time taken for the bubble to rise the same distance in both the tubes.
    - ➢ In a falling ball viscometer, time taken for the ball to travel vertically downward for a specified distance in the liquid in a tube will give viscosity from the equation for Stokes' law.
  - Rotational viscometers (Figure 1.3) use the principle that the force required to turn an object in a fluid depends on the viscosity of the liquid. Absolute viscosity can be directly obtained from a rotational viscometer that measures the force needed to rotate a spindle in the fluid.
  - *Brookfield viscometer* belongs to this class and determines the required force for rotating a disk or bob in a fluid at a known speed.

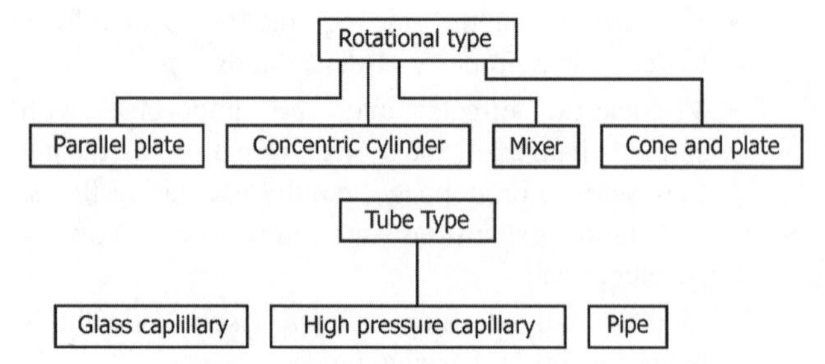

FIGURE 1.2 Classification of viscometers.

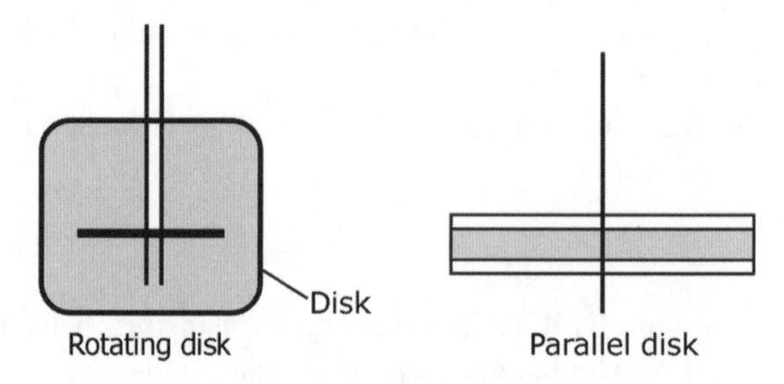

FIGURE 1.3 Rotating disk and parallel plate viscometers.

- In this device, a concentric rotating cylinder (spindle) spins at a constant rotational speed inside another cylinder. In general, there is a very small gap between the walls. This annulus is filled with the fluid. The torque needed to maintain this constant rotation rate of the inner spindle is measured by means of a torsion wire from which the spindle is suspended. This device is sometimes called Couette viscometer, named after its discoverer. Some types of viscometers rotate the outer cylinder.

- *Cup and bob* viscometers (Figure 1.4) work by defining the exact volume of sample that is to be sheared within a test cell; the torque required to achieve a certain rotational speed is measured and plotted. There are two classical geometries in *cup and bob* viscometers known as either the *Couette* or *Searle* systems—distinguished by whether the cup or bob rotates.
- A Cambridge moving piston viscometer is illustrated in Figure 1.5.
- Cone and plate viscometers (Figure 1.6) use a cone of very shallow angle in bare contact with a flat plate. With this system, the shear rate beneath the plate is constant. A graph of shear stress (torque) versus shear rate (angular velocity) gives the viscosity.
- Controlled flow and controlled pressure viscometers are illustrated in Figure 1.7.
  - In controlled flow rate viscometers, a piston forces the liquid through a horizontal or vertical tube and the resultant pressure drop is measured.
  - In controlled pressure viscometers, compressed air drives the liquid through a horizontal or vertical tube and the resultant volumetric flow rate is measured.
  - A *Stormer viscometer* is a rotation-type viscometer used to determine viscosity of paints. It consists of a paddle-type rotor that is rotated by an internal motor, submerged into a cylinder of the viscous material. The rotor speed can be adjusted by changing the magnitude of load applied onto the rotor. For

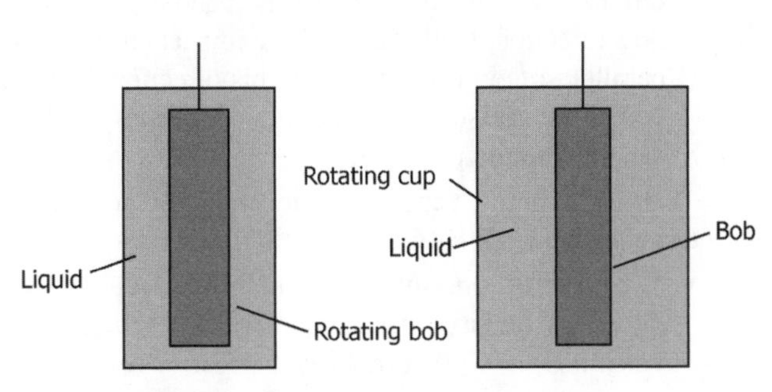

FIGURE 1.4 Cup and bob coaxial viscometers.

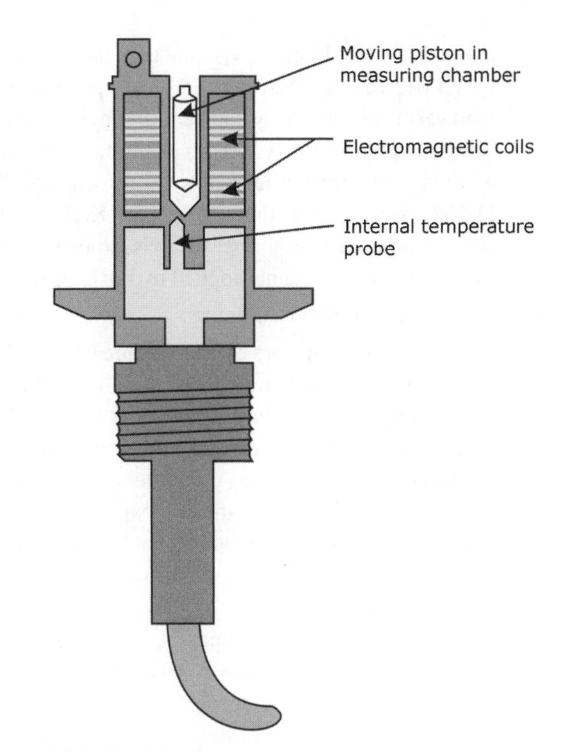

FIGURE 1.5 Cambridge moving piston viscometer.

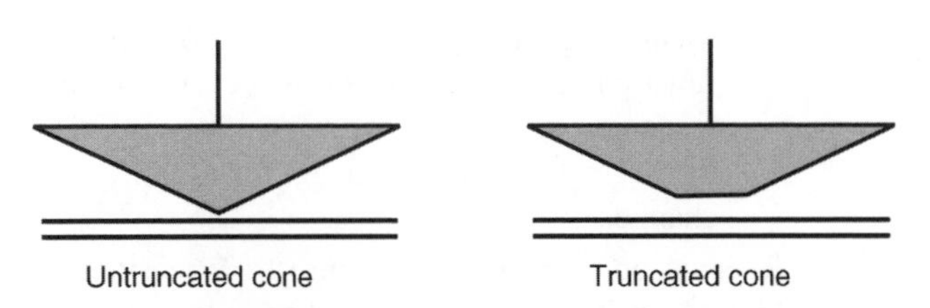

FIGURE 1.6 Cone and plate viscometers.

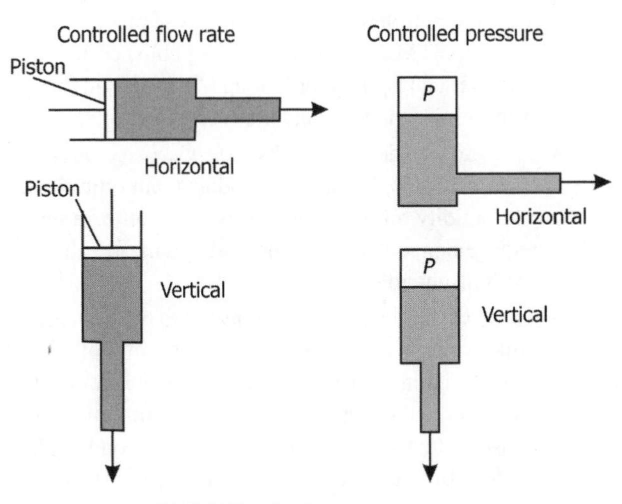

FIGURE 1.7 Tube viscometers.

example, in one brand of viscometers, the load and speed decrease by pushing the level upward, and vice versa. The viscosity can be found by adjusting the load until the rpm is 200. By examining the load applied and comparing tables found on ASTM D-562, one can find the viscosity in Krebs units (KU), unique only to Stormer-type viscometers. This method is used for paints applied by brush or roller.

- Capillary glass tube viscometers.
  - ➢ Vibrational viscometers operate by measuring the damping of an oscillating electromechanical resonator immersed in a fluid whose viscosity is to be determined. The resonator generally oscillates in torsion or transversely (as a cantilever beam or a tuning fork). The higher the viscosity, the larger the damping imposed on the resonator. The damping of the resonator may be measured by one of the several methods:
    - Measuring the power input necessary to keep the oscillator vibrating at a constant amplitude. The higher the viscosity, the more the power needed to maintain the amplitude of oscillation.
    - Measuring the decay time of the oscillation once the excitation is switched off. The higher the viscosity, the faster the signal decays.
    - Measuring the frequency of the resonator as a function of phase angle between excitation and response waveforms. The higher the viscosity, the larger the frequency change for a given phase change.
- Rheoviscometers are used to measure viscosities of high-viscosity fluids such as molten polymers.

- Name some petroleum testing viscometers and describe their characteristics.
  - These are Saybolt, Redwood, and Engler viscometers, used for petroleum testing using ASTM or IP standards. (ASTM and IP standards are published annually by American Society for Testing Materials and Energy Institute of United Kingdom, respectively.)
  - They do not determine absolute viscosity of petroleum products, tar or fluid products, but empirically test fluidity of petroleum oils, by conforming to requirements of the test methods giving results that are reproducible.
  - These viscometers basically involve a tube of certain dimensions to accommodate fixed volumes of test oil and precision orifices of specified bores, attached to the tube at the bottom. The tube is immersed in a constant temperature bath. Time, in seconds, taken for flow of fixed quantity of oil into a calibrated glass bottle is measured.
  - Saybolt viscometers are of two types, namely, Saybolt Universal and Saybolt Furol, based on the diameter of the orifice used. Universal orifice, with smaller bore, is intended to measure viscosities of lighter oils. Furol orifice is of larger bore for measuring viscosities of high-viscosity oils such as furnace oil. For the Saybolt viscometer, the amount of oil to be measured is 60 mL. Viscosity is expressed as Saybolt Seconds Universal (SSU) or Saybolt Seconds Furol (SSF).
  - Redwood viscometers are of two types, Redwood No. 1 or Redwood No. 2. The former having a smaller bore orifice and the latter a larger bore orifice are intended to measure lower and higher viscosity oil samples, respectively. When the flow time exceeds 2000 s, Redwood No. 2 will be more convenient to use.
  - Redwood seconds refer to the number of seconds required for 50 mL of the oil to flow out of the device at a predefined temperature.
  - For Engler viscometers, the reading is the time, in seconds, required for 200 mL of the oil to flow through the device at a predefined temperature.
  - Conversion charts are used to convert viscosities from these instruments into kinematic viscosities.
- What type of viscometer is used for viscosity measurement for non-Newtonian liquids?
  - Rheoviscometers.
- Give selected approximate viscosity conversions.
  - Table 1.2 gives viscosity conversions among kinematic, Saybolt, Redwood, and Engler viscosities.

1.5 FLUID STATICS

- What are static and dynamic pressures?
  - Static pressure is the pressure exerted by a fluid at rest. Dynamic pressure is the pressure exerted by a fluid in motion.
  - *Static pressure* is uniform in all directions, so pressure measurements are independent of direction in an immobile (static) fluid. Flow, however, applies additional pressure on surfaces perpendicular to the flow direction, while having little impact on surfaces parallel to the flow direction. This directional component of pressure in a moving (dynamic) fluid is called dynamic pressure.
- "Values of high vacuum are generally expressed in terms of *Torr*." What is Torr?
  - The unit of pressure in mmHg is often called *Torr*, particularly used in vacuum applications. 760 mmHg = 760 Torr. 1 kPa = 1 Torr × 0.13332.
- What is a barometer? How does it work?

TABLE 1.2 Viscosity Conversions

Kinematic Viscosity (cSt)	Saybolt Viscosity		Redwood Viscosity (s)		Engler Viscosity (s)
	Universal (SSU)	Furol (SSF)	No. 1	No. 2	
1	31	–	29	–	1
2	32.6	–	30.2	–	1.1
3	36	–	32.9	–	1.2
4	39.2	–	35.5	–	1.3
5	42.4	–	38.6	5.28	1.37
6	45.6	–	41.8	5.51	1.43
7	46.8	–	43.1	5.6	1.48
8	52.1	–	46	6.03	1.64
9	55.4	–	48.6	6.34	1.74
10	58.8	–	51.3	6.6	1.83
20.52	100	–	85.6	10.12	3.02
42.95	200	–	170	18.9	5.92
108	500	52.3	423	46.2	14.6
215.8	1000	102.1	846	92.3	29.2
431.7	2000	204	1693	185	58.4
1078.8	5000	509	4230	461	146
1510.3	7000	712	5922	646	204
2157.6	10,000	1018	8461	922	292
3236.5	15,000	1526	12,692	–	438
4315.3	20,000	2035	16,923	–	584

- Barometer is used for measuring the pressure of the air, due to the weight of the column of air above it. As the Earth's atmosphere gets thinner with increasing height, it follows that at higher altitudes above sea level, the weight of air will decrease. In other words, pressure decreases.
- Barometer consists of a long glass tube closed at one end and filled with a liquid, usually mercury, and inverted upside down in an open container having the same liquid, taking care that no air is entrapped in the tube. Pressure of air acting on the liquid in the container forces the liquid up the tube. Height of the liquid column in the tube above the liquid surface in the container indicates atmospheric pressure at the place. The space above the liquid column is not absolute vacuum as it contains vapors of the liquid used as barometric fluid, exerting its vapor pressure, which is a function of temperature.
- Figure 1.8 shows a barometer setup.
- At sea level, height of mercury column will be 760 mm. Mercury is normally used as the barometric liquid as its vapor pressure is very low.

• If water is used as a barometric fluid instead of mercury, what will be height of the water column at sea level?

- The height of the water column above sea level, corrected to vapor pressure exerted above the column, is 10.33 m.

• What is the working principle of an aneroid barometer? What are its plus points?

- An aneroid is a flexible metal bellows that has been tightly sealed after having some of the air removed. Higher atmospheric pressures squeeze the metal bellows while lower pressures allow it to expand.
- Aneroid barometer has certain advantages over a mercury barometer because it is much smaller and compact and can also record a week's worth of data.

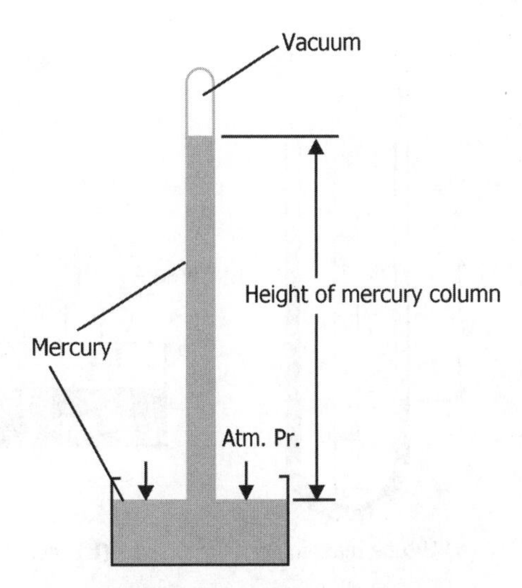

FIGURE 1.8 Mercury barometer.

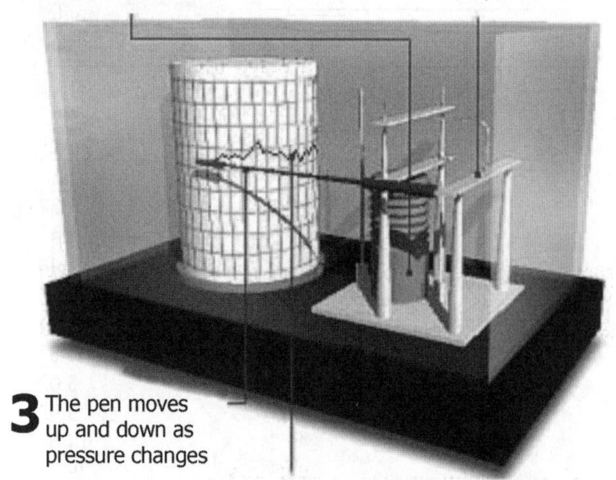

FIGURE 1.9 Aneroid barograph.

- Aneroid barometers are the heart of altimeters used in modern aviation.
- What is an aneroid barograph? Illustrate by means of a diagram.
  - A pen at the end of a lever attached to the aneroid moves up and down according to pressure changes and records the pressure on the graph paper wrapped around the cylinder, which slowly rotates.
  - Figure 1.9 illustrates an aneroid barograph.
- Name and illustrate different types of manometers.
  - U-tube, inclined, and two-liquid types of manometers are in common use. These are illustrated in Figure 1.10.
  - Closed U-tube manometers are used with mercury as manometric fluid, to directly measure absolute pressure P of a fluid, provided that the space between the closed end and mercury is perfect vacuum.
- Name some fluids used as manometric liquids.
  - Low freezing point fluids are used for cold weather climates and outdoor use. High-temperature fluids are used for higher temperature environments. Indicating fluid specific gravity (density), options range from 0.827 Red Oil to 13.54 Mercury. Higher specific gravity liquids can be used to increase the measurement range of any glass tube manometer.
  - Mercury is used for larger ΔP measurements and oils for small pressure differences.
- What is a two-liquid manometer? What are its applications?
  - A two-liquid manometer, a highly sensitive device, is used for measuring small pressure differences. Figure 1.11 shows a two-liquid manometer.
- What is a differential U-tube manometer?
  - A differential U-tube manometer is used for measuring pressure difference between two points of a closed system such as a head-type flow meter.
- How does Bourdon tube (pressure gauge) work? Does it indicate absolute pressure?
  - Bourdon tubes are hollow, cross-sectional beryllium, copper, or steel tubes, shaped into a three-quarter circle.
  - It is a thin-walled tube with wall thicknesses in the range of 0.25–1.25 mm (0.01–0.05 in.) that is flattened diametrically on opposite sides to produce a cross-sectional area elliptical in shape, having two long flat sides and two short round sides. The tube is bent lengthwise into an arc of a circle of 270–300°.

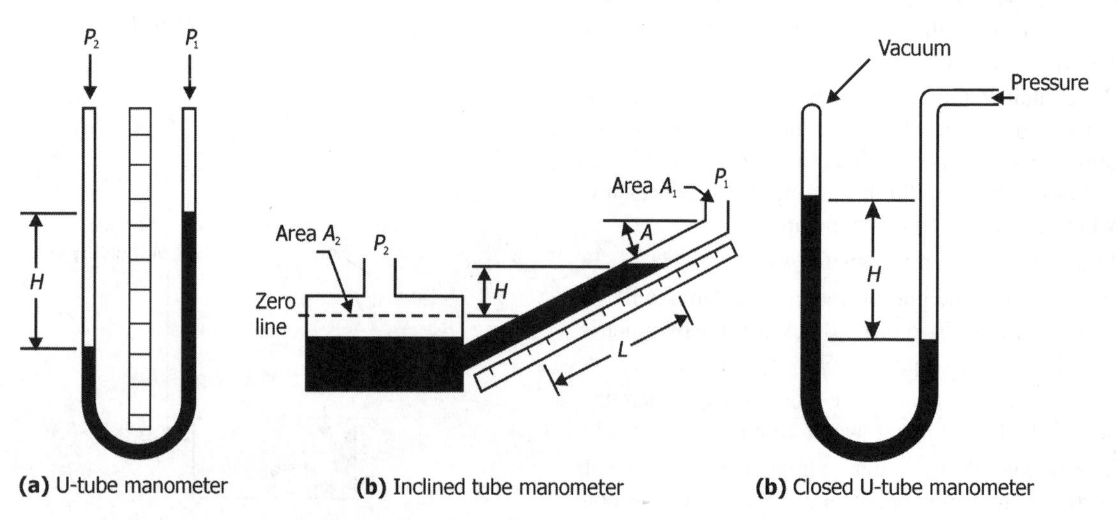

FIGURE 1.10 Different types of manometers.

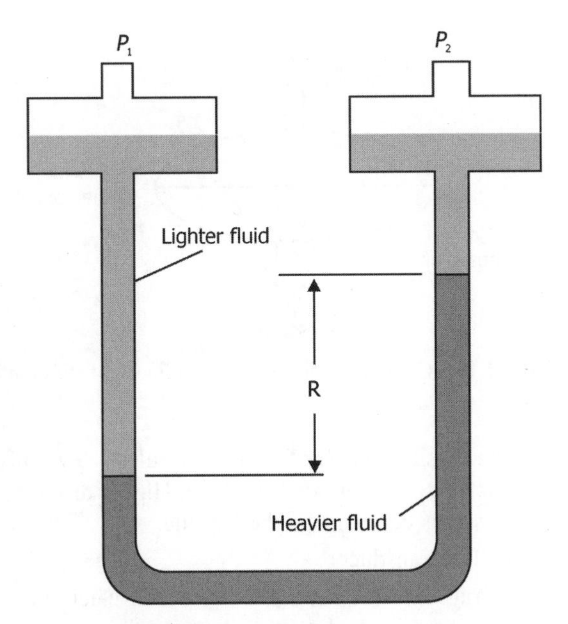

FIGURE 1.11 Two-liquid manometer.

- Pressure applied to the inside of the tube causes change of the flat sections and tends to restore its original round cross section.
- This change in cross section causes the tube to straighten slightly. Since the tube is permanently fastened at one end, the tip of the tube traces a curve that is the result of the change in angular position with respect to the center.
- As the gauge pressure increases the tube will tend to uncoil, while a reduced gauge pressure will cause the tube to coil more tightly. This motion is transferred through a linkage to a gear train connected to an indicating needle. Movement of the tip of the tube can then be used to position a pointer on a circular calibrated dial or to develop an equivalent electrical signal to indicate the value of the applied internal pressure.
- Typical shapes of tubes are spiral or helical.
- This instrument indicates gauge pressure.
- Figures 1.12 and 1.13 illustrate spiral and helical types of pressure gauges.
- The helical configuration is more sensitive than the circular Bourdon tube.
- Due to their robust construction, Bourdon tubes are often used in harsh environments and high pressures, but can also be used for very low pressures.

- Name other types of pressure sensing devices.
  - Bellows.
  - Diaphragms.
  - Capsules.
- Describe a bellows-type pressure sensing device? What are its characteristic features?
  - Metallic bellows pressure sensing detectors are developed to provide a device that is extremely sensitive to low pressures.
  - The metallic bellows is most accurate when measuring pressures from 3.4 to 39 kPa gauge (0.5–75 psig). However, when used in conjunction with a heavy range spring, some bellows can be used to measure pressures of over 6890 kPa gauge (1000 psig).
  - Figure 1.14 shows a basic metallic bellows pressure sensing element.
  - The bellows is a one-piece, collapsible, seamless metallic unit that has deep folds formed from very

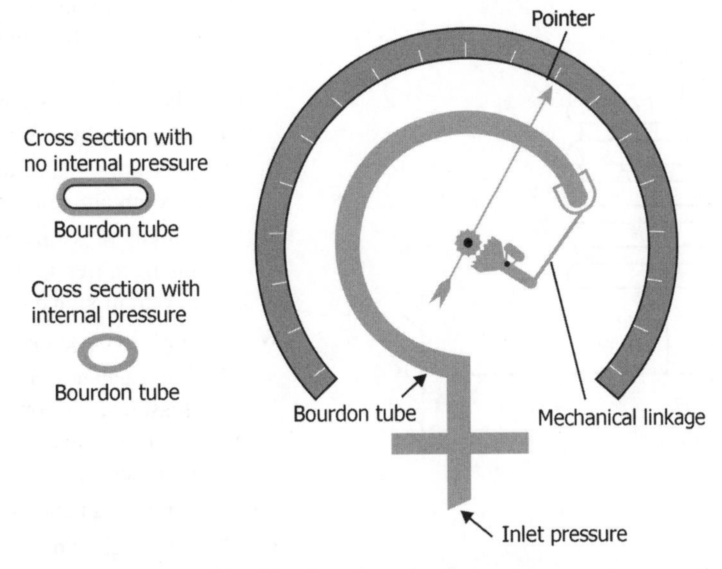

FIGURE 1.12 Bourdon tube pressure gauge.

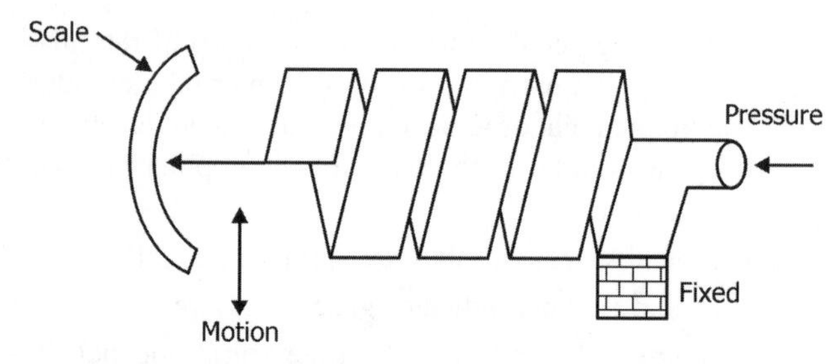

FIGURE 1.13 Helical Bourdon tube.

thin-walled tubing. System pressure is applied to the internal volume of the bellows.

  - As the inlet pressure to the instrument varies, the bellows will expand or contract. The moving end of the bellows is connected to a mechanical linkage assembly.
  - As the bellows and linkage assembly moves, either an electrical signal is generated or a direct pressure indication is provided.
  - The flexibility of a metallic bellows is similar in character to that of a helical, coiled compression spring.

- How does a capsule device for pressure measurement work? What are its characteristics? Illustrate.
  - The capsule consists of two circular shaped, convoluted membranes (usually stainless steel) sealed tight around the circumference.
  - Pressure is applied to the inside of the capsule.
  - Figure 1.15 illustrates the principle.
  - Unlike a diaphragm, which is fixed on one side, a capsule expands like a balloon in both directions.

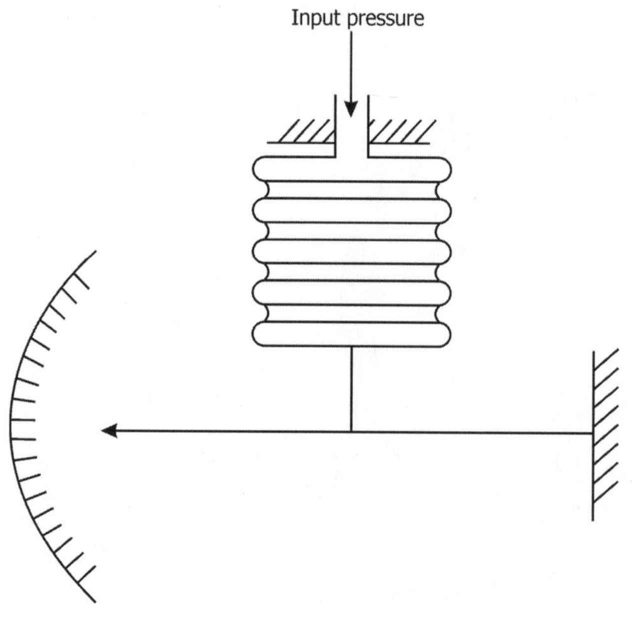

FIGURE 1.14 Basic metallic bellows.

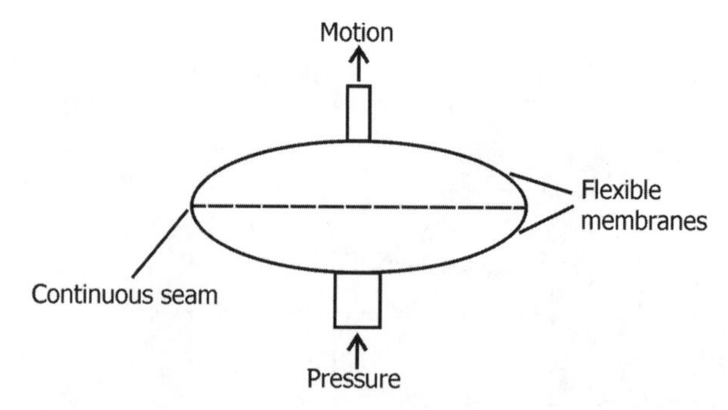

FIGURE 1.15 Capsule device for measurement of differential pressure.

  - The sensitivity range for most capsules is low, only a few kPa of differential pressure. Higher differential pressures can damage the capsule.

- What is a transducer?
  - A transducer is a device, electrical, electronic, or electromechanical, that converts one type of energy to another for various purposes including measurements or information transfer.
  - In a broader sense, a transducer is sometimes defined as any device that converts a signal from one form to another.
  - A pressure transducer converts pressure into an analog electrical signal. Although there are various types of pressure transducers, one of the most common types is the strain gauge-based transducer.
  - The conversion of pressure into electrical signal is achieved by the physical deformation of strain gauges that are bonded into the plastic diaphragm of the pressure transducer and wired into a Wheatstone bridge configuration.
  - The pressure applied to the pressure transducer produces a deflection of the diaphragm, which introduces strain to the gauges. The strain produces an electrical resistance change proportional to the pressure. The electrical signal can get altered even if the transducer is touched by hand.
  - Most pressure signals transmitted from the field to the control room are generated by pressure transducers.
  - Differential pressure indicators give differential readings from two transducers and generate a milliampere output signal.

- What is a McLeod gauge?
  - A McLeod gauge is an instrument to measure very low pressures, as low as 10^{-7} Torr. It isolates a sample of gas and compresses it in a modified mercury manometer until the pressure is few mmHg.

- What is a Pirani gauge? Where is it used?
  - A Pirani gauge consists of a metal wire open to the pressure to be measured. The wire is heated by the

current flowing through it and cooled by the gas surrounding it. If the gas pressure is reduced, the cooling effect will decrease and hence the temperature of the wire will increase. The resistance of the wire is a function of its temperature. By measuring the voltage across the wire and the current flowing through it, the resistance and hence the gas pressure can be determined.

- Name other methods for measuring low pressures.
  - At pressures below 1 Torr, forces exerted by gaseous molecules are not sufficient to measure pressures directly with an absolute pressure gauge. Pressure must be inferred by assessing a pressure-dependent physical property of the gas, such as thermal conductivity, ionizability, or viscosity.
  - Ionization gauges (IGs) are one class of instruments for indirect measurement of gas density and pressure. IGs generally work by ionizing neutral gas molecules and then determining their number by measuring an electric current.
    - One type of IG can be classified as emitting cathode IGs (ECGs) (also known as hot cathode IGs). These are characterized by a heated filament that serves as a source of electrons.
    - Another type of ionization gauges can be classified as crossed electromagnetic field (EMF) IGs (also called cold cathode IGs). This IG type is characterized by the generation of a discharge between cathode and anode that is maintained with a magnetic field.

1.5.1 Liquid Level

- What is the importance of control of liquid level in process equipment?
  - Liquid level in a vessel should be maintained above the exit nozzle as otherwise outflow from the vessel will stop, starving downstream equipment such as pumps, process heaters, and numerous other equipment, creating damaging situations for downstream equipment. Loss of coolant is one of the worst scenarios giving rise to catastrophic accidents in nuclear power plants.
  - If liquid level increases beyond certain limits, overflow and spillages might occur, causing hazardous conditions such as toxic exposures or fires and explosions. Spillages have been the causes for major accidents in industry in the past and continue to be so at present times.
  - Also, high liquid levels might increase entrainment in a vapor–liquid separator and can lead to liquid flowing into vapor lines leading to operational upsets.
  - Level can influence the performance of a process, the most common example being a liquid-phase chemical reactor.
  - Reliability of level control systems should be of high order.
- What are the types of liquid level monitoring techniques used in industry?
  - There are two techniques for liquid level monitoring sensors, point and continuous level detection.
  - Point level detection is mainly conducted by liquid level switches. These liquid level switches consist of mechanical switches or liquid level gauges that can detect the presence of only a liquid and not the level.
  - The main application for point level detection consists of controlling the maximum or the minimum (or both) liquid level allowed in a container. Point level detection is less expensive because it uses mechanical switches or simple gauges.
  - The main drawbacks of point level detection systems are the need for an empty tank to add low cutoff sensors, the need to add mounting fixtures, and the problems associated with contact sensing.
  - Continuous level detection is used when it is necessary to know the specific level at all times. The need to know the actual level of the liquid is necessary when transferring liquids, mixing liquids, and determining production levels, and for many other company- and industry-specific reasons.
  - Monitoring of the liquid level is performed by continuous liquid level gauges. Continuous monitoring is typically achieved through entering existing openings at the top of the tank and employing both contact and noncontact sensors.
  - Continuous liquid level monitoring can also be used as point detection.
  - The main criterion in using point detection or continuous monitoring is the users' need to know the exact level of the liquid or the need for knowing when the level is high or low.
- How are level sensors located?
  - Level sensors can be located in the vessel holding the liquid or in an external leg that acts as a manometer.
  - When in the vessel, float and displacement sensors are usually placed in a chamber, free from flow disturbances, which reduces the effects of flows in the vessel.
- What are the methods used for liquid level measurement?
  - *Float*: The float material that is lighter than the fluid follows the movement of the liquid level. The position of the float, perhaps attached to a rod, can be

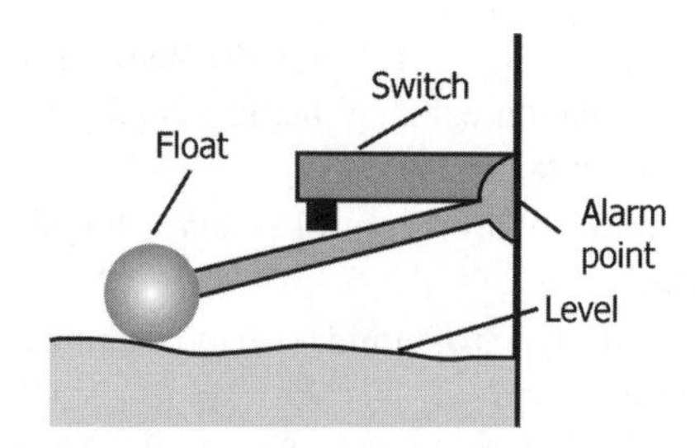

FIGURE 1.16 Float-type level measurement.

determined to measure the level. It cannot be used with sticky fluids that coat the float.

- Figure 1.16 illustrates float-type level measurement.
- *Displacement*. By Archimedes' principle, a body immersed in a liquid is buoyed by a force equal to the weight of the liquid displaced by the body. Thus, a body that is denser than the liquid can be placed in the vessel, and the amount of liquid displaced by the body, measured by the weight of the body when in the liquid, can be used to determine the level.
- Mechanical floats and displacers measure on the simple principle that the buoyant force on an immersed object equals the weight of fluid displaced. Therefore, when liquid level rises, the weight of a displacer decreases. A low-density float is attached to a horizontal rod that is mounted to a tank wall and linked to a switch. As the level rises and falls, the switch opens and closes.
- Mechanical floats and displacers are inexpensive, easy to install, and work well in a variety of fluid densities.
- However, the float or displacer is calibrated to the density of the liquid it measures, so if the density is changed, the float must be recalibrated.
- Mechanical floats operate at temperatures up to 650°C and mechanical displacers can handle temperatures up to 370°C.
- Another limiting factor for mechanical technology is buildup of material on the float or displacer. Buildup causes changes in weight displacement and the displacer will require cleaning or recalibration.
- Other disadvantages include suitability only for non-freezing liquids, difficulty of calibration, and susceptibility to changes in specific gravity.
- Mechanical floats and displacers can work well for backup level measurement.
- *Differential Pressure*: The difference in pressures between two points in a vessel depends on the fluids between these two points.
- If the difference in densities between the fluids is significant, which is certainly true for a vapor and liquid and can be true for two different liquids, the difference in pressure can be used to determine the interface level between the fluids.
- Usually, a seal liquid is used in the two connecting pipes (legs) to prevent plugging at the sensing points.
- Differential pressure devices are a common means of continuous level measurement in industry because of their ease of use. The high-pressure side of a differential pressure instrument is connected to the vapor space at the top of the vessel. The measured pressure differential is the pressure of the liquid column in the tank. This provides a true level reading *if the fluid density is constant*. If not, changes in liquid composition or temperature will change the specific gravity and create a false reading.
- Figure 1.17 shows a typical open tank level measurement installation using a pressure capsule level transmitter (LT).
- Any change in density, such as that caused by a change in temperature, necessitates recalibration.
- Differential pressure devices offer the advantage of easy installation in liquid applications that are relatively clean and free of suspended solids.
- Drawbacks include the requirement of seal fluid in pressurized vessels, difficulty in calibration, and technical difficulties related to density and temperature.
- *Electrical Methods*: Two electrical characteristics of fluids, namely, conductivity and dielectric constant, are frequently used to distinguish between two phases for level measurement purposes.
  - ➢ *Electromechanical devices* have a motor operated paddle that is submerged into a vessel. The paddle stops rotating when its sensor is covered by liquid or solid material. The stalled motor activates a switch, signaling the level of material to a control

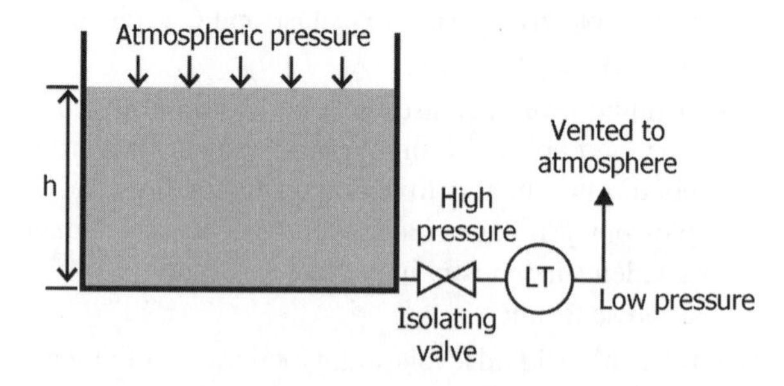

FIGURE 1.17 Liquid level measurement in an open tank.

device. The higher the liquid density, the smaller the paddle required.

- The cost-effective and low-maintenance electromechanical designs are well suited for solids such as plastic pellets, carbon black, fertilizers, Styrofoam, and rubber chips and beads.
- The paddle switch can handle bulk densities as low as 35 g/L and the technology is completely independent of dielectric properties of the material.

- *Capacitance*: A capacitance probe can be immersed in the liquid in the tank and the capacitance between the probe and the vessel wall depends on the level. By measuring the capacitance of the liquid, the level in the tank can be determined. Capacitance can be affected by density variations.
  - Capacitance technology is widely used in many industries and can handle a wide range of applications from simple storage of acids in small tanks to high-temperature and high-pressure characteristics of fine chemicals in a turbulent process reactor. This technology also produces highly accurate and repeatable results.
  - Because capacitance is a contacting technology, chemical compatibility with the device must be taken into account, as well as the potential for buildup problems.
  - Also, the chemistry of liquids must remain constant or homogeneous in nonconductive or insulating hydrocarbon fluids such as oil and methane.
  - Any change due to temperature or chemical composition causes the dielectric property of the material to change, resulting in errors and the requirement to recalibrate.
  - Capacitance technology requires only one opening in a vessel, making it easy to install, and has no moving parts that may wear out over time.
- *Ultrasonic Methods*: Ultrasonic technology uses a piezoelectric crystal stored inside a transducer to convert an electrical signal into sound energy. The sound energy is fired toward the material and is reflected back to the transducer (Figure 1.18).
  - The transducer then acts as a receiving device and converts the sonic energy back into an electrical signal.
  - An electronic signal processor analyzes the return echo and calculates the distance between the transducer and the target. The time lapse between the sound burst and the return echo is proportional to the distance between the transducer and the material in a vessel.

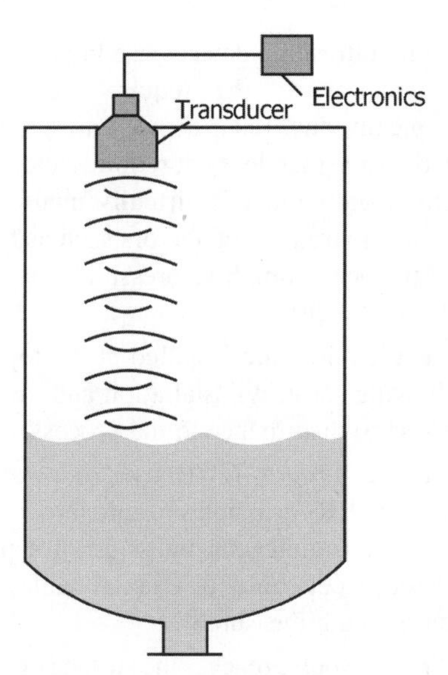

FIGURE 1.18 Ultrasonic level measurement.

- *Thermal methods* are based on the difference in thermal characteristics between the fluids, such as temperature or thermal conductivity.
- *Radar*: Radar devices transmit an electromagnetic wave traveling at the speed of light toward a material. Once the electromagnetic wave comes into contact with the material, it is reflected back to the source.
  - The total transit time to and from the target is calculated and is directly related to the distance.
  - There are two techniques of transmitting radar, namely, pulse- or frequency-modulated continuous wave (FMCW).
  - Pulse radar is similar to ultrasonic noncontact method in that fixed-frequency pulses are transmitted to a material and then reflected back to the source where the time of flight is calculated.
  - FMCW radar devices continuously transmit a range of frequencies, known as a frequency sweep. The receiver continuously monitors the received frequencies and the difference between the transmitter and receiver frequency is directly proportional to the distance to the target.
  - High-frequency radar is suited for low-dielectric media due to the narrower, more focused signal, which improves the reflection from material.
  - Low-frequency radar has antenna deposit resistance, a longer wavelength, and physical wave reflection properties.

        - While ultrasonic devices use high-frequency mechanical waves that require a carrier medium (typically air), radar devices use an electromagnetic wave that does not require a carrier medium. This means radar is virtually unaffected by extreme environmental factors such as temperature, turbulence, humidity, pressure, vacuum, vapor, steam, or dust.
        - Radar devices are installed at the top of vessels, allowing for easy installation and, in most situations, no disturbance to the process.
    - *Guided wave radar (GWR)* uses *time domain reflectometry* (TDR) principles to measure level by guiding an electromagnetic pulse down a probe (solid steel rod, steel cable, or coaxial cable) toward the material being measured.
        - When the pulse reaches the surface of material, the change in dielectric value between air and the material causes a portion of the pulse to reflect back toward the transmitter.
        - The signal strength of the reflected pulse is proportional to the dielectric constant of the material. The higher the dielectric value of the material, the stronger the signal will be.
        - Once the return pulse reaches the transmitter, the device uses time-of-flight principles to calculate the distance to the material surface.
        - Guided wave radar is unaffected by vapor, density, foam, dielectric, temperature, and pressure, and thus tends to work well for short- and medium-range measurements.
        - Guided wave radar is not suited for agitated vessels or vessels containing abrasive materials.
        - The technology is commonly used to measure liquid, plastic pellets, slurries, and displacer replacement.
    - *Other methods* include radiation, neutron backscatter technique, and X-ray methods.
- What are the advantages and disadvantages of radar technology for level measurement?
    - Because radar technology is noncontacting, it is ideal for applications where deposits or encrustation may be a problem.
    - The other advantages include use of a microprocessor to process the signal providing numerous monitoring, control, communication, setup, and diagnostic capabilities. It can detect the level under a layer of light dust or airy foam. Radar is effective in solids applications with extreme levels of dust such as plastic powders, washing powders, fly ash, and lime powder.
        - If the dust particle size increases, or if the foam or dust gets thick, the sensors will no longer detect the liquid level, and instead, the level of the dust or foam will be measured.
        - Radar is also very effective in applications with turbulence or extreme levels of condensation found in chemical process applications.
        - The first disadvantage is the high price of the radar technology.
        - The second disadvantage is the need to know the empty level, the full level, and application parameters in order to program the sensor.
        - The third disadvantage is the interference of echoes from multiple reflections, the tank walls, and improper orientation of the radar sensor.
        - Another disadvantage is the loss of accuracy due to gas layering in the tank that varies the speed of the radar signal.
        - In some applications, the antenna can become contaminated, which can hinder the emission and reception of microwaves.
        - It requires application-specific antennas and mounting conditions.
- What are bubbler sensors for level measurement? How do they work? What are their applications?
    - The air bubbler system works by introducing air into a pipe. The pressure created in the pipe returns to the sensor where it can be displayed visually or converted into an electrical signal.
    - The pressure in the pipe is equal to the pressure exerted by the water and effectively measures the same as that by the hydrostatic pressure sensors. If the tank is not vented, then another line needs to be added so that the bubbler does not build pressure inside the tank.
    - The pressure in the pipe has to be considered in the level calculation depending on the air pressure above the liquid, the size and shape of the tank, and the distance from the bottom to the pipe.
    - Air bubbler systems are a good choice for open tanks at atmospheric pressure and can be built so that high-pressure air is routed through a bypass valve to dislodge solids that may clog the bubble tube.
    - The technique is inherently *self-cleaning*. It is highly recommended for liquid level measurement applications where ultrasonic, float, or microwave techniques have proved undependable.
    - Air bubbler systems contain no moving parts, making them suitable for measuring the level of sewage, drainage water, night soil, or water with large quantities of suspended solids.

TABLE 1.3 Modern Level Measuring Methods and Their Applications

Method	Applications
Electromechanical	Plastic pellets, Styrofoam beads and chips, carbon black, fertilizers
Capacitance	Acids, alkalis, styrene and other aromatics, adhesives, and so on
Ultrasonic	Chemical storage tanks, wastewater effluents, plastic pellets
Radar	Bulk storage vessels for chemicals, agitated and reaction process vessels, sulfur storage
Guided wave radar	Liquids and slurries, plastic pellets, displacer replacement

- The only part of the sensor that contacts the liquid is a bubble tube, which is chemically compatible with the material level to be measured. Since the point of measurement has no electrical components, the technique is a good choice for electrically classified hazardous areas.
- The control portion of the system can be safely located, with the pneumatic plumbing isolating the hazardous areas from the safe areas.

• Give an overview of modern level measurement methods with suitable examples of applications.

- Table 1.3 gives a summary of modern level measuring methods.

• Illustrate the causes of discrepancy between external level and internal level (inside an equipment).

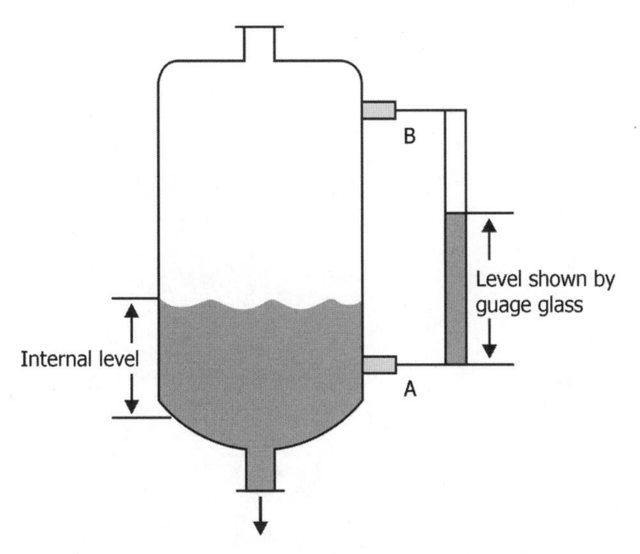

FIGURE 1.19 Liquid level measurement.

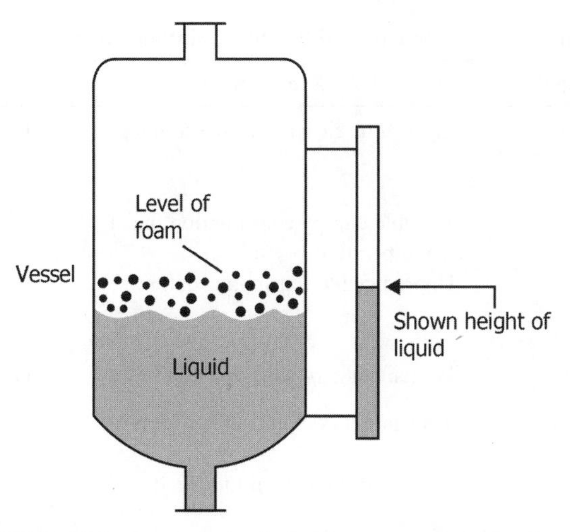

FIGURE 1.20 Inaccuracies in level measurements for foaming liquids.

- Figure 1.19 shows the discrepancy between internal and external levels for a vessel.
- Ambient heat loss from external heat loss.
- The density of the liquid in the glass tube (lower temperature) is more than the density of the liquid inside the process vessel (higher temperature). Exception is for the refrigeration process where refrigerated liquid inside the vessel is at lower temperature than liquid in the gauge glass. The density differences are due to ambient heat losses or heat gain from the gauge glass.
- Rule of thumb for hydrocarbon liquids: For every 40°C decrease in liquid temperature in the gauge glass, density of the liquid increases by 5%.
- *Example*:

  Height of the liquid in the gauge glass: 1.2 m.

  Liquid temperature in the gauge glass: 15°C.

  Liquid temperature inside the vessel: 290°C.

  Level difference: ≈30 cm.
- Plugged taps give rise to false high level readings.
- Foaming liquids cause inaccurate level readings as illustrated in Figure 1.20.

• What are differential pressure (DP) cells? How are they used for liquid level measurement?

- DP cells measure pressure difference between two points and send a differential pressure reading to the control system.
- The control system converts the DP cell reading into a liquid level based on an assumed specific gravity inside the vessel. Levels based on DP cells can be in error for different reasons. Some reasons are given in the above question.

TABLE 1.4 Problems and Possible Solutions in Pressure and Level Measurements

Variable	Symptom	Problem Source	Possible Solution
Pressure	Zero shift, air leaks in signal lines	Excessive vibration from positive displacement equipment	Use independent transmitter, flexible connection lines. Use liquid-filled gauge
	Variable energy consumption under temperature control	Change in atmospheric pressure	Use absolute pressure transmitter
	Unpredictable transmitter output	Wet instrument air	Mount local dryer, use regulator with sump, slope air lines away from transmitter
	Permanent zero shift	Overpressure	Install pressure snubber for spikes
Level	Transmitter does not agree with level	Liquid gravity changes	Gravity compensate measurement or recalibrate
	Zero shift, high level indicated	Water in the process absorbed by glycol seal liquid	Use transmitter with integral remote seals
	Zero shift, low level indicated	Condensable gas above liquid	Heat trace vapor leg. Mount trans mitter above connections and slope vapor line away from transmitter
	Noisy measurements, high level indicated	Liquid boils at ambient temperature	Insulate liquid leg

- What is the optimum liquid level in a stirred tank reactor?
  - Liquid level = tank diameter. It should be noted that tank diameters are smaller at high pressures.
- What are the common problems involved in pressure and level measurements? Give suitable solutions.
  - Table 1.4 briefly gives the problems and solutions in pressure and level measurements.

2

FLUID FLOW

2.1	Flow Phenomena	21
2.2	Water Hammer	30
2.3	Compressible Fluids	32

2.1 FLOW PHENOMENA

- How is Reynolds number useful in interpreting flow conditions?
  - Reynolds number serves as a criterion to describe laminar, transitional, and turbulent flow conditions.
    - $N_{Re} > 4000$, flow will be turbulent with inertial forces dominating.
    - $N_{Re} < 2100$, flow will be laminar in nature with viscous forces dominating the flow.
- What are eddies? Where do you find these?
  - An eddy is the swirling of a fluid and the reverse current created when the fluid flows past an obstacle. The moving fluid creates a space devoid of downstream flowing fluid on the downstream side of the object. Fluid behind the obstacle flows into the void creating a swirl of fluid on each edge of the obstacle, followed by a short reverse flow of fluid behind the obstacle flowing upstream, toward the back of the obstacle.
  - Eddies are areas of still or reverse (upstream) moving fluid. They are formed, for example, in rivers when water comes up against an obstacle (most commonly rocks) and has to pass around it, creating a void in the flow directly downstream. They are also formed in any flowing fluid under turbulent flow conditions in a conduit such as a pipe.
  - Recirculating eddies are formed during flow through weirs or where the main flow runs up against a wall or obstruction and part of the flow moves back upstream. The flow is often erratic.
- What is turbulence? Explain.
  - Turbulence is irregular and seemingly random (chaotic) movement of flowing fluid elements in all directions, the net flow being unidirectional. Turbulence manifests itself at high flow velocities at Reynolds numbers above 10,000.
  - The origin of turbulence is rooted in the instability of shear flows. It is also derived from buoyancy-driven flows.
  - Turbulence is rotational and three-dimensional motion.
  - Large-scale turbulent motion is roughly independent of viscosity. In other words, at high Reynolds numbers, viscous forces, which contribute to flow stability, are insignificant compared to inertial forces that contribute to flow instability.
  - Turbulence is associated with high levels of vorticity fluctuation. Smaller scales are generated by the vortex stretching mechanism.
  - Turbulence is highly dissipative and requires a source of energy to maintain it.
  - Rapid mixing involved in turbulence increases momentum, heat, and mass transfer processes.
  - Turbulence can be generated by contact between two layers of fluid moving at different velocities or by a flowing stream in contact with a solid boundary.
  - Turbulence can arise when a jet of fluid from an orifice flows into a mass of fluid.
  - In turbulent flow at a given place and time, large eddies continually form and break up into smaller eddies and finally disappear. Eddies can be as small

Fluid Mechanics, Heat Transfer, and Mass Transfer Chemical Engineering Practice, By K. S. N. Raju

as 0.1–1 mm and can be as large as the smallest dimension of the turbulent stream.

- Flow inside an eddy is laminar because of its large size.
- In turbulent flow, velocity is fluctuating in all directions.

• What is Prandtl mixing length?

- Prandtl, in his work on turbulent shear stresses, developed the concept of *mixing length*, which is defined by assuming that eddies move in a fluid in a manner similar to the movement of molecules in a gas. The eddies move a distance, called *mixing length* (*L*), before they lose their identity.
- Actually, the moving eddy or *lump* of fluid will *gradually* lose its identity. However, in the definition of Prandtl mixing length *L*, this small lump of fluid is assumed to retain its identity while traveling the entire length *L* and then lose its identity or get absorbed in the host region.

• What is eddy viscosity?

- Eddy viscosity refers to the internal friction generated within the fluid as laminar flow becomes irregular and turbulent as it passes over irregular solid objects. Turbulent transfer of momentum by eddies gives rise to the internal fluid friction, in a manner analogous to the action of molecular viscosity in laminar flow, but taking place on a much larger scale.
- Eddies are very efficient at transferring high-energy, high-momentum fluid from the higher velocity parts of the flow to the lower velocity parts of the boundary layer. This energy transfer between the boundary layer and the main flow results in an increase in apparent viscosity. The existence of eddies effectively acts as a resistance to shear.
- It is important to remember that eddy viscosity is neither a true viscosity nor a constant quantity, but that it is a useful concept helpful in solving problems related to sediment movement in turbulent flow.
- In turbulent flows, there are local velocity gradients that cause local stresses that work against the mean velocity gradient. This results in a loss of energy from the flow.
- Newton's equation for laminar flow is

$$\tau = \mu(\mathrm{d}U/\mathrm{d}y), \tag{2.1}$$

where the shear stress is equal to the rate of change of velocity times a proportionality factor, which is viscosity. For turbulent flows, the equation is written as

$$\tau = (\mu + \eta)(\mathrm{d}U/\mathrm{d}y), \tag{2.2}$$

where η is the eddy viscosity. The value of η is variable for different flows, depending on the size and velocity of the eddies, and for air and water it is much larger than μ.

• What is meant by laminar and turbulent flows? Explain with illustrations.

- In laminar flow, the fluid moves in layers called laminas, in streamlines. Laminar flow need not be in a straight line. For laminar flow, the flow follows the curved surface of an airfoil smoothly, in layers. The closer the fluid layers are to the airfoil surface, the slower they move. Moreover, the fluid layers slide over one another without fluid being exchanged between the layers, that is, without lateral mixing of the layers.
- In turbulent flow, secondary random motions are superimposed on the principal flow and there is an exchange of fluid from one adjacent sector to another. Also, there is an exchange of momentum such that slow moving fluid particles speed up and fast moving particles give up their momentum to the slower moving particles and slow down themselves. Flow moves instantaneously in many directions, but mean flow is in the downstream direction.
- Figure 2.1 illustrates change over from laminar to turbulent flow when an obstruction comes in its path.

• Show the velocity profiles in a pipe for laminar and turbulent flows by means of diagrams.

- *Laminar Flow*:

$$v = 2V[1-(r^2/R^2)], \tag{2.3}$$

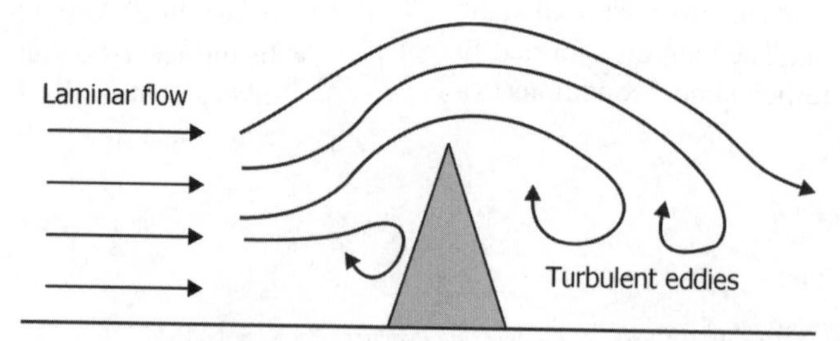

FIGURE 2.1 Changeover of laminar flow into turbulent eddies.

where v is the velocity at a pipe radius of r, V is the average velocity (Q/A), which is the equation for a parabola, and R is the radius of the pipe.

- Average velocity for laminar flow is 0.5 of maximum velocity in the centerline. The velocity profile is parabolic, the gradient of which depends on the viscosity of the fluid for a set flow rate.
- Average velocity for turbulent flow depends on N_{Re}. For N_{Re} of 5000, it is equal to 0.77 and for N_{Re} of 3×10^6 it is equal to 0.87 of maximum velocity in the centerline (logarithmic profile).
- *Turbulent Flow*: For rough pipes in turbulent flow, velocity profile is much more flattened than in the case of laminar flow and is given by

$$u_+ = 2.5 \ln(y/\varepsilon) + 8.5 \quad \text{for} \quad y_+ > 30, \qquad (2.4)$$

where u_+ is v/u_* (dimensionless), $u_* = \sqrt{(\tau_w/\rho)}$, which is called *friction velocity*, and τ_w is the wall stress given by $f\rho V^2/2 = \Delta P/4L$ for a circular pipe. y_+ is the dimensionless distance from the wall ($=yu_*\rho/\mu$).

$$v_{av} = 0.5 v_{max}. \qquad (2.5)$$

- As the liquid enters the pipe, its profile will be more *blunt* due to a smaller difference in velocity between its outer layers and those toward the center.
- This flow pattern, and the resulting friction, is due to the rather complex interactions of the forces of adhesion and cohesion and the momentum of the moving liquid.
- Figure 2.2 illustrates the velocity profiles for laminar and turbulent flows.

- What types of velocity profiles develop for liquid metal flow in pipes?
  - Velocity profiles for liquid metal flow will be flatter than those for turbulent flow and velocities are practically constant along the radius of the pipe from center to the wall.
- "Flow far from the surface of a solid object is inviscid." *True/False?*
  - *True*. Effects of viscosity are manifest only in a thin layer near the surface where steep velocity gradients occur.
- "Velocity profile of Newtonian fluids in laminar flow is flatter than that in turbulent flow." *True/False?*
  - *False*.
- "Maximum velocity of Newtonian fluids in laminar flow inside a circular pipe is twice the average velocity." *True/False?*
  - *True*.
- What criteria can be adopted for pressure drop calculations in laminar to turbulent flow conditions in pipeline flow?
  - In transition flow range, frictional losses should be calculated based on both laminar and turbulent conditions and the highest resulting loss should be used in subsequent system calculations.
- What is potential flow?
  - Flow behavior strongly depends on the influence of solid boundaries.
  - If the influence of solid boundary is negligible or small, shear stress may be negligible and the fluid behavior approaches zero viscosity, that is, inviscid.
  - This type of flow is called potential flow.
- Why the discharge ends of the pipes are elevated compared to the test pipes in a pipe friction experiment?
  - This arrangement ensures that the pipe is full, water occupying the entire cross section.
- What is superficial velocity?
  - Superficial velocity is defined as volumetric flow rate divided by cross-sectional area for flow.
- What is mass velocity?
  - Mass velocity, G, is defined as mass flow rate divided by cross-sectional area for flow, $G = V\rho$.
- How and why a Reynolds number experiment is carried out?
  - A Reynolds number experiment (Figure 2.3) is carried out in pipe flows with different diameter pipes, smooth and rough. Flow is gradually varied, visually observing flow streamlines by injecting a dye or ink

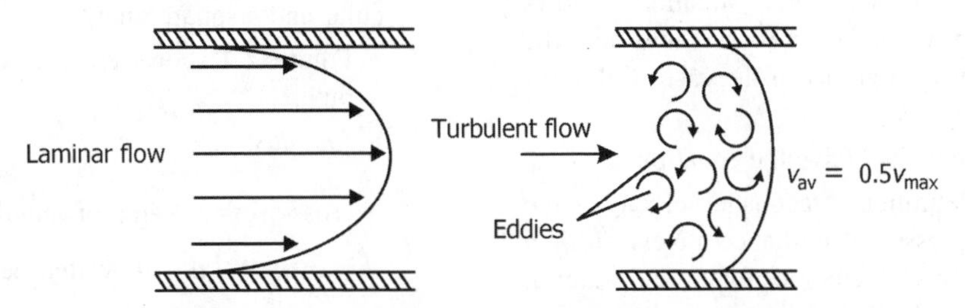

FIGURE 2.2 Velocity profiles for laminar and turbulent flows.

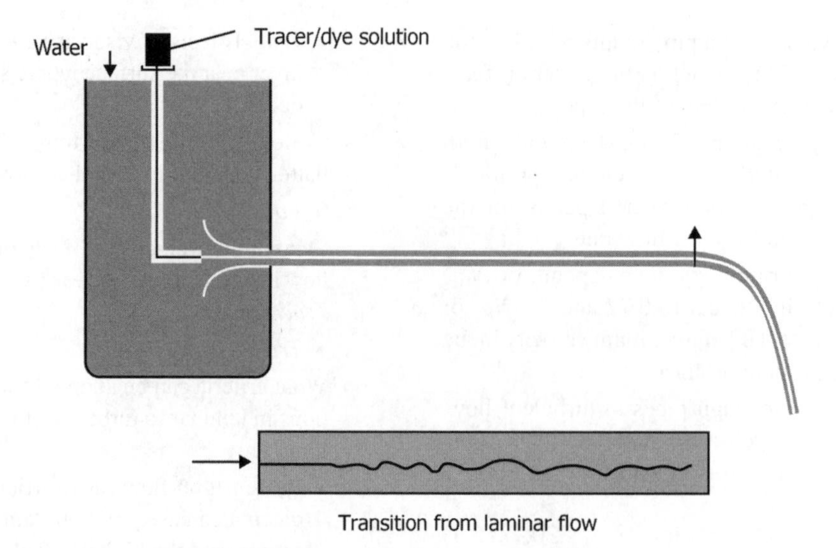

FIGURE 2.3 Reynolds number demonstration experiment.

or other tracer near the inlet at the center of the pipe by means of a hypodermic needle. Flow rates are measured by weighing liquid, noting reading of a stopwatch. For each flow rate, Reynolds number is calculated.

- The onset of turbulent flow is identified and Reynolds number is determined at the onset. Observations are made repeating the experiment for all the pipes. Transition Reynolds numbers are calculated. Results show different onsets based on diameters and roughness.

- What is meant by fully developed flow? What is transition length?
  - When a fluid is flowing through a pipe, obstructions and cross-section and direction changes can result in variations in flow velocities and flow profiles. The laminar or turbulent flow criteria in terms of N_{Re} can be in error. In fully developed flow, the distortions will be absent.
  - The length of obstructionless pipe to restore fully developed flow is called the transition length.
- What is Stokes flow?
  - Stokes flow is a type of flow where inertial forces are small compared to viscous forces. The Reynolds number is low. This is a typical situation in flows where the fluid velocities are very low and the viscosities are very high, as in the case of flow of viscous polymers.
- What is acoustic velocity? Give the equation.
  - The technical definition of acoustic velocity is the rate at which a pressure disturbance travels within a fluid. Typical acoustic velocities for air and water at 20°C are 343.1 and 1481 m/s, respectively.
  - The equation for acoustic velocity is

$$c = (k_\beta/\rho)^{0.5}, \tag{2.6}$$

where c is the acoustic velocity (m/s), k_β is the bulk modulus ($\times 10^9$ N/m^2), and ρ is the density (kg/m^3).

  - This equation is valid for liquids, solids, and gases.
  - Sound travels faster through media with higher elasticity and/or lower density.
- What is equivalent diameter?
  - Equivalent diameter for flow through a conduit is defined as 4(cross-sectional area for flow through the conduit)/wetted perimeter. For a circular conduit,

$$\text{equivalent diameter} = 4(\pi D^2/4)/\pi D = D, \tag{2.7}$$

that is, the diameter of the pipe.

  - Equivalent diameter is used for N_{Re} calculations in noncircular conduits, annular flow such as flow through the annulus of a double pipe heat exchange, shell side flow in shell and tube heat exchanger, packed bed flows, and so on.
- Calculate the equivalent diameter for the annulus of a pipe. What are the equivalent diameters for a rectangular and a square duct?
  - Figure 2.4 shows annulus of a concentric double pipe.
  - *Annulus*:

$$\text{cross-sectional area of annulus} = (\pi D^2/4) - (\pi d^2/4) = \pi(D^2 - d^2)/4, \text{ wetted perimeter} = \pi(D+d), \tag{2.8}$$

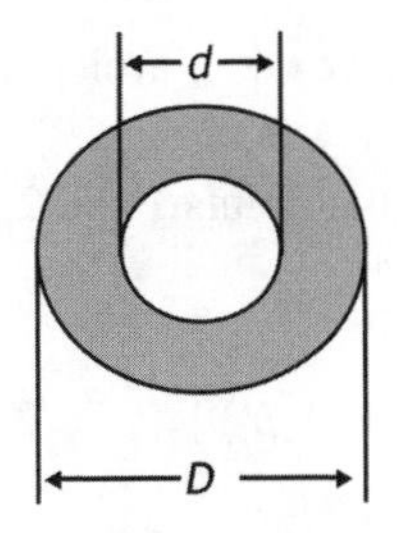

FIGURE 2.4 Equivalent diameter for annulus.

$$\text{equivalent diameter}, D_e = 4[\pi(D^2 - d^2)/4]/\pi(D + d) = D - d. \qquad (2.9)$$

Rectangular duct:

$$D_e = 4LW/2(L + W), \qquad (2.10)$$

where L is the length and W is the width of the duct.

Square duct:

$$D_e = L. \qquad (2.11)$$

- Give the equation for hydraulic radius of a packed bed.

$$r_H = (\text{volume of voids})/(\text{volume of bed})/(\text{wetted surface})/(\text{volume of bed}) = \varepsilon/a, \qquad (2.12)$$

where a is $6(1-\varepsilon)/D_p = [\varepsilon/6(1-\varepsilon)]D_p$.

- How is Reynolds number defined for (i) noncircular conduits, (ii) packed beds, and (iii) mixing?
  - (i) $N_{Re} = D_e\rho/\mu$, (2.13)

    where D_e is the equivalent diameter of the noncircular flow conduit.

    (ii) $N_{Re,p} = D_p V'_\rho/[(1-\varepsilon)\mu]$ (Ergun's definition), (2.14)

    where D_p is the particle/packing diameter, V' is the superficial velocity, ε is the void fraction, and ρ and μ are fluid density and viscosity, respectively.

$$\text{Superficial velocity}, V' = \varepsilon v, \qquad (2.15)$$

    where v is the actual velocity.

    (iii) Impeller Reynolds number,

$$N_{Re} = D^2 N\rho/\mu, \qquad (2.16)$$

    where D is the impeller diameter and N is rpm.

- What is boundary layer? Illustrate.
  - Flow far from the surface of a solid object is inviscid and effects of viscosity are manifest only in a thin layer near the surface where steep velocity gradients occur.
  - Boundary layer is where the fluid is influenced by friction with its boundaries. Flow is zero at the boundary and increases away from the boundary until it reaches the mean or maximum velocity of the flow. The zone of flow velocity increase is called boundary layer.
  - In other words, the thin layer where velocity decreases from the inviscid potential flow velocity to zero at solid surfaces is called *boundary layer*.
  - Development of boundary layer is illustrated in Figure 2.5.
- State Bernoulli's principle.
  - Bernoulli's principle states that in an ideal fluid with no work being performed on the fluid, an increase in velocity occurs simultaneously with decrease in pressure or a change in the gravitational potential energy of the fluid.
  - This principle is a simplification of Bernoulli's equation, which states that the sum of all forms of energy in a fluid flowing along an enclosed path (a streamline) is the same at any two points in that path.
  - In fluid flow with no viscosity, and, therefore, one in which a pressure difference is the only accelerating force, it is equivalent to Newton's laws of motion. It

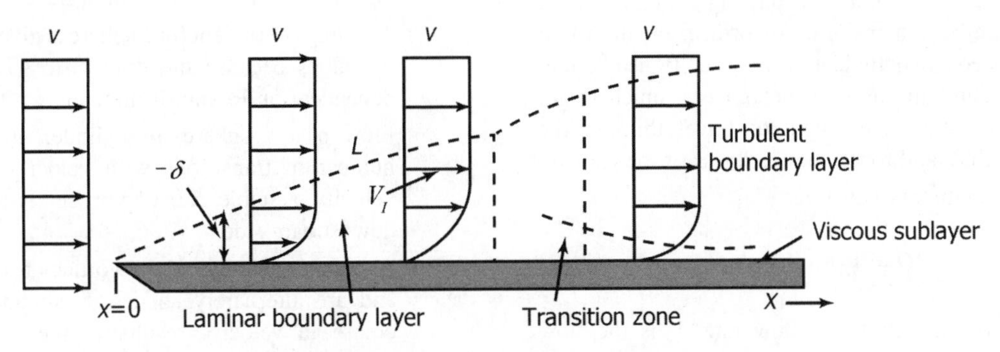

FIGURE 2.5 Development of boundary layer.

is important to note that the only cause of the change in fluid velocity is the difference in balanced pressure on either side of it.

- Write Bernoulli's equation for unit mass, unit length, and unit volume of the fluid.
  - For incompressible flow in a uniform gravitational field, Bernoulli's equation can be written as

$$v^2/2 + gh + p/\rho = \text{constant}, \qquad (2.17)$$

  where v is the fluid velocity along the streamline, g is the acceleration due to gravity, h is the height of the fluid, p is the pressure along the streamline, and ρ is the density of the fluid.
- What are the assumptions involved in Bernoulli's equation?
  - Flow is inviscid, that is, viscosity of the fluid is zero.
  - Steady-state incompressible flow.
  - In general, the equation applies along a streamline. For constant density potential flow, it applies throughout the entire flow field.
  - Density, ρ, is constant, though it may vary from streamline to streamline.
- What are velocity head and pressure head?
  - The term $v^2/2g$ is called the velocity head and $p/\rho g$ is called the pressure head.
- Give an example of the utility of the velocity head concept.
  - It is used, for example, for sizing the holes in a sparger, calculating leakage through a small hole, sizing a restriction orifice, and calculating the flow with a pitot tube and the like.
  - With a coefficient, it is used for orifice calculations, relating fitting losses, relief valve sizing, and heat exchanger tube leak calculations.
- "The velocity head concept is used in sizing holes in a sparger." *True/False*?
  - *True*. For a sparger consisting of a large pipe having small holes drilled along its length, the velocity head concept applies directly, because the hole diameter and the length of fluid travel passing through the hole have similar dimensions. An orifice, on the other hand, needs a coefficient in the velocity head equation because the hole diameter has a much larger dimension than the length of travel through the orifice, that is, through the thickness of the orifice.
- Write the continuity equation.

$$Q = V_1 A_1 = V_2 A_2, \qquad (2.18)$$

where Q is the volumetric flow rate, V is the flow velocity, and A_1 and A_2 are cross-sectional areas of the conduit at points 1 and 2, between which flow continuity exists.

  - If the fluid density is different at points 1 and 2, the mass flow rate is given by

$$m = V_1 A_1 \rho_1 = V_2 A_2 \rho_2. \qquad (2.19)$$

- What is Venturi effect?
  - The Venturi effect is an example involving application of Bernoulli's principle, in the case of a fluid flowing through a tube or pipe with a constriction in it. The fluid velocity must increase through the constriction to satisfy the equation of continuity, while its pressure must decrease due to conservation of energy. The gain in kinetic energy is supplied by a drop in pressure or a pressure gradient force.
- Define skin and form friction.
  - Skin friction is due to viscous drag.
  - Form friction is the drag due to pressure distribution.
- What is drag? Explain.
  - Drag is a force on a body due to a moving fluid interacting with it. Drag force slows down the flowing fluid, causes push of the object directly downstream, and transfers downstream momentum to the object.
  - Drag is a function of the body shape over which fluid is flowing. Different shapes will cause the flow to accelerate around them differently, that is, *streamlined shapes*. Low drag shapes: gentle curves; continuous surfaces prevent flow separation (reduce wake). The shape of an object has a very large effect on the magnitude of drag.
  - Figure 2.6 gives drag coefficients for different shapes of objects.
  - A quick comparison of the shapes shown in Figure 2.6 shows that a flat plate gives the highest drag and a streamlined symmetric airfoil gives the lowest drag, by a factor of almost 30. *Shape has a very large effect on the amount of drag produced.*
  - The drag coefficient for a sphere is given with a range of values because the drag on a sphere is highly dependent on Reynolds number.
  - Flow past a sphere, or cylinder, goes through a number of transitions with velocity. At very low velocity, a stable pair of vortices is formed on the downstream side.
  - As velocity increases, the vortices become unstable and are alternately shed downstream. Further increase in velocity results in the boundary layer transitions to chaotic turbulent flow with vortices of

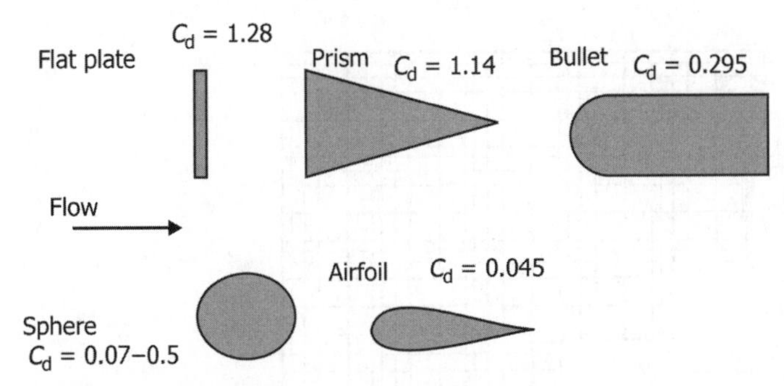

FIGURE 2.6 Drag coefficients for different shaped objects. *Note*: All objects have the same projected (frontal) area. Drag coefficient $C_d = F_D/0.5\rho AV^2$, where A is the projected area.

many different scales being shed in a turbulent wake from the body.

- Each of these flow regimes produce a different amount of drag on the sphere.
- Comparing different downstream shapes, it can be noted that the downstream shape can be modified to reduce drag.

• What are the different types of drag and lift?

- There are two basic types of drag as explained below.
- For the case of fluid friction inside conduits, the transfer of momentum perpendicular to the surface results in a tangential shear stress or drag on the smooth surface parallel to the direction of flow. This force exerted by the fluid on the solid in the direction of flow is called *skin friction* or *skin* or *wall drag*.
- For any surface in contact with a flowing fluid, skin friction will exist. In addition to skin friction, if the fluid is not flowing parallel to the surface but must change direction to pass around a solid body such as a sphere, significant additional frictional losses will occur. This phenomenon, called *form drag*, is in addition to the skin drag in the boundary layer.
- Drag is also classified into the following types:
  - Surface friction causes *viscous drag*.
  - Turbulence causes *pressure drag*.
  - Deflected flow causes *lift*.
  - Deflected flow causes *induced drag*.

• Define drag coefficient.

- Drag coefficient, C_d, is a dimensionless number given by

$$C_d = F_D/0.5\rho AV^2$$
$$= \text{drag force}/(\text{projected area} \times \text{velocity head}), \quad (2.20)$$

where F_D is the drag force, ρ is the fluid density, V is the fluid velocity, and A is the projected area of the solid object.

• Above what Reynolds numbers (for flow past a sphere), flow separation takes place?

- $N_{Re} \geq 20$.

• What is a stagnation point?

- In flow past an immersed body, boundary layer separation starts at the front center of the body where the fluid velocity will be zero. This point is called the *stagnation point*. Boundary layer growth begins at this point and continues over the surface until the layer separates.

• What is stagnation pressure?

- *Stagnation pressure* is the pressure at a stagnation point in a fluid flow, where the kinetic energy is converted into pressure energy. It is the sum of the dynamic pressure and static pressure at the stagnation point.

• Illustrate the relationship between drag coefficient and Reynolds number.

- At low Reynolds numbers, involving laminar flow, drag coefficients are high. With increase in N_{Re} drag coefficients decrease sharply and under fully turbulent conditions they remain constant with further increases in N_{Re}.
- Figure 2.7 illustrates this point.

• What are the common additives used to reduce drag in pipeline flow?

- Carboxymethyl cellulose (CMC), methyl methacrylate, or other polymer solutions.

• What is Coanda effect?

- The Coanda effect is the tendency of a stream of fluid to follow the contour of a surface.

• What is magnetohydrodynamics (MHD)?

- The study of flow of electrically conducting fluids in a magnetic field is called magnetohydrodynamics.
- The phenomenon arising from a coupling between either electric fields or electric discharges and velocity fields is called MHD.
- For example, gas–liquid flows can be altered by electric field effects.
- The simplest example of an electrically conducting fluid is a liquid metal, for example, mercury or liquid sodium. However, the major use of MHD is in plasma physics. (A plasma is a hot, ionized gas containing electrons and ions.)
- There are two serious technological applications of MHD that may both become very important in future. First, strong magnetic fields may be used to confine rings or columns of hot plasma that might be held in place long enough for thermonuclear fusion to occur and for net power to be generated. In the second application, which is directed toward a similar goal,

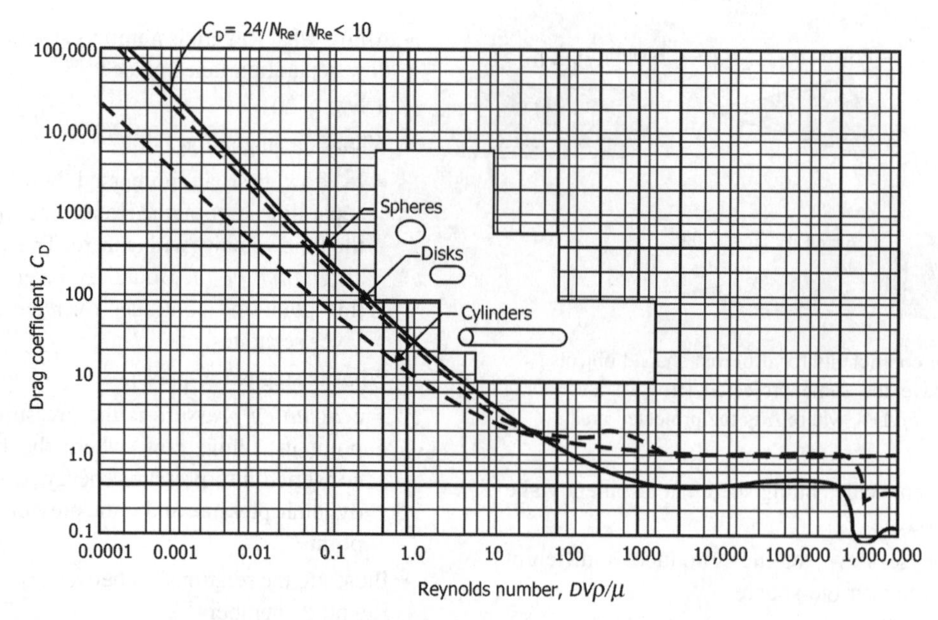

FIGURE 2.7 Drag coefficients for spheres, disks, and cylinders.

liquid metals are driven through a magnetic field in order to generate electricity.

- The study of magnetohydrodynamics is also motivated by its widespread application to the description of space (within the solar system) and astrophysical plasmas (beyond the solar system).

- What is creeping flow? Under what circumstances it occurs? Give examples of systems in which creeping flow is likely to be present.
  - Fluid flow at very small values of Reynolds number, that is, at very low velocities relative to a solid object, in general a very viscous flow, is called *creeping flow* or *Stokes flow*.
  - Basic assumption for creeping flow: The inertia terms are negligible in the momentum equation if $Re \ll 1$.
  - Flow over immersed bodies, usually small particles.
  - Creeping flow can occur in lubrication systems involving narrow and variable passages, for example, flow through porous media such as ground water movement.
- Above what Reynolds numbers, for flow past a sphere, flow separation takes place?
  - $N_{Re} > 20$.
- What is vortex shedding?
  - Vortex shedding is an unsteady flow that takes place at special flow velocities (according to the size and shape of the cylindrical body). In this flow, vortices are created at the back of the body and periodically from both sides of the body.
  - Vortex shedding is caused when air flows past a blunt structure. The air flow past the object creates alternating low-pressure vortices on the downwind side of the object. The object will tend to move toward the low-pressure zone.
  - Eventually, if the frequency of vortex shedding matches the resonant frequency of the structure, the structure will begin to resonate and the movement of the structure can become self-sustaining.
  - As Reynolds number (for flow over spherical bodies) increases above 20, vortices form in the wake of the sphere.
  - For Reynolds numbers in the range of 100–200, instabilities in flow break up vortices. This phenomenon is called *vortex shedding*.
- What are the undesirable effects of vortex shedding?
  - It causes severe vibration and mechanical failure of cylindrical elements such as heat exchanger tubes, columns, stacks, transmission lines, suspended piping, and so on.
- What is Von Kármán vortex street? Explain.
  - A Von Kármán *vortex street* is a repeating pattern of swirling vortices caused by the unsteady separation of flow over bluff bodies. A vortex street will only be observed over a given range of Reynolds numbers, typically above a limiting N_{Re} value of about 90. The range of N_{Re} values will vary with the size and shape of the body from which the eddies are being shed, as well as with the kinematic viscosity of the fluid. The governing parameter, the Reynolds number, is

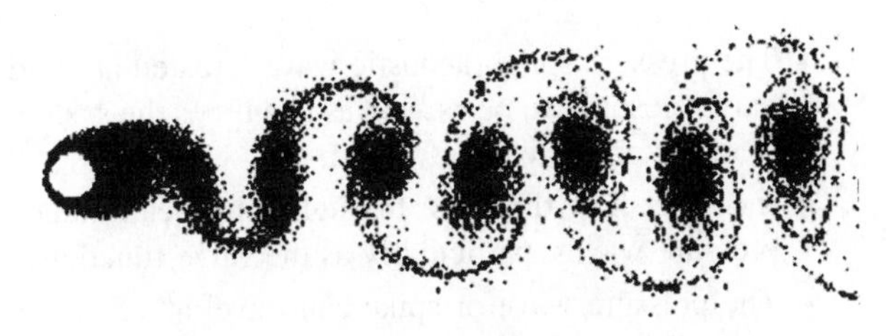

FIGURE 2.8 Vortex street phenomenon.

essentially a measure of the ratio of inertial to viscous forces in the flow.

- Over a large N_{Re} range ($47 < N_{Re} < 10^7$ for circular cylinders), eddies are shed continuously from each side of the body, forming rows of vortices in its wake. The alternation leads to the core of a vortex in one row being opposite to the point midway between two vortex cores in the other row, giving rise to the distinctive pattern shown in Figure 2.8. Ultimately, the energy of the vortices is consumed by viscosity as they move further downstream and the regular pattern disappears.
- When a vortex is shed, an unsymmetrical flow pattern forms around the body, which therefore changes the pressure distribution. This means that the alternate shedding of vortices can create periodic lateral forces on the body in question, causing it to vibrate. If the vortex shedding frequency is similar to the natural frequency of a body or structure, it causes resonance. It is this forced vibration that, when at the correct frequency, causes telephone or power lines to *hum*, the antennae on a car to vibrate more strongly at certain speeds, and it is also responsible for the fluttering of Venetian blinds as the wind passes through them.
- Periodic forcing set up in this way can be highly undesirable and hence it is important for engineers to account for the possible effects of vortex shedding when designing a wide range of structures, from submarine periscopes to industrial chimneys. In order to prevent the unwanted vibration of such cylindrical bodies, a longitudinal fin can be fitted on the downstream side, which, as long as it is longer than the diameter of the cylinder, will prevent the eddies from interacting, and consequently they remain attached. Obviously for a tall building or mast, the relative wind could come from any direction. For this reason, helical projections that look like large screw threads are sometimes placed at the top, which effectively create unsymmetrical three-dimensional flow, thereby discouraging the alternate shedding of vortices.
- A vortex street phenomenon is illustrated in Figure 2.8.
- When considering a long circular cylinder, the frequency of vortex shedding is given by the empirical formula

$$f_{vs}D_O/V_c = 0.198[1-(19.7/N_{Re})], \qquad (2.21)$$

where f_{vs} is the vortex shedding frequency (Hz), D_O is the diameter of the cylinder, and V_c is the flow velocity.

- This formula will generally hold good for the range $250 < N_{Re} < 2 \times 10^5$.

- What is Strouhal number?
  - The dimensionless parameter, $f_{vs}D_O/V_c$, is known as the Strouhal number.
  - Studies have shown that insects such as bees borrow energy from the vortices that form around their wings during flight. Vortices inherently create drag. Insects can recapture some of this energy and use it to improve speed and maneuverability. Insects rotate their wings before starting the return stroke, and the wings are lifted by the eddies of air created on the downstroke. The high-frequency oscillation of insect wings means that many hundreds of vortices are shed every second. However, this leads to a symmetric vortex street pattern, unlike the ones shown above.
  - Flow across a tube produces a series of vortices in the downstream wake formed as the flow separates alternately from the opposite sides of the tube.
  - This alternate shedding of vortices produces alternating forces that occur more frequently as the velocity of flow increases.
  - For a single cylinder, the tube diameter, the flow velocity, and the frequency of vortex shedding can be described by the dimensionless Strouhal number.
  - For single cylinders, the vortex shedding Strouhal number is a constant with a value of about 0.2.
  - Vortex shedding occurs in the range of Reynolds numbers 100 to 5×10^5 and greater than 2×10^6. The gap is due to a shift of the flow separation point of the vortices in this intermediate transcritical Reynolds number range.
  - Vortex shedding also occurs for flow across tube banks. The Strouhal number is no longer a constant, but varies with the arrangement and spacing of the tubes. There is less certainty that there is a gap in the vortex shedding Reynolds number.
  - Vortex shedding is fluid mechanical in nature and does not depend on any movement of the tubes. For a given arrangement and tube size, the frequency of the vortex shedding increases as the velocity increases.

- The vortex shedding frequency can be an exciting frequency when it matches the natural frequency of the tube and vibration results.
- With tube motion the flow areas between the tubes are being expanded and contracted in concert with the frequency of vibration. This in turn changes the flow velocity that controls the frequency of the vortex shedding. Since tubes vibrate only at unique frequencies, the vortex shedding frequency can become *locked in* with a natural frequency.

2.2 WATER HAMMER

- What is water hammer?
  - Water hammer is generally defined as pressure wave or surge or momentary increase in pressure or hydraulic shock, caused by the kinetic energy of a nonviscous fluid in motion when it is forced to stop or change direction suddenly.
  - Kinetic energy of the moving mass of liquid upon sudden stoppage or abrupt change of direction is transformed into pressure energy.
  - Slow or abrupt start-up or shutdown of a pump system, pipe breakage, turbine failure, or electric power interruption to the motor of a pump are some examples. Surge is a slow motion mass oscillation of water caused by internal pressure fluctuations in the system.
  - Water hammer occurs during the start-up or energizing of a steam system. If the steam line is energized too quickly without proper warm up time and the condensate created during the start-up is not properly removed, water hammer will be the result.
  - When a rapidly closed valve suddenly stops water flowing in a pipeline, pressure energy is transferred to the valve and pipe wall. Shock waves are set up within the system. Pressure waves travel backward until encountering the next solid obstacle, then forward, and then back again. The velocity of the pressure wave is equal to the speed of sound. It generates damaging noise as it travels back and forth, until dissipated by frictional losses.
  - Successive reflections of the pressure wave between pipe inlet and closed valve resulting in alternating pressure increases and decreases often result in severe mechanical damage.
  - A valve closing in 1.5 s or less depending on valve size and system conditions causes an abrupt stoppage of flow.
  - The pressure spike (acoustic wave) created at rapid valve closure can be as high as five times the system working pressure.
  - Incorrect flow direction through valves can induce pressure waves, particularly as the valve functions.
  - The pressure wave or spike can travel at velocities exceeding 1400 m/s for steel tubes.
  - Rapid changes in flow, operation of positive displacement pumps, entrained or separated gases, and increased temperatures induce damaging pressure changes.
  - When a vapor bubble collapses as in the case of cavitation, water rushes into this space from all directions causing hammer.
  - Lack of proper drainage ahead of a steam control valve. When the valve is opened, a slug of condensate enters the equipment at high velocity impinging on the interior walls, causing hammer.
  - Improper operation or incorporation of surge protection devices can do more harm than good. An example is oversizing the surge relief valve or improperly selecting the vacuum breaker–air release valve.
  - Mixing of steam with relatively cool water in a pipe or confined space also results in water hammer.
  - If not controlled, the transients in pressures cause damage to pipes, fittings, valves, and steam traps, causing leaks and shortening the life of the system. Liquid for all practical purposes is not compressible and any energy that is applied to it is instantly transmitted. Neither the pipe walls nor the liquid will absorb the created shocks. Such transients can also lead to blown diaphragms, destruction of seals, gaskets, meters, gauges, and steam trap bodies, heat exchanger tube failures, and injuries to personnel.
- What are the conditions that cause water hammer?
  - *Hydraulic Shock:* Start, stop, abrupt change in the speed of a pump, power failure, or rapid closure of a valve (usually a control valve, which can slam shutdown in 1 or 2 s). Sudden stoppage generates a shock wave similar to that of stopping a hammer with a bang. Pressures generated can be as high as 40 bar (600 psi).
  - *Differential Shock:* Develops in two-phase systems, for example, whenever steam and condensate flow in the same line with different velocities (steam velocities will be $\geq$10 times the liquid velocities). When condensate fills the line, steam flow is stopped and on the downstream side pressure drops. Steam pushes the condensate slug like a piston/hammer.
  - *Thermal Shock:* Steam bubbles may become trapped in pools of condensate in a flooded main,

branch, or tracer line or pumped condensate lines or heat exchanger tubing. Steam will collapse, since condensate temperature is mostly subcooled in the lines.

  - This collapse will release energy in a short period of time.
  - Vacuum is created due to sudden volume reduction by about 1600 times.
  - Water rushes into this space from all directions.

- *Flow Shock:* Caused by lack of proper drainage ahead of a steam line isolation valve or steam control valve.
  - Changes in power demand from turbines.
  - Operation of reciprocating pumps.
  - Changing elevation of a reservoir/storage tank.
  - Waves on a reservoir.
  - Unstable fan or pump characteristics.
  - Tube failure in a heat exchanger.

- What are the important phenomenon parameters useful in the analysis of water hammer?
  - Velocity of the pressure wave.
  - Critical time of the phenomenon.
  - Maximum head developed in the maximum pressure time.
  - Minimum head developed in the critical time.
- What are the precautions to prevent or reduce water hammer?
  - Designing the system with low velocities.
  - Water hammer often damages centrifugal pumps when electric power fails. In this situation, the best form of prevention is to have automatically controlled valves, which close slowly. (These valves work without electricity or batteries. The direction of the flow controls them.) Closing the valve slowly can moderate the rise in the pressure when the downsurge wave, resulting from the valve closing, returns from the reservoir.
  - Providing slow closing bypasses around fast closing valves.
  - Ensuring correct flow directions in the process plant can reduce or even prevent pressure wave problems.
  - With air-operated seat valves, incorrect direction of flow can cause the valve plug to close rapidly against the valve seat inducing pressure waves.
  - As rapid changes in operating conditions of valves and pumps are the major reasons for pressure waves, it is important to reduce the speed of these changes. For example, valve plug movement can be damped by means of special dampers, such as the one shown in Figure 2.9.
  - A pressure wave as a result of a pump stopping is more damaging than a pump starting due to the large change in pressure that will continue much longer after the pump is stopped, compared to start of the pump.
  - Entrained air or temperature changes of the water can be controlled by pressure relief valves, which are set to open with excess pressure in the line and then closed when pressure drops. Relief valves are commonly used in pump stations to control pressure surges and to protect the pump station. Positively controlled relief valves act as surge suppressors. These valves can be an effective method of controlling transients.
  - However, they must be properly sized and selected.
  - Where pressure drops due to condensation or unloading operations, giving rise to vacuum in the system, properly selected and sized vacuum breaker–air release valves can be the least expensive means of protecting a piping system.
  - Pump start-up problems can usually be avoided by increasing the flow slowly to collapse or flush out the voids gently. Also, a simple means of reducing hydraulic surge pressure is to keep pipeline velocities low. This not only results in lower surge pressures, but also results in lower drive horsepower and, thus, maximum operating economy.
  - Providing surge tanks in long pipelines to relieve pressures generated, storing excess liquid. These tanks can serve as dampeners for both positive and negative pressures. However, surge tanks for surge control are expensive compared to other devices.
  - Since severity of water hammer is highly dependent on the acoustic velocity, a reduction in acoustic velocity will reduce hammer. This can be accomplished by adding a small quantity of air to a liquid system or switching to noncircular piping.
  - Installing air chambers in systems with frequent water hammer problems: shaped like thin, upside-down bottles with a small orifice connection to the pipe, they are air-filled. The air compresses to absorb the shock, protecting the fixture and piping.

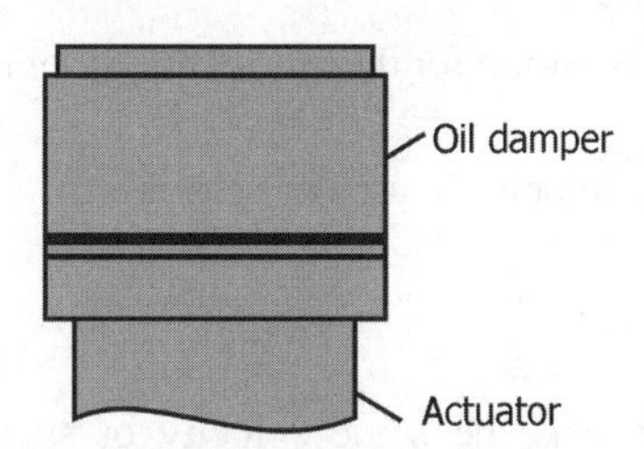

FIGURE 2.9 Oil damper to cushion valve closure.

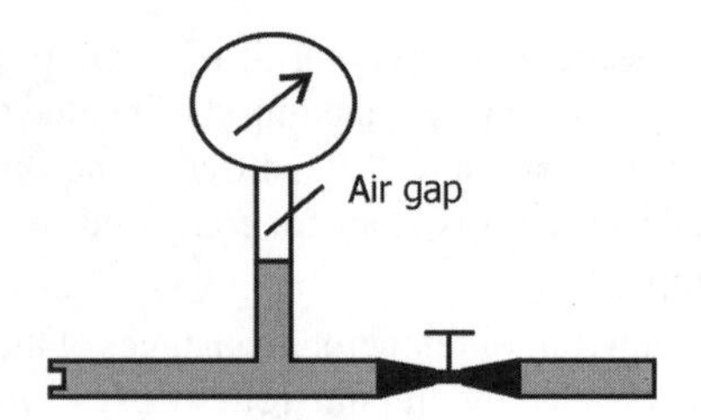

FIGURE 2.10 Air gap cushions shock waves when valve is suddenly closed.

- Installing a riser tube with a pressure gauge on the pipe on the upstream of valve on the piping: When the valve is closed, the hammer pulse generated is absorbed by the air column between the gauge and piping, as illustrated in Figure 2.10. Air is compressed absorbing the shock wave with the pressure rise indicated by the pressure gauge.
- Since the pressure spike is proportional to the initial flow velocity, doubling the pipe diameter can reduce the spike by a factor of 4.
- When water flows are split or combined, using vacuum breakers to admit air that cushions the shock resulting from sudden opening or closing of the second split stream.
- Minimizing number of elbows, where elbows are needed, using bends or two 45° elbows in place of one 90° elbow, and using rigid supports with clamps in place of hangers are some remedies to reduce pipe vibration problems.

2.3 COMPRESSIBLE FLUIDS

- What is the maximum velocity that a compressible fluid can attain in a pipe?
  - Critical/sonic velocity in the fluid.
  - Maximum velocity in a pipe is limited by the velocity of propagation of a pressure wave that travels at the speed of sound in the fluid.
  - Shock waves travel at supersonic velocities and exhibit a near discontinuity in pressure, density, and temperature.
  - Normally velocities are limited to 30% of sonic velocity.
- Give an equation for the estimation of sonic velocity in a gas.
  - Sonic velocity in an ideal gas is

$$c = \sqrt{[kRT/M_w]}. \qquad (2.22)$$

where c is the sonic velocity or speed of sound or acoustic velocity, k is the ratio of specific heats (C_p/C_v), M_w is the molecular weight, and T is the absolute temperature.

- What is Mach number? Explain its significance and usefulness.
  - Mach number is defined as the speed of an object relative to a fluid medium, divided by the speed of sound in that medium.
  - It is the number of times the speed of sound an object or a duct, or the fluid medium itself, moves relative to the other.
  - Mach number is commonly used both with objects traveling at high speed in a fluid and with high-speed fluid flows inside channels such as nozzles, diffusers, or wind tunnels.
  - At a temperature of 15 °C and at sea level, Mach 1 is 340.3 m/s (1225 km/h or 761.2 mph) in the Earth's atmosphere. The speed represented by Mach 1 is not a constant but temperature dependent.
  - Since the speed of sound increases as the temperature increases, the actual speed of an object traveling at Mach 1 will depend on the fluid temperature around it.
  - Mach number is useful because the fluid behaves in a similar way at the same Mach number. For example, an aircraft traveling at Mach 1 at sea level (340.3 m/s, 1225.08 km/h) will experience shock waves in much the same manner as when it is traveling at Mach 1 at 11,000 m (36,000 ft), even though it is traveling at 295 m/s (654.632 mph, 1062 km/h, 86% of its speed at sea level).
  - It can be shown that Mach number is also the ratio of inertial forces (also referred to as aerodynamic forces) to elastic forces.
- How are high-speed flows classified?
  - High-speed flows can be classified into five categories:
    - Subsonic: $M < 1$.
    - Sonic: $M = 1$.
    - Transonic: $0.8 < M < 1.2$.
    - Supersonic: $1.2 < M < 5$.
    - Hypersonic: $M > 5$.
  - At transonic speeds, the flow field around the object includes both subsonic and supersonic components. The transonic regime begins when first zones of $M > 1$ flow appear around the object. In case of an airfoil (such as the wing of an aircraft), this is typically above the wing. Supersonic flow can decelerate back to subsonic only in a normal shock, which typically happens before the trailing edge.
  - As the velocity increases, the zone of $M > 1$ flow increases toward both leading and trailing edges. As

$M = 1$ is reached and passed, the normal shock reaches the trailing edge and becomes a weak oblique shock, the flow decelerating over the shock but remaining supersonic. A normal shock is created ahead of the object and only subsonic zone in the flow field is a small area around the leading edge of the object.

- How does a convergent–divergent nozzle work for compressible flows?
  - As flow in a channel crosses $M = 1$, it becomes supersonic, one significant change takes place. Common sense would lead one to expect that contracting the flow channel would increase the velocity (i.e., making the channel narrower) resulting in faster air flow and at subsonic velocities this holds true. However, once the flow becomes supersonic, the relationship between flow area and velocity is reversed; that is, expanding the channel actually increases the velocity.
  - The obvious result is that in order to accelerate a flow to supersonic, one needs a convergent–divergent nozzle, where the converging section accelerates the flow to $M = 1$, sonic speeds, and the diverging section continues the acceleration. Such nozzles are called *de Laval* nozzles, and in extreme cases, they are able to reach incredible hypersonic velocities (M of 13 at sea level).
- "In a converging nozzle, the velocity of the gas stream will never exceed the sonic velocity, though in a converging–diverging nozzle supersonic velocities may be obtained in the diverging section." *True/False*?
  - *True*.
- Give equations for calculating Mach numbers at subsonic and supersonic velocities in air.
  - *Subsonic Velocities:* Assuming air to be an ideal gas, the formula to compute Mach number in a subsonic compressible flow is derived from the Bernoulli equation for $M < 1$:

$$M = \sqrt{[5(q_c/P + 1)^{2/7} - 1]}, \tag{2.23}$$

    where M is the Mach number, q_c is the impact pressure, and P is the static pressure.
  - *Supersonic Velocities:* The formula to compute Mach number in a supersonic compressible flow is derived from the Rayleigh supersonic pitot equation:

$$M = 0.88128485\sqrt{[(q_c/P + 1)\{1 - (1/7M^2)\}^{2.5}]}, \tag{2.24}$$

    where q_c is the impact pressure measured behind a normal shock. M is obtained by an iterative process. First determine if $M > 1.0$ by calculating it from the subsonic equation. If $M > 1.0$ at that point, then use the value of M from the subsonic equation as the initial condition in the supersonic equation.
- What is choked flow?
  - Choked flow of a fluid is a fluid dynamic condition caused by the Venturi effect. When a flowing fluid at a certain pressure and temperature flows through a restriction (such as the hole in an orifice plate or a valve in a pipe) into a lower pressure environment, under the conservation of mass, the fluid velocity must increase for initially subsonic upstream conditions as it flows through the smaller cross-sectional area of the restriction. At the same time, the Venturi effect causes the pressure to decrease.
  - Choked flow is a limiting condition that occurs when the mass flux does not increase with a further decrease in the downstream pressure environment.
  - For homogenous fluids, the physical point at which choking occurs for adiabatic conditions is when the exit plane velocity is under sonic conditions or at a Mach number of 1. For isothermal flow of an ideal gas, choking occurs when the Mach number is equal to the square root of C_v/C_p.
  - The choked flow of gases is useful in many engineering applications because the mass flow rate is independent of the downstream pressure, depending only on the temperature and pressure on the upstream side of the restriction. Under choked conditions, valves and calibrated orifice plates can be used to produce a particular mass flow rate.
  - If the fluid is a liquid, choked flow occurs when the Venturi effect acting on the liquid flow through the restriction decreases the liquid pressure to below that of the liquid vapor pressure at the prevailing liquid temperature. At that point, the liquid will partially boil into bubbles of vapor and the subsequent collapse of the bubbles causes cavitation. Cavitation is quite noisy and can be sufficiently violent to physically damage valves, pipes, and associated equipment. In effect, the vapor bubble formation in the restriction limits the flow from increasing any further.
  - Chocked flow (or critical flow) for a liquid occurs when the mass flux through a restricted area is at its maximum (velocity is sonic). If the downstream pressure is decreased, the mass flow will not increase. Besides the possibility of physical equipment damage due to flashing or cavitation, formation of vapor bubbles in the liquid flow stream causes a crowding

condition at the vena contracta that tends to limit flow through the restriction, for example, a valve.

- "In adiabatic compressible flow in a conduit of constant cross section, velocity can exceed sonic velocity." *True/False*?
  - *False*.
- "For an ideal gas, temperature increases with increase in velocity." *True/False?*
  - *False*. Temperature decreases. For real gases, temperature may increase.

3

PIPING, SEALS, AND VALVES

3.1 Friction and Piping 35
3.2 Gaskets and Mechanical Seals 42
3.3 Valves 46

3.1 FRICTION AND PIPING

- What is friction factor? What is its significance?
  - Friction factor is a dimensionless quantity signifying resistance offered by the wall of a pipe when a fluid flows through it.
- Give the equation for the estimation of friction factor for laminar flow in a pipe.

$$f = 16/N_{Re} \quad \text{(Fanning friction factor)}. \tag{3.1}$$

$$f' = 64/N_{Re} \quad \text{(D' Archy friction factor)}. \tag{3.2}$$

- What are the effects of temperature changes on friction factors?
  - Sieder–Tate gives corrections in equations for determining friction factors inside heat exchanger tubes with varying temperatures: First, the average, bulk mean temperature in the processing line is determined to estimate the physical properties and friction factors.
  - *Laminar Flow*:
    - Cooling of the liquid: The friction factor obtained from the mean temperature and bulk properties is divided by (bulk viscosity/wall viscosity)$^{0.23}$.
    - Heating of the liquid: The friction factor obtained from the mean temperature and bulk properties is divided by (bulk viscosity/wall viscosity)$^{0.38}$.
    - In laminar flow, the bulk and the wall viscosities are determined at the mean temperature over the length of the line.
  - *Turbulent Flow*:
    - Cooling of the liquid: The friction factor obtained from the mean temperature and bulk properties is *divided* by (bulk viscosity/wall viscosity)$^{0.11}$.
    - Heating of the liquid: The friction factor obtained from the mean temperature and bulk properties is *divided* by (bulk viscosity/wall viscosity)$^{0.17}$.
    - In turbulent flow, the bulk and the wall viscosities are determined at the mean temperature over the length of the line.
- Give the equation for the estimation of pressure drop in pipeline flow of an incompressible fluid.
  - *Fanning Equation*:

$$\Delta P = 4fLV^2/2D, \tag{3.3}$$

  where f is the Fanning friction factor.
  - *D'Arcy Equation*:

$$\Delta P = f'LV^2/2\text{D}, \tag{3.4}$$

  where D'Arcy friction factor is equal to four times the Fanning friction factor, f.
- What is roughness factor/relative roughness?
  - Roughness factor or relative roughness is a dimensionless number, defined as ε/D, where ε is the average length of surface projections, which constitute roughness of inner surface of the pipe in which a fluid is flowing and D is the inside diameter of the pipe.
- When ε/D is more, are the friction losses more or less for the same Reynolds number? Why?
  - More. Because increased roughness induces increased turbulence, which results in higher frictional pressure losses for the flow.
- Give equation(s) and Moody chart for the estimation of friction factors as function of N_{Re} for turbulent flow in smooth and rough pipes.

Fluid Mechanics, Heat Transfer, and Mass Transfer: Chemical Engineering Practice, By K. S. N. Raju

- Smooth pipes:

$$1/\sqrt{f} = 4\log_{10}(N_{Re}\sqrt{f}) - 0.4. \quad (3.5)$$

This equation requires iterative techniques.

➢ Blasius equation:

$$f = 0.079/N_{Re}^{0.25},\ 4000 < N_{Re} < 10^5. \quad (3.6)$$

- Rough pipes:

$$1/\sqrt{f} = -2\log_{10}[\{\varepsilon/D/3.7\} + 2.51/N_{Re}(1/\sqrt{f})]. \quad (3.7)$$

This equation also requires an iterative technique such as Newton–Raphson method for estimating *f*.

➢ Rough pipes—noniterative equation:

$$1/\sqrt{f} = 0.8686\ \ln\,[0.4587N_{Re}/(s/s^{s+1})], \quad (3.8)$$

where $s = 0.1240(\in/D)N_{Re} + \ln(0.4587N_{Re})$. Maximum error of 1% is claimed in the estimation of *f*.

- Churchill equation for smooth and rough pipes:

$$1/\sqrt{f} = -4\log_{10}[0.27(\varepsilon/D) + (7/N_{Re})^{0.9}],\quad N_{Re} > 4000. \quad (3.9)$$

This equation is noniterative and applies to both laminar and turbulent flows, by putting $\varepsilon/D = 1$ for laminar flow.

- *Colbrook equation*:

$$1/\sqrt{f} = -4\log_{10}[(\varepsilon/3.7D) + (1.256/N_{Re}f)],\quad N_{Re} > 4000. \quad (3.10)$$

This equation formed the basis for the Moody chart.

- Approximate relationships for Fanning friction factors are as follows:

➢ For laminar flow

$$f = 16/N_{Re}.$$

➢ For commercial pipes

$$f = 0.054/N_{Re}^{0.2}. \quad (3.11)$$

➢ Smooth pipes

$$f = 0.46/N_{Re}^{0.2}. \quad (3.12)$$

➢ For extremely rough pipes

$$f = 0.013. \quad (3.13)$$

- Use of equations has become more convenient in the present scenario of computerization compared to the past when use of charts has been the practice.
- Figure 3.1 represents Moody diagram.
- The values of roughness for different materials are listed in Table 3.1.

• What are the recommended pipe surface roughness values for steel piping at different stages of its life?

- New pipe: 0.04 mm.
- After longer use: 0.2 mm.
- Slightly rusted condition: 0.4 mm.
- Severely rusted condition: 4.0 mm.

• Arrange the following pipe materials in the order of increasing surface roughness: Polyethylene, stainless steel, concrete, GI, copper, commercial steel, cast iron, Teflon-coated carbon steel, severely rusted steel, and glass.

- Glass, Teflon-coated carbon steel, copper, polyethylene, stainless steel, GI, cast iron, concrete, and severely rusted steel.

• Name different types of fittings used in piping.

- Flanges, elbows (90° and 45°), bends, tees, reducers, crosses, nipples, unions, valves, and so on.

• Friction losses are more in a bend or in an elbow?

- In an elbow, as it involves sudden change in direction, promoting energy losses through eddy formation.

• "Pressure drop in a fitting for sudden enlargement (e.g., tank inlet) is higher than for sudden contraction (e.g., tank outlet) for the same size ratio." *True/False*?

- *True*. The number of diameters equivalent to straight pipe for sudden enlargement is 50, while it is 25 for sudden contraction.

• "Friction factor has a linear relationship with Reynolds Number under turbulent conditions." *True/False*?

- *False*.

• "In laminar flow friction factor *f* is dependent on roughness factor ε/D." *True/False*?

- *False*.

• Is there any effect of temperature on friction factors for flow through a pipe?

- *Yes*. As temperature decreases, for example, fluid viscosity increases and also fluid density. But viscosity increase is much more in magnitude than density increase, with the net result that Reynolds number decreases. A look at Moody diagram

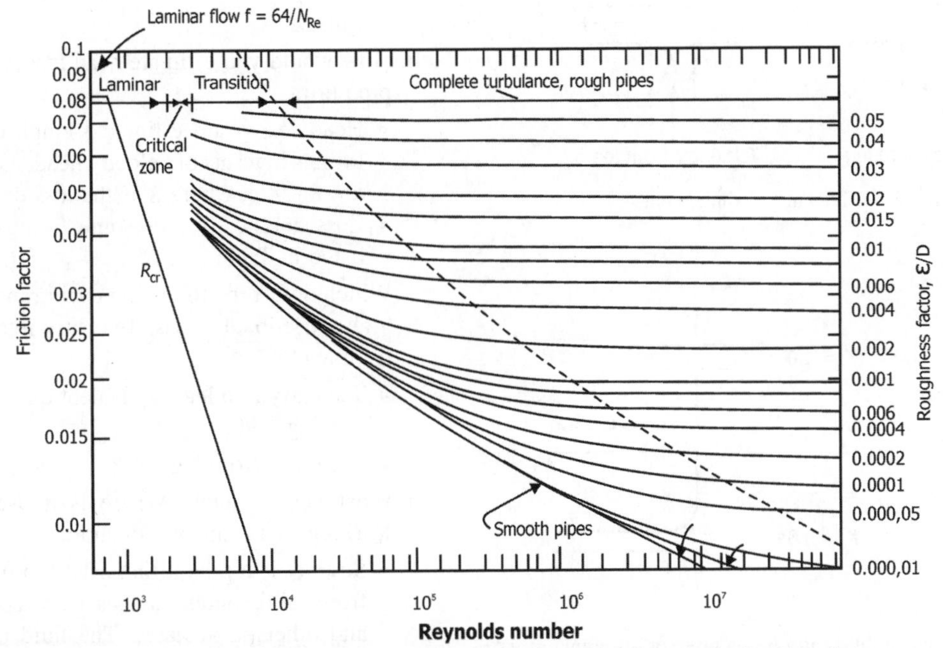

FIGURE 3.1 D'Arcy friction factors as function of N_{Re}.

TABLE 3.1 Values of Roughness for Different Materials

Material	ε (mm)
Glass, drawn brass, and copper tubes	0.0015
Commercial steel or wrought iron	0.045
Asphalted cast iron	0.12
Galvanized iron	0.15
Cast iron	0.26
Concrete	0.3–3.0
Corrugated metal	45
PVC	0.12

Note: Roughness will change with use of the piping.

suggests that decreased Reynolds number results in increased friction factors.

- "Pressure drop due to friction for a flowing fluid in a straight pipe is proportional to the velocity of the fluid." *True/False*?
  - *False*. It is proportional to the square of velocity.
- "In turbulent flow, the higher the surface roughness of the pipe the higher the influence of the Reynolds number on the friction factor." *True/False*?
  - *False*.
- "Friction factor f in laminar flow depends on the Reynolds number and the surface roughness of the pipe." *True/False*?
  - *False*. In laminar flow, f is independent of Reynolds number.
- Which one of the arrangements shown in Figure 3.2 involves higher frictional losses?
  - Tee entry into leg. (Number of diameters equivalent to straight pipe for tee entry into leg are 90, while tee entry from leg are 60.
- What are the equivalent lengths for fittings with sudden enlargement and sudden contraction?
  - Sudden enlargement: 50.
  - Sudden contraction: 25.
- "Equivalent lengths for welded fittings are more than those for threaded fittings with the same diameters." *True/False*?
  - *False*. It is the other way round.
- Illustrate how entrance losses for flow can be minimized.
  - Losses can be reduced by accelerating the flow gradually and eliminating the vena contracta. Figure 3.3 illustrates different entrance configurations, showing loss coefficients.
- Give Hagen–Poiseuille equation for head losses for flow throw a pipe. What is the condition for its applicability?

$$h_f = 32\mu Lu/\rho g D^2, \tag{3.14}$$

where u is the average flow velocity, D is the pipe diameter, and L is the length of the pipe. The equation is applicable for laminar flow conditions.

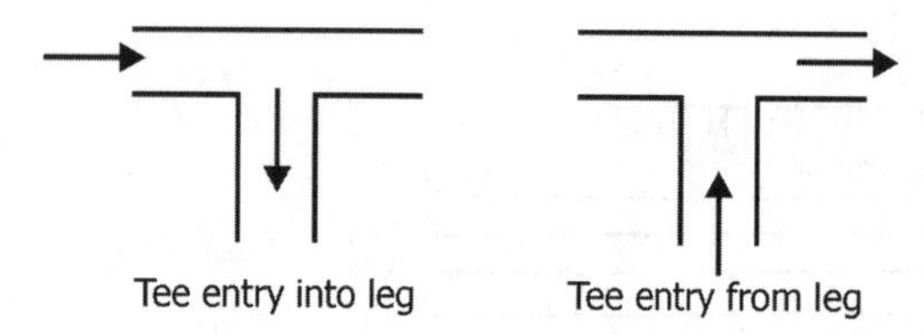

FIGURE 3.2 Tee entry arrangements.

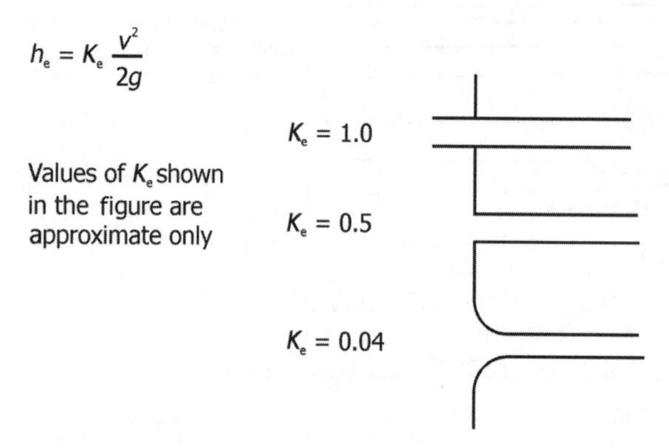

FIGURE 3.3 Losses on fluid entry into a pipe for different entry configurations.

- Give a procedure for estimating ΔP for the turbulent flow inside commercial pipes.
  - Find the effective length, L, of a pipe adding equivalent lengths of valves and fittings in the line.
  - Measure the flow rate of the fluid by means of a flow meter or by direct measurement and calculate the average velocity of the fluid, knowing the inside diameter of the pipe.
  - Calculate N_{Re} from density and viscosity data, using calculated velocity and known inside diameter of pipe.
  - Take values of ε, pipe roughness, and obtain the roughness factor.
  - Using Colbrook or Churchill equations or Moody diagram, obtain friction factor.
  - Using Fanning or D'Archy equations, calculate head losses and hence ΔP, using the appropriate friction factor (Fanning or D'Archy values).
- Give an equation for head losses for turbulent flow through pipes.

$$h_f = fLu^2/2gD$$

where f is Darcy Weisbach friction factor. It differs from Fanning friction in that it is *four* times the Fanning friction factor, the rest of the terms in Fanning equation being the same as in the above equation.

- "Hagen–Poiseuille equation gives the pressure drop as a function of the average velocity for turbulent flow in a horizontal pipe." *True/False*?
  - *False.*
- How would you estimate head losses for flow through pipe fittings?
  - Head losses are estimated empirically for various fittings in terms of velocity head, using the equation, $h = KV^2/2g$, where K is the loss coefficient.
  - Loss coefficients for some fittings and valves are listed in Table 3.2.
- Which one of the following two arrangements involves higher frictional losses: Tee entry into leg or tee entry from leg?
  - Tee entry into leg. Equivalent diameter for tee entry into leg = 90.
  - Tee entry from leg = 60.
- What is a strainer? Where is it used? Illustrate its working by means of a diagram.
  - Strainer is a pipe fitting used to filter flowing fluids from solid contaminants such as corrosion products and other particulates. The fluid passes through a screen and solids remain in the leg of the screen basket, which are to be removed occasionally by opening a plug fitted at the bottom of the strainer. Suction side of centrifugal and reciprocating pumps is normally provided with strainers. Figure 3.4 shows a typical strainer.
- What is the difference in the specifications for a pipe and a tube?
  - Tube is specified by its outside diameter and wall thickness in terms of SWG (standard wire gauge) or BWG (Birmingham wire gauge).
  - Pipe is specified by nominal diameter and schedule number.
- Define schedule number.

$$\text{Schedule number} = P_S \times 1000/\sigma_S, \qquad (3.15)$$

where P_S is safe working pressure and σ_S is safe working stress.

  - The higher the schedule number, the thicker the pipe is.

TABLE 3.2 Loss Coefficients for Some Pipe Fittings

Fitting/Valve	K
Pipe inlets	0.5–0.9
90° Elbows (short radius, $r/d = 1$)	0.24
90° Elbows (long radius, $r/d = 1.5$)	0.19
Fully open gate valve	0.1–0.3
Fully open globe valve	3–10
Fully open butterfly valve	0.2–0.6
Swing check valve	0.29–2.2
Lift check valve	0.85–9.1

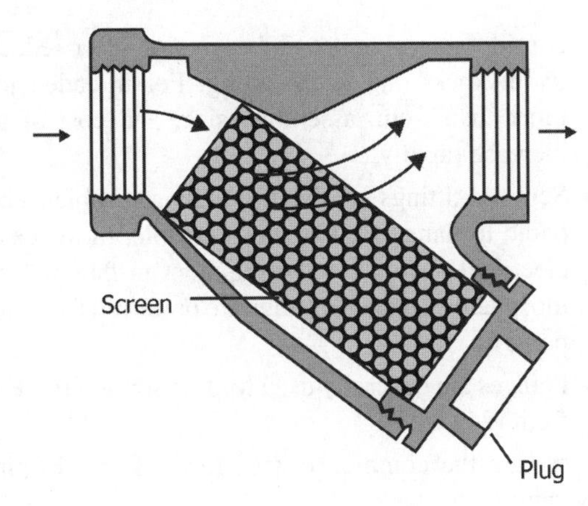

FIGURE 3.4 Strainer.

- The outside diameter of each pipe size is standardized.
- Therefore, a particular nominal pipe size will have a different inside diameter depending on the schedule number specified.
- Schedule 40, 60, and 80 are examples while Schedule 40 is most common.

• Which of the following has higher wall thickness, 12 SWG or 18 SWG?
- 12 SWG.

• Some times, piping is specified as Class A, Class B, and Class C. Which class of piping has the highest wall thickness?
- Class C.

• Give an equation for the estimation of internal design pressure (IDP) for a pipe.

$$\mathrm{IDP} = (P_{\max} + P_{\mathrm{s}})S_{\mathrm{F}}, \qquad (3.16)$$

where $P_{\max}$ is the maximum steady-state operating pressure, P_s is the surge or water hammer pressure, and S_F is the safety factor to take care of unknowns such as external earth loads or live loads; variation of pipe properties with temperature; damage that could result from handling, shipping, or installing; and corrosion, erosion, and other effects associated with long use. S_F is usually in the range of 3–4.

• What is the maximum allowable working pressure of a material?
- The maximum allowable working pressure of the material selected should significantly exceed the maximum expected operating pressure; 20% is a reasonable minimum safety factor but 200% is a desirable target.

• What is a GI pipe? What is its application?
- When a steel pipe is coated with zinc, it is called a GI pipe. Zinc coating protects the pipe from corrosion. It is commonly used for water lines.

• What type of pipe is used for steam lines?
- Class C pipe, which is some times called black pipe.

• What are the generally recommended considerations in determining utility gas piping size?
- Piping sizing should be selected so that pressure drop at the furthest point in the system does not exceed 10% of the inlet pressure under actual flow conditions.
- The source and the pipeline should be sized itself with an adequate but affordable safety factor to allow for peak demand and future growth. It is not unusual to apply a safety factor of two times the current anticipated flow and from 1.2 to two times the required pressures. This provides a margin so future additions or changes in process requirements do not necessarily demand a totally new piping system.

• What are the common errors in compressed air distribution systems and how are they addressed?
- A common error in compressed air systems is too small line sizes for the desired air flow. This includes the interconnecting piping from compressor discharge, air dryer, and header. It also applies to the distribution lines conveying air to production areas and within the equipment found there. Undersized piping restricts the flow and reduces the discharge pressure, thereby preventing the user of expensive compressed air power. Poor systems not only consume significant energy but also degrade productivity and quality.
- Use of charts showing standard pressure drop as a function of pipe size and fittings, which sizes the line for the so-called *acceptable pressure drop*. This practice can be misleading because the charts cannot accommodate velocity- and flow-induced turbulence.
- The interconnecting piping is a critical element that must deliver air to the distribution headers with small pressure drop.
- The objective in sizing interconnecting piping is to transport the maximum expected volumetric flow from the compressor discharge, through the dryers, filters, and receivers, to the main distribution header with minimum pressure drop (about 20 kPa or less).

• Excessive air velocity can be a root cause of back pressure, erratic control signals, turbulence, and turbulence-driven pressure drop. The recommended design pipeline velocity for interconnecting piping and main headers is 6 m/s or less, and never to exceed 9 m/s.

- What are the ways by which pipe sections are joined?
  - Welding (no leakage problems)
    - Welded joints: Used wherever feasible to eliminate leaks.
    - Screwed joints.
    - Flanged joints.
  - What are the different types of fittings used in joining piping?
  - Butt welded fittings: Most resistant to both vibration and fatigue. Limitations are requirement on site welding equipment, time taken for installation, and accessibility for maintenance.
  - Threaded fittings: Threaded or screwed fittings are very common. NPT (National Pipe Thread) fittings have a tapered thread on both male and female ends. Lubricant or sealant such as PTFE tape is used on the male threads to prevent damage.
  - SAE (Society of Automotive Engineers) straight thread fittings: Unlike in NPT fittings, seal is not used. Installation or accessibility is much easier.
  - ISO (International Standards Organization) parallel and tapered thread fittings.
  - NPTF (National Pipe Tapered Dry Seal Fittings).
  - 37° AN (Army–Navy) flare fittings.
- What are the different types of tube fittings?
  - Compression fittings: Made up of nut, body, and gasket ring or ferrule. Utilizes a friction grip. Unlike with pipe fittings, no special tools are required in assembling. Cannot withstand high pressures. Not good for systems involving thermal stresses, vibration, and other dynamic forces.
  - Flare fittings: Made up of nut, sleeve, and body with a flare or coed end. Can handle higher pressures and have larger seal area providing ease of maintenance.
  - Bite-type fittings: Comprised of a fitting with a nut, body, and ferrule(s) having a sharp leading edge, which bites into the skin of the tubing to achieve holding ability.
  - Mechanical grip-type fittings: Disassembly and remake after installation can be more successfully accomplished without damage than bite type.
- How are flanges rated? What are the normally used flange ratings?
  - Flanges are rated in terms of design pressures of piping.
  - Flange ratings include 10, 20, 40, 103, and 175 bar (150, 300, 600, 1500, and 2500 psig).
- When are screwed fittings and flanges used in joining pipe sections?
  - Screwed fittings for small pipe sizes [≤5 cm (2″)]: The walls should be thick enough to withstand considerable pressure and corrosion after reduction in thickness due to threading. For threaded joints more than 5 cm, assembly, size, and cost of tools increase rapidly.
  - Screwed fittings involve less leakage problems compared to flange fittings. For example, in the case of electrical conduits and equipment in flammable atmospheres, screwed fittings are preferred from safety point of view.
  - Flanges are generally used for line sizes of more than 5 cm (2 in.).
- What are the common causes for leakages in piping systems?
  - Process piping, analytical instrumentation, and utility lines involve leakages, which add significantly to process plant operating costs.
  - Common causes include system vibration, pulsation, and thermal cycling for leakages.
    - Vibration fatigue is a factor that can be aggravated by poor metallurgical consistency within the fitting material, undue stress imposed on the connection from side load or other system design characteristics, or simply improper installation practices.
    - High amplitude of alternating stress on a piping component is primarily responsible for early failure.
    - Deepness of the groove or notch made in the pipe or tubing line by the fitting as it is tightened adds to intensification for the leakage.
  - Types of connecting devices used in joining process pipe.
  - The level of knowledge and the practical experience of those installing and maintaining the system.
- How are leakages minimized/prevented?
  - Proper choice of type of fitting, including compatibility of material of construction with piping and fluids handled.
  - The fitting connection most resistant to both vibration and fatigue is a Butt welded fitting. Its ability to resist vibration and fatigue is determined by the strength and integrity of the connection made.
  - Proper installation and taping procedures for threaded connections.
  - Periodic inspection and maintenance.
- What is the desirable information that is to be incorporated in specifications of piping systems?
  - Line size.
  - Fluid handled.
  - Fluid service category.

- Material of construction.
- Pipe specification.
- Insulation specification.
- Piping and instrument diagram (P&ID).
- Line sequence number.
- From and to information.
- Pipe code.
- Heat tracing.
- Operating pressure.
- Design pressure.
- Operating temperature.
- Design temperature.
- Type of cleaning.
- Piping test pressure test fluid and type of test.

• What is a duplex tube?
  - Duplex tube is a bimetallic tube involving a combination of two different materials. Used where corrosive environments are involved with respect to one of the fluids in a heat exchanger.

• Can copper be used for piping and fittings for handing acetylene? Give reason.
  - Copper can never be used for an acetylene piping system because the acetylene will react with the copper to form hazardous and explosive compounds such as copper acetalide that explodes on contact with copper.

• What are the options available in selection of materials for modern piping systems, with special reference to gaseous systems?
  - New alloys have greatly extended the life of lines in corrosive service.
  - Improvements in welding processes, joining techniques, and fittings have reduced the maintenance tasks on many types of installations while reducing chances of leaks and purity of the resulting systems.
  - Consideration of properties of the fluid and the process or instrumentation.
  - For process and instrumentation gases, maintaining purity is a key requirement.
  - It is of no good to purchase high purity grades of instrumentation gases if the delivery pipeline contaminates them.
  - All materials selected should be cleaned to very high levels, delivered to the site capped and installed by persons using processes capable of maintaining that level of purity.
  - Copper tubing or pipe should be of a standard suitable at least *cleaned for oxygen service* that guarantees it as free of hydrocarbon contamination as possible and safe for use in oxygen piping systems.
  - Type 316L and other stainless steels or better alloys make them suitable from electropolishing.
  - For inert or nonreactive gases, using copper is as expensive as 316L stainless steel, making replacement of copper by the later as an attractive choice, due to its many advantages: it is generally suitable for even corrosive or reactive gases, the leakproof characteristics of compression fittings or welded joints is far superior, with its less installation costs.
  - For extremely corrosive or reactive sample or process lines, high nickel alloys may be a good choice.
  - Piping materials and installation for oxygen deserve special care. At concentrations above 23%, oxygen can be extremely hazardous. Even a small amount of a contaminant that is ignition proof in air will ignite or explode when exposed to a pure oxygen atmosphere. Therefore, the piping system and components must be certified as cleaned and suitably installed for oxygen service.
  - Many facilities require piping systems carrying high-pressure oxygen (20 bar) to be made of Monel, which strongly resists promoted ignition in pure oxygen.
  - Fluorine, fluorine mixtures, and certain fluorine compounds such as hydrogen fluoride require high nickel alloys or, for some components, pure nickel.
  - All components and piping must be passivated for fluorine service. This involves slowly increasing the concentration of fluorine in the system until the internal surface reaction sites of the piping and its components have a metal fluoride layer that prevents further reaction with fluorine. Any system modification or addition necessitates repeating this process.
  - Glass-coated stainless steel often is an attractive option for highly reactive samples, which imparts inertness and corrosion resistance.
  - Besides tubing, many components including valves, pressure control regulators, and different other parts can be glass coated.

• What are the advantages and applications of plastic piping?
  - In general, one of the greatest advantages of plastic pipe is its corrosion resistance.
  - Different types of plastic piping can be buried in alkaline or acidic soils without requiring any paint or special coating.
  - Plastics containing titanium oxide for ultraviolet protection strongly resist weathering.
  - Most plastic pipe also is not susceptible to scaling. Such piping systems maintain their full fluid handling capability throughout their entire service life. This means that it is often possible to reduce the

diameter of the pipe when converting from metal, reducing material costs, and to the use of smaller pumps, saving energy.

- What are the commonly used plastic piping in process industry?
  - Polyvinyl chloride (PVC):
    - Polyvinyl chloride (PVC) contains a good combination of long-term strength and high stiffness. These characteristics have made PVC the principal plastic pipe material for both pressure and nonpressure applications. An example of a nonpressure application may be using PVC as an electrical conduit.
    - PVC cannot handle high-temperature applications.
  - Chlorinated polyvinyl chloride (CPVC): CPVC provides superior chemical resistance as well as a high heat distortion temperature, due to its molecular structure. Large chlorine atoms surround the carbon backbone to protect it like armor plating.
    - Due to its high heat distortion temperature; chemical inertness; and outstanding mechanical, dielectric, flame, and smoke properties, CPVC can serve any chemical plant. Applications include processing operations, cleaning systems involving high temperatures and harsh cleaning agents.
    - CPVC piping can handle chemicals that cause process leaks, flow restrictions and, ultimately, premature failure in metal systems. It withstands most mineral acids, bases, and salts, as well as aliphatic hydrocarbons.
  - Temperature and pressure bearing capabilities of CPVC can be increased by wrapping it with fiberglass.
  - Polyvinylidene fluoride (PVDF).
  - Polyethylene:
    - Polyethylene retains its strength and flexibility even at subfreezing temperatures.
    - Coiling is even possible with small diameter polyethylene pipe; therefore, it is used for gas distribution and water services.
    - Polyethylene is also abrasion resistant.
    - Common uses also include chemical transfer, power ducts, and sewage mains. In some instances, the polyethylene molecules are cross-linked in order to raise the maximum operating temperature up to 95°C.
    - Common applications include underfloor heating systems, melting ice and snow, and hot or cold water systems.
  - Polybutylene (PB):
    - Polybutylene is flexible, yet stronger than even high-density polyethylene. Its strength increases at higher temperatures.
    - Its temperature limits are 95°C for pressure applications and slightly higher for nonpressure applications.
    - Polybutylene is mostly used for hot effluent lines and slurry transportation.
  - Polypropylene (PP):
    - Polypropylene shares similar properties with polyethylene. One key difference is polypropylene's excellent chemical resistance to organic solvents and some other chemicals.
    - Due to the chemical resistance of polypropylene, good rigidity, good strength, and high-temperature limits, it is used primarily for chemical waste movement.
    - Polypropylene's key limitation is its moderate impact resistance.
  - Acrylonitrile-butadiene-styrene (ABS):
    - ABS is formed from three distinct monomer building blocks.
    - Substances of these types are usually referred to as copolymers. The proportions of each substance will determine the physical properties of the final product.
    - In this case, acrylonitrile contributes rigidity, strength, hardness, and chemical and heat resistance.
    - Butadiene contributes impact resistance and styrene is added to increase the ease of processing.
    - ABS is primarily used for drain, waste, and venting applications.
    - However, one formulation of ABS has shown to be particularly useful for aboveground compressed air applications.

3.2 GASKETS AND MECHANICAL SEALS

- What are the methods used to prevent leaks from pipe joints and pumps?
  - By the use of gaskets and seals wherever frequent disassembly is required.
  - Bell and spigot joints, used for brittle materials, involve pouring a low melting material as seal between two pipe sections. The pouring compound may be molten, or chemical setting, or merely compacted.
  - Push-on joints require diametral control of the end of the pipe. They are used for brittle materials, usually for underground piping.

- Others include grooved joints, expanded joints, V-clamp joints, seal ring joints, pressure seal joints, soldered, brazed, cemented joints, and so on.
- Welding is another way to join piping, wherever the joints need not be disassembled for general maintenance purposes.

- What is a gasket?
  - Gaskets (made out of comparatively softer materials than the material of the pipe or section) are used to secure leakproof conditions in flanged joints in piping or between two surfaces of different sections of an equipment that must be assembled together. Tightening the bolts causes the gasket material to flow into the minor machining imperfections, resulting in a fluid-tight seal.
- What are the important factors that go into the selection of gaskets?
  - Compatibility with process fluid so that the gasket will not swell or dissolve (physical effects).
  - Should withstand thermal cycling, vibration, aging, and should not harden.
  - Should not be corrosive to flange surfaces and/or bolting.
  - Should be uniformly covering the flange surfaces.
  - Must have high coefficient of friction to be held properly between the flanges.
- What are spiral wound gaskets? What are their characteristics and applications?
  - Combination of metal and filler like asbestos or other material, spirally wound under high pressures.
  - Metals used include stainless steel, nickel, Monel, Inconel, titanium, Hastelloy, and so on.
  - Fillers used for spiral wound gaskets are asbestos, PTFE, ceramic, graphite, and so on.
  - In view of the cancer risks involved in the use of asbestos, other metal–nonmetal combinations, such as metal–TFE and metal–graphite spiral wound gaskets are marketed as substitutes to asbestos and may seal better than metal–asbestos gaskets.
  - Used for higher temperature (up to 600°C) and pressure services, including cyclic or difficultly contained fluids.
  - Density of the gasket is determined by the compactness of windings and fillers.
  - Spiral wound gaskets are also used widely in high-pressure steam services.
  - Spiral wound gaskets should preferably be used with a smooth flange finish.
  - The spiral wound gaskets are furnished with a solid metallic ring on the outside, to limit gasket compression, provides protection against blowout when used with raised facing.
- "Life of a gasket depends on the operating temperature and halves for every 10°C rise in temperature." *True/False*?
  - *True*.
- What is an O-Ring? For what purpose it is used?
  - O-ring seal joints are used for applications requiring heavy wall tubing. The outside of the tubing must be clean and smooth. The joint may be assembled repeatedly and as long as the tubing is not damaged, leaks can usually be corrected by replacing the O-ring and the anti-extrusion washer. This joint is used extensively in oil-filled hydraulic systems.
  - Viton, fluorinated ethylene–propylene polymers (FEP), vulcanized rubber, perfluoro-elastomers, and so on go into the manufacture of O-rings.
- Name some materials that go into the manufacture of gaskets, O-rings, and other seal materials for valves and other components used in piping systems.
  - CAS (compressed asbestos sheet).
  - Rubber-based compounds (usually compounded with a binder).
  - SBR (styrene-butadiene rubbers) or NBR (nitrile), FPM/Viton (fluorinated rubber), and other synthetic rubbers.
  - EPDM (ethylene–propylene).
  - TEFLON/PTFE (poly-tetra-fluoro-ethylene).
  - Viton.
- "Rubber gaskets are used to prevent oil leakages from a pipe joint." *True/False*?
  - *False*. Rubber is not compatible with oil.
- Name some of the most commonly used gasket materials used for mechanical seals for pumps in modern chemical plants and give their characteristics.
  - EPDM (ethylene–propylene copolymer): Good resistance for some ketones, alcohols, food products, ozone, hot water, and radiation. Not compatible with hydrocarbons, organic and nonorganic oils, and fats.
  - Nitrile/NBR (acrylonitrile-butadiene copolymer): Good general capabilities and excellent low temperature properties.
  - PTFE: Near universal chemical compatibility but not elastic.
  - Fluorinated rubber (FPM)/Viton): Resistance to most chemicals and ozone. Not suitable for fluids such as hot water, alcohol, steam, lye, and acid.
  - Fluoro-elastomers (FEP, fluorinated ethylene–propylene): Best standard duty elastomer with

excellent overall chemical compatibility and resistant to ozone. More elastic than PTFE.

- Perfluoro-elastomers (Kalrez and Chemraz): A wide range of specific elastomer compounds with superior chemical compatibility and higher temperature properties. Resistant to ozone. Elastic.
- Silicone (MVQ): Resistant to ozone, alcohols, glycols, and many products used in food industry. Not resistant to steam, mineral oils, organic solvents and inorganic acids.
- Flexible graphite: Near universal chemical compatibility and high-temperature capability but no resiliency.

• What are the temperature limits for use of common gasketing/mechanical seal materials?

- Asbestos: Up to 550°C.
- NBR: −40–100°C
- EPDM: −40–150°C
- Teflon: −30–200°C.
- FPM/Viton: −20–200°C.
- FEP: Suitable for temperatures up to 200°C.
- MVQ: −50–230°.
- Kalrez and Chemraz: −20–250°C.
- CAS: Up to 400°C.
- Grafoil: Up to 3000°. Graphite and metallic gasket materials are used for high temperatures.

• What is the effect of increase in temperature on the life of a gasket?

- Gasket strength deteriorates with increase in operating temperature.
- Rule of thumb is that gasket life becomes half for every 10°C increase in temperature above the minimum permissible temperatures.

• What are the requirements for seal faces? Name the commonly used face materials.

- In most seal applications, the face materials consist of a hard face and soft face. This combination has a proven record of providing a low coefficient of friction and best tolerance of face contact. The soft face can wear to match the profile of the hard face, resulting a thin fluid film and low leakage.
- Hard faces used are almost exclusively of ceramic materials. In some aggressive services, some seals use both faces made from hard materials.
- Soft faces are blends of amorphous carbon, graphite, and impregnants.
- Most hard face materials are tungsten carbide with cobalt or nickel binders, reaction bonded silicon carbide with free silicon, alumina with homogeneous silicon materials and direct sintered silicon carbide with homogeneous ceramic silicon carbide materials.
- Most common soft face materials are metalized carbon with carbon grade with metal phase, resin-impregnated carbon with general duty carbon and acid grade carbon with low impregnant and ash content.

• What are the types of mechanical seals?

- Pusher seals.
- Bellows seals.

• What are the reasons, other than loss of the fluid material, that require special attention for eliminating leakages in piping systems?

- Environmental concerns requiring minimization/elimination of stream contamination/fugitive emissions to meet increasingly stringent regulations.
- Safety issues while handling hazardous materials such as flammable, toxic, and radioactive fluids.

• What are the causes of leakage in piping systems?

- Primarily leakages occur from pipe fittings.
- Mechanical vibration leading to *vibration fatigue*, which can be aggravated by poor metallurgical consistency within the fitting material, undue stress imposed on the connection from side load or other system design characteristics, or improper installation practices. The greater the amplitude of alternating stress on the piping/fitting material, the sooner it will fail.
- The most critical areas contributing to leakage are as follows:
  - Types of connecting devices used in joining piping.
  - The level of knowledge and practical experience of those installing and maintaining the piping.

• What are the causes of flange leakage?

- Improper flange facing.
- Improper flange alignment.
- Unclean or damaged flange facings.
- Excessive loads at flange locations.
- Thermal shock.
- Differential expansion between bolts and flanges (due to temperature gradients or use of different materials for flanges and bolts, e.g., steel flanges and aluminum bolts).
- Uneven bolt stress.
- Flange bowing, bolt hole distortion or nonparallelism of flanges.
- Improper gasket material or size.
- Gasket failure that is attributed to extrusion, crushing, and creep relaxation. Creep is the cold flow or thinning of a material due to applied pressure while relaxation is the loss of spring stiffness resiliency of a material.

- What are the different methods of minimizing flange leakages?
  - One can minimize flange leakages by the following methods:
    - Reducing the distance between bolt centers.
    - Increasing the flange thickness.
    - Increasing the elastic modulus of the flange material.
    - Increasing the width of the flange sealing surface.
    - Using flat or conical washers under the bolt head.
    - Selecting a gasket material with a lower sealing stress. Using less compressible gasket material.
    - Decreasing the gasket area, which is one way to increase flange pressure.
    - Reducing gasket thickness.
    - Reducing the initial tightening torque if possible.
    - Eliminating the lubricants on the flange face and the gasket.
    - Roughening the sealing surface.
    - Use of flat washers.
    - Tightening all bolts at the same time, for example, by using a multiple head torque wrench.
- What are the materials used in pipes and tubes?
  - Cast iron, carbon and low alloy steels, high alloy stainless steels, nickel and nickel-based alloys, aluminum and its alloys, copper and its alloys, titanium and its alloys, tantalum, zirconium and its alloys, plastic and plastic-lined/coated steels, FRP, rubber and rubber lined, stoneware, cement and concrete, glass and glass-lined steel, and so on.
- What are the general considerations involved in the selection of materials for piping?
  - Considerations to be evaluated when selecting the piping materials include the following:
    1. Possible exposure to fire with respect to the loss of strength, degradation temperature, melting point, or combustibility of the pipe or support material.
    2. Ability of thermal insulation to protect the pipe from fire.
    3. Susceptibility of the pipe to brittle failure, possibly resulting in fragmentation hazards, or failure from thermal shock when exposed to fire or fire-fighting measures.
    4. Susceptibility of the piping material to crevice corrosion in stagnant confined areas (screwed joints) or adverse electrolytic effects if the metal is subjected to contact a dissimilar metal.
    5. Suitability of packing, seals, gaskets, and lubricants or sealants used on threads as well as compatibility with the fluid handled.
    6. Refrigerating or cryogenic effect of a sudden loss of pressure on volatile fluids in determining the lowest expected service temperature.
- What is the hydrostatic test pressure employed in testing piping?
  - Usually 1.5 times or more of the design pressure.
- What are the recommended practices in selecting pipe sizes?
  - Pipe sizes are normally selected by optimizing cost of energy losses due to friction that will increase with decrease in pipe diameter and increased costs of pipe that will increase with increase in pipe diameter.
  - If the velocity is too low, suspended solids may settle and air may not be swept out but collect at high points of the pipe. Velocities of over 1 m/s will be adequate for this purpose.
  - The safe lower velocity limit to avoid collecting air and sediment depends on the amount and type of solid contaminants and on the pipe diameter and pipe profile. Velocities greater than about 1 m/s are usually sufficient to move trapped air to air release valves and keep the solid contaminants in suspension.
  - Problems associated with high velocities include erosion of the pipe wall or lining/coating, cavitation at control valves and other restrictions, increased pumping costs, and increased risk of hydraulic transients.
  - A typical upper velocity for many applications involving nonviscous liquids is 6 m/s.
  - Other considerations include problems of erosion, corrosion, noise, vibration, hammer, and cavitation.
- "Sizing piping is done based on recommended velocities."
  - (a) Among dry gas and wet gas, which possesses higher recommended velocities?
    - Dry gas.
  - (b) What velocity is normally recommended for air/dry gas?
    - 30 m/s (100 ft/s).
  - (c) What velocity is normally recommended for water flow?
    - 1.2–2.4 m/s (4–8 ft/s).
  - (d) What velocity is normally recommended for pump suction line?
    - 0.15–1 m/s (0.5–3 ft/s).
- "Vapor/gas velocities to be used while sizing (designing) piping depend on line sizes and increase with increased line sizes." *True/False*?
  - *True*.
- Suggest suitable velocities for line/duct sizing for steam lines.

- Maximum velocities for saturated steam lines should be 37 m/s (120 ft/s) to avoid erosion.
- Maximum velocities for superheated, dry steam or gas lines should be 61 m/s (200 ft/s) and a pressure drop of 10 kPa/100 m (or 0.5 psi/100 ft) of pipe.
- Steam or gas lines can be sized for $6D$ m/s and pressure drops of 10 kPa/100 m of pipe.
- Liquid lines should be sized for a velocity of $(1.5 + D/3)$ m/s and a pressure drop of 40 kPa/100 m (2 psi/100 ft) of pipe at pump discharges. Generally, velocities are in the range of 1–3 m/s, depending on the nature of liquids.
- In preliminary estimates, line pressure drops may be set for an equivalent length of 30 m of pipe between each piece of equipment.

• "For the same line size, design velocities are more for saturated vapors compared to gas/superheated vapors." *True/False*?
- *False* (other way round).

• "Recommended optimum velocities in sizing piping increase with increase in fluid densities." *True/False*?
- *False.*

• Summarize typical approximate design velocities for process system applications.
- The typical design velocities for different applications are listed in Table 3.3.

• What are the recommended maximum velocities to avoid static electricity generation in pipelines handling flammable hydrocarbons?

TABLE 3.3 Design Velocities for Different Applications

Service	Velocity m/s	Velocity ft/s
Average liquid (process)	1.2–2.0	4–6.5
Pump suction (except boiling)	0.3–1.5	1–5
Pump suction (boiling)	0.25–0.9	0.5–3
Boiler feed water (discharge pressure)	1.2–2.4	4–8
Drain lines	0.46–1.2	1.5–4
Liquid to reboiler (no pump); downcomer	0.6–2.1	2–7
Two-phase flow	10–23	35–75
Vapor–liquid mixture from reboiler	4.6–9.1	15–30
Vapor to condenser	4.6–24	15–80
Vapor lines	≤0.3 Mach	
Compressor suction	23–60	75–200
Compressor discharge	30–75	100–250
Steam turbine inlet	35–100	120–320
Gas turbine inlet	45–100	150–350
Relief valve discharge	0.5 Mach	

Note: For heavy and viscous liquids, the above values should be reduced by half. The above values are for fluids with no suspended solids.

TABLE 3.4 Maximum Hydrocarbon Flow Velocities to Avoid Static Electricity Problems

Pipe Size (mm)	Maximum Velocity (m/s)
10	8.0
25	4.9
50	3.5
100	2.5
200	1.8
400	1.3
600 +	1.0

Source: Australian Standard.
Note: (1) Static electricity generation and accumulation increases with decrease in electrical conductivity of the fluids. (2) The conducting pipeline wall, well grounded, provides the path of static discharge. (3) It should be noticed that the velocities are to be reduced with increased pipe sizes as increased pipe sizes increases the path of discharge of static electricity thereby decreasing the discharge rate of static accumulation.

- The maximum velocities as function of pipe size are listed in Table 3.4.

3.3 VALVES

• What are the functions of valves?
- Blocking of flow.
- Throttling of flow.
- Prevent flow reversal.

• Illustrate the important types of valves by a suitable diagram.
- Some important types of valves are illustrated in Figure 3.5.

• Name different types of valves, stating briefly their applications and their good and bad aspects.
Gate Valves:
- Variety of designs are available, for example, plain wedge, flexible wedge, split wedge, and double disk types.
- Used for blocking/stopping flow. Operate either in fully open or in fully closed condition. Used only with clean fluids and infrequent operation.
- These bodies are used primary for hand-operated valves and valves automated for emergency shutoff. Infrequent operation.
- Low ΔP in open position as it will make full bore available for flow.
- Minimum amount of fluid trapped in the line. Cheap.
- Not suitable for flow control or slurry service.
- Flow control is very poor, particularly at low flow rates. When partially open, the crescent-type opening for flow and flow area greatly increases even with slight movement. For example, 5–10% opening results in 85–95% of full flow.

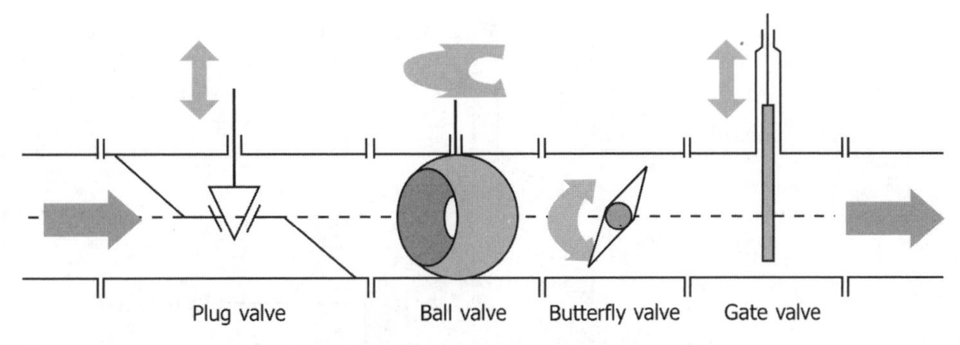

FIGURE 3.5 Important types of valves.

- High flow velocities, particularly at low openings, may lead to erosion of the disk.
- Tight shutoff is not possible when handling fluids with particulates as particles can get lodged in the valve seating preventing full closer, which can give rise to leakages. For this reason, not suitable for flammable and toxic materials.
- Cavitate at low pressure drops.
- Not quick opening type.
- A relief valve may be required in cases where pressure build up is likely due to expansion of fluid between two gate valves in closed position.A typical gate valve is illustrated in Figure 3.6.

Plug Cocks:

- These valves are used in fully open or in fully closed position.
- These are quick opening type, with a quarter turn, that is, 90° rotation of the plug is required to open or to close.

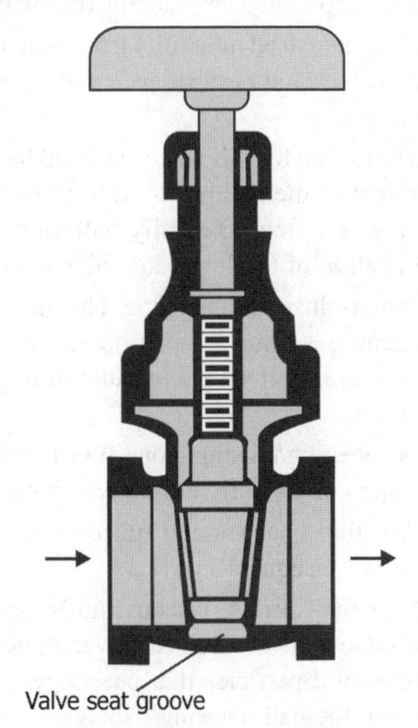

FIGURE 3.6 Gate valve.

- In plug cocks, the contact area between plug and body is large and because of this feature, there is little likelihood of leakage when closed.
- Lubricated plug cocks may use straight or tapered plugs. Lubrication permits easy and quick operation and removal of contaminants from the plug surfaces when rotated.
- For lubricated plug cocks, the lubricant must have limited viscosity change over the range of operating temperature, must have low solubility in the fluid handled, and must be applied regularly. There must be no chemical reaction between the lubricant and the fluid, which would harden or soften the lubricant or contaminate the fluid.
- These are limited to temperatures below 260°C since differential expansion between the plug and the body results in seizure.
- An elastomer is used in nonlubricated plug cocks as a body liner, a plug coating, or port seals on the body or on the plug. Provide a low-friction surface for plug rotation. Eliminate the need for frequent lubrication.
- Temperature limitation is equally relevant for plastic-coated valves.
- Because there is a large flow change near shutoff at high velocity, these valves are not normally used for throttling service.
- In control applications, it can be used for block and bypass service.

Globe Valves:

- As the valve plug changes, the area for flow between the plug and the seat (opening) changes (Figure 3.7).
- The name globe refers to the external shape of the valve, not the internal flow area.
- In cast-iron globe valves, disk and seat rings are usually made of bronze.
- Most economic for throttling flow. Can be hand controlled.
- Provides tight shutoff.
- Suitable for frequent operation.

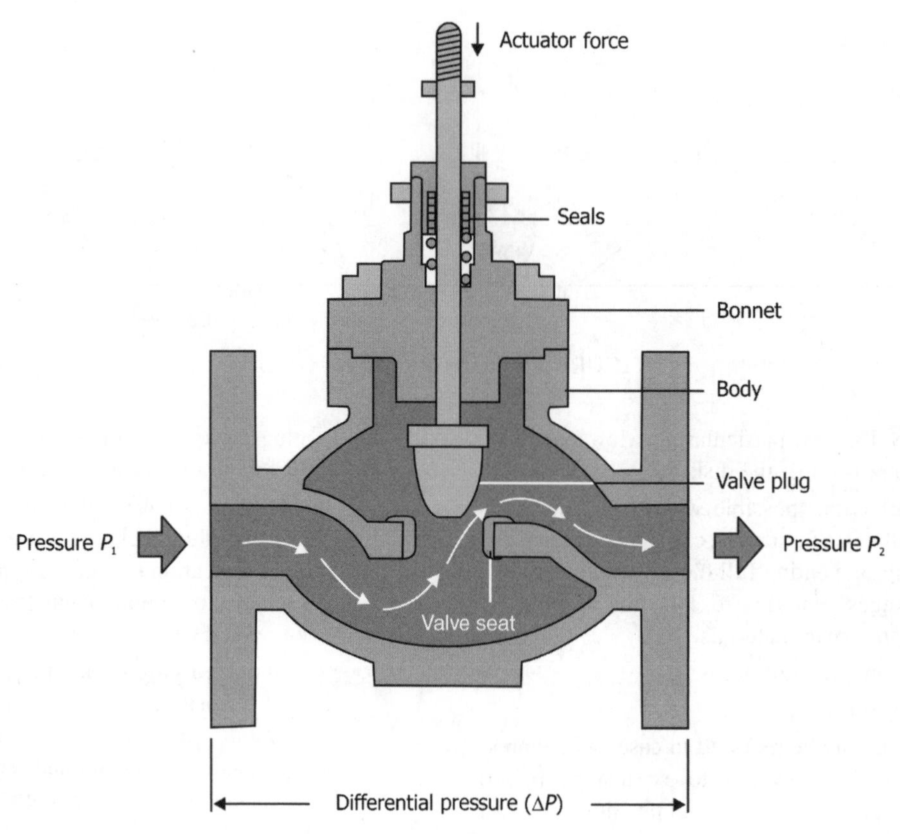

FIGURE 3.7 Globe valve.

- Best applications are for clean fluids including gases.
- It can be jacketed for heating or cooling and can be built as a diverting or blending valve.
- With disk not attached to the stem, valve can be used as a stop-check valve.
- Not good for slurries.
- Not suitable for scraping or rodding.
- Too costly for on/off block operations.
- High ΔP due to changes in flow direction.
- Frequent lubrication is necessary when used for on–off service.
- Due to the process fluid being dragged through the packing by the stem, it is susceptible to leaks causing emissions or corrosion.

Needle Valves:

- Needle valves generally are used for instrument, gauge, and meter line service.
- Very accurate throttling is possible with needle valves.
- In needle valves the valve plug is tapered with the end being a needlepoint.
- Not recommended for high temperature or slurry service.

Rotary Globe Valves:

- It is a transitional valve between the traditional rotary ball valve and the rising stem globe valve.
- It has the capability approaching that of a globe valve.
- Provides increased reliability from stem leakage over rotary valves and performance over globe valves.

Ball Valves:

- The restriction for this valve is a solid ball, which has some part of the ball removed to provide an adjustable area for flow. Basically ball valve design is a modification of nonlubricated plug cock.
- Opening is flush with the pipe. The inner diameter of the opening is same as inner diameter of the pipe in which it is installed. This results in negligible flow restriction.
- Quick opening/closing—one-fourth of a turn for fully open or for fully closed condition.
- No direction changes and full bore is available for flow and consequently low ΔP.
- Suitable for slurries, viscous fluids, and cryogenic liquids due to its low ΔP. However, there are chances for fine solid particles that get lodged between the seats and the ball, causing damage to the valve and increasing the potential for leaks. For such service, if

the valve is to be opened and closed frequently, leaks are possible.

- Tight sealing and least maintenance problems. Low cost.
- Widely used for a range of services and outsells the rotary type by a factor of 2.
- Materials of construction, such as ceramics, perfluoro-alkoxy copolymer (PFA) and stainless steel coated with corrosion and abrasive resistant materials increase its range of applications.
- For control applications, it is used in sizes less than 10 cm (4 in.).
- Not as good as globe valve for modulating control.
- Poor throttling capability.
- Not good option for corrosives or processes where cavitation may be a problem.
- Fluid can get trapped under closed condition increasing inventory of fluids, which gives rise to hazardous conditions.
- Special ball valves are available for three way applications and with vented balls.
- Figure 3.8 illustrates a typical ball valve, showing different parts.

Diaphragm Valves:

- Diaphragm valve has one surface, which is deformed by the force from the valve stem to vary the resistance to flow.
- Does not have pockets to trap solids, slurries, and other impurities.
- These valves are excellent for slurries, viscous fluids, and other unclean fluids such as sewage sludges and gritty fluids because of their simple construction and ease of maintenance and replacement of diaphragms.
- Well suited for throttling service where precise flow control is required. Several turns (four to six) are required to open and closed the valve, making fine-tuning possible to achieve the desired flow rate.
- Flow control is not good at low flow rates.
- These valves do not permit contamination of flow medium; thus, they are used extensively in food processing, pharmaceutical, brewing, and other applications that cannot tolerate any contamination.
- Suitable for corrosive fluids.
- Range is small.
- Not suitable for high-pressure and high-temperature applications.
- Small movement of diaphragm is enough for opening or closing.
- ΔP across the valve is higher than that for full bore valves.
- Diaphragms may be made from rubber with fabric reinforcement and coated/lined with Teflon. Thin corrosion-resistant metal wall diaphragms are also used for some applications.
- Requires no packing and is leakproof.
- Can be installed in any position.
- The only maintenance required is frequent replacement of the diaphragm, which can be done very quickly without removing the valve from the line.
- As the diaphragm flexes forward and backward on a frame to which it is fixed, there will be fatigue on the diaphragm, leading to loss of strength, development of cracks, and ultimate failure, requiring frequent replacement.
- Figure 3.9 illustrates a typical diaphragm valve.

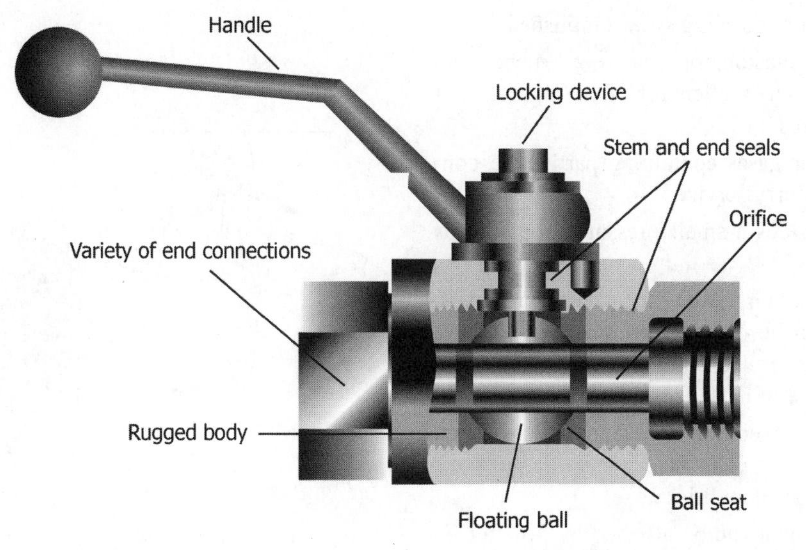

FIGURE 3.8 Ball valve. (*Courtesy*: Vaishnavi Engineering.)

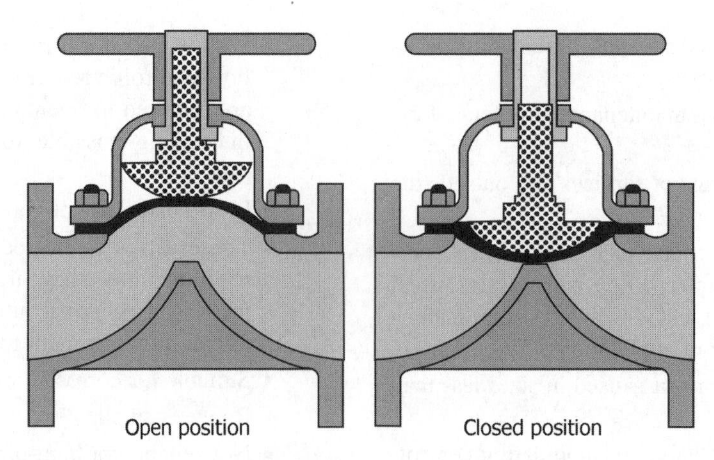

FIGURE 3.9 Diaphragm valve.

Butterfly Valves:

- Butterfly valve provides a damper that is rotated to adjust the resistance to flow. Used as dampers for control of gas/vapor flow.
- Quick opening type. Quarter turn is required for fully open or closed condition.
- These valves occupy less space in the line than any other valves, which is an advantage in large sizes (>10 cm), coupled with its lighter weight and cost.
- The main advantage of butterfly valve is that it can be installed in very large piping, as large as 2 m size, but not the best choice for small pipes (<75 mm).
- These are also available for line sizes over 20 cm, as high-performance type, in which case as ANSI rated, metal seated variety. Can operate at higher temperatures.
- Suitable for throttling service. Good for control as long as the upstream shutoff pressure is not high. Have a narrow range of control.
- Comparatively tight sealing is accomplished.
- It can have an elastomeric seat (e.g., rubber and PTFE) that makes it excellent for bubble-tight shutoff at lower pressures.
- Can be used for gases containing particulate contaminants and slurry service.
- This valve provides a small pressure drop for gas flows.
- Compared with other valves for low pressure drops, these valves can be operated by smaller hydraulic cylinders. In this service, butterfly valves are the first lowest cost valve in pipe sizes 25 cm and larger.
- Cavitation and choked flow are the main drawbacks. Where cavitation might take place, large sizes (>15 cm) are considered risky.
- The disk movement can be affected by flow turbulence as it is unguided.
- Fluid pressure distribution tends to close the valve. For this reason, the smaller manually operated valves have a latching device on the handle, and the larger manually operated valves use worm gearing on the stem.
- High torque.
- Not satisfactory for fine flow control.
- Available in lug or wafer styles.
- Figure 3.10 illustrates a butterfly valve.

Check Valves:

- Prevent flow reversal. Different types—swing check valve, lift check valve, ball check valves, restrained

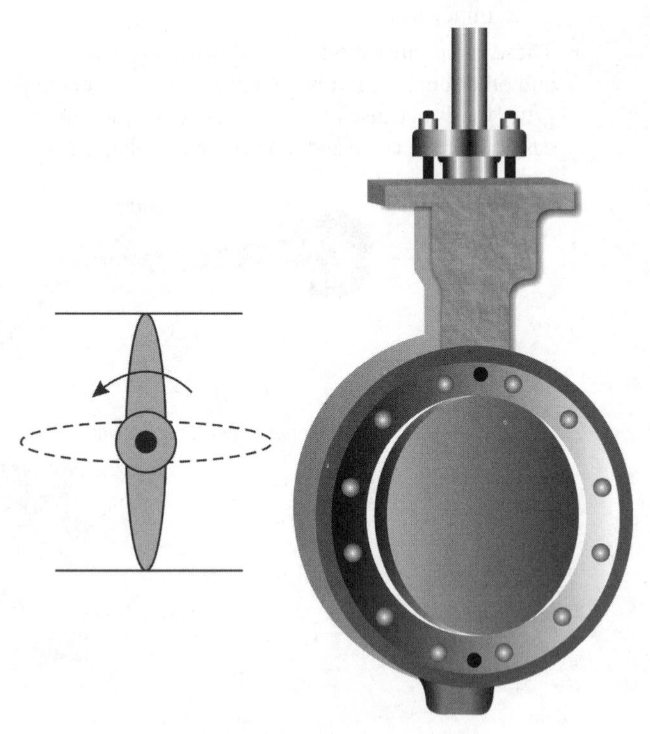

FIGURE 3.10 Butterfly valve.

check valve, tilting disk check valve, diaphragm check valve, and foot valve.

- Swing check valves are often used in piping systems with gate valves. Can be installed in horizontal piping or in vertical piping with upward flow. One difficulty with these valves is in piping systems with pulsating flow. Does not assure tight shutoff and back flow leakage may cause contamination of upstream fluid, which might also create hazards.
  - Swing check valves are commonly used in larger pipes.
- Typical swing, ball, and lift check valves are illustrated in Figures 3.11, 3.12 and 3.13.
- Ball check valve is rugged in construction and can stand repeated or cyclic operation. The ball revolves in the flowing fluid so that a different part of its surface rests in the seat each time the valve closes, thus distributing the wear on the ball. The heavy ball ensures reliability of seating in clean service. Involves high ΔP to the piping system. With low viscosity fluids, it can give rise to water hammer.
  - Ball check valves are frequently installed on tanks and prevents back flow when pump is stopped.
  - One limitation of ball check valves is its flow restriction feature. The ball is directly in the flow stream and the fluid flows around it.
  - Susceptible for damage from abrasive solids.
- Lift check valve—globe and vertical types. It is a type of check valve used on pump suction lines to prevent liquid draining out of the suction lines. Not satisfactory for unclean or viscous fluids.
- Other types of check valves include tilting disk, restrained, butterfly, and spring-loaded types.

• What is the other name for a check valve?
  - Nonreturn valve.

• What are the considerations involved in the selection of check valves?

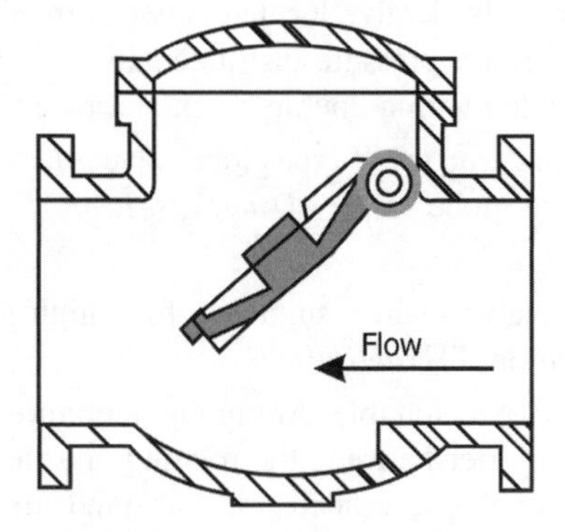

FIGURE 3.11 Swing check valve.

- Opening characteristics, that is, velocity versus disk position data.
- Velocity required to fully open and firmly backseat the disk. For most applications, it is preferable to size the check valve so that the disc is fully open and firmly backseated at normal flow rates.
- Pressure drop at maximum flow.
- Stability of the disk at partial openings. Disk stability varies with flow rate, disk position, and upstream disturbances and is an important factor in determining the useful life of a check valve.
- Sensitivity of disk flutter to upstream disturbances.
- One of the worst design errors is to oversize a check valve that is located just downstream from a disturbance such as a pump, elbow, or control valve. If the disk does not fully open, it will be subjected to severe motion that will accelerate wear. To avoid this problem, it may be necessary to select a check valve that is smaller than the pipe size.
- Speed of valve closure compared with the rate of flow reversal of the system.
- The transient pressure rise generated at check valve closure is another important consideration. The pressure rise is a function of how fast the valve disk closes compared with how fast the flow in the system reverses. The speed that the flow in a system reverses depends on the system. In systems where rapid flow reversals occur, the disk can slam shut causing a pressure transient.
- The closing speed of a valve is determined by the mass of the disk, the forces closing the disk, and the distance of travel from full open to closed.
- Fast closing valves have the following properties: the disk (including all moving parts) is lightweight, closure is assisted by springs, and the full stroke of the disk is short.
- Swing check valves are the slowest closing valves because they have heavy disks, no springs, and long disk travel.
- The nozzle check valve is one of the fastest closing valves because the closing element is light, spring loaded, and has a short stroke.
- The silent, duo, double door, and lift check valves with springs are similar to nozzle valves in their closing times, mainly because of the closing force of the spring.
- Systems where rapid flow reversals occur include parallel pumps, where one pump is stopped while the others are still operating, and systems that have air chambers or surge tanks close to the check valve. For these systems, there is a high-energy source

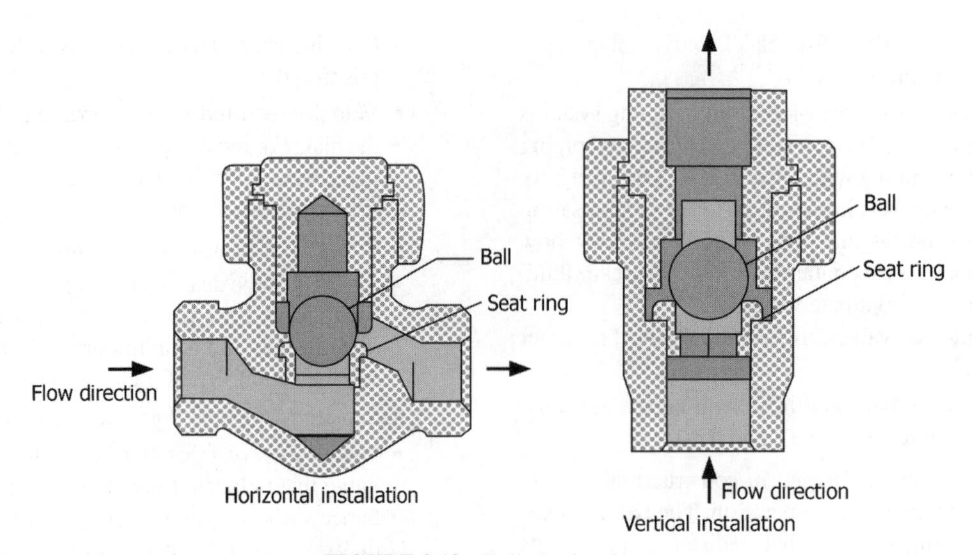

FIGURE 3.12 Ball check valves.

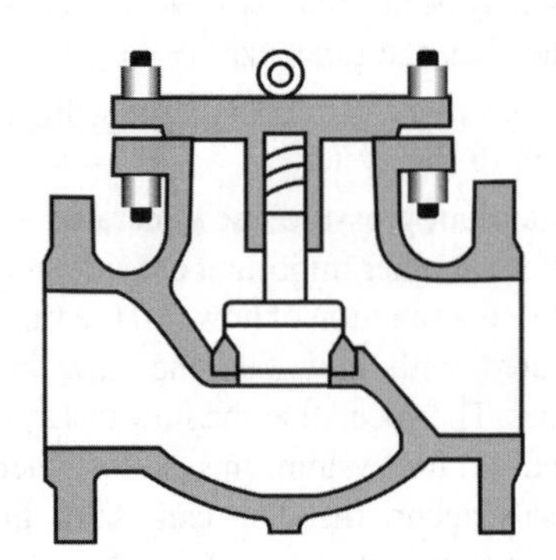

FIGURE 3.13 Lift check valve.

downstream from the check valve to cause the flow to quickly reverse. As the disk nears its seat, it starts to restrict the reverse flow. This builds up the pressure, accelerates the disk, and slams it into the seat.

- Great reductions in the transient pressures at disk closure can be achieved by replacing a slow-closing swing check valve with a fast-acting check valve. For example, in a system containing parallel pumps where the transient was generated by stopping one of the pumps, the peak transient pressure was reduced from 745 to 76 kPa when a swing check valve was replaced with a nozzle check valve.

- What are the important considerations involved in the selection of a control valve?
  - Fluid temperature, viscosity, and density.
  - Flow capacity (maximum and minimum).
  - Inlet and outlet pressures (maximum and minimum.).
  - Pressure drop during normal flow and shutoff conditions.
  - Degree of superheat or occurrence of flashing if known.
  - Noise, hammer, and cavitation problems.
  - Inlet and outlet pipe sizes and specifications.
  - Special tagging requirements.
  - Body material.
  - End connections and valve rating (screw or flange connections, etc.). Desired action (valve to open, to close, or to retain last control position).
  - Available instrument air supply pressure and instrument signal.
  - Valve type, size, and other body and packing details.
  - Valve plug action (push down to close or open).
  - Actuator size, and so on.
- What are (i) block, (ii) bleed, and (iii) double block and bleed valves?
  - Block valves: Valves that are used to shut off a separate system and applies to any valve used for shutoff service rather than for throttling.
  - Bleed valves: Small valves used to draw off liquids.
  - Double block and bleed valves: Two block valves with a bleed valve located between them. Bleed off valving for pressure instrumentation on hot lines is provided with a double block and bleed valve.
- "Pressure drop in a Y-type globe valve is less than that in a normal globe valve." *True/False*?
  - *True.*
- "Ball valves are suitable for high-temperature applications." *True/False*?
  - *False.* Not suitable. At high temperatures, the polymer barrier between the rotating and the stationary part of the valve will not withstand high temperatures. Use of graphite seating may permit operation at higher temperatures.

- "Ball valves can create hazardous conditions in flammable atmospheres." Comment.
  - Liquid is trapped in the valve bore in closed condition increasing inventory of flammable/hazardous materials. Also such trapped liquid can give rise to corrosion problems. Some designs of ball valves, in larger sizes, have provision to drain out trapped liquid on shutdown to overcome these problems.
  - Polymeric seats/packing used in ball valves can electrically insulate ball and stem from valve body. Movement of ball relative to the polymeric film between the ball and the stationary surface can give rise to static electricity generation that can act as ignition source for flammables.
  - Use of conducting polymeric films or use of antistatic device to discharge static electricity reduce these problems.
  - If left open partially for extended periods under high pressure drop conditions, soft seat will flow around edge of the ball opening and possibly lock the ball in that position.
- "Gate valve is not normally recommended for handling unclean fluids/slurries." Comment.
  - Particulates can get embedded in the valve seat preventing complete shutoff resulting in leakages.
- "It is not desirable to keep a gate valve partially open." Comment.
  - Partially open condition can lead to distortion and bending caused by increased velocities of the flowing fluid with attending turbulence.
- What is a flush bottom valve? What is its application?
  - Fitted at the bottom of a process vessel for complete drainage of the vessel contents through the valve.
  - Its distinguishing feature is that the closure is flush with the bottom of the vessel so that there is no holdup of contents in the valve body. The outlet of the valve discharges at a 45° angle from the vertical.
  - Its disadvantage is that it requires considerable vertical distance under the bottom of the vessel for installation and operation.
- What are the salient features of a flush bottom valve?
  - For controlling flow of liquid/slurry from bottom of a vessel.
  - No holdup in the valve body.
  - Complete drainage of vessel contents through the valve.
  - No stagnant liquid pockets in outlet nozzle.
  - No lubricant is required on internal working parts of the valve and therefore no contamination of process fluid.
- What is a foot valve? Where is it used?
  - Foot valve is used in pump suction lines to prevent back flow of liquids from the pump discharge side, reversing the direction of movement of pump impeller, which might damage the pump, apart from draining the pumped liquid back. It is a type of check valve.
  - In addition, when the pump is restarted, it will run dry without pumping liquid due to air entering the pump casing when it is drained.
- What is a flexible valve? What are its plus and minus points?
  - Flexible valves consist of a flexible tube/hose with a clamping device for squeezing the tube for closure, for example, pinch clamps on plastic tubes.
  - Flow is controlled by squeezing the tube.
  - The simplest form of the device is the spring pinchcock.
  - Fluid will not come in contact with any moving parts of the valve.
  - It is leakproof and requires no packing.
  - Low pressure drop.
  - Flow control is good for moderate to full flow.
  - Used for sanitary handling of biofluids and food- and pharmaceutical-related fluids for which contamination is detrimental.
  - Can be used for abrasive/corrosive fluids, slurries, and powders/granules.
- What are the limitations of a flexible valve?
  - Repeated squeezing/pressing of the tube/hose results in loss of flexibility and weakening of the tube/hose requiring frequent replacements.
  - Not suitable for high-pressure and high-temperature applications.
  - Vacuum service requires use of heavy-walled hose, which is expensive and has low levels of flexibility.
- Name some applications for pinch valves in slurry service.
  - Lime slurries, which are associated with abrasion and scaling problems.
  - Abrasive copper concentrate and metal minerals. Requires changes in rubber sleeve once in 1–3 years.
  - Plastic beads with abrasive characteristics.
  - Chemical pulp stock, which is slightly abrasive.
  - Cement and sand, which are abrasive in nature. Requires changes in rubber sleeve once in 1–2 years.
  - Titanium dioxide. Abrasive.
  - Pulp stock rejects. Abrasive. Requires changes in rubber sleeve once in 6 months to 1 year.

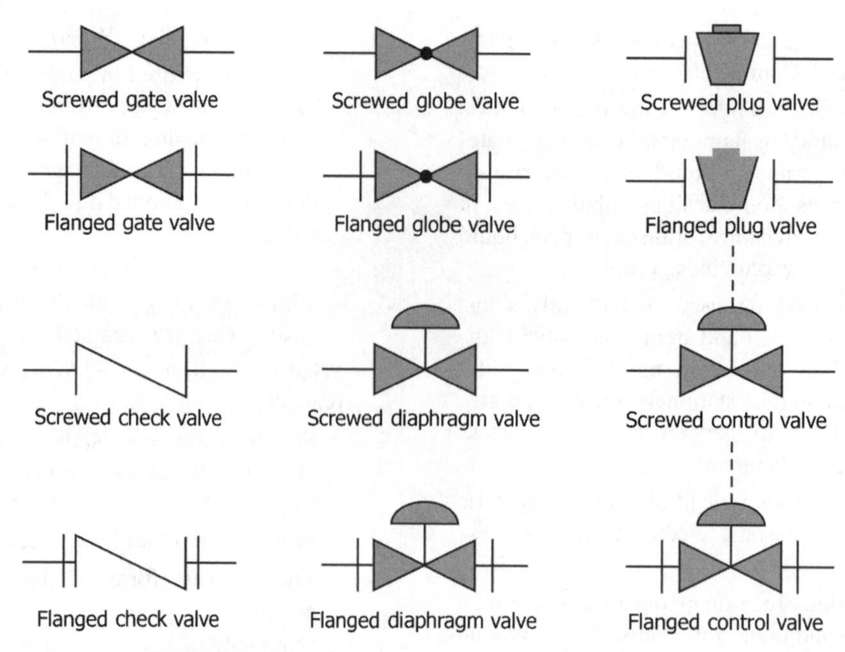

FIGURE 3.14 Symbols for some common types of valves.

- What are the symbols used for different types of valves? Illustrate with diagrams.
  - Symbols of some commonly used valves are given in Figure 3.14.
- What type of valves do you recommend for gas lines requiring tight shutoff?
  - Globe valves are most commonly used for gases for control and when tight shutoff is required.
  - Where the valve is to function for fully open or fully closed conditions requiring no flow control, gate valves are used.
- What types of valves do you recommend for use on slurry lines?
  - Pinch valve is claimed to be one of the best valves available for slurry service. This type of valve can handle abrasive slurries such as lime, minerals such as alumina, rejects from pulp stocks, titanium dioxide, cement, and sand.
  - Diaphragm valves are used for slurries for a long time. Diaphragms require frequent replacement, especially for abrasive slurries.
  - Ball valves are used for nonabrasive slurries, as ball moves across seats with a wiping action.
- "Friction losses are more for a gate valve compared to globe valve both being under fully opened conditions." *True/False*?
  - *False.* With a gate valve, flow direction will not change whereas in a globe valve flow direction changes increasing ΔP.
- "A ball cock is used to prevent overflow from an overhead tank." Illustrate how it works.
  - Ball cock is connected to a float, which is a plastic ball. As liquid level rises, the float moves up over the liquid surface. When the tank fills up to its predetermined level, the lever attached to the ball closes the valve stopping the flow.
- "Pressure drop in a Y-type globe valve is less than that for a standard globe valve." *True/False*?
  - *True.*
- "In a laboratory experimental setup, gate valves are installed for varying water flow rates." Is it the correct arrangement? Explain.
  - *No.* Not suitable for flow control. Reason is explained elsewhere. Gate valves are operated either in fully open or in fully closed condition.
- What are safety, relief, and safety relief valves?
  - Safety valve is an automatic pressure-relieving device actuated by the static pressure upstream of the valve and characterized by full opening pop up action.
    - It is used for air, steam, or vapor service.
    - Rated capacity is reached at 3%, 10%, or 20% overpressure, depending on applicable code.
  - Relief valve is an automatic pressure-relieving device actuated by the static pressure upstream of the valve that opens further with the increase in pressure over the opening pressure.
    - It is used primarily for liquid service.

- Rated capacity is usually attained at 25% overpressure.

- Safety relief valve is an automatic pressure actuated relieving device suitable for use either as a safety valve or as a relief valve, depending on the application.
  - It is characterized by an adjustment to allow reclosure, either a *popup* or a *nonpopup* action and a nozzle-type entrance.
  - Opens in proportion to increase in internal pressure.
  - It reseats as pressure drops.
  - Used on steam, gas, vapor and liquid (with adjustments).
  - Most general type in petroleum and petrochemical plants.
  - Rated capacity is reached at 3% or 10% overpressure, depending on code and/or process conditions.
- A typical relief valve assembly is shown in Figure 3.15.

- What is an air/vacuum valve?
  - The air/vacuum valve is designed for releasing air while the pipe is being filled and for admitting air when the pipe is being drained.
  - The valve must be large enough that it can admit and expel large quantities of air at a low ΔP. The outlet orifice is generally of the same diameter as the inlet pipe.
- How does an air/vacuum valve operate?
  - These valves typically contain a float that rises and closes the orifice as the valve body fills with liquid.

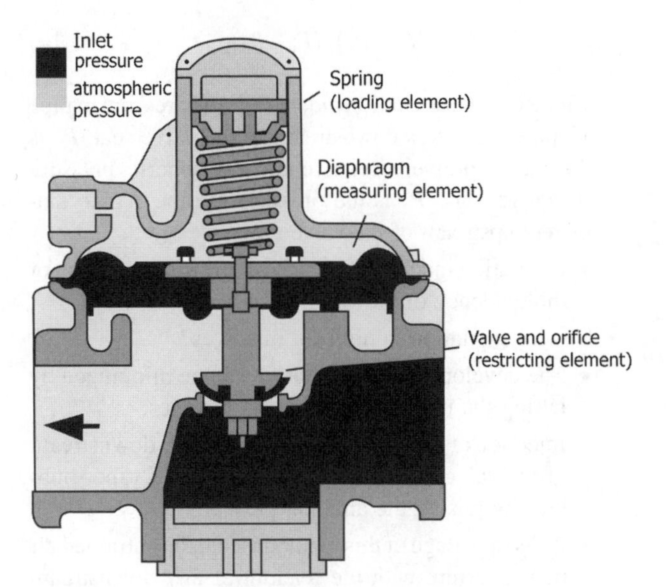

FIGURE 3.15 Relief valve.

Once the line is pressurized, this type of valve cannot reopen to remove air that may subsequently accumulate until the pressure becomes negative, allowing the float to drop. If the pressure becomes negative during a transient or while draining, the float drops and admits air into the line.

  - Air release valves contain a small orifice and are designed to release small quantities of pressurized air trapped during filling the small orifice is controlled by a plunger activated by a float at the end of a lever arm. As air accumulates in the valve body, the float drops and opens the orifice. As the air is expelled, the float rises and closes off the orifice.
- What are the desirable characteristics for the selection of a valve for flow control?
  - For many flow control applications, it is desirable to select a valve that has linear control characteristics. This means that if you close the valve 10%, the flow reduces to about 10%.
  - Selecting the proper flow control valve should consider the following criteria:
    - The valve should not produce excessive pressure drop when fully open.
    - It should control at least 50% of its movement.
    - At maximum flow, the operating torque must not exceed the capacity of the operator or valve shaft and connections.
    - The valve should not be subjected to excessive cavitation.
    - Pressure transients should not exceed the safe limits of the system.
    - Some valves should not be operated at very small openings. Other valves should be operated near full open.
- What are the solutions available to overcome transient problems encountered in valves meant for flow control?
  - Increasing the closing time of control valves.
  - Using a smaller valve to provide better control.
  - Designing special facilities for filling, flushing, and removing air from pipelines.
  - Increasing the pressure class of the pipeline.
  - Limiting the flow velocity.
  - Using pressure relief valves, surge tanks, air chambers, and so on.
- What is the operating range for a valve to get good control?
  - It is much easier to control a valve in the 10–80% stroke range. Using the lower 10% and upper 20% of the valve stroke should be avoided.

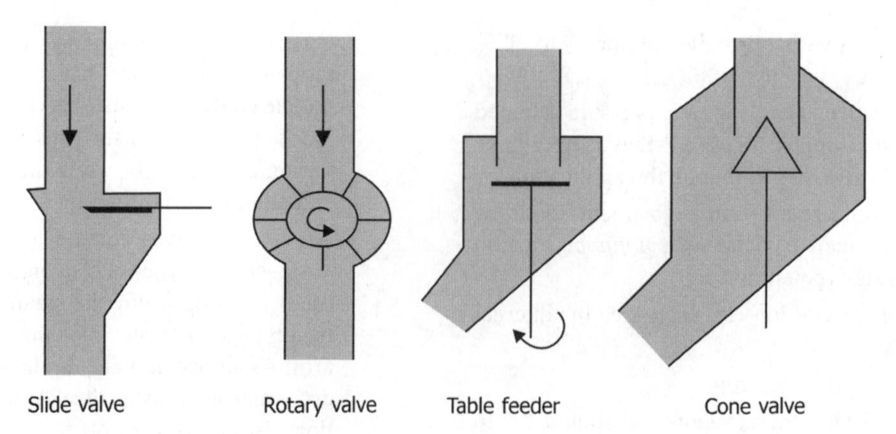

FIGURE 3.16 Different types of valves for solids flow.

- Name and illustrate different types of valves used for solids flow.
  - Typical solids flow valves are shown in Figure 3.16.
- What are the different problems associated with valves?
  - A high ΔP through a valve creates a number of problems, such as cavitation, hammer effects, flashing, choked flow, high noise levels and vibration, leading to erosion or cavitation damage, malfunction, or poor performance.
  - Blockages and leakages.
  - Seizure, that is, getting stuck.
- Describe cavitation in valves.
  - When liquid passes through partially closed valve, static pressure in the region of increasing velocity and in the wake of the closure member, drops and may reach vapor pressure of the liquid in the low pressure region may vaporize and form vapor-filled cavities that grow around minute gas bubbles and impurities in the liquid.
  - When the liquid reaches a region of high static pressure, the bubbles collapse suddenly.
  - Impingement of opposing liquid elements in the form of microjets on the collapsing vapor bubbles produces locally very high pressures that are short lived. The microjets can reach velocities of 50–100 m/s.
  - For water, pressure surges with amplitudes between 750 and 1500 N/mm^2 are reached with the jet velocities mentioned above.
  - The surge lasts between several microseconds and several milliseconds.
  - The effect of one single surge is limited to an area of only a few micrometers in diameter.
  - If the collapse occurs near the boundaries of the valve body/pipe wall, local fatigue failures that can cause boundary surfaces roughened until large cavities are formed.

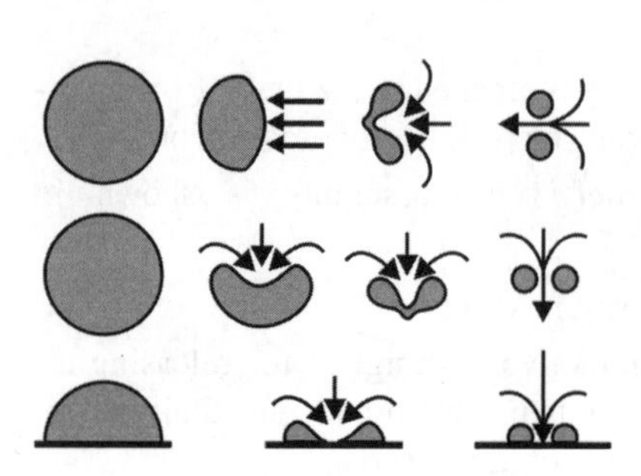

FIGURE 3.17 Collapse of a bubble. (*Source:* Samson AG, Frankfurt.)

  - Figure 3.17 illustrates the collapse of a bubble.
- What is cavitation index with respect to valves? What is its significance?

$$C = (P_d - P_v)/(P_u - P_d), \tag{3.17}$$

where C is the cavitation index, P_d is the pressure in pipe 12 pipe diameters downstream of the valve seat, P_v is the vapor pressure relative to atmospheric pressure (negative), and P_u is the pressure in pipe 3 pipe diameters upstream of the valve seat.

  - Cavitation index indicates the degree of cavitation or the tendency of the valve to cavitate.
- How is cavitation minimized in valves?
  - The development of cavitation can be minimized by letting the pressure drop occur in stages.
  - Injection of compressed air immediately downstream of the valve minimizes the formation of vapor bubbles by raising the ambient pressure.
  - A disadvantage of this method is that the entrained air may interfere with the reading of any downstream instrumentation.

TABLE 3.5 Recommended Equivalent Lengths of Valves and Fittings

Valve/Fitting	Description	Equivalent Length (*L/D*)	Remarks
Globe valve	Fully open	340–450	Depending on the type of seat and disk
Y-pattern	Fully open	175	Stem 60° from pipe run
Y-pattern	Fully open	145	Stem 45° from pipe run
Angle valve	Fully open	145–200	Depending on the type of seat and disk
Gate valve	Fully open	13	—
	Three-fourth open	35	—
	Half open	160	—
	One-fourth open	900	—
Check valve	Fully open	50–150	Depending on the type
Butterfly valve	Fully open	40	8 in. and larger sizes
Foot valves with strainer	Fully open	75–420	Depending on the type
90° Standard elbow	—	30	—
45° Standard elbow	—	16	—
90° Long-radius elbow	—	20	—
Standard tee	—	20	With flow through run
Standard tee	—	60	With flow through branch

- Sudden enlargement of the flow passage just downstream of the valve seat can protect the boundaries of valve body and pipe from cavitation damage. A chamber with a diameter 1.5 times the pipe diameter and a length of 8 times the pipe diameter including the exit taper has proved satisfactory for needle valves used in waterworks.
- Give an example of severe noise generation involving valves.
  - Letting down of gas by valves from high pressure to low pressure can produce excessive noise, which is mainly due to the turbulence generated by the high velocity jet shearing the relatively still medium downstream.
- How is valve noise attenuated?
  - By the use of perforated diffuser-type silencers in which gas is made to flow through numerous orifices.
- How is hammer minimized/avoided in valves?
  - By providing one or more surge protection devices at strategic locations in piping.
  - A stand pipe containing gas in direct contact with the liquid or separated from the liquid by a flexible wall/membrane or a pressure relief valve.
  - By changing acoustic properties of the fluid, for example, by introducing insoluble gas bubbles into liquid stream.
  - Letting down of gas by valves from a high pressure to low pressure can produce unbearable noise, which is mainly due to the turbulence generated by the high velocity jet shearing the relatively still medium downstream.

TABLE 3.6 Causes and Consequences of Operational Deviations of Piping Systems

Deviation	Causes and Consequences
Overpressure	Blockage of piping, valves, or flame arresters due to solid deposition.
	Rapid closure of valve in the line resulting in liquid hammer and pipe rupture.
	Thermal expansion of liquid in blocked line leading to line rupture.
	Automatic control valve opens inadvertently leading to high-pressure downstream of the valve.
	Block valve upstream or downstream of relief device accidentally closed resulting in loss of relief capability.
	Blockage of relief device by solids deposition (polymerization, solidification).
	Deflagration and detonation in pipelines causing failure and loss of containment.
High temperature	Faulty heat tracing or jacketing of line leading to hot spots resulting in exothermic reactions.
	External fire leading to undesired process reactions (e.g., acetylene decomposition).
Low temperature	Cold weather conditions causing freezing of accumulated water or solidification of product in line or dead ends.
	Condensation in steam lines due to cold ambient conditions resulting in steam hammer.
High flow	High fluid velocity in piping causing erosion especially if two phase flow or abrasive solids are present leading to loss of containment.
	High pressure drop across control valve causing flashing/vibration leading to loss of containment.

(*continued*)

TABLE 3.6 (*Continued*)

Deviation	Causes and Consequences
Reverse flow	Differential pressure on joining lines, drains or temporary connections causing back flow of fluid resulting in undesirable reaction or over-filling, and so on.
Loss of containment	Failure to isolate flow from sample connection, drain and other fittings resulting in discharge to environment.
	Breakage of sight glasses and glass rotameters due to overpressure, thermal stress, or physical impact.
	Loss of containment due to leakages from piping, flanges, valves, Hoses, pipe rupture, collision, or improper support.
	Pipe failure due to excessive thermal stress.
	Breakdown of pipe/hose lining.

- Use of perforated diffuser-type silencers in which the gas is made to flow through numerous small orifices (as small a possible, minimum diameter of 5 mm, dictated by considerations of blockage). Shapes can be perforated plate, cone, or bucket.
- Hammer in valves will be reduced if valve opening/closing is made gradual than sudden. Dampening these operations considerably reduces hammer.

• Give recommended equivalent lengths for valves and pipe fittings for the estimation of pressure drops.

- The recommended equivalent lengths for valves and fittings are listed in Table 3.5.

• Summarize operational deviations and their causes and consequences in piping systems.

- A summary is presented in Table 3.6.

4

FLOW MEASUREMENT

4.1 Flow Measurement 59
- 4.1.1 Differential Pressure Flow Meters 60
  - 4.1.1.1 Orifice and Venturi Meters 60
- 4.1.2 Other Differential Pressure Flow Meters 68
  - 4.1.2.1 Pitot Tubes 71
- 4.1.3 Variable Area Flow Meters 73
  - 4.1.3.1 Rotameters 73
- 4.1.4 Mechanical Flow Meters 78
  - 4.1.4.1 Turbine and Paddle Wheel Flow Meters 78
- 4.1.5 Electronic Flow Meters 81
  - 4.1.5.1 Ultrasonic Flow Meters 81
  - 4.1.5.2 Magnetic Flow Meters 83
- 4.1.6 Mass Flow Meters 85
  - 4.1.6.1 Coriolis Flow Meters 85
- 4.1.7 Positive Displacement Flow Meters 88
- 4.1.8 Miscellaneous Types of Flow Meters 94
  - 4.1.8.1 Target Flow Meters 94
  - 4.1.8.2 Vortex Shedding Flow Meters 94
  - 4.1.8.3 Anemometers 96
  - 4.1.8.4 Bubble Flow Meters 97
- 4.1.9 Weirs 98

4.1 FLOW MEASUREMENT

- What are the requirements for reliable flow measurement in piping systems?
  - The basic requirement for accurate flow measurement is to ensure that the flow is fully developed.
  - Uncertainty values for metering devices are always based on the steady-state flow of the single-phase homogenous Newtonian fluids with a meter entrance velocity profile that resembles that obtained in a long straight run of pipe. Departure from these flow conditions will result in metering errors due to installation effects. The fluid flow condition at the flow meter entrance necessary for accurate measurement is referred to as the fully developed velocity profile.
  - ISO 5167 refers to a *fully developed velocity flow profile* as a state of flow where the actual fluid velocity flow profile resembles the fully developed state within 5%, and the swirl angle is within 2° of zero. This requirement is deemed to be acceptable for all types of metering applications, yet not imposing impossible restrictions on fluid flow conditions in the pipe.
  - Ninety-nine percent of flow meter applications do not exhibit the fully developed flow state. Piping lengths of up to 150–200 internal pipe diameters are required to create this essential pipe flow condition. The alternative is to install flow conditioners into the pipe. This is the most economical and accurate method to create a fully developed flow profile in the pipe with meter run lengths as short as 13 pipe diameters.
- How are pulsations eliminated in gas flow upstream of a metering device?
  - For gas flow, a combination of a surge chamber located close to pulsation source and a constriction in the line between surge chamber and metering element are used to eliminate pulsations.
- What is *turndown ratio* for a flow meter? Illustrate.
  - Ratio of the highest to lowest flow rates measurable by the flow meter is called turndown ratio.
  - Turndown ratio is often used to compare the span (the range) of flow measuring devices.
  - Turndown ratio can be expressed as

$$\mathrm{TR} = q_{\max}/q_{\min}, \tag{4.1}$$

where TR is the turndown ratio, $q_{\max}$ is the maximum flow, and $q_{\min}$ is the minimum flow.

  - Maximum and minimum flows are stated within a specified accuracy and repeatability for the device.
  - Turndown ratios are illustrated in Figure 4.1.

Fluid Mechanics, Heat Transfer, and Mass Transfer: Chemical Engineering Practice, By K. S. N. Raju

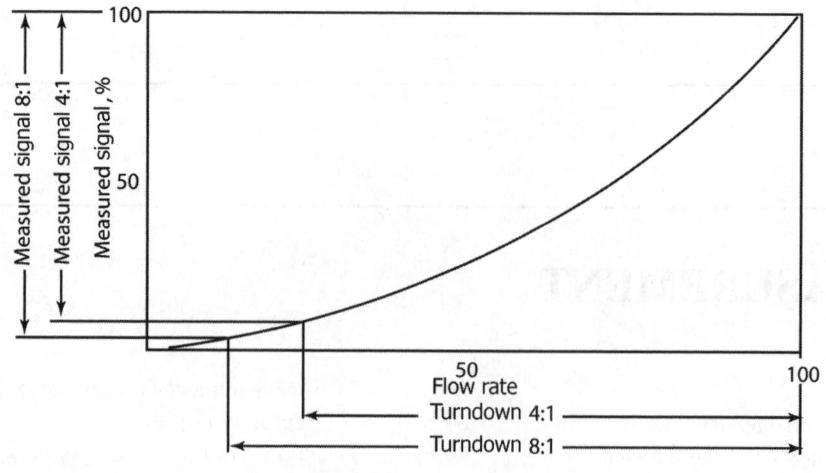

FIGURE 4.1 Turndown ratios.

- How is full flow through a flow meter ensured during flow measurement in a pipe?
  - Full conduit flow is important in liquid systems. The flowing pipe must run full, or measurements made will be in error. This can be a problem if piping design does not keep the meter below the rest of the piping.
  - If the meter is at the high point, vapor can collect and create a void in the meter so any velocity or volume displacement measured will be in error.
  - This point is illustrated in Figure 4.2.

4.1.1 Differential Pressure Flow Meters

4.1.1.1 Orifice and Venturi Meters

- What are obstruction flow meters? Are they same as differential pressure meters? What is the principle involved in these meters?
  - Obstruction flow meters, also called differential pressure (DP) flow meters, are based on the differential pressure measurements.

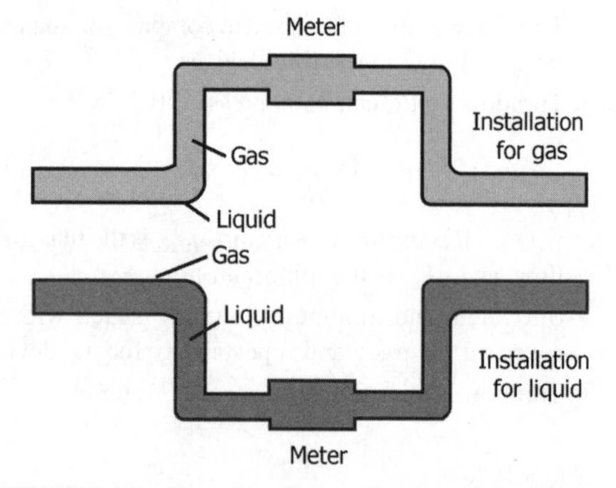

FIGURE 4.2 Meter installations for gas and liquid flows.

> All DP primary elements restrict the flow in some way. A restriction in a pipe results in an increase in the fluid velocity, according to Bernoulli's law of conservation of energy. The ensuing conversion to kinetic energy reduces the static pressure. This pressure drop, the measured DP, is proportional to the square root of the flow rate and thus provides a means to measure flow.

  - Orifice meter, venturi meter, and flow nozzle are examples of obstruction flow meters. They have no moving parts and can be fabricated in a wide selection of materials. Their purchase cost is relatively low, even for large pipe sizes.
  - Accuracy is moderate, ranging from 1% to 5%; compensation techniques can improve these values to better than 1%. DP meters are generally easy to select for a specific application.
  - When the flow area changes abruptly, the effective flow area immediately downstream of the alteration will not necessarily be the same as the pipe flow area. This effect is brought about by an inability of a fluid to expand immediately upon encountering an expansion as a result of the inertia of each fluid particle. This forms a central core flow bounded by regions of slower moving recirculating eddies.
  - As a consequence, the pressure sensed with pipe wall taps located within the vena contracta region will correspond to the higher moving velocity within the vena contracta of unknown flow area, A_2.
  - The unknown vena contracta area will be accounted for by introducing a contraction coefficient, C_c.
  - The low pressure at the point of highest velocity creates the possibility for the liquid to partially vaporize (called *flashing*). It might remain partially vaporized after the meter or it might return to a liquid as the pressure increases after the lowest pressure

point (called *cavitation*). Any vaporization should be avoided to ensure proper operation of the meter and to retain the relationship between pressure difference and flow. Vaporization can be prevented by maintaining the inlet pressure sufficiently high and the inlet temperature sufficiently low.

- What is an orifice meter?
  - An orifice meter consists of a circular plate, containing a hole (orifice), which is inserted into a pipe such that the orifice is concentric with the pipe inside diameter.
- What is the principle of operation of orifice meter?
  - As the fluid approaches the orifice, the pressure increases slightly and then drops suddenly as the orifice is passed. It continues to drop until *vena contracta* is reached and then gradually increases until at approximately 5–8 diameters downstream a maximum pressure point is reached that will be lower than the pressure upstream of the orifice. The decrease in pressure as the fluid passes through the orifice is a result of the increased velocity of the fluid passing through the reduced area of the orifice. When the velocity decreases as the fluid leaves the orifice the pressure increases and tends to return to its original level. All of the pressure loss is not recovered because of friction and turbulence losses in the stream. The pressure drop, ΔP, across the orifice increases when the rate of flow increases.
  - With the orifice plate or any flow measurement based on differential pressure, pressure drop is at zero at no flow and then increases with the square of the flow. Thus, if the application requiring a 5-to-1 turndown in flow, a differential pressure transmitter with a turndown of 25-to-1 is required. This can create limitations with the application of orifice meters.
  - Similarly, when sizing an orifice to handle the maximum flow rate, it may not be possible to also measure the flow at the lower end of the range, due to loss of signal.
  - Pressure drop for a gas flow is based on the flow rate at the gas density at the actual operating conditions. To get a mass flow measurement at standard conditions of temperature and pressure, it is also necessary to have temperature and pressure transmitters and a flow computer or multivariable transmitter. As a result, while the cost of the orifice plate itself is relatively inexpensive, the installed cost of the complete system becomes substantially more expensive due to the additional instrumentation that is required to obtain an accurate mass flow measurement.
- What is vena contracta?
  - Vena contracta is the position on the downstream side of an orifice plate or a nozzle from which a fluid is flowing out at which the pressure becomes the lowest with fluid velocity attaining its maximum value.

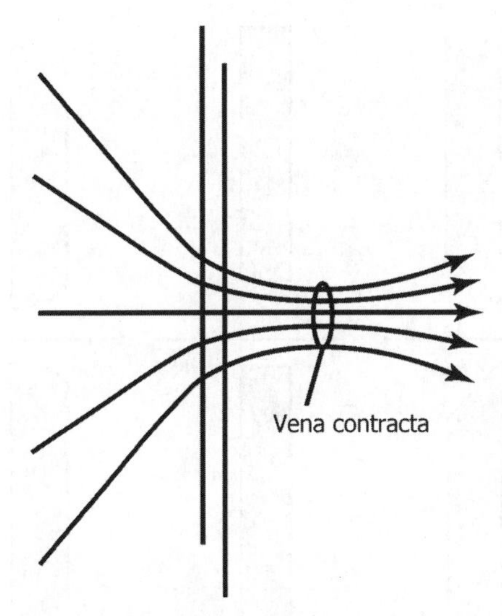

FIGURE 4.3 Vena contracta.

  - Vena contracta is the point in a fluid stream where the diameter of the stream is the least (Figure 4.3).
  - The maximum contraction takes place at a section slightly on the downstream side of the orifice, where the jet is more or less horizontal.
  - The reason for this phenomenon is that fluid streamlines cannot abruptly change direction. In the case of both the free jet and the sudden pipe diameter change, the streamlines are unable to closely follow the sharp angle in the pipe/tank wall. The converging streamlines follow a smooth path, which results in the narrowing of the jet (or primary pipe flow) observed.
- What is sharp-edged orifice? What is its advantage?
  - With sharp-edged orifice, the cross section of discharging liquid, contracts as it leaves the opening. Figure 4.4 illustrates a sharp-edged orifice along with square and rounded orifices. Sharp-edged orifices are used as standard orifices. Flow through the other two orifices is affected by the thickness of the plate and the roughness of the surface, and for the rounded type, the radius of curvature.
  - Velocity attainable will be maximum and consequently pressure will reach its lowest value at vena contracta and pressure differential developed will be highest, making accuracy of measurements by a manometer more precise.
  - If the orifice is blunt edged, for a given flow rate, head will be less than that for sharp-edged orifice.
- For a sharp-edged orifice, how is the area of vena contracta related to downstream pressure?
  - Vena contracta area increases with decrease in downstream pressure.

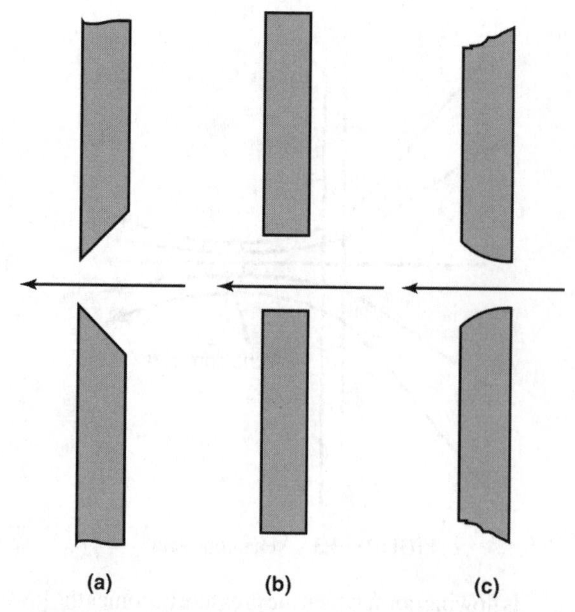

FIGURE 4.4 Types of orifice plates. (a) Sharp-edged, (b) thick plate, and (c) thick plate with curved radius.

- What is beta-ratio?
  - Beta ratio (β) is the ratio of orifice plate bore divided by pipe internal diameter, that is, d/D where d is the plate bore and D is the inside diameter of the pipe.
- What are the different types of taps used in orifice meters?
  - Flange taps: These taps are located 1 in. (25.4 mm) from the upstream face of the orifice plate and 1 in. (25.4 mm) from the downstream face with a $+1/64$ to $+1/32$ tolerance. These are more commonly used.
  - Figure 4.5 illustrates installation and flow pattern for a sharp-edged orifice with flange taps.
  - Radius taps: Radius taps are similar to vena contracta taps, except the downstream tap is fixed at $0.5D$ from the orifice plate. On the upstream side, these are located at a distance of 1 pipe diameter from the orifice plate.
  - Pipe taps: These taps are located at 2.5 nominal pipe diameters upstream and at 8 nominal pipe diameters downstream (point of maximum pressure recovery). They detect the smallest pressure difference and because of the tap distance from the orifice, the effects of pipe roughness, dimensional inconsistencies, and, therefore, measurement errors are the greatest.
  - Vena contracta taps: Upstream static hole is 0.5–2 pipe diameters from the orifice plate. Downstream tap is located at the position of minimum pressure, called vena contracta. This point, however, varies with the β-ratio. Usually, upstream tap is located 1 pipe diameter (inside) and 0.3–0.8 pipe diameters downstream from the face of the orifice plate. They are not commonly used in other than plant measurements, where flows are relatively constant and plates are not changed. If the plate is changed, it may require a change in the tap location. Exact dimensions are given in appropriate tables.
    - The vena contracta taps provide not only the maximum pressure differential but also give rise to noise. Also, in small pipes, the vena contracta might lie under a flange. Therefore, vena contracta taps normally are used only in pipe sizes exceeding 15 cm (6 in.).
  - Corner taps: Static holes are drilled, one in the upstream and one in the downstream flange, with openings as close as possible to the orifice plate.
  - Corner taps are widely used in Europe. In-line sizes less than 5 cm, they are used with special honed flow meter tubes for low flow rates.
  - Figure 4.6 illustrates orifice tap location alternatives.
- Name some methods of reducing upstream calming section requirements for a head flow meter.

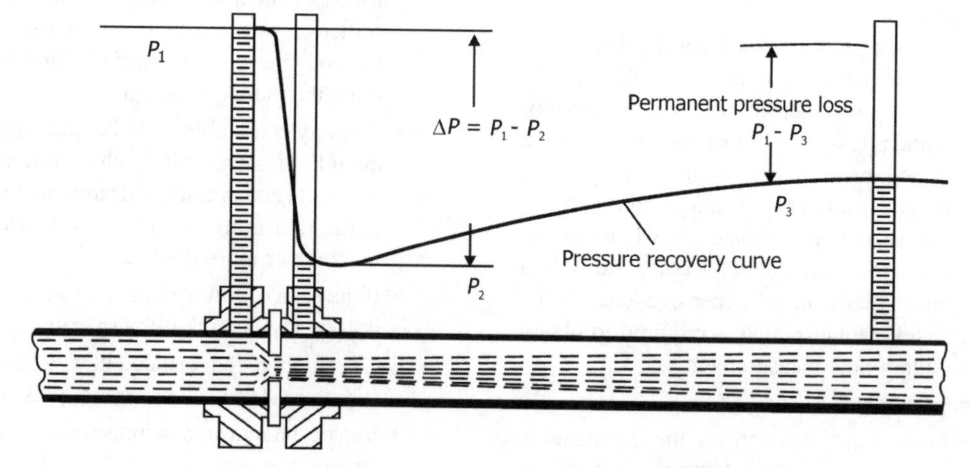

FIGURE 4.5 Sharp-edged orifice meter showing flow pattern with flange taps.

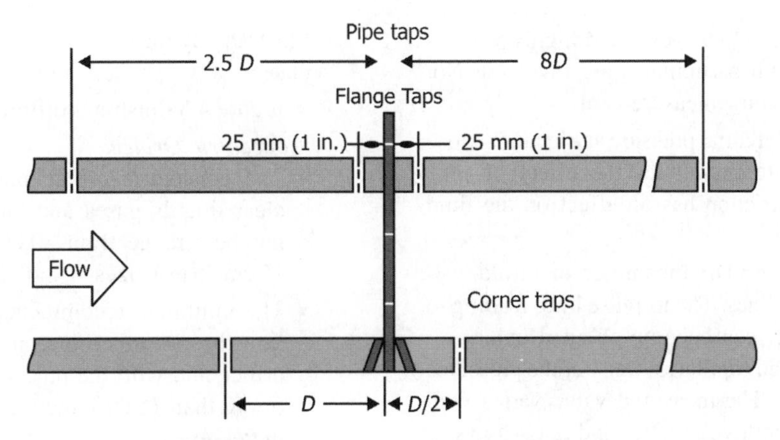

FIGURE 4.6 Pressure tap location alternatives.

- In order to fully develop the flow (and the pressure drop to be predictable), straight pipe runs are required both upstream and downstream of the differential pressure measuring element.
- The length of straight run required depends on both the β-ratio of the installation and on the nature of the upstream components in the piping.
- For example, when a single 90° elbow precedes an orifice plate, the straight pipe requirement ranges from 6 to 20 pipe diameters as the diameter ratio is increased from 0.2 to 0.8.
- In order to reduce the straight run requirement, flow straighteners such as tube bundles, perforated plates, or internal tabs can be installed upstream of the primary element. Flow straighteners are illustrated in Figure 4.7.

- What are the good points in using orifice meters for flow measurement?
  - Orifice meters are simple, rugged, widely accepted, reliable flow measuring devices.
  - Can be used for clean and unclean liquids and for some slurries.
  - Do not require direct fluid flow calibration.
  - They have no moving parts.
  - Standards for construction and calibration are readily available.
  - Relatively cheap to fabricate and install.
- What are the bad points in the use of orifice meters?
  - Not good for viscous liquids.
  - Accuracy is ±1% with calibration and less if uncalibrated.
  - They involve high pressure losses. Large part of ΔP (50–80%) is not recovered.
  - Long, straight run of pipe is required to avoid effects of upstream conditions.
  - ΔP versus flow rate relationship is not linear so that the range is limited to a 4:1 ratio between maximum flow and minimum flow.
    - The square root relationship limits the range of flow rates that realistically can be measured in a particular application.
  - Orifice plates can be knocked out of position by impurities in the flow stream, and they are subject to wear.

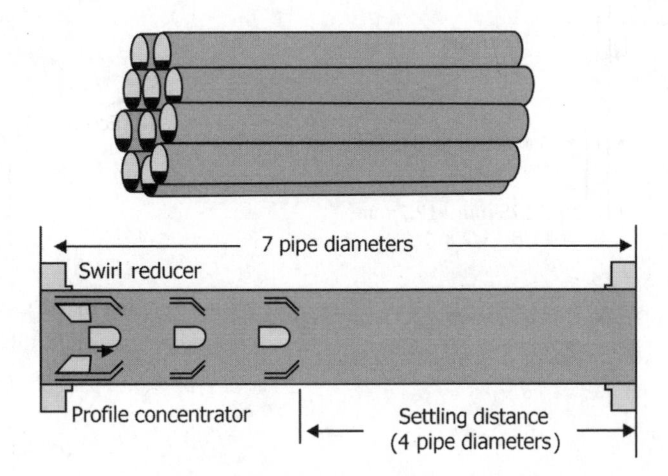

FIGURE 4.7 Flow straighteners.

- They are subject to blockages when highly viscous fluids or fluids with particulates are involved. Not suitable for slurry flow measurement.
- Measuring the differential pressure at a location close to the orifice plate minimizes the effect of pipe roughness, since friction has an effect on the fluid and the pipe wall.
- Installation requires a DP transmitter, manifold, valving, and impulse lines. The impulse lines leading to the DP transmitter can become plugged unless remote seals and filled capillaries transfer the pressures to the transmitter. The measured value varies with fluid density for both volumetric and mass flows.
- Additionally, flow elements tend to be sensitive to flow profiles within the pipe, requiring long upstream pipe runs or flow straightening devices.

• On what factors the recommended minimum distances upstream and downstream of a head meter depend?

- β-ratio.
- Type of fitting upstream, for example, 90° elbow, tee or cross, short or long radius elbows, contraction/enlargement, gate/globe/check valve, etc.

• "With an orifice meter, accuracy of flow measurement depends on its range." Comment.

- Near top end of its range, indicated flow is likely to be accurate.
- Near the lower end, flow measurement cannot be accurate.

• Illustrate, by means of diagrams, some types of orifice plates.

- Figure 4.8 illustrates different types of orifice plates.

Concentric Orifice:

- The concentric orifice plate is recommended for clean liquids, gases, and steam flows when Reynolds numbers range from 20,000 to 10^7 in pipes under 15 cm (6 in.) sizes.
- The minimum recommended Reynolds number for flow through an orifice varies with the β-ratio of the orifice and with the pipe size. For larger size pipes (more than 15 cm), the minimum Reynolds number increases.
- Concentric orifice plates can be provided with drain holes to prevent buildup of entrained liquids in gas streams, or with vent holes for venting entrained gases from liquids. The unmeasured flow passing through the vent or drain hole is usually less than 1% of the total flow if the hole diameter is less than 10% of the orifice bore.
- The effectiveness of vent/drain holes is, however, limited because they often plug-up.
- Concentric orifice plates are not recommended for multiphase fluids in horizontal lines because the secondary phase can buildup around the upstream edge of the plate. In extreme cases, this can clog the opening, or it can change the flow pattern, creating measurement error.

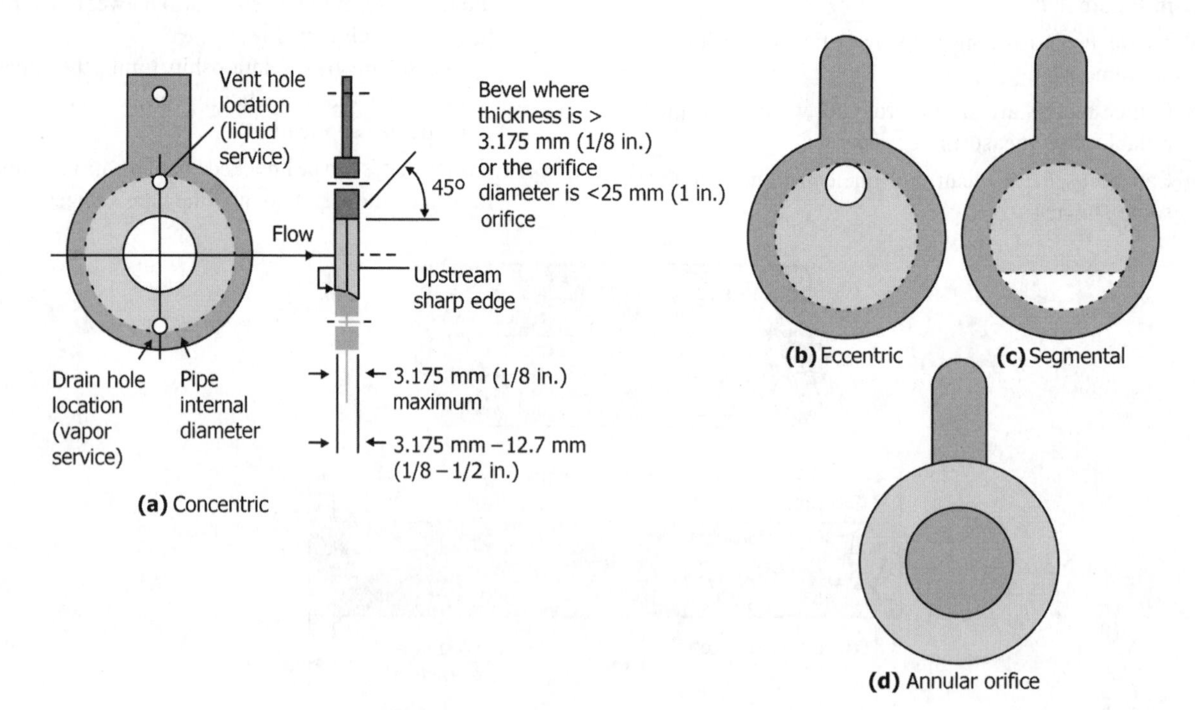

FIGURE 4.8 Different types of orifice plates.

- Concentric orifices are, however, preferred for multiphase flows in vertical lines because accumulation of material is less likely and the sizing data for these plates is more reliable.
- The concentric orifice plate has a sharp (square-edged) concentric bore that provides an almost pure line contact between the plate and the fluid, with negligible friction drag at the boundary.
- The β-ratios of concentric orifice plates range from 0.25 to 0.75.
- The maximum velocity and minimum static pressure occurs at some 0.35 to 0.85 pipe diameters downstream from the orifice plate, that is, vena contracta.

Eccentric Orifice:

- The eccentric plate has a round opening (bore) tangent to the inside wall of the pipe. This type of plate is most commonly used to measure fluids that carry a small amount of nonabrasive solids, or gases with small amounts of liquid, since with the opening at the bottom of the pipe, the solids and liquids will flow through, rather than collect at the orifice plate.
- For liquid metering with entrained gas, the opening is placed on the upper side of the orifice plate.
- For gas metering in which entrained liquids or a slurry accumulate in front of a concentric circular orifice, the opening is placed on the lower side of the orifice plate.

Segmental Orifice:

- The opening in a segmental orifice plate is comparable to a partially opened gate valve. This plate is generally used for measuring liquids or gases taht carry nonabrasive impurities such as dilute slurries or exceptionally unclean gases. Predictable accuracy of both the eccentric and segmental plates is not as good as the concentric plate.
- Segmental orifice plates are usually used in pipe sizes exceeding 10 cm (4 in.) in diameter.
- These must be carefully installed to make sure that no portion of the flange or gasket interferes with the opening.
- Flange taps are used with eccentric and segmental types of plates and are located in the quadrant opposite the opening for the eccentric orifice, in line with the maximum dam height for the segmental orifice.
- The drainage area of the segmental orifice is larger than that of the eccentric orifice and, therefore, it is preferred in applications with high proportions of the secondary phase.

Annular Orifice:

- Flow is through the annulus.
- Used for gas metering with entrained liquids or solids.
- Used for liquid metering with entrained gas in small quantities.

Other Types:

- Quadrant edge plate:
  - The quarter-circle orifice or the quadrant orifice is used for fluids of high viscosity. The orifice incorporates a rounded edge of definite radius, which is a particular function of the orifice diameter.
- Conical edge plate:
  - The conical edge plate has a 45° bevel facing upstream into the flowing stream. It is useful for even lower Reynolds numbers than the quadrant edge.
  - Quadrant-edged and conical orifice plates are recommended when the Reynolds number is less than 10,000.
  - Quadrant-edged and conical orifice plates, used for viscous flows are illustrated in Figure 4.9.
  - Flange taps, corner taps, and radius taps can all be used with quadrant-edged orifices, but only corner taps should be used with a conical orifice.
  - Conical and quadrant orifices are relatively new. The units were developed primarily to measure liquids with low Reynolds numbers. Essentially, constant flow coefficients can be maintained at N_{Re} values below 5000. Conical orifice plates have an upstream bevel, the depth and the angle of which must be calculated and machined for each application.
  - The segmental wedge is a variation of the segmental orifice. It is a restriction orifice primarily designed to measure the flow of liquids containing solids. The unit has the ability to measure flows at low Reynolds numbers and still maintain the desired square root relationship. Its design is simple, and there is only

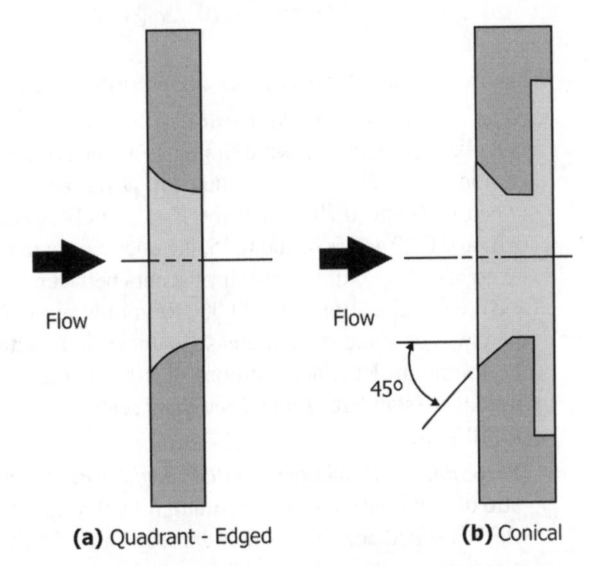

FIGURE 4.9 Orifices for viscous flows.

one critical dimension, the wedge gap. Pressure drop through the unit is only about half of that of conventional orifices.
- Integral wedge assemblies combine the wedge element and the pressure taps into a one-piece pipe coupling bolted to a conventional pressure transmitter. No special piping or fittings are needed to install the device in a pipeline.

- What are the considerations in the selection of orifice plates?
  - Concentric square edge orifice plates are recommended for general applications in clean fluid with relatively high Reynolds number.
  - Concentric quarter-circle orifice plates are recommended for applications when low Reynolds number values occur, that is with low rate liquid flow or highly viscous streams.
  - Eccentric orifice plates are recommended for applications where the drainage of extraneous matters is required.
  - Segmental orifice plates are recommended for applications where heavier or lighter components are mixed in a given fluid as in the case of two-phase flow measurement.
- Give the equation for volumetric flow through an orifice.

$$Q = C_0 A_0 [(2\Delta P/\rho)/(1-\beta^4)]^{0.5}. \tag{4.2}$$

where A_0 is the orifice area and β is the ratio of diameters of orifice to pipe.

  - For corner taps

$$C_0 \approx 0.5969 + 0.0312\beta^{2.1} - 0.184\beta^8 + (0.0029\beta^{2.5})(10^6/N_{ReD})^{0.75}. \tag{4.3}$$

- What are the three "R"s in assessing an orifice meter?
  - *Reliability/Uncertainty/Accuracy*: The coefficients calculated for flange taps are subject to an uncertainty of approximately +0.5% when the β-ratio is between 0.20 and 0.70. When the β-ratio is between 0.10 and 0.20 and 0.70 and 0.75, the uncertainty may be greater. Minimum uncertainty occurs between 0.2 and 0.6 β-ratios. Below 1,000,000 Reynolds number, there will be some small increase in uncertainty with the minimum Reynolds number of 4000 being the limit of the standard. Typical accuracy is about 2–4% of full scale.
  - *Rangeability*: Sometimes called *turndown*, is the ratio of maximum flow to minimum flow throughout which a stated accuracy is maintained. This is in the range of 4–1.
  - *Repeatability*: The ability of a flow meter to indicate the same readings each time the same flow conditions exist. These readings may or may not be accurate, but will repeat. This capability is important when a flow meter is used for flow control.
  - In general, turndown ratios for orifice meters are between 3:1 and 5:1.
- What is a venturi meter?
  - A venturi meter consists of a smooth converging contraction to a narrow throat followed by a shallow diverging section.
  - The meter is installed between two flanges intended for this purpose.
  - Pressure is sensed between a location upstream of the throat and a location at the throat.
- What are the recommended *proportions* (structural details) of a venturi meter?
  - Entrance cone angle = 21 ± 2°.
  - Exit cone angle = 5–15°.
  - Throat length = 1 throat diameter.
  - Upstream tap is located at 0.25–0.5 pipe diameter upstream of the entrance cone.
  - Venturi tubes are available in sizes up to 183 cm (72 in.) and can pass 25–50% more flow than an orifice with the same ΔP.
  - Pipes of up to 25 cm diameter usually utilize machined constrictions. They can be installed in large diameter pipes using flanged, welded, or threaded end fittings.
  - Venturi accuracy is best for N_{Re} between 10^5 and 10^6.
  - Because of the cone and the gradual reduction in the cross-sectional area, there is no vena contracta. The flow area is minimum at the throat.
  - Figure 4.10 gives the constructional details of a venturi meter.
- Illustrate the effect of Reynolds number on discharge coefficients for different types head flow meters.
  - Figure 4.11 gives the effect of Reynolds numbers on different flow meters.
- What are the average values of orifice and venturi coefficients?
  - Orifice coefficient: 0.61. This is true if N_{Re} at the orifice is above 20,000 and d/D is less than about 0.5, the value of the coefficient being nearly constant at 0.61. This value is adequate for design for liquids and gases. The coefficient is determined experimentally. For $N_{Re} < 20{,}000$, the coefficient increases sharply and then drops.
  - Venturi coefficient: 0.9.
- What is the significance of discharge coefficient, C_d?

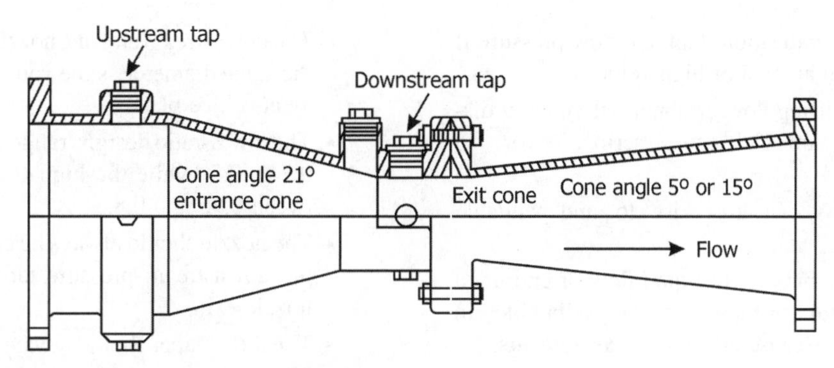

FIGURE 4.10 Venturi meter.

- Discharge coefficient signifies the quantum of energy recovery or permanent energy loss in a flow meter.

- Why C_d of venturi is higher than C_d for orifice?
  - In a venturi meter, gradual change in cross-sectional area of the convergent and divergent cone ensures smooth flow profiles without formation of eddies that represent large energy losses. Because of the non-turbulent streamline conditions, discharge coefficient—a measure of energy losses—will be high.
- Why the angle of discharge cone (diffuser) is smaller in a venturi meter?
  - The purpose of the diffuser is to ensure steady and gradual deceleration of the fluid, with pressure rise to nearly to that of original stream pressure, after the throat. The angle of the diffuser is normally in the range of 6–8°. Wider angles cause separation of flow from the walls, creating eddies that increase frictional losses between the walls and the fluid. If the angle is less, the meter becomes very long with pressure losses increasing. The objective is to convert back the increased kinetic energy due to increased velocity back to pressure energy.

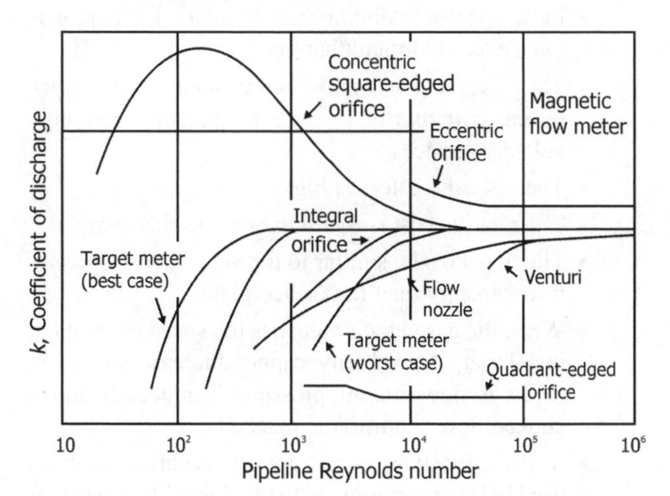

FIGURE 4.11 Effect of Reynolds numbers on different flow meters.

- What is the maximum flow velocity attainable in a venturi meter?
  - Sonic velocity at the throat, which corresponds to minimum cross-sectional area.
- What are the characteristic features and applications of a venturi meter?
  - High recovery of pressure and energy recovery makes a venturi meter particularly suitable where only small pressure heads are available.
  - Suitable for clean, unclean, slurry, and viscous flow services. Plugging problems are less with slurries.
  - The rangeability is 4:1.
  - Typical accuracy is 1% of full range.
  - Required upstream pipe length is 5–20 diameters.
  - Both meter and installation costs are comparatively higher for the venturi meter and generally used for line sizes less than 15 cm.
  - Venturi tubes have the advantage of being able to handle large flow volumes at low pressure drops.
  - Have no moving parts and can be used for fluids with high solids content.
  - Four or more pressure taps are usually installed with the unit to average the measured pressure.
  - The initial cost of venturi tubes is high, so they are primarily used on larger flows or on more difficult or demanding flow applications.
  - Venturi meters are insensitive to velocity profile effects and therefore require less straight pipe run than orifice meters.
  - Their contoured nature, combined with the self-scouring action of the flow through the tube, makes them immune to corrosion, erosion, and internal scale buildup.
  - In spite of their high initial cost, the total cost can still be favorable because of savings in installation, operating, and maintenance costs.
- What are the sources of error for pressure measurement in a differential pressure meter?

- Lower pressure indication than real flow pressure if tap is located in an area of high velocity.
- Presence of swirling flow or abnormal velocity distribution upstream can cause serious errors in measurements.
- Flow pulsations can give rise to undependable measurements.
- Presence of gas bubbles in liquid flow or entrained liquid in gas flow in piping and/or gas bubbles in manometer leads cause errors in measurements.
- Use of purge gas to keep the tap from plugging can cause a high pressure reading if too much of purge gas is used.
- If tap is located below a liquid level, reading will be too high.
- For gases, due to the low ΔP values (low head differentials), measuring accuracy is poor. At low velocities, differentials are extremely low.

4.1.2 Other Differential Pressure Flow Meters

- What is a flow nozzle? Give a diagram and describe its features.
  - A flow nozzle is essentially a short cylinder with the approach being elliptical in shape to a narrow throat (Figure 4.12). It is typically installed in-line, but can also be used at the inlet to and the outlet of a pipe or outlet from a tank.
  - It can be installed in any position, although horizontal orientation is preferred. Vertical down flow is preferred for wet steam, gases, or liquids containing solids.
  - Pressure taps are usually located at one pipe diameter upstream of the nozzle inlet and at the nozzle throat by using either wall taps or throat taps.
  - Required upstream straight pipe length is 10–30 diameters, similar to those of orifice plates.

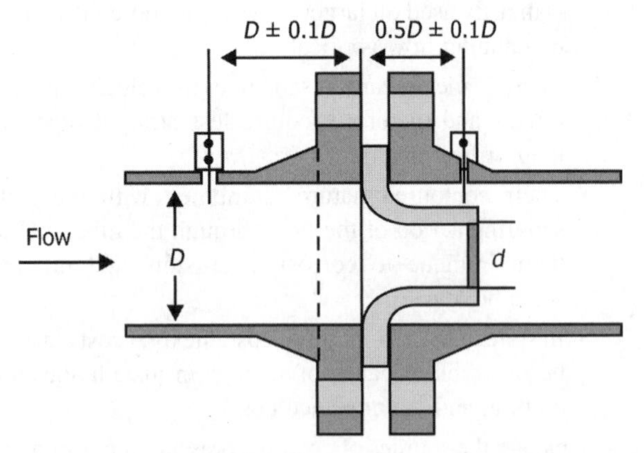

FIGURE 4.12 Flow nozzle.

  - The downstream end of a nozzle is a short tube having the same diameter as the vena contracta of an equivalent orifice plate.
  - The low β-ratio designs range in diameter ratios from 0.2 to 0.5, while the high β-ratio designs vary between 0.45 and 0.8.
  - The nozzle should always be centered in the pipe and the downstream pressure tap should be inside the nozzle exit.
  - The throat taper should always decrease the diameter toward the exit.
  - The most common flow nozzle is the flange type. Taps are commonly located 1 pipe diameter upstream and 0.5 pipe diameter downstream from the inlet face.
  - A major disadvantage of the nozzle is that it is more difficult to replace than the orifice.
- What are the characteristics and applications of flow nozzles?
  - Flow nozzle has characteristics similar to those of the venturi meter, but is shorter and much less expensive.
  - It is recommended generally for clean liquids and gases. Applicable to some slurry systems, but normally not recommended for slurries or unclean fluids.
  - It has often been used to measure high flow rates of superheated steam.
  - The rangeability is 4–1.
  - The relative pressure loss is medium, less than that for a venturi.
  - ΔP lies between orifice plates and venturi tubes.
  - Typical accuracy is 1–2% of full range.
  - Flow nozzles are highly accurate in measuring gas flows and are often used in laboratories as standards for calibrating other gas flow meters.
  - Flow nozzles maintain their accuracy for long periods, even in difficult service.
  - They are dimensionally more stable than orifice plates, particularly in high temperature and high velocity services.
  - The viscosity effect is high.
  - The relative cost is medium and less than a venturi.
  - The flow nozzle, similar to the venturi, has a greater flow capacity than the orifice plate.
  - When the gas velocity reaches the speed of sound in the throat, the velocity cannot increase any more (even if downstream pressure is reduced) and a choked flow condition is reached.
  - Such *critical flow nozzles* are very accurate and often used in laboratories as standards for calibrating other gas flow metering devices.

- What is a critical flow nozzle?
  - For a given set of upstream conditions, rate of discharge of a gas from a nozzle will increase for a decrease in absolute pressure ratio, $P_{nozzle}/P_{upstream}$ (say, P_2/P_1), until linear velocity in the throat reaches sonic velocity at that location.
  - Value of P_2/P_1 for which acoustic velocity is just attained is called critical pressure ratio, r_c.
  - Actual pressure in the throat will not fall below $P_1 r_c$, even if much lower pressure exists downstream.
- What is a sonic nozzle?
  - Sonic nozzle is used to measure and to control the flow rate of compressible gases. It may take the form of any of the previously described obstruction meters. If the gas flow rate through and obstruction meter becomes sufficiently high, the sonic condition will be achieved at the meter throat.
  - At the sonic condition, the gas velocity will equal the acoustic wave speed (speed of sound) of the gas. At that point, the throat is considered to be choked and the mass flow rate through the throat will be at a maximum for the given inlet conditions regardless of any further increase in pressure drop across the meter.
- On what factors do the recommended minimum distances upstream and downstream of orifice and venturi meters depend?
  - The main consideration for accurate is that the flow should be fully developed before entry into the meter. In other words, sufficient straight section of the upstream pipe is required.
  - On the downstream side, disturbance created in flow by the meter must be removed for maximum recovery of pressure to minimize energy losses.
- Illustrate, graphically, permanent pressure losses with respect to orifice meter, venturi meter, and flow nozzles.
  - From Figure 4.13, it is evident that venturi meter involves lowest permanent pressure losses compared to orifice meter or flow nozzle, the later having marginally higher losses.
  - Losses decrease with increased diameter ratios in all the three cases. The decrease is uniformly steeper with increased diameter ratios for both flow nozzle and orifice where as for the venturi, it is steeper at low diameter ratios and becomes marginal at increased ratios.
  - For an area ratio of 0.5, the pressure loss is about 65–70% of the orifice pressure differential.

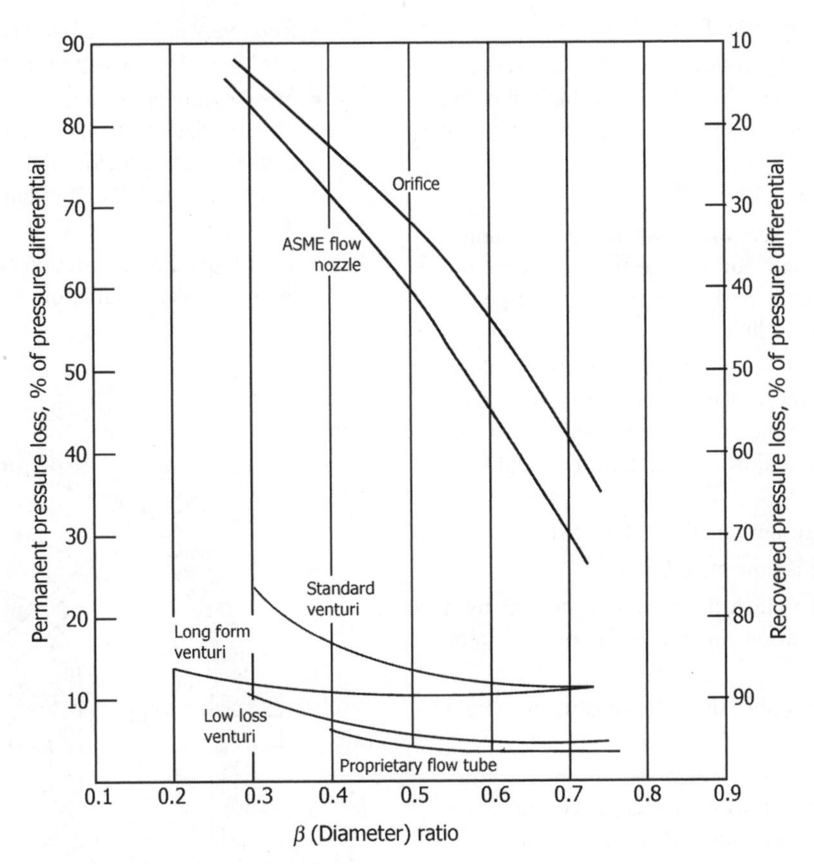

FIGURE 4.13 Pressure losses for different head flow meters as function of β-ratio.

TABLE 4.1 Comparison of Head Flow Meters

Primary Element	Recommended Service	Minimum N_{Re}	Advantages	Limitations
Square-edge concentric orifice	Clean liquids, gases, and steam	2,000	Easy to install, easy to replace, low cost	Calming piping length required, high head loss, accuracy is affected by conditions of orifice and installation
Conical quadrant-edge concentric orifice plate	Viscous liquids	500	Same as above	Same as above
Eccentric/segmental orifice	Two-phase fluids	10,000	Same as above	Same as above, higher uncertainties in discharge coefficient data
Venturi/flow tube	Clean and unclean fluids, slurries, steam	75,000	Low head loss, two to nine times less calming section for same ΔP, accuracy is less affected by wear and installation conditions than orifice	High initial cost
Flow nozzle	Clean fluids, steam	50,000	Higher capacity than orifice, accuracy less affected by wear and installation, good for high temperature and high velocity	High head loss, harder to replace
Segmental-cost wedge	Unclean and viscous liquids, slurries, gases	500	40% less head loss than orifice, minimum calming piping	Requires calibration and a differential transmitter

- Give a comparison of head flow meters.
  - A comparison of head flow meters with respect to their recommended service along with their plus and minus points for their choice is presented in Table 4.1.
- What is an elbow meter?
  - A differential pressure exists when a fluid changes direction due to an elbow or pipe turn. The pressure difference results from the centrifugal force generated by the flowing fluid.
  - Since pipe elbows exist in plants, the cost for these meters is very low, but the accuracy is very poor. These are only applied when reproducibility is sufficient and other flow measurements would be expensive.
  - Figure 4.14 illustrates an elbow flow meter.
- What is a V-element flow meter?
  - A flow meter in which the flow is restricted by a V-shaped indentation in the side of the pipe (Figure 4.15).
  - The pipe requires only 5 diameters upstream straight pipe section.
  - The segmental-wedge element is a proprietary designed for use in services such as slurry, corrosive, erosive, viscous, gas flow with liquid droplets or liquid flow with gas bubbles.
  - Relatively expensive but has high accuracy of about ±0.5% of the measured flow rate.
  - The segmental wedge has a V-shaped restriction characterized by the *H*/*D* ratio, where *H* is the height of the opening below the restriction and *D* is the diameter. The *H*/*D* ratio can be varied to match the flow range.
  - The unique flow restriction can be designed to match the life of the installation without any deterioration.

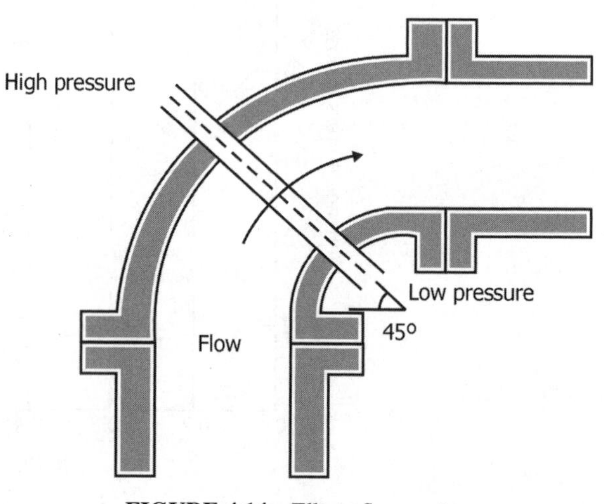

FIGURE 4.14 Elbow flow meter.

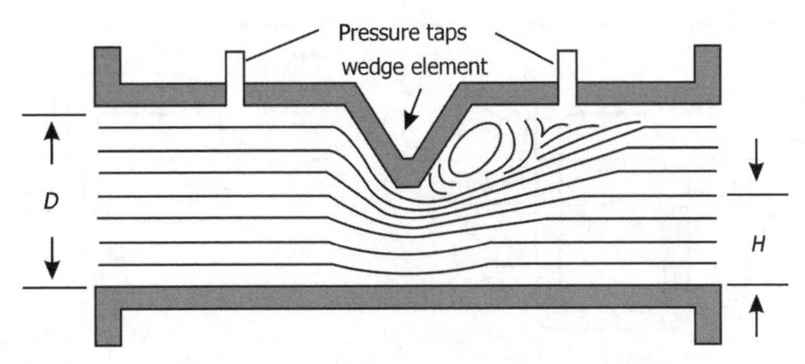

FIGURE 4.15 Segmental-wedge V-element flow meter.

- Unlike with an orifice meter, the meter coefficient is constant at a value of about 0.8 at low flow rates with N_{Re} as low as 500.
- Wedge elements are used with 7.6 cm (3 in.) diameter chemical seals, eliminating both the lead lines and any dead-end cavities. The seals are attached to the meter immediately upstream and downstream of the restriction. They rarely require any cleaning even in services like dewatered sludge, black liquor, coal and fly ash slurries, and crude oils.
- The oncoming flow creates a sweeping action through the meter. This provides a scouring effect on both faces of the restriction, helping to keep it clean and free of buildup.
- Segmental wedges can measure flow in both directions, but the DP transmitter must be calibrated for a split range, or the flow element should be provided with two sets of connections for two DP transmitters (one for forward flow and another for reverse flow).
- The venturi-cone (V-cone) element is another proprietary design that promises consistent performance at low Reynolds numbers and is insensitive to velocity profile distortion or swirl effects. However, it is relatively expensive. The V-cone restriction has a unique geometry due to wear, making it a good choice for high velocity flows and erosive/corrosive applications.
- The V-cone creates a controlled turbulence region that flattens the incoming irregular velocity profile and induces a stable differential pressure that is sensed by a downstream tap.
- The β-ratio of a V-cone is so defined that an orifice and a V-cone, with equal β-ratios, will have equal opening areas

$$\beta\text{-ratio} = (D^2 - d^2)^{0.05}/D, \qquad (4.4)$$

where d is the cone diameter and D is the inside diameter of the pipe.

- Figure 4.16 illustrates the V-cone flow meter.

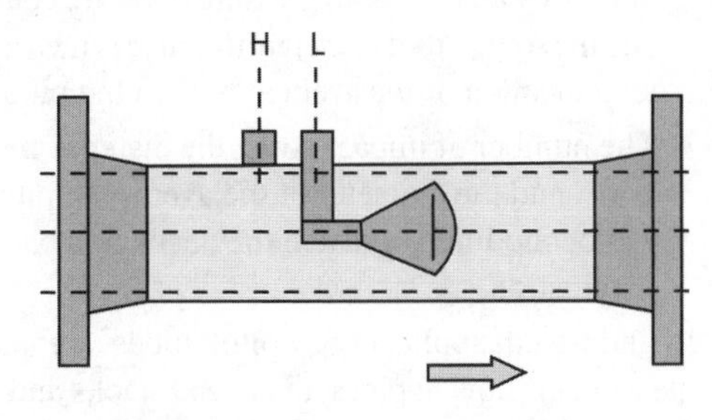

FIGURE 4.16 V-cone flow meter.

4.1.2.1 Pitot Tubes

- What is a pitot tube? Is it used to measure velocity or pressure? Illustrate by means of a diagram.
  - Pitot tube is used to measure the local velocity at a given point in the flow stream and not the average velocity in the pipe or conduit. A diagram of pitot tube is shown in Figure 4.17.
  - One tube, the impact tube, has its opening normal to the direction of flow, while the static tube has its opening parallel to the direction of flow.
  - The fluid flows into the opening at point 2, pressure builds up and then remains stationary at this point, called the *stagnation point.* The difference in the stagnation pressure at point 2 and the static pressure measured by the static tube represents the pressure rise associated with deceleration of the fluid. The manometer measures this small pressure rise (Δh in Figure 4.17). Δp is obtained from the manometer reading from the equation

$$\Delta p = \Delta h(\rho_{\text{m}} - \rho), \qquad (4.5)$$

where ρ_{m} is the density of manometric fluid and ρ is the density of flowing fluid.

  - For incompressible fluids, velocity can be obtained from the following equation, by applying Bernoulli equation between point 1, where the velocity v_1 is undisturbed before the fluid decelerates, and point 2,

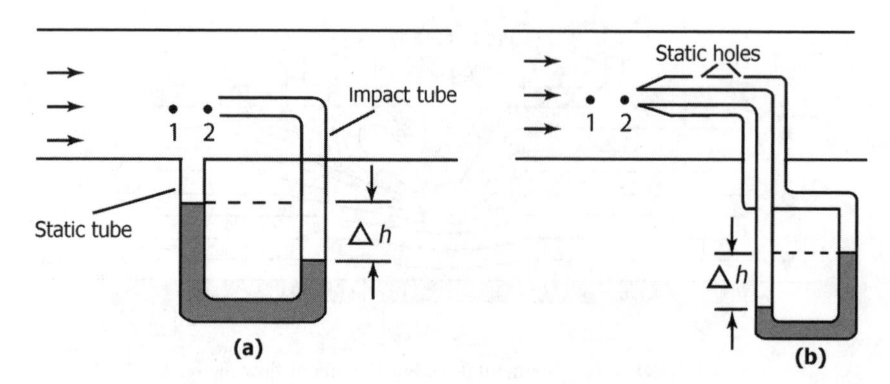

FIGURE 4.17 Pitot tube arrangements for flow measurement.

where the velocity v_2 is zero:

$$v = C_P[2(p_2 - p_1)/\rho]^{0.5}, \quad (4.6)$$

where v is the velocity v_1 in the tube at point 1, in meters per second; p_2 is the stagnation pressure; ρ is the density of the flowing fluid at the static pressure p_1; and C_p is a dimensionless coefficient that generally varies between about 0.98 and 1.0. For accurate use, the coefficient should be determined by calibration of the pitot tube. Average velocity can be found from v_{max}, obtained by velocity measurement at center of the pipe and then using appropriate velocity profile expressions.

- When the flow rate is obtained by multiplying the point velocity by the cross-sectional area of the pipe or duct, it is critical that the velocity measurement be made at an insertion depth, which corresponds to the average velocity. As the flow velocity increases, the velocity profile in the pipe changes from laminar (elongated) to turbulent (more flat). This changes the point of average velocity and requires an adjustment of the insertion depth.
- Pitot tubes are recommended only for highly turbulent flows ($N_{Re} > 20{,}000$) and, under these conditions, the velocity profile tends to be flat enough so that the insertion depth is not critical.
- An averaging pitot tube is provided with multiple impact and static pressure ports and is designed to extend across the entire diameter of the pipe.
  - The pressures detected by all the impact (and separately by all the static) pressure ports are combined and the square root of their difference is measured as an indication of the average flow in the pipe.
  - The number of impact ports, the distance between ports, and the diameter of the averaging pitot tube can be modified to match the needs of a particular application.
- In industrial applications, pitot tubes are used to measure air flow in pipes; ducts and stacks and liquid flow in pipes; and weirs and open channels.
- Pitot tubes also can be used in square, rectangular, or circular air ducts. A pitot tube also can be used to measure water velocity in open channels, at drops, chutes, or over fall crests.
- While accuracy and rangeability are relatively low, pitot tubes are simple, reliable, inexpensive, and suited for a variety of environmental conditions, including extremely high temperatures and a wide range of pressures.
- The pitot tube is an inexpensive alternative to an orifice plate.
- Accuracy is comparable to that of an orifice. Its flow rangeability of 3:1 (some operate at 4:1) is also similar to the capability of the orifice plate. The main difference is that, while an orifice measures the full flow stream, the pitot tube detects the flow velocity at only one point in the flow stream.
- Its advantage is that it can be inserted into existing and pressurized pipelines without requiring a shutdown.

• What are the sources of error in a pitot tube?
  - Angle of tip with respect to flow direction influences ΔP measured.
  - Wall effects in small flow lines influence readings.
  - Positioning the probe with respect to cross section of the pipe gives different readings (with respect to the center line of the pipe gives maximum velocity).
  - For gases, the pressure change is often quite low and hence accurate measurement of velocities is difficult.
  - Care should be taken to have the pitot tube at least 100 diameters downstream from any pipe obstruction as flow disturbances upstream of the probe can cause large errors.
  - Suitable for flow measurements involving clean fluids as presence of solid particles give rise to inaccurate readings and the tip might get blocked.
  - Pulsations in the pipe need to be dampened.

> Specially designed pitot probes have been developed for use with pulsating flows.
> One design uses a pitot probe filled with silicone oil to transmit the process pressures to the DP cell. At high-frequency pulsating applications, the oil serves as a pulsation dampening and pressure averaging medium.

- Errors in detecting static pressure arise from fluid viscosity, velocity, and fluid compressibility.
- The key to accurate static pressure detection is to minimize the kinetic component in the pressure measurement.

• What is a pitot-venturi flow element? What is its advantage?
  - A pitot-venturi element consists of a pair of concentric venturi elements in place of a pitot tube.
  - Low-pressure tap is connected to throat of inner venturi, which in turn discharges into throat of outer venturi.
  - It is capable of developing a pressure differential of five to ten times that of a standard pitot tube.

• What is a reverse pitot tube or pitometer?
  - Pitot tube with one pressure opening facing upstream and the other downstream is called reverse tube/pitometer.

• What is the advantage of reverse pitot tube over standard pitot tube?
  - Coefficient C for this type is of the order of 0.85. This gives about 40% increase in pressure differential as compared with standard pitot tubes and is an advantage at low velocities. Compact types of pitometers are available, which require comparatively small openings for their insertion into a duct.

• What is an annubar?
  - An annubar consists of several pitot tubes placed across a pipe to provide an approximation to the velocity profile and the total flow can be determined based on the multiple measurements (Figure 4.18).

• "Averaging pitot tubes are a good choice for the measurement of high flows." *True/False*?
  - *True*.

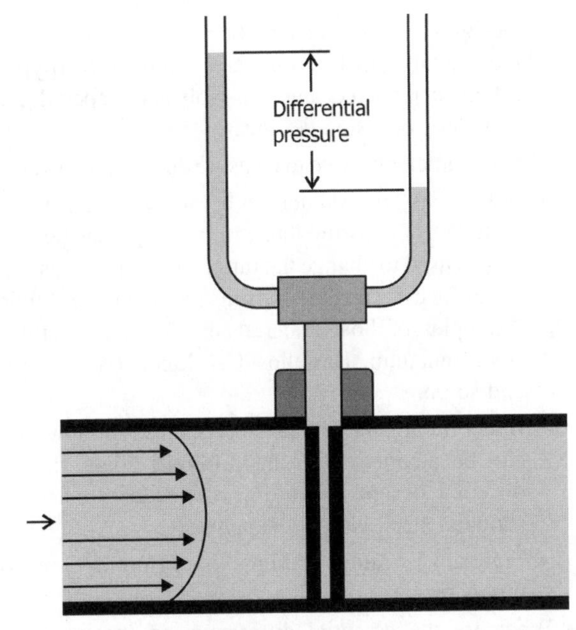

FIGURE 4.18 Annubar.

4.1.3 Variable Area Flow Meters

4.1.3.1 Rotameters

• What is a rotameter? Has the glass tube comprising the rotameter is of constant cross section?
  - Rotameters, or variable area flow meters, operate on the principle that the variation in area, of flow stream required to produce a constant pressure differential, is proportional to the flow rate. The flowing fluid enters the bottom of the meter, passes upward through a tapered metering tube, and around a float, exiting at the top. The float is forced upward until the force is balanced by gravitational forces.

> Either the force of gravity or a spring is used to return the flow element to its resting position when the flow decreases.
> Gravity operated meters (rotameters) must be installed in a vertical position, whereas spring operated ones can be mounted in any position.

  - Flow is through the annular area between float and tapered tube whose area is variable depending on flow rate.
  - The height of rise of the float in the tube is calibrated with the flow rate, and the relation between the meter reading, that is, position of the float, and the flow rate is approximately linear.
  - Floats have a sharp edge at the point where the reading should be observed on the tube-mounted scale.
  - The reading is taken at the center of the float. It is recommended that the float be at eye level to minimize reading errors.
  - The scale can be calibrated for direct reading of air or water, or can read percentage of range.
  - Most rotameters are made of glass with etched scale on the outside so that flow readings can be taken visually.
  - Used to measure the flow of liquids, gases, and steam.

- Design variations include the rotameter (a float in a tapered tube), orifice/rotameter combination (bypass rotameter), open channel variable gate, tapered plug and vane, or piston designs.

- Can a rotameter be used to measure different flow rates?
  - *Yes*. Floats are available in a variety of shapes and materials of construction, with varying densities that can be used to change the range of the meter, as well as to resist corrosion from the measured fluid. Examples of float materials include 316 stainless steel, tantalum, Hastelloy C, Monel, PTFE, PVC, and so on.
  - If a correlated flow tube is used, different flow rates can be attained by using different floats, that is, different designs and materials made of carboloy, stainless steel, glass, or sapphire.
  - Figure 4.19 shows designs of different types of rotameters.
- What are the essential differences in the operating principle of a rotameter from a venturi or orifice meter?
  - Orifice and venturi meters use constant restriction/flow area and a variable pressure differential. Rotameters use a variable restriction/flow area and a constant pressure differential to measure flow.
  - Typically, rotameters are used to measure smaller flows and the reading is usually done locally, although transmission of the readings is possible.
- What are the forces acting on the float in a rotameter?
  - Buoyancy force acting upward.
  - Gravity acting downward.
  - Drag force acting downward.
- What is the normal size ranges of rotameter tubes?
  - Tube diameters vary from 0.5 to >15 cm.
- What are the advantages of rotameters?
  - Relatively cheap.
  - Somewhat self-cleaning because as the fluid flows between the tube wall and the float, it produces a scouring action that tends to prevent the buildup of foreign matter. Nevertheless, rotameters should be used only on clean fluids that do not coat the float or the tube. Liquids with fibrous materials, abrasives, and large particles should also be avoided.
  - Rotameter accuracy is not affected by the upstream piping configuration. The meter also can be installed directly after a pipe elbow without adverse effect on metering accuracy.
  - Available in different materials for chemical compatibility.
  - Used for measuring volumetric flow rates of many liquids and gases.
  - With a glass tube meter, the main metering elements and the fluid are clearly visible. The user can immediately see any change in flow rate.
  - Wide rangeability: From 5:1 to 12:1.
  - Rangeability of 12:1, as compared to 4:1, for differential pressure meters that are commonly used with orifice plates.
  - Viscosity compensation: The float can be designed to compensate for increased viscosities.
    - Most rotameters are relatively insensitive to viscosity variations. The most sensitive are very small rotameters with ball floats, while

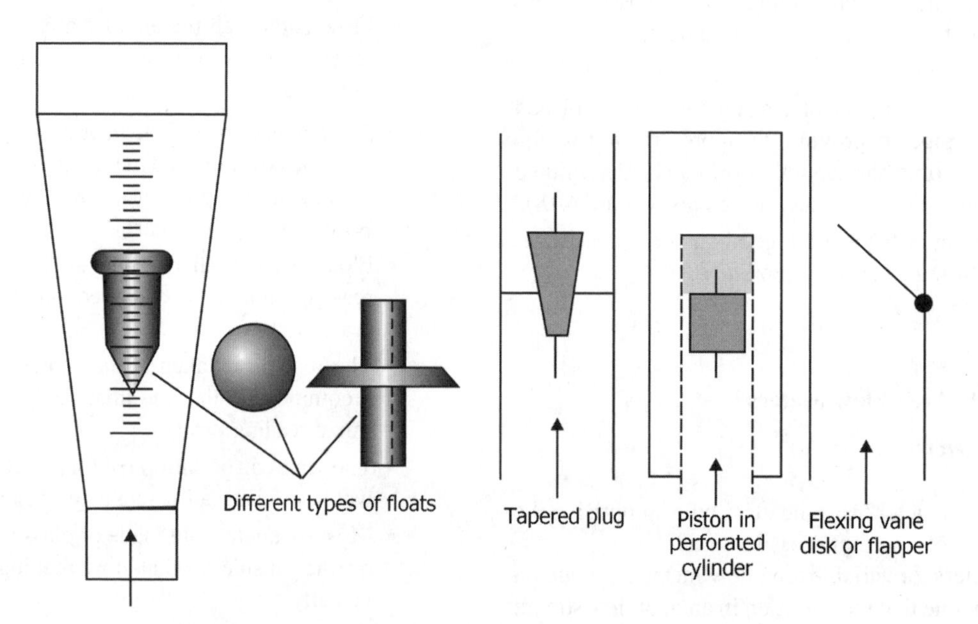

FIGURE 4.19 Rotameter showing different types of float designs.

larger rotameters are less sensitive to viscosity effects.

  - The float shape does affect the viscosity limit. If the viscosity limit is exceeded, the indicated flow must be corrected for viscosity.

- Sustained high repeatability: The float moves freely in the metering tube, experiencing no friction or hysteresis. The rotameter thus attains the ideal design goal of having high repeatability and being easily readable. Repeatability is 0.5–1% of full scale.
- Low pressure drop: Because the area between the float and the tapered tube increases with flow rate, ΔP across the float is low and relatively constant. This reduces pumping costs. Also, a meter can be selected to provide lower ΔP by using an oversized tube with a light float.
- Linear scale: Because area variation is the measure of flow rate (rather than head or differential pressure), the calibration curve is practically a straight line. This means that the meter can have an indicating linear scale with evenly spaced divisions, comparing better for readability than any differential pressure flow meter. Rotameter scale readings are not compressed (harder to read) at low flow rates, so the same degree of accuracy holds throughout the entire flow range.
- Lack of need for electricity: Rotameters used as local indicators require no connection to an electric power source; hence, there is no need for explosion-proofing when flammables may be present. Useful where the flow measurement must be made in a hazardous area where electric power is unavailable, or would be potentially dangerous.
- Special applications: Used as a bypass meter around a mainline orifice for larger pipe sizes. Very high flow rates of liquids or gases can be measured by installing a rotameter in a bypass line around an orifice plate in the main line.
- Signal: Visual or electronic.
- Sight glass functionality: A glass tube rotameter can also serve as a sight glass, eliminating the need for a separate device to show that the process fluid is flowing.
- Familiarity: Most users are familiar with rotameters and can readily understand their operation.
- Ease of corrosion-proofing: Because of its design simplicity, a rotameter can be economically constructed of high corrosion-resistant materials.
- Ease of conversion to measure a different fluid: A model installed for service on one fluid can be recalibrated for measurement of another with ease.
- Ease of installation and maintenance: Rotameters mount vertically in the pipe without need for pipe taps, connecting lines, seal pots, valves, or requirements for a straight run pipe upstream or downstream, as is the case for orifice meters.
- Design temperature: Metal tube up to about 450°C. Glass tube up to 120°C. Plastic tube up to 60°C.
- Design pressure: Metal tube up to 100 bar. Glass tube up to 20 bar. Plastic tube up to about 5 bar.
- Ability to handle low flow rates: The small rotameters can measure liquid flow rates down to 0.65 mL/min and gas flow rates down to 47 mL/min.
- Viscosity: Liquids up to 200 cP.

- How is the effect of fluid density variations taken care of in rotameters to variations in operation?
  - Because the float is sensitive to changes in fluid density, a rotameter can be furnished with two floats (one sensitive to density, the other to velocity) and used to approximate the mass flow rate.
  - The more closely the float density matches the fluid density, the greater the effect of a fluid density change will be on the float position.
  - Mass flow rotameters work best with low viscosity fluids such as raw sugar juice, gasoline, jet fuel, and light hydrocarbons.
- What type of rotameter is used for purge gas flow measurements?
  - Low-capacity glass tube rotameters are extensively used in purge systems in which service they are known as *purge meters*.
    - Instrument Society of America (ISA) defines purge meter as a *device designed to measure small flow rates of liquids and gases used for purging measurement piping*.
  - With the help of a needle valve purge meters facilitate accurate control of low flow rates. For example, for water, these rates can be as low as 3.8 L/min (1 gpm) and for air 56 L/min (<2 ft^3/min).
- What is a purge flow regulator?
  - If a needle valve is placed at the inlet or outlet of a rotameter and a DP regulator controls the pressure difference across this combination, the result is a purge flow regulator.
  - Such instrumentation packages are used as self-contained purge flow meters.
  - These are among the least expensive and most widely used flow meters.
  - Their main application is to control small gas or liquid purge streams.
  - They are used to protect instruments from contacting hot and corrosive fluids, to protect pressure taps from

plugging, to protect the cleanliness of optical devices, and to protect electrical devices from igniting upon contact with combustibles.

- Purge meters are quite useful in adding nitrogen gas to the vapor spaces of tanks and other equipment. Purging with nitrogen gas reduces the possibility of developing a flammable mixture because it displaces flammable gases.
- The purge flow regulator is reliable, *intrinsically safe* and inexpensive.
- Figure 4.20 illustrates its working:

• What is the difference between correlated and direct reading rotameters?

- A direct reading flow meter indicates the flow rate on its scale in specific engineering units (e.g., mL/min or scf/h). Direct reading scales are designed for a specific gas or liquid at a given temperature and pressure. While it is more convenient than a correlated flow meter, a direct reading flow meter is less accurate and limited in its applications.
- A correlated flow meter is scaled along either a 65 mm or a 150 mm length, from which a reading is taken. The reading is then compared to a correlation table for a specific gas or liquid. This will give the actual flow in engineering units. One correlated flow meter can be used with a variety of fluids or gases.

• " Pressure drop in a rotameter is constant" *True/False*?

- *True.* Pressure drop (head) will be constant across the float only if flow area remains constant.

• "A rotameter must not constrict flow." *True/False*?

- *True.*

• What are the disadvantages of a rotameter?

- Glass is breakable which could lead to hazardous situations when handling toxic or flammable liquids.
- With opaque or cloudy liquids, float will not be visible (special rotameters are available, which are more expensive).
- No output for data transmission.
- Sensitive to differing gas types and changes in temperature and pressure. In other words, as these meters measure volumetric flow rates, gas density variations will affect measurements.
- ΔP is high.
- Accuracy is only 2–3% unless expensive high precision rotameter is employed.
- Not suitable for high capacities.
- Accumulation of dirt or other foreign matter deposited on the float or tube walls can cause wrong readings.

• What type of rotameter is used for opaque or cloudy liquids?

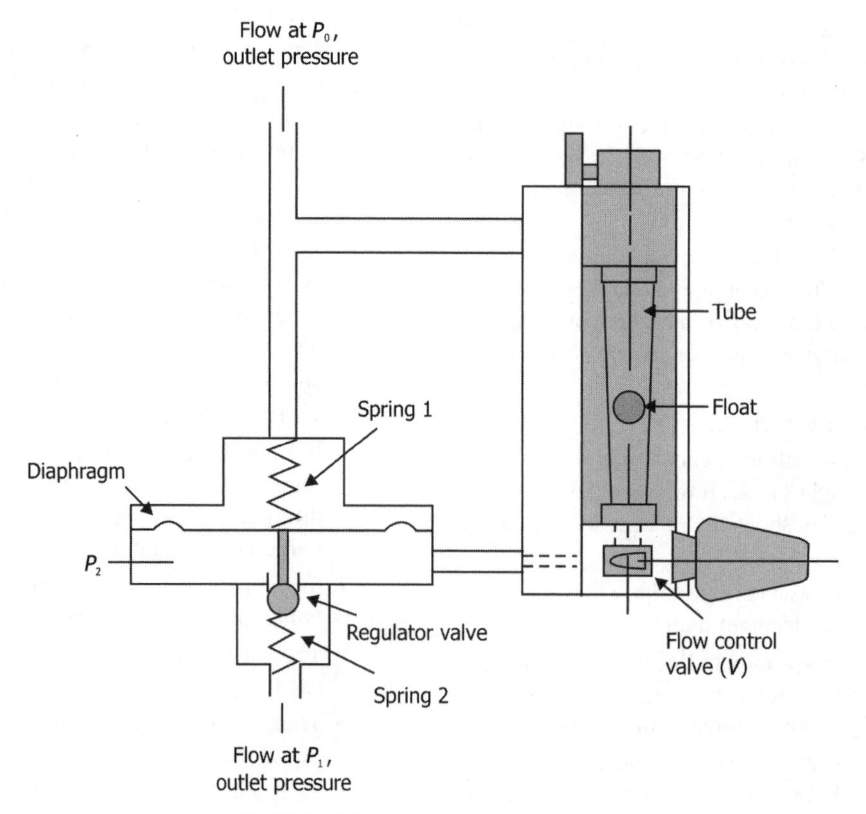

FIGURE 4.20 Purge flow meter design.

- Metal tube rotameters, which are also called, armored tube rotameters, in which indication is achieved by means of a magnet inside the float and an external follower magnet linked to the pointer. No external electric power is involved.
- Alternatively, armored (metal tube) rotameters can be set up with an electrical output taken off a converter. Then, metal tubes can be specified in applications that require remote transmission of the measured flow rate. This feature is not generally available with glass tube meters.

• What are the temperature and pressure limits for glass tube rotameters?

- 120°C and 20 bar.

• What type of rotameter one would recommend for flow measurements involving hot alkalis or hydrofluoric acid?

- Transparent polymeric rotameters with floats made of plastic. All-Teflon meters are available to resist corrosive damage by aggressive chemicals.

• Can a rotameter be used in a vacuum application or with back pressure?

- *Yes*, but if there is a valve, which must be placed at the outlet (top of the flow meter). This is done by inverting the tube inside the frame and then turning over the frame. At this position, the tube should read correctly from the original perspective and the valve should be at the outlet, or top of the flow meter. This allows for proper control of the vacuum.

• What are the differences between a 150 and a 65 mm flow meters?

- A 150 mm flow meter has a 150 mm scale length and is graduated accordingly. It provides better resolution than the more economical 65 mm flow meter.

• Must a rotameter be mounted vertically?

- In general, rotameters must be mounted vertically, because the float must center itself in the fluid stream. At high flow rates, the float assumes a position toward the tip of the metering tube and at low flow rates positions itself lower in the tube.
- Some rotameters have spring loaded floats and therefore may be mounted in any orientation.

• Illustrate how orifice meter–rotameter combination can be used to measure very high flow rates of liquids or gases.

- Major disadvantages of the rotameter are its relatively high cost in larger sizes and the requirement that it be installed vertically (there may not be enough head room).
- The cost of a large rotameter installation can be reduced by using an orifice bypass or a pitot tube in combination with a smaller rotameter.
- Rotameter can be installed in a bypass line around an orifice plate in the main line as shown in Figure 4.21.
- The differential pressure developed by orifice plate in the main line causes a corresponding flow through the rotameter. An integral ranging orifice proportions the bypass flow to the mainline flow, permitting use of a 1 in. (1.25 cm) rotameter regardless of the size of the main line. Since the float position in a rotameter linearly with flow rate, the rotameter can be used to indicate mainline flow rate in direct flow units on a linear scale.
- Benefits of such a bypass arrangement include increased rangeability of 12:1 compared to rangeability of 4:1 for orifice meters; rotameter scale readings can be graduated in direct flow units for flow in the main line; and easy range change or cleaning of the rotameter tube without disassembly or removal of the meter from the bypass line simply by isolation by valves in the bypass line.
- The same size bypass rotameter can be used to measure a variety of flows, with the only difference between applications being the orifice plate and the differential it produces.

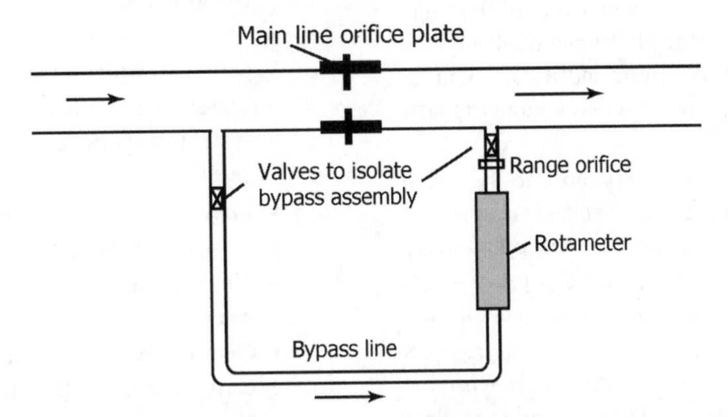

FIGURE 4.21 Rotameter installed in a bypass line around an orifice plate in the main line.

- Advantages of a bypass rotameter include low cost.
- Its major disadvantage is inaccuracy and sensitivity to material buildup.

4.1.4 Mechanical Flow Meters

- What are mechanical flow meters?
  - Mechanical flow meters measure flow using an arrangement of moving parts, either by passing isolated, known volumes of a fluid through a series of gears or chambers (positive displacement) or by means of a rotating turbine or rotor elements.
  - All positive displacement (PD) flow meters operate by isolating and counting known volumes of a fluid (gas or liquid) while making it flow through the meter. By counting the number of passed isolated volumes, flow measurement is obtained.
  - Each PD design uses a different means of isolating and counting these volumes.
  - The frequency of the resulting pulse train is a measure of flow rate, while the total number of pulses gives the size of the batch.
  - While PD meters are operated by the kinetic energy of the flowing fluid, metering pumps determine the flow rate while also adding kinetic energy to the fluid.
  - Turbine flow meter belongs to the category of mechanical flow meters.
  - Other types of rotary element flow meters include the propeller (impeller), shunt, and paddle wheel designs.

4.1.4.1 Turbine and Paddle Wheel Flow Meters

- What is a turbine flow meter and how does it operate?
  - A turbine flow meter is a straight flow tube equipped with rotating vanes/blades (turbine) mounted at right angles to flow direction, suspended in the fluid stream free to rotate on a shaft, supported by one or more bearings and located on the center line of the tube. Flowing fluid will strike the blades at a small angle to balance the forces on the rotor and it rotates at an angular speed that is proportional to volumetric flow rate.
  - The diameter of the rotor is very close to the inside diameter of the metering chamber and its speed of rotation is proportional to the volumetric flow rate.
  - Turbine rotation can be detected by solid-state devices or by mechanical sensors that include infrared beams, photoelectric sensors, or very small magnets are embedded in the tips of the rotor blades sense the frequency of rotation, an electrical pulse is then generated and converted to a frequency output proportional to the flow rate. Rotors are typically made of a nonmagnetic material, like polypropylene, Ryton, or PVDF (Kynar).
  - In the water distribution industry, mechanical-drive turbine flow meters continue to be the standard.
  - These turbine meters use a gear train to convert the rotation of the rotor into the rotation of a vertical shaft. The shaft passes between the metering tube and the register section through a mechanical stuffing box, turning a geared mechanical register assembly to indicate flow rate and actuate a mechanical totalizer counter.
  - The water distribution industry, in recent years, has adopted a magnetic drive as an improvement over high-maintenance mechanical drive turbine meters. This type of meter has a sealing disk between the measuring chamber and the register. On the measuring chamber side, the vertical shaft turns a magnet instead of a gear.
  - On the register side, an opposing magnet is mounted to turn the gear. This permits a completely sealed register to be used with a mechanical drive mechanism.
  - Used for clean and noncorrosive fluids.
- What are the applications and constraints in use of turbine meters?
  - Turbine meters have a low ΔP and are very accurate ($\pm 0.25\%$ or better).
  - Residential and industrial gas and water meters are often of the rotary wheel type, that is, turbine type.
  - They are exceptionally repeatable, but are restricted to clean and low viscosity fluids because of possible fouling of their rotating parts.
  - They have fast response rate (few milliseconds); high pressure and high temperature capabilities, up to 345 bar (5000 psi) and 220°C (800°F) with high temperature pick coils); and compact rugged construction.
  - They have high pressure and high temperature capabilities with compact and rugged construction.
  - These meters should not be used for slurries or systems experiencing large, rapid flow or pressure variation.
  - They are sensitive to temperature changes that affect fluid viscosity.
  - Turbine meters have moving parts that are subject to wear.
  - One disadvantage in some designs is a loss of linearity at the low flow rates.
  - Tend to be fragile with high maintenance costs.

- Relatively expensive.

- Does a turbine flow meter require a minimum straight distance before its installation?
  - To maintain an even cross-sectional flow, it is recommended that there be a straight pipe length of at least 10 inner diameters of the meter upstream and at least 5 inner diameters downstream of the sensor.
- Can a turbine meter be used for flow of air or gases?
  - Some turbine flow meters can be used with air or gases. However, if there are air bubbles or vapor pockets in the liquid, the reading will be inaccurate. There should be a laminar (stable) flow through the cross section of the pipe.
- What are the plus points in using a turbine flow meter?
  - The turbine flow meter is an accurate and reliable flow meter for both liquids and gases.
  - Good accuracy with liquids. Used as a calibration standard.
  - Range of flow rates is 10:1 and response is very fast.
  - Compact and has the same diameter as the pipe.
  - Signal output available for totalizing flows.
  - Available for low flow rates.
  - Easy to install and easy to maintain.
  - Unrecoverable ΔP is moderate.
- What type of flow measuring device is used for domestic water supply lines?
  - Turbine flow meter.
- What are the limitations of using a turbine flow meter?
  - Sensitive to viscosity changes. Calibration is dependent on viscosity.
  - Straight pipeline required upstream.
  - Good for clean liquids and gases.
  - High cost.
  - Problems with bearings.
- What are the materials that are normally used the construction of turbine flow meters?
  - Most industrial turbine flow meters are manufactured from austenitic stainless steel (301, 303, 304 SS), whereas turbine meters intended for municipal water service are bronze or cast iron.
  - The rotor and bearing materials are selected to match the process fluid and the service.
  - Rotors are often made from stainless steel and bearings of graphite, tungsten carbide, ceramics, or in special cases of synthetic ruby or sapphire combined with tungsten carbide.
  - In all cases, bearings and shafts are designed to provide minimum friction and maximum resistance to wear.
  - Some corrosion-resistant designs are made from plastic materials such as PVC.
- What are the operating flow ranges of turbine flow meters?
  - Turbine meters should be sized so that the expected average flow is between 60% and 75% of the maximum capacity of the meter.
  - Flow velocities under 0.3 m/s (1 ft/s) can be insufficient, while velocities in excess of 3 m/s (10 ft/s) can result in excessive wear.
  - Most turbine meters are designed for maximum velocities of 9 m/s (30 ft/s).
- What is a dual-rotor turbine flow meter?
  - A dual-rotor turbine flow meter involves two rotors turning in opposite directions
    - ➢ The front one acts as a conditioner, directing the flow to the back rotor.
    - ➢ The rotors lock hydraulically and continue to turn as the flow decreases even to very low rates.
  - Dual-rotor liquid turbines increase the operating range in small line size, under 5 cm (2 in.) applications.
- What are the fluid parameters that affect performance of turbine flow meters?
  - Viscosity affects the accuracy and linearity of turbine meters. It is therefore important to calibrate the meter for the specific fluid it is intended to measure.
  - Repeatability is generally not greatly affected by changes in viscosity and turbine meters often are used to control the flow of viscous fluids.
  - In general, turbine meters perform well if the Reynolds number is greater than 4000 and less than or equal to 20,000.
  - Because it affects viscosity, temperature variation can also adversely affect accuracy and must be compensated for or controlled.
  - The turbine meter's operating temperature ranges from −200 to 450°C.
  - Density changes do not greatly affect turbine meters. On low-density fluids (sp. gr. < 0.7), the minimum flow rate is increased due to the reduced torque, but the meter's accuracy usually is not affected.
  - The linearity of a turbine meter is affected by the velocity profile (often dictated by the installation), the viscosity, and the temperature.
- What are the features incorporated in turbine flow meters for gas service?
  - Gas meters compensate for the lower driving torque produced by the relatively low density of gases.

  - This compensation is obtained by very large rotor hubs, very light rotor assemblies and larger numbers of rotor blades.
- What are the installation guidelines for turbine flow meters?
  - Turbine meters are sensitive to upstream piping geometry that can cause vortices and swirling flow.
  - Specifications require 10–15 diameters of straight pipe run upstream and 5 diameters of straight run downstream of the meter.
  - However, the presence of any of the following obstructions upstream would necessitate that there be more than 15 diameters of upstream straight pipe runs:
    - 20 diameters for 90° elbow, tee, filter, strainer, or thermowell.
    - 25 diameters for a partially open valve.
    - 50 or more diameters if there are two elbows in different planes or if the flow is spiraling or corkscrewing.
  - In order to reduce this straight run requirement, straightening vanes are installed. Tube bundles or radial vane elements are used as external flow straighteners located at least 5 diameters upstream of the meter.
  - Under certain conditions, the pressure drop across the turbine can cause flashing or cavitation. The first causes the meter to read high, the second results in rotor damage. In order to protect against this, the downstream pressure must be held at a value equaling 1.25 times the vapor pressure plus twice the pressure drop.
  - Small amounts of air entrainment (100 mg/L or less) will make the meter to read only a bit high, while large quantities can destroy the rotor.
  - Turbine meters also can be damaged by solids entrained in the fluid.
  - If the amount of suspended solids exceeds 100 mg/L of + 75 μm size, a flushing Y-strainer or a motorized cartridge filter must be installed at least 20 diameters of straight run upstream of the flow meter.
- What is a shunt flow meter?
  - Shunt flow meters consist of an orifice in the main line and a rotor assembly in the bypass and are used in gas and steam service.
- What is a paddle wheel flow meter and how does it work?
  - A paddle flow meter consists of rotating paddles with magnets installed on each paddle of the sensor, which is inserted into the liquid. As the paddle turns, an electrical frequency output proportional to the flow velocity is generated.

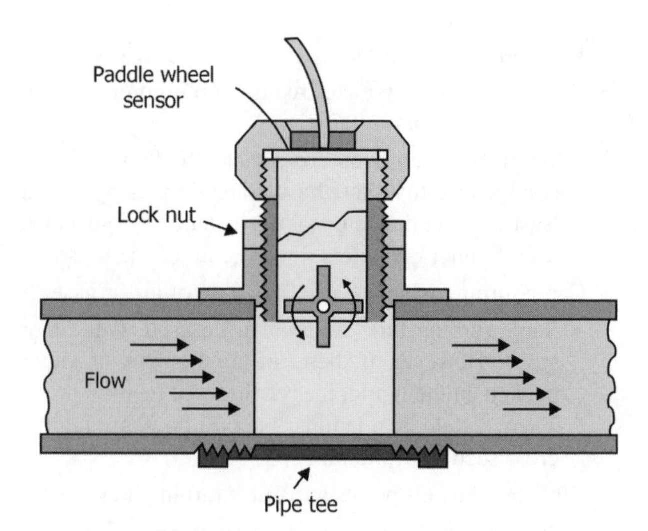

FIGURE 4.22 Paddle wheel flow meter.

  - Paddle wheel flow meters use a rotor whose axis of rotation is parallel to the direction of flow.
  - Most paddle wheel meters have flat-bladed rotors and are generally bidirectional.
  - Figure 4.22 illustrates a paddle wheel type flow meter.
- Are the paddle meters suitable for turbulent flows or foamy liquids?
  - Because the sensors use laminar flow characteristics, turbulent or foamy liquids will not be read accurately. The sensors must also be installed in a full flowing, straight section of pipe.
- What are straight pipe requirements upstream and downstream of the meter?
  - For systems with no bends or restrictions, a minimum of 15 pipe diameters upstream and 5 pipe diameters downstream can be provided.
- What are the advantages of using a paddle wheel flow meter?
  - Good repeatability.
  - Low pressure drop.
  - Easy maintenance.
- What is a propeller flow meter?
  - Propeller meters are commonly used in large diameter over 10 cm (4 in.) irrigation and water distribution systems.
  - They are low cost and low accuracy flow meters. Accuracy is generally is 2% of reading.
  - Propeller meters have a rangeability of about 4:1 and exhibit very poor performance if the velocity drops below 0.5 m/s (1.5 ft/s).
  - Most propeller meters are equipped with mechanical registers.

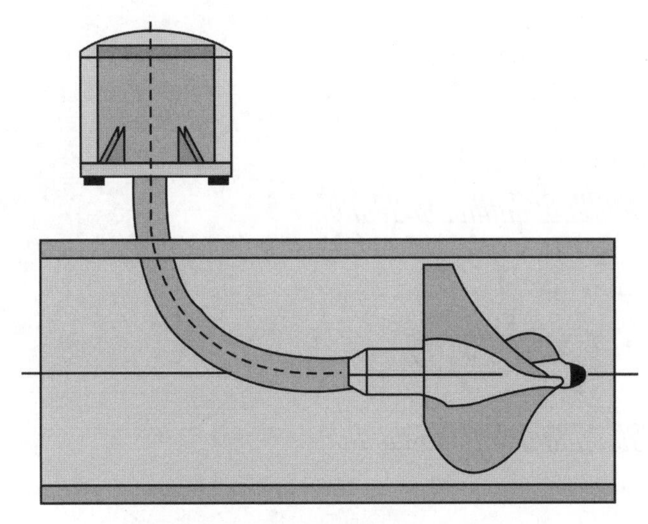

FIGURE 4.23 Propeller flow meter.

- Mechanical wear, straightening, and conditioning requirements are the same as for turbine meters.
- Figure 4.23 illustrates a propeller flow meter.

• Describe a flow meter suitable for line sizes less than 5 cm (2 in.) for measuring low flows.

- Low-flow meters have a small jet orifice that projects the fluid onto a Pelton wheel.
- Varying the diameter and the shape of the jet orifice matches the required flow range.
- Because of the small size of the jet orifice, these meters can only be used for clean fluids.

4.1.5 Electronic Flow Meters

4.1.5.1 Ultrasonic Flow Meters

• What are the operating principles that are used to form the basis for the operation of ultrasonic flow meters?

- There are two types of ultrasonic flow meters, namely, *transit time* and *Doppler* meters.
- Transit time meters have both a sending transducer and a receiving transducer. The sending transducer sends an ultrasonic signal from one side of to other side a pipe. A signal is then sent in the reverse direction.
- When an ultrasonic signal travels with the flow, it travels faster than when it travels against the flow. The flow meter measures both transit times. The difference between the two transit times (across the pipe and back again) is proportional to flow rate.
- Figure 4.24 illustrates transit time flow meter.
- When the flow is zero, the time for the signal T_1 to get to T_2 is the same as that required to get from T_2 to T_1.
- When there is flow, the effect is to boost the speed of the signal in the downstream direction, while decreasing it in the upstream direction. The flowing velocity, V can be determined by Equation 4.5.

$$V = K\mathrm{d}t/T_\mathrm{L}, \tag{4.7}$$

where K is a calibration factor for the volume and the time units used, $\mathrm{d}t$ is the time differential between the upstream and the downstream transit times, and T_L is the zero flow transit time.

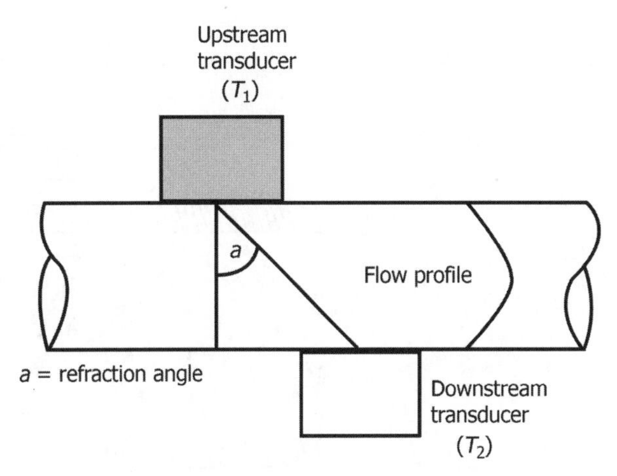

FIGURE 4.24 Transit time flow meter.

- Transit time meters can be used to measure both very hot (e.g., liquid sulfur, molten metals) and cryogenic liquids (liquid nitrogen, liquid argon, and liquid helium) down to −300°C, and also to detect very low flows.
- Transit time flow meters are often used to measure the flow of crude oils and simple fractions in the petroleum industry. They also work well with viscous liquids, provided that the Reynolds number at minimum flow is either less than 4000 (laminar flow) or more than 10,000 (turbulent flow).
- In a Doppler flow meter, a high-frequency signal is projected through the wall of the pipe and into the liquid. The signal is reflected off impurities in the liquid such as air bubbles or particles, and sent back to the receiver. The frequency difference between the transmitted and the received signal is directly proportional to the flow velocity of the fluid.
- Like transit time flow meters, Doppler meters send an ultrasonic signal across a pipe. However, the signal is reflected off moving particles in the flow stream, instead of being sent to a receiver on the other side. The moving particles are traveling at the same speed as the flow. As the signal passes through the stream, its frequency shifts in proportion to the average velocity of the fluid. A receiver detects the reflected signal and measures its frequency. The meter calculates flow by comparing the generated and detected frequencies.
- Figure 4.25 illustrates a Doppler flow meter.

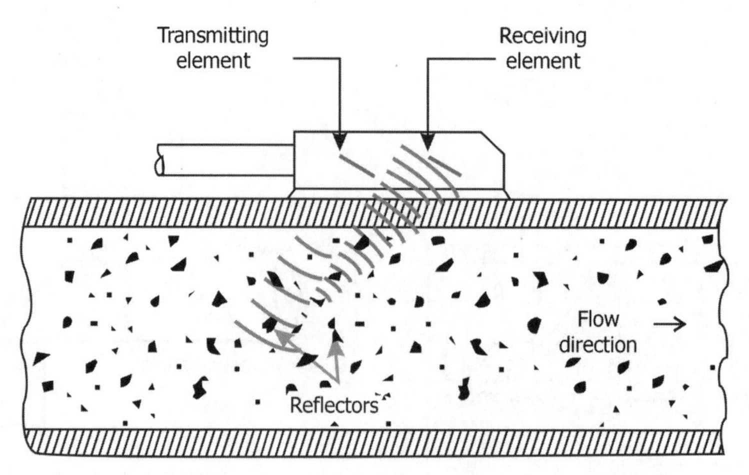

FIGURE 4.25 Doppler flow meter.

- Multipath ultrasonic flow meters use several sets of sending and receiving transducers placed in different paths placed across the flow section, to determine the flow rate. Most multipath flow meters use four to six different paths or ultrasonic signals to determine the velocity profile across the entire cross section of the pipe. The transducers alternate sending and receiving a signal over the same path length. Flow rate is determined by averaging the values given by the different paths, yielding greater accuracy than provided by single-path meters.
- Multipath instruments are used in large diameter conduits, such as stack gas flows in power plant scrubbers and in other applications where nonuniform flow velocity profiles exist.

• What are the applications of ultrasonic flow meters?
  - Ultrasonic flow meters are used to measure the flow of both liquids and gases.
  - Transit time flow meters are mainly used for clean, swirl-free liquids and gases of known profile. Having clean fluid is the most important constraint on ultrasonic flow meters although transit time meters can handle some impurities.
  - They can be affected by swirl.
  - It is important to take into consideration both pipe and fluid characteristics into account.
  - Doppler ultrasonic flow meter requires the presence of particulates or bubbles in the flow stream so the signal can bounce off them. Hence, they are used with unclean liquids and slurries and not recommended for clean fluids. Minimum size of particulates in the stream should be 30 μm and concentration of particulates should be >25 ppm.
  - Raw wastewater-related applications are problematic, as the solids concentration can be too high for either transit time or Doppler flow meters to work properly.
  - The use of multipath flow meters in raw wastewater and storm water applications is common, while Doppler or cross-correlation hybrid designs are most often used to measure activated sludge and digested sludge flows.
  - For mining slurries, Doppler flow meters work well.
  - Among the few problem applications are those in HDPE pipe because the pipe wall flexes enough to change the diameter of the measurement area. This affects the accuracy of the meter.
  - Another problem area is the measurement of slurries that are acoustically absorbent, such as lime or kaolin slurries. These applications fail because the highly absorbent solids attenuate the signal below usable strength.
  - These meters must be calibrated for the fluid flowing through them.

• Will pipe thickness or insulation affect readings of a Doppler flow meter?
  - *Yes.* Insulation should be removed before mounting the sensor.

• Does a Doppler flow meter require a minimum upstream straight pipe length?
  - *Yes.* Doppler flow meters require 10 pipe diameters from any valve, tee, bend, and so on.
  - Doppler flow meters also require a full pipe flow.

• What are the advantages and the disadvantages of using a Doppler flow meter?

  Advantages
  - The main advantage of the Doppler ultrasonic meter is its nonintrusive design. An acoustic coupling compound is used on the surface of the pipe and the

sensors are simply held in place to take a measurement or, for a more permanent installation, they are strapped around the pipe. Other advantages include the following:

- Easy installation and removal. No process downtime during installation.
- No moving parts to wear out.
- Zero pressure drop.
- No process contamination.
- Works well with slurries or corrosive fluids.
- Works with a wide range of pipe sizes.
- No leakage potential.
- Meters are available that work with laminar, turbulent, or transitional flow characteristics.
- Portable. Battery powered units are available for remote or field applications.
- Sensors are available for pulsating flows.
- Insensitive to liquid temperature, viscosity, density, or pressure variations.

Disadvantages

- The main disadvantage is the fact that the liquid stream must have particulates, bubbles, or other types of solids in order to reflect the ultrasonic signal. This means that the Doppler meter is not a good choice for distilled water or very clean fluids. A good rule of thumb is to have a bare minimum of 25 ppm at roughly 30 μm in order for the ultrasonic signal to be reflected efficiently.
- If the solids content is too high (around 50% and higher by weight), the ultrasonic signal may attenuate beyond the limits of measurability. Another disadvantage is that the accuracy can depend on particle size distribution and concentration and also on any relative velocity that may exist between the particulates and the fluid. If there are not enough particulates available, the repeatability will also degrade.
- It can give rise to troubles when operated at very low flow velocities. *Applications*:
- Doppler meters, being nonintrusive, have a wide variety of applications in water, waste water, heating, ventilation and air conditioning, and petroleum and general process markets. Possible applications are as follows: Waste water, potable water, cooling water, makeup water, hot and chilled water, crude oil flow, mining slurries, acids, caustic, liquefied gases, and so on.

- What is the important limitation of using a Doppler flow meter?
  - Not suitable for clean liquids.
  - Requires straight tube run upstream of the meter.
- "Normally Doppler flow meters are calibrated for water flow measurements." Do they require recalibration for other fluids?
  - *Yes*. Because the flow rate depends on the velocity of sound through the fluid, which is dependent on fluid properties.
- Give examples of typical applications of Doppler flow meters.
  - Waste water.
  - Cooling water.
  - Clarifier monitoring.
  - Digester feed control.
  - Hot and chilled water.
  - Crude oil flow.
  - Mining slurries.
  - Acids.
  - Alkalis.
  - Liquefied gases.
- What type of flow measuring devices are best for slurries?
  - Doppler flow meter.
  - Venturi meters are also suitable for some slurries.
- How does a laser velocity meter work?
  - Small particles in a flowing fluid will follow the flow and can be detected by observing the scattered light from a focused laser beam.
  - Seed particles are naturally present in the fluids. Some times, seeding is necessary.

4.1.5.2 Magnetic Flow Meters

- What is the principle of operation of an electromagnetic flow meter?
  - Magnetic flow meters operate on the principle of Faraday's law of electromagnetic induction; an electrical voltage is induced in a conductor that is moving through a magnetic field and at right angles to the field. The faster the conductor moves through the magnetic field, the greater the voltage induced in the conductor in proportion to speed of movement of the conductor.
  - The magnetic flow meter consists of a nonmagnetic pipe lined with an insulating material. As current passes through the coils, generally mounted outside of the pipe through which fluid flows, a magnetic field is generated inside the pipe. As conductive fluid passes through the pipe, a voltage is generated and detected by electrodes that are mounted on either side of the pipe. The flow meter uses this voltage value to calculate flow rate.

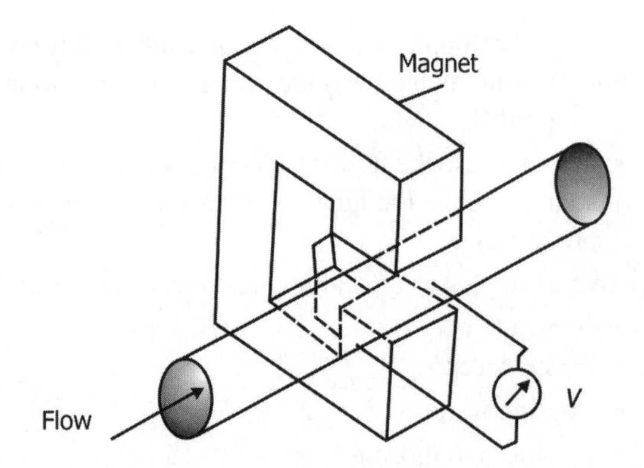

FIGURE 4.26 Elements of an electromagnetic flow meter.

- Figure 4.26 shows the elements of an electronic flow meter.
- When velocity profile is symmetrical and magnetic flux density, B, is uniform, velocity is given by

$$V = E/BD, \qquad (4.8)$$

where V is the velocity, E the is voltage, B is the magnetic flux density, and D is pipe diameter.

- A pair of magnetic coils is situated as shown in Figure 4.27 and a pair of electrodes penetrates the pipe and its lining.
- Because the magnetic field density and the pipe diameter are fixed values, they can be combined into a calibration factor K and the equation reduces to

$$E = KV. \qquad (4.9)$$

- Manufacturers determine K factor for each meter, by water calibration of each flow tube. The K value thus obtained is valid for any other conductive liquid and is linear over the entire flow meter range. For this reason, flow tubes are usually calibrated at only one velocity.
- Magnetic meters can measure flow in both directions, as reversing direction will change the polarity but not the magnitude of the signal.
- The K value obtained by water testing might not be valid for non-Newtonian fluids (with velocity-dependent viscosity) or magnetic slurries (those containing magnetic particles). These types of fluids can affect the density of the magnetic field in the tube.
- In-line calibration and special compensating designs should be considered for both of these fluids.
- The velocity differences at different points of the flow profile are compensated for by a signal-weighing factor. Compensation is also provided by shaping the magnetic coils such that the magnetic flux will be greatest where the signal-weighing factor is lowest, and vice versa.
- The voltage that develops at the electrodes is a millivolt signal. This signal is typically converted into a standard current (4–20 mA) or frequency output (0–10,000 Hz) at or near the flow tube.
- Intelligent magnetic transmitters with digital outputs allow direct connection to a distributed control system.
- Because the magnetic flow meter signal is a weak one, the lead wire should be shielded and twisted if the transmitter is remote.
- The electromagnetic flow meter comes commercially as a packaged flow device, which is installed directly in-line and connected to an external electronic output unit.

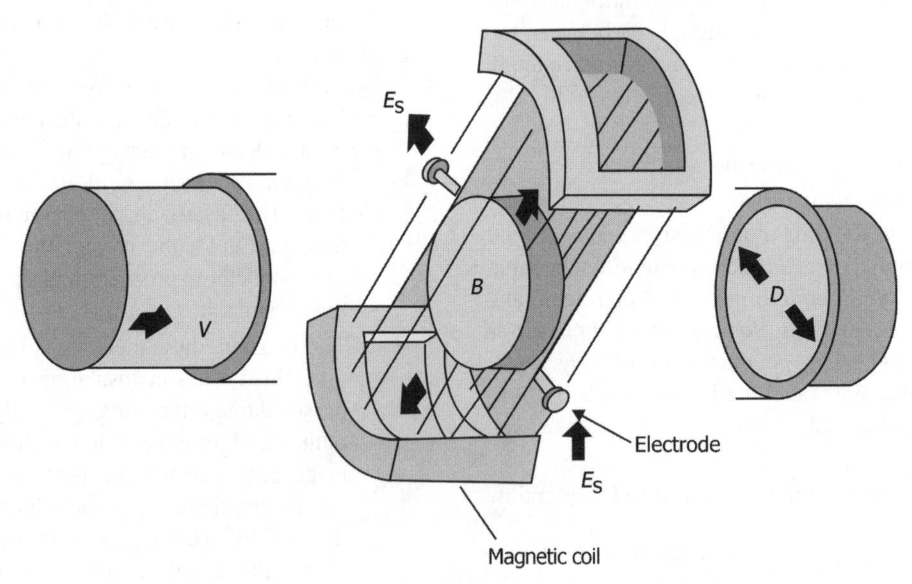

FIGURE 4.27 Magnetic flow meter and its components.

- Units are available using either permanent magnets, called DC units, or variable flux strength electromagnets, called AC units.

- State the advantages and disadvantages of a magnetic flow meter.
  - In general, magnetic flow meters are accurate, reliable, measurement devices that do not intrude into the system.
  - Range of velocity (flow rate) can be increased beyond 10:1 by increasing sensitivity for voltage detection.
  - The electromagnetic flow meter has a very low pressure drop associated due to its open tube, no obstruction design, and it is suitable for installations that can tolerate only a small pressure drop.
  - This absence of internal parts is very attractive for metering corrosive and *unclean* fluids.
  - The operating principle is independent of fluid density and viscosity, responding only to average velocity and there is no difficulty with measurements in either laminar or turbulent flows, provided that the velocity profile is reasonably symmetrical.
  - Magnetic flow meters are widely used to measure the flow rate of *conductive* liquids and slurries in process applications.
  - Magnetic flow meters can detect the flow of conductive fluids only. They cannot be used to measure flows of hydrocarbons, which are nonconductive. As a result, they are not widely used in the petroleum industry.
  - Since gases and steam are nonconductive, these meters cannot be used to measure them.
  - The pipe has to be full of liquid, since these meters compute flow rate based on velocity times area.
  - Liner damage and electrode coating can affect the accuracy of magnetic flow meters.
  - These meters are expensive. While their initial purchase cost is relatively high, most of these meters are priced lower than equivalent Coriolis meters.
  - Small meters are bulky in size.

4.1.6 Mass Flow Meters

- How does a mass flow meter work?
  - A volume of gas has a known mass at standard conditions. As pressure and temperature are applied, the volume will change, but the mass remains constant. Mass flow meters measure flow based on the molecular mass of the gas. This measurement is independent of temperature and pressure. One technique to measure mass flow is to send a part of the flow through a sensor tube. In the tube, the gas is heated in a coil and then measured downstream. The temperature differential is directly related to the mass flow.
- Can a mass flow meter give a total accumulation of gas?
  - *Yes.*
- Why is it necessary to have a filter upstream of a mass flow meter?
  - Mass flow meters require clean gases. In general, any particle larger than 50 μm require a filter upstream of the meter.
- What are the advantages of using a mass flow meter?
  - It measures mass directly.
  - It can handle applications whose stream temperature and line pressures fluctuate.
- What are the limitations of using a mass flow meter?
  - A mass flow meter calibrated for air will not work on other gases or gas mixtures without factory recalibration. When the gas is changed, the calibration must be updated.
  - The gas must be dry.
- Give typical applications of mass flow meters.
  - Nitrogen delivery and control for tank blanketing.
  - Control of methane or argon to gas burners.
  - Gas delivery and control for fermenters and bioreactors.
  - Leak testing.
  - Regulating CO_2 injected into bottles during beverage production.
  - Monitoring and controlling air flow during gas chromatography.
- What is a thermal mass flow meter?
  - Thermal mass flow meter includes a heated sensor element that is cooled by the gas flow. The result of the temperature difference is in proportion with the gas mass flow and therefore a standard for the flow rate. A flow-proportional frequency or analog signal provides the flow rate.

4.1.6.1 Coriolis Flow Meters

- What is a Coriolis flow meter? What is the principle involved in its operation?
  - Coriolis flow meters measure *mass flow* by taking advantage of the Coriolis effect, which states that if a particle inside a rotating body moves in a direction toward or away from the center of rotation, the particle generates internal forces that act on the body.
  - Simply stated, the inertial effects that arise as a fluid flows through a tube are directly proportional to the mass flow of the fluid.

- If a tube is rotated around a point while liquid is flowing through it (toward or away from the center of rotation), that fluid will generate an inertial force (acting on the pipe) that will be at right angles to the direction of the flow.
- Naturally, rotating a tube is not practical when building a commercial flow meter, but oscillating or vibrating the tube can achieve the same effect.
- Coriolis flow meters are made up of one or more vibrating tubes, usually bent. The fluid to be measured travels through the vibrating tubes. The fluid accelerates as it approaches the point of maximum vibration, and decelerates as it leaves this point. As a result, the tubes take on a twisting motion. The amount of twisting motion is directly proportional to mass flow.
- In a Coriolis flow meter, vibration is induced in the process fluid filled flow tube(s), then the mass flow rate is captured by measuring the difference in the phase of vibration between the one end and the other end of the flow tube.
- Coriolis flow meters can measure flow through the tube in either the forward or the reverse directions.
- In most designs, the tube is anchored at two points and vibrated between these anchors. This configuration can be envisioned as vibrating a spring and mass assembly. Once placed in motion, a spring and mass assembly will vibrate at its resonant frequency, which is a function of the mass of that assembly.
- This resonant frequency is selected because the smallest driving force is needed to keep the filled tube in constant vibration.
- Position detectors are used to sense the positions of the vibrating tubes.
- Most Coriolis flow meter tubes are bent, but straight tube meters are also in use.
- A tube can be of a curved or straight form and some designs can also be self-draining when mounted vertically.
- When the design consists of two parallel tubes, flow is divided into two streams by a splitter near the meter's inlet and is recombined at the exit. In the single continuous tube design (or in two tubes joined in series), the flow is not split inside the meter.
- In either case, drivers vibrate the tubes. These drivers consist of a coil connected to one tube and a magnet connected to other tube.
- The transmitter applies an alternating current to the coil, which causes the magnet to be attracted and repelled by turns, thereby forcing the tubes toward and away from one another.

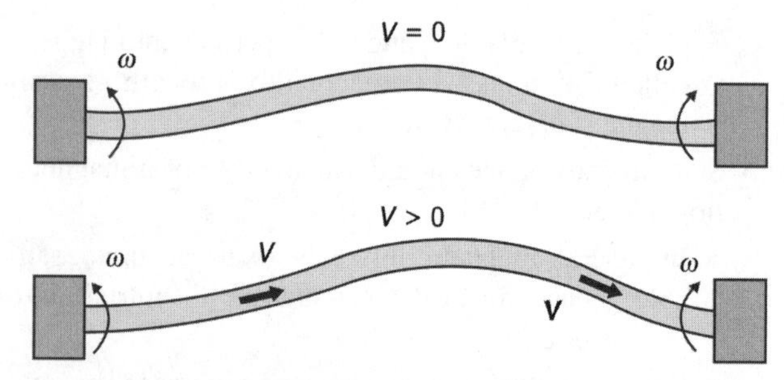

FIGURE 4.28 Single straight tube Coriolis mass flow meter.

- The sensor can detect the position, velocity, or acceleration of the tubes.
- If electromagnetic sensors are used, the magnet and coil in the sensor change their relative positions as the tubes vibrate, causing a change in the magnetic field of the coil.
- Therefore, the sinusoidal voltage output from the coil represents the motion of the tubes.
- When there is no flow in a two tube design, the vibration caused by the coil and the magnet drive results in identical displacements at the two sensing points B_1 and B_2 as shown in Figure 4.30 (double U-tube design).
- When flow is present, Coriolis forces act to produce a secondary twisting vibration, resulting in a small phase difference in the relative motions. This is detected at the sensing points. The deflection of the tubes caused by the Coriolis force only exists when both axial fluid flow and tube vibration are present.
- Vibration at zero flow, or flow without vibration, does not produce an output from the meter.
- The natural resonance frequency of the tube structure is a function of its geometry, materials of construction, and the mass of the tube assembly (mass of the tube plus the mass of the fluid inside the tube).
- Figures 4.28–4.30 illustrate single- and two-tube Coriolis flow meters:

- For what types of applications bent tube and straight tube Coriolis flow meters are preferred?

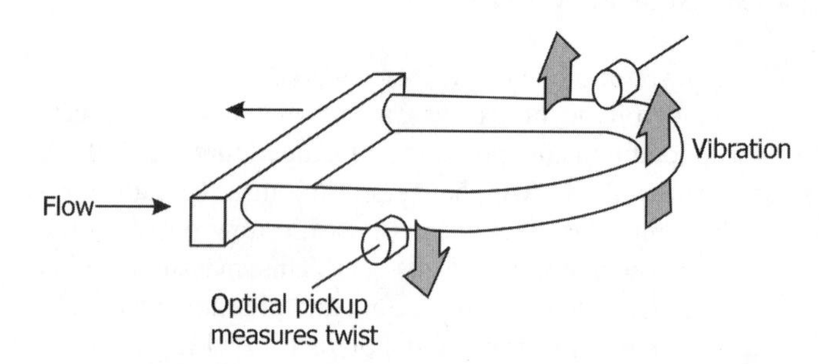

FIGURE 4.29 Single U-tube Coriolis mass flow meter.

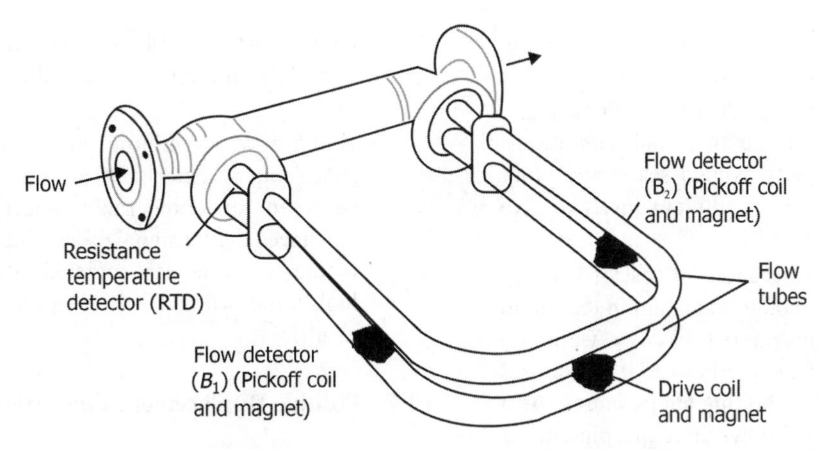

Note: Driver or drive coil and magnet assembly, which is used to vibrate the sensor flow tubes.

FIGURE 4.30 Double U-tube Coriolis mass flow meter designs.

- Bent tube Coriolis flow meters are preferred wherever highly accurate, reliable, rangeable mass flow is required.
- Straight tube meters are preferred when raw material contamination is to be avoided. Between runs of the same material or between runs of different materials it may be necessary to avoid cross-contamination

- What are the applications of Coriolis flow meters?
  - Used for clean liquids and gases flowing fast enough to operate the meter.
  - Coriolis meters can be used to measure flow of both liquids and gases.
  - These meters can be used to measure the flow of some fluids with varying densities that cannot be measured by other flow meters.
- What are the limitations of Coriolis flow meters?
  - These meters are sensitive to pipe and building vibrations.
  - Not suitable for flow measurement involving slug flow of two-phase gas–liquid systems. Entrained air or gases in liquid flow measurement could cause significant errors in measurements. They under predict the actual flow rates at low void fractions and over predict flows at high void fractions of entrained gases. Such errors could be as high as 20% or more when entrained gases or low flow rates are involved.
  - Over sized meters could result in low flow rates which reduce Coriolis effect and hence accuracy of the readings. They must be properly sized for the flow rates involved.
  - Temperature is constantly measured because the oscillatory properties of the tube vary with temperature.
  - Use of these meters is limited for smaller pipe sizes (usually 7–10 cm) as they become expensive for larger pipe sizes.
  - Some low-pressure gases do not have sufficient density to operate the meter.
  - The main limitation of the Coriolis meter is that pressure drop can become large as fluid viscosity increases.
  - Coriolis flow meters have a relatively high initial cost, which is offset by normally low maintenance costs.
- What type of application Coriolis flow meters is particularly advantageous over conventional ones?
  - General-purpose gas or liquid flow.
  - Blending ingredients and additives.
  - Conducting a primary check on secondary flow meters.
  - Metering natural gas consumption.
  - Monitoring such fluids as syrups, oils, suspensions, and pharmaceuticals.
  - *True* mass flow capability of Coriolis meters provide accurate flow requirement of expensive raw materials without wastage, for example, detergents, shampoos, and toothpastes.

- Well suited for cryogenic applications involving LNG, liquid helium, and so on.
- With normal flow measuring devices, equipment and components are affected by extreme cold conditions. Moving mechanical parts, wetted seals, and other parts can cease to function, fail prematurely or give rise to other problems.
- A complication arises because the cost of keeping cryogenic fluids cold enough to remain in the liquid state goes up as the temperature goes down. Common practice is to keep cryogenic fluids at temperatures only slightly below their boiling temperatures. As a consequence, as the fluid flows through a pipeline or past an obstacle like valve, ΔP, even small, can cause flashing to take place. Small or large pockets of gas form in the liquid, which makes flow measurement erroneous.

• Summarize the plus and minus points of Coriolis flow meters.
  - Plus points:
    - Higher accuracy than most flow meters ($\pm 0.1\%$ for mass flow).
    - Can be used in a wide range of liquid flow conditions.
    - Measures the mass flow rate directly, which eliminates the need to compensate for changing temperature, viscosity and pressure conditions.
    - Ideally suited to products that are accounted for on a mass basis, such as LPG, NGL, ethylene, and CO_2.
    - Capable of measuring hot (e.g., molten sulfur and liquid toffee) and cold (e.g., cryogenic helium and liquid nitrogen) fluid flow.
    - Low maintenance, because there are no parts that wear with time.
    - The meters can be significantly over ranged without causing damage to the sensor.
    - Low ΔP.
    - Suitable for bidirectional flow.
  - Minus points:
    - High initial setup cost.
    - Clogging may occur and difficult to clean.
    - Larger in overall size compared to other flow meters.
    - Limited line size availability.

• How does flashing is avoided or minimized in cryogenic applications of Coriolis flow meters?
  - A good rule of thumb for avoiding or minimizing flashing is that the difference between the discharge pressure and the liquid vapor pressure at the fluid temperature should be maintained at a factor of at least three times the value of ΔP across the meter.
  - A logical way to minimize flashing is to limit ΔP by specifying a meter that is larger than that might be used in comparable noncryogenic service. Advanced digital signal processing makes it possible to filter out the noise associated with a large meter at a high turndown and still achieves a highly accurate result.

4.1.7 Positive Displacement Flow Meters

• What are positive displacement flow meters?
  - Operation of positive displacement meters consists of separating liquids into accurately measured increments and moving them on. Each segment is counted by a connecting register. Because every increment represents a discrete volume, positive-displacement units are popular for automatic batching and accounting applications. Positive displacement meters are good for measuring the flows of viscous liquids or for use where a simple mechanical meter system is needed.

• What are the characteristic features and applications of positive displacement flow meters?
  - Positive displacement meters provide high accuracy ($\pm 0.1\%$ of actual flow rate in some cases) and good repeatability (as high as 0.05% of reading). Accuracy is not affected by pulsating flow unless it entrains air or gas in the fluid.
  - PD meters do not require a power supply for their operation and do not require straight upstream and downstream pipe runs for their installation.
  - PD meters are available in sizes from 6 to 300 mm (one-fourth in to 12 in.) and can operate with turndown ratios as high as 100:1 although ranges of 15:1 or lower are common.
  - PD meters operate with small clearances between their precision-machined parts.
  - As fluid viscosity increases, slippage between the flow meter components is reduced, and metering accuracy is therefore increased. Although slippage through the meter decreases with increased fluid viscosity (i.e., accuracy increases), pressure drop through the meter also increases.
  - The maximum (and minimum) flow capacity of the flow meter is decreased as viscosity increases. The higher the viscosity is, the lesser the slippage and the lower the measurable flow rate are.
  - As viscosity decreases, the low flow performance of the meter deteriorates.

- The maximum allowable pressure drop across the meter constrains the maximum operating flow in high-viscosity services.
- Wear rapidly destroys their accuracy. For this reason, PD meters are generally not recommended for measuring slurries or abrasive fluids.
- The process fluid must be clean. Particles greater than 100 μm in size must be removed by filtering.
- In clean fluid services, their precision and wide rangeability make them ideal for custody transfer and batch charging.
- They are most widely used as household water meters.
- In industrial and petrochemical applications, PD meters are commonly used for batch charging of both liquids and gases.

• What is a nutating disk flow meter? Illustrate with a diagram.

- Nutating disk meters are the most common PD meters. They are used as residential water meters.
- As water flows through the metering chamber, it causes a disk to wobble (nutate), turning a spindle, which rotates a magnet. This magnet is coupled to a mechanical register or a pulse transmitter.
- Because the flow meter entraps a fixed quantity of fluid each time the spindle is rotated, the rate of flow is proportional to the rotational velocity of the spindle.
- Figure 4.31 illustrates nutating disk flow meter.
- The meter housing is usually made of bronze but can be made from plastic for corrosion resistance or cost savings.
- The wetted parts such as the disk and spindle are usually bronze, rubber, aluminum, neoprene, Buna-N, or a fluoroelastomer such as Viton.
- Nutating disk meters are designed for water service and the materials of which they are made must be checked for compatibility with other fluids.
- Meters with rubber disks give better accuracy than metal disks due to the better sealing they provide.
- Nutating disk meters are available in sizes from 1.6 to 5 cm (five-eighth to 2 in.). They are suited for 10 barg (150 psig) operating pressures with overpressure to a maximum of 20 barg (300 psig). Cold-water service units are temperature limited to 50°C. Hot-water units are available up to 120°C.
- The accuracy of these meters is required to be ±2% of actual flow rate.
- Higher viscosity can produce higher accuracy, while lower viscosity and wear over time will reduce accuracy.

• Describe a rotating vane PD flow meter and illustrate it.

- Rotating vane meters (Figure 4.32) have spring loaded vanes that entrap increments of liquid between the eccentrically mounted rotor and the casing.
- The rotation of the vanes moves the flow increment from inlet to outlet and discharge.
- Accuracy of ±0.1% of actual rate is normal, and larger size meters on higher viscosity services can achieve accuracy to within 0.05% of rate.
- Rotating vane meters are regularly used in the petroleum industry and are capable of metering solids-laden crude oils at flow rates as high as 4000 m^3/h (17,500 gpm).
- Pressure and temperature limits depend on the materials of construction and can be as high as 175°C and 70 barg (1000-psig).
- Viscosity limits are 1–25,000 cP.
- In the rotary displacement meter, a fluted central rotor operates in constant relationship with two wiper rotors in a six-phase cycle.
- Its applications and features are similar to those of the rotary vane meter.

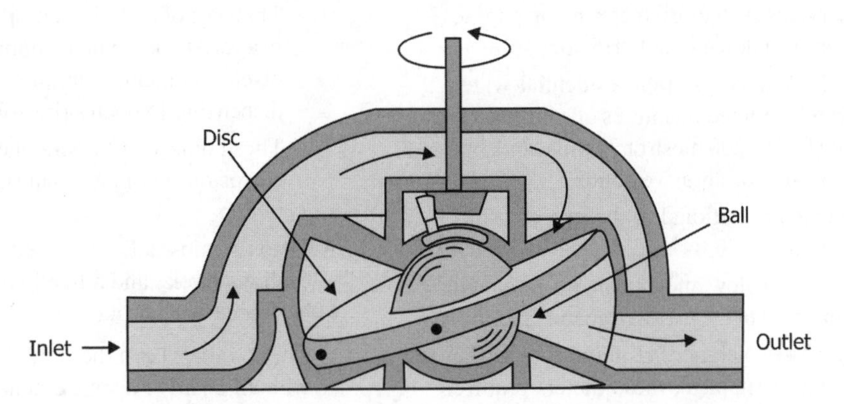

FIGURE 4.31 Nutating disk positive displacement flow meter.

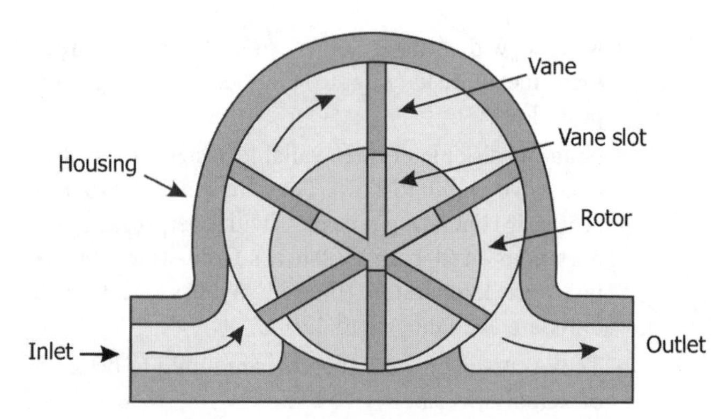

FIGURE 4.32 Rotating vane positive displacement flow meter.

- What is an oscillating piston positive displacement flow meter?
  - Figure 4.33 illustrates an oscillating piston positive displacement flow meter.
  - The measurement chamber is cylindrical with a partition plate separating its inlet port from its outlet port.
  - The piston is also cylindrical and is punctured by numerous openings to allow free flow on both sides of the piston and the post.
  - The piston is guided by a control roller within the measuring chamber, and the motion of the piston is transferred to a follower magnet that is external to the flow stream. The follower magnet can be used to drive either a transmitter, a register, or both. The motion of the piston is oscillatory (not rotary) since it is constrained to move in one plane.
  - When a quantity of fluid enters the chamber it causes a piston to rotate on its shaft. As it does so, a specific volume of fluid is moved through the meter and discharged at the outlet port. Each revolution of the piston corresponds to the movement of a fixed volume of fluid through the meter. A sensing system, typically magnetic or optical, senses a pulse each time a portion of a revolution occurs.
  - Oscillating piston flow meters typically are used in viscous fluid services such as oil metering on engine test stands where turndown is not critical.
  - These meters also can be used on residential water service and can pass limited quantities of solids, such as pipe scale and fine (−200 mesh or 74 μm) sand, but not large particle size or abrasive solids.
  - The rate of flow is proportional to the rate of oscillation of the piston.
  - The internals of this flow meter can be removed without disconnection of the meter from the pipeline.
  - Because of the close tolerances required to seal the piston and to reduce slippage, these meters require regular maintenance.
  - Oscillating piston flow meters are available in sizes from 11.25 to 75 mm (half in to 3 in.) sizes, and can generally be used between 6.9 and 10.3 barg (100 and 150 psig).
  - Some industrial versions are rated to 103 barg (1500 psig). They can meter flow rates from 0.2273 to 15 m^3/h (1–65 gpm) in continuous service with intermittent excursions to 22.73 m^3/h (100 gpm).
  - Meters are sized so that pressure drop is below 2.5 barg (35 psig) at maximum flow rate.
  - Accuracy ranges from ±0.5 % for viscous fluids to ±2% for nonviscous applications. Upper limit on viscosity is 10,000 cP.
  - Reciprocating piston meters are probably the oldest positive displacement meter designs.
  - They are available with multiple pistons, double-acting pistons, or rotary pistons. As in a reciprocating piston engine, fluid is drawn into one piston chamber as it is discharged from the opposed piston in the meter.
  - Typically, either a crankshaft or a horizontal slide is used to control the opening and closing of the proper orifices in the meter.
  - These meters are usually smaller (available in sizes down to 2.5 mm (0.1 in.)) diameter and are used for measuring very low flows of viscous liquids.
  - A single piston reciprocating positive displacement flow meter is shown in Figure 4.34.
- What are metering pumps?
  - Metering pumps are PD meters that also impart kinetic energy to the process fluid.
  - There are three basic designs: Peristaltic, piston, and diaphragm.
  - Peristaltic pumps operate by having fingers or a cam systematically squeeze a plastic tubing against the housing, which also serves to position the tubing.
    - This type of metering pump is used in laboratories, in a variety of medical applications, in the majority of environmental sampling systems, and also in dispensing hypochlorite solutions.
    - The tubing can be silicone rubber or, if a more corrosion-resistant material is desired, PTFE tubing.
  - Piston pumps deliver a fixed volume of liquid with each *out* stroke and a fixed volume enters the chamber on each *in* stroke.
  - Check valves keep the fluid flow from reversing. As with all positive displacement pumps, piston pumps generate a pulsating flow. To minimize the pulsation,

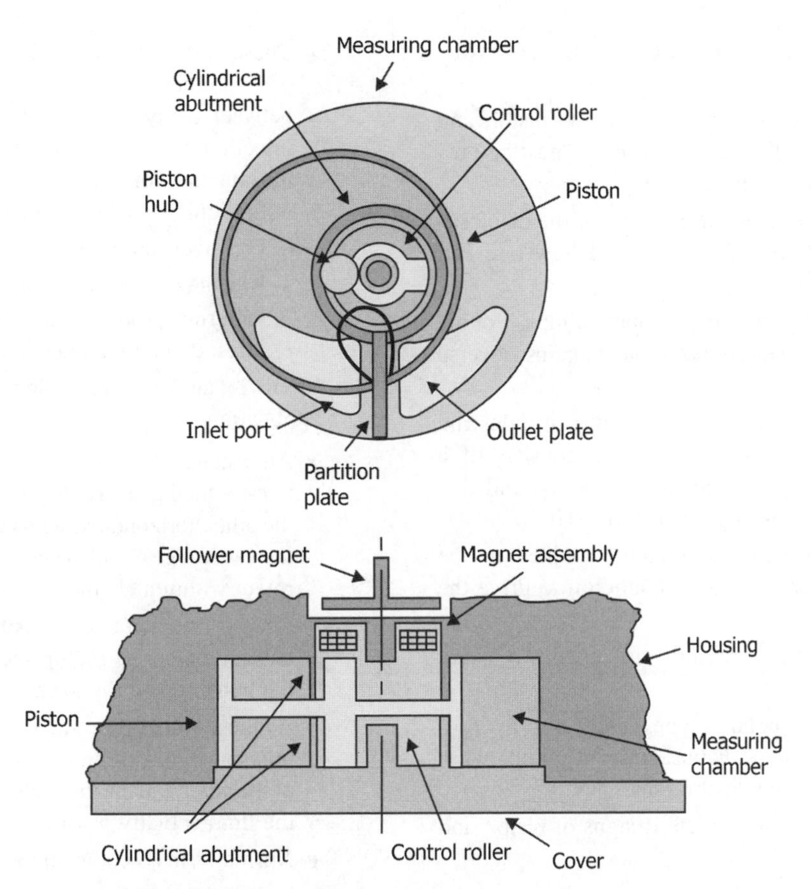

FIGURE 4.33 Oscillating piston positive displacement flow meter.

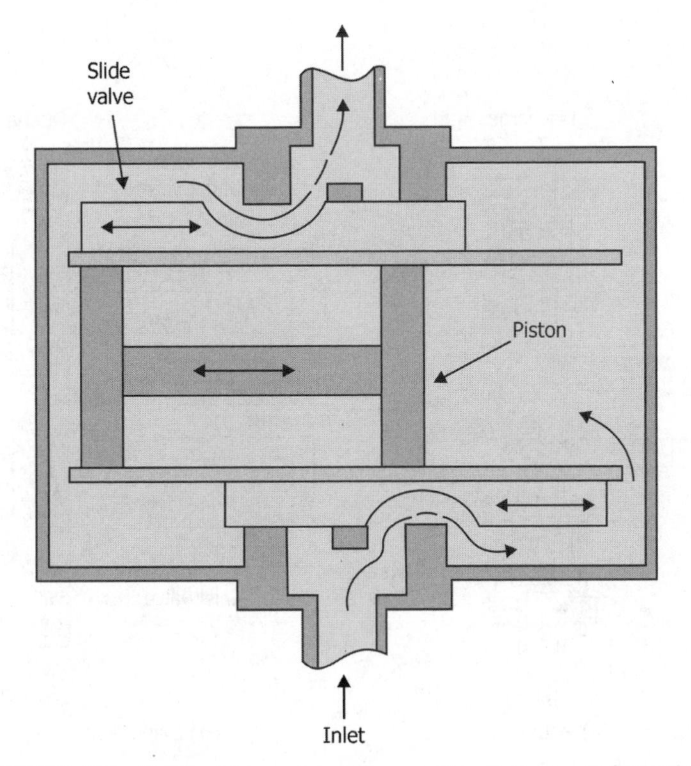

FIGURE 4.34 Single piston reciprocating positive displacement flow meter.

multiple pistons or pulsation dampening reservoirs are installed.

- Because of the close tolerances of the piston and cylinder sleeve, a flushing mechanism must be provided in abrasive applications.
- Piston pumps are sized on the basis of the displacement of the piston and the required flow rate and discharge pressure.
- Check valves (or, on critical applications, double check valves) are selected to protect against reverse flow.
- Diaphragm pumps are the most common industrial PD pumps. A typical configuration consists of a single diaphragm, a chamber, and suction and discharge check valves to prevent reverse flow.
- The piston can either be directly coupled to the diaphragm or can force a hydraulic oil to drive the diaphragm.
- Maximum output pressure is about 8.6 barg (125 psig).
- Variations include bellows type diaphragms, hydraulically actuated double diaphragms, and air operated, reciprocating double diaphragms.
- Figure 4.35 illustrates typical designs of piston and diaphragm pumps.

• Give examples of applications of chemical *proportionating pumps* as flow meters.

- These applications involve accurate measurement of volumetric flow ratios to be maintained between the process liquid volume and the chemical injection volume. Examples include injection, or introduction, of
  - Sodium hypochlorite disinfectant into water.
  - Fertilizer into irrigation water.
  - Bioengineered organisms into liquids for manufacturing products such as soaps for large laundries, defoaming chemicals, and insecticides.

• What is an oval gear flow meter? What are its applications?

- An oval gear flow meter consists of two oval-shaped, fine-toothed gear rotors, one mounted vertically and the other horizontally, with gears meshing at the tip of the vertical gear and the center of horizontal gear that rotate within a chamber of specified geometry.
- The two rotors rotate opposite to each other, creating an entrapment in the crescent-shaped gap between the housing and the gear.
- As these rotors turn, they sweep out and trap a very precise volume of fluid between the outer oval shape of the gears and the inner chamber walls, with none of the fluid actually passing through the gear teeth.
- Normally, magnets are embedded in the rotors, which then can actuate a reed switch or provide a pulse output via a specialized, designated sensor.

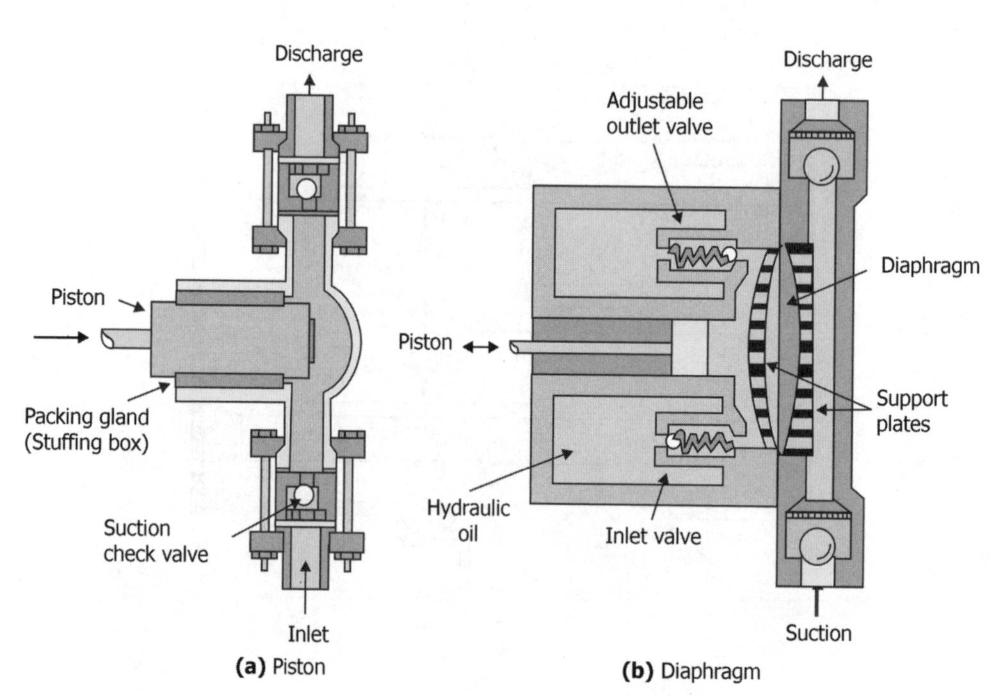

FIGURE 4.35 Piston and diaphragm positive displacement metering pumps.

FIGURE 4.36 Oval gear lobe flow meter.

- Each pulse or switch closure then represents a precise increment of liquid volume that passes through the meter. The result is a high accuracy (usually ±0.5% of reading) and resolution if slippage between the housing and the gears is kept small.
- These meters involve almost negligible effects for varying fluid viscosity, density, and temperature.
- Accuracy of these meters will be more with liquids involving high viscosities (>10 cP) as with high viscosities, liquid slippage will become less.
- At lower flows and at lower viscosity, slippage increases and accuracy decreases.
- Figure 4.36 illustrates an oval gear lobe flow meter.
- Pressure drop is the only limiting factor when the application requires the metering of highly viscous liquids.
- The lubricating characteristics of the process fluid also affect the turndown ratio of an oval gear meter. With liquids that do not lubricate well, maximum rotor speed must be kept low to limit wear.
- Another way to limit wear is to keep the pressure drop across the meter low. The pressure drop across the meter limits the allowable maximum flow in high-viscosity service.
- Oval gear meters are generally not recommended for water or water-like fluids, because the increased chances of fluid slippage between the gears and the chamber walls.
- Range of fluids that can be used with oval gear flow meters include sugar solutions, syrups, cooking oils, sauces, beverages, honey, molasses, milk products, juices, chocolate, coatings, shampoos, gels, perfumes, creams, fuel oils, lubricants, latex-based paints, wax finishes, and so on.
- Rotating lobe and impeller-type PD meters are variations of the oval gear flow meter that do not share its precise gearing.
- In the rotating lobe design, two impellers rotate in opposite directions within the ovoid housing. As they rotate, a fixed volume of liquid is entrapped and then transported toward the outlet.
- Because the lobe gears remain in a fixed relative position, it is only necessary to measure the rotational velocity of one of them.
- The impeller is either geared to a register or is magnetically coupled to a transmitter.
- The lobe gear meter is available in a wide range of materials of construction, from thermoplastics to highly corrosion-resistant metals.
- Disadvantages of this design include loss of accuracy at low flows. Also, the maximum flow through this meter is less than for the same size oscillatory piston or nutating disk meter.
- In the rotating impeller meter, very coarse gears entrap the fluid and pass a fixed volume of fluid with each rotation.
- These meters are accurate to 0.5% of rate if the viscosity of the process fluid is both high and constant, or varies only within a narrow band.
- These meters are used to measure paints, and, because they are available sanitary designs, also used for milk, juices, and chocolate.

- What is a helical gear flow meter? What are its applications?
  - The helix (helical gear) meter is a positive displacement device that uses two radially pitched helical gears to continuously entrap the process fluid as it flows. The flow forces the helical gears to rotate in the plane of the pipeline.
  - Optical or magnetic sensors are used to encode a pulse train proportional to the rotational speed of the helical gears.
  - The forces required to make the helices rotate are relatively small and therefore, in comparison to other PD meters, the pressure drop is relatively low.
  - Helical gear meters can measure the flow of highly viscous fluids (from 3 to 300,000 cP), making them ideal for extremely thick fluids such as glues and very viscous polymers.
  - The maximum rated flow through the meter is reduced as the fluid viscosity increases.
  - If the process fluid has good lubricating characteristics, the meter turndown can be as high as 100:1, but lower (10:1) turndowns are more typical.

4.1.8 Miscellaneous Types of Flow Meters

4.1.8.1 Target Flow Meters

- What are target meters? How do they work? What is their plus point?
  - In a target meter, a sharp-edged disk is set at right angles to direction of flow and drag force exerted on the disk by the fluid is measured.
  - Figure 4.37 illustrates a target flow meter.
  - Flow rate is proportional to the square root of drag force and the fluid density.
  - Target meters sense and measure forces caused by liquid impacting on a target or drag disk suspended in the liquid stream.
  - A direct indication of the liquid flow rate is achieved by measuring the force exerted on the target.
  - In its simplest form, the meter consists only of a hinged, swinging plate that moves outward, along with the liquid stream. In such cases, the device serves as a flow indicator.
  - These meters are rugged and inexpensive.
  - A more sophisticated version uses a precision, low-level force transducer-sensing element. The force of the target caused by the liquid flow is sensed by a strain gauge. The output signal from the gauge is indicative of the flow rate.
  - Target meters are useful for measuring flows of viscous liquids, slurries, or corrosive liquids.

4.1.8.2 Vortex Shedding Flow Meters

- What is the principle of vortex shedding flow meter? Illustrate by means of a sketch.
  - Vortex shedding is a natural phenomenon where alternating vortices are shed in the wake of a body at a frequency that depends on the flow velocity past the body.
  - The vortices formed on opposite sides of the body are carried downstream in the body's wake, forming a *vortex street*, each vortex having an opposite sign of rotation.

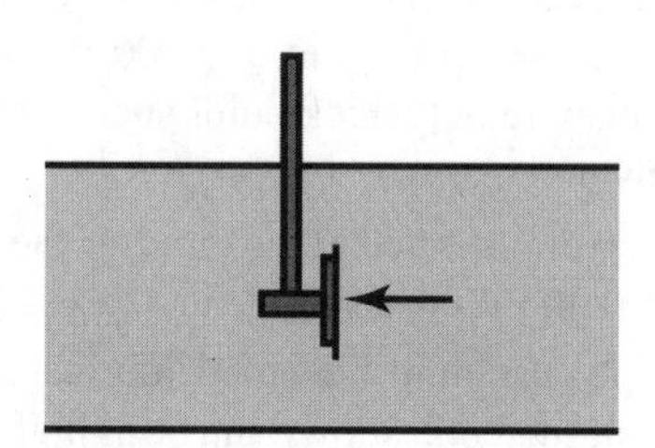

FIGURE 4.37 Target meter.

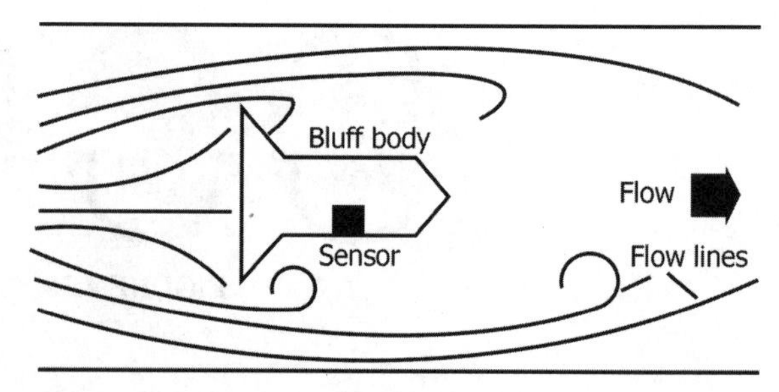

FIGURE 4.38 Major components in a vortex shedding flow meter.

  - Vortex flow meters make use of von Karman effect. According to this principle, the presence of an obstruction in the flow stream causes the fluid to generate alternating vortices. In a vortex meter, this obstruction is called a bluff body. It consists of a piece of material with a broad, flat front that is mounted at right angles to the surface of the flow stream.
  - Vortex shedding frequency is directly proportional to the velocity of the fluid in the pipe and therefore to volumetric flow rate.
  - Flow rate is determined by multiplying flow velocity times the area of the pipe.
  - The vortex shedding phenomenon is used to sense average velocity in pipe flows in a vortex flow meter.
  - The shedding frequency is independent of fluid properties, such as density, viscosity, conductivity, and so on, except that the flow must be turbulent for vortex shedding.
  - The basic components of a vortex shedding flow meter are illustrated in Figures 4.38 and 4.39.
  - A vortex flow meter is typically made of 316 stainless steel or Hastelloy and includes a bluff body, a vortex sensor assembly, and the transmitter electronics.
  - The installed cost of vortex meters is competitive with that of orifice meters in sizes under 15 cm (6 in.). Wafer body meters (flangeless) have the lowest cost, while flanged meters are preferred if the process fluid is hazardous or is at a high temperature.
  - Bluff body shapes (square, rectangular, t-shaped, and trapezoidal) and dimensions have been used to achieve the desired characteristics.
  - Linearity, low Reynolds number limitation, and sensitivity to velocity profile distortion vary only slightly with bluff body shape.
    - In size, the bluff body must have a width that is a large enough fraction of the pipe diameter that the entire flow participates in the shedding.
    - It must have protruding edges on the upstream face to fix the lines of flow separation, regardless of the flow rate.

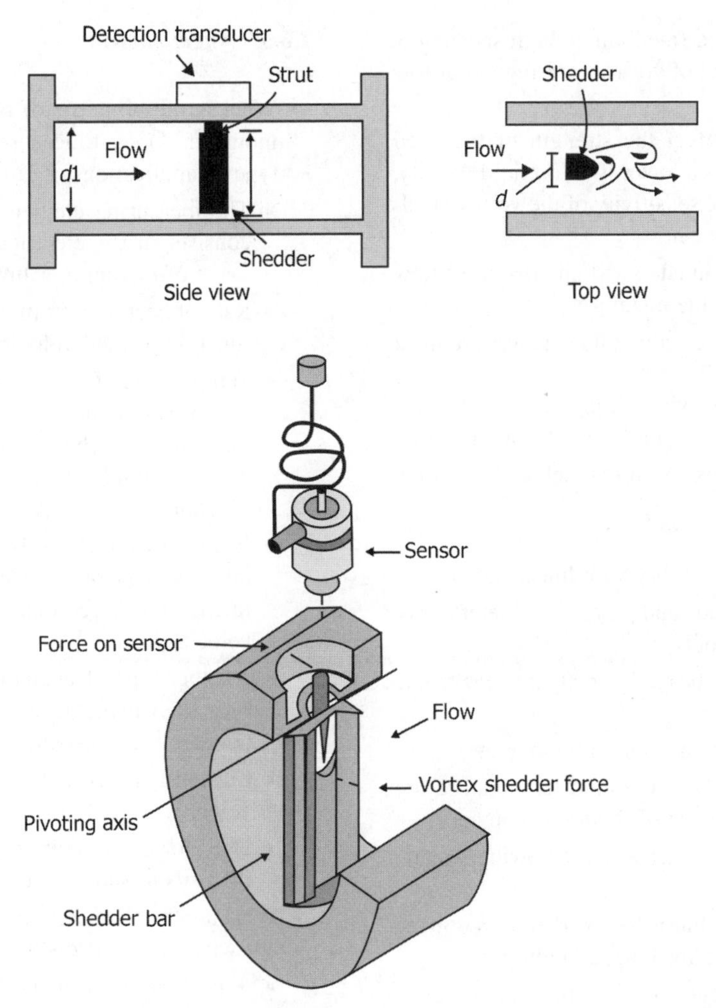

FIGURE 4.39 Vortex shedding flow meters. *Note:* Different shapes of shedders are used in the designs for getting the desired flow rate measurements.

        - ➢ Its length in the direction of the flow must be a certain multiple of its width.
    - ▪ The majority of vortex meters use piezoelectric- or capacitance-type sensors to detect the pressure oscillation around the bluff body. These detectors respond to the pressure oscillation with a low voltage output signal that has the same frequency as the oscillation. Such sensors are modular, inexpensive, easily replaced, and can operate over a wide range of temperature ranges—from cryogenic liquids to superheated steam.
    - ▪ Sensors can be located inside the meter body or outside.
- • What are the characteristics and applications of vortex shedding flow meters?
    - ▪ Vortex meters are among the most versatile of meters, and can be used to measure liquid, gas, and steam flows.
    - ▪ Used for clean, low-viscosity, and swirl-free fluids flowing at medium to high speed. Vortex formation in high-viscosity fluids may not be dependable.
    - ▪ Any erosion, corrosion, or deposits that change the shape of the bluff body can shift flow meter calibration.
    - ▪ The meter has no moving parts and involve relatively low pressure drop compared to obstruction meters.
    - ▪ Vortex meter accuracy is from medium to high, depending on the manufacturer and the model.
    - ▪ Under low flow conditions, vortices are formed irregularly, so low flow conditions present a problem for vortex meters.
    - ▪ The lower flow rate limit on vortex meters appears to be at Reynolds numbers near 10,000. This can be a problem in metering high-viscosity pipe flows.

- Ideal conditions include medium- to high-speed flow because the formation of vortices is irregular at low flow rates.
- Density variations affect the strength of the shed vortex and this places a lower limit on fluid density, which is based on the sensitivity of the vortex shedding detection equipment.
- It is necessary to eliminate swirl and distorted flow patterns upstream of the meter.

• Under what circumstances vortex flow meters are most accurate?
  - High-speed flows involving clean and low viscosity liquids, free from swirl and distorted flow patterns.

• What are the advantages and disadvantages of vortex shedding flow meters?

  Advantages
  - Relatively wide rangeability with linear output.
  - On clean fluids (liquids and gases), the meters have long-term stable proofs.
  - Frequency output can be read directly into electronic readout systems.
  - Installation is simple and costs are moderate.
  - Effects of viscosity, pressure, and temperature are minimal over wide range of Reynolds numbers.
  - No moving parts in contact with the flowing stream.

  Disadvantages
  - Flow into a meter must be swirl-free requiring straightening vane and/or long, straight piping.
  - Output may have frequency instability and/or fade in certain areas of operation, which affect readout requirements.
  - Not available in sizes above 20 cm.
  - Pulse train is irregular and proving requires a long test time to obtain a representative average pulse rate.
  - Pulse resolution is the same for all meter sizes, meaning a low pulse rate with larger meters yields low volume resolution.
  - Subject to range limitations at lower Reynolds numbers.

• Give typical applications of vortex meters.
  - Cryogenic fluids.
  - Steam measurement.
  - Condensate measurement.
  - Ultra pure and deionized water.
  - Acids.
  - Solvents.

• "Vortex meters with integral reducing flanges are a good choice for measuring low flows". *True/False*?
  - *True*. These provide flexibility after installation.

4.1.8.3 Anemometers

• What is the principle of operation of a hot-wire anemometer? Does it measure velocity or flow? For what type of applications, it is well suited?
  - The measuring element of a hot-wire anemometer consists of an electrically heated wire, generally made of platinum or tungsten.
  - Rate of heat flow from the wire to the fluid (gas) is a function of fluid velocity.
  - Temperature of the wire is a function of its resistance (cf. principle involved in platinum-resistant thermometer), which in turn is a function of fluid velocity. Thus, fluid velocity is calibrated in terms of wire resistance.
  - It measures velocity of gases/air in conduits/ducts into which the probe is inserted with the wire element facing flow direction. Velocity measured is point velocity.
  - Another type of anemometer is constant resistance type, in which the current is varied to keep wire temperature constant.
  - Current is measured and expressed as a function of velocity.
  - Used for gas velocity measurements in the range of 0.15 m/s to supersonic velocities.

• What are the characteristic features and applications of hot-wire anemometers?
  - A positive consequence of their small mass is fast speed of response. They are widely used in HVAC and ventilation applications.
  - Larger and more rugged anemometers are also available for more demanding industrial applications.
  - To ensure the proper formation of the velocity profile, a straight duct section is usually provided upstream of the anemometer station (usually 10 diameters long).
  - A conditioning nozzle is used to eliminate boundary layer effects.
  - If there is no room for the straight pipe section, a honeycomb flow straightener can be incorporated into the sensor assembly.
  - Typically, the anemometer wire is made of platinum or tungsten and is 4–10 μm in diameter and 1 mm in length.
  - Because of the small size and fragility of the wire, hot-wire anemometers are susceptible to dirt buildup and breakage.

• What is a heated thermocouple anemometer?
  - Heated thermocouple anemometer measures gas velocity from cooling effect of the fluid (gas) stream

flowing across hot junctions of a thermopile supplied with constant electric power input.

- The measured frequency shift is related to true average velocity through a calibration technique.

• What is a hot-film anemometer?
  - A hot-film anemometer consists of a platinum-sensing element deposited on a glass substrate.
  - It is less susceptible to fouling by bubbles or dirt when used for liquids.

• "A thermal anemometer for gas flow measurement is more universally applicable than a pitot tube for measuring velocity because it does not need to be adjusted for differences in the molecular weight of the gas." *True/False*?
  - *False*. The signal from a thermal anemometer varies with heat loss and gases with different heat absorbing properties (i.e., heat capacity and thermal conductivity) will remove different amounts of heat. A pitot tube outputs a simpler property (velocity head) than a thermal anemometer (relative heat loss).

4.1.8.4 Bubble Flow Meters

• What is a bubble flow meter? What are its applications, advantages, and limitations?
  - The bubble flow meters are of two designs, namely, for liquids and for gases. The design for liquids makes use of a timed measurement of a meniscus rising between two optical sensors.
  - The fluid enters the inlet and moves up inside a glass tube, past the sensor block and around the tube toward the outlet. As this happens, a solenoid valve is timed to periodically open and close, thereby sucking a small amount of air into the tube. This creates separate columns of liquid that move upward inside the tube, and toward the optical sensor block. The meniscus that is formed by these columns of fluid against the glass capillary tube walls is measured by the optical sensors. Since the meniscus travels at the same rate as the column of fluid, measuring the rate of meniscus travel gives a direct correlation to the liquid flow.
  - Two infrared sensors located within the sensor block time the rate of rise of the meniscus and this volume-over-time measurement is then converted to a flow rate and displayed on a digital readout. As the fluid moves around the top of the tube, air is vented at the top while the liquid continues around and exits at the overflow tube. The process then repeats itself as the solenoid valve opens to create another air gap.
  - By comparison, the bubble design for gas flow works somewhat differently although the same basic concept remains. For the gas bubble flow meters, a soapy solution is used to fill the lower reservoir of the glass flow tube. The gas flow source is then connected to a point above the bubble solution reservoir and gas travels around to the glass flow tube. At this point, a rubber bulb is either manually squeezed or a clamp is used to continuously generate bubbles that travel at the same speed as the gas.
  - When the bubble passes the lower optical sensor within the sensor block, an internal timer is automatically started, and when the bubble passes the upper optical sensor, the timer is stopped. The total elapsed time is correlated to a gas flow rate and displayed on a digital readout. The small amount of liquid soap left over from the process collects in a flow trap for disposal.
  - *Advantages*: The major advantage of the bubble meter for gases is that it is not affected by the gas composition. By contrast, most electronic meters must be calibrated for a specific gas or gas mixture. Whether one is measuring ordinary gases such as N_2, O_2, H_2, CO_2, and Ar, or measuring a unique gas mixture, one bubble meter can do it all. This versatility helps to lower equipment costs and can save recalibration time. However, it should be kept in mind that some gases may have a chemical reaction with the water used to make the bubble solution and the user should be careful when specifying bubble flow meters for such compounds.
  - Another useful advantage of the bubble design is that the calibration does not drift over time. The main electrical parts of the system are the optical sensors for detecting the presence or absence of a bubble or meniscus layer. These noncontact sensors do not wear out or experience a drift in accuracy. The glass tube is fixed in diameter and will not change with time. In the gas chromatography market, bubble meters can be qualified as a primary flow standard. Earlier, bubble flow meters were available only for very low flow rates, but currently they are available also for expanded flow rate ranges. While gas flows ranging from 0.1 to 25 L/min can be accurately measured, liquid bubble meters do not have quite the range as the gas versions and are available in sizes ranging from roughly 1 –30 mL/min.
  - *Disadvantages*: In order to make an in-line measurement with a bubble flow meter, one needs to make a break in the line where the flow reading is desired, then make measurement and finally restore the line to its original condition. Bubble meters are therefore adequate for *end-of-line* readings, but are not well suited for continuous, in-line monitoring. In some applications, the use of a bubble solution could be a

TABLE 4.2 Summary of Plus and Minus Points of Different Types of Flow Meters

	Liquids						
Flow Meter Type	Low μ	High μ	Slurries	Gases	Turndown Ratio	Accuracy ($\pm$%)	Cost
Head type							
Orifice	Yes	Limited	No	Yes	20:1	2–3	Low
Venturi	Yes	Limited	Yes	Yes	20:1	2–3	High
Rotometers	Yes	Limited	Limited	Yes	10:1	2–4	Low
Target Meters	Yes	Limited	Limited	No	—	—	Medium
Velocity type							
Magnetic	Yes	Yes	Yes	No	20:1	0.5–1	High
Vortex	Yes	Limited	No	Yes	20:1	1	Medium
Displacement							
Turbine	Yes	Limited	Limited	Yes	10:1	0.25–1	Medium
Oval Gear	Yes	Yes (>10 cP)	No	No	25:1	0.1–0.5	Medium
Mass flow type							
Coriolis	Yes	Yes	—	Yes	100:1	0.05–0.1	High
Gas mass flow	—	—	—	Yes	50:1	1.5–2	Medium

minor inconvenience, since it needs to be cleaned up after the measurement.

- *Applications*: Bubble meters are used in laboratory and low-flow research applications. Their use in more industrial applications is extremely limited. Some of the applications for a bubble flow meter include the following:
  - Supercritical fluid extraction.
  - Chromatography column, detector and carrier gas measurement.
  - Monitoring post-detector flow volumes in HPLC systems.
  - Calibration and flow verification for variable area and electronic flow meters.
  - Accurate flow measurement of gas mixtures without recalibration.
  - Accurate flow measurement of changing gas
  - concentrations.
  - Calibration of air sampling pumps.

- Give a brief summary of plus and minus points of different types of flow meters.
  - Table 4.2 summarizes the plus and minus points of different types of flow meters.
- Name some best practices that aid in the selection of flow measurements.
  - Measurement of mass flow in gases and steam.
  - Elimination of impulse lines.
  - Minimization of permanent pressure loss.
  - Selection of in-line meters for small lines and insertion meters for large lines to reduce the overall installation cost of the flow meter.
  - Elimination of mechanical meters to reduce maintenance costs. Moving parts in mechanical meters require regular maintenance.
- What are the common flow measurement problems and possible solutions?
  - Table 4.3 gives a summary of flow measurement problems.

4.1.9 Weirs

- What are weirs? Name different types of weirs and give their applications?
  - A *weir* is a dam over which the liquid flows.
  - Weirs operate on the principle that an obstruction in a channel will cause water to back up, creating a high

TABLE 4.3 Summary of Flow Measurement Problems

Symptom	Problem Source	Solution
Low mass flow indicated	Liquid droplets in gas	Install demister upstream. Heat gas upstream of sensor.
Mass flow error	Static pressure change in gas	Add pressure recording pen
Transmitter zero shift	Free water in fluid	Mount transmitter above taps
Measurement is high	Pulsation in flow	Add process pulsation dampener
Measurement error	Nonstandard pipe runs	Estimate limits of error

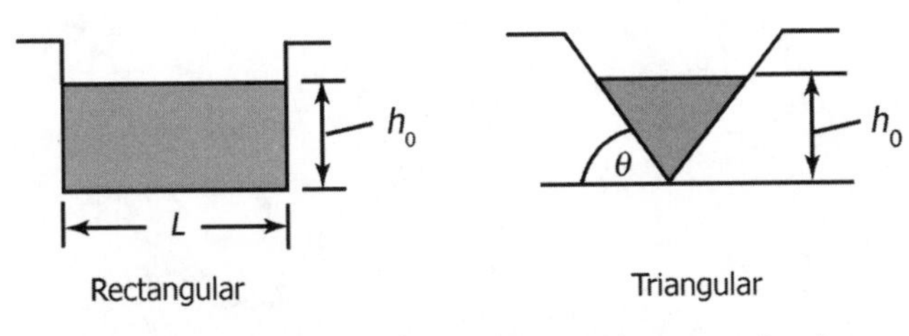

FIGURE 4.40 Rectangular and triangular types of weirs.

level (head) behind the barrier. The head is a function of flow velocity, and, therefore, the flow rate through the device.

- Weirs consist of vertical plates with sharp crests. The top of the plate can be straight or notched. Weirs are classified in accordance with the shape of the notch. The basic types are rectangular, V-notch, and trapezoidal.
- The two types of weirs commonly used are the rectangular weir and the triangular weir as shown in Figure 4.40.
- The liquid flows over the weir, and the height, h_0 (weir head), in m is measured above the flat base or the notch as shown. This head should be measured at a distance of about $3h_0$ m upstream of the weir by a level or float gauge.
- Weirs are used for open channel flow measurement, as liquids flow in open channels in process plants, rivers, canals, drains, and so on. An example in process equipment is in plate distillation columns.
- The equation for the volumetric flow rate q in m^3/s for a rectangular weir is given by

$$q = 0.415(L - 0.2h_0)h_0^{1.5}(2g)^{0.5}. \tag{4.10}$$

For a triangular notch,

$$q = [0.31h_0^{2.5}](2g)^{0.5}/\tan\theta, \tag{4.11}$$

where q is measured in m^3/s; L is the crest length in meters; $g = 9.80665$ m/s^2; h_0 = weir height in meters; and Ø is the angle as shown in Figure 4.40. These formulas were developed for flow of water and require corrections for other liquids.

- What is the difference between a sharp-edged weir and a rounded upstream-edged weir?
  - With sharp-edged weir, the sheet of discharging liquid, contracts as it leaves the opening and free discharge occurs.
  - Rounding upstream edge will reduce the contraction and increase the flow rate for a given head.
  - That is, for a given flow rate, head will be less than that for sharp-edged weir.
  - Result is that accuracy of head readings will decrease for a blunt-edged weirs.
- Give the equations for flow measurement in open channels using rectangular and V-notches.
  - Flow over wide rectangular weirs:

$$q_{\text{weir}} = C_d b g^{1/2} H^{3/2}. \tag{4.12}$$

where q_{weir} is the volumetric flow rate measured in m^3/s and C_d is the dimensionless weir coefficient. For turbulent upstream flow, C_d depends on weir geometry, b is the width of the weir in meters. For narrow weirs with side walls, b is to be replaced by $b - 0.1$H, where H is the height of the weir in meters, and $g = 9.81$.

- C_d can be computed from the following empirical equations:

For sharp-crested weirs,

$$C_d \approx 0.564 + 0.0846\text{L/H}, \quad \text{for } L/H < 0.07. \tag{4.13}$$

For broad-crested weirs,

$$C_d \approx 0.462 \quad \text{for } 0.08 < H/L < 0.33. \tag{4.14}$$

where L is the length of the weir in meters and H is the height of the weir in meters.

- Flow over triangular notches with the angle, 2θ,

$$Q \approx 0.44 \tan\theta\, g^{1/2} H^{5.2}, \quad \text{for} \quad 10^\circ < \theta \leq 50^\circ \tag{4.15}$$

- The V-notch is more sensitive at low flow rates (large H for a small Q) and thus is popular in laboratory measurements of channel flow rates.

5

PUMPS, EJECTORS, BLOWERS, AND COMPRESSORS

5.1 Pumps 101
  5.1.1 Centrifugal Pumps 101
  5.1.2 Cavitation 112
  5.1.3 Centrifugal Pump Performance and Operational Issues 118
  5.1.4 Positive Displacement Pumps 127
  5.1.5 Miscellaneous Pumps 137
  5.1.6 Vacuum Pumps 141
5.2 Ejectors 145
5.3 Fans, Blowers, and Compressors 152

5.1 PUMPS

- Classify pumps.
  - Pumps are divided into two fundamental types based on the manner in which they transmit energy to the pumped media, namely, kinetic displacement or positive displacement.
  - In kinetic displacement, a centrifugal force of the rotating element, called an impeller, imparts kinetic energy to the fluid, moving the fluid from pump suction to the discharge.
  - On the other hand, positive displacement uses the reciprocating action of one or several pistons, or a squeezing action of meshing gears, lobes, or other moving bodies, to displace the media from one area into another (i.e., moving the material from suction to discharge).
  - Figure 5.1 presents a broad classification of pumps.
  - Hydraulic Institute classifies pumps by type, not by application, into two basic types, namely, kinetic types and positive displacement types, which are detailed in Figures 5.2 and 5.3.

5.1.1 Centrifugal Pumps

- What is the principle of operation of a centrifugal pump? How does it work?
  - The purpose of a centrifugal pump is to convert the energy of the motor or other prime mover, first into the kinetic energy (velocity) of the fluid being pumped and then into pressure energy. The velocity is developed in the rotating impeller, whereas the conversion to pressure takes place within the stationary volute or diffuser.
  - The process liquid enters the pump near the impeller axis, that is, eye of the impeller and the rotating impeller vanes. As a result of this rotation, impeller vanes transfer mechanical work to the fluid in the impeller channel, which is formed by the impeller vanes. The rotating impeller channel sweeps the liquid out the ends of the impeller blades at high velocity. As liquid leaves the impeller eye, a low-pressure area is created, causing more liquid to flow toward the inlet. Because the blades are curved, the fluid is pushed in a tangential and radial direction by the centrifugal force.
  - Energy added to the pump by centrifugal force is *kinetic energy*, which is given to the liquid proportionate to velocity at the edge or vane tip of the impeller. The faster the impeller rotates or the bigger the impeller is, the higher will be the velocity of the liquid at the vane tip and the greater the energy imparted to the liquid.
- What are the salient features of centrifugal pumps? What type of applications a centrifugal pump is suitable?
  - Very high capacity and low head.
  - Smooth nonpulsating flow.
  - No damage on closure of discharge valve, unlike positive displacement pumps.

Fluid Mechanics, Heat Transfer, and Mass Transfer: Chemical Engineering Practice, By K. S. N. Raju

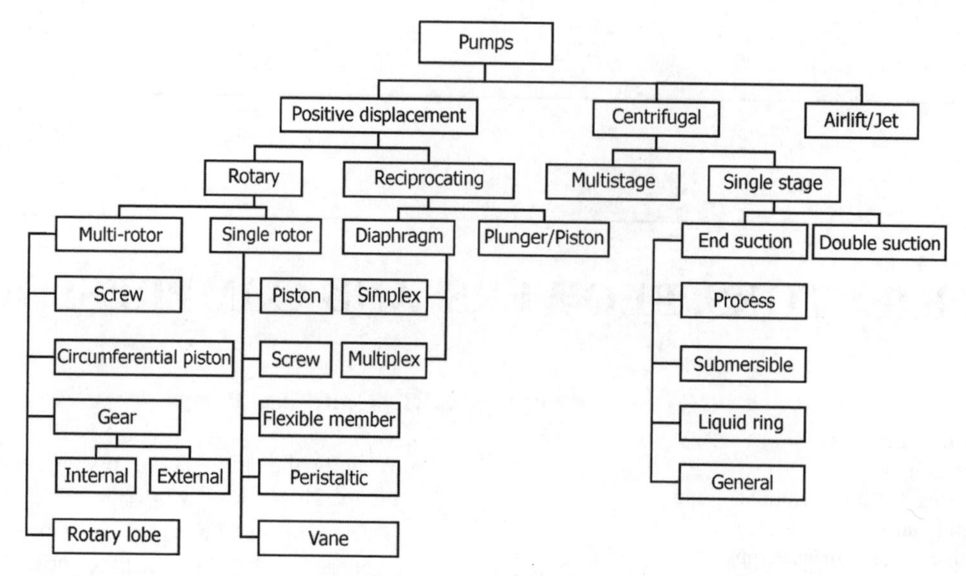

FIGURE 5.1 Pump classification.

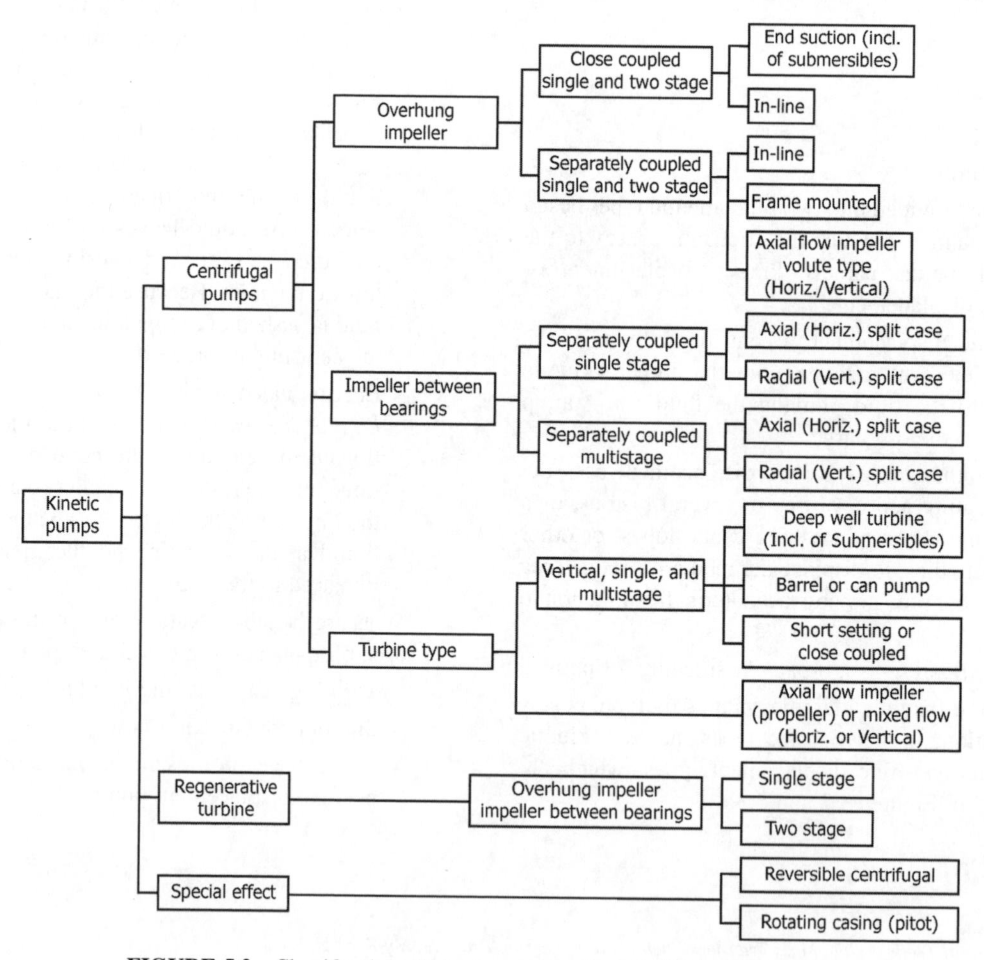

FIGURE 5.2 Classification of kinetic pumps. (*Source*: Hydraulic Institute.)

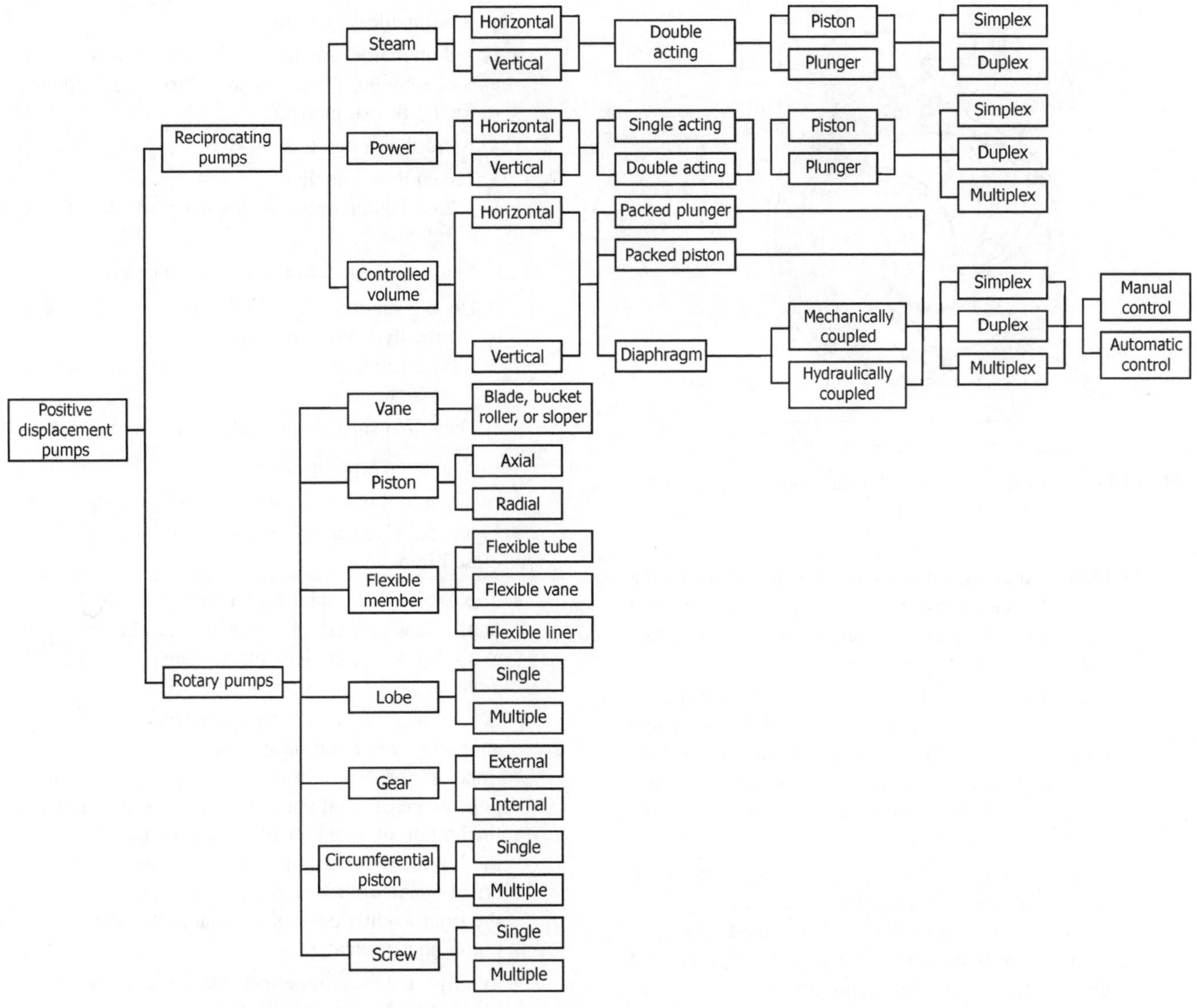

FIGURE 5.3 Classification of positive displacement pumps. (*Source*: Hydraulic Institute.)

- Only moving parts are shaft and impeller.
- No close clearances.
- No valves/reciprocating parts as integral parts of the pump.
- Ease of maintenance.
- Less noise.
- Low first cost.
- Small floor space requirement.

• What are overhung impeller-type pumps?
  - The impeller or impellers are mounted on the extreme end of the shaft in a cantilevered condition hanging from the support bearings.
  - This type of pumps is also subdivided into a class known as motor pumps or close-coupled pumps where the impeller is directly mounted onto the motor shaft and supported by the bearings of the motor.

• What are the features of pumps with impellers mounted between bearings?
  - In this group of centrifugal pumps, the impeller or the impellers are mounted onto the shaft with the bearings on both ends. The impellers are mounted between the bearings.
  - These types of pumps are further divided into single-stage (one impeller) or multistage pumps (multiple impellers).

• What are the parameters required in specifying a centrifugal pump?
  - *Required flow rate*, which is determined by material and energy balances.

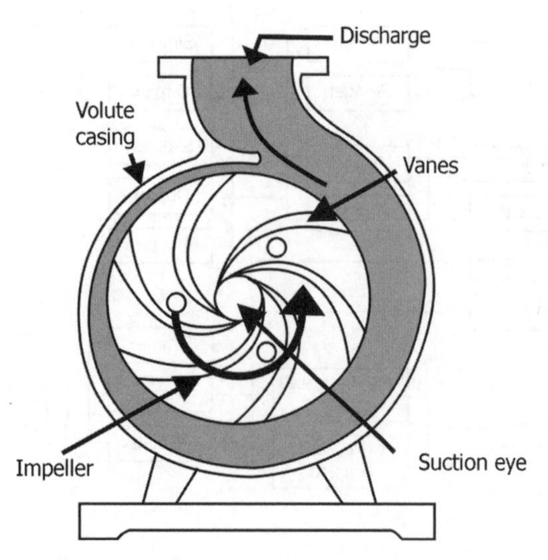

FIGURE 5.4 Centrifugal pump showing important parameters. (*Source*: www.cheresources.com.)

- Design margins, up to about 25%, are added to the material balance flow rate to account for unexpected variations in properties or conditions or to ensure that the plant meets its performance criteria.
- Minimum flow protection is added as continuous circulation. If the required flow rate falls in the low range for centrifugal pumps, a minimum size pump, rated for continuous service, is specified and the extra pump capacity is consumed by circulation from the discharge to the source.
- The maximum pressure a pump will develop during any aspect of operation, including start-up, shutdown, and upset conditions is determined. The shutoff pressure is the maximum pressure a pump will develop under zero flow conditions.
- Variables to consider when determining the design pressure include the following:
  - Maximum static head of the source at pump suction level.
  - Maximum head developed by the pump, that is, shutoff head.
  - Maximum pump operating speed (for variable speed drives).
  - Possibility of operator intervention during an upset.
- Other requirements, apart from normal values of density and viscosity, the maximum values that a pump is likely to encounter during extreme conditions, such as start-up, shutdown, process upsets, or weather effects, as these property variations influence the work needed to drive the pump.
- Maximum and minimum operating temperatures, presence of suspended solids, dissolved gases, and avoidance of cavitational problems.

- How are impellers classified?
  - Based on major direction of flow in reference to the axis of rotation, impellers are classified as follows:
    - Radial flow impellers.
    - Axial flow impellers.
    - Mixed flow impellers.
      - Based on suction type, impellers are classified as follows:
    - Single suction: Liquid inlet on one side.
    - Double suction: Liquid inlet to the impeller symmetrically from both sides.
      - Based on mechanical construction, impellers are classified as follows:
    - Open: No shrouds or wall to enclose the vanes.
    - Semiopen type or vortex type.
    - Closed: Shrouds or sidewall enclosing the vanes.
  - Figure 5.5 gives photographs of open, semiopen, and closed impellers.
  - The totally open axial flow impeller has high volumetric flow capacity, but not high head or pressure. With its open tolerances for moving solids, they are generally not high-efficiency devices.
  - A semiopen impeller has exposed blades, but with a support plate or shroud on one side. These types of impeller are generally used for liquids with a small percentage of solid particles such as sediment from the bottom of a tank or river, or crystals mixed with the liquid. The efficiency of these impellers is governed by the limited free space or tolerance between the front leading edge of the blades and the internal pump housing wall.
  - Totally enclosed impellers are designed with the blades between two support plates. These impellers are used for clean liquids because tolerances are tight at the eye and the housing and there is no space for suspended solids, crystals or sediment. Solid contamination will destroy the tolerance between the outside of the eye and the bore of the pump housing. This specific tolerance governs the efficiency of the pump.
  - The number of impellers determines the number of stages of the pump. A single-stage pump has one impeller only and is best for low head service. A two-stage pump has two impellers in series for medium head service. A multistage pump has three or more impellers in series for high head service.
- What are the applications of open, semienclosed, and enclosed impellers?
  - *Open*: Used for low heads, small flows, and liquids with suspended solids. May be preferable for higher viscosity liquids. Easier to clean as well as relatively self-cleaning. Axial flow-type impellers are effective than open impeller type.

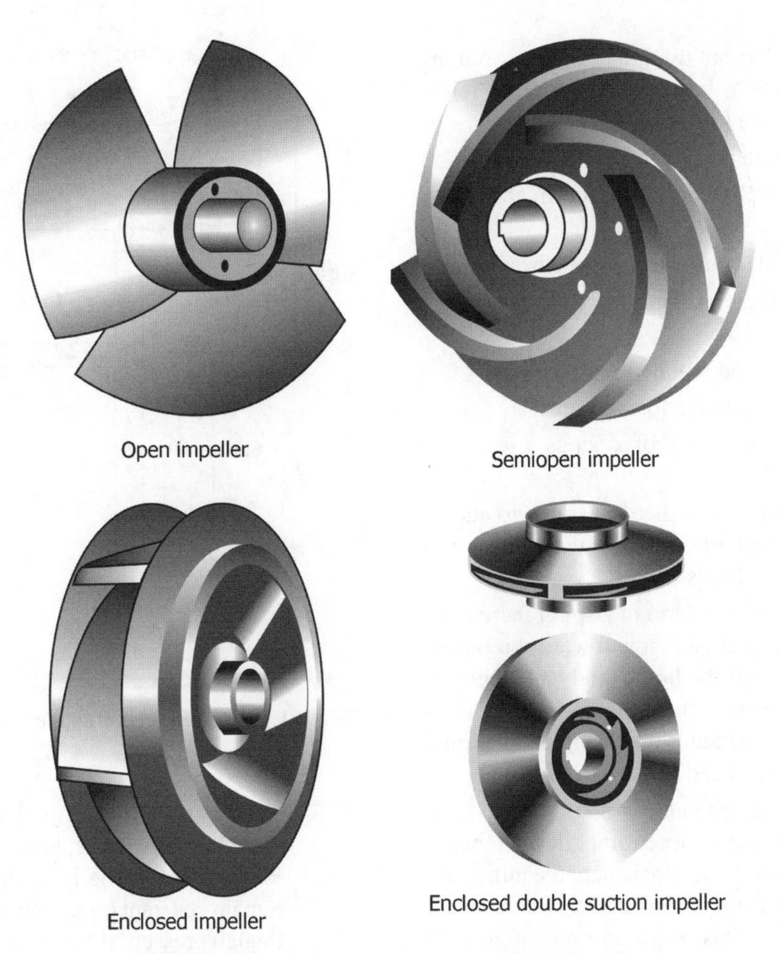

FIGURE 5.5 Open, semiopen, and enclosed impellers.

- *Semienclosed*: Used for general purpose. For higher heads. Have open vane tips at entrance to break up suspended solid particles and prevent clogging.
- *Enclosed*: Used for higher heads and higher pressure applications. Higher efficiencies because liquid flow is close to radial conditions with minimum lateral movement. NPSH requirements are lowest.

• What type of impeller is used for pumping water from a sump/pit?
  - Open impeller as water pumping from sumps/pits involves contamination with solid particles and any other type creates deposits on internal surfaces and it would be difficult to clean them.

• What is meant by a double-suction impeller for a centrifugal pump? What are its merits and limitations?
  - Fluid enters from both sides of the impeller eye.
  - Useful for high flows.
  - Lowers suction specific speed and maintain hydraulic balance. This lessens NPSH requirements and reduces thrust bearing loads.
  - Used in many horizontal services with more than two stages to avoid a high-pressure seal.
  - Impractical to use on most vertical pumps and for mixed flow or axial flow impellers.
  - Compared to single-suction impellers, double-suction impellers are less compact, have two stuffing boxes to maintain, less flexible in piping arrangement, handle solids poorly and less resistant to abrasion, corrosive, high temperatures, and expensive.
  - Figure 5.6 shows a double-suction centrifugal pump.

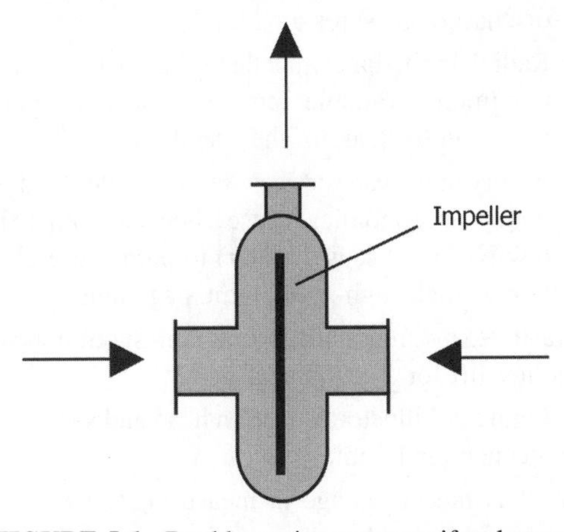

FIGURE 5.6 Double suction to a centrifugal pump.

- What types of impellers are used in centrifugal pumps (based on flow direction) for (i) high heads, (ii) intermediate heads, and (iii) low heads?
  i. Radial flow impellers are used for high heads.
  ii. Mixed flow impellers are used for intermediate heads.
  iii. Axial flow impellers are used for low heads.
- What is the main application of an axial flow (propeller) pump?
  - Very high capacity and low head applications.
  - Used in closed loop circulation systems.
- What are the different types of casings used in a centrifugal pump?
  - *Circular casing*: Involve higher losses due to eddies and shock when liquid leaves impeller at high velocity. Used for low heads and high capacity.
  - *Volute casing*: It is in the form of a spiral increasing uniformly in cross-sectional area toward the outlet. The volute decelerates the liquid, and its velocity is converted to pressure, efficiently as in a venturi meter. The volute constitutes the casing of the pump. Used for higher heads and lower capacities.
  - *Diffuser casing*: Turbine pumps—Guide vanes (diffusers) are interposed between impeller discharge and casing chamber; losses are kept to the minimum and improved efficiency is obtained over a wider range of capacities. This construction is often used in multistage high head pumps. These pumps are relatively inflexible and often unsatisfactory for low flows.
- Are centrifugal pumps available as multistage units? What is their principal advantage over positive displacement pumps?
  - *Yes.* Multistaging increases delivery pressures, which is a limitation for single-stage units, while retaining their high-capacity capability. Capacities of positive displacement pumps are generally low although their discharge pressures are high.
  - Radial multistage impellers should be operated at maximum available rpm, because head per stage varies with square of the speed.
  - In very high head services, such as boiler feed pumps, it may be economical to use a booster pump to handle low NPSH (discussed later) to permit use of a relatively small high-speed multistage unit.
- Illustrate by means of a diagram suction head and suction lift for a centrifugal pump.
  - Figure 5.7 illustrates suction head and suction lift for a centrifugal pump.
- What is the advantage in measuring the energy of a pump as head rather than pressure?

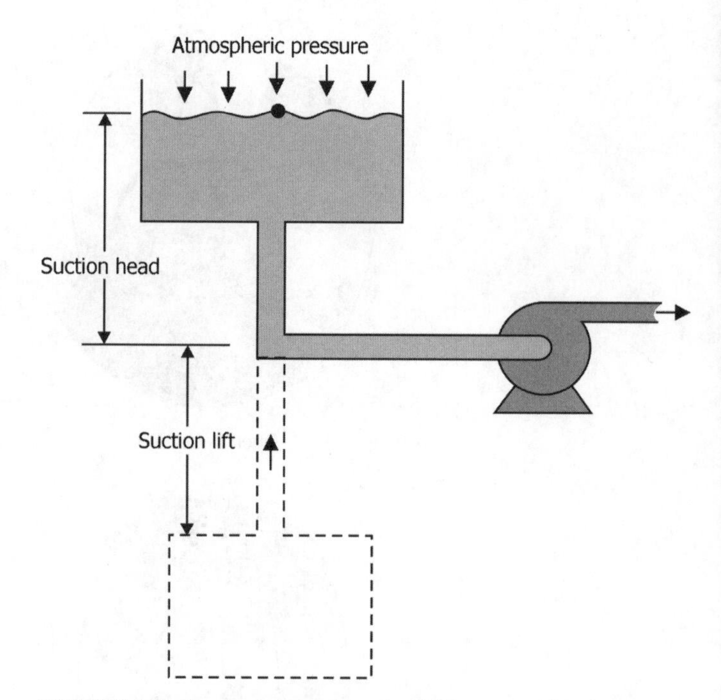

FIGURE 5.7 Suction head and suction lift for a centrifugal pump.

  - Head applies to any liquid that is pumped at the same rated capacity and speed, as long as the viscosity is low (<10 cP). The head developed by a pump will remain constant (at a given flow rate and speed), even though pressure differential and power requirements vary.
- What is TDH?
  - TDH is the *total dynamic head*, the energy that the pump imparts to the liquid. TDH takes into account differences in pressure, liquid elevation and velocity between the source and the destination, line frictional losses, and pressure drop through the instrumentation and other items in the flow path of the liquid.
  - It is composed of up to four heads or pressures:
    - *Static head* or the change in elevation of the liquid across the system. It is the difference in the liquid surface level at the suction source or vessel, subtracted from the liquid surface level where the pump delivers the liquid. Some systems do not have static head or elevation change.
    - *Pressure head* or the change in pressure across the system. It also may or may not exist in every system.
    - *Velocity head* or the energy lost into the system due to the velocity of the liquid moving through the pipes. It has normally an insignificant value.
    - *Friction head* is the friction losses in the system.

- What is the function of suction piping?
  - The suction side piping of a pump is much more important than the piping on the discharge side.
  - The suction piping is to supply an evenly distributed liquid flow to pump suction with sufficient pressure to avoid cavitation (detailed later) and the consequent damage to the impeller.
  - An uneven flow distribution is characterized by strong local currents, swirls, and/or excessive entrained air.
  - The suction piping should preferably straight with no turns or flow-disturbing fittings close to the pump inlet.
  - Isolation valves, strainers, and other devices used on the suction side of the pump should be sized and located to minimize flow disturbance into the pump.
  - The suction piping should at least have the same diameter as the pump inlet nozzle and sized to ensure that the maximum liquid velocity at any point in the inlet piping does not exceed 2.5 m/s (8 ft/s).
  - The minimum straight pipe length before pump suction, to ensure laminar flow to the eye of the impeller, should be
    - 1–8 pipe diameters (for low suction energy/low specific speeds)
    - 3–16 pipe diameters (for high suction energy/high specific speeds)
    - smaller multiplier would be used on the larger pipe diameters and vice versa.
- What problems can result from using improper suction piping?
  - Noisy operation.
  - Random axial load oscillations.
  - Premature bearing/seal failure.
  - Cavitation damage to the impeller and inlet portions of the casing.
    - Cavitation occurs when the suction pressure of the pumped fluid drops below its vapor pressure, leading to the formation of vapor bubbles. As the fluid becomes pressurized again in the pump, these bubbles implode, leading to pitting of the impeller and other pump components.
    - In addition, since vapor has a lower density than liquid, cavitation leads to a reduction in the pump capacity and efficiency.
  - Occasional damage from liquid separation on the discharge side.
- What is the effect of installing an elbow close to suction side of a centrifugal pump?
  - There is always an uneven flow in an elbow, and when one is installed on the suction of any pump, it introduces that uneven flow into the eye of the impeller.
  - This can create turbulence and air entrainment, which may result in impeller damage and vibration.
  - When the elbow is installed in a horizontal plane on the inlet of a double suction pump, uneven flows are introduced into the opposing eyes of the impeller, upsetting the hydraulic balance of the rotating element.
  - Under these conditions, the overloaded bearing or mechanical seal will fail prematurely.
- What is NPSH?
  - When the absolute suction pressure is low, cavitation may be the result. To avoid cavitation, it is necessary to maintain a *net positive suction head,* NPSH, which is the total head of the liquid at the pump center line minus vapor pressure of the liquid at the prevailing temperature, expressed in equivalent of head units.
  - Some times NPSH is written as NPIP, *net positive inlet pressure.* This terminology is used for positive displacement pumps for which the inlet conditions are traditionally defined in *pressure terms* rather than the head terms. NPIP is inlet pressure minus fluid vapor pressure.
- What are $NPSH_A$ and $NPSH_R$? How is $NPSH_A$ calculated?
  - $NPSH_A$: Available NPSH.
  - $NPSH_R$: NPSH recommended by the pump manufacturers.
  - $NPSH_A$ must be equal to or greater than $NPSH_R$ for the desired capacity.
  - $NPSH_A$ can be calculated as

$$NPSH_A = P_a \pm h_s - h_f - P^0 \tag{5.1}$$

where P_a is the absolute pressure above liquid level in the source tank/vessel, expressed as units of liquid head (m), h_s is the static suction head (m), h_f is the head loss due to friction in suction line (m), and P^0 is the vapor pressure of the liquid at the prevailing temperature expressed as units of liquid head (m).

  - Petroleum products are not pure components but mixtures of hydrocarbons, which do not have finite vapor pressures. In such cases, Reid vapor pressures (RVP) may be used in place of P^0 in the equation for $NPSH_A$ for estimation purposes. RVP determinations are to be made at the liquid temperatures, particularly from the process streams being pumped directly from condensers without cooling.

- For low temperature applications, especially for liquids involving high boiling points, P^0 is not critical and assumed to be negligible.
- On what factors NPSH of a centrifugal pump depend?
  - Speed of rotation.
  - Inlet area of the impeller.
  - Number and types of vanes.
- What are the ways by which $NPSH_A$ can be increased?
  - Raise liquid level on the suction side by either raising level in the tank or raising suction side tank itself.
  - Lower the pump (install it in a sump or trench).
  - Pressurize the suction side tank.
  - Reduce friction losses in suction piping by increasing pipe diameter, reducing valves (e.g., replace globe valves by gate or butterfly valves by ball valves) and fittings (e.g., eliminate some elbows or change standard elbow by long radius elbow) or changing pipe material with that with smooth surface (e.g., plastic or stainless steel pipe).
  - Use a booster pump.
  - Subcool the liquid.
- What is the effect of altitude of a pump installation above sea level on $NPSH_A$?
  - $NPSH_A$ decreases with altitude as barometric pressure decreases.
- What are the factors on which $NPSH_R$ for a given application depends?
  - $NPSH_R$ is dependent on pump design, for example, on suction area of the impeller including double suction, shape and number of vanes, area between vanes, eye diameter, specific speed, shaft diameter, and so on.
  - Normally $NPSH_R$ is provided by the pump manufacturers.
- How is $NPSH_R$ reduced?
  - Use lower speeds. This needs installing a larger size impeller to handle a given pumping task.
  - Use double-suction impeller. A double-suction impeller design needs only two-thirds as much NPSH as a similarly rated single-suction design.
  - Use large impeller eye area or increase suction nozzle size.
  - Use oversize pump.
  - Use several smaller pumps in parallel.
  - Use a booster pump.
  - Use inducers ahead of conventional impellers. Inducer is a low head axial-type impeller with few blades located in front of conventional impeller.
  - Polish suction throat and pathway to the impeller.
- If $NPSH_A$ is low (<3 m), what modifications may be recommended for the pump system to prevent cavitation problems?
  - Oversizing the pump suction line and the impeller eye.
- "$NPSH_A$ should always be $>NPSH_R$." *True/False?*
  - *True.*
- What is $NPSH_{R3}$?
  - $NPSH_{R3}$ is defined as the cold water pump head drop of 3%, if $NPSH_A = NPSH_R$, that is, pump head will be reduced by 3% in case of pumping cold water over that obtainable theoretically, which is the accepted industry standard.
- What is the effect of excessive increase in $NPSH_A$ over $NPSH_R$ on impeller life?
  - Maximum pump damage rate due to erosion to the impeller occurs if $NPSH_A$ is in the range of two to three times the $NPSH_{R3}$.
- What are $NPIP_R$ and $NPIP_A$?
  - $NPIP_R$ is the required NPIP and is the difference between the inlet pressure and the vapor pressure (corrected to the center line of the pump inlet port) necessary for the pump to operate without a reduction of flow.
  - The Hydraulic Institute defines the minimum required pressure (or equivalent $NPIP_R$) as the pressure where 5% of the flow reduction occurs due to cavitation.
  - $NPIP_A$ is the available NPIP.
- What are the effects of low $NPIP_A$?
  - $NPIP_A$ should always be more than $NPIP_R$.
  - For positive displacement pumps, an increase in $NPIP_A$ has no effect on volumetric efficiency as long as $NPIP_A$ is greater than $NPIP_R$.
  - Low values of $NPIP_A$ may result not only in flow reduction but also in significant pressure spikes, noise, vibrations, and possible damage to the pump.
- "NPSH for hot/boiling liquids is high." *True/False?* Comment.
  - *False.* Hot/boiling liquids have high vapor pressures that promote vaporization of the liquid in the suction line, especially in the eye of the impeller where pressure is very low.
- What is the common recommended range for $NPSH_A$?
  - 1.2–6.1 m of liquid.
- What is priming in a pump operation?
  - The process of introducing fluid into the pump casing to improve sealing of pump parts on starting and expelling air from it is called priming.

- What is a self-priming pump?
  - A self-priming pump is a pump that draws liquid up from below pump inlet (suction lift), as opposed to a pump requiring flooded suction.
- "Centrifugal pumps are usually self-primed pumps." *True/False?*
  - *False.*
- Define specific speed for a pump.
  - Specific speed, N_s, is a dimensionless design index that identifies the geometric similarity of pumps. It is used to classify pump impellers as to their type and proportions. Pumps of the same N_s but of different size are considered to be geometrically similar, one pump being a size factor of the other.
  - Specific speed is defined as the speed in revolutions per minute at which a geometrically similar impeller would operate if it were of such a size as to deliver 1 gpm flow against 1 ft head.
  - Specific speed,

$$N_S = NQ^{1/2}/H^{3/4} \quad (5.2)$$

where N is the number of rotations per minute and Q is the flow rate at or near BEP measured in gpm. (BEP is defined and discussed later). H is the head developed per stage at BEP at maximum impeller diameter in ft.

  - In general, pumps with a low specific speed have a low capacity and high specific speed, high capacity.
  - The understanding of this definition is of design engineering significance only, however, and specific speed should be thought of only as an index used to predict certain pump characteristics.
  - Specific speed determines the general shape or class of the impellers. As the specific speed increases, the ratio of the impeller outlet diameter, d_2, to the inlet or eye diameter, d_1, decreases. This ratio becomes 1.0 for a true axial flow impeller.
  - *Radial flow impellers* develop head principally through centrifugal force. Radial impellers are generally low-flow high head designs.
  - Pumps of higher specific speeds develop head partly by centrifugal force and partly by axial force. A higher specific speed indicates a pump design with head generation more by axial forces and less by centrifugal forces.
  - An axial flow or propeller pump with a specific speed of 10,000 or greater generates its head exclusively through axial forces. Axial flow impellers are high flow low head designs.
  - Specific speed identifies the approximate acceptable ratio of the impeller eye diameter (d_1) to the impeller maximum diameter (d_2) in designing a good impeller.
  - N_s is 500–5000, $d_1/d_2 > 1.5$ for radial flow pump; 5000–10,000, $d_1/d_2 < 1.5$ for mixed flow pump; and 10,000–15,000, $d_1/d_2 = 1$ for axial flow pump.
  - Specific speed is also used in designing a new pump by size-factoring a smaller pump of the same specific speed. The performance and construction of the smaller pump are used to predict the performance and model the construction of the new pump.
  - Pump may be damaged if certain limits of N_s are exceeded and efficiency is best in some ranges.
- How does specific speed indicate pump type to be selected?
  - If $N_S > 500$, centrifugal pump is the choice.
  - If $N_S < 500$, positive displacement pump is a likely candidate.
  - Some kinetic pump types can generate very high head (such as regenerative turbine pumps), and the specific speed criterion, as suggested above would not hold true for them.
- What is suction specific speed?
  - Suction specific speed is expressed as

$$N_{SS} = NQ^{1/2}/(\text{NPSH})^{3/4}. \quad (5.3)$$

  - The formula for N_{SS} is same as for N_S, by substituting NPSH_R for H.
  - Suction specific speed, N_{ss}, is a dimensionless number or index that defines the suction characteristics of a pump. It is calculated from the same formula as N_s by substituting H by NPSH_R.
  - N_{ss} is commonly used as a basis for estimating the safe operating range of capacity for a pump. The higher the N_{ss} is, the narrower is its safe operating range from its BEP. The numbers range between 3000 and 20,000. Most users prefer that their pumps have N_{ss} in the range of 8000–11,000 for optimum and trouble-free operation.
  - In multistage pump the NPSH_R is based on the first-stage impeller NPSH_R.
- What are the desirable suction-specific speeds for a centrifugal pump?
  - For single-suction pumps, $N_{SS} > 11{,}000$: excellent; $N_{SS} = 7000$–9000: average; and $N_{SS} < 7000$: poor.
  - For double-suction pumps, $N_{SS} > 14{,}000$: excellent; $N_{SS} = 9000$–$11{,}000$: average; $N_{SS} < 7000$: poor.
- What are the ranges of specific speeds for centrifugal pumps? Are these ranges more or less for reciprocating pumps?

- For centrifugal pumps, specific speeds are in the range of 400–10,000 or over.
- For reciprocating pumps specific speeds are much lower than those for centrifugal pumps.

- What are the ranges of specific speeds of centrifugal pumps in relation to the type of impeller used?
  - Radial flow impeller: 400–1000.
  - Mixed flow impeller: 1500–7000.
  - Axial flow impeller: >7000.
- "Axial flow centrifugal pumps are of (i) high/low flow and (ii) high/low head type, and have a (iii) low/high specific speed." Give correct answers for each of the cases sited above.
  i. High flow.
  ii. Low head.
  iii. High specific speed.
- "Purely radial pumps are of (i) high/low flow and (ii) high/low head and have a (iii) high/low specific speed." Give correct answers for each of the above cases.
  i. Low flow.
  ii. High head.
  iii. Low specific speed.
- "The head developed by a centrifugal pump is proportional to the speed of the impeller." *True/False*?
  - *False*. It is proportional to the square of speed.
- What are the normal ranges of capacities, heads developed and efficiencies of different types of pumps?
  - Table 5.1 lists the normal ranges of capacities, the heads developed, and the efficiencies of different types of pumps.
- What happens if a pump pumping water is switched to pumping a petroleum product under the same piping and tank arrangement?
  - There will be no change in head developed.
  - Power required to pump a lower specific gravity liquid such as a petroleum product will be reduced.
- What are BHP and FHP?
  - The gross power delivered to the pump by the driver is known as *break horsepower* (BHP).
  - The net power delivered to the fluid by the pump is the *fluid or hydraulic horsepower* (FHP).
  - The difference between BHP and FHP is due to internal mechanical and volumetric losses in the pump.
- Give the three affinity laws/similarity relationships among impeller speed, N, capacity, Q, head developed, H, and BHP for a centrifugal pump.

$$Q \propto N;\ H \propto N^2;\ \text{and BHP} \propto N^3. \tag{5.4}$$

$$\text{Capacity: } Q_1/Q_2 = N_1/N_2 = D_1/D_2. \tag{5.5}$$

$$\text{Head: } H_1/H_2 = (N_1/N_2)^2 = (D_1/D_2)^2. \tag{5.6}$$

$$\text{BHP: } (\text{BHP})_1/(\text{BHP})_2 = (N_1/N_2)^3 = (P_1/P_2)^3. \tag{5.7}$$

$$\text{Head-capacity: } (Q_2/Q_1)^2 = H_2/H_1. \tag{5.8}$$

- What is required power/brake horsepower?
  - Required power or brake horsepower (BHP) is the power needed at the pump shaft. This is always higher than the hydraulic power due to energy losses in the pump.
  - Required power is expressed as

$$\omega \times T, \tag{5.9}$$

where ω is shaft angular velocity and T is shaft torque.

- What is hydraulic power for a pump?
  - The theoretical energy required to pump a given quantity of fluid against a given total head is known as hydraulic power, hydraulic horsepower, or water horsepower.

TABLE 5.1 Capacities, Heads Developed, and Efficiencies of Different Types of Pumps

Pump Type	Capacity m^3/min (gpm)	Maximum Head m (ft)	Efficiency (%)
Centrifugal, single stage	0.057–18.9 (15–5000)	152 (500)	[a]
Centrifugal, multistage	0.076-41.6 (20–11,000)	1675 (5500)	
Axial flow	0.076–378 (20–100,000)	12 (40)	65–85
Rotary	0.00378–18.9 (1–5000)	15,200 (50,000)	50–80
Reciprocating	0.0378–37.8 (10–10,000)	300 km (1×10^6)	[b]

[a] 45% at 0.378 m^3/min (100 gpm); 70% at 1.89 m^3/min (500 gpm); and 80% at 37.8 m^3/min (10,000 gpm).
[b] 70% at 7.46 kW (10 HP); 85% at 37.3 kW (50 HP); and 90% at 373 kW (500 HP).

- What is power output for a pump? Give the equation for its calculation.
  - Power output is the product of the total dynamic head and the mass of liquid pumped in a given time.
  - In SI units:

$$\text{Power output in kW} = HQ\rho/3.67 \times 10^5. \quad (5.10)$$

  where H is the total dynamic head in N-m/kg (column of liquid); Q is the capacity in m^3/h; and ρ is the liquid density in kg/m^3. When the total dynamic head H is expressed in Pascals, the power output in kW is

$$HQ/3.599 \times 10^6. \quad (5.11)$$

- What is volumetric efficiency of a pump?
  - Ratio of actual volume of the fluid delivered to the theoretical volume that can be delivered by the pump.
- What is *BEP*? What is its significance?
  - *Best efficiency point* (BEP) is the capacity at maximum impeller diameter at which the efficiency is highest. All points to the right or the left of BEP have a lower efficiency.
  - BEP is a measure of optimum energy conversion. BEP is the area on the performance curve where the change of velocity energy into pressure energy at a given capacity is optimum. In other words, the point where the pump is most efficient.
  - When sizing and selecting centrifugal pumps for a given application, the pump efficiency at design should be taken into consideration. The efficiency of centrifugal pumps is stated as a percentage and represents a unit of measure describing the change of centrifugal force (expressed as the velocity of the fluid) into pressure energy.
  - The BEP is the area on the curve where the change of velocity energy into pressure energy at a given flow rate is optimum, that is, the point where the pump is most efficient.
  - The impeller is subject to nonsymmetrical forces when operating to the right or left of the BEP. These forces give rise to many mechanically unstable conditions such as vibration, excessive hydraulic thrust, temperature rise, cavitation, and erosion.
  - The operation of a centrifugal pump should not be outside the farthest left or right efficiency curves published by the manufacturer.
  - Performance in these areas induces premature bearing and mechanical seal failures due to shaft deflection, and an increase in temperature of the process fluid in the pump casing causing seizure of close tolerance parts and cavitation.
  - BEP is an important parameter in that many parametric calculations such as specific speed, suction specific speed, hydrodynamic size, viscosity correction, head rise to shutoff, and so on are based on capacity at BEP. Many users prefer that pumps operate within 80–110% of BEP for optimum performance.
- What is specific energy for a pumping system?
  - Specific energy is the power consumed per unit volume of fluid pumped. It is determined by measuring the flow delivered into the system over a period of time and calculating the power consumed during the same period of time.
  - This measure takes into account all of the factors that will influence the efficiency of an installation, not just pump efficiency.
  - Specific energy is a useful measure to consider when evaluating combinations of pump type, model, and system. Another benefit of using specific energy as a measure is that it allows some approximate comparisons between similar pumping installations.
- "Generally, the larger the pump, the higher the efficiency." *True/False*?
  - *True*.
- Discuss the factors that influence pump efficiencies.
  - The important causes of pump inefficiencies are hydraulic effects, mechanical losses, and internal leakage.
  - *Hydraulic losses* may be caused by boundary layer effects, disruptions of the velocity profile and the flow separation.
    - Boundary layer losses can be minimized by making pumps with clean, smooth, and uniform hydraulic passages.
    - Mechanical grinding and polishing of hydraulic surfaces, or modern casting techniques, can be used to improve the surface finish, decrease vane thickness and improve efficiency.
    - Generally, the improvements in surface finishes are economically justifiable in pumps with low specific speeds and for pumps handling pharmaceutical, biological, and food products.
    - Shell molds, ceramic cores, and special sands produce castings with smoother and more uniform hydraulic passages.
    - Separation of flow occurs when a pump is operated well away from the BEP.
    - The flow separation occurs because the incidence angle of the fluid entering the hydraulic passage is significantly different from the angle of the blade.

- Voided areas increase the amount of energy required to force the fluid through the passage.

- *Mechanical losses* in a pump are caused by viscous disk friction, bearing losses, seal or packing losses, and recirculation devices.
  - Bearings, lip seals, mechanical seals, and packings, all consume energy and reduce the efficiency of the pump. Small pumps are particularly susceptible.
- Close tolerances on the wear rings have a tremendous effect on the efficiency of a pump, particularly for pumps with a low specific speed ($N_s < 1500$).
  - If the clearance between the impeller and the casing sidewall is too large, disk friction can increase, reducing the efficiency.
  - Bearings, thrust balancing devices, seals and packing all contribute to frictional losses. Most modern bearing and seal designs generate full fluid film lubrication to minimize frictional losses and wear.
  - Frequently, recirculation devices such as auxiliary impellers or pumping rings are used to provide cooling and lubrication to bearings and seals. Similar to the main impeller, these devices pump fluid and can have significant power requirements.
- *Internal leakage* occurs as the result of flow between the rotating and the stationary parts of the pump, from the discharge of the impeller back to the suction.
  - The rate of leakage is a function of the clearances in the pump. Reducing the clearances will decrease the leakage but can result in reliability problems if mating materials are not properly selected.
  - Some designs bleed off flows from the discharge to balance thrust, provide bearing lubrication, or to cool the seal.
- Impeller diameter influences efficiency. There will be an efficiency reduction with a reduction in the impeller diameter. For this reason, it is not recommended to reduce the impeller size by more than 20%.
- Viscous liquids generally affect efficiency. As the viscosity of the fluid goes up, generally the efficiency of most pumps goes down.
- Slurries with low solids concentrations (less than 10% average) classified by size and material generally exhibit no adverse effect to pump efficiency. Sanitary and wastewater pumps that handle high solids, have two or three blades on a specially designed impeller that gives lower efficiency.

5.1.2 Cavitation

- What is cavitation? What are the causes and the effects of cavitation?
  - Cavitation occurs in liquid when bubbles form and implode in pump systems or around impellers. Pumps put liquid under pressure, but if the pressure of the liquid drops or its temperature increases, it begins to vaporize, just as boiling water. The bubbles that form inside the liquid are vapor bubbles, gas bubbles, or a mixture of both.
  - Vapor bubbles are formed due to the vaporization of the liquid being pumped, at a point inside the pump where the local static pressure is less than the vapor pressure of the liquid. A cavitation condition induced by formation and collapse of vapor bubbles is commonly referred to as *vapor cavitation.*
  - Gas bubbles, by contrast, are formed due to the presence of dissolved gases in the liquid that is being pumped. In many situations, the gas dissolved is air.
  - *Vapor cavitation* bubbles get carried in the liquid as it flows from the impeller eye to the impeller tip, along the trailing edge of the blade. Due to the rotation of the impeller, the bubbles first attain very high velocity, then reach the regions of higher pressure. The pressure around the bubbles begins to increase until they collapse. This process is an implosion (inward bursting). Hundreds of bubbles implode at approximately the same point on each impeller blade.
  - The bubbles collapse in such a way that the surrounding liquid rushes to fill the void, forming a liquid microjet. The microjet subsequently ruptures the bubbles with such a force that a hammering action occurs.
  - Figure 3.17 illustrates bubble collapse.
  - After bubbles collapse, a choke wave emanates outward from the point of collapse. This choke wave is what is actually heard and called cavitation.
  - The collapse of the bubbles also ejects destructive microjets of extremely high velocity, up to 100 m/s, causing abnormal sounds, vibrations, and extreme erosion of the pump parts, pitting, and denting in the metal of the casing and the impeller blades, as illustrated in Figure 5.8.
  - Figure 5.9 shows the potential cavitation damage areas of a pump impeller.
  - It has been estimated that during collapse of bubbles pressures of the order of 10^4 bar develops.
  - Apart from erosion of pump parts, cavitation can also result in imbalance of radial and axial thrusts on the impeller, due to lack of symmetry in the bubble formation and collapse.

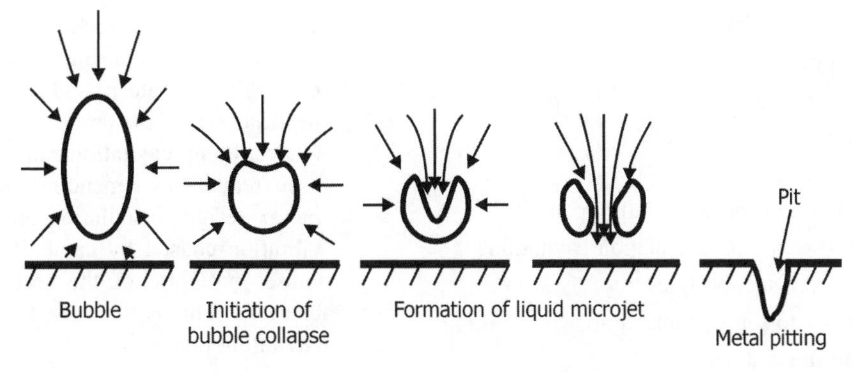

FIGURE 5.8 Bubble formation, collapse, and metal damage. (*Source*: www.cheresources.com.)

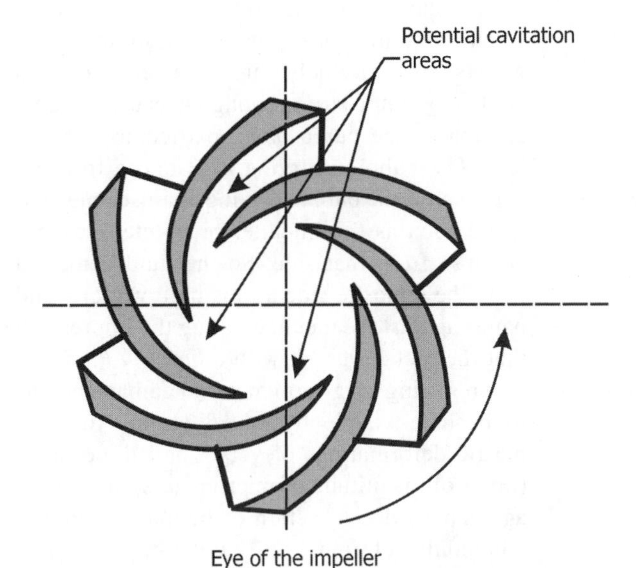

FIGURE 5.9 Potential cavitation damage areas in the eye of pump impeller.

- In many cases, cavitation also leads to corrosion. The implosion of bubbles destroys protective layers on the metal surface, making the metal permanently activated for the chemical attack.
- *Gas bubble cavitation* that amounts to two-phase pumping occurs when any gas (most commonly air) enters a centrifugal pump along with liquid.
- Unlike vapor cavitation, gas bubbles seldom cause damage to the impeller or casing. Instead, their main effects are surge and choke phenomena that cause deterioration of pump head and capacity. Separation of the gas phase from the liquid phase, accompanied by a tendency of the gas to coalesce in large pockets at the impeller blade entry, results in a sonic choke effect.
- The pump performance decreases continuously as the gas volume increases, until at a certain critical gas content the pump stops delivering liquid.
- Cavitation occurs not only in centrifugal pumps but also in orifices, valves, and similar equipment involving flow restrictions.

- What are the indications of cavitation during pump operation?
  - Indications of the cavitation during pump operation include loud noises, vibrations, and an unsteadily working pump. Fluctuations in flow and discharge pressure take place with a sudden and drastic reduction in head rise and pump capacity. Depending upon the size and the quantum of the bubbles formed and the severity of their collapse, the pump faces problems ranging from a partial loss in capacity and head to total failure in pumping along with severe damages to the internal parts.
- What are the types of cavitation in fluid pumping?
  - There are two types of cavitation that can occur in the different stages of pumping, but both are the results of the same phenomenon.
    - Suction cavitation occurs around the impeller as it is drawing liquid through the chamber. The movement of the impeller creates the changes in pressure necessary for vaporization.
    - Discharge or recirculation cavitation is the result of changing pressure at the point of exit, the discharge valve. The valve is not able to let all the liquid through, as fast as it should, so the different velocities of the current create miniature changes in the uniform pressure. Even such small variations are enough to create the ideal circumstances for cavitation.
  - Suction cavitation occurs when the NPSH available to the pump is less than what is required, that is, $NPSH_A < NPSH_R$.
- What are the symptoms for pump suction cavitation?
  - The pump gives rise to high levels of noise.
  - Suction line shows a high vacuum reading.
  - Results in low discharge pressure and high flow.

- What are the causes for pump suction cavitation?
  - Clogged suction pipe.
  - Suction line is too long.
  - Suction line diameter is too small.
  - Suction lift is too high.
  - Valve on suction line is only partially open.
- What are the remedies to prevent suction cavitation?
  - Remove debris from suction line.
  - Move pump closer to source tank/sump.
  - Increase suction line diameter.
  - Decrease suction lift requirement.
  - Install larger pump running slower that will decrease the $NPSH_R$ by the pump.
  - Increase discharge pressure.
  - Fully open suction line valve.
- Under what circumstances cavitation occurs on the discharge side of a pump?
  - Discharge cavitation occurs when the pump discharge head is too high where the pump runs at or near shutoff.
- What are the symptoms for pump discharge cavitation?
  - High noise levels.
  - High discharge gauge reading.
  - Low flow.
- What are the causes for pump discharge cavitation?
  - Clogged discharge pipe.
  - Discharge line is too long.
  - Discharge line diameter is too small.
  - Discharge static head is too high.
  - Discharge line valve is only partially open.
- What are the remedies to prevent pump discharge cavitation?
  - Remove debris from discharge line.
  - Decrease discharge line length.
  - Increase discharge line diameter.
  - Decrease discharge static head requirement.
  - Install larger pump that will maintain the required flow without discharge cavitating.
  - Fully open discharge line valve.
- How can one determine whether the noise produced by a pump is due to cavitation or bearing problem?
  - To distinguish between the noise due to a bad bearing or cavitation, operate the pump with no flow. The disappearance of noise will be an indication of cavitation.
- What condition causes vibration during pump operation? Explain.
  - Vibration is due to the uneven loading of the impeller as the mixture of vapor and liquid passes through it, and to the local shock wave that occurs as each bubble collapses. Formation and collapsing of bubbles will alternate periodically with the frequency resulting out of the product of speed and number of blades. Pump cavitation can produce various vibration frequencies depending on the cavitation type, pump design, installation and use. The excessive vibration caused by cavitation often subsequently causes a failure of the seal of the pump and/or bearings. This is the most likely failure mode of a cavitating pump.
- What types of pump damage occurs due to cavitation?
  - Cavitation erosion or pitting:
    - During cavitation, the collapse of the bubbles occurs at sonic speed ejecting destructive microjets of extremely high velocity (up to 1000 m/s) liquid strong enough to cause extreme erosion of the pump parts, particularly impellers. The bubble is trying to collapse from all sides, but if the bubble is lying against a piece of metal such as the impeller or volute it cannot collapse from that side. So the fluid comes in from the opposite side at this high velocity and bangs against the metal, creating the impression that the metal was hit with a *ball pin hammer*. The resulting long-term material damage begins to become visible by so-called pits that are plastic deformations of very small dimensions (order of magnitude of micrometers). The damage caused due to action of bubble collapse is commonly referred to as cavitation erosion or pitting.
    - Cavitation erosion from bubble collapse occurs primarily by fatigue fracture due to repeated bubble implosions on the cavitating surface, if the implosions have sufficient impact force. The erosion or pitting effect is quite similar to sand blasting. High head pumps are more likely to suffer from cavitation erosion, making cavitation a *high-energy* pump phenomenon.
    - The most sensitive areas where cavitation erosion has been observed are the low pressure sides of the impeller vanes near the inlet edge. The cavitation erosion damages at the impeller are more or less spread out. The pitting has also been observed on impeller vanes, diffuser vanes, impeller tips, and so on. In some instances, cavitation has been severe enough to wear holes in the impeller and damage the vanes to such a degree that the impeller becomes completely ineffective.
    - Near the outside edge of the impeller, pressure builds to its highest point. This pressure im-

plodes the gas bubbles, changing the state of the liquid from gas into liquid. When cavitation is less severe, the damage can occur further down toward the eye of the impeller. A careful investigation and diagnosis of point of the impeller erosion on impeller, volute, diffuser, and so on can help predict the type and cause of cavitation.

  - The extent of cavitation erosion or pitting depends on a number of factors such as presence of foreign materials in the liquid, liquid temperature, age of equipment, and velocity of the collapsing bubble.
  - Figure 5.10 is a photograph of blade damage due to cavitation of a mixed flow pump impeller.

- Mechanical deformations: Apart from erosion of pump parts, in larger pumps, longer duration of cavitation condition can result in unbalancing (due to unequal distribution in bubble formation and collapse) of radial and axial thrusts on the impeller. This unbalancing often leads to following mechanical problems:
  - Bending and deflection of shafts.
  - Bearing damage and rubs from radial vibration.
  - Thrust bearing damage from axial movement.
  - Breaking of impeller check-nuts.
  - Seal faces damage, and so on.
- These mechanical deformations can completely wreck the pump and require replacement of parts. The cost of such replacements can be huge.
- Cavitation corrosion: Frequently, cavitation is combined with corrosion. The implosion of bubbles destroys existing protective layers, as stated under an earlier question, making the metal surface permanently activated for the chemical attack. Thus, in this way even in case of slight cavitation it may lead to considerable damage to the materials. The rate of erosion may be accentuated if the liquid itself has corrosive tendencies such as water with large amounts of dissolved oxygen to acids.

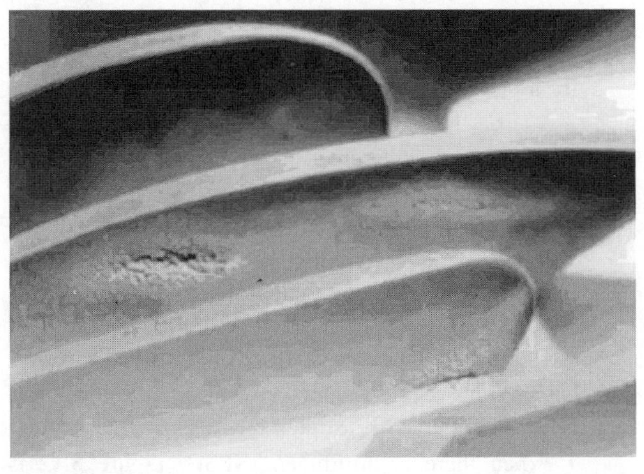

FIGURE 5.10 Typical blade damage due to cavitation of a mixed flow pump.

- Summarize the most severe damaging effects of pump cavitation.
  - Pitting marks on the impeller blades and on the internal volute casing wall of the pump.
  - Premature bearing failure.
  - Premature mechanical seal failure.
  - Shaft breakage and other fatigue failures in the pump.
- Suggest a remedy to suppress cavitation in a centrifugal pump.
  - Add a bleed system: An effective method to suppress cavitation in centrifugal pumps is to bleed some of the high-pressure fluid from the pump discharge back to the suction side, through essentially tangential-entry nozzles that give the fluid a rotating motion before entry into the impeller eye.
- "Cavitation problems are more severe for water compared to hydrocarbon liquids." *True/False?* Comment.
  - *True.* Specific volume of hydrocarbon vapors is very small compared to that of water vapor. Water occupies 1600 times its volume on vaporization whereas hydrocarbon liquids on vaporization occupy much less volumes. Volume reduction on implosion of vapor bubbles is very large for water creating much higher destructive forces.
- "Flow restrictions can cause cavitation damage." *True/False?* Illustrate with an example.
  - *True.* In heat exchanger tubes, internal tube cavitation erosion can occur in the area of sudden contraction at the tube sheet as shown in Figure 5.11.
- What is pump vapor locking? How is it detected?
  - When there is no flow from a running pump, but flow resumes after it is shut down and restarted, there is more likelihood of vapor lock in the pump. This occurs when a quantity of vapor or gas, as little as 10% by volume, enters the pump casing and becomes trapped in the eye of the impeller, thereby blocking the inflow of the fluid.
- The reason for this accumulation of the gas in the eye is because the centrifugal forces generated by the impeller vanes cause the fluid with higher density, that is, liquid to be thrown outward from the eye while the gas remaining concentrated in the center of the eye. When

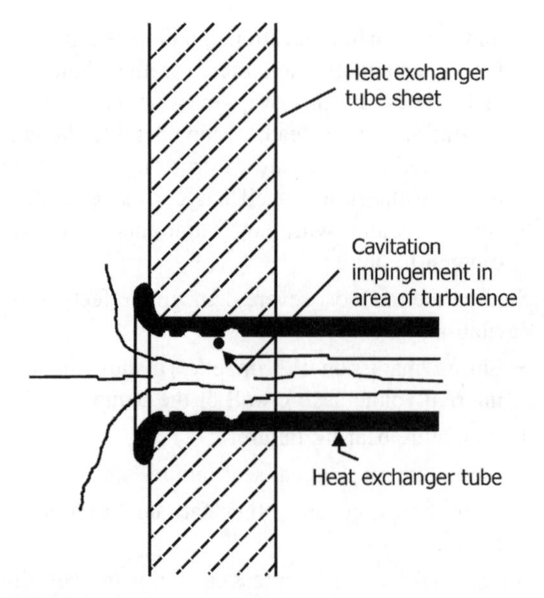

FIGURE 5.11 Heat exchanger tube damage at the entrance area into the tube due to cavitation.

the pump is shut off, the gas floats to the top and when the pump is restarted, the gas is pushed by the liquid that enters out of the discharge nozzle.

- Vapor locking occurs in a pump when a critical amount of a gas or air enters the pump and gets trapped in the eye of the impeller, blocking inflow from the suction line.

• How does air/gas enter a pump casing?

- It can enter through a vortex that can form when liquid is entering suction pipe from a tank or sump.
- Vortex formation is helped if the suction head is low or the liquid velocity is high.
  - To prevent vortexing, the suction head may be increased, that is, increase submergence, liquid flow rate may be decreased or suction pipe diameter increased. Vortex breakers may be used in cases where it is not practicable to increase submergence.
- It can also enter due to leaks in packing of pump shaft, suction side valve stem, joint rings on suction piping, flange face sheet gaskets at pipe joints, O-rings and threaded fittings on instrumentation in the suction piping, O-rings and other secondary seals on single mechanical seals, and the faces of single mechanical seals.
- It can also enter into the pump from bubbles and air/gas pockets in the suction piping and products that foam can introduce air into the pump.
- *Gas pockets.* Presence of a gas pocket in the suction line can cause vapor lock.
- Sources for air/gas can be release of dissolved gas, formation of slugs by entrained gas or through vaporization of the liquid.
- *Vaporization.* Vaporization in the suction line can also cause vapor lock. If the vapor formed is 10% or more, vapor lock can occur.
  - Vapor formation can be avoided by increasing pressure through increase in level in the supply tank, reducing frictional losses in suction line through use of larger pipe diameters and minimum number of fittings, valves, and so on, decreasing liquid temperature by cooling or using a pump with a lower $NPSH_R$.

• Explain how pump capacity reduces if cavitation is occurring.

- The formation of bubbles causes a volume increase, decreasing the space available in the pump casing for the liquid and thus diminishes pumping capacity. For example, when water changes state from liquid to gas its volume increases by approximately 1600 times. If the bubbles get large enough at the eye of the impeller, the pump *chokes*, that is, loses all suction resulting in loss of flow. The unequal and uneven formation and collapse of bubbles causes fluctuations in the flow and the pumping of liquid occurs in spurts. This symptom is common to all types of cavitations.

• Explain how head developed by a centrifugal pump decreases under cavitation.

- Bubbles unlike liquid are compressible. The head developed diminishes drastically because energy has to be expended to increase the velocity of the liquid used to fill up the cavities, as the bubbles collapse. According to one definition, cavitation is occurring in a pump when there is a drop of head developed by 3% of the normal value. Like reduction in capacity, this symptom is also common to all types of cavitations.

• What is the difference between NPSH and suction head for a pump?

- Suction head is pressure, in terms of head, above atmospheric pressure.
- NPSH: To ensure sufficient head of liquid at the entrance of pump impeller to overcome internal flow losses of pump, making the liquid essentially free from flashing vapor bubbles due to boiling action of the liquid. Liquid must not vaporize in the eye/entrance of the impeller where pressure is the lowest.

• Under what circumstances cavitation develops in a centrifugal pump? Illustrate.

- When there is insufficient NPSH. Figure 5.12 is illustrative of cavitation conditions due to insufficient $NPSH_A$.

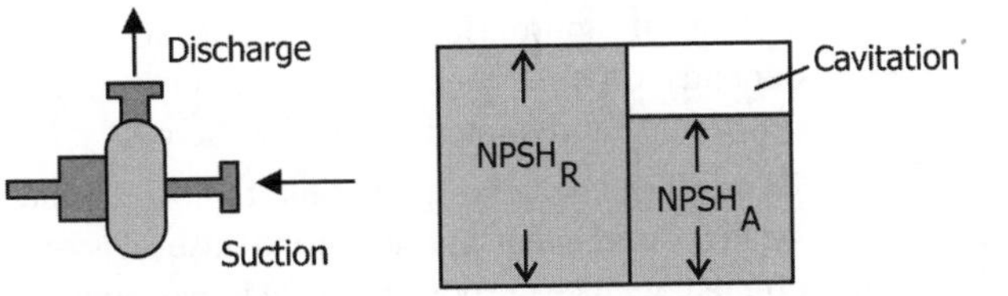

Discharge

Suction

Cavitation

$NPSH_R$

$NPSH_A$

FIGURE 5.12 Illustrative diagram for cavitation conditions.

- What are the pump arrangements that can lead to cavitation problems?
  - When operating only one pump out of pumps operating in parallel, the operating pump can develop cavitation.
  - Pumps that perform more than one duty through a valve manifold tend to suffer cavitation.
  - Pumps that fill and drain tanks from the bottom tend to cavitate.
  - The last pump drawing on a suction header tends to cavitate.
  - Vacuum pumps and pumps in a high suction lift suffer cavitation.
- Does cavitation occur in all types of pumps?
  - *Yes*. In centrifugal, reciprocating, or rotary pumps.
- What are the remedies for trouble-free operation of a centrifugal pump, with reference to avoidance of cavitation problems?
  - Keep $NPSH_A > NPSH_R$.
  - $NPSH_A > NPSH_R$ + A minimum of 0.60 m (2 ft) or preferably 1 m or more of excess liquid head. $NPSH_A$ should be at least 10% more than $NPSH_R$ for hydrocarbons and 20% more for water.
  - Internal clearances between the impeller and the pump casing must be kept low to enlarge the flow passages. This requires use of lesser number of blades in the impeller design.
  - Obstructions such as nozzles, screens, fittings, valves, and blockages in suction piping should be kept minimum.
  - Entry of noncondensable gases should be avoided. However, small amounts of entrained gas (1–2%) can actually cushion the forces from the collapsing cavitation bubbles and can reduce the resulting noise, vibration, and erosion damage.
  - Deviations/fluctuations in suction side pressures (decreases), temperatures (increases), and liquid level (decreases) must be avoided/corrected.
  - Attention for suction piping layout to minimize frictional losses: Tee intersections, globe valves, baffles, long lines with numerous elbows, and so on must receive special attention.
  - Avoiding liquid vortex formation in suction side vessel/tank: Installing sturdy vortex breakers firmly anchored to the vessel should be considered.
  - Keeping liquid velocities in suction line below 1–2 m/s.
  - Reducing liquid temperatures by cooling before admitting into pump suction side. Caution is to be exercised that liquid viscosities do not appreciably increase, thereby increasing pressure drop.
  - Using a booster pump, having the same capacity range, on the suction side to increase suction pressure.
  - Using a number of smaller flow pumps in parallel in place of a large flow pump. Require less NPSH.
  - An effective method to suppress cavitation in centrifugal pumps is to bleed some of the high-pressure fluid from the pump discharge back to the suction side, through essentially tangential-entry nozzles that give the fluid a rotating motion before entry into the impeller eye.
  - A thumb rule is to keep pump suction line at least 1 pipe size larger than the pump suction nozzle.
- What are the issues in steam condensate pumping and how are they met?
  - Removal of condensate from a condenser hot well requires a high-suction lift of a near boiling liquid with a wide range of capacities under varying loads.
  - Typical horizontal condensate pumps have only 0.5–1.5 m of $NPSH_A$.
  - Usually, 1750 rpm is used for low capacities and 880 rpm or even less for high flows.
  - Double suction is sometimes used to reduce $NPSH_R$. Sometimes, a double-suction first stage is used with a split into two equal single-suction second stages.
  - Vertical canned turbine pumps with adequate submergence by locating the first-stage impeller well below the floor level is one of the solutions for such applications.
- What are the effects of using very short suction pipe lengths to a centrifugal pump?
  - While shorter pipe lengths reduce friction losses thereby decreasing available NPSH, the flow path of the liquid into the impeller eye might be compromised.
- What is *critical cavitation number*?
  - Critical cavitation number is

$$\sigma_I = (P - P^0)/\rho(V^2/2) \qquad (5.12)$$

where σ_I is the cavitation number at the inception/start of cavitation, P is the static pressure in undisturbed flow, P^0 is the vapor pressure of the liquid, V is the free stream velocity of the liquid, and ρ is the liquid density.

- The ratio between the static pressure and the vaporization pressure, an indication of the possibility of vaporization, is often expressed by the cavitation number.
- Cavitation number depends on the nature of the equipment.
- As a guide, σ_I for blunt forms is 1.0–2.5 and for streamlined forms σ_I is 0.2–0.5.

• What is *cavitation coefficient*?

- Cavitation coefficient is defined as

$$\sum = \text{NPSH}/H. \tag{5.13}$$

where Σ is the cavitation coefficient and H is the pump differential head (head per stage in the case of multistage pumps) at design flow (i.e., optimum efficiency).

- Cavitation coefficient, a dimensionless number, is useful in correlating specific speeds with other pump parameters. Increased values of pump cavitation coefficients are associated with increased pump suction specific speeds.

5.1.3 Centrifugal Pump Performance and Operational Issues

• What are the major components that influence efficiency of a pumping system?

- The major components of a typical pumping system that have a large effect on the system efficiencies include the selection of efficient and properly sized electric motors, use of variable speed drives when appropriate, proper piping inlet and outlet configurations, appropriate selection and operation of valves, especially any throttling or bypass valves, pump speed control, multiple pump arrangements, and bypass and throttling valves, which are the primary methods for controlling flow rates in pumping systems.
  - ➢ The most appropriate type of speed control depends on the system size and layout, fluid properties, and sometimes other factors.
  - ➢ Bypass arrangements allow fluid to flow around a system component but at the expense of system efficiency, since the power used to bypass any fluid is wasted.
  - ➢ Throttling valves restrict fluid flow at the expense of pressure drops across the valves.

• Give a checklist of symptoms for inefficient pump system operation.

- The pump is oversized and has to be throttled to deliver the right amount of flow. Energy is lost in the valve.
- Flow control valves that are highly throttled.
- Pumps that are not running close to their best efficiency points operate at lower efficiency. Throttled pumps usually fall into this category.
- Pumps are running with bypass, or recirculation lines open.
- The pump is worn and the efficiency has deteriorated.
- Batch type processes in which one or more pumps operate continuously during a batch.
- Frequent on–off cycling of a pump in a continuous process.
- Presence of noise, due to cavitation, either at the pump or elsewhere in the system.
- A parallel pump system with the same number of pumps always operating.
- A pump system that has undergone a change in function, without modification.
- A pump system with no means of measuring flow, pressure, or power.
- The pump/system was installed or designed incorrectly (piping, base plate, etc.)

• What are the common centrifugal pump problems? Illustrate.

- Figure 5.13 sums up the common centrifugal pump problems.
- Pump not properly primed or suction pipe not completely filled.

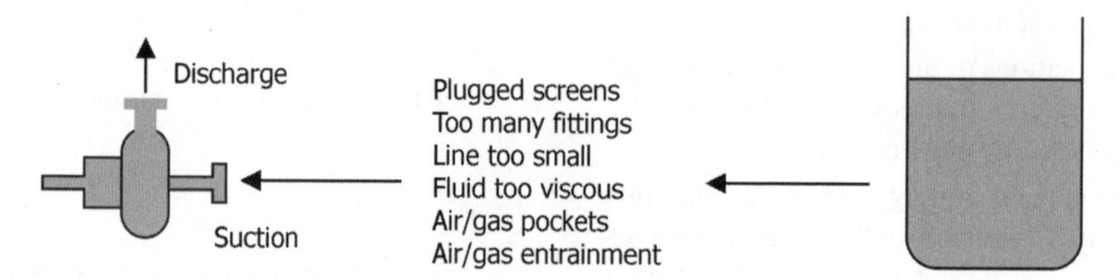

FIGURE 5.13 Pump suction side problems.

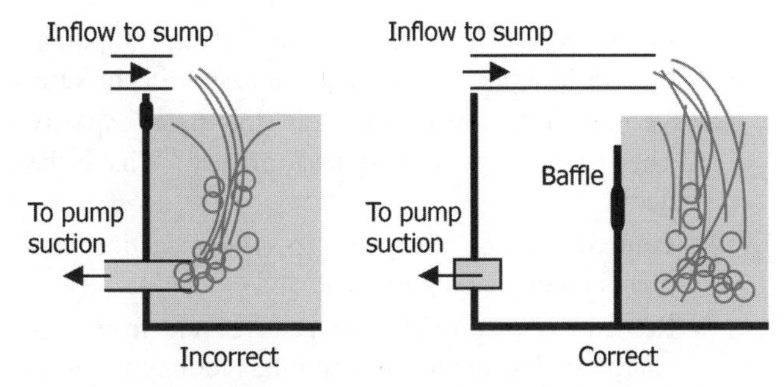

FIGURE 5.14 Pump suction through a sump.

- Suction lift is too high or insufficient margin between suction pressure and vapor pressure.
- Air/vapor present or entrained in suction liquid. Air entrained in the inflow should not pass into the suction opening. This can usually be prevented by providing sufficient submergence of liquid over the suction opening.
- An alternative sump design is to provide a baffle in the sump as illustrated in Figure 5.14.
- Wrong direction of rotation.
- Too low or too high speed.
- Total system head is higher/lower than pump design head.
- Specific gravity/viscosity of the liquid is far from design value.
- Very low capacity operation.
- Foreign matter in the impeller. Installing a filter on the suction side, such as a wire mesh strainer with a mesh size of about 120 μm, is advisable. This provision becomes all the more a necessity at the time of plant start-up to avoid debris entering the pump casing.
- Damaged impeller/wrong size impeller installed.
- Misalignment of shaft or bent shaft.
- Lack or loss of lubrication. Contamination of lubricant.
- Abnormal noise and vibration levels.
- Driver HP not adequate.
- Seal-related problems such as leakages, loss of flushing, cooling, and quenching systems.
- Pump problems can be summed up into three categories, namely, *design errors*, *operational errors*, and *maintenance errors*.
- Pump failures generally are due to three reasons, namely, mechanical problems within the pump, driver problems and hydraulic limits in the system.
  - Suction recirculation imposes mechanical limits and hydraulic requirements. Operation with suction recirculation damages the pump, increases failure rates and raises pumping system costs.
  - A rule of thumb is that reliable operation requires using pumps with N_{SS} under 8500. As N_{SS} increases, the reliable operating range of the pump usually decreases. Some low N_{SS} pumps continue to have problems. Some high N_{SS} pumps appear to operate with widely varying flow rates without problems.
  - Experience has shown two moves can improve reliability, namely, running with lower rpm pumps and raising the NPSH available, $NPSH_A$, at the pump suction. Reducing rpm increases the diameter of the pump needed or mandates more stages in a multistage pump.
  - Boosting $NPSH_A$ at the pump suction usually requires elevating vessels higher above the pump. Both steps can be expensive. No clear guidelines detail the trade-offs. One solution is to always operate close to the BEP by using pump recirculation.

- What are the effects of high-viscosity liquids and entrained gases on centrifugal pump performance?
  - As viscosity increases, head and capacity deteriorates and markedly when viscosity reaches 65 cP. BHP increases.
  - Above a viscosity of 65 cP, it would be better to use a positive displacement pump.
  - Entrained gases decrease performance in terms of efficiency, head, and power leading to high levels of severe effects when air/gases increase beyond 6% by volume, measured under suction conditions. 0.5% is tolerable limit.
- What is the effect of high pump speed while pumping highly viscous liquids?
  - While handling viscous liquids, selecting too high a pump speed results in insufficient inflows, causing cavitation problems.
- How can reliability in the operation of a centrifugal pump be improved?
  - Once a unit is built, often the only option to improve pump reliability is to replace the pump with a lower speed one. It is found that dropping from 3600 to 1800 rpm nearly solves most N_{SS}-related problems.
  - It has been useful to use the concept of suction energy, $E_{Suction,}$ in predicting pump reliability as related to suction conditions by using a *suction recirculation factor* that includes peripheral velocity in the impeller eye at the vane edge maximum radius in its calculation:

$$E_{\text{Suction}} = D_{\text{Eye}} \times N \times N_{\text{SS}} \times (\text{sp.gr.})_{\text{C}} \quad (5.14)$$

where D_{Eye} is the impeller eye diameter and $(\text{sp. gr.})_{\text{C}}$ is the specific gravity of the fluid at operations conditions. If the D_{Eye} is not available, it may be approximated by using 0.9 times the suction nozzle diameter for end-suction pumps and 0.75 times the suction nozzle diameter for double-suction pumps.

- Name an important contributing factor for lower efficiency of a centrifugal pump.
  - Recirculation of the fluid in the casing of a centrifugal pump is considerable, resulting unproductive energy consumption.
- How is recirculation of the liquid inside pump casing caused and how it is minimized?
  - Recirculation is caused by oversized flow channels that allow liquid to turn around or reverse flow. This reversal causes a vortex that attaches itself to the pressure side of the vane.
  - If there is enough energy available and the velocities are high enough, damage will occur.
  - Suction recirculation is reduced by lowering the peripheral velocity, which in turn, increases NPSH. To avoid this, it is better to opt for a lower speed pump, two smaller pumps, or an increase in NPSH_{A}.
  - Discharge recirculation is caused by flow reversal and high velocities producing damaging vortices on the pressure side of the vane at the outlet.
  - The solution to this problem lies in the impeller design. The problem is the result of a mismatched casing and impeller, too little vane overlap in the impeller design, or trimming the impeller below the minimum diameter for which it was designed.
- What is shut-in pressure for a centrifugal pump?
  - Shut-in pressure is the pressure developed on the discharge side at zero flow rate, that is, when the discharge valve is closed while the pump is in operation.
  - Pressure developed at shutoff may be of interest particularly when the pump and associated equipment can be isolated by valving, so that maximum design pressure of the equipment should reflect shutoff head.
  - Relatively high head pumps have little increase in discharge pressure at shutoff.
  - Typical shutoff head is about 25 m for mixed flow pumps and about 350% of the head at the maximum efficiency flow for axial flow pumps.
  - Codes require that all pump downstream equipment must be designed to withstand pump shut-in pressure.
- What is the effect of near shutoff (low-flow) capacity and head conditions on pump performance? What is the remedy?
  - Pump overheats, giving rise to serious suction problems involving vaporization and cavitation.
  - Remedy is to provide a bypass/recycle from discharge side to suction side through cooling arrangement to artificially keep a minimum safe flow through the pump.
- What is the difference between performance curve and system curve for a centrifugal pump?
  - The interdependency of the system and the centrifugal pump can be explained with the use of the pump performance curve and the system curve.
  - A centrifugal pump *performance curve* shows that the pressure the pump can develop is reduced as the capacity increases. Conversely, as the capacity drops, the pressure it can achieve is gradually increased until it reaches a maximum where no liquid can pass through the pump.
  - The *system curve* represents the pressures needed at different flow rates to move the product through the system.
  - To simplify a comparison with the centrifugal pump performance curve, *head* measurement is used.
  - The system head consists of the following three factors:
    - Static head or the vertical elevation through which the liquid must be lifted.
    - Friction head or the head required to overcome the friction losses in the pipe, the valves, and all the fittings and equipment.
    - Velocity head, which is the head required to accelerate the flow of liquid through the pump (velocity head is generally quite small and often ignored).
  - As the static head does not vary simply because of a change in flow rate, the graph would show a straight line. However, both the friction head and the velocity head will always vary directly with the capacity.
  - The combination of all three creates the system curve.
  - When the pump curve is superimposed on the system curve, the point of intersection represents the conditions (head and capacity) at which the pump will operate, as illustrated in Figure 5.15.
- What are the conditions that will cause the pump performance curve change its position and shape?

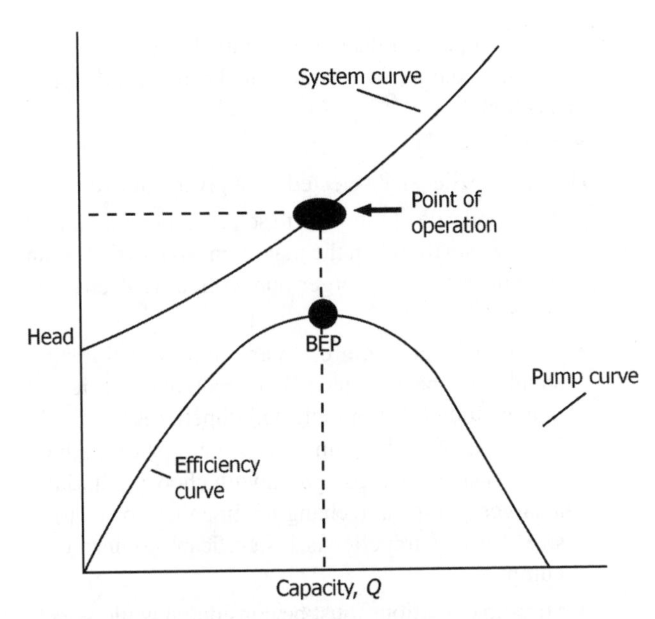

FIGURE 5.15 Pump performance curve and system curve illustrated.

- Conditions which will cause the pump performance curve to change its position and shape are as follows:
  - Wear of the impeller.
  - Change in rotational speed.
  - Change in impeller diameter.
  - Change in liquid viscosity.

- What are the conditions that will cause a change in system curve? What is the importance of changes in system curve?
  - Static head, velocity head, and friction head.
  - As velocity head is generally small, changes in static head and friction head will influence changes in system head.
    - A change in the static head is normally a result of a change in tank level.
    - An increase in friction head can be caused by a wide variety of conditions such as the change in a valve setting or buildup of solids in a strainer. This will give the system curve a new slope.
  - Regardless of the rated capacity of the centrifugal pump, it will only provide what the system requires.
  - It is important to understand the conditions under which system changes occur, the acceptability of the new operating point on the pump curve, and the manner in which it can be moved.
  - When the operating conditions of a system fitted with a centrifugal pump change, it is helpful to consider these curves, focus on how the system is controlling the operation of the pump and then control the system in the appropriate way.

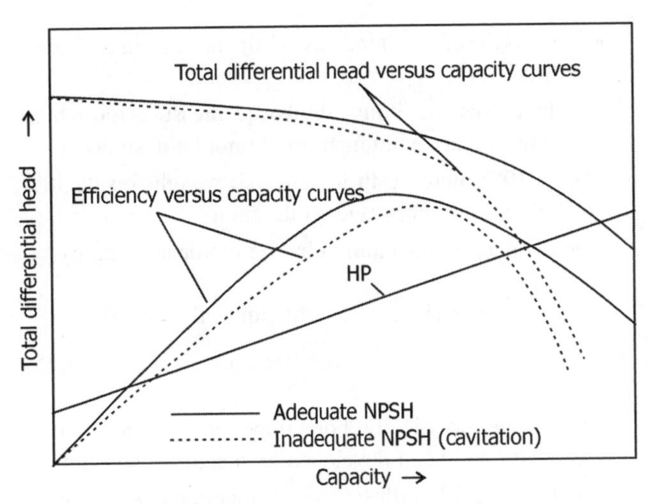

FIGURE 5.16 Pump performance curves.

- Give typical performance curves for a centrifugal pump indicating cavitation effects.
  - Figure 5.16 gives typical performance curves. The solid line curves represent condition of adequate $NPSH_A$ whereas the dotted lines represent condition of inadequate $NPSH_A$, that is, cavitation effect.
  - As pump capacity increases head developed marginally decreases in the lower capacity ranges and rapidly deteriorates at high capacities. The deterioration leads to breakdown conditions under cavitation conditions.
  - Efficiencies are low at low-flow conditions and increase rapidly with increased capacities and fall rapidly at very high capacities.
  - Cavitation effects are noticeable as shown by the dotted lines.

- What are the reasons of drop in head versus capacity curve (dotted lines in Figure 5.16) originally provided by the pump manufacturer?
  - Pumping liquids other than the one for which the manufacturer supplied the performance curves, for example, a higher viscosity liquid.
  - Wear resulting increased clearance between the impeller and the casing, that is, worn out impeller.
  - If pump driver is changed to steam turbine-driven pump, speed could be lower than the design speed.

- How is a centrifugal pump tested for cavitation?
  - When the head drops 3% or less for a given speed and capacity from design curves, that is, curves provided by the manufacturer, it is an indication that cavitation is occurring.

- What are the ways by which capacity of a centrifugal can be increased?
  - Reduce downstream ΔP.
  - Increase impeller size.

- What is the effect of increased impeller size in the pump casing?
  - Increases maximum discharge pressure for which downstream equipment might not be designed.
  - A 10% increase in impeller size might result 30% increase in amperage on the motor.
  - Amperage, A, required by the motor is given by

$$A \propto (\text{Head})(\text{sp.gr.})(\text{Liquid flow rate})/\text{motor efficiency}. \quad (5.15)$$

- Illustrate graphically how impeller size determines capacity and head developed by a centrifugal pump.
  - Figure 5.17 is illustrative of impeller size, capacity, and head developed relationships for a centrifugal pump.
  - For the same flow rate, head developed increases with increased diameter of the impeller.

$$h_2 = h_1(d_1/d_2)^2. \quad (5.16)$$

  - For the same head developed, flow rate (capacity) increases with impeller size.

$$Q_2 = Q_1(d_2/d_1). \quad (5.17)$$

  - Amperage for the motor increases with increased impeller diameter.

$$A_2 = A_1(d_2/d_1)^3. \quad (5.18)$$

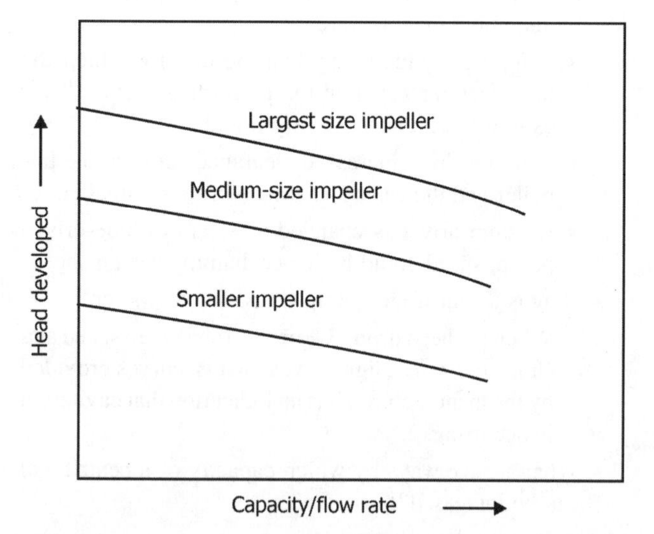

FIGURE 5.17 Effect of impeller size on capacity versus head developed for a centrifugal pump.

- Up to what percent decrease in impeller diameter for a centrifugal pump, is tolerable to keep the efficiency levels high?
  - $\leq$20%.
- How is impeller size selected for a given pump casing?
  - A general rule of thumb is to select an impeller that is one size smaller than the maximum size so that it can be replaced with a larger one without replacing the casing.
  - A pump that requires maximum size impeller should not be selected. When put into service, it may be found that the selected impeller is too small. In such a case, the pump has to be replaced, which is expensive when compared with choosing initially a larger pump and changing impeller to a larger size. Cost of impeller is insignificant compared to pump.
  - Start-up conditions must be considered while selecting the impeller. May be the pump has to put up 50% more head to establish *initial* circulation than normal.
  - While increasing impeller size, the consequences of increase in shut-in pressure must be examined and taken care of.
  - Motor selection should be based on the largest size impeller that can fit into the pump and not on size of actual impeller used.
- "Speed control for a centrifugal pump will not affect its efficiency." Comment.
  - Changes in speed up to 20% will not adversely affect efficiency.
- How are pulsations reduced through design of a centrifugal pump?
  - Number of vanes: Increasing number of vanes.
  - Radial clearance between tip of the impeller and the stationary collector.
  - Type of impeller: Pulsations tend to increase with increase in specific speed of the impeller. Thus, the radial impeller has the lowest pulsations, followed by Francis and axial-type impellers.
  - Impeller tip speed: Higher tip speeds generally mean larger pulsations.
  - Staggering impellers in multistage machines: Staggering reduces pulsations.
- What is the difference in operating centrifugal pumps in series and parallel?
  - Options for handling high head or high flow applications include using pumps in series or parallel.
  - When running in series, the heads are added and the total capacity is equal to that of the pump with the smallest capacity.

- In parallel, the capacities of the pumps are added and the head of all pumps will be equal at the point where the discharged liquids recombine.

- Give reasons for operation of pumps in parallel.
  - Parallel pumps are used for a variety of reasons:
    - Cost (two smaller pumps may cost less than a larger one).
    - Increase in the size of an existing plant.
    - To compensate for a process with varying capacity.
  - It must be noted that pumps operated in parallel must have similar head characteristics to avoid potential operating problems.

- What are the problems involved in the operation of pumps in parallel? How are they overcome?
  - Pumps are operated in parallel to increase pumping capacity and introduce flexibility in operation when flow capacity changes are expected.
  - Parallel operation is most effective with identical pumps. However, they do not have to be identical, nor have the same shutoff head or capacity to be paralleled.
  - Each pump must have a check valve to prevent reverse flow. When one of the pumps is turned off, the flow reverses almost immediately in that line because of the high manifold pressure supplied by the operating pumps. This causes the check valve to close.
  - If a slow-closing check valve is installed, the flow can attain a high reverse velocity before the valve closes, generating high-pressure transients. Proper selection of the check valve is necessary.
  - If one of the pumps is relatively weaker than the other, the weaker pump must be started first and then the stronger one.
  - When draining a tank with two pumps, one should not use a "T" with two connections. The dominant pump may asphyxiate the other pump. Each pump needs its own supply pipe as illustrated in Figure 5.18.

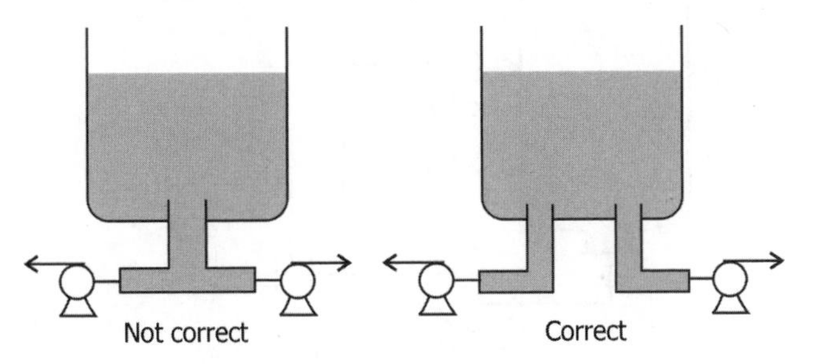

FIGURE 5.18 Two pumps draining liquid from a tank.

  - One pump running in a parallel system tends to suffer from cavitation because $NPSH_R$ of the pump increases drastically. Parallel pumps should have dual or double seals installed to withstand cavitation when only one pump is running. Also changes in operating conditions may warrant increased power requirements for the motor when only one pump is running.
  - Pumps in series with one of the pumps running may warrant selection of double seals for them to avoid operational problems.

- Give pump performance curves for pumps in parallel and series.
  - Figures 5.19 and 5.20 show centrifugal pump performance curves for parallel and series operations.

- What are the general guidelines for the selection of a centrifugal pump?
  - The pump should be selected based on rated conditions.
  - The BEP should be between the rated point and the normal operating point.
  - The head/capacity characteristic curve should continuously rise as flow is reduced to shut off (or zero flow).
  - The pump should be capable of a head increase at rated conditions by installing a larger impeller.
  - The pump should not be operated below the minimum continuous flow rate specified by the manufacturer.

- What are the causes of high flows in a pumping system?
  - When an oversized pump is used for a specific application.
  - When two or more pumps are used in parallel and one of them is taken out of service due to decrease in demand or for maintenance.

- What are the effects of high flows in a pumping system?
  - In the case of oversized pump, unless sufficient $NPSH_A$ is provided, pump may suffer cavitation damage and power consumption will be excessive.
  - In case of parallel pumps, flows in excess of a single-pump capacity may result in cavitation and high-power consumption.

- What are the causes of low flows in pumping?
  - Due to reduced demand by the process served by the pump.
  - Two pumps operating in parallel may be unsuitable for service at reduced flows and one of the pumps on the line may have its check valve closed by the higher pressure developed by the stronger pump.

- What are the effects of low flows or no flows due to closing of discharge valve in a pumping system and what are the remedies?

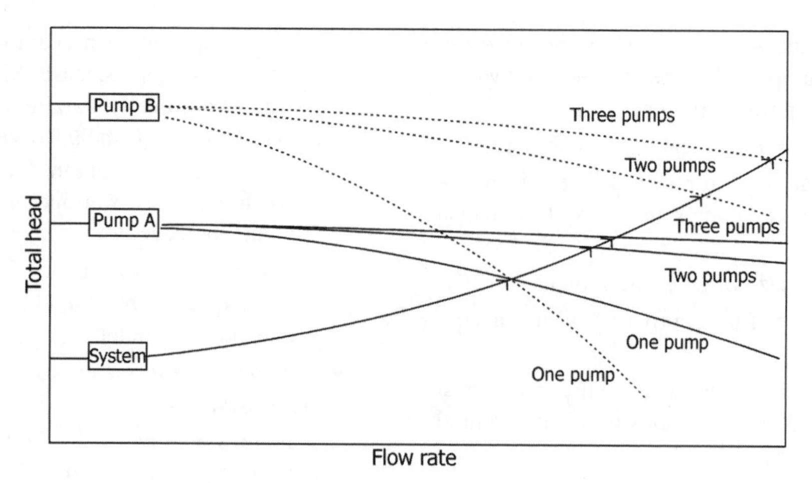

FIGURE 5.19 Performance curves for centrifugal pumps operating in parallel. (*Courtesy*: GPSA Engineering Data Book, 12th ed.)

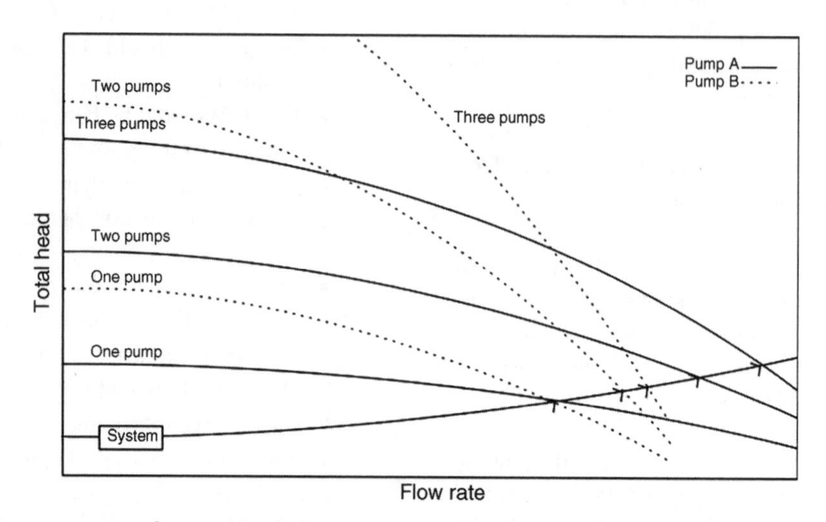

FIGURE 5.20 Performance curves for centrifugal pumps operating in series. (*Courtesy*: GPSA Engineering Data Book, 12th ed.)

Effects

- Decreased efficiencies.
- Higher bearing loads.
- Increased temperature.
- Internal recirculation causing surging, intense vortices and damage to impeller.

Remedies

- Use of variable speed drives or use of several pumps for total required capacity and shutting down pumps sequentially as demand reduces.
- Appropriate design to take care of higher bearing loads.
- Use of minimum flow bypass working automatically (Figure 5.21). This also protects against accidental closure of check valve while pump is running.
- Installing coolers on pump suction side to reduce liquid temperature and thereby increasing NPSH.
- Providing adequate liquid holdup in vessels acting as surge tanks (5–15 min holdup time is typical).
- Providing vortex breakers in the bottom of the vessel.
- Suggested suction-specific speeds to reduce low-flow problems: 8500–9500.

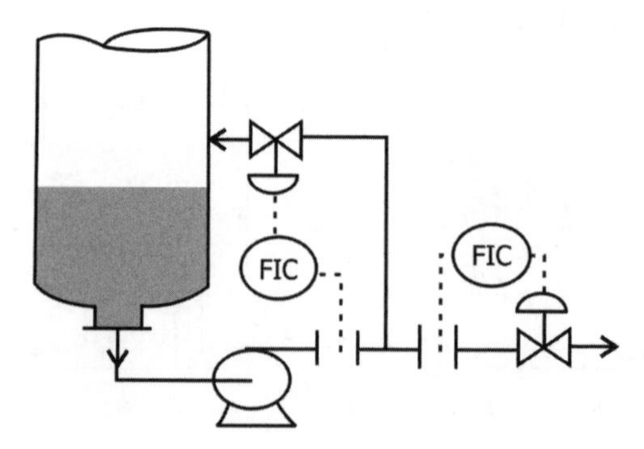

FIGURE 5.21 Minimum flow bypass.

- *Note*: Pumps handling hydrocarbons can tolerate lower flows than cold water, as specific volume of water is much higher than that for hydrocarbons, (cf. see cavitation issues).

- What is the significance of the minimum flow required by a pump?
  - Centrifugal pumps have a required minimum flow below which severe operational problems and damage can occur. Below minimum flow, pump may cavitate drawing air into its suction. Its casing gets filled and looses its prime.

- What are design requirements to handle minimum flows in a centrifugal pump system?
  - If the primary pump discharge path can be blocked, a recirculation line is required to accommodate the minimum flow of the pump. Such lines can be designed either for continuous or for controlled flows.
    - Continuous flows are more common in smaller pumps and use a flow restriction orifice in the recirculation line to bleed off the pressure gained from the pump.
    - Controlled flows are normally used in higher capacity systems where a continuous, minimum flow would lead to significantly higher operating costs or the selection of a larger pump. Controlled flows are most often accomplished with a self-regulated pressure control valve on the pump discharge or a flow control valve and controller with the minimum flow configured as the set point.
  - Excess pumping energy increases the temperature of the fluid being pumped. Generally, the temperature rise is negligible, except when the pump is operating near shutoff.
    - For this reason, the minimum flow recirculation line should be placed in such a way that the outlet of the line connects to a vessel instead of the pump suction line, thus giving the fluid time to shed its excess heat. If the fluid is added to the suction line, the temperature increase will not be dissipated. A small heat exchanger may be placed ahead of the pump suction to achieve the same result for relatively high head pumps (~150 m) or when handling heat-sensitive fluids.
    - Low suction pressure switches interlocked into the plant control system sometimes trip the pump or prevent its start-up when the suction pressure is below a set value. High-pressure switches can be installed on the pump discharge so as not to exceed the maximum allowable output pressure.
    - The motor is usually tripped or not allowed to start under these conditions. Similarly, low-flow switches on the discharge are used to alarm and trip the motor when a low-flow condition has occurred.
    - This interlock is often bypassed (manually or with a timer) during pump start-up to allow the flow to reach its normal operating point, at which time the bypass is removed. Changing from switches to transmitters for increased accuracy and easier failure diagnosis is presently favored.

- For a given size of already installed pump motor, how is the maximum possible size of an impeller in the pump casing decided?
  1. Place pump discharge control valve in wide-open position.
  2. Measure the amperage drawn by the pump?
  3. Determine maximum possible impeller size that can be used with the existing motor, using the relationship

$$d_2 = d_1[A_r/A_1]^{1/3} \tag{5.19}$$

  where A_r is the rated capacity of the motor in amperes (service factor, which is 10–15% of the rated capacity); A_1 is the actual amperage as measured under step 2; and d_1 is the existing impeller diameter.

- Which of the following two arrangements in the control of centrifugal pumps involves better energy conservation?: Throttling the flow with a valve on pump discharge or using a variable speed control on the pump drive?
  - Using a variable speed control conserves energy better. It is used when flow rates are large to justify additional costs in installing variable speed control.
  - For low flow capacities, throttling is cheaper.

- What are the additional requirements for the following pump options?
  - Table 5.2 lists the additional requirements for different pump options.

- "Centrifugal pumps are used as metering pumps." *True/False?*
  - *False.*

- Under what circumstances use of inducers is desirable on pump inlet?
  - Inducers are used to improve pump suction conditions.
  - Inducers improve NPSH requirements for difficult applications.
  - They assist flow of viscous fluids into pump casing.

TABLE 5.2 Additional Requirements for Different Pump Options

Option Details	Additional Requirements
Impeller replacement with higher/lower diameter impeller of same or different design.	Possible change in drive; base frame; foundation requiring additional space; change in speed; possible change in pump control system.
Addition of identical/lower/higher capacity pump parallel to existing pump(s)/standby pumps of higher/lower capacities for parallel operation.	Additional space; changes in drives, base frame, foundation; control system change; piping modifications; pump operation at different speed or with different impeller sizes; adding individual minimum bypass system for each parallel pump.
Replacing existing pump(s) with pump(s) of higher capacity.	Additional space; piping modifications; adding individual minimum bypass system for each parallel pump; change in control system.
Replacing positive displacement pump with centrifugal pump.	Additional space; change in control system.

- What are the maintenance problems involved in pump performance and life?
  - Bent shafts and misalignment of the motor and pump shafts.
  - If a bent shaft is installed into a pump and run, it will fail prematurely, leaving evidence and specific signs on the circumference of close tolerance stationary parts around the volute circle of the pump. The shaft will exhibit a wear spot on its surface where the close tolerance parts were rubbing. A deflected shaft is absolutely straight when rotated in a lathe or dynamic balancer. The deflection is the result of a problem induced either by operation or by system design. The deflected shaft also will fail prematurely in the pump, leaving similar, but different evidence on the close tolerance rubbing parts in the pump.
  - Pump shaft and driver shaft alignments are very important for long useful equipment life, and to extend the running time between repairs.
  - Besides, good alignment reduces the progressive degradation of the pump.
  - One of the most important and least considered points of correct alignment is the relationship with the power transmitted from the motor to the pump. An almost perfect alignment with an adequate and new coupling transmits nearly 100% of the power of the motor, with some small losses.
- What are the symptoms and possible root causes for centrifugal pump problems?
  - Table 5.3 lists the symptoms and the possible causes for centrifugal pump problems.
- Give a summary of good and bad qualities of centrifugal pumps.

Good Qualities

- Simple in construction, inexpensive, available in a large variety of materials and have low maintenance cost.
- Operate at high speed so that they can be driven directly by electrical motors.
- Give steady delivery, can handle slurries and take up little floor space.

Drawbacks

- Single-stage pumps cannot develop high pressures except at very high speeds (10,000 rpm for instance).
- Multistage pumps for high pressures are expensive, particularly in corrosion-resistant materials.
- Efficiencies drop off rapidly at flow rates much different from those at peak efficiency.
- They are not self-priming and their performance drops off rapidly with increasing viscosity.

- "The discharge line of a centrifugal pump can be completely closed without damaging the pump." *True/False?*
  - *True.* When discharge valve is closed, the fluid churns/recirculates inside the casing without rupturing the discharge line although the pump overheats.
- What are the difficult pumping applications that require highly demanding mechanical seals?
  - Crystallizing liquids.
  - Solidifying liquids.
  - Vaporizing liquids.
  - Film forming liquids.
  - High-temperature liquids.
  - Nonlubricating liquids.
  - Dry-running the pump.
  - Hazardous liquids.
  - Gases and liquids that phase to gas.
  - Slurries.
  - Cryogenic liquids.
  - High-pressure liquids.
  - Vacuum conditions.
  - High-speed pumps.

TABLE 5.3 Symptoms and Possible Causes for Centrifugal Pump Problems

Symptom	Possible Causes
Pump does not deliver rated capacity.	Excessive discharge head, not enough $NPSH_A$, suction line not filled, air or vapor pocket in suction line, plugged impeller or piping, worn or damaged impeller or wearing rings, inadequate foot valve size, foot valve clogged with debris, internal leakage due to defective gaskets, wrong direction of rotation, viscosity of liquid more than that for which pump is designed, inlet to suction line not sufficiently submerged, and pump not up to rated speed.
Failure to deliver liquid.	Wrong direction of rotation, pump not primed, available NPSH not sufficient, suction line not filled with liquid, air or vapor pocket in suction line, inlet to suction pipe not sufficiently submerged, pump not up to rated speed, and total head required greater than head which pump is capable of delivering.
No discharge pressure.	Pump improperly primed, not enough $NPSH_A$, plugged impeller or piping, inlet to suction line not sufficiently submerged, inadequate speed, suction line not filled, air ingress into suction line, total head required is greater than head pump is capable of delivering, incorrect direction of rotation, and closed discharge valve.
Pressure surge.	Not enough $NPSH_A$, air ingress into suction line, entrained air, plugged impeller.
Inadequate discharge pressure.	Not enough velocity, air/gases in pumped liquid, too small impeller diameter, pump not up to rated speed, worn or damaged impeller or wearing rings, incorrect rotation, discharge pressure more than pressure for which pump is designed, viscosity of liquid more than that for which pump is designed, and internal leakage due to defective gaskets.
Pump loses liquid after starting.	Suction line not filled with liquid, air leaks in suction line or stuffing boxes, two-phase flow, slug flow, inlet to suction line not sufficiently submerged, not enough $NPSH_A$, liquid seal piping to lantern ring plugged, and lantern ring not properly located in stuffing box.
Excessive power consumption.	Head too small, excessive recirculation, high sp. gr. or high μ liquid, misalignment, bent shaft, bound shaft, packing too tight, rotating element dragging, and incorrect rotation.
Vibration and noise.	Cavitation, excessive suction lift, starved suction, not enough $NPSH_A$, air ingress, inlet to suction line not sufficiently submerged, worn or loose bearings, bent shaft, misalignment, bound rotor, improper location of control valve in discharge line, and foundation not rigid.
Stuffing boxes overheat.	Packing too tight and/or not lubricated, wrong grade of packing, box insufficiently packed, and insufficient cooling water to jackets.
Bearings overheat.	Oil level too low, improper or poor grade oil, dirty bearings and oil, moisture in oil, oil cooler clogged or scaled, oiling system failure, insufficient cooling water circulation, insufficient cooling air, bearings too tight, oil seals too close fit on shaft, and misalignment.
Bearing wear rapidly.	Misalignment, bent shaft, vibration, bearings improperly installed, dirt in bearings, moisture in oil, excessive cooling of bearings, excessive thrust resulting from mechanical failure inside the pump, and lack of lubrication.

Courtesy: GPSA Engineering Data Book, 12th ed.

- Can the suction size of a centrifugal pump be larger than the pipeline?
  - The piping leading to a centrifugal pump must be at least one size larger than the inlet flange on the pump. Similarly, the discharge piping leading from the pump should be at least one size larger than the flange size on the discharge of the pump.

5.1.4 Positive Displacement Pumps

- "Positive displacement pumps generally operate at low inlet velocities." *True/False?*
  - *True.* In these cases, generally velocity head portion is ignored.
- "Positive displacement pumps are usually self-primed pumps." *True/False*?
  - *True.*
- What is meant by *Simplex*, *duplex*, *triplex*, and *multiplex* in pump terminology?
  - *Simplex* refers to single-piston type of reciprocal piston pump. A *Duplex* pump has two pistons; one is on suction mode while the other is on delivery mode. Similarly, *triplex* and *multiplex* mean three and multiple number of pistons, respectively, in different modes of operation.
- Illustrate valve action for a double-acting reciprocating piston pump and discharge curves for different flow configurations.

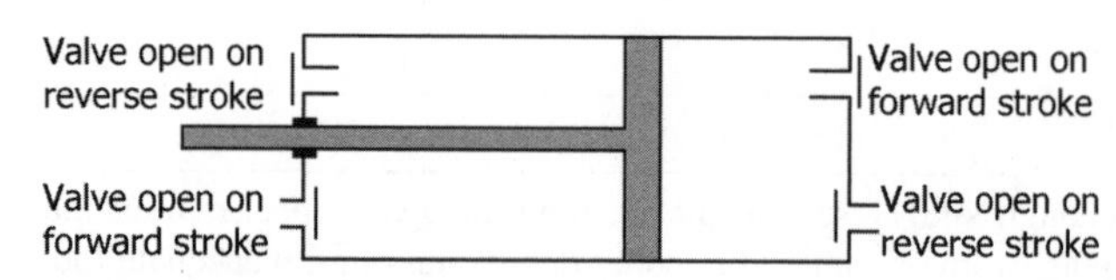

FIGURE 5.22 Valve action for a double-acting reciprocating piston pump.

- Figure 5.22 shows valve action for a double-acting reciprocating piston pump.
- Figure 5.23 shows discharge curves for different flow configurations for a reciprocating piston pump.
- Flow fluctuations are reduced, as illustrated in the above figures, from single-acting piston type, through simplex double-acting-type, to duplex double-acting-type pumps.

• What are the variations/arrangements involved in reciprocating pump systems?
- *Main Pump Components*: (i) Piston is used for low-pressure light duty or intermittent service, which is less expensive than the plunger design, but cannot handle gritty liquids. (ii) Plunger is used for high-pressure heavy duty or continuous service, which is suitable for gritty and foreign material service. Expensive than the piston design.
- *Types*:
  - ➢ Simplex: One piston.
  - ➢ Duplex: Two pistons.
  - ➢ Triplex: Three pistons. Not used as steam-driven pumps.
- *Piston/Plunger Action*:
  - ➢ Single acting: One stroke per revolution.
  - ➢ Double acting: Two strokes per revolution.
  - ➢ Cylinder fills and discharges per each stroke.
- *Packing for Piston/Plunger*:
  - ➢ Piston type: Piston packed in which packing is mounted on the piston and moves with it.
  - ➢ Plunger type: Cylinder packed in which packing is stationary and plunger moves; Expensive than piston packed.

• What is a duplex pump?
- Reciprocating pump with two cylinders is called duplex pump.

• What are the applications of (i) piston pumps and (ii) plunger pumps?
- Piston pumps: Used for comparatively lower pressures, light duty and intermittent service; less expensive than plunger pumps; and cannot handle gritty liquids.
- Plunger pumps: Used for high pressures, heavy duty, and a continuous operation; expensive than piston type; and can handle liquids with gritty/foreign material.

• "The flow rate in a positive displacement pump decreases significantly as the head increases." *True/False?*
- *False.*

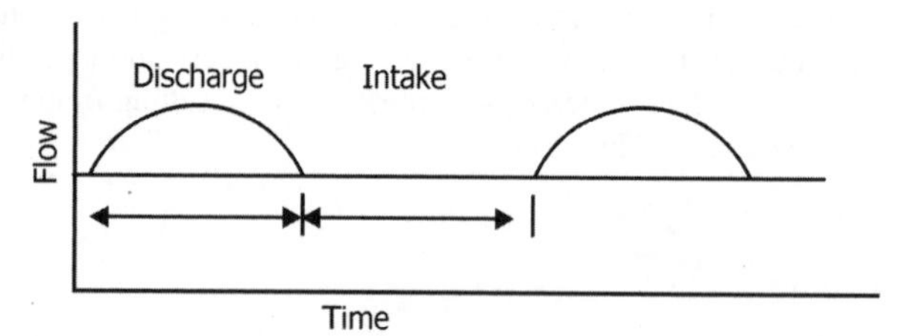

Discharge curve of a single-acting piston pump operated by a crank, Half-sine curve

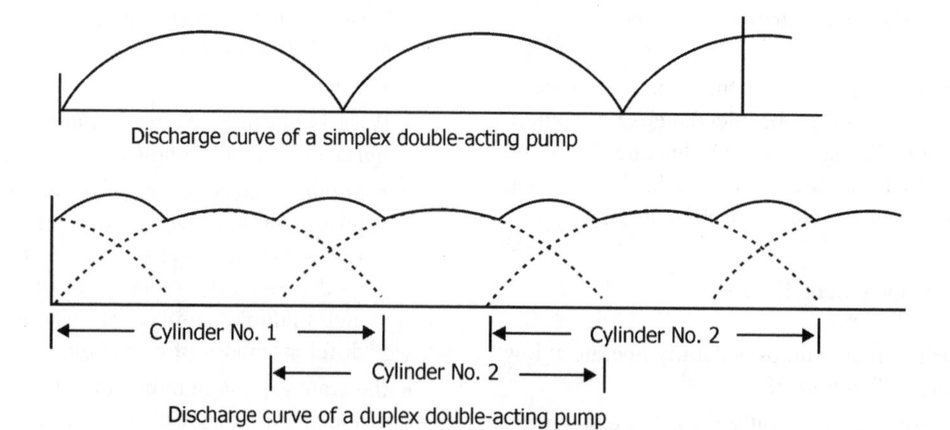

Discharge curve of a simplex double-acting pump

Discharge curve of a duplex double-acting pump

FIGURE 5.23 Discharge curves for different reciprocating flow configurations.

- What are the ranges of operating pressures and capacities of reciprocating pumps?
  - Pressures: 7×10^4 kPa or higher.
  - Capacities: up to 250 m^3/h.
- "The discharge line of a positive displacement pump can be closed without damaging the pump." *True/False*?
  - *False*. A positive displacement pump delivers a fixed quantity of the fluid with each stroke. If discharge line is closed, the delivered fluid increases pressure with each stroke and finally damages the discharge line so that the fluid escapes through the ruptured line.
  - As a precaution, a relief valve is installed on the discharge line so that when pressure increases beyond the set pressure of the valve, the valve opens and relieves the pressure, thus preventing damage to the pipe.
- What is overall or mechanical efficiency of a positive displacement pump?
  - It is the ratio of the useful hydraulic power transmitted to the fluid leaving the pump to the total power absorbed by the pump:

$$\eta = \text{FHP}/\text{BHP}. \tag{5.20}$$

- Define (i) volumetric efficiency and (ii) mechanical efficiency for a reciprocating pump.
  - Volumetric efficiency: Ratio of actual discharge to that based on piston displacement.
  - Mechanical efficiency: Ratio of energy supplied to the fluid to the energy supplied to the pump.
- Which of the two pumps, centrifugal or positive displacement, has higher efficiency? Why?
  - There will be considerable internal recirculation in a centrifugal pump leading to unproductive work, whereas in a positive displacement pump, there will not be any recirculation, and once fluid enters pump casing it will be discharged, barring small amount of slip through clearances that are very small compared to those in a centrifugal pump.
- What are the overall efficiencies obtainable in a positive displacement pump?
  - 85–94%.
- "Plunger pumps are always single acting." *True/False?*
  - *True.*
- What are the sources of pulsations in flow?
  - In discharge lines in reciprocating pumps/compressors because of flow (during discharge) and no flow (during suction) strokes, pulsations in flow occur.
  - In lines supplying steam to reciprocating machinery.
- How are flow pulsations eliminated/reduced?
  - Liquid, being incompressible, cannot dampen pulsations that are characteristic with positive displacement pumps. Therefore, air or gas, being compressible, is filled in a surge chamber on the discharge side of a pump/compressor to cushion out/dampen pulsations.
  - For gas flow, a combination of a surge chamber located close to pulsation source and a constriction in the line between surge chamber and metering element.
  - A liquid level gauge is desirable to permit checking the amount of air/gas in the chamber.
  - Use of multiple cylinders, for example, duplex and double-acting pumps reduces amplitude and frequency of pulsations in discharge lines.
- What are the differences in amplitude and frequency of pulsations between positive displacement pumps/compressors and centrifugal pumps/compressors?
  - In general, pulsations generated by positive displacement pumps/compressors are of low frequency and high amplitude whereas those generated by centrifugal pumps/compressors are of high frequency and low amplitude.
- What is slip of a pump?
  - Fraction/% loss of capacity relative to the theoretical capacity that is based on the displacement volume of the cylinder due to the stroke of piston/plunger.

$$\text{Slip} = (1 - \eta_{\text{vol}}) \tag{5.21}$$

  where

$$\eta_{\text{vol}} = \frac{\text{Actual amount of liquid pumped}}{\text{Theoretical capacity based on piston movement.}}$$

  - Slip varies 2–10% of displacement, average being 3%.
- Why is it necessary to use a relief valve set to open at a safe pressure on the discharge side of a positive displacement pump?
  - Each stroke of the pump discharges a finite quantity of fluid to the discharge line, unlike in centrifugal pumps in which considerable fluid recirculation occurs in the pump casing.
  - If a valve on the discharge side is closed, pressure goes on increasing with each stroke in a positive displacement pump.
  - This pressure increase leads to discharge pipe rupture if pressure increases beyond design pressure of the pipe.

TABLE 5.4 Comparison of Capacities, Heads, and Efficiencies for Positive Displacement Pumps

Pump Type	Capacities	Heads	Efficiencies
Axial flow	0.076–376 m^3/min (20–100,000 gpm)	Up to 12 m (40 ft)	65–85
Rotary	0.00378 m^3/min (20–100,000 gpm)	Up to 15,200 m (50,000 ft)	50–80
Reciprocating	0.0378–37.8 m^3/min (10–100,000 gpm)	Up to 300,000 m (1,000,000 ft)	70% at 7.46 kW (10 HP) 85% at 37.3 kW (50 HP) 90% at 373 kW (500 HP)

264 gpm = 1 m^3/min; 3.28 ft = 1 m.

  - A relief valve, set to open at a pressure below design pressure, avoids pipe rupture.
- Give a comparison of axial flow, rotary, and reciprocating pumps with respect to ranges of capacities, heads developed, and efficiencies.
  - Table 5.4 presents a comparison of capacities, heads, and efficiencies for axial flow, rotary, and reciprocating pumps.
- What type of pump is normally recommended for hydraulic testing of vessels? Why?
  - Direct-acting plunger pump is suitable as it stalls at a set pressure and pumps only when pressure falls.
- What type of pump is suitable for pumping light hydrocarbons from refinery knockout drums?
  - Direct-acting positive displacement pump because of low NPSH and variable speed requirements.
- "Plunger pumps are suitable for homogenization of food products." *True/False*?
  - *True*. Milk and food products are pumped to high pressures and throttled through special valves.
- What are the characteristics and applications of rotary pumps? Give a diagram of a typical rotary pump.
  - Flow is proportional to speed and is independent of ΔP.
  - BHP varies directly with speed and pressure.
  - With speed and pressure constant, BHP is proportional to viscosity of liquid.
  - Have close clearances.
  - Sometimes used as metering pumps.
  - Some designs involve rotation in either direction.
  - Internal slip reduces efficiency; increases with pressure and decreases with viscosity.
  - Self-priming.
  - Can be used for liquids of any viscosity but effective for high-viscosity liquids, greases, paints, and so on.
  - Majority of these pumps handle clean liquids with wide range of viscosities up to 10^6 cP and pressures of over 70 bar.
  - Used for small and medium capacities in the range of 0.2–115 m^3/h (1–500 gpm).
  - It is sometimes better to select a larger pump running at low speeds than a smaller pump at high speeds when dealing with viscous liquids.
  - As a general guide, speed is reduced 25–35% below rating for each 10-fold increase in viscosity above 1000 SSU. Also, generally, the mechanical efficiency of the pump is decreased 10% for each 10-fold increase in viscosity above 1000 SSU.
  - Not suitable for liquids with abrasive solids.
  - They are usually self-priming, able to handle very low NPSH and reversible.
  - Low cost.
  - Require small space for installation.
  - Low volumetric efficiency.
  - Require overpressure relief protection.
  - Entrained gases reduce liquid capacity and cause pulsations.
  - Figure 5.24 illustrates a sliding vane rotary pump.
    - Vane pumps are effective for low-viscosity liquids and when dry-priming is required. They are not ideal for abrasive liquids.
- What are the factors that influence the amount of slip in a rotary pump?
  - Clearance: Increased clearances result in increase in slip.

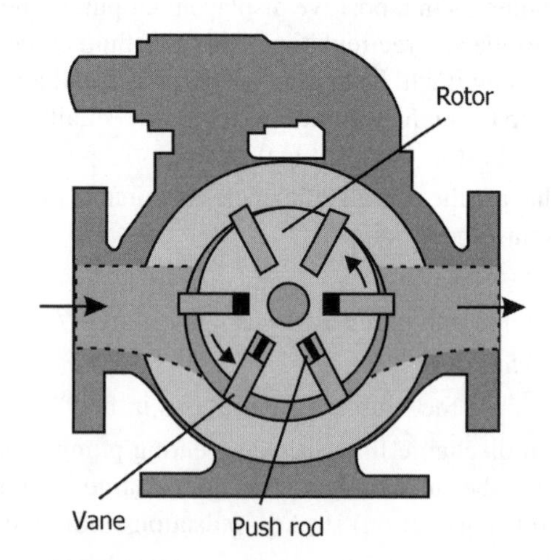

FIGURE 5.24 Sliding vane rotary pump.

- Size and shape of the rotor also influences slip.
- Pressure: Increased pressure increases slip.
- Liquid viscosity: Increased viscosity decreases slip.
- Pump speed: Slip is independent of speed.

• "For a rotary pump, flow is proportional to speed and almost independent of pressure differential." *True/False*?

- *True.*

• " In a rotary pump, entrained gases will not have any effect on liquid capacity." *True/False*?

- *False.*

• What are the attributes of a liquid ring pump?

- A liquid ring pump is primarily a high efficiency centrifugal pump.
- Self-priming when pump casing is half filled.
- Suitable for aerated fluids.
- Impeller is radial type with straight vanes.
- For each type of liquid ring pump there is only one impeller size.

• What is dead head speed?

- Pump speed required to overcome slip is known as dead head speed.

• What are the features/characteristics/applications of rotary gear pumps? Give a sketch of a rotary gear pump.

- Gear pumps may be of internal or external gear type as illustrated in Figure 5.25.
- Fluid entering the pump casing fills the cavities between the gear teeth and the casing moves circumferentially to the outer port and from there it discharges.
- These pumps are self-priming.
- They give a constant flow of liquid for a given rotor speed.
- Pulsations are slight and may be eliminated by the use of helical gears.
- They are valveless and can reverse flow. They are relatively simple to maintain and rebuild.
- They are less sensitive to cavitation problems compared to centrifugal pumps.
- Major disadvantage is their inability to resist erosion, abrasion, and corrosion. Disadvantages include limitations in the selection of materials of construction due to tight tolerances required. Since the gears touch, the materials of construction should be dissimilar, especially for low-viscosity liquids or poorly lubricated applications.
- High shear is placed on the liquid.
- The fluid must be free from abrasives.
- Gear pumps must be controlled through the control of motor speed. Throttling the discharge is not an acceptable means of control.
- They can be damaged if run dry.
- Uneconomical for large flows.
- These pumps are appropriate for use for high-pressure and low-capacity applications.
- Even with limitations on operating speed, a relatively small pump can have a large capacity. They have the ability to handle a wide range of viscosities. Used for liquid viscosities normally in the range of 60–300 cS, but at low speeds have been used with liquids up to 10,000 cS. These pumps can be used as metering pumps.

• Compare characteristic features of internal and external gear pumps with respect to their working and applications.

Internal Gear Pumps:

- Internal gear pumps have an outer gear called the rotor that is used to drive a smaller inner gear called

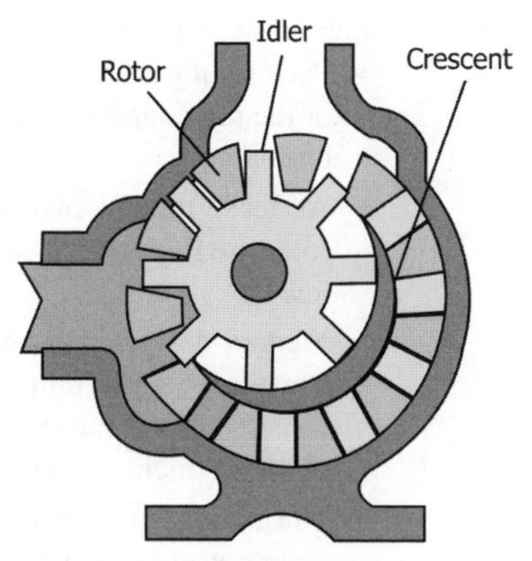

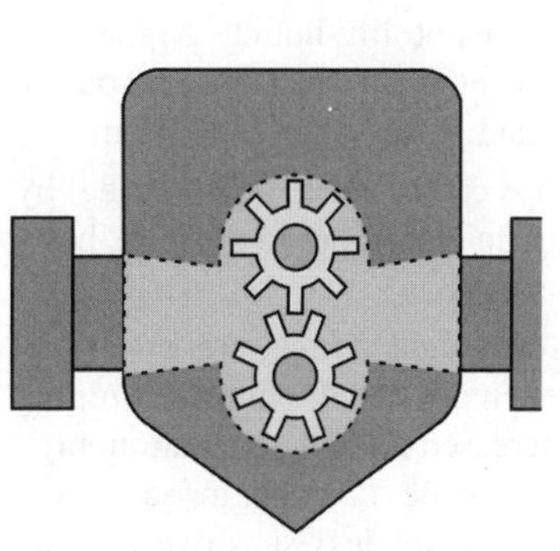

FIGURE 5.25 Gear pumps.

the idler. The idler gear rotates on a stationary pin and operates inside the rotor gear. As the two gears come out of mesh, they create voids into which the liquid flows.

- When the gears come back into mesh, there will be a reduction in void spaces and liquid is forced out of the discharge port. A *crescent* is formed between the two gears that functions as a seal between the suction and the discharge by trapping the liquid carried between the teeth of the rotor and the idler.
- Internal gear pumps are effective with viscous liquids, but do not perform well for liquids containing solid particles.

External Gear Pumps:

- External gear pumps have a similar pumping action to internal gear pumps in that two gears come into and out of mesh to give rise to flow. The external gear pumps have two identical gears rotating against each other. Each gear is supported by a shaft with bearings on both sides of each gear.
- External gear pumps work well in high-pressure applications, such as hydraulics, but are not effective in applications requiring critical suction conditions.

- What are the advantages of rotary pumps?
  - The advantages of rotary pumps are the following:
    - Efficiency: 20–50% (depending on pressure) higher efficiency for most typically pumped liquids, with high viscosities.
    - Viscous liquids handling: Above approximately 300–1000 SSU, a centrifugal pump simply cannot be used, as viscous drag reduces efficiency to nearly zero. Positive displacement pumps continue to pump at high efficiency, with no problems.
    - Pressure versatility: Practically unaffected by pressure, within a wide range.
    - Self-priming: A typical ANSI-dimensioned centrifugal pump that is commonly found in many chemical plants cannot lift liquid. A standard positive displacement pump, such as a gear pump, can easily lift liquid in the range of 0.3–7 m.
    - Centrifugal pumps can be made self-priming, by adding a priming fluid chamber in the inlet, however, this adds expense.
    - Inlet piping: Centrifugal pumps are extremely sensitive to inlet piping details. Improper piping may cause an increased $NPSH_R$, cavitation, high vibrations and possible damage to seals and bearings. PD pumps are less sensitive to inlet piping and can be a real solution for many difficult installations with space constraints, since piping modifications to the existing setups are very costly.
    - Bidirectional: By simply reversing the direction of motor rotation, many PD pumps will pump in reverse, which can be advantageous in many processes. Centrifugal pumps can pump in only one direction. In some installations, two centrifugal pumps are used: one for loading, and another for unloading, which doubles the piping runs, valves, and auxiliaries. A *single* rotary pump would do the same job. (*Note*: Relief valves should be installed in both directions in such cases.)
    - Flow maintainability: PD pumps produce almost constant flow, regardless of properties of the fluid and conditions (viscosity, pressure, and temperature). For centrifugal pumps, a change in fluid properties and external conditions would result in a definite change in performance.
    - Metering capability: PD pumps can be used as convenient and simple metering devices. Centrifugal pumps have no such capabilities.
    - Inventory reduction: Since PD pumps can pump a wide variety of fluids in an extreme range of viscosities, the same pump parts inventory is required for a wide range of applications throughout the plant. Centrifugal pumps require a greater multitude of sizes for different applications, which results in increased inventory of parts.
- What are the salient features of a rotary screw pump?
  - Rotary screw pump makes use of two long helical rotors in parallel, which rotate in opposite directions. The rotors do not touch each other.
  - Helical timing gears are used to synchronize the rotation.
  - When used for moving gases as a vacuum pump, it moves the gas axially along the screw without any compression from suction to discharge.
    - Pockets of gas are trapped within the convolutions of the rotors and the casing and transported to the discharge.
    - At least three convoluted gas pockets in the rotor are required to achieve acceptable vacuum levels.
    - Modern designs use a variable pitch (number of convolutions per unit length), which essentially consists of two individual short rotors, each with a different pitch, connected in series.
    - The gas at the inlet is first transported by the lower pitch portion of the rotors and then by the higher pitch portion, which results in internal compression of the trapped gas.

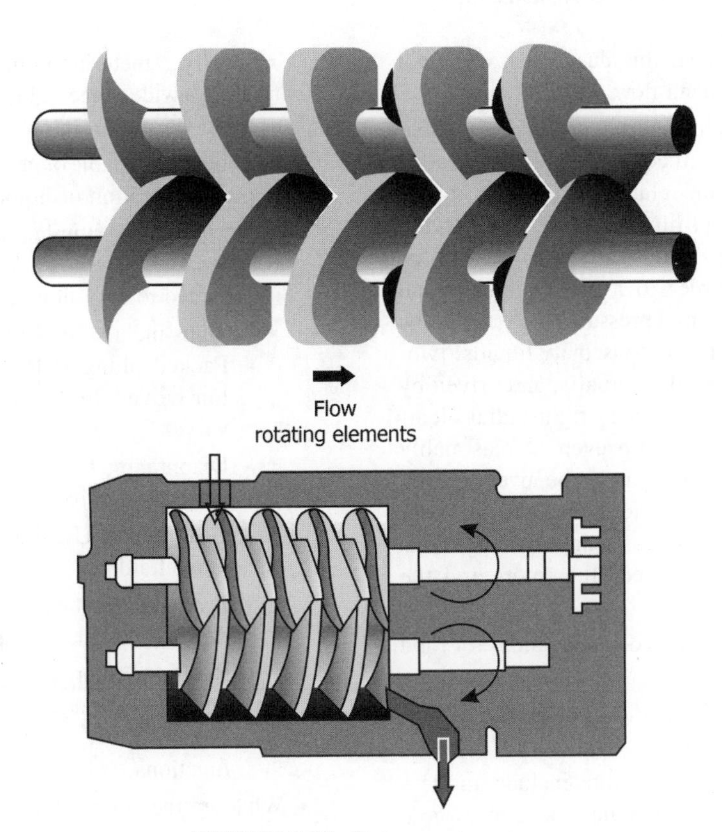

FIGURE 5.26 Rotary screw pump.

- ➢ The pump is unique in that it uses a single stage rather than the multistage design used by other types of pumps. Because of this design, flow path is simple without any volumes in which material can accumulate.

- "A screw pump is a special case of a rotary pump." *True/False?*
  - *True.*
- What are the applications and limitations of a screw pump? Give a diagram.
  - Used for high increases in pressures, especially while handling viscous liquids such as heavy oils. Can handle extremely high viscosities.
  - Not suitable for handling liquids with abrasive solid particles.
  - Because these pumps produce immense heat, temperature-sensitive materials, flammable gases, and severe service applications will cause problems for this design. Discharge temperatures of 350C are possible.
  - Polymerization can occur because of the high operating temperatures.
  - If any wear occurs on the screws, they need to be replaced in matched sets. Other dry pump designs allow for individual part replacement.
  - Figure 5.26 shows the rotating elements and the assembly of a rotary screw pump.
- "Gear and screw pumps are well suited to handle high-viscosity liquids." *True/False?*
  - *True.*
- What is the effect of fluid viscosity on the operation of a positive displacement pump?
  - Power required for the pump increases with increase in fluid viscosity to overcome the internal hydraulic viscous drag.
  - Suction conditions become more demanding with increased viscosity, reflecting the ability of the fluid to get to the pump suction port and fill its displacement mechanism (screws, gears, etc.) in the positive displacement pumps.
- How does viscosity determine/affect selection of a pump? Explain.
  - Centrifugal pumps are not recommended above approximately 30 cP. The viscous friction becomes high and the flow and pressure reduction is dramatic. For example, above 65–130 cP, there is practically no flow through the pump.
  - On the other hand, rotary displacement pumps may not be the first choice at very low viscosities. They are seldom applied below 20–30 cP. (Reciprocating

pumps are the exception to this, due to their very tight clearances. They maintain flow, with low slip).

- There are also some exceptions to this rule. Gear pumps are sometimes made with pressure loaded end plates to minimize lateral clearance to the thickness of the liquid film. However, their radial clearances (between gears and case) cannot be made too tight in order to avoid metal-to-metal contact, unless differential pressures are low. Screw pumps are available for low-viscosity liquids: two-screw designs, supported externally, and driven by the timing gears, can have very tight radial clearances and maintain low slip even at reasonably high pressures. The shafts must be oversized in such cases; otherwise, the long span of the rotor between bearings would cause sagging deflections and potential contact between the rotors and the casing.

- What are the considerations involved in pumps for food applications?
  - Specific standards and specifications govern pumps for the food industry.
  - Allowable materials of construction include stainless steel, certain plastics, and certain grades of carbon. Another specification covers many design issues such as self-cleaning capabilities, absence of internal crevices where bacteria may spread, and seals and seal chamber dimensions.
  - Lobe pumps are often used in these applications. Lobes do not contact (in theory) and are driven by the external timing gears. Absence of contact prevents any contaminant passing through the pump.
  - Shafts are robust, as is the rest of the pump, to ensure low deflections of rotors, resistance to piping loads, and so on.
- What is a proportioning/metering pump?
  - Proportionating/metering pump is a controlled volume, positive displacement pump, which will accurately transfer a predetermined volume of fluid (liquid, gas, or slurry) at a specified amount or rate into a process or system.
  - These pumps are capable of both continuous flow metering and dispensing.
  - Flow rates are predetermined accurately and are repeatable within $\pm 1\%$ (*within the specified range and turndown ratio*).
  - Adjustable volume control is typically inherent in the design.
  - Traditionally, they were reciprocating (back and forth action) designs although rotary configurations (gear and peristaltic) are increasingly used for metering applications.
  - Ideally, a metering pump should be capable of handling a wide range of liquids, including those that are toxic, corrosive, hazardous, volatile, and abrasive.
  - Should be capable of generating sufficient pressure to permit injection of liquids into processes.
  - Reciprocating simplex plunger pumps with variable speed drive or a stroke-adjusting mechanism are provided to vary flow as per requirement.
  - Designs include the following:
    - Packed plunger: Reciprocating plunger or piston moves fluid through inlet and outlet check valves.
    - Diaphragm: Uses a flexible diaphragm to move fluid through inlet and outlet check valves.
    - Gear pumps: Use the spacing between gear sets to move fluid.
    - Peristaltic tubing pumps: Use rollers to move fluid through flexible tubing.
    - Rotating/reciprocating: *Valveless pump* uses only one moving part, a rotating reciprocating piston, both to move fluid and to accomplish valving functions.
- What are the applications of metering pumps?
  - Metering pumps are used in virtually every segment of industry to inject, transfer, dispense, or proportion fluids. The chemical process industries have by far the greatest diversity of applications.
  - In addition to a broad range of industrial applications, metering pumps and dispensers are used extensively in laboratory, analytical instrumentation, and automated medical diagnostic equipment:
    - Industrial applications cover industries such as chemical, petroleum, electronics, water and wastewater treatment, food processing, agriculture, metal finishing, aerospace, automotive and mining.
    - Analytical instrumentation includes chemical analyzers and environmental monitors.
    - Medical applications are in the areas of diagnostic systems, dialysis, disposable component assembly, and so on.
- What are the considerations involved in the selection of metering pumps?
  - Volume: Flow rate or volume per dispense.
  - Pressure: Operating pressure, system pressure, and differential pressure (pressure across the inlet and the discharge ports).
  - Accuracy: Accuracy and precision (repeatability).
  - Temperature: Process temperature, ambient temperature, and nonoperating temperature (component sterilization).

- Plumbing: Connection size, material, and fitting. Suction side tubing size is critical to prevent cavitation, which may occur when the tubing size is too small to allow adequate flow of fluid to enter the pump. Cavitation can cause problems including hammer, flow variation, bubble formation, loss of prime, and eventually pump damage.
- Drive power: AC, DC, pneumatic, stepper, and mechanical.
- Flow control: Fixed, variable, and reversing.
- Control source: Manual, electronic, and mechanical.
- Special requirements: Hazardous, sanitary, sterile, outdoor, and so on.
- In addition to the above, fluid characteristics, chemical compatibility (with pump wetted parts or fluid path), viscosity, corrosive application, specific gravity, suspended solids (percent and particle size), air sensitivity, crystallizing characteristics, shear sensitivity, size, and price are to be given due consideration.

- What precaution is required for pumping shear-sensitive fluids such as yeast?
  - Use lower speeds (100–400 rpm depending upon fluid being pumped).
- Give diagrams of lobe-type rotary pumps and explain their features.
  - Figure 5.27 illustrates lobe-type rotary pumps.
  - Lobe pumps resemble external gear pumps in operation, except that the pumping elements do not make contact. Lobe contact is prevented by external timing gears.
  - Lobe pumps perform well with liquids that contain solid materials, but do not perform well with low-viscosity liquids.
- What are the applications of rotary lobe pumps?
  - *Rotary lobe.* The rotary lobe pump is typically used as a mechanical booster operating in series with an oil-sealed piston or vane pump to boost pumping capacity at low pressures.
  - This pump consists of two symmetrical two-lobe rotors mounted on separate shafts in parallel, which rotate in opposite directions to each other at high speeds.
  - As with a rotary claw pump, timing gears are used to synchronize the rotation of the lobes to provide constant clearance between the two.
  - The number of lobes varies between one and five.
- What are the salient features of a rotary bilobe pump?
  - Rotary bilobe pump consists of two symmetrical two-lobe rotors, each mounted on a separate shaft in parallel, which rotate in opposite direction to each other at high rotational speeds without making any contact or using any seal liquid.
  - It uses timing gears to synchronize the rotation of the lobes to provide constant clearance between the two. The clearances are kept to low levels to minimize back slip of the fluid.
- What are the applications of rotary pumps with bilobe rotors?
  - These are mainly used for high-viscosity products containing solid particulates. Using such rotors keeps the volumetric efficiency high.
  - Optimum performance is obtained by using large slow running pumps.
  - Bilobe rotors give decreased efficiencies on water-like low-viscosity liquids for which trilobed pumps are preferred.

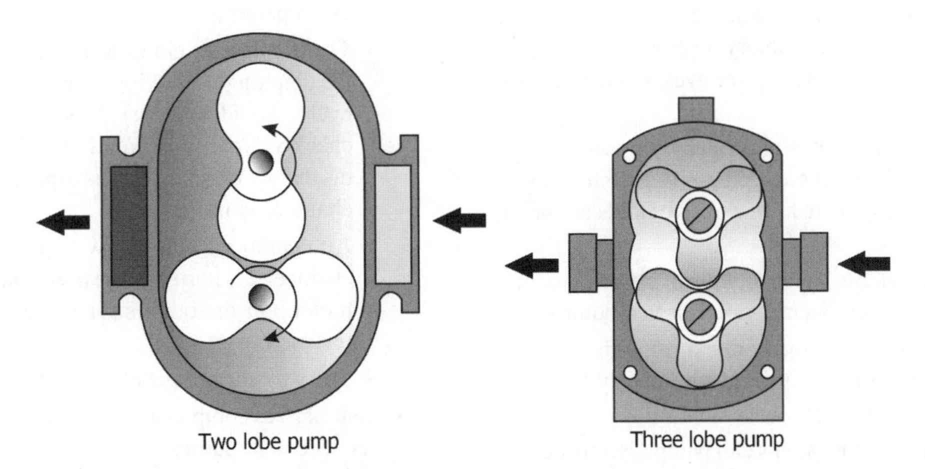

FIGURE 5.27 Rotary lobe pumps.

- These are used for liquids with delicate suspended solids requiring minimum damage. Examples include jams with fruit pieces, sausage meat filling, pet foods, soups, and sausages containing solids.
- These pumps are used as boosters in series with a vacuum pump, discharging gases to the vacuum pump. The booster traps a pocket of gas and transfers it from low pressure to higher pressure, limited to less than about 100 Torr.

- Name the important criteria to be considered for desirable optimum conditions for lobe pumps handling solids in suspension.
  - Desirable properties of solids and operating characteristics:
    - Spherical and soft solid particles possessing resilience and shear strength. Small solid concentrations in the liquid.
    - Smooth solid surfaces.
    - Flow velocities sufficient to keep solids in suspension but not damage them. For example, while pumping yeast, yoghurt, and other food products, cell structure can be damaged if velocities are high.
    - Bilobe pumps are suited for the above applications.
- Does a rotary lobe pump require installation of pressure relief valves or other protective devices on discharge line of a rotary lobe pump?
  - *Yes*. These pumps are positive displacement pumps and pressure may build up if the line on the discharge side is closed/blanked.
- What is the highest viscosity of a liquid that can be handled by a rotary lobe pump?
  - 10^6 cP.
- For what type of applications use of heating/cooling jackets are desirable on pumps?
  - Heating: To reduce viscosity of the liquid being pumped to give satisfactory operation. Examples include products such as adhesives, chocolate, gelatin, resins, and jams.
  - Cooling: Where heat is generated by means of fluid being recirculated through pumps (e.g., throttling of liquid) in cases where heat sensitive fluids are being pumped.
- What are the merits of air-operated piston and diaphragm pumps over electric motor-driven pumps?
  - The pumps can be turned on and off with a control valve. They do not overheat, nor do they need bypassing or power input when shut off.
  - Air-operated piston pumps can pump viscous liquids of more than 1,000,000 cP at pressures of up to about 50 times the supply air pressure.

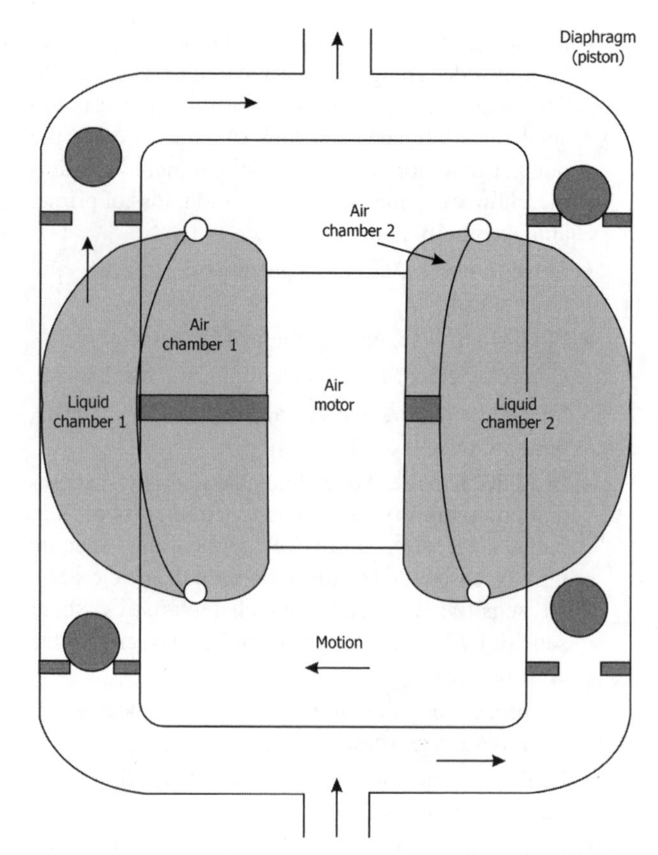

FIGURE 5.28 Air-operated double diaphragm pump.

  - Air-operated diaphragm pumps are normally used with liquids of lower viscosity (up to about 10,000 cP) and pressures equal to the supply air pressure.
  - Liquid velocities are relatively low through the pump and there is less degradation of shear-sensitive liquids when compared to rotary pumps.
- What is an air-operated double-diaphragm pump? How does it operate?
  - The diaphragm pump in Figure 5.28 is shown with the diaphragms moving to the left. On this part of the cycle, liquid chamber 1 is discharging. The inlet check valve in this liquid chamber is closed and its discharge check valve is open, while the second chamber is filling.
  - Air chamber 1 is filling with pressurized air, and air chamber 2 is ported to the atmosphere through the air motor. On the other stroke, the valve positions are reversed.
  - The two chambers help smoothen flow pulsations.
- Summarize comparison of centrifugal and positive displacement pumps.
  - Table 5.5 presents a comparison between centrifugal and positive displacement pumps.

TABLE 5.5 Comparison between Centrifugal and Positive Displacement Pumps

Variable	Centrifugal Pump	Positive Displacement Pump
Principle	The pump imparts velocity to the liquid, which is converted back to pressure at the outlet. Pressure is created and flow results.	The pump takes in specific quantities of liquid and transfers them from the suction to discharge port. Flow is created and pressure results.
Viscosity	Efficiency increases with increasing viscosity reaching maximum at the BEP. At higher or lower pressures, efficiency decreases.	Efficiency increases with increasing viscosity.
Inlet conditions	There should be liquid in the pump to create a pressure differential. A dry pump will not prime on its own.	A dry pump will prime on its own.
Performance	Flow varies with changing pressure.	Flow is constant with changing pressure.

5.1.5 Miscellaneous Pumps

- What are seal-less pumps? What is their importance?
  - Pumps in which mechanical seals, which can give rise to fugitive emissions, are eliminated in their design.
  - Stringent environmental and safety regulations force process industry to adopt methods that reduce/eliminate fugitive emissions in the work place. Most emissions are contributions from pumps used by the industry.
  - The two most common types of seal-less pumps increasingly used by industry are magnetic drive and canned motor types.
  - Both the pumps feature a fully contained pump housing with no break needed in this containment shell to admit a motor drive shaft. This eliminates the need for a shaft seal, and, with it, the potential for seal leakage.
  - They differ in how they get power from the motor to the pump impeller.
  - Magnetic drive pumps have separate, standard motors coupled to the internal pump drive via synchronous magnetic couplings, usually made of rare earth neodymium iron or samarium cobalt (for higher temperatures).
  - The drive in canned motor pumps, on the other hand, is integral within the pump housing, with the stator and rotor of the AC electric motor separated from the pumped liquid by a nonmagnetic membrane.
  - The membrane is in the air gap, that is, the space between the rotor and the stator of the motor.
- What is a canned motor pump?
  - It is a close-coupled unit in which the cavity housing the motor, rotor, and pump casing are interconnected.
  - Motor bearings run in the process liquid and all seals are eliminated. Process liquid acts as lubricant.
  - Used for clean liquids as otherwise due to abrasion from solid particles in the fluid, bearings are affected.
  - These pumps are developed to meet environmental and safety norms for reduction of fugitive emissions.
  - Developments in mechanical seal technology for conventional pumps keep them competitive to canned motor pumps.
- What are the applications of canned motor pumps?
  - Used to eliminate fugitive emissions, meeting environmental regulations.
  - Examples include handling organic solvents, organic heat transfer liquids, light oils, and many clean toxic/hazardous liquids or for plants where leakages are economic problems.
  - Organic heat transfer liquids.
- What is an electromagnetic pump? How does it work?
  - Electromagnetic pump utilizes the principle of operation of an electric motor. A conductor in a magnetic field, carrying a current that flows at right angles to the direction of field, has a force exerted on it, the force being mutually perpendicular to both the field and the current.
  - Fluid is the conductor.
  - Magnetic drive pumps were originally designed to pump toxic and other hazardous fluids without the use of mechanical seals. This is achieved by retaining the fluid inside the pump casing and a containment shell, while the impeller shaft is supported on sleeve bearings lubricated by the fluid. The impeller shaft is driven by a magnetic field passing through the containment shell from the driver shaft.
  - When a metal containment shell is used, adequate removal of the heat generated by eddy currents is vital, particularly when the liquid being pumped is heat sensitive.
  - A major difference between the magnetic drive pumps and the conventional pumps is the location and type of bearings. In conventional pumps, the bearings are usually located well away from the pumped liquid and a wide choice of lubricants can be used.

- With magnetic drive pumps, however, the bearings on the impeller shaft are lubricated by the fluid that may not be an appropriate lubricant. In addition, when the pump runs dry, or operates at very low flows, the lubricant tends to disappear and the bearings will overheat. Any solids in the fluid will be detrimental to the bearing reliability.
- Magnets are also temperature sensitive and will demagnetize if exposed to temperatures exceeding their upper limit. Any upset conditions, such as running the pump dry or against a closed discharge valve.
- To provide some degree of protection against this problem, the material of the magnets should be selected to be able to handle 12–25°C above the expected maximum operating temperature.
- One application for such pumps is use in liquid metal pumping in nuclear power plants.

• What is the disadvantage of magnetic drive pumps?
- Low efficiencies due to internal heat generation.
- Load for refrigeration applications will be high because of heat generation characteristics.

• Give a comparison of magnetic drive pumps and canned motor pumps.
- Table 5.6 presents a comparison of magnetic drive and canned motor pumps.

• What is a vertical pump?
- Pump with a vertical shaft, having a minimum length of 20 m, from drive end to impeller.

• What are the advantages of vertical pumps?
- Liquid level is above the impeller and thus pump is self-priming.
- Shaft seal is above liquid level and so not wetted by pumped liquid with sealing task simplified.
- Used for large capacity applications.
- Vertical submersible turbine pumps are used as pipeline boosters as well as for deep well service.

• What are the disadvantages of vertical pumps?
- Require intermediate or line bearings when shaft length is more than three meters, to avoid shaft resonance problems.
- These bearings must be lubricated.
- Since all wetted parts must be corrosion-resistant, low-cost materials may not be suitable for shaft, column, and so on.
- Costly maintenance since pump is larger and more difficult to handle.

• What is a turbine pump?
- This uses mixed flow-type impellers, giving partly axial and partly centrifugal flows.
- These pumps are usually vertical pumps.
- Turbine pumps are characterized as having bearings lubricated with the pumped liquid.
- These pumps are popular in multistage construction.
- The impellers discharge into a vertical support column housing the rotating shaft.

TABLE 5.6 Comparison of Magnetic Drive and Canned Motor Pumps

Selection Parameter	Magnetic Drive Pump	Canned Motor Pump
Safety on failure	Has only one sealed liner and if it ruptures, fluid escapes into the atmosphere	Two boundaries exist, the can and the motor housing. If the can ruptures, the motor housing takes over as a barrier
Viscosity	Has limitation for use with high fluid viscosities	Fluid is warmed inside the motor section allowing the pumping of higher viscosity liquid
Temperature	Limited to applications between 35 and 400°C	Can be used at temperatures between 130 and 540°C
Maximum operating pressures	Can thickness limits the maximum pressure	Pressures are independent of the can thickness because support can be furnished outside of the gap
$NPSH_R$	Better NPSH characteristics	Poorer NPSH characteristics due to warm up of liquid inside pump
Sensitivity to solids	More sensitive to solids, especially with ferrous solids in the fluid	Available with slurry designs, which isolate the motor section from the pumped fluids
Noise levels	Higher due to fan on the motor and additional bearings in the couplings	Lower. Especially quiet operation
Starting problems	Extreme care must be taken in applying the torque requirements to these units	Easy starting
Cost	Higher	Lower
Cost of repair	Less	More
Foundation	Required	Not required

- These pumps are often installed into deep well water applications.
- The impellers are commonly mixed flow types, where one stage feeds the next stage through a bell shaped vertical diffuser.

- What type of application a turbine pump is suitable?
  - For low flow rates of low-viscosity liquids requiring high delivery pressures than normal centrifugal pumps.
- What type of flow patterns are obtainable for turbine pumps?
  - Mixed flow: Partly radial (centrifugal) and partly axial.
- What type of pump may be appropriate for pumping liquids near saturation, low flow rates and very limited $NPSH_A$?
  - Regenerative turbine pump. The regenerative turbine is specifically developed for these conditions and also high discharge pressures. The regenerative turbine can give a $NPSH_R$ of 15 cm (0.5 ft) with ease. They are particularly suited to saturated boiler feed water.
- What is an airlift pump? How does it operate?
  - In an airlift pump, liquid is raised by means of compressed air, as shown in Figure 5.29.
  - Operates by introducing compressed air into the liquid near the bottom of a well (widely used for pumping water from wells).
  - Air–liquid mixture, being lighter than the liquid, rises in the well casing.

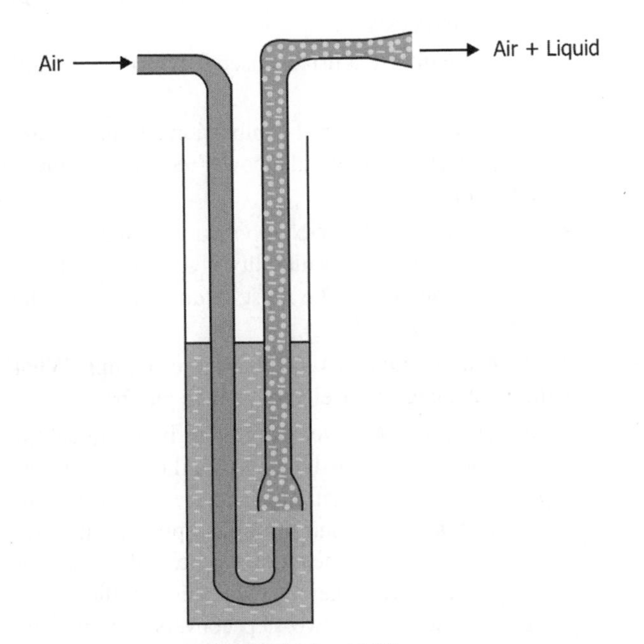

FIGURE 5.29 Airlift pump.

  - No moving parts in the well. Compressor is located on top/surface.
  - Airlift pumps serve two functions. First, they can eliminate fluid stratification in large unagitated tanks. Second, they can lift a liquid above its normal surface elevation without dilution and without introducing shear that damages fragile suspended solids.
  - Gas-lift pumps are used to lift oil from oil wells.
  - Its construction consists of a vertical pipe, open on both ends, that serves as a fluid pipe. A separate, smaller line, extends to the bottom of the fluid pipe and makes a U-turn. The open end extends a short distance into the bottom of the fluid line. When compressed air starts flowing, rising bubbles transport slugs of fluid upward. The process is enhanced by the fact that bubbles also expand as they rise because the hydrostatic head is decreasing.
  - When used for agitating a stratified tank, the fluid pipe should be completely submerged to move material from the lower parts of the vessel to the upper sections. When used to lift fluid above its normal surface, the fluid pipe should be submerged only partially. As the two-phase flow reaches the top of the pipe, it overflows and the liquid disengages from the gaseous phase. The resulting highly nonuniform rate of overflow is collected to provide a more uniform flow for downstream gravity feeding or pumping.
  - Other applications include moving gritty and corrosive fluids that destroy metallic pumps, dewatering mine shafts, removing sediment from vessels and collecting samples from boreholes.
  - Because it is easier to remove contamination from air than it is to treat contaminated water, airlift pumps find a role in environmental remediation. As a result of the mass transfer and intimate, turbulent mixing of air and water, volatile organic materials in the aqueous phase are transferred to the gaseous phase as the material rises in the pipe.
- How is an airlift pump sized?
  - The flow rate through an airlift pump is proportional to the flow rate of the air powering it. Airlift pump flow rates can be in the range of 4.5–455 m^3/h (20–2000 gpm) and lifts more than 200 m.
  - The following empirical correlation gives the flow of air with that of water.

$$V_a = 0.8\,L_l/[C\log_{10}\{(L_s+34)/34\}] \qquad (5.22)$$

where V_a is the volume of free air (ft^3) needed per gallon of water; L_s is the length of the submerged section (ft); L_l is the length of the lift section (ft); and C is the constant that depends on L_l.

- What is an *acid egg*?
  - Egg-shaped container filled with a liquid that is pumped out by the use of compressed air, as in the case of airlift pump.
  - Used for pumping corrosive liquids such as acids.
  - Can be hand operated or arranged for semiautomatic or automatic operation.
- What is a regenerative pump?
  - Regenerative pumps are essentially turbine pumps that employ a combination of mechanical impulse and centrifugal force to produce high heads at low flows.
  - Impellers used are high-speed type with close clearances.
  - Have many short radial passages milled on each side at the periphery.
  - Similar channels are milled in the mating surfaces of the casing.
  - Liquid is directed into the impeller passages and proceeds in a spiral pattern around the periphery, passing alternatively from impeller to casing and receiving successive impulses as it does so.
  - Used for clean low-viscosity liquids (due to close clearances).
- What is a peristaltic pump? What are its characteristics and applications?
  - Peristaltic pump is positive displacement pump with the fluid not coming in contact with pump parts. It involves movement of the fluid through a flexible tube, which is alternately squeezed and released synonymous to suction and discharge of positive displacement pumps. Figure 5.30 illustrates working of a peristaltic pump.
  - Flow can be varied, providing flexibility to handle a variety of containers, fluid viscosities, and process capacities.

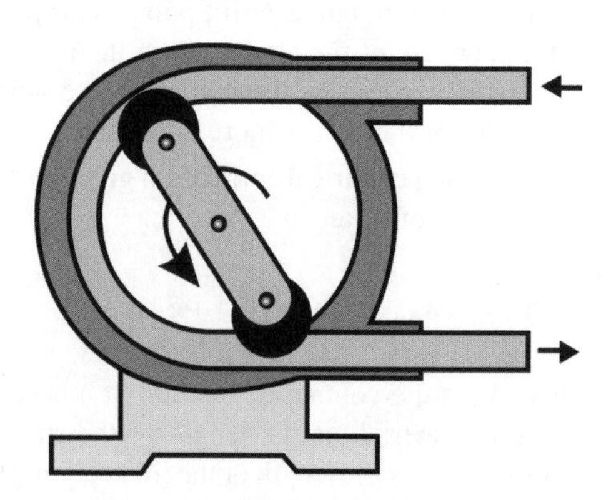

FIGURE 5.30 Peristaltic pump.

  - They meet aseptic production requirements and require a minimum of maintenance-related downtime. They provide for sterile, precision filling of liquid volumes with accuracies of 0.5% or better, drive with variable speeds of 1–650 rpm and dispense volumes from 10 μL to 10,000 L.
  - Well suited to handle a wide range of fluid viscosities and also particularly gentle on shear-sensitive products such as cell cultures and bioreactor constituents.
  - Also find applications in handling abrasive and corrosive fluids.
  - Other applications include dialysis machines, open heart bypass pump machines, and sewage sludge handling.
  - Fluid changeover is accomplished with a simple tubing change. No other parts of the pump come in contact of the fluid.
  - Suited for low-pressure operation (up to about 2 bar).
- What are the desirable performance parameters for the tubing of peristaltic pumps?
  - Chemical compatibility: The tubing material must be compatible with the pumped fluid to provide good pumping performance, as well as safety. Many tubing suppliers provide chemical compatibility charts. While using such compatibility charts, care should be exercised as tubing that gets an acceptable rating for general contact with a given chemical might be influenced by factors such as physical stresses of pumping, concentration of solutions, contaminants in the fluids, fluid permeability through tube walls, regulatory requirements, temperature and pressure effects, and costs.
- What are the different drivers used for the operation of pumps?
  - Electric motors, internal combustion engines, steam and gas turbines, hydraulic power recovery turbines, and so on.
  - Driver selection depends upon many factors including type of service, availability of steam or fuel (for turbines), and cost. The most common driver is the electric motor.
- What are the applications of steam-driven pumps? What is their advantage over electric-driven pumps?
  - Large size steam-driven pumps are in use in chemical process industry due to (i) availability of high-pressure steam internally generated for different power generation and process applications and (ii) direct use of steam eliminates the step of steam-to-electricity thereby giving higher efficiencies than electric driven (two-step conversions, namely, steam to power generation and power to motor).

- Such use is justified if large pump sizes are involved for providing steam lines and condensate handling facilities that are expensive than power supply.
- Steam turbines are used on larger vital spare pumps in some plants during power failure and consume excess steam generated in the process or to provide low-pressure exhaust steam, from the turbines, required for heating in the process.

• What are the different types of variable speed drives used for pumps?

- Two-speed type: For simple pumping not requiring high accuracy.
- Direct current type: These are among the first variable speed drives using wound motors.
- Variable voltage type.
- Variable frequency type: Most reliable and efficient.
- Wound rotor regenerative type: Gives one of the best efficiencies.

• Give a summary checklist for troubleshooting a pump.

- Has a temporary suction strainer been provided for pump start-up?
- Has adequate valving been provided for pump maintenance or removal? Is there a check valve to prevent excess reverse flow?
- Does the pump have a valved drain connection for maintenance work?
- Is there provision for venting noncondensables?
- Has all auxiliary instrumentation been provided as specified?
- Is there any discharge pressure gauge for monitoring pump performance?
- Is there a relief valve to prevent overpressurization of a positive displacement pump?
- Have proper measures provided to tackle weather problems?

• What are the essential variables involved in pump specifications?

- Based on the application in which a pump will be used, the pump type, service, and operating conditions, the specifications of a pump can be determined.
- Casing connection: Volute casing efficiently converts velocity energy impacted to the liquid from the impeller into pressure energy. A casing with guide vanes reduce loses and improve efficiency over a wide range of capacities and are best for multistage high head pumps.
- Impeller details: Closed-type impellers are most efficient. Open-type impellers are best for viscous liquids, liquids containing solid matter, and general purposes.
- Seals: Rotating shafts must have proper sealing methods to prevent leakage without affecting process efficiency negatively.
  - Seals can be grouped into noncontacting seals and mechanical face seals. Noncontacting seals are often used for gas service in high-speed rotating equipment. Mechanical face seals provide excellent sealing for high leakage protection.
- Bearings: Factors to take into consideration while choosing a bearing type include shaft speed range, maximum tolerable shaft misalignment, critical speed analysis, and loading of impellers.
  - Bearing styles include cylindrical bore, cylindrical bore with dammed groove, lemon bore, three lobe, offset halves, tilting pad, plain washer, and taper land.
- Materials: Pump material is often stainless steel. Material should be chosen to reduce costs and maintain personnel safety while avoiding materials that will react with the process liquid to create corrosion, erosion, or liquid contamination.

5.1.6 Vacuum Pumps

• "Subatmospheric pressures can be divided into four regions." Name them, giving pressure ranges involved.

- Rough vacuum 760–1 Torr.
- Medium vacuum 1–10^{-3} Torr.
- High vacuum from 10^{-3} to 10^{-7} Torr.
- Ultrahigh vacuum 10^{-7} Torr and below (*Note*: 1 Torr = 1 mmHg; 1 μm = 0.001 Torr.)

• Give examples of applications of rough, medium, high, and ultrahigh vacuum in chemical process industry.

- Rough vacuum: Polymer reactors, vacuum distillation columns, vacuum dryers, and so on.
- Medium vacuum: Degassing of molten metals, molecular distillation, freeze drying, and so on.
- High and ultrahigh vacuum: Production of thin films, mass spectrometry, low-temperature research, surface physics research, nuclear research, and space simulation.
- Rough to ultrahigh vacuum: Semiconductor applications.

• What are the ranges of vacuum obtainable with different types of vacuum producing equipment?

- Table 5.7 lists ranges of pressures obtainable with different vacuum producing equipment.

• Which of the following piston vacuum pumps are capable of producing higher vacuum than the other? Give the ranges of vacuum obtainable for each type?

TABLE 5.7 Pressures Obtainable with Different Vacuum Producing Equipment

Type of Equipment	Pressure (Torr)
Reciprocating piston type	Up to 1
Rotary piston type	Up to 0.001
Two-lobe rotary	Up to 0.0001
Steam jet ejectors	
Single stage	100
Two stage	10
Three stage	1
Five stage	0.05

- Reciprocating piston vacuum pumps are generally capable of vacuum to 1 Torr abs, rotary piston types can achieve vacuums of 0.001 Torr.

• "A two-lobe rotary vacuum pump is capable of producing much higher vacuum compared to a rotary piston vacuum pump." *True/False?*

- *True.* Pressures attainable for a two-lobe rotary pumps are down to 0.0001 Torr while rotary piston pumps can produce pressures down to 0.001 Torr.

• What is a liquid ring vacuum pump? Give a diagram.

- A liquid ring vacuum pump consists of a multibladed impeller with straight radial vanes, eccentrically positioned within a cylindrical casing, in such a way that at top dead center the clearance between the casing and the impeller blade tip is at a minimum and at bottom dead center the clearance is at a maximum. Figure 5.31 illustrates a liquid ring vacuum pump.

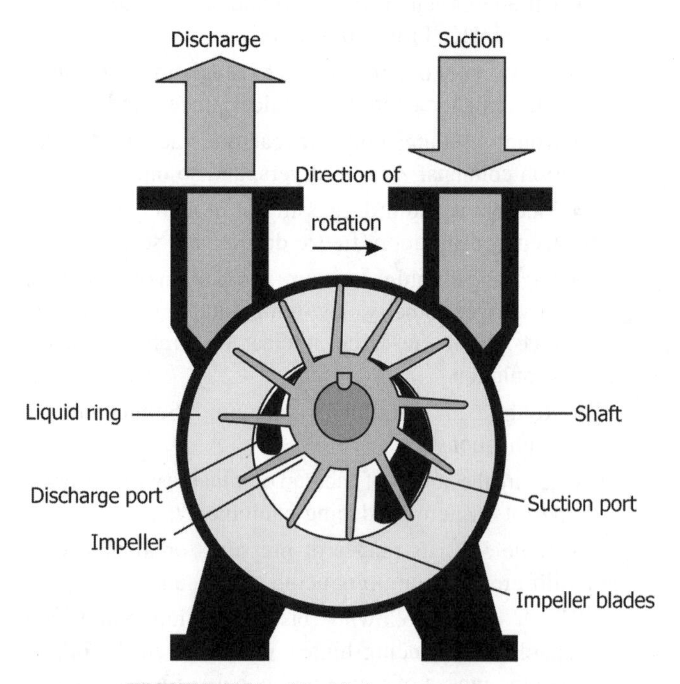

FIGURE 5.31 Liquid ring vacuum pump.

- The compression of the gas occurs in one or two impeller stages depending on the ultimate vacuum required although for this service two-stage pumps are generally required.
- Water or some other incompressible fluid, called sealant liquid, is used as the sealant liquid. The sealant liquid, used for both sealing and cooling, can be any liquid that is compatible with the process having the following ranges of physical properties:
  - Specific gravity: 0.5–1.5.
  - Specific heat: 0.35–1.0, relative to that for water.
  - Viscosity: 1 to 32 cP.
  - Vapor pressure at operating temperature: Less than or equal to vapor pressure of water at 15°C.
  - Low-viscosity oils, glycols, and many solvents such as toluene, methanol, ethanol, propanol, butanol, and ethylbenzene can be used.
- As the impeller rotates, a ring of liquid is formed inside the pump casing developing centrifugal force that throws the liquid against the inside walls of the casing, where it forms a rotating ring of liquid. This action draws the gaseous stream into the pump through the inlet port.
- The gas is compressed by the liquid ring, exiting the first stage through a smaller area discharge port and into the second stage of the pump. The second stage is volumetrically smaller doing the final compression of the gas. The gas then exits the pump usually at atmospheric pressure, along with the service liquid.
- Single-stage pumps are capable of achieving 100 mmHg abs, while two-stage pumps can achieve 30 mmHg abs. Pumping capacities are up to and more than 500 m^3/min.

• What are the advantages of a liquid ring pump?

- Simpler design and operation than most other vacuum pumps.
- It does not contain many parts. Employs only one rotating assembly.
- Self-priming when half filled with liquid.
- Suitable for aerated liquids.
- Can use any type of liquid for the sealant fluid in situations where mingling with the process vapor is permissible.
- No lubricating liquid in the vacuum chamber to be contaminated.
- Small tolerances between the impeller and the pump casing.
- Can handle condensable loads.

- Ability to handle small liquid streams along with the gas from the process or precondensers.
- No damage from liquid or small particulates in the process fluid.
- High efficiency.
- Inherently slow rotational speed (1800 rpm or less), which maximizes operating life.
- Very small increase in gas outlet temperature. This is an advantage while handling flammable or temperature-sensitive process fluids.
- Minimal noise, vibration, and maintenance. Rebuilding is simple.
- Can be fabricated from any castable metal.

• What are the disadvantages of liquid ring vacuum pumps?
  - Mixing of the evacuated gas with the sealant liquid.
  - Risk of cavitation, which requires that portion of the process load be noncondensable under the pump operating conditions.
  - High power requirement to form and maintain the liquid ring, resulting in larger motors than for other types of pumps.
  - Achievable vacuum is limited by the vapor pressure of sealant liquid at the operating temperature.
  - All liquid ring pumps must cope with cavitation when operating at low inlet pressures. Sealant liquid, sealant temperature, impeller rpm, blade angle, and inlet pressure influence cavitation.

• What are the typical applications of liquid ring pumps?
  - Used for vacuum distillation, vacuum drying, and evaporator service because the pump handles noncondensables saturated with process vapors.

• What is a dry vacuum pump? Name some dry vacuum pumps used in industry.
  - It is a positive displacement vacuum pump that discharges continuously to atmospheric pressure and in which the swept volume is free of lubricants or sealing liquids.
  - Rotary claw, rotary lobe, and rotary screw pumps are examples of dry pumps in the CPI, particularly in larger size pump applications.

• What is the operating principle of dry vacuum pumps developed from rotary lobe Roots blower?
  - Rotary lobe Roots blower is a positive displacement machine that normally operates as a dry compressor.
  - Two interlocking rotors on two parallel shafts synchronized by timing gears and rotating in opposite directions trap and transport gases.
  - Gears and bearings are oil lubricated, but are external to the pump; the rotors run dry.
  - Clearances between the rotors and the casing are generally 0.1–0.5 mm (0.004–0.02 in.).
  - Back leakages across these clearances, which reduce pump capacity, increase with increase in ΔP between intake and exhaust.
  - Therefore, these machines are limited for use across relatively small ΔP.

• What are the plus points of using dry vacuum pumps?
  - Dry pumps are compact and energy efficient and do not contribute to air pollution, a problem with oil-sealed pumps, or water pollution, a problem with steam jets and water-sealed liquid ring pumps.
  - Dry pumps are unique among vacuum pumps, because they do not require a working fluid to produce vacuum, so nothing contacts the load being pumped.
  - Solvents or products aspirated from the process can be discharged to an aftercondenser. Contamination is not a concern and the condensate can be recycled directly to the process.
  - For these reasons, dry vacuum pumps have found applications in semiconductor industry. The dry pumps for the semiconductor industry are medium vacuum pumps.
  - Dry vacuum pumps are generally rotary lobe Roots blower type. Roots blowers have limited application as process vacuum pumps discharging against high pressure differentials to the atmosphere, but they are used extensively as vacuum boosters in the 0.001–50 Torr range.
  - The potential for eliminating process contamination is the main reason for specifying dry vacuum pumps for fine chemicals and good manufacturing practices plants.

• What is a rotary claw vacuum pump? Illustrate.
  - Rotary claw: The geometric shape of this pump allows for a greater compression ratio across the rotors at higher pressures.
  - Two claw rotors rotate in opposite directions of rotation without touching, using timing gears to synchronize the rotation.
  - The gas enters through an inlet port opened to the gas during its rotation and fills the pump housing.
  - On the next rotation, the same trapped gas is compressed and discharged as the discharge port opens.
  - Figure 5.32 illustrates rotary claw vacuum pump.
  - A minimum of three stages in series is required to achieve pressures comparable to those of an oil-sealed mechanical pump.
  - Some dry designs use two technologies in combination, for example, a rotary lobe as a booster for a claw pump.

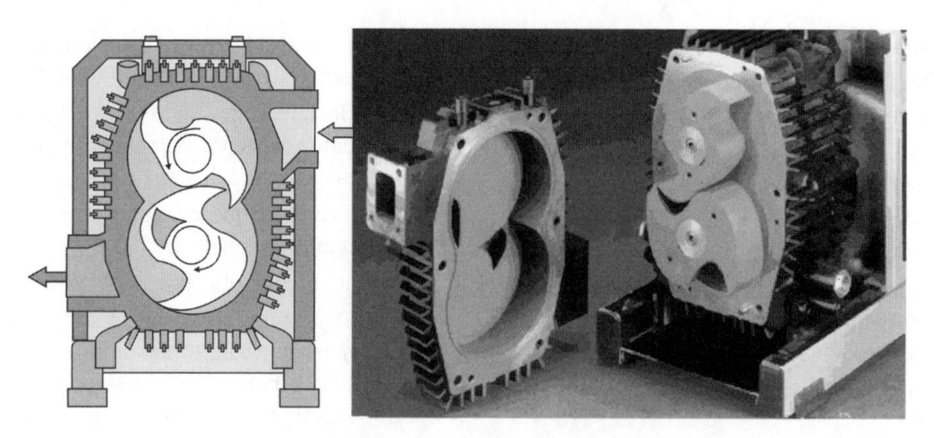

FIGURE 5.32 Rotary claw pump.

- Describe the working of rotary screw pumps for vacuum service?
  - Two long helical rotors in parallel rotate in opposite directions without touching, synchronized by helical timing gears.
  - Gas flow moves axially along the screw without any internal compression from suction to discharge.
  - Pockets of gas are trapped within the convolutions of the rotors and the casing and transported to the discharge.
  - Compression occurs at the discharge port, where the trapped gas must be discharged against atmospheric pressure.
  - Each convolution of the rotor acts similarly to a stage in series with the one behind it.
  - At least three convoluted gas pockets in the rotor are required to achieve acceptable vacuum levels.
  - The rotary screw type vacuum pump can give extremely high vacuum levels in the range of 10–100 μm (low absolute pressures).
  - These pumps move gas along the length of the screws. Gas compression and mechanical compression does not occur until the final half-turn of the screws.
- What are the advantages and the disadvantages of dry vacuum pumps?

  Advantages
  - Rugged rotor design, constructed of sturdy cast or ductile iron without any flimsy rotating components. Due to lack of condensation, pump can be fabricated of standard, inexpensive cast iron.
  - Dry pumps are compact and energy efficient, and do not contribute to air pollution, a problem with oil-sealed pumps, or water pollution, a problem with steam jet ejectors and water-sealed liquid ring pumps.
  - These pumps do not require a working fluid to produce vacuum, so nothing contacts the load being pumped.
  - Noncontact design facilitated by timing gears.
  - High rotational speed reduces the ratio of gas slip to displacement, increases net pumping capacity and reduces ultimate pressure.
  - No contamination of evacuated gas.
  - Can discharge to the atmosphere.
  - Multiple staging provides inlet pressures below 1 mmHg abs while discharging to atmosphere.
  - May discharge gases at high temperatures; in some situations, as high as 300°C or more. Liquid in ring pumps acts as a cooling and flushing medium for cleaning the pump internals of process material.

  Disadvantages
  - Dry pumps have difficulty in handling slugs of liquids. Knockout pots are generally required. A liquid ring pump is the only type of vacuum pump that can safely handle slugs of liquids.
  - Cannot handle particulates. The tight clearances make some of the dry pumps sensitive to particulate buildup.
  - Inlet filters and knockout pots, along with discharge silencers are normally required for a complete operating system.
  - Eliminating a liquid within the pumping chamber also eliminates a method of sealing between the pump clearances.
  - Due to the high operating temperature, some process gases may have a tendency to polymerize.
  - Internal clearances are to be kept tight, as tight as 0.1 mm to reduce gas slippage, allowing for thermal expansion of the rotors.
  - May require an inert gas purge for cooling or to protect the bearings and seals from the process gas.

The inert gas acts as a diluent for flammable or corrosive gases, besides helping to control internal temperatures.

- Dry pumps are sensitive to process upsets and should be carefully selected to ensure maximum run time.
- In most models, difficult to repair or rebuild.
- Expensive than either ejectors or liquid ring pumps.

• What is a scroll pump?

- Scroll pump uses a rotating plate shaped into a spiral (involute curve), which moves within a second stationary plate, shaped as a similar spiral. Gas moves within the spiral passage, from the outside of the spiral to the center where it is exhausted through a valve.
- Multiple stages can be used to provide pressures down to 0.01 mmHg abs, with pumping capacities limited to less than 1.4 m^3/min (50 ft^3/min).
- This type of pump is limited for applications limited to clean gases because of the tortuous flow path.

• What are the applications of dry vacuum pumps?

- Used when contamination of the process fluid is to be avoided, when solvent recovery is the main objective or when emissions must be kept low.
- In cases where there is risk of liquid carryover from the evacuated vessel, use of knockout pots, filters, or other similar devices are required to avoid pump damage.
- Used in semiconductor industry for etching of wafers and chemical vapor deposition requiring pressures of 0.01–1 mmHg abs.. Oil-sealed pumps require lubrication with inert, fluorine-based fluids for protection from corrosive gases (e.g., HCl) and harsh conditions (condensation/precipitation of hard solids such as $AlCl_3$ and SiO_2) of fabrication. These involve high costs for liquid ring pumps for such applications.

• What are the most important parameters that affect vacuum pump selection?

- Suction pressure.
- Capacity required for the process.

5.2 EJECTORS

• What are the incentives for the use of ejectors?

- They develop any reasonable vacuum needed for industrial operations.
- All sizes are available to match any small- or large-capacity requirements.
- Largest throughput capacities, more than 30,000 m^3/min, handled in relatively small equipment sizes.
- Can handle wet, dry, or corrosive vapor mixtures.
- Can handle condensable loads.
- Suitable for handling practically any type of gas or vapor or for handling wet or dry mixtures or gases containing sticky or solid matter.
- Tolerance for entrained liquids, even in slug form and intermittency.
- Their efficiencies are reasonable to good and sustainable.
- No special start-up or shutdown procedures required.
- Simple and easy operation. Quiet and stable operation within design range.
- Higher reliability in severe services.
- Lowest capital cost among vacuum producing devices.
- Installation costs are relatively low when compared to mechanical vacuum pumps.
- They have no moving parts, hence, maintenance is low and operation is fairly constant when corrosion is not a factor.
- Long life.
- Corrosion damage is easily repaired at low cost. Ejectors are available in many materials of construction to suit process requirements.
- No stuffing box sealing is required.
- Space requirements are small.
- Can be mounted in any orientation.

• What are the disadvantages of ejectors?

- Very low efficiency (1–20%).
- Efficiency is high in its application for pumping boiler feed water by steam in an injector, where heat of steam used for pumping is added to steam space in the boiler, thus recovering the heat.
- Head developed is low.
- Dilutes fluid pumped through mixing.
- Requires steam at medium pressures (2–4 barg) as the motivating fluid.
- The effluents (condensates) from the ejector are contaminated with process fluids and require to be treated to meet pollution control requirements.
- Can be noisy. May require discharge silencers or sound insulation.

• What are the applications of ejectors?

- Steam jet ejectors are used on all kinds of distillation units, vacuum deaerators, evaporators, crystallizers, oil deodorizers, steam vacuum refrigeration, flash coolers, condensers, vacuum pan dryers, dehydrators, vacuum impregnators, freeze dryers, vacuum filters, and stream degassing and vacuum melting of metals.

- Steam vacuum refrigeration systems produce chilled water in the range of 415°C for cooling process heat exchangers, air conditioning and many other applications where chilled water is required. Flash cooling and concentration of liquid solutions are other applications for the steam vacuum refrigeration system.
- Steam jet ejectors can be designed to pump liquids and even finely divided solids for pneumatic conveying systems. In the latter service, they are used to transfer fluidized catalyst.
- Liquid jet ejectors can be used for applications involving agitation, mixing or metering liquid solutions. Water operated ejectors are used for fume scrubbing in which large volumes of gases and vapors can be entrained with good water economy. In these applications, obnoxious, corrosive, or poisonous fumes can be sucked out of an enclosed area by means of a simple water jet ejector.
- Ejectors have found many applications for compressing fluids to pressures above atmospheric.
- Steam jet thermocompressors are a very efficient means of boosting low-pressure waste steam to a pressure where it can serve a useful purpose. High-pressure motive steam that might otherwise be throttled for use in an evaporator or heat exchanger can be used to actuate a steam jet thermocompressor. The thermocompressor can serves two purposes, namely, throttling the high-pressure steam to the desired pressure and compressing low-pressure waste steam to a pressure where it is again useful.
- Steam jet air compressors are useful for supplying compressed air in an explosion hazardous area where electrical equipment would have to be of explosion proof construction and relatively expensive.
- Compressed air at 140 kPag (20 psig) for pneumatic controls is a typical application for steam jet air compressors.
- Handling slurries and granular solids.
- Heating liquids by direct contact.

• What is the principle of operation of an ejector? What is its general name?
  - An ejector uses momentum of one fluid to move another.
  - Its general name is jet pump.
  - The operating principle involves conversion of pressure energy of the motive fluid, generally steam, into kinetic energy (velocity). Pressure of the steam used as motive fluid is in the range of 7–14 barg.
  - The motive steam expands adiabatically through the steam nozzle into the suction chamber. This adiabatic expansion occurs across a converging–diverging nozzle (Figure 5.32).
  - This results in supersonic velocity off the motive fluid nozzle, typically in the range of Mach 3–4.
  - This creates a low-pressure zone for pulling the gases from the process equipment into the ejector. High-velocity motive fluid entrains and mixes with the suction gas.
  - The mixture passes through the diffuser (venturi), in the converging section of which kinetic energy (velocity) is converted to pressure as cross-sectional flow area is reduced. During the mixing, much of the kinetic energy of the motive steam is converted into heat energy. At the throat section, a shock wave is established, boosting pressure energy at the expense of kinetic energy across the shock wave.
  - Flow across the shock wave goes from supersonic ahead of the shock wave, to sonic at the shock wave and subsonic after the shock wave.
  - In the diverging section, velocity is further reduced and converted into pressure.
  - The recompression into pressure is achieved with an efficiency of about 80%.
  - Motive pressure, temperature, and quality of the motive steam are critical variables for proper ejector performance.
  - Calculation of a required motive nozzle throat diameter is based on the necessary amount of motive steam, its pressure, and specific volume.
  - Motive steam quality is important because moisture droplets affect the amount of steam passing through the nozzle. High-velocity liquid droplets also erode ejector internals, reducing performance.

• What are the components that may be present in the vapor load to an ejector?
  - Air, noncondensable gases, condensable vapors, and steam from distillation overheads and water vapor.
  - Air leakage, which is common to subatmospheric systems, is usually the basic component of the load to an ejector and the quantities of water vapor and/or condensable vapor are normally directly proportional to the air load.
  - Entrainment.

• Give approximate estimates for leakage of air and noncondensable and condensable vapors into ejectors from process systems, to be used in ejector selection and design.
  - For very tight, small process systems, an air leakage of 1–2.5 kg/h may be taken.
  - For moderately tight, small process systems (about 15 m^3 volume), air leakage may be of the order of 4–5 kg/h may be a reasonable estimate.

- For large systems, 8–10 kg/h may be a reasonable figure for air leakage.
- Solubility of dissolved gases plus any gases produced in chemical reactions in the feed streams and the evacuated equipment can be a rough guide in estimation of noncondensables.
- Condensables may be estimated from a knowledge of saturation conditions of such vapors in the noncondensables.

• What are the operating conditions that must be supplied to the vendor while ordering an ejector system?

- Flows of all components to be purged from the system (often air plus water vapor).
- Temperature and pressure entering the ejectors and pressure leaving if not atmospheric.
- Temperature and pressure of steam available to drive the ejectors.
- Temperature and quantity of cooling water available for the intercondensers and cooling water allowable pressure drop for the intercondensers.
- The vendor will convert the component flow data into an *air equivalent*. Because ejectors are rated on air handling ability, the vendor can then build up a system from his standard hardware.
- The vendor should provide air equivalent capability data with the equipment he supplies.
- Air equivalent can be determined with the following equation:

$$e_R = F\sqrt{(0.0345 \times M_W)} \qquad (5.23)$$

where e_R is the entrainment ratio (or air equivalent). It is the ratio of the weight of gas handled to the weight of air that would be handled by the same ejector operating under the same conditions; M_W is the gas molecular weight.

$$\begin{aligned} &F \text{ is } 1.00, \text{ for } M_W\ 1-30; \\ &F = 1.076-0.0026, \text{ for } M_W\ 31-140. \end{aligned} \qquad (5.24)$$

• Give the Heat Exchanger Institute Standard equation for steam jet ejectors.

$$W = 892.4\, C_d D_n^2 (P_s/V_g)^{0.5} \qquad (5.25)$$

where W is the motive steam (lb/h); C_d is the nozzle discharge coefficient; D_n is the nozzle throat diameter (in.); P_s is the motive steam pressure at ejector, psia; V_g is the motive steam specific volume (ft^3/lb).

• What is ejector compression ratio?

- The ratio of discharge pressure to suction pressure is the ejector compression ratio. The values normally vary from 3 to 15.
- The individual compression ratio of an ejector is a function of cooling water temperature, steam use and condensation profile of vapors handled.
- The first-stage ejector, tied directly to column discharge, will have a compression ratio set primarily by intercondenser cooling water temperature. Intercondensers are positioned between ejectors.

• What is the recommended pressure for ejection of noncondensables into the atmosphere from the last stage of the ejector system?

- About 800 mmHg abs.

• Classify ejectors stating their relative characteristics and usefulness.

- Ejectors can be classified as *single stage* or *multistage*. Multistage ejectors may be further divided into condensing or noncondensing types.
- The single-stage ejector, the simplest and the most common type, is generally recommended for pressures ranging from atmospheric pressure (101.3 kPa abs) to 13.3 kPa (100 Torr) abs. Discharge is typically at or near atmospheric pressure.
- A three-stage ejector requires 100 kg steam/kg air to maintain a pressure of 133.3 Pa (1 Torr). A five-stage ejector produces a pressure of 6.7 Pa (0.05 Torr).
- Multistage noncondensing ejectors are usually of two-stage type although six-stage units have been used successfully.
- Multistage noncondensing ejectors are used to produce suction pressures lower than 10–15 kPa (3 or 4 in.Hg) abs.
- Steam consumption in a multistage noncondensing ejector is relatively high. Each successive stage is required to handle the load plus the motive steam from the previous stage.
- Multistage noncondensing ejectors are frequently used when low first cost is more important than long-range economy. They are also used for intermittent service or when condensing water is not available.
- Multistage condensing ejectors are available in two through six stages. Intercondensers (surface or direct contact) between stages condense steam from the preceding stage, reducing the load to be compressed in the succeeding stage.
- Four, five, and six-stage ejectors are used to achieve suction pressures as low as 5 μmHg abs. Under such vacuum conditions, pressure between the preliminary stages is too low to permit condensation of ejector steam, and only the final two stages are fitted with condensers.
- Multistage condensing ejectors remove condensable vapor ahead of a given ejector stage. They also permit

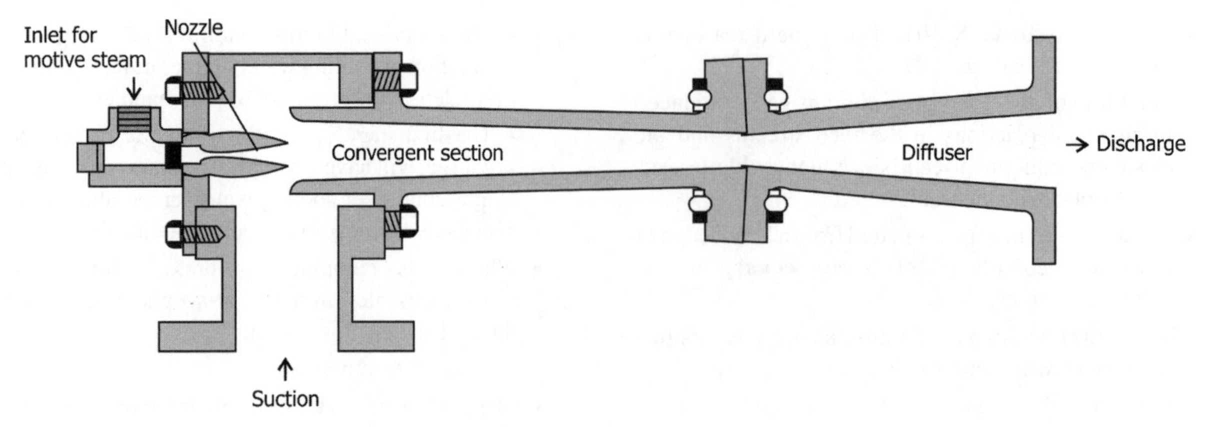

FIGURE 5.33 Steam jet ejector.

use of a smaller ejector, and a reduction in the amount of steam required.

- *Precondensers* are used when the absolute pressure of the process is sufficiently high to allow condensation at the temperature of the available water supply. Noncondensables are removed from the precondenser by one or more ejector stages.

• Give diagrams showing parts of a steam jet ejector and arrangement of a Two-stage ejector.

- Figure 5.33 illustrates a single-stage steam jet ejector and Figure 5.34 illustrates a two-stage ejector.

• Illustrate, with suitable diagrams, condensate drain leg layouts from vacuum condensers. Discuss the considerations involved.

- Figure 5.35 gives typical condensate drain leg layouts.
- Length of drain leg should be at least 10.33 m water and 13.7 m for hydrocarbons.
- Accumulation of trapped bubbles is a common hazard in barometric or shell and tube condenser tailpipes.
- Condensate from a shell and tube condenser, or cooling water plus condensed steam or hydrocarbons from a direct contact barometric condenser, always contain air or other noncondensable gases.
- A horizontal or slightly downward sloped line is vulnerable for clinging of these gases to upper pipe surfaces.

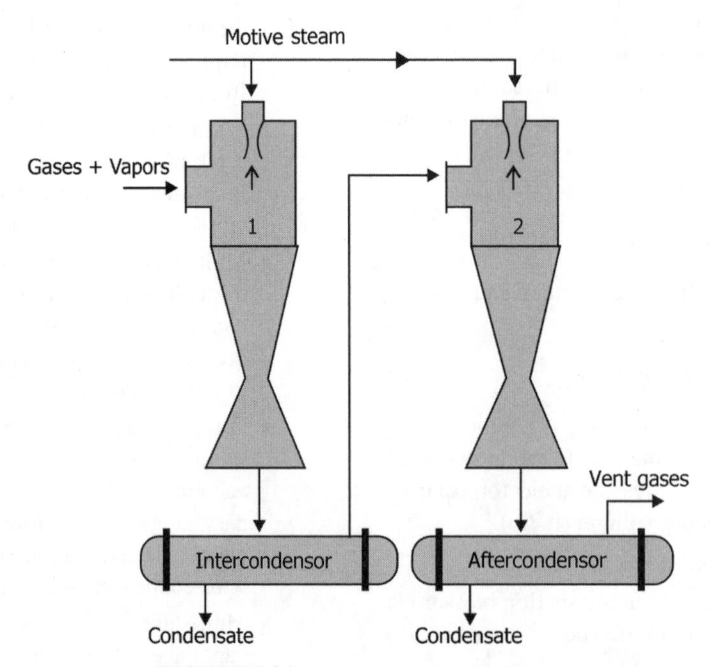

FIGURE 5.34 Two-stage ejector system.

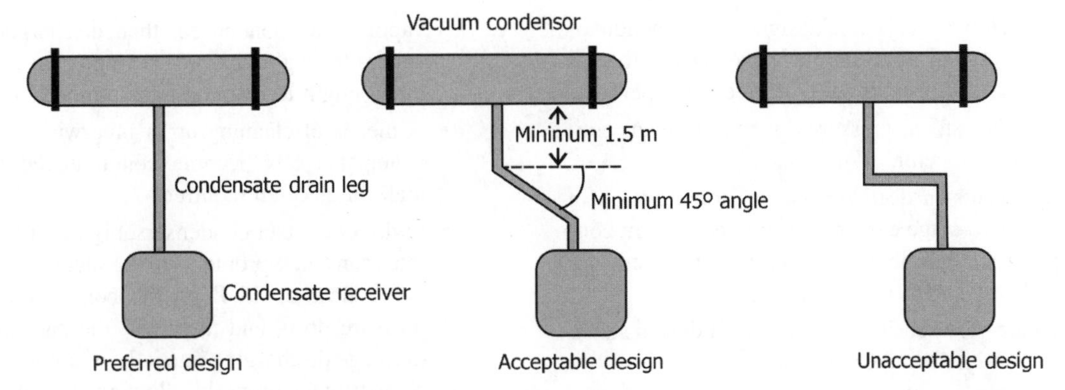

FIGURE 5.35 Condensate drain leg layouts.

- Pipe roughness can cause gases tending to building up in the crevices.
- Flanged joints are other places for gases to collect.
- As these gases accumulate, they form tiny bubbles, growing into larger ones that eventually become large enough to partially or completely block off piping at that point. The condensate cannot flow downward and its level rises, flooding the condenser.
- If piping changes direction, it must form at least a 45° angle from the horizontal (Figure 5.34). With this amount of sloping, gases will either rise back up the pipe or continue downward with the thrust of the flowing water.
- When a change in direction is required, there must always be a vertical straight distance of five pipe diameters or 1 m minimum between each change. This allows flowing liquid to develop a minimum velocity head and a straight downward pattern before the first change in direction. There should not be any valves in the tailpipes.

• What is a booster ejector?

- Ejector used for high-volume, low-pressure stages of a multistage system, up to the pressure level where condensers are effective with available cooling water temperatures.

• What are the factors that are to be considered in the selection of an ejector?

- The selection of an ejector for a given application depends upon available steam pressure, maximum water temperature, highest vacuum, temperature, and capacity required.

• What are the factors that influence performance of ejectors?

- The performance of an ejector is a function of the area of the motive fluid nozzle and venturi throat, pressure of the motive fluid, suction and discharge pressures, ratios of specific heats, molecular weights, and temperatures.
- Wet steam causes decrease in efficiency and erosion (due to water droplets moving fast with steam) of steam inlet nozzle, increasing steam flow area and hence steam rate, if erosion is uniform over the inner cross section. If design steam rate is exceeded, ejector performance deteriorates. Periodic replacement of the nozzle is a solution for this problem, apart from installing a steam filter on steam inlet line. Some times, 2–8°C superheat in steam is recommended.

• How does motive steam influence operation of an ejector?

- Motive steam plays probably the most important role in the operation of a steam jet ejector. Since internal dimensions are fixed, the ejector is designed for only one steam condition. When the steam condition changes, there will be a change in the operation and the efficiency of the ejector.
- In a critical flow ejector, a very small decrease in steam pressure will result in a broken or unstable vacuum. An increase in steam pressure above design will not have a noticeable effect in the operation of an ejector unless the increase is significant (>25%).
- Besides wasting steam, excess motive pressure tends to choke the venturi with steam, thereby decreasing the suction capacity of the ejector.
- The performance of ejectors designed for saturated steam will be adversely affected if operated with superheated steam.
- The specific volume of steam increases with increasing temperature, which may require an increase in pressure to maintain required steam flow.
- Excess moisture in steam is one of the most common problems found in ejectors. Wet steam causes poor performance and, depending on the degree of wetness, can permanently damage an ejector in a very short period of time. A steam quality of less than 2% moisture is tolerable with most moderate vacuum

systems. However, ejectors designed for a vacuum of 5 mmHg abs or less should have steam that is completely dry or with a few degrees of superheat.

- "Use of higher steam pressures increases steam consumption in an ejector." *True/False?*
  - *False.* Results in decreased steam consumption in single or two-stage ejectors. Decrease in steam consumption is negligible if steam pressures are *more than* 24 barg (>350 psig).
- "Ejectors are very sensitive to lower than design steam pressures." *True/False?* Comment.
  - *True.* Even a small decrease in steam pressures (by few units) from design values lead to improper operation.
- How can one detect steam used in an ejector is wet and what are the remedies to prevent wet steam entering an ejector?
  - Fluttering of the needle on a steam pressure gauge during operation is a sign that steam flowing is wet.
  - Testing the steam with a throttling calorimeter, which is a constant enthalpy device that measures steam pressure and temperature. When used in combination with a Mollier Chart, a reading of steam quality is obtained.
  - Maintaining few degrees of superheat in the steam supply.
  - Installing a steam separator in the steam line as close to the ejector as possible.
  - Keeping all steam piping, and the steam separator, completely insulated to help prevent the formation of wet steam.
- What are the effects of changes in process loads on the performance of ejectors?
  - A change in process load will have a direct effect on the ejector system. Any increase in load will result in a higher absolute pressure.
  - Any increased noncondensables will travel through the system, affecting the following ejector stages. Discharge pressures of each stage will increase to the point of a breakdown in operation.
- What is the effect of rise in temperature of cooling water used in condensing-type ejectors?
  - When inlet water temperature increases above design maximum, loads to the following stage increase, resulting in a poorer vacuum at that stage. If the affected stage is the last stage, then the vacuums of all the preceding stages could also be affected.
- Evaluate direct and indirect contact condensers in a condensing ejector system.
  - In direct contact (countercurrent, barometric design) condensers, cooling water is mixed directly with the vapor to be condensed, then discharged to atmosphere through a barometric leg or tailpipe of sufficient length to overcome the atmospheric pressure.
  - A means of cleaning up or otherwise disposing the water that has become contaminated by process material is often required.
  - In direct contact condensers, high or low water flow rates can cause problems in the vacuum system. High water flow could flood the condenser, increasing pressure drop, and therefore, the back pressure on the stage discharging into it. Low water flow may not be distributed properly, allowing condensable load to bypass into the following ejector and resulting in a poor vacuum.
  - The surface contact condenser permits main condenser cooling water to be used as cooling water through inter- and after-condensers, for energy and process-water conservation.
- How does a steam jet water chiller work?
  - Steam jet water chiller works on the principle of evaporative cooling.
  - To cool water quickly (chilling) evaporation process must be performed under vacuum.
  - The process is generally completed in a multistage ejector system, which normally involves two to four chilling stages.
  - A typical chiller employs several booster ejectors, one or more per chilling stage.
  - The booster section pulls a vacuum, evaporating water and entraining dissolved noncondensables and leakage air and then discharging into a condenser, which while operating under vacuum, condenses motive steam and water vapor, using cooling water.
- What is compression ratio for a steam jet ejector? What is the range of compression ratios for these units?
  - Compression ratio of an ejector is defined as the ratio of (discharge pressure of the jet)/(suction pressure).
  - Compression ratios are normally in the range of 3:1–4:1. Some times, these values can be as high as 8:1.
  - Serious problems can result in the operation of a steam jet ejector if the compression ratio falls below 2:1.
- What is overall compression ratio for a multistage ejector?
  - Overall compression ratio is the product of compression ratios of individual stages working in series.
- What are the tolerable operational limits with respect to deviations from design capacities of steam jet ejectors?
  - A reasonable range of operation for good efficiency would be from 50% to 115% of the design capacity.

An ejector may operate well in excess of these figures but the efficiency will usually decrease at points beyond these limits.

- Is it desirable to use *higher pressure steam than the design values* in the operation of an ejector?
  - *No*. It does not increase suction capacity. It just uses more steam.
  - At pressures more than 25% above design pressure, capacity will actually fall due to *choking* in the venturi.
- What is the effect of using higher *design* steam pressures in ejectors?
  - Decreased steam consumption (single and two-stage ejectors).
  - Decrease is negligible if steam pressure is >25 bar.
- "Ejectors are very sensitive to lower than design steam pressures." *True/False*? Comment.
  - *True*. Even a small decrease (few millimeters of mercury gauge pressure) leads to improper operation.
- What are motive steam *pickup* and *break* pressures for an ejector? What are their significances?
  - If the steam pressure is being increased from a region of unstable operation, the point at which the ejector first becomes stable is called the *motive steam pickup pressure*.
  - The pickup pressure is a direct function of the discharge pressure. At the higher discharge pressure, the ejector will regain its stability once the motive steam pressure is increased to the pickup pressure.
  - For every discharge pressure in an ejector there is also a minimum steam flow below which the operation will be unstable.
  - If the steam pressure is being decreased from a region of stable operation, the point at which the ejector becomes unstable is called its *motive steam break pressure*.
  - The motive steam break pressure is below the motive steam pickup pressure for any given discharge pressure and load. For this reason, the ejector operating with steam pressure between the *break* and the *pickup* points may be stable or unstable depending on the direction of the steam pressure change.
- What are the common causes of problems arising in a vacuum system operating with the use of ejectors?
  - Air leakage: Air leakage often occurs through joints or flanged connections, or through corroded or eroded parts.
    - A hole measuring just a few millimeters of diameter will cause sufficient air leakage, which damages the vacuum, consequently increasing the need for ejector steam.
    - Estimate air or other gas leakage into the system. Every effort should be made to keep it as tight as possible. Possible leak points can be sealed with polystyrene, which produces an excellent seal.
    - In-leakage of air to evacuated equipment depends on the absolute pressure (in Torr) and the volume of the equipment, V (in m^3), according to

$$W = kV^{2/3}\ \text{kg/h}, \tag{5.26}$$

where $k = 0.98$, when $P > 90$ Torr; $k = 0.39$, when P is between 0.4 and 2.67 kPa (3 and 20 Torr); and $k = 0.12$ at $p < 133.3$ Pa (1 Torr).

  - *Blockage or Pressure Drop in the Lines*
    - A shifted gasket can cause a pressure drop in the vacuum system.
    - If the final atmospheric discharge piping becomes blocked, it could result in a pressure drop that affects the entire vacuum system.
  - *Insufficient Steam Pressure*
    - Under normal conditions, an ejector needs its full design motive pressure to operate.
    - In practice, an ejector works with steam that is fed at a pressure up to 10% below the design pressure of the ejector.
  - *High-Pressure Steam*
    - Running an ejector at a pressure substantially above the design pressure decreases capacity, mainly because of choking in the ejector throat.
    - It is safe to run an ejector at a pressure of about 25% above its design pressure.
    - At higher pressures, the steam to the ejector should be throttled back to a pressure that is closure to the design pressure.
  - *Superheat in the Steam*
    - Superheat in the steam will increase specific volume of steam.
    - Since an ejector nozzle has a fixed orifice, more or less steam will be passed, depending on the degree of superheat.
    - This can have the same effect as increasing or decreasing the motive steam pressure.
    - Too little superheat is not normally a problem, unless the ejector was designed for a large degree of superheat.
  - *Excessive Moisture in the Steam*
    - Wet steam alters performance of the ejector. Efficiency decreases.

    - It accelerates erosion on the system (e.g., changes critical dimensions on nozzles and venturis) and leads to ejector failure.
    - Erosion is due to fast movement of water droplets with the steam. About 3–8°C superheat is sometimes recommended.
    - Erosion of steam inlet nozzle increases its diameter permitting more steam to pass through. This increases downstream condenser pressure.
    - Diffuser operation suffers as diameter of the diffuser is designed to operate at a certain steam flow.

- Give a troubleshooting checklist for vacuum systems based on ejectors.
  - Steam lines from the utility side should be well insulated to prevent condensation. Insulation thickness is a function of steam temperature as well as pipe diameter.
  - A steam separator is installed before the ejector to ensure that the steam is dry. If steam quality is questionable, a calorimeter is used to measure the percentage of moisture in the steam.
  - The maximum allowable level of moisture in the steam is 3%.
  - The steam pressure on each ejector is checked to ensure that it is within limits. Checking the pressure of only the main steam line will not be enough.
  - Water pressure should be within 15% of the design pressure and its temperature should be less than or equal to design temperature.
  - Water flow rate must be adequate for proper condensation and cooling. About 8–15°C temperature rise in the condenser is acceptable.
  - The system must be checked for excessive air leakage. Excess air leakage is based on volume and time taken to build up pressure from vacuum after the ejectors are turned off.
  - The system for corroded/eroded/cracked parts, especially ejector nozzles, must be checked.
  - Blocked nozzles/lines/diffusers/strainers/barometric legs/fouled condenser should be checked.
  - Steam leaks between the steam chest and the nozzle, which will increase the suction load to ejector, should be checked.
  - Always spares for critical items should be maintained as inventory.
- Give some general rules of thumb for ejectors.
  - To determine number of stages required, assume a maximum compression ratio of 7:1 per stage.
  - The supply steam conditions should not be allowed to vary greatly. Pressure below design can lower capacity. Pressure above design usually does not increase capacity and can even lower it.
  - Use *Stellite* or other hard surface material in the jet nozzle. For example, 316 s/s is insufficient.
  - Always provide a suitable knockout drum ahead of the ejectors. Water droplets can quickly damage an ejector. The steam should enter the drum tangentially. Any condensate leaves through a steam trap at the bottom. It is a good idea to provide a donut baffle near the top to knock back any water creeping up the vessel walls.
  - The ejector barometric legs should go in a straight line to the seal tank. A 60–90° slope from horizontal is best.
  - The suction to the ejector must be designed for half the absolute pressure of the evacuated system.
  - Screwed fittings should never be used in any vacuum system regardless of size.
  - For large size vapor lines and condensers (frequently possible in vacuum systems), the line, condenser, and top of the column should always be insulated; otherwise, rain or sudden weather variations will change column control. It is possible to have more surface in the overhead line than in the condenser.
  - Liquid traps in vacuum system piping should be avoided by never going vertical after having gone to horizontal.
  - The vacuum system control valves should be installed at the highest point of a horizontal run with the control valve bypass in the same horizontal plane.
- Summarize troubleshooting of steam jet ejectors.
  - Table 5.8 presents troubleshooting guidelines for steam jet ejectors.
- Compare different types of vacuum producing equipment with respect to their capacities and operating ranges.
  - Table 5.9 presents a comparison of different types of vacuum producing equipment on pressures and capacities obtainable.
  - Dry pumps work in the range from 0.05 to 760 Torr with capacities of 1.4–40 m^3/min.
  - Liquid ring pumps have capacities of more than 600 m^3/min.

5.3 FANS, BLOWERS, AND COMPRESSORS

- What are the different types of fans and blowers?
  - Centrifugal or axial flow types.

TABLE 5.8 Troubleshooting Guidelines for Steam Jet Ejectors

Problem	Effect	Corrective Action
Lower than design motive steam pressure	Poor ejector performance	Raise steam pressure or bore steam nozzles
Higher than design motive steam pressure	Reduced ejector capacity and wasted steam	Reduce motive steam pressure
Higher than design temperature	Poor ejector performance	Raise steam pressure or bore steam nozzles
Higher than design discharge pressure	Poor ejector performance	Downstream condenser or ejector problems, restrictions in discharge piping
Low ejector discharge temperature	Reduced capacity/poor performance	Insulate steam lines/add moisture separator to steam lines
Higher than design suction pressure	Higher than design load or mechanical problems, for example, worn parts and internal steam leaks.	Inspect internal dimensions and replace if necessary. Tighten steam nozzle to steam chest.

- Fans are classified according to the direction of air flow. In a centrifugal fan, air/gas flows along the fan shaft, turns 90° by the impeller, which imparts kinetic energy to the air/gas as it flows radially outward. The kinetic energy is converted to pressure as it leaves the fan parallel to the shaft.

TABLE 5.9 Pressures and Capacities Obtainable for Vacuum Equipment

Type	Ultimate Pressure	Lower Limit for Process Applications	Single Unit Capacity range (m^3/min)
Steam jet ejectors			
Single stage	50 Torr	75 Torr	0.3–30,000
Two stage	4 Torr	10 Torr	
Three stage	0.8 Torr	1.5 Torr	
Four stage	0.1 Torr	0.25 Torr	
Five stage	10 μm	50 μm	
Six stage	1 μm	3 μm	
Liquid ring pumps			1–500
Water sealed (15°C)			
Single stage	50 Torr	50 Torr	
Two stage	20 Torr	25 Torr	
Oil sealed	1 Torr	10 Torr	
Air ejector first stage	1 Torr	10 Torr	
Dry vacuum pumps			
Three-stage rotary lobe	0.5 Torr	1.5 Torr	1.7–6.8
Three-stage claw	0.1 Torr	0.3 Torr	1.4–7.6
Screw compressor	50 μm	0.1 Torr	1.4–40
Integrated systems			
Booster–liquid ring pump	1 Torr	5 Torr	2.8–42.5
Booster–rotary lobe dry pump	25 μm	0.25 Torr	2.8–42.5
Booster–claw compressor	10 μm	0.1 Torr	2.8–70
Booster–screw compressor	<0.1 μm	1 μm	2.8–140

- In an axial flow fan, air/gas enters and leaves the fan parallel to the shaft.
- Centrifugal fans are classified according to their blade geometry—radial, forward curved, backward curved, and airfoil. The major characteristic of radial fan is its ability to compress air/gases to a higher pressure but delivers lower flow rates than the other fan types.
- The blades are self-cleaning, tending to fling off particles and thus can be used to pneumatically convey solids.
- Figure 5.36 illustrates different types of fans.
- The backward incline fan design consists of the single thickness blade and the airfoil blade. The single thickness blade can be used for pneumatic conveying of solids.
- The airfoil type has aerodynamically shaped blades to reduce flow resistance, resulting in a high efficiency. Entrained particles will damage the blades, and thus this fan is not suitable for pneumatic conveying.
- For both types as the flow rate increases, the required power increases, reaches a maximum, and then decreases instead of continuously increasing. When operating under conditions where the flow rate varies, this characteristic is an asset.
- The forward curved blade fan is designed for low to medium flow rates at low pressures. Because of the cup-shaped blades, solids tend to be held in the fan, and thus this fan is also not suitable for pneumatic conveying of solids.
- Axial fans consist of the tube axial fan and the vane axial fan, which are designed for a wide range of flow rates at low pressures. These fans consist of a propeller enclosed in a duct. They are limited to applications where the gas/air does not contain entrained solids.
- In a tube axial fan, the discharged flow follows a helical path creating turbulence. To reduce

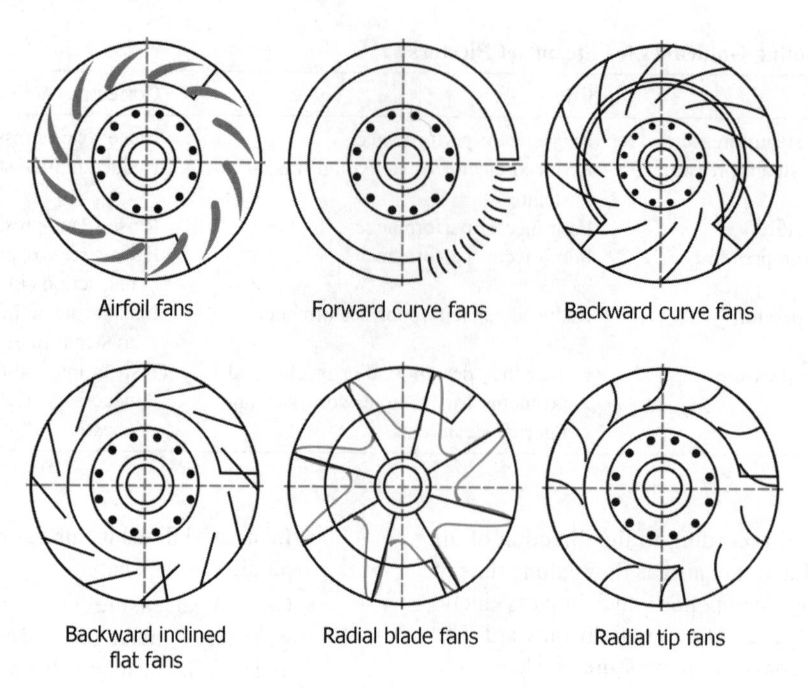

FIGURE 5.36 Types of fans.

turbulence and increase the fan efficiency, the vane axial fan contains flow-straightening vanes.

- What are the characteristics and applications of fans?
  - Fans are used for high flow and low-pressure applications to move large volumes of air or gas through ducts at near atmospheric pressures, supplying air for drying, cooling towers, removing fumes, and so on.
    - Because the clearances between the impeller and the casing are large, the pressure developed is low, between 1.01 and 1.15 bar.
  - In large buildings, blowers are often used due to the high delivery pressures needed to overcome the pressure drop in the ventilation system. Most of these blowers are of the centrifugal type. Blowers are also used to supply draft air to boilers and furnaces.
- What are the pressure increases involved in fans, blowers, and compressors?
  - Fans: Low delivery pressures: 3% (<40 kPa) (30 cm H_2O).
  - Blowers: <300 kPa (<2.75 barg).
  - Compressors: Higher pressures: >300 kPa (>2.75 barg).
- What is a jet compressor?
  - Jet compressors utilize a high-pressure gas to raise other gases at low pressure to some intermediate value by mixing with them.
- What are the factors that affect performance of centrifugal fans?
  - The performance of a centrifugal fan varies with changes in conditions such as temperature, speed, and density of the gas being handled.
- "Axial flow fans and blowers have high specific speeds compared to centrifugal compressors." *True/False?*
  - *True.*
- "Mixed flow fans and blowers have higher specific speeds than axial flow types." *True/False?*
  - *False.*
- What are the advantages and disadvantages of straight lobe blowers?
  - Low cost.
  - Low maintenance requirements.
  - Applicable for low-pressure operations.
  - Poor efficiency compared to screw compressors because of slip and high-frequency flow direction reversal, which takes place at the discharge port.
  - Solid contaminants deteriorate critical internal clearances.
- Classify different types of compressors.
  - Figure 5.37 gives classification of different types of compressors.
  - *Positive Displacement Compressors*:
    - *Reciprocating compressors* are used mainly when high-pressure head is required at a low flow.
    - Reciprocating compressor ratings vary from fractional to more than 30,000 kW (40,000 hp) per

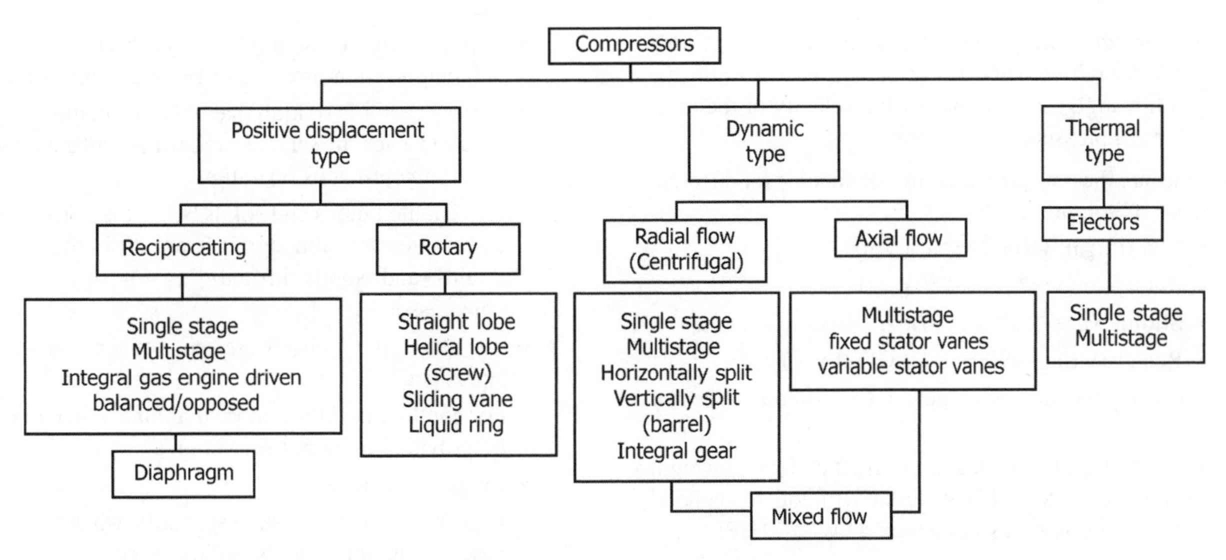

FIGURE 5.37 Types of compressors. (*Courtesy*: GPSA Engineering Data Book, 12th ed.)

unit. Pressures range from low vacuum at suction to 200 mPa (30,000 psi) and higher at discharge for special process compressors.

- Reciprocating compressors are furnished either single stage or multistage. The number of stages is determined by the overall compression ratio. The compression ratio per stage (and valve life) is generally limited by the discharge temperature and usually does not exceed 4 although small-sized units (intermittent duty) are furnished with a compression ratio as high as 8.
- On multistage machines, intercoolers are normally provided between stages to remove heat of compression and reduce the temperature to approximately the temperature existing at the compressor intake.

- Such cooling (1) reduces the actual volume of gas going to the high-pressure downstream equipment (2) reduces the power required for compression, and (3) keeps the temperature within safe operating limits.

- Reciprocating compressors should be supplied with clean gas as they cannot satisfactorily handle liquids and solid particles that may be entrained in the gas. Liquids and solid particles tend to destroy cylinder lubrication and cause excessive wear. Liquids are noncompressible and their presence could rupture the compressor cylinder or cause other major damage.
- Rotary compressors are very close to centrifugal and axial flow compressors. Like reciprocating compressors, they are positive displacement type; however, unlike the reciprocating compressors, they do not have very high vibration problem. They have a casing with one or more rotating elements that either mesh with each other such as lobes or screws or displace a fixed volume with each rotation.
- Screw type of rotary compressor is generally appropriate for a flow range of 85–170 m^3/h and capable of handling capacities up to about 4×10^4 m^3/h at pressure range of 2070– 2760 kPa (300–400 psig) at pressure ratios of 4:1 and higher. Relatively small diameter rotors allow rotary speeds of several thousand rpm.
- Screw compressors are of two types, namely, dry screw compressors and oil-flooded screw compressors. Oil-flooded screw compressors use oil for bearing lubrication as well as to seal the compression chamber. They also remove heat generated during compression.

- *Dynamic Compressors*: The dynamic types include radial flow centrifugal, axial flow, and mixed flow machines. They are rotary continuous flow compressors in which the rotating element (impeller or bladed rotor) accelerates the gas as it passes through the element, converting the velocity head into static pressure, partially in the rotating element and partially in stationary diffusers or blades.
  - Centrifugal compressors are generally used for higher pressure ratios and lower flow rates compared to lower stage pressure ratios and higher flow rates in axial flow compressors.
  - Axial flow compressors are used mainly as compressors for gas turbines.

- *Thermal Compressors*: These are the ejectors that use a high-velocity gas or steam jet to entrain the inflowing gas, then convert the velocity of the mixture to pressure in a diffuser.

- What are the ranges of pressures obtainable for different types of compressors?
  - Centrifugal: Over 345 bar (>5000 psig).
  - Rotary screw: 20 bar (250 psig).
  - Sliding vane: 10 bar (150 psig).
  - Reciprocating: 6900 bar (100,000 psig).
- What are the ranges of power for different types of compressors?
  - Centrifugal (multistage): For light molecular weight gases more than 15 kW (2000 HP) and for general hydrocarbon gases more than 4 kW (500 HP).
  - Centrifugal (single stage): Low power and low head applications, for example, blowers.
  - Rotary screw: From 74.6×10^{-3} to 45 kW (10–6000 HP).
  - Sliding vane: Up to 3 kW (400 HP).
  - Reciprocating: Up to 110 kW (15000 HP).
- What are the factors to be considered while choosing between dry and oil-flooded screw compressors?

  Dry Screw Compressors
  - Dry type is the choice if oil is not compatible with oil or it should be free from oil. Also if the gas contains small particulates, dry compressor is preferred.

  Oil-Flooded Compressors
  - Used if oil is compatible with gas or if gas carries droplets of oil as in natural gas processing systems. Also oil-flooded compressors are good to minimize or avoid shaft seal leakages.
- "Centrifugal compressors are classed as dynamic machines" Comment.
  - These do their work by using inertial forces applied to gas by means of rotating bladed impellers.
- What are the advantages of centrifugal compressors?
  - Large capacities: 30–4500 atm m^3/min) and pressures up to about 350 bar.
  - Efficiencies are in the range of 68–76%.
  - No contamination by lubricating oil, as they do not have rubbing surfaces.
  - Balanced machines and do not require heavy foundations.
  - Low maintenance costs.
  - Within their operating range, initial cost is less than that for reciprocating compressors within the same operating range.
  - Capacity can be controlled by speed variations, reducing suction pressure or by inlet vane control.
  - Service factor is so high that only one compressor is required even in services requiring three or more years in continuous operation.
  - Can handle liquids and solids better than other types of compressors, although it is not desirable to have liquids and solids in the gas for any type of compressor.
- What are the disadvantages of reciprocating compressors?
  - For continuous duties, more than one machine must be provided to permit servicing.
  - Large and expensive.
  - High maintenance costs, especially when handling gases containing liquids, solids, or corrosives.
  - Require large foundations.
- How does a diaphragm compressor work? What are its applications?
  - In a diaphragm compressor, a piston acts indirectly by applying pressure to hydraulic oil, which flexes a thin metal diaphragm to compress the gas.
  - It is used for small flow rates, below the range for reciprocating compressors, and is limited by the construction of the diaphragm.
  - An advantage of the diaphragm compressor is that leakage of either the gas or the oil into the gas is prevented. Thus, the diaphragm compressor is ideal for compressing flammable, corrosive, or toxic gases at high pressures.
  - A disadvantage is its high maintenance cost, mainly because the diaphragm has to be replaced after about 2000 h of operation.
- What are the plus and minus points for rotary screw compressors?

  Plus Points
  - Balanced machines and require light foundations.
  - As there are no rubbing surfaces, do not contaminate compressed gas with lube oil.
  - Low maintenance costs.
  - Cheaper than centrifugal and reciprocating compressors.
  - Can be used to compress gases with contaminants such as liquid droplets, tars, and polymers, with good levels of efficiency.

  Minus Points
  - Noisy.
  - Range of capacity variation at constant speed is very small.
  - Designed for a specified gas and compression ratio.

- What are the advantages and disadvantages of vane type rotary compressors compared to other rotary compressors?

  Advantages
  - Low cost.
  - Highest compression efficiency and overall efficiency over all types of rotary compressors.
  - Very high volumetric efficiency and hence high flow rate with respect to machine size.
  - Low starting torque requirement.

  Disadvantages
  - Continuous lubrication of internal rubbing parts is required (10 times more than reciprocating machines).
  - High maintenance due to wear.
  - Sensitive to solids in gas stream and so require filters in air service.
- What are the advantages and the disadvantages of liquid ring rotary compressor?

  Advantages
  - Low discharge temperature capability. Useful for polymerizing gases such as acetylene and gas mixtures with exothermic reactions.
  - Selection and processing of liquid allows unit operations such as absorption, scrubbing, and cooling, besides compression.
  - Insensitive to liquid and solids fines carry over.
  - Oil-free operation.
  - Simple with only one moving part.
  - Maintenance in corrosive environments is lower than for other types.
  - Nearly five times as efficient as ejectors in vacuum service.
  - Pulsations in discharge pressure are low due to circulation of liquid inside.

  Disadvantages
  - Low overall efficiency is 35–50%.
- What is the advantage of a dry screw compressor over an oil-flooded screw compressor?
  - Oil-flooded screw compressor contaminates with oil the product gas and instrument air handled by it.
  - Oil contamination in air compressors resulted in several explosions involving oil and air.
  - Such risks are eliminated in dry screw compressors.
- Summarize characteristics of different types of compressors.
  - Table 5.10 presents a summary of the characteristics of different types of compressors.
- Give equations for theoretical adiabatic and polytropic power for compression.
  - Adiabatic power:

$$mz_1RT_1[(P_2/P_1)^a-1)]/a, \quad (5.27)$$

  where T_1 is the inlet temperature (K), R is the gas constant (8.314 J/mol K), z_1 is the compressibility factor, m is the molar flow rate, $a=(k-1)/k$, and $k=C_p/C_v$.
  - The outlet for the adiabatic reversible flow is

$$T_2 = T_1(P_2/P_1a]. \quad (5.28)$$

    - Exit temperatures should not exceed 204°C (400°F).
    - For diatomic gases ($C_p/C_v=1.4$), this corresponds to a compression ratio of about 4.

TABLE 5.10 Summary of Characteristics of Different Types of Compressors

Type	Inlet Flow Rate (1000 m^3/h)	Compression Ratio	Maximum Temperature (K)	Overall Efficiency
Positive Displacement				
Reciprocating	—	3–4	450–510	0.75–0.85
Diaphragm	0.0051–0.051	20	—	—
Rotary				
Helical screw	34	2–4	450–510	0.75
Spiral–axial	22	3	450–510	0.70
Straight lobe	52	1.7	450–510	0.68
Sliding vane	10	2–4	450–510	0.72
Liquid ring	22	5	450–510	0.50
Dynamic				
Centrifugal	85–340	6–8	111–505	—
Axial	1.3–1000	12–24	590	—

- Compression ratios should be about the same in each stage for a multistage unit.

$$\text{Compression ratio} = (P_n/P_1)^{1/n}, \text{ with } n \text{ stages.} \tag{5.29}$$

- Polytropic power:

$$mz_1RT_1[P_2/P_1)_p^{a/\eta}-1]/(a/\eta_p), \tag{5.30}$$

where η_p is the polytropic efficiency (use 75% for preliminary work).

- What is the outlet temperature for reversible adiabatic process?

$$T_2 = T_1(P_2/P_1)^a \tag{5.31}$$

- How are the outlet temperatures estimated for multistage compressors?

$$\text{Outlet temperature } T_2 = T_1[P_2/P_1]^n. \tag{5.32}$$

- What is the maximum exit temperature for compressors handling diatomic gases?
  - 167–204°C (350–400°F).
- What is the normally recommended maximum outlet temperatures for multistage compressors?
  - 50–60°C.
  - Intercooling can be used to hold desired temperatures for high overall compression ratio applications. This can be done between stages in a single compressor frame or between series frames.
    - Sometimes for high compression ratio applications, the job cannot be done in a single compressor frame. Usually, a frame will not contain more than about eight stages (wheels).
  - Sometimes, economics rather than a temperature limit dictate intercooling.
- Define specific speed for a compressor.
  - Speed in rpm at which an impeller would rotate if reduced proportionately in size so as to deliver 1 cft of gas per minute against a total head of 30.48 cm (1 ft) of fluid.
- What is the range of specific speeds for a centrifugal compressor?
  - 1500–3000 rpm at the high-efficiency point.
- "Axial flow blowers and fans have high specific speeds compared to centrifugal compressors." *True/False?*
  - *True.*
- "Mixed flow blowers and fans have higher specific speeds than axial flow types." *True/False?*
  - *False.*
- What is the reason that specific speeds of centrifugal compressors are the lowest compared to fans and blowers?
  - Because they have narrow impellers.
- What is percent clearance for a reciprocating compressor?
  - Clearance volume is usually expressed as a percent of piston displacement and referred to as percent clearance, or cylinder clearance, *C*.
  - Percent clearance = 100(clearance volume)/(volume of piston displacement).
  - For double-acting cylinders, the percent clearance is based on the total clearance volume for both the head end and the crank end of a cylinder. These two clearance volumes are not the same due to the presence of the piston rod in the crank end of the cylinder.
  - Sometimes, additional clearance volume (external) is intentionally added to reduce cylinder capacity.
- Define volumetric efficiency for a reciprocating compressor.
  - The term *volumetric efficiency* refers to the actual pumping capacity of a cylinder compared to the piston displacement.
  - Without a clearance volume for the gas to expand and delay the opening of the suction valve(s), the cylinder could deliver its entire piston displacement as gas capacity.
  - When a nonlubricated compressor is used, the volumetric efficiency should be corrected by subtracting an additional 5% for slippage of gas.
- "Efficiencies for pumps are generally more than for compressors." *True/False?*
  - *False.* Efficiencies for pumps are in the range of 40–60% and for compressors the efficiencies are in the range of 60–80%.
- What are the usual efficiencies for large (170–3000 atm. m^3/min) at suction condition) centrifugal compressors? Illustrate.
  - 76–78%.
- What is the range of efficiencies of rotary compressors?
  - 50–70%.
- What are the normally used compression ratios for different compressors? What are the considerations involved in limiting compression ratios?
  - Typical compression ratios of reciprocating compressors:
    - Large pipeline compressors: 1.2–2.0.
    - Process compressors: 1.5–4.0.
    - Small compressors: up to 6.0.

- To save on equipment costs, it is desirable to use as few stages of compression as possible.
- As a rule, compression ratio is limited by a practical desirability to keep outlet temperatures below 150°C or so to minimize the possibility of ignition of lubricants, as well as the effect that power requirements go up as the outlet temperatures go up.
- For minimum equipment cost, the work requirement should be the same for each stage.
- For ideal gases with no friction losses between stages, this implies equal compression ratios.

• Does compression ratio differ from stage to stage for a multistage compressor? Give the equation.

- Normally compression ratio should be about the same in each stage of a multistage compressor.
- Compression ratio can be represented as

$$\text{Ratio} = [P_n/P_1]^{1/n}, \text{ with } n \text{ stages.} \tag{5.33}$$

• How are efficiencies related to compression ratios for reciprocating compressors?

- Efficiencies increase with increase in compression ratios as illustrated in Table 5.11.

• What is meant by surge in a compressor? Characterize surge.

- Surging is defined as a self-oscillation of the discharge pressure and flow rate, including a flow reversal. Every centrifugal or axial compressor has a characteristic combination of maximum head and minimum flow. Beyond this point, surging will occur. During surging, a fast flow reversal through a compressor is often accompanied by a pressure drop. This flow reversal is accompanied with a very violent change in energy.
- If the demand for gas is reduced, the operating point will move toward the surge point. If the load is sufficiently reduced, the compressor loses the ability to increase the discharge pressure and *flow reversal occurs*.
- The compressor surge limit is a complex function that is dependent on the gas composition, rpm, suction temperature, and pressure.
- Symptoms of surge are excessive vibration and a large audible sound.

TABLE 5.11 Compression Ratio Versus Efficiency

Compression Ratio	Efficiency (%)
1.5	65
2.0	75
3–6	80–85

- Surge happens at low flow rates, often when the downstream demand decreases.
- >When the flow decreases below a certain minimum point, flow patterns within the compressor become unstable and fluid can move back through the compressor from the high-pressure side to the low-pressure side.
- Figure 5.38 illustrates surge point on the centrifugal pump operating curve.
- In a centrifugal compressor, surge is usually initiated at the exit of the impeller or at the diffuser entrance for impellers producing a pressure ratio of less than 3:1. For higher pressure ratios, the initiation of surge can occur in the inducer.
- Flow reversals are temporary and forward flow is quickly reestablished once the compressor discharge pressure drops.
- The pattern starts again, setting up a large-scale flow oscillation in the system.
- Thus, surge process is cyclic in nature and if allowed to cycle for some time, irreparable damage can occur to the compressor.
- The set of operating conditions under which surge begins, at a given rotational speed, is referred to as the *surge point*. Surge normally occurs at about 50% of the design inlet capacity at design speed.
- Because surge is fast, high-energy phenomenon, it can introduce excessive dynamic loads on internal components, such as thrust bearings, seals and blades, and introduce pipe vibrations.
- Over a period of time, surge can lead to fatigue failures that can damage the compressor.

• What is surge limit line (SLL)? Illustrate the difference between surge limit line and surge control line (SCL).

- Surge limit line is the locus of surge points at different values of rpm.

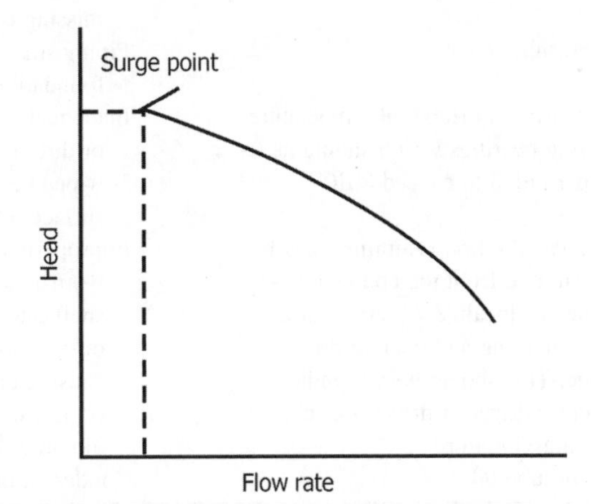

FIGURE 5.38 Centrifugal compressor operating curve.

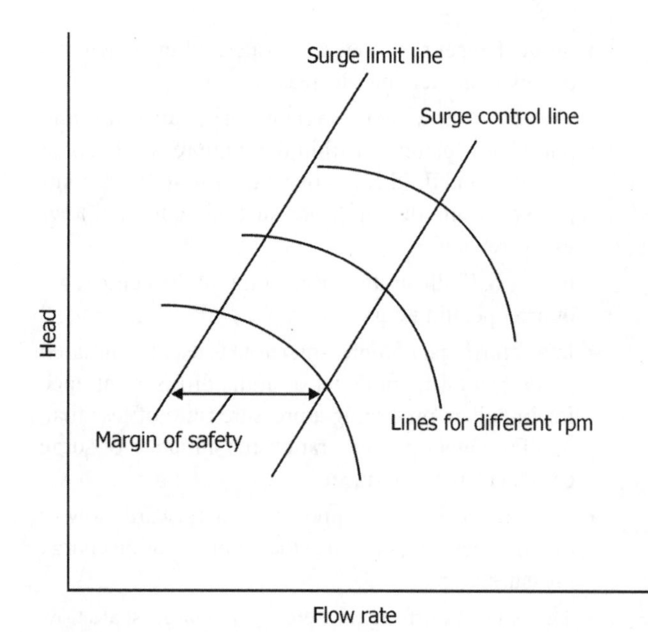

FIGURE 5.39 Illustration of operable range for the compressor with surge-free conditions.

- Figure 5.39 illustrates the difference between surge limit line and surge control line, below which inoperable conditions exist and the margin of safety for operation of the compressor, with small drop in head over a range of capacities.
- A surge limit line (SLL) is the line connecting the various surge points of a compressor at varying rpms. The set point of the antisurge controller is represented on the compressor map shown in Figure 5.38 by a line that runs parallel to the surge limit line. This line is called the surge controller line (SCL).

- What are the consequences of compressor surging?
  - Consequences of surging can include the following:
    - Rapid flow and pressure oscillations cause process instabilities.
    - Rising temperatures inside the compressor.
    - Tripping of the compressor.
    - Mechanical damage.

TABLE 5.12 Summary of Centrifugal Compressor Problems

Trouble	Probable Cause(s)
Low discharge pressure	Compressor not up to speed; excessive compressor inlet temperature; low inlet pressure, leak in discharge piping; excessive system demand from compressor.
High pressure in low pressure stages	Backflow via recycle loop due to control system failure.
Compressor surge	Inadequate flow through the compressor; change in system resistance due to obstruction in the discharge piping or improper valve position; deposit buildup on rotor or diffusers restricting gas flow.
High temperature in subsequent stages	Loss of interstage cooling.
Overspeed	Due to speed control system failure and loss of containment.
Wrong gas composition	Liquid in suction leading to damage to rotor.
Low lube oil pressure	Faulty lube oil pressure gauge or switch; low level in oil reservoir; oil pump suction plugged; leak in oil pump suction piping; clogged oil strainers or filters; failure of both main and auxiliary oil pumps; operation at a low speed without the auxiliary oil pump running (if main oil pump is shaft driven); leaks in the oil system; relief valve improperly set or stuck open; incorrect pressure control valve setting or operation; bearing lube oil orifices missing or plugged.
Shaft misalignment	Piping strain; warped bedplate; compressor or driver; warped foundation; loose or broken foundation bolts; defective grouting.
High bearing oil (lube oil temperature leaving bearings which should never be permitted to exceed 80°C)	Inadequate or restricted flow of lube oil to bearings; poor temperature conditions of lube oil or dirt or gummy deposits in bearings; inadequate cooling water flow lube oil cooler; wiped bearing; fouled lube oil cooler; high oil viscosity; water in lube oil; rough journal surface; excessive vibration.
Excessive vibration (vibration may be transmitted from the coupled machine. To localize vibration, disconnect coupling and operate driver alone. This should help to indicate whether driver or driven machine is causing vibration)	Improperly assembled parts; loose or broken bolting; piping strain; shaft misalignment; worn or damaged coupling; dry coupling (if continuously lubricated type is used); warped shaft caused by uneven heating or cooling; damaged rotor or bent shaft; unbalanced rotor or warped shaft due to severe rubbing; uneven buildup of deposits on rotor wheels, causing unbalance; excessive bearing clearance; loose wheel(s) (rare case); operating at or near critical speed, operating in surge region; liquid slugs striking wheels; excessive vibration of adjacent machinery (sympathetic vibration).
Water in lube oil	Condensation in oil reservoir; leak in lube oil cooler tubes or tube sheet.

Courtesy: GPSA Engineering Data Book, 12th ed.

– Mechanical damage can include the following:

1. Radial bearing load during the initial phase of surging. A side load is placed on the rotor that acts perpendicular to the axis.
2. Thrust bearing load due to loading and unloading.
3. Seal rubbing.
4. Stationary and rotating part contact if thrust bearing is overloaded.

- How is surge controlled?
  - The only way to prevent surging is to recycle or blowdown a portion of the flow to keep the compressor away from its surge limit.
  - Unfortunately, compressing extra flow results in a severe economic penalty.
  - The control system must be able to accurately determine the operating point of the compressor as to provide adequate, but not excessive, recycle flow.
- What types of compressors experience surge problems?
  - The surge problem is inherent in dynamic compressors—centrifugal and axial, as distinguished from positive displacement types.
- What is the effect of wetness of inlet gas on compressor performance? How is the wetness minimized/eliminated?
  - Results in damage to the compressor in the form of pitting corrosion, salt deposits and diluted lubricants.
  - Use of liquid knockout drums on suction side fitted with mist eliminators, which are installed in the knockout drum.
  - The mist eliminators should have appropriate specifications free from incorrect installation, overloading, uneven velocity profiles, high liquid viscosity, waxy deposits, liquid slugs, foaming, and several other possibilities.
- "Efficiencies for compressors are generally more than for pumps." *True/False?*
  - *True.* Compressor efficiencies are in the range of 60–80%, whereas for pumps, efficiencies are in the range of 40–60%.
- What are the possible causes of centrifugal compressors problems?
  - Table 5.12 presents a summary of possible centrifugal compressor problems.

TABLE 5.13 Summary of Reciprocating Compressor Problems

Trouble	Probable Cause(s)
Compressor will not start	Power supply failure; switchgear or starting panel; low oil pressure shut down switch; control panel.
Motor will not synchronize	Low voltage; excessive starting torque; incorrect power factor; excitation voltage failure.
Low oil pressure	Oil pump failure; oil foaming from counterweights striking oil surface; cold oil; dirty oil filter; interior frame oil leaks; excessive leakage at bearing shim tabs and/or bearings; improper low oil pressure switch setting; low gear oil pump bypass/relief valve setting; defective pressure gauge; plugged oil sump strainer; defective oil relief valve.
Noise in cylinder	Loose piston; piston striking outer head or frame end of cylinder; loose crosshead lock nut; broken or leaking valve(s); worn or broken piston rings or expanders; improper valve position.
Excessive packing leakage (blue rings).	Worn packing rings; improper lube oil and/or insufficient lube rate; dirt in packing; excessive rate of pressure increase; packing rings assembled incorrectly; improper ring side or end gap clearance; plugged packing vent system, scored piston rod; excessive piston rod runout.
Packing overheating	Lubrication failure; improper lube oil and/or insufficient lube rate; insufficient cooling.
Excessive carbon on valves	Excessive lube oil; improper lube oil (too light, high carbon residue); oil carryover from inlet system or previous stage; broken or leaking valves causing high temperature; excessive temperature due to high-pressure ratio across cylinders.
Relief valve popping	Faulty relief valve, Leaking suction valves or rings on next higher stage, Obstruction (foreign material, rags), blind or valve closed in discharge line.
High discharge temperature	Excessive ratio on cylinder due to leaking inlet valves or rings on next higher stage; fouled intercooler/piping; leaking discharge valves or piston rings; high inlet temperature; fouled water jackets on cylinder improper lube oil and/or lube rate.
Frame knocks	Loose crosshead pin; pin caps or crosshead shoes; loose/worn main; crankpin or crosshead bearings; low oil pressure; cold oil; incorrect oil; knock is actually from cylinder end.
Crankshaft oil seal leaks	Faulty seal installation; clogged drain hole.
Piston rod oil scraper leaks	Worn scraper rings; scrapers incorrectly assembled. Worn/scored rod; improper fit of rings to rod/side clearance.

Courtesy: GPSA Engineering Data Book, 12th ed.

- What are the possible causes of reciprocating compressors problems?
  - Table 5.13 presents a summary of possible reciprocating compressor problems.
- What are the differences between the *utility* and the *instrument grades* of compressed air used in process industry?
  - *Utility grade* compressed air is typically used as a source to power air tools, air motors, hoists, and for other tasks within a facility that require the air to simply be of sufficient pressure to do the intended work.
  - The quality of this air stream is not normally monitored for the presence of water, oil aerosols or particulate matter, as it is not used in or consumed by the actual manufacturing process itself.
  - A typical utility grade compressed air system consists of a compressor, aftercooler, receiver, and water separator. The compressor is usually of an oil-flooded design, equipped with an air–oil separator that is intended to prevent the passage of oil downstream. This is done primarily to maintain the volume of oil required in the lubrication of internal components and to ensure proper levels for introduction into the compression cycle.
  - The aftercooler is also normally integrated into the compressor package and may utilize a heat exchanger as a means of removing the heat of compression from the compressed air stream.
  - The receiver vessel (if used) will be sized according to consumption trends and as a surge vessel to control pressure fluctuations in the header.
  - Water separators are installed at different points in the air header system to remove condensate as the air continuously cools on its movement through the air distribution system.
  - *Instrument grade* compressed air is used for both task-specific functions and some times, as a part of the manufacturing process.
  - These applications require air free from particulate contamination, oil aerosols, and water vapor.
  - Examples of instrument quality air being used as a component of the process include oil-free air that is injected into a blending process as a means of introducing oxygen into the process and use of instrument air in injection molding process for plastics.
  - In order to achieve instrument grade air, the compressed air system should be equipped with an oil-free compressor, such as a dry rotary screw or a centrifugal compressor along with a refrigerated or regenerative desiccant dryer.
  - The items in an instrument air system that are often common to both utility grade and instrument grade systems include the aftercooler and the receiver vessel.
  - The compressed air cycle in both utility grade and instrument grade air are equipped with removal system for the heat of compression, particulate contamination and water vapor.
  - The major difference between the two systems ensures the oil-free compressor module and the use of an air dryer.

6

MIXING

6.1	Mixing	163
6.2	Mixing Equipment	169
6.2.1	Baffled Vessels	169
6.2.2	Impellers	172
6.2.3	Turbine and Propeller Mixers	173
6.2.4	Paddle Mixers	176
6.2.5	Draft Tubes	177
6.2.6	Other Types of Mixers	177
6.2.6.1	Static Mixers	183
6.2.7	Solids Mixing	186
6.2.8	Mixer Seals	194

6.1 MIXING

- What is mixing? What are its objectives and characteristics?
  - Mixing is a unit operation in which a relatively uniform mixture is obtained from two or more components. In some cases, mixing is required to promote mass transfer or chemical reaction.
  - Typically, the objectives include creating a suspension of solid particles, blending miscible liquids, dispersing gases through liquids, blending or dispersing immiscible liquids in each other, and promoting heat transfer.
  - The degree of uniformity achievable varies widely. It is easy to achieve virtually complete homogeneity when mixing miscible liquids or mixing soluble solids into a liquid, but it can be difficult to achieve a homogeneous result when mixing two solids, mixing two highly viscous liquids, or mixing items with widely varying densities, especially if the amount of one component is very small compared to the amounts of the others.
  - The efficiency of mixing depends on the efficient use of energy to generate flow of the components.
  - Mixing reduces concentration and temperature gradients in a mixture, thus exerting a favorable effect on the overall rates of mass and heat transfer.
  - Important aspects in the design of a mixer include the following:
    - Provision of adequate input energy (for an appropriate time).
    - Design of the mechanism for introducing the energy.
    - Properties of the components.
- How is mixing achieved? What are its objectives? What are its applications?
  - Mixing processes use mechanical means to forcefully circulate materials of separate phases to yield a random distribution of the components.
  - Mixing objectives can encompass blending, gas induction, homogenization, particle reduction, emulsification, and so on. A homogeneous mixture is one in which the separate molecules being combined are interspersed. A heterogeneous mixture contains a random dispersion of materials having distinguishable phases.
  - Mixing applications involve contacting and interfacing of a variety of substances, including gas–gas, liquid–liquid, solid–solid, gas–liquid, solid–liquid, and gas–paste.
- What is the difference between mixing and agitation?
  - Mixing refers to any operation used to change a nonuniform system into a uniform one (the random distribution of two or more initially separated phases).
  - Agitation implies forcing a fluid by mechanical means to flow in a circulatory or other pattern inside a vessel.

Fluid Mechanics, Heat Transfer, and Mass Transfer: Chemical Engineering Practice, By K. S. N. Raju

- The words mixing and agitation are often used interchangeably.

- What are macro- and mesomixing?
  - Mixing at nanometer level involves high shear stresses and turbulence. The energy dissipation rate determines whether mixing processes are at macro, meso, or micro levels. The overall mixing process occurs within a flow field continuum that covers a wide range of time- and length scales that are indicative of these mixing scales.
  - For two miscible liquids, the initial large-scale distribution by flow patterns that cause gross dispersion is considered as *macromixing*.
  - The breakdown of large eddies into smaller ones via an *eddy cascade* is termed *mesomixing*. This is followed by fluid engulfment in small eddies and subsequent laminar stretching of these eddies, with molecular diffusion as the final mechanism to obtain uniform composition; this process is termed as *micromixing*. The length scale for this final diffusion process is determined by the size of the smaller eddies formed, which is referred to as *Kolmogorov length scale*. These hydrodynamic processes are important to understand nanoscale phenomena.
- Classify process criteria for mixing processes.
  - *Physical Processing:* suspension, dispersion, emulsion, blending, pumping, and so on.
  - *Application Processing:* liquid–solid, gas–liquid, immiscible liquids, miscible liquids, and so on.
  - *Chemical Processing:* dissolving, absorption, extraction, reactions, heat transfer, and so on.
- Give examples of common processes that require agitation.
  - Blending of two miscible or immiscible liquids, such as hydrogen peroxide and water.
  - Dissolving solids in liquids, such as salt in water.
  - Dispersing a gas in a liquid as fine bubbles, such as oxygen in a suspension of microorganisms for waste treatment.
  - Dispersion of droplets of one immiscible liquid in another in liquid–liquid extraction or in heterogeneous reaction processes.
  - Suspension of fine solid particles, such as metallic pigments in car paint.
  - Agitation of a fluid to eliminate temperature gradients arising from poor heat transfer or to increase heat transfer between the fluid and a heating element in the vessel.
- What are the basic flow patterns in mixing processes?
  - Tangential, axial, and radial flows. For a cylindrical vessel without baffles and any type of impeller on the center vertical axis, all turbulent flow will be tangential.
  - Axial flow impellers generate flow parallel to the axis (or shaft), either upward or downward.
  - Radial flow impellers create flow perpendicular to the axis or in a radial direction.
  - Some impellers, such as pitched blade turbine, create a mixed flow pattern, that is, partially axial and partially radial.
- What are primary and secondary flows in a mixing process?
  - The primary flow is directly induced by the rotation of the impeller. The secondary (or entrained) flows are due to the flowing liquid dragging adjacent liquid and entraining it.
  - Secondary flows allow the entire contents of a vessel to circulate even when a small impeller is inducing the primary flow. The secondary flow component is smaller relative to primary flow for larger impellers.
- What are the factors that influence efficiency of mixing processes?
  - Design of impeller, diameter of impeller, speed, and baffles.
- What are the variables that influence power requirements of an agitator?
  - Fluid circulation rate (pumping capacity) and fluid shear rate.
  - The pumping capacity of an impeller is defined as the volumetric flow rate normal to the impeller discharge area.
  - The fluid shear rate is based on the velocity of the fluid leaving the tips of the impeller.
  - A large impeller running at slow speed has a high pumping capacity and a low fluid shear rate, while a small impeller running at high speed has a high shear rate and a low pumping capacity. This means the power input may be distributed in different ways by selecting the appropriate ratio of impeller to tank size.
- What are the characteristic features of batch and continuous mixing processes?
  - The mixers commonly used in batch processes (often referred to as *dynamic mixers*) are freestanding machines in which the ingredients are loaded, agitated, and then discharged, typically into holding bins, smaller containers, or packaging equipment.
  - In general, batch mixing processes are energy intensive. Energy is imparted to an impeller via an electric motor.
  - The mixers associated with continuous processes are generally integral parts of large production lines and

continuously discharge the finished product into the next piece of equipment in the process. A continuous mixer is constantly fed with the correct amounts and proportions of ingredients at one end. As the materials are being mixed, they are also conveyed to the opposite end where the product is discharged.

- In comparison to batch mixers, continuous mixers (also known as *in-line* or *static mixers*) have fairly low energy requirements. The energy of the flowing fluid completes the mixing and that energy is derived from a pump upstream of the mixer.
- Continuous processing utilizes smaller vessels that reduce capital costs. Improved micromixing and reaction selectivity reduce unwanted by-products.

• What are the effects of mixing on different types of fluids (Newtonian and non-Newtonian)?

- *Newtonian:* Unaffected by magnitude and kind of fluid motion.
- *Dilatant:* Viscosity increases with increase in mixing intensity.
- *Bingham Plastics:* Have definite yield values that must be exceeded before flow starts. Once flow starts, viscosity decreases with increase in mixing intensity.
- *Pseudoplastics:* Do not have a yield value, but viscosity decreases with increase in mixing intensity.
- *Thixotropic:* Viscosity normally decreases with increased mixing, but this depends on duration of agitation, viscosity of fluid, and rate of motion before agitation.

• How does non-Newtonian behavior affect agitation in a mixer?

- The shear rate, γ, in an agitated vessel is a function of position and time within the vessel.
- The highest values of γ occur near the impeller at points of highest local velocity.
- For open turbine impellers, velocity and shear rate are highest at the tip of the impeller.
- γ decreases with increasing distance from the impeller and is lowest at the vessel wall. Therefore, shear thinning liquids will have their lowest apparent viscosity, μ_a, in the impeller region of an open turbine.
- With close wall clearance impellers, such as helical ribbons, shear thinning liquids will have the lowest μ_a near the vessel wall.
- As the shear rate varies throughout the vessel, μ_a also varies. Therefore, it is necessary to estimate effective impeller shear rate and effective μ_a in order to quantify performance characteristics of an impeller. Experimental data are required for such quantification.
- Several dimensionless numbers, which are influenced by geometric factors, such as ratio of impeller to vessel diameter and ratio of clearance from vessel bottom to vessel diameter, are used in the design of a mixer.
- Databases are available in the literature in order to obtain impeller performance parameters for scale-up. From these databases, the dimensionless numbers are obtained for a variety of common impeller types. Impeller N_{Re} for a power law fluid is centric to different dimensionless numbers of significance.

• Define impeller Reynolds number.

$$N_{Re} = D_0^2 N\rho/\mu, \tag{6.1}$$

where N is revolutions per second and D_0 is the impeller swept (maximum) diameter.

- N_{Re} for power law fluids measures the ratio of inertial to viscous forces within the mixer and is given by

$$N_{Re} = D_0^2 N\rho/(k_s n)^{n'-1}, \tag{6.2}$$

where k_s is the effective shear rate constant, n is the rotational shaft speed (rpm), and n' is the flow behavior index of a power law fluid.

- For Newtonian fluids, laminar flow regime is for $N_{Re} \leq 10$, transition flow is up to N_{Re} of 300, and above 300 it will be turbulent flow. For N_{Re} above 10,000, there will not be any effect of Reynolds number on power consumption, since viscosity is not a dominant factor.
- Some references state that turbulence is ensured throughout the mixing vessel if $N_{Re} > 20{,}000$ stating that $10 < N_{Re} < 20{,}000$ is the transition region.

• Define Froude number for an agitator.

$$N_{Fr} = DN^2/g, \tag{6.3}$$

where D is the agitator diameter.

- Froude number is the ratio of applied forces to gravitational forces and is considered only if there are significant gravitational effects present, such as in vortexing systems. For systems with top-entering or bottom-entering agitators in baffled tanks or applications employing side-entering propellers and in general for all types of systems where $N_{Re} < 300$, there is no vortex formation, Froude number has no effect.

• Define power number for an impeller of a mixer?

$$P_o = P/N^3 D^5 \rho. \tag{6.4}$$

- Impeller power number is a dimensionless number expressing the power drawn by an impeller and is a form of drag coefficient.
- Power number expresses the ratio of energy causing local turbulence to that providing bulk flow and is

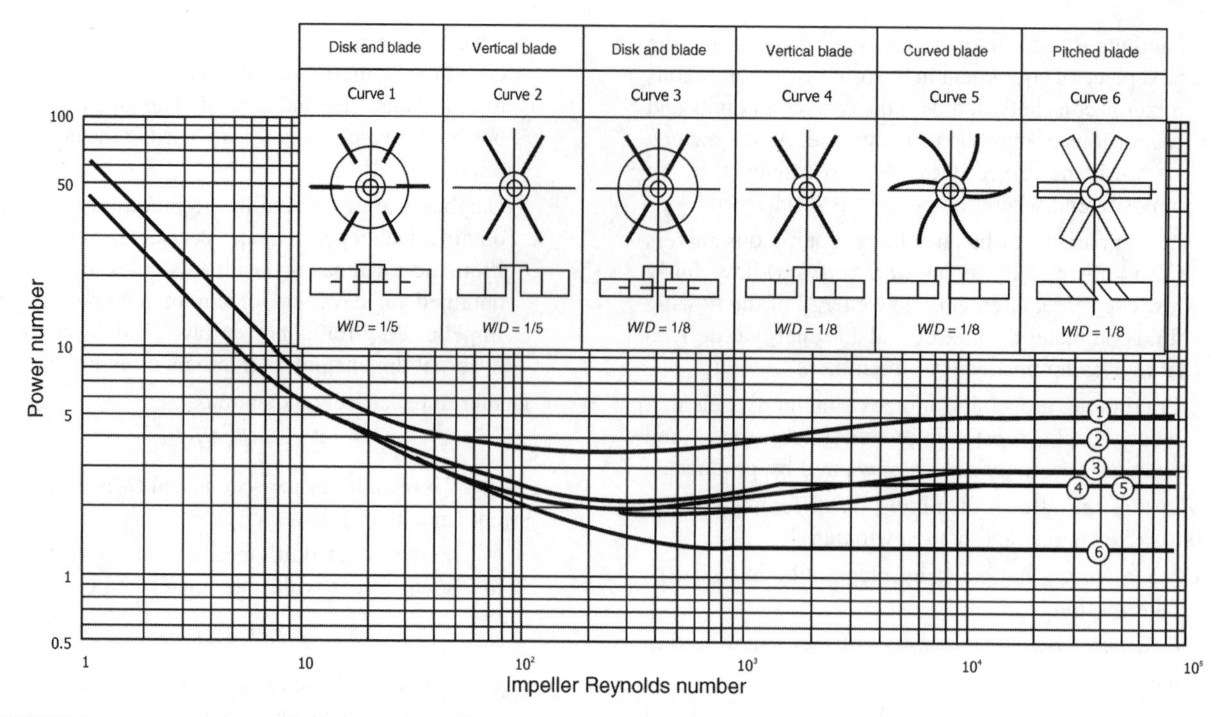

FIGURE 6.1 Power number as a function of Reynolds number for different turbine impellers. *D*: impeller diameter; *W*: blade width.

sensitive to anything that changes flow patterns produced by the impeller.

- Power number is inversely proportional to N_{Re} in the laminar regime, has weaker dependence on N_{Re} in the transition regime, and is almost independent of N_{Re} under highly turbulent conditions.
- Power number will be low for impellers that produce a uniform flow with relatively low levels of local turbulence (e.g., narrow blade hydrofoils, $P_o \cong 0.3$) and much higher for impellers that produce a highly nonuniform flow with high local turbulence intensity (e.g., the disk turbine, $P_o \cong 5$).
- It is used to predict impeller power directly and torque indirectly.

• How is power number related to Reynolds number for mixing?

$$P_o = K(N_{Re})^n (N_{Fr})^m. \quad (6.5)$$

- $K, n,$ and m are constants, whose values depend on the type of impeller. For example, for a propeller with a pitch equal to the diameter, $n = -1$ and $K = 41$, according to one reference, based on experimental data.
- The Froude number correlates the effects of gravitational forces and it becomes significant only when the propeller disturbs the liquid surface. Below Reynolds numbers of about 300, the Froude number is found to have little or no effect and

$$P_o \approx K(N_{Re})^n. \quad (6.6)$$

- Figure 6.1 illustrates the relationship between N_{Re} and P_o.
- In most industrial situations, power number of an impeller is constant under turbulent conditions. This means that under turbulent conditions, power is proportional to fluid density and independent of viscosity.
- Under laminar conditions, power number is inversely proportional to Reynolds number, which means that power is proportional to viscosity and independent of density.
- Under transitional conditions, impeller power is influenced by both viscosity and density.
- Under turbulent conditions, typical power numbers are listed in Table 6.1 for different impeller designs.

• What is pumping number/flow number?

$$N_Q = Q/ND^3. \quad (6.7)$$

- It is a dimensionless number used to represent actual flow during mixing in a vessel.

TABLE 6.1 Power Numbers for Different Impeller Designs

Impeller Type	Number of Blades	Power Number
45° pitched blade turbine	4	1.37
Straight blade turbine	4	3.96
Straight blade turbine	6	4.92
Disk turbine	6	5.0
Hydrofoil impeller	3	0.31

- Define and explain *agitation intensity number*.
  - Tank turnover rate has been used as a criterion for agitation intensity.
  - Agitation intensity is directly proportional to the rate at which tank contents are turned over.
  - Agitation intensity is assumed to vary linearly with tank fluid velocities.
  - Range of agitation intensity (scale) is defined by bulk fluid velocities, V_b, ranging from 1.8 to 18 m/min (6–60 ft/min).
  - A 1–10 scale is established for the above range.
  - Agitation intensity number

$$N_I = V_b/(6\,\text{ft/min}). \tag{6.8}$$

$V_b = q/A$, where q is the capacity at volumetric flow rate and A is the cross-sectional area of the vessel.

- Give examples of applications of *flow-sensitive* mixing systems in process industry, indicating the ranges of agitation scales for each of these applications.
  - Table 6.2 classifies mixing in terms of mixing intensity along with applications in industry.
- How is mixing intensity related to impeller tip speeds?
  - Table 6.3 gives the relationship between mixing intensity and impeller tip speed.
- Illustrate how solids settling velocities and intensity of agitation are related.
  - For settling velocities around 0.9 cm/s, solids suspension can be accomplished with turbine or propeller impellers.

TABLE 6.2 Examples of Mixing Processes Based on Mixing Intensity Scales

Agitation Intensity, N_i	Applications
Mild, ≤1	Noncritical blending operations; blending to prevent concentration surges; storage or holding tanks; feed tanks; equalization basins; water treatment; flocculation; and so on (surface barely in motion)
Moderate, 2–3	Make up tanks; reaction tanks; blend tanks; pigment suspension (paint); maintaining suspension (surface in strong motion)
Vigorous, 3–8	Critical mixing operations; most heat transfer; pH control; reactors; blend tank for adhesives
Violent, 6–10	Special critical applications; high shear requirements; critical heat transfer; emulsion polymerization; monomer emulsions in water with stabilizers; bulk polymerization; polymer in molten form or solution in monomer; reactors; surface boiling; splashing; vortexing

TABLE 6.3 Mixing Intensity Versus Impeller Tip Speed

		Tip Speed	
Intensity	N_I	(m/min)	(ft/min)
Low	5.3–6.8	150–200	500–650
Medium	6.8–8.4	200–245	650–800
Very high	8.4–11.6	245–335	800–1100

  - For settling velocities above 4.5 cm/s, intense propeller agitation is needed.
- Define impeller blend number.
  - Impeller blend number is used to predict the blend time in a mixing system:

$$N_B = n\theta(D/T)^{2.3}, \tag{6.9}$$

where n is the rotational shaft speed (rpm) and θ is the blend time (s).

- Define impeller force number.
  - Impeller force number, N_F, correlates the axial force, F_{ax}, or the thrust generated by an impeller. F_{ax} is used in the correlation to predict cavern dimensions and is also important for mechanical design considerations.

$$N_F = F_{ax}/\rho n^2 D^4. \tag{6.10}$$

  - Cavern is used to describe the well-mixed, turbulent region around the impeller. Equations are developed to predict cavern diameters.
- Give an equation for blend time to achieve the desired concentration?
  - Grenville gives the blend time to achieve within 5% of the desired concentration as

$$(\text{blend time})_{95\%} = 5.4(T/D)^2/N_p^{1/3}N. \tag{6.11}$$

  - This correlation is claimed to be valid for turbulent conditions with $N_{Re} > 10{,}000$ and a liquid depth equal to the vessel diameter, T. D is the impeller diameter, N_p is the power number, and N is rpm.
- While mixing low to moderate viscosity fluids, how turbulence and vortex formation are to be handled?
  - Turbulence should be induced to entrain slow moving parts within faster moving parts. Turbulence is highest near the impeller and liquid should be circulated through this region as much as possible.
  - A vortex should be avoided because adjoining layers of circulating liquids travel at a similar speed and entrainment does not take place—the liquids simply rotate around in the mixer.
- What are the superficial fluid velocity ranges in a mixer for mild and intense mixing?
  - Mild agitation results from superficial fluid velocities of 0.03–0.06 m/s (0.10–0.20 ft/s).

- Intense agitation results from velocities of 0.21–0.30 m/s (0.70–1.0 ft/s).
- Describe agitation achievable in relation to fluid velocities for different fluid systems.
  - Table 6.4 gives agitation achievable for different fluid velocities.
- Give approximate ranges of power requirements for different applications for baffled agitated tanks.
  - Table 6.5 gives power requirements for baffled vessels for different applications.
- "In a mixer, high pumping capacity is obtained by using large diameter impellers at slow speeds compared to higher shear rates obtained by using smaller impellers and higher speeds." *True/False*?
  - *True*. The pumping capacity of mixing impellers is proportional to ND^3, as indicated by the relationship

$$Q \propto ND^3, \tag{6.12}$$

  where Q is the pumping capacity, N is the impeller speed, and D is the impeller diameter.
- "Power requirements for mixing a gas with a liquid are more than those for mixing the liquid alone." *True/False*?
  - *False*. Power requirements for gas–liquid systems can be 25–50% *less*, depending on gas/liquid ratios, than those for liquid systems alone.
- What are the ranges of relative power requirements for mixing a gas with a liquid and the liquid alone?
  - Power to mix a gas and a liquid can be 25–50% less than the power to mix the liquid alone.
- What are the disadvantages of vortex formation in mixing operations?
  - Once vortex reaches impeller, severe air entrainment may occur.
  - In addition to air entrainment, swirling mass of liquid may generate an oscillating surge in the tank, which, coupled with the deep vortex, may create a large fluctuating force acting on the shaft.
- What are the undesirable effects of vortex shedding?
  - Severe vibration.
  - Mechanical failure of cylindrical elements such as suspended piping, transmission lines, heat exchanger tubes, columns, stacks, and so on.
- How are vortices broken/eliminated?
  - By the use of baffled tanks and vortex breakers. For axial flow impellers, the effect of full baffling can be achieved in an unbaffled vessel with an off-center and angled impeller shaft location. Off-center angled shaft eliminates vortexing and swirling to a large extent.

TABLE 6.4 Agitation Achievable for Different Fluid Velocities

Agitation Results Corresponding to Specific Superficial Velocities

Superficial Velocity (cm/s)	Description
Liquid systems	
3–6	Low degree of agitation; a velocity of 6 cm/s will blend miscible liquids to uniformity when $\Delta\rho < 0.1$; blend miscible liquids to uniformity if $\mu_1/\mu_2 < 100$; establish liquid movement throughout the vessel; produce a flat but moving surface
9–18	Characteristic of most agitations used in chemical processing; a velocity of 18 cm/s will blend miscible liquids to uniformity if $\Delta\rho < 0.6$; blend miscible liquids to uniformity if $\mu_1/\mu_2 < 10{,}000$; suspend trace solids (<2%) with settling rates of 0.60–1.2 m/min; produce surface rippling at low viscosities
21–30	High degree of agitation; a velocity of 30 cm/s velocity will blend miscible liquids to uniformity if $\Delta\rho < 1.0$; blend miscible liquids to uniformity if $\mu_1/\mu_2 < 100{,}000$; suspend trace solids (<2%) with settling rates of 1.2–1.8 m/min; produce surging surface at low μ
Solids suspension	
3–6	Minimal solids suspension; a velocity of 3 cm/s will produce motion of all solids with the design settling velocity; move fillets of solids on the tank bottom and suspend them intermittently
9–15	Characteristic of most applications of solids suspension and dissolution; a velocity of 9 cm/s will suspend all solids with the design settling velocity completely off the bottom of the vessel; provide slurry uniformity to at least one-third of the liquid level; be suitable for slurry draw-off at low exit nozzle locations
Gas dispersions	
3–6	Used when degree of dispersion is not critical to the process; a velocity of 6 cm/s will provide nonflooded impeller conditions for coarse dispersion; be typical of situations that are not mass transfer limited
9–15	Used where moderate degree of dispersion is needed; a velocity of 15 m/s will drive fine bubbles completely to the wall of the vessel; provide recirculation of dispersed bubbles back into the impeller
18–30	Used where rapid mass transfer is needed; a velocity of 30 cm/s will maximize interfacial area and recirculation of dispersed bubbles through the impeller

TABLE 6.5 Power Requirements for Baffled Vessels

Agitation	Applications	Power (kW/m^3)
Mild	Blending, mixing	0.04–0.10
	Homogeneous reactions	0.01–0.03
Medium	Heat transfer	0.03–1.0
	Liquid–liquid mixing	1.0–1.5
Vigorous	Slurry suspension	1.5–2.0
	Gas absorption	1.5–2.0
	Emulsions	1.5–2.0
Intense	Fine slurry suspension	>2.0

- What are the means normally used to increase mixing in a reactor?
  - By adding/improving baffles.
  - Using a different mixer blade design.
  - Installing higher rpm motor on the agitator.
  - Using multiple impellers.
  - Pumped recirculation: addition and/or increasing the quantity.
  - Using in-line static mixers for mixing entering streams.
- How is a mixing operation scaled up?
  - When changing the scale, that is, size of equipment, with the goal of obtaining the same mixing performance at the new scale:
    (i) Geometric similarity should be preserved: dimensional ratios should be the same in the large tank as in the small one.
    (ii) Dynamic similarity should be preserved: Reynolds numbers should be the same in the large tank as in the small one.
  - It must be noted that the above analysis is correct for Newtonian fluids only: it becomes more complex when the non-Newtonian behavior of materials is taken into account.

6.2 MIXING EQUIPMENT

6.2.1 Baffled Vessels

- What are the generally used dimensions for a mixing vessel?
  - Generally circular with rounded bottoms to minimize dead spots.
  - Depth to diameter ratio is normally 0.5–1.5 (1.0 is often recommended).
  - If tall vessels are used, impellers should be installed for each vessel diameter equal to height.
- What is the purpose of using baffles in a mixer?
  - During agitation of a low-viscosity liquid, the rotating impeller imparts tangential motion to the liquid. Without baffling, this swirling motion approximates solid body rotation in which little mixing actually occurs.
  - For example, while stirring a cup of coffee or a bowl of soup, the majority of the mixing occurs when the spoon is stopped or the direction of stirring is reversed.
  - The primary purpose of baffling is to convert swirling motion into a preferred flow pattern to accomplish process objectives.
  - The most common flow patterns are axial flow, typically used for blending and solids suspension, and radial flow, used for dispersion.
  - The presence of baffles produces axial flow, in which the discharge flow produced by the impeller impinges on the base of the vessel, flows radially to the vessel wall, then up the wall, returning to the impeller from above.
  - Baffling has some other effects as well, such as suppressing vortex formation, increasing the power input, and improving mechanical stability.
- Give commonly adopted geometries for an agitated tank relative to its diameter (D).
  - Various geometries of an agitated tank relative to diameter (D) of the vessel include the following:
    - Liquid level $= D$.
    - Turbine impeller diameter $= D/3$.
    - Impeller level above bottom $= D/3$.
    - Impeller blade width $= D/15$.
    - Vertical baffle width $= D/10$ (four baffles).
- What is standard baffling for an agitated vessel?
  - Standard baffling consists of three or four flat vertical plates, each baffle occupying about 1/10th of tank diameter and radially directed (i.e., normal to the vessel wall), spaced at 90° around the vessel periphery and running the length of the straight side of the vessel. They prevent tangential flow.
  - Flat plate baffles are common because of their ease of manufacture and installation and the associated economy.
  - Standard baffle width is 1/10th or 1/12th of the vessel diameter.
  - Sometimes, baffles are flushed with the vessel wall and base, but, more often, gaps are left to permit the flow to clean the baffles. Recommended gaps are equal to 1/72nd of the vessel diameter between the baffles and the vessel wall and 1/4th to one full baffle width between the bottom of the baffles and the vessel base.
- Give some rules of thumb for mixing.
  - For fluid with viscosities less than 10,000 cP, baffles are highly recommended. There should be four

baffles, 90° apart. The baffles should be 1/12th the tank diameter in width and should be spaced off the wall by 1/5th the baffle width. The off-wall spacing helps eliminate dead zones.

- If baffles are used, the mixer should be mounted in the vertical position in the center of the tank. If baffles are *not used*, the mixer should be mounted at an angle of about 15° to the right and positioned off-center. This breaks up the symmetry of the tank and simulates baffles although not nearly as good as baffles.
- The purpose of baffles is to prevent solid body rotation, that is, all points in the tank move at the same angular velocity and there is no top-to-bottom turnover. The formation of a large central vortex is a characteristic of solid body rotation.
- However, small vortices, which travel around the fluid surface, collapse, and reform, are more a function of the level of agitation. Violent and vigorous agitation will have these vortices present. In fact, they are desired for processes that require solids addition from the liquid surface.
- The impellers are located at different positions depending on the design.
- Axial flow impellers should be positioned between 0.5 and 1.5 impeller diameters off the bottom of the tank. Radial flow impellers can be positioned just few centimeters off the bottom.
- If multiple impellers are used, the spacing will depend on the liquid height to tank diameter ratio. Care must be taken to prevent impeller spacings of one impeller diameter. This can lead to a cancellation of flow.
- Torque is one of the most important factors. A large diameter slow spinning impeller is much better for blending than a small diameter fast spinning impeller at equal power levels.
- Impeller diameters for relatively low viscosities should be between 0.25 and 0.45 times the tank diameter with some exceptions.
- For turbulent mixing, large diameter, low-speed, and axial flow impellers are preferable.

• "It is easier to mix a high-viscosity liquid with a low-viscosity liquid than the other way round." *True/False*?
  - *True*.

• What are the speed ranges for high- and low-speed impellers? Give the applications with respect to speeds.
  - High speeds: 1150–1750 rpm for liquids such as aqueous solutions.
  - Medium speeds: 450–1150 rpm for liquids of medium viscosity such as varnishes, medium oils, syrups, and so on.
  - Low speeds: 35–420 rpm for liquids under the following conditions:
    - When the mixture is thick, viscous, or slippery, so that small high-speed propellers tend to channel rather than propel the entire mass into circulation.
    - When the mixture contains particulates such as crystals, sliced fruit, or other foods that will be broken up at high speeds.
    - When the mixture is of foamy nature and the foam is undesirable.
    - When colloids such as milk and cream are damaged at high speeds.

• Give a diagram of a baffled draft tube mixer showing standard dimensions.
  - Figure 6.2 illustrates standard recommended dimensions for a baffled draft tube mixer.
  - Arrangement of heat transfer surfaces is also shown.
  - A dished bottom requires less power than a flat one.
  - When a single impeller is to be used, a liquid level equal to the diameter is optimum, with the impeller located at the center for an all-liquid system. Economic and manufacturing considerations, however, often dictate higher ratios of depth to diameter.
  - Except at very high Reynolds numbers, baffles are needed to prevent vortexing and rotation of the liquid mass as a whole.
  - A baffle width $w = D_t/12$ is used.
  - Baffle length extending from one-half the impeller diameter, $d/2$, from the tangent line at the bottom to the liquid level is normally used, but sometimes terminated just above the level of the eye of the uppermost impeller.
  - When solids are present or when a heat transfer jacket is used, the baffles are offset from the wall a distance equal to one-sixth the baffle width. Four radial baffles at equal spacing are used in standard designs. Heat transfer surfaces are often installed inside the vessel and jackets (both sidewall and bottom head) so that the vessel wall and bottom head can be used as heat transfer surfaces.
  - Helical coils are attached to wall baffles, as shown in Figure 6.2, with tube diameters typically equal to $Z/30$ and tube row spacing equal to one tube diameter.
  - Six baffles are only slightly more effective and three are appreciably less effective.
  - When the mixer shaft is located off-center (one-fourth to one-half the tank radius), the resulting flow pattern has less swirl, and baffles may not be needed, particularly at low viscosities.

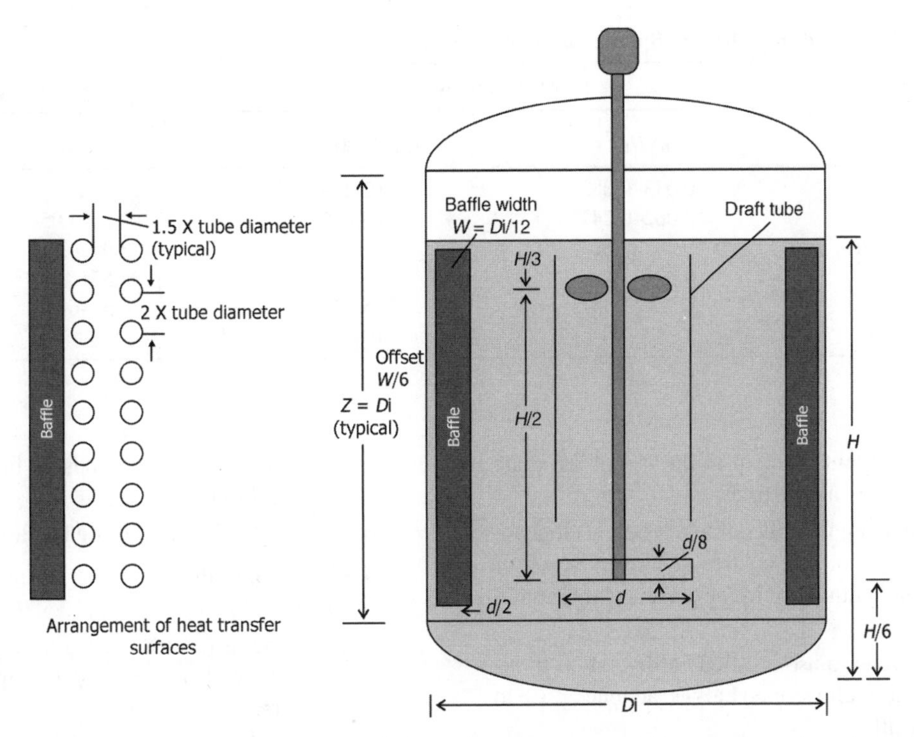

FIGURE 6.2 A basic stirred tank design showing a lower radial impeller and an upper axial impeller housed in a draft tube (not to scale). Four equally spaced baffles are standard. *H*: height of liquid level; D_i: tank diameter; *Z*: height of the vessel (excluding heights of dished heads) *d*: impeller diameter. For radial impellers, $0.3 \leq d/D_i \leq 0.6$.

- An impeller–draft tube system behaves as an axial flow pump of somewhat low efficiency. Its top-to-bottom circulation behavior is of particular value in deep tanks for suspension of solids and for dispersion of gases.

- What are the effects of reducing/removing baffles in a mixing vessel?
  - Significantly increases tangential (swirling) flow of the liquid, rather than the radial and axial flow generated by the well-baffled system.
  - This reduces or even inhibits top-to-bottom mixing and entrains surface gas, even at low impeller speeds.
- "For standard baffles, power and pumping numbers are essentially independent of the impeller Reynolds number." *True/False*?
  - *True*.
- Above what impeller Reynolds numbers, for standard baffles, fully turbulent agitation occurs?
  - Above 10,000.
- What are the effects of increasing number of baffles on (a) power requirements and (b) time of agitation?
  - (a) Power requirements increase, particularly for larger impeller submergence and (b) time of agitation decreases.
- For different applications of baffled tanks, illustrate how power inputs and impeller tip speeds are related.
  - For baffled tanks, agitation intensity is measured by power input and impeller tip speeds, as illustrated in Table 6.6.
- Under what circumstances, baffles are not recommended/not used in agitated vessels?
  - Side-entering agitators or with close clearance impellers, such as gates, anchors, and helical ribbons, for which impeller to tank diameter ratio is more than 90%.
  - Rectangular or square tanks that prevent swirling by providing some natural baffling in their sharp corners.
  - No or narrow baffles are used for impeller Reynolds numbers <50 because of the viscous effects eliminating near-solid body rotation of the liquid.
  - Small agitated vessels also do not generally use baffles, and instead, use of angled and/or off-center mounting of agitator eliminates excessive swirling.
    - This arrangement involves high costs and mechanical complications in the case of large agitators.
  - Baffles are not favored for vessels that require sterility or in which material hang-up during draining is problematic.
  - Vessels containing heating/cooling tubes also do not use baffles (e.g., highly exothermic/endothermic reaction vessels).

TABLE 6.6 Power Input and Impeller Tip Speeds for Baffled Tanks

	Power Requirements		Tip Speeds	
Process	(kW/m^3)	(hp/1000 gal)	(m/s)	(ft/s)
Blending	0.033–0.082	0.2–0.5	–	–
Homogeneous reaction	0.082–0.247	0.5–1.5	2.29–3.05	7.5–10
Reaction with heat transfer	0.247–0.824	1.5–5.0	3.05–4.57	10–15
Liquid–liquid mixtures	0.824	5.0	4.57–6.09	15–20
Liquid–gas mixtures	0.824–1.647	5–10	4.57–6.09	15–20
Slurries	1.647	10	–	–

- What is the important problem in the use of flat plate baffles? What is the alternative?
  - Material can hang up or become trapped in stagnant regions near them, particularly in more viscous or non-Newtonian liquids or in the presence of filamentous materials.
  - An alternative is to use profiled baffles, often triangular or semicircular shaped baffles, attached flush to the vessel wall.
    - Use is limited to critical applications such as polymerization reactors and clean-in-place (CIP) reactors that are commonly used in pharmaceutical industry.
- How are baffles fitted in glass-lined vessels?
  - The baffles in these vessels hang from flanges in the top head of the vessel.
    - Not more than two baffles are fitted this way due to space limitations.

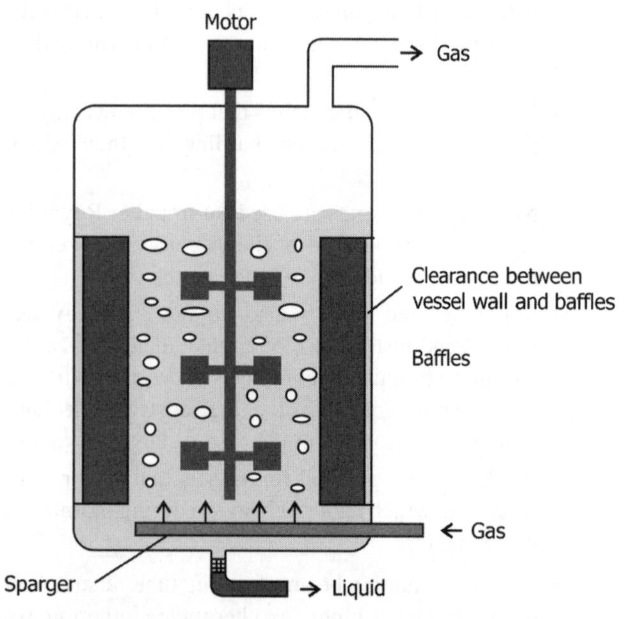

FIGURE 6.3 Multistage agitator with baffles and sparger for gas–liquid reactors.

- Illustrate a baffled vessel used as a gas–liquid reactor.
  - Figure 6.3 illustrates the essential features of a gas–liquid reactor.
  - This type of reactor is used as a gas–liquid reactor, equipped with baffles and multiple turbine/paddle agitators.
  - A sparger supplies the gas at the bottom of the vessel below the mixer.
  - Gas bubbles from the sparger will result in mild agitation at a superficial gas velocity of 0.3 m/min and intense agitation at over 0.3 m/min.
  - Notice that there is a clearance between the baffles and vessel walls to facilitate movement of the fluids close to the vessel wall to prevent any deposits forming on the wall and baffle surfaces.
- "Power requirements in a mixer increase with increased number of baffles." *True/False*?
  - *True.*
- Name different types of devices that can induce mixing.
  - Flow/in-line mixers: small time of contact.
  - Jet mixers: impingement of one liquid on another liquid.
  - Injectors: flow of one liquid is induced by flow of majority liquid pumped at high velocity.
  - Orifices and mixing nozzles.
  - Valves.
  - Pumps.
  - Packed tubes.
  - Agitated line mixer.

6.2.2 Impellers

- What are the general types of impellers used in mixing processes?

- Impellers can be divided into two general classes, axial flow/hydrofoil or radial flow, depending on the angle the blade makes with the axis of the drive shaft.
- Axial flow and hydrofoil impellers make an angle of less than 90° with the mixer shaft. Axial flow impellers are used at high speeds to promote rapid dispersion and used at low speeds for keeping solids in suspension. The impellers tend to have high flow rates and low shear rates. Often used with low-viscosity fluids (<1000 cP). In most cases, baffling is required for the optimal performance of this impeller design. Examples of axial flow impellers include the marine propeller, fan turbine, and pitched paddle.
- Radial flow impellers have blades that are parallel to the axis of the mixer shaft. The smaller, multiblade types are called turbines. Larger, slow-speed impellers, with two to four blades, are known as paddles, while other larger two-sided types are known as gate or anchor impellers.
- Radial flow impellers are used in turbines, which come in a variety of types, and provide excellent circulation of fluid throughout the mixing vessel. These impellers tend to provide low flow rates and higher shear rates. The diameter of a turbine impeller is normally between 0.3 and 0.6 of the tank diameter.
- Paddles are used at slower speeds and normally have a diameter greater than 0.6 of the tank diameter.
- Turbine impellers and paddles are the two types most often used for large-scale mixing of solid/liquid suspensions.
- Gate and anchor impellers are used to sweep the entire peripheral area of the tank, both walls and bottom.

• Name different types of mixers.
  - Pitched blade turbine, flat blade turbine, propeller, paddle, rotating disk, anchor, and helical ribbon types.
• Name some axial flow impellers.
  - Propeller and pitched blade turbine impellers.
• Name some radial flow impellers.
  - Turbines and paddles. Turbines are of two types:
    - Curved blade: Advantageous in starting the impeller in settled solids.
    - Flat blade.

6.2.3 Turbine and Propeller Mixers

• Give diagrams of turbine and propeller mixers showing flow lines.

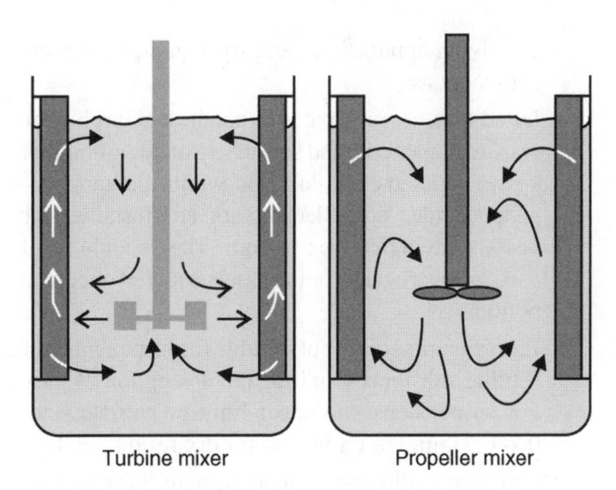

FIGURE 6.4 Turbine and propeller mixers.

  - Figure 6.4 illustrates flow lines for turbine and propeller mixers.

• What are the characteristic features of propeller mixers?
  - These are short bladed impellers (usually less than 25% of vessel diameter) rotating at high speeds of 500 to several thousand rpm.
  - They are capable of high-speed operation (up to 1800 rpm) without the use of a gear box and hence provide a more cost-effective operation because there are no mechanical losses in transmission.
  - Produce primarily longitudinal (axial flow) and rotational velocities.
  - Can be effective in quite large vessels with low-viscosity liquids (up to about 4000 cP), but generally not mounted centrally. Inclined mounting at an angle or even horizontal mounting through the vessel wall.
  - Due to their more streamlined shape, power requirements for a propeller mixer are less than those for the other types of mixers at the same Reynolds number.
  - Depending on the height of the liquid layer, one shaft may carry one to three propellers. Single impellers are used for tank height to diameter ratios of ≤1.0. Dual impellers are used for tank height to diameter ratios of ≥1.5. Duel impellers are also used for mixing solids in liquids with difficult wettability. The upper impeller is located on the shaft just below liquid level to create slight vortex, which draws the solids in, while the lower impeller is used for thorough mixing of the solids with the liquid.
  - Propeller mixers are used for mixing liquids with viscosities up to 2000 cP. They are suitable for the formation of low-viscosity emulsions, for

dissolving applications, and for liquid-phase chemical reactions.

- Disadvantages, compared to paddle and turbine mixers, are higher cost and sensitivity of operation to the vessel geometry and location within the tank. As a general rule, propeller mixers are installed with vessels having convex bottom. They should not be used in square tanks or in vessels with flat or concave bottoms.
- Propeller mixers are not suitable for suspending rapid settling substances and for the absorption of gases. For suspensions, the upper limit of particle size is 0.1–0.5 mm, with a maximum dry residue of 10%.
- Figure 6.5 illustrates three designs of propellers, namely, marine, saw-toothed, and perforated types.
- Marine type circulates by axial flow parallel to the shaft and its flow pattern is modified by baffles, normally a downward flow.
  - Operates over wide speed range.
  - Can be pitched at various angles, most common is the three blades on square pitch (pitch equal to diameter).
  - Shearing action is very good at high speed, but not generally used for this purpose.
  - Power consumption is economical.
  - Generally self-cleaning.
  - Relatively difficult to locate in vessels to obtain optimum performance.
  - Not effective for viscous liquids.
- Sawtooth design displaces a large amount of liquid and combines cutting and tearing action. Suitable for fibrous materials.
- Perforated propeller is sometimes recommended for wetting dry powders, especially those that form lumps.

• What are the characteristic features of turbine mixers?

- Turbine impellers have a constant blade angle with respect to a vertical plane, over its entire length or over finite sections, having blades either vertical or set at an angle less than 90° with the vertical plane, usually 45°.
- Blades may have different configurations, curved or flat. Impeller has generally more than four blades in the same plane of rotation. Number of blades is in the range of 2–8, with 6 blades being common.
- The predominantly radial flow from the impeller impinges onto the vessel walls, where it splits into two streams.
- Generally smaller than paddles, measuring 30–50% of the vessel diameter. Speeds of rotation are 30–500 rpm.
- Baffles and pitched blades may be used in a similar way as with paddles.
- Velocities of the liquid are relatively high and fluid currents travel throughout the vessel, with vertical flow often being set up by the deflection of currents from the vessel walls.
- When operated at high rotational speeds, both radial and tangential flows become pronounced, along with vortex formation. In such situations, it is preferable to use baffles to ensure a more uniform flow distribution throughout the mixing vessel.
- Vaned disk impellers may be used to disperse gases in liquids. Suited for liquids up to 100,000 cP at high pumping capacity and particularly effective in moderately viscous liquids. Used for fairly high shear and turbulence. Better than axial flow units for tanks with conical bottom of >15° angle, to lift material from bottom of the cone and mix with bulk of the liquid.
- Curved blade turbines effectively disperse fibrous materials without fouling. The swept back blades have a lower starting torque than straight ones, which is important when starting up settled slurries.
- Shrouded turbines consisting of a rotor and a stator ensure a high degree of radial flow and shearing action and are well adapted to emulsification and dispersion.

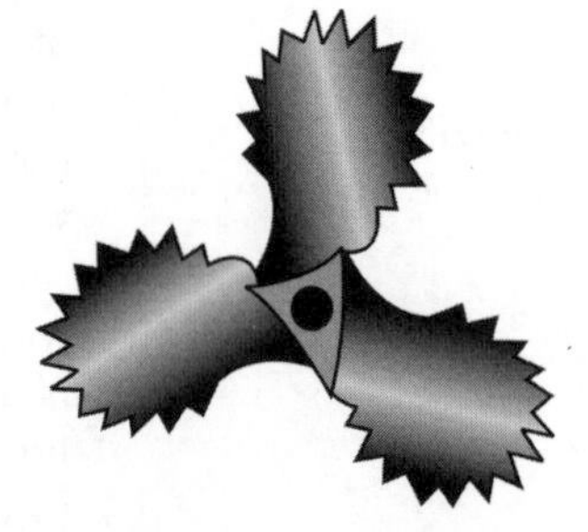

FIGURE 6.5 Marine, saw-toothed, and perforated propellers.

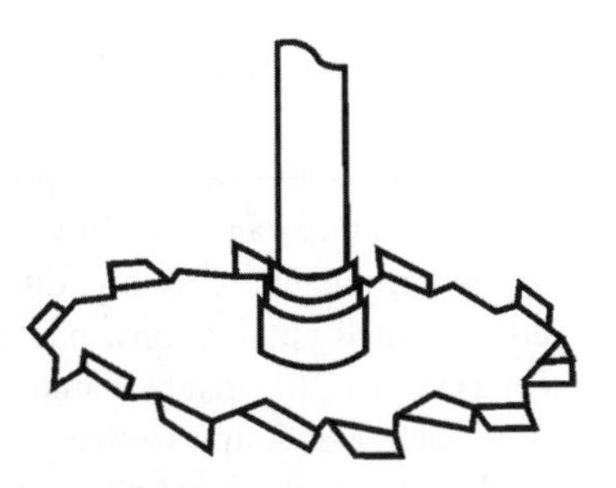

FIGURE 6.6 Flat plate impeller with sawtooth edges.

- Flat plate impellers with sawtooth edges are suited to emulsification and dispersion. Since the shearing action is localized, baffles are not required. Figure 6.6 illustrates the design.

- Illustrate different types of turbine impellers.
  - Figure 6.7 illustrates some turbine impellers.
  - A standard six-blade (vertical curved) turbine gives good efficiency per unit of power for suspensions and mixing fibrous materials. Gives high pumping capacity.
  - A flat blade disk turbine is used for gas dispersion at low and intermediate gas flow rates.
  - A straight blade turbine involves local liquid motion and is used for blending, dispersion, and low liquid level solids suspension, keeping outlets clear of solids.
  - A pitched blade turbine gives mainly axial flow design (with significant radial component) for wide changes in process viscosity. Used for blending, liquid dispersion, and solids suspension where increased shear is required.
  - A narrow blade turbine is used for liquid–liquid and solid–liquid dispersions.
- What are the recommended dimension proportions used in turbine agitated vessels?
  - Tank to impeller diameter ratio = 2–4.
  - Impeller diameter = ≤ 5 m.
  - Impeller blade width = $D/15$.
  - Maximum tank diameter = 20 m.
  - Liquid level = one tank diameter, D.
  - Impeller level above tank bottom = one-third of the tank diameter.
  - Four vertical baffles with baffle width = $D/10$.
- Name the types of impellers used for mixing low-viscosity liquids.
  - Propellers, axial, and radial flow turbines.
- What are the considerations involved in the selection of mixers used in tank mixing, with respect to their installation and orientation?
  - Impellers used in small portable mixers are often inclined at an angle and are located off-center to give a good top-to-bottom flow pattern in the system.
  - Large top-entering drives usually use either the axial flow turbine or the radial flow flat blade turbine.
  - Bottom-entering drives have the advantage of keeping the mixer off the top of tanks and required superstructure, but have the disadvantage that if the sealing mechanism fails, the mixer is in a vulnerable location for damage and loss of product by leakage.
  - Side-entering mixers are used for many types of blending and storage applications.

Standard six-blade (vertical curved) turbine

Flat blade disk turbine (Ruston turbine)

Standard blade turbine

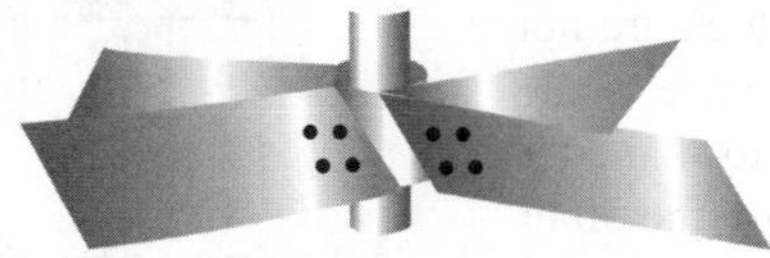

Pitched blade turbine

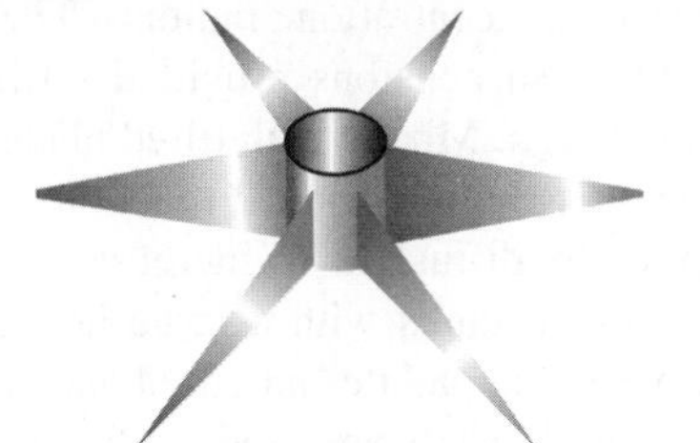

Narrow blade turbine

FIGURE 6.7 Different types of turbine impellers.

6.2.4 Paddle Mixers

- What are the types and characteristic features of paddle mixers?
  - Paddle mixers consist of two or more blades mounted on a vertical or inclined shaft.
  - Speeds normally are in the range of 20–50 rpm. Under these conditions, mixing action is small and vortex formation is prevented.
  - Paddle mixers are often used in vessels without baffles. In broad blade paddles, which operate at speeds up to 120 rpm, baffles are incorporated into the design to minimize vortex formation.
  - Commonly used paddles are of half to three quarters the diameter of the vessel with width of the blades 1/10th to 1/6th of their length.
  - Consist of high radial and rotational components. Little or no vertical flow, unless baffles are incorporated.
  - Pitched blades may be used to increase vertical flow.
  - Multivane or gate agitators are used for more viscous liquids and anchor paddles are used to just clear vessel walls to promote heat transfer and minimize deposits; counterrotating multiblade paddles may be used to develop high localized shear.
  - Single blade paddles are often used for gentle mixing action with sensitive materials. Flow capacity can be high for multiple blades.
  - Baffling is often used to reduce swirling and vortexing.
  - Easy to fabricate and are of low cost.
  - One of the disadvantages is slow axial flow, which does not provide good mixing of the tank volume. Good mixing is attained only in a relatively thin stratum of liquid in the immediate vicinity of the blades. Therefore, paddle mixers are used for liquids with viscosities only up to about 1000 cP.
  - Because of a concentration gradient that is often created in the liquid, they are not suitable for continuous operation.
  - Tilting the paddle blades 30–45° to the axis of the shaft produces increased axial flow, thereby decreasing concentration gradients. These mixers can maintain suspensions provided settling velocities are not high. Mixers with tilted blades are used for processing slow chemical reactions, which are not limited by diffusion. To increase turbulence of the medium in tanks with a large height to diameter ratio, several paddles mounted one above the other on a single shaft are used. Individual paddles are separated by distances between them in the range of 0.3–0.8d, where d is the diameter of the paddle. These distances are based on the viscosity of the mixture.
  - For mixing liquids with viscosities up to 1000 cP, as well as for heated tanks in cases in which sedimentation can occur, anchor or gate paddle mixers are employed. For such applications, paddle diameters are almost as large as the inside diameter of the tank so that the outer and bottom edges of the paddle scrape or clean the walls and bottom.
  - Leaf-shaped (broad blade) paddle mixers provide a predominant tangential flow of liquid, but there is also turbulence at the upper and lower edges of the blade. These are employed for mixing low-viscosity liquids, intensifying heat transfer processes, promoting chemical reactions in a reactor, and dissolving materials.
  - Not easily fouled.
  - Figure 6.8 illustrates gate and leaf impellers.
  - For dissolving applications, leaf blades are usually perforated. Jets formed at the exits from the holes promote the dissolution of materials.
- Speed of rotation of paddle mixers is around (i) 80–150 m/s; (ii) 1000 m/s; (iii) 500 m/s. Give the correct answer.
  - 80–150 m/s.
- What are the applications of single shaft ribbon and paddle mixers?
  - Used for simple applications, such as light blending, easy backmixing for flash dryers and the like, and attenuating fluctuating feeder outputs.
  - Simple ribbon mixers allow a degree of backmixing that causes light mixing and axial diffusion. The blades can be cut and folded, or pegs fitted, to give extra disturbance to the contents but only for relatively free-flowing bulk materials.
  - Paddles may be quadrant or ribbon shaped and sometimes sharpened to separate the bulk easier and reduce buildup on the ribs.
- Summarize advantages and disadvantages of important liquid mixers.
  - Table 6.7 gives advantages and disadvantages of major liquid mixers.

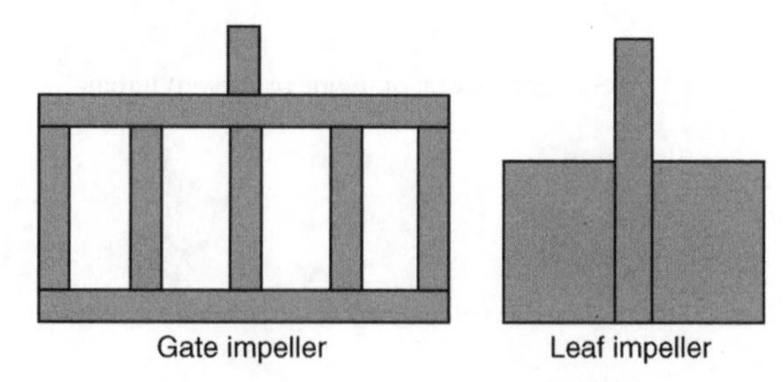

FIGURE 6.8 Gate and leaf impellers.

TABLE 6.7 Advantages and Disadvantages of Selected Liquid Mixers

Type of Mixer	Advantages	Disadvantages
Paddle mixer	Good radial and rotational flow; cheap	Poor perpendicular flow; high vortex formation risk at higher speeds
Multiple paddle mixer	Good flow in all three directions	More expensive; higher energy requirements
Propeller mixer	Good flow in all three directions	More expensive than the paddle mixer
Turbine mixer	Very good mixing	Expensive; blockage risks

6.2.5 Draft Tubes

- What is a draft tube? What is its purpose in an agitator?
  - The draft tube directs the flow to the regions of the vessel that otherwise would not be agitated by the liquid stream.
  - It is a tube or shell around the shaft of a mixer, enclosing the usual axial flow impeller, which allows a special or top-to-bottom fixed flow pattern to be set up in the fluid system.
  - Draft tubes are employed to improve the mixing of large quantities of liquids by directing the motion of the liquid.
  - Size and location of the tube are related to both the mechanical and mixing performance characteristics.
  - Usually these tubes are used to ensure a mixing flow pattern that cannot or will not develop in the system otherwise. These are favorable for large ratios of liquid depth to mixer diameter.
  - With a draft tube inserted in the tank, no sidewall baffles would be required.
  - Flow into the axial impeller mounted inside the tube is flooded to give a uniform and high flow pattern into the inlet to the impeller.

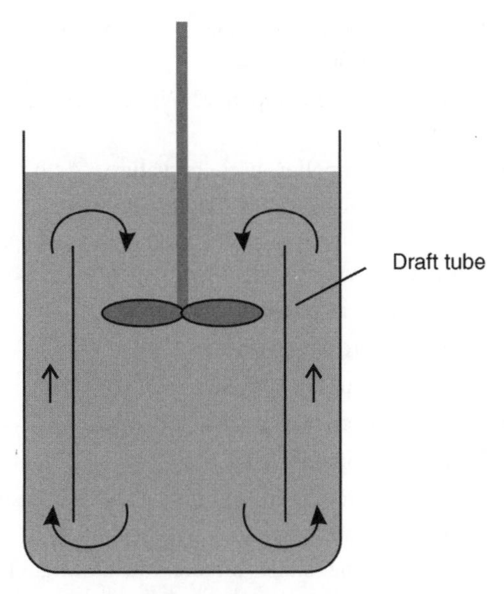

FIGURE 6.9 Draft tube agitator.

  - The upflow in the annulus around the tube has sufficient velocity to keep particles in suspension.
  - Figure 6.9 illustrates a draft tube agitator.
- What are the advantages of a draft tube?
  - The advantages of a draft tube are as follows:
    - Increases mixing efficiency by preventing short-circuiting of the fluid.
    - Minimizes areas of inadequate turbulence.
    - Amplifies mixing action by effectively increasing ratio of mixer to container diameter.

6.2.6 Other Types of Mixers

- What is a vortex mixer? What are its applications?
  - A vortex mixer is used for turbulent processes.
  - These mixers consist of a series of baffles or tabs inclined relative to one another and at an angle relative to the pipe axis. The mixer elements are rotated by 90° and arranged successively in the pipe.
  - Mixing is achieved by controlled vortex structures generated by the baffle geometry that requires a mixer length less than two pipe diameters.
  - This design can be used in all turbulent flow mixing applications regardless of line shape or size and has pressure losses 75% less than conventional static mixers.
  - The vortex created by the impeller travels to the bottom of the tank and into the nooks and corners of the container. Such an impeller and flow field are excellent for solids suspension.
  - Water-borne paints and many other solids are difficult to suspend in solution and keep from stratifying. The V-impeller and its associative flow field are excellent for solids suspension.
  - Typical applications include low-viscosity liquid–liquid blending processes, as well as gas–gas mixing.
- What is a rotor–stator mixer?
  - A *rotor–stator mixer* is similar in concept to an electric motor and consists of a high-speed rotor surrounded in close proximity by a stator.
  - Rotor is a rapidly spinning device nested inside a stationary stator. Typical rotor tip speeds range from 10 to 50 m/s.

- Modern rotor–stator mixers involve numerous concentric rows of intermeshing teeth and other proprietary designs that achieve the tight tolerances needed to impart high mechanical and hydraulic shear.
- As the liquids or solid slurries are drawn from the vessel into the rotor inlet, they are accelerated and expelled either radially or axially at high velocity, depending on the design. The materials are subject to intense shear as they are forced through the stator slots and the narrow gap between the ends of the rotor blades and the stator, through the perforations in the stator.
- As fluid enters the center of the stator, pumping vanes on the rotor, which spin at 55 m/s, accelerate the product through grooves in the respective parts, but in opposite directions, resulting in opposed flow collision that imparts tremendous shear forces upon the product.
- This provides much higher shear rates than the normal mixers.
- In many cases, fluids are forced to decelerate from 300 m/s or more to zero across the extremely narrow gaps.

- What are the applications of rotor–stator mixers?
  - Homogenization, dispersion, emulsification, grinding, dissolving, chemical reaction, cell disruption, coagulation (due to shear), and so on.
  - Used in the manufacture of latexes, adhesives, personal care and cleaning products, dispersion of chemicals, agricultural formulations, and so on.
  - Used in the fluid viscosity ranges up to 150 Pa s (150,000 cP). Above these values, extruders are used.
- What are high shear mixers? What are their applications?
  - High shear mixers are basically stator–rotor devices with a wide variety of designs. The rotor is equipped with blades or teeth on the inner and outer surfaces. The stator is patterned with slots, holes, and other perforations.
  - High shear mixing is an energy-intensive process that combines immiscible liquids to produce emulsions and dispersions through the controlled formation and integration of droplets.
  - These are also used to improve deagglomeration and improve blending of solid ingredients into liquids.
  - Applications include manufacture of mayonnaise, toothpaste, lubricants, paints, salad dressing, pharmaceuticals, and a wide variety of other products.
- Name different types of impellers used in mixing equipment for the following applications:

  (a) Low-viscosity liquids (high Reynolds number applications).

  (b) High-viscosity liquids/slurries/suspensions (low Reynolds number applications).

  - (a) Relatively small, high-speed impellers, for example, propellers, turbines, and paddles.
  - (b) Larger, lower speed impellers, for example, anchor, helical ribbons, and so on.
- What are the essential differences between impellers used for low-viscosity liquids and high-viscosity liquids, with respect to their constructional features?
  - Low-viscosity liquids:
    - ➢ Diameters of impellers are about one-fourth to half of the vessel diameter.
    - ➢ Impellers are placed at about one-third of the liquid height above the base.
    - ➢ Baffles are used to prevent vortex formation.
    - ➢ Baffle width is about 1/10th of the vessel diameter.
  - High-viscosity liquids:
    - ➢ Heavy duty impellers with diameters just less than vessel diameter. Small clearance between impeller and vessel wall.
    - ➢ No baffles.
    - ➢ Use of helical coils for heat transfer is difficult and expensive.
    - ➢ Jacketed vessels are used for such heat transfer applications.
- What type(s) of impeller(s) is recommended for solid suspensions involving low and high settling velocities?
  - For settling velocities around 0.9 cm/s, solids suspension can be accomplished with turbine or propeller impellers.
  - For settling velocities above 4.5 cm/s, intense propeller agitation is needed.
- What is a fluid foil impeller? What is its advantage/application?
  - Fluid foil impeller uses curvature of the blades instead of usual turbine blades.
  - It has improved blending and solids suspension characteristics.
- How is mixing in a reactor increased?
  - Adding/improving baffles.
  - Installing a higher rpm motor on the mixer.
  - Using a different mixer blade design.
  - Using multiple impellers.
  - Pumped recirculation: addition and/or increasing the quantity.
  - Use of in-line static mixers for mixing entering streams.

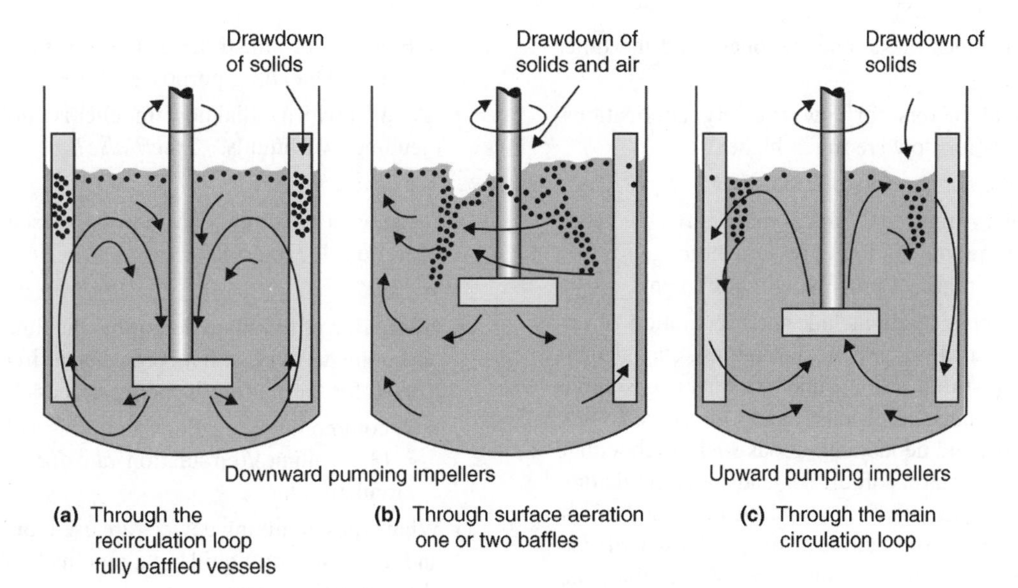

FIGURE 6.10 Incorporating floating solids into liquids.

- How are fine floating solids dispersed into liquids? Illustrate and explain the considerations involved.
  - Figure 6.10 is a schematic illustration of how solids can be entrapped from the surface, depending on the impeller pumping mode, position, and number of baffles.
  - With a reduced number of baffles and the impeller mounted close to the liquid surface, floating solids are entrapped into the liquid along with air through vortices, as shown in Figure 6.10b.
  - It is also possible to incorporate solids into the liquid without vortex formation as shown in Figure 6.10a and c. This can be achieved in fully baffled vessels at different locations on the liquid surface depending on the impeller pumping mode.
  - Fully baffled vessels ensure a more stable operation. Vortex formation, which is not desirable, can be avoided and a more flexible operation can be ensured.
  - The key to achieving drawdown of floating solids at the lowest possible power input is to ensure that the impeller discharge (where liquid velocities and turbulence are highest) is closest to the liquid surface.
  - Therefore, operating in the up pumping mode requires less power input for drawdown from the surface, in a similar way as down pumping impellers are recommended for suspending heavy solids from the vessel base.
  - When micron or submicron size particles are introduced into a liquid, they tend to form macroscale clusters.
  - Due to the air entrapped in these clusters, which reduces the apparent density, they will float even though the particles themselves are denser than the liquid phase.
  - Narrow blade hydrofoils pumping up require low power input for drawdown.
  - Once the particles are in the liquid phase, they may settle on account of their high density. Careful positioning of the impeller may solve this issue.
  - Alternatively, an additional impeller may be needed to achieve off-bottom suspension—this second one pumping down and mounted closer to the vessel base.
  - The large clusters must be broken up that require high power input.
  - In practice, such processes are carried out in a series of equipment, the stirred tank being often used as the first stage in the process to generate a predispersion, which is then fed into a more energy-intensive device for breakup.
  - The use of the stirred tank is not only to generate a suspension, but also to reduce the initial size. A sawtooth impeller can be used.
  - Particles that are difficult to wet pose additional challenges with very low drawdown rates. Such particles may rise to the surface as soon as the impeller is switched off or the speed decreased slightly.
  - While the use of an appropriate wetting agent will facilitate particle incorporation, the process can also be improved further if drawdown and breakup can be achieved simultaneously.
- What is the basic difference between an emulsifier and a normal mixer?
  - Emulsifiers predominantly use shearing action to achieve emulsification. They typically consist of

close clearance disks, one stationary and the other rotating.

- In normal mixers for low-viscosity applications, clearances for flow are much higher.

• What are the applications of emulsifiers?
  - Emulsifiers are used as an alternative to slow speed impeller mixing or high-pressure homogenization for a wide range of processing requirements.
  - Typical applications include the preparation of adhesives, asphalts, carbon dispersions, clay dispersions, dyestuffs, paints, inks, lacquers, cosmetics such as creams, emulsions, hand lotions, perfumes, shampoos, and deodorants, foods such as chocolate coatings, mustard, soft drinks, and sugar emulsions, pharmaceuticals such as antibiotics and ointments, plastics (e.g., cold cutting resins, polyester dispersions, resin solutions), and various miscellaneous mixtures such as floor polishes, gum dispersions, lubricants, petroleum emulsions, and so on.
  - These types of mixers are normally used in dished or conical bottom vessels.

• What is blending? What type of impeller is used in blending processes?
  - Though the terms *mixing* and *blending* are often used interchangeably, they are technically considered to be slightly different.
  - Blending involves mixing of different liquids to obtain homogeneous composition for the resultant liquid. Examples include mixing of low-viscosity petroleum products in a refinery, alcohol blending with gasoline to produce *gasohol*, biofuel blending, and so on.
  - For solids, the term blending is used for *gentle* mixing of multiple dissimilar materials such as plastic pellets, talc, plastic powder with impact modifiers, colorants, and flame retardants to give a homogeneous product. Sometimes materials of same chemical nature are also blended to give uniform particle size distribution.
  - Side-entry propeller mixers are generally used for blending liquids in tanks.
  - Figure 6.11 illustrates side-entry propeller mixers used for blending purposes.

• "A mixer with a radial flow impeller is a good device for blending two liquids." *True/False*?
  - *False.*

• "Side-entering mixers are normally used in blending operations." *True/False*?
  - *True.*

• Give a rule of thumb to determine the time required for tank contents to be well mixed when a circulating pump is used for the mixing/blending process?
  - A commonly used rule of thumb is as follows: time $= (3 \times \text{volume})/(\text{circulation rate for mixing with a circulation tank})$.

• What types of mixing devices are used for mixing gases and low-viscosity liquids in a pipeline? What are their characteristic features?
  - Orifice plate mixers:
    - ➢ Induce turbulence and recirculation similar to those of fluid jets.
    - ➢ Ratios of orifice to pipe diameters are in the range of 0.5–0.2.
    - ➢ ΔP is in the range of 5–30 kPa.
  - Spargers:
    - ➢ Used for gas–liquid contact with mild agitation.
    - ➢ Consist of perforated tubes or porous elements.
    - ➢ Recommended gas rates, m^3/s per m^2 of tank cross section:

    Mild agitation: 0.004.

    Moderate agitation: 0.008.

    Intense agitation: 0.02

• What are aerators?
  - Aerators work at the air–liquid interfaces. These are of two types. One type uses a relatively high-speed propeller unit pumping liquid up and spraying it out through the air. The second type uses slower speed, larger diameter turbine impellers that operate at the interface causing spraying and entraining action. Both types can be either fixed mounted or float

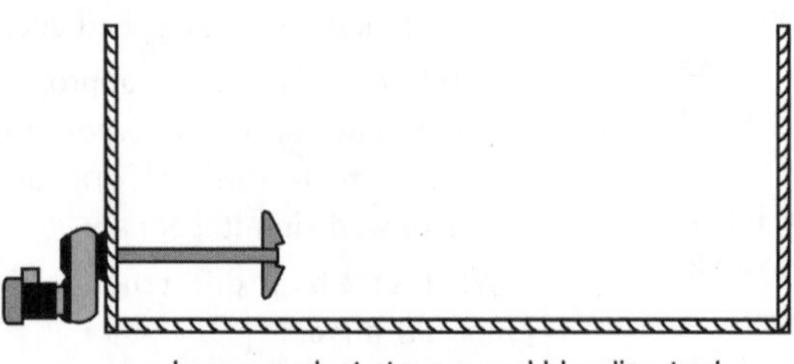
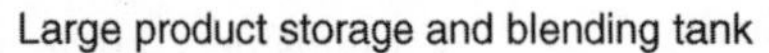

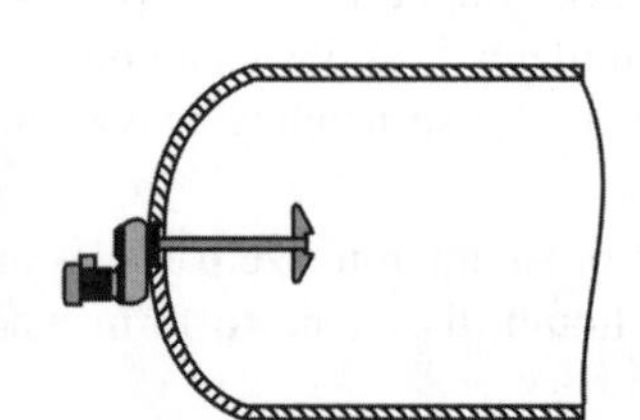

FIGURE 6.11 Side-entry propeller mixers used for blending liquids.

mounted in the aeration basin. These are used in wastewater treatment and increasing oxygen supply to reduce eutrophication of stationary ponds, bioreactors, and so on.

- What are planetary or pan mixers? What are their characteristic features and applications?
  - Planetary mixers are used for a variety of liquid and solid mixing applications, from simple mixtures to sophisticated reactions, involving high temperature, vacuum, or internal pressure. This type of mixer is employed in batch operations.
  - During the mix cycle, two rectangular shaped stirrer blades revolve around the tank on a central axis. Each blade revolves on its own axis simultaneously, at approximately the speed of the central rotation.
  - With each revolution on its own axis, the mixing elements move in a planetary path, visiting all parts of the mixing pan, normally with only small clearance from the pan wall.
  - This movement provides homogeneity of the material being mixed and does not depend on the flow characteristics of the mix. Instead, the stirrers cover every point within the mix tank.
  - Double planetary mixers have no packing glands or bearings in the product zone, and hence, cleaning between batches is minimized.
  - In a rotating pan mixer, the mixing vessel is mounted on a rotating turntable. The mixing elements rotate in a fixed position near the pan wall.
  - Various designs of mixing elements may be used.
  - The mixer is equipped with a hydraulic lift that permits the stirrer blades to be lowered and raised in and out of the mixing tank.
  - The mixer is capable of handling low-viscosity fluids to very high-viscosity pastes and dough-like materials. Typical capacities of these units range from 38 to more than 1200 L (10 to more than 300 gallons).
- What is the principle of operation of continuous paste mixers?
  - A common principle is to force material through obstructions such as perforated plates, meshes, grids, and so on, by means of screw conveyors. The material is kneaded and sheared between the screws and the walls of a trough and further acted on mechanically by being forced through or past obstructions.
- What are the objectives involved in mixing of pastes and high-viscosity materials?
  - Uniform mixing may not be the only objective; for example, mechanical action is required for dough development.
  - The general principle is that mixing performance depends on direct contact between the mixing element and the material.
  - Flow inside the material is laminar, not turbulent.
  - The material must be brought to the mixing elements or the mixing elements must travel to all parts of the vessel.
  - Mixing occurs by kneading the material against the vessel wall or against other material, folding unmixed food or other material into the mixed part, and shearing to stretch the material.
- Name mixers used for pastes and viscous materials.
  - Anchor impeller: The anchor impeller design is best suited for mixing of high-viscosity fluids. The design produces radial flow at low speeds. These types of impellers often incorporate wipers that remove material from the vessel walls during mixing, which enhances heat transfer.
  - Change-can mixers: The container is a separate unit easily placed in or removed from the frame of the mixer.
  - Helical blade mixers.
  - Double arm kneading mixers: Consist of two counterrotating blades in a rectangular trough curved at the bottom to form two longitudinal half cylinders and a saddle section. Mixing action is a combination of bulk movement, smearing, stretching, folding, dividing, and recombining as the material is pulled and squeezed against blades. Different blade designs are used for specific applications. Sigma blade is most widely used for mixing foods, epoxy resins, adhesives, and other plastic mass.
  - Figure 6.12 illustrates common designs of double arm kneader mixers.
  - Screw mixers and extruders: Extensively used in plastics industry for uniform mixing of various additives, stabilizers, fillers, and so on.
  - Mixing rolls.
  - Muller mixer.
  - Banbury mixer: Used for mixing rubbers and plastics. It is a high-intensity mixer with power input up to 6000 kW/m^3. Speeds are up to 40 rpm.
- What are the features and good points of twin shaft ribbon and paddle mixers?
  - These are most commonly used continuous mixers, being highly flexible and capable of high output rates.
  - Twin overlapping screws rotating down at the outer walls bring the material into a compression triangle in the center of the casing. This confined region allows relatively high shear loads to develop at high

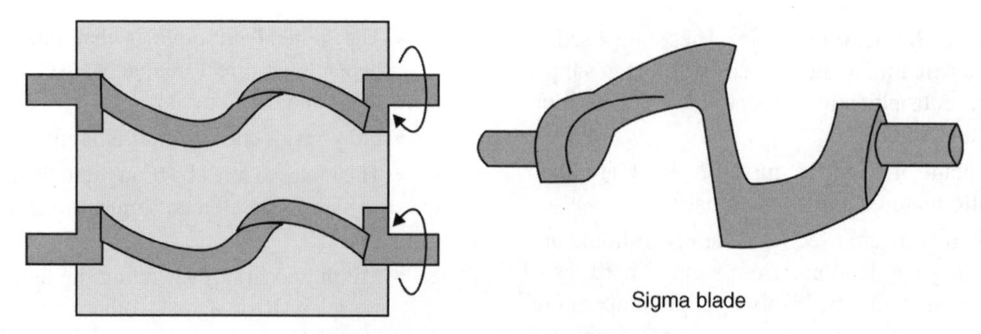

FIGURE 6.12 Double arm kneader.

cross-sectional loading of the machine as the submerged blades converge within the mass.

- Ribbon-type screws allow product to be expressed and introduce a degree of backmixing in the mixer.
- The two shafts might be oriented with the mixing blades in either an *even-space* mode or a *close-space* mode. Close spacing of the blades increases the amount of shear imparted to the product. The alignment of the blades is usually set by the factory according to the degree of work input considered appropriate for the duty.
- The angle of the blades is set at around 70° to the shaft for general use. Flatter blades are fitted for more agitation and finer settings for dealing with difficult flow materials.

• Where are helical ribbon impellers used? Give a diagram.

- The helical ribbon impellers are used when turbines and anchors cannot provide the necessary fluid movement to prevent stratification in the vessel. Used for liquid viscosities above 100,000 cP.
- Figure 6.13 illustrates a helical ribbon impeller.

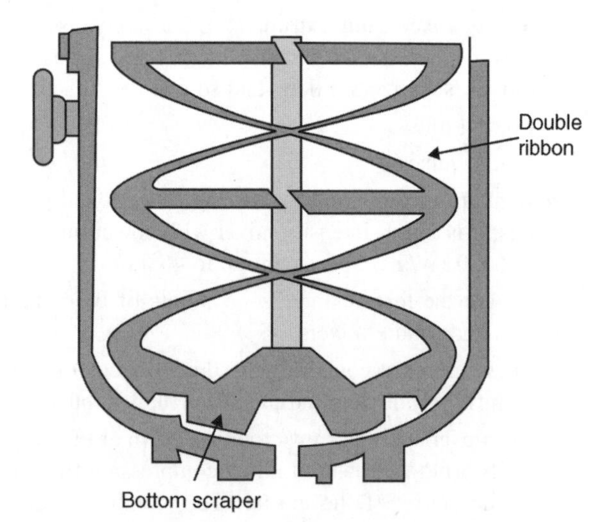

FIGURE 6.13 Helical ribbon impeller.

• What are helical mixers? For what type of applications these are used?

- Helical mixers comprise a series of mixing elements with the leading edge of one element being perpendicular to the trailing edge of the previous. Each mixing element is a metal or polymeric ribbon with a 180° helical twist that measures approximately one and a half pipe diameters in length. The mixing elements can be permanently mounted inside a tube or removable so as to allow for frequent cleaning and inspection.
- The helical mixer is used primarily for laminar flow operations. Applications for the helical mixer include liquid–liquid and gas–liquid mixing. These mixers can be used for laminar, transitional, and turbulent flow applications and are suitable for most blending and dispersion processes involving liquids and gases.

• What is the important advantage of a helical coil impeller over an anchor impeller for a mixer? What is its disadvantage?

- With helical coil type there will be better top-to-bottom mixing than with anchor type.
- Helical coil type is more expensive.

• How mixing rolls operate?

- Material is forced through a narrow space between two or more rotating rollers that may rotate at different speeds to create shear as well as compression.

• How does a Muller mixer operate?

- A Muller mixer operates with wheels rolling over the material, crushing and rubbing the material similar in action to that of a mortar and pestle.

• Name close clearance impellers and state their working and applications.

- Helical coil and anchor impellers (Figure 6.14).
- Operate near the tank wall and are particularly effective in pseudoplastic fluids in which mixing energy is concentrated near the tank wall.
- These are slow mixers with speeds ranging from 5 to 20 rpm.

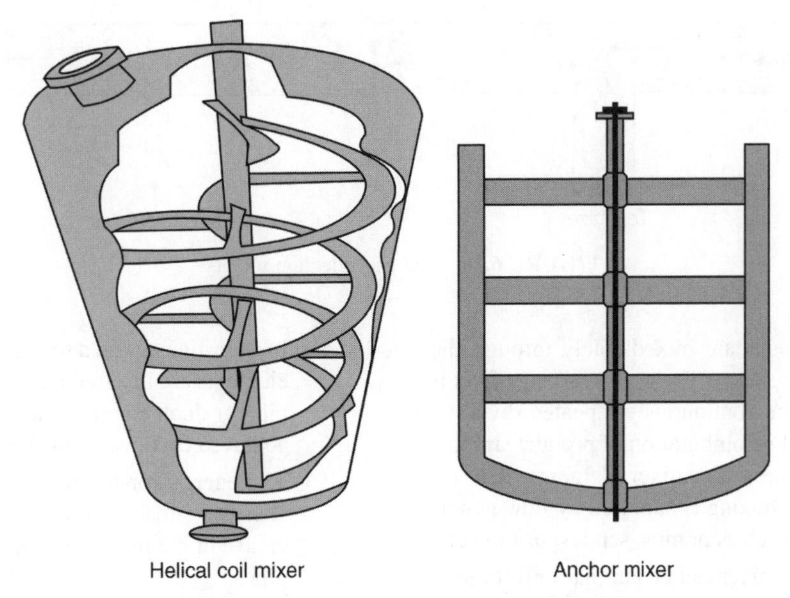

FIGURE 6.14 Helical coil and anchor mixers.

- Anchor impellers are used for an intermediate range of $0.5 > N_{Re} > 10$, because they are much less expensive than helical ribbons and they sweep the entire vessel volume, whereas a turbine leaves stagnant areas near the vessel walls for $N_{Re} < 10$.

- "For high-viscosity liquids propeller mixers are better suited than turbine mixers." *True/False*?
  - *False.*
- What type of mixer is used for very high-viscosity applications (40,000–50,000 cP)?
  - Anchor type.
- "Slow speed close clearance impellers (helical coil or anchor type) are used when mixing high-viscosity liquids." *True/False*?
  - *True.*
- How does liquid viscosity influence impeller selection for a mixer?
  - Table 6.8 and Figure 6.15 give impeller selection for different liquid viscosity ranges.

TABLE 6.8 Impellers Used for Different Liquid Viscosity Ranges

Type of Impeller	Viscosity Range of Liquid (cP)
Anchor	10^2 to 2×10^3
Propeller	10–10^4
Flat blade turbine	10 to 3×10^4
Paddle	30–100
Gate	10^3–10^5
Helical screw	3×10^3 to 3×10^5
Helical ribbon	10^4 to 2×10^3
Extruders	$>10^6$

6.2.6.1 Static Mixers

- What are static mixers? Give examples.
  - Static mixing involves homogenization without using moving parts. In contrast to dynamic mixing,

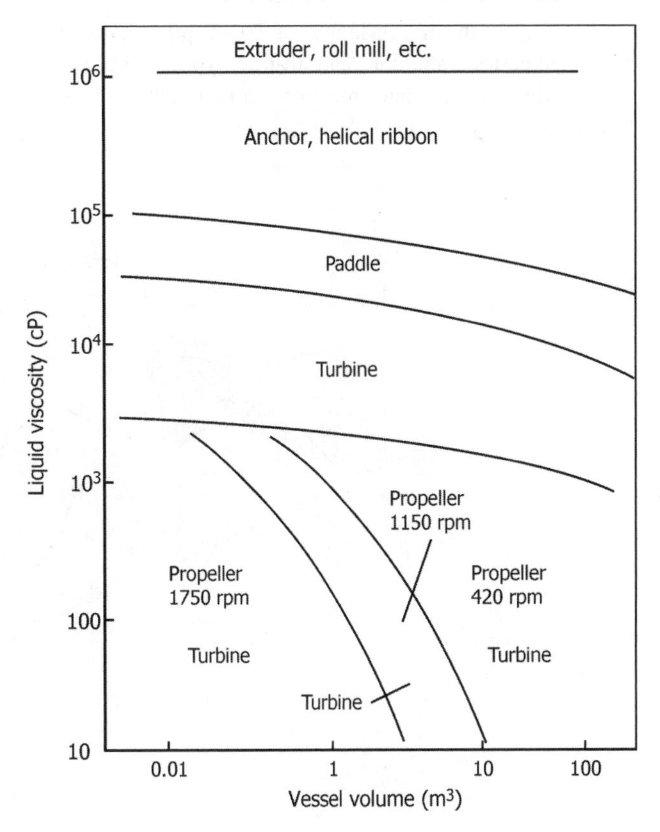

FIGURE 6.15 Impeller selection.

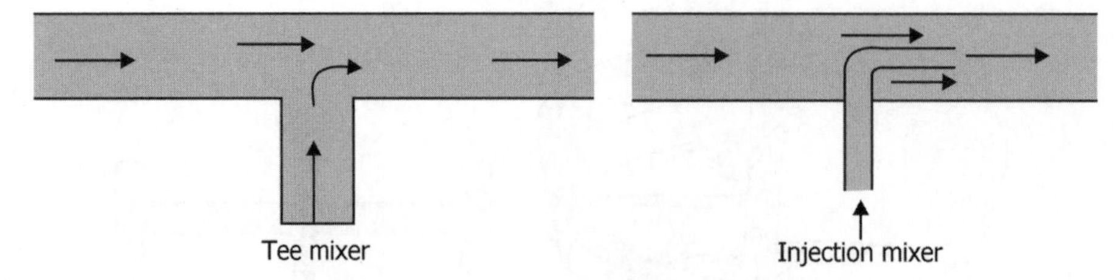

FIGURE 6.16 Tee and injection mixers.

different components are mixed solely through the utilization of flow energy. The actual mixing effect is generated from the continuously repeated division, transposition, and recombination of product streams flowing around the static mixer elements. The required energy for mixing is supplied by flow generation equipment such as pumps, screws, or blowers.

- Mixing in a static mixer can be laminar or turbulent.
- Laminar mixing applications include handling of liquids with large viscosity differences, homogenization and dispersing of viscous liquids, blending melts and homogenization of fibers, injection molding, and the like.
- Turbulent mixing applications include dispersing and contacting for mass transfer and reactions, mixing and homogenization of low-viscosity liquids, additives, and gases, and the like.
- Tee (with mixing length of 10–20 pipe diameters), injection, annular, and in-line types. Figure 6.16 illustrates tee and injection static mixers.

• What is an in-line static mixer?

- In an in-line mixer, a series of stationary rigid mixing elements are inserted axially into a straight length of pipe or duct. Elements are stationary and mixing is achieved by flow of fluid over mixing elements.
- The energy for mixing is derived from the kinetic energy of the fluid stream. Hence, there is an increase in the pressure drop relative to that of empty pipes.
- The unit becomes more efficient by adding additional static elements causing the flowing fluid elements to split, rearrange, recombine, and again split, thus repeating the process until a uniform/homogeneous flowing stream is produced at the discharge.
- The shear on the fluid carries the mixing process forward.
- Figure 6.17 illustrates a typical in-line static mixer.

• What are the broad areas in which static mixers are employed?

- Polymerization reactors.
- Heat transfer.
- Thermal homogenization.

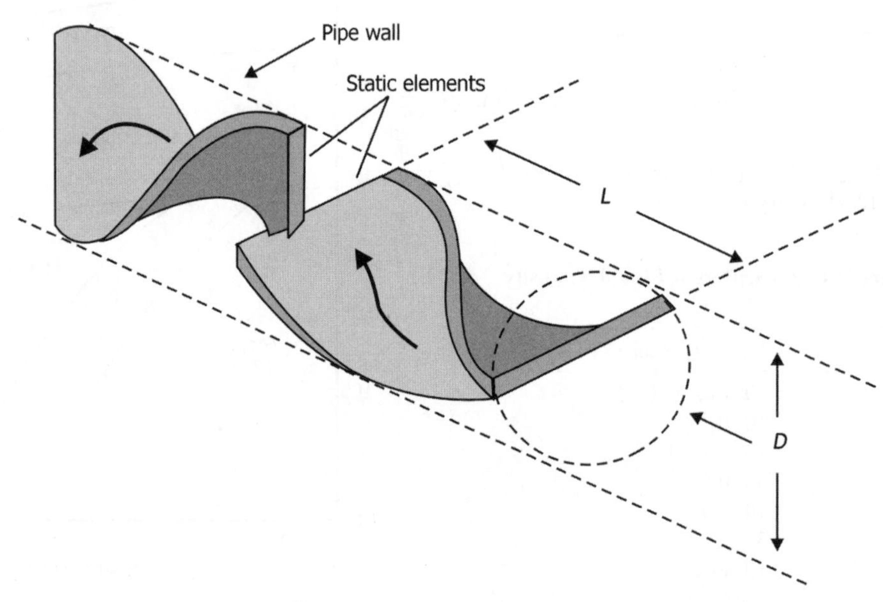

FIGURE 6.17 An in-line static mixer.

- Mixing of similar and varying viscosity liquids.
- Dispersion of immiscible liquids and gases into polymer melts.
- Devolatilization.

- What are the types of applications of in-line mixers?
  - Mixing of low-viscosity liquids.
  - Laminar blending of high-viscosity liquids.
  - Turbulent mixing of miscible liquids.
  - Blending, for example, blending of gasoline of different grades.
  - Blending catalyst, dye, or additive into a viscous liquid.
  - Homogenization of gas flows with regard to concentration or temperature.
  - Intimate and intensive contacting of liquids with gases for increased mass transfer.
  - Dispersion of liquids into liquids (immiscible) in liquid–liquid extraction operations.
  - pH control. Neutralization of waste streams.
  - Gas–liquid reactors.
  - Controlled reactions over a narrow residence time distribution.
  - Mixing of dry powders.
  - Enhanced heat transfer in pipes, especially while dealing with viscous liquids at low Reynolds numbers.
    - Promote significant radial motion in a pipe and thereby reduce temperature gradients across the pipe cross section, especially for viscous liquids, giving enhanced heat transfer rates.
  - Gentle heating of temperature-sensitive products, as well as cooling with mixing action.
  - Solid–solid mixing, for example, food products, dry powders, and so on.
  - Mixing additives such as fragrances, flavors, ice creams, chocolate, dyestuffs, minerals, trace elements, vitamins, emulsifiers, and preservatives, pasteurization and sterilization, and many other areas.

- What are the advantages of static mixers?
  - The absence of moving parts reduces downtime for servicing and repairs to a minimum. Low maintenance costs.
  - Simple construction. No mechanical seals. Requires small space.
  - Lower capital and operating costs.
  - Low shear forces to prevent excessive stressing of the product.
  - Low energy requirements (by 10–100 times) compared to agitated vessels or extruders.
  - Controlled mixing quality. Narrow residence time distributions.
  - Plug flow behavior.
  - The closed system permits hygienic product processing.
  - Semibatch processes can also be operated continuously, even when volumes are low resulting in maximum flexibility, no additional reservoir, and products that are always fresh.
  - Self-cleaning, requiring short purge/changeover time.

- What are the disadvantages of static mixer?
  - The main disadvantages of static mixers are increased pressure drop and fouling problems.

- What are the specific advantages of static mixers for use in food processing industry?
  - The specific advantages of using static mixers in food processing are the following:
    - Good clean-in-place characteristics.
    - No production losses.
    - No unwanted air ingress.
    - Small volumes, short residence times.
    - Low space requirements.
    - Easy installation in existing plant.
    - Low investment costs.

- What are the circumstances that favor use of in-line blenders?
  - When contact time requirements are very small, of the order of 1 or 2 s.

- What are the essential differences between rotor–stator and in-line mixers?
  - Rotor–stator mixers are typically introduced into the top of the vessel and operated in batch mode. This allows moving a given mixer from vessel to vessel.
  - Rotor–stator mixers can slow down overall mixing process and result in off-specification batches. As vessel size and viscosity increase, self-pumping capabilities of these mixers may not be sufficient. Dead zones may be created resulting in off-specification products due to not only insufficient mixing but also thermal degradation of the product due to hot spots developed.
  - In an in-line rotor–stator mixer, the rotor–stator head is installed in a closed housing, like in a pump casing. These are used for continuous operation or for operation between two or more vessels. More uniformity is achieved as the fluid particles flowing experience the same residence characteristics.

6.2.7 Solids Mixing

- Give examples of solids mixing processes in pharmaceutical, food, and chemical process industries.
  - In the pharmaceutical industry, small amounts of a powdered active drug are carefully blended with materials such as sugar, starch, cellulose, lactose, or lubricants.
  - In the food industry, many powdered consumer products result from custom mixed batches. Examples include cake mix, ice tea, or Indian curry, a blend of many fine spices.
  - Thousands of processes in the chemical process industries involve mixing or blending of specialty chemicals, explosives, fertilizers, dry powdered detergents, glass or ceramics, and rubber compounds.
- What is the importance of uniformity in solids mixing on industrial scale?
  - The costs to business are increased substantially with poor mixing process implementation, when the requirements are to address larger batch sizes, faster mixing times, energy conservation, and minimization of segregation.
  - The goals of producing an acceptable mix, maintaining that mix through additional handling steps, and verifying that both the mix and the finished product are sufficiently homogeneous are often difficult to achieve on the first attempt.
  - In many cases, the costs attributed to troubleshooting a poorly performing mixing system can far outweigh the initial investment costs. For example, an inadequate mix or segregation of a pharmaceutical drug can cause the batch to fail, wasting millions of dollars involved, even though the equipment used to mix and transfer the powder can be a small percentage of this cost.
- Compare batch and continuous mixing processes for solids mixing.
  - A batch mixing process typically consists of three sequential steps, namely, weighing and loading the components, mixing, and discharge of the mixed product.
  - In a batch mixer, solids motion is confined only by the vessel, and directional changes are frequent and critical with most units.
  - The retention time in a batch mixer is normally controlled, while for a continuous mixer, this is not the case.
  - Mixing cycles can last from a few seconds with high-intensity units to 30 min or more where additional processing such as heating or cooling may be involved.
  - Mixer discharge may be rapid, or it may take substantial time, particularly if the mixer is used as a surge vessel to feed a downstream process.
  - Ideally, a mixer should not be used for storage capacity, because this can create a process bottleneck, as the mixer cannot perform operations of storage and mixing concurrently.
  - Batch mixers are often used where
    - Quality control requires strict batch control.
    - Ingredient properties change over time and compensation must be on a batch-by-batch basis.
    - The mixer cannot be dedicated to a specific product line.
    - Production quantities are small.
    - Many formulations are produced on the same production line.
  - Major advantages of batch over continuous mixing include the following:
    - Lower installed and operating costs for small to medium capacities.
    - Lower cleaning costs when product changes are frequent.
    - Production flexibility.
    - Premixing of minor ingredients is easily accomplished.
    - Control of mixing time.
  - In a continuous mixing process, the weighing, loading, mixing, and discharge steps occur continuously and simultaneously. Product motion is generally directed from the feed point toward the outlet.
  - Unlike batch mixers where product retention time is carefully controlled, with continuous mixers, material retention time is not uniform and can be directly affected by mixer speed, feed rate, mixer geometry, and design of internals.
  - Continuous mixing is typically used when
    - A continuous, high production rate is required.
    - Strict batch integrity is not essential.
    - Combining several process streams.
    - Smoothing out product variations.
  - Some of the advantages of a continuous mixing system are the following:
    - Ease of equipment integration into continuous processes.
    - Less opportunity for batch-to-batch variation caused by loading errors.
    - Automation can improve quality and reduce labor costs.
    - Higher throughputs are often possible.

- What are the mechanisms of particulate mixing? What are mix structures?
  - The three primary mechanisms of mixing are convection, diffusion, and shear.
    - ➢ Convective mixing involves gross movement of particles through the mixer, either by a force action from a paddle or by gentle cascading or tumbling under rotational effects.
    - ➢ Diffusion is a slow mixing mechanism and will pace a mixing process in certain tumbler mixers if proper equipment fill order and method are not utilized.
    - ➢ The shear mechanism of mixing involves thorough incorporation of material passing along high-intensity, forced slip planes in a mixer. Often, these mixers will infuse a liquid or powdered binder into the mix components to achieve a special consistency, such as granulates.
  - There are two types of mix structures, namely, random and ordered.
  - A random mix occurs when the mix components do not adhere or bind with each other during motion through the mixing vessel. In this case, dissimilar particles can readily separate from each other and collect in zones of similar particles when forces such as gravity, air flow, or vibration act on the blend.
  - More commonly, ordered or structured blends result in most industrial processes. This occurs when the mix components interact with one another by physical, chemical, or molecular means and some form of agglomeration or coating takes place. The process of granulation involves this approach, whereby larger particles are created from smaller building block ingredient particles and each *super*particle has ideally the correct mix uniformity.
  - A mix of perfect superparticles of identical size will not segregate after discharge from the mixer, which is clearly an advantage over a random mix.
  - However, if these particles are not monosized, then segregation by size may occur and induce problems with bulk density, reactivity, or solubility in postmix processing.
  - With regard to mix structure, there are cases where some ingredients have a tendency to adhere only to themselves, without adhering to dissimilar ingredients. This often happens with fine materials, such as fumed silica, titanium dioxide, and carbon black.
  - Sometimes, a mix can reach saturation, where minor fine components will no longer coat larger particles and concentrations of the fine component will build and segregate from the mix.
- Name and illustrate, with suitable diagrams, some commonly used mixers for solids mixing.
  - Ribbon blenders and double cone mixers are typical examples of mixers used for solids mixing.
  - The ribbon blender, shown in Figure 6.18, consists of a trough in which a shaft rotates with two open helical screws attached to it, one screw being right handed and the other left handed. As the shaft rotates, sections of the powder move in opposite directions, so particles are vigorously displaced relative to each other.
  - A double cone blender consists of two cones mounted with their open ends fastened together. They are rotated about an axis through their common base. This mixer is shown in Figure 6.18.
  - Figure 6.19, a photographic diagram, shows a vertical mixer, employed for mixing solid food products.
- What are the different types of solids mixing equipment?

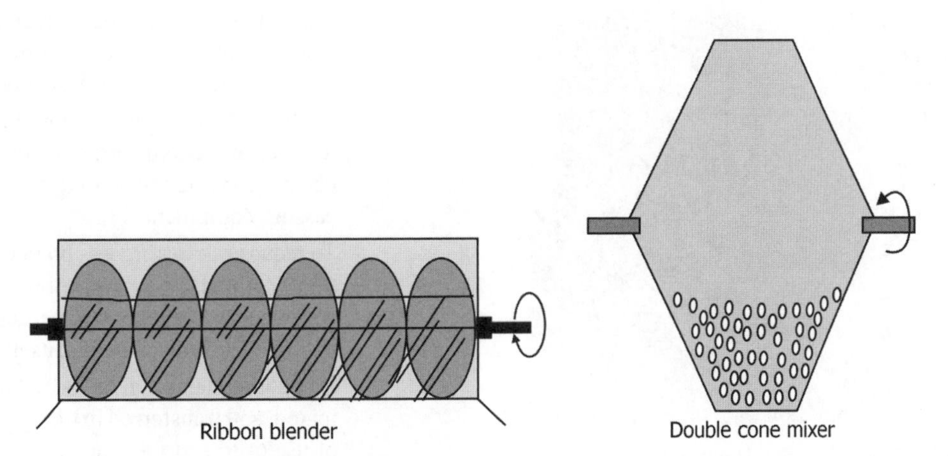

FIGURE 6.18 Ribbon blender and double cone mixer.

  - There are four main types of mixing equipment, namely, tumbler, convective, hopper, and fluidization.
- What are tumbler mixers? What are their characteristics?
  - Tumbler mixers operate by tumbling the solids inside a revolving vessel.
  - May be fitted with baffles to assist mixing or with internal rotating devices to break up agglomerates.
  - The tumbler mixer is a mainstay in the pharmaceutical and food industries because of its features of close quality control in batch operation, effective convective and diffusive mechanisms of mixing, and gentle mixing for friable particles. This type of rotating mixer comes in double cone or V-shaped configurations, and in some cases, these geometries are given asymmetric features to reduce mix times and improve mix uniformity.

FIGURE 6.19 Solids mixer.

  - Operate at speeds up to about 100 rpm (about half the critical speed, at which the centrifugal force on the particles exceeds the pull of gravity).
    - Rotational speed is generally not as important of a factor in achieving uniformity as loading method and mix time, based on the number of rotations.
  - Working capacity is about 50–60% of volume. Capacities are in the range of 0.014–5.66 m^3 (0.5–200 ft^3).
  - Best suited for gentle blending of particles with similar physical characteristics; segregation can be a problem.
  - Equilibrium is generally reached in about 10–15 min.
  - Typically, a top-to-bottom component loading is better than a side-to-side loading with tumbler mixers.
    - With top-to-bottom loading, ingredients are allowed to cascade into one another with diffusive effects occurring perpendicular to the main flow. This approach yields far faster mixing times than side-to-side loading, whereby diffusive mixing paces, and greatly increases, the mixing time.
    - It is also important to avoid ingredient adherence to the walls of the mixer. This is common with fine additives, such as pigments and fumed silica, and component loss can occur if the material does not leave the wall surface.
    - In some cases, the sticky ingredient can be premixed into another component, a practice known as master batching, to help predisperse the material and prevent wall adherence.
    - Mix cohesiveness, which directly correlates to the tendency of a material to form a bridge over the outlet of the mixer, must also be considered.
    - Highly cohesive blends should not be handled in tumbler mixers if bridging or *rat holing* flow obstructions have been experienced in past processing equipment.
    - To reduce mixing process bottlenecks and segregation potential, tumbling in-bin mixers have been developed where the storage container, called an intermediate bulk container, itself becomes a mixer.
    - Mix components can be loaded into the container, mixed, and transferred in the container to the point of use or to a storage area.

    - This process leads to highly flexible production and has been popular in the pharmaceutical, food, and powdered metal industries.
    - The greatest benefit of in-built mixing technology is the elimination of a transfer step from a mixer to a container, by which segregation through various mechanisms can result.
    - No cleaning between batches is required and the mix is stored in a sealed container until use.
    - Optimum in-bin tumbler mixers incorporate mass flow technology, where all of the material is in motion whenever any is discharged, to ensure that the mix does not segregate during container discharge.

- What are convection mixers? What are their operational characteristics?
  - *Convection Mixers*: These mixers use a fixed U-shaped or cylindrical shell with an internal rotating element (impeller) such as a ribbon, paddle, or plow. Due to the action of the impeller, the particles are moved rapidly from one location to another within the bulk of the mixture.
    - The mixing action can range from relatively gentle to aggressive, depending on the agitator design and speed and the use of intensifiers, which are known as choppers.
  - These mixers work well with cohesive materials, which normally need substantially longer mixing times in tumbler mixers.
  - They require less headroom, allowing liquid addition and heating or cooling and potential for continuous operation instead of only batch mixing as with tumbler mixers.
  - These mixers are less likely to experience mix segregation during discharge, since the impellers typically operate during this process.
  - Total capacity for these mixers generally ranges from 28 to 28×10^3 L (1–1000 ft^3), and working capacity is normally 50–70% of the total capacity.
- What are hopper mixers?
  - *Hopper mixers* are usually cone-in-cone to tube-type units, where particles flow under the influence of gravity in a contact bed fashion without moving parts, making them attractive for highly abrasive bulk materials, given their wear potential.
    - With the former unit, the inner cone produces a pronounced faster flow through the inner hopper compared to the outer annulus section, thereby allowing moderate mixing of material.
  - These hoppers typically require two to four passes with a recirculation system to achieve proper uniformity.
  - Tube mixers utilize open pipes within a bin. The pipes have notches in them to allow pellet or granular material to partially flow in and out of the tubes over the height of the bin, or for reintroduction into a lower portion of the bin, such as in a mixing chamber.
  - These mixers can handle much larger volumes of material than tumbler or convective mixers, since no free board space is required and their technology can be applied to storage bins or silos.
  - Another type of hopper mixer, called a planetary or conical screw mixer, is commonly used for cohesive powder mixing.
  - The planetary screw is composed of a screw conveyor inside a conical hopper. The screw is located so that one end is near the apex of the cone and the other end is near the top of the hopper, with the tip of the flights near the wall of the hopper. The screw rotates while revolving around the walls of the hopper, pulling material up from the bottom.
  - It has the ability to handle a wide range of materials, from free flowing to cohesive.
  - Potential disadvantages include sifting segregation within the mixer, possible segregation during mix discharge, and a dead region at the bottom of the cone during mixing.
  - These mixers are commonly jacketed for heating or cooling of a material during the mixing cycle.
  - The Nauta-type mixer can also be fitted with a vertically oriented ribbon mixer, although there are limitations on its capacity, given the high level of operating torque and horsepower.
- What are the characteristics of fluid bed mixers?
  - Effective for materials that will fluidize and those with similar settling characteristics.
  - These mixers use high flow rates of gas to fully fluidize powders in order to rapidly mix components. The gas can also be used to process (e.g., heat or cool) the mix.
  - Particulates should be fine, free-flowing powders that have a narrow size distribution and are close in particle density.
  - Fluidized mixers combine fluidization and convective features, yielding rapid mixing times with a high degree of blend uniformity.
  - Added jets of air that produce *spouting* are said to be effective in decreasing the time required to achieve good mixing.
  - Very rapid: 1–2 min compared to around 15 min for a tumbler mixer.
  - This type of mixer consists of twin counterrotating paddled agitators that mechanically fluidize the

TABLE 6.9 Solids Mixer Selection for Different Applications

Type	Range of Materials	Handle Cohesive Materials	Handle Segregative Materials	Mixing Time	Ease to Clean
Ribbon	Wide	Yes	Yes	Moderate	Easy
Tumbler	Moderate	Moderately	Moderately	Long	Moderate
In-bin tumbler	Moderate	Moderately	Yes	Long	Yes
Planetary	Moderate	Yes	Moderately	Moderate	Moderate
Fluidized	Narrow	No	No	Short	Yes
High shear	Moderate	Yes	Moderately	Short	Moderate

ingredients. Rotation is such that the mix is lifted in the center, between the rotors.

- Mixing is quite intensive, producing intimate blends in a short period of time.
- Mixing cycles are often less than a minute and *bomb bay* doors allow rapid discharge of the entire mix.
- These features combine to give this mixer a high throughput capacity relative to its batch size and highly cohesive materials can be readily mixed.

- What are horizontal trough mixers?
  - Horizontal trough mixers consist of semicylindrical horizontal vessels in which one or more rotating devices (such as screw conveyors or a ribbon mixer) are located.
  - In a typical ribbon mixer, one ribbon moves the material slowly in one direction, while the other moves it quickly in the opposite direction, so there is a net movement of material and the system can be used as a continuous mixer.
  - Particle damage can occur due to the small clearance between the ribbon and the vessel wall and the mixer has a high power requirement.
  - Segregation is less of a problem.
- What are vertical screw mixers?
  - A vertical screw mixer consists of a rotating screw located in a cylindrical or cone-shaped vessel.
  - The screw may be fixed centrally or may rotate around the vessel near the wall.
  - Quick and efficient and good for mixing a small quantity into a larger one.
  - Good for materials prone to segregation.
- What are the properties of different types of mixers for mixing different solid materials?
  - Table 6.9 gives selection of mixers for different applications of solids mixing.
- What are the types of mixing action and characteristics of different types of mixers?
  - Table 6.10 presents different variables that influence selection of mixers for solids.
- Explain the phenomena of segregation of bulk solid materials during the mixing process.
  - Mixing and segregation (demixing) are competing processes.
  - Segregation is separation of particles into distinct zones by particle size, shape, density, resiliency, or other physical attributes such as static charge.
  - A general rule of thumb is that every time a transfer step is added to a process, the powder mix can segregate.
  - *Mechanisms*: Common particle segregation mechanisms include sifting, fluidization (air entrainment), and dusting (particle entrainment).
    - *Sifting Segregation*: This results when fine particles concentrate in the center of a bin or drum during filling, while more coarse particles roll to the pile's periphery. In this case, smaller particles move through a matrix of larger ones. If discharge

TABLE 6.10 Selection of Solids Mixers Based on Operational Requirements

	Gentle Mixing	Lump Breaking	Jacketed Vessel	Ability to Add Liquid	Equipment Height	Dead Spots	Power Required
Ribbon	Moderate	Good	Yes	Yes	Short	Likely	High
Tumbler	Yes	Poor	Difficult	Difficult	Tall	Possible	Moderate
In-bin tumbler	Yes	Poor	Difficult	Difficult	Tall	Possible	Moderate
Planetary	Yes	Good	Yes	Yes	Tall	Likely	Moderate
Fluidized	No	Poor	Yes	Yes	Short	Possible	Low
High shear	No	Excellent	Yes	Yes	Short	Unlikely	High

from this segregated pile occurs from the central core, then a concentration of fine particles will occur, eventually followed by the coarse material.

- This mechanism requires four conditions, namely, difference in particle sizes, relatively large particles (average size greater than 100 μm), free-flowing material, and interparticle motion.
- If any one of the four is absent, the mixture will not segregate by this mechanism.

➢ *Fluidization Segregation*: This type of segregation results when finer, lighter particles rise to the top surface of a fluidized mix of powder, while the larger, heavier particles concentrate at the bottom of the bed. In this case, the fluidizing air entrains the lower permeability fines and carries them to the top surface.

- This mechanism generally occurs only with powders with an average particle size smaller than 100 μm.
- Fluidization segregation is likely to occur when fine materials are pneumatically conveyed, when they are filled or discharged at high rates, or if gas counterflow occurs. As with most segregation mechanisms, the more cohesive the material, the less likely it will segregate by this mechanism.

➢ *Dusting Segregation*: Finally, this form of segregation concentrates the ultrafine and fine particles at container's walls or at points farthest from the incoming stream of material.

- Dusting segregation is a common problem with fine pharmaceutical and food powders being discharged from mixers into drums, tableting press hoppers, and packaging equipment surge hoppers. This effect starts to become prevalent around 50 μm and is very common below 10 μm.

▪ By knowing which segregation mechanisms, if any, might dominate, precautions can be taken in the design and equipment selection process to avert problems.

- How segregation effects are reduced?
  - There are three general approaches for the reduction of segregation problems, namely, changing the material, changing the process, or changing the equipment design.
  - *Changing the Material*: Highly segregating materials are free flowing, and thus prone to sifting segregation.
    - ➢ For example, many pharmaceutical dry blends must be highly free flowing to allow flow through small orifices in a tableting press or encapsulation equipment. However, this feature commonly allows segregation of the formulation.
    - ➢ Product formulators may be capable of altering the cohesiveness of the mix using liquid binders or changing particle size distribution, although a significant change could induce other flow problems, such as bridging or rat holing, which immediately creates a process bottleneck.
  - *Changing the Process or Equipment*: The following general practices can help to prevent or minimize the effects of segregation:
    - ➢ Mixing process should be located as far downstream in the process as possible.
    - ➢ Post-blend handling of the material should be minimized.
    - ➢ Surge and storage bins should be designed for mass flow (no stagnant regions in hopper).
    - ➢ Velocity gradients within bins should be minimized.
    - ➢ Asymmetric hoppers should be avoided due to their large velocity gradients.
    - ➢ A mass flow bin with a tall, narrow cylinder will minimize the potential for segregation compared to that of a short, wide bin. It is preferred to keep the material level in the bin high.
    - ➢ Design symmetry should be utilized whenever possible. For example, if designing a bin with multiple outlets, all of the outlets should be located at the same distance from the bin centerline.
    - ➢ Ensuring that any splitting of a process stream does not result in differences between the streams.
    - ➢ Venting should be used to avoid air counterflow (which induces fluidization segregation).
    - ➢ Generation of dust should be minimized.
- What are the parameters involved in solids mixing?
  - It is not possible to achieve a completely uniform mixture of dry powders or particulate solids. The degree of mixing achieved depends on the following:
    - ➢ The relative particle size, shape, and density.
    - ➢ The efficiency of the particular mixer for the components being mixed.
    - ➢ The tendency of the materials to aggregate.
    - ➢ The moisture content, surface characteristics, and flow characteristics of each component.
  - Generally, materials similar in size, shape, and density are able to form the most uniform mixtures.

Differences in these properties can also cause unmixing or segregation during mixing or mechanical jiggling of the mixture.

- Experience shows that materials with a size greater than 75 μm will segregate readily during mechanical jiggling of the mixture, but those below 10 μm will not segregate appreciably.
- Means of overcoming segregation and poor mixing include the following:
  - Comminution to smaller sizes.
  - Use of powders with a narrow size distribution.
  - Use of the same volume-average diameter for all components.
  - Granulation.
  - Coating processes.
  - Controlled continuous mixing.

- What is the criterion for selection of mixers for solids?
  - Selection of mixers must take into account any tendency toward segregation.
  - This may be evaluated from a *heap test*, in which a well-mixed material is poured through a funnel to form a heap.
  - If the composition of samples taken from the outside varies significantly from compositions of samples taken from the center of the heap, the material is likely to segregate during mixing or later processing.
- How are solids mixers classified with respect to segregation?
  - Mixers can be classed into two groups with respect to segregation:
    - *Segregating mixers* have mainly diffusive mechanisms, encouraging the movement of individual particles, making segregation more significant. Non-impeller mixers tend to be of this type.
    - *Less segregating mixers* have mainly convective mixing mechanisms. These are typically impeller types in which blades, screws, ploughs, and so on sweep groups of particles through the mixing zone.
- Illustrate the working of a solids static mixer for solids mixing.
  - Figure 6.20 illustrates a static mixer for solids mixing.
  - A solids static mixer employs gravity as the driving force for promoting the flow and mixing process. It is characterized by multiple element separations and recollation of the flow stream. As the mechanics of the process is limited by the free-flowing behavior of the bulk material, the operation is generally gently repetitive, and hence more suitable for handling delicate products than a mechanical mixer that normally exerts high local and individual forces on the constituent particles and bulk regions.
- What is acoustic/sonic mixing?
  - Sonic waves cause powders to fluidize, turn over, and mix rapidly. Sonic mixers use acoustic fluidization. Acoustic fluidization creates bubbling through the powder and a high-intensity mixing zone. Mixing chamber shape helps to focus sonic energy into mixing regions. Sonic mixing/blending is said to

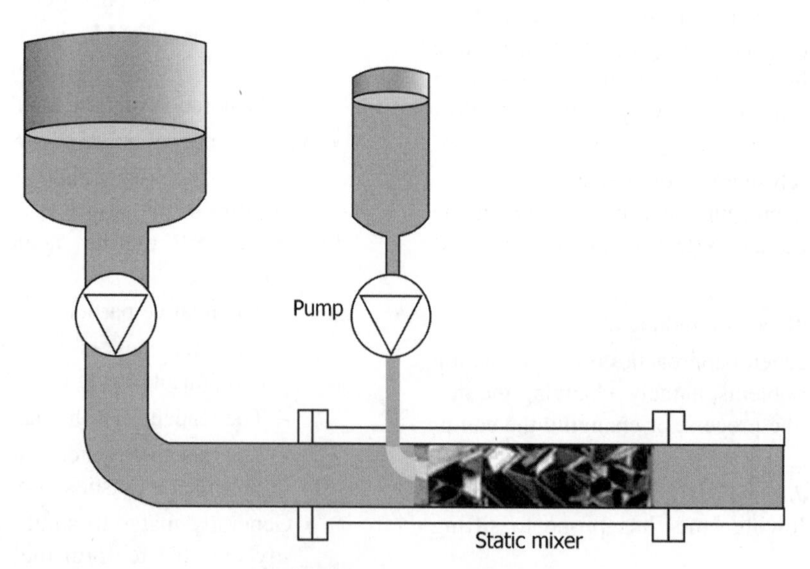

FIGURE 6.20 Static mixer.

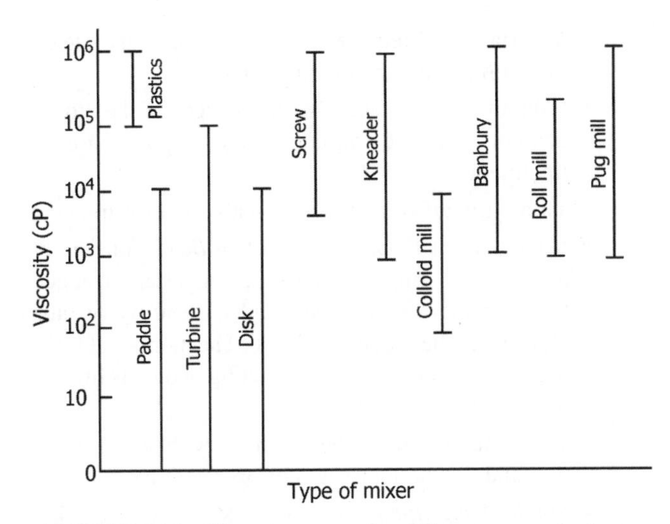

FIGURE 6.21 Viscosity ranges for different types of mixers.

work well and quickly with all sorts of mixtures, including particles of dissimilar size, shape, and specific gravity. And unlike mechanical blending, sonic devices are more efficient with dense powders than with fluffy materials.

- The technology has its limitations, the most obvious being a capacity limit of 200 L.
- Sonic blending also does not work well with moistened blends, probably because of sonic damping by water.

• Summarize, graphically, selection of mixers based on viscosity ranges.

- Figure 6.21 gives a graphical representation of viscosity ranges of materials for different mixing equipment.

• Summarize selection of solids mixing equipment for different applications.

- Table 6.11 gives a summary of mixing action and applications for different types of mixers.

• What is the minimum agitator speed to keep solids under suspension in a vessel? What other considerations influence the settling process?

- The most common method for estimating the minimum suspension condition is the following Zwietering correlation for the minimum agitator speed:

$$N_{\min} = \Psi(D_T/D_A)^{\alpha} d_p^{0.2} \mu_l^{0.1} (g\Delta\rho)^{0.45} W_s^{0.13} / \rho_l^{0.55} D_A^{0.85}, \tag{6.13}$$

where D_A is the impeller diameter (m), D_T is the tank diameter (m), d_p is the particle diameter (m), W_s is the ratio of weight of solids to weight of liquid (%), ρ_l is the liquid density (kg/m^3), $\Delta\rho$ is the density difference between solids and liquid (kg/m^3), μ_l is the liquid viscosity (Pa s), g is the gravitational constant (m/s^2), and Ψ and α depend on the characteristics of the agitator. For propeller, turbine, and flat blade agitators, Ψ and α are 1.4 and 1.5, respectively.

- Dished tank bottoms and baffles help prevent settling in cylindrical tanks. Baffles impart a high vertical component of velocity and help eliminate dead zones. If baffles are undesirable due to slime buildup or other reasons, the agitator should be offset and mounted at an angle to reduce swirl and increase vertical motion (10–15° from vertical + offset).

TABLE 6.11 Types of Mixers and Their Mixing Action and Applications

Type of Mixer	Mixing Action	Applications
Rotating: cone, double cone, drum	Tumbling action	Blending dry, free-flowing powders, granules, and crystals. For example, pharmaceuticals, food, and chemicals
Air blast fluidization	Air blast lifts and mixes particles	Dry powders and granules. For example, milk powder, detergents, and chemicals
Horizontal trough mixer with ribbon blades, paddles, or beaters	Rotating element produces contraflow movement of materials	Dry and moist powders. For example, tablet granulation, food, chemicals, and pigments
Z-blade mixers	Shearing and kneading by the specially shaped blades	Mixing heavy pastes, creams, and doughs. For example, bakeries, rubber doughs, and plastic dispersions
Pan mixers	Vertical, rotating paddles, often with planetary motion	Mixing, whipping, and kneading of materials ranging from low-viscosity pastes to stiff doughs. For example, food, pharmaceuticals, chemicals, printing inks, and ceramics
Cylinder mixers, single and double	Shearing and kneading action	Compounding of rubbers and plastics. For example, rubbers, plastics, and pigment dispersions

Jones R L. Mixing equipment for powders and pastes. *Chemical Engineering (London)* 1985;419:41.

  - Another option is use of rectangular tanks since the corners act as baffles.
  - The most common impellers for slurries are marine propellers, four-bladed/45° pitched turbines, and hydrofoils of proprietary designs. Two or more impellers may be required, either on separate shafts or on the same shaft, to provide suspension of solids, as well as mixing or blending.
- What are the ways to resuspend solids after a power failure?
  - Agitators are often unable to resuspend the solids once they have settled.
  - Most pumps are designed for light slurry loads and cannot handle the settled solids, usually due to high viscosities.
  - Sparging with a liquid is the best solution for agitation. Liquid jets can recover the solids suspension, provided the solids are not extremely cohesive.
  - Nozzles should have the capability of either backflow or solids-free liquid injection after power failures.
  - Positive displacement pumps can handle very high solids concentrations (maybe as high as 95%), but add considerable costs.

6.2.8 Mixer Seals

- What are the functions of agitator seals?
  - Containment of vessel contents to save valuable materials, from safety considerations involved with toxic and flammable materials.
  - Avoidance of contamination and corrosion problems.
  - Static and dynamic sealing to maintain vessel pressure.
- What are the entry aspects that influence seal selection?
  - Entry of the shaft of an agitator can be top entry (most common), side entry, or bottom entry.
  - In top entry, seal is not wetted by the liquid contents and agitator is in contact with only vapor.
  - Bottom-entry seal is immersed in the liquid contents. This may provide seal lubrication. For abrasive materials, seal requires external flushing and cleaning with product compatible fluid.
  - Side-entry seals are also immersed in the liquid contents of the agitator and require similar treatment.
- What are the different types of seals used for mixers?
  - *Compression Packing or Stuffing Box*: Consists of a series of rings of packing material arranged between the rotating shaft and the stationary tube attached to the mounting flange. The number of rings used determines the pressure rating, which is limited to 6 bar gauge.
  - This type of seal can be sterilized with steam injection and can be lubricated or operated dry.
  - *Single Mechanical Seal*: Consists of two sealing faces, one rotating with the shaft and the other stationary. These seals give lower rates of leakage, are more expensive, and have longer life. Should not be used on sensitive or pathogenic cultures or toxic contents of the vessel.
  - *Single Fully Split Mechanical Seal*: Similar to single mechanical seal but with the addition of a longitudinal split line for ease of maintenance. Disadvantage is lower pressure rating with higher leakage rates.
  - *Dual Pressurized (Double) Mechanical Seal*: A barrier liquid is introduced into the sealing chamber for lubrication, which isolates process fluid and vapors from the environment.
  - *Gas-Lubricated Seal*: A thin gas film separates the sealing faces. Designed to run dry without face contact and involve no wear. Dry N_2 is used as the separating gas.
  - *Combination of Contact Dry Running and Gas-Lubricated Seals*: Can reduce problems of N_2 accumulation in unvented closed building spaces.
  - *Modular Cartridge Seals*: Preassembled self-contained seals.
  - *Magnetic Drives*: High levels of containment and isolation. High initial and long-term maintenance costs.

7

TWO-PHASE FLOW SYSTEMS

7.1 Two-Phase Flow 195
7.1.1 Gas/Vapor–Liquid Flow 195
7.1.2 Fluid–Solid Systems 199
7.1.2.1 Packed Beds 199
7.1.2.2 Fluidization 200
7.1.2.3 Gas–Solids Transport 208
7.1.2.3.1 Pneumatic Conveyance 208
7.1.2.4 Solid–Liquid (Slurry) Flow 217

7.1 TWO-PHASE FLOW

7.1.1 Gas/Vapor–Liquid Flow

- Give examples of industrially important applications of gas/vapor–liquid flows.
  - Gas bubbles in oil, wet steam, vapor–liquid flow in refrigeration systems, steam–water flows in boilers and condensers, vapor–liquid flows in heat exchangers, evaporators, distillation columns, and reactors, and so on.
- What are the various types of flow patterns in gas–liquid flow in vertical pipes? Illustrate.
  - Bubble flow, slug flow, churn flow, annular flow, and mist flow. The flow regime depends on the interaction of two forces, namely, gravity and vapor shear, acting in different directions. At low vapor flow rates, gravity dominates, giving rise to stratified, slug–plug, or bubble flow, depending on the relative amounts of liquid present. At high vapor velocities, vapor shear dominates, giving rise to wavy, annular, or annular–mist flow.
  - Figure 7.1 illustrates the flow patterns.
- What is bubble flow?
  - When vapor/gas bubbles, distributed in the liquid, move along at about the same velocity as the liquid, the flow is described as bubble flow.
  - Bubble flow describes the flow of distinct, roughly spherical vapor regions surrounded by continuous liquid. The diameter of the bubbles is generally considerably smaller than that of the container through which they flow.
- Under what conditions bubble flow can be expected?
  - Bubble flow usually occurs at low vapor concentrations. When vapor content is less than 30% of total weight flow rate, bubble flow is possible.
- "Bubble flow in a horizontal pipe is prevalent at high ratios of gas to liquid flow rates." *True/False*?
  - *False.*
- Explain what is meant by slug flow?
  - *Slug Flow*: Slugs of gas bubbles flow through the liquid. If the vapor and liquid are flowing through a pipe, bubbles may coalesce into long vapor regions that have almost the same diameter as the pipe. This is called slug flow. Slug flow develops when high waves of liquid develop with progressive increase of gas or vapor content in wavy flow.
- What are the undesirable effects of slug flow?
  - Slug flow can cause vibrations in equipment because of high-velocity slugs of liquid impinging against fittings and bends.
  - Can cause flooding in gas–liquid separators.
  - Pressure fluctuations develop in the piping, which can upset process conditions and cause inconsistent instrument sensing.
- How is slug flow avoided in process piping?
  - Reducing line sizes to minimum.
  - Designing parallel pipe runs that will increase flow capacity without increasing overall friction losses.

Fluid Mechanics, Heat Transfer, and Mass Transfer: Chemical Engineering Practice, By K. S. N. Raju

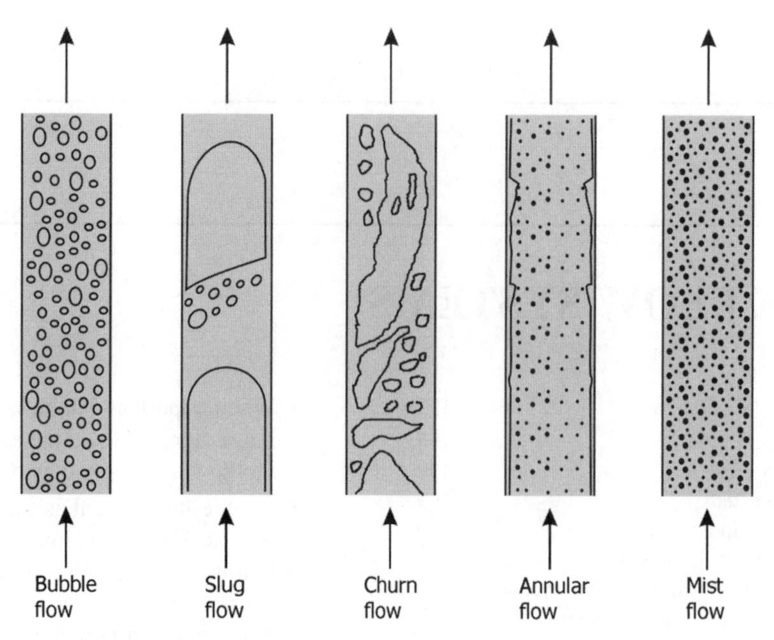

FIGURE 7.1 Flow patterns in gas–liquid flow in vertical pipes.

- Using valved auxiliary pipe runs to regulate alternative flow rates, avoiding slug flow.
- Using a low point effluent drain or bypass.
- Arranging pipe configurations to protect against slug flow; for example, in a pocketed line where liquid can collect, slug flow may develop.

• What is churn flow?

- At moderate to high flow velocities and roughly equal proportions of vapor and liquid, the flow pattern is often very irregular and chaotic. If the flow contains no distinct entities with spherical or, in a pipe, cylindrical symmetry, it is called churn flow.

• Under what circumstances churn or froth flow can occur in two-phase flow systems?

- Churn flows, sometimes called froth flows, are characterized by strong intermittency and intense mixing, with neither phase easily described as continuous or dispersed.

• What is annular flow?

- When liquid forms a film or ring around the inside wall of the pipe and gas flows at higher velocity as a central core, the flow is called annular flow.

• Under what circumstances annular flow develops?

- At high vapor content, the liquid may form a thin film or ring around the inner wall of the pipe, wetting the pipe wall, and gas or vapor flows at higher velocity as a central core. This type of flow is called annular flow.

• What are *annular droplet* and *droplet* flows?

- In *annular droplet* flow, liquid, besides forming annular film along the walls of the conduit, may get entrained as small roughly spherical droplets in the vapor or gas stream. In *droplet flow*, only entrained droplets flow with the absence of liquid film. If the droplet sizes are very small, they can be treated as mists. These flows are possible at high-velocity vapor or gas flows.
- *Spray/Dispersed/Mist Flow*: Gas and liquid flow in dispersed form. All the liquid is carried as fine droplets in the gas phase. In such flows, gas velocities involved are high, of the order of 20–30 m/s.
- Figure 7.2 illustrates annular droplet flow.

• "In two-phase flow, the flow is dispersed when more than 30 wt% of flow rate is vapor." *True/False*?

- *True.*

• Illustrate by means of a diagram different two-phase flow patterns in a vertical tube heated from outside.

- Figure 7.3 illustrates patterns of two-phase flow in a heated vertical pipe.

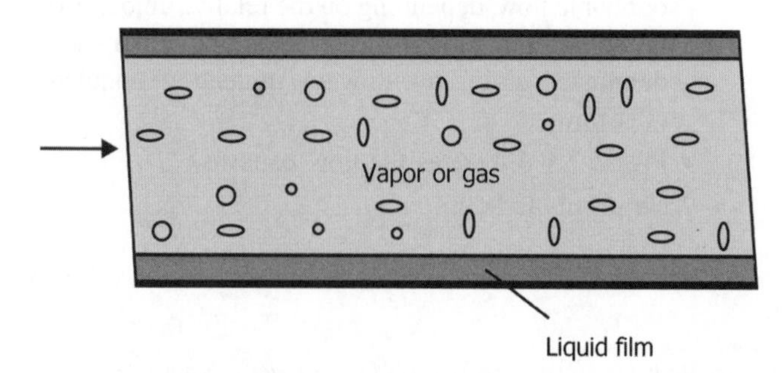

FIGURE 7.2 Annular droplet or mist flows.

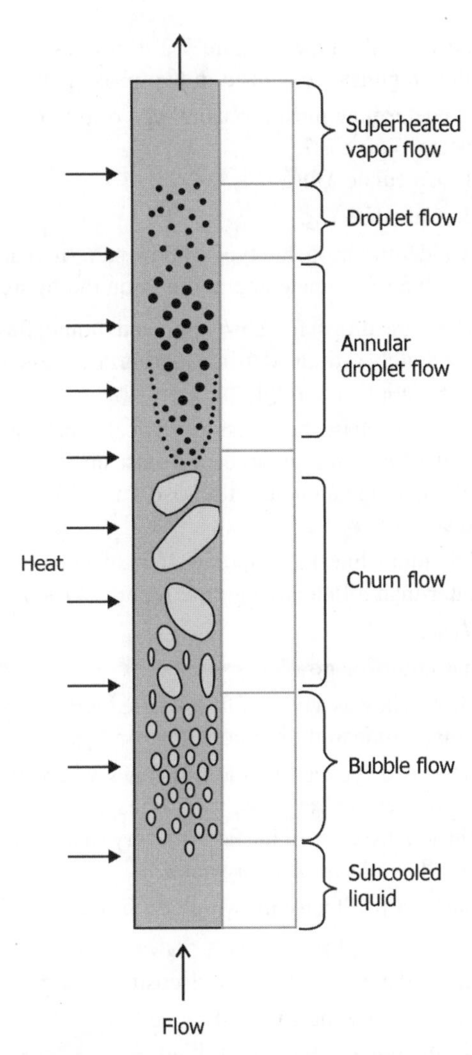

FIGURE 7.3 Two-phase flow patterns in a vertical evaporator tube.

- Vapor fraction in the two-phase mixture determines the flow patterns as illustrated in Figure 7.4.

• What is flashing flow?

- Flashing flow and condensing flow are two examples of multiphase flow with phase change.

• Under what circumstances flashing flow occurs?

- Flashing flow occurs when pressure drops below bubble point pressure of a flowing fluid.

• Illustrate gas–liquid two-phase flow patterns in horizontal pipes.

- Figure 7.5 illustrates gas/vapor–liquid flow patterns in horizontal pipes.
- *Bubble Flow*: Bubbles dispersed in liquid.
- *Plug Flow*: Plugs of liquid flow followed by plugs of gas/vapor flow.
- *Stratified Flow*: Liquid and gas/vapor flow in stratified layers.

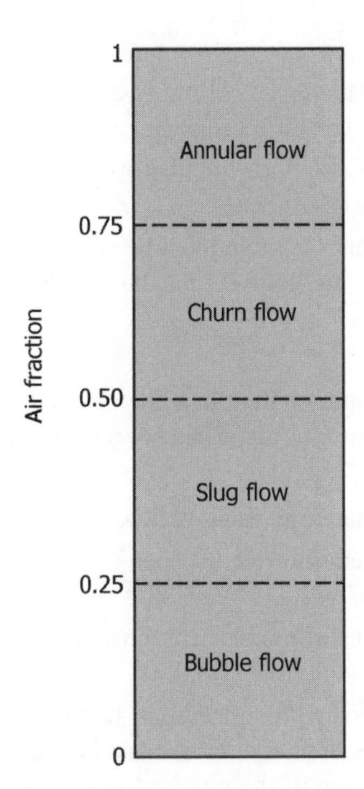

FIGURE 7.4 Flow pattern as a function of fraction of air in air–water flow in a vertical pipe.

- *Wavy Flow*: Gas/vapor flows in the upper pipe section and liquid as waves in the lower section.
- *Slug Flow*: Slugs of gas/vapor bubbles flowing through the liquid.

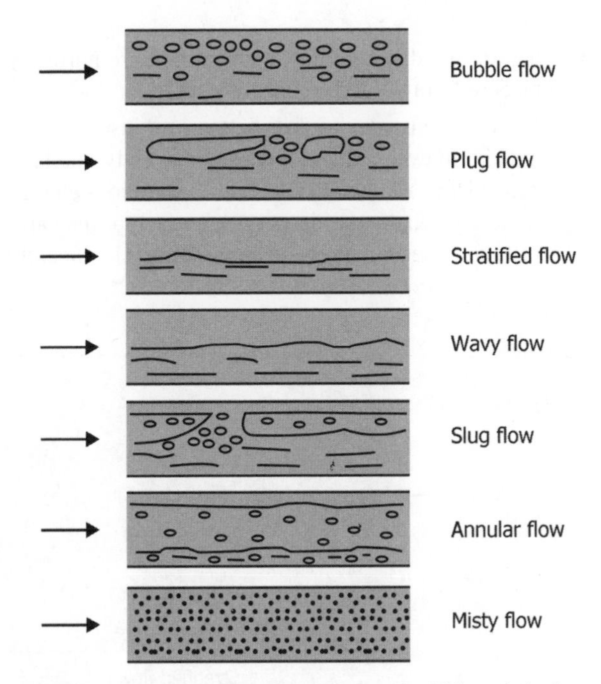

FIGURE 7.5 Flow patterns in gas/vapor–liquid flows in horizontal pipes.

- *Annular Flow*: Liquid flows in continuous annular ring on the pipe wall and gas/vapor flows through the central part of the pipe cross section.
- *Misty Flow*: Fine liquid droplets are dispersed in the flowing gas/vapor.

- "When vapor content in a line is more than about 30% of total weight flow rate, bubble flow is possible." *True/False*?
  - *False.*
- "When vapor content in a line is less than 30% of total weight flow rate, dispersed flow can occur." *True/False*?
  - *False.*
- Under what conditions stratified flow occurs?
  - Stratified flow occurs when the ratio of liquid to gas flow is low.
- What is the difference between stratified flow and wavy flow?
  - *Stratified Flow*: Occurs in horizontal lines. Liquid and gas flow in stratified layers. Liquid flows along the bottom of the pipe and gas flows over a smooth gas–liquid interface.
  - *Wavy Flow*: Gas flows in the top part of the pipe and liquid as waves in the lower section over a smooth gas–liquid interface. The interface has waves moving in the flow direction.
  - Figure 7.6 illustrates these flow patterns.
  - As gas flow relative to liquid flow increases, stratified flow transforms into wavy stratified flow that progressively produces high waves and transforms into slug flow.
- What are the different flow patterns that are normally considered in two-phase flow systems?
  - Six or seven types of flow patterns are usually considered in evaluating two-phase flow. Only one type can exist in a line at a time, but as conditions change (velocity, roughness, elevation, etc.) the type may also change. The unit pressure drop varies significantly between the types. Figure 7.7 illustrates the typical flow regimes recognized in two-phase flow.
  - To determine most probable type of two-phase flow using Figure 7.7,
    1. Calculate $\lambda \Phi G_L/G_G$.
    2. Calculate G_G.
  - To identify probable type of flow pattern, read intersection of ordinate and abscissa on the figure.
  - Pressure drops for gas/vapor and liquid flows are separately estimated using appropriate pressure drop correlations for single-phase flows.
  - For two-phase flow inside vertical pipes, stratified and wavy flow regimes do not exist, and bubble, slug, annular, and annular with mist are the flow regimes that exist.
- "If a liquid line has vapor content, frictional loss is greater than a line having only liquid." *True/False*?
  - *True.*
- What is Stokes flow?
  - Stokes flow is a type of flow where inertial forces are small compared to viscous forces: $N_{Re} \ll 1$.
  - This is a typical situation in flows where the fluid velocities are very low, viscosities are very high, or the length scales of the flow are very small as found in the flow of viscous polymers.
- What is slip velocity in two-phase flow?

$$\text{Slip velocity} = V_{gas} - V_{liquid}. \tag{7.1}$$

- What is the effect of liquid viscosity on slip?
  - As viscosity increases, slip decreases.
- Does the volume fraction or holdup of each phase in a two-phase system same as the volumetric flow rate of that fraction?
  - *No*. Because of velocity differences or slip between the phases.
- "If a liquid line has vapor content, frictional losses are greater than a line having only liquid." *True/False*?

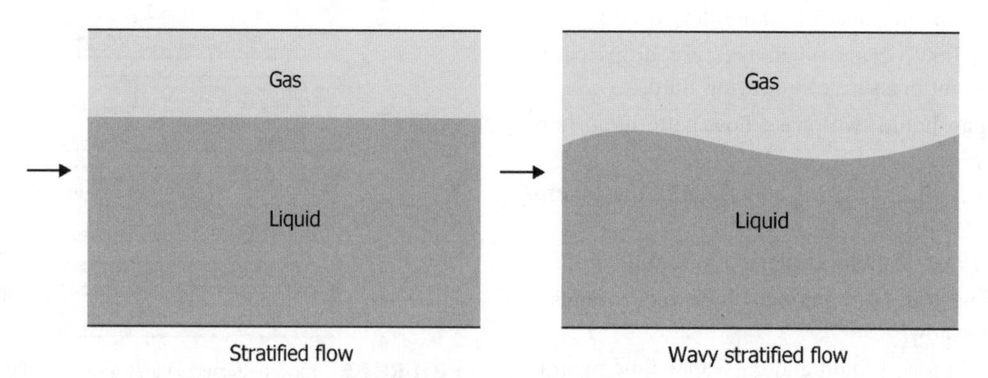

FIGURE 7.6 Stratified and wavy flows.

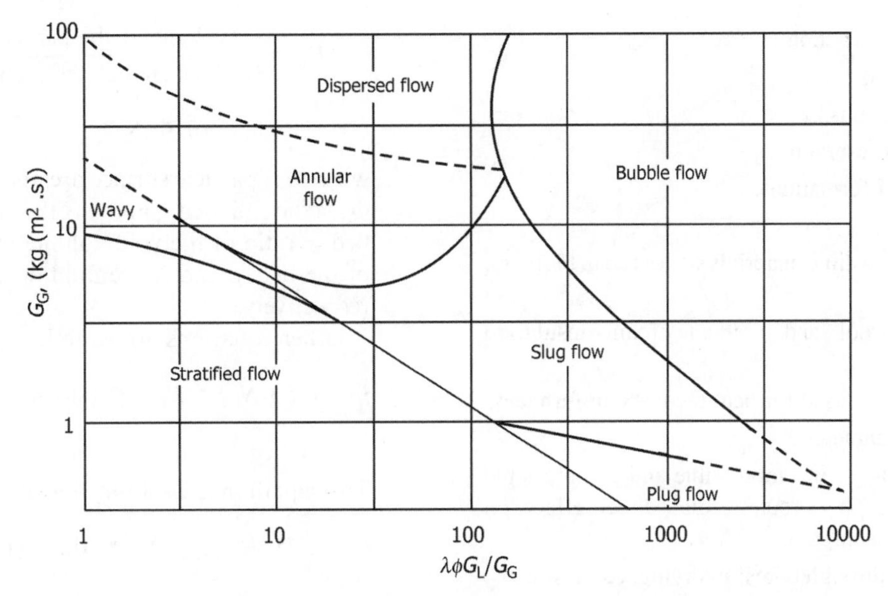

FIGURE 7.7 Flow patterns for two-phase flow in horizontal pipes. *Notation*: G_G: mass flux of gas phase; G_L: mass flux of liquid phase; $\Phi = \sigma_w/\sigma_L[\mu_L/\mu_w(\rho_w/\rho_L)^2]^{1/2}$; $\lambda = (\rho_G/\rho_A)(\rho_L/\rho_w)^{1/2}$; σ: surface tension; Φ and λ are fluid property correction factors. Subscripts w and A refer to water and air at 20°C.

- *True*. Flow velocities increase as a result of vapors and hence increase pressure drop due to friction.

- What is the function of sphering in pipelines?
  - Some gas pipelines are designed to transport condensate as two-phase flow.
  - This liquid formation can cause a number of adverse effects such as corrosion, reduced efficiency, and increased operating cost of overloading of downstream processing plants.
  - Using spheres and developing a routine sphering operation can greatly reduce or control these adverse effects.
  - Maintaining and cleaning pipelines with spheres reduces pressure loss and increases efficiency.
  - Spheres are used for product separation during batching of multiple products in pipelines to minimize interface mixing.
  - In natural gas pipelines, variations in temperature and pressure can cause liquid dropout. This dropout varies with the process conditions and whether the gas is raw or processed.
  - Pipeline spheres are used in the various stages of pipeline construction, acceptance testing, commissioning, batch separation, and pipeline maintenance/cleaning operations.
  - Controlling wax buildup on the pipe walls of crude oil pipelines is effective using spheres in a routine maintenance program.
  - Batching of corrosion inhibitors in the pipeline is another ideal application for spheres.

- What are the normally used materials for the spheres?
  - Neoprene rubber, polyurethane compounds, and so on.

- Explain the differences in characteristics of settling of solid particles, gas bubbles, liquid drops in two-phase liquid systems, and liquid drops in gases.
  - Gas bubbles may undergo deformation and internal circulation unlike rigid solid particles.
  - Very small liquid drops in two-liquid phase systems behave like rigid solid particles.
  - Large liquid drops in liquid–liquid or gas–liquid systems may undergo deformation and internal circulation.

- Why inclined piping is not recommended for two-phase gas–liquid flow?
  - Gas–liquid separation is poor in inclined pipes compared to vertical pipes.

7.1.2 Fluid–Solid Systems

7.1.2.1 Packed Beds

- Give examples of porous media of engineering importance.
  - Packed beds include the following:
    - ➢ Packed columns for mass transfer.
    - ➢ Porous catalysts.
    - ➢ Adsorbents.
    - ➢ Filtration equipment.

   - Fibers and filtration media.
   - Membranes.
   - Sintered metals.
   - Petroleum reservoirs.
   - Geological formations.
   - Aquifers.
   - Solid combustible materials such as coal, biomass, and so on.
   - Mineral wool and urethane foam insulation materials.
   - Regenerators used for heat recovery in furnaces.
- Important parameters:
   - ΔP for flow of incompressible and compressible fluids through porous media: direct effect on energy consumption.
   - Diffusion through pores: in drying, catalyst beds, and adsorbents.
   - Diffusion through membranes.

     Note: Applications involving packed beds are discussed under mass transfer and other relevant topics.

- Explain what is meant by wall effect in a packed bed.
   - Particles will not pack as closely in the region near the wall as in the center of the bed.
   - Therefore, flow resistance in small diameter beds is less than it would be in an infinite container for the same flow rate per unit area of bed cross section.
   - This is called the wall effect.
- Give equations for friction factor, f_p, for flow through a packed bed.

$$f_p = (150/N_{Rep}) + 1.75 \text{ (Ergun equation)}. \quad (7.2)$$

   - Here, the Reynolds number N_{Rep} and the friction factor f_p for the packed bed are defined as follows:

$$N_{Rep} = D_p V_s \rho/(1-\varepsilon)\mu, \quad (7.3)$$

$$f_p = (\Delta P/L)(D_p/\rho V_s^2)[\varepsilon^3/(1-\varepsilon)], \quad (7.4)$$

where $\Delta P/L$ is the pressure drop per unit bed height and V_s is the superficial velocity, defined as Q/A, where Q is the volumetric flow rate of the fluid and A is the cross-sectional area of the bed. Actual velocities vary inside the bed. Sometimes the concept of interstitial velocity, V_i, which is the velocity that prevails in the pores of the bed, is used where

$$V_i = V_s/\varepsilon. \quad (7.5)$$

D_p is the equivalent spherical diameter of the particle defined as

$$D_p = \frac{(\text{volume of the particle})}{(\text{surface area of the particle})} = 6(1-\varepsilon)/\Phi_S S, \quad (7.6)$$

where S is particle surface area per unit bed volume, Φ_S is the sphericity, and ε is the void fraction of the bed = ratio of the void volume to the total volume of the bed. ρ and μ are fluid density and viscosity, respectively.

Other equations are as follows:

$$f_p = (150/N_{Rep}),\ N_{Rep} \leq 1 \text{(Kozeny-Karman equation)}. \quad (7.7)$$

This equation is used for flow of very viscous fluids.

$$f_p = 1.75,\ N_{Rep} \geq 10,000 \text{(Burke-Plummer equation)}. \quad (7.8)$$

- How is ΔP influenced by N_{Re} in a packed bed?
   - At low Reynolds numbers, ΔP is dominated by viscous forces and is proportional to fluid viscosity and superficial velocity.
   - At high Reynolds numbers, ΔP is proportional to fluid density and square of superficial velocity.
- What are *coating* flows?
   - In coating flows, liquid films are entrained on moving solid surfaces.
   - In dip coating flows, or free withdrawal coating, a solid surface is withdrawn from a liquid pool.
- Give typical examples of porosities.
   - Table 7.1 gives typical porosities for different materials.

7.1.2.2 Fluidization

- What is fluidization?
   - Fluidization involves fluid–solid systems in which solid particles are suspended in a fluid, maybe liquid or gas/vapor, the system as a whole behaving as if it is a fluid taking the shape of the container.
   - The upward velocity of the fluid balances the gravitational pull exerted by the particles, keeping the particles in suspension without allowing them to

TABLE 7.1 Examples of Porosities of Different Materials

Loose sand beds	35–50%
Salt	45–55%
Brick	12–35%
Fiber glass	88–93%
Limestone	4–19%

settle down (sedimentation) or allowing the fluid to carry away the particles (*pneumatic conveyance* if the fluid is air, *hydraulic transport* if it is water, or *carryover* in the general sense of the term if the fluid is a gas or liquid other than water). Initially, particles are at rest over a porous support such as a screen or a distributor plate with openings whose size is less than the size of the particles.

- As the velocity of the fluid from the bottom of the distributor is increased, the particles start getting disturbed and begin to rise, first as aggregates and then as individual particles. At this point, the force of the pressure drop times the cross-sectional area equals the gravitational force on the mass of particles. This is the *onset* of fluidization.
- The equation for the onset of fluidization is

$$V_f = [(\rho_p - \rho_f) g D_p^2 / 150\mu][\varepsilon^3/(1-\varepsilon)]. \quad (7.9)$$

- When the superficial velocity $V_s = V_f$, the condition of the bed is referred as one of *incipient fluidization.*
- At a certain velocity of the fluid, called *minimum fluidization velocity*, the particles get *fluidized.* Minimum fluidization velocity is based on empty cross section of the container (superficial velocity). Liquid fluidized beds are generally characterized by the regular expansion of the bed, which takes place as the velocity increases from the minimum fluidization velocity to terminal falling/settling velocity of the particles. This phenomenon is termed *bed expansion.*
- The pressure drop increases as the fluid velocity is increased until the onset of minimum fluidization. Then, as the velocity is further increased, the pressure drop decreases very slightly and then it remains practically unchanged as the bed continues to expand or increase in porosity with increase in velocity.

- On what factors minimum fluidization velocity depends?
  - The factors include particle diameter, its density, void fraction of the bed at the onset of fluidization, and viscosity of the fluid. Void fraction of the fluidized bed is not accurately known but is normally taken as about 0.4 at the onset of fluidization.
  - Commercial gaseous fluidized beds are usually operated at flow rates many times those at minimum fluidization requirements. Typically, 5–20 times the minimum values, taking care that carryover will not occur.
  - Liquid fluidized beds operate at values closer to minimum values.
- What is the relationship between bed height and porosity?

$$L_1/L_2 = (1-\varepsilon_1)/(1-\varepsilon_2), \quad (7.10)$$

where L is the bed height and ε is the bed porosity.

- What are the different regimes of fluidization? Illustrate.
  - With increased fluid velocities, the following transformations, ranging from fixed bed to pneumatic conveyance, take place: fixed bed → particulate regime → bubbling regime → slug flow → turbulent regime (bubbling bed, slug flow, and turbulent regimes constitute aggregative fluidization) → fast fluidization → pneumatic conveyance.
  - Figure 7.8 illustrates the regimes of fluidization.
  - Six different regimes of fluidization are identified in Figure 7.8.
  - Particulate fluidization, class (b) of the figure, is desirable for most processing since it affords intimate contacting of phases.
  - Fluidization depends primarily on the sizes and densities of the particles, but also on their roughness and the temperature, pressure, and humidity of the gas.

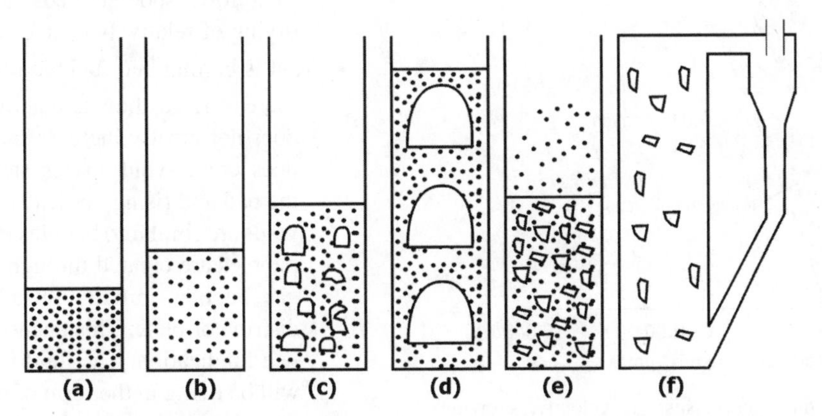

FIGURE 7.8 Six regimes of fluidization identified with increasing gas superficial velocities: (a) fixed bed; (b) particulate fluidization; (c) bubbling fluidization; (d) slugging fluidization; (e) turbulent regime.

- Small particles are subject to electrostatic and interparticle forces, which increase nonuniformity in fluidization.

- Show the *range band* for good fluidization conditions on a plot of particle diameter versus gas velocity in a fluidized bed.
  - Figure 7.9 illustrates the range of band for good fluidization.

- Give a correlation for predicting minimum fluidization velocity in a gas–solid system.

$$u_{mf} = 0.0093 d_p^{1.82} (\rho_p - \rho_f)^{0.94} / (\mu^{0.88} \rho_f^{0.06}), \quad (7.11)$$

where d_p is the particle diameter (μm), ρ_p and ρ_f are particle and fluid densities (kg/m^3), μ is the fluid viscosity, and u_{mf} is the minimum fluidization velocity (m/s).

- What is *incipient fluidization*?
  - As the gas velocity is increased from zero, gas pressure drop across the bed will reach a maximum. At this point, friction between gas and solids equals solids gravity pull and solids separate. This is called *incipient fluidization.*

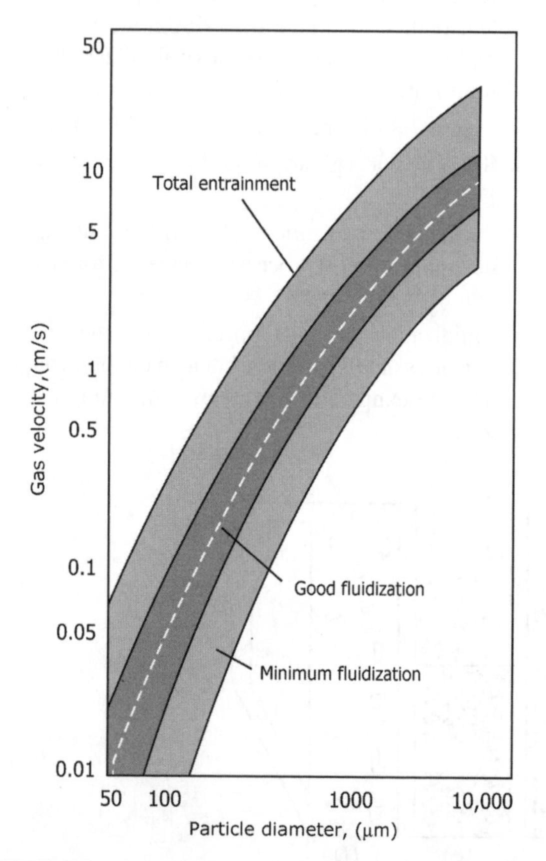

FIGURE 7.9 Particle diameter versus gas velocity, showing minimum fluidization velocity for good fluidization and total carryover bands.

- What is a spouted bed?
  - Gas entering into solids at the conical base in the bed makes/drills a passage through the granular solids inside the vessel (through a central core) and on reaching the top of the solid layer forms a *spout* (relatively narrow channel) and fountains the carried solids on the top of the spout.
  - It is an intermediate stage between fixed and fluidized beds.
  - Fairly coarse, uniformly sized particles when fluidized from a single nozzle will behave as if they are spouting out.
  - In spouting, a central stable gas jet carrying solids penetrates the bed. In the surrounding annulus, solids move downward like a packed bed in plug flow (piston-/rodlike flow). The solids form a wall for the central spout of the gas and are gradually entrained, disengaging at the top due to the decrease in velocity as a result of increase in flow area compared to that of the vessel.
  - The solids circulation pattern consists of rapid upward movement in the spout and slow downward movement in the surrounding bed. Spouted bed is an intermediate stage between fixed bed and fluidized bed.
  - The pressure drop over the spouted bed is normally lower than that for a fluidized bed because part of the weight of solids is supported by the frictional force between the solids and the wall of the vessel.
  - Good contact is achieved between the solids and the gas and high rates of heat and mass transfer occur because the relative velocity between the gas and the particles at the wall of the spout is considerably higher than that in a fluidized bed.
  - Furthermore, spouted beds can handle sticky materials in which the adhesion between particles is too strong for the possibility of fluidization taking place. Therefore, spouted beds are frequently used for drying of relatively coarse solids.

- What is boiling bed (bubble bed) fluidization?
  - Increasing gas flow rate above incipient fluidization does not greatly increase interparticle distance, but does create void spaces starting at the bottom of the bed and rising vertically through the bed. These voids are similar to bubbles in a boiling liquid. These bubbles are created through coalescence of smaller gas bubbles initially introduced at the base gas distributor plate. This phenomenon is called boiling bed fluidization. The fluid in some parts of the bed will be rising in the form of bubbles with practically no solids inside, at velocities higher than the rest of the fluid in which particles are in a fluidized state.

The bubbles may be carrying some fines within them. The whole bed resembles that of a boiling liquid. This bubbling bed results at low velocities above minimum fluidization velocity.

- What are the differences between particulate and aggregative fluidization?
  - Above minimum fluidization velocity, the bed can expand in two ways:
    i. *Particulate Fluidization*: As the fluid velocity is increased, the bed continues to expand and remains homogeneous, with the particles moving farther apart more rapidly. The average bed density at a given velocity is the same in all regions of the bed. The bed voidage increases with fluid velocity, but remains uniform. This type of fluidization is very desirable in promoting intimate contact between the gas and solids. Liquids often give particulate fluidization. Particulates in this type of fluidization are classified, based on their density and size range, as follows:

$$\Delta\rho = (\rho_p - \rho) = 2000\,\text{kg/m}^3,\ D_p = 20\text{–}125\,\mu\text{m};$$

$$\Delta\rho = 1000\,\text{kg/m}^3,\ D_p = 25\text{–}250\,\mu\text{m};$$

$$\Delta\rho = 500\,\text{kg/m}^3,\ D_p = 40\text{–}450\,\mu\text{m};$$

$$\Delta\rho = 200\,\text{kg/m}^3,\ D_p = 100\text{–}1000\,\mu\text{m}.$$

    ii. *Aggregative Fluidization*: The bed is very nonuniform. Most of the fluid (usually gas or vapor) passes through the bed in fast moving bubbles that are nearly free from particles, while the rest of the bed, the emulsion phase, remains close to minimum fluidization conditions. It may become quite rough/violent at high fluid velocities and bed structure and flow patterns are much more complex than those in particulate fluidization. The expansion of the bed is small as gas velocity is increased. Sand and glass beads provide examples of this behavior. This behavior is sometimes described as *bubbling fluidization.* Particulates in this type of fluidization are classified as follows:

$$\Delta\rho = 2000\,\text{kg/m}^3,\ D_p = 125\text{–}700\,\mu\text{m}.$$

$$\Delta\rho = 1000\,\text{kg/m}^3,\ D_p = 250\text{–}1000\,\mu\text{m}.$$

$$\Delta\rho = 500\,\text{kg/m}^3,\ D_p = 450\text{–}1500\,\mu\text{m}.$$

$$\Delta\rho = 200\,\text{kg/m}^3,\ D_p = 1000\text{–}2000\,\mu\text{m}.$$

  - Generally liquid fluidized beds exhibit particulate behavior while gas fluidized beds tend to be aggregative in character. Very dense, coarse particles can fluidize aggregatively while fine particles in dense gas may exhibit particulate fluidization tendencies.
- What is slugging in a fluidized bed?
  - Particulates will be distributed nonuniformly forming aggregates of slugs in the gas–solid systems.
  - *Slugging* can occur in bubbling fluidization since the bubbles tend to coalesce and grow as they rise in the bed. If the column is small in diameter with a deep bed, bubbles can become large and fill the entire cross section and travel up the column separated by slugs of solids.
- What is a *tapered bed*? What are its advantages?
  - In a tapered bed, cross-sectional area of the bed increases from a minimum at the bottom to a uniform cross section (that of the vessel) at some height.
  - When there is a wide range of particle sizes in a powder, fluidization will be more even in a bed that is tapered so as to provide minimum cross-sectional area at the bottom. If pressure gradient is low and the *gas* does not significantly *expand*, velocity will decrease in the direction of flow.
  - Coarse particles that tend to fluidize at the bottom (high velocity) assist in the dispersion of the fluidized gas. Carryover of fines from top will be reduced because of the lower velocity at the exit.
- What is a *centrifugal fluidized bed*?
  - In the centrifugal fluidized bed, the solids are rotated in a basket and the gravitational field is replaced by a centrifugal field that is usually sufficiently strong for gravitational effects to be neglected. Gas is fed in at the periphery and travels inward through the bed.
  - The equipment required for a centrifugal fluidized bed is much more complex than that needed for conventional fluidized bed, and therefore, it has found use only in highly specialized situations. One such application is in zero gravity situations, such as spacecraft, for the absorption of carbon dioxide from the atmosphere inside.
- What is a recirculation fluidized bed? Explain with an example.
  - The recirculating fluidized bed with a draft tube concept is briefly illustrated in Figure 7.10.
  - In application as a coal devolatilizer, dry coal is introduced into the devolatilizer below the bottom of the draft tube through a coal feeding tube concentric with the draft tube gas supply.
  - The coal feed and recycled char at up to 100 times the coal feed rate are mixed inside the draft tube and

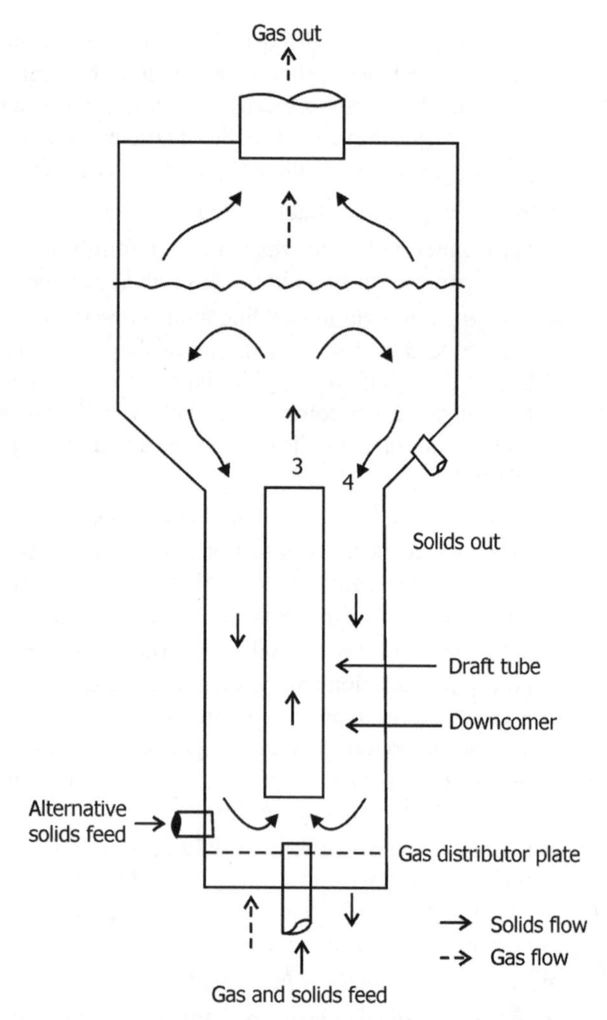

FIGURE 7.10 Recirculating fluidized bed concept. The draft tube operates as a dilute phase pneumatic transport tube.

carried upward pneumatically in dilute phase at velocities greater than 4.6 m/s.

- The solids disengage in a fluidized bed above the top of the draft tube and then descend in an annular downcomer surrounding the draft tube as a packed bed at close to minimum fluidization velocity.
- Gas is introduced at the base of the downcomer at a rate permitting the downward flow of the solids.
- The recirculating solids effectively prevent agglomeration of the caking coal as it devolatilizes and passes through the plastic stage.

- What is the influence of Froude number on fluidization characteristics of particulates?
  - Froude number for a fluidized bed is defined by Wilhelm and Kwauk as V_m^2/gd, where V_m is the minimum fluidization velocity, d is the diameter of the particles, and g is the acceleration due to gravity.
  - For $N_{Fr} < 1$, particulate fluidization normally occurs, and for $N_{Fr} > 1$, aggregative fluidization takes place.
  - For fluidization involving liquid–solid systems, much lower values of Froude numbers are involved because much lower values of minimum velocities are involved for fluidization. Uniformity for fluidization in the case of liquids is much more compared to gas-phase fluidization.
- How does pressure drop vary with velocity in fluid–solid beds? Illustrate.
  - As the superficial velocity approaches the *minimum* fluidization velocity, V_{mf}, the bed starts to expand, and when the particles are no longer in physical contact with one another, the bed *is fluidized.* ΔP then becomes lower because of the increased voidage, and consequently, the weight of particles per unit height of bed is smaller. This drop continues until the velocity is high enough for transport of the material to take place and ΔP then starts to increase again because the frictional drag of the fluid at the walls of the pipe starts to become significant.
  - Figure 7.11 illustrates the history of transformation of pressure drop from fixed bed to fluidized bed to pneumatic conveyance as superficial velocity increases.
- Give the equation for minimum fluidization velocity.

$$V_{mf} = 0.0055[\varepsilon_{mf}^3/(1-\varepsilon_{mf})]d^2(\rho_s-\rho)g/\mu, \quad (7.12)$$

where ε_{mf} is the void fraction.

  - This equation is based on the Kozeny–Karman equation for pressure drop for fixed beds and is applicable for laminar flow conditions.
- What is the function of a seal leg in a fluid bed process?
  - A seal leg equalizes pressures and strips trapped or adsorbed gases from the solids.
- What are the advantages and disadvantages of fluidized beds?

Advantages

  - Liquid-like behavior of solids, permitting easy addition and withdrawal of solids.
  - Rapid solids mixing within the bed, which when coupled with high thermal capacity (heat capacity) of solids compared to gas provides near-isothermal conditions within the bed. *Hot spots* will be absent. This benefits temperature-sensitive reactions, physical processing such as drying heat-sensitive solids, and the like.
  - Heat and mass transfer rates between gases and particles are very high, providing intimate contact between the phases.
  - High heat transfer rates between the bed and immersed heat transfer surfaces are well suited for

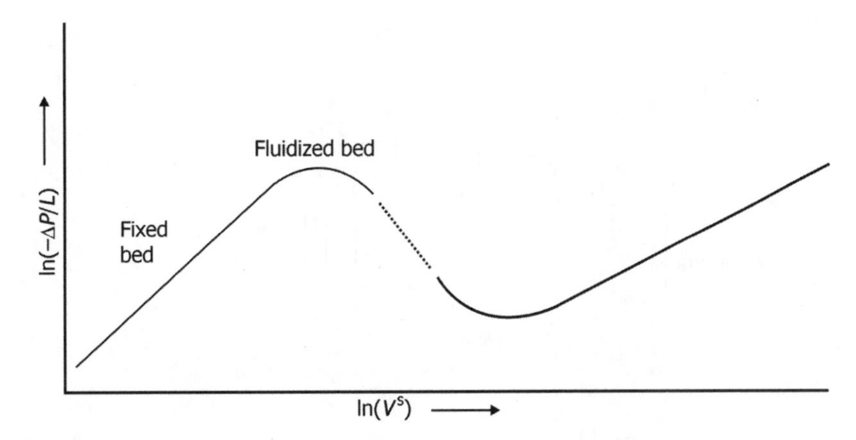

FIGURE 7.11 Pressure drop in a fluid–solid bed as a function of fluid superficial velocity.

applications involving heat removal from the reactors carrying out highly exothermic reactions or heat addition to endothermic processes.

Disadvantages

- Tendency for attrition of breakable bed solids with subsequent entrainment losses of solids. Dust control and treatment equipment requirement might be more expensive than the capital and running cost of fluidization equipment.
- Erosion of immersed surfaces creating maintenance problems.
- Single beds are generally unsuitable for processes that require plug flow of solids because of the well-mixed behavior of the solids.

- Name some important applications of fluidization.
  - Cooling/heating of granular solids.
  - Drying.
  - Drying of suspensions and solutions.
  - Blending of solids.
  - Treatment of heat-sensitive materials in a well-controlled environment due to uniformity of bed temperatures without hot spots.
  - Agglomeration, granulation, coating of tablets and granular solids, and so on.
  - Communition of solid particles.
  - Evaporative crystallization.
  - Fluidized leaching.
  - Catalytic cracking, hydrocracking, hydrodesulfurization, and reforming of petroleum fractions.
  - Fluidized coal combustion.
  - Fischer–Tropsch synthesis.
  - Catalytic chemical reactions.
  - Fluid bed bioreactors.
  - Iron ore reduction.
  - Application of heat to decompose a solid like calcination of calcium carbonate to CaO or gypsum to plaster.
- Classify applications of fluidization.
  - Chemical reactions:
    - Catalytic.
    - Noncatalytic.
    - Combustion/incineration of solids.
  - Physical contacting:
    - Solids mixing/blending.
    - Gas mixing.
    - Solids classification.
    - Size reduction.
    - Size enlargement.
    - Coating.
    - Granulation.
    - Heat transfer.
    - Drying
      - Solids.
      - Gases.
      - Suspensions and solutions.
    - Heat treatment.
    - Adsorption/desorption.
- Illustrate, with flow diagrams, fluidized bed catalytic cracking processes.
  - Figure 7.12 illustrates a typical catalytic cracking unit.
  - Gas oil is fed to the reactor and contacts the hot finely divided catalyst in the reactor and is vaporized and catalytically cracked into smaller molecules.
  - The product gases leave the fluid bed and, after passing through cyclones that remove entrained catalyst, leave the reactor for fractionation.

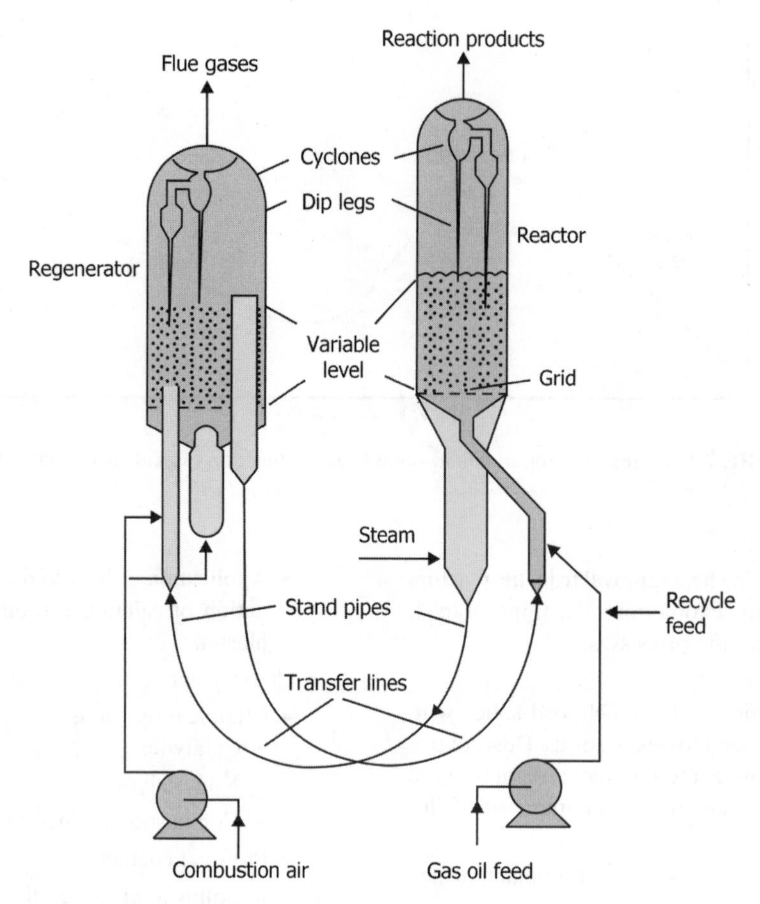

FIGURE 7.12 Typical fluid bed catalytic cracking unit.

- The reaction is endothermic and carbonaceous residue is deposited on the catalyst particles, reducing its activity. The catalyst is continuously removed for regeneration and reheating. It passes down through the reactor stripper in which unconverted hydrocarbons are removed and then through a standpipe and transfer line to the regenerator vessel.
- In the regenerator, air is used as the fluidizing gas and the coke deposits are burned off the catalyst particles. The burning is highly exothermic and provides the heat necessary for the cracking reaction. The hot regenerated catalyst is then returned to the reactor.
- Steam is normally used to strip the catalyst particles from the interstices of the solids as well as gases adsorbed by the particles.
- There are a number of variations in the design of catalytic cracking and reforming units. Figure 7.13 gives another variation used in fluid catalytic cracking units.

• What is the function of standpipes/seal legs in fluid bed solids transfer lines?

- Seal legs are frequently used in conjunction with solids flow control valves to equalize pressures and to strip trapped or adsorbed gases from the solids.
- Catalyst is transferred between the reactor and the regenerator through standpipes/seal legs and risers.
- Seal and/or stripping gas is introduced near the bottom of the leg. This gas flows both upward and downward. The transfer line is aerated with a high gas flow, thereby reducing the density there to a much lower value than that in the standpipe. This resulting pressure imbalance causes the desired catalyst flow.
- Bulk density in the standpipe is fairly high and the hydrostatic pressure produced at the bottom is substantial.
- Accurate pressure balances between vessels across the standpipe/riser loops are essential for proper design of a fluid–solids unit.

• What is vibrofluidization?

- When a fluidized bed is given vibrations, the process is called vibrofluidization. The bed of solids is

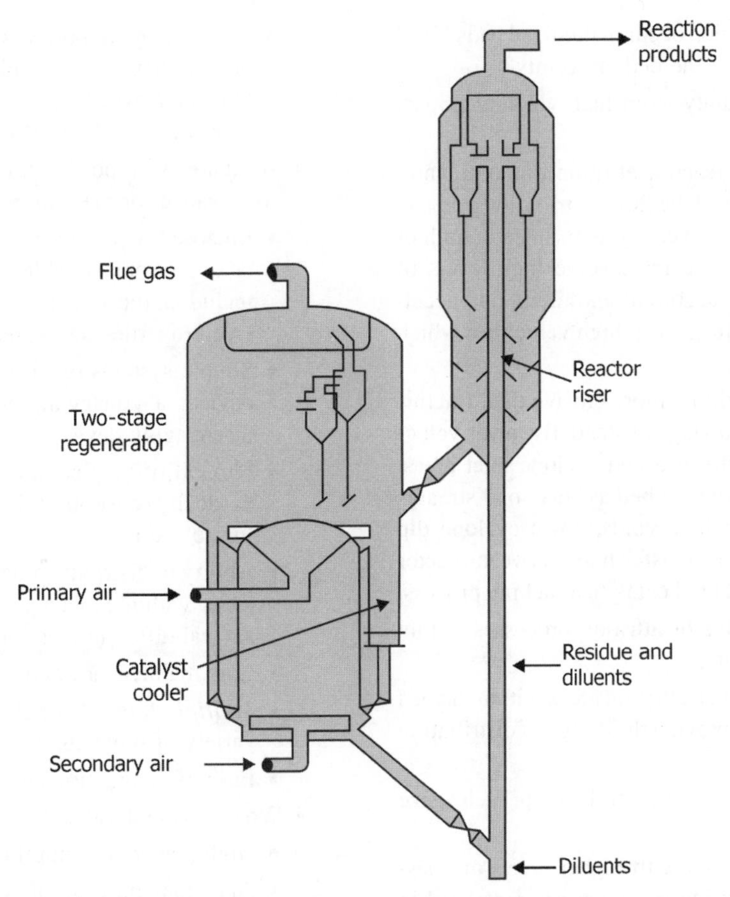

FIGURE 7.13 Fluid catalytic cracking unit with a two-stage regenerator (UOP).

fluidized mechanically by imposing vibration to throw particles vertically upward cyclically. It can be accomplished by vibrating the gas distributor plate or by a combination of vibration and gas pulsation. Bed mixing will be more vigorous. Lateral mixing will reduce channeling of different size particles.

- Vibrofluidization enables the bed to operate with either no gas upward velocity or reduced gas flow. Entrainment can also be greatly reduced compared to unaided fluidization.
- The technique is used commercially in drying and other applications.
- Chemical reaction applications are possible.

• What are the desirable characteristics of particles for smooth fluidization?

- Rounded, smooth shape (sand and glass beads).
- Enough toughness (abrasion) to resist attrition.
- Size range: 50–500 μm.
- Ratio of largest to smallest sizes: 10–25.
- Cohesive and large (>1 mm) particles are not good for fluidization.

• What are the characteristics of catalysts used in fluid beds?

- Diameters range from 30 to 150 μm.
- Density is about 1.5 g/mL.
- Bed expansion occurs to an appreciable extent before fluidization sets in.
- Minimum bubbling velocity is greater than minimum fluidization velocity.
- Rapid disengagement of bubbles takes place.

• What are the types of particles that are not suitable for fluid bed operations?

- Cohesive particles and particles larger than 1 mm do not fluidize well and are processed in other ways.
- What are the effects of particle attrition in fluidized beds?
- Reduction in uniformity in fluidization of the particles within the bed.
- Increased loads on cyclones and loss of cyclone efficiency.
- Loss of valuable catalyst requiring increased fresh makeup.

- How does particle attrition occur in fluidized beds? Will it take place throughout the bed uniformly?
  - Attrition occurs mainly from high-velocity impact within the bed.
  - Within a fluid bed reactor, attrition predominantly occurs in the vicinity of the fluid distribution grid and within the cyclone recovery system, since in both of these locations gas streams involve high levels of turbulence and are capable of entraining and accelerating particles to their local high velocities, which can be over 100 m/s.
  - Some degree of attrition, more erosive than fracturing in character involving substantially lower velocities than those at the grid and cyclone inlet areas, also takes place within the bed as bulk solid streams flow into random bubble voids, down cyclone dip legs, and in standpipe transfer lines between reactor and regenerator of a fluid catalytic cracking process.
- In what way knowledge of attrition processes within fluidized beds is useful?
  - Understanding the high attrition areas within the bed will be helpful in improving the design of distribution grids and cyclones.
  - Better recognition and control of particle size analysis.
  - Preparation of catalysts with higher levels of resistance to attrition, which is sometimes measured in terms of *attrition index*.

7.1.2.3 Gas–Solids Transport

7.1.2.3.1 Pneumatic Conveyance

- What is pneumatic conveyance?
  - Transport of solid particles with air as the carrier is called *pneumatic transport*. As velocity of air is increased beyond velocities at which fluidization takes place, the entrained particles get carried away by the air. Fluidization process gets transformed into pneumatic transport. The velocities involved will be such to support particulates in the fluid overcoming gravitational pull on the particles plus overcoming frictional losses in the conduit as well as maintaining bulk movement of the fluid and solids.
  - When a fluid other than air is used, the process might be termed as gas–solids transport, hydraulic transport, slurry transport, and so on.
- "Pneumatic conveying implies use of air for conveying solids." Is it always true? Under what circumstances a gas other than air is used?
  - *No*. Conveying powders that are susceptible to oxidation and that form explosive mixtures with air need to be conveyed by gases other than air, for example, inert gases such as nitrogen.
- What are the important reasons for adopting pneumatic conveyance for the transport of solids?
  - Enclosed, safe, and environmentally attractive method of transport suitable for a wide variety of products, including those with bacteria-prone, toxic, or explosive properties. Dust-free transport.
  - Simple systems requiring a prime mover, a feeding device, a conveying pipeline, and a cleaning or disengaging device.
  - Flexibility in pipeline layout. Can be transported vertically (over 300 m in a single lift) or horizontally (up to 3 km).
  - Ability to distribute product to a number of different areas within a plant and/or pick up material from several different locations.
  - Low maintenance and manpower costs.
  - *Multiple Uses*: A single pipeline can be used for a variety of products.
  - Ease of automation and control.
- What are the disadvantages of pneumatic conveying?
  - High power consumption.
  - Abrasion and wear of equipment.
  - Incorrect design can lead to attrition or degradation of the bulk solid particles.
  - Requires high skills in designing, operation, and maintenance.
  - Pneumatic conveying systems are limited in overall conveying distance and conveying capacity, compared to hydraulic transport.
  - Cohesive or sticky materials are often difficult to convey pneumatically. Moist substances that are wet enough to stick to the walls of the pipeline usually cannot be handled successfully. Materials with high oil or fat contents can also cause severe buildup in pipelines such that conveying is not practical, although this can sometimes be overcome with temperature control or flexible pipelines.
- Give examples of applications of pneumatic conveyors in industry?
  - Cement transport from grinders to silos and silos to loading points.
  - Conveying finely ground powders such as flours, pigments, and catalyst powders.
  - Unloading trucks, railcars, and barges, transferring materials to and from storage vessels, injecting solids into reactors and combustion chambers, and collecting fugitive dust by vacuum.

TABLE 7.2 Materials Transported by Pneumatic Conveyance

ABS powder	Acetyl salicylic acid	Adipic acid	Activated earth	Alumina
Ammonium sulfate	Ammonium nitrate	Granulated cattle feed	Anthracite	Cement
Bagasse fines	Cellulose acetate	Charcoal granules	Chalk	Clays
Ground coffee	Coke dust	Bleaching powder	Cotton seeds	Copra
Calcium chloride	Limestone	Magnesia	Flour	Bran
Milk powder	Glass fibers	Nitrocellulose	Sawdust	Soot
Graphite powder	Gypsum	Paper mill wood waste	Peanut shells	Starch
Phthalic anhydride	Plaster	Polyethylene granules	Granulated sugar	Sulfur
Polystyrene beads	Polyester fibers	PVC powder and granules	Sodium sulfate	Talc
Terephthalic acid	Titanium dioxide	Soap flakes	Silica gel	Yeast
Sodium carbonate	Zinc oxide	Wood shavings	Tobacco	Fly ash

- Give a list of some materials conveyed by pneumatic conveyance.
  - Table 7.2 provides a list of materials that are conveyed by pneumatic conveyance.
- What are the important guidelines with respect to selection, specification, and design of pneumatic conveyance systems?
  - During the project stage, current conveying rate requirements and plant layout along with plans for future expansions or potential rate enhancements due to improvements in process technologies need to be considered.
  - Minimum and maximum conveying rates should be carefully defined.
  - A reasonable horizontal conveying length (15–20 pipe diameters), with minimum flexible hose length, should be allowed before the first bend to allow the bulk material to accelerate.
  - Placing bends/elbows back to back in the line should be avoided. The number of bends should be kept to minimum.
  - Pipe diameter should be increased at the end of the line to prevent excessive velocity and reentrainment of the particulates.
  - Proper venting of the rotary air lock and feeder in a positive pressure system is critical for reliable operation. Gas leakage at feeder must be considered in design calculations and adequately compensated.
  - The gas flow control system for multiproduct and multidestination systems must ensure that the operating point is maintained within the stable operating zone.
  - It must be noted that product damage and wear at pipe bends depend on the nature of the material.
  - A proper purge control sequence need to be designed and tested to avoid product degradation or blockage.
  - Minimum conveying velocity is a function of flow rate.
  - Relative humidity of conveying air should be more than 70%, to avoid electrostatic effects.
- What are the characteristics of particulates that are handled by pneumatic conveying?
  - Pneumatic conveying can be used for particles ranging from fine powders to pellets with bulk densities of 16–3200 kg/m^3 (1–200 lb/ft^3).
  - Among the important particle characteristics are particle size and size distribution, shape, density, hardness, and friability.
    - Particle size, distribution, and shape are known to be among the most significant variables affecting pneumatic conveying. For example, uniformly sized round and smooth particles are easier to convey than angular, rough ones having a wide size distribution.
    - Another rule of thumb is that, to prevent mechanical plugging of conveying lines, particularly when conveying materials containing large particles, the pipe diameter must be at least three to five times the maximum particle dimension.
    - Particle density affects the minimum conveying velocity and pressure drop required for transport.
  - Bulk properties that are important include bulk density and compressibility, permeability, cohesive strength, segregation tendency, explosibility, toxicity, reactivity, and electrostatic effects.
    - The compressibility and permeability of a bulk solid determine how readily the material will deaerate.
    - The direct effect of the cohesiveness of a bulk solid on pneumatic conveying can be buildup in the lines. Flow stoppages, or erratic flow through equipment upstream of a conveying line, such as in a bin, feeder, or chute, can be detrimental to achieving the desired transfer rates.

  - Other bulk properties that must be considered during design include explosivity and toxicity. Materials that may contain residual hydrocarbons, such as newly reacted polyethylene or polypropylene powders, may have to be conveyed using nitrogen to limit exposure to oxygen.
  - Static electricity may be a source of ignition for materials prone to explosion, in which case the charge must be either dissipated by proper grounding or neutralized.
  - Materials that are toxic or need strict containment may require a vacuum system, in which any leak will be into the line, rather than out to the environment.
  - Double walled pipelines under positive pressure have been used to convey materials such as contaminated soils.
  - Hygroscopic materials exposed to humid air often become more difficult to handle.

- What is *saltation velocity* and how is it used in designing pneumatic conveying systems?
  - *Saltation velocity* is defined as the actual gas velocity (in a horizontal pipe run) at which the particles of a homogeneous solid flow will start to fall out of the gas stream.
  - In designing, the saltation velocity is used as a basis for choosing the design gas velocity in a pneumatic conveying system.
  - Usually, the saltation gas velocity is multiplied by a factor that depends on the nature of the solids to arrive at a design gas velocity.
  - For example, the saltation velocity factor for fine particles may be of the order of 2.5 while the factor could be as high as 5 for coarse particles such as soya beans.
- What is *dilute phase* conveying? What are its characteristic features?
  - *Dilute phase* conveying takes place when particles are conveyed at a gas/air velocity that is greater than saltation and choking velocities. Particles are freely suspended in the gas stream.
  - Adverse effects of dilute phase conveying include pipeline wear and attrition among the particles due to the high velocities involved.
- What is *dense phase* conveying? What are its characteristics?
  - In *dense phase* conveying, velocities involved are below saltation velocity and plug or piston flow or moving bed flow occurs.
  - With plug flow, coarse and permeable bulk solids such as pellets or beans can be reliably conveyed.
  - With moving bed flow, impermeable fine solids such as cement and fly ash can be effectively conveyed.
  - Because of the lower velocities involved in dense phase conveying, pipeline wear and damage to the solids due to attrition are reduced. Gas and energy requirements are lower with decreased operating costs. Long conveying distances are possible.
- What is *choking velocity* in pneumatic conveying?
  - The gas velocity at which particles that are conveyed upward within a vertical pipeline approach their free fall velocity.
  - Partial settling out of solids from a flowing gas stream and other instabilities may develop below certain linear velocities of the gas called *choking velocities*.
  - Normal pneumatic transport of solids accordingly is conducted above such a calculated rate by a factor of 2 or more because the best correlations are not that accurate.
  - Above choking velocities, the process is called *dilute phase* transport, and below, *dense phase* transport.
  - Transition from dense phase flow to dilute phase flow characterizes choking point.
- Classify gas–solids transport in horizontal piping.
  - The flow of gases and solids in horizontal pipe is usually classified as either *dilute phase* or *dense phase* flow.
  - For *dilute phase* flow, achieved at low solids loadings, high volumes, and high gas velocities, the solids may be fully suspended and fairly uniformly dispersed over the pipe cross section, particularly for low-density or small particle solids. The gas stream carries the materials as discrete particles by means of lift and drag forces acting on the individual particles.
  - At lower gas velocities, the conveying process takes place with a certain proportion of the solids moving through the upper part of the pipe cross section together with a highly loaded stream in the lower part of the cross section. Depending on the characteristics of the solids, gas velocity, solids flow rate, and other factors such as pipe size and roughness, the flow patterns in the *dense phase* mode can vary from being unstable to stable or an intermediate stable/unstable regime.
  - The stable regime involves smooth flow while the unstable regime is characterized by violent pressure surges as the moving layer breaks up.
  - With higher loadings and lower gas velocities, the flow patterns in the *dense phase* mode can vary from conditions in which the particles may pack the pipe and move as slugs to situations in which the particles may settle to the bottom of the pipe, forming

aggregates that move along the bottom section of the pipe, similar to movement of sand dunes in a desert, with a dilute phase layer of solids moving above the dunes.

- The main principle of a *dense phase* conveying system is to slow down the velocity of the product in the pipe to a point that is below the speed at which the product breaks or degrades. In a *dense phase* system, the velocity range at the source can be as low as 5–8 km/h for the majority of products. The product velocity at the destination is always a function of the system ΔP.
- Because of the compressibility of the air or gas used as the conveying medium, the gas will expand as it moves to the end of the pipeline and the product velocity will increase accordingly. The terminal velocities of a dense phase system can vary, but in most cases they rarely exceed 30 km/h. In a typical dilute phase system, the starting velocities begin at about 65 km/h while the terminal velocities can reach 160 km/h.

- What are the advantages of *dense phase* conveying over *dilute phase* conveying?

  Dense Phase Conveying

  - Decreased energy usage due to significantly reduced volumes of air.
  - Reduced material breakage or degradation and reduced pipeline wear due to lower conveying velocities. Best suited for handling friable products.
    - Many products within the food industry fall into this category. As an example, if a consumer opens a bag of cheese puffs and finds that half of them are broken into little pieces, he/she will quickly change product brand to that with whole cheese puffs. Degradation of the final product is a very important factor for any manufacturer of goods.
  - Advantageous for handling abrasive solids as in dilute phase conveying velocities are higher resulting in increased wear of pipe and associated elements. Many minerals and chemicals fall into this category. The pipe wear can also result in a contamination problem because the pipe material will get mixed with the product.
  - Another advantage of using the dense phase system for plastics transport is the following:
    - Some of the softer plastics, such as polypropylene and polyethylene, smear onto the pipe wall when the product slides along the outer wall of an elbow in a dilute phase transport system. The plastics actually melt due to the frictional contact with the pipe wall and leave a long thin layer of material. The layers are peeled off into strips and reentrained into the system. Dense phase transportation eliminates these problems that are commonly associated with dilute phase conveying.
  - Smaller conveying pipeline sizes due to heavier line loading capabilities.
  - Smaller dust collection requirements at the material destination due to lower conveying air volumes.

- What are the disadvantages of *dense phase* conveying systems?
  - *Dense phase* systems are more complicated than dilute phase systems to control properly. The narrow operating range for stable operation gives rise to many control problems. It is necessary to constantly change the gas amounts for varying conditions in the system. The gas control device must be flexible and accurate.
    - Varying product transfer capacities, varying products, or varying product grades that result in dissimilar physical product characteristics pose problems in operation. Even product temperature variations need to be known at the design stage and incorporated into the system design.
    - As temperature changes, velocity changes, in addition to product characteristics.

- What are the different systems that are used in *dense phase* conveying?
  - *Gatty System*: Air is injected from a subpipe mounted inside the conveying line. Used for conveying plastic pellets.
  - *Buhler Fluidstat System*: A series of bypass lines are used every half a meter or so. Solids having good air retention properties are well handled by this system.
  - *Tracer Air System*: Uses a series of boosters employing a pressure regulator and a check valve.
  - *Pulsed Conveying System*: Consists of a flow vessel and an air knife downstream from the discharge valve. The plug lengths are controlled by the air knife operation time.
  - *Takt-Schub System*: Incorporates a double pulse system employing two gas inlet valves whose timing can be controlled. The system is operated in an alternating pulsing manner.
  - *Plug Pulse System*: The discharge valve is alternately operated with a gas pulse. It works well with free-flowing materials.
  - *Molerus–Siebenhaar System*: This system relies on vibrations being put into the system by an unbalanced motor. These vibrations cause the materials to move with a low velocity throughout the system. This system is particularly useful for fragile materials.

- For upflow of gas–solid systems in vertical pipes, what is the recommended minimum conveying velocity for the solids?
  - The minimum recommended conveying velocity for low loadings may be estimated as *twice* the terminal settling velocity of the largest particles.
  - Choking occurs as the velocity drops below the minimum conveying velocity and the solids are no longer transported, collapsing into solid plugs.
- How does conveying distance affect capacity and conveying pressure?
  - Conveying capacity is inversely proportional to conveying distance.
  - Conveying pressure is directly proportional to conveying distance.
- What are the general types of pneumatic conveying systems? State briefly their uses.
  - *Negative (Vacuum) System:* Normally used when conveying from several pickup points to one discharge point.
    - The negative system usually sucks on a cyclone and/or filter/receiver mounted above the receiving storage hopper or bin. Solids are usually sent down to the hopper with a rotary air lock feeder. Air is sucked into the transfer pipe or duct at the pickup end of the system.
    - A variety of feeders can be used to introduce solids into the flowing air stream such as rotary air lock feeders, pan-type manifolds under railcar hoppers, paddle-type railcar unloaders, screw conveyors, and so on.
  - *Positive Pressure System*: Normally used when conveying from one pickup point to several discharge points.
    - In the positive pressure system, air is blown into the pickup duct often up to a cyclone with atmospheric vent.
    - Usually solids are introduced to the conveying air stream with a rotary air lock feeder.
    - For sending the solids down to the storage bin from the cyclone, a simple spout connection can be used instead of the rotary air lock feeder required in the negative pressure system. The positive system does not need a vacuum vessel at each receiving location.
    - Also in the positive pressure system conveying to a number of hoppers, a simple bag-type cloth can serve as the filter.
    - So for conveying from one pickup location to several receiving locations, the pressure system is often cheaper than the negative system.
    - Positive system is most widely used.
  - *Pressure Negative (Push–Pull) Combination System*: Normally used when conveying from several pickup points to several discharge points.
    - The pressure negative system is ideal for unloading railcars and also for conveying light dusty materials because the suction at the inlet helps the product enter the conveying line.
    - Positive systems are poorer than negative systems for handling such materials because feeding into the pressure line can be difficult and can present a dust collection problem at the pickup location because of blowback or leakage air through the rotary valve.
    - Also the negative system works better if there are lumps at the pickup end.
    - The positive pressure system tends to pack at the lump, while the negative system will often keep solids moving around the lump and gradually wear it away.
    - For the pressure negative system, a single blower can be used for both the negative and positive sides of the system.
    - Lack of flexibility with a single blower usually dictates the need for separate blowers for the negative and positive sides.
  - Venturi, product systems, or blow tanks.
- When is a negative (vacuum) system recommended for conveying solids? Illustrate.
  - When conveying solids from several pickup points as illustrated in Figure 7.14.
  - Vacuum systems may require feeders with a good seal to minimize leakage of air/gas into the pipeline.
  - An acceleration zone is required after the feed point to obtain steady velocity for transport.
- When is a positive system (pressure) recommended for conveying solids?
  - When conveying solids from one pickup point to several discharge points (Figure 7.15).
  - Positive pressure systems require devices that can feed material from atmospheric pressure into a pressurized pipeline.
  - Positive systems are most common. Diverter valves deliver to several receiving vessels.
- What is a closed loop system? Where is it used?
  - Closed systems are used whenever hazardous solids are pneumatically conveyed. Reason is environmental regulations and safety prohibiting atmospheric discharge of entrained solids. Examples include toxic and combustible solids.

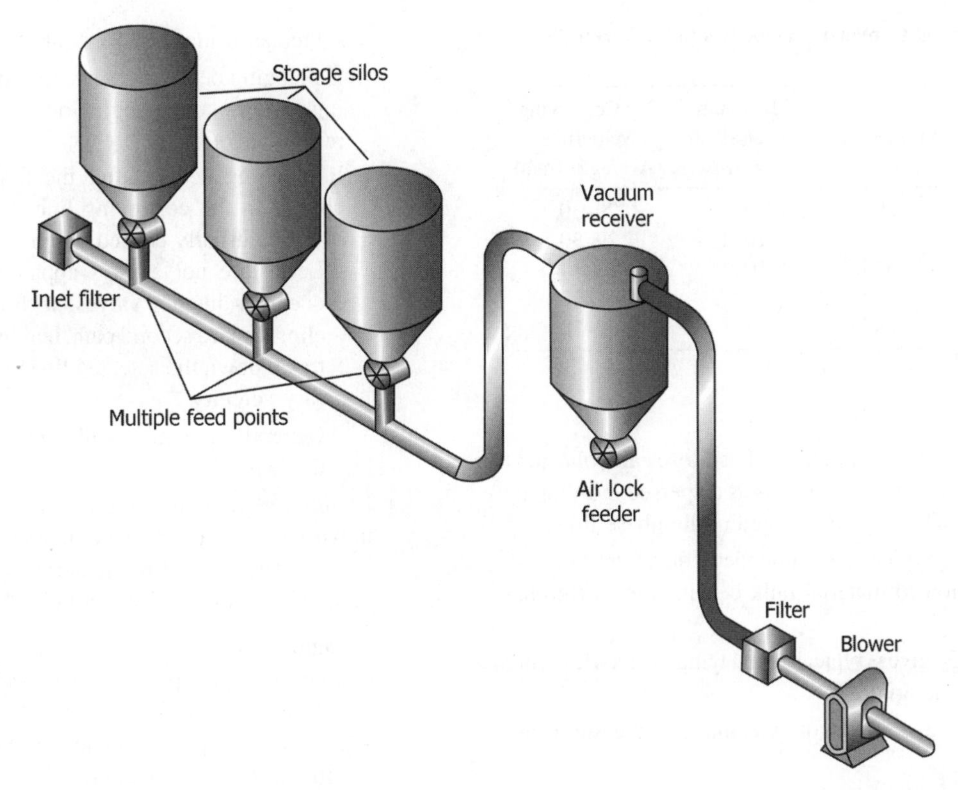

FIGURE 7.14 Pneumatic conveyance: negative system.

- Nitrogen might be used to avoid explosions if air is used with combustible solids.
- The gas is recycled instead of releasing to atmosphere.

• Describe different flow conditions that occur as gas velocity is increased in a pneumatic conveyor.

- Beyond minimum fluidization velocity, there will be moving bed flow.
- At higher gas rates, solids are carried in gas as slugs. This is called slugging dense phase flow.

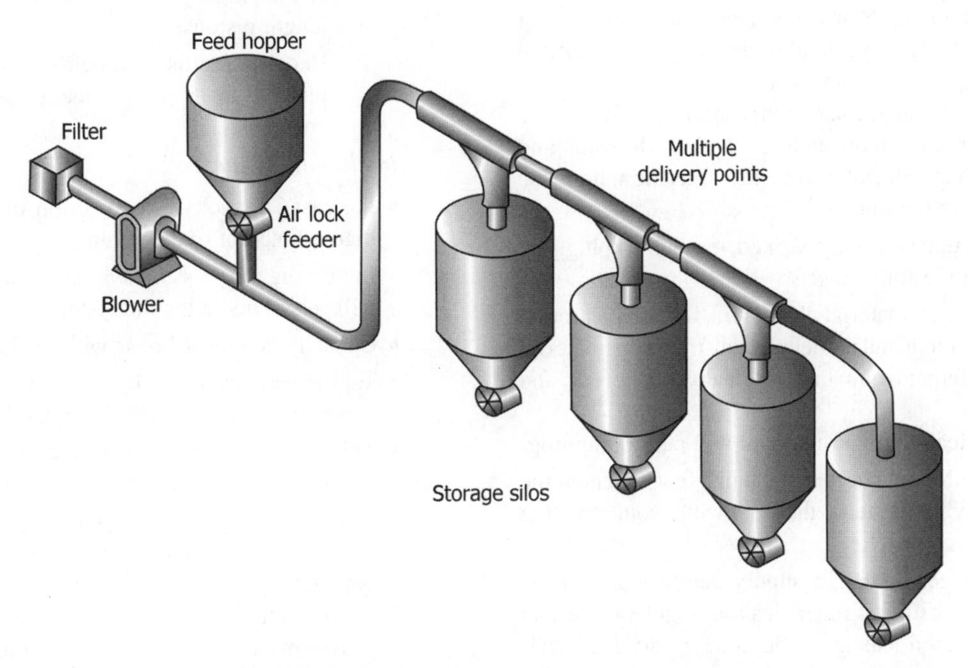

FIGURE 7.15 Pneumatic conveyor: positive system.

TABLE 7.3 Typical Conveying Velocities for Different Materials

Material	Bulk Density (kg/m^3)	Maximum Material/Air Ratio (wt/wt)	Air Conveying Velocities (Average) (m/s)
Flour	593	10 : 1	29–30
Polyolefin pellets	529	10 : 1	29–30
Wheat	513	10 : 1	33–34
Coarse sand	1522	6 : 1	41–42
Lime (pebble)	897	7 : 1	37–38

- At still higher gas rates, solid bed loses all cohesion and particles stream with gas as dispersed individual particles. This region is called dilute phase flow.

- What are the typical recommended conveying velocities in relation to material bulk densities for different materials?
  - Table 7.3 gives typical conveying velocities for different materials.
- What are the most common pneumatic conveying problems?

  Plugging of Lines
  - Plugging is caused when the conveying pipe is filled for a substantial length and the conveying air builds up pressure behind the accumulated material.
  - If the accumulation is short enough, ΔP will increase, the force exerted against the accumulation will overcome friction, and the material will slide along in the pipe.
  - If the material forms a longer piston, as the air pressure builds up behind it, the plug is forced more tightly against the pipe wall.
    - ➢ The material becomes nonpenetrable, such that the higher the frictional force against the wall and the plug will not move, no matter how high the pressure buildup.
  - If the system originally worked, reasons for plugging could be the following:
    - ➢ Change of material characteristics such as moisture content and particle size distribution, material is different from that originally conveyed, and so on.
      - Typically, a finer material will cause plugging.
    - ➢ If the plugging occurs in the first few meters of the conveying system, the following could be the reasons:
      - Changes in the air supply due to worn equipment, dirty air filters, leaking relief valves, and increased leakage through the rotary feeder/air lock.
      - Increased material feed rate.
    - ➢ If plugging occurs at the beginning of the conveying system, conveying velocity or air to material ratio is incorrect.
    - ➢ If plugging occurs after the first few meters, a change in the conveying line itself is indicated, which is usually caused by an air leak.
      - Leaks are not always apparent as they often occur in diverter valves, with the lost air traveling down a second path, leaving the material to travel down the selected line without the necessary velocity.
      - Generally plugging will occur about 12 m after the leak.
  - In *dilute phase* conveying (more than enough air to move the material), the air velocity and material to air loading determine if the material will be moved by drag effect on individual particles.
  - In *dense phase* conveying, material is fed into the system to form a piston that entirely fills the pipe cross section. Drag has no effect on conveying.
    - ➢ For a coarse, permeable material having no fines, little problem is encountered.
    - ➢ The length of nonpermeable fine materials is first controlled by the way they are introduced into the dense phase system.
      - In most systems, the material is discharged through a reducing elbow that compacts it into a solid plug.
      - Injecting air downstream of feed controls length of the plug by cutting the extrusion into proper length pistons.
      - Boosters are used to help control length of the pistons and keep them separate during conveying.

  Inadequate Capacity
  - Increased capacity is a function of pressure rather than a function of air volume.
  - Increasing the pressure available from the air supply will permit operating the system at a higher pressure.
  - Capacity is almost linear with operating pressure.
  - To increase capacity by 10%, for example, 10% increase in pressure may be required.
  - This is true for both dilute and dense phase conveying.
  - Sometimes increasing the operating pressure may be expensive, so reducing the *required* pressure of the system may be more attractive.
    - ➢ For example, eliminating some of the bends in the system and taking a more direct line with a reduced line length will lower the operating pressure.

- Capacity can then be increased by the same amount as the reduction in line restriction.
- Line length and capacity are inversely proportional.
- Reducing line length by 10% will increase capacity by 10%.

- If the system is a dilute phase and the conveying velocity is higher than that required, reducing air volume will reduce *required* pressure making increase in capacity possible.
  - There are systems in which slowing down the blower reduces pressure and increases conveying capacity.
- Increasing air flow by speeding up the blower: it is observed that the capacity, instead of increasing, reduced.
  - A given material requires a minimum air flow to be conveyed in dilute phase, but increasing the velocity above that point only raises the *required* pressure and has little effect on capacity.

Line Wear and Maintenance

- Line wear is caused by the impact of an abrasive material on the pipe and is most noticeable at changes in direction.
  - Abrasive wear depends on particle characteristics such as hardness, particle size and shape, and particle density.
  - Hardness is measured on the Mohs scale with diamond having a value of 10 and talc having a value of 1.
    - For example, a No. 6 material will scratch a No. 5 material, but a No. 1 material will have little effect on a No. 6 material.
- Particle size affects the rate of wear, since wear is a function of the energy of impact between the material and the pipe wall.
  - Larger particles produce greater attrition on the pipe wall than smaller particles. For example, pneumatic conveying of broken glass pieces will have a much greater impact on the line than finely pulverized glass.
- Shape of the particle influences the abrasive wear.
  - If the particle is spherical, it will impact the wall surface over a larger area than if it is jagged and contacts it at a small point.
- Velocity of conveying determines the energy of impact that influences wear rate.
  - Impact is a function of $(\text{velocity})^2$.
  - To improve life of the line, velocity must be kept to the minimum.
- Another way is to use abrasion-resistant materials for the line.
  - The higher the material is on the Mohs scale, the better resistive it is for abrasion with lower Mohs materials.
- The fourth method is to use bends that have a different configuration than radius bends.
  - Any bend that forms a pocket to allow material accumulate in it will have a greatly increased life as the conveyed material now impacts itself instead of the pipe wall.

Venting of Feeder/Air Locks

- Improper or lack of venting is the primary cause for most feeding problems of conveying systems.
  - If material is to flow into a vessel, chamber, feeder, conveying line, and so on, the air that is filling the volume must be vented.
  - When feeding into a pressure conveying system, some air tries to escape from the pressure line, both through clearances in the feeder and through the displacement volume of the rotor. If the feeder is under a silo or storage bin, the air must be vented out for smooth flow.
  - With coarse materials that are not fluidizable, venting is simple.
  - A channel installed inside the bin allows the leakage air to pass upward a small distance, where an opening allows it to escape through a filter bag or continue upward to the top of the bin.
- If the above method is used for fine fluidizable material, leakage air fluidizes the material and then flows like water, seeking its own level, up into the venting area, and quickly plugs the vent line.
- The approaches that are used fairly successfully are venting the feeder body, separating the feeder from the air lock, and venting the air by venturi.
  - Venting of the feeder body is done through ports either on the side of the feeder or on the end plates. There are problems with this arrangement:
    - There must be an adequate number of blades on the rotor to assure separation from the feed bin to the vent and from the vent to the conveying line, regardless of the position of the rotor.
    - This is often not considered and, after a short period, vent line plugs and feed problems resume.
    - If the material does not drop out of the rotor at the bottom, maybe due to too high speed of operation, the material is carried along with the rotor past the dropout point and is then carried into the vent line. This also plugs the vent over time. Injecting air into the rotor in the discharge

area helps material to move out of the rotor during the limited time available.

- Separating the feeder from the air lock allows for a free space between the two, in which venting can be accomplished.
- In venting by venturi, a small venturi using plant air from the feeder either aspirates material and air from the feeder and introduces it into the conveying line or vents it to the top of the bin/silo.

Material Degradation

- Frequently, the material being conveyed has a change in physical characteristics making it unsuitable for marketing.
  - Pneumatic conveying can create fines.
  - Altered appearance due to scuffing of the surface, which would change the luster of the product.
- Conveying with as high a loading (solids to air ratio) as possible and as low a velocity as possible will minimize the above problems.
  - Pressure conveying systems are better suited for this purpose than vacuum conveying systems as the latter involve dilute phase with low solids content and lower velocities.
  - Avoiding cyclones and slowing down the material at the end of the system.
    - More material degradation takes place in one cyclone than in the entire conveying system. In a cyclone, material rubs against walls causing accelerated degradation. Another type of separator may be considered for such cases.
    - To minimize material degradation at the end of the line, diameter of the line may be increased by one pipe diameter for a distance of 6 m. This reduces material impact on the receiver and hence degradation.

- Summarize the important parameters that affect particle attrition during pneumatic conveyance of solids.

  Process-Related Factors
  - Mode of conveying (dense versus dilute phase).
  - Gas velocity or particle velocity.
  - Solids loading (or concentration).
  - Temperature of gas and solids (coupled with material properties).
  - Conveying distance.
  - Materials of construction of straight pipeline sections and bends.
  - Surface finish of pipeline and bends.
  - Number of bends (frequent change in direction).
  - Bend geometry and flow pattern at the bend.

  Material-Related Factors
  - Particle size.
  - Particle shape.
  - Particle strength or modulus or hardness.
  - Elasticity of particles.
  - Breakage function of material.
- What are the principal measures that help reduce wear in existing pneumatic conveying systems?
  - Reducing conveying velocity or increasing the solids loading ratio.
  - Reducing the number of bends by simplifying the line layout wherever possible.
  - Replacing bends with designs that are less prone to attrition.
- What are the commonly used elbows or bends in pneumatic conveying? How are they classified?
  - Elbows or bends are classified based on their radius of curvature and diameter of the conveyor line.
  - *Elbow*: Ratio of radius of curvature R and line size D in the range of 1–2.5.
  - *Short Radius Bend*: R/D in the range of 3–7.
  - *Long Radius Bend*: R/D in the range of 8–14.
  - *Long Sweep Bend*: R/D in the range of 15–24.
- Compare a blind tee with a blind bend used in pneumatic conveyor lines. Illustrate.
  - Figure 7.16 illustrates a blind tee and a blind bend showing solids impact points by means of arrows.
  - *Blind Tee*:

    Advantages
    - In a blind tee, one of the outlets is plugged allowing conveyed solids to accumulate in the pocket. The accumulated pocket of material cushions the impact of the incoming material, significantly reducing the potential for wear and product attrition.
    - Low cost.
    - Erosion/wear resistant.
    - Short turn radius; compact design.

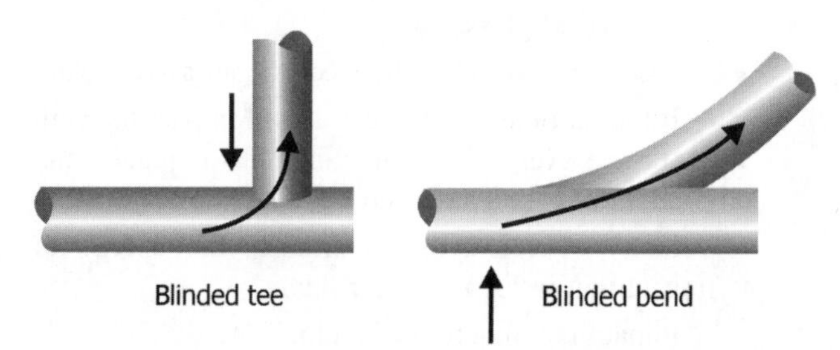

FIGURE 7.16 Blind tee and blind bend with arrows showing impact points.

- ➢ Easy to retrofit.
- ➢ Low particle attrition (no chipping or surface abrasion).

Disadvantages

- ➢ High pressure drop. The conveyed solids lose most of their momentum during the impact and must be reaccelerated downstream of the bend. As a result, pressure drop across a blind tee can be as much as three times that of a long radius bend.
- ➢ Not suitable for moist, cohesive, or sticky materials.
- ➢ May result in cross-contamination if the pocket does not self-clean.

- *Blind Bend*:

Advantage

- ➢ Better erosion resistance.

Disadvantages

- ➢ Same as those for blind tee.
- ➢ Secondary impact wear zone on the inner radius.

- "In pneumatic conveying of abrasive solids, pipe wear is a problem reducing its life, especially severe wear effects at points of change in direction." How is this problem solved?
  - Use of elbows with a wear-resistant ceramic liner or a replaceable impact plate insert. Examples include high-density alumina ceramics, zirconium corundum, hardened cast iron, silicon carbide, and tungsten carbide.
  - Use of a blind tee elbow.
  - Use of a spherical chamber elbow.
  - Use of proprietary elbow designs.
- What are the minimum conveying velocities for various types of materials for good ventilation?
  - Table 7.4 gives minimum conveying velocities for different types of materials for good ventilation requirements.

TABLE 7.4 Minimum Recommended Conveying Velocities for Good Ventilation

Nature of Material	Industrial Examples	V_{fmin} (m/s)
Vapors, gases, smoke	All vapors, gases, and smoke	5–10
Fumes	Welding	10–12
Very light fine dust	Cotton lint, wood flour	12–15
Dry dusts and powders	Fine rubber dust, bakelite molding powder dust, cotton dust, light shavings, soap dust, leather shavings	15–20
Average industrial dust	Grinding dust, dry buffing lint, coffee beans, granite dust, silica flour, general materials handling, brick cutting, clay dust, foundry, limestone dust, asbestos dust	18–20
Heavy dusts	Heavy/wet saw dust, metal turnings, foundry tumbling barrels and shake-out, sand blast dust, wood blocks, brass turnings, cast iron boring dust, lead dust	20–23
Heavy or moist	Lead dust with small chips, moist cement dust, asbestos chunks from machines, sticky buffing lint	>23

Source: ACGIH. *Industrial Ventilation: A Manual of Recommended Practice*, 21st edition, American Conference of Governmental Industrial Hygienists, Inc., Cincinnati, OH, 1992.

7.1.2.4 Solid–Liquid (Slurry) Flow

- Give important examples of slurries that are transported in pipelines.
  - A coal–oil slurry used as fuel and a lime slurry used for acid waste neutralization are two examples of slurries involved in process applications.
  - Long-distance movement of coal, limestone, ores, and others. A few such installations have been made with lengths ranging from several kilometers to several hundred kilometers.
- What are the general considerations involved in the design of a slurry pipeline?
  - Whenever possible, piping should be designed to be self-draining.
  - Manual draining should be installed to drain sections of the piping when self-draining is not possible.
  - Blowout or rod-out connections should be provided to clear lines in places where plugging is likely to occur.
  - Access flanges should be provided at T-connections.
  - Manifolds should have flanged rather than capped connections to allow for easy access.
  - Clean-out connections should be provided on both sides of main line valves so that flushing can take place in either direction.
  - Break flanges should be provided every 6 m of horizontal pipe or after every two changes in direction.
- What type of pump(s) is suitable for slurry handling?
  - In short process lines, slurries are readily handled by centrifugal pumps with open impellers and large clearances.
  - Positive displacement pumps, notably diaphragm pumps, are another type of pumps that are widely used in slurry transport in long-distance lines, in pumping stages. Plunger or piston-type pumps are

also common for use in conventional slurry systems because of their high-pressure capability. Positive displacement pumps are limited by the maximum particle size, approximately 2.4 mm, which will pass through the valves.

- Line sizes of 45 cm, handling 50–60 wt% coal particles with particle sizes up to 14 mesh, using velocities of about 1.5 m/s, have been used successfully in the past.

- Give reasons why particle settling velocities in slurries are *lower* than free settling velocities of single particles, other factors remaining the same for both cases.
  - Hydrodynamic interaction between particles and upward motion of displaced liquid.
  - Increased viscosity of suspension over the liquid without particles.
- What are the two categories of slurries handled in slurry flow? What are the issues involved in piping design for slurry transport for these two categories?
  - *Nonsettling* slurries are made up of very fine, highly concentrated, or neutrally buoyant particles. These slurries are normally treated as pseudohomogeneous fluids. They may be quite viscous and are frequently non-Newtonian.
  - Slurries of particles that tend to settle out rapidly are called *settling* slurries or fast settling slurries.
  - For fast settling slurries, ensuring conveyance is usually the key design issue while pressure drop is somewhat less important.
  - For nonsettling slurries conveyance is not an issue, because the particles do not separate from the liquid. Here, viscous and rheological behaviors, which control pressure drop, take on critical importance.
  - Fine particles, often at high concentration, form nonsettling slurries for which useful design equations can be developed by treating them as homogeneous fluids. These fluids are usually very viscous and often non-Newtonian. Shear thinning and Bingham plastic behaviors are common. Dilatancy is sometimes observed. Rheology of such fluids must in general be empirically determined, although theoretical results are available for some very limited circumstances. Sewage flow is an example of such slurries.
- Describe settling characteristics of particulates in slurry flow in horizontal piping in relation to particle size ranges.
  - *Ultrafine particles*, 10 μm or smaller, are generally fully suspended and the particle distributions are not influenced by gravity.
  - *Fine particles*, 10–100 μm, are usually fully suspended, but gravity causes concentration gradients.
  - *Medium size particles*, 100–1000 μm, may be fully suspended at high velocity, but often form a moving deposit at the bottom of the pipe.
  - *Coarse particles*, 1000–10,000 ×m, are rarely fully suspended and are usually conveyed as a moving deposit.
  - *Ultracoarse particles* larger than 10,000 ×m are not suspended at normal velocities unless they are unusually light.
  - Turbulence in the line helps to keep particles in suspension. It is essential, however, to avoid dead spaces in which solids could accumulate and also to make provisions for periodic cleaning of the line.
- What is *deposition velocity*?
  - The velocity below which particles tend to settle out and form a deposit in the pipe is called the *deposition velocity*. The pipe diameter should be selected such that the velocity in the pipeline is maintained above the deposition velocity over the operating range of flow rates.
- What are the flow conditions in slurry flow in horizontal pipes?
  - Flow conditions can be classified into four categories:
    - *Homogeneous Flow*: Homogeneous flow implies that the solid particles are uniformly distributed across the pipeline cross section. Examples include sewage sludge, drilling muds, paper pulp, and many finely ground materials.
    - *Heterogeneous Flow*: Slurries at low concentration with rapidly settling (coarse particles) solids generally exhibit heterogeneous flow. Typical examples are sand and gravel slurries and coarse coal slurries.
    - *Intermediate Regime*: This type of flow occurs when some of the particles are homogeneously distributed while others are heterogeneously distributed. Examples are tailings slurry from mineral processing plants and transportation of coal–water slurries.
    - *Saltation Regime*: The fluid turbulence may not be sufficient to keep fast settling particles in suspension. The particles travel by discontinuous jumps or roll along a sliding or stationary bed at the bottom of the pipe. This type of flow will occur with coarse sand and gravel slurries.
    - The regimes are illustrated in Figures 7.17 and 7.18.
- What are the common problems in slurry handling systems?

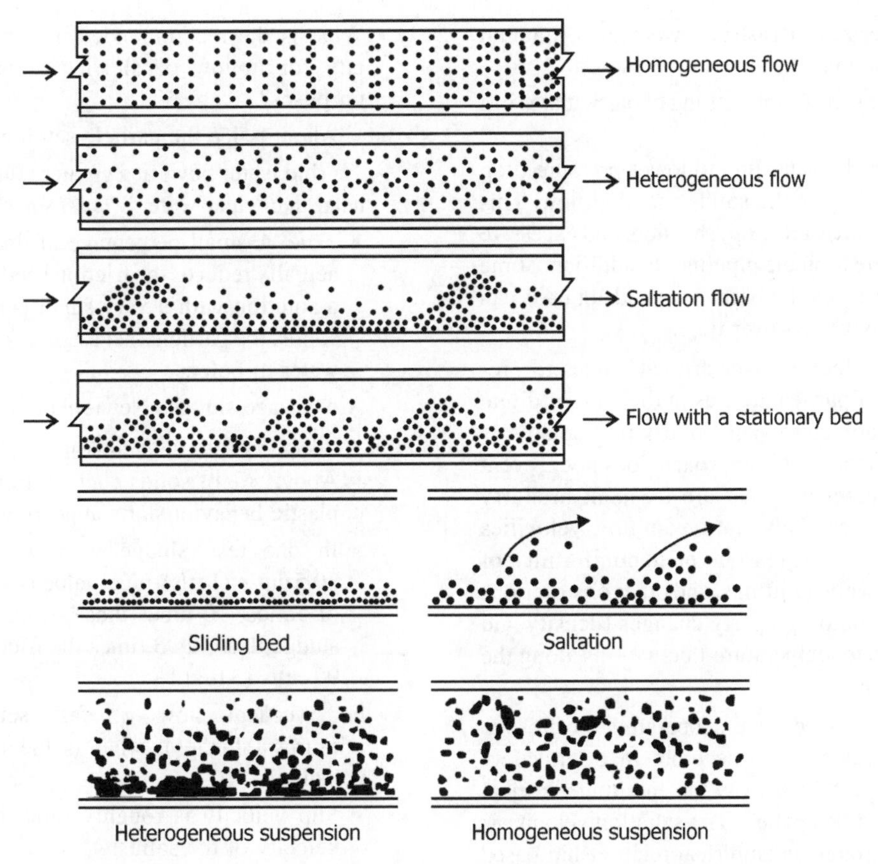

FIGURE 7.17 Regimes of flow for settling slurries in horizontal pipelines.

  - Scale buildup on valves and fittings.
  - Abrasion and wear.
  - Leakages from valves and fittings.
- How is ΔP reduced in pipeline flow of slurries?
  - One way is to use additives to prevent flocculation and improve dispersion of solids. Additives are soluble ionic compounds used to disperse flocculated slurries. The ionic compounds break up the flocculated particles that give rise to higher shear stresses in pipeline flow. Another method to reduce ΔP is to inject air into the pipeline. This is effective for non-Newtonian shear thinning slurry flow involving laminar flow. Boundary layer injection systems involve liquids at three or four points along the pipe wall to generate an annulus that lubricates the flow. Water, aqueous polymer solutions, waste oils, or polyelectrolytes are some of the liquids used. This works well for concentrated viscous slurries as core flowing slurry, since there is little mixing between the annular wall layer of the injected liquid and the slurry.

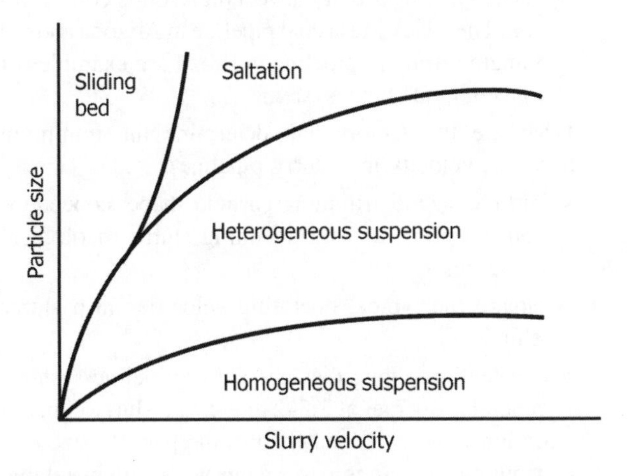

FIGURE 7.18 Schematic representation of the boundaries between the flow regimes for settling slurries in horizontal pipelines.

- What are the advantages of air injection downstream of slurry pump discharge?
  - Reduction of pump discharge pressure requirement.
  - Increase in capacity of existing pipeline retaining the same pump.
  - Application of same pump–pipeline combination for handling higher viscosity shear thinning slurry while maintaining the same discharge pressure.
  - Examples of air injection technology include pumping waste molasses, red mud waste in alumina pro-

duction, pulverized coal ash at power plants, titanium dioxide, chalk and clay slurries, and so on.

- Name the methods to prevent settling of particulates in a slurry pipeline.
  - Using a larger diameter line to lower pressure drop will cause settling of the solids and slug flow. This leads to higher pressure drop, abrasion, and excessive mechanical stress on the pipeline. In addition, slime may form on the interior of the pipe and abrasion can reduce wall roughness over time.
  - Maintaining velocities over critical values to prevent settling of particulates, as in the case of dilute phase pneumatic conveyance of dry particulates, is the normal method. This approach does not prevent settling and blockages that are frequent in slurry piping as there can be variations in flow velocities along the piping for reasons of nonuniformity of particle sizes, valves, fittings, and other obstructions in the piping, liquid property changes (density and viscosity) due to temperature fluctuations along the line, and so on.
  - As with pneumatic conveyors, operating too close to the saltation velocity can be problematic, a safety factor of at least 25% over the minimum critical velocity should be applied. The saltation velocity is determined through an empirical relationship based on the solids loading, particle size, and physical properties of the system.
  - While the saltation velocity for the largest particle in the suspension is usually used, particle size distribution may be more important. For instance, a slurry with a large percentage of very fine particles may behave as a single larger particle and overshadow the effect of a few large particles. The fine particles may cluster and cause the larger particles to settle at an even higher rate, which would require a higher slurry velocity.
  - One technology involves swirl inducement in the flow at strategic points along the piping, which causes rotational flow along the pipe axis, thereby lifting the particulates concentrated in the lower part of the pipe to the upper part creating more homogeneous flow pattern. Aspen technologies are used in computer simulation of these techniques.
  - Swirl inducement is claimed to operate the piping at reduced average flow velocities to prevent settling, lower pumping power and pipe erosion, and reduce settling at pipe bends and fittings. It has to be realized that the cost of maintaining pressures required to induce swirl is understandably higher. Efforts are being aimed at improving efficient swirl inducing piping technologies.
- What is the recommended minimum transport velocity for upflow of slurries to be used for design purposes?
  - About twice the particle settling velocity.
- "Turbid water flows at a greater volume under the same head than clear water." *True/False*?
  - *True*. A small percentage of fine particles in water actually reduces the friction head loss. Apparently, in a solution with a very small percentage of flowing solids, the particles act as guide vanes and reduce the water turbulence.
- Are sewage sludges Newtonian or non-Newtonian?
  - Sludges below 3% solids are usually near Newtonian.
  - Above 3% of solids content, thixotropic or pseudoplastic behavior starts appearing.
  - In one test, sludge with 9% solids flowing at 30.5 cm/s (1 ft/s) pipe velocity gave friction losses of almost 10 times those of water, while the same sludge had only 3 times the friction loss of water at 90 cm/s (3 ft/s).
- For vertical flow of fast settling solid–liquid mixtures (slurries), what is the thumb rule of slip velocity?
  - Slip velocity is roughly same as terminal settling velocity of the solids.
- What are the two types of slurry pipelines?
  - *Brute Force*: These are usually short, high-velocity systems that operate at relatively low solids concentrations and usually carry relatively large particles. Most dredging and tailings pipelines are in this category.
  - *Conventional*: These are generally well-designed systems in which particle size distribution and solids concentration are closely controlled in order to maintain an economical yet stable operating velocity. These systems usually have a high solids concentration. The Black Mesa coal pipeline in Arizona and the Samarco iron ore pipeline in Brazil are examples of conventional slurry systems.
- What are the factors that determine the minimum transport velocity in a slurry pipeline?
  - Particle size distribution, particle shape, concentration of solids, and the resulting slurry rheological properties.
- What are the typical operating velocities in a slurry pipeline?
  - Conventional long-distance slurry pipeline systems generally operate at 1.5–2 m/s. If the slurry concentration is properly matched to the particle size distribution, velocities in this range will result in a stable operating system that may be easily restarted after

shutdown and in which pipe wall abrasion is minimized.

- Brute force systems, on the other hand, may require operating velocities as high as 5.5–6 m/s.

- "It is a good practice to give gentle slope to horizontal slurry lines, wherever possible." *True/False*?
  - *True*. Giving a slope of about 12.5 mm for every 3 m pipe run is a good practice.
- What are the considerations involved in the design of slurry pipelines with respect to pipeline wear due to abrasion?
  - Abrasion may be considered as the rubbing together of the solids in the slurry and the pipe. This needs to be considered in the design of brute force pipelines such as dredging lines. In these systems, pipe wear is a major consideration. These lines may utilize different linings or different types of pipes such as hardened steel, concrete, and so on.
  - Conventional slurry systems are designed to operate at velocities low enough that abrasion is minimal and pipeline wall can be designed for long-term operation. These systems use standard steel pipe, the same as that used for oil and gas pipelines.

SECTION II

HEAT TRANSFER

8

DIMENSIONLESS NUMBERS, TEMPERATURE MEASUREMENT, AND CONDUCTION HEAT TRANSFER

8.1 Important Dimensionless Groups in Heat Transfer 225
8.2 Temperature Measurement 226
8.3 Conduction Heat Transfer 232
8.3.1 Thermal Insulation 236

8.1 IMPORTANT DIMENSIONLESS GROUPS IN HEAT TRANSFER

- Name dimensionless groups of importance in heat transfer and state their physical significance (other than those mentioned in Chapter 1).
  - Fourier number (N_{Fo}): $k\theta/\rho C_p L^2$ is the ratio of the rate of heat transfer by conduction to the rate of energy storage in the system.
  - Prandtl number (N_{Pr}): $C_p\mu/k$ is the ratio of momentum diffusivity (μ) and thermal diffusivity. Since it is a material property, Prandtl number depends only on the physical conditions (temperature and pressure) that a material is held at, not on the system in which it is placed.
  - Peclet number ($N_{Re} \times N_{Pr}$): $DV\rho C_p/k$ is the convective transport/diffusive transport.
  - Nusselt number (N_{Nu}): hD/k is the ratio of convective heat transfer to conductive heat transfer in the fluid perpendicular to flow direction. Measures enhancement of heat transfer from a surface that occurs in a real situation compared to heat transferred if only conduction occurred. Nusselt number conveys how important convection is compared to conduction. It involves a heat transfer coefficient and a characteristic length, both of which depend on the type of system one is using.
  - Grashof number (N_{Gr}): $L^3\rho^2\beta g\Delta t/\mu^2$ is the ratio of buoyancy force to viscous force acting on a fluid. L is characteristic length. It is generally used to model natural convection.
  - Rayleigh number (N_{Ra}): $N_{Gr} \times N_{Pr} = L^3\rho^2 g\beta C_p \Delta T/\mu\alpha$ is the quantity that governs natural convection heat transfer.
  - Graetz number (N_{Gz}): $WC_p/kL = D_i/LN_{Re} \times N_{Pr}$ characterizes laminar flow in a conduit. Ratio of the sensible heat change of the flowing fluid to the rate of heat conduction through a film of thickness D or L.
  - Biot number (N_{Bi}): hL/k is the ratio of the internal thermal resistance to the external thermal resistance. It represents the relative importance of the thermal resistance within a solid body.
  - Stanton number (N_{St}): $h/C_p V\rho = N_{Nu}/N_{Re} \times N_{Pr}$ measures the ratio of the heat transferred into a fluid to the thermal capacity of the fluid. Used in forced convection.
  - Condensation number (N_{Co}): $(h/k)(\mu^2/\rho^2 g)^{1/3}$.
  - Vapor condensation number (N_{Cv}): $L^3\rho^2 g\lambda/k\Delta t$.
  - Bond number, B_d, is the ratio of gravity forces to surface tension forces and is significant in drainage of condensate from heat exchanger surfaces and is useful in the design of finned surfaces for efficient drainage of condensate.
  - Bond number at the base of a fin can be approximated by $B_d = (\rho_l - \rho_g)ge^2/\sigma\theta_m$, where ρ_l and ρ_g are the liquid and vapor densities, e is the fin height, σ is surface tension, and θ_m is fin angle.
    - If the surface tension forces are dominant over gravity forces, the condensate drainage is determined by surface tension.
    - $B_d = 1$ implies that surface tension forces are equal to gravity forces at the end of the fin and that

Fluid Mechanics, Heat Transfer, and Mass Transfer: Chemical Engineering Practice, By K. S. N. Raju

surface tension forces are greater than gravity forces for the remainder of the fin.

- "Biot number (B_i) expresses the relative importance of the thermal resistance of a body to that of the convection resistance at its surface". *True/False*?
  - *True*.
- "If $B_i > 40$, the surface temperature of a solid may be assumed to be equal to the temperature of the surroundings." *True/False*?
  - *True*.

8.2 TEMPERATURE MEASUREMENT

- What are reference temperatures? Give some reference temperatures.
  - Temperatures established by physical phenomena that are easily observed and consistent in nature.
  - The International Temperature Scale (ITS) is based on such phenomena. It establishes 17 fixed points and corresponding temperatures. A sampling is given in Table 8.1.
- Give a typical pressure–temperature phase diagram.
  - If matter is heated to a high enough temperature, it becomes gaseous. If matter is subjected to a high enough pressure, it becomes a solid. At combinations of pressure and temperature in between these limits, matter can exist as a liquid.
  - The boundaries that separate these states of matter are called the melting (or freezing) curve, the vaporization (or condensation) curve, and the sublimation curve. The intersection of all three curves is called the triple point. All three states of matter can coexist at that pressure and temperature. Triple point is used as a fixed temperature point in the International Temperature Scale.

TABLE 8.1 Reference Temperatures

Element	Type	Temperature (°C)
Hydrogen (H_2)	Triple point	−259.3467
Neon (Ne)	Triple point	−248.5939
Oxygen (O_2)	Triple point	−218.7916
Argon (Ar)	Triple point	−189.3442
Mercury (Hg)	Triple point	−38.8344
Water (H_2O)	Triple point	+0.01 (273.16K)
Gallium (Ga)	Melting point	29.7646
Indium (In)	Freezing point	156.5985
Tin (Sn)	Freezing point	231.928
Zinc (Zn)	Freezing point	419.527
Aluminium (Al)	Freezing point	660.323
Silver (Ag)	Freezing point	961.78
Gild (Au)	Freezing point	1064.18

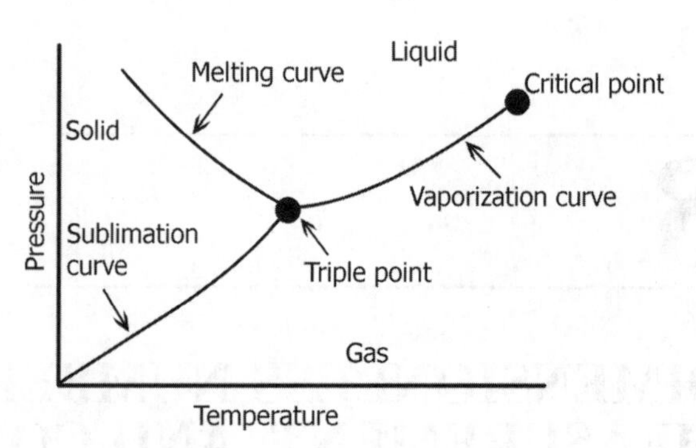

FIGURE 8.1 Illustration for triple point.

  - Figure 8.1 gives the phase diagram illustrating the triple point.
- Name important types of temperature measuring devices.
  - Thermocouples.
  - Resistance temperature detectors (RTDs).
  - Thermistor and integrated circuit temperature sensors (ICTS).
  - Other temperature-sensing devices: Liquid-in-glass thermometers, bimetallic strip thermometers, pressure thermometers, pyrometers, and infrared thermometers.
- What are the effects associated with temperature measurement by thermocouples?
  - Seebeck effect: It states that the voltage produced in a thermocouple is proportional to the temperature between the two junctions.
  - Peltier effect: It states that if a current flows through a thermocouple, one junction is heated (puts out energy) and the other junction is cooled (absorbs energy).
  - Thompson effect: It states that when a current flows in a conductor along which there is a temperature difference, heat is produced or absorbed, depending upon the direction of the current and the variation of temperature.
  - In practice, the Seebeck voltage is the sum of the electromotive forces generated by the Peltier and Thompson effects.
- What is a thermocouple? How does it work?
  - Thermocouples consist essentially of two strips or wires made of different metals and joined at one end. Changes in the temperature at that junction induce a change in electromotive force (emf) between the other ends. As temperature goes up, this output emf of the thermocouple rises although not necessarily linearly. Temperature ranges for thermocouples are from −270 to 2300°C.

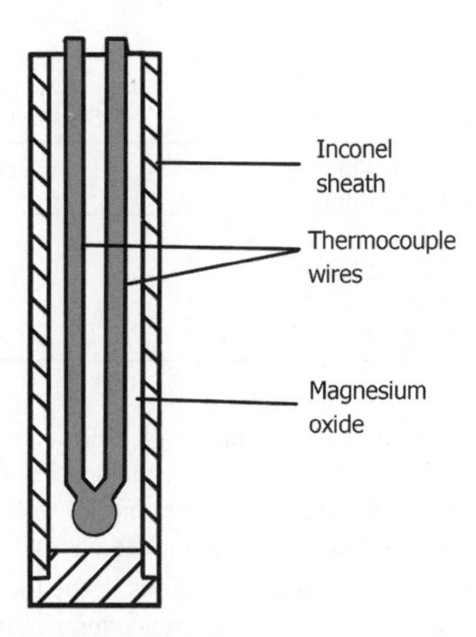

FIGURE 8.2 Internal construction of a typical thermocouple.

- Figure 8.2 shows the internal construction of a typical thermocouple. The leads of the thermocouple are encased in a rigid metal sheath. The measuring junction is normally formed at the bottom of the thermocouple housing. Magnesium oxide surrounds the thermocouple wires to prevent vibration that could damage the fine wires and to enhance heat transfer between the measuring junction and the medium surrounding the thermocouple.
- Thermocouples cause an electric current to flow in the attached circuit when subjected to changes in temperature.
- The voltage generated is dependent on the temperature difference between the measuring and the reference junctions, the characteristics of the two metals/alloys used, and the characteristics of the attached circuit.
- Figure 8.3 illustrates a simple thermocouple circuit.

• What are the different metal combinations used in thermocouples?
  - Iron–constantan
  - Copper–constantan
  - Chromel–constantan
  - Chromel–alumel
  - Platinum–platinum + 13% rhodium
  - Platinum–platinum + 10% rhodium

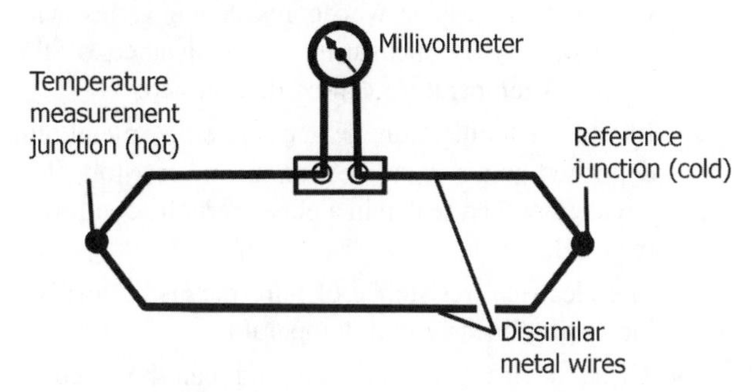

FIGURE 8.3 Simple thermocouple circuit.

• What are the advantages and the disadvantages of thermocouples?

Advantages

- A thermocouple is capable of measuring a wider temperature range than an RTD.
- Simple, rugged, easy to use, inexpensive, self-powered, wide temperature ranges.

Disadvantages

- If the thermocouple is located at some distance away from the measuring device, expensive extension grade thermocouple wires or compensating cables have to be used.
- Thermocouples are not used in areas where high radiation fields are present. Radioactive radiation (e.g., beta radiation from neutron activation) will induce a voltage in the thermocouple wires. Since the signal from thermocouple is also a voltage, the induced voltage will cause an error in the temperature transmitter output.
- Thermocouples are slower in response than RTDs.
- If the control logic is remotely located and temperature transmitters (millivolt to milliampere transducers) are used, a power supply failure will cause faulty readings.
- Other disadvantages include nonlinearity, low-voltage output, requirement of reference junction, poor stability, and poor sensitivity; connections result in additional junctions and voltage depends on composition of metals in the wires.

• Give the characteristics of different types of standard thermocouples.
  - Table 8.2 gives the characteristics of thermocouples.
  - Types R and S are expensive and are resistant to corrosion and high temperatures. Not very sensitive.
  - Type T thermocouples are not expensive, and very sensitive but not suitable for corrosive and high-temperature environments. Neither wire is magnetic. Copper–constantan thermocouples are very susceptible to conduction error due to the high thermal conductivity of the copper and should not be used unless long runs of wire (100–200 wire diameters) can be laid along an isotherm.
  - Type K thermocouples are more widely used for their moderate cost and have reasonable corrosion and high-temperature resistance. The alumel wire is magnetic. Chromel–alumel thermocouples generate electrical signals, while the wires are being bent and

TABLE 8.2 Characteristics of Thermocouples

Type	Materials	Temperature Range (°C)	Approximate Sensitivity (mV/°C)
T	Copper–constantan	From −250 to 400	0.052
E	Chromel–constantan	From −270 to 1000	0.076
J	Iron–constantan	From −210 to 760	0.050
K	Chromel–alumel	From −270 to 1372	0.039
R	Platinum/platinum–13% rhodium	From −50 to 1768	0.011
S	Platinum/platinum–10% rhodium	From −50 to 1768	0.012

should not be used on vibrating systems, unless strain relief loops can be provided.

- Iron–constantan thermocouples can generate a galvanic emf between the two wires and should not be used in applications where they might get wet. The iron wire is magnetic.

- What are the different reference temperature systems that are used with thermocouples?
  - The signal from a thermocouple depends as much on the reference junction temperature as it does on the measuring junction temperature.
  - There are several different systems for establishing a reference temperature.
    - Ice baths: Ice baths are widely used, because they are accurate and inexpensive. Any potable water freezes within about 0.01°C of zero.
    - Electronically controlled references: Electronically controlled reference temperature devices are available, both high temperature and ice point. These devices require periodic calibration and generally are not as stable as ice baths, but are more convenient.
    - Compensated reference temperature systems: Dedicated temperature indicators terminate each thermocouple at a connection panel inside the chassis and use a compensation network to inject a signal that compensates for the temperature of the panel before calculating the temperature.
    - Zone boxes: A zone box is a region of uniform temperature used to ensure that all connections made within it are at the same temperature. The temperature needs to be neither controlled nor measured. It needs only to be uniform.
- How are thermocouple sheaths protected from deterioration from oxygen-rich atmospheres?
  - A thin platinum film is used to cover the sheath to prevent corrosive destruction by oxygen.
  - Platinum-coated sheaths can be used in blast furnaces, glass furnaces, and high-temperature oxygen producing solid electrolytic cells.
- What is a thermopile?
  - A thermopile is a number of thermocouples connected in series, to increase the sensitivity and the accuracy by increasing the output voltage when measuring low temperature differences.
  - Each of the reference junctions in the thermopile is returned to a common reference temperature.
- What are resistance temperature devices (RTDs)?
  - Every type of metal has a unique composition and has a different resistance to the flow of electrical current. This is termed the resistivity *constant* for that metal.
  - For most metals, the change in electrical resistance is directly proportional to its change in temperature and is linear over a range of temperatures. This constant factor called the temperature coefficient of electrical resistance is the basis of resistance temperature detectors.
  - Several different pure metals (such as platinum, nickel, and copper) can be used in the manufacture of an RTD. A typical RTD probe contains a coil of very fine metal wire, allowing for a large resistance change accommodated in a small space.
  - Usually, platinum RTDs are used as process temperature monitors because of their accuracy and linearity.
  - RTDs rely on resistance change in a metal, with the resistance rising more or less linearly with temperature. Temperature ranges are from −250 to 850°C. Metallic devices are commonly made from platinum.
  - The RTD can actually be regarded as a high precision wire-wound resistor whose resistance varies with temperature. By measuring the resistance of the metal, its temperature can be determined.
  - Resistance temperature devices are either metal film deposited on a surface or wire-wound resistors. The devices are then sealed in a glass–ceramic composite material.
  - The electrical resistance of pure metals is positive, increasing linearly with temperature.
  - These devices are accurate and can be used to measure temperatures from −170 to 780°C.

- What are the advantages and disadvantages of resistance temperature devices?

  Advantages
  - The response time compared to thermocouples is very fast, in the order of fractions of a second.
  - Within its range, it is more accurate, more linear, and has higher sensitivity than a thermocouple.
  - Unlike thermocouples, radioactive radiation (beta, gamma, and neutrons) has minimal effect on RTDs since the parameter measured is resistance, not voltage.
  - An RTD will not experience drift problems because it is not self-powered.
  - In an installation where long leads are required, the RTD does not require special extension cable.

  Disadvantages
  - Because the metal used for an RTD must be in its purest form, they are much more expensive than thermocouples.
  - In general, an RTD is not capable of measuring as wide a temperature range as that of a thermocouple.
  - Current source is required. A power supply failure can cause erroneous readings.
  - Small changes in resistance are being measured, thus all connections must be tight and free of corrosion, which will create errors.
  - Among the many uses in a nuclear station, RTDs can be found in the reactor area temperature measurement and fuel channel coolant temperature.
- What is a thermistor?
  - Thermistor is a thermally sensitive resistor, whose primary function is to exhibit a change in electrical resistance with a change in temperature.
  - It is made from compressed and sintered metal oxides (semiconductor materials).
  - Metals used are Ni, Co, Mn, Fe, Cu, Mg, and Ti.
  - By changing the oxide proportions, basic resistance of the thermistor can be varied.
  - Thermistors are a class of metal oxide (semiconductor material), which typically have a high negative temperature coefficient of resistance, but can also be positive.
  - Thermistors are based on resistance change in a ceramic semiconductor. The resistance drops nonlinearly with temperature rise as shown in Figure 8.4.
  - Temperature ranges for thermistors are from −50 to 300°C. Devices are available with the temperature range extended to 500°C.
  - Thermistors have high sensitivity that can be up to 10% change per °C, making them the most sensitive temperature elements available, but with very nonlinear characteristics.

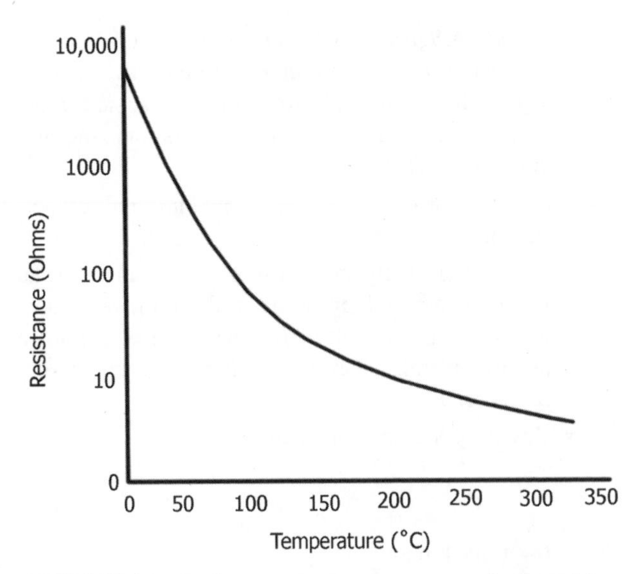

FIGURE 8.4 Resistance—temperature curve of a thermistor.

  - These are low-cost devices and manufactured in a wide range of shapes, sizes, and values.
  - When in use, care has to be taken to minimize the effects of internal heating.
  - Not suitable for wide span temperature measurement as resistance changes are too large to be conveniently measured by a single instrument. Maximum spans are about 85°C.
  - Particularly suitable for narrow temperature spans due to large resistance changes (i.e., good accuracy) involving high sensitivity, which can be up to 10% change per °C, making them the most sensitive temperature measuring devices available.
  - The typical response times are 0.5–5 s with an operating range from −50°C to typically 300°C. Devices are available with the temperature range extended to 500°C.
  - When in use, care has to be taken to minimize the effects of internal heating.
  - The nonlinear characteristics make the device difficult to use as an accurate measuring device without compensation, but its sensitivity and low cost makes it useful in many applications.
  - The nonlinear characteristics are illustrated in Figure 8.4.
- What are the advantages and the disadvantages of thermistors?
  - *Advantages*: Fast and high output. Being low in heat capacity, more accurate than liquid thermometer for measurements involving small volumes.

- *Disadvantages*: Nonlinear, limited temperature range, fragile, current source required, and self-heating. Being semiconductors, thermistors are more susceptible to permanent decalibration than thermocouples or RTDs.

- What is the principle of operation of infrared temperature sensors?
  - They obtain temperature by measuring the thermal radiation emitted by a material. Infrared sensors are noncontacting devices. These are also known as pyrometers. Used for furnace temperature measurements.
  - Based on Stefan–Boltzmann equation,

$$q = \sigma T^4 \tag{8.1}$$

for a black body.

- What are the advantages and disadvantages of pyrometers?
  - *Advantages*: Pyrometers can measure high temperatures without melting or oxidation. They can also be used for measuring low temperatures.
  - *Disadvantages*: Pyrometers are not as accurate as other methods and readings are based on black body radiation. Many bodies are not black in practice.
- How do bimetallic thermometers work?
  - Bimetallic devices take advantage of the difference in rate of thermal expansion between different metals. Strips of two metals are bonded together. When heated or cooled, one side will expand or contract more than the other and the resulting bending is translated into a temperature reading by mechanical linkage to a pointer.
  - If two strips of dissimilar metals such as brass and invar (copper–nickel alloy) are joined together along their length, they will flex to form an arc as the temperature changes, as shown in Figure 8.5a.

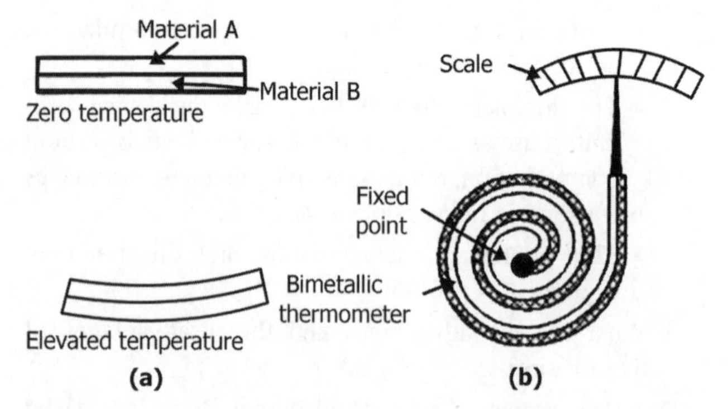

FIGURE 8.5 (a) Effect of temperature change on a bimetallic strip. (b) Bimetallic strip thermometer.

  - Bimetallic strips are usually configured as a spiral or helix for compactness and can then be used with a pointer to make a cheap compact rugged thermometer as shown in Figure 8.5b.
  - These are portable and do not require a power supply, but are usually not as accurate as thermocouples or RTDs. Temperature recording is not possible.
  - Their operating range is from −180 to 430°C and can be used in applications from oven thermometers to home and industrial control thermostats.
  - The bimetallic strip is extensively used in ON/OFF applications, and is not requiring high accuracy as it is rugged and cost-effective.
  - Bimetallic thermometers are relatively inaccurate, slow to respond, not normally used in analog applications to give remote indication and have hysteresis.
- What are the characteristics of fluid expansion thermometers?
  - Fluid expansion devices are liquid-in-glass thermometers. The liquid may be mercury or organic liquid (typically alcohol). Versions employing gas instead of liquid are also available.
  - Mercury is considered an environmental hazard, so there are regulations governing the shipment of devices that contain it.
  - Fluid expansion sensors do not require electric power, do not pose explosion hazards, and are stable even after repeated cycling.
  - Useful for calibrating other temperature measuring devices. For this purpose, standard short-range thermometer sets, called Anschutz thermometers, are available.
  - On the other hand, they do not generate data that are easily recorded or transmitted, and cannot make spot or point measurements.
  - Glass being breakable, the liquid may pose problems on spillage.
  - Need to make sure liquid is continuous in the capillary column.
  - Readings are subject to human error in reading.
- What are change-of-state temperature measuring devices?
  - Change-of-state temperature sensors consist of labels, pellets, crayons, lacquers, or liquid crystals whose appearance changes once a certain temperature is reached. They are used, for instance, with steam traps—when a trap exceeds a certain temperature, a white dot on a sensor label attached to the trap will turn black. Response time typically takes minutes, so these devices often do not respond to transient temperature changes. Accuracy is lower than

that with other types of sensors. Furthermore, the change in state is irreversible, except in the case of liquid crystal displays. Even so, change-of-state sensors can be handy when one needs confirmation that the temperature of a piece of equipment or a material has not exceeded a certain level, for instance, for technical or legal reasons during product shipment.

- What are vapor pressure thermometers? How do they work? Illustrate.
  - Figure 8.6 illustrates vapor pressure thermometer.
  - System is filled with gas, liquid, or vapor–liquid mixture.
    - Gas: Pressure is proportional to temperature.
    - Liquid: Differential thermal expansion.
    - Vapor–liquid mixture: Gauge reads vapor pressure, which is a function of temperature.
  - Vapor pressure thermometer system is partially filled with liquid and vapor such as methyl chloride, ethyl alcohol, ether, toluene, and so on.
  - In this system, the lowest operating temperature must be above the boiling point of the liquid and the maximum temperature is limited by the critical temperature of the liquid. The response time of the system is slow, being of the order of 20 s.
  - The temperature–pressure characteristic of the thermometer is nonlinear as shown in the vapor pressure curve for methyl chloride in Figure 8.7.
  - Gas thermometer is filled with a gas such as nitrogen at a pressure range of 1000–3350 kPa at room temperature.
- What are the problems involved in temperature measuring instruments in industrial environments?
  - Fouling of the exterior of thermowell results in a lower reading.
  - Radiation losses from external cap of a thermowell to the atmosphere give low readings, giving rise to significant errors, when measuring temperatures are high, for example, temperatures more than 300°C.

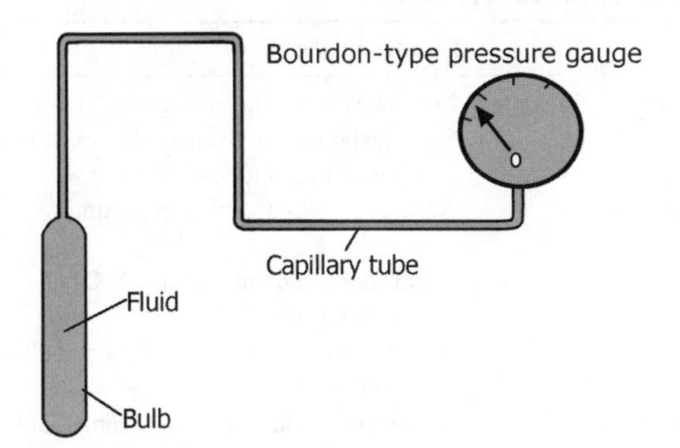

FIGURE 8.6 Vapor pressure thermometer.

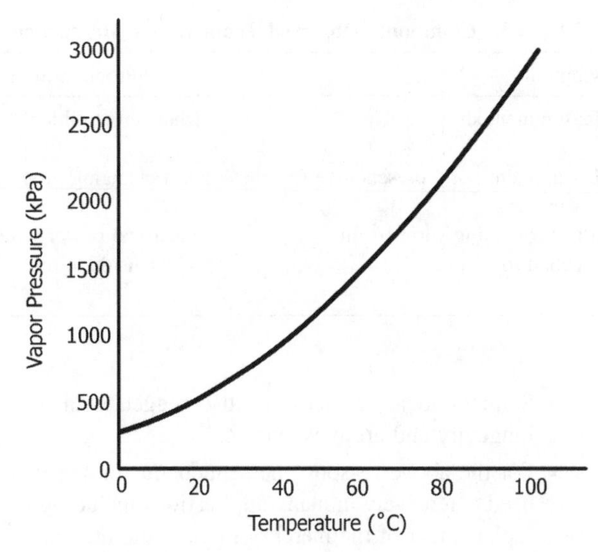

FIGURE 8.7 Vapor pressure curve for methyl chloride.

Sometimes, values of temperatures read can be 20–25°C below the true values.

  - Thermowell should be fully inserted several centimeters into the process fluid. If process fluid is a gas or vapor, depth of thermowell into the fluid should be not less than 15 cm, this requirement being due to poorer heat transfer with vapors and gases compared to liquids.
- What are pyrometers?
  - *Pyrometers* are devices that measure temperature by sensing the heat radiated from a hot body through a fixed lens that focuses the heat energy on to a thermopile.
  - These are noncontact devices. Furnace temperatures, for instance, are normally measured through a small hole in the furnace wall.
  - The distance from the source to the pyrometer can be fixed and the radiation should fill the field of view of the sensor.
- What are the characteristics of semiconductors as temperature measuring devices?
  - Semiconductors have a number of parameters that vary linearly with temperature.
  - Normally the reference voltage of a Zener diode or the junction voltage variations is used for temperature sensing.
  - Semiconductor temperature sensors have a limited operating range from −50 to 150°C but are very linear with accuracies of ±1°C or better.
  - Other advantages are that electronics can be integrated onto the same die as the sensor giving high sensitivity, easy interfacing to control systems, and making different digital output configurations possible.

TABLE 8.3 Commonly Observed Temperature Measurement Problems

Symptom	Problem Source	Solution
Measurement shift	Change in ambient temperature	Increase immersion depth, insulate surface
Measurement not representative of process	Fast changing process temperature	Use quick response or low thermal time constant device
Indicator reading varies from second to second	Electrical power wires near thermocouple extension wires	Use shielded, twisted pair, thermocouple extension wire, and/or install in conduit

  - Semiconductor devices are also rugged with good longevity and are inexpensive.
  - For the above reasons, the semiconductor sensor is used extensively in many applications including the replacement of the mercury in glass thermometer.
- What are the commonly observed temperature measurement problems and remedies?
  - Table 8.3 gives possible problems associated with temperature measurements.
- Summarize the characteristics of different temperature measuring devices.
  - Table 8.4 summarizes characteristics, advantages, and disadvantages of commonly used temperature measuring devices.

8.3 CONDUCTION HEAT TRANSFER

- Briefly explain the mechanism of heat conduction.
  - Thermal energy is transported within a solid by the electrons and the phonons (lattice vibrations) inside the material. The transport of energy is hindered by the presence of imperfections or by any kind of scattering sites.
  - If there is macroscopic transport of matter (e.g., fluid flow) inside the body, the mass flow makes an additional contribution to the transport of energy (convective heat transfer). This contribution is disregarded when studying conduction heat transfer.
  - In gases and liquids, conduction is due to collisions and diffusion of the molecules during their random motion.
  - In solids, conduction is due to a combination of vibrations of molecules in a lattice and energy transport is by free electrons.
- State Fourier's law and explain why there is negative sign.
  - According to Fourier's law

$$Q_{\text{conduction}} = -kA\, \mathrm{d}T/\mathrm{d}x. \tag{8.2}$$

  - Rate of conduction $\propto$ (area)(ΔT/thickness).
  - k is proportionality constant, designated as thermal conductivity, which is a measure of ability of the material to conduct heat. It is one of the transport properties, such as viscosity.
  - According to Fourier's law, k is independent of temperature gradient, ΔT.
  - It is dependent on temperature, but not strongly. For small temperature ranges, k might be considered independent of T. For large temperature ranges, k may be approximated as a function of temperature by

TABLE 8.4 Characteristics, Advantages, and Disadvantages of Temperature Measuring Devices

Type	Linearity	Advantages	Disadvantages
Thermocouple	Good	Low cost, rugged, and very wide range	Low sensitivity and reference needed
Resistance	Very good	Stable, wide range and accurate	Slow response, low sensitivity, expensive, self-heating, and limited range
Thermistor	Poor	Low cost, small, high sensitivity, and fast response	Nonlinear, range, and self-heating
Bimetallic	Good	Low cost, rugged, and wide range	Local measurement or for ON/OFF switching only
Pressure	Medium	Accurate and wide range	Needs temperature compensation and vapor is nonlinear
Semiconductor	Excellent	Low cost, sensitive, and easy to interface	Self-heating, slow response, range, and power source

the equation of the form,

$$k = a + bT, \quad (8.3)$$

where a and b are constants.

- dT/dx is temperature gradient.
- The negative sign on the right-hand side signifies that conduction is in the direction of decreasing temperature. It indicates that thermal energy flows from hot regions to cold regions.
- k, thermal conductivity, is the rate of thermal energy transfer per unit area and per unit temperature gradient. Units are W/(m-°C).
- A is the area of the surface that is perpendicular to the flow direction for heat energy.

- Define similar laws for (i) momentum transfer, (ii) mass transfer, and (iii) electrical energy transfer.

(i) *Momentum Transfer*:

- Momentum transfer is described by Newton's law that relates shear stress to velocity gradient, employing a proportionality constant called viscosity.

$$\text{Shear stress} = -\mu \, du/dy. \quad (1.2)$$

(ii) *Mass Transfer*:

- Fick's first law relates flux of a component to its composition gradient, employing a constant of proportionality called diffusivity.
- Rate of mass transfer is

$$N_A = -D_{AB} \, dc_A/dx, \quad (8.4)$$

where D_{AB} is the proportionality constant, designated as diffusivity, which is a measure of ability of transfer of mass.

- Mass transfer is in the direction decreasing concentration, which explains the negative sign.

(iii) *Electrical Energy Transfer*:

- Ohm's law is expressed as

$$I = V/R, \quad (8.5)$$

where I is the current in amperes, V is the voltage, and R is the resistance.

- Write three-dimensional conduction equation.
  - Three-dimensional conduction equation is

$$\partial^2T/\partial x^2 + \partial^2T/\partial y^2 + \partial^2T/\partial z^2 + q/k = (1/\alpha)(\partial T/\partial t), \quad (8.6)$$

where T is the absolute temperature; x, y, and z are the directions of flow in the three dimensions; q is the conductive heat transfer rate; k is the thermal conductivity; α is the thermal diffusivity, $k/\rho c_p$, where ρ is the density and c_p is the heat capacity; and t is the time.

- Give an example of heat transfer by conduction with internal heat generation.
  - Electrical resistance heaters.
- Define steady-state heat conduction.
  - Steady-state conduction is said to exist when the temperature at all locations in a substance is constant with time, as in the case of heat flow through a uniform wall.
  - In other words, temperature is a function of position only and rate of heat transfer at any point is constant.
- What happens in unsteady-state heat conduction?
  - In unsteady-state heat conduction, temperature varies with both time and location.
- Give examples of materials having (i) high thermal conductivity and (ii) low thermal conductivity.

(i) Diamond (900–2320), silver (429–415), gold (318), and copper (401).

(ii) Air (0.025), polyurethane foam (0.026), and glass fiber (0.043) (values in brackets are in W/(m-°C)).

- "Thermal conductivities of metals can significantly be affected by the presence of impurities in them." *True/False*?
  - *True*. Impurities in metals can give rise to variations in thermal conductivity by as much as 50–75%.
- "Thermal conductivity of an alloy is usually much lower than that of either metal of which it is composed." *True/False*? Give examples.
  - *True*. For example, k for copper is 401, for nickel is 91; for constantan (55% Cu + 45% Ni) is 23, and for aluminium is 237; and for bronze (90% Cu + 10% Al) is 52.
- "Ice has a thermal conductivity much higher than water." *True/False*?
  - *True*.
- What is the effect of temperature on thermal conductivity of solids?
  - The conductivity of solids changes mildly with temperature except at very low temperatures where it can acquire very large values. For instance, pure copper at 10K has a conductivity of about 20,000 W/(m-°C), whereas its conductivity at normal temperatures is 401 W/(m-°C).
- How does thermal conductivity vary with temperature for (i) liquids and (ii) gases?
  - For most liquids, k is lower than that for solids, typical values being about 0.17 W/(m-°C).

- For most liquids, thermal conductivity decreases with increase in temperature. Simple liquids are more sensitive than complex liquids such as highly polar liquids for which the decrease is slower.
- k decreases by about 3–4% for a 10°C rise in temperature.
- Thermal conductivity for gases increases with increase in temperature. For monatomic gases, for example, it is approximately proportional to $T^{1/2}$. For small temperature range, k increases nearly linearly with temperature but for wide ranges, the increase is more rapid than linear increases.
- Increase in molecular weight increases k values.
- Gases have very low values of k, as low as 0.007. For air at 0°C, k is 0.024 W/(m-°C).
- Figure 8.8 illustrates ranges of thermal conductivities of different materials.

• Arrange the following materials in the order of increasing thermal conductivity: magnesium, silver, mild steel, copper, aluminium powder, glass wool, rubber, and cardboard.
- Glass wool, cardboard, rubber, mild steel, magnesium, aluminium, copper, and silver.

• "For the same heat transfer rate, the slope of the temperature gradient in insulating materials is smaller than in noninsulating materials." *True/False*?
- *False.*

• Why is it necessary to use the concept of *apparent thermal conductivity* to describe thermal conductivity? Explain.
- While thermal conductivity is a property of the material, apparent thermal conductivity depends not only on temperature but also on its impurities present and bulk density. With increase in bulk density, thermal conductivity increases.

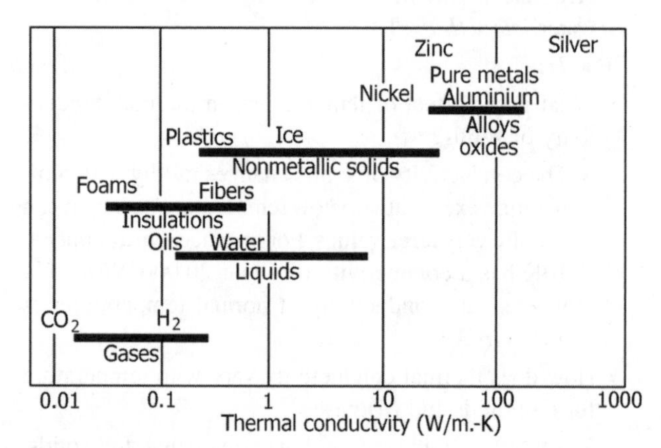

FIGURE 8.8 Approximate ranges of thermal conductivities of materials at normal temperatures and pressures.

- For granular solids, presence of air pockets can cause heat transfer not only by the mechanism of conduction but also by convection and by radiation from surface to surface of the individual particles. This enhances apparent thermal conductivities by as much as 1.5–2.5 times those for the case of still air pockets in which no convection effects are induced.

• What is thermal diffusivity? What is its significance?
- Thermal diffusivity describes the speed of penetration of heat into the body of a material under applied thermal load at its surface.
- Thermal diffusivity, α, is defined as

$$\alpha = \frac{\text{heat conducted}}{\text{heat stored}} = \frac{k}{\rho c_p} \ \text{m}^2/\text{s}. \tag{8.7}$$

- Values of thermal diffusivity, α, range from 0.1×10^{-6} for cork to 300×10^{-6} for potassium.
- Large value of α denotes that heat is propagated fast into the medium.
- Small value of α means heat is mostly absorbed by material and small amount conducted further.

• Name similar terms in momentum and mass transfer.
- Kinematic viscosity, ν, is known as momentum diffusivity, which is the ratio of absolute viscosity to density.

$$\nu = \mu/\rho. \tag{8.8}$$

- The mass diffusivity is the rate of mass transfer per unit area and per unit concentration gradient. Mass diffusivity is measured in terms of diffusivity coefficients.

• "Thermal diffusivity is a measure of the ability of a material to transfer thermal energy by conduction compared to the ability of the material to store thermal energy." *True/False*?
- *True.*

• "Materials with high thermal diffusivity will need more time to reach equilibrium with their surroundings." *True/False*?
- *False.*

• What is heat flux? Give its units.
- Heat transfer per unit area is called heat flux, Q/A.
- Units are kJ/(s)(m² c.s. area).

• What is a composite wall? Why is it necessary to use layers of different materials for insulation?
- A composite wall is a wall with flat surfaces constructed of slabs of different materials in series with same or different thicknesses. Furnace walls are examples, consisting of refractory bricks for the inner

wall(s) and other insulating materials such as magnesia and slag wool for the outer wall(s).

- Furnace temperatures are very high compared to the outside environment, requiring a high-temperature material for the inner walls that is also expected to serve as a structural wall apart from its insulating properties. Acid or basic bricks are used as refractory bricks depending on the nature of the environments they come in contact, that is, acidic or alkaline atmospheres.
- Outer walls serve more as insulating walls to serve much lower temperatures. Outer insulation is provided by slag wool or similar insulating material.

• Give expressions for heat conduction through a flat slab and hollow cylinder.

- *Flat Slab*:

$$q = kA(t_1 - t_2)/\Delta x, \tag{8.9}$$

where A is the surface area of the slab, t_1 and t_2 are temperatures on the two surfaces of the slab, and Δx is the thickness of the slab.

- *Hollow Cylinder*:

$$q = kA_L(t_i - t_o)/(r_o - r_i), \tag{8.10}$$

where

$$A = 2\pi L(r_o - r_i)/\ln(r_o/r_i), \tag{8.11}$$

where the subscripts o and i represent outer and inner, respectively.

• When are arithmetic mean and logarithmic mean radii used in heat conduction to a hollow cylinder?

- Arithmetic mean can be used for thin-walled tubes where the ratio of outer to inner radii is about 1.0.

• Write down equations for heat conduction for (i) multilayer flat wall, (ii) multilayer hollow cylinder, and (iii) multilayer hollow sphere.

- Multilayer flat slab: One-dimensional heat conduction—Figure 8.9 shows a series of three flat slabs in thermal contact with each other.

$$\Delta T = \Delta T_1 + \Delta T_2 + \Delta T_3. \tag{8.12}$$

- Overall thermal resistance,

$$R = R_1 + R_2 + R_3. \tag{8.13}$$

$$q = \Delta T/R = \Delta T_1/R_1 = \Delta T_2/R_2 = \Delta T_3/R_3 = (\Delta T_1 + \Delta T_2 + \Delta T_3)/(R_1 + R_2 + R_3). \tag{8.14}$$

- Multilayer hollow cylinder: One-dimensional heat conduction (Figure 8.10).

$$q = 2\pi L(T_1 - T_4)/[\ln(r_2/r_1)/k_a + \ln(r_3/r_2)/k_b + \ln(r_4/r_3)/k_c]. \tag{8.15}$$

- In the above equation, contact resistance between the adjacent layers is neglected. This is true if the layers of the composite cylinder are bonded together to provide intimate contact between adjacent layers, excluding any air gaps or gaps containing any contaminants.Multilayer hollow sphere: One-dimensional heat conduction ((Figure 8.11).

$$q = 4\pi(T_1 - T_4)/[(r_2 - r_1)/r_2 r_1/k_a + (r_3 - r_2)/r_3 r_2/k_b + (r_4 - r_3)/r_4 r_3/k_c]. \tag{8.16}$$

• What are heat conduction shape factors?

- Two-dimensional heat transfer in a medium bounded by two isothermal surfaces at T_1 and T_2 may be represented in terms of a conduction shape factor S (Figure 8.12).

$$q = Sk(T_1 - T_2). \tag{8.17}$$

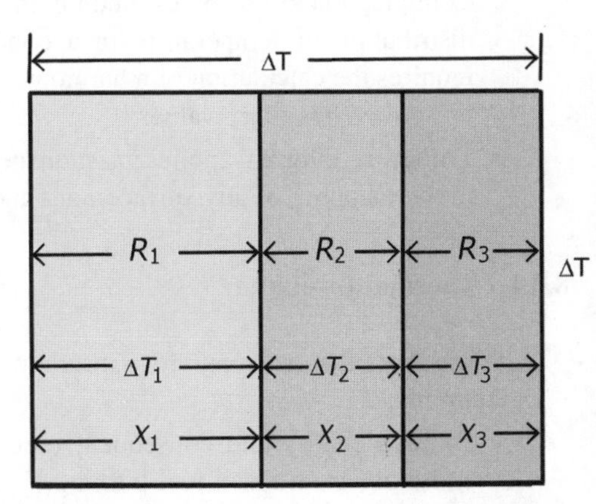

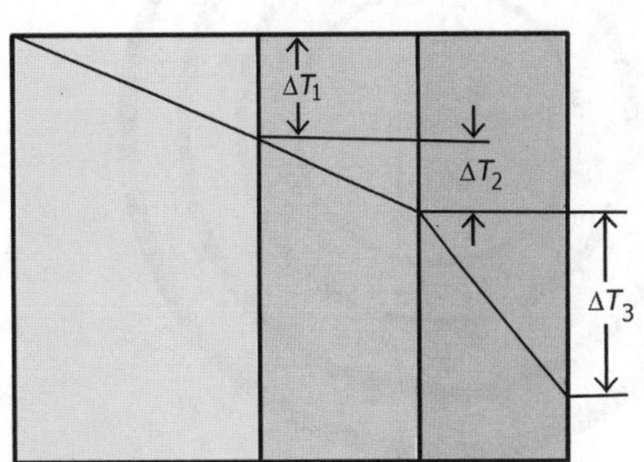

FIGURE 8.9 Multilayer slab.

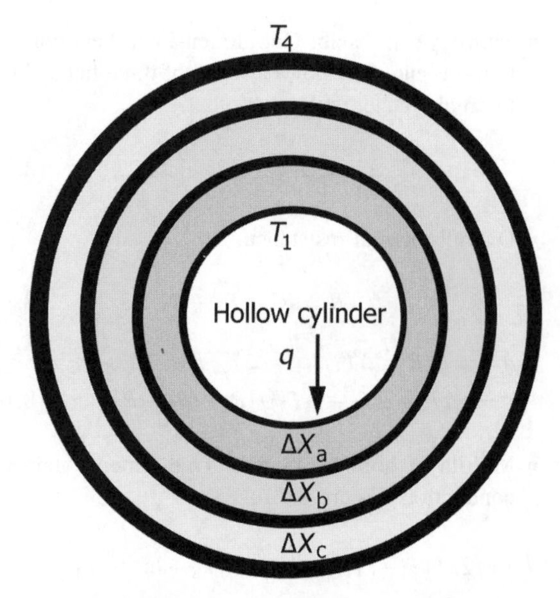

FIGURE 8.10 Multilayer hollow cylinder.

- For example, for the case of a long ($L \gg w$) circular cylinder centered in square solid of equal length,

$$S = 2\pi L / \ln(1.08\, w/D). \tag{8.18}$$

- What are Heisler charts?
  - Heisler charts are a series of charts that depict unsteady-state temperature distributions in different shaped bodies such as slabs, cylinders, and spheres.
- What is conformal mapping?
  - A special operation in mathematics in which a set of points in one coordinate system is mapped or transformed into a corresponding set in another coordinate system, preserving the angle of intersection between pairs of curves.

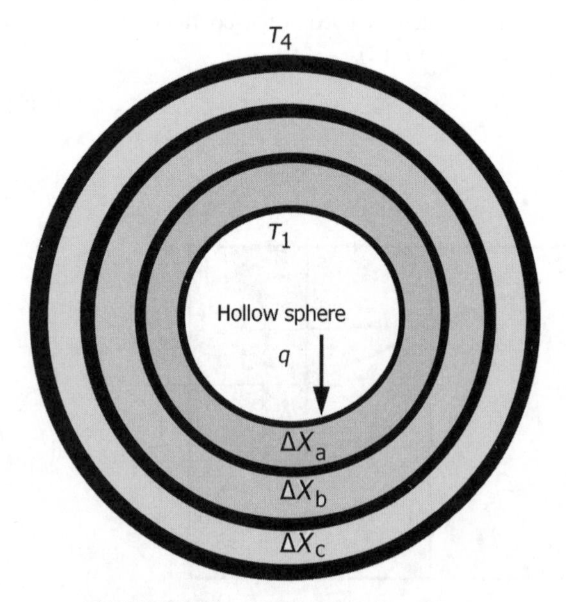

FIGURE 8.11 Multilayer hollow sphere.

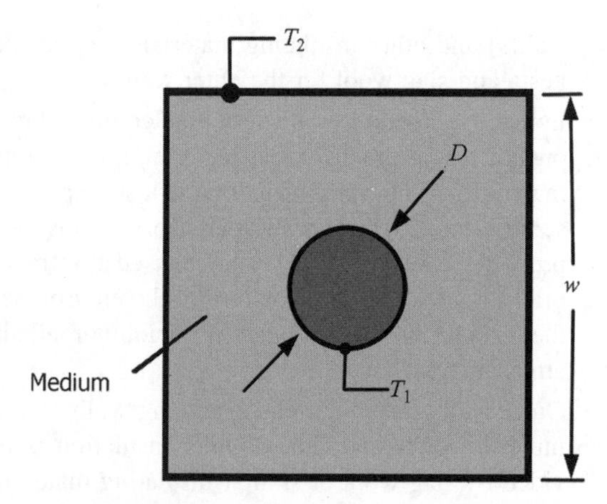

FIGURE 8.12 Heat transfer shape factors.

  - A mapping or transformation of a set E of points in the xy plane onto a set F in the uv plane is a correspondence that is defined for each point (x, y) in E and sends it to a point (u, v) in F, so that each point in F is the image of some point in E.
  - A mapping is one to one if distinct points in E are transformed to distinct points in F.
  - A mapping is conformal if it is one to one and preserves the magnitudes and orientations of the angles between curves. Conformal mappings preserve the shape but not the size of small figures.
  - If the points (x, y) and (u, v) are viewed as the complex numbers $z = x + iy$ and $w = u + iv$, the mapping becomes a function of a complex variable: $w = f(z)$. It is an important fact that a one-to-one mapping is conformal if and only if the function f is analytic and its derivative $f'(z)$ is never equal to zero.
  - Conformal mappings are important in two-dimensional problems of fluid flow, heat conduction, and potential theory. They provide suitable changes of coordinates for the analysis of difficult problems. For example, the problem of finding the steady-state distribution of temperature in a conducting plate requires the calculation of a harmonic function with prescribed boundary values.
  - The term conformal applies in a more general context to the mapping of any surface onto another.

8.3.1 Thermal Insulation

- What are the reasons for insulating plant and equipment?
  - To save energy and consequent reduction in CO_2 emissions from fossil fuel burning.
  - To help maintain process temperatures.

- To protect site personnel against burns. Temperatures of outer surfaces of insulation should not be allowed to rise above 60°C to prevent burns.
- To prevent condensation.
- To protect against frost.
- To protect equipment from corrosion.
- To provide fire protection.
- To provide acoustic insulation.

• What are the desirable characteristics for the selection of an insulating material?

- Low thermal conductivity.
- Resistance to moisture absorption.
- High porosity.
- Ease of application on surfaces.
- Nontoxic and nonflammable.
- Noncorrosive.
- Good strength.

• "Air has very low thermal conductivity and therefore is one of the best insulators." Why then heat transfer surfaces are provided with various types of solid insulators when these surfaces are surrounded by atmospheric air?

- Atmospheric air involves convection currents that remove heat from the hot surfaces or add heat to cold surfaces through convection process. Air, in the absence of convection currents, will only act as a good insulating material. Therefore, in order to make air free from convection currents, it has to be broken down into tiny globules by entrapping it by nonconducting cellular solid structures to make an effective insulation.

• Name some thermal insulating materials for low- and high-temperature insulation applications.

- *Low-Temperature Applications*:
  - ➢ Cork board.
  - ➢ Fiberglass with asphalt coating.
  - ➢ Expanded foam glass, glass blocks.
  - ➢ Polyurethane, polyisocyanurate, phenolic, polyimide, and polystyrene foams and elastomeric cellular plastic.
  - ➢ Vegetable fiberboard with asphalt coating.
  - ➢ Mineral wool board or *rock wool.*
  - ➢ Expanded rubber board or *Rubatex.*
  - ➢ Silica aerogel powder or *Santocel.*
  - ➢ Perlite: This involves fine pores of trapped air. Attributes include low thermal conductivity, nonflammable, ease of application, and low moisture retention and relatively low cost. Expanded perlite is used as a loose-filling material in the annulus of a double-walled vessel for storage of cryogenic gases at −130°C. It can also be used for high-temperature applications up to 750°C.
- *High-Temperature Applications*:
  - ➢ Mineral/slag wool, glass wool.
  - ➢ Diatomaceous earth, silica powder (up to 820°C).
  - ➢ Diatomaceous earth, asbestos, and bonding material (up to 870°C).
  - ➢ Glass blocks (up to 650°C).
  - ➢ 85% magnesia (up to 340°C).
  - ➢ Ceramic refractories (for higher temperatures).
  - ➢ Insulating firebrick.
  - ➢ Zirconia grain (up to 1650°C).
  - ➢ Microquartz fibers (up to 1650°C).
  - ➢ Calcium silicate products are known for exceptional strength and durability in both intermediate- and high-temperature applications.

• What is a Supertherm ceramic insulation? What are its characteristics?

- Supertherm is essentially a ceramic coating type of insulation consisting of different types of ceramic powders with combinations of acrylic and urethane polymers to provide elasticity, adhesion, toughness, and moisture repulsion. The ceramic components repel radiation with resistance for heat conduction.
- It can be used as spray-on coating on irregular surfaces.
- Ceramic is more expensive than fiberglass insulation. Therefore, where fiberglass or other *conventional* insulation can be effectively used, ceramic is not an economically attractive insulation material.
- Ceramic is also a good corrosion inhibitor and more resistant to impact damage and moisture.
- There are types of ceramic that can be applied to hot surfaces, reducing or eliminating downtime application.
- Supertherm is also a good soundproofing material.

• "Asbestos is a good insulating material." What is the important limitation involved in its application?

- Asbestos is found to be a carcinogenic material and many countries banned its manufacture and use. Asbestos dust resulting from demolition of structures with asbestos-containing materials is particularly harmful to exposed workers.
- Due to this reason, asbestos is discontinued as an insulating material.

• What is the utility of reusable insulation pads? What are their applications?

- Reusable insulating pads are commonly used in industrial facilities for insulating flanges, valves, expansion joints, heat exchangers, pumps, turbines, tanks, and other irregular surfaces. The pads are flexible and vibration resistant and can be used with equipment that is horizontally or vertically mounted or that is difficult to access.

- The pads are made of a noncombustible inside cover, insulation material, and a noncombustible outside cover that resists tears and abrasion.

- Polyurethane foams can be used as insulating materials. Name some applications for these materials.
  - Polyurethane foams are good insulating materials that combine low thermal conductivities of polyurethanes with low thermal conductivity of air that is entrapped in the micropores of polyurethane.
  - These foams are available as rigid foams in molded form and in liquid form for spraying on surfaces for subsequent setting into highly porous solid, adhering on to the solid surfaces on which they are sprayed.
  - Rigid foams are molds made into panels in different shapes to suit the application to different types of equipment operating at ambient and low temperatures.
  - Foams in liquid form are sprayed on external walls of houses in cold climates, cold storages, and on other surfaces to prevent heat ingress and conserve energy.
- What is mineral/slag wool? Compare this with glass wool for insulating purposes.
  - Mineral/slag wool is a raw glass fiber insulation made out of molten blast furnace slag, which is forced through multiple hole nozzles made out of heat resistant materials, to give rise to solidified fibers. Blast furnace slag, which is a waste mineral matter from steel plants, provides a cheap insulating material entrapping air to give high-efficiency insulation. Slag wool insulation is widely used in industrial insulation applications.
  - Glass wool, on the other hand, is made out of molten glass that is equally good insulation but expensive compared to slag wool.
- What is the important application of ceramic fiber insulation?
  - High-temperature systems, for example, ceramic fiber blankets are used to line fired heaters.
- Cork is a good cold insulating material. Is it being used for the present day cold insulation applications? Comment.
  - Cork has become a scarce material with consequent increases in its price. Equivalent insulating materials, namely, plastic foams, have replaced cork. These materials are cheaper and easily available and last longer in the moist atmospheres where cork has been used earlier.
- What type of insulating material is used as insulation in domestic refrigerators? What are the other applications of rigid insulation?
  - Rigid polyurethane foams, cellular polystyrene, and other rigid insulations are widely used as insulation in domestic refrigerators.
- What are the applications of rigid insulations?
  - Rigid insulation is resistant to deformation when subjected to foot traffic.
  - Areas that experience loads or repetitive personnel access/use will require a firmer rigid insulation than inaccessible areas. Piping used as ladders/walkways and horizontal surfaces subject to vibration/loads are examples where rigid insulation is required.
- Name some commonly used rigid insulations and give their characteristics.
  - Calcium silicate insulation is a rigid, dense material used for above ambient to 650°C. Commonly used for high-temperature applications. It has good compressive strength and is noncombustible.
  - Calcium silicate block and pipe insulation is used on surfaces with temperatures up to 870°C.
  - Expanded or exfoliated *vermiculite* thermal insulating cement in the form of dry cement or plaster is mixed with a suitable proportion of water, applied as a plastic mass, and dried in place, for use as insulation on surfaces operating at temperatures 35–1000°C. The cement is not suitable where it will be exposed to combustion conditions, such as the hot face lining of a furnace. Vermiculite cement is also used as loose-fill insulation.
  - Cellular glass is also rigid, dense material normally used in the temperature range from −230 to +200°C. Its closed cell structure makes it preferable for low-temperature applications and for use on services where fluid absorption into the insulation could be a problem. This is particularly useful where leaks can introduce fluids into the insulation.
    - Fires due to leaked organic fluids that *self-heat* through slow oxidation inside the insulation are common phenomena. The insulation prevents the dissipation of heat from such oxidation, increasing temperatures inside the insulation to the point where self-ignition occurs, leading to the so-called *lagging fires*.
- What are the measures that are useful in preventing fires in insulations?
  - Use of cellular glass insulation that resists fluid absorption by the insulation.
  - If a leak develops, the insulation must be removed and replaced by new insulation, after the leaks are contained.

- On vertical runs of pipe where occasional leaks can develop at flanges, protective tight fitting caps should be installed to divert any fluid leakage outside the insulation.
- Valve stems should be installed horizontally or in a downward position so that any stem leakage does not enter the insulation.
- Applying impervious protecting layer over the cellular insulation to prevent air ingress into the insulation. Weather barriers, vapor barriers, rigid and soft jackets, and a multitude of coatings exist for all types of applications.
- The possibility of leakage through joints and fittings is characteristic of most organic fluids unless the fittings are extremely tight. System design should minimize the number of connections in the piping layout. The best way to prevent piping leakage is to weld all connections. Use of threaded fittings should be strongly discouraged due to their tendency to leak. Where access is necessary, raised face flanges with weld neck joints or equivalent raised face flanges should be used.

- What are the applications of compressible insulation?
  - Compressible insulation is used for filling voids and closing gaps in insulation, which allows expansion, contraction, or movement of rigid insulation.
- Name materials used for insulation of electrical systems.
  - PVC, rubber, and other polymer and ceramic composites.
- What is a vapor barrier over insulation? Why is it necessary to use it?
  - Vapor barrier is a coating or metal shield applied over the surface of the insulation, especially when ambient temperatures are below 0°C, to prevent ingress of water into the pores of the insulation.
  - Particular attention should be paid to ensure integrity and continuity of the vapor barrier. Care should be taken at termination points, where any exposed insulation edges should be covered properly.
  - Metal cladding applied over insulation in hot service should have drain holes located at the lowest point to allow any entrapped water to escape.
  - Where a liquid-applied (*mastic*) vapor barrier is used (e.g., on cryogenic processes), joints should be filled with a suitable sealant to effectively isolate each section and hence localize water (and water vapor) ingress in case of any mechanical damage.
  - It should be noted that liquid-applied finishes are divided into water vapor barriers for cold applications and vapor breathers for hot service. Each type may or may not be self-weatherproofing.
- What is the common vapor barrier wrapping in thermal insulations?
  - Aluminium foil wrapping is commonly used. Typical thickness of aluminium used ranges from 0.2 to 0.4 mm. In case of low-temperature insulations, such a wrapping provides reflecting surface to reduce heat leak into the system. End sections of the insulation and where insulation abuts supports the vapor barrier should be taped by aluminium foil to all inserts. Any stickpins protruding through the foil facing should be securely taped with a minimum of 100 mm wide.
- What are the reasons for wrapping insulated surfaces with metallic sheets?
  - To protect insulation from peeling off.
  - To prevent heat ingress into insulated cold systems with reflective wrapping sheets.
  - To prevent moisture ingress.
  - To provide stagnant air film between insulation and wrapper to increase effectiveness, that is, resistance for heat flow.
  - To prevent crumbling of insulation by erosion by wind, contact by plant personnel that can also be a health hazard to them and fire exposure?
- "For high fire risk plants, stainless steel shielding is recommended to protect the insulation even though stainless steel is an expensive material compared to normally used shielding material." When such a consideration does become attractive?
  - For high fire risk areas, the finished insulation system should not get dislodged when subjected to the fire water stream used for fire fighting, either by hand lines or monitor nozzles. Most insulation systems used in fire protection are metal with stainless steel jackets and bands, which meet these criteria.
  - Stainless steel shield not only provides a vapor barrier to the insulation but also protects insulation from crumbling on fire exposure, taking the role of fireproofing material, thus protecting the structural integrity of piping/equipment from collapse.
- "In cold insulation system design, vapor barriers are extremely important compared to hot insulation." Comment.
  - Usually, the cost of removing heat (heat gain) by refrigeration is greater than that of producing process heat (heat loss) by heat-generating equipment.
  - Therefore, the heat gain in cold processes must be kept to a minimum. The rule of thumb is to provide sufficient insulation to maintain heat gain of 25–30 W/m^2 to the cold process.

- What is the relationship between dew point temperature of ambient air and insulation thickness?
  - It is crucial that sufficient insulation is added so that the outer temperature of the insulation remains *above* the dew point temperature. At the dew point temperature, moisture in the air will condense onto the insulation and damages it.
- Why the surface of the insulating material is treated to give smooth and impervious surface finish, blocking the pores inside the insulation?
  - The surface is made impervious to prevent ingress of moisture, which lowers effectiveness of insulation, acting as vapor barrier as explained earlier.
  - Leaked flammable liquids into the insulation catch fire due to self-heating, and continued oxygen supply feeds such fires. Self-heating occurs as oxidation of leaked flammable liquids into the insulation releases heat that accumulates inside the insulation as the surrounding insulation prevents heat dissipation, raising temperatures to self-ignition values of the leaked fuel, causing insulation fires, as explained before.
  - The impervious surface prevents oxygen supply to the interior of the insulation and *kills* insulation fires that are precursors to major fire and explosion hazards in process industry.
- What are the effects of water ingress into insulation?
  - Water ingress reduces ability of the material to insulate.
  - Just 4% moisture by volume can reduce thermal efficiency by 70%, since water has a thermal conductivity of up to 20 times greater than most insulation materials.
  - The risk of plant corrosion increases.
  - If the plant is operating at temperatures below 0°C, then water vapor may pass through the insulation and condense and freeze. This may cause mechanical breakage of the insulation, corrosion of the plant, and total breakdown of the insulation system or the equipment itself.
- What are the solutions to handle wet insulation?
  - If the insulation is wetted by leaked flammable liquids, it should be replaced by new insulation.
  - If the insulation gets wet due to moisture ingress, replacement may not be the best option as it involves high costs—material costs, labor costs, and disposal costs of defective insulation due to its environmental pollution effects.
  - Applying new insulation directly over the existing insulation, known as *overfitting*. Overfitting is a good option, because it is a cost-effective and immediate solution. An overfitted system will result in reduced energy costs. It also restores existing deteriorated insulation to its former thermal values, reduces system downtime, and improves plant safety by properly maintaining surface temperatures.
  - Even though pipes and equipment are at high temperatures, it is impossible to completely dry out *permeable* insulation once it becomes wet. The fuel loads required to maintain the operating temperature of steam systems with wet *permeable* insulation can double and, in some cases, triple.
  - Overfitting pipes and equipment with cellular glass insulation with a minimal amount of cellular glass, the insulation can become self-drying by creating an interface temperature that is high enough to vaporize the existing moisture in the permeable insulation.
  - A typical overfit system consists of a 3–5 cm layer of closed cell glass insulation, the thickness of which is determined by the resultant interface temperature between the existing insulation and the new overfit system. This interface temperature must be kept at or above 100°C to drive out the moisture in the existing insulation. The vaporized moisture is forced to escape through the interface and out of the system through the joints of the cellular insulation.
  - After the system is overfitted with cellular glass, it has an outer layer of impermeable insulation that eliminates the penetration of moisture (in both liquid and vapor forms) into the permeable insulation.
- What are the precautions required while applying mineral wool insulation to process equipment?
  - Mineral wool insulation, when comes in contact with skin, gives rise to irritation and skin rashes. Skilled labor using protective clothing and gloves are required.
  - The wool insulation requires protective sheathing in the form of metal sheets wrapped over it to prevent peeling off due to wind or contact with moving objects.
  - Alternative to wrapping, magnesia insulation in the form of paste may be applied to cover surface pores in order to prevent entry of air into it in the event of lagging fires.
- What are the common maintenance snags with respect to insulation to process equipment and piping in industry?
  - During maintenance work in a plant, commonly observed snags include the following:
    - Damage to wrappings and insulation.
    - During valve/fitting maintenance, omissions in restoring back molded insulation blocks in place.

  - ➢ Repairs giving rise to gaps for moisture entry.
- What type of insulation one would recommend for insulating cold storages?
  - Spray-on foam insulations.
- What insulation is recommended for insulating hot oil storage tanks?
  - Spray-on rigid urethane foam.
- Explain how a thermos flask maintains temperature of its contents.
  - Thermos flask is a double-walled glass vessel, with vacuum inside the sealed outer annulus and highly reflective surface on the outer glass wall.
  - Vacuum and the reflective surface are the effective means of maintaining inside temperature of the vessel.
  - Glass, with reasonably low thermal conductivity, is used to enhance insulating power of the flask.
  - Additionally, the flask is protected by a plastic or a metal container that also provides *nearly stagnant* air gap between the glass and the outer containers, adding to the insulation effectiveness.
- What type of construction is used for storage tanks for liquid ammonia with the objective of preventing heat leakages into the system?
  - Double-walled construction is used with the outside of the external wall coated with alumina paint to reflect back the heat incident on it.
- Why is it necessary to insulate inside steel surface of an ammonia reactor?
  - Ammonia reactors operate at temperatures of over 500°C and 250 bar pressures.
    - ➢ Inside surfaces of the steel wall must be kept at temperatures below 370°C, temperature at which steels begin to decline in strength and also to prevent hydrogen access to steel shell through which it diffuses causing hydrogen embrittlement.
  - Providing an air gap (about 2 cm) between the outer shell and the insulating liner significantly improves overall insulating quality.
- "Entrapping air within an insulation will substantially increase insulation power. Convection currents within entrapped air will decrease the insulating power." *True/False*? Comment.
  - *True.* Dry *still* air is a very good insulator. Any convection currents increase heat transfer reducing insulating power of the insulation.
- "It is said that insulation of surfaces reduces noise." Comment.
  - Sound attenuation is a natural by-product of the insulation design.
  - Because of their sound absorption characteristics, some insulations provide greater sound attenuation than others.
  - Mineral fiber products are among the best thermal insulation materials for sound attenuation.
  - The jacketing material used to cover the insulation can play an important role in sound attenuation.
  - A fabric-reinforced mastic finish over insulation has better sound absorption properties than metal jacketing.
- What is *critical radius*? If a thicker insulation exceeding the *critical radius* is used, what happens to heat losses?
  - As thickness of insulation increases, resistance for heat conduction through the insulation increases and hence heat losses decrease.
  - Also, as insulation thickness increases on cylindrical surfaces, exposed surface area increases.
    - ➢ Heat losses are a direct function of exposed surface.
    - ➢ So, effect of increased insulation thickness is to decrease heat losses due to increased resistance and increased heat losses due to increased surface area.
    - ➢ There will be a maximum insulation thickness beyond which losses due to increased surface will overtake the reduction of heat losses due to increased resistance.
    - ➢ This maximum thickness is called *critical thickness* and the radius of the cylinder at this thickness is called *critical radius*.
- Explain why heat losses increase when thermal insulation is applied to thin wires.
  - Surface area increases rapidly as insulation is applied to wires, increasing heat losses.
- What is optimum insulation thickness?
  - As insulation thickness increases, savings in energy costs (operating costs) will increase. Exposed surface area of the insulation on piping or walls with a curvature, such as in the case of tanks, increases contributing to heat losses.
  - Also as insulation thickness increases, cost of insulation material as well as costs involved in applying the material increases (fixed costs).
  - Therefore, there will be a maximum thickness beyond which fixed costs overtake the costs involved in energy savings. This thickness is called optimum thickness.
- "Optimum insulation thickness varies with temperature." *True/False*? Illustrate.

  - *True*. For example, at 100°C, the thickness is 1.27 cm (0.5 in.); at 200°C, the thickness is 2.54 cm (1.0 in.); and at 300°C, the thickness is 3.17 cm (1.25 in.)
- What is the effect of windy conditions on insulation requirements?
  - Where windy conditions (12 km/h) prevail, 10–20% greater insulation thickness is justified.
  - What is the minimum thickness suitable for using multilayer insulation? What precautions are taken in applying multilayer insulation?
  - It is generally recommended that insulation thicknesses greater than 75 mm are applied in multilayers.
  - Where insulation is applied in two or more layers, all joints, both longitudinal and circumferential, should be staggered.
  - Metal cladding used outdoors should have joints overlapped and be sealed with a suitable *mastic*. If securing screws penetrate the metal finish, these should be sealed.
- Give an example where the purpose of using insulation is to increase rather than decrease heat losses.
  - Electrical transmission lines are often lagged to increase heat dissipation rate.
    - In this case, as wires are of very small diameter compared to pipes, thickness of the insulation exceeds the critical thickness resulting in increased heat transfer rates due to surface increase compared to reduction of heat transfer rates due to increased thickness, that is, increased resistance to heat flow. Net effect is increased heat losses.
  - Heat pickup by insulated small diameter refrigerated lines can be more than that by the bare pipes.
- How does an ablative material work as a thermal insulation/shield?
  - Ablative materials, coated on the surfaces to be insulated, will expand into a foam structure on exposure to high temperatures, for example, fire exposure, acting as effective insulation, protecting the surfaces from yielding due to softening or melting before remedial measures take over the job of protection. These materials, initially developed for spacecraft applications, are frequently used to protect equipment on fire exposure. Once such an exposure occurs, the material must be replaced.
- What are refractories?
  - Refractories are generally ceramic materials that are capable of withstanding high temperatures. The bulk of refractory materials consist of single or mixed high-melting point oxides of elements such as silicon, aluminium, magnesium, calcium, and zirconium. Nonoxide refractories also exist and include materials such as carbides, nitrides, borides, and graphite. The actual composition of a refractory material is dependent on operating factors such as temperature, atmosphere, and the materials it will be in contact with.
- What is refractoriness? How is refractoriness of a material measured?
  - The refractoriness of a material is a measure of its ability to withstand exposure to elevated temperatures without undergoing appreciable deformation.
  - It is generally measured using what are known as *Sager cones*, which are cone-shaped ceramic objects of differing compositions. When heated to different temperatures, these cones will slump as they soften in response to the temperature, with the degree of slumping being dependent on the composition.
  - A similar cone-shaped object is made from the material to be measured and heated along with standard Sager cones. After the conclusion of the heating cycle, the sample material is compared to the Sager cones to gain a comparative measure of its refractoriness.
  - However, it should be noted that when a material is subject to a mechanical load, it may well soften at well below the temperature indicated by the Sager cone test.
- What types of refractories are used in steel making?
  - In steel making, basic refractories are used because the refractories often come into contact with basic slags containing magnesium and calcium oxides. If the refractory lining was made from acidic refractories, it would be eroded quickly by the chemical interaction of the basic slag and the acidic lining (e.g., silica) forming low melting point compounds.
- Where are (i) magnesia (basic) and (ii) silica bricks (acidic) used?
  - Due to its excellent corrosion resistance, refractory grade fused magnesia is used in high wear areas in steel making, for example, basic oxygen and electric arc furnaces, converters, and ladles.
  - Silica bricks are widely used in glass furnaces, glass kiln vaults, and so on.
- What are the applications of graphite refractories?
  - Graphite refractories can operate at temperatures of up to several thousand degrees Celsius under reducing conditions, or oxygen-free conditions, such as vacuum. However, they may begin to sublime at approximately 1000°C under oxidizing conditions.
- What is the effect of porosity of refractories on their insulating and high-temperature characteristics?

- High-porosity refractories have low thermal conductivities due to entrapment of air in the pores.
- High-porosity refractories, on the other hand, are structurally weak and cannot cope with flame impingement in the furnaces and therefore not desirable to be used in hotter zones. They shrink on exposure to high temperatures, which adds to their structural instability.
- Dense refractories are used in hotter zones and high-porosity refractories are used as backup lining layer to serve insulation requirements.

- How are refractory layers organized in furnace construction?
  - Furnaces may use a number of layers of different refractories in their construction. For instance, a dense hot face material may be used on the inner surfaces that are exposed to the highest temperatures.
  - On the outside, a low-density highly insulating refractory layer may be employed to conserve energy.
  - Between the hot face refractory and the low-density insulation, a number of different layers of intermediate materials may also be employed.
  - Each successive layer (moving away from the hot face) would have an increasingly lower density and more than likely a lower refractoriness compared to the previous layer.
- What type of insulation is used in modern furnaces?
  - Ceramic fiber insulation panels. Ceramic fibers are replacing customary refractory insulation. Ceramic fiber insulation has lower thermal conductivities than refractories. Consequently, the insulation thicknesses have reduced considerably for the same duty, making the furnace construction lighter and more compact. Construction times have reduced and maintenance has become easier.
- What are geopolymers? What are their characteristics and applications?
  - Inorganic polymers are different and show promise for use in elevated temperature insulation applications. Inorganic polymers made from aluminosilicates are termed geopolymers. They are amorphous to semicrystalline and consist of two- or three-dimensional aluminosilicate networks, dependent on the composition. Geopolymers can be formed using a relatively low-temperature processing techniques.
  - Physical behavior of geopolymers is similar to those of Portland cement. Consequently, they have been considered as a possible improvement on conventional cements with respect to compressive strength, resistance to fire, heat, and acidity as well as a medium for the encapsulation of hazardous or low/intermediate level radioactive wastes.
  - These are suitable as refractory coatings, heat insulation materials for continuous use at temperatures in the range from 1000 to 1400°C.

9

CONVECTIVE HEAT TRANSFER BASICS

9.1 Convective Heat Transfer 245
9.1.1 Heat Transfer Coefficients 246
9.1.2 Condensation 251
9.1.3 Boiling and Vaporization 254
9.1.4 Heat Transfer Coefficients: Magnitudes and Data 259
9.1.5 Heat Transfer Fluids 262
9.2 Annexure (Heat Transfer Coefficients for Refinery Services) 269

9.1 CONVECTIVE HEAT TRANSFER

- What is convection? Explain.
  - Convection heat transfer can be defined as transport of heat from one point to another in a fluid as a result of macroscopic motions of the fluid, the heat being carried as internal energy.
  - Convection between a solid surface and adjacent fluid that is in motion involves combined effects of conduction and fluid motion.
  - Density differences between adjacent layers of fluids induce relative motion between fluid layers.
  - The density differences may be due to temperature differences between two layers of the same fluid or two different fluids with different densities at the same temperature.
  - External means of energy input could also induce relative motion.
- What is the difference between natural and forced convection?
  - *Natural or Free Convection:* If fluid motion is caused by buoyancy forces that are induced by density differences due to variation in fluid temperature or due to variation in densities of two different fluids adjacent to each other or solid–fluid interfaces are at different temperatures. Convection currents are induced by temperature differences only without any external energy inputs such as pump energy. Local velocities are generally low, giving rise to near laminar flow conditions.
  - *Forced Convection:* Fluid is forced over the surface by external means such as a fan, pump, mixer, or wind, in addition to natural temperature difference-induced convection currents, to increase localized velocities to increase turbulence and hence heat transfer coefficients.
- Give examples for natural convection processes.
  - Air circulation in the atmosphere and water circulation in oceans and lakes.
  - Storms, including dust storms, are propelled by density differences in the atmosphere caused by temperature differences.
  - Example of a burning candle: Air next to the flame is hot and rises up while air away from the flame is cooler and moves downward to fill the region from which hot air moved up.
  - Stacks from which hot combustion gases move upward.
  - Light gases like hydrogen rising in open environment when leaked from a container due to density difference between air and hydrogen.
- "For large pipe diameters and large ΔT between pipe wall and bulk of the fluid, natural convection effects are more." *True/False*?
  - *True.* In the case of small pipe diameters, scope for natural convection currents will be very poor.
- What is *conjugate heat transfer*?
  - Conjugate heat transfer refers to the ability to compute conduction heat transfer through solids coupled by convective heat transfer in a fluid.
  - Either the solid zone or fluid zone or both may contain heat sources, for example, coolant flowing over fuel rods that generate heat.

Fluid Mechanics, Heat Transfer, and Mass Transfer: Chemical Engineering Practice, By K. S. N. Raju

9.1.1 Heat Transfer Coefficients

- What is Newton's law of cooling?
  - Newton postulated that heat transfer Q is proportional to surface area of the object and temperature difference, ΔT.
  - The proportionality *constant* is called heat transfer coefficient, h, which lumps together several factors that govern the heat transfer process.

$$Q_{\text{convection}} = hA(T_S - T_\infty) = hA\,\Delta T, \tag{9.1}$$

    where h is the heat transfer coefficient W/(m^2 °C), T_S is the surface temperature (°C), and T_∞ is the temperature of the fluid sufficiently far from the surface, that is, bulk fluid (°C). $\Delta T = T_S - T_\infty$ is the temperature difference.
- What are the units for (i) thermal conductivity and (ii) heat transfer coefficient?
  - Thermal conductivity: W/(m °C).
  - Heat transfer coefficient: W/(m^2 °C).
- What is the significance of heat transfer coefficient?
  - Heat transfer coefficient is a parameter whose value depends on all the variables that influence convection, for example, nature of the fluid in motion and its properties, bulk fluid velocity, surface geometry, and so on.
  - h is not an intrinsic property of the fluid.
- "Prandtl number is the ratio of momentum diffusivity to thermal diffusivity." *True/False*?
  - *True*.
- "Prandtl number for a gas is nearly independent of temperature." *True/False*?
  - *True*. Prandtl number varies little over a wide range of temperatures, approximately 3% from 300 to 2000 K.
- "Prandtl numbers for liquid metals are far below Prandtl numbers for water." *True/False*?
  - *True*. Values are of the order of 0.01.
- Give typical values of Prandtl numbers for gases, water, oil, and mercury.
  - Most gases: 0.7.
  - Water: 7.
  - Oil: 100–40,000.
  - Mercury: 0.015.
  - Liquid metals: 0.01.
  - For mercury, heat conduction is very effective compared to convection, with thermal diffusivity being dominant. For oil, convection is very effective compared to conduction, with momentum diffusivity being dominant.
- For what type of fluids Prandtl numbers are of the order of 0.01 or less?
  - For liquid metals.
- Illustrate how Prandtl numbers vary with temperature.
  - Figure 9.1 illustrates variation of Prandtl numbers with temperature for light hydrocarbon gases.
    - From the figure, it could be noticed that Prandtl numbers decrease with increase in temperature, though the decrease is not linear.
    - Also, Prandtl numbers are higher for heavier hydrocarbons.
- What is Graetz number?

$$\text{Gz} = N_{\text{Re}} \cdot N_{\text{Pr}} \cdot D/L. \tag{9.2}$$

- What are the applications of liquid metals as heat transfer media?
  - Liquid metals are used as heat transfer fluids in nuclear reactors. Their high temperatures and high heat capacities are attractive for removing heat from the core of nuclear reactors.
- What is a thermal boundary layer?
  - A layer across which there is a significant temperature difference and the heat transfer is primarily via heat conduction.
  - Thermal boundary layer thickness is different from hydrodynamic boundary layer (Figure 9.2).
- "A thermal boundary layer for gases is thinner than a hydrodynamic boundary layer." *True/False*?
  - *False*. Prandtl number controls the relative thickness of the momentum and thermal boundary layers.
  - Thermal boundary layer thickness for a high Prandtl number fluid, like water, is much less than hydrodynamic boundary layer thickness.

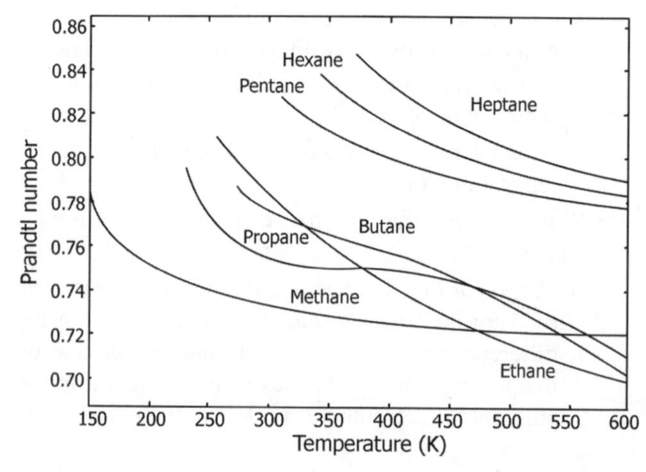

FIGURE 9.1 Prandtl numbers of light hydrocarbon gases at 1 bar.

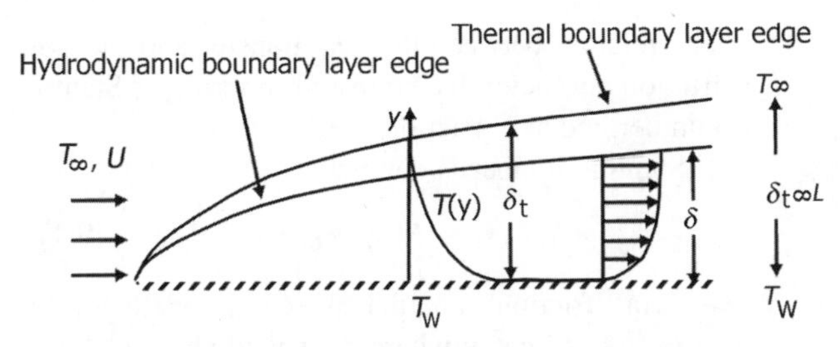

FIGURE 9.2 Hydrodynamic and thermal boundary layers.

- If the Prandtl number is less than 1 (low Prandtl number fluids), which is the case for air and many other gases at standard conditions, the thermal boundary layer is thicker than the hydrodynamic boundary layer. Mercury and liquid metals have far less values of Prandtl numbers than gases and boundary layer thicknesses are much higher.
- When Prandtl number $\cong$ 1, the boundary layer thicknesses are nearly same.

- "Prandtl number relates the thickness of the hydrodynamic boundary layer to the thickness of the thermal boundary layer." *True/False*?
  - *True.*
- What are the dimensionless groups that are significant in natural convection heat transfer?
  - Grashof number.
  - Rayleigh number.
- Define Grashof number. For what type of heat transfer, this number is of significance?
  - Grashof number is given by N_{Gr},

$$L^3\rho^2\beta g\,\Delta t/\mu^2 = \text{ratio of buoyancy force to viscous force acting on the fluid.} \tag{9.3}$$

  - Natural convection heat transfer.
- Define Rayleigh number.
  - Rayleigh number is the product of Grashof and Prandtl numbers.

$$N_{Ra} = N_{Gr}\cdot N_{Pr} = L^3\rho^2 g\beta C_p\,\Delta T/\mu\alpha. \tag{9.4}$$

  - It governs natural convection heat transfer.
- What are the equations that are used in correlating natural convection heat transfer?
  - Natural convection about *vertical cylinders and planes*:

$$N_{Nu} = 0.53\,(N_{Pr}\cdot N_{Gr})^{0.33}, \quad \text{for } 10^4 < N_{Pr}\cdot N_{Gr} < 10^9 \tag{9.5}$$

and

$$N_{Nu} = 0.12\,(N_{Pr}\cdot N_{Gr})^{0.33} \quad \text{for } 10^9 < N_{Pr}\cdot N_{Gr} < 10^{12}. \tag{9.6}$$

For air, these equations can be approximated respectively by

$$h_c = 1.3(\Delta T/L)^{0.25} \tag{9.7}$$

and

$$h_c = 1.8(\Delta T)^{0.25}. \tag{9.8}$$

  - ➢ The equations for air are not dimensionless where ΔT is in °C, L or D in m, and h_c in J/(m^2 s °C). The characteristic dimension to be used in the calculation of N_{Nu} and N_{Gr} in these equations is the height of the plane or cylinder.
  - Natural convection about *horizontal cylinders* such as a steam pipe:

$$N_{Nu} = 0.54(N_{Pr}\cdot N_{Gr})^{0.25} \text{ for laminar flow in range } 10^3 < N_{Pr}\cdot N_{Gr} < 10^9. \tag{9.9}$$

  - ➢ For air the approximate equations are the same as for vertical surfaces, except for higher ranges of $N_{Pr}\cdot N_{Gr}$, the exponent for ΔT is 0.33 instead of 0.25.
  - For *horizontal planes*, the equations for cylinders may be used, using L in place of D of the cylinder wherever D is used in N_{Nu} and N_{Gr}.
  - In the case of *horizontal planes*, cooled when facing upward, or heated when facing downward, which appear to be working against natural convection circulation, half of the value of h_c, found in the given correlations, may be used.
- What is Colburn J_H factor? What is its significance? Give the appropriate equations.
  - *J*-factor is evolved in observing analogous conditions between momentum and heat transfer.
  - The analogy between heat and momentum is evolved by assuming that diffusion of momentum and heat occurs by essentially the same mechanism so that a relatively simple relationship exists between the momentum and heat diffusion coefficients.
  - Colburn has defined *J*-factor for momentum as

$$J_M = f/2, \tag{9.10}$$

where f is the friction factor. *J*-factor for heat transfer was assumed to be equal to *J*-factor for momentum transfer and designated as J_H.
  - As per Colburn's analogy,

$$J_M = f/2 = J_H = h/C_pG(N_{Pr})^{2/3}, \tag{9.11}$$

where h is the heat transfer coefficient, G is the mass velocity $= V\rho$, velocity multiplied by density of the fluid.

- For viscous fluids,

$$J_{\mathrm{H}} = h/C_{\mathrm{p}}G(N_{\mathrm{Pr}})^{2/3}(\mu_{\mathrm{w}}/\mu_{\mathrm{b}})^{0.14}. \quad (9.12)$$

- Friction factor, f, is related to N_{Re} by the following equation, applicable in the range of $5000 < N_{\mathrm{Re}} < 200{,}000$:

$$f = 0.046(N_{\mathrm{Re}})^{-0.2}. \quad (9.13)$$

- For transition region to turbulent flow, that is, $2100 < N_{\mathrm{Re}} < 6000$:

$$J_{\mathrm{H}} = 1.86(D/L)^{1/3}(N_{\mathrm{Re}})^{-2/3}, \quad (9.14)$$

which is shown on a plot of N_{Re} versus J_{H} (See Figure 9.3)

- What is the basis for the analogy between momentum and heat transfer?
  - Mechanism of diffusion of heat and diffusion of momentum are essentially the same.
- Give the equations involved in Colburn analogy for momentum and heat transfer.

$$J_{\mathrm{M}} = J_{\mathrm{H}} \text{ or } f/2 = [h/(C_{\mathrm{p}}G)](N_{\mathrm{Pr}})^{2/3}. \quad (9.15)$$

  - Where viscosity variation is considerable between wall and bulk fluid, the equation can be written as

$$J_{\mathrm{H}} = [h/(C_{\mathrm{p}}G)](N_{\mathrm{Pr}})^{2/3}(\mu_{\mathrm{b}}/\mu_{\mathrm{w}})^{0.14}. \quad (9.16)$$

  - The viscosity ratio factor can be neglected if fluid properties are evaluated at the mean temperature between bulk fluid and fluid at wall.
- Write Reynolds analogy equation between transfer of momentum and heat and name other analogies.
  - The relation between the heat transfer and the skin friction coefficient is expressed in terms of Stanton number and skin friction coefficient:
    - Stanton number is defined as

$$h/C_{\mathrm{p}}G = N_{\mathrm{Nu}}/(N_{\mathrm{Re}}N_{\mathrm{Pr}}). \quad (9.17)$$

    - Skin friction coefficient C_{f} is defined by $\tau_{\mathrm{w}}/[(1/2)(\rho v^2)]$, where τ_{w} is *wall shear stress.*
  - Thus, heat transfer coefficient can be related to skin friction coefficient by

$$N_{\mathrm{St}} \approx 1/2(C_{\mathrm{f}}) \text{ or } h \approx C_{\mathrm{p}}GC_{\mathrm{f}}/2. \quad (9.18)$$

  - This equation provides a useful estimate of h, based on knowing the skin friction, or drag.
  - This relation is known as the *Reynolds analogy* between shear stress and heat transfer. The Reynolds analogy is extremely useful in obtaining a first approximation for heat transfer in situations in which the shear stress is known.
  - This can be written as

$$\frac{\text{heat flux to wall}}{\text{convected heat flux}} = \frac{\text{momentum flux to wall}}{\text{convected momentum flux.}} \quad (9.19)$$

  - A modified Reynolds analogy has been obtained to take into consideration the fact that Prandtl number Pr is usually not equal to 1:

$$1/2(C_{\mathrm{f}}) = N_{\mathrm{St}} \cdot N_{\mathrm{Pr}}^{2/3}, \; 0.6 < N_{\mathrm{Pr}} < 60. \quad (9.20)$$

  - The following equation is often recommended for heat transfer in tubes:

$$N_{\mathrm{Nu}} = 0.023(N_{\mathrm{Re}})^{0.8}(N_{\mathrm{Pr}})^{0.33}(\mu_{\mathrm{b}}/\mu_{\mathrm{w}})^{0.14}. \quad (9.21)$$

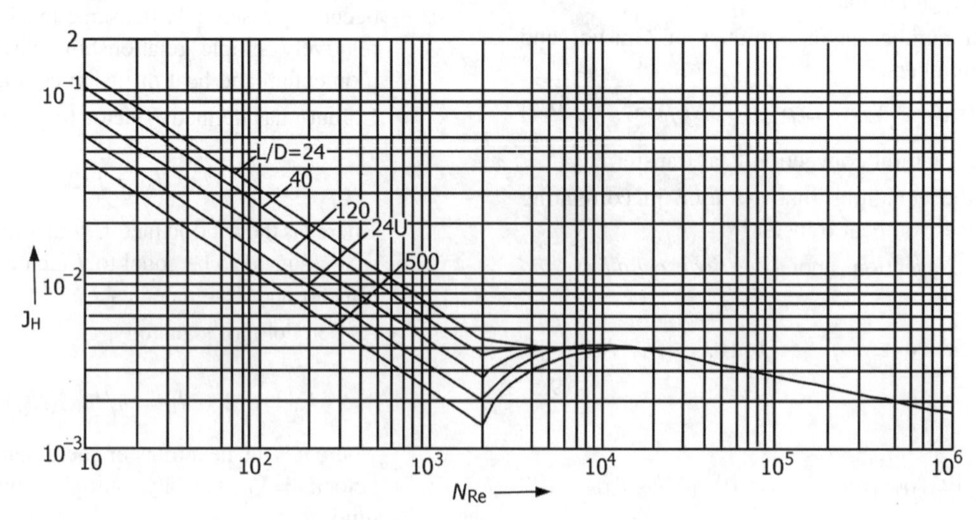

FIGURE 9.3 J_{H} factors as function of N_{Re}.

- Under what circumstances the term $(\mu_b/\mu_w)^{0.14}$ is of significance?
  - Viscosity of liquids decreases with increase in temperature. When viscosity of the liquids is high, for example, oils, there will be considerable variation in temperature of the bulk of the liquid and that of the wall. In other words, viscosities differ significantly with this temperature variation from *bulk* of the liquid and tube wall and the ratio differs significantly from unity. For low-viscosity liquids like water, this temperature gradient will be very small and therefore viscosity ratio tends to unity.
- Is the above equation applicable for laminar flow conditions?
  - No. This equation, called Dittus–Boelter equation, is for turbulent flow inside pipes. For laminar flow conditions, similar equation with different constant and exponent terms, Seider–Tate equation, is used.
  - Dittus–Boelter equation is applicable for $N_{Re} > 6000$. It should not be used for flow of liquid metals, which have abnormally low values of N_{Pr}.
  - It is applicable for fully developed flow conditions. Local values of h near entrance to tubes are much higher than those for fully developed flow.
  - It must be recognized that local values of h differ from average values given by the equation as temperatures and hence fluid properties differ from point to point along the length of the tube.
  - For gases effect of temperature on fluid properties are much less than for liquids as increase in k and C_p with temperature offset the increase in μ, giving a slight increase in h.
  - For liquids effect of temperature is much greater as variation in μ is much more rapid than k and C_p.
  - In practice unless the tube is very long with variation of local value of h is more than 2:1, using average value of h in the calculation of overall heat transfer coefficient, U, is adequate.
- Why higher coefficients are found when a liquid is being heated rather than cooled?
  - Reason is because of viscosity increases while cooling. For low-viscosity fluids, the ratio of μ/μ_w is not very important. However, for viscous fluids, for example, oils, μ_w and μ_{bulk} may differ by 10-fold.
- "Local heat transfer coefficients in flow through tubes are higher near the entrance region than those in regions of fully developed flow." *True/False*?
  - *True.* Reason is that turbulence effects are more near the entrance than inside tubes.
- Give an equation for Nusselt number for turbulent flow through tubes as function of friction factors, N_{Re} and N_{Pr}.
  - Petukhov–Kirillov equation for turbulent flow through tubes:

$$N_{Nu} = (f/2)\, N_{Re} \cdot N_{Pr}/[1.07 + 12.7(f/2)^{1/2}(N_{Pr}^{2/3} - 1)]. \quad (9.22)$$

  - Friction factor, f, can be calculated from

$$f = (1.58 \ln N_{Re} - 3.28)^{-2}. \quad (9.23)$$

  - This equation is claimed to predict heat transfer coefficients with 5–6% error in the range between $10^4 < N_{Re} < 5 \times 10^6$ and $0.5 < N_{Pr} < 200$ and in the range $0.5 < N_{Pr} < 2000$, with 10% error.
- What are the dimensionless groups involved in Seider–Tate equation? State its applications.
  - Gz and μ/μ_W.
  - Laminar flow inside pipes/tubes.
- Write Seider–Tate equation.

$$N_{Nu} = 1.86(N_{Gz})^{1/3}(\mu_b/\mu_w)^{0.14}, \quad (9.24)$$

  where N_{Gz}, Graetz number $= N_{Re}N_{Pr}D/L$ as given in Equation 9.2.
  - Seider–Tate equation is satisfactory for small diameters and small temperature differences.
- Give Hausen correlation for heat transfer in laminar flow.

$$hD/k = [3.65 + \{0.0668 N_{Re}N_{Pr}(D/L)\}/ \{1 + 0.04 + N_{Re}N_{Pr}(D/L)^{2/3}\}](\mu/\mu_w)^{0.14}. \quad (9.25)$$

  - This equation is one of the widely recommended equations for laminar flow inside tubes.
  - h is the mean coefficient for the entire length of the tube. Examination of the equation shows that the mean coefficient decreases with increasing length of the tube, L. This is a consequence of the build up of an adverse temperature gradient in laminar flow.
  - The value of L to be used is the length of a single pass, or in a U-tube bundle, the length of the straight tube from the tube sheet to the tangent point of the bend. In other words, the adverse temperature gradient is assumed to be completely destroyed by the strong secondary flow induced in the U-bend.
- Give an equation for liquid natural convection outside single horizontal tubes.

$$N_{Nu} = hD_o/k = 0.53[(D_o^3\rho^2 g\beta\,\Delta T/\mu^2)(C_p\mu/k)]^{1/4}, \quad (9.26)$$

where $\Delta T = T_w - T_b$ and β is the coefficient of volumetric expansion.

- Applicable for tube wall temperatures that are too low to initiate nucleate boiling.

- Is the following correlation for forced or for natural convention? How do you know?

$$N_{Nu} = 0.2\,[(N_{Gr}N_{Pr})^{1/4}/(L/\delta)]. \quad (9.27)$$

  - The above equation is for natural convection. N_{Gr}, Grashof number involves buoyancy forces that are based on density differences. In forced convection, buoyancy forces have limited influence.

- If the velocity of the fluid that is flowing in laminar flow in a pipe is doubled, how will the heat transfer coefficient change?
  - Heat transfer coefficient is proportional to $v^{0.4}$ for turbulent flow and to $v^{1/3}$ for laminar flow. In both the cases, h will increase by doubling velocity.
  - If flow changes over to turbulence conditions, h will be higher than change over, falling within the range of laminar conditions.

- Give an equation for the estimation of heat transfer coefficients to banks of tubes.

$$hD/k = C(DV_{max}\rho/\mu)^n \cdot N_{Pr}^{1/3}. \quad (9.28)$$

  - Values of the constants C and n depend on whether tube banks are arranged in-line or staggered.
  - In-line: $0.05 < C < 0.5$ and $0.55 < n < 0.8$.
  - Staggered: $0.2 < C < 0.6$ and $0.55 < n < 0.65$.
  - V_{max} = maximum fluid velocity and μ = viscosity.
  - In-line and staggered arrangements are illustrated in Figure 9.4.

- Heat transfer coefficient for turbulent flow is somewhat greater for a pipe than for a smooth tube. Explain why?
  - For a smooth tube like copper, the fluid film will have nearly the same thickness, gets least disturbed, and near laminar flow conditions exist.
  - For a pipe, surface roughness is higher and hence fluid flowing near the wall gets disturbed by the local surface projections, that is, roughness. This increases local turbulence, thereby increasing heat transfer coefficients. It should be noted that pressure drop also increases.

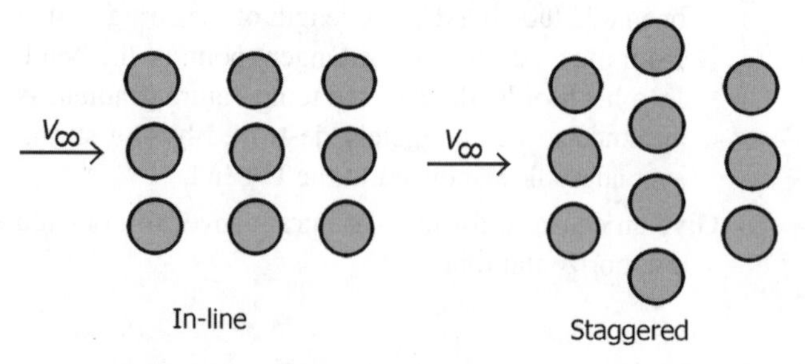

FIGURE 9.4 In-line and staggered tube arrangements.

- What is the effect of vibrations and pulsations on heat transfer coefficients?
  - Vibrations/pulsations increase turbulence that results in increased heat transfer coefficients.

- What is the difference between film coefficients and overall coefficients?
  - Film coefficient involves the resistance for heat transfer offered by the individual fluid *film*. It is assumed that all the resistance offered by the fluid lies in a fictitious film of the fluid. It is a direct measure of heat transfer rate per unit time per unit area per unit temperature difference.
  - Overall coefficient includes all the resistances for heat transfer between two fluids separated by a solid barrier that includes the metal wall and the solid deposits on both sides of the metal wall in a heat exchanger. In other words, overall coefficient gives a measure of the combined heat transfer rate through a series of barriers consisting of two fluid films, metal wall, and two solid deposits on both sides of the metal wall.

- Under what circumstances overall heat transfer coefficient can be approximated to an individual film coefficient?
  - When one of the film coefficients is very large compared to the other and resistances offered by the metal and other solid boundaries (deposits) are negligible, overall coefficient can be approximated to the smaller of the two individual film coefficients.
  - For example, when one of the fluids is a viscous liquid that is heated by condensing steam and the metal wall is clean and made out of a high-conductivity metal or alloy, overall coefficient can be approximated to the coefficient on the viscous fluid side.

- Under what circumstances the ratio of inside and outside film coefficients for tubular heat exchangers is approximately equal to the ratio of inside and outside surface areas of tubes?
  - When the heat transfer resistances are approximately equal, that is,

$$1/A_i h_i \approx 1/A_o h_o \text{ or } A_o/A_i \approx h_i/h_o. \quad (9.29)$$

  - Example is in applications where water–water or same fluid heat exchange is involved.

9.1.2 Condensation

- What are the different modes of condensation? Explain.
  - *Dropwise Condensation*: In dropwise condensation, drops of liquid are formed from vapor at particularly favored locations, called nucleation sites, on a solid surface. These locations can be in the form of pits or any other surface irregularities that may be thousands in number per unit surface area.
  - The drops grow by continued condensation and coalesce with adjacent drops, forming larger drops that detach from the solid surface by the action of gravity or by sweeping action by the moving vapors, clearing the surface and exposing it to vapor.
  - Dropwise condensation occurs on surfaces that are not readily wettable, such as oily or extremely smooth surfaces and the condensate does not spread on the surfaces and detach once they grow to be influenced by gravity or by sweeping action by the vapors.
  - There will not be any film of liquid on the surface and drops occupy very small proportion of the surface, leaving most of the surface available for direct contact with the vapor. This, coupled with the sweeping action of the detaching drops, gives very high heat transfer coefficients. The average value of h for this process may be 5–10 times more than the values obtainable for film condensation under similar conditions. For condensation of pure vapors, h will be as high as 114 W/(m^2 °C).
  - Although dropwise condensation gives high heat transfer coefficients, its formation is unstable and unpredictable under process conditions.
  - *Film Condensation*: In film condensation, the initially formed drops quickly coalesce and spread as a film on the solid surface. The film acts as a barrier for heat transfer to condense more liquid. Thus, it offers more resistance for heat transfer than dropwise condensation. Consequently, heat transfer coefficients are far less for film condensation than for dropwise condensation. Film condensation is the normal mode in practice and therefore assumed in heat transfer estimations for design purposes.
  - Comparison between dropwise and film-type boiling is illustrated by Figure 9.5.
  - *Direct Contact Condensation:* In direct contact condensation, the coolant liquid is sprayed into the vapor, which directly condenses on the sprayed coolant droplets. Solid surface for condensation is eliminated making heat transfer highly efficient, but results in mixing of the condensate and coolant. It is used where the coolant and condensate are of the same material or they become immiscible layers that are separated by settling processes.

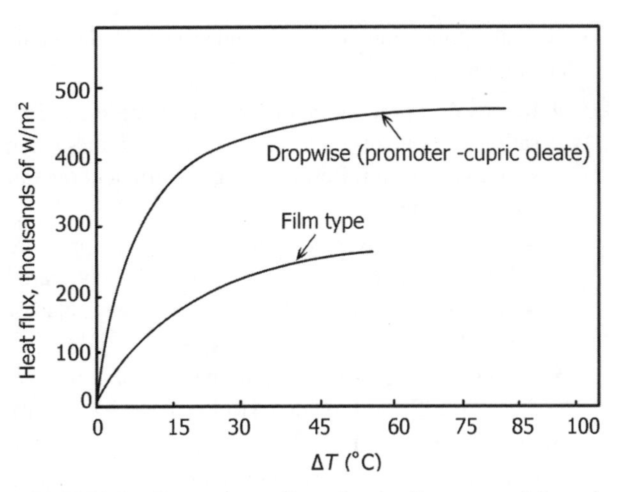

FIGURE 9.5 Comparison of heat flux for film-type and dropwise condensation.

  - *Homogeneous Condensation:* In homogeneous condensation, the liquid phase forms directly from supersaturated vapors. For condensation to occur this way, microscopic contaminants should be present in the vapors to act as nuclei for initiating condensation process as in the case of rain. In industrial equipment, this process is important in fog formation in condensers.
- How is dropwise condensation achieved?
  1. By adding a promoting chemical into the vapor.
  2. Treating the surface with a promoting chemical. Promoters include waxes, fatty acids like oleic, stearic, and so on.
  3. Coating the surface with a polymer like Teflon, or a noble metal like gold, silver, rhodium, palladium, or platinum.
- What is freeze condensation? Explain.
  - Heat transfer surfaces in a condenser are maintained below the freezing point of the vapors.
  - Process vapors or steam solidify on the heat transfer surfaces through (a) condensation followed by freezing or (b) directly by means of deposition.
    - Ethylene glycol, for example, freezes at approximately −13°C, so if the tube wall temperature is −12°C, ethylene glycol will condense and then freeze onto the tube wall. Most applications involve condensation followed by freezing.
    - Deposition occurs when vapors change directly to ice without passing through the liquid phase (opposite of sublimation). For example, freeze condensers for edible oil deodorization operate on the deposition principle.

- Give an application of freeze condensation and state its advantages.
  - Removal of process vapors by either freeze condensation or condensation ahead of a vacuum system such as an ejector. Results in a more efficient overall vacuum system.
  - Substantially lower consumption of high-pressure motive steam (e.g., 500 kg/h (1100 lb/h) compared to 4536 kg/h (10,000 lb/h) for conventional ejector system).
  - Lower cooling water requirements, for example, $28\ m^3/h$ (125 gpm) compared to $455\ m^3/h$ (2000 gpm).
  - Much smaller and easier to maintain ejector system (largest ejector is 3–4 m long compared to 12 m long one).
  - Easy maintenance and accessibility as ejectors are mounted horizontally unlike conventional ones.
  - Less environmental impact because far less wastewater is produced, for example, 500 kg/h (1100 lb/h) compared to 4536 kg/h (10,000 lb/h).
  - Flexible operation making possibilities for future expansion.
- What is Γ? What are the units of Γ appearing in Nusselt equation for condensation?
  - The Reynolds number of the condensate film (falling film) is $4\Gamma/\mu$, where Γ is the weight rate of flow (loading rate) of condensate per unit perimeter kg/(s m).
  - The thickness of the condensate film for Reynolds number less than 2100 is $(3\mu\Gamma/\rho^2 g)^{1/3}$.
- Define Reynolds number for condensate flow.

$$N_{\text{Re}} = D_{\text{h}} u_{\text{av}} \rho_{\text{l}} / \mu_{\text{l}}, \tag{9.30}$$

  where D_{h} is the hydraulic diameter for condensate flow = 4(cross-sectional area for condensate flow)/wetted perimeter = $4A/P$.

$$\begin{aligned} P &= \pi D \text{ (wetted perimeter), for vertical tube of outside diameter } D \\ &= 2L, \text{ for horizontal tube of length } L \\ &= W, \text{ for vertical or inclined plate of width } W. \end{aligned} \tag{9.31}$$

- Give Nusselt equations for condensation on horizontal single tubes and banks of tubes.
  - Horizontal single tubes:

$$hd/k = 0.725[D^3 \rho^2 g/\mu\Gamma]^{1/3}. \tag{9.32}$$

  - Condensation is in filmwise mode and the film flows under the influence of gravity to the bottom of the tube and drips off on to the next tube below it. The flow is almost always laminar as the flow path is too short for turbulence to develop, especially so for viscous fluids. For low-viscosity fluids, turbulent flow is possible.
  - If the vapor flow rate is high, vapor shear action in cross-flow on the condensate film becomes significant in blowing off the condensate, carrying it downstream as a spray. This shear effect makes the film turbulent even earlier than it would have under the influence of gravity alone.
  - Heat transfer coefficients for dropwise condensation may be as much as 10 times as high as the rates for film condensation.
  - Horizontal tube banks with N tubes directly falling one below the other so that condensate from tube above falls on tube below without splashing under laminar conditions:

$$h_{\text{N}} = [h(_{\text{top tube}})]^{-1/4}, \tag{9.33}$$

  where h_{N} is average coefficient.
    - h for top tube will be higher than that for tube below and so on, as liquid film thickness increases due to liquid from top tube falling on tube below creating an *accumulating effect*, theoretically speaking as shown in Figure 9.6.
    - There is a limit to increase in film thickness as gravity effects take over, facilitating drainage of liquid from tube surface, after about 3–4 rows of tubes, directly falling one below the other.
- Give the corresponding Colburn-type equation for horizontal tubes.

FIGURE 9.6 Condensation on horizontal tube banks.

$$(h/C_pG)(N_{Pr}) = 4.4/(4\Gamma/\mu), \tag{9.34}$$

where $G = \Gamma/(3\mu\Gamma/\rho^2 g)^{1/3}$.

- Write Nusselt and corresponding Colburn-type equations for condensation on vertical tubes.
  - *Nusselt*:

$$hL/k = 0.943[L^3\rho^2 g\lambda/k\mu\,\Delta T]^{1/4}. \tag{9.35}$$

  - *Colburn Type*:

$$(h/C_pG)(N_{Pr}) = 5.35/(4\Gamma/\mu). \tag{9.36}$$

- Explain the applicability of the above equation.
  - The equation for vertical tubes is derived on the assumption that the condensate flows under laminar conditions.
  - For long tubes, the condensate film becomes progressively thick and flows at much higher velocities, creating turbulent conditions in the lower part of the tube.
  - It has been observed that experimental values of h were found to be about 20% more than those predicted from the equation. This is attributed to the formation of ripples on the falling film surface.
- Under what circumstances vertical tubes are preferred over horizontal tubes for condensation?
  - Vertical tubes are preferred over horizontal tubes when the condensate must be subcooled below its condensation temperature.
  - Another reason involves layout considerations. Floor space requirements are much more for horizontal exchanger than for vertical one. On the other hand, vertical layout increases headroom with reduced accessibility for maintenance, if the tube lengths are large.
- Illustrate by means of a diagram the flow of condensate on vertical plates.
  - Figure 9.7 shows transition of condensate flow from laminar to turbulent, through wavy laminar intermediate flow.
  - The condensate film begins at the top and flows down under the influence of gravity, adding more condensate as it flows.
  - The flow is laminar and wave-free up to a certain distance, turns wavy, remaining laminar as shown, and finally turns turbulent at an approximate Reynolds number of about 2000. Figure 9.7 illustrates this point.
  - The profile is based on the assumptions that the vapor temperature is uniform and at the saturation value, gravity flow is the only external force acting on the film and there is no drag by the vapor on it.

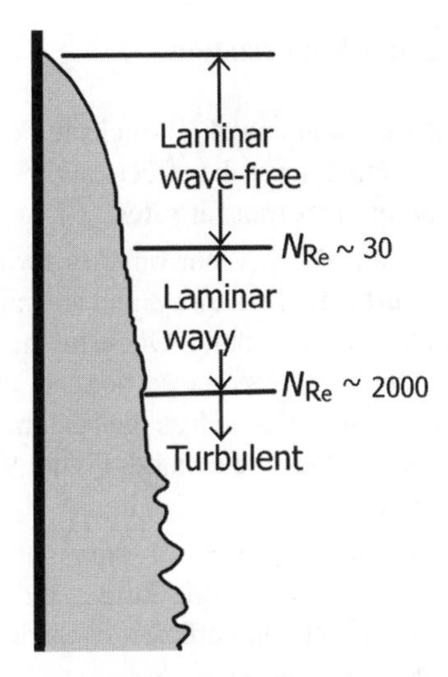

FIGURE 9.7 Film condensation on a vertical plate.

- What is the effect of ripples formed on the condensate film on heat transfer rates?
  - The ripples increase heat transfer, increasing interfacial area and reducing film thickness.
- "Condensation is a constant pressure process." *True/False*? Comment.
  - *True.* Friction losses are small therefore it can be assumed that pressure remains constant on the condensing side of a tube.
- What is the order of h_i/h_o in cases involving heat transfer in film condensation outside tubes with cooling water circulating inside tubes?
  - Normally 5–10.
- What is the effect of contamination of noncondensable gases in condensable vapors on condensing film coefficients?
  - The contamination of the condensing vapor by gases or vapors, which do not condense under the condenser conditions, can have a profound effect on overall coefficients. Presence of air in condensing steam is a common example. Noncondensable gases affect condensing heat transfer rates adversely. Condensing film coefficients are of the order of 200–2000 times those of natural convection or low-velocity gas film coefficients.
  - If noncondensable gases accumulate as a film on condensing surfaces, there will be drastic reduction in condensing film coefficients.
- "Heat transfer coefficients for dropwise condensation are more than those for film condensation by a factor of 10." *True/False*?
  - *True.*

9.1.3 Boiling and Vaporization

- Explain the phenomena of (i) nucleate boiling and (ii) film boiling and discuss the effects of these two boiling phenomena on heat transfer rates.
  - In *nucleate boiling*, vapor bubbles form at the heat transfer surface, break away, and are carried into the mainstream of the fluid. Once in the fluid mainstream, the bubbles collapse because the bulk fluid temperature is not as high as the heat transfer surface where the bubbles are created. Figure 9.8 illustrates nucleate boiling.
  - This heat transfer process is sometimes desirable as the energy imparted to the fluid at the heat transfer surface is quickly and efficiently carried away.
  - When the pressure of a system drops or the flow decreases, the bubbles cannot escape as quickly as they are formed at the heat transfer surface.
  - Similarly, if the temperature of the heat transfer surface increases, more bubbles are formed than can be efficiently carried away. The bubbles then grow and coalesce with the neighboring bubbles covering small areas of the surface with vapor film. This is known as *partial film boiling*.
  - Since vapor has lower heat transfer coefficient than liquid, the vapor patches on the heat transfer surface act as insulation, making heat transfer more difficult.
  - As the area of the heat transfer surface covered with vapor increases, temperature of the surface rapidly increases, while heat flux from the surface decreases.
  - This unstable situation continues until the surface is covered by a stable blanket of vapor, preventing contact between the surface and the liquid.
  - This condition after the stable vapor blanket has formed is called *film boiling*.

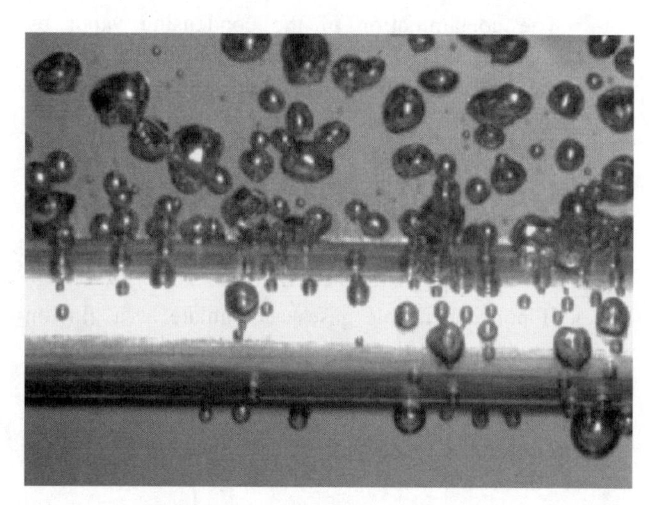

FIGURE 9.8 Nucleate boiling illustrated.

  - Although film boiling is not desirable because of low heat transfer coefficients involved, it is unavoidable in some situations as in the vaporization of low boiling liquids or in cryogenic vaporizers.
  - Film boiling might appear attractive from fouling and corrosion point of view but is impractical due to variation in operating conditions during start-up or shutdown and the fluctuations in the condition of the film in the transition region.
- Give reasons why a rough surface gives higher boiling heat transfer coefficients than a smooth surface.
  - Rough surfaces provide cavities for bubble formation and give increased turbulence.
  - Sometimes the surfaces are made rough by etching, scratching, sand blasting, or other means.
  - Heat transfer coefficients for a rough surface may be 10–50 times higher than for a smooth surface at a given ΔT.
- What is a Linde boiling surface?
  - The Linde surface consists of a thin layer of porous metal bonded to the heat transfer substrate.
  - Bubble nucleation and growth are promoted within a porous layer that provides a large number of stable nucleation sites of a predesigned shape and size.
  - Figure 9.9 illustrates Linde surface.
  - Linde porous surface in industrial boiling practice can be traced to the use of pumice stones in a beaker in laboratory practice while boiling a liquid.
  - Microscopic vapor nuclei in the form of bubbles entrapped on the heat transfer surface must exist in order for nucleate boiling to occur.
  - Surface tension at the vapor–liquid interface of the bubbles exerts a pressure above that of the liquid. This excess pressure requires that the liquid be superheated in order for the bubble to exist and grow.
  - The porous surface substantially reduces the superheat required to generate vapor.

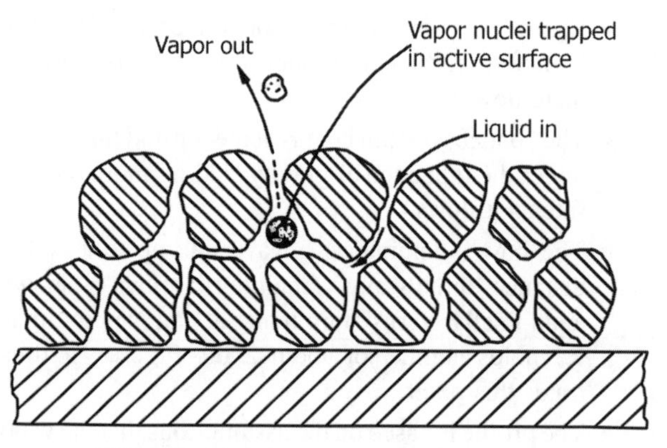

FIGURE 9.9 Linde porous boiling surface.

- The entrances to the many nucleation sites are restricted in order to retain part of the vapor in the form of a bubble and to prevent flooding of the site when liquid replaces the escaping bubble.
- Many individual sites are also interconnected so that fresh liquid is continually supplied.
- Boiling heat transfer coefficients, which are functions of pore size, fluid properties, and heat flux, are 10–50 times greater than smooth surface values at a given temperature difference.
- Nucleate boiling exists at much lower temperature differences than required for smooth surfaces.

• Illustrate boiling regimes by means of the boiling curve, ΔT versus heat flux.

- Figures 9.10 and 9.11 illustrate pool boiling curve and stages involved in pool boiling curve.
- *Natural Convection Boiling* (AB): As the temperature of the heater is increased from the initial bulk value, the liquid close to the heated surface rises and is replaced by cooler fluid. Since no vapor is formed in the heating process, the heat transfer is governed by natural convection. Transport by natural convection occurs until the heater temperature is slightly greater than the boiling/saturation temperature (2–6°C for water). Liquid is slightly superheated and vaporizes when it reaches the surface.
- *Nucleate Boiling* (B′C): When the surface temperature is increased beyond the boiling point, small vapor bubbles form at points along the heater surface. In Figure 9.10, this temperature difference corresponds to point B′, which represents the *onset of nucleate boiling*. The bubbles formed along the heater surface collapse as they move and grow into the colder subcooled liquid. The growth and collapse of the bubbles causes increased turbulence from that found in the natural convection process. When bubble collapses, liquid rushes to occupy the space originally occupied by the bubble. This increased turbulence along with the latent heat transported by the vapor bubbles increases surface heat transfer. Therefore, the nucleate boiling region is accompanied by a substantial increase in heat transfer from that for natural convection as shown in Figure 9.10.
- Between B′ and C, the nucleate boiling, that is, hot surface temperature further increases, increasing rate of bubble formation. Bubbles rise as numerous continuous columns of vapor in liquid, reach free surface and break up, releasing vapor from liquid. The curve has a very steep slope, ranging from 2 to 4 and heat

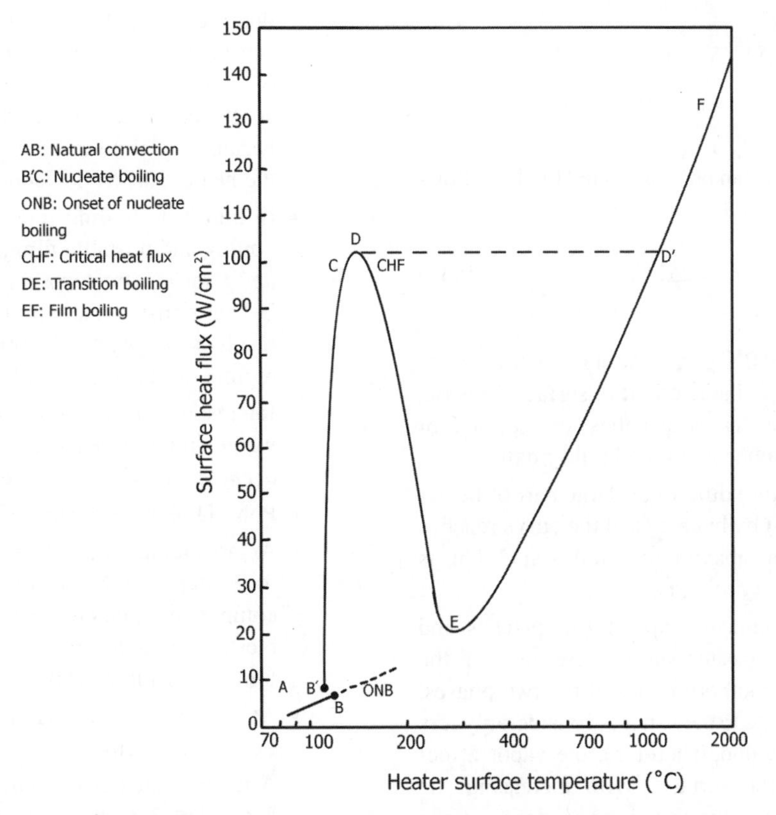

FIGURE 9.10 Pool boiling curve for water at atmospheric pressure.

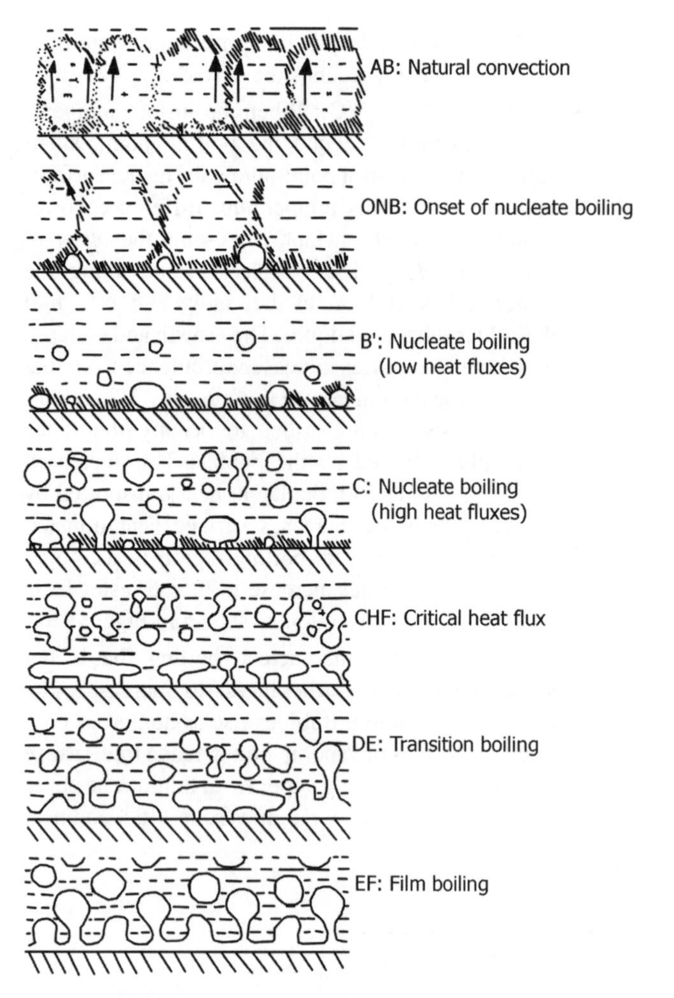

FIGURE 9.11 Stages in pool boiling curve.

transfer coefficient can be represented by the following equation:

$$h = a\Delta T^{m}, \quad (9.37)$$

where $\Delta T = T_w - T_{sat}$.

- The position of the B′C part of the curve is affected by the solid surface characteristics, surface tension, pressure, dissolved gases or solids, or presence of high boiling components in the liquid mixture.
- At large temperature difference, large part of heater surface is covered by bubbles and the curve reaches C, which represents maximum heat flux at C. This is called *critical heat flux* (CHF).
- At the critical point, the vapor forms patches and columns near the heater surface. Because of the differences in thermal properties of the two phases, the heat transfer rate to the vapor is considerably less than that to the liquid. Therefore, the vapor effectively insulates the surface. With increasing the surface temperature, the vapor covers more of the heating surface until a maximum value for the heat transfer rate is reached.
- Beyond the critical heat flux, two possibilities exist depending on the heating conditions. If the surface heat flux is controlled and increased beyond critical point, the surface temperature increases dramatically as shown by the dashed line from point D to D′ in Figure 9.10.
- The temperature at D′ is often higher than the maximum temperature that heater surface can maintain, and thus this heat flux is referred to as the *burn out point*. Surface temperature at E is beyond melting point of most materials.
- If the surface temperature is controlled and increased beyond the temperature at D′, the insulating effect continues and the heat transfer rate decreases. This regime, called *transition boiling*, is characterized by the unstable vapor blanket that covers the surface. The vapor blanket collapses periodically and allows the fluid to contact the surface. This periodic motion results in large variations in surface temperature and a highly unstable flow.
- The periodic contacts between liquid and heated surface in the transition boiling region of the boiling curve result in the formation of both large amounts of vapor, which forces liquid away from the surface, and creates an unstable vapor film or blanket. Because of this, the surface heat flux and the surface temperature can experience variations both with time and position on a heater. However, the average heat transfer coefficient decreases as the temperature increases, because the time of contact between the liquid and the heater surface is decreased.
- In transition boiling region, both unstable nucleate boiling and unstable film boiling alternately exist at any given location on a heating surface. The variation in heat transfer rate with temperature is primarily a result of a change in the fraction of time each boiling regime exists at a given location.
- Interest in transition boiling regime arises because of its potential importance during a *loss of coolant accident* (LOCA) in nuclear reactors.
- Point D represents start of film boiling.
- At large temperature differences, greater than 250°C for water at atmospheric pressure, the vapor film completely blankets the heated surface. Vapor bubbles are released regularly from the surface and the film is considered stable.
- At E, heat flux reaches minimum. This point is called Leidenfrost point.
- After E, heat transfer coefficient increases mainly due to the effect of radiation particularly at low flows,

low void fractions, and surface temperatures in excess of 700°C.

- What is critical heat flux in boiling heat transfer?
  - Maximum heat flux achievable with nucleate boiling is called critical heat flux.
- Give an equation for heat transfer at critical heat flux conditions.

$$Q_{cr} = 0.18\lambda\rho_v[g\sigma g_c(\rho_l-\rho_v)/\rho_v^2]^{1/4}. \qquad (9.38)$$

  - The above equation is for flat plates and, depending on geometry, the constant varies between 0.12 and 0.2, depending on the dimensionless parameter, $L[g(\rho_l - \rho_v)/g_c\sigma]$, where L is radius of surface or length of plate.
  - This equation is due to Cichelli and Bonilla.
- Explain why heat transfer rates fall off above critical heat flux in boiling.
  - Above critical flux, high rate of vapor generation leads to film formation (blanketing).
  - Dry patches develop, lowering heat transfer rates.
- Define *pool boiling*, *subcooled or local boiling*, and *saturated or bulk boiling*.
  - *Pool boiling* takes place when the heated surface is submerged below a free surface of liquid.
  - *Subcooled or local boiling* takes place under conditions of temperature of the liquid is below saturation temperature.
  - *Saturated or bulk boiling* involves boiling when the liquid is maintained at saturation temperature.
- Explain the heat transfer mechanisms involved in nucleate pool boiling. Illustrate.
  - *Bubble Agitation:* The systematic pumping motion of the growing and departing bubbles agitates the liquid, pushing it back and forth across the heater surface, which in effect transforms the otherwise natural convection process into a localized forced convection process. Sensible heat is transported away in the form of superheated liquid and depends on the intensity of the boiling process. Figure 9.12 illustrates bubble agitation process.

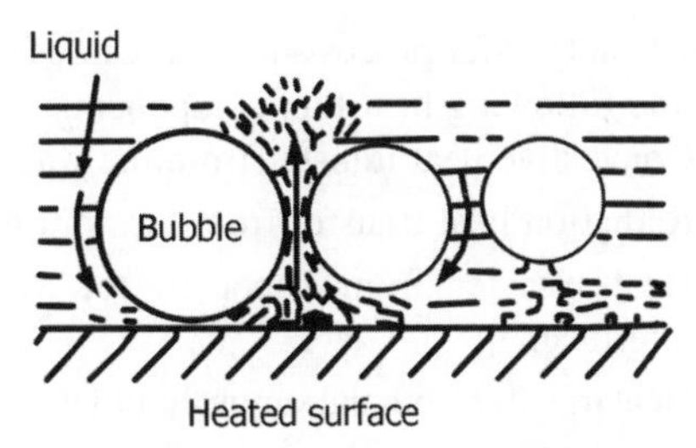

FIGURE 9.12 Bubble agitation.

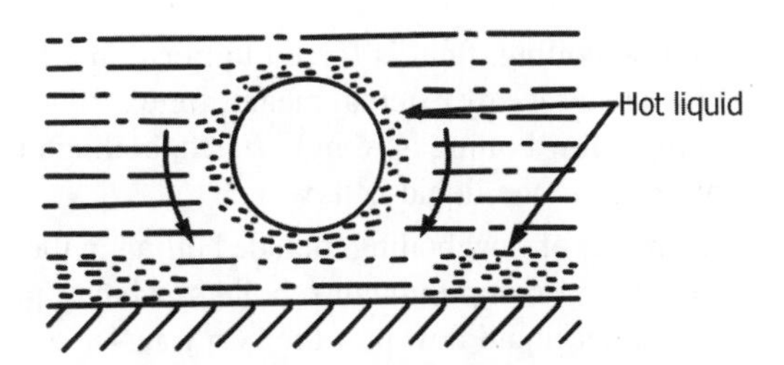

FIGURE 9.13 Vapor–liquid exchange.

  - *Vapor–Liquid Exchange:* The wakes of departing bubbles remove the thermal boundary layer from the heated surface and create a cyclic thermal boundary layer stripping process. Sensible heat is transported away in the form of superheated liquid, whose rate of removal is proportional to the thickness of the layer, its mean temperature, the area of the boundary layer removed by a departing bubble, the bubble departure frequency and the density of active boiling sites. Figure 9.13 illustrates vapor–liquid exchange.
  - *Vaporization* Illustrated in Figure 9.14. Heat is conducted into the thermal boundary layer and then to the bubble interface, where it is converted to latent heat. Macrovaporization occurs over the top of the bubble while microvaporization occurs underneath the bubble across the thin liquid layer trapped between the bubble and the surface, the later often referred to as *micro-layer vaporization*. Since bubbles rise much faster than the liquid natural convection currents and contain a large quantity of energy due to the latent heat absorbed by the bubble making this heat transfer mechanism very efficient. The rate of latent heat transport depends on the volumetric flow of vapor away from the surface per unit area.
- What is the difference between *pool boiling* and *flow boiling*?
  - Pool boiling involves boiling of stationary liquids. Fluid is not forced to flow by means of a pump. Any motion of the fluid is by natural convection currents and motion of bubbles under the influence of buoyancy. Heat transfer coefficient for pool boiling increases with increase of surface roughness.

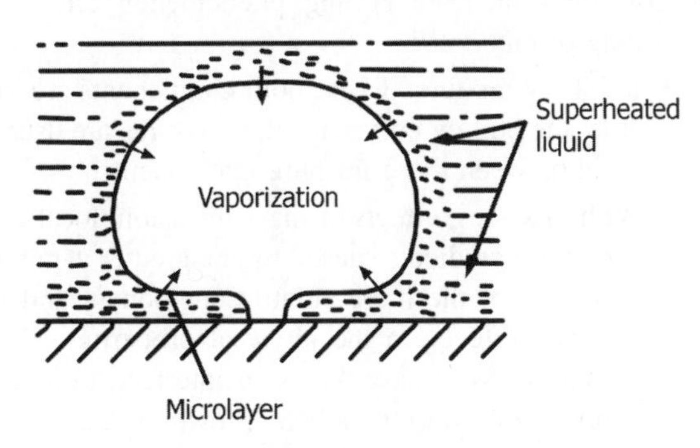

FIGURE 9.14 Vaporization.

- In flow boiling, fluid is forced to move in a heated pipe or surface by external means, for example, by a pump. Flow boiling is, simply stated, boiling taking place while the liquid is flowing.
- Examples of flow boiling include boiling in thermosiphon and pump-through reboilers, which operate with a net liquid flow past the heat transfer surface. Under these conditions, shear stress on the layer of liquid immediately adjacent to the heat transfer surface will influence the boiling process.
- In flow boiling, nucleation processes are suppressed and liquid gets superheated and transported from the tube wall by turbulent eddies to the vapor–liquid interface, where vaporization takes place.
- Heat transfer coefficients are much higher in flow boiling processes compared to nucleate boiling processes.
- Film boiling is also possible under forced convection vaporization if the tube wall temperatures are high. In these cases, sometimes mist flow is encountered in which vaporization takes place from suspended droplets in the superheated vapor.
- In such cases, heat transfer coefficients become very low and design of thermosiphon and pump-through reboilers avoid conditions of high tube wall temperatures.

- What is Leidenfrost phenomenon?
  - Leidenfrost noted that when liquids were spilled/placed on very hot surfaces, drops were formed that did not contact the surface but floated above it and slowly vaporized. When surface temperature was reduced below a certain value, the drops contacted the surface and rapidly vaporized. This phenomenon is called Leidenfrost phenomenon and point E on the boiling curve (Figure 9.10) is called Leidenfrost point.
  - Temperature at point E is known by different terms that include Leidenfrost temperature, minimum film boiling temperature (T_{MFB}), rewetting temperature, quench temperature, and film boiling collapse temperature.
- How does nucleate boiling phenomenon affect the boiling of mixtures?
  - In binary mixtures where both components are volatile, the boiling curves for the mixtures are usually lying between those for pure components.
  - With mixtures, effects of mass diffusion, local concentration gradients caused by the greater vaporization rates of the more volatile component and the resultant effects on the physical properties of the mixture as well as changes in interface saturation temperatures during bubble growth influence the boiling curves.
  - If one of the components has very high boiling point, essentially making it nonvolatile, the effect of increasing its concentration is to shift the curve B′C to the right reducing heat transfer rates. Accumulation of this heavy component in the nuclei cavities can make them inactive for boiling. For this reason, sometimes, special surfaces are used to provide washing effect of the circulating fluid on the heavy components from the cavities.
- What are the additional factors involved in nucleate and film boiling phenomena inside tubes?
  - The vapor and liquid inside a tube must travel together. The pattern of the resulting two-phase flow affects both heat transfer and pressure drop, because of the changing vaporization and relative vapor–liquid loads.
  - In film boiling, while the liquid does not contact the tube surface, it will be in one of the following forms:
    - A dispersed spray of droplets, normally encountered at void fractions in excess of 80% (liquid-deficient or dispersed flow film boiling regime).
    - A continuous liquid core (surrounded by a vapor annulus that may contain entrained droplets) usually encountered at void fractions below 40% (inverted annular film boiling or IAFB regime).
    - A transition between the above two cases, which can be in the form of an inverted slug flow for low to medium flow.
  - Changes in pressure drop along the tubes cause changes in boiling point of the liquid.
- What is *inverted annular film boiling* (IAFB)? Describe heat transfer processes during inverted annular film boiling.
  - IAFB refers to the film boiling type characterized by a vapor layer separating the continuous liquid core from the heated surface.
  - In the inverted annular flow regime, few entrained droplets are present while the bulk of the liquid is in the form of a continuous liquid core that may contain entrained bubbles.
  - At dry out, the continuous liquid core becomes separated from the wall by a low-viscosity vapor layer.
  - The heat transfer process in IAFB can be considered by the following heat flux components:
    - Convective heat transfer from the wall to vapor.
    - Radiation heat transfer from the wall to liquid.
    - Heat transfer from vapor to the vapor–liquid interface.
    - Heat transfer from the vapor–liquid interface to the liquid core.

- What is *slug flow film boiling*?
  - *Slug flow film boiling* is usually encountered at low flows and void fractions that are too high to maintain inverted annular film boiling but too low to maintain dispersed flow film boiling.
- What is *dispersed flow film boiling*?
  - The *dispersed flow film boiling* regime is characterized by the existence of discrete liquid drops entrained in a continuous vapor flow.
  - This flow regime may be defined as dispersed flow film boiling, liquid-deficient heat transfer, or mist flow.
  - It is of importance in nuclear reactor cores for off-normal conditions as well as in steam generators.
  - The dispersed flow film boiling regime usually occurs at void fractions in excess of 40%.
- What are the mechanisms involved in convective boiling in a vertical tube? Illustrate.
  - The mechanisms occurring in convective boiling are quite different from those occurring in pool boiling. The following conditions occur in different parts of the tube as shown in Figure 9.15.
    - Subcooled single-phase liquid region.
    - Subcooled boiling region.
    - Saturated boiling region.
    - Dry wall region.
  - The saturated boiling region is the most important region encountered in design.
- What are the types of special surfaces used to enhance boiling side heat transfer coefficients?
  - Finned surfaces, specially plated surfaces, and porous surfaces produced by electroplating, sintering, or machining.
- Is there any advantage in using finned tubes for boiling heat transfer?
  - Yes. Finned surfaces provide sites for nucleate boiling, enhancing nucleation process, besides giving enhancement of surface area per unit length, which makes the equipment more compact. Shorter tube lengths can be used with decrease in pressure drop inside tubes. Layout costs will be less due to savings in space requirements.
  - Such surfaces typically have 16–40 fins/in. with approximately 1/16 in. height. Increase in area ratio of about 2.2–6.7 over plain tubes.
  - Main application of enhanced tubes is in boiling clean liquids at low temperature differences. Not advantageous for wide boiling range mixtures. At high ΔT, the relative performance of enhanced tubes and plain tubes is less compared to small ΔT.

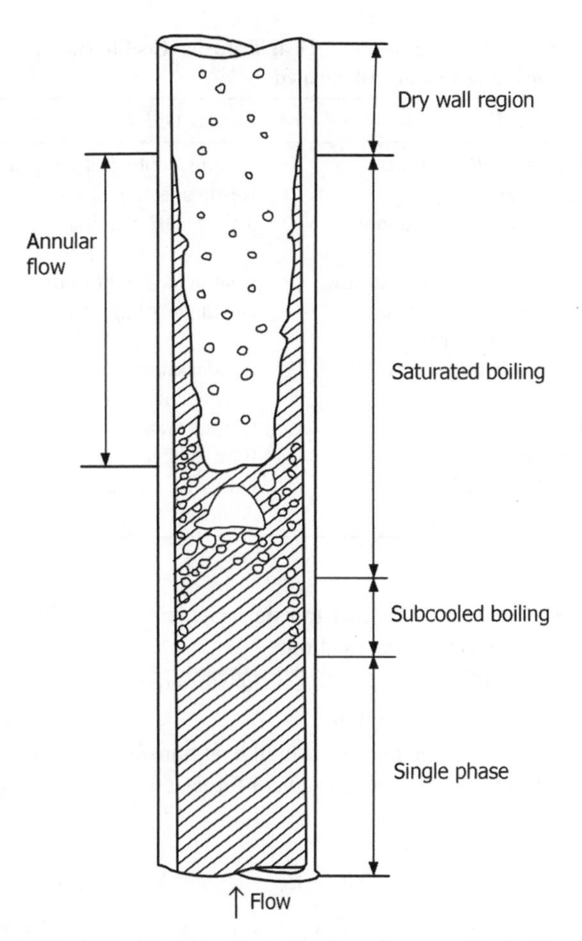

FIGURE 9.15 Heat transfer mechanisms in convective boiling in a vertical tube.

  - For a given duty, the required ΔT will be smaller than for plain tubes.
- What are the considerations involved in selecting enhanced tubes for boiling heat transfer?
  - These are expensive and evaluation of cost comparisons must be made.
  - As finned tubes are available in limited materials of construction, the corrosion characteristics of the fluids need to be evaluated.
  - The boiling range and fouling characteristics of the liquids influence the selection. Ability to clean the surfaces is of primary concern.

9.1.4 Heat Transfer Coefficients: Magnitudes and Data

- Arrange the following cases in the order of increasing heat transfer coefficients.
  - The given order and correct order are shown in Table 9.1.

TABLE 9.1 Magnitude of Heat Transfer Coefficients in Increasing Order are Illustrated

Given Order	Correct Order
Heating/cooling of oils	Heating/cooling of air/gases
Boiling water	Desuperheating steam
Dropwise condensation of steam	30% NaOH solution
Film condensation of steam	Heating/cooling of oils
Condensation of steam containing air	Heating/cooling of water
Desuperheating steam	Condensation of steam containing air
Heating/cooling of water	Boiling water
Heating/cooling of air/gases	Film condensation of steam
30% NaOH solution	Dropwise condensation of steam

- Give ranges of heat transfer coefficients for different categories of applications.
  - Table 9.2 presents heat transfer coefficients for different applications.
- Give the equation for overall heat transfer coefficient in terms of different resistances for heat transfer in tubular heat exchangers.

$$U_o = 1/[1/h_o + R_{fo} + R_{fins} + \Delta x_w A_o/k_w A_m + R_{fi} A_o/A_i + 1/h_i A_o/A_i], \quad (9.39)$$

where h_o and h_i are outside and inside film coefficients, respectively; R_{fo} and R_{fi} are outside and inside fouling resistances, respectively; Δx_w and k_w are wall thickness and wall thermal conductivity, respectively; R_{fins} is resistance due to fins (if any); and A_o and A_i are outside and inside tube surface areas, respectively.

TABLE 9.2 Heat Transfer Coefficients for Different Applications

Application	h (W/(m^2 °C))
Free convection in air/gases	1–50
Forced convection in gases	25–250
Superheated steam	30–100
Free convection in liquids	10–1,000
Forced convection in liquids	50–20,000
Oils: heating or cooling	50–1,500
Water: heating or cooling	300–20,000
Liquid metals	5,000–250,000
Steam: film condensation	6,000–20,000
Steam: dropwise condensation	30,000–100,000
Boiling water	1,700–50,000
Condensing organic vapors	1,000–2000
Boiling/condensation	2,500–100,000

Note: Ranges taken from different sources.

  - U is commonly referred to A_o, the total outside tube heat transfer area, including fins (if any).
  - Mean wall heat transfer area A_m is given by

$$A_m = \pi L(D_i + D_o)/2. \quad (9.40)$$

  - If U_i, based on inside heat transfer area, is preferred over U_o, the following relationship holds:

$$U_o A_o = U_i A_i \text{ or } U_o/U_i = A_i/A_o = D_i/D_o. \quad (9.41)$$

  - Identification of the reference area, that is, whether the heat transfer coefficient is based on the outside or inside surface area, should be specified.

$$U_o \neq U_i, \text{ unless } A_o = A_i. \quad (9.42)$$

  - Temperature gradients through hot fluid, metal wall, and cold fluid are illustrated below, signifying the resistances offered by the metal wall and the two fluid films on both sides of the metal wall. The boundaries of steep temperature changes on either side of the metal wall are indicated in Figure 9.16.
- Under what circumstances overall heat transfer coefficient can be approximated to an individual film coefficient?
  - Overall heat transfer coefficient represents the overall heat transfer rate that is contributed by the two fluid films, and the solid barriers separating the fluids.
  - It is given by the equation in approximate simplified form:

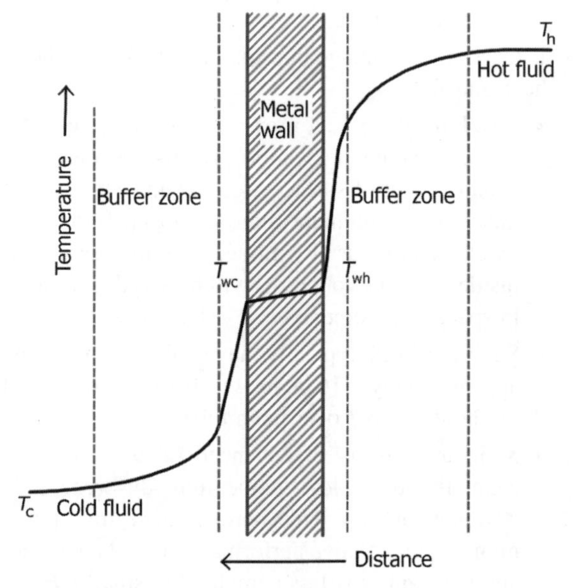

FIGURE 9.16 Temperature gradients in forced convection.

$$1/U = 1/h_o + 1/h_i + \sum(\text{metal wall and fouling resistances on both sides of the metal wall}). \quad (9.43)$$

- If the inside and outside surfaces are clean, fouling resistances will be zero. Generally, metal wall resistance can be assumed to be negligible, metals being good conductors of heat. In such a case, the equation reduces to the form

$$1/U = 1/h_o + 1/h_i. \quad (9.44)$$

- If the outside fluid is steam as in the case of steam heaters and inside fluid is oil, liquid film offers much higher resistance as condensing steam will have very high heat transfer coefficients compared to the heat transfer coefficients for the oil. This can be represented by

$$h_o \gg h_i \text{ or } 1/h_o = 1/h_i.$$

- In such cases,

$$1/U \approx 1/h_o \text{ or } U \approx h_o. \quad (9.45)$$

- What are fouling factors? How do they influence heat transfer rates?
  - Resistances to heat transfer in a tubular heat exchanger are due to inside fluid film, deposits of foreign solid material (fouling deposit) inside the tube, metal wall, deposits of foreign solid material outside the tube, and outside fluid film. Resistances due to deposits are significantly high due to low thermal conductivities of these deposit materials. These resistances are lumped into what are known as *fouling factors* that depend on nature, thicknesses, average thermal conductivities, structure, and porosities of the deposits.
  - Fouling factors contribute to decrease of heat transfer rates.
- Give typical ranges of overall heat transfer coefficients for the cases given in Table 9.3.
- What are the normally recommended values of overall heat transfer coefficients to be used for estimation purposes?
  - Table 9.4 gives recommended overall heat transfer coefficients for broad categories of applications.
- What are the sources of getting heat transfer coefficients for estimation purposes for shell and tube exchangers?
  - Shell and tube heat transfer coefficient for estimation purposes can be found in many reference books or an online list can be found at one of the two following addresses:
    - http://www.cheresources.com/uexchangers.shtml
    - http://www.processassociates.com/process/heat/uvalues1.htm

TABLE 9.3 Typical Ranges of Overall Heat Transfer Coefficients

Type	Application	U (W/(m^2 K))
Shell and tube exchanger: heating/cooling	Gases at atmospheric pressure inside and outside tubes	5–35
	Gases at high pressure inside and outside tubes	150–500
	Liquid inside/outside and gas outside/inside at atmospheric pressure	15–70
	Gas inside at high pressure and liquid outside tubes	200–400
	Liquid inside and outside tubes	150–1200
	Steam outside and liquid inside tubes	300–1200
Shell and tube exchanger: condensation	Steam outside and cooling water inside tubes	1500–4000
	Organic vapors or NH_3 outside and cooling water inside tubes	300–1200
Shell and tube exchanger: vaporization	Steam outside and high-viscosity liquid inside tubes: natural circulation	300–900
	Steam outside and low-viscosity liquid inside tubes: natural circulation	600–1700
	Steam outside and liquid inside tubes: forced circulation	900–3000
Air-cooled exchanger	Cooling water	600–750
	Cooling light hydrocarbon liquids	400–550
	Cooling of tarry liquids	30–60
	Cooling of air or flue gases	60–180
	Cooling hydrocarbon gases	200–450
	Condensation of low-pressure steam	700–850
	Condensation of organic vapors	350–500
Plate heat exchanger	Liquid to liquid	1000–4000
Spiral heat exchanger	Liquid to liquid	700–2500
	Condensing vapors	900–3500

Source: cheresources.com.

TABLE 9.4 Recommended *U* for Broad Categories of Design Applications

Category	W/(m^2 °C)
Water–liquid	850
Liquid–liquid	280
Liquid–gas	60
Gas–gas	30
Reboilers	1140

  - *Annexure* gives typical ranges of heat transfer coefficients that can be used in the preliminary design of heat transfer equipment in different refinery units.

- Give an estimation method for overall heat transfer coefficients for use in preliminary design of heat exchangers.
  - Figure 9.17 gives a nomograph for the estimation of overall heat transfer coefficients for preliminary design of heat exchangers.

9.1.5 Heat Transfer Fluids

- What are heat transfer fluids? What are the desirable characteristics of heat transfer fluids?
  - A gas or liquid used to move heat energy from one place to another is called a heat transfer fluid.
  - Desirable properties include
    - ➢ Low coefficient of expansion.
    - ➢ Low viscosity for better heat transfer and the need for low pumping power.
    - ➢ High heat capacity.
    - ➢ Low thermal diffusivity.
    - ➢ Low freezing point (at least 20°C lower than the lowest operating temperature) to avoid freezing on the heat transfer surfaces.
    - ➢ High flash point and high self-ignition temperatures making it less susceptible to ignition.
    - ➢ From ignition point of view, not only flash point and self-ignition temperature but also tendency for formation of aerosols during use of heat transfer fluids is an important criterion. The following four properties/conditions of the heat transfer fluids influence this tendency:
      - Higher density fluids tend to form smaller droplets upon leaking.
      - Higher viscosity fluids are less likely to form an aerosol.
      - Higher surface tension fluids will form larger droplets upon leaking.
      - Higher operating pressures will produce smaller droplet sizes closer to the point of leak.
    - ➢ Low vapor pressure or high boiling point to avoid the need to pressurize the system at elevated temperatures.
    - ➢ Low corrosiveness.
    - ➢ Should be nontoxic and environment friendly. Its vapor should neither contribute to greenhouse effect nor depletion of the ozone layer.
    - ➢ High thermal and oxidation stability. Most organic fluids oxidize at high temperatures in the presence of air and can form acidic and polymerization products in the system that can initiate corrosion and fouling.

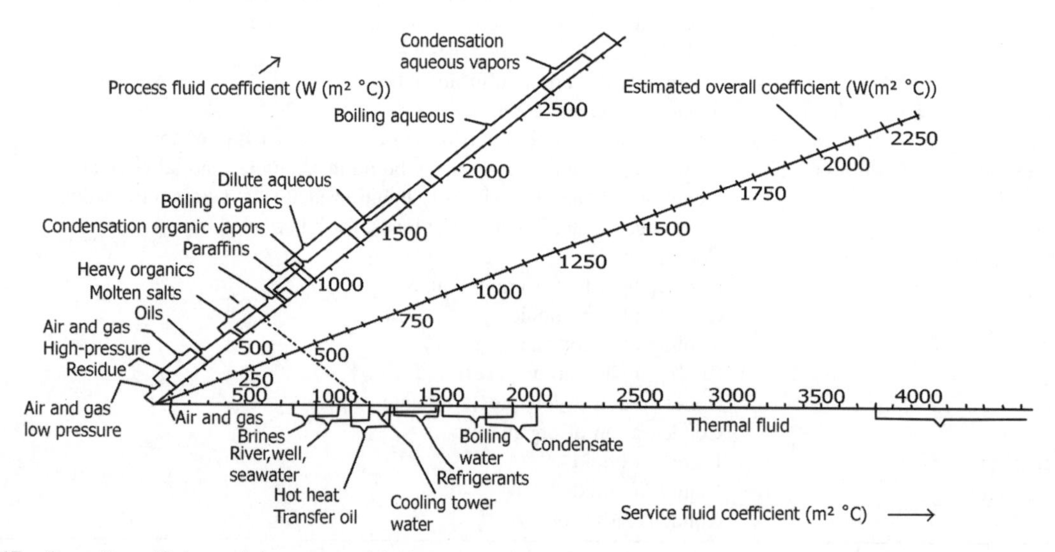

FIGURE 9.17 Overall coefficients. Join process side duty to service side and read U from center scale (see dotted line for illustration).

  - No regulatory constraints. Preferably it should be of food grade quality and satisfy the criteria of *incidental food contact*.
  - Favorable economics.

- What are the commonly used heat transfer fluids in the chemical process industry, other than steam?
  - Heat transfer fluids include glycol–water solutions, hydrocarbon oils, silicone oils, molten salts, and liquid metals.
  - Hydrocarbon oils are very commonly used fluids in the chemical process industries, either petroleum-based or synthetic, which are available for use over a wide range of operating temperatures. Some are suitable for low-temperature operation and others for good thermal stability at temperatures up to 400°C.
  - Proprietary heat transfer fluids like Dowtherm and Therminol fluid products.
- What are the particularly important requirements for low-temperature heat transfer fluids?
  - Low freezing point and low viscosity.
- Name some low-temperature heat transfer fluids?
  - *Air:* Air will not boil or freeze and is noncorrosive. However, it has a very low heat capacity and tends to leak out of equipment.
  - *Water:* Water is nontoxic and inexpensive. With a high specific heat, and a very low viscosity, it is easy to pump. Unfortunately, water has a relatively low boiling point and a high freezing point. It can also be corrosive if the pH (acidity/alkalinity level) is not maintained at a neutral level. Water with a high mineral content (i.e., *hard* water) can cause mineral deposits to form on heat transfer surfaces and reduce heat transfer rates.
  - *Glycol–Water Mixtures:* Ethylene and propylene glycol are antifreezes and used as heat transfer fluids in the temperature range of 121–40°C.
  - *Trichloroethylene (TCE):* Has also been used extensively in the past because of its low viscosity at low temperatures and nonflammability. Glycols and TCE are inexpensive, but they have several limitations. The most important drawback is the environmental concerns associated with their use. Ethylene glycol and TCE are considered toxic, so they cannot be used in many processes. Propylene glycol, although nontoxic, has very high viscosity at low temperatures, so it is normally used at temperatures above about −23°C. Glycols also degrade over time and form glycolic acid, which can contribute to the corrosion of metals in the system.
  - Glycols are usually added to water to control freezing point and viscosity of water. Glycol/water mixtures have a 50/50 or 60/40 glycol to water ratio. Ethylene glycol is extremely toxic. Food grade propylene glycol/water mixtures can be used in a single-walled heat exchanger. Most glycols deteriorate at very high temperatures. One must check the pH value, freezing point, and concentration of inhibitors annually to determine whether the mixture needs any adjustments or replacements to maintain its stability and effectiveness.
  - *Hydrocarbon Oils:* Hydrocarbon oils have a higher viscosity and lower specific heat than water. These oils are relatively inexpensive and have a low freezing point. The basic categories of hydrocarbon oils are synthetic hydrocarbons, paraffin hydrocarbons, and aromatic refined mineral oils.
  - Synthetic hydrocarbons are relatively nontoxic and require little maintenance. Paraffin hydrocarbons have a wider temperature range between freezing and boiling points than water, but they are toxic. Aromatic oils are the least viscous of the hydrocarbon oils.
  - *Aromatics:* Aromatic hydrocarbons such as diethylbenzene are very common low-temperature heat transfer fluids in the temperature range from −70 to 260°C. Their low-temperature heat transfer characteristics and thermal stability are excellent. However, these are not nontoxic. They have a strong odor that can be irritating to the personnel handling them, and very few aromatic compounds have freezing points lower than −80°C. Hence, these are used above −70°C in closed airtight systems in chemical processing and industrial refrigeration.
  - *Aliphatics:* Paraffinic and isoparaffinic aliphatic hydrocarbons are used in some low-temperature systems. Many petroleum-based aliphatic compounds meet the criteria for *incidental food contact*. These fluids do not form hazardous degradation by-products, and most have an insignificant odor and are nontoxic in case of contact with skin or ingestion.
  - Because of their high viscosity at low temperatures, these fluids are not very common in low-temperature applications. Also, the thermal stability of aliphatic compounds is not as good as aromatic compounds.
  - Some of the isoparaffinic fluids (with 12–14 carbons) can be used from −60 up to 150°C. These fluids are preferred in food and pharmaceutical applications where toxicity is a major issue.
  - *Refrigerants/Phase Change Fluids:* Refrigerants are well-known examples of heat transfer fluids. These are commonly used as the heat transfer fluid in refrigerators, air conditioners, and heat pumps. They generally have a low boiling point and a high

heat capacity. Chlorofluorocarbon (CFC) refrigerants, such as Freon, were the primary fluids used by refrigerator, air conditioner, and heat pump manufacturers because they are nonflammable, low in toxicity, stable, noncorrosive, and do not freeze. However, due to the negative effect of CFCs on the ozone layer of the Earth, they are being phased out. Ammonia can also be used as a refrigerant. It is commonly used in industrial applications. These can be aqueous ammonia or a calcium chloride ammonia mixture.

- *Silicones:* Dimethylpolysiloxane is an example of silicone oils. Its molecular weight and thermophysical properties can be adjusted by varying the chain length. Silicones have a very low freezing point, and a very high boiling point. They can be used at temperatures as low as −100°C and as high as 260°C. They are noncorrosive and stable. They have excellent service life in closed systems in the absence of oxygen. Also, they have essentially no odor and very low toxicity. Since silicone fluids are virtually nontoxic, their potential for applications in the pharmaceutical industry is good. They have a high viscosity and low heat capacities. They also leak easily through pipe fittings and microscopic holes because of their low surface tension, although this improves wettability. Silicones are more expensive than the aromatic- and aliphatic-based heat transfer fluids.
- *Fluorocarbons:* Fluorinated compounds such as hydrofluoroethers and perfluorocarbon ethers have a low freezing point and low viscosity at low temperatures. Typical applications of fluorocarbon-based fluids are in the pharmaceutical and semiconductor industries within the temperature range from −100 to 150°C. They are nonflammable and nontoxic. Some fluorinated compounds have no ozone depleting potential and other favorable environmental properties. However, these fluids are very expensive. Due to the extremely low surface tension, leaks can develop around fittings.

- Name some Dowtherm fluids for heat transfer applications, giving temperature ranges for which they are suitable.
  - Dowtherm synthetic organic and Syltherm silicone heat transfer fluids (See Figure 9.18).
  - Dowtherm, Dowfrost, and Dowcal (See Figure 9.19).
- What are the temperature ranges for different types of heat transfer fluids?
  - Petroleum oils: <315°C (600°F).
  - Dowtherm or other synthetics: <400°C (750°F).
  - Fused salts (molten salts): <600°C (1100°F).
  - Direct firing/electricity: >230°C (450°F).

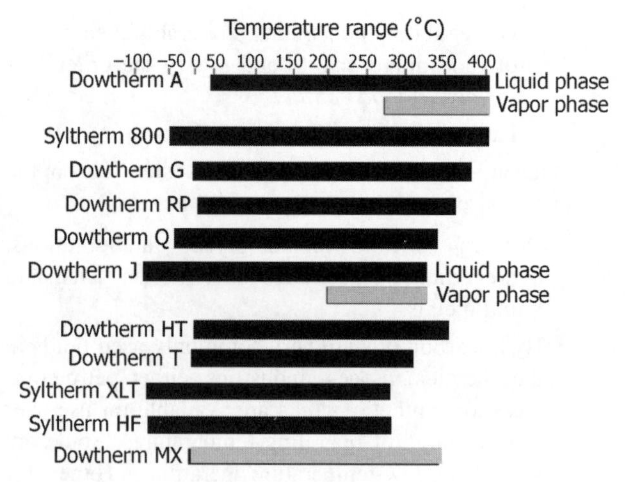

FIGURE 9.18 Dowtherm and Syltherm heat transfer fluids.

- Name some heat transfer fluids used in nuclear power plants.
  - *Liquid Metals:* These are used in nuclear reactors as heat transfer fluids.
- Give an example of heat transfer fluids used for concentration of caustic soda solutions in high-viscosity range.
  - Heat transfer salts. For example, sodium chloride is used as molten salt at temperatures around 500°C.
- What are the operating temperatures in high-temperature heat transfer equipment that are most important in the selection of heat transfer fluids?
  - Fluid film temperatures, rather than bulk fluid temperatures, are most important in the selection of a suitable heat transfer fluid, as fluid film temperatures

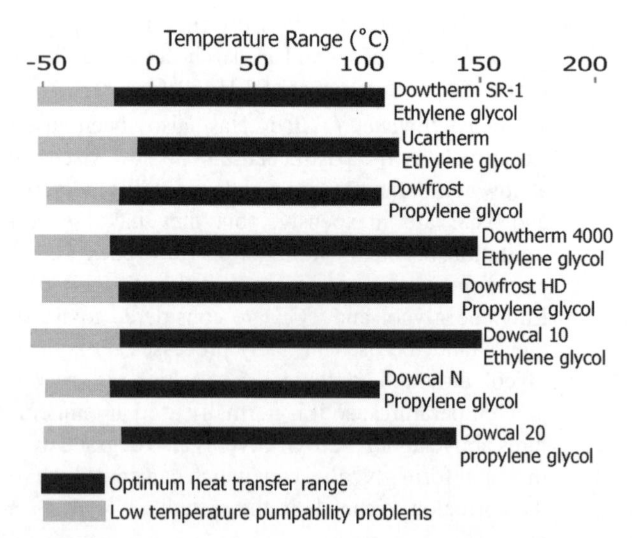

FIGURE 9.19 Dowtherm, Dowfrost, and Dowcal inhibited glycol-based heat transfer fluids.

are the highest in the heat transfer equipment operating at high temperatures.

- What are the performance problems involved in the use of high-temperature heat transfer fluids?
  - *Fluid Degradation:* Fluid degradation starts in the fluid film, as the fluid is hottest. Fluid degradation occurs due to oxidation of the heat transfer oils giving rise to formation of solids and high-viscosity products.
  - For typical heat transfer oils, the fluid decomposition rate doubles for roughly every 10°C.
    - Increase in degradation rates in the film can exceed those in the bulk fluid by a factor of 20 or more.
  - *Fouling:* Overheating of the fluid film may result in fouling of the heat transfer surface.
  - *Coke Formation:* Formation of abrasive coke particles due to fluid cracking reduces the life of pump seals, bearing surfaces, and valve seats.
  - *Contamination:* Accumulation of dirt, contamination from the process systems, and corrosion products from piping and equipment.
  - *Low Boiling Fluid Formation:* Formation of low boiling fluid or gas formation due to decomposition at high temperatures requires vapor venting that contributes to fluid loss apart from lowering heat transfer coefficients.
- What are the reasons for heat transfer fluid degradation?
  - When a heat transfer fluid reacts with oxygen, organic acids are formed and oxidation occurs.
  - Coke deposits, sludge, and increased viscosity are the symptoms of oxidation and are the most common reasons for severe fluid degradation.
  - Unless water is present, these acids are not corrosive.
  - The problem is because the acids, which have poor thermal stability, break down inside the heat transfer equipment.
  - One of the consequences of this acid deterioration is the extremely fine carbon particles (soot) that can agglomerate to form sludge deposits.
    - These particles will remain suspended in the fluid while it is flowing, but can stick together and form blockages where excessive turbulence will pack the particles together.
    - These carbon particles will also settle out of the suspension and form sediment when the fluid is stagnant and cool.
  - Another way that acids cause problems in heat transfer systems is that they increase the viscosity of the heat transfer fluid *when they polymerize.* If localized hot surfaces exist inside the tubes or on electrical elements, high acid levels will cause hard deposits (coke) to form on the hot spot.
  - Acid number and viscosity measurements of the fluid in use give an indication of the degradation of the fluid and replacement requirements.
- What are the consequences of contamination of heat transfer fluid?
  - Reduced heat transfer rates.
  - Decreased fuel efficiency.
  - Flow blockages in small diameter or low-velocity areas.
  - Extended start-up times at low temperatures.
  - Fouling of heat transfer surfaces.
  - Overheating and damage, or even complete failure of heater tubes.
- Steam is the most commonly used heat transfer fluid. Would you recommend high-pressure steam or low-pressure steam for this purpose? Give reasons for your choice.
  - High-pressure steam is normally not a good choice for heating, though it gives larger ΔT, for the following reasons:
    - It is more valuable for power generation, the exhaust from the turbines being normally used for heating purposes.
    - It has lower λ_v, heat of condensation, thus releasing less energy per unit quantity compared to low-pressure steam.
    - Equipment and distribution lines must be designed and fabricated to withstand higher pressures, an expensive proposition.
    - Where higher ΔT is required, some other heat transfer fluid such as Dowtherm with high λ_v or other high boiling fluids should be selected.
- To get high ΔT, high-pressure steam is preferred as a heating medium than low-pressure steam. Explain the pros and cons in supporting such a proposition.
  - While high-pressure steam increases ΔT for heat transfer, it is not preferred as a heating medium for reasons given under the previous question.
- Use of superheated steam as a heating medium increases ΔT for heat transfer. Is it advisable to use superheated steam for such an application?
  - Superheated steam behaves as a gas while it is giving up its superheat. Heat transfer coefficients are very low for gases compared to heat transfer coefficients for condensation of vapors. Condensing film coefficients are over 800–1000 times higher than gas film coefficients.
  - Area requirements for heat transfer will be very high compared to the benefits obtainable due to getting higher ΔT. Thus, equipment costs will increase.

- Use of superheated steam is not advisable as an option for obtaining higher ΔT as the amount of heat obtainable from the superheat is very small compared to latent heat obtainable on condensation of the saturated steam.
- Heat exchangers operating at higher temperature levels involve increased heat losses or require more insulation to offset the increased heat losses.
- However, sometimes, superheated steam is used as a heating medium in desuperheaters for energy recovery or while heating highly viscous liquids where high ΔT is desirable.

• What are the practices that involve wasteful energy losses in steam heating systems?

- Many refineries and petrochemical plants use expensive medium-pressure (MP) steam for low-temperature heating duties when they could instead be using cheaper and more readily available low-pressure (LP) steam. A common reason to use MP steam is that if there is significant back pressure in the condensate return system, the pressure of the LP steam is too low to drive condensate through a steam trap and into the condensate main, resulting in condensate back up in the equipment impeding heat transfer.
- This MP steam demand for heating purposes is met often by letting down high-pressure steam, which is a wasteful operation.
- LP steam is a versatile heat source and is frequently vented to atmosphere, or condensate from heat transfer equipment is sent to drain, both involving wastage of energy.
- The above practices add to ever increasing energy costs.

• How can the energy losses by the above practices be reduced?

- Combined steam trap and pump systems are available that avoid the pressure drop problems by returning LP condensate to the boiler house even against significant back pressure, thus avoiding use of MP steam.

• If steam used as a heat transfer fluid contains some air, what is its effect on heat transfer rates?

- Presence of air reduces heat transfer.

• What are the factors that are to be considered in evaluating heat transfer fluids?

- *Working Temperature Range:* This is the primary consideration to match process temperature requirements with the selected fluid.
  - ➢ Steam is normally considered first for high temperatures, but above 180°C steam pressure increases rapidly with increasing temperature.
  - ➢ Consequently, piping and vessel costs will also rise rapidly. Thus, other high-temperature heat transfer fluids must be considered.
  - ➢ A low vapor pressure at a high temperature is the major reason for choosing an organic fluid over steam.
  - ➢ Pressurized water could be used from 300 to 400°C, but high pressures are required to maintain the water in the liquid state.
  - ➢ Above 500°C, combustion gases and liquid metals are possibilities.
  - ➢ Liquid metals are used for cooling nuclear reactors.
  - ➢ Temperatures from 50 to 1000°C can also be achieved by electrical heating.
- *Environmental Effects:* Since accidental chemical spills occur occasionally, the effect of the heat transfer fluid on the environment and health must be considered.
  - ➢ Since the use of chemicals may be governed by laws, the process engineer must comply. For example, EPA has banned use of polychlorinated biphenyls (PCBs) because of the concern over environmental contamination.
- *Viscosity:* High viscosity reduces heat transfer rates, defeating the very purpose of heat transfer.
- *Thermal Stability:* Organic heat transfer fluids can degrade somewhat, either by oxidation or thermal cracking.
  - ➢ The primary cause is thermal degradation. In thermal degradation, chemical bonds are broken forming new smaller compounds that lower the flash point of the fluid.
  - ➢ At the flash point, flammable fluids will momentarily ignite on application of a flame or spark.
  - ➢ Organic fluids will also degrade to form active compounds. These compounds will then polymerize to form large molecules thereby increasing the fluid viscosity, which reduces heat transfer.
- *Corrosivity:* Generally, a heat transfer fluid should be noncorrosive to carbon steel because of its low cost.
  - ➢ Carbon steel may be used with all the organic fluids and with molten salts up to 450°C. With the sodium–potassium alloys, carbon and low-alloy steels can be used up to 540°C, but above 540°C stainless steels should be used. Cryogenic fluids require special steels. For example, liquid methane requires steels containing 9% nickel.
- *Toxicity and Flammability:* Organic heat transfer fluids require stringent leakage control because they

are all flammable from 180 to 540°C. Most of these fluids have toxic effects and irritate eyes and skin.

- Summarize fluids used as heat transfer media and their working temperature ranges.
  - Table 9.5 gives a summary of heat transfer fluids and the temperature ranges involved in their use.
- What are the capacities and steam pressures involved in generation of steam for use as heating medium in process plants?
  - Medium-pressure boilers for generating steam for use as a heat transfer fluid generally have capacities up to about 10,000 kg/h at an operating pressure of around 1000 kPa.
- What are the quality requirements for the water used for steam generation?
  - Deaeration of boiler feed water and blowdown control.

TABLE 9.5 Operating Temperature Ranges of Different Heat Transfer Fluids

Heat Transfer Fluid	Operating Temperature Range (°C)
Refrigerants	
Ethane and ethylene	−60 to −115
Propane and propylene	5 to −46
Butanes	−12 to 16
Ammonia	−32 to 27
Fluorocarbon (R-12) (dichlorodifluoromethane)	−29 to 27
Water + ethylene glycol (50%/0%)	−50 to 90
Water	
Water (wells, rivers, and lakes)	32–49
Chilled water	1.7–16
Cooling tower water	30
High-temperature water	300–400
Air	65–260
Steam	
Low pressure (2.7 bar)	126
Low pressure (4.6 bar)	148
Organic oils[a]	−50 to 430
Silicone oils	−23 to 399
Molten salts	
25% $AlCl_3$, 75% $AlBr_3$	75–500
40% $NaNO_2$, 7% $NaNO_3$, 53% KNO_3	204–454
Liquid metals	
56% Na, 44% K or 22% Na, 78% K	204–454
Mercury (not in current use due to toxicity)	316–538
Combustion gases	>500

[a] For example, diphenyl-diphynyl oxide, hydrogenated terphenyl, aliphatic, and aromatic oils.

- Why is it necessary to deaerate water used for steam generation?
  - Dissolved air in water gets released on vaporization of water for steam generation. Thus, steam will be contaminated by air, which is a noncondensable gas.
  - When steam is used as a heating medium, it condenses in heat transfer equipment with accumulation of air on heat transfer surfaces.
  - Air has very low heat transfer coefficient, being a bad conductor, thus retarding heat transfer.
  - Even 1% air in steam will reduce heat transfer coefficients for steam very appreciable by as much as over 50–60%.
  - Due to this reason, deaeration of boiler feed water becomes a necessity for maintaining high heat transfer rates with steam as a heating medium.
  - Presence of dissolved gases in process streams will have similar effect on heat transfer as presence of air in steam.
- What are the other effects of presence of dissolved gases in process streams? How are they removed?
  - Apart from reducing heat transfer rates in heat exchangers, dissolved gases on release inside equipment or pumping systems can create operational problems such as vapor lock, cavitation in pumps, increased entrainment problems inside equipment and increased losses of liquids as presence of dissolved gases increases vaporization rates by their partial pressure effect on the liquid, interference in measurements, corrosion problems in downstream equipment, environmental pollution, and safety problems.
  - Water containing dissolved oxygen can be highly corrosive to boiler system components. Oxygen in water gives rise to localized pitting corrosion that can very rapidly lead to catastrophic failure of steel boiler components.
  - Besides power plants, breweries require deaerated water for acceptable quality of beer.
  - If dissolved gases are present in the liquid, the liquid should be passed through a separation vessel to allow for vapor disengagement prior to transferring heat to or from the stream or sending for further processing the stream. Dissolved gases may be present in significant amounts in solvents recycled in absorption and other gas processing units.
- What is the maximum dissolved oxygen content permissible in steam condenser systems in power plants for efficient heat transfer?
  - 7 ppb of dissolved oxygen. 5 ppb is recommended for nuclear power plants.

- How dissolved oxygen buildup can occur in power plant condenser systems?
  - Cold makeup water with high solubility for oxygen is a major source.
  - Air leakages into the system under lower condenser pressures.
  - High vapor loads in vent systems resulting in insufficient vent capacity.
  - High air ingress occurs at start-up and reduced loads. It can occur between hot condensate well and the condensate pump.
- How is air ingress into the condensate system prevented/reduced? Illustrate.
  - By preheating makeup water with turbine exhaust steam and passing the preheated water through a packed bed or other deaerator of suitable design before it enters the condenser and mixes with the condensate.
  - Prevention of oxygen absorption can be accomplished by designing the condenser internals to permit easy flow of noncondensable gases and associated water vapor within the condenser. This requires adequate provision of venting system capacity taking care of fluctuating operating vapor loads. At part load operating conditions, for example, the venting equipment capacity decreases as the condenser pressure decreases.
  - In the following illustration, Figure 9.20, oxygen removal from makeup water is accomplished by distributing the water over a packed bed in a vented column countercurrent to turbine exhaust steam drawn from the steam surface condenser that operates at the vacuum produced in the condenser. Vacuum is created by the use of steam jet ejectors or liquid ring vacuum pumps.
- Why it is necessary to remove/control dissolved salts from boiler feed water?
  - Dissolved salts in the feed water get concentrated in the water inside the boiler leading to scale formation and corrosion of metal surfaces. The net result will be lowering the efficiency of steam generation, besides damage to boiler due to corrosion.
  - The feed water is demineralized by reverse osmosis or other methods and concentrations of remaining solids in the water inside the boiler are controlled from excessive increases by blowdown. High dissolved solids content can also increase foaming tendency and carry over of the solids with the steam, which will increase scale formation in the heat transfer equipment.
  - The total dissolved solids (TDS) in the water in the boiler should be kept below 3000 mg/L, which is achieved by blowdown.
  - The quantity of blowdown needed to ensure that the TDS level is kept below the recommended level is determined by the TDS in the feed water, TDS to be maintained in the boiler and the evaporation rate.
  - There are two types of blowdown, namely, intermittent and continuous.
  - Intermittent blowdown is manually operated by means of a valve. This involves large short-term increases in feed water and substantial heat losses.
  - Continuous blowdown ensures constant TDS and steam purity. Heat lost can be recovered. This practice is common in high-pressure boilers.

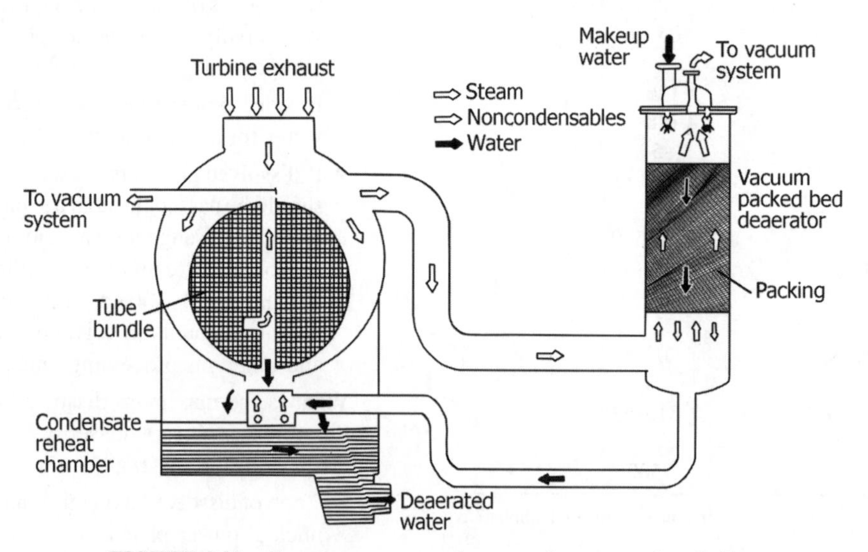

FIGURE 9.20 Condenser vacuum deaerator for power plant boiler.

- The required blowdown rate is estimated by the following formula:

$$\text{Blowdown rate} = F \times S/(B-F), \quad (9.46)$$

where F is the TDS in feed water (mg/L), S is the steam generation rate (kg/h), and B is the maximum TDS in the boiler (mg/L).

9.2 ANNEXURE (HEAT TRANSFER COEFFICIENTS FOR REFINERY SERVICES)

- Table 9.6 gives typical ranges of heat transfer coefficients for design in refinery services.

TABLE 9.6 Heat Transfer Coefficients for Preliminary Design of Heat Exchangers in Refinery Service

Refinery Unit	Range (W/(m^2 °C)) (Btu/(h ft^2 °F))
Crude distillation	
Crude/heavy gas oil	227 (40)–284 (50)
Crude/kerosene	199 (35)–255 (45)
Crude/light gas oil	227 (40)–284 (50)
Crude/naphtha	227 (40)–284 (50)
Crude/reduced crude	142 (25)–199 (35)
Crude/vacuum tar	142 (25)–170 (30)
Heavy gas oil cooler	284 (50)–284 (50)
Heavy gas oil–steam generation	568 (100)–681 (120)
Kerosene cooler	312 (55)–369 (65)
Light gas oil cooler	255 (45)–312 (55)
Lube distillation/crude	227 (40)–284 (50)
Naphtha cooler	340 (60)–398 (70)
Reduced crude/gas oil	199 (35)–255 (45)
Reduced crude/naphtha	227 (40)–284 (50)
Reduced crude cooler	114 (20)–170 (30)
Column overhead/crude	170 (30)–227 (40)
Column condenser	255 (45)–312 (55)
Vacuum tar/steam generation	255 (45)–312 (55)
Vacuum tar cooler	57 (10)–114 (20)
Catalytic cracking	
Naphtha cooler	312 (55)–369 (65)
Feed/DFO	199 (35)–255 (45)
DFO cooler	255 (45)–312 (55)
DFO cooler (150–200°C)	454 (80)–568 (100)
Gas oil/DFO	227 (40)–284 (50)
Gas oil/quench	199 (35)–284 (50)
Gas oil/tar	170 (30)–227 (40)
Quench or STB/feed	199 (35)–255 (45)
Quench/BFW	341 (60)–454 (80)
Quench or STB cooler	170 (30)–284 (50)
Quench and/or STB cooler (150–200°C)	284 (50)–398 (70)
Quench steam generation	398 (70)–511 (90)
Reduced crude/quench	199 (35)–255 (45)
Reduced crude/tar separator bottoms	170 (30)–227 (40)
SYN column condenser	170 (30)–227 (40)

TABLE 9.6 *(Continued)*

Refinery Unit	Range (W/(m^2 °C)) (Btu/(h ft^2 °F))
Thermal cracking	
Coker combination TWR condenser	227 (40)–284 (50)
Gas oil/gas oil	170 (30)–227 (40)
Gas oil cracker overhead condenser	255 (45)–369 (65)
Gas oil reflux steam generation	454 (80)–511 (90)
Gasoline/naphtha charge	341 (60)–454 (80)
Thermal tar cooler (box)	57 (10)–114 (20)
Thermal tar steam generation	227 (40)–284 (50)
Top reflux/BFW	341 (60)–454 (80)
Top reflux/naphtha	341 (60)–454 (80)
Top reflux/naphtha charge	284 (50)–341 (60)
Top reflux cooler	312 (55)–369 (65)
Lube and wax processing/extraction	
Solvent vapor/extract mix (75% solvent)	1278 (225)
Treating column intercooler	341 (60)–568 (100)
Solvent/charge oil	114 (20)–227 (40)
Solvent water cooler	511 (90)
Raffinate oil/raffinate oil mix	114 (20)–341 (60)
Raffinate oil/water cooler	142 (25)–341 (60)
Solvent/raffinate oil mix	170 (30)
Raffinate oil mix/steam	199 (35)
Raffinate oil mix/steam evaporator (80% solvent)	511 (90)
Raffinate oil mix/steam evaporator (45% solvent)	398 (70)
Raffinate oil mix/steam evaporator (13% solvent)	227 (40)
Atmospheric solvent vapors/extract mix	341 (60)
Extract/water cooler (submerged coil)	57 (10)–85 (15)
Dry solvent vapor (vacuum)/water condenser (+ subcooling)	341 (60)
CBM/water condenser (subcooling to 150°F)	681 (120)
Vacuum pump aftercooler	57 (10)
Steam/steam superheater	85 (15)
Extract mix/steam evaporator (80% solvent)	511 (90)
Extract mix/steam evaporator (45% solvent)	398 (70)
Extract mix/steam evaporator (13% solvent)	227 (40)
Extract mix/solvent vapor (80% solvent)	511 (90)–653 (115)
Extract mix/solvent vapor (45% solvent)	312 (55)
Extract mix/solvent vapor (13% solvent)	199 (35)
Wet solvent vapor (vacuum)/water condenser (+ subcooling)	596 (105)
Extract mix (10% solvent)/extract mix (80% solvent)	85 (15)
Dewaxing	
Cold pressed oil mix/solvent	170 (30)
Cold slack wax mix/steam	255 (45)
Flue gas/ammonia chiller	34 (6)–57 (10)
Flue gas/water cooler	57 (10)–114 (20)
Hot solvent/wax oil	369 (65)

TABLE 9.6 *(Continued)*

Refinery Unit	Range (W/(m^2 °C)) (Btu/(h ft^2 °F))
Pressed oil/pressed oil mix	170 (30)–227 (40)
Pressed oil/water cooler (to 150°F)	142 (25)
Slack wax/slack wax mix	170 (30)
Solvent/ammonia chiller	341 (60)
Solvent/cold slack wax mix	227 (10)
Solvent/water cooler (5 # ΔP)	511 (90)
Solvent/water cooler (1 # ΔP)	398 (70)
Solvent vapor/pressed oil mix	255 (45)–568 (100)
Solvent vapor/slack wax mix	369 (65)
Solvent vapor/pressed oil	454 (80)
Solvent vapor/pressed oil	511 (90)
Solvent vapor/slack wax mix (3:1 solvent ratio)	454 (80)
Solvent vapor/slack wax mix (4:1 solvent ratio)	511 (90)
Solvent vapor/water condenser (no subcooling)	568 (100)
Steam/steam superheater	85 (15)
Steam/pressed oil (2:1 solvent ratio)	85 (15)–398 (70)
Steam/pressed oil (3:1 solvent ratio)	199 (35)–568 (100)
Steam/pressed oil (4:1 solvent ratio)	710 (125)
Steam/slack wax mix (1:2 solvent ratio)	85 (15)–398 (70)
Steam/slack wax mix (2:1 solvent ratio)	341 (60)
Steam/slack wax mix (3:1 solvent ratio)	398 (70)
Steam/slack wax mix (4:1 solvent ratio)	710 (125)
Warm wash heater	568 (100)
Wax oil mix/ammonia DP chiller (with scrapers)	170 (30)
Wax oil mix/ammonia DP chiller (without scrapers)	114 (20)
Wax oil mix/pressed oil mix DP exchanger	85 (15)–114 (20)
Wax oil mix/water cooler	170 (30)
Wet solvent vapor condenser (with subcooling)	425 (75)
Naphtha hydrotreating and reforming	
Pretreater reactor effluent/charge (cold end)	369 (65)–425 (75)
Pretreater reactor effluent/charge (hot end)	398 (70)–454 (80)
Pretreater reactor effluent condenser	454 (80)–511 (90)
Naphtha splitter feed/bottoms	369 (65)–454 (80)
Naphtha splitter condenser	369 (65)–425 (75)
Reactor effluent condenser	454 (80)–568 (100)
Reactor effluent/feed (cold end)	398 (70)–483 (85)
Reactor effluent/feed (hot end)	454 (80)–511 (90)
Splitter or stripper feed/pretreater effluent	425 (75)–483 (85)
Stabilizer reboiler (hot oil)	425 (75)–540 (95)
Hydrodesulfurization	
Charge/reactor effluent	341 (60)–398 (70)
Charge/gas oil product	312 (55)–369 (65)

TABLE 9.6 *(Continued)*

Refinery Unit	Range (W/(m^2 °C)) (Btu/(h ft^2 °F))
Charge/HTS offgas	284 (50)–341 (60)
HTS offgas/water condenser	425 (75)–483 (85)
Stripper feed/gas oil product	312 (55)–369 (65)
Stripper bottoms cooler	369 (65)–425 (75)
Stripper condenser	312 (55)–369 (65)
Total gas/gas oil product	312 (55)–369 (65)
Light ends processing	
Absorber intercooler	341 (60)–398 (70)
Compressor contactor cooler	425 (75)–511 (90)
Deethanizer condenser	54 (80)–568 (100)
Deethanizer reboiler	454 (80)–568 (100)
Debutanizer feed/bottoms	341 (60)–398 (70)
Debutanizer feed preheater (STM)	454 (80)–568 (100)
Debutanizer condenser	75 (425)–511 (90)
Debutanizer condenser aftercooler	227 (40)–284 (50)
Debutanizer reboiler (STM)	454 (80)–568 (100)
Depropanizer feed/bottoms	398 (70)–511 (90)
Depropanizer condenser	511 (90)–568 (100)
Depropanizer reboiler (hot oil)	312 (55)–369 (65)
Depropanizer reboiler (STM)	454 (80)–568 (100)
Fractionating reboiler	425 (75)–511 (90)
Gasoline cooler	398 (70)–454 (80)
Lean oil/rich oil exchanger	256 (45)–312 (55)
Lean oil cooler	341 (60)–398 (70)
Propane and/or butane cooler	425 (75)–483 (85)
Stabilizer bottoms/feed	369 (65)–425 (75)
Stabilizer condenser	398 (70)–511 (90)
Stabilizer reboiler	425 (75)–568 (100)
Alkylation	
Debutanizer condenser	425 (75)–511 (90)
Debutanizer reboiler (STM)	483 (85)–568 (100)
Deisobutanizer feed preheater (STM)	425 (75)–568 (100)
Deisobutanizer condenser	454 (80)–511 (90)
Deisobutanizer reboiler (STM)	511 (90)–568 (100)
Depropanizer feed preheater (STM)	511 (90)–568 (100)
Depropanizer condenser	511 (90)–568 (100)
Depropanizer reboiler (STM)	425 (75)–540 (95)
Depropanizer feed/bottoms exchanger	341 (60)–454 (80)
Olefin feed chiller	398 (70)–454 (80)
Refrigeration condenser	511 (90)–568 (100)
Rerun tower preheater (STM)	341 (60)–454 (80)
Rerun column condenser	369 (65)–454 (80)
Rerun column reboiler (STM)	454 (80)–568 (100)
Rerun column bottoms cooler	284 (50)–398 (70)
Amine treating	
Rich/lean amine exchanger	425 (75)–511 (90)
Regenerator condenser	398 (70)–511 (90)
Regenerator reboiler (STM)	568 (100)–681 (120)
Lean amine cooler	454 (80)–511 (90)

Note: Values in brackets are in Btu/(ft^2 h °F). Conversion: 1 Btu/(ft^2 h °F) = 5.6785 W/(m^2 °C).

10

SHELL AND TUBE HEAT EXCHANGERS

10.1 Heat Exchangers 271
 10.1.1 Tube Pitch 285
 10.1.2 Baffles 287
 10.1.3 Fouling 296
 10.1.4 Pressure Drop 308
 10.1.5 Shell Side Versus Tube Side 309
 10.1.6 Specification Sheet 311
 10.1.7 Log Mean Temperature Difference 311
 10.1.8 Performance 317
10.2 Thermal Design of Shell and Tube Heat Exchangers 321
 10.2.1 Kern Method 321
 10.2.2 ε-NTU Method 322
 10.2.3 Bell Delaware Method 324
10.3 Miscellaneous Design Equations 327
10.4 Annexure (TEMA Fouling Factors) 328

10.1 HEAT EXCHANGERS

- Name different types and names of heat transfer equipment and state their functions/applications.
  - *Shell and Tube Heat Exchanger:* General name for a commonly used heat transfer equipment involving exchange of heat between two fluids, one flowing on the tube side and the other on the shell side, with no direct contact between them. This is the workhorse of process industry.
  - *Condenser:* To condense steam or process vapors, using in most cases, cooling water for the purpose.
  - *Cooler:* To cool hot streams/products from process, before sending to storage or other applications requiring colder fluids for exampling, recycling in reactors or distillation units. On one side is process stream and on the other usually cooling water.
  - *Air-Cooled Exchanger:* Has an integral electric motor or fluid-powered fan for the cooling or heat removal.
  - *Heater:* Supplying heat energy to a process stream, for example, preheating a reactor or distillation column feed. Steam or a hot heat transfer fluid supplies the required energy.
  - *Chiller:* One stream, a process fluid being condensed at subatmospheric temperatures and the other a boiling refrigerant or process stream.
  - *Plate Heat Exchanger:* An alternative for shell and tube exchanger replacing shell and tube system by alternating channels created by thinner plate partitions for the flow of two fluids exchanging heat between them. Plate heat exchangers are often used in low-viscosity applications with moderate demands on operating temperatures and pressures, typically below 150°C. Gasket material is chosen to withstand both the operating temperature and the properties of the processing fluid. Gasketed, brazed, welded, semiwelded, or hybrid types are some variations.
  - *Finned Tube/Extended Surface Heat Exchanger:* This is a class of compact heat exchangers in which surface area (one side) of heat transfer equipment is increased by attaching fins or creating grooves/threaded surfaces or by other configurations. Used for applications involving poor heat transfer that requires large surface area for a given duty, as in exchangers involving air or gases.
  - *Cross-Flow Heat Exchanger:* Shell side flow in a heat exchanger is in the perpendicular direction instead of the customary counter or parallel flow, used in most exchangers.
  - *Spiral Heat Exchanger:* Heat transfer surfaces are arranged in the form of a spiral to increase turbulence through increased shear by continuous direction change. Used in special applications justifying increased costs.
  - *Scraped Surface Heat Exchanger:* Heat transfer surface involving highly viscous liquids is subjected

Fluid Mechanics, Heat Transfer, and Mass Transfer: Chemical Engineering Practice, By K. S. N. Raju

to mechanical scraping as in ladles used, for example, in the kitchen for making sweets or other products involving solidification or thickening on being heated. Surfaces are constantly renewed to prevent charring of products with increased residence times and consequent quality deterioration of heat-sensitive materials. It typically consists of a jacketed cylinder with rotating rows of scraper blades. The product is pumped through the cylinder while a heating or cooling medium is circulated between the cylinder and the insulating jacket. Typically used for freezing of ice cream and in the cooling of fats during margarine manufacture.

- *Reboiler:* One stream, a bottoms stream from a distillation column and the other a hot utility (steam or hot oil) or a process stream.
- *Kettle Reboiler:* Heat exchanger in which boiling of a liquid takes place on the shell side, outside tubes. Steam or other heating medium flows through tubes. Used for energy supply to distillation process by partial boiling of the bottom stream. This is a shell and tube heat exchanger in which tube bundle is submerged in the shell side boiling liquid, as in electric kettles with the enclosed heater element submerged in the liquid.
- *Thermosiphon Reboiler:* In thermosiphon reboiler, boiling/vaporizing liquid is inside tubes and heating medium is on the shell side. Two-phase flow is involved in the tubes.
- *Barometric Condenser:* Vapors produced in an evaporator are condensed in a direct contact condenser in which vacuum is created to assist vaporization in the evaporator at relatively lower temperatures. Principle of a barometer is involved with the liquid leg acting as the liquid column in a barometer. Above the liquid column, there will be vacuum in a closed system involving vapor space in the evaporator. Vapor from the evaporator is drawn by the vacuum into the condenser into which water is introduced through nozzles for condensing the vapors with direct contact between the two fluids.
- *Jacketed Vessel:* An outer jacket is provided on a cylindrical vessel through which heating or cooling medium is circulated while the liquid in the vessel is heated or cooled. Such vessels are normally equipped with mixers to mix the fluids inside the vessel, for example, to carry out batch chemical reactions.
- *Coiled Vessels:* Vessels in which heating or cooling is required are equipped with tubular coils, through which heating or cooling fluid circulated. This is an alternative to a jacketed vessel.
- *Recuperators and Regenerators:* There are two types of air preheaters in furnace systems, namely, recuperators and regenerators.
  - Recuperators are gas–gas heat exchangers placed on the furnace stack. Internal tubes or plates transfer heat from the outgoing exhaust gas to the incoming combustion air. The two streams are prevented from mixing by the tubes or plates.
  - Regenerators include two or more separate heat storage sections. Flue gases and combustion air take turns flowing through each regenerator, alternately heating the storage medium and then withdrawing heat from it. Refractory walls act as storage media for the heat.

- What are the recommended selection criteria, based on area, among shell and tube, double pipe, and coil types of heat exchangers?
  - $A < 2\,m^2$, select a coiled heat exchanger.
  - $2\,m^2 < A < 50\,m^2$, select a double pipe heat exchanger.
  - $A > 50\,m^2$, select a shell and tube heat exchanger.
    - The coiled heat exchanger is very compact and is frequently used when space is limited. The decision between the heat exchanger types is not as distinct as indicated.
- Name different types of shell and tube heat exchangers.
  - Double pipe heat exchanger: Annulus formed by the outside tube may be considered to be shell side space and the exchanger may be treated as a single tube exchanger for small duties.
  - Single and multipass straight tube bundle exchangers: Fixed tube sheet and floating head types.
  - U-bundle heat exchangers.
  - Reboilers: Kettle and thermosiphon types.
  - Finned tube heat exchangers.
- Can we use polymers for heat exchanger construction? If yes, what are their benefits?
  - Polymeric materials have a number of features that make them suitable for use in heat exchangers.
  - Low cost and corrosion free. When streams contain chlorides and other corrosives, polymeric exchangers have an advantage.
  - Thin structures are possible with reduced weight.
  - Their smooth and noncrystalline surface finish results in low ΔP and low fouling and scaling.
  - Surface properties can be changed for improved wettability.
  - Their thermal conductivity can be improved by incorporating heat conducting fillers and fibers made from graphite, boron nitride, and carbon black. Using

such fillers with liquid crystal polymers, their thermal conductivities can be improved to the level of stainless steels.

- Their hydrophobic surfaces enhance dropwise condensation.
- Easy to join and seal. Good moldability.
- A broad range of polymers is available to choose for different temperature ranges. Polyolefin materials are cost effective up to 90°C. Poly-ether-ether-ketone (PEEK) has a maximum operating temperature of 220°C.
- Polymeric exchangers are developed for applications involving concentration of acids and mechanical vapor compression evaporators.
- Lower operating temperatures can be achieved to make polymers suitable for heat transfer applications, for example, in evaporators by reducing operating pressures.

• What are the heat transfer areas for which a double pipe heat exchanger is suitable?

- Double pipe heat exchangers may be a good choice for areas from 9.3 to 18.6 m^2 (100–200 ft^2).

• "Double pipe heat exchanger is especially suitable for small heat transfer area surfaces and high tube pressures." *True/False*?

- *True.*

• What are the applications of double pipe heat exchangers? What are their plus points?

- Double pipe exchangers are extremely flexible with respect to configuration of hairpin arrangements, since both the inner pipes and annuli can be connected either in series or in parallel.
- In order to meet pressure drop constraints, it is sometimes convenient to divide one stream into two or more parallel branches while leaving the other stream intact. Such a case is shown in Figure 10.1, where the inner pipes are connected in parallel, while the annuli are connected in series.

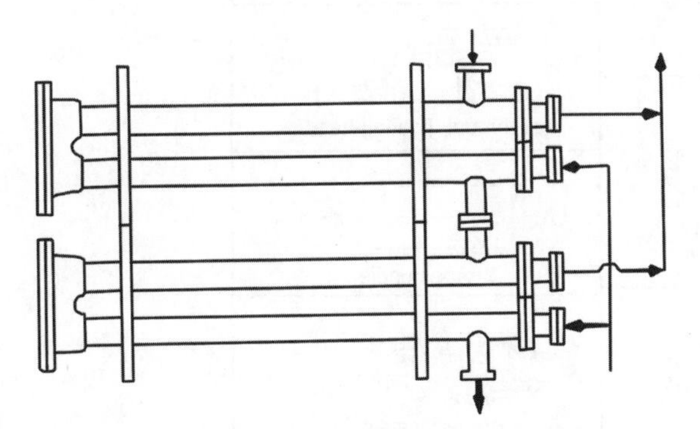

FIGURE 10.1 Double pipe (hairpin) heat exchanger with annuli connected in series and inner pipes connected in parallel.

• What is meant by 1–2 and 2–4 exchangers? Show flow arrangements for the above cases.

- 1–2 shell and tube heat exchanger means that it has one shell pass and two tube passes (see Figure 10.4). The tube passes will be in multiples of two (1–2, 1–4, 2–4, etc.).
- Odd numbers of tube passes have more complicated mechanical stresses and so on. An exception is 1–1 exchangers that are sometimes used for vaporizers and condensers (Figure 10.3).
- 2–4 exchanger means that it has two shell passes and four tube passes (Figure 10.5).

• Under what circumstances, a 1–2 heat exchanger is recommended?

- 1–2 exchangers involve moderate pressure drops with reasonable fluid velocities and hence good heat transfer coefficients on the tube side.

• What are the important limitations of 1–1 exchangers?

- Low tube side fluid velocities and consequently not suitable for fouling fluids.
- Normally, being of fixed tube sheet construction, not suitable for large temperature differences.

• What is meant by TEMA? What are its functions?

- Tubular Exchanger Manufacturers Association (United States).
- TEMA standards are widely followed in the design and fabrication of shell and tube exchangers. These standards cover the following main topics:
  - ➢ Nomenclature.
  - ➢ Fabrication tolerances.
  - ➢ Mechanical practices.
    - Examples include required tube thicknesses for different pressures, recommended gaskets, types of end covers, and expected effects of fouling on heat transfer.
  - ➢ Recommended practices.
    - Examples include design practices for supports, lifting lugs, and wind and seismic considerations.
- An advantage of TEMA standards is that end users recognize that TEMA specifications comprise industry standards that directly relate to recognized quality practices for manufacturing. Fabricators who build to TEMA standards can be competitively compared because tolerances and construction methods should be very similar for a given design.

• Illustrate the TEMA nomenclature for different constructional features of shell and tube heat exchangers.

- Figure 10.2 illustrates TEMA patterns for shell and tube heat exchangers.
- Shell patterns are defined as E, F, G, H, J, K, and X by TEMA based on flow through the shell.
  - E: Single-pass shell. Fluid enters at one end and leaves from the other end.
  - F: Two-pass shell in which a longitudinal baffle divides the shell into two passes. Fluid enters at one end, flows through one-half of shell area and returns, flowing in the reverse direction through the other half, exiting at the end of the second pass. Used for temperature cross situations, that is, where cold stream leaves at a higher temperature than the outlet temperature of exiting hot stream. Such a shell gives true countercurrent flow if there are two tube passes involved.
  - G: Split flow shell. There is only a central support plate and no baffles. Used as a horizontal thermosiphon reboiler. Tube lengths are <3 m, as larger tube lengths exceed TEMA-specified unsupported tubes leading to sagging of tubes.
  - H: Double split flow. Used when a larger tube length is required. This is equivalent to two G-shells placed side by side so that there are two support plates. Also used as a horizontal thermosiphon reboiler. G- and H-shells involve lower pressure drops because of absence of baffles.
  - J: Divided flow shell. Fluid enters the shell at the center and divided into two streams flowing in opposite directions and leave separately. Such a shell can be used for fluid entering as two streams at both ends and leaving as a combined stream.

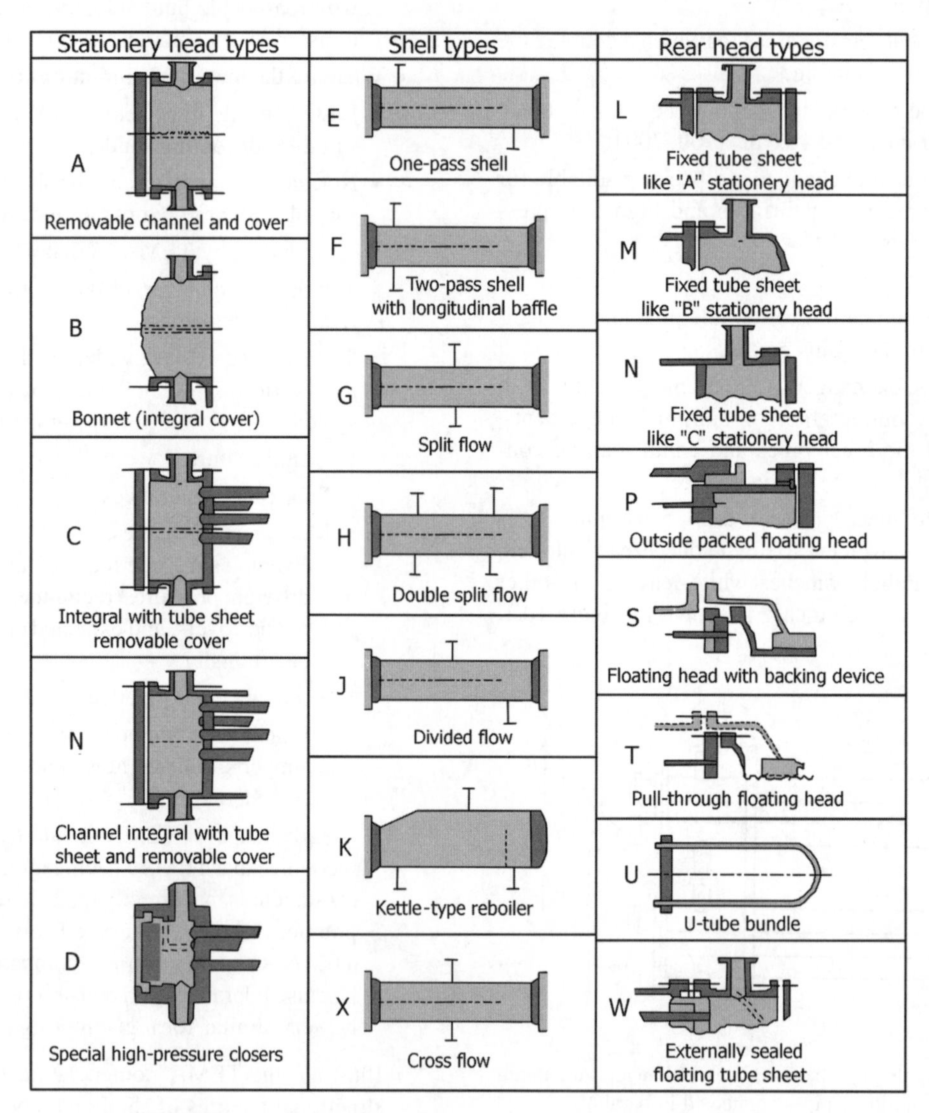

FIGURE 10.2 TEMA designations for shell and tube heat exchangers.

> K: Special cross-flow shell used as kettle reboiler, with an enlarged shell to facilitate vapor disengagement.

> X: Cross-flow shell. The shell side fluid enters at the top (or bottom) of the shell, flows across the tubes, and exits from the opposite side of the shell. The flow may be introduced through multiple nozzles located strategically along the length of the shell for better distribution. The pressure drop will be extremely low. Used for cooling and condensing vapors at low pressures.

- Give diagrams for different shell and tube heat exchangers, showing the tube and shell passes.
  - Figures 10.3–10.7 illustrate different shell and tube heat exchangers.
- Give a schematic diagram showing two heat exchangers connected in series. When is such an arrangement useful?
  - Figure 10.8 illustrates flow arrangement for two exchangers arranged in series.
  - Usually one designs for the least number of shells for a system. Two or more shells are connected in series when there is a relatively low flow in the shell side and the shell stream has the lowest heat transfer coefficient. This happens when the baffle spacing is close to the minimum, which is equal to shell I.D./5. Then adding a shell in series gives a higher velocity and hence heat transfer because of the smaller flow area in the smaller exchangers that are required.
- Compare arranging two shells in series and a two-pass shell.
  - Two-pass shell is cheaper than two-shell arrangement for the same duty.
  - A two-pass shell has improved heat transfer efficiency.
- What are the features and plus and minus points of a fixed tube sheet heat exchanger?
  - A fixed tube sheet heat exchanger (Figure 10.3) has straight tubes that are secured at both ends to tube sheets welded to the shell. The construction may have removable channel covers, bonnet-type channel covers, or integral tube sheets.
  - It is simple in construction and least expensive when temperature differences are small, requiring no expansion joints. Provides maximum amount of surface for a given shell and tube diameter.
  - For large temperature differences, expansion joints are to be incorporated making the exchanger more expensive. Use of fixed tube sheet exchangers is limited for ΔT values <95°C.
  - Tube can be cleaned by removing the channel cover.
  - Shell side cannot be mechanically cleaned because of the fixed tube sheets and therefore its application is

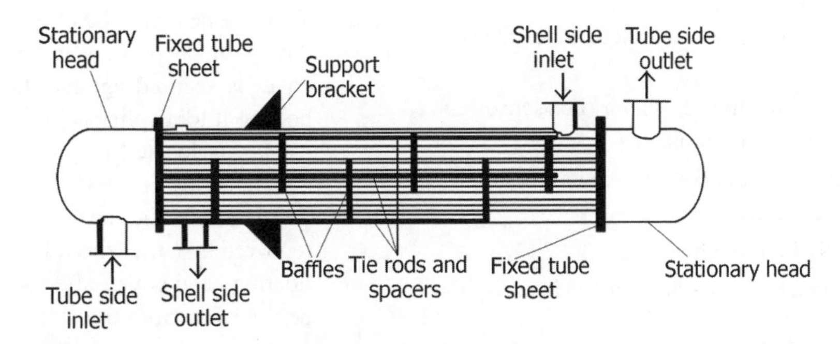

FIGURE 10.3 1–1 Fixed tube sheet shell and tube heat exchanger with baffles.

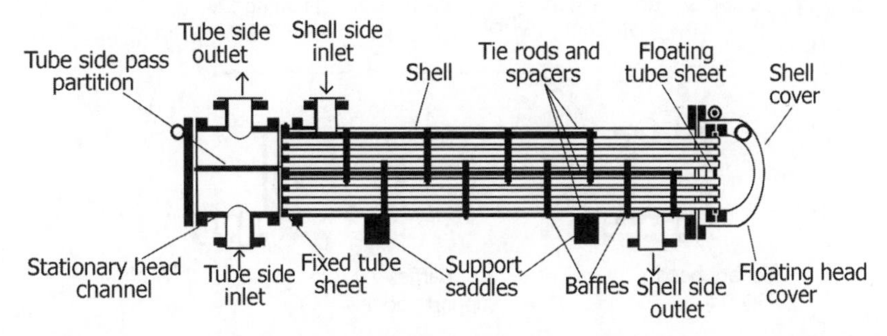

FIGURE 10.4 Pull-through 1–2 floating head heat exchanger with baffles (TEMA S).

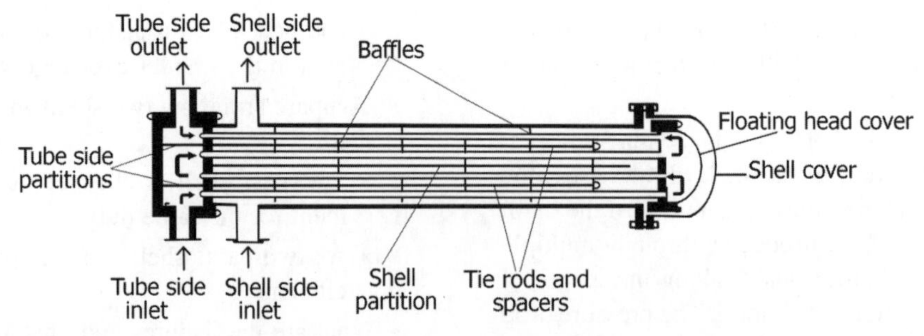

FIGURE 10.5 2–4 Floating head heat exchanger with baffles.

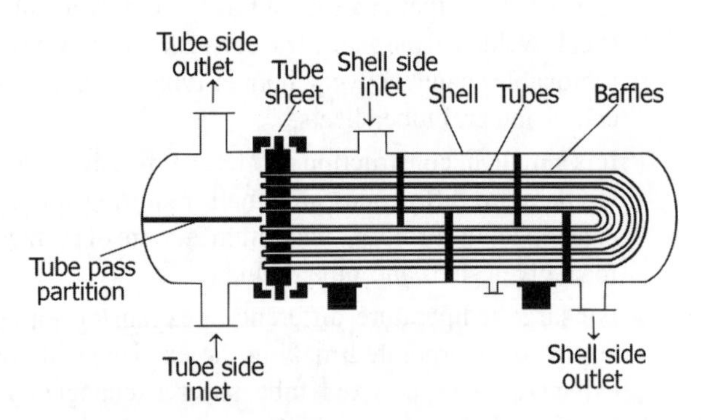

FIGURE 10.6 U-bundle heat exchanger with baffles.

limited to clean fluids. Chemical cleaning may be used.

- Can be designed for single and multiple tube passes to assure proper velocity.

- What are the applications of a fixed tube sheet exchanger?
  - Oil coolers, liquid to liquid, vapor condensers, reboilers, gas coolers, and so on.
- What type of heat exchanger is normally recommended for large temperature differences?
  - Floating head or U-bundle type.
- What is a floating head heat exchanger? Where is it used?
  - In a floating head heat exchanger, one tube sheet is fixed relative to the shell and the other is free to *float* within the shell. This permits free expansion of the tube bundle (Figures 10.4 and 10.5).
  - Floating head heat exchanger is the most versatile type of shell and tube heat exchanger and also the costliest.
  - Shell circuit can be inspected and cleaned with steam or mechanically.
  - The tube bundle can be repaired or replaced without disturbing the shell.
  - Provides maximum surface for a given shell diameter for removable bundle design.
  - There are various types of floating head construction. The most common ones are the pull-through type with backing device, TEMA S (Figure 10.4) and pull-through TEMA T (Figure 10.7) designs.
  - TEMA S design is the most common configuration in the chemical process industries. The floating head cover is secured against the floating tube sheet by bolting it to a split-backing ring. This floating head closure is located beyond the end of the shell and contained by a shell cover of a larger diameter. To dismantle the heat exchanger, the shell cover is removed first, then the split-backing ring and then the floating head cover, after which the tube bundle can be removed from the stationary end.

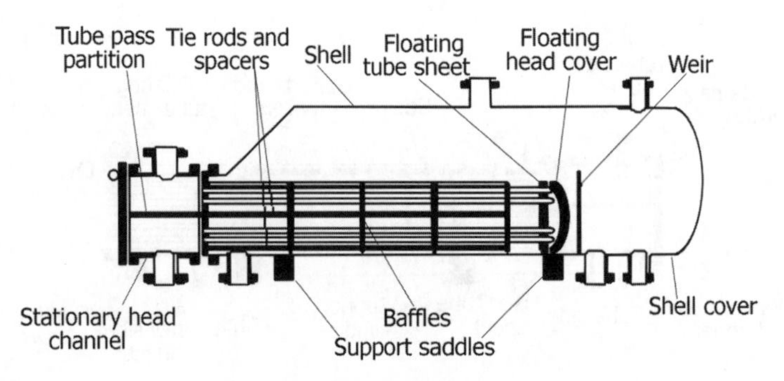

FIGURE 10.7 Pull-through floating head heat exchanger, suitable for kettle reboilers (TEMA T).

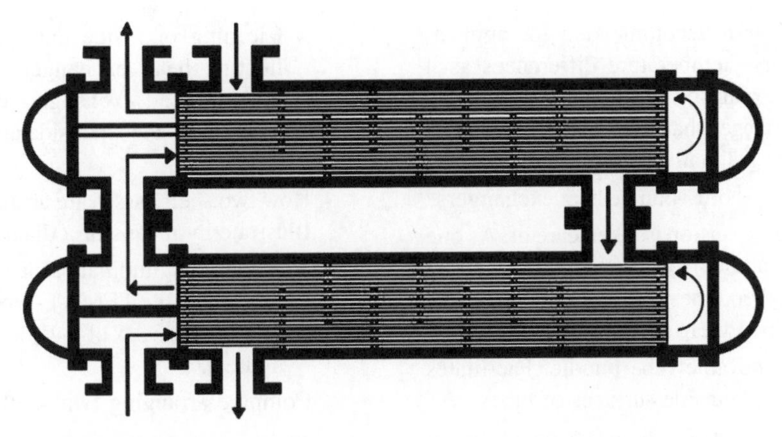

FIGURE 10.8 Flow arrangement for two heat exchangers in series.

- In the TEMA T construction (Figure 10.7), the entire tube bundle, including the floating head assembly, can be removed from the stationary end, since the shell diameter is larger than the floating head flange. The floating head cover is bolted directly to the floating tube sheet so that a split-backing ring is not required.
- The advantage of this construction is that the tube bundle may be removed from the shell without removing either the shell or the floating head cover, thus reducing maintenance time.
- This design is particularly suited to kettle reboilers having a dirty heating medium where U-tubes cannot be employed.
- Due to the enlarged shell, this construction has the highest cost of all exchanger types.

- What are the limitations of a floating head heat exchanger?
  - More expensive than normal type.
  - Tube side passes limited to single- or two-pass design.
  - All tubes are attached to two tube sheets. Individual tubes cannot expand independently so large local thermal shock applications should be avoided.
  - Involve internal gaskets with possibilities of leakages.
  - Packing materials produce limits on design pressure and temperature.
  - Corrosion possibilities of shell side floating parts by fluids.
- What are the differences in the construction of fixed tube sheet and floating head-type heat exchangers with respect to shell sizes and tube lengths?
  - Floating head heat exchangers are often limited to a shell inside diameter of 1.4–1.5 m and a tube length of 6–9 m, whereas fixed tube sheet heat exchangers can have shells as large as 3 m and tube lengths up to 12 m or more.
  - Considerations for these differences are largely based on tube bundle removal requirements for floating head exchangers and fabrication limitations and availability of components such as dished ends and flanges for fixed tube sheet exchangers.
- For what types of applications, a floating head heat exchanger is best suited?
  - For $\Delta T > 95°C$.
  - For service involving dirty fluids requiring cleaning inside and outside tubes.
- Name the shell and tube heat exchanger that has only one tube sheet. Comment on its utility.
  - U-bundle exchanger (Figure 10.6). Single tube sheet tends to save on costs, but costs involved in bending the tubes, coupled with the requirement of larger diameter shell (due to the minimum U-bend radius), offsets these savings.
  - As one end of the bundle is free, the bundle or individual tubes can expand or contract in response to stress differentials.
  - Capable of withstanding thermal shock applications.
  - The outside surfaces of the tubes can be cleaned, as the bundle can be removed.
  - To ease manufacturing and service, it is common to use a removable tube bundle design.
  - The disadvantage of the U-tube construction is that the insides of the tubes cannot be cleaned effectively, since the U-bends would require flexible-end drill shafts for cleaning. Thus, U-tube heat exchangers should not be used for services with a dirty fluid inside tubes.
- For what type of application, a U-bundle heat exchanger is recommended?

- U-bundle exchanger is recommended for applications involving large temperature differences as it provides for tube expansion in a simpler way than a floating head exchanger that involves internal parts that are difficult to maintain.

- What are the advantages of U-bundle heat exchangers?
  - Possibility of tube expansion is taken care of. As one end is free, each tube is free to expand or contract independent of other tubes, in response to stress differences developed due to temperature differences.
  - Replaceable or removable tube bundle. Facilitates cleaning of shell and outside surfaces of tubes.
  - Use of hydraulic cleaners (water jets forced through spray nozzles at high pressure for cleaning deposits in tube interiors reduces cleaning problems).
  - Minimum clearance between outer tube limit and inside of shell.
  - Number of joints required is reduced, making it well suited for high-pressure applications, with low initial and maintenance costs.
  - No internal gaskets. (Required for floating head type.)
  - Replaces floating head exchanger for applications involving large temperature differences.
- What are the negative points in the use of U-bundle heat exchangers?
  - Difficult to clean inside of tubes, especially in the curved ends, as the U-bends require flexible-end drill shafts for cleaning.
  - Therefore, U-tube heat exchangers should not be used for services with a dirty fluid inside tubes, as mentioned earlier.
  - U-tube bundles do not have as much tube surface as straight tube bundles due to the bending radius.
  - Because of U-tube nesting, it is difficult to replace a tube, particularly the interior tubes, and often requires the removal of outer layers. Such tubes are often plugged, reducing heat transfer area.
  - No single tube pass or true countercurrent flow is possible.
  - Tube wall thickness at the U-bend is thinner than at straight portion of tubes.
  - Draining of tube circuit is difficult when mounted with vertical position with the head side up.
- Compare U-bundle and floating head heat exchangers.
  - Floating head design involves more complexity than a U-bundle design with the consequence that a floating head exchanger is costly. Also its maintenance costs are higher.
  - Cleaning of inside tubes is difficult in a U-bundle exchanger.
  - Cleaning of inside and outside of the tubes in a floating head exchanger is easier and, therefore, floating head exchangers can be used for services where both the shell side and the tube side fluids are dirty.
- How two shell passes are arranged in a heat exchanger? Illustrate by means of a diagram?
  - A solid longitudinal baffle is provided to form a two-pass shell (see TEMA F-type shell in Figure 10.2 and also Figure 10.5 of 2–4 floating head exchanger for figures).
- Compare arranging two shells in series and two-pass shell.
  - Two-pass shell is cheaper than two-shell arrangement for the same duty and also has improved heat transfer efficiency.
- What are the different classes of heat exchangers according to TEMA? Give applications of each class.
  - *Class R:* For generally severe requirements of refineries and related process applications. Equipment fabricated in this class is designed for safety and durability under rigorous service and maintenance conditions.
  - *Class C:* For generally moderate requirements of commercial and general process applications.
  - *Class B:* For chemical process service.
  - TEMA R is the most restrictive and TEMA C is the least stringent. TEMA B and R are very similar in scope. TEMA R includes the requirement for confined joints where recesses must be machined in the flanges and tube sheets. Spiral-wound gaskets with a ring construction also meet this TEMA R requirement. TEMA R also requires a greater minimum thickness for some components.
- "TEMA Class C heat exchanger is used for heavier duty applications compared to TEMA Class R exchanger." *True/False*?
  - *False.*
- List the important heat exchanger parts and connections.
  - Shell, tubes, shell cover, channel, channel cover, tube sheets, baffles, nozzles, tie rods and spacers, pass partition plates, impingement baffle plate, longitudinal baffle plate, expansion joints, supports, drain connections, vents and relief valve connections, instrumentation connections, and so on.
- Give a summary for the selection of heat exchangers, with respect to their types.
  - Table 10.1 gives selection of heat exchangers with respect to their types.

TABLE 10.1 Selection of Shell and Tube Heat Exchangers

Type of Design	U-Bundle	Fixed Tube Sheet	Floating Head Outside Packed	Floating Head Split-Backing Ring	Floating Head Pull-Through Bundle
Provision for differential expansion	Individual tubes free to expand	Expansion joint in shell	Floating head	Floating head	Floating head
Removable bundle	Yes	No	Yes	Yes	Yes
Replacement bundle possible	Yes	Not practical	Yes	Yes	Yes
Individual tubes replaceable	Only those in outside row	Yes	Yes	Yes	Yes
Tube side cleanability	Mechanical	Yes	Yes	Yes	Yes
Shell side cleanability	Cleaning not easy, chemical cleaning possible	Mechanical or chemical	Mechanical or chemical	Mechanical or chemical	Mechanical or chemical
Δ-Pitch	Chemical only	Chemical only	Chemical only	Chemical only	Chemical only
Square pitch	Mechanical and chemical	Mechanical and chemical	Mechanical and chemical	Mechanical and chemical	Mechanical and chemical
No. of tube passes	Any practical even number	Normally no limitations	Normally no limitations	Normally no limitations	Normally no limitations
Internal gaskets eliminated	Yes	Yes	Yes	No	No

Courtesy: GPSA Engineering Data Book, 12th edition.

- "A shell and tube heat exchanger cannot be used in high-pressure applications." *True/False*?
  - *False.*
- What is the function of tie rods and spacers in a heat exchanger?
  - To hold baffles in place.
- What is the function of a header in a shell and tube heat exchanger?
  - To distribute and control the flow of tube side fluid in the tube circuit.
- What are the common nozzle connections to shell and heads of a heat exchanger?
  - Inlet and outlet nozzles for the two fluids (shell and tube side).
  - Safety valves normally on shell side and sometimes on tube side.
  - Drains on shell and on bottom of most heads.
    - Sometimes several drains are necessary on the shell side to facilitate drainage between baffles, when flushing is a part of operation.
  - Vents on shell and tube side headers to allow venting noncondensables.
  - Couplings are handy on process inlet and outlet nozzles on both shell and tube side and may be required for flushing, sampling, pressure gauges, thermowells, and so on.
  - It is desirable to match nozzle sizes with line sizes to avoid expanders or reducers.
  - However, sizing criteria for nozzles are usually more stringent than for lines, especially for the shell side inlet. Consequently, nozzle sizes must sometimes be one size (or more) larger than the corresponding line sizes, especially for small lines.
- Why is it necessary to use vacuum breakers on the shell side (steam side) of a shell and tube heat exchanger?
  - Vacuum breakers are often installed on the shell side (steam side) of shell and tube exchangers to allow air to enter the shell in case of vacuum conditions developing inside the shell. For an exchanger such as this, the shell side should already be rated for full vacuum so the vacuum breaker is not a pressure (vacuum) relief device. Development of vacuum in the shell could allow condensate to build in the unit and water hammer may result.
- What are the practices for the fabrication of heat exchanger shells?
  - Commonly available steel pipe is generally used up to 60 cm (24 in.) in diameter.
  - Above 60 cm (24 in.), manufactures use rolled and welded steel plate, which is more costly and obtaining roundness is a critical issue.
  - Roundness and consistent shell inside diameter is necessary to minimize the space between the baffle

outside edge and the shell as excessive space allows fluid bypass and reduced performance.
- Roundness can be increased by expanding the shell around a mandrel or double rolling after welding the longitudinal seam. In extreme cases, the shell can be cast and then bored to the correct inside diameter.

- What are the constructional features of tube sheets?
  - Tube sheets are usually made from a round, flat piece of metal. Holes are drilled for the tube ends in a precise location and pattern relative to one another.
  - Tube sheets are manufactured from the same range of materials as tubes.
  - The tube sheet is in contact with both fluids, so it must have corrosion resistance allowances and metallurgical and electrochemical properties appropriate for the fluids and velocities.
  - Low carbon steel tube sheets can include a layer of a higher alloy metal bonded to the surface to provide more effective corrosion resistance without the expense of using the solid alloy.
  - In cases where it is critical to avoid fluid intermixing, a double tube sheet can be provided. In this design, the outer tube sheet is outside the shell circuit, virtually eliminating the chance of fluid intermixing. The inner tube sheet is vented to atmosphere, so any fluid leak is detected easily.
- How are tubes fixed to tube sheets? Describe the practices and give a typical illustration.
  - Tube ends are expanded onto thick tube sheets by the use of rollers or pneumatic or hydraulic pressure for a length of two tube diameters, or 50 mm or tube sheet thickness minus 3 mm. Holes are drilled in the tube sheets, which can be machined with one or more grooves for increased joint strength, for inserting tube ends (Figure 10.9).

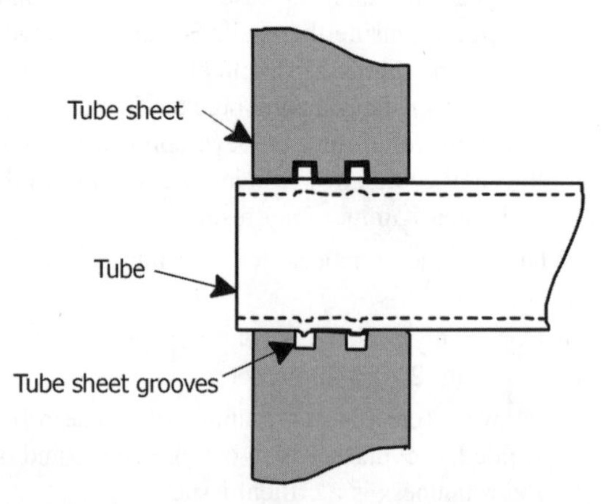

FIGURE 10.9 Grooves are made in the tube sheet for increased jointing between tube and tube sheet.

  - Properly rolled joints have uniform tightness to minimize tube fractures, stress corrosion, and tube sheet deformation.
  - Tube holes are typically drilled in the tube sheet and then reamed and can be machined with one or more grooves. This greatly increases the strength of the tube joint (Figure 10.9).
  - For moderate general process requirements at gauge pressures less than 2058 kPa and less than 177°C, tube sheet holes without grooves are standard. For all other services with expanded tubes at least two grooves in each tube hole are common. The number of grooves is sometimes changed to one or three in proportion to tube sheet thickness.
  - Expanding the tube into the *grooved tube holes* provides a stronger joint but results in greater difficulties during tube removal.
  - Figure 10.10 illustrates different tube-to-tube joints.
  - Some manufacturers offer a low-cost design that brazes the tubes to a thin tube sheet.
  - Failed tubes are either plugged or replaced, depending on the design.
- Illustrate different types of expansion joints used in heat exchangers.
  - Figure 10.11 illustrates some typical expansion joints used in heat exchangers.
  - Internal bellows designs are also used for applications such as waste heat thermosiphon reboilers, where only one pass is permitted on the tube side. These bellows have been designed to successfully operate with high-pressure boiling water on the tube side and high-temperature reactor effluent gas on the shell side.
- How can provision be made for thermal expansion of the shell of a heat exchanger? Illustrate.
  - Shell is designed and fabricated with an expansion joint (Figure 10.12).
- Under what circumstances, an expansion joint is needed on the shell of a heat exchanger?
  - A fixed tube sheet exchanger does not have provision for expansion of the tubing when there is a difference in metal temperature between the shell and tubing. When this temperature difference reaches a certain point, an expansion joint in the shell is required to relieve the stress.
  - It takes a much lower metal temperature difference when the tube metal wall temperature is higher than the shell metal wall temperature requiring an expansion joint. Typically an all steel exchanger can take a maximum of approximately 25°C metal temperature difference when the tube side is the hottest. When the

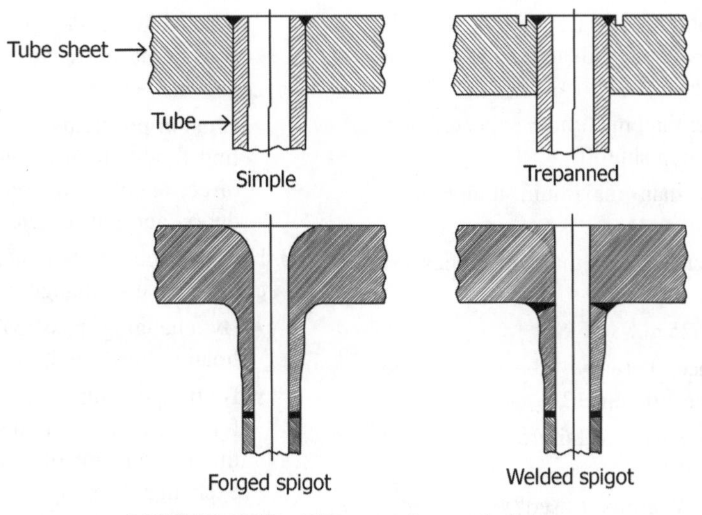

FIGURE 10.10 Different tube-to-tube joints.

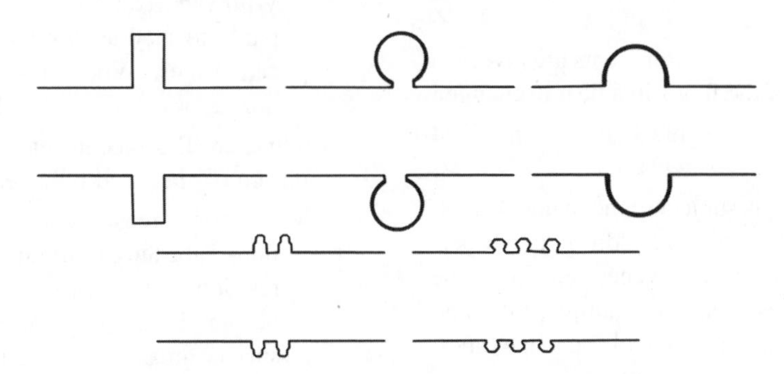

FIGURE 10.11 Different types of expansion joints.

shell side is the hottest, the maximum is typically 85°C.

- Usually if an expansion joint is required, it is because the maximum allowable tube compressive stress has been exceeded. According to the TEMA procedure for evaluating this stress, the compressive stress is a strong function of the unsupported tube span. This is normally twice the baffle spacing.

• What is a double tube sheet design? When is it used? Illustrate.

- If mixing of the shell side and tube side fluids cannot be tolerated, the double tube sheet design, shown in Figure 10.13, is used for extra protection. Because leaks could occur at the tube sheets, either the shell or tube side fluid will collect in the space between both tube sheets. It is unlikely that both tube sheets will leak simultaneously.

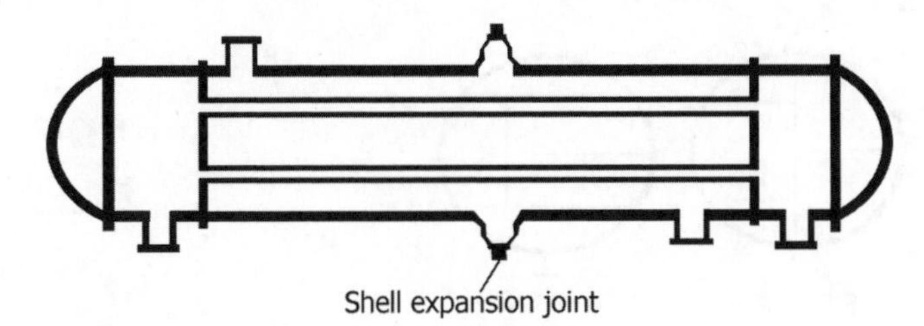

FIGURE 10.12 Illustration showing shell expansion joint.

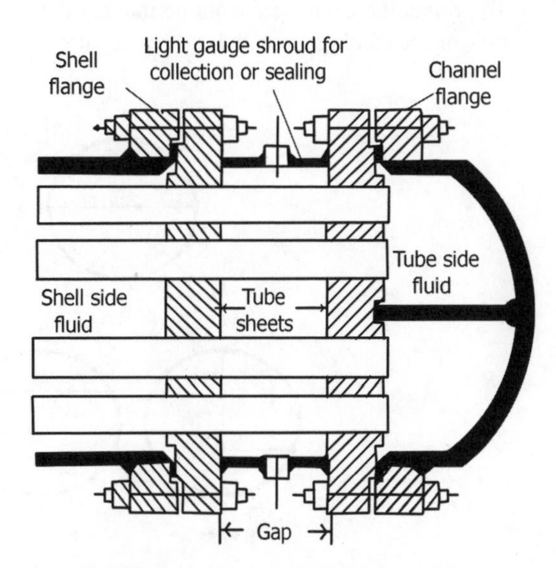

FIGURE 10.13 Double tube sheet design.

- What are the normal margins of safety adopted while specifying design pressures and temperatures for TEMA-specified exchangers?
  - About 200 kPa greater than maximum expected during operation or at pump shutoff.
  - About 14°C greater than maximum temperature while in service.
- What are the recommended maximum distances for tube supports?
  - Not more than 1.0–1.25 m.
- Why cannot the distances between tube supports be more than the recommended values?
  - Tubes can sag and mechanical failure can occur if distances are increased.
- What is a duplex tube? Where is it used?
  - It is a bimetallic tube involving a combination of two different materials.
  - It is used where corrosive environments are involved with respect to one of the fluids in a heat exchanger.
- What are the sources of mechanical stresses in a heat exchanger?
  - Every heat exchanger is subject to mechanical stresses from a variety of sources, in addition to temperature gradients. Stresses are generated from the fabrication techniques used, for example, tube and tube sheet stresses resulting from rolling in the tubes.
  - During fabrication, shipping, and installation of the exchanger, many stresses could develop. These could be on account of inadequate support structure, stresses from the connecting piping, stresses occurring during normal operation, process stream conditions such as pressures and pressure fluctuations, start-up, and shutdown, vibration, process upsets, and so on.
  - To protect the exchanger from permanent deformation or weakening from these stresses, it becomes necessary to design the exchanger, taking into consideration such developments as mentioned above.
  - Erosion problems should be controlled by controlling flow velocities, especially in bends and other direction change areas, filtering the fluids from debris and particulate matter and other measures.
- How are the number of tube passes increased for an existing heat exchanger? Illustrate.
  - By changing pass partitions in the channel and floating heads as illustrated in Figure 10.14.
  - B arrangements accommodate less number of tubes for a given shell diameter than A arrangements. For this reason, use of B arrangements is restricted to exchangers having appreciable pass-to-pass temperature differences or to other special cases.
  - With vaporization or condensation, nozzles are normally located as close as possible, to top or bottom centerlines. Where no phase change is involved, nozzle orientations may be rotated by 90°.
- What are the mechanical constraints involved in the design of shell and tube heat exchangers?
  - For some reasons, for example, layout considerations, tube lengths might have to be restricted. Such a restriction can have important implications for the design. In the case of exchangers requiring large surface areas, the restriction drives the design toward large tube counts.
  - If such large tube counts lead to low tube side velocities, the designer is tempted to increase the number of tube passes in order to maintain a reasonable tube side heat transfer coefficient.
  - Thermal expansion considerations can also lead the designer to opt for multiple tube passes, because the cost of floating head is generally lower than the cost

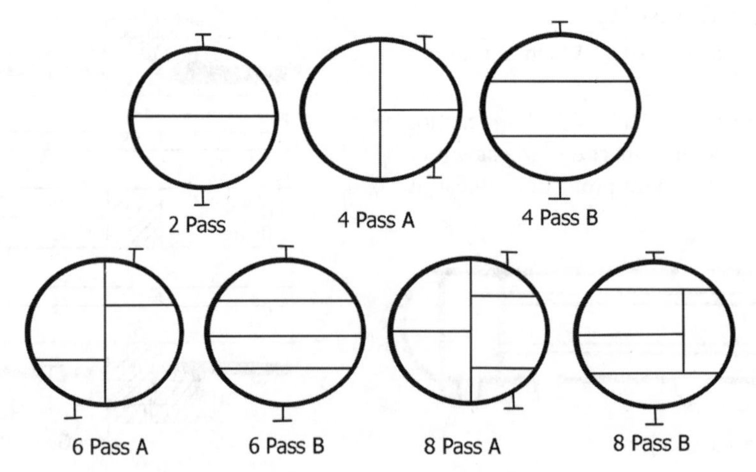

FIGURE 10.14 Typical pass partitions for two to eight tube passes.

of installing an expansion bellows in the exchanger shell.

- What are the relationships between velocity and (i) heat transfer coefficient and (ii) pressure drop through tubes in a heat exchanger? How are tube side velocities optimized in a heat exchanger?
  - Velocity strongly influences heat transfer coefficient. For turbulent flow, the tube side heat transfer coefficient varies to the 0.8 power of tube side velocity, whereas tube side pressure drop varies to the *square* of velocity.
  - Thus, with increasing velocity, pressure drop increases more rapidly than does the heat transfer coefficient.
  - Consequently, there will be an optimum velocity above which it will not be useful to increase it any further.
  - Furthermore, very high velocities lead to tube vibration and erosion.
  - Tube side pressure drop rises steeply with an increase in the number of tube passes. Consequently, it often happens that for a given number of tubes and two passes, the pressure drop is much lower than the allowable value, but with four passes it exceeds the allowable pressure drop.
  - If, in such circumstances, a standard tube has to be employed, the designer may be forced to accept a rather low velocity. However, if the tube diameter and length varied, the allowable pressure drop can be better utilized and a higher tube side velocity realized.
- Compare pressure drops in 1–1 and 1–2 heat exchangers.
  - In a 1–1 heat exchanger, the flow cross-sectional area for the fluid is the sum of the cross-sectional areas of *all* the tubes, whereas in a 1–2 exchanger only half the total tube cross-sectional area is available for the flow. Therefore, tube velocity doubles in a 1–2 exchanger over that in a 1–1 exchanger.
  - As pressure drop is proportional to the square of velocity, doubling the velocity results in four times increase in pressure drop in a 1–2 exchanger, compared to that for a 1–1 exchanger. Allowance to this is to be given for the decrease in flow cross section due to increased fouling in a 1–1 exchanger, on account of lower velocities, which increases pressure drop due to fouling over that in a clean exchanger.
- What are the considerations involved in deciding the value of allowable pressure drop for a heat exchanger?
  - Allowable pressure drop for a given heat exchanger cannot be arrived at in isolation.
  - If pressure drop requirements are low, using four or more tube passes should be avoided as this will drastically increase pressure drop.
  - It is important to realize that the total pressure drop for a given stream must be met.
  - The distribution of pressure drop in the various heat exchangers for a given stream in a particular circuit may be varied to obtain good heat transfer in all the heat exchangers. Low ΔP exchangers perform poorly for effective heat transfer, as these involve low fluid velocities.
  - High fouling fluids require higher velocities to prevent deposits, with the consequence of higher ΔP requirements.
  - Also it must be noted that higher viscosity liquids require higher ΔP for good heat transfer.
  - Normally, a pressure drop of 70 kPa per shell is permitted for liquid streams. If, for example, there are five such exchangers through which a particular stream is flowing, a total pressure drop of 340 kPa for the circuit would be permitted.
  - If the pressure drop through two of these exchangers turns out to be only 80 kPa, the balance of 260 kPa would be available for the other three.
- What are the typical pressure drops in a shell and tube heat exchanger?
  - In most applications, pressure drops are higher on the tube side compared to the pressure drops on the shell side.
  - Typical pressure drops are 30–60 kPa (5–8 psi) on the tube side and 20–30 kPa (3–5 psi) on the shell side.
- Give equations for estimation of tube side pressure drops in shell and tube heat exchangers.
  - Table 10.2 gives tube side pressure drops in shell and tube heat exchangers.
- "Maximum recommended velocities through nozzle connections and piping associated with shell and tube heat exchangers will be more for low-viscosity liquids compared to high-viscosity liquids." *True/False*?
  - *True.*
- What are the consequences of low fluid velocities in a heat exchanger?
  - Poor heat transfer coefficients.
  - Increased fouling rates.
  - Tube plugging for high-viscosity liquids.
- What is the normal recommended range of liquid velocities in heat exchanger tubes?
  - Not less than 0.9 m/s and not more than 4 m/s. Generally 1–2 m/s.

TABLE 10.2 Tube Side Pressure Drops in Shell and Tube Heat Exchangers

Section	ΔP in Terms of No. of Velocity Heads	Equation
Entrance and exit of the exchanger	1.6	$\Delta h = 1.6v_p^2/2g$ (10.1)
Entrance and exit of the tubes	1.5	$\Delta h = 1.5(v_t^2/2g)N$ (10.2)
End losses in tube side bonnets and channels	1.0	$\Delta h = 1.0(v_t^2/2g)N$ (10.3)
Straight tube losses	Calculated using equation for pipe ΔP	

Δh is the head loss in feet of flowing fluid, v_p is the velocity in the pipe leading to and from the exchanger (ft/s), v_t is the tube velocity, and N is the number of tube passes.

- Typical velocities in the tubes should be 1–3 m/s (3–10 ft/s) for liquids and 15–30 m/s (50–100 ft/s) for gases.
- The number of tubes is selected such that the tube side velocity for water and similar liquids are in the above ranges.
- The lower velocity limit corresponds to limiting the *fouling* and the upper velocity limit corresponds to limiting the rate of *erosion*.
- When sand and silt are present, the velocity is kept high enough to prevent settling.

• What is the minimum recommended velocity for liquid flow through heat exchanger tubes to prevent solids deposition?
  - 1.5 m/s.

• What is the effect of using excessively high fluid velocities in heat exchangers?
  - Excessive erosion rates.

• What are the maximum recommended design fluid velocities for flow inside tubes to minimize erosion problems with different materials of construction?
  - *Water:*

	m/s
Low carbon steel	3
Stainless steel	4.5
Aluminum	2
Copper	2
90–10 Cupronickel	3
70–30 Cupronickel	4.5
Titanium	15

  - *Other Liquids:*

$$\text{Recommended velocity for the given liquid} = \text{recommended velocity for water} \times (\rho_{water}/\rho_{liquid})^{1/2}. \quad (10.4)$$

  - *Gases and Dry Vapors for Steel Tubing:*

$$V(\text{ft/s}) = 1800/[\sqrt{(\text{absolute pressure in psia})}(\text{molecular weight}). \quad (10.5)$$

For other materials, maximum recommended velocities may be taken in the same ratio as for water.

• For a given flow rate, how can one increase tube side velocities in a heat exchanger?
  - By increasing number of tube passes and by decreasing number of tubes per pass.
  - Tube side velocity $\propto$ (tube side flow)/(number of tubes per pass).

• In which of the following cases, velocities can be controlled more closely: (a) tube side and (b) shell side?
  - Shell side. Any design velocity can be achieved by changing baffle spacing.
  - On tube side, velocity changes in larger increments with change in number of tube passes:
    - Two tube passes: 4 cm/s.
    - Four tube passes: 8 cm/s.
    - Six tube passes: 12 cm/s.
  - There is no way to design for a tube velocity of, say, 9 cm/s.

• What is the normal range of shell side liquid velocity?
  - 0.6–1.5 m/s (2–5 ft/s).
  - For water flow on shell side, cross-flow velocities of the order of about 1.0–1.5 m/s (3–5 ft/s) are usually employed.
  - For other fluids, shell side cross-flow velocities may be estimated from the following equation:

$$V(\text{ft/s}) = 30/(\rho)^{0.5}, \quad (10.6)$$

where ρ is in lb/ft^3.

• What are the tube sizes used in heat exchangers? What are the commonly used tube sizes and tube pitch arrangements for a shell and tube heat exchanger?
  - 12.7, 19, 25.4, 31.7, 38, 51 mm (0.5, 0.75, 1.0, 1.25, 1.5, 2.0 in.).
  - Most commonly used tubes are 19 mm (0.75 in.) in outer diameter. Tubes smaller than 19 mm (0.75 in.)

should not be used for fouling service as cleaning becomes difficult. Therefore, small tube diameters are used for clean fluids. Tubes with diameter 12.7 mm (0.5 in.) are used for small exchangers with heat transfer areas less than 20–30 m^2.

  - Tubes with diameter 19 mm (0.75 in.) and 4.9 m (16 ft) length on a 25.4 mm (1 in.) triangular pitch are common. Tubes with diameter 25.4 mm (1 in.) are also common.
  - Normally, tube lengths are 2.44, 3.66, 4.88, and 6.10 m (8, 12, 16, and 20 ft). Tube length of 4.9 m (16 ft) is common.

- Under what circumstances, use of 19 mm (0.75 in.) tubes is cheaper than use of 25.4 mm (1 in.)?
  - An exchanger using 19 mm (0.75 in) tubes will be cheaper than that using 25.4 mm tubes when the tube side has a much lower heat transfer coefficient than the outside of the tubes, under conditions of laminar flow in a two-tube pass exchanger. If four tube passes are used, the 19 mm tubes will have to be significantly shorter than allowed in order to meet pressure drop requirements. On the other hand, the 25.4 mm (1 in.) tube design uses the full allowable tube length.
- How much tube surface can be accommodated in shells of (i) 30 cm (1 ft) diameter, (ii) 60 cm (2 ft) diameter, and (iii) 90 cm (3 ft) diameter?
  - A 1 ft (30 cm) shell will contain about 9.3 m^2 (100 ft^2).
  - A 2 ft (60 cm) shell will contain about 37.2 m^2 (400 ft^2).
  - A 3 ft (90 cm) shell will contain about 102 m^2 (1100 ft^2).

10.1.1 Tube Pitch

- What is meant by tube pitch? Define.
  - Pitch is the shortest center-to-center distance between two adjacent tubes in a heat exchanger.
- What are the different ways by which tube pitch is arranged in a heat exchanger? Illustrate.
  - (i) Square pitch (90°), (ii) rotated square pitch (45°), (iii) triangular pitch (30°), and (iv) rotated triangular pitch (60°).
  - Figures 10.15 and 10.16 illustrate types of tube pitch.
- "Square or rotated square pitch accommodates more tubes than triangular or rotated triangular pitch." *True/False*?
  - *False:* It is the other way round. Triangular or rotated triangular pitch accommodates more tubes.

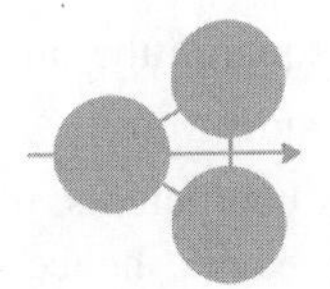

Triangular pitch (30°)

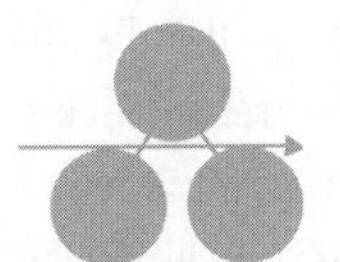

Rotated triangular pitch (60°)

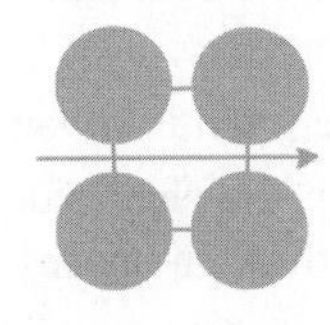

Square pitch (90°)

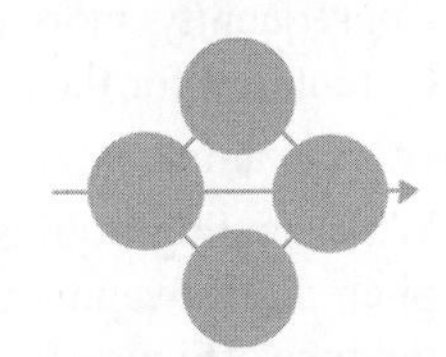

Rotated square pitch (45°)

FIGURE 10.15 Types of tube pitch.

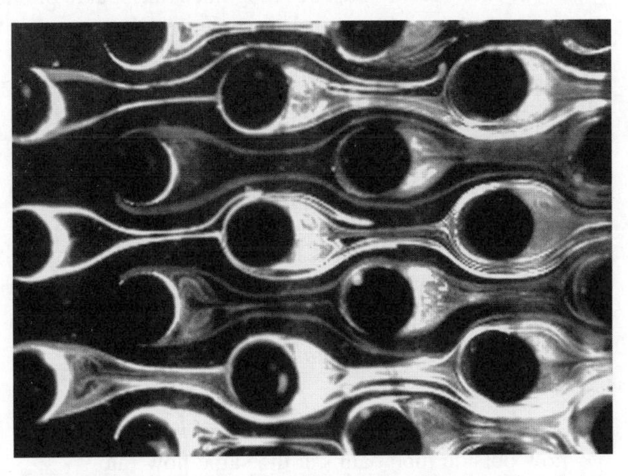

FIGURE 10.16 Shell side flow patterns in triangular pitch.

- Compare heat transfer and pressure drop characteristics of arranging tubes in square pitch and triangular pitch.
  - *Triangular Pitch:* Higher turbulence and hence higher h and higher ΔP.
  - *Square Pitch:* Less turbulence and hence poorer heat transfer and lower ΔP.
- Arrange the following in the order of increasing ΔP for the case of flow outside tube banks, flow rate remaining the same.
  1. Tube arrangement with triangular pitch (staggered).
  2. Tube arrangement for square pitch (in-line).
  3. Use of finned tubes with staggered arrangement.
     - 2, 1, 3.
- What are the advantages of square tube pitch in a heat exchanger?
  - Lower ΔP.
  - Ease of mechanical cleaning.

- What are the disadvantages of square tube pitch?
  - Poor turbulence on the shell side.
  - Lower shell side heat transfer coefficients, requiring larger heat transfer surface area and hence number tubes for the same duty.
  - Consequently, more expensive larger size shell will be required for the same heat duty.
- Give an example where square pitch is advantageous?
  - When boiling liquids are on the shell side. Square pitch allows vapors to rise through spaces between the tubes (Figure 10.17).
- What type of tube pitch arrangement is normally recommended for fouling fluids?
  - Square pitch is amenable for mechanical cleaning on the shell side. It must be recognized that square pitch does not produce much turbulence that promotes greater fouling deposit rates. When shell side Reynolds numbers are low (<2000), a rotated square pitch is advantageous as it produces much higher turbulence.
  - However, triangular pitch can be used if chemical cleaning is suitable and effective, as it does not require straight access lanes.
- "In U-bundle heat exchangers, ease of cleaning of tube exteriors is more if triangular tube pitch is used instead of square pitch." *True/False*?
  - *False.*
- Arrange the following in the order of *increasing* ΔP for flow outside tube banks at the same flow rate.
  1. Tube arrangement with triangular pitch (staggered).
  2. Tube arrangement with square pitch (in-line).

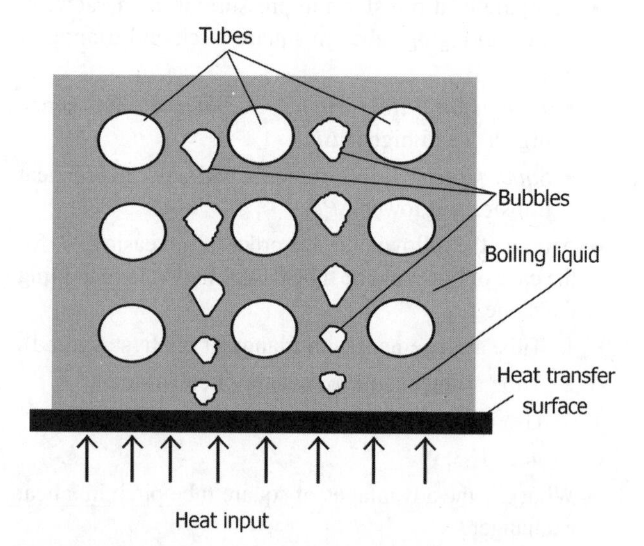

FIGURE 10.17 Vapor bubbles rising through boiling liquid inside a heat exchanger with square tube pitch.

  3. Use of finned tubes with staggered arrangement.
  - 2. In-line square pitch.
    1. Triangular staggered pitch.
    3. Finned tubes with staggered arrangement.
- "Arrangement of tubes in a heat exchanger with square pitch is better suited for access for cleaning rather than triangular pitch." *True/False*?
  - *True.*
- "Shell side heat transfer coefficients are higher for square pitch compared to triangular pitch." *True/False*?
  - *False.*
- What are the normally recommended values of tube pitch?
  - Values of tube pitch depend on the tube diameter. For example, TEMA specifies a minimum tube pitch of 1.25 times the tube O.D. for triangular pattern. Thus, a 25 mm pitch is usually used for 20 mm O.D. tubes.
  - For square patterns, TEMA additionally recommends a minimum cleaning lane of 6 mm (4 in.) between adjacent tubes. Thus, the minimum tube pitch for square patterns is either 1.25 times the tube O.D. or the tube O.D. plus 6 mm, whichever is larger. For example, 20 mm tubes should be laid on a 26 mm (20 mm + 6 mm) square pitch, but 25 mm tubes should be laid on a 31.25 mm (25 mm × 1.25 mm) square pitch.
  - Normally the minimum specified values of tube pitch are used in the design as it leads to the smallest size shell for a given number of tubes.
  - However, in exceptional circumstances, the tube pitch may be increased to a higher value, for example, to reduce shell side pressure drop. This is particularly true in the case of cross-flow shell.
  - Normally used values of tube pitch are 25, 30, and 38 mm.
- What are the advantages and disadvantages of using closer tube pitch?
  - *Advantages*
    - Larger number of tubes is fitted in a given shell or for a given number of tubes; a smaller diameter shell will be required.
    - Reduces initial cost.
  - *Disadvantages*
    - Closer pitch.
    - Sagged tubes, particularly in hot service, touch each other after some use.
      - When the tubes are arranged too close to each other, while more tubes can be accommodated

for a given shell diameter, tube sheet becomes too weak.

- Access for maintenance of tube bundles becomes more difficult.

- What is the recommended range of optimum tube pitch to tube diameter ratio for conversion of pressure drop to heat transfer?
  - Pressure drop–heat transfer relationships: Higher turbulence is achieved at the expense of higher pressure drops.
  - Increased turbulence increases heat transfer rates.
  - In the design of exchangers, these two variables are to be optimized in order to achieve the best possible energy economy in terms of heat transferred and energy expended on pumping to overcome pressure drops.
  - The optimum tube pitch to tube diameter ratio for conversion of pressure drop to heat transfer is typically 1.25–1.35 for turbulent flow and around 1.4 for laminar flow.
- What are the reasons for not increasing tube pitch to reduce pressure drop?
  - There are two reasons:
    (i) It increases shell diameter that is the single variable for increased equipment costs.
    (ii) Reducing pressure drop by modifying the baffle spacing, baffle cut, or shell type will result in a cheaper design.

10.1.2 Baffles

- What are the functions of baffles used in a heat exchanger?
  - *First Function:*
    - ➢ To support tubes.
    - ➢ Must be supported at intervals of not more than 1.5 m (5 ft) depending on diameter and material.
    - ➢ Shorter intervals where flow-induced vibrations occur.
    - ➢ To prevent mechanical vibrations, sagging that results in tubes touching each other and giving rise to leakages and failures especially near tube sheets.
    - ➢ Supporting distances depend on outside diameter of the tubes and materials of construction as illustrated in Table 10.3.
  - *Second Function:*
    - ➢ To direct flow through the shell in the desired pattern.
    - ➢ To enable a desirable velocity to be maintained for the shell side fluid.
    - ➢ To increase turbulence and reduce stagnant pockets in the exchanger.

TABLE 10.3 Maximum Unsupported Span for the Tubes

	Maximum Unsupported Span (mm)	
Approximate O.D. of Tube (mm)	Group A Materials	Group B Materials
19	1520	1321
25	1880	1626
32	2240	1930
38	2540	2210
50	3175	2794

Group A: Carbon and high alloy steel, low alloy steel, nickel–copper, nickel, nickel–chromium–iron. Group B: Aluminum and aluminum alloys, copper and copper alloys, titanium and zirconium.

- Show by means of a diagram how a baffled shell and tube heat exchanger look like.
  - Figure 10.18 is a cut section view of a baffled shell and tube heat exchanger.
- Why is it necessary to use baffles in a condenser?
  - In a condenser, necessity to increase turbulence on the condensing side is not there as heat transfer coefficients for condensing vapors are quite large. But it will be necessary to support tubes.
- Name different types of baffles used in a heat exchanger.
  - (i) Segmental, (ii) disk and doughnut, (iii) orifice, and (iv) rod baffles.
  - Segmental, disk and doughnut, and orifice baffles are plate-type baffles.

FIGURE 10.18 Photograph of a cutaway of a baffled shell and tube heat exchanger.

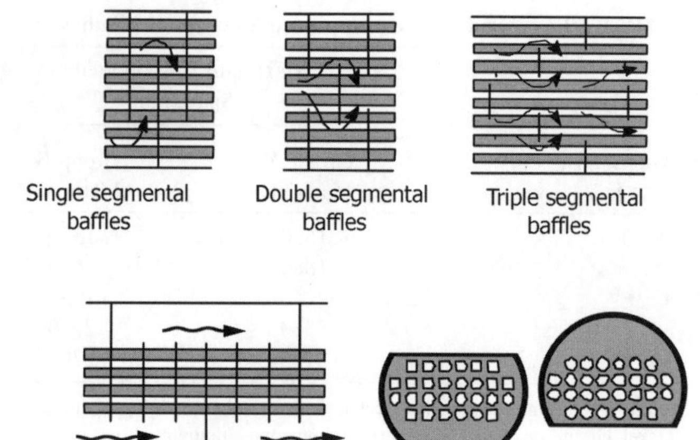

FIGURE 10.19 Different arrangements for segmental baffles.

- What is the purpose of a segmental baffle?
  - Segmental baffles direct flow across tube bundle, perpendicular to tube axis.
- Show the arrangements for different types of baffles in a heat exchanger.
  - Plate baffles may be single segmental, double segmental, or triple segmental, as shown in Figures 10.19 and 10.20.
  - *Segmental Baffles:*
    - ➢ The single and double segmental baffles are most frequently used. They divert the flow most effectively across the tubes.
    - ➢ The triple segmental baffles are used for low-pressure applications.
  - *Disk and Doughnut Baffles:* These baffles are composed of alternating outer rings and inner disks,

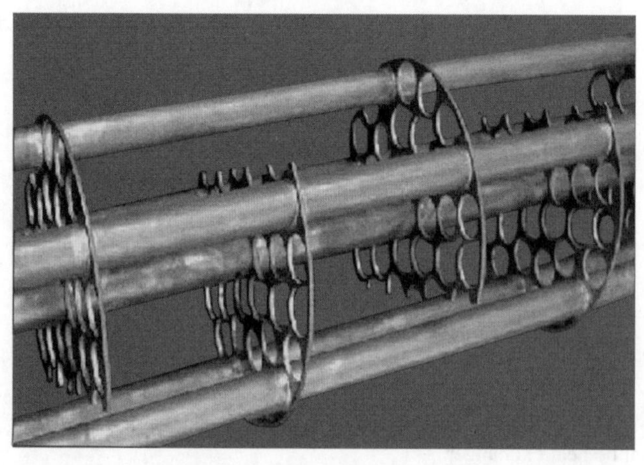

FIGURE 10.20 Segmental baffles.

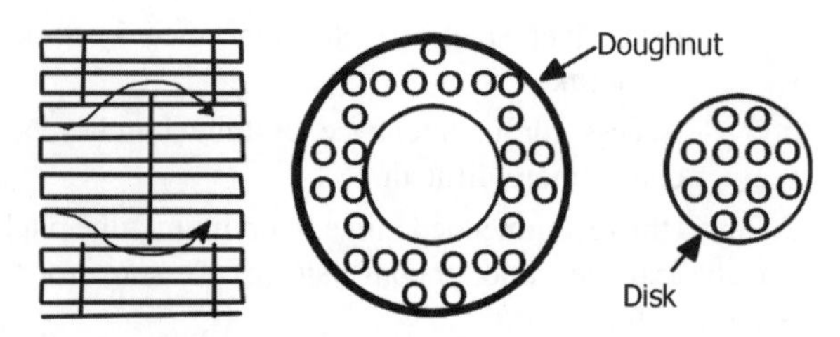

FIGURE 10.21 Disk and doughnut baffle.

which direct the flow radially across the tube field (Figure 10.21).
    - ➢ The potential bundle-to-shell bypass stream is eliminated.
    - ➢ This baffle type is very effective in pressure drop to heat transfer conversion.
  - *Orifice Baffles:* In an orifice baffle, shell side fluid flows through the clearance between tube outside diameter and baffle hole diameter (Figure 10.22).
  - *Rod Baffles:* Metal rods rather than sheet metal baffles are used to support tubes and flow in the shell is mainly parallel to tube axis (Figure 10.23).
  - Eliminate tube vibrations, which occur with plate baffles when fluid velocities are high and used wherever very low pressure drop is required.
    - ➢ Flow in the shell is mainly parallel to the tube axis. Flow across the rods leads to vortex formation, giving heat transfer coefficients for turbulent flow about 1.5 times those predicted by the Dittus–Boelter equation for the same Reynolds number, but are not as high as those for a segmentally baffled exchanger with close baffle spacing.
    - ➢ Lower ΔP and reduced vibration make the rod baffle exchanger preferable for many applications.
    - ➢ Tubes are arranged with square pitch. Rods, with a diameter equal to the clearance between tube rows, are attached to ring supports and placed between alternate tubes in both horizontal and vertical directions.

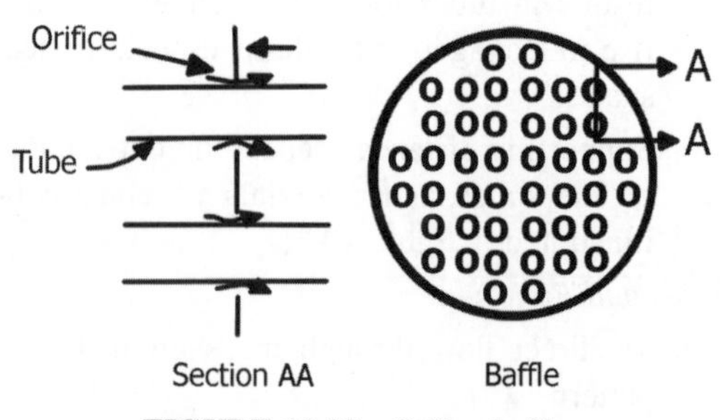

FIGURE 10.22 Orifice baffle.

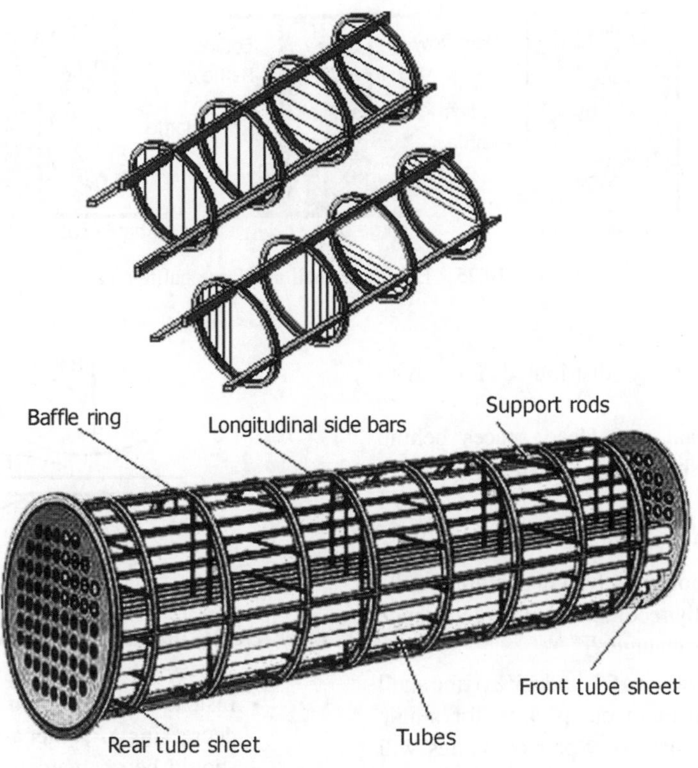

FIGURE 10.23 Rod baffles.

  - ➢ The normal rod diameter is 6.35 mm and each tube is supported on all four sides at several points along the exchanger.
  - ➢ Not good for large diameter shells.
- ▪ *Twisted Tubes:* Tubes are twisted into a helical shape with an oval cross section, so that each tube is supported over its entire length by multiple contact points with adjacent tubes. The end sections are kept circular to permit mounting in standard tube sheets.
  - ➢ Due to higher turbulence both inside and outside of the tubes, higher heat transfer rates are achieved, reducing heat transfer area requirements for the same duty, thus offsetting increased costs.
  - ➢ Fouling rates are reduced and tube vibrations are eliminated.
  - ➢ Flow distribution will not be even in both rod baffle and twisted tube designs for large diameter shells. With a single inlet nozzle, tubes near the inlet have more than the average flow and those opposite to inlet will have low flows for an appreciable distance down the exchanger.

- Under what circumstances, disk and doughnut baffles are used in a heat exchanger?
  - ▪ Where ΔP available is very small, that is, where there is pressure drop problem.

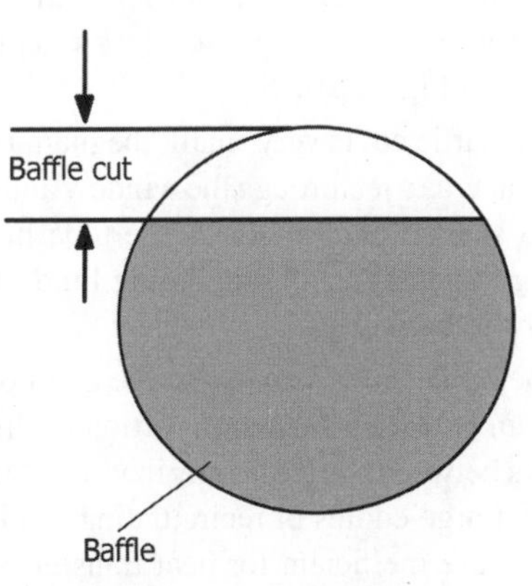

FIGURE 10.24 Baffle cut.

- What is meant by baffle cut? Illustrate.
  - ▪ Baffle cut is the height of the segment that is cut in each baffle to permit the shell side fluid to flow across the baffle. Baffle cut is expressed as the percentage of shell inside diameter not covered by the baffle (Figure 10.24).
- What is the advantage and disadvantage of using (i) higher baffle cut and (ii) lower baffle cut than the normally recommended value of 20–25% of shell diameter? Give brief answers.
  - (i) *Larger Cut*: Less ΔP.

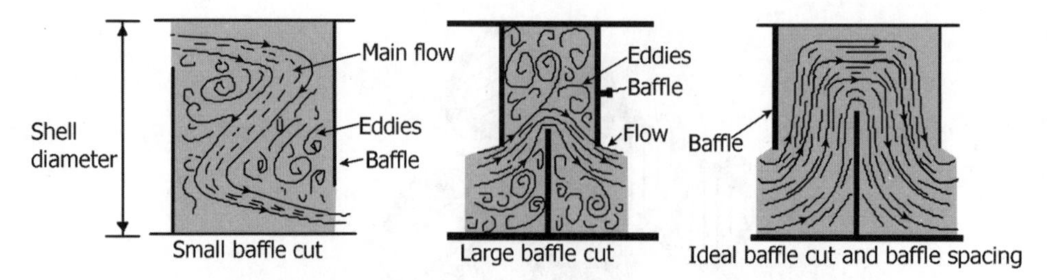

FIGURE 10.25 Effect of small and large baffle cuts.

➢ *Disadvantage:* Poorly distributed flow with large eddies.
  - Creates stagnant areas/dead spaces behind baffles in the shell.
  - Lower h.

(ii) *Smaller Cut*: High h and high ΔP.

- What are the considerations involved in deciding baffle cut? What is the normally recommended percent range of baffle cut in a heat exchanger?
  - Baffle cut can vary between 15% and 45% of the shell inside diameter. Maximum cut is 45% for single segmental baffles so that every pair of baffles will support each tube.
  - Both very small and very large baffle cuts are detrimental to efficient heat transfer on the shell side due to large deviation from an ideal situation, as illustrated in Figure 10.25.
  - If the baffle cut is very small, the main portion of the flow acts as a jet through the window and then follows an S-shaped pattern across the tube bundle, generating large eddies of circulating fluid in the regions near the baffle tips.
  - If the baffle cut is very large, the main portion of the stream bypasses the major portion of the bundle and flows between the baffle tips in virtually longitudinal flow. Large eddies of recirculating fluid are created, which are inefficient for heat transfer.
  - The ideal flow pattern on the shell side is cross flow. However, the baffles that are needed to increase the shell side velocity have the negative effect of altering the ideal cross-flow pattern. Therefore, a suitable correction has to be employed to the heat transfer coefficient for the ideal tube bundle. This correction may be significant for very small and very large baffle cuts.
  - The normally recommended range of baffle cuts are between 20% and 25%.
  - Reducing baffle cut below 20% to increase the shell side heat transfer coefficient or increasing the baffle cut beyond 35% to decrease the shell side pressure drop usually leads to poor designs.

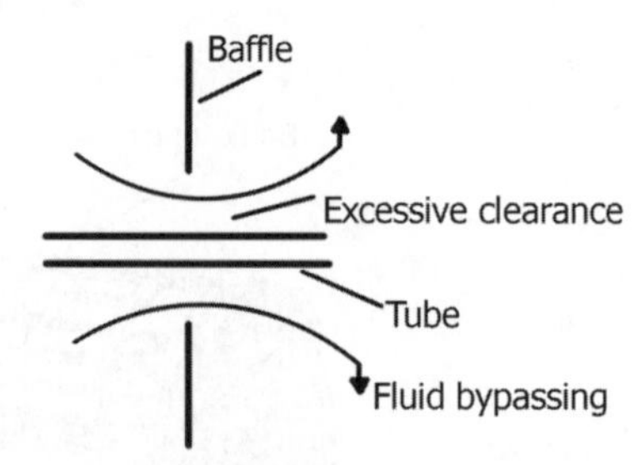

FIGURE 10.26 Large clearance between baffle and tube.

  - Instead of increasing baffle cut beyond 35% for decreasing ΔP, other aspects of tube bundle geometry should be changed instead to achieve those goals. For example, double segmental baffles or a divided flow shell (TEMA J), or even a cross-flow shell (TEMA X), may be used to reduce the shell side pressure drop.
- TEMA specifies clearances between baffles and tubes. What will be the consequences of using larger clearances?
  - Large clearances result in fluid bypassing as shown in Figure 10.26.
  - The tubes are to be inserted or withdrawn through baffle holes with ease.
  - Spacing should provide for differential expansion and contraction between baffle plate and tubes that normally have different wall thicknesses and might be of different materials of construction.
  - Certain amount of leakage through the clearances is to be permitted in order to reduce stagnant regions and make turbulent conditions more uniform and reduce shell side pressure drop.
  - Such leakage reduces deposits in stagnant areas.
  - TEMA gives recommendations on the optimum clearance requirements. Excessive bypassing occurs if these clearances are large, resulting in fluid starving regions in the shell with low fluid velocities in the shell.
- Illustrate different orientations used for baffle cuts in a shell and tube exchanger.
  - Figure 10.27 illustrates different baffle cut orientations.

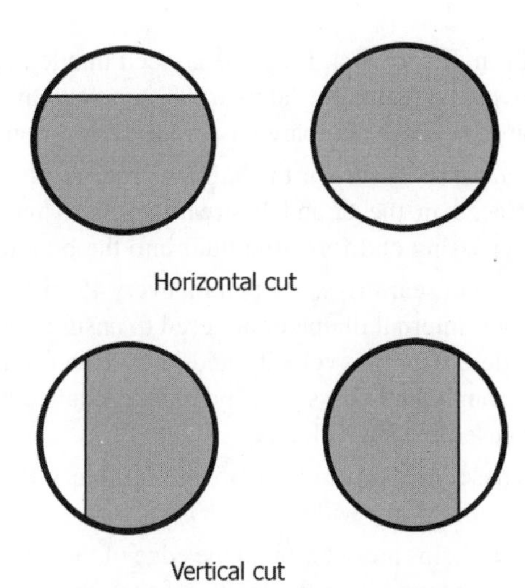

FIGURE 10.27 Baffle cut orientations.

- Give examples when horizontal and vertical baffle cut orientations are used.
  - For single-phase fluids on the shell side, a horizontal baffle cut (Figure 10.27) is recommended, because this minimizes accumulation of deposits at the bottom of the shell and also prevents stratification.
  - However, in the case of a two-pass shell (TEMA F), a vertical cut is preferred for ease of fabrication and bundle assembly.
  - For condensers and vaporizers, vertically cut baffles are usually employed to enable condensed or vaporized fluid to separate from the uncondensed vapor or unvaporized liquid.
  - If the condensation or vaporization is in the shear controlled regime, separation of phases will not take place and horizontally cut baffles should be used.
  - This is particularly true for partial condensers or partial vaporizers.
- What is baffle spacing/pitch?
  - Baffle spacing is the centerline-to-centerline distance between adjacent baffles.
- What are the considerations involved in fixing baffle spacing?
  - Baffle spacing is the most vital parameter in shell and tube heat exchanger design.
  - The TEMA standards specify the minimum baffle spacing as one-fifth of the shell inside diameter or 5 cm (2 in.), whichever is greater.
  - Closer spacing will result in poor bundle penetration by the shell side fluid and difficulty in mechanically cleaning the outside of the tubes.
  - Low baffle spacing results in a poor stream distribution.
  - The maximum baffle spacing is the shell inside diameter.
  - Higher baffle spacing will lead to predominantly longitudinal flow, which is less efficient than cross flow and large unsupported tube spans, which will make the exchanger prone to tube sagging and failure due to flow-induced vibration.
- What is the recommended optimum baffle spacing?
  - *Optimum Baffle Spacing:* For turbulent flow on the shell side ($N_{Re} > 1000$), the heat transfer coefficient varies 0.6–0.7 power of velocity.
  - However, pressure drop varies 1.7–2.0 power.
  - For laminar flow ($N_{Re} < 100$), the exponents are 0.33 for the heat transfer coefficient and 1.0 for pressure drop.
  - Thus, as baffle spacing is reduced, pressure drop increases at a much faster rate than does the heat transfer coefficient.
  - This means that there will be an optimum ratio of baffle spacing to shell inside diameter that will result in the highest efficiency of conversion of pressure drop to heat transfer.
  - This optimum ratio is normally between 0.3 and 0.6.
- Illustrate leaking paths for flow bypassing on the shell side and discuss the effects of different leakage streams on exchanger performance.
  - Figure 10.28 illustrates different leakage streams in a baffled heat exchanger.

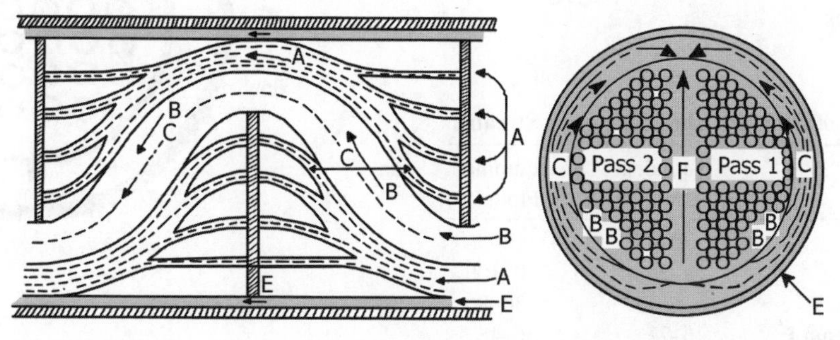

FIGURE 10.28 Leaking paths for flow bypassing the tube matrix. Both through baffle clearances between the tube matrix and the shell.

- On the shell side, there is not just one stream, as inside tube side. There are many bypass and leakage streams that reduce the baffle effectiveness.
- *Stream B* is the main effective cross-flow stream, which can be related to flow across ideal tube banks.
- *Stream A* is the leakage stream in the orifice formed by the clearance between the baffle tube hole and the tube wall.
- *Stream C* is the tube bundle bypass stream that flows in the space between the outside of tube bundle and the inside of the shell wall.
- *Stream E* that leaks through the clearance between the baffle edge and shell wall.
- *Stream F* is the bypass stream in flow channel partitions due to omissions of tubes in tube pass partitions. It occurs only in tube side multipass bundles and where the pass divider lane is oriented parallel to the cross-flow B-stream.
- The leakage and bypass streams are less efficient for heat transfer than main cross-flow stream (B-stream).
- In fact, the baffle-to-shell leakage stream (E-stream) is totally ineffective, since it does not encounter any heat transfer surface.
- Due to the leakage streams, the temperature of the main cross-flow stream (B-stream) changes rapidly and its final temperature prior to mixing before the outlet nozzle may be substantially different from the mixed outlet temperature.
- This results in a temperature profile distortion and a consequent reduction in the mean temperature difference (MTD).
- The MTD is more pronounced when the leakage/bypass streams are high, especially the baffle-to-shell leakage stream (E-stream) and/or the ratio of shell side temperature difference to the minimum approach temperature difference is high.
- The leakage/bypass streams tend to be high when the shell side viscosity is high and the baffle spacing is very low. Thus, care must be exercised in the design of coolers for viscous liquids.
- Table 10.4 illustrates the fractions of different shell side flow streams.

TABLE 10.4 Fractions of Different Shell Side Flow Streams

Flow Stream	Turbulent Flow (%)	Laminar Flow (%)
Cross-flow stream B	30–65	10–50
Leakage stream A	9–23	0–100
Tube bundle bypass stream C	15–35	30–80
Baffle-to-shell leakage stream E	6–21	6–48

- From Table 10.4, it can be noticed that leakage and bypass streams for laminar/viscous flow, the effects are very high compared to turbulent flow conditions.
- *Side strip baffles* or the baffle tie rods are frequently placed in the C and F stream paths to reduce the bypassing and force the fluid into the bundle.
- Pairs of seal strips, one pair for every 45 cm (18 in.) of shell internal diameter are used to ensure good shell side cross-flow velocity and help reduce fluid flow through the bypass area and reduce localized fouling caused by low velocity.
- These are typically 6.35 mm (1/4 in.) thick and 10 mm (4 in.) wide.
- Seal strips are set around the edge of the tube bundle and extend along the length of the tubes.
- They are inserted in grooves cut in the tube support baffles.
- Seal strips increase heat transfer efficiency by 5–10%.
- Figure 10.29 illustrates use of seal strips.
- Close manufacturing tolerances are the only means of controlling the A and E leakage streams.
- In spite of all the measures taken in manufacturing to reduce leakage streams, the leakage and bypass rates are high and as low as half the total flow goes through the bundle.

• What is the effect of baffle pitch/spacing (distance between two adjacent baffles) on h and ΔP?
  - Smaller pitch: Higher h and higher ΔP.
  - Pitch is normally set to give highest h within allowable ΔP.

• What is the minimum baffle pitch? What is the normally used value?

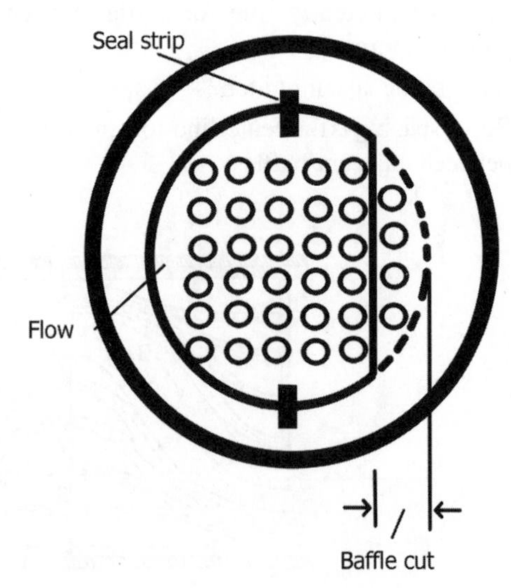

FIGURE 10.29 Seal strips reduce bypassing around tube bundle.

- Not less than 16% of shell diameter.
- Normal pitch is 20% of shell diameter.

• What is the effect of increasing number of baffles in a heat exchanger?
- Velocity and ΔP on the shell side increase.
- Doubling number of baffles doubles velocity across tube bundle and increase ΔP by four times.

• On what consideration maximum baffle spacing is based?
- Adequate support requirement for the tubes.
- Maximum unsupported tube span in inches is given by

$$74d^{0.75}, \quad (10.7)$$

where d is the outside diameter of tube, in inches.

- For copper, aluminum, and their alloys, unsupported tube span is reduced by 12% (because tubes of these materials sag more easily).

• What is the generally recommended minimum baffle spacing?
- 0.3–0.5 × (shell diameter). Not less than 2 in. (50.8 mm).

• Give an equation for the estimation of number of baffles in a heat exchanger.

$$\text{Number of baffles} = 10 \times \text{tube length}/(\%\ \text{baffle pitch})(\text{shell diameter}). \quad (10.8)$$

• Give the equations for space available for flow on the shell side and free area of flow between baffles.

$$W = D_i - (d_o \times T_{cl}), \quad (10.9)$$

where W is the space for flow, D_i is the I.D. of shell, d_o is the O.D. of tube, and T_{cl} is the number of tubes across centerline.

- Free area of flow between baffles,

$$A_f = W(B_p - 0.187), \quad (10.10)$$

where A_f is in inch^2 and B_p is baffle pitch, inch.

• Give equations for calculation of number of tubes across centerline of tube bundle.
- For square pitch,

$$T_{cl} = 1.19(\text{number of tubes})^{0.5}. \quad (10.11)$$

- For triangular pitch,

$$T_{cl} = 1.1(\text{number of tubes})^{0.5}. \quad (10.12)$$

• Summarize design guidelines for baffles.
- Baffle spacing/shell diameter = 0.2–1.
- Baffle spacing/tube diameter = 30–40.
- Baffle height/shell diameter = 0.75.

• What are the disadvantages of using baffles in the shell side and on the passes in a heat exchanger?
- Pressure drop will increase.

• What is a *Helixchanger* heat exchanger? How does it reduce fouling rates compared to conventional shell and tube exchanger?
- A *Helixchanger* heat exchanger uses quadrant-shaped baffle plates that are arranged at an angle to the tube axis in a sequential pattern, creating a helical flow path through the tube bundle as shown in Figure 10.30.
- Baffle plates act as guide vanes rather than forming a flow channel as in conventionally baffled heat exchanger. Uniformly higher flow velocities achieved in a Helixchanger heat exchanger offers enhanced convective heat transfer coefficients.
- A well-designed *Helixchanger* heat exchanger involves the following characteristics:
  - ➢ Uniform flow velocities through the tube bundle offering uniform film and metal temperatures.
  - ➢ Elimination of backflow and eddies.
  - ➢ Shell side flow approaches plug flow conditions improving the temperature driving force.
  - ➢ Higher flow velocities are achieved with correspondingly lower pressure drops.
  - ➢ Reduced shell size achieved with the *Helixchanger* heat exchanger offers higher tube side velocities as a secondary benefit in reducing the tube side fouling rates.
  - ➢ Helical baffles are recommended for fluids with viscosities over 5 cP.

• What is the purpose of an impingement plate just below the inlet nozzle on the shell side of a heat exchanger? Illustrate schematically.

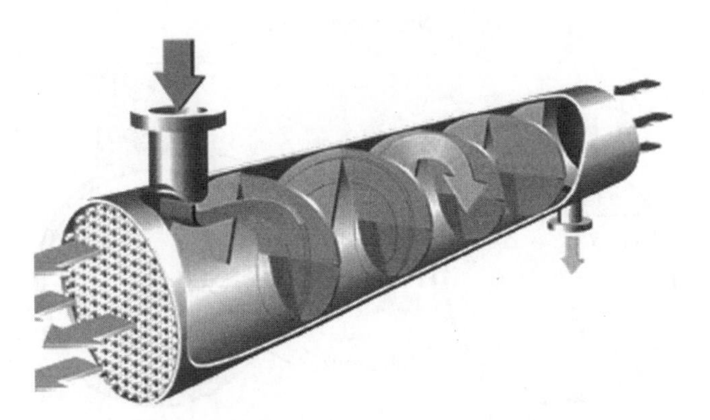

FIGURE 10.30 Helixchanger heat exchanger.

- Direct impingement of fluid jet onto the tubes from inlet nozzle into the shell can cause cavitation, vibration, and erosion.
- Impingement plate/baffle is sometimes used to divert incoming fluid jet into the shell from impacting directly at high velocity onto the top row of tubes. This ensures distribution of the fluid evenly and prevents fluid-induced erosion, cavitation, and vibration.
- Installing the impingement baffle inside the shell prevents installing a full tube bundle, resulting in less available surface.
- It can alternately be installed in a domed area above the shell. The domed area can either be reducing coupling or a fabricated dome. This style allows a full tube count and therefore maximizes the utilization of shell space.
- Figure 10.31 illustrates alternative arrangements of impingement baffle.
- Sometimes slotted distributor plate is used in place of a baffle plate for the above purpose.
- In order to accommodate such plates, it may sometimes be necessary to omit some tubes near the plate to avoid excessive pressure drop.

• What are the considerations involved for the selection of materials of construction for heat exchangers?
  - Corrosion and/or operation at elevated temperatures are the main functional considerations in material selection.
  - Requirement for low-cost, lightweight, high-conductivity, and good joining characteristics often leads to the selection of aluminum for the heat transfer surface.
  - Stainless steel is used for food processing or fluids that require corrosion resistance.
  - Commonly used tubes are seamless or welded with materials of construction usually alloy steels; copper and alloys of nickel. Titanium or aluminum may also be used for special purposes.
  - Table 10.5 gives briefly, material selection for corrosive and noncorrosive service for heat transfer surfaces.

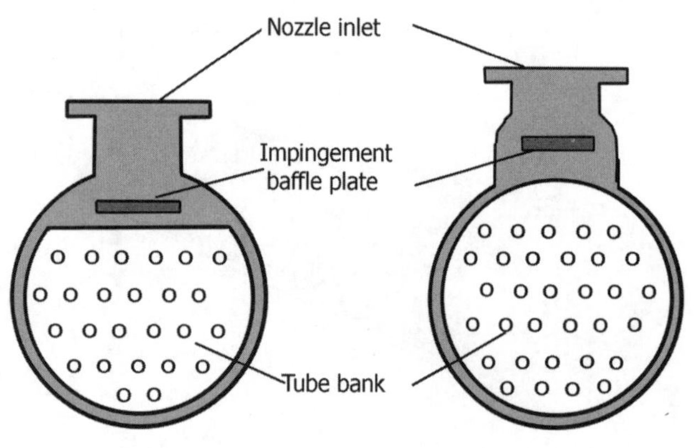

FIGURE 10.31 Use of impingement baffle.

• What are the different types of tube inserts in heat exchanger tubes? How do they work? What are their benefits and applications?
  - Twisted tapes, rotating helical coils, and wire matrix inserts are used in shell and tube heat exchanger tubes to enhance heat transfer through increased turbulence and control fouling inside the tubes.
  - Twisted tapes tend to promote intensity of mixing and turbulence. The partitioning and blockage of the tube flow cross section by the tape results in higher flow

TABLE 10.5 Selection of Materials of Construction for Heat Exchangers

Material	Exchanger Type/Typical Service
Noncorrosive Service	
Al and austenitic chromium–nickel steel	Any type, $T < -100°C$
Nickel steel (3.5% Ni)	Any type, $-100 < T < -45°C$
Carbon steel (impact tested)	Any type, $0 < T < 500°C$
Refractory-lined steel	Shell and tube, $T > 500°C$
Corrosive Service	
Carbon steel	Mildly corrosive fluids; tempered cooling water
Ferritic carbon–molybdenum and chromium–molybdenum alloys	Sulfur-bearing oils above 300°C; H_2 at elevated temperatures
Ferritic chromium steel	Tubes for moderately corrosive service; cladding for shell or channels in contact with sulfur-bearing oils
Austenitic chromium–nickel steel	Corrosion-resistant service
Aluminum	Mildly corrosive fluids
Copper alloys: admiralty, aluminum brass, cupro–nickel	Freshwater cooling in surface condensers; brackish and seawater cooling
High Ni–Cr–Mo alloys	Resistance to mineral acids and chlorine-containing acids
Titanium	Seawater coolers and condensers including plate heat exchangers
Glass	Air preheaters for large furnaces
Carbon	Severely corrosive service
Coatings: Al, epoxy resins	Exposure to sea and brackish water
Linings: lead, rubber	Channels for seawater cooling
Linings: austenitic Cr–Ni steel	General corrosion resistance

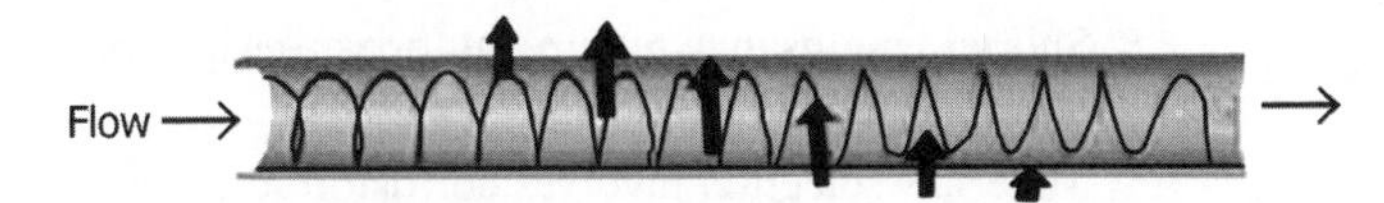

FIGURE 10.32 Rotating helical coil tube insert.

velocities. Secondary fluid motion is generated by the tape twist, and the resulting twist mixing improves heat transfer.

- Helical metal elements inside the tubes rotate making use of the energy of the flowing fluid. This rotation causes a high turbulence in the flow and thus improves the internal heat transfer coefficient. As the boundary layer is continuously renewed, the wall temperature is lowered and fouling is reduced. By proper design and installation, rotating elements can increase heat transfer coefficients by over 50% and reduce fouling to a significant effect.
- Applications include exchangers for crude oil preheating, petroleum products, and the like.
- Figure 10.32 illustrates a typical rotating helical coil insert.
  - The inserts are held in place by straight wires at each end (inlet and outlet).
- Inserts are also used in U-bundle exchangers.
- Some inserts operate with much lower shear than that occurs in plain tubes. These allow fuller use of ΔP without resulting higher erosion problems when tube velocities are increased due to the insert.
- The inserts are used for fluids in the temperature ranges of 200–350°C with flow velocities in the range of 0.6–3 m/s. Tube sizes are between 12 and 50 mm and lengths of 3–10 m.
- Figure 10.33 illustrates twisted tape inserts.
- Wire matrix inserts tend to create well-distributed flow conditions within individual tubes by continuously removing stagnant fluid from the tube wall and replacing it with fluid from the center of the tubes.
- In viscous, single-phase flow applications, wire matrix inserts are widely used to improve heat transfer characteristics. These devices have also been successfully applied for two-phase flow applications,

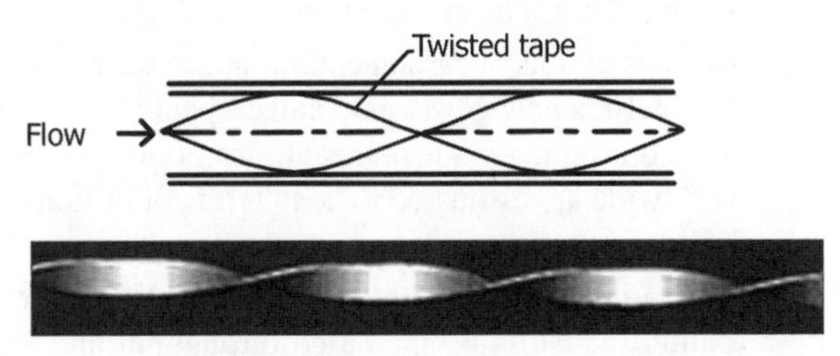

FIGURE 10.33 Twisted tape tube insert.

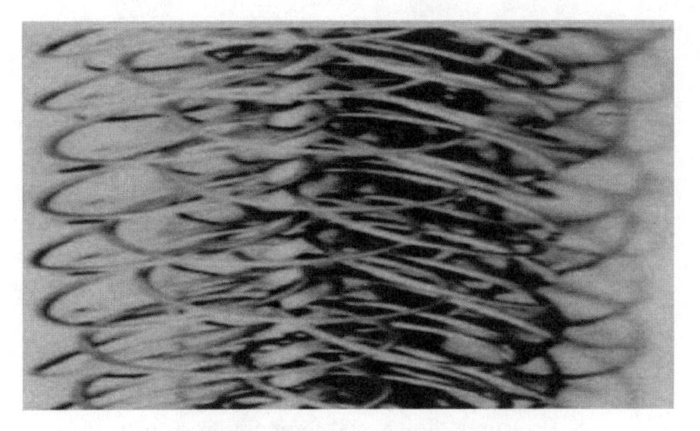

FIGURE 10.34 Wire mesh insert.

such as reflux condensers and thermosiphon reboilers. In all these applications, the boundary layer fluid is continuously displaced from the tube inner surface and remixed with the bulk fluid, thereby improving the convective heat transfer.

- Figure 10.34 illustrates wire mesh tube insert.
- The use of a full-length helically coiled insert can increase the swirl and pressure gradient in the radial direction. The boundary layer along the tube wall is thinner with the increase of radial swirl and pressure, resulting in better heat flow through the fluid.

• What are the negative factors in the use of inserts in heat transfer tubes?
  - Inserts introduce tube cleaning problems and might cause plugging of the tubes with fouling fluids.
  - Some inserts introduce higher tube side ΔP.

• What is the important information that should be provided to manufacturers of heat exchangers for proper selection and sizing?
  - *Choice of Exchanger Type:*
    - This directly affects process performance and also influences plant size and layout, the length of pipe runs and the strength and size of supporting structures.
    - Other than shell and tube options such as plate exchangers, finned tube, and other compact and microexchanger designs might be examined.
    - Enhancing features within the shell and tube exchanger such as tube inserts, helical baffles, twisted tubes, and rod baffles should also be included in the consideration.
  - *Process Conditions:*
    - Maximum, minimum, and normal flows of process fluids.
    - Design pressure and maximum allowable ΔP.
    - Heat duty expressed in terms of inlet and outlet temperatures and maximum deviations that can

occur for heating and cooling of single-phase fluids.

- *Fouling Nature of the Fluid:*
  - ➢ How much fouling allowance is to be provided and what type of fouling-resistant materials is to be specified.
    - *Fluid properties such as C_p, μ, k, and ρ at the anticipated range of operating temperatures and pressures, for both tube side and shell side fluids.*
- *Condition of Steam or Other Condensing Medium*:
  - ➢ Pressures, flows, and temperatures upstream and downstream of control valve.
  - ➢ Design conditions, maximum, minimum, and normal operating conditions.
- *Expected Turndown Ratio:* This information is required for designing proper control strategies and should be known to avoid oversizing the exchanger.
- *Materials of Construction:* Some applications stipulate special materials to ensure compatibility with the fluids to be handled.
- *Special Requirements:* Occasionally, applications require special construction requirements such as removable tube bundles or double tube sheets.

• Under what conditions, ASME standards are necessary for a heat exchanger?

- When operating pressures exceed 1 barg.

• What are the maintenance, service, layout, and economic considerations involved in the use of heat transfer equipment?

- *Maintenance, Servicing, Layout, and Economic Considerations:*
  - ➢ Design and layout must permit access to heat transfer area (tube bundle) for cleaning and tube replacement on failure. Cleaning is easier for a 1–1 and plate exchangers.
  - ➢ For high-pressure applications (above 8 barg), shell and tube exchangers are preferred over plate heat exchangers.
  - ➢ Higher steam pressures reduce area requirements due to larger ΔT obtainable. However, higher pressures involve higher costs in terms of equipment, control, and safety.
  - ➢ Use of highest possible ΔP reduces exchanger area requirements.
  - ➢ Using long and small diameter shell is cost effective, provided horizontal/vertical space requirements accommodate such design.
  - ➢ Overall space requirement should be at least double the length of the shell.
  - ➢ Shorter lengths may require multipass designs but cleaning becomes difficult.
  - ➢ Materials selection involves corrosion resistance, initial cost, expected life, maintenance, and performance. A 10-year life should be considered.
  - ➢ Proper condensate drainage, preferably by gravity, is essential while considering installation. Condensate drainage must have sufficient pipe length, typically 0.5 m vertical and *not more than* 0.2 m to steam trap inlet is required.
  - ➢ Sizing and length of pipe from control valve outlet to inlet nozzle to heat exchanger is crucial. Minimum 10 pipe diameters length from control valve to inlet connection is required.
  - ➢ Air venting and vacuum breaking provisions must be made for the heat transfer equipment.

• Give a checklist incorporating installation guidelines for heat transfer equipment.

- Provide a condensate drip leg with steam trap before steam control valve.
- Use ball valves with locking handles for all pipe sizes less than 5 cm in diameter. This provides the best lock out/tag out safety procedure.
- Install a strainer in front of the control valve.
- Select a proper control valve based on the application.
- Install pressure gauges before and after the control valve.
- Control valve outlet piping must be more than or equal to inlet connection to the exchanger.
- Use a steam trap for condensate capacities of $\leq$5000 kg/h.
- Use a control valve with a level controller for condensate capacities of $\geq$5000 kg/h.

10.1.3 Fouling

• What are the general effects of heat exchanger fouling in the chemical industry?

- Estimates have been made of fouling costs primarily due to wasted energy through excess fuel burnt that are as high as 0.25% of the gross national product (GNP) of the industrialized countries.
- Millions of tons of carbon emissions are the result of this inefficiency. Costs associated specifically with crude oil fouling in the preheat trains of oil refineries worldwide are estimated to be in terms of billions of dollars.
- With oil prices at record highs, the payback from fouling reduction by increased throughput and less wasted fuel increases year after year.

- What are the major cost increasing factors due to fouling in the operation of a plant?
  - Reduced flow: Due to decreased flow areas.
  - Increased pressure: Due to increased pressure drops.
  - Reduced production: Due to shutdowns consequent to fouling.
  - Cleaning and disposing of toxic wastes resulting from cleaning operations.
  - Use of costly chemical additives to prevent fouling deposits.
  - Increased safety hazards: Due to particularly, flow and pressure variations with consequent equipment leaks and failures leading to fires and toxic releases.
  - Increased investment: Use of switchable exchangers.
    - Providing excess heat transfer surfaces.
    - Costs of antifouling equipment.
    - Costs of cleaning equipment, chemicals, and waste disposal.
    - Increased costs of maintenance.
  - Equipment replacement.
- List out streams that normally *do not foul* and *foul heavily*.
  - *Streams that Normally do not Foul:*
    - Refrigerants.
    - Demineralized water.
    - LNG.
    - Olefin-free nonpolymerizing condensing gases.
  - *Streams that Normally Foul Heavily:*
    - Crude oil.
    - Crude distillation overheads.
    - Amines.
    - Hydrogen fluoride.
    - Coal gasification fluids.
    - Cooling water that is not properly maintained.
- Give an example of the consequences of fouling in the operation of a refinery.
  - Most fouling arises from asphaltene deposition from the crude oil onto the metal surfaces of the preheated train of heat exchangers. This fouling leads to a decline in furnace inlet temperature, by perhaps as much as 30°C, and a subsequent need to burn extra fuel in the furnace to make up the temperature necessary for efficient distillation.
  - Generally, the fouling deposits that occur as high molecular weight polymers are formed in the crude preheat systems. Products of corrosion and inorganic salts mix with the polymers and increase the volume of the fouling deposits.
  - Factors affecting fouling in crude oil preheat exchangers include process conditions (temperature, pressure, and flow rate), exchanger and piping configuration, crude oil composition, and inorganic contaminants such as salts.
  - More often, the fouling mechanism responsible for the deterioration of heat exchanger performance is flow velocity dependent. Maldistribution of flow, wakes, and eddies caused by poor heat exchanger geometry can have detrimental effect on fouling.
  - Uneven velocity profiles, backflows, and eddies generated on the shell side of a segmentally baffled heat exchanger results in higher fouling and shorter run lengths between periodic cleaning and maintenance of tube bundles.
- What are the differences in the resistances offered for heat transfer between a new heat exchanger and an exchanger in operation for a period of time for the same process application?
  - Resistances get added due to fouling caused by deposits.
  - Resistances get added due to formation of oxide films due to corrosion.
- What is the overall effect of fouling of a heat exchanger tube on ΔP?
  - A tube with fouling deposits will increase ΔP in two ways:
    (i) Roughness resulting from fouling will offer more resistance for flow.
    (ii) Fouled tube will decrease effective flow cross section and offer more resistance for flow.
- What are the different types of fouling on heat transfer surfaces?
  - *Precipitation or Crystallization Fouling:* Dissolved solids get supersaturated at heat transfer surfaces and crystallization takes place from solution. Crystals adhere to surfaces.
    - Crystalline ionic salts deposit on heat transfer surfaces.
    - Normal solubility salts (inorganic/organic) precipitate on colder surfaces.
    - Inverse solubility salts precipitate on hot surfaces. Examples are calcium salts such as calcium carbonates and sulfates. These are less soluble in hot water than in cold water. Buildup of such salts on heat transfer surfaces start on nucleation sites such as scratches and pits. Such scales are hard and require vigorous mechanical or chemical means for their removal.
  - *Particulate Fouling:* Accumulation of finely divided particulates (silt, mud, sand, insoluble products,

etc.), suspended in process fluid, onto heat transfer surfaces where velocities are low.

- *Sedimentation Fouling:* Due to gravity settling. Deposits formed through sedimentation processes are generally loose and do not adhere strongly to the heat transfer surfaces. These deposits are self-limiting, that is, as they grow they get washed off by the flowing fluid. Sedimentation fouling is strongly affected by fluid velocity and less so by temperature. However, a deposit can get *baked on* to a hot wall and become very difficult to remove.
- *Chemical Reaction Fouling:* Formation of insoluble products by chemical reaction and subsequent deposit formation on heat transfer surfaces, for example, petroleum coke deposits in cracking processes and asphaltenes, polymer, and food derived products.
  - For example, coking and polymerization reactions take place on hot metal surfaces, producing an adhering solid product of reaction.
- *Corrosion Fouling:* Accumulation of corrosion products.
  - Corrosion layer (less conducting than metal) is produced by reaction of heat transfer surface with flowing fluids.
  - Sometimes a corrosion layer acts as a shield for further corrosion, in which case attempts to clean the surface may only result in accelerated corrosion and eventual failure of the exchanger.
- *Biofouling:* Attachment of macroorganisms (*macrobiofouling*) and/or microorganisms (microbial or *microbiofouling*) present in fluid stream to warm heat transfer surfaces, where they adhere, grow, and reproduce. These biofouling processes can be further classified as *aerobic* and *nonaerobic* types.
  - Microbial slime and algae are examples of microbiofoulants.
  - Snails, barnacles, and mussels are examples of macrobiofoulants.
  - For tackling biofouling, the usual solution is to kill the life forms by chlorination or to discourage their settling on the surfaces by using 90–10 copper--nickel alloy or other high copper alloy tubes. Intermittent *shock* chlorination is an alternative solution than continuous chlorination.
- *Solidification Fouling:* Overcooling of a fluid below freezing point at heat transfer surface resulting in solidification of the process fluid and coating of heat transfer surface.
  - Higher melting components of a multicomponent solution freeze on colder surfaces.
  - This type of fouling occurs in refrigeration and cryogenic systems and those involving high melting point components, for example, waxy components.
- *Combination Mechanisms:* In most cases, fouling occurs involving a combination of the above mechanisms. A common example is fouling of cooling water involving sedimentation and inverse solubility, as most surface waters used for cooling contain sediment and calcium carbonates. Macrofouling cuts down on cooling water flow and thus allows more sedimentary deposits, which in turn, can lead to microfouling and corrosion.
  - Cooling waters are not normally once-through flows but involve recirculation. Evaporative cooling reduces quantities of waters, thus increasing concentration of solids and, in addition, contains treatment chemicals. Use of makeup water in cooling water systems is to be carefully planned. Blowdown requirements should also be properly calculated and implemented.

- What are the different contributors for scale formation in equipment?
  - *Oxidation Products:*
    - Metal oxide scales. Oxides are more stable than metals themselves.
    - Mill scales.
    - Magnetite scales (Fe_3O_4) are formed during forging, hot rolling, and other high-temperature manufacturing operations.
  - *Fluid oxidation* due to air infiltration mainly causes solids formation and fouling in organic thermal liquid heat transfer systems. Oxidation of the organic fluid creates insoluble solids. The combination of corrosion products and oxidized heat transfer fluid insolubles usually deposits on heat transfer surfaces. Rapid thermal decomposition of a heat transfer fluid occurs when the fluid is exposed to excessively high temperatures. Off-design operating conditions (e.g., high heat flux or reduced flow conditions) are common causes of excessive fluid temperatures. These conditions may result in surface fouling and, in severe cases, rapid coking.
  - *Corrosion:*
    - Magnetite scales crack, producing fissures in the scales. Exposed metal acts as anode and mill scales act as cathodes, accelerating corrosion process.
    - *Rust or Red Iron Scales (Fe_2O_3) on Metal Surfaces*: In water transfer lines, heat exchangers on water cooling side, boilers, and so on.

- Waters with sulfide contaminants give rise to metallic sulfide + iron scales.
- *Water Insolubles*: Precipitation of insolubles on heating of water, for example,

$$Ca(HCO_3)_2 + heat \rightarrow CaCO_3 + H_2O + CO_2. \tag{10.13}$$

$$Mg(HCO_3)_2 + heat \rightarrow MgCO_3 + H_2O + CO_2. \tag{10.14}$$

- $CaCO_3$ forms thick scales in boilers, cooling engine jackets, steam-heated kettles, and so on.
- Sulfates in water give rise to calcium sulfate mixed with $CaCO_3$ scales (inverse solubility at higher temperatures).

- *Conversion Products:*
  Additives used in water treatment, for example, phosphates used to precipitate calcium as soft sludge.
  - Removal of sludge by blowdown prevents accumulation, but with time deposits of Ca or Mg hydroxylapatite form.
  - Dispersants used thermally degrade giving rise to residues of heavy chalk like accumulations.
- *Silicate Scales:* Thin, tightly adherent, and very hard scales.
  - Require special consideration in descaling.
  - Formed by a combination or rearrangement of ions in water to form highly complex insoluble deposits.
  - For example, water with Mg, Na, SO_4^-, and silicate ions in solution. If pH is raised by addition of alkali, hard glassy scales will form.

$$3MgSO_4 + 2Na_2SiO_3 + 2NaOH + H_2O \rightarrow 3MgO \cdot 2SiO_2 \cdot 2H_2O + 3Na_2SO_4 \tag{10.15}$$

- *Process Deposits:* Due to degradation of products in flow lines, coolers, evaporators, and so on.
  - Polymers may form from processing hydrocarbons. These range from thin oily substances to thick black gummy matrices.
  - Coke and rubber-like scales.
- *Electrochemical Action:* Water with traces of copper is stripped of copper content when passed through steel piping due to electrolytic action. Copper is deposited electrolytically on steel.

- What are the most significant process variables that affect the fouling process? Explain.
  - *Flow Velocity:* As velocity increases, entrainment and transport of the fouling species increases, providing more chances for deposition to occur on the heat transfer surface.
    - Simultaneously, the shear forces acting at the fluid–heat transfer surface interface increase, which aids deposit removal.
    - The actual amount of fouling is a balance between these two opposing effects.
    - Effect of flow velocity on fouling mechanisms is given in Table 10.6.
  - *Temperature:* The prevailing temperature of a fluid passing through a heat exchanger and temperature of the heat transfer surface can have a profound influence on the fouling mechanisms.
    - Table 10.7 gives the influence of temperature on fouling mechanisms.
    - The presence of a deposit will affect temperature distribution across the exchanger, which, in turn, changes temperature at the point of deposition, thereby influencing rate of temperature-dependent fouling.
  - *Conceptual Design and Geometry of the Exchanger:* It is known that shell and tube heat exchangers are more sensitive to fouling than, for example, a plate and frame or double pipe heat exchangers. This is mostly because velocities and turbulence levels are higher for the latter one.

TABLE 10.6 Influence of Flow Velocity on Fouling Mechanisms

Particulate fouling decreases with flow and stops above 1.7 m/s
Crystalline fouling slightly decreases with flow over 0.5 m/s
Corrosion fouling decreases very slightly with flows above 1.2 m/s
Biological fouling decreases with flow over 0.7 m/s and stops over 2.2 m/s

TABLE 10.7 Influence of Temperature on Fouling Mechanisms

Mechanism	Temperature Dependence
Particulate deposition	Little effect except physical conditions are affected
Precipitation or crystallization	Solubility; crystalline fouling involving inverse solubility salts greatly increases
Chemical reaction	Reaction rate
Corrosion	Corrosion rate; corrosion fouling slightly decreases above 50°C
Biofouling	Metabolic activity; biofouling decreases above 50°C and absent above 140°C
Freezing	Solidification

- *Material of Construction:* There will be decreased fouling on alloy steel surfaces due to their relative smoothness compared to mild steel surfaces. Surface material may also influence biological fouling, for example, copper is more sensitive to biological fouling than most other materials.
- *Surface Roughness:* Rough surfaces promote fouling rates not only due to tendency of particulates to deposit in the microcavities of the surface but also due to enhanced corrosion and chemical reaction rates.
- *Other Factors:* Rate of deposit formation is also influenced by the following:
  - The concentration and nature of the foulant (or its precursor).
  - The process fluid in which they are carried:
    - Shape and size of particles.
    - Chemical composition.
    - pH.
    - Availability of nutrients for biological growth.

- Under what circumstances, one should focus on tube wall temperatures on the cooling water side of a shell and tube exchanger?
  - Where $CaCO_3$ deposits on heat exchanger surfaces are possible from circulating cooling water, tube wall temperatures in excess of 60°C should be avoided.
  - Corrosion effects should also be considered at hot tube walls. As a rough rule of thumb, this check may be made if the inlet process temperature is above 90°C for light hydrocarbon liquids and 150–200°C for heavy hydrocarbons. Using air coolers may be considered to bring the process fluid temperature down before it enters the water-cooled exchanger.

- Draw qualitatively idealized fouling curve and show the different regions on it.
  - There are basically three regions for fouling deposit formation on heat transfer surfaces with respect to time:
    - Initiation of fouling on fresh surfaces takes place relatively slowly that provides the necessary sites for further deposition of particulates.
    - Further deposition proceeds with rapid increase in thickness in an exponential way as shown by the steep curve (Figure 10.35).
    - This buildup gradually tapers off with the ultimate rates of deposition and detachment coming to an equilibrium condition, the thickness remaining nearly constant, as illustrated by the asymptotic part of the curve.

- What are the sequential events in fouling?
  - Initiation (delay, nucleation, induction, incubation, surface conditioning).
  - Transport (mass transfer).
  - Attachment (sticking, adhesion, bonding, surface integration).
  - Removal (release, reentrainment, detachment, scouring, erosion, spalling, sloughing off).
  - Aging (changes in crystal or chemical structure, for example, dehydration, polymerization, chemical degradation, for example, hydrocarbon gums converted to coke; thermal stresses can cause chemical degradation).

- Discuss fouling problems associated with condensers.
  - Typical cooling waters contain the hardness contributing ions, calcium and magnesium, and other cations such as sodium and potassium. These are

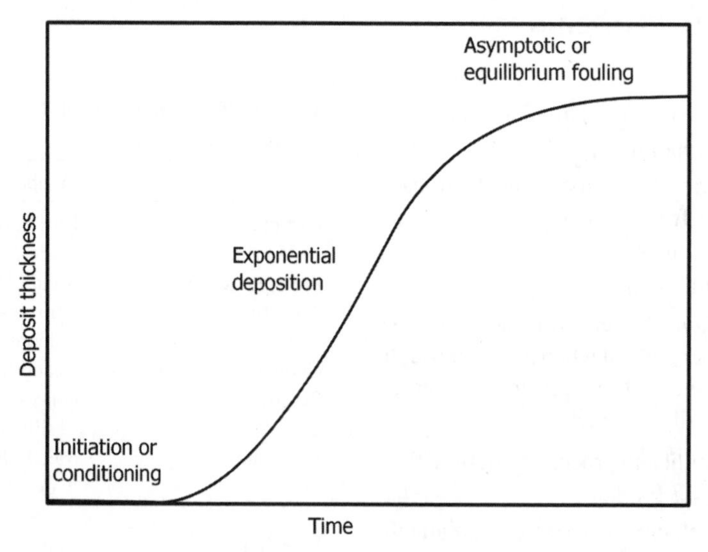

FIGURE 10.35 Idealized fouling curve.

counterbalanced by anions such as bicarbonates, chlorides, and sulfates.

- Groundwaters usually contain higher concentrations of dissolved ions.
- Surface waters are softer but contain suspended solids, silica, organics, and microorganisms.
- Inverse solubility of many compounds formed from different ions ($CaCO_3$, silicates including $MgSiO_3$) contained in cooling waters precipitate on temperature increases on heat transfer surfaces.
- These give rise to scaling in condensers, which gets aggravated due to recirculation. Phosphate-based corrosion inhibitors give rise to scale forming $Ca_3(PO_4)_2$.
- Microbial activity in cooling waters gives rise to protective sticky films on tube walls, which absorbs silt and sediment, further increasing deposit volume. Biofouling may also induce corrosion with the dissolved oxygen in cooling water diffusing toward the oxygen-deficient zone under the deposit near the tube walls. Many microbes produce acids in the presence of oxygen as part of their metabolic processes. These acids attack the metal walls.

- What are the effects of fouling on finned surfaces?
  - Finned surfaces are on outside of tubes in tubular exchangers. Effects of fouling on these surfaces are more complicated than inside tubes, because in extreme situations there is a possibility that the finite thickness of the fouling layer can effectively block off flow through the fins, negating the usefulness of the fins.
  - On the other hand, finned surfaces are sometimes found to be more resistant to fouling than normal plain surfaces. One reason for this could be the expansion and contraction of the surface during normal operational cycles that tend to break off the brittle deposits.
  - Normally, it is not recommended to have finned surfaces for heat transfer involving dirty liquids as cleaning becomes a constraint.
  - High finned surfaces are commonly used with air and other low pressure and relatively clean gases. Dust is the main fouling agent in such cases and is easily removed by blowing.
- How does selection of materials influence fouling?
  - Corrosion-type fouling is minimized by proper selection of material for heat transfer surfaces. Such material should also withstand corrosive attack on fluids used for chemical cleaning.
  - Biological fouling can be largely eliminated by the use of copper-based alloys containing over 70% of copper.
  - It is normal to consider high flow velocities to minimize fouling deposits. High flow velocities give rise to erosion problems on equipment, particularly wherever direction changes for flow occurs. Heat exchanger failures occur in U-bends and where impingements are involved. Materials such as titanium or stainless steels resist erosion better. A cupronickel alloy with 83% copper, 17% nickel, and 0.4% chromium is claimed to be having much greater resistance to erosion attack, being capable of operating with seawater at velocities of 7–8 m/s.
- "For a particular service, use of carbon steel tubes serves the purpose from corrosion point of view." Is there any advantage of using more expensive alloy tubes or smooth surfaces?
  - Yes.
  - Alloy tubes are smoother and result in less fouling.
  - Highly polished surfaces (mechanical or electrochemical polishing) are used in food industry.
  - Surface treatment by coating the surface chemically or modification of the surface itself to reduce adhesion of deposits.
  - Easy to clean.
  - Better heat transfer rates, although equipment is more expensive.
  - Longer life.
- Name the conditions in the operation of plant equipment that change fouling rates.
  - When plant operating conditions change too fast.
  - When a process upset occurs.
  - When pumps are switched on.
  - During unloading tank cars.
  - During addition of new catalysts.
  - When surfaces become rough due to corrosion or due to roughness from earlier fouling.
  - Due to excessive cooling water temperatures causing scale formation on the waterside surfaces.
- Name some of the considerations involved for the choice of heat exchanger design with respect to fouling.
  - Table 10.8 gives considerations involved in the selection of a heat exchanger with respect to fouling.
- What are the indications that an exchanger has fouled and how maintenance handles this?
  - A light sludge or scale coating on the tube greatly reduces its efficiency. A marked increase in pressure drop and/or reduction in performance usually indicates cleaning is necessary.
  - The unit should first be checked for air or vapor binding to confirm that this is not the cause for the

TABLE 10.8 Some Considerations for the Choice of Heat Exchanger Design

Exchanger Type	Materials	Cleaning	Comments
Shell and tube	Most materials	Tubes easier to clean; shell more difficult to clean	Widely used
Plate	Usually stainless steel	Easy to clean	Compact
Plate-fin	Al, stainless steel, titanium	Only chemical cleaning possible	Highly compact
Double pipe	Carbon steel common	Inner tube easier to clean; annulus difficult or impossible if welded	Used for small area requirements
Immersion coils	Most materials	Inside tubes only through chemical cleaning	Limited application
Spiral	Most materials	Easy access	Compact; useful for fouling fluids and slurries
Air-cooled	Al fins on carbon steel tubes common, other combinations possible	Inside tubes easier, finned surfaces difficult	Large area required for installation
Scraped surface	Most materials	Generally self-cleaning	Have moving parts
Graphite block	Graphite	Mechanical cleaning not possible, chemical cleaning possible	Useful for corrosives

reduction in performance. Since the difficulty in cleaning increases rapidly as the scale thickness or deposit increases, the intervals between cleanings should not be excessive.

- Fouling can be detected through pressure drop monitoring, as nonuniform rough fouled surfaces increase pressure drops.
- With water-cooled exchangers, performance degradation could also be due to decreased flow rates. Therefore, all the other possibilities, namely, air or vapor binding (accumulation) on the surfaces and decreased flow rates should be examined before concluding on the fouling process.
- Buildup of noncondensables could be checked by monitoring the temperature of the vent gases at constant condensing pressure. If the temperature is close to that of the cooling water, flow rate of cooling water may be increased.

• Under what conditions, fouling deposits will not be uniform on the heat transfer surfaces?

- Under two-phase conditions, the pattern of fouling will not be uniform and significant localized fouling can occur because part of the surfaces may become dry because of contact with vapor bubbles.
- Evaporators and reboilers suffer from maldistribution of vapor–liquid phases.
- Use of high-pressure steam as heating medium promotes film boiling inside tubes because of large temperature differences, with the consequent increase in fouling rates.
- Less than satisfactory performance in fired heaters is possible under unexpected two-phase conditions. If, under normal operating conditions, two-phase flow lies between two regimes (e.g., transition from slug flow to annular flow to misty flow), slight changes in process conditions (e.g., flow rate, vapor quality, or pressure) can alter the local axial and circumferential heat transfer coefficient that alters the tube wall temperatures with consequent fouling rates.

• What are the remedies for making vapor–liquid distribution more uniform under two-phase flow conditions?

- Use of tube inserts and alternative baffle designs may improve vapor and liquid phase distribution inside and outside tubes, respectively.
- Avoiding film boiling through use of lower pressure steam as heating medium in tubular heaters and adjustments in firing rates and proper temperature control through burner management and tube side flow rates in fired heaters are some measures for controlling fouling.

• Suggest suitable guidelines for mitigating fouling problems.

- *Evaluation of Causes and Effects of Fouling:* During start-up, flow rates might be low leading high tube wall temperatures. Alternative start-up procedures involving stabilizing flow rates with, for example, careful firing rates in fired heaters.
  - ➢ In coolers, cooling of hot process streams might cause scaling of tube walls. Use of higher tube velocities (>2 m/s) and use of additives might be used to keep tube wall temperatures low. It must be recognized that in such situations, ΔP might increase. Swirling flow might be created by use of tube inserts, particularly in transition regions from annular to mist flow.

- *Identifying the Fouling Mechanism(s):* Eliminating some types of corrosion (e.g., naphthenic acid corrosion) may reduce fouling while minimizing fouling might reduce some types of corrosion (e.g., sulfide corrosion in hydrotreating of hydrocarbon streams). Use of dispersant-type chemical additives might keep insolubles from forming scales on tube walls.
  - Fouling due to soluble corrosion products can be curbed by preventing the chemical reaction(s) involved in the formation of fouling products.
  - Biofouling is dependent on the tube wall temperatures, rather than cooling water temperatures. As tube wall temperatures in the range of 25–35°C give rise to biofouling, these temperatures should be avoided by adjusting water flow rates in summer and winter conditions. Biofouling decreases with increased tube wall temperatures above 50°C.
- *Developing a Fouling Mitigation Strategy:* This should involve some of the solutions discussed earlier as well as determining cleaning schedules, alternating blending of feed stocks, changing operating conditions, use of appropriate additives, adopting alternative tube bundle designs, changing number of tube passes, changing baffle designs, use of tube inserts, changes in process conditions, modifying piping networks, and use of alternative heat exchangers.
  - Guidelines are available from different organizations such as Heat Exchanger Institute Standards, Engineering Sciences Data Unit Guide data, and others. Best practice is the organizations themselves developing their own performance models. Many organizations follow this practice.
  - The simplest way to develop a model is to compare heat transfer just before and just after cleaning and then calculating the increase in efficiency due to the gain in heat transfer, assigning costs to the heat transfer gain and costs involved in cleaning, process interruption (if any), maintenance, and so on.
  - An example is provided by a 500 MW power plant condenser, which was not cleaned for a year, amounting to $265,000 in 1 year as costs due to fouling. One cleaning would have paid off seven times the direct fuel costs. This example involves only the direct losses due to fouling and not the costs of corrosion of materials due to fouling and loss of equipment life due to corrosion.

- What are the techniques used to minimize fouling heat transfer surfaces?
  - *Corrosion Control:*
    - Minimization of corrosion through selection of appropriate materials of construction.
    - Growth of oxide scale offers resistance to heat transfer that is equivalent to fouling.
    - Use of coatings on tube surfaces. For example, epoxy coatings are of great success in seawater cooling exchangers in preventing corrosion and reducing fouling.
  - *Design Steps:*
    - Pretreatment of feed streams to a heat exchanger by filtration, softening (water), desalting, and so on.
    - Use of high fluid velocities (>1.8 m/s) to prevent settling of particles on heat transfer surfaces.
      - For high-viscosity liquids, at above velocities, ΔP can be prohibitive.
    - Minimizing areas of lower velocities, that is, stagnant regions, for example, on shell side of a heat exchanger where stagnant areas are possible, especially behind baffles, where deposition can occur.
    - Regions of recirculation are also possible, which can result in extended residence times, with a chance for chemical reactions to take place influencing incidence of fouling.
    - Use of helical baffles, rod baffles, or twisted tubes can minimize or eliminate shell side fouling.
    - Improvements on the tube side may be made by using vibrating or fixed inserts to promote turbulence and enhance heat transfer, which, in turn, may also reduce incidence of fouling.
    - Polymerization, precipitation, and freezing fouling are direct results of temperature extremes at heat transfer surface.
      - Reduced by lowering ΔT between the surface and bulk flow region.
      - Lowering ΔT is done by increasing velocity that increases h.
      - Increased h in turn results in decrease in ΔT.
      - There is a limit for increasing velocities.
      - Better approach would be use of adequate velocities + some form of extended surface.
      - Extended surface is used only when construction permits access for cleaning.
    - Shell side fouling is greatest for segmentally baffled bundles in the regions of low velocities.
    - Keeping heat transfer surface temperatures as low as possible (for heating applications).
    - Use of cocurrent flow where rise in temperature is detrimental to keeping surfaces clean of fouling.

- ➢ Use of low fouling exchangers that make use of elimination of cross-flow plate baffle.
  - Examples are plate heat exchangers, spiral plate exchangers, twisted tube exchangers, and so on. These involve no baffles and provide 25–30% excess surface to compensate for fouling.
- ▪ *Other Designs:* To develop easily removable fouling layers on which further deposition occurs.
- ▪ *Use of Additives:* Periodic or continuous injection of small amounts of additives.
  - ➢ Different additives reduce fouling problem by chemical or physicochemical effects.
  - ➢ Controlled acid dosing (H_2SO_4, HCl taking care that there is no corrosion) of hard waters to combat calcium carbonate deposits.
  - ➢ Antiscalants for aqueous systems, for example, chelates such as ethylene diamine tetra acetic acid, which combine with scaling material and give rise to soluble compounds.
  - ➢ Threshold agents, for example, polyphosphates that retard precipitation.
  - ➢ Crystal habit modifiers, for example, polycarbolic acids, which distort crystal habits of scaling compounds that prevent adherence to hot surfaces.
  - ➢ Dispersants, for example, polyelectrolytes to keep scaling materials in suspension.
- ▪ Biofouling is controlled by injection of (1) biocides that kill microorganisms or (2) biostats that arrest growth of microorganisms.
  - ➢ Chlorine is a common biocide but gives rise to corrosion. Less corrosive substitutes are used.
  - ➢ Use of copper alloys and/or chemical treatment of fluid to prevent organism growth and reproduction.

- What are the ways by which fouling deposits are removed?
  - ▪ *Cleaning:* Periodic cleaning is necessary.
    - ➢ Off-line cleaning methods: Mechanical methods (rodding, drilling, brushing, etc.), hydraulic or pneumatic methods, or chemical cleaning methods.
    - ➢ Onstream methods: Cleaning of inside tubes/pipes by the use of recirculating sponge rubber balls or by flow-driven brushes, wherever high degree of cleanliness is desirable. The moving balls remove deposits and corrosion products as they pass through the tubes. A mesh basket in the outlet piping collects the balls and a ball pump reinjects them into the inlet fluid.
    - ➢ Use of tube inserts, ultrasound, and circulation of polymer fibers.
    - ➢ Use of an additive in conjunction with a physical method may be more effective than either technique alone.
    - ➢ Use of sonic vibrations.
  - ▪ *Chemical Cleaning:* Weak acids, solvents, and so on.
  - ▪ *Loose Deposits:* Flushing with high-velocity steam or water jets.
    - ➢ Circulating hot wash oil or light distillate through tubes or shell at high velocity.
    - ➢ Sand blasting or use of sand–water slurry.
    - ➢ Periodic use of high fluid velocities.
    - ➢ Some salt deposits may be washed out by circulating hot freshwater.
  - ▪ *Scales:* Mechanical action.
    - ➢ Rodding.
    - ➢ Turbining.
    - ➢ Scraping and rotating wire brushing. These methods are limited to surfaces that can be reached by the tools. In shell and tube exchangers, use of large clearances between tubes and/or the use of rotated square tube pitch. Scraping should not be used on finned tubes.
    - ➢ Powered rotating drills cut and propel their way through hard, dense deposits.
    - ➢ Water jets + drills are used for hard, dense/thick deposits.
    - ➢ Hydraulic jets at pressures up to 700 bar. For shell side, the jets will not be very effective deep inside a large tube bank.
  - ▪ Mechanical cleaning sometimes requires pulling out tube bundle for tube outside scale removal.
  - ▪ Inside scales need not require removal of bundle.
- What are the precautions to be taken while cleaning deposits?
  - ▪ Tubes should not be cleaned by blowing steam through individual tubes since this heats the tube and may result in severe expansion strain, deformation of the tube, or loosening of the tube-to-tube sheet joint.
  - ▪ When mechanically cleaning a tube bundle, care should be exercised to avoid damaging the tubes.
  - ▪ Cleaning compounds must be compatible with the metallurgy of the exchanger.
- What are the precautions taken while cleaning heat transfer surfaces on which pyrophoric iron sulfide scales are formed?
  - ▪ Such scales, which might form on petroleum crude heat exchangers, should *never* be removed in dry condition as it creates fire hazards.

- These scales should be scraped off by ensuring wetting such surfaces by water.

- What is *rubber ball cleaning* of heat exchanger tubes? What are its advantages?
  - Sponge rubber balls are used to clean inside of tubes while the unit is in operation.
  - Balls are slightly larger in diameter than the tube and are compressed as they travel the length of the tube. This constant rubbing action keeps tube walls clean and virtually free from all types of deposits.
  - Suspended solids are kept moving and not allowed to settle, while bacterial fouling is wiped quickly.
  - Balls are forced to move due to ΔP between inlet and outlet.
  - Abrasive-coated balls are available to remove heavy fouling.
  - Balls are circulated in a closed loop.
  - This technique is used in natural gas pipelines to flush out condensates, keeping pipeline capacities for gas flow high.
- When dirty liquids like waste streams are involved on the tube side of a heat exchanger, what are the ways to maintain clean tube surfaces?
  - Use of recirculated sponge balls or reversing brushes without shutting down the unit, that is, online cleaning.
- What are the various chemicals employed in chemical cleaning?
  - *Acids:*
    - Strong acids damage equipment.
    - HCl with 0.25–1% ammonium bifluoride and copper complexing agents as additives. Effective for siliceous deposits. Not generally recommended for austenitic alloys.
    - H_2SO_4 for austenitic alloys.
    - Weak acids such as citric, formic, sulfamic, and so on are less effective and more expensive but less corrosive compared to HCl.
      - Citric acid is sometimes preferred for preoperational cleaning, where iron oxides constitute the bulk of deposits.
      - Hydroxyacetic acid and formic acid mixtures (typically 2% hydroxyacetic acid + 1% formic acid).
    - Primarily used to clean once-through boilers.
      - EDTA (ethylene diamine tetra acetic acid) formulations (ammoniacal solutions).
      - Ammonium citrate.
  - *Chlorine:* Being discouraged/eliminated by government regulations. Use of chlorine-containing solvents should be avoided as residual solvents can create severe corrosion problems. Also disposal of halogenated compounds is more expensive.
  - *Baking Soda:* Aqueous solution is forced at high pressure through nozzles as an alternative to high-pressure water jets.
    - Cleans oil, grease, polymers, dirt, carbon soot, and so on.
    - Foulant does not redeposit.
- How does EDTA work as a cleaning fluid?
  - Many chemical cleanings are performed with EDTA.
  - EDTA belongs to a class of compounds known as chelants.
  - EDTA is particularly effective in complexing divalent and trivalent cations.
  - The most popular procedure uses tetraammonium EDTA, in which the two hydrogen atoms at the end of each molecule (four total) are replaced with ammonium (NH_4^+) ions.
  - Ammonia is alkaline and a typical EDTA cleaning is performed at a pH of 9–9.5.
  - Tetraammonium EDTA is not as aggressive toward deposits as HCl, so common practice calls for the boiler to be filled with a 5% solution and then fired until liquid temperatures reach about 135°C. This increases reactivity of the chemical.
  - In natural circulation units, following the initial firing, the boiler must be alternately cooled to about 135°C or so and then refired to 135°C to circulate the chemical.
  - The iron removal stage of an EDTA cleaning may take from 12 to 36 h.
    - Free EDTA concentrations should not be allowed to fall below 0.4%.
    - Once iron removal is complete, the system is allowed to cool to about 65°C and then an oxidant (such as air with sodium nitrite, oxygen, or hydrogen peroxide) is injected into the solution to effect copper removal.
    - This may take 4–8 h.
  - The oxidant converts copper to the +2 oxidation state, whereupon it is complexed by the EDTA.
- What are the advantages and disadvantages of tetraammonium EDTA?
  - *Advantages:*
    - EDTA is much less corrosive than HCl.
    - The process is performed at an alkaline pH, so if a bit of residual remains in the boiler after the cleaning and rinses, it will not attack tubes unlike HCl.

  - EDTA is not as hazardous as HCl, although ammonia smell will be evident.
  - Copper removal often does not require a preliminary stage or a completely separate second step.
  - The spent solvent may be evaporated in the boiler, greatly reducing project costs.
- *Disadvantages:*
  - Boiler must be fired during the cleaning process.
  - Chemical must be manually circulated because soaking does not work well.
  - Although EDTA is not as hazardous as HCl, safety issues regarding the high temperature of the liquid must be considered.
    - If boiler temperatures exceed 150°C, EDTA decomposes.
- During firing, the boiler water volume swells, which may require partial draining of the solution, which should be done under nitrogen blanketing.
- EDTA is more expensive than HCl.

- "Organic heat transfer fluids give rise to oxidation and decomposition products on usage." How are such fluids cleaned to prevent fouling deposits arising from them?
  - For in-system filtration, disposable glass fiber-wound filter cartridges are generally the most satisfactory since they withstand system temperatures up to 400°C and have adequate solids handling capacity. Sintered metal filters, while satisfying high-temperature capability, are more expensive and difficult to clean.
- What are *fouling factors*? How do they influence heat transfer rates?
  - For making an allowance for fouling in the design of heat exchangers, it is customary to assign a value to the anticipated heat transfer resistance, which is often referred to as *fouling factor*.
  - At the start of commissioning, there will not be fouling resistance and hence, the exchanger will initially overperform, which has implications for control:
    - Depending on the duty of the exchanger, one may be tempted to use a bypass so that the mixed exit temperature from the exchanger system has the desired value.
    - If the flow rate of the stream is not critical (e.g., cooling water), this variable may be reduced.
    - Both the above alternatives effectively reduce the stream velocity through the exchanger, with the potential to accelerate fouling process.
- Are fouling factors affected for steam contaminated with traces of oil? If so, compare fouling factors for oil-free steam and steam with traces of oil.
  - Yes. Fouling factors for oil-free steam are much less than for oil contaminated steam.
- What are the sources for obtaining fouling factors?
  - TEMA standards.
  - Published literature.
  - Company generated data or in-house experience for use in specific situations.
  - Other sources.
- What are the uncertainties involved in specifying fouling factors while designing or evaluating the performance of a heat exchanger?
  - TEMA recommended values are based on the fouling propensity of the service fluid and the resistance caused by that fouling. They do not account for the type of material of construction and the smoothness of its finished surfaces. Other sources of error include differences between the process fluid and ostensibly *identical* fluid on which fouling data are based and any additional *out of context* application arising from plant revamping. Microbial activity, temperature, and fluid velocity might be different.
  - Published fouling factors do not reflect true performance, as for some services they are too high while for others they are too low.
  - Specified fouling factors are static values, while in actual situations, they vary based on the operational variations during service. During the initial stages of the operational life of the exchanger, it will *overperform*, which may have process control implications, as stated earlier.
  - Temperature and velocity are two very significant variables that affect fouling rates.
  - Crude oils, for example, foul by different mechanisms including sedimentation, crystallization/precipitation, corrosion, asphaltene flocculation, and coking. It will foul by one mechanism, or combination of mechanisms, at one point in a refinery preheat train but will foul via a different mechanism at another point. Many crude oils foul very heavily whereas others foul hardly at all. Level of desalting of the crude influences fouling rates.
    - Specific crude oil, wall temperature, shear stress, run time, exchanger design, and many other variables influence fouling rates. Therefore, alternatives to fouling factors are needed.
  - Despite the constraints mentioned above, TEMA and other fouling factor data used do provide qualitative

guidance on the likely severity of the fouling problem in a particular application.

- Advanced design methods aim at reducing the margin of errors associated with fouling by using iterative mathematical techniques, considering thermal requirements and maximum ΔP on both tube and shell side. Maximum ΔP is related to maximum velocity. The complexity of shell side flow patterns makes it difficult to specify velocity constraints compared to tube side flow patterns. As low velocities are conducive to fouling process, particular attention must be paid to low-velocity constraints. The minimum velocity could be defined as the velocity for threshold fouling rate in applying these design techniques.

• Give typical fouling factors for different streams in heat exchangers.

- Table 10.9 gives fouling factor data for various process streams.

• What are the common errors in the design of heat exchangers with respect to specifying fouling factors?

- One of the most common errors made in specifying a new heat exchanger is overdesign. Specifying too large a fouling factor will often result in more tube or parallel channels. This will lower the velocity in the exchanger and actually promote fouling.
- Another common mistake is to apply fouling factor information from one type of equipment to a completely different type of equipment. For example, what is true for fouling of a shell and tube exchanger is not true with a compact heat exchanger. Specifying same fouling factors for both result in different levels of overdesign. Moreover, TEMA-specified fouling factors are based on experience gathered for shell and tube exchangers.
- While shell and tube exchangers have long used fouling factors, compact heat exchangers generally utilize a *heat transfer margin* that is typically 10–25% over the clean heat transfer coefficient.
- Also it must be realized that overdesigning in compact heat exchangers is even more detrimental to performance than in a shell and tube heat exchanger.

• What are the limitations in using TEMA-specified fouling factors in the design of shell and tube exchangers?

- TEMA tables for fouling factors are largely based on experience with water and provide incomplete coverage to the large variety of possible process fluids and exchanger configurations.
- These tables barely give recognition to such factors as fluid velocities, bulk temperatures, and composition and surface temperatures.

TABLE 10.9 Fouling Factors for Different Streams (10^4 m^2 K/W)

Stream	Fouling Factor
Water Streams	
Boiler blowdown, brackish, river waters	3.5–5.3
Seawater, treated cooling tower water, spray pond water	1.75–3.5
Distilled water, closed-cycle condensate	0.7–1.75
Closed loop treated water, engine jacket water	1.75
Treated boiler feed water	0.9
Industrial Liquid Streams	
Ammonia (oil bearing)	5.25
Engine lube oil, hydraulic oil, transformer oil, refrigerants	1.75
Methanol, ethanol, ethylene glycol	3.5
Industrial organic fluids	1.75–3.5
Cracking and Coking Unit Streams	
Heavy cycle oil, light coker gas oil	5.3–7
Bottom slurry oils	5.3
Heavy coker gas oil	7–9
Light cycle oil	3.5–5.3
Light liquid products, overhead vapors	3.5
Light-End Processing Streams	
Overhead gas, vapor, and overhead liquid products	1.75
Alkylation trace acid streams	3.5
Absorber oils	3.5–5.3
Reboiler streams	3–5.5
Chemical Process Streams	
Solvent vapor, stable overhead products	1.75
Natural gas	1.75–3.5
Acid gas	3.5–5.3
Crude Oil Refinery Streams	
120°C	3.5–7
120–180°C	5.25–7
180–230°C	7–9
>230°C	9–10.5
Petroleum Streams	
LPG	1.75–3
Natural gasoline, rich oil	1.75–3.5
Lean oil	3.5
Process Liquid Streams	
Bottom products	1.75–3.5
MEA and DEA solutions, DEG and TEG solutions, caustic solutions	3.5
Crude and Vacuum Liquid Streams	
Gasoline	3.5
Kerosene, light distillates, gas oil, naphtha	3.5–5.3
Heavy fuel oil	5.3–12.3
Atmospheric tower bottoms	12.3
Vacuum tower bottoms	17.6
Industrial Gas or Vapor Streams	
Compressed air, ammonia	1.75
Oily exhaust steam	2.6–3.5
CO_2, oily refrigerant	3.5
Nonoily steam, natural gas originated flue gas	9
Coal originated flue gas	17.5

Source: Chenoweth JM. Final Report of the HTRI/TEMA Joint Committee to Review the Fouling Section of the TEMA Standards. *Heat Transfer Engineering* 1990;11(1):73–107.

- Whether the fouling factors specified denote asymptotic values or values after a fixed operating time is not clearly stated.
- TEMA-specified fouling factors do not account for the effect of material of construction, which could result in inaccurate design. For example, copper alloy tubes in exchangers used in crude oil processing units are highly susceptible to coking resulting in higher fouling resistances than TEMA-specified values. Stainless tubes have been found to offer lower fouling resistances than copper alloy tubes. In one study on fuel oil, it was found that fouling resistances for copper tubes were 0.005 whereas for 316 stainless steel tubes were found to be 0.0025. The implications are less cleaning and maintenance requirements for stainless steel than copper, although copper having higher thermal conductivity offers less resistance than stainless steel.
- Another aspect is that since the exchanger does not exhibit the specified fouling resistance at the start of its operational life, it will initially overperform, which may involve implications on process control. For example, the overdesign, with respect to initial operation, may require lower initial fluid velocities, which in turn gives rise to higher surface temperatures than prescribed by design, thus resulting in more or faster fouling than would be the case otherwise.
- In spite of the above limitations, no clear and effective alternative to TEMA-specified fouling factors is available and these are continued to be used.

• What are the effects of specifying larger fouling factors than necessary in the design of a heat exchanger?
  - Results in oversize exchanger:
    - *Low Fluid Velocities*: More fouling on two counts, namely, increased settling rates of solids due to decreased velocities and increased surface temperatures that can give rise to decomposition and polymerization reactions of the fluids and also higher corrosion rates giving rise to more corrosion products.
  - Net effect is decreased heat transfer in spite of using larger and more expensive exchanger.
    - For example, if overall coefficient, U, is 100 W/(m^2 °C) for a clean exchanger and fouling factors of 0.001 are specified on either side, required area of the exchanger increases by 20%. If clean exchanger U is 100 and total fouling factor (both sides combined) specified as 0.01, required A will double.

• "Overdesign of a heat exchanger leads to good performance." *True/False*?
  - *False* (*See previous question.*)

• What is a self-cleaning heat exchanger? State its operating principle.
  - Self-cleaning heat exchanger is essentially a vertical shell and tube heat exchanger through which a fouling fluid flows upward inside the tubes charged with solid particles that are swept upward along with the fluid, producing a scouring action on the walls of the tubes as they move up.
  - A distributing system in the inlet channel provides uniform distribution of the fluid and the particles into the tubes.
  - From the outlet channel, the particles and fluid are carried into a separator. From the separator, the particles are returned to the inlet channel through a control system.
  - Existing heat exchangers such as evaporators and reboilers can be retrofitted with the solids circulation system, reducing fouling problems.
  - The disadvantage with such systems is erosion problems for the tubes and separation of fouling solids from the circulating particles. Certain percentages of the fouled solids are removed and fresh solids introduced into the system as practiced, for example, in boiler blowdown.

10.1.4 Pressure Drop

• What are the normally assumed pressure drops in the design of heat exchangers?
  - 0.2–0.62 bar (3–9 psi) for most services and 0.1 bar (1.5 psi) for boiling.
  - For low-viscosity liquids, allowable pressure drops are lower compared to high-viscosity liquids. This is because flow velocities should be kept higher to increase heat transfer coefficients for high-viscosity liquids, which are otherwise low and sometimes flow blockages occur. When a high pressure drop is used in the design, care should be taken to prevent erosion of equipment surfaces.
  - For gases and vapors, allowable pressure drops for design purposes depend on system pressure. Very low pressure drops are a necessity for vacuum systems compared to atmospheric and high-pressure exchangers.

• Give equations for estimation of ΔP in the tube side of a heat exchanger.
  - For straight circular tubes, Fanning equation can be used for ΔP:

$$P_{in} - P_{out} = 2fG^2L/g_c\rho D_i. \quad (10.16)$$

  - Including expansion, contraction, or changes in direction (nonisothermal, multipass exchangers):

$$P_{in} - P_{out} = K_P[2N_P f G^2 L / g_c \rho D_i \Phi]. \quad (10.17)$$

$$\Phi = 1.02(\mu_b/\mu_w)^{0.14}. \quad (10.18)$$

Where K_P is the correction factor for contraction/expansion/reversal and N_P is the number of tube passes.

- Will there be any possibility of ΔP changing during the operation of a heat exchanger when flow rates remain constant? If so, why?
  - Fouling reduces flow areas and hence increases ΔP.
- What are the considerations involved in deciding the value of allowable pressure drop for a heat exchanger?
  - Allowable ΔP for an exchanger depends on the *combined* pressure drops for the network of equipment and associated piping in the circuit, which must be overcome by the pump pumping the fluid through the network.
  - Depending on the nature of each of the equipment plus the piping and fittings, the designer distributes the available ΔP among the individual pieces of equipment and piping.
  - Pressure drop considerations are crucial for certain types of equipment like, for example, vacuum distillation columns for which very low pressure drops are required.
  - Excessive increases in pump discharge pressures depend on the nature of the pump and its input power that also affects mechanical design of the individual equipment in the first few stages of equipment.
  - Therefore, the designer has to carefully examine the options and *distribute* the available overall ΔP, including considerations for use of intermediate pumping stages.
- Arrange the following in the order of increasing pressure drop for the case of flow outside tube banks (for the same flow rate):
  (ii) Tube arrangement with triangular pitch (staggered).
  (iii) Tube arrangement with square pitch (in-line).
  (iv) Use of finned tubes with staggered arrangement.
  - Square pitch (ii).
  - Triangular pitch (i).
  - Finned tubes (iii).
- What is the normally recommended ΔP specified for shell and tube exchangers handling heavy hydrocarbons?
  - Frequently 35–70 kPa (5 or 10 psi) is specified for allowable pressure drop inside heat exchanger tubing. For heavy liquids that have fouling characteristics, this is usually not enough. There are cases where the fouling excludes using turbulence promoters and using more than the customary tube pressure drop is cost effective. This is especially true if there is a relatively higher outside heat transfer coefficient.
- How can shell side ΔP be reduced while designing a shell and tube exchanger?
  - Rod baffles, instead of plate baffles, give lower ΔP on the shell side.
  - Use of rods or tube protectors in top rows instead of plates reduces ΔP. These create less pressure drop and better distribution than an impingement plate. An impingement plate causes an abrupt 90° turn of the shell stream, which causes extra pressure drop.
- Compare exchangers arranged in parallel and in series.
  - *Parallel:* Low ΔP, low velocities; poorer thermal performance; savings in pumping costs.
  - *Series:* High ΔP, high velocities; good heat transfer; higher pumping costs (may require a booster pump).
- How is ΔP reduced while retrofitting an existing heat exchanger for increase in its capacity?
  - When an increase in capacity will cause excessive pressure drop, a relatively inexpensive alteration is to reduce the number of tube passes. Other possibilities are arranging the exchangers in parallel or using low fins or other special tubing.
- Give an expression for a quick estimate of additional ΔP resulting from increased number of tube passes in a shell and tube exchanger.
  - When the calculated pressure drop inside the tubes is underutilized, the estimated pressure drop with increased number of tube passes can be estimated from the relationship,

$$(\Delta P)_{New} = (\Delta P)_{Old} \times \left(\frac{\text{New number of passes}}{\text{Old number of passes}}\right)^3. \quad (10.19)$$

  - This would be a good estimate if advantage is not taken of the increase in heat transfer. Since the increased number of tube passes gives a higher velocity and increases the calculated heat transfer coefficient, the number of tubes to be used will decrease. For a better estimate of the new pressure drop, add 25% if the heat transfer is all due to sensible heat transfer.

10.1.5 Shell Side Versus Tube Side

- In the following cases, how do you select which of the fluids is to be taken on the shell side and which one on the tube side of a shell and tube heat exchanger? Explain.
  - *Gas–Liquid:* Gas on tube side as its velocity and hence h can be more easily increased.

- *Viscous Liquid–Nonviscous Liquid:* The decision of whether the more viscous fluid should be taken on the tube side or shell side involves the evaluation of many, often contradictory factors. It will be easier to increase turbulence on the shell side by the adjustment of baffle spacing, baffle cut, and staggered tube pitch than on the tube side. Also, to maintain flow of viscous liquids through tubes will involve prohibitively high pressure drops. However, if the high-viscosity fluid results in shell side $N_{Re} < 200$ and is being cooled, unstable operation becomes a possibility, and it would be better to pass such a fluid on the tube side. This has additional advantages such as ease of cleaning and increasing tube velocities with increase of tube passes more easily than shell passes. The overall heat transfer coefficient is controlled by the heat transfer coefficient for the viscous fluid.
- *High-Pressure Fluid and Low-Pressure Fluid:* High-pressure fluids are usually placed in the tubes. Placing the high-pressure fluid in the tubes will minimize the cost associated with the exchanger because the cost of thicker tube walls is generally less expensive than a thick shell. With their small diameter and nominal wall thicknesses, tubes are better able to withstand high pressures. This approach avoids having to design more expensive, larger diameter components for high pressure. If it is necessary to put the higher pressure fluid stream in the shell, it should be placed in a small diameter, long shell.
- *Corrosive and Noncorrosive Fluids:* Corrosive fluids that require a higher alloy are best placed in the tubes so that the shell does not have to be cladded with or fabricated from an expensive material. It is much less expensive to use the special alloys suitable to resist corrosion for the tubes than for the shell. Other tube side components can be clad with corrosion-resistant materials or epoxy coated.
- *Slurries/Fouling Liquids and Nonfouling Liquids:* Fouling/scaling liquids inside tubes. Tubes are easier to clean using common mechanical methods than the shell, the later being very difficult to clean.
- *Air and Flue Gas:* Air on shell side and flue gas on tube side. Flue gas is dirty and cleaning is easy on tube side.
- Many possible designs and configurations, affecting tube pitch, baffle use and spacing, and multiple nozzles, to name a few, can be used when laying out the shell circuit. Because of this, it is best to place fluids requiring low pressure drops in the shell circuit.
- The fluid with the lower heat transfer coefficient is admitted in the shell circuit, if low-fin tubing, which will increase available surface area, can be used to offset the low heat transfer rate.
- *Large Volumetric Flow Rate and Small Volumetric Flow Rate Fluids:* Large volume flows on shell side as shell side volume is more than tube side volume.
- Condensing fluids are typically placed on the shell side.
- *Steam and Cooling Water:* Steam on shell side as it occupies higher volume.
- *High-Temperature and Normal-Temperature Liquid:* High-temperature fluid is on tube side as otherwise heat losses will be more.
- Toxic and hazardous fluids are taken on tube side, as containment is easier on tube side than on shell side.
- For applications where complete drainability for servicing or process fluid changes is required, the critical fluid should be placed on the tube side.

• Why generally a fouling fluid is placed on the tube side of a heat exchanger?
  - For better control over design, fluid velocity and higher allowable velocity in tubes will reduce fouling.

• What are the questions to be addressed before preparing a specification sheet for a shell and tube exchanger?
  - Any phase changes expected?
    - In liquid–liquid exchangers, no vaporization must be ensured by use of proper operating pressures for the fluids.
  - Any dissolved gases in either stream?
    - Dissolved gases reduce heat transfer, these must be removed in a separation vessel prior to admitting to the exchanger. Example is in gas absorption systems.
    - Any dissolved or suspended solids in either stream? For example, cooling water should not be allowed to exceed 50°C to prevent reverse solubility salts to deposit on the heat transfer surfaces. Amounts and nature (particle size distribution and relative hardness) of suspended solids are to be ascertained to take appropriate measures such as velocities, erosion aspects, and so on.
  - Operating pressure and available ΔP in the exchanger (for existing pumps)?
    - The design pressure must be a certain factor above the highest operating pressure. For example, if the highest operating pressure for a heat exchanger is to be 700 kPa, then a reasonable design pressure may be 1000 or 1500 kPa, considering the higher the design pressure, the more expensive the exchanger will be.
    - Leakage considerations, with the consequences of contamination of the two fluids with one another and safety aspects, determine which fluid should be at higher pressure.

    - Available ΔP depends on the system considerations (discussed elsewhere). Most heat exchangers should need between 35 and 100 kPa of pressure loss to operate effectively.
  - Fouling tendencies of the fluids involved?
  - Either of the fluids non-Newtonian?
  - Non-Newtonian fluids have flow characteristics that dictate viscosity characteristics depending on the forces acting on the fluid. This has implications on selection and design.
  - Either of the fluids corrosive? Required material of constructions?
    - While selecting material of construction, considerations of relative corrosive characteristics of the fluids, temperatures, and pH must be addressed. For example, fluids with corrosive nature should be on the tube side.
    - Where requirements are for expensive alloys, a compact exchanger might be considered as they involve thinner gauge plates. Care should be exercised. Another point to consider is that just because a fluid is compatible with a stainless steel tube, for example, it may not be compatible with a stainless steel plate that has been prestressed (during the pressing process). Prestressing of metals can make them susceptible to pitting corrosion such as chloride attack.
  - Compatibility of elastomers and/or compression gaskets with the fluids?
    - Elastomer gaskets are commonly offered in materials such as EPDM, nitrile, PTFE, and FKMG (a generic form of Viton-G from Dupont). Elastomer gaskets can seldom be rated for temperatures in excess of 150°C.
    - Selection of nonmetallic or metallic gaskets is to be evaluated. Nonmetallic gaskets are seldom used at pressures in excess of 8000 kPa and temperatures in excess of 450°C.
  - Toxicity of either of the fluids?
    - The ASME pressure vessel code stipulates very specific pressure vessel requirements for heat transfer service that are qualified as *lethal.* If the service requires an ASME *L* stamp, it must be communicated to the heat exchanger manufacturer.
  - Any mechanical cleaning expected for one or both fluids?
    - Floating head shell and tube heat exchangers, gasketed plate exchangers, spiral heat exchangers, and some welded plate heat exchangers allow good access for mechanical cleaning.
  - Suitability of any particular cleaning solution?
  - Space requirements for proper maintenance?

10.1.6 Specification Sheet

- Prepare a typical specification sheet for a heat exchanger.
  - Table 10.10 gives a typical specification sheet for a shell and tube heat exchanger.

10.1.7 Log Mean Temperature Difference

- What is LMTD?
  - LMTD stands for log mean temperature difference, which is used in the heat exchanger design equation,

$$Q = UA\Delta T_m, \tag{10.20}$$

where ΔT_m is the mean temperature difference expressed as logarithmic mean value.

$$\text{LMTD} = \Delta T_{lm} = (\Delta T_1 - \Delta T_2)/\ln(\Delta T_1/(\Delta T_2). \tag{10.21}$$

  - Sometimes ΔT_{lm} can be approximated to arithmetic mean temperature difference, $(\Delta T_1 + \Delta T_2)/2$, when $1 \leq (\Delta T_1/\Delta T_2) \leq 2.2$, the error introduced by considering the arithmetic mean instead of logarithmic mean, is within 5%, that is,

$$\Delta T_{am}/\Delta T_{lm} = 1.05 \tag{10.22}$$

where $\Delta T_{am} = (\Delta T_1 + \Delta T_2)/2$.

- Give expressions for LMTD for (i) countercurrent flow and (ii) cocurrent flow in a heat exchanger.
  - *Countercurrent Flow:* The two fluids flow in opposite directions, each entering the exchanger at opposite ends. Because the cooler fluid leaves the exchanger at the end where the hot fluid enters, the cooler fluid will approach the inlet temperature of the hot fluid. Counterflow heat exchanger can have the hottest cold fluid temperature greater than the coldest hot fluid temperature.
  - Temperature profiles for countercurrent flow are given in Figure 10.36.

$$\text{LMTD} = [(T_1 - t_2) - (T_2 - t_1)]/\ln[(T_1 - t_2)/(T_2 - t_1)]. \tag{10.23}$$

  - *Cocurrent or Parallel Flow:* The two fluids enter the heat exchanger from the same end with a large temperature difference (Figure 10.37). As the fluids transfer heat, hotter to cooler, the temperatures of the two fluids approach each other. It should be noted that

TABLE 10.10 Specification Sheet for a Shell and Tube Heat Exchanger

1	Company/Customer. Reference No. Proposal No. Date:				
2	Plant Location:				
3	Service: *To remove heat from water in closed circuit with pond water*				
5	Size: Type: *TEMA NEN* Connected In Parallel Series				
6	Surface/Unit (Gross/Effective) m^2 No. of Shells/Unit Surface/Shell (Gross/Effective) m^2				
7	*Performance of One Unit*				
8	Fluid Allocation	(In) Shell (Out)		(In) Tube (Out)	
9	Fluid Name	Water		Pond Water	
10	Fluid Quantity, kg/h				
11	Temperature,°C				
12	Density, kg/m^3				
13	Viscosity, cP				
14	Specific Heat, J/(kg °C)				
15	Thermal Conductivity, W/(m °C)				
16	Inlet Pressure, kPa (abs)				
17	Fluid Velocity, m/s				
18	Pressure Drop (Allowable/Calculated), kPa	/		/	
19	Fouling Resistance (min) (m^2 °C/W)				
20	Heat Exchanged, J/h				
21	Heat Transfer Rate for Clean Service, W/(m^2 °C)				
22	*Construction of One Shell*				

the highest cold fluid temperature is always less than the lowest hot fluid temperature.

$$\text{LMTD} = [(T_1 - t_1) - (T_2 - t_2)]/\ln[(T_1 - t_1)/(T_2 - t_2)]. \quad (10.24)$$

- *Cross Flow:* Exists when one fluid flows perpendicular to the second fluid, that is, one fluid flows through tubes and the second fluid passes around the tubes at 90° angle.
- Cross-flow heat exchangers are usually found in applications where one of the fluids changes state (two-phase flow). An example is a condenser, in which vapor enters the shell side and cooling water flows through the tubes condensing the vapor.
- Large volumes of vapor may be condensed using cross flow.
- In actuality, most large heat exchangers are not purely parallel flow, counterflow, or cross flow. They are usually a combination of the two or all three types of heat exchangers.
- The rate of heat transfer in any of the types of flow varies along the length of the exchanger tubes because its value depends upon the temperature

TABLE 10.10 ***(Continued)***

23	Fluid Allocation			Shell Side	Tube/Channel Side
24	Design/Test Pressure (min/max), kPa (gauge)			/	/
25	Design Temperature, (max/min) (°C)				
26	Number of Passes			*1*	*2*
27	Corrosion Allowance (mm)				
28	Connection Input Size and Rating				
29	Connection Output Size and Rating				
30	Tube No.	O.D.	Wall Thickness	Length	
31	Tube Type	Material:		Pattern/Pitch:	
32	Shell I.D.	O.D.	Shell Cover		
33	Channel Material:		Channel Cover Material:		
34	Tube Sheet — Fixed:		Tube Sheet — Floating:		
35	Floating Head Cover		Impingement Protection	Yes	
37	Baffles — Cross Material	Type	%Cut (Dia/Area)	Spacing	Inlet
38	Baffles — Long		Seal Type		
39	Supports — Tube	U-Bend		Type	
40	Bypass Seal Arrangement		Tube-to-Tube Sheet Joint *Double Grooved, Rolled, and Seal Welded*		
41	Expansion Joint		Type		
42	Pv^2 — Inlet Nozzle	Bundle Entrance		Bundle Exit	
43	Gaskets Shell Side		Tube Side		
44	Floating Head				
45	Code Requirements — *ASME Section VIII, Division 1*	*TEMA Class C — Type NEN*			
46	Weight/Shell	Filled with Water		Bundle	

difference between the hot and the cold fluid at the point being viewed.

- What are the assumptions involved in the applicability of the LMTD equations given in the previous questions?
  - All elements of a given stream that enter an exchanger have the same thermal history, that is, follow paths through the exchanger that have the same heat transfer characteristics and have same exposure to heat transfer. In actual situation, there will be some flow paths that offer less flow resistance and less heat transfer surface than others, resulting in less heat transfer.
  - The heat exchanger operates under steady-state conditions.
  - Heat capacity of each stream is constant along the path of the exchanger.
  - Overall heat transfer coefficient is constant.

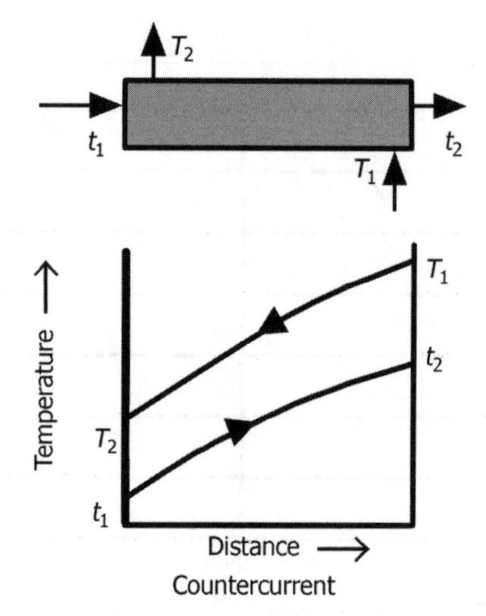

FIGURE 10.36 Temperature profiles for countercurrent flow.

- There will be no heat losses from the exchanger.
- There is no longitudinal heat transfer within a given stream. That is, plug flow conditions prevail without backmixing.
- The flow is either countercurrent or cocurrent.

- What are the important advantages of countercurrent flow arrangement over cocurrent arrangement?
  - In countercurrent flow arrangement, the maximum temperature change is limited by one of the outlet temperatures equilibrating with the inlet temperature of the other stream, giving a basically more efficient heat transfer for otherwise identical inlet conditions compared to the cocurrent arrangement. Due to this reason, countercurrent flow arrangement is chosen wherever possible.

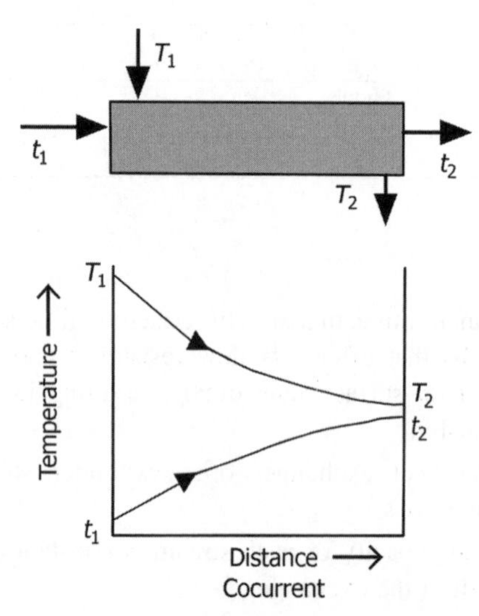

FIGURE 10.37 Temperature profiles for cocurrent flow.

  - The outlet temperature of the cold fluid can approach the highest temperature of the hot fluid (the inlet temperature).
  - The more uniform temperature difference between the two fluids minimizes the thermal stresses throughout the exchanger and produces a more uniform rate of heat transfer throughout the heat exchanger.
  - The efficiency of a counterflow heat exchanger is due to the fact that the average ΔT between the two fluids, over the length of the heat exchanger, is maximized resulting in larger LMTD for a counterflow heat exchanger than for a similar parallel or cross-flow heat exchanger.
  - If one stream is isothermal, the two cases are equivalent and the choice of countercurrent or cocurrent is immaterial.

- What are the disadvantages of parallel flow arrangement?
  - The large temperature difference at the ends causes large thermal stresses. The opposing expansion and contraction of the materials due to diverse fluid temperatures can lead to eventual material failure.
  - The temperature of the cold fluid exiting the heat exchanger never exceeds the lowest temperature of the hot fluid. This relationship is a distinct disadvantage if the design purpose is to raise the temperature of the cold fluid.

- Under what circumstances, parallel flow is advantageous?
  - The design of a parallel flow heat exchanger is advantageous when the two fluids are required to be brought to nearly the same temperature.

- Give examples of cases where LMTD is not the correct mean temperature difference to be used.
  - LMTD is equal to true or effective mean temperature difference, sometimes simply referred to as mean temperature difference (MTD), only if flow of the two streams is truly countercurrent or parallel.
  - For all *other* flow arrangements, LMTD $\neq$ MTD. For example, if there is any cross flow, partial or otherwise, is involved,

$$\text{MTD} = F(\text{LMTD}), \qquad (10.25)$$

    that is, $\Delta T_m = F\Delta T_{lm}$, where F is a nondimensional correction factor.
  - LMTD is not correct if U changes appreciably or when ΔT is not a linear function of Q. Examples are as follows:
    - The exchanger is required to cool and condense a superheated vapor or subcool condensate.

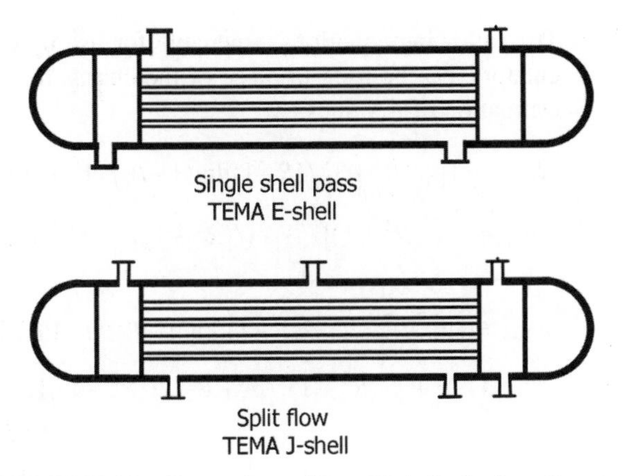

FIGURE 10.38 Comparison of E and J shells for flow directions.

> Heat transfer to or from a reacting fluid in a jacketed reactor.

- Is there a difference in MTD between E and J (divided flow) type shell and tube heat exchangers?
  - Figure 10.38 compares E and J shells for flow directions.
  - In the single-pass E-shell, MTD is same as LMTD, whether the flow is countercurrent or parallel.
  - In the split flow J-shell, the shell length is divided into two halves, one flow is in one direction and the other in the opposite direction, giving rise to cross-flow components to significant lengths, with countercurrent or parallel flows limited to shorter sections. Thus, MTD is less than LMTD, as the correction factor, F, needs to be applied to LMTD, reducing the effectiveness.
- "In a condenser, there is no difference between countercurrent or parallel flows." *True/False*?
  - *True*. One of the fluids, that is, condensing fluid, is at constant temperature, provided the exchanger is functioning only as a condenser and neither desuperheating nor subcooling is involved.
- Show temperature profiles for a 1–2 heat exchanger.
  - Figure 10.39 illustrates temperature profiles for a 1–2 heat exchanger.
- Why is it necessary to apply correction factors to LMTD for multipass heat exchanger?
  - In practice, flows through a heat exchanger, especially multipass exchangers, are not truly countercurrent but there will be a cross-flow element, reducing effective LMTD and hence heat transfer rates. For such cases, a correction factor needs to be applied to the heat transfer equation, making the equation,

$$Q = UAF\Delta T_{\text{lm}}, \tag{10.26}$$

where F being the correction factor to account for the extent of deviation from counterflow conditions. ΔT_{lm} is LMTD. F is always less than 1.0 and typically more than 0.8.
  - Mean temperature difference is normally represented by the symbol ΔT_{m} that is equal to $F\Delta T_{\text{lm}}$.
- What is the significance of F?
  - The limiting significance of F should be clearly understood. It does not represent the effectiveness of the heat exchanger, but represents a degree of departure for the true mean temperature difference from the countercurrent flow LMTD. In other words, F is a gauge of the exchanger performance in comparison to that of a counterflow exchanger.
- On what factors, the value of F depends?
  - F depends on the exact arrangement of the two streams in the exchanger, number of exchangers in

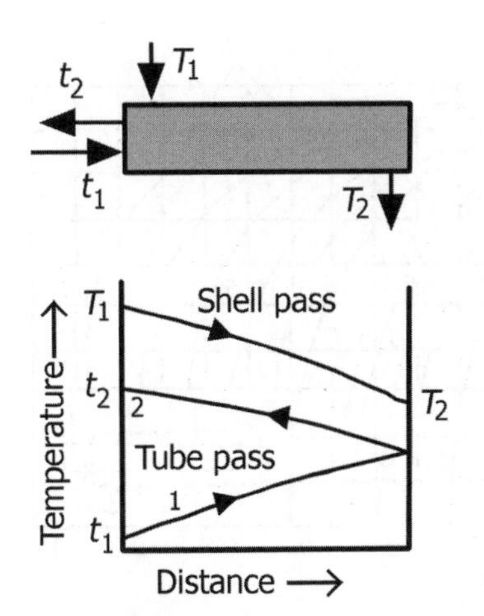

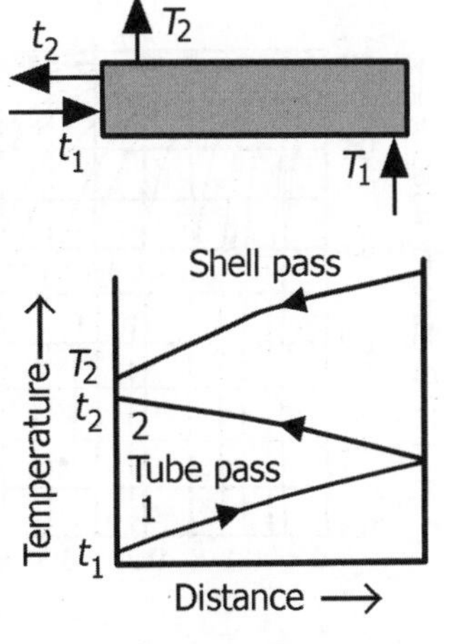

FIGURE 10.39 Temperature profiles for a 1–2 heat exchanger.

series, and the parameters, R and P, as defined in the following relationships:

$$R = (T_1 - T_2)/(t_2 - t_1)$$
$$= \text{(temperature range of the shell side fluid)/(temperature range of the tube side fluid)}. \quad (10.27)$$

- In other words, R is the ratio of the fall in temperature of the hot fluid to the rise in temperature of the cold fluid.
- R is called *heat capacity rate ratio* and its value ranges from 0 to ∞, zero being for pure vapor condensation and infinity being for pure liquid vaporization.

$$P = (t_2 - t_1)/(T_1 - t_1)$$
$$= \text{(temperature range of tube side fluid)/(maximum temperature difference)}. \quad (10.28)$$

- P, in other words, is the heat transfer (or thermal) effectiveness, which is the ratio of actual temperature rise of the cold fluid to the maximum temperature rise obtainable, that is, if the warm end approach is zero, based on countercurrent flow.
- The value of P ranges from 0 to 1.

- The relationships among F, R, and P are given in the form of charts for different exchanger configurations as illustrated for 1–2 and 2–4 heat exchangers (Figures 10.40 and 10.41).
- Similar charts are given in standard texts for other exchanger configurations.
- Once the terminal temperatures of both streams of the exchanger are specified/determined, R, P, and LMTD can be calculated.
- The following equations, amenable for use by calculators, can be used in place of the charts, for the estimation of F values:

$$F_{1-2} = \{[\sqrt{(R^2+1)}/(R-1)]\ln[(1-P)/(1-PR)]\}/\ln\{[A+\sqrt{(R^2+1)}]/[A-\sqrt{(R^2+1)}\}. \quad (10.29)$$

$$F_{2-4} = \{[\sqrt{(R^2+1)}/2(R-1)]\ln[(1-P)/(1-PR)]\}/\ln\{[A+B+\sqrt{(R^2+1)}]/[A+B-\sqrt{(R^2+1)}]\}, \quad (10.30)$$

where

$$A = (2/P)-1-R \text{ and } B = (2/P)\sqrt{(1-P)(1-PR)}. \quad (10.31)$$

F_{1-2} stands for F value for a 1–2 exchanger and F_{2-4} stands for F for a 2–4 exchanger.

- The equation given for F_{1-2} also applies to one shell pass and two, four, or any multiple of two tube passes.
- Similarly, the equation for F_{2-4} also applies to two shell passes and four, eight, or any multiple of four tube passes.

- What value of LMTD correction factor is generally assumed in the absence of LMTD correction charts?
  - For the heat exchanger equation, $Q = UAF\Delta T_{lm}$, use $F = 0.9$ when charts for the LMTD correction factor are not available.
- What are the negative aspects of using F values below 0.8 or 0.75 in the design of a heat exchanger? What are the alternatives in such cases?
  - Low value of F means use of substantially large surface area for the exchanger to compensate for the

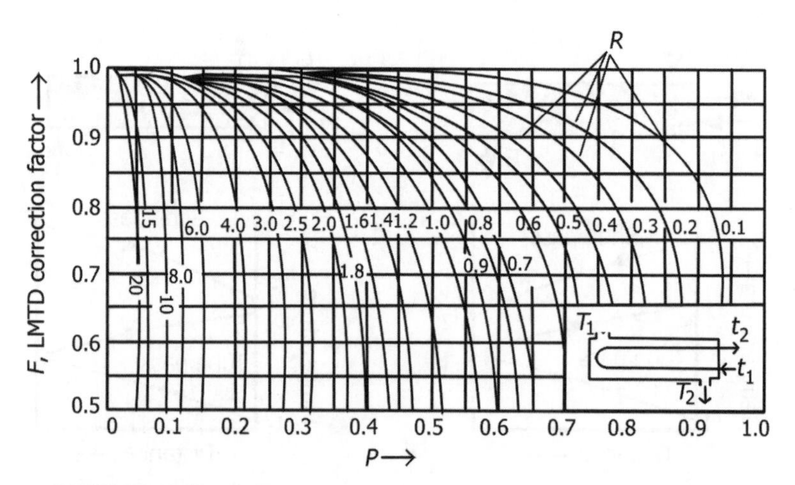

FIGURE 10.40 LMTD correction factors, F, for a 1–2 heat exchanger.

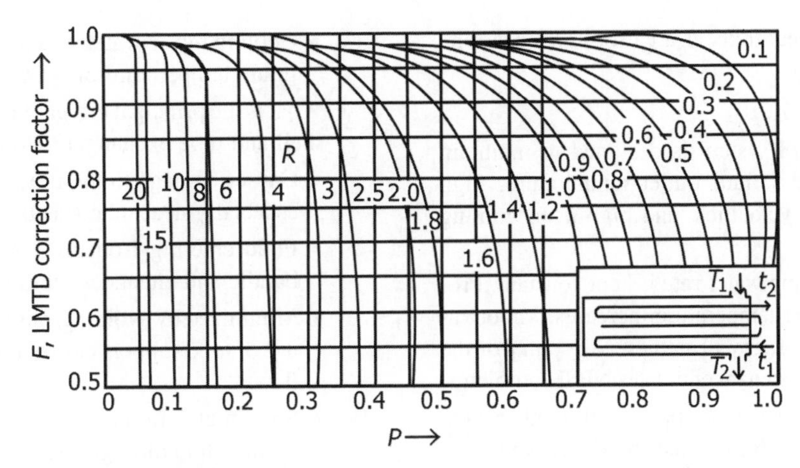

FIGURE 10.41 LMTD correction factors, F, for a 2–4 heat exchanger.

inefficient temperature profile. This increases exchanger costs.

- The charts cannot be relied upon for accuracy in the steep portions of the curves.
- Design in or near the steep portion of the curves indicates that thermodynamically limiting configuration is approached, even if all the assumptions are satisfied. Violation of even one assumption, for example, excessive bypassing, may result in an exchanger that is thermodynamically incapable of meeting the specified temperatures.
- If the value of F determined is too low for the proposed configuration, after considering all options of shell and tube passes by using relevant charts, use of additional shells in series might be considered for improvement. Alternatively, it may be possible to redesign the exchangers to permit the use of fixed tube sheet exchangers and purely countercurrent flow, for which F is unity.

- Does an increase in LMTD increase the overall heat transfer coefficient in a shell and tube heat exchanger?
  - The overall heat transfer coefficient is generally weakly dependent on temperature. As the temperatures of the fluids change, the degree to which the overall heat transfer coefficient will be affected depends on the sensitivity of the viscosity of the fluids to temperature. If both fluids are water, for example, the overall heat transfer coefficient will not vary much with temperature because viscosity of water does not change greatly with temperature.
  - If, however, one of the fluids is oil that may have a viscosity of 1000 cP at 10°C and 5 cP at 200°C, then the overall heat transfer coefficient would be much better at higher temperatures since the oil side would be the controlling factor.

- What is temperature approach in a heat exchanger?
  - Difference between outlet temperature of one stream and inlet temperature of the other stream. In other words, it is the temperature difference between the hot side outlet temperature and the cold side outlet temperature.
  - A good rule of thumb is that a single shell and tube heat exchanger should be designed with a minimum temperature approach of 5°C.
- What are the minimum recommended temperature approaches for shell and tube heat exchangers?
  - The minimum approach temperature is about 10°C for fluids and 5°C for refrigerants.

10.1.8 Performance

- What are the effects of weather conditions on the performance of a cooler?
  - *Winter Conditions:*
    - *Case I*: If the exchanger control system is designed to maintain constant process fluid outlet temperature, winter conditions reduce the cooling water flows, bringing cooling water velocities to low levels. Decreased turbulence and increased fouling resulting in decreased heat transfer coefficient on cooling waterside.
    - *Case II*: If cooling water rate is kept constant, process fluid outlet temperature decreases. If it is liquid, for example, oil, its viscosity increases, heat transfer coefficient on process fluid side decreases. Increased viscosity also increases ΔP as well as the chances for blockages. If it is a condenser, subcooling of the condensate will occur.
      - Condensate backup.
      - More tubes get submerged in condensate.

- Decreased heat transfer area for condensation.
- Decreased Q.

- *Summer Conditions.*
  - *Case I*: Control system is designed to maintain constant process fluid outlet temperature, high cooling water velocities, and high h on cooling waterside.
  - *Case II*: Cooling water rate is kept constant. Process fluid outlet temperature increases. Vaporization losses on storage will increase. Any pump on the exit side of the exchanger creates NPSH problems.

- "Temperature-sensitive fluids passing through a heat exchanger can be a source of waste products. Reducing tube wall temperatures is one way to prevent formation of such wastes." What are the ways to reduce tube wall temperatures?
  - Use of lower pressure steam for heating.
  - Desuperheating plant steam reduces tube wall temperature and increases effective area for heat transfer by increasing heat transfer coefficient (h for condensing steam is about 10 times that for superheated steam).
  - Use of staged heating minimizes degradation of heat-sensitive fluids.
  - Using air-fin cooling instead of water cooling reduces cooling water requirements.
  - Using online cleaning techniques keeps tube surfaces clean that makes it possible use of lower pressure steam.
  - Controlling cooling water temperature reduces scale formation.
- "Maximum recommended temperatures for cooling water in heat exchanger outlets are about 45°C." If temperatures exceed this value, what are the disadvantages?
  - More fouling.
  - More evaporation losses.
- "Maximum recommended velocities through nozzle connections and piping associated with shell and tube heat exchangers will be more for low-viscosity liquids compared to high-viscosity liquids." *True/False*?
  - *True.*
- Water cooling has been normal practice in industry, but of late air cooling is acquiring importance. Give reasons for this shift.
  - Water shortages.
  - Costs involved in treatment and disposal of wastewater requirements originating from water cooling systems to meet pollution control/regulations.
- What are the maximum temperatures at which cooling water is normally available?
  - Cooling tower water is typically available at a maximum temperature of 30°C.
- What is the principal effect of tube bundle vibration in a shell and tube exchanger? How is it minimized?
  - Causes damage to the tubes and results in leakages, especially near tube sheets.
  - Fluid entering from inlet nozzle impinges on tube bundle and enhances vibration.
  - Minimized by avoiding/minimizing cross flow by the use of tube support baffles that promote longitudinal flow.
  - Special attention is given to minimize impingement of inlet fluid through shell inlet nozzle (by the use of impingement baffle at inlet nozzle).
- If a tube breaks within a shell (might occur while removing a tube from bundle), what is the suggested course of action?
  - When tube breaks, it is very difficult to remove or replace it.
  - In such cases, the holes on both tube sheets are plugged to avoid mix-up of the two fluids. This involves less loss of downtime compared to the replacement of the tube.
  - On turn around for maintenance of the plant or unit, the plugged tube(s) may be replaced.
- When a pinhole or holes develop in a tube, what are the likely consequences?
  - The consequences depend on the nature of fluids, their temperature and pressure levels, and the type of service.
    - In a heat exchanger, shell side fluid gets contaminated with tube side fluid or vice versa, depending upon which side fluid is at higher pressure. Ingress of one fluid into the other causes contamination, flow/pressure variations, fouling, corrosion, reaction, and other problems.
    - If shell side fluid is a condensing organic vapor, for example, like a petroleum product and tube side is having cooling water, the condensate forms two layers and mutual solubilities contaminate both the fluids that require treatment as moisture in the oil can be detrimental for the oil product and oily waters create environmental problems. If the condensing vapor is miscible with water, there will be concentration changes that will affect quality.
    - If the tube side fluid is at a high pressure, the shell side gets pressurized with the consequence of failure/explosion of the shell as it is not normally designed to withstand high pressures.
    - In a tube still heater, leakage of oil into the firebox alters combustion process, with incomplete

combustion and formation of CO, soot, and other products, which can be hazardous leading to the need to additional soot blowing to prevent deposits on heat transfer surfaces, furnace explosions, and creating air pollution problems.

 ➢ Plugging or replacing such a tube, preventing flow, becomes a priority to avoid any intolerable consequence.

- What are the consequences of tube vibration in a heat exchanger? What type of heat transfer application is prone to tube vibrations?
  - Bundle vibration can cause leaks due to tubes being cut at the baffle holes or being loosened at the tube sheet joint.
  - There are services that are more likely to cause bundle vibration than others. The most likely service to cause vibration is a single-phase gas operating at a pressure of 700–2000 kPa. This is especially true if the baffle spacing is greater than 46 cm (18 in.) and single segmental.
  - Most flow-induced vibration occurs with the tubes that pass through the baffle window of the inlet zone. The unsupported lengths in the end zones are normally longer than those in the rest of the bundle.
  - For 19 mm (3/4 in.) tubes, the unsupported length can be 1.2–1.5 m (4–5 ft). The cure for removable bundles, where the vibration is not severe, is to stiffen the bundle. This can be done by inserting metal slats or rods between the tubes. Normally this only needs to be done with the first few tube rows.
  - Another solution is to add a shell nozzle opposite the inlet so as to cut the inlet fluid velocity by half. For nonremovable bundles, this is the best solution. Adding a distributor belt on the shell would be a very good solution if it is not very expensive.
- Write the equation for rate of heat transfer in terms of overall heat transfer coefficients.

$$Q = UAF\Delta T_{\text{lm}}. \quad (10.32a)$$

- Give an equation for heat transfer rate for varying overall heat transfer coefficient. Give the assumption involved in the applicability of the equation.

$$Q = A[U_2\Delta T_1 - U_1\Delta T_2]/\ln[U_2\Delta T_1/U_1\Delta T_2]. \quad (10.32b)$$

The subscripts refer to the ends of the heat exchanger.

- "According to the equation, $Q = UAF\Delta T_{\text{lm}}$, increase in A by increasing number of tubes in a shell and tube heat exchanger should result in increase in Q." Comment and explain.
  - Exchanger size and hence fixed costs will increase.
  - Larger exchanger sizes result in decrease in flow velocities.
    - ➢ Decreased flow velocities result in decreased heat transfer coefficients.
    - ➢ Also at low velocities, fouling deposits increase, further reducing heat transfer coefficients.
    - ➢ Plugging of tubes can result, especially when liquids are viscous.
    - ➢ Second aspect, (ii), is very critical.
    - ➢ Net result can be increased A might result in reduced Q, instead of increased Q.
    - ➢ *Another Aspect for High-Viscosity Liquids*: Decreased velocity reduces h and hence Q, which means *lower tube wall temperatures*, thereby increasing μ_W. This has a further decreasing effect on h as $h \propto (\mu/\mu_W)^{0.14}$. This effect is very considerable for very high-viscosity liquids.
- What alternatives can be suggested for overcoming the problem mentioned in the above question in cases where A is to be increased.
  - Velocities should be increased by increasing number of tube passes. It is always better to use a higher ΔP exchanger, although it involves higher pumping costs that are much less than loss of energy recovery.
  - Use of larger tube lengths increases A for the same number of tubes, with marginal increase in shell costs compared to use of larger diameter shells with more number of tubes. While there are limitations in increasing tube lengths beyond certain limits, use of multiple shells in series can be an alternative.
  - Another way to increase Q by increasing A will be through the use of finned tubes in place of plain tubes. Surface areas of finned tubes will be higher compared to plain tubes, involving no flow area increase, thereby not affecting flow velocities.
- "When cooling a highly viscous liquid with water in a tubular heat exchanger, tube wall temperature is approximately equal to cooling water temperature." *True/False*?
  - *True*. Combined heat transfer resistance for water film + metal wall will be negligible compared to that for viscous liquid film.
- What are the detrimental effects of using multiple tube passes?
  - Multiple tube passes can lead to a reduction of number of tubes that can be accommodated in a given shell. In other words, increasing the shell diameter will be more for a given number of tubes resulting in increased equipment cost.
  - Use of multiple tube passes results in the thermal contacting of the streams not being pure counterflow,

which reduces the effective mean temperature driving force and possibly can lead to a temperature cross, where the outlet temperature of the cold stream is higher than the inlet temperature of the hot stream. This reduces the effective mean temperature difference, necessitating use of two or more exchangers operating in series.

- Tube bundles having more than four tube passes, introducing pass partition lanes into the bundles, result in increased bundle bypassing and a reduction in shell side heat transfer coefficient.
- Part of tube side pressure drop is wasted in the return headers, as no heat transfer will take place in the headers.

- "Increasing ΔT in a heat exchanger increases heat transfer rate according to the equation, $Q = UAF\Delta T$." What are the limitations involved in this approach?
  - Instead of using ambient cooling water supply to condensers, refrigeration could be used. But it very much increases process costs.
  - Use of high steam temperatures in heaters involves high steam pressures.
    - *High Steam Pressures*: Low λ_V and therefore high steam consumption compared to using low-pressure steam.
    - *High Pressure Steam*:
      - More valuable for power generation.
      - Turbine exhaust steam is at low pressure with high λ_V.
      - Exchanger is to be designed for high pressure, which is expensive with poorer safety.
    - Increased ΔT will also result from using superheated steam as heating medium but it is not normally preferred for the following reasons:
      - Superheated steam is more valuable for power generation involving steam turbines.
      - Heat transfer coefficients are very low in the desuperheating section of the heat exchanger, steam acting as a gas during this cooling process.
      - Heat given out by the steam during desuperheating process is sensible heat that is very low compared to latent heat liberated during condensation. For example, steam at 1 bar absolute with a 50°C superheat transfers about 101 kJ/kg while cooling from the superheat to the saturation temperature. But it transfers about 2485 kJ/kg of latent heat on complete condensation.
    - Use of Dowtherm or other high-temperature heat transfer fluids requires special boilers and piping, which is expensive for normal applications.

- What are the means by which one can increase ΔT in a heat exchanger involving a heating medium?
  - By using heat transfer fluids.
- "Increasing U increases Q according the equation, $Q = UAF\Delta T_{lm}$." Comment.
  - Increase U by increasing V ($U \propto V^{0.8}$).
  - V is increased on tube side by increasing number of tube passes.
  - V is increased on shell side by decreasing baffle spacing.
  - Negative aspect in the above approaches is increased pressure drop.
  - But pumping energy is much less than energy transfer in an exchanger.
- What are the ways by which U can be increased in a heat exchanger?
  - U can be increased by increasing turbulence in the flowing fluids on tube and/or shell side. It must be understood that pressure drop also increases and whether such an increase is considered to be under allowable ΔP limits.
  - By reducing fouling rates through pretreatment of the fluids or increasing cleaning schedules.
  - By eliminating stagnant areas inside the exchanger by judicious design.
- What are the causes for overpressure development and failure of heat exchangers?
  - Corrosion/erosion of exchanger internals resulting in a heat transfer surface leak or rupture leading to possible overpressure of the low-pressure side.
  - Differential thermal expansion/contraction between tubes and shell resulting in tube leak/rupture (fixed tube sheet exchanger).
  - Excessive tube vibration resulting in tube leak/rupture and possible overpressure of the low-pressure side.
  - Excessive heat input resulting in vaporization of the cold side fluid.
  - Loss of heat transfer due to fouling, accumulation of noncondensables, or loss of cooling medium.
  - Ambient temperature increase resulting in higher vaporization rate in air heated exchanger.
  - Cold side fluid blocked in while heating medium continues to flow.
- A process liquid at high pressure is being heated by steam with pressure lower than the process liquid. A pinhole develops on a tube due to corrosion. What happens to the operation of the exchanger? Explain and discuss.

- The process fluid leaks into the shell increasing shell side pressure.
- The shell, being much weaker than the tubes, might not be able to withstand the increased pressure and might fail spilling the steam and the tube side fluid if adequate safety relief is not provided on shell side.
- There will be possibilities for explosion.
- Shell side condensate gets contaminated and results in pollution problems with discharged condensate.
- Flow/pressure variations on tube/shell side might result.
- Corrosion, fouling, and other problems can arise.

10.2 THERMAL DESIGN OF SHELL AND TUBE HEAT EXCHANGERS

- What are the criteria to be considered for designing shell and tube heat exchangers?
  - The process requirements for heat transfer within the allowable pressure drops, effects of fouling and cleaning requirements, must be fulfilled.
  - The heat exchanger must withstand the service conditions of the plant environment.
  - Minimum or maximum flow velocities.
  - The exchanger must be maintainable. In other words, a configuration that permits replacement of any component that is especially vulnerable to thermal expansion, corrosion, erosion, vibration, or aging must be chosen.
  - Consideration of the advantages of a multishell arrangement with flexible piping and valving provided to allow one unit to be taken out of service for maintenance without disturbing the rest of the plant.
  - Cost considerations while satisfying the above criteria.
  - Limitations on the heat exchanger length, diameter, weight, and/or tube specifications due to site requirements, lifting, and servicing capabilities must be all taken into consideration in the design.
- What are the reasons for adding margins while designing a heat exchanger?
  - To account for
    - Fouling.
    - Uncertainties in the methods used and fluid properties. Uncertainties in estimation of heat transfer coefficients depend on exchanger geometry and phase condition of the fluids.
      - For a shell and tube exchanger, on the tube side, normally variation is within ±10%, while on shell side it can be as high as ±20–50%.
      - For a plate heat exchanger, the uncertainty can be ±10–30%.
      - For a plate-fin exchanger, it can be ±20%.
    - Variations in process and ambient conditions.
      - *Process Conditions*: Capacity increases can be the primary cause for increased erosion and vibration problems that can cut down the life of an exchanger. Flow decreases increase fouling rates.
      - *Ambient Conditions:* Results in variations in MTD and hence performance of an exchanger. Effects are particularly high with air-cooled exchangers.
    - Lessons from previous experience.
    - Risks associated with an exchanger that does not meet the process requirements.
- Define heat exchanger design margin.
  - Exchanger design margin is defined as any heat transfer area exceeding the required area for a clean exchanger to satisfy a specified duty.
  - The following equations complete the definition:
    - Percent excess area from fouling

$$= 100[(U_{\text{clean}}/U_{\text{actual}})-1]. \tag{10.33}$$

    - Percent over design

$$= 100[(U_{\text{actual}}/U_{\text{required}})-1]. \tag{10.34}$$

    - Percent total excess area

$$= 100[(U_{\text{clean}}/U_{\text{required}})-1]. \tag{10.35}$$

$$U_{\text{clean}} \geq U_{\text{actual}} \geq U_{\text{required}}. \tag{10.36}$$

- Name the important methods used in designing shell and tube heat exchangers.
  - *Kern Method:* Simplest and long established. Restricted to a fixed baffle cut (25%) and cannot adequately account for baffle-to-shell and tube-to-baffle leakages. However, the Kern equation is not particularly accurate, it does allow a very simple and rapid calculation of shell side coefficients and pressure drop to be carried out and has been successfully used since its inception.
  - *Bell Delaware Method:* The most widely accepted method. Used as a rating method.

10.2.1 Kern Method

- Describe Kern method for the design of shell and tube heat exchanger.

- *Equations for Shell Diameter and Shell Equivalent Diameter:*

$$N = \pi(\text{CTP}) \cdot D_s^2/4(\text{CL})(\text{PR})^2\,(d_o^2), \quad (10.37)$$

where N is the total number of tubes. CTP is the tube count calculation constant that accounts for the incomplete coverage of the shell by the tubes due to necessary clearances between the shell and the outer tube circle and tube omissions due to tube pass lanes for multitude pass design. CTP values for different tube passes are 0.93 for one tube pass, 0.90 for two tube passes, and 0.85 for three tube passes. D_s is the shell inside diameter and CL is the tube layout constant (1.0 for 90° and 45° layouts and 0.87 for 30° and 60° layouts). PR $= P_T/d_o$, where P_T is the tube pitch and d_o is the outside tube diameter.

- *Equation for Shell Inside Diameter, D_s:*

$$D_s = 0.637(\text{CL}/\text{CTP})^{1/2}[A_o(\text{PR})^2 d_o/L]^{1/2}, \quad (10.38)$$

where A_o is outside heat transfer surface area based on tube outside diameter $= \pi d_o NL$, L being tube length.

- *Shell Equivalent Diameter, D_e:* The equivalent diameter is calculated along (instead of across) the long axes of the shell and therefore is taken as four times the net flow area as layout on the tube sheet (for any pitch layout) divided by the wetted perimeter.
- $D_e = 4$ (free flow area/wetted perimeter).
- For square pitch,

$$D_e = 4(P_T^2 - \pi d_o^2/4)/\pi d_o \quad (10.39)$$

- For triangular pitch,

$$D_e = 4\{[(P_T^2\sqrt{3})/4] - (\pi d_o^2/8)\}/\pi d_o/2 \quad (10.40)$$

- *Equation for the Number of Tubes at the Centerline of the Shell:*

$$\text{Number of tubes, } N_t = D_s/P_T. \quad (10.41)$$

- *Equation for Shell Side N_{Re}:*

$$N_{\text{Re s}} = (m_s/A_s)\,(D_e/\mu_s). \quad (10.42)$$

where m_s is shell side fluid flow rate and A_s is the cross-flow area at the shell diameter.

$$A_s = (D_s/P_T)CB, \quad (10.43)$$

where A_s is the bundle cross-flow area, C is the clearance between adjacent tubes, $(P_T - d_o)$, and B is the baffle spacing.

- *Equations for the Estimation of Shell Side and Overall Heat Transfer Coefficients, Heat Transfer Area, and Exchanger Tube Length and Shell Diameter:*
  - Shell side

$$N_{\text{Nu}} = h_o D_e/k_s = 0.36(N_{\text{Re s}})^{0.55}(N_{\text{Pr s}})^{1/3}(\mu_b/\mu_w)^{0.14} \quad (10.44)$$

$$2 \times 10^3 < N_{\text{Re s}} = (G_s D_e/\mu) < 1 \times 10^6. \quad (10.45)$$

  - Overall coefficient for clean surfaces, U_c:

$$1/U_c = 1/h_o + 1/h_i\,(d_o/d_i) + r_o \ln(r_o/r_i)/k, \quad (10.46)$$

  where r refers to radius and the subscripts c, o, and i refer to clean, outside, and inside, respectively.

  - Considering fouling resistances, overall coefficient for fouled surface is

$$1/U_f = 1/U_c + R_{\text{ft}}, \quad (10.47)$$

  where R_{ft} is total fouling resistance.

  - Heat transfer area A_f, for fouled exchanger, is given by

$$A_f = Q/U_f(F)(\text{LMTD}). \quad (10.48)$$

  - Exchanger length is given by

$$L = A_f/N_t \pi d_o. \quad (10.49)$$

  - Shell diameter, D_s, is calculated by the equation given earlier.
- Method for finding LMTD is given earlier along with sample plots for finding LMTD correction factor, F.
- Kern method involves estimation of LMTD, which assumes that both inlet and outlet temperatures are known. When this is not the case, the solution to a heat exchanger problem becomes somewhat tedious.

10.2.2 ε-NTU Method

- Define effectiveness of a heat exchanger.
  - Effectiveness is the ratio of actual heat transfer rate from hot to cold fluid in a given exchanger of any flow arrangement to the thermodynamically limited maximum possible heat transfer rate.

$$\varepsilon = Q/Q_{\max} = \text{actual heat transfer rate/maximum possible heat transfer rate from one stream to the other.} \quad (10.50)$$

  - Equations for effectiveness for parallel and countercurrent flows are given below:

➢ For countercurrent flow,

$$\varepsilon = [1-\exp\{-(UA/C_{\min})(1-C_{\min}/C_{\max})\}]/[1-(C_{\min}/C_{\max})\exp\{-(UA/C_{\min})\times(1-C_{\min}/C_{\max})\}]. \quad (10.51)$$

➢ For parallel flow,

$$\varepsilon = [1-\exp\{-(UA/C_{\min})(1+C_{\min}/C_{\max})\}]/[1+C_{\min}/C_{\max}], \quad (10.52)$$

where U is the overall heat transfer coefficient, A is the heat transfer area, $C_{\min}$ is the lower of the two fluids heat capacities, $mC_{P\min}$, and $C_{\max}$ is the higher of the two fluids heat capacities, $mC_{P\max}$.

- How is $Q_{\max}$ determined?
  - $Q_{\max}$ is obtainable in a counterflow heat exchanger of infinite surface area operating with fluid flow rates and inlet temperatures same as those of an actual exchanger. In other words, it is obtainable with a counterflow exchanger, if temperature change of the fluid having minimum value of mC_p equals difference in inlet temperatures of hot and cold fluids.
- Define NTU for heat transfer.

$$\text{NTU} = |(t_i - t_o)|/\Delta t_m = [\text{total temperature change for process fluid}]/[\Delta t_m \text{ for the exchanger}] = U_{Av}\cdot A_t/WC_P. \quad (10.53)$$

  - It can also be written as

$$\text{NTU} = UA/C_{\min} = \text{ratio of overall heat transfer to smaller heat capacity} = (1/C_{\min})\int U dA = U_{Av}A/C_m. \quad (10.54)$$

$$C_{\min} = \text{smaller of } (mC_P)_{hot} \text{ and } (mC_P)_{cold}. \quad (10.55)$$

  - NTU designates nondimensional *heat transfer size* or *thermal size* of a heat exchanger. Physical size is heat transfer area.
  - Physical significance of NTU: Ratio of heat capacity of exchanger, W/°C divided by heat capacity of flow, W/°C.
- Give the equivalent plots, based on the equations given earlier for the estimation of effectiveness, in terms of NTU.
  - Figures 10.42 and 10.43 give relationships for NTU and effectiveness of the heat exchangers for countercurrent and parallel flow conditions, respectively.
- How is heat load and outlet temperature estimated from the effectiveness and input temperatures for a heat exchanger?

$$Q = (\text{effectiveness}, \varepsilon)(T_{\text{hot in}} - T_{\text{cold in}}). \quad (10.56)$$

$$T_{\text{hot out}} = T_{\text{hot in}} - Q/(m_{\text{hot}}\cdot C_{P\,\text{hot}}). \quad (10.57)$$

$$T_{\text{cold out}} = T_{\text{cold in}} + Q/(m_{\text{cold}}\cdot C_{P\,\text{cold}}). \quad (10.58)$$

- What is the difference between LMTD and NTU design methods?
  - *LMTD Method:* Temperatures of inlet hot and cold fluids are known.
    - ➢ Type of heat exchanger is selected and exchanger size (heat transfer area) is estimated.
  - *NTU Method:* Heat exchanger type and size are known.

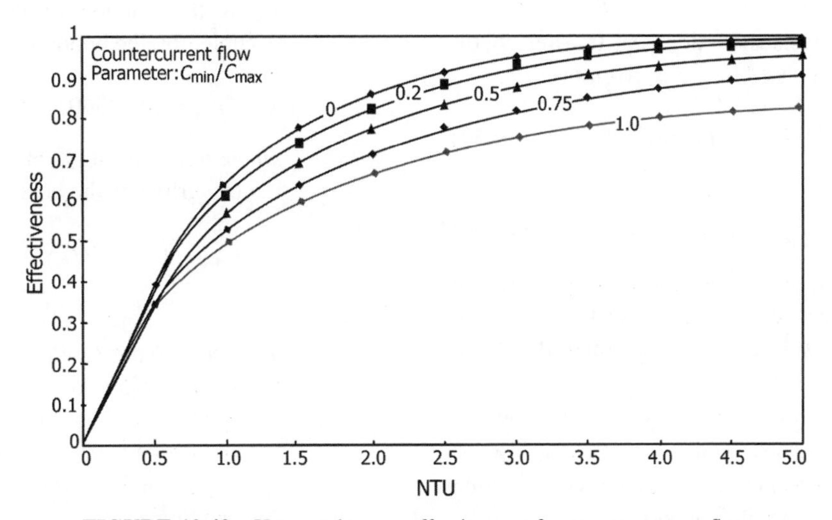

FIGURE 10.42 Heat exchanger effectiveness for countercurrent flow.

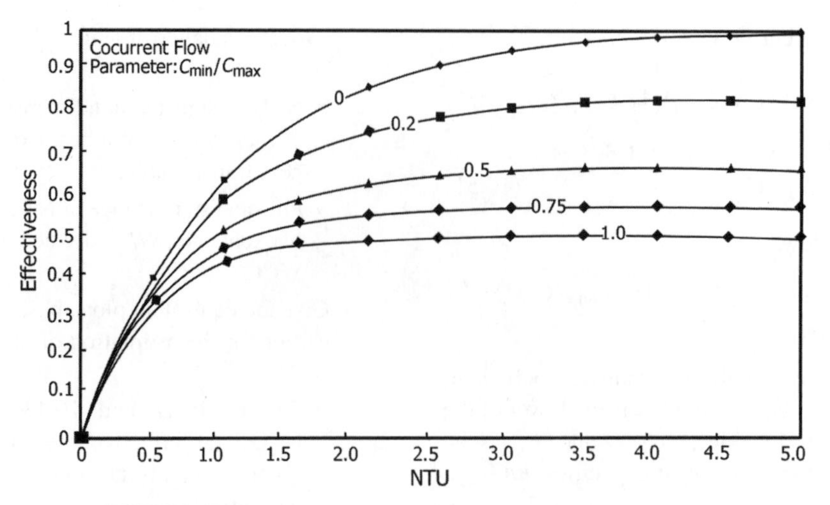

FIGURE 10.43 Heat exchanger effectiveness for cocurrent flow.

  - Heat transfer rate and fluid outlet temperatures are to be estimated, knowing flow rates and inlet temperatures.
- One type of calculation where the ε-NTU method may be used to clear advantage would be cases in which neither fluid outlet temperature is known.

10.2.3 Bell Delaware Method

- What are the essential elements in Bell Delaware method in the design of shell and tube heat exchangers?
  - In the Delaware method, the fluid flow in the shell is divided into a number of individual streams A through F as discussed before (Figure 10.28).
  - Correction factors to heat transfer correlation for cross flow across tube banks for each of the streams are introduced. Note that in the case of LMTD method, only one correction factor, F, is involved.
  - The stream analysis shell side heat transfer coefficient for single-phase flow h_o is given by

$$h_o = (J_c J_l J_b J_s J_r J_\mu) h_{ideal}. \quad (10.59)$$

$$h_{ideal} = J_H C_{Ps} (m_s/A_s)(k_s/C_{Ps}\mu_s)^{2/3}. \quad (10.60)$$

  - h_{ideal} is the ideal tube bank heat transfer coefficient for all the flow across the tube bundle, that is, as if all the flow in the exchanger were in stream B without any bypass flows.
  - J_H: Colburn J-factor. It is a function of the following:
    - Shell side Reynolds number based on the outside tube diameter and on the minimum cross-section flow area at the shell diameter.
    - Tube layout.
    - Pitch size.
  - J_c: Correction factor for baffle cut and spacing. This factor accounts for the nonideal flow effects of window flow on heat transfer since the velocity through the window (that of the baffle cut) is not the same as that for cross flow over the bundle. The window flow velocity can be larger or smaller than for cross flow depending on the size of the cut and spacing. In addition, the window flow is partially longitudinal to the tubes, which is less effective than cross flow.
    - It is a function of the baffle cut, the outer tube limit diameter, and the window flow area and is calculated by the following procedure:

$$J_c = 0.55 + 0.72F_c, \text{ where } F_c = 1 - F_w. \quad (10.61)$$

F_w is the fraction of the cross-sectional area occupied by the window.

$$F_w = (\theta_{ctl}/360) - (\sin\theta_{ctl}/2\pi), \quad (10.62)$$

where θ_{ctl} is the angle of the baffle cut relative to the centerline of the heat exchanger, in degrees:

$$\theta_{ctl} = 2\cos^{-1}[(D_s/D_{ctl})\{1 - 2(B_c/100)\}]. \quad (10.63)$$

    - The above expression is valid for baffle cuts in the range of 15–45% of the shell diameter. Use of baffle cuts outside this range is not normally recommended because of flow maldistribution.
    - J_c typically ranges from 0.65 to 1.175 in a well-designed unit.

- J_l: Correction factor for baffle leakage effects including tube-to-baffle and shell-to-baffle leakage (A- and E-streams).
  - ➢ The pressure differences between neighboring baffle compartments force a fraction of the flow through the baffle-to-tube hole gaps in the baffle (A-stream) and through the annular space between the shell and the baffle edge (E-stream).
  - ➢ These streams reduce the part of the flow that passes over the tube bundle as cross flow (B-stream), reducing both the heat transfer coefficient and pressure drop.
  - ➢ E-stream is very detrimental to thermal design as it is not effective for heat transfer.
  - ➢ If the baffles are put too close together, then the fraction of the flow in the leakage streams increases compared with the cross flow.
    - J_l is a function of the ratio of total leakage area per baffle to the cross-flow area between adjacent baffles, the ratio of the shell-to-baffle leakage area to the tube-to-baffle leakage area, and is estimated from

$$J_l = 0.44(1-r_s) + [1-0.44(1-r_s)]\exp[-2.2\, r_{lm}]. \tag{10.64}$$

$$r_s = S_{sb}/(S_{sb} + S_{tb}). \tag{10.65}$$

$$r_{lm} = (S_{sb} + S_{tb})/S_m. \tag{10.66}$$

- The shell-to-baffle leakage area, S_{sb}, the tube-to-baffle hole leakage area, S_{tb}, for $N_{tt}(1 - F_w)$ tube holes, and the cross-flow area at the bundle centerline S_m are determined by the following equations:

$$S_{sb} = 0.00436 D_s L_{sb}(360-\theta_{ds}), \tag{10.67}$$

where L_{sb} is the diametrical shell-to-baffle clearance and the baffle cut angle θ_{ds} in degrees is

$$\theta_{ds} = 2\cos^{-1}[1-2(B_c/100)]. \tag{10.68}$$

$$S_{tb} = [\pi/4\{(D_t + L_{tb})^2 - D_t^2\} N_{tt}(1-F_w)]. \tag{10.69}$$

$$S_m = L_{bc}[L_{bb} + (D_{ctl}/L_{tp,\,eff})(L_{tp}-D_t)]. \tag{10.70}$$

L_{bc} is the central baffle spacing, L_{bb} is the bypass channel diametrical gap, N_{tt} is the total number of tubes in the bundle. The effective tube pitch, $L_{tp,\,eff}$, which is equal to L_{tp} for 30° and 90° tube layouts while for 45° staggered layouts,

$$L_{tp,\,eff} = 0.707 L_{tp}. \tag{10.71}$$

  - ➢ A typical value of J_l is in the range of 0.7–0.9. $J_l < 0.6$ should be avoided. The maximum value of J_l is 1.0.
  - ➢ For refrigeration chillers and water-cooled condensers, a value of 0.85–0.9 is achievable because of their tighter constructional tolerances and smaller clearances than TEMA standards.
- J_b: Correction factor for bundle bypassing effects due to the clearance between the tube and the inner wall of the shell (C-stream) and the bypass lane created by any pass partition lanes.
  - ➢ (F-stream) in the flow direction. F-stream is not always present and can be eliminated completely by placing dummy tubes in the pass partition lanes. C-stream can be reduced by a tighter fit of the tube bundle into the shell and also by placing sealing strips (in pairs) around the bundle perimeter, up to a maximum of one pair of strips for every two tube rows passed by the flow between the baffle cuts. The sealing strips can increase the value of J_b.

$$J_b = \exp[-C_{bh} F_{sbp}\{1-(2r_{ss})^{3/2}\}]. \tag{10.72}$$

C_{bh} is an empirical factor with a value of 1.35 for laminar flow ($100 \geq N_{Re}$), 1.25 for transition and turbulent flows ($N_{Re} > 100$).

$$F_{sbp} = S_b/S_m, \tag{10.73}$$

where S_m is as given earlier and

$$S_b = L_{bc}[D_s - D_{0tl}] + L_{pl}]. \tag{10.74}$$

L_{pl} is the width of bypass lane between tubes, L_{pl} is 0, for situations without a pass partition lane or such a lane normal to the flow direction. L_{pl} is 1/2 the actual dimension of the lane, for a pass partition lane parallel to the flow direction. It can be assumed to be equal to D_t, tube diameter.

$$r_{ss} = N_{ss}/N_{tcc}, \tag{10.75}$$

where r_{ss} is the number of sealing strips and N_{ss} is the number of pairs, if any, passed by the flow to the number of tube rows crossed between baffle tips in one baffle section N_{tcc}.

$$N_{tcc} = (D_s/L_{pp})[1-2(B_c/100)]. \tag{10.76}$$

$L_{pp} = 0.866L_{tp}$ for a 30° layout, $L_{tp} = L_{tp}$ for a 90° layout, and $L_{pp} = 0.707L_{tp}$ for a 45° layout.

- There is a maximum limit of $J_b = 1$, at $r_{ss} \geq 1/2$.
- For relatively small clearance between the outermost tubes and the shell for fixed tube sheet construction, $J_b = 0.90$.
- For a pull-through floating head, requiring larger clearance, $J_b = 0.7$.

- J_s: Correction factor for variable baffle spacing at the inlet and outlet. Because of the nozzle spacing at the inlet and outlet and the changes (decrease) in local velocities, the average heat transfer coefficient on the shell side will be adversely influenced.

$$J_s = [(N_b - 1) + (L_{bi}/L_{bc})^{1-n} + (L_{bo}/L_{bc})^{1-n}]/[(N_b - 1) + (L_{bi}/L_{bc}) + (L_{bo}/L_{bc})]. \quad (10.77)$$

$n = 0.6$ for turbulent flow and 1/3 for laminar flow. L_{bi} and L_{bo} are inlet and/or outlet baffle spacings, larger than the central baffle spacing L_{bc}. $J_s < 1$, for larger inlet and outlet baffle spacings than the central baffle spacing and for inlet and outlet equal to the central baffle spacing, $J_s = 1$, that is, no correction is required. J_s value will usually be between 0.85 and 1.00. N_b is the number of baffle compartments, determined from the effective tube length and baffle spacings.

- J_r: Laminar flow correction factor. In laminar flows, heat transfer is reduced by the adverse temperature gradient formed in the boundary layer as the flow thermally develops along the flow channel. J_r accounts for this effect.
  - $J_r < 1$ for laminar flow, that is, $100 \geq N_{Re}$.
  - $J_r = 1$, for $N_{Re} > 100$.
  - For $N_{Res} < 20$,

$$J_r = (J_r)_{20} = (10/N_c)^{0.18}. \quad (10.78)$$

  - N_c is the total number of tube rows crossed by the flow over the entire heat exchanger and is given by

$$N_c = (N_{tcc} + N_{tcw})(N_b + 1). \quad (10.79)$$

  - N_{tcc} is the number of tube rows crossed between baffle tips and is given under J_b.
  - N_{tcw} is the number of tube rows crossed in the window area and is given by

$$N_{tcw} = (0.8/L_{pp})[D_s(B_c/100) - (D_s - D_{ctl})/2]. \quad (10.80)$$

  - For $20 > N_{Re} < 100$, the value of J_r is prorated as

$$J_r = (J_r)_{20} + [(20 - N_{Re})/80][(J_r)_{20} - 1]. \quad (10.81)$$

  - The minimum value of J_r in all cases is 0.4.

- J_μ is viscosity correction factor. For heating and cooling of liquids, the effect of variation of the fluid properties between bulk fluid temperature and the wall temperature is corrected by $J_\mu = (\mu/\mu_w)^{0.14}$.
  - For gases and low-viscosity liquids, no such correction is required as temperature variation between the bulk fluid and the wall will be negligible.
  - For heating of gases, correction based on temperature rather than viscosity is applied.

$$J_\mu = (T/T_w)^{0.25}, \quad (10.82)$$

where T is in K.

- A_s cross-flow area at the centerline of shell for one cross flow between two baffles.
- Subscripts s and w stand for shell and wall temperature, respectively.
- *The combined effects of all these correction factors for a reasonably well-designed shell and tube heat exchanger are of the order of 0.60.*
- Taborek gave the following equation for the estimation of h_{ideal}:

$$h_{ideal} = J_i C_P \dot{m} \, (N_{Pr})^{-2/3}, \quad (10.83)$$

where $\dot{m}$ is the mass velocity of the fluid based on the total flow through the minimum flow area normal to the flow, $kg/m^2s = M/S_m$.

- M is shell side flow rate (kg/s) and S_m is as defined earlier.

$$J_i = a_1[1.33/(L_{tp}/D_t)^a (N_{Re})^{a_2}. \quad (10.84)$$

$$a = a_3/(1 + 0.14(N_{Re})^{a_4}. \quad (10.85)$$

- The values of the empirical constants a_1, a_2, a_3, and a_4 are given by Taborek in tabular form.
- Values of a_1 and a_2 are functions of N_{Re} and tube layout angles, that is, whether 30°, 45°, or 90°.
- Values of a_3 and a_4 are dependant on the tube layout angles.
- Equations for ΔP are given by empirical equations by Taborek.
- Solutions for Bell Delaware equations with Taborek's empirical equations are best handled by available software.

• Compare Kern and Bell Delaware methods for the design of shell and tube heat exchangers.
  - Kern method is commonly used giving satisfactory results. Its main advantage is that it requires little knowledge of the geometrical parameters of the exchanger. It only requires tube diameter and pitch, physical properties of the fluid entering shell side, and assumes a 25% baffle cut in all cases.

- When more accurate results are desirable, Bell Delaware method is a better choice. This method requires awareness of the configuration of the exchanger, as it is based on correction factors of the configuration of the exchanger, leakage and bypass flow corrections. It requires baffle configuration, leakage through the gaps between tubes and baffle and baffles and shell, bypassing of the flow between shell and tube bundle and baffle configuration.

10.3 MISCELLANEOUS DESIGN EQUATIONS

- What are the recommended design guidelines for determining shell diameter and tube counts for a heat exchanger?
  - *Shell Diameter:* The design process is to fit the number of tubes into a suitable shell to achieve the desired shell side velocity of 1.219 m/s (4 ft/s), subject to pressure drop constraints.
  - Most efficient conditions for heat transfer are to have the maximum number of tubes possible in the shell to maximize turbulence.
  - Preferred tube length to shell diameter ratio is in the range 5–10.
  - Criteria for tube count data are as follows:
    - Tubes are eliminated to provide entrance area for a nozzle equal to 0.2 times shell diameter.
    - Tube layouts are symmetrical about both the horizontal and vertical axes.
    - Distance from tube O.D. to centerline of pass partition 7.9 mm (5/16 in.) for shell I.D. <559 mm (22 in.) and 9.5 mm (3/8 in.) for larger shells.
- Give an equation for the estimation of number of baffles.

$$\text{Number of baffles} = 10 \times T_l/(B_P,\ \%\text{ of } D_i)(D_i), \tag{10.86}$$

where T_l is the tube length, B_P is the baffle pitch, D_i is the shell I.D.

- Give equations for (i) space available for flow on shell side and (ii) free area for flow between baffles.

(i) $$W = D_i - (d_o \times T_c), \tag{10.87}$$

where W is the space for flow (in.^{2}), D_i is the shell I. D. (in.), d_o is the tube O.D. (in.), and T_c is the number of tubes across centerline.

(ii) $$A_f = W(B_P - 0.187), \tag{10.88}$$

where A_f is free area between baffles and B_P baffle pitch, in.

- Give equations for calculation of number of tubes across the centerline of tube bundle.
  - For square pitch,

$$T_c \cong 1.19 \quad (\text{number of tubes})^{0.5}. \tag{10.89}$$

  - For triangular pitch,

$$T_c \cong 1.1 \quad (\text{number of tubes})^{0.5}. \tag{10.90}$$

- Give Taborek equation for the estimation of number of tubes and state its applicability.

$$N_t = 0.7854 D_{ctl}^2 / C_1 L_{tp}^2, \tag{10.91}$$

where N_t is the number of tubes, D_{ctl} is the centerline tube limit diameter, L_{tp} is the tube pitch, and C_1 is the constant (1.0 for square and rotated square tube layouts and 0.866 for triangular tube layouts).

- Give an equation for the estimation of number of tubes based on maximum allowable pressure drop for a liquid–liquid exchanger.

$$N = (m/1.111)[\sqrt{f_t L N_t^3 / \rho D_i^2 \Delta P_t}], \tag{10.92}$$

where ΔP_t is in Pa, N is the number of tubes, m is the mass flow rate on the shell side, f_t is the tube side friction factor, L is the length of tube bundle, N_t is the number of tube side passes, D_i is the inside tube diameter, and ΔP_t is the maximum allowable ΔP.

- Give an equation for ΔP across a tube bank.
  - For turbulent flow on shell side,

$$\Delta P = K_s[2 N_R f' G_s^2 / g_c \rho \Phi]. \tag{10.93}$$

$$f' = b(D_o G_s/\mu)^{-0.15}. \tag{10.94}$$

$$b = 0.23 + 0.11/(x_T - 1)^{1.08} \tag{10.95}$$

for staggered arrangement.

$$b = 0.044 + 0.08 x_L/(x_T - 1)^{(0.43 + 1.13/x_L)} \tag{10.96}$$

for inline arrangement.

$$K_s = \text{correction factor } (1 + \text{number of baffles}). \tag{10.97}$$

N_R = number of tube rows across which shell side fluid flows.

f' = modified friction factor.

x_T, x_L = ratio of pitch transverse, parallel to flow-to-tube outside diameter.

$$\Phi = 1.02(\mu_b/\mu_w)^{0.14}. \tag{10.98}$$

- Give an equation for shell side ΔP.

$$\Delta P_s = f G_s^2 [\{(L/B) - 1\} + 1] D_s / 2 \rho D_e (\mu_b/\mu_w)^{0.14}. \tag{10.99}$$

$$f = \exp[0.576 - 0.19 \ln(N_{Re\ s})]. \tag{10.100}$$

10.4 ANNEXURE (TEMA FOULING FACTORS)

TABLE 10.11 TEMA Fouling Factors

I. *Water*

Temperature of heating medium	≤240°F	240–400°F
Temperature of water	≤125°F	>125°F

Type of Water	Water Velocity (ft/s)			
↓	≤3	>3	≤3	>3
Seawater	0.005	0.001	0.003	0.002
Cooling Tower/Spray Pond				
Treated makeup	0.001	0.001	0.002	0.002
Untreated	0.003	0.003	0.005	0.005
River water (minimum)	0.002	0.001	0.003	0.002
Muddy/silty water	0.003	0.003	0.005	0.005
Hard water	0.003	0.003	0.005	0.005
Distilled water	0.0005	0.0005	0.0005	0.0005
Treated boiler feed water	0.001	0.0005	0.001	0.001
Boiler blowdown water	0.002	0.002	0.002	0.002

Ratings in the last two columns are based on a temperature of the heating medium of 240–400°F. If heating medium temperatures are >400°F and the cooling medium is known to give rise to scaling, these ratings should be modified accordingly.

II. *General Industrial Fluids*

Gases and Vapors	
Steam (nonoily)	0.0005
Steam (oily)	0.001
Refrigerant vapors (oily)	0.002
Compressed air	0.001
Industrial organic heat transfer fluids	0.001
Liquids	
Refrigerant liquids	0.001
Hydraulic liquids	0.001
Industrial organic heat transfer liquids	0.001
Transformer oil	0.001
Engine lube oil	0.001
Fuel oil	0.005
Quench oil	0.004
Molten heat transfer salts	0. 0005

III. *Chemical Process Streams*

Gases and Vapors	
Acid gases	0.001
Solvent vapors	0.001
Stable overhead products	0.001
Liquids	
Ethanol amine solutions	0.002
Glycol solutions	0.002
Stable side draw and bottom products	0.001
Caustic solutions	0.002
Vegetable oils	0.003

IV. *Natural Gas/Gasoline Process Streams*

Gases and Vapors	
Natural gas	0.001
Overhead products	0.001
Liquids	
Lean oil	0.002
Rich oil	0.001
Natural gasoline and LPG	0.001

V. *Refinery Streams*

Crude and Vacuum Unit Vapors	
Atmospheric column overhead vapors	0.001
Light naphthas	0.001
Vacuum column overhead vapors	0.002

Crude Oil

Temp. (°F)	0–199			200–299			300–399			400–499		
Vel. (ft/s)	<2	2–4	>4	<2	2–4	>4	<2	2–4	>4	<2	2–4	>4
Dry crude	0.003	0.002	0.002	0.003	0.002	0.002	0.004	0.002	0.002	0.005	0.004	0.003
Salty crude	0.003	0.002	0.002	0.005	0.004	0.004	0.006	0.005	0.004	0.007	0.006	0.005

TABLE 10.11 *(Continued)*

Refinery Straight Run Products	
Gasoline	0.001
Naphtha and light distillates	0.001
Kerosene	0.001
Light gas oil	0.002
Heavy gas oil	0.003
Heavy fuel oils	0.005
Asphalt and residues	0.010
Cracking and Coking Unit Streams	
Overhead vapors	0.002
Light cycle oil	0.002
Heavy cycle oil	0.003
Light coker gas oil	0.003
Heavy coker gas oil	0.004
Bottoms slurry oil (vel. min. 4.5 ft/s)	0.003
Catalytic Reforming, Hydrocracking, and Hydrodesulfurization Streams	
Reformer charge	0.002
Reformer effluent	0.001
Hydrocracking charge and effluent	0.002
Recycle gas	0.001
Hydrodesulfurization charge and effluent	0.002
Overhead vapors	0.001
Liquid product (>50° API)	0.001
Liquid product (30–50° API)	0.002

TABLE 10.11 *(Continued)*

Light Ends Processing Streams	
Overhead gases and vapors	0.001
Liquid products	0.001
Absorption oils	0.002
Alkylation trace acid streams	0.003
Reboiler streams	0.003
Lube oil processing streams	0.003
Feed stock	0.002
Solvent feed mix	0.002
Solvent	0.001
Extract	0.003
Raffinate	0.001
Asphalt	0.005
Wax slurries (wax deposition to be prevented on cold tube walls)	0.003
Refined lube oil	0.001

Note: Reported fouling factors in the above tables are in ft^2 (°F Btu). 1 Btu/(ft^2 °F h) = 5.6785 W/(m^2 K).

11

HEAT TRANSFER EQUIPMENT INVOLVING PHASE TRANSFER

11.1 Condensers 331
11.2 Reboilers 336
11.3 Evaporation and Evaporators 347
  11.3.1 Multiple Effect Evaporators 361
  11.3.2 Evaporator Performance 364
  11.3.3 Auxiliary Equipment 368
  11.3.4 Utilities 370

11.1 CONDENSERS

- What are the advantages and limitations of direct contact condensers?
  - The main advantage of a direct contact condenser is that there are no metal wall and fouling resistances involved with the consequence of high heat transfer rates compared to indirect contact condensers.
  - Environmental considerations preclude use of direct contact condensers as the effluent needs to be freed from contaminants, requirement of treatment of effluents before discharge to permissible levels.
  - Because of the above limitation these are not favored any more, except in cases of vapors containing solids that would foul heat transfer surfaces of surface condensers very quickly.
- In condensation of vapors from a noncondensable gas, is fog formation a desirable phenomena? Explain.
  - If bulk temperature of the gas falls below dew point of vapors, fog formation can take place. Droplet sizes in a fog are in the range of 1–20 μm.
  - Fog can be formed in a condenser when the ratio of noncondensable to condensable vapor is high and the temperature difference is high.
  - External nuclei in the inlet vapor will enhance fog formation.
  - Not desirable. Fog formation hinders condensation process. The tiny droplets in a fog hinder movement of vapors toward the condensing surfaces. These droplets tend not to collect on the condensing surfaces and may be lost through the condenser vent unless special precautions are taken for removal. Usual demisters will not capture fog droplets.
  - Carry over of liquid droplets out of the condenser may give rise to plumes of fog outside the vent systems with environmentally objectionable effects.
- What is the basic reason for fog formation in a condenser?
  - Fogging occurs in a condenser when mass transfer is slower than heat transfer.
  - The design must provide sufficient time for mass transfer to occur.
  - A high ΔT with noncondensables present or a wide range of molecular weights can produce a fog.
  - The high ΔT gives a high driving force for heat transfer. The driving force for mass transfer, however, is limited to the concentration driving force (ΔY) between the composition of the condensable component in the gas phase and the composition in equilibrium with the liquid at the tube wall temperature.
  - The mass transfer driving force (ΔY) thus has a limit. The ΔT driving force can, under certain conditions, increase to the point where heat transfer completely outstrips mass transfer, which produces fogging.
- "In condensers, external nuclei in the inlet vapors will enhance fog formation." What are the sources of such nuclei?
  - Solids in air.

Fluid Mechanics, Heat Transfer, and Mass Transfer: Chemical Engineering Practice, By K. S. N. Raju

  - Ions.
  - Solids produced by combustion in upstream processing.
  - Upstream reactions.
  - Entrained liquids.
- What are the ways to prevent/reduce fogging in condensers?
  - By reducing ΔT, thereby providing more surface area to help mass transfer.
  - By minimizing $\Delta T/\Delta Y$.
  - Increasing inlet fluid superheat.
  - Filtering inlet gases to remove foreign nuclei.
  - Seeding gas stream with condensation nuclei to produce drops, which will be captured by conventional demisters.
  - Heating the vent to reduce fog.
  - Allowing fog to form and then removing with electrostatic precipitators or special demisters which first coalesce the droplets before removal.
- For what type of fluids, condensation could give rise to fog formation?
  - Acid gases.
- What are the ways by which condenser capacity can be increased for an existing distillation unit?
  - Adding additional condensers in parallel to existing ones.
  - Adding a low ΔP partial condenser upstream in series with the existing total condenser.
- Compare adding condensers in parallel to adding in series to existing condensers for an existing distillation unit.
  - *Parallel*:
    - If new condensers have larger surface and higher ΔP, vapors take path of least resistance and mostly pass through old condensers, wasting additional area provided in the form of new added condensers.
    - If new condensers have smaller surface and lower ΔP, vapors flow mostly through new condensers. Surface of new condensers is not adequate, resulting in partial condensation. Uncondensed vapors enter reflux drum.
    - In short, careful balancing hydraulics between new and old condensers becomes a necessity.
  - *Series*: Reduce ΔP in old condenser since vapor flow to old condenser is decreased due to condensation in the new one. (ΔP is mainly due to vapor flow and not liquid flow.)
- What is the effect of subcooling in a condenser and why so?
  - Condenser capacity is reduced.
  - Heat transfer coefficient for condensation is 20 to 30 times more than for cooling of liquids.
  - Subcooling leads to creation of a pool of stagnant liquid in condenser bottom (Figure 11.1), submerging some tubes whose contact with vapors is thereby cut off. Condensate backup can be due to inadequate drainage of condensate or lack of provision of a condensate well below the exchanger or blockage of condensate outlet or any other reason.
  - Therefore, these tube surfaces act as liquid coolers rather than as vapor condensers.
- How is subcooling detected in a condenser?
  - Cannot be detected during shutdown.
  - By observing condenser outlet temperature going down below liquid bubble point (subcooling).
  - By observing ΔP in condenser outlet line (increased ΔP for condensate flow).
  - For low boiling liquids, by observing indications for moisture condensation on lower outside surfaces of the condenser.
- How does the adverse effects of subcooling are prevented or reduced?
  - By reducing cooling water rate/velocity.
  - By providing proper drainage of condensate to prevent backup.
  - To prevent the condensate level from rising to the lower tubes of the condenser, a hot well level control system may be employed.
  - Varying the pumping of the condensate through use of variable speed pumps with suitable controls is one method used to accomplish hot well level control. A level sensing network controls the condensate pump speed or pump discharge flow control valve position.
  - Another method employs an overflow system that discharges water from the hot well when a high level is reached.
- Give an instance where subcooling is helpful.

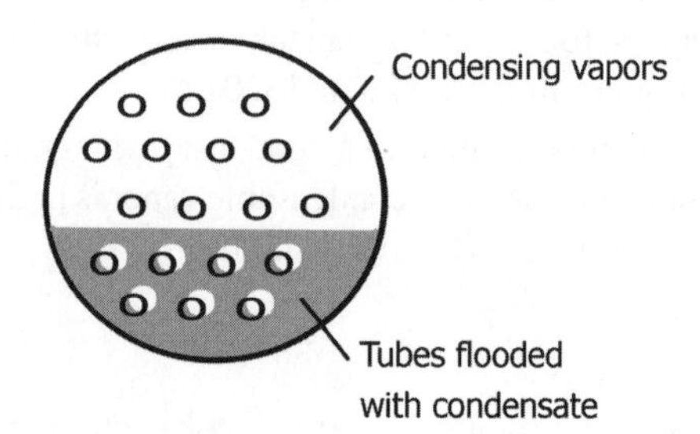

FIGURE 11.1 Illustration showing condensate backup.

- Condensate at bubble point may create NPSH problems for the condensate outlet pump, that is, enough NPSH may not be there, resulting in cavitation, etc.

- What is *condensate depression*? How does it decrease efficiency of a power plant?
  - The difference between the saturation temperature at the existing condenser vacuum and the temperature of the condensate is termed *condensate depression*. This is expressed as a number of degrees condensate depression or degrees subcooled.
  - Excessive condensate depression decreases the operating efficiency of the plant because the subcooled condensate must be reheated in the boiler, which in turn requires more heat from the fuel or other heat source.
- Suggest a suitable way to remove condensate from a reboiler involving high flows of condensate.
  - A typical diagram of a typical system adoptable to large condensate removable capacities is given as an illustration in Figure 11.2.
  - It must be ensured that the condensate receiver is properly vapor-balanced with the top of the reboiler tube bundle. This is to allow unrestricted condensate gravity flow. Proper line sizes must be ensured to minimize pressure drop and permit for venting. The vapor balance line ensures that no noncondensables will collect at the top of the condensate receiver and stop the gravity flow of condensate. Steam chest around the tube bundle must be provided for purging of all noncondensables, by the use of adequate line sizes of not less than 2.5 cm to provide enough vent capacity and mechanical rigidity.
  - Design of condensate control valve should take into consideration the amount of flashing that is likely to occur, giving rise to two-phase flow.
  - Condensate receiver should be located well below the level of bottom tube sheet, in case of a vertical thermosiphon reboiler, or below the lowest possible point of the reboiler, making condensate outlet line as large as possible to assist gravity flow. It must be ensured that no condensate accumulates around the tube bundle of the reboiler.
  - The condensate receiver should have adequate condensate inventory to provide for positive control, operability, maintenance and space requirements. The vessel is normally sized for an L/D ratio of 2.5–3.0.
  - Illustrate stratified and annular flows for the case of vapors condensing in a horizontal tube.
  - Stratified and annular flows for vapors condensing inside horizontal tubes are shown in Figure 11.3.
- What are the most serious factors that affect condenser performance? Explain.
  - Condenser scaling and fouling. Scaling normally occurs on the cooling waterside and is possible on the vapor condensing side also due to water infiltration into, for example, steam side.
    - ➢ Cooling waters contain hardness ions, calcium and magnesium, and other cations such as sodium and potassium, all of which are counter-balanced by bicarbonate, chloride, and sulfate anions.
    - ➢ Groundwaters usually contain higher concentrations of dissolved ions, while surface waters are often softer but typically contain suspended solids, organic compounds, and microorganisms. Silica is another problematic contaminant, which in some

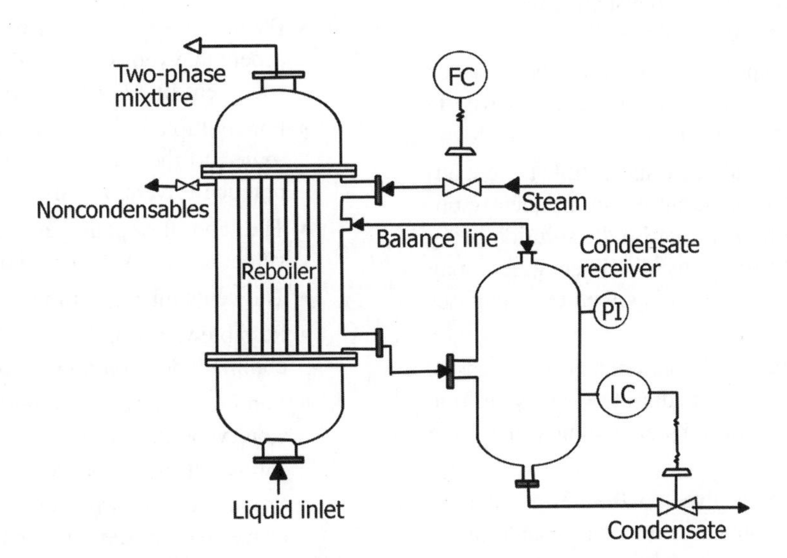

FIGURE 11.2 Condensate removal system for a reboiler.

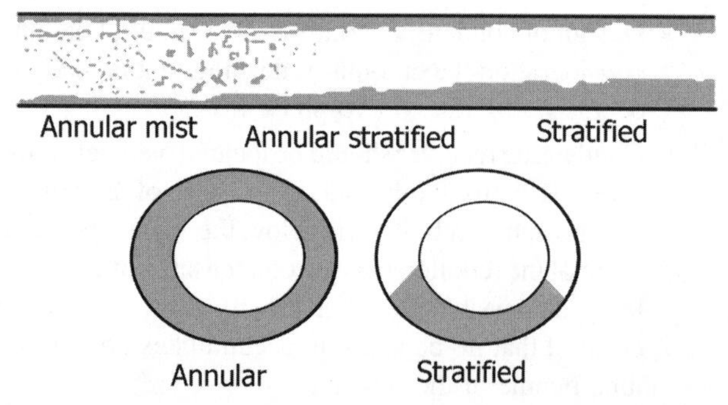

FIGURE 11.3 Annular and stratified flows inside horizontal condensers.

places exceed concentrations of 50 ppm in groundwaters. Calcium phosphate is a major scale-forming material due to the break down of phosphate-based scale or corrosion inhibitors.

  - The general problems of fouling are discussed elsewhere.
  - Because cooling water temperature rises in condensers, condensers are susceptible to scaling. Scaling problems may be more severe in open recirculating systems, for example, cooling towers, where the concentrations of impurities increase, the quantum of which depends on treatment, blowdown, and other factors.
  - Microorganisms attach to tube walls, secrete a sticky, protective film that increases the deposit mass by absorbing silt, etc. Many microbes produce acids, and the acids trapped beneath the deposit can directly attack the tube surface.
- Air leakage into a condenser. Air leakage occurs at points around the condenser due to the strong vacuum generated within the system.

- What are the effects of noncondensable gases in condensing vapors on heat transfer rates?
  - Condensation of vapors is impeded by the presence of noncondensable gases. This is especially true in condensers operating under vacuum.
  - Evaporators, in particular, are susceptible to noncondensable gas problems because they frequently operate under vacuum and because the solution being evaporated usually contains dissolved gases, may contain entrained gases and may liberate other gases on being concentrated.
  - Although air is the usual contaminant, other gases, as long as they do not dissolve easily in the condensate, affect heat transfer in almost the same manner as air.
  - Two problems are involved in the design of condensers for vapors containing noncondensable gases:
    - Effect of noncondensables on heat transfer rates.
    - Physical arrangement of the heating surface to properly move the vapors past the heating surface and the condensables out through the vent.
  - Noncondensable gases reduce heat transfer rates in two ways:
    - Heat transfer coefficient is reduced because heat must be transferred across a gas film if noncondensables are present in significant concentration, the available temperature difference is reduced because the condensation is no longer isothermal.
    - There are two significant effects of noncondensables on the condensation of vapors. The first is the effect on the temperature-composition profile, which determines the fraction condensed as a function of temperature. As material condenses, the vapor phase composition becomes richer in air and it becomes more difficult to condense the remaining vapors. The second effect is to add a mass transfer resistance as the vapor near the liquid film is depleted of the condensing component, resulting in a radial composition profile.
  - The reduction of heat transfer coefficient can be treated as an empirical *fouling factor*.
  - *Apparent fouling factor* for condensing steam in the presence of noncondensables is expressed by the following equation:

$$h_f = 24.7/C, \quad (11.1)$$

where h_f is in KW/(m^2°C) and C is concentration of air, wt%. ($h_f = 4350/C$, where h_f is in BTU/(h ft^2°F)).

    - The fouling factor would be $1/h_f$.

- What are the venting provisions to be considered while designing condensers involving noncondensables?
  - Each condenser should be vented separately to an acceptable vent system. If a vent header is provided, each vent line must be separately controlled.
  - For multiple effect evaporators, the vent should be routed to the condenser. It is not generally a good practice to vent an earlier effect to a later effect.
  - The vapor flow path preferably should result in vapor velocities as constant as possible.
  - The vents must be at the end of the vapor flow path.
  - Possible variations in heat transfer conditions on the cooling side should be considered.
  - Venting is especially important for a vapor compression evaporator because reduced evaporator capacity can result in compressor surge.
  - Proper venting arrangements serve not only to improve heat transfer but also to reduce corrosion. High concentrations of noncondensables result in some

redissolving in the condensate which accelerates corrosion either in the evaporator or of the condensate return system.

- Equipment with tube side condensation is more easily vented than on shell side condensation, provided condensate is not permitted to flood the heat transfer surface.

- Give examples of condensation involving presence of noncondensable gases.
  - Dehumidification in an air conditioning system in which water vapor is partially condensed from the humid air.
  - Refinery and chemical plant distillation units, overhead condensers often involve condensable vapors in the presence of gases.
  - Power plant condensers involve noncondensable air derived from feed waters as dissolved gas.
  - Condensers used in evaporator systems.
- How is corrosion of boiler tubes minimized?
  - By removal of dissolved oxygen in feed water and maintaining alkaline conditions.
  - By keeping surfaces clean.
  - By counteracting effect of corrosive gases in steam and condensate systems with chemical treatment.
  - By avoiding absorption of CO_2 in water as it forms corrosive carbonic acid.
- In a steam-generating system, what are the possibilities for air leakage into it? How are the air leakage effects minimized?
  - Air enters a steam-generating system at points around the condenser due to the vacuum generated within the system. Typical air entry points are the expansion joints between the turbine and condenser, turbine and pump seals, etc.
  - Structural failures in the condenser shell or other mechanical problems can increase air leakage, sometimes very rapidly, which cannot be handled by the normal air removal systems. In such cases air forms a film on the heat transfer surfaces resulting drastic reduction in heat transfer rates.
  - Most condensers in power plants are equipped with one or more air removal compartments, in which mechanical vacuum is applied to remove the gases that enter the condenser.
- What are the considerations involved in the layout of condensers?
  - Space considerations often influence the cost of a condenser.
  - For example, if the surface condenser is limited in tube length, a multipass design may be required in place of a single-pass design. This results in a short, but larger diameter unit, which is generally more expensive than a long, thin unit.
  - Headroom limitations and permissible condenser tube length should take into consideration the distance beyond the tube sheets of the condenser required to permit removing tubes for replacement.
  - Condensers must be mounted so that liquid level on the hot well is sufficiently high to permit removal by condensate pumps, without cavitation effects.
  - For air conditioning installations, the liquid level should be ≥ 1.5 m above the inlet to the condensate pump and preferably 0.3 or 0.6 m higher.
  - This permits an economical pump selection.
- In a well-insulated steam heater, process fluid being heated is on the tube side and steam is on the shell side. The insulation on the shell is taken out during maintenance work and the plant restarted without insulation? What happens?
  - Condensation of steam occurs due to (i) heat transfer to tube side fluid and (ii) heat losses to atmosphere as a consequence of insulation being removed.
  - Additional condensation due to loss of insulation might result in flooding of the condenser as provision for condensate drainage may not be enough to remove condensate at the rate at which it forms (e.g., insufficient steam trap capacity).
    - Some of the tubes are submerged in the accumulated condensate.
    - The submerged tube surfaces are not available for condensation heat transfer.
    - Net effect is reduction of heat transfer area for condensation.
  - Heat transfer coefficients for these tubes will be far less than those for condensation.
  - Tube fluid outlet temperatures will come down.
  - In other words, performance of the heater will be adversely affected.
- Why a vacuum breaker is used on shell and tube heat exchangers that are utilizing steam as the heating medium?
  - During plant shutdown, absence of a vacuum breaker on the condensing side of a heat exchanger can create vacuum on account of condensation of vapors in the vapor space. Vacuum breaker avoids such a situation and prevents accidental ingress of any undesirable fluids into the space, contaminating the condensates, necessitating treatment before disposal.
  - In the absence of a vacuum breaker, condensation on shutdown creates vacuum in the exchanger. The exchanger might not have been designed to withstand

vacuum and exchanger failure becomes a possibility. This problem gets aggravated, especially for large diameter shells when condensing fluid is on the shell side.

- What is the generally recommended temperature approach in steam condensers in air conditioning and power plants?
  - Lowest feasible approach (steam condensing temperature *minus* entering cooling water temperature) is 3°C.
  - Generally the approach used in most air conditioning installations is 10°C but not less than 5°C, to save on costs involved in large heat transfer surface requirements.
- What is meant by *cleanliness factor*?
  - Cleanliness factor is defined as $100 \times (U_{Design}/U_{Clean})$.
- What is the normally recommended cleanliness factor for the design of condensers for air conditioning and power plants?
  - Most surface condensers for air conditioning and power plant applications are designed with a cleanliness factor of 85%. This means that the heat transfer rate used in designing the condenser is 85% of the clean heat transfer coefficient.
  - For most applications using clean cooling tower water, refrigeration condensers should be specified with a fouling factor of 0.0005 and steam surface condensers designed for an 85% clean tube coefficient.
  - If the refrigerant fouling factor is increased, the surface condenser cleanliness factor should be increased by the same percentage. Thus, if the refrigerant condenser is specified as 0.001 versus 0.0005, the corresponding cleanliness factor for the surface condenser is 70% clean versus 85% clean.
  - For those installations where cooling water sources produce rapid fouling of the tubes, a higher factor must be used.
- What are the effects of weather on condenser duties?
  - Can limit production in summer.
  - Capable of condensing more vapor at night than during day. This situation will be more pronounced in plants located in desert atmospheres.
- "When light components accumulate, condensation is impeded." *True/False*? Explain.
  - *True*. Presence of light components and/or noncondensables reduce condenser capacity due to reduced heat transfer coefficients.
  - Solution for this problem is to vent noncondensables.
- "Condensation inside tubes is not appropriate for vacuum column overhead condenser because ...." Comment.
  - This statement is true because of the high tube side ΔP and the difficulty in piping and supporting a vertically mounted condenser. Therefore, condensation is the best choice.
- Summarize troubleshooting of vacuum condensers.
  - Table 11.1 gives condenser troubleshooting.

11.2 REBOILERS

- Name common types of reboilers.
  - Natural circulation reboilers
    - ➢ Once-through reboiler.
    - ➢ Recirculation reboiler.
  - Forced circulation reboilers.
  - Vertical thermosiphon reboilers.
  - Horizontal thermosiphon reboilers.
  - Flooded bundle reboilers.

TABLE 11.1 Condenser Troubleshooting

Problem	Effect	Corrective Action
High shell side ΔP	Shell/tube side fouling	Clean tubes; reduce temp.
Should be $\leq$5% of design operating pressure	Cooling water temp. > design value	Increase cooling water flow
	Low cooling water flows > design condensables (>20–30% design)	Reduce vapor flow
		Use larger condenser
		Use ejector downstream
>Design tube side ΔP	Tube side fouling; >design cooling water flow	Clean tubes; higher flow is not a problem
>Design ΔT	Low cooling water flows	Increase fluid flows
	Higher than design duty	Increase cooling water flows/replace condenser
High vapor outlet temp.	Tube fouling; low cooling water flows/high inlet temp.	Clean tubes; increase cooling water; reduce inlet temp.

- Recirculating, baffled bottom reboilers.
- Fired heaters carrying out reboiler duties.

• What are the factors that must be considered in the selection of reboilers?

- Total duty required.
- Fraction of distillation column liquid vaporized.
- *Cleanability (Fouling):* Tube side is easier to clean than shell side.
- *Corrosion:* Corrosion may dictate use of expensive alloys and to save costs, corrosive fluid must be placed inside tubes, avoiding an expensive shell, in addition to alloy tubes.
- *Pressure:* High-pressure fluids are placed inside tubes to avoid an expensive thick-walled shell. For low pressures (vacuum), other factors involved determine the tube side fluid.
- *Temperatures:* Very hot fluid is placed on tube side to reduce shell costs. The lower stress limits at high temperatures affects shell design as is the case for high pressures. Shell insulation requirements also increase.
- Heating medium requirements are more important than the boiling liquid requirements.
- *Boiling Fluid Characteristics:* Heat-sensitive liquids require low holdup design. Boiling range and mixture concentration, together with available ΔT, affect circulation requirements to avoid stagnation. Foaming liquids can best be handled inside tubes.
- ΔT and type of boiling (nucleate or film) affect selection.
- Available and required temperature approach.
- *Space Constraints:* For example, if headroom is limited, horizontal units should be the choice. Limitation of available space dictates use of internal reboilers.
- Extended surfaces are suitable only for some types.

• What are the important configuration selection factors for reboilers?

- Forced versus natural circulation.
- Tube side versus shell side vaporization.
- Once-through versus process recirculation.
- Single versus multiple shell systems.
- Vertical versus horizontal orientation.
- Column internal bundles.
- Other types.

• Illustrate, by means of diagrams, different configurations for the operation of reboilers and discuss their characteristics.

- Figures 11.4–11.7 are illustrative of common types of reboiler configurations, with horizontal or vertical arrangements.
- Horizontal thermosiphon reboilers are regular baffled exchangers with boiling taking place in the shell side.
- By proper piping arrangement a driving force for circulation is established by the density differences between the liquid in the column and the two-phase mixture in the exit piping.
- Advantages include higher circulation rates (can give a better ΔT) than a kettle reboiler, less skirt height for the column and less headroom requirements. High velocities and low exit vapor fractions reduce fouling and high residual boiling fractions. Larger areas can be built.
- Disadvantages include difficult cleaning on the shell side, creation of vapor blanketing and local dry-outs, poor understanding of hydrodynamic problems, more complicated piping and requirement of large plot area.

• "Once-through reboiler functions as the bottom theoretical tray of the distillation column whereas a circulating reboiler does not." Comment.

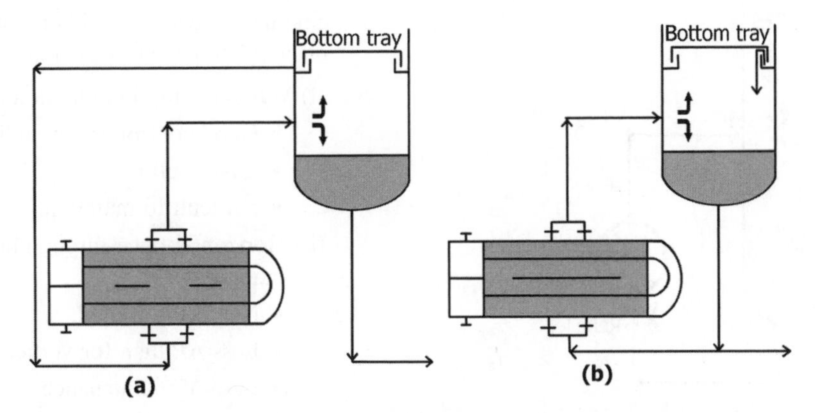

FIGURE 11.4 (a) Horizontal once-through reboiler with shell side boiling. (b) Horizontal recirculating reboiler with shell side boiling.

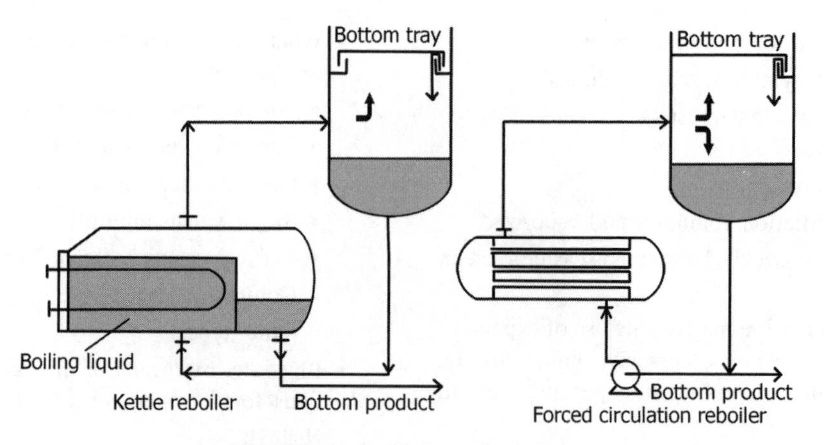

FIGURE 11.5 Horizontal reboilers.

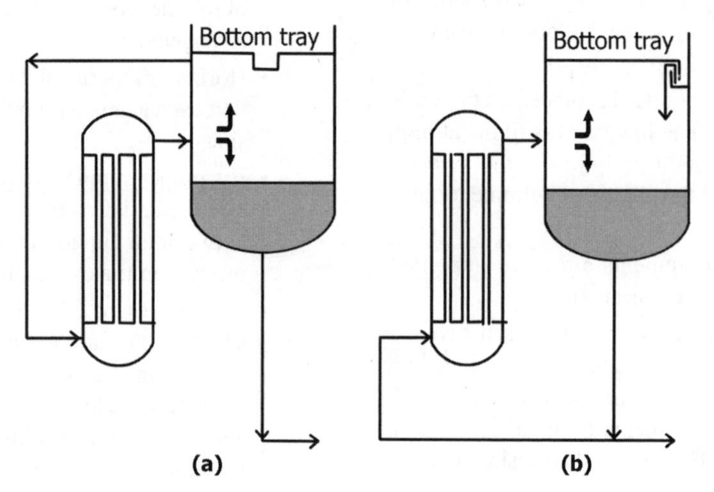

FIGURE 11.6 (a) Vertical single pass, once-through with tube side boiling. (b) Vertical, recirculating with shell side boiling.

- All the liquid from the bottom tray passes to the once-through reboiler where partial vaporization takes place, establishing equilibrium between the two phases due to intimate contact between the phases before they are separated on admission into the column.
- With a circulating reboiler, because the liquid from the *bottom tray* of a column is of a composition similar to that of the bottoms product, one can assume that it *does not* act as a theoretical tray.

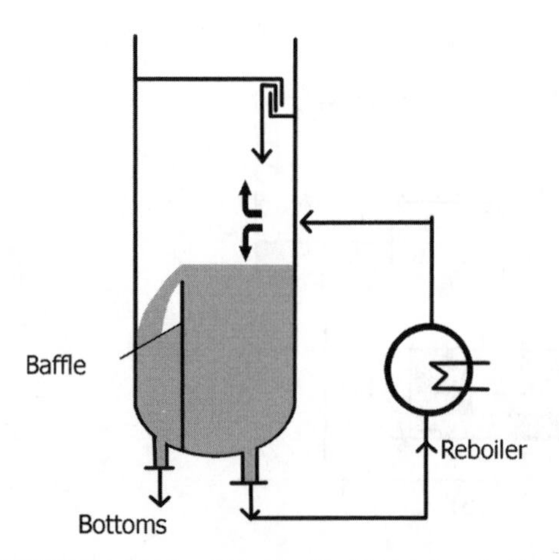

FIGURE 11.7 Recirculating baffled bottoms reboiler system.

- What are the plus and minus points of (i) vertical natural circulation reboiler, (ii) horizontal once-through/natural circulation type, and (iii) forced circulation reboiler?

(i) Vertical natural circulation type:
- Good controllability unlike once-through type.
- Easy fouling.
- Difficult to maintain.

(ii) Horizontal once-through/horizontal natural circulation type:
- Not easily fouled.
- Less ΔP than for vertical type.
- Ease of maintenance.

(iii) Forced circulation type:

- Suitable for high viscosity and fouling liquids or when fraction vaporized must be kept low.
- Force recirculation reboilers may be designed so that boiling occurs inside vertical tubes, inside horizontal tubes, or on the shell side.
- Normally used in fired heaters as reduction or loss of circulation results in tube failures. Forced circulation ensures higher flow velocities.
- Since the feed liquid is at its bubble point, adequate NPSH must be assured for the pump if it is a centrifugal type.
- Linear velocities in the tubes of 4.6–6 m/s (15–20 ft/s) usually are adequate.
- Very high equipment, maintenance, and operating costs with additional piping and pumps.
- Larger space requirements.

- What are the recommended maximum heat flux for a (i) natural circulation reboiler and (ii) forced circulation reboiler?

  (i) Natural circulation reboiler: 37.85 kW/m^2 (12,000 Btu/(h ft^2)).

  (ii) Forced circulation reboiler: 94.64 kW/m^2 (30,000 Btu/(h ft^2)).

- What is the normally recommended ratio of liquid circulated to vapor generated in a circulation reboiler?
  - 4:1.
- What are the consequences of high vaporization rates in a reboiler?
  - Large temperature differences in vaporizers/reboilers are not conducive to good heat transfer rates on account of film boiling, which prevent liquid reaching heat transfer surfaces as a result of vapor blanketing. In this condition, the heat transfer rates decrease because of the lower thermal conductivity of the vapor.
  - If a design analysis shows that the temperature difference is close to causing film boiling, the vaporizer should be started with the boiling side full of relatively cooler liquid. This way, flashing of the liquid is avoided. If the vaporizer is steam heated, the steam pressure should be reduced which will reduce the temperature difference. With steam heating, the design should consider keeping the mean temperature be kept below 50°C, to avoid film boiling.
- What are the issues involved in the selection of forced circulation reboilers?
  - Since the feed liquid is at its bubble point, adequate NPSH must be assured for the pump if it is a centrifugal type.
  - Linear velocities in the tubes of 3–5 m/s usually are adequate. The main disadvantages are the costs of pump and power and possibly severe maintenance. This mode of operation is a last resort with viscous or fouling materials, or when the fraction vaporized must be kept low.
- How is a kettle reboiler different from a U-bundle heat exchanger?
  - Shell diameter for a kettle reboiler is much larger than that for a U-bundle heat exchanger on account of provision of vapor disengaging space in a kettle reboiler.
  - In a kettle reboiler, shell side liquid submerges the tube bundle with provision of an overflow baffle/weir at the end of the bundle to ensure submergence of the tube bundle. Liquid product overflows into a reservoir section from which it is withdrawn.
- "In a kettle reboiler, liquid vaporizes inside tubes." *True/False*?
  - *False*. Heating medium is admitted into the tubes.
- What are the advantages of kettle reboilers?
  - Vapor disengaging space with reduced entrainment.
  - Can operate at low ΔP.
  - High heat fluxes are possible.
  - Can handle high vaporization rates (up to 80%).
  - Low skirt height.
  - Ease of maintenance.
  - Ease of control.
  - Handles higher viscosity liquids.
- What are the disadvantages of kettle reboilers?
  - Difficult to determine degree of mixing.
  - High residence time and hence not suitable for heat-sensitive liquids.
  - Liquid distribution along the entire length of the shell is not uniform within the shell, resulting in excessive local vaporization and vapor binding.
  - Fouling problems. Reboiler shell side fouling may lead to tray flooding in the column, as fouling can cause ΔP buildup on the shell side, leading to liquid backup in the column.
  - Shell side is difficult to clean.
  - Excessive pressure drop in kettle reboiler circuits is the prominent kettle malfunction, causing liquid to backup in the column base beyond the reboiler return elevation. This high liquid level leads to premature flooding and capacity loss.
  - Requires extra space and piping.
  - High cost, mainly due to large shell size.
- What is a column internal reboiler and what are its characteristic features? Illustrate.
  - Column internal reboiler is a U-bundle inserted into the side of the column (Figure 11.8). It works as a

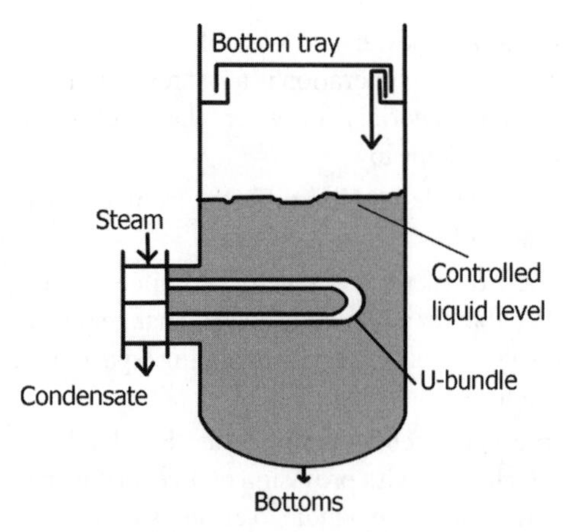

FIGURE 11.8 Column internal reboiler.

kettle reboiler, except it does not have shell and the connecting piping.

- Hydraulic problems are few, unlike with a kettle reboiler for which sizing and location of feed and vapor lines can give rise to operating problems.
- Disadvantages are limited heat transfer area can be accommodated, requires large flange and internal supports, tubes are short and hence bundle costs are high and column has to be shut down if the reboiler is to be serviced as no alternative operation is possible.

- What are the advantages of using low-fin tubes in a kettle reboiler?
  - In a kettle reboiler with low ΔT and in fouling media, low-fin tube gives rise to more nucleation sites per unit of surface than a bare tube.
  - This is because finning operation involving alteration of tube surface with tools with tiny imperfections being introduced on finished fin surface.
  - Therefore for the same ΔTs, there will be more bubbles generated from a finned surface than for a comparable length of bare tube surface, not only because of ratio of surface areas is about 2.5, finned to bare.
- When a 1–2 heat exchanger is used as a vaporizer, is it advisable to vaporize all the feed? If not, explain why.
  - No. It results in formation of scales.
- What is the normally recommended ratio of liquid circulated to vapor generated in a reboiler?
  - 4:1.
- What is the advantage in using recirculation of liquid in a vaporizer/reboiler/evaporator?
  - Less fouling of heat transfer surfaces.
- How is the submergence of tube bundle ensured in a kettle reboiler?
  - By using a vertical partition baffle plate acting as a weir between the tube bundle and the liquid outlet, its height being more than the bundle diameter as illustrated in Figure 11.9.
- What are the factors that are to be considered in the design of the vapor space in a kettle reboiler?
  - Entrainment from kettle reboilers occurs when the disengagement space above the bundle is insufficient to deentrain liquid droplets. One important factor is to provide enough space to decrease the vapor velocity sufficiently for most of the liquid droplets to fall back by gravity to the boiling surface.
  - Also, vapor outlet nozzle should be sized adequately to avoid excessive velocities and consequent entrainment into the column.
  - The extent of entrainment separation depends on the nature of the vapor destination. A distillation column with a large disengaging space, low column efficiency, and high reflux rate does not require as much kettle vapor space as a normal reboiler.
    - Several criteria and rules of thumb for preventing entrainment are available, the most common of which is keeping the top tube row not higher than

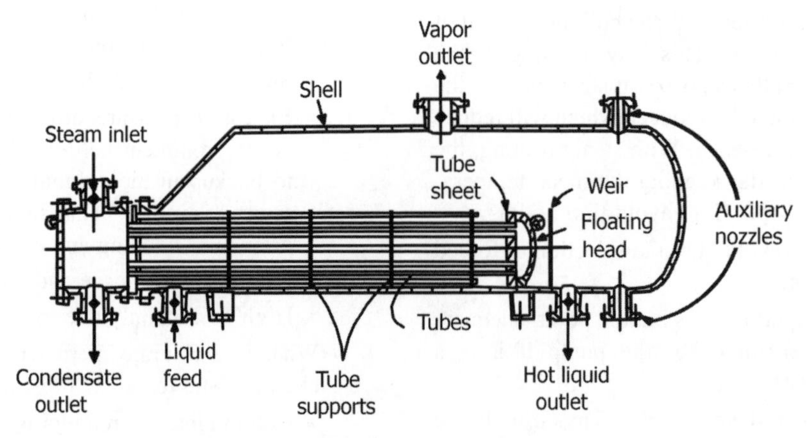

FIGURE 11.9 Kettle reboiler.

60% of the shell diameter. Other criteria include keeping a minimum height of 0.3 m or 1.3 to 1.6 times the bundle diameter (whichever is greater) above the liquid level.

- Normally the vapor outlet is in the center above the bundle. Then the vapor comes from two different directions as it approaches the outlet nozzle. Only in rare cases are these two vapor streams equal in quantity. A simplification that has been extensively used is to assume the highest vapor flow is 60% of the total. One case where this would cause an undersized vapor space is when there is a much larger temperature difference at one end of the kettle than the other. The minimum height of the vapor space is typically 18 cm (8 in.). It is higher for high heat flux kettles.
- Using internal baffles or demisters to reduce entrainment will entail maintenance problems and frequent blockages.

• What are the effects of an undersized kettle reboiler?

- The effect will be a decrease in the boiling coefficient. A boiling coefficient depends on a nucleate boiling component and a two-phase component that depends on the recirculation rate. An undersized kettle will not have enough space at the sides of the bundle for good recirculation. Another effect is high entrainment or even a two-phase mixture going back to the column.

• If reboiler heating medium is steam, a desuperheater is provided if superheat is more than 4–15°C. Why?

- Heat transfer coefficient is very low for superheated steam; desuperheating increases h by 100 to 200 times.

• What is a thermosiphon reboiler?

- *Thermosiphon:* Convective movement of liquid starts when liquid in the loop is heated, causing it to expand and become less dense and thus more buoyant than the cooler liquid in the bottom of the loop. Convection moves heated liquid upwards in the system as it is simultaneously replaced by cooler liquid returning by gravity. The liquid may be heated beyond its boiling temperature, causing vaporization of part of the liquid. Since the vapor is much more buoyant than the hot liquid, the convective process is increased considerably.
- Thermosiphon reboiler operates making use of the buoyancy effect. Typically 25–30% of the liquid is vaporized in the reboiler. The two-phase flow is sent back to the distillation column where the vapor is released to move up through the trays and the liquid drops back to the bottom of the column.

• "Thermosiphon reboilers can be horizontal or vertical." *True/False*?

- *True*.

• "In horizontal thermosiphon reboilers, fluid is not taken on the tube side." Comment.

- Due to their inability to establish a vertical density gradient inside horizontally aligned tubes.

• What is the maximum percent vaporization that may be acceptable in a vertical thermosiphon vaporizer illustrated in Figure 11.10. Give reasons for this choice.

- About 30% of the feed. For higher fouling fluids, a lower percent vaporization should be used to avoid fouling deposits.
- Experience dictates that a 30% vapor fraction helps keep a 1.8–2.4 m (6–8 ft.) long tube wet along its entire length (or nearly the entire length) as when the tube begins to dry out, fouling occurs.

• What is the maximum recommended ΔT between the fluids in a thermosiphon reboiler?

- About 50°C.

• Is mist flow a desirable phenomenon in a reboiler?

- No. In this type of flow, the tube wall is almost dry and the liquid droplets are carried along in a vapor core. Therefore, the heat transfer is much lower because the much higher thermal conductivity of the liquid is in very little contact with the tube wall. The higher the percent vaporization, the lower should be the velocity to avoid mist flow.
- If the mist flow region cannot be avoided, then use of twisted tape turbulence enhancers can be used to increase heat transfer. They will throw the liquid in the vapor core toward the tube wall.

• What are the reasons for loss of circulation through a once-through thermosiphon reboiler?

- Reboiler outlet temperature is higher than column bottoms temperature.

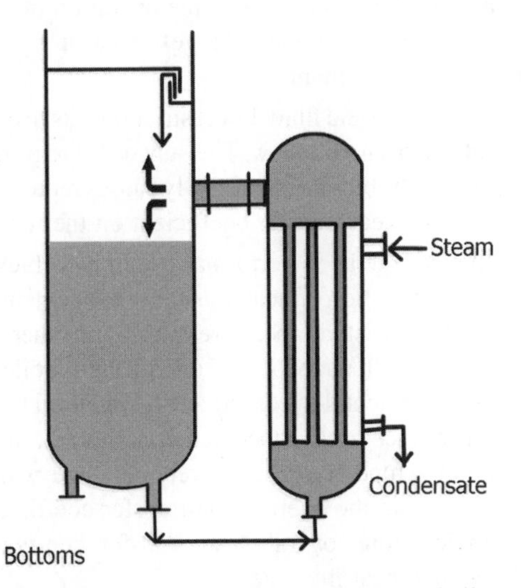

FIGURE 11.10 Vertical thermosiphon reboiler.

- Inability to achieve normal reboiler duty.
- Bottoms product is very light.
- Steam inlet valve does not function well, resulting inadequate heat supply to the column.
- Low reflux drum level, accompanied by low column pressure, even at low reflux rate.
- *Causes* include leaking bottom tray, bottom tray seal pan, or drawoff pan damaged, partially plugged reboiler, restricted reboiler feed line, excessive reboiler design ΔP, and submerged reboiler return line in column bottoms.

• "In a circulating thermosiphon reboiler, the reboiler outlet temperature is lower than the column bottoms temperature." *True/False*?

- *False.* The reboiler outlet temperature is always higher than the column bottoms temperature. Part of the bottom product is recirculated back into the column through the reboiler, thus further heating the circulating part of the liquid and hence the higher temperature.

• "Chances of flooding are more in a circulating reboiler than in a once-through reboiler." Comment.

- In a once-through reboiler, liquid flow is limited to amount of liquid overflowing bottom tray.
- In a circulating reboiler, velocities are higher as liquid flow can be very high making it more susceptible for flooding of the column. Due to this reason, fouling will be low making it better suited for dirty bottom product.

• Illustrate by means of a diagram, flow arrangements with a vertical thermosiphon reboiler with respect to a distillation column.

- Figure 11.10 illustrates flow arrangements with a vertical thermosiphon reboiler.

• What are the bottlenecks in the operation of a vertical thermosiphon reboiler? Suggest measures in performance improvements.

- Due to size and fluid flow restrictions, its heat transfer rate is often quite low. The subcooled region formed at the tube bundle base usually causes reduction in the average heat transfer coefficient on the tube side.
- Improvements in performance can be achieved with the application of heat transfer enhancement devices such as twisted tape, wire matrix, or other types of inserts in the subcooled zone of the reboiler. These enhancement devices reduce the required heat transfer surface area, subcooled zone length, and maximum temperature of the reboiler tube wall, while increasing the overall heat transfer coefficient, subcooled zone overall heat transfer coefficient and average heat flux rate.
- These benefits result in a reduction in the size of the exchanger, the space requirements, and the initial investment.
- However, use of inserts increases pressure drop in the exchangers and increase cleaning and maintenance costs, especially for viscous and dirty fluids.

• "For thermosiphon reboilers, hydraulic aspects are as important as heat transfer aspects." *True/False*?

- *True.* A single pass TEMA E-type shell is generally used as shown in Figure 11.10.
- General characteristics of this type of reboiler are a large diameter vapor exit pipe with a cross-sectional area equal to the total cross-sectional area of the tubes arranged to minimize the vertical distance between the top tube sheet and the column nozzle.
- The liquid level in the column is usually kept near the top tube sheet level in order to provide for maximum circulation.
- The driving force for circulation is the density difference between the liquid in the column and the two-phase mixture in the tubes.
- The exit vapor weight fraction for hydrocarbons is in the range of 0.1–0.35 and for water it is about 0.02–0.1.
- The E-type shell and associated piping is relatively inexpensive, easy to support and compact.

• What is the effect of choking in a vertical thermosiphon reboiler on heat transfer rates?

- Choking on the outlet nozzle and piping reduces the circulation rate through a thermosiphon reboiler. Since the tube side heat transfer rate depends on velocity, the heat transfer is lower at reduced recirculation rates. A rule of thumb is that the inside flow area of the channel outlet nozzle and piping should be the same as the flow area inside the tubing. For example, a ratio of 0.7 in nozzle flow area/tube flow area reduces the heat flux by about 10%. A ratio of 0.4 reduces the heat flux by almost 50%.

• What are the advantages and disadvantages of thermosiphon reboilers?

Advantages:

- High circulation rates.
- Compact with simple piping.
- Reliable operation.
- No/low maintenance requirements during normal operation.
- Low investment relative to a pumped reboiler system.
- No auxiliary equipment such as a pump for recirculation and flow control system are required, saving

pump energy. Pumping in reboiler systems involve handling hot liquids with pump NPSH, leakage and other pump maintenance problems.

- Savings in energy which is required if pump is used.
- No leakage problems as there is no pump.
- Not easily fouled because of the relatively high two-phase velocities obtained in the tubes. For a fluid that is considered to have average fouling tendencies, using a vertical thermosiphon reboiler is quite common.
- Heavy components are not likely to accumulate.
- Less cost than kettle type, the later requiring large diameter shell and more space for installation.

Disadvantages:

- Requires more headroom and column skirt height.
- Its main disadvantage is the problem of hydrodynamic instability, as the upward flow of the vapor should be capable to support the flow of liquid slugs. Experience has shown that the instability at high heat fluxes is very sensitive to the size and hence to the ΔP in the exit piping.
- More difficult to design satisfactorily than kettle reboilers, especially in vacuum operation. Lack of understanding of design issues which include integration of complex columns, reboiler systems and control schemes, temperature gradient analysis, and imbalances in multiexchanger thermosiphon reboiler circuits.
- Control over circulation rates is not easy with upsets in operation due to energy input variations.
- Poor thermosiphoning for high viscosity liquids.
- Steam being on the shell side, which involves low velocities, fouling tendencies are high on the shell side.
- Suspended solids in the process fluid are recirculated between the column and the reboiler.
- Large surface area requirements.

• Give general guidelines for thermosiphon reboilers.
  - Inclined piping should never be used for two-phase flow in a process plant. This is particularly true for reboiler return piping. Only horizontal or vertical runs should be used.
  - If the reboiler heating medium is condensing steam, a desuperheater should be provided, if the superheat is more than 4–15°C.
• "For a horizontal thermosiphon reboiler, process fluid is taken on tube side." *True/False*?
  - *False.* Process fluid is on shell side, unless forced circulation is employed.
• "Land area required for a horizontal thermosiphon reboiler is more than that for a vertical one." *True/False*?
  - *True.*
• "Maintenance for a vertical thermosiphon reboiler is easy." *True/False*?
  - *False.*
• Briefly explain the impact of relative volatility on the operation of thermosiphon reboilers.
  - High relative volatility gives rise to large temperature gradients.
  - Large variations in reboiler feed temperatures result from compositional differences in the reboiler feed arising from different configurations. Tendency to create foaming will be more with high relative volatility fluids. Foaming creates control, operability, product quality, and stability problems.
• What are the symptoms for loss of circulation in a once-through thermosiphon reboiler?
  - Inability to achieve normal reboiler duty.
  - Low reflux drum level, accompanied by low column pressure even at low reflux ratio.
  - Too light bottoms, that is, bottoms composition with respect to low boiling component(s) increases.
  - Reboiler outlet temperature is higher than column bottoms temperature.
  - On opening heat source (steam/hot oil) inlet valve will not increase heat input.
• What are the causes for loss of circulation? (cf. above question)
  - Leaking column bottom tray due to low dry ΔP.
  - Damaged bottoms tray seal pan/drawoff pan.
  - Partially plugged reboiler.
  - Restricted flow to reboiler feed line.
  - Excessive design ΔP for the reboiler.
  - Bottoms liquid level covering reboiler vapor return nozzle.
• How would one detect malfunctioning of thermosiphon circulation in a once-through thermosiphon reboiler?
  - By observing temperatures of column bottoms and reboiler outlet. If column bottoms are cooler than the reboiler outlet liquid, it means that some thing has gone wrong with circulation and remedial steps are to be taken.
• What are the consequences of submerged reboiler return inlet below the column base liquid?
  - The high velocity vapor entering the column from the reboiler outlet, when below the liquid level in the column, gives rise to entrainment and carry-over large amounts of liquid on to the trays or packing above. This liquid gives rise to flooding the column to considerable heights.

- There have been cases where the flooding led to excessive liquid discharges from the column overhead.
- Often this flooding occurs with formation of violent slugs which uplift trays/packing and cause damage with deformation and dislocation. Liquid levels rising above the reboiler return inlet is one of the most common causes of this damage.

• What are the causes for the liquid levels rising above reboiler return nozzle?

- Restriction in the bottoms outlet line, which includes loss of bottoms pumping, obstruction by sediment and debris and undersized outlet lines.
- Faulty level measurement and control.
- Excessive ΔP in kettle reboiler circuits.
- Other operational problems of the reboilers.
- These level rising problems are frequent with kettle reboilers.

• What are the remedial measures for the above problem(s)?

- Reboiler return flows should not impinge on the bottom liquid surface, bottom seal pan, seal pan overflow, or the bottom downcomer.
- Reboiler return lines should be properly sized and excessive velocities should be avoided.
- The vapor disengaging region above the reboiler return should be of adequate geometry and height.
- Liquid overflowing the bottom seal pan should descend without interference from the reboiler return vapor.
- If pipes are used to bring liquid from the bottom downcomer into the column base, they must be designed for self-venting flow and must be adequately sealed and submerged.
- Any baffling in the column base should be simple and of sound hydraulic design, adequately allowing for possible vaporization where bottom tray liquid contacts reboiler return liquid.
- If reboiler return line is to be submerged by design intent, the return should be through a well designed submerged sparger, which releases the vapor into the liquid as bubbles so that slugging is avoided. In such a case adequate level monitoring becomes important because the bubbles and froth lowers the density of the base liquid, which can give rise to incorrect level transmission. This type of arrangement is not normally used because of the above-cited reason.

• What are the ways to increase heat input capacity of existing distillation units? Comment on each approach.

(a) Installing additional reboiler in parallel (Figure 11.11).

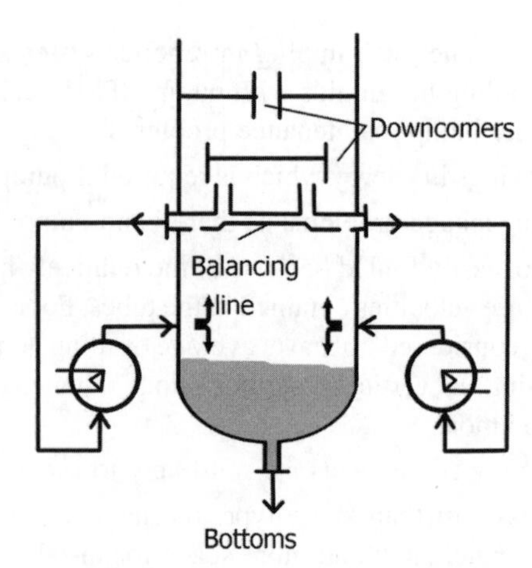

FIGURE 11.11 Two reboilers in parallel.

- Difficult to balance flows to new and old reboilers (similar to condensers—See Question on condensers).
- If flows are not balanced, liquid flows to one of the reboilers in preference to the other, making the later run dry.
  - ➢ Lack of flow/reduced flow promotes fouling, which in turn reduces effective flow area thereby further decreasing flows.
- Adjusting flows by means of providing inlet valves creates operational complexities and increased fouling and blockages of partially closed valves.

(b) Providing separate drawoff nozzles from the column to each reboiler with proper balancing.

(c) Installing second reboiler in series

- Requires sufficient liquid head at column bottom for the flow overcoming ΔP in both the reboilers (Figure 11.12).

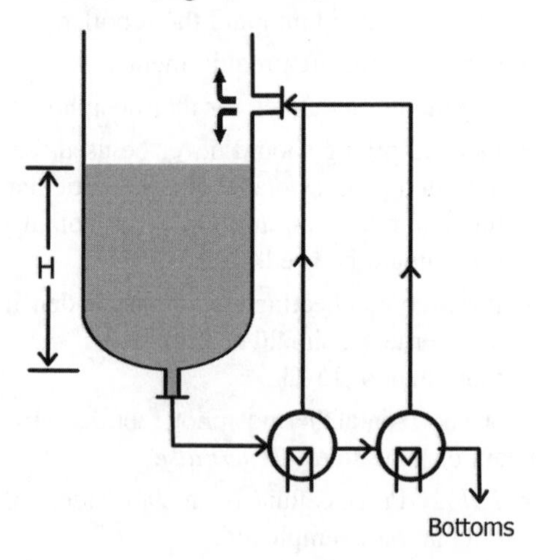

FIGURE 11.12 Two reboilers in series.

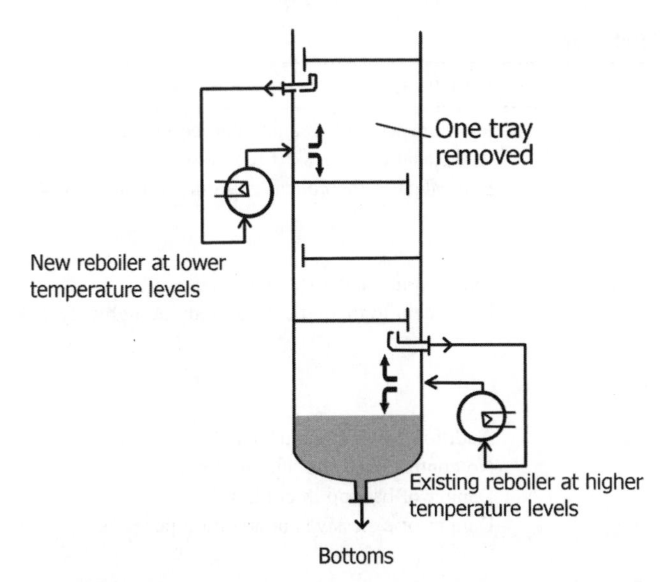

FIGURE 11.13 New reboiler installed at an upper section of the column.

- To make the above arrangement to an existing unit is very difficult. For a new unit this contingency can be taken care of.

(d) Installing a new reboiler at an upper section of the column, instead of withdrawing liquid from the same tray and admitting vapor at the same level (Figure 11.13).

- This is very convenient to do so by installing liquid drawoff and vapor inlet nozzles and liquid trap-out pan.
- Requires removal of one or two trays.
- Lower energy levels are involved as illustrated below.

Advantages:

- Circulation through the existing reboiler is not adversely affected.
- Load (vapor and liquid flows), which is usually maximum in the lower section, is reduced on the lower trays.
- Energy saving: Lower pressure steam (waste steam) may be used in upper reboiler.

Disadvantage:

- Reduction in vapor flows: Decreases separation obtainable as a stripper (lower section).
  - ➢ Summarize comparison of reboiler types.
  - ➢ Table 11.2 gives a summary of comparison of reboiler type.

- Summarize advantages and disadvantages of different types of reboilers.
  - Table 11.3 summarizes advantages and disadvantages of different reboiler types.
- What are the factors that must be considered while selecting a reboiler?

TABLE 11.2 Summary of Reboiler Types and their Characteristic Features

	Thermosiphon				
Features	Vertical	Horizontal	Kettle	Forced Circulation	Internal Reboiler
Boiling side	Tube	Shell	Shell	Tube	Shell
Heat transfer rate	High	Moderately high	Low to moderate	High	Low to moderate
Required space	Small	Large	Large	Small (if vertical)	None
Piping	Simple	Extra required	Extra required	Extra required	None
Pump	None	None	None	Yes	None
Skirt	Required	Required but less	Small	Small (depends on NPSH or residence time)	Small
Residence time	Low	Low	High	Low	High
Fouling	Low	Moderate	High	Very low	Moderate
High μ-liquids	Poor	Poor (better than vertical)	Poor	Good	Poor
Instability	High	High	Low	Low	Low
Controllability	Good	Good	Poor	Excellent	Poor
Maintenance and cleaning	Can be difficult	Relatively easy	Relatively easy	Can be difficult	Difficult
Surface area	For large area >1 shell required	Easier than vertical	Large surface in single shell	>1 shell required	Limited by column size
ΔT required	High	Moderate	Low	High	Moderate to high
Reliability	High	Medium	Excellent	Good to excellent	Medium
Capital cost	Low	Moderate	High	Moderate	Very low

TABLE 11.3 Advantages and Disadvantages of Different Types of Reboilers

Type	Advantages	Disadvantages
Vertical thermosiphon	Capable of very high heat transfer rates Compact, simple piping required Low residence time in heated zone Not easily fouled Good controllability	Maintenance and cleaning can be inconvenient Additional column skirt required Equivalent to theoretical tray only at high recycle
Horizontal thermosiphon	Capable of moderately high heat transfer rates Low residence time in heated zone Not easily fouled Good controllability Easy maintenance and cleaning	Extra piping and space required Equivalent to theoretical tray only at high recycle
Once-through natural circulation	Capable of moderately high heat transfer rates Compact, simple piping required Low residence time in heated zone Not easily fouled Equivalent to theoretical plate	Additional column skirt height required No control over circulation rate Danger of backup in column Danger of excessive vaporization per pass
Kettle (flooded bundle)	Easy maintenance and cleaning Convenient when heating medium is dirty Equivalent to one theoretical plate Contains vapor disengaging space Low skirt height Easy control	Lower heat transfer rates Extra piping and space required High residence time in heated zone, leading to degradation of heat-sensitive fluids Easily fouled
Forced circulation[a]	Viscous and solid-containing liquids can be circulated Enables an erosion-fouling balance Circulation rate can be controlled Higher heat transfer coefficients	Relatively expensive due to extra shell volume Cost of pump and pumping Leakage of material at stuffing box

[a] Advantages and disadvantages will in general correspond to the type of reboiler to which forced circulation is applied. The advantages and disadvantages shown are in addition.

- Temperature difference and type of boiling, namely, nucleate or film affects the selection.
- *Boiling Fluid Characteristics:* Heat-sensitive liquids require low holdup design. Boiling range and mixture composition together with available ΔT affect circulation requirements to avoid stagnation.
- Foaming can be better handled inside tubes.
- *Temperatures:* Very hot fluids are placed inside tubes to reduce shell costs.
- Heating medium requirements may be more important than the boiling liquid requirements.
- *Pressure:* High-pressure fluids are placed inside tubes to avoid requirement of thick-walled shells. For very low pressures (vacuum), other factors involved in the selection determine the tube side fluid.
- Enhanced surfaces are only suitable for some types of reboilers.
- *Cleanability:* Tube side is easier to clean than shell side.
- *Corrosion:* May dictate the use of expensive alloys and therefore corrosive fluids are placed in the tube side, avoiding expensive alloy shell.
- *Space Constraints:* For example, if headroom is limited, vertical units are inappropriate.

- Give a summary of reboiler selection.
  - Table 11.4 is a reboiler selection guide.
- Give a quick guide for reboiler selection.
  - Quick Selection Guide for reboilers (Figure 11.14).
    1. Preferable over recirculating type where low vaporization rates (<25—30%) are alright. Also preferable where minimum exposure of degradable and/or fouling materials to high temperatures.
    2. For large duties, dirty process and where frequent cleaning is required. Used for most refinery thermosiphon applications. Require less headroom than for vertical type. More complex piping is needed. More easily maintainable as tube bundle can be more easily pulled out.
    3. For small duties, clean process and where infrequent cleaning is alright. Vaporization rates <30% and >15%, if operating pressures are below 350 kPa gauge (50 psig). Viscosity of feed is <0.5 cP. Used for most chemical and petrochemical plant thermosiphon applications. Put a

TABLE 11.4 Reboiler Selection Guide

Process Conditions	Kettle or Internal	Horizontal Shell Side Thermosiphon	Vertical Tube Side Thermosiphon	Forced Circulation
		Operating Pressure		
Moderate	E	G	B	E
Near critical	B-E	R	Rd	E
Deep vacuum	B	R	Rd	E
		Design ΔT		
Moderate	E	G	B	E
Large	B	R	G-Rd	E
Small (mixture)	F	F	Rd	P
Very small (pure component)	B	F	P	P
		Fouling		
Clean	G	G	G	E
Moderate	Rd	B	G	E
Heavy	P	Rd	B	G
Very heavy	P	P	Rd	B
		Mixture Boiling Range		
Pure component	G	G	G	E
Narrow	G	G	B	E
Wide	F	G	B	E
Very wide with viscous liquid	F-P	G-Rd	P	B

Category Abbreviations: B, best; G, good operation; F, fair operation, but better choice is possible; Rd, risky unless carefully designed, but could be best choice in some cases; R, risky because of insufficient data; P, poor operation; E, operable but expensive.

butterfly valve in the reboiler inlet piping. Disadvantage is that the base must be elevated to provide hydrostatic head required for thermosiphon effect. Increases column support costs.

4. Greater stability than unbaffled.
5. Used where ΔP_{Piping} is high and natural circulation is impracticable. Especially suitable for viscous and fouling liquids. Also suitable for low vacuum and low vaporization applications. Main disadvantage is pump is required. As pump is to operate at high temperatures, leakage problems at pump seals, and pumping is expensive.
6. Very stable and easy to control. Has no two-phase flow. Permits low skirt height. Expensive as a larger shell, than that for thermosiphon type, is required. It is gravity fed from the distillation column. Lower heat transfer coefficient. Not suitable for fouling liquids. High residence time. No separate vapor–liquid disengagement vessel is required.

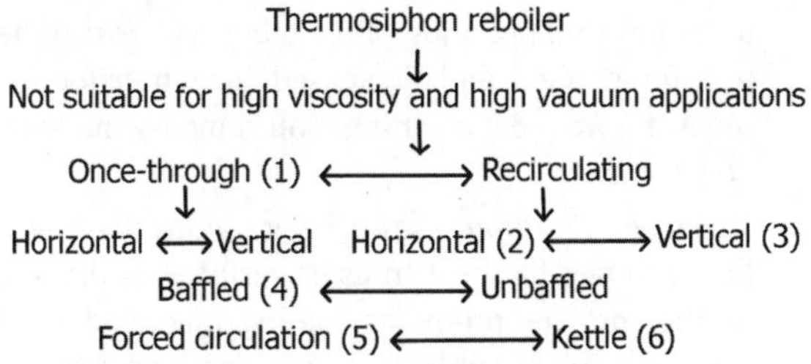

➢ Suitable for vacuum operation. High vaporization rates (up to 80% of feed).

Thermosiphon reboiler
↓
Not suitable for high viscosity and high vacuum applications
↓
Once-through (1) ⟷ Recirculating
↓ ↓
Horizontal ⟷ Vertical Horizontal (2) ⟷ Vertical (3)
Baffled (4) ⟷ Unbaffled
Forced circulation (5) ⟷ Kettle (6)

FIGURE 11.14 Quick selection guide for reboilers.

11.3 EVAPORATION AND EVAPORATORS

- What is an evaporator? What are its applications?
  - An evaporator is used to vaporize a volatile solvent, usually water, from a solution. Its purpose is to concentrate nonvolatile solutes such as organic compounds, inorganic salts, acids, or bases. Typical solutes include phosphoric acid, sulfuric acid, acid sulfate liquor, Kraft liquor, caustic soda, sodium chloride, sodium sulfate, sugars, gelatin, radioactive wastes, syrups, urea, etc.
  - In many applications, evaporation results in the precipitation of solutes in the form of crystals, which are usually separated from the solution with settlers, cyclones, wash columns, elutriating legs, filters, or centrifuges. Examples of precipitates are sodium chloride, sodium sulfate, sodium carbonate, and

calcium sulfate. The desired product can be the concentrated solution, precipitated solids or both.

  - In some applications, the evaporator is used primarily to recover a solvent, such as potable water from saline water (desalination). In any case, the relatively pure condensate from many evaporators is recovered for boiler feed makeup, salt washing, salt dissolving, pump seals, instrument purges equipment and line washing, and other uses.

- Name different types of evaporators.
  - *Short Tube/Calandria Type Evaporators*: Generally used in sugar industry. Short tube evaporators can be of vertical or horizontal tube types. These are less popular for present day applications. The calandria can handle both salting and scaling applications as well as those where no precipitates are formed.
  - *Most Popular Types Are Long Tube Vertical with Natural or Forced Circulation:*
    - High capacities and for low viscosity solutions.
    - Forced circulation type, usually with a submerged inlet, are advantageous for fouling or crystallizing solutions.
  - *Falling Film and Rising Film Evaporators*: Less holdup and boiling takes place from a film of liquid. Shorter residence times. Suitable for viscous and heat-sensitive liquids. Long tube vertical rising film evaporators are commonly used to concentrate many nonsalting liquors and falling type are preferred for liquors requiring evaporation at extremely low ΔT and for critically heat-sensitive liquors. Both types provide maximum evaporative performance for the least capital investment.
  - *Agitated Film/Wiped Film Evaporators*: For very high viscosity solutions.
  - Plate and spiral type evaporators.
  - Scraped surface evaporators.
  - *Direct-Heated Evaporators*: Solar pans, submerged combustion type, etc.
  - Jacketed and coiled vessels for small-scale evaporation applications. Jacketed evaporators are used when the product is very viscous, the batches are small, good mixing is required, ease of cleaning is important, or glass-lined equipment is required.
  - Coils for evaporator heating surfaces come in an almost unlimited variety of shapes and sizes. The most common application is to provide coils inside a tank with the evaporation process outside the coils. Evaporation can also occur inside coils with the heating medium outside the coil.

- What are the factors that are to be considered for the selection of an evaporator?
  - *Capacity*: High heat transfer rates reduce equipment sizes for a given capacity.
  - *Viscosity of Feed and Increase in Viscosity During Evaporation*: The density and viscosity may increase with solid content until the heat transfer performance is reduced or the solution becomes saturated.
  - *Nature and Quality of the Product Required*: Solid, slurry, or concentrated solution.
  - *Crystallization and Salting During Concentration of the Solutions*: Continued boiling of a saturated solution may cause crystals to form, which often must be removed to prevent plugging or fouling of the heat transfer surface.
  - *Heat-Sensitive Fluid Characteristics*: Food products, pharmaceuticals, resins, and temperature-sensitive products are to be handled at lower temperatures and short residence times. Requirements for handling such materials include a combination of the following:
    - Minimizing volume of the liquid at any time in the evaporator.
    - Minimizing time of contact of the liquid with the heat transfer surface.
    - Reducing product's bulk boiling temperature by the use of vacuum.
  - *Fouling or Nonfouling Solutions*: Whether any hard scales form, for example, sugar evaporators.
  - *Foaming or Nonfoaming Liquids*: Stable foams may cause excessive entrainment. Foaming may be caused by dissolved gases in the liquor, by an air leak below the liquid level, special designs for feed inlet, that is, separation of feed liquid from vapor stream, reducing boiling intensity of the liquid on the heat transfer surface by operating at lower temperatures, that is, under vacuum or at a higher pressure and by the presence of surface active agents or finely divided particles in the liquor. Foams may be reduced/suppressed by operating at low liquid levels, reducing vapor velocities in the tubes, by mechanical or hydraulic methods and use of antifoaming agents.
  - *Liquids with Boiling Point Rise*.
  - *Solids*: The properties of the concentrate might change with increase in solids concentration leading to fouling and plugging of the tubes with loss of heat transfer surface and decreased heat transfer rates requiring frequent cleaning requirements and loss of downtime.
  - *Vapor to Solution Ratio*: There must be enough liquid to wet the heat transfer surfaces as dry spots on the surfaces promotes fouling rates and salting problems and possible deterioration of product qual-

ity. Where high vapor to liquid ratios are required, recirculation might be employed.

- *Vapor Velocity (Pressure Drop and Entrainment)*: Requirements of high vapor velocities to give high heat transfer coefficients must be balanced with considerations of entrainment, pressure drop, and erosion.
- *Entrainment Separation*: Effective separation of vapor produced and the liquid. Careful attention must be paid to the selection and design of entrainment separators.
- *Energy Efficiency*.
- *Heat Transfer Medium*: Selection of heat transfer medium (steam, high-temperature heat transfer fluids such as Dowtherm vapors, hot oil may influence the selection of the evaporator. Liquid-heated evaporators typically have lower heat transfer coefficients, requiring larger heat transfer surfaces. Hot oil may give higher temperatures, provided the fluid being evaporated is not heat sensitive.
- *Whether Direct Heating can be Used.*
- *Choice of Corrosion-Resistant Materials of Construction*: Corrosion may influence the selection of evaporator. Corrosion and erosion are frequently more severe in evaporators than in other types of equipment because of the high liquid and vapor velocities, the frequent presence of suspended solids, and the concentrations required.
- *Product Quality may Require Low Holdup and Low Temperatures*: Low holdup time requirements may eliminate application of some evaporator types. Product quality may also dictate special materials of construction.
- *Other Fluid Properties must also be Considered*: These include viscosity increases of the product, heat of solution, toxicity, explosion hazards, radioactivity, and ease of cleaning. Salting, scaling, and fouling result in steadily diminishing heat transfer rates until the evaporator must be shut down and cleaned. Some deposits may be difficult and expensive to remove.

- Summarize evaporator selection through the use of a chart.
  - Figure 11.15 gives an evaporator selection guide.
- "Fruit juices are concentrated in falling film evaporators." *True/False*?
  - *True*.
- "Gelatin is concentrated in agitated film evaporators." *True/False*?
  - *True*.
- "Falling film tubular evaporators are more compact than plate evaporators." *True/False*?
  - *False*.
- What are the ways by which energy efficiency of an evaporation system can be improved?
  - There are three basic possibilities to save energy:
    - Multiple effect evaporation.
    - Thermal vapor recompression.
    - Mechanical vapor recompression.
  - Application of one of these techniques will considerably decrease the energy consumption. Often it is feasible to combine two of these possibilities to minimize capital and operating costs. In highly sophisticated evaporation plants all three techniques may be applied.

	Feed Conditions							Suitability for Heat-Sensitive Materials
	Viscosity (mN s/m²)							
Evaporator Type	High Viscosity >1000	Medium Viscosity <1000 Max	Low Viscosity < 100	Foaming	Scaling or Fouling	Crystals Produced	Solids in Suspension	
Recirculating calandria (short vertical tube)		←	—	—	—	—	→	No
Forced circulation		←	—	—	—	→		Yes
Falling film			←→					No
Natural circulation			←	→				No
Single pass wiped film	←	—	—	—	—	—	→	Yes
Tubular (long tube) falling film			←→					Yes
Tubular (long tube) rising film			←	→				Yes

FIGURE 11.15 Evaporator selection guide.

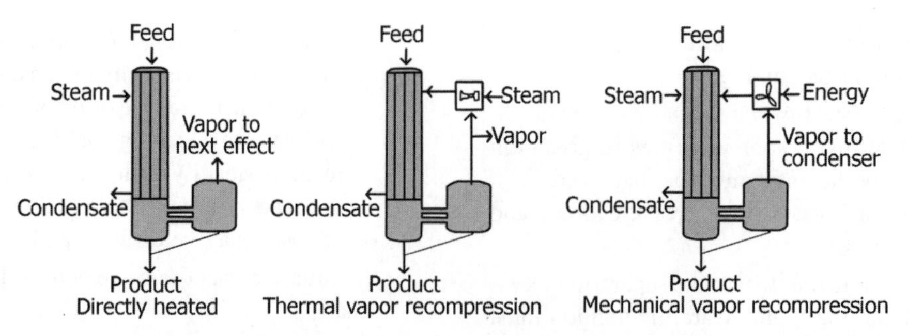

FIGURE 11.16 Energy-efficient evaporation systems.

- The principles are illustrated in Figure 11.16.

- What are the important problems arising in an evaporation process?
  - High viscosity.
  - Heat-sensitivity.
  - Scaling.
  - Entrainment problems.
  - Problems involved with vacuum-generating equipment.
  - Handling crystallized solids.
  - Formation on heat transfer surfaces an insulating blanket of air through which steam must diffuse before it can condense. This is due to accumulation of noncondensables on the surface.
- Illustrate with a diagram short vertical tube calandria-type evaporator and describe its characteristic features.

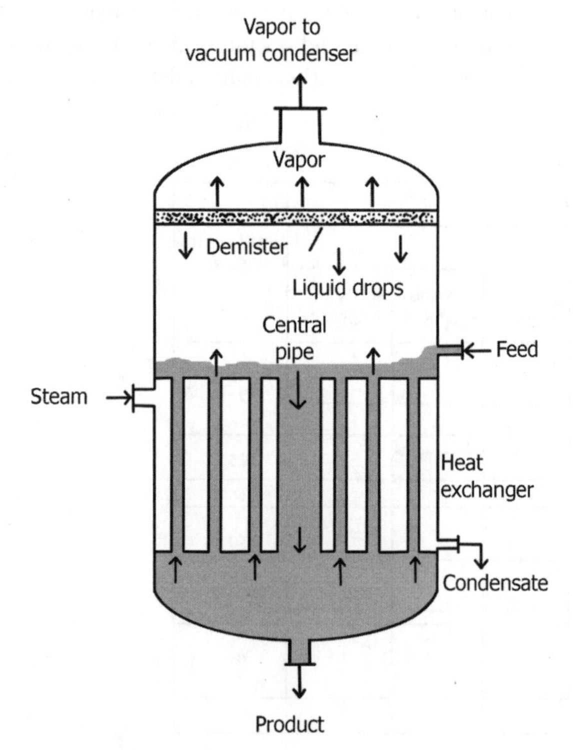

FIGURE 11.17 Short tube vertical calandria-type evaporator.

  - Figure 11.17 illustrates calandria-type short tube vertical evaporator.
  - This is one of the earliest types and its main use is in the evaporation of cane sugar juice.
  - Circulation over the heating surface is induced by boiling in the tubes, which are usually 5–7.5 cm (2–3 in.) in diameter by 1.2–1.8 m (4–6 ft) long. The circulation rate through the tubes is many times the feed rate. Due to this reason, a return passage from above the top tube sheet to below the bottom tube sheet is provided in the form of a large diameter central pipe as shown in Figure 11.19, so that friction losses through the central pipe do not appreciably impede circulation up through the tubes. The area of the central pipe is about of the same order of magnitude as the combined cross-sectional area of the tubes. This results in a central pipe almost half of the diameter of the tube sheet.
  - Circulation and heat transfer in this type of evaporator are strongly affected by the liquid level. Highest heat transfer coefficients are achieved when the level is about half the height of the tubes. Any reduction in level, below the optimum, result in incomplete wetting of the tube walls with a consequent increased tendency to foul and a rapid reduction in capacity.
  - The calandria can handle both salting and scaling-type applications as well as those where no precipitates are formed.
  - When this type of evaporator is used with a liquid that can deposit salt or give rise to scaling, it is normal to operate with the liquid level appreciably higher than the top tube sheet.
  - Circulation in the standard short tube vertical evaporator is dependent entirely on boiling, and when boiling stops, any solids present settle out of suspension. Consequently, this type is rarely used as a crystallizing evaporator. To avoid salting when the evaporator is used for crystallizing solutions, the liquid level must be kept well above the top tube sheet.

- A knitted wire mesh pad (demister) serves as an effective entrainment separator when it cannot easily be fouled by solids in the liquor. The mesh is available in woven metal wire of most alloys and is installed across the top of the evaporator as shown in Figure 11.19.
- Demisters have low pressure drops, usually of the order of 13 mm of water with a collection efficiency of above 99.8% at vapor velocities ranging from 2.5 to 6 m/s.

• What are the advantages, disadvantages, and applications of a standard short tube vertical evaporator?

- *Advantages*:
  - ➢ High heat transfer coefficients at high-temperature differences.
  - ➢ Easy mechanical descaling.
  - ➢ Low headroom.
  - ➢ Relatively of low cost.
- *Disadvantages*:
  - ➢ Poor heat transfer at low-temperature differences and low temperature.
  - ➢ Poor heat transfer with viscous liquids.
  - ➢ Relatively high holdup.
  - ➢ High floor space and weight.
- *Applications*:
  - ➢ Clear liquids.
  - ➢ Relatively noncorrosive liquids, since body is large and expensive if built of materials other than mild steel or cast iron.
  - ➢ Mild scaling solutions requiring mechanical cleaning, since tubes are short and large in diameter.

• Compare vertical tube and horizontal tube evaporators.

- Vertical units provide more hydraulic head and higher circulation rates.
- However, such units have boiling point elevations due to the hydrostatic head.
- To offset this, they can be inclined, but not less than 15° from horizontal. Inclined units result in different flow patterns and velocity in various tubes because elevation is not constant, especially for large units.
- Horizontal units frequently have poor distribution and vapor binding, especially when boiling on the shell side.
- The vertical boiling-in-tube unit is generally the most economical choice, both in initial investment and operating costs. The greater boiling point elevation is seldom a serious penalty when all the advantages are weighed.

• What is the range of tube sizes and lengths normally used in evaporators?

- Tubes range from 19 to 63 mm (3/4—2.5 in.) in diameter and 3.6 to 9.1 m (12–30 ft) in length. For short tube vertical evaporators, tube lengths are in the range of 1–3 m.

• What are the advantages, disadvantages, and applications of horizontal tube evaporators?

- In these evaporators, liquid is outside the tubes and steam is passed inside tubes.
- *Advantages*:
  - ➢ Large vapor–liquid disengaging area (submerged tube type), with low entrainment losses.
  - ➢ Very low headroom requirements.
  - ➢ Relatively low cost in small capacity (straight tube type).
  - ➢ Good heat transfer coefficients for a properly designed evaporator.
  - ➢ They are well adapted for nonscaling, low viscosity liquids.
- *Disadvantages*:
  - ➢ Not suitable for salting and scaling liquids.
  - ➢ Not suitable for foaming liquids.
  - ➢ Maintaining liquid distribution is difficult in film type.
- *Applications*:
  - ➢ Wherever limited headroom is required.
  - ➢ For small capacities.
  - ➢ For concentrating nonscaling liquids (straight tube type).

• What are the characteristic features of long tube vertical evaporators?

- Long tube vertical (LTV) evaporator consists of a single-pass vertical shell and tube heat exchanger discharging into a relatively small vapor head. Normally no liquid level is maintained in the vapor head and residence time of liquid is only a few seconds.
- The tubes are usually about 5 cm (2 in.) in diameter but may be smaller than 2.5 cm (1 in.) with lengths varying from less than 6 to 10.7 m (20–35 ft) in the rising film version and to as large as 20 m (65 ft) in the falling film version.
- The evaporator concentrates the liquid from the feed to discharge density in a once-through flow. Because of the long tubes and relatively high heat transfer coefficients, it is possible to achieve higher single unit capacities in this type of evaporator than in any other.
- LTV evaporators are used for concentrating black liquor in the pulp and paper industry and also used for caustic soda evaporation.

- *Falling film* version of LTV evaporator is widely used for concentrating heat-sensitive materials, such as fruit juices, because the holdup time is very small, the liquid is not overheated during passage through the evaporator, and heat transfer coefficients are high even at low boiling temperatures.
  - ➢ The liquid to be concentrated is supplied to the top of the heating tubes and distributed in such a way as to flow down the inside of the tube walls as a thin film. The liquid film starts to boil due to the external heating of the tubes and is partially evaporated as a result. The downward flow, caused initially by gravity, is enhanced by the parallel, downward flow of the vapor formed.
  - ➢ Residual film liquid and vapor is separated in the lower part of the calandria and in the downstream centrifugal droplet separator.
  - ➢ It is essential that the entire film heating surface, especially in the lower regions, be evenly and sufficiently wetted with liquid. Where this is not the case, dry spots will result that will lead to encrustation and buildup of deposits.
  - ➢ For complete wetting, it is important that a suitable distribution system is selected for the head of the evaporator.
  - ➢ Wetting rates are increased by using longer heating tubes, dividing the evaporator into several compartments or by recirculating the product.
  - ➢ For some applications, it is necessary to supplement an insufficient quantity of feed liquor with product liquor pumped to the top liquor chamber to avoid vapor blanketing of the inside tube surface.
- *Rising film evaporators* involve the following characteristics:
  - ➢ *High-Temperature Difference between Heating Chamber and Boiling Chamber*: In order to ensure sufficient liquid transfer in tubes of a length of 5–7 m and to cause the film to rise.
  - ➢ *High Turbulence in the Liquid*: Due to the upward movement against gravity. For this reason, rising film evaporators are also suited for products of high viscosity and those with the tendency to foul on the heating surface.
  - ➢ *Stable High Performance Operation*: Based on product recirculation within a wide range of conditions.
    - If the vapor pressure of the feed equals or exceeds the system pressure at the bottom tube sheet, vaporization will occur immediately. For colder feed, the lower portion of the tubes is used to preheat the liquor to its boiling point. Vaporization then begins at that height within the tubes where the vapor pressure of the feed liquor equals the system pressure.
    - Heat transfer rates are enhanced in the nonboiling section by surface or local boiling and in the boiling section by nucleate boiling.
    - In the nucleate boiling zone, heat transfer rates are several times higher than those in the nonboiling zone and therefore, it is important to reduce the nonboiling zone to the minimum.
    - Different two-phase regimes are created in the boiling zone, ranging from *slug flow*, *annular flow*, and *mist flow*.
    - *Mist flow* should be avoided because poor heat transfer results when there is not enough liquid present to wet the tube walls.
    - To avoid mist flow, it is sometimes necessary to recycle concentrated product from the vapor body to the bottom liquor chamber to supplement the feed liquor.
- In *rising film evaporators*, liquid is fed into the bottom liquid chamber and then into the tubes. As a result of the upward movement of the vapor bubbles, the liquid rises to the top (Figure 11.18).
- Due to external heating, the liquid film starts to boil on the inside walls of the tubes and is partially vaporized during this process.
- During the ascent more and more vapors form. The film starts to move along the wall. As the liquor climbs up the inside of the tubes, additional vapor is generated and the velocity of the liquid–vapor mixture increases to a maximum at the tube exit.
- As the liquor climbs up the inside of the tubes, additional vapor is generated and the velocity of the liquid–vapor mixture increases to a maximum at the tube exit.
- Heat transfer rates are enhanced in the nonboiling section by surface or local boiling and in the boiling section by nucleate boiling.
- The exit mixture impinges upon a deflector baffle, mounted above the top tube sheet of the heat exchanger, where most of the entrainment separation of the liquid from the vapor occurs.
- Additional entrainment is separated from the vapor by gravity as the vapor rises in the vapor body. A mesh type or centrifugal entrainment separator can be installed near the top of the vapor body to remove most of the remaining traces of liquid from the vapor.
- To provide for good heat transfer rates, ΔT between the heat transfer medium and the liquid should preferably be $>10°C$.

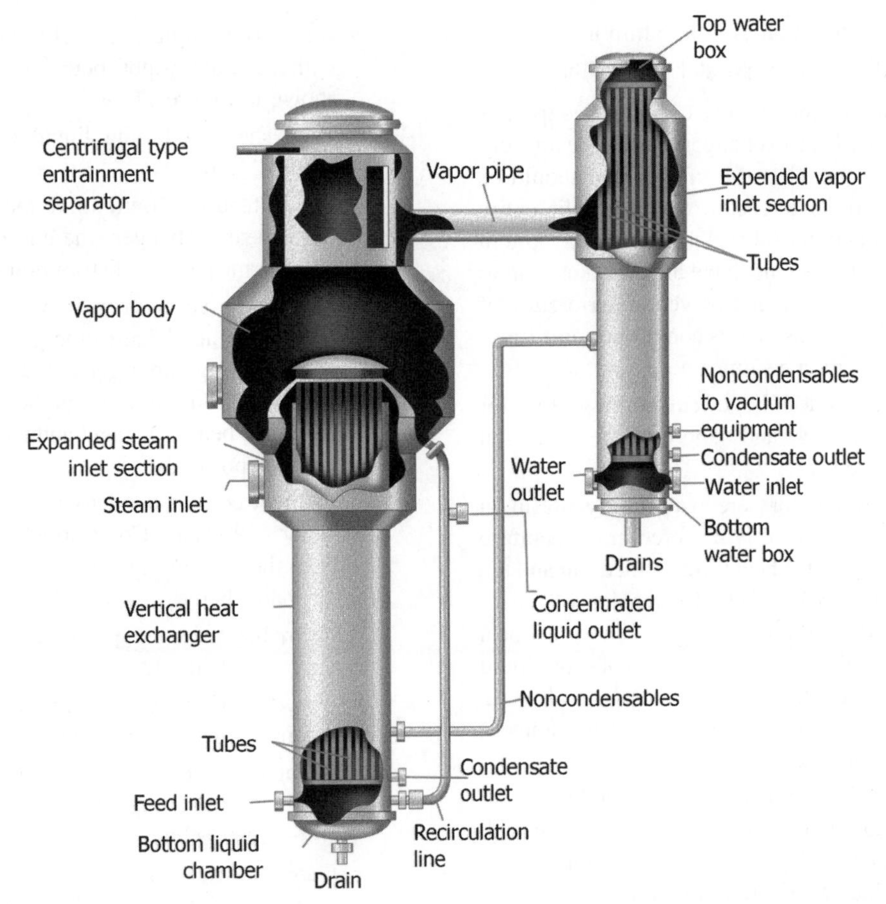

FIGURE 11.18 Swenson rising film evaporator unit. (*Courtesy*: Swenson Technology, Inc.)

- Condensing steam or any other suitable heat transfer fluid, such as Dowtherm or hot liquid may be used.
- Feed distribution to the tubes is the main problem in these evaporators.
- These evaporators are normally not suitable for salting and scale-forming liquids. These are used primarily to concentrate nonsalting liquids.
- Because of their simplicity of construction, compactness, and generally high heat transfer coefficients, LTV evaporators are well suited to service with corrosive liquids.
- Polished stainless-steel LTV evaporators are widely used for food products.

• What are the advantages, disadvantages, applications, and performance problems of LTV evaporators?

- *Advantages*:
  - Good heat transfer coefficients at reasonable temperature differences (rising film).
  - Good heat-transfer coefficients at all temperature differences (falling film).
  - Low holdup.
  - Large heating surface in one body.
  - Low cost.
  - Small floor space.
- *Disadvantages*:
  - High headroom.
  - Generally unsuitable for salting and severely scaling liquids.
  - Poor heat transfer coefficients of rising film version at low-temperature differences.
  - Recirculation usually required for falling-film version.
- *Applications*:
  - Clear liquids.
  - Foaming liquids. High velocities attainable in long tube evaporators break up foams.
  - Corrosive solutions.
  - Large evaporation loads.
  - High-temperature differences (rising film), low-temperature differences (falling film).
  - Low-temperature operation: Falling film.
  - Vapor compression operation: Falling film.
- *Performance Problems*:
  - Sensitivity of rising film units to changes in operating conditions.

- ➢ Poor feed distribution to falling film units.

- How does a falling film evaporator work? Illustrate.
  - A thin film of the product to be concentrated trickles down inside of heat exchanger tubes. Steam condenses outside the tubes. Minimum flow should be maintained in order to ensure wetting of the entire internal surfaces of the tubes. Particularly the bottom sections of the tubes must have an unbroken film to avoid solids deposition and prevent deterioration of product quality. To ensure this condition, some times product recirculation becomes necessary.
  - Used on the shell side of large *heat pump* systems, for example, in absorbers and vaporizers in absorption heat pump systems.
  - Falling film evaporators are particularly useful in applications where the driving force in temperature difference between the heat transfer medium and the liquid is small, with $\Delta T < 8°C$. The residence time for the liquid in this evaporator is less than that for a rising film evaporator, the combination of short liquid residence time, and the ability to operate at a low ΔT makes the falling film evaporator ideal for concentrating highly heat-sensitive liquids.
  - High heat transfer coefficients are attained in falling film evaporators when a continuous film of liquid, preferably at its boiling point, flows down the inside tube wall with a vapor core in the center of the tube.
  - Falling film evaporators are especially popular in the food industry where many materials are heat sensitive.
  - A Swenson single effect, LTV falling film evaporator with separate vapor body and heat exchanger is shown in Figure 11.19.
  - Tube showing falling liquid film is illustrated in Figure 11.20.
  - Liquid feed is admitted into the top liquid chamber of the heat exchanger where it is distributed to each tube by means of a distribution device.
  - The liquor accelerates in velocity as it descends inside the tubes due to the gravity and drag of the down flowing vapor generated by boiling. Liquid is separated from the vapor in the bottom liquid chamber of the heat exchanger and with a skirt-type baffle in the vapor body.
  - A direct contact condenser is used to condense the vapor with water. Concentrated liquid is discharged from the bottom liquid chamber and cone bottom of the vapor body.

- What are the various services for which falling film heat exchangers are used?
  - *Liquid Coolers and Condensers*:
    - ➢ Dirty water can be used for cooling.
    - ➢ Top of cooler is open to atmosphere for access to tubes for cleaning while in operation by removing distributors one at a time and scrubbing the tubes.
  - *Evaporators*:
    - ➢ Heat-sensitive materials such as ammonium nitrate, urea, etc., are handled.

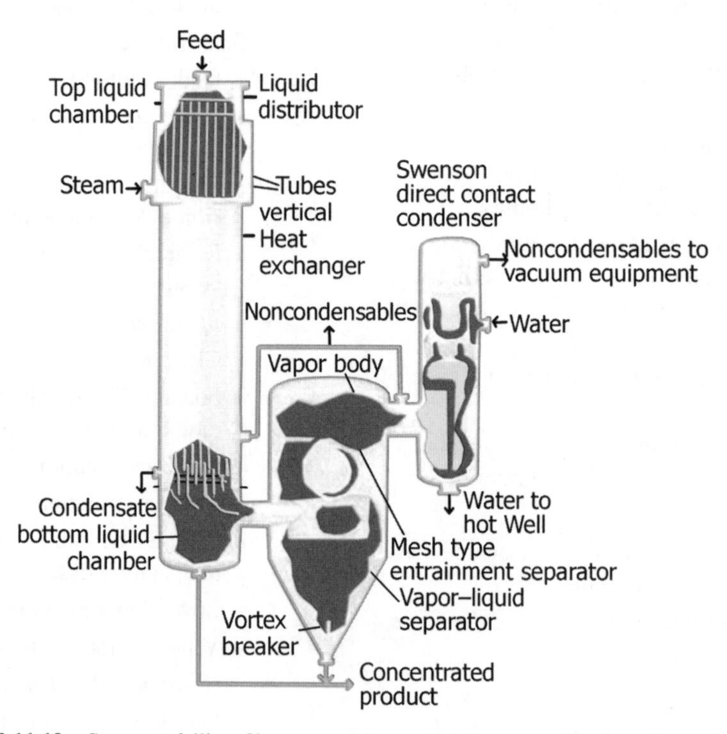

FIGURE 11.19 Swenson falling film evaporation unit. (*Courtesy*: Swenson Technology, Inc.)

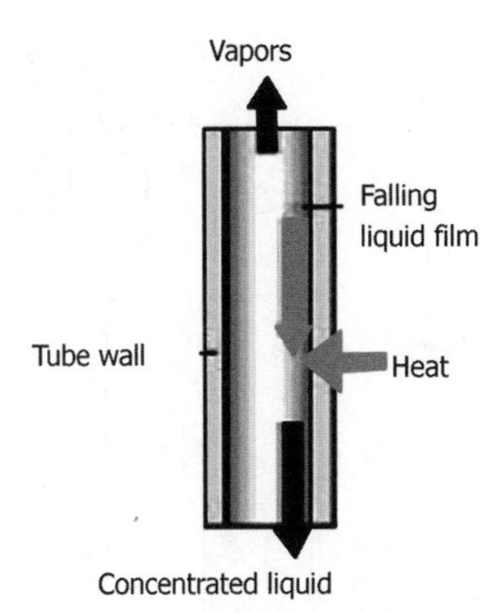

FIGURE 11.20 Tube showing falling liquid film.

- ➢ Air is introduced sometimes into tubes to reduce partial pressure of solutions whose boiling points are high.
- ▪ *Absorbers*:
  - ➢ Particularly those involved with heat effects.
  - ➢ Liquid film inside the tubes will be falling while gas rising through the tubes.
  - ➢ Cooling water circulates outside the tubes.
- ▪ *Freezers*:
  - ➢ By cooling the falling film to its freezing point, these exchangers convert a variety of chemicals to solid phase.
  - ➢ Employed for the production of sized ice and *p*-dichlorobenzene.
  - ➢ Selective freezing is used for isomer separation.

- What are the principal advantages of falling film evaporators?
  - ▪ High heat-transfer rates.
  - ▪ More uniformity in the overall heat-transfer coefficients.
  - ▪ Short residence times (important for heat-sensitive materials).
  - ▪ Large reduction in liquid charge to the evaporator. This reduction of inventory also helps increasing safety in the evaporation of hazardous liquids.
  - ▪ Lower values of ΔT (closer temperature approaches) and also lower temperatures are involved for the evaporation process.
  - ▪ No internal ΔP.
  - ▪ More compactness in the evaporator design. Relatively small floor space requirement.
  - ▪ Easy accessibility to tubes for cleaning.
  - ▪ In some cases, prevention of leakages from one side to the other.

- What are the applications of falling film evaporators?
  - ▪ Falling film evaporators have virtually replaced other evaporator types in many fields. These evaporators are especially popular in the food industry where many materials are heat sensitive.
  - ▪ Evaporating fruit and vegetable juices presents a special challenge. Juices are heat sensitive and their viscosities increase significantly as they are concentrated. Small solids in the juices tend to stick to the heat transfer surface thus causing spoilage and burning.
  - ▪ Juice evaporations are usually performed in a vacuum to reduce boiling temperatures (due to heat sensitivity).
  - ▪ High flow circulation rates help avoid buildups on the tube walls.
  - ▪ For some juices (e.g., orange juice), it is unavoidable that the flavor changes as concentration increases. Some of the volatile, flavor-containing components are lost during evaporation. In such cases, some of the raw juice is mixed with the concentrate to replace the lost flavors.
  - ▪ Sometimes these are applied to desalination units, large capacity air separation plants and ocean thermal energy conversion (OTEC) plants, in which closer temperature approaches are realized achieving high-energy efficiencies.
  - ▪ For liquids which contain small quantities of solids and have a low-to-moderate tendency to form incrustations.
  - ▪ Some heat-sensitive materials concentrated in falling film evaporators include coffee extract, syrups, tomato juice, whey, pectin, glue, gelatin, animal blood, and the like.

- "In a falling film heat transfer equipment, the tubes must be strictly vertical." Comment.
  - ▪ Tilted tubes
    - ➢ Film thickness along the tube perimeter will not be uniform.
    - ➢ Lower heat transfer rates.
  - ▪ Tube areas where film thickness is very small, result in enhanced surface fouling by creating dry out conditions.
  - ▪ Long tubes must be kept free from vibrations by rigid supports.

- Give the heat transfer equations applicable for falling film evaporators.
  - ▪ Falling film evaporators are characterized by having thin liquid films in contact with the heat transfer surfaces. Therefore, the heat transfer equations are basically the same as those for condensation, the only

difference being the direction of heat flow. These equations do not consider nucleation and are conservative because nucleation effects are to increase heat transfer coefficients. Uniformity in liquid distribution is essential for in-tube falling films.

- In the preheating section, the applicable heat transfer equations are:
  - For the laminar region:

$$h = 4.71k/[3\Gamma\mu/gv^2]^{1/3}. \quad (11.2)$$

  - For the turbulent region:

$$h[\mu^2/k^3v^2g]^{1/3} = 5.7(10^{-3})(4\Gamma/\mu)^{0.4}(C_p\mu/k)^{0.34}. \quad (11.3)$$

- For surface vaporization, the respective equations for heat transfer coefficients are:
  - Laminar flow:

$$h = 0.821[\mu^2/k^3v^2g]^{-1/3}(4\Gamma/\mu)^{-0.22}. \quad (11.4)$$

- This equation includes the effect of waves and ripples.
  - Turbulent flow:

$$h = 3.8(10^{-3})[\mu^2/k^3v^2g]^{-1/3}(4\Gamma/\mu)^{0.4}(C_p\mu/k)^{0.34}. \quad (11.5)$$

- The transition N_{Re} is just the value from the intersection of these equations, not the transition of the flow regimes and is given by:

$$(4\Gamma/\mu)_{trans} = 5800(C_p\mu/k)^{-1.06}. \quad (11.6)$$

- These equations provide the local heat transfer coefficients, which are used in the design of the vaporizer. All the properties are evaluated for the liquid phase.

- "Rising film evaporators require smaller temperature differences than falling film evaporators." *True/False*?
  - *False.*
- "Rising film evaporators are short tube evaporators." *True/False*?
  - *False.*
- "If the cooling water temperature in the condenser is decreased, the final product concentration will increase." *True/False*?
  - *True.*
- "Residence time in a falling film evaporator is smaller than in a forced circulation evaporator." *True/False*?
  - *True.*
- Illustrate, by means of a diagram, a vertical forced circulation evaporator showing the essential parts. What are its essential features?

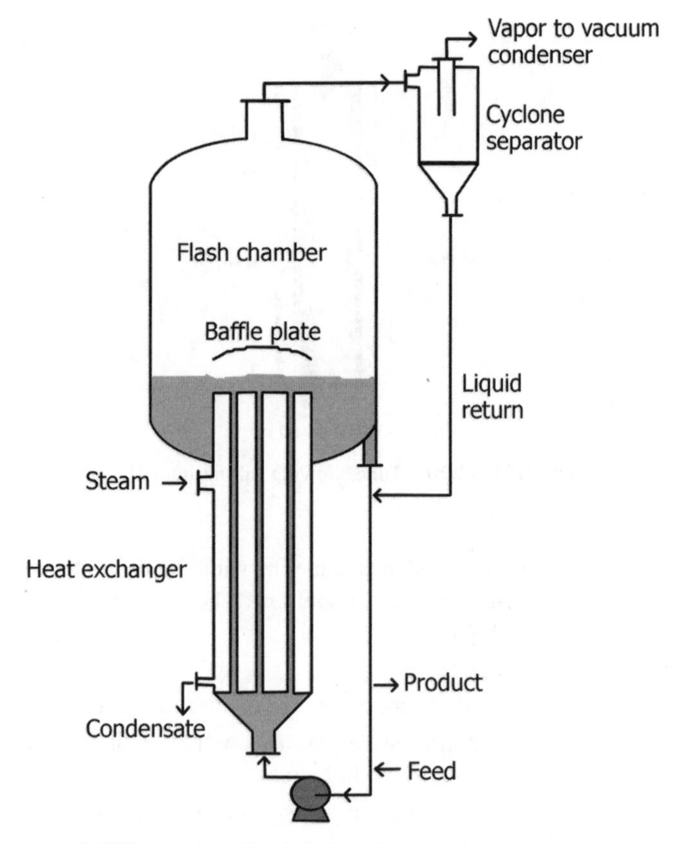

FIGURE 11.21 Vertical forced circulation evaporator.

- Vertical shell and tube heat exchanger or plate heat exchanger as the calandria, with flash vessel/separator arranged above the calandria, circulation pump. Exchanger can be horizontal also (Figure 11.21).
- The liquid is circulated through the calandria by means of a circulation pump, where it is superheated at an elevated pressure, higher than its normal boiling pressure. Upon entering the separator, the pressure in the liquid is rapidly reduced resulting in some of the liquid being flashed, or rapidly boiled off. Since liquid circulation is maintained, the flow velocity in the tubes and the liquid temperature can be controlled to suit the product requirements independently of the preselected temperature difference.
- Boiling/evaporation will not take place on the heating surfaces, but in the separator. Fouling due to incrustation and precipitation in the calandria is therefore minimized.
- A pump is used to ensure circulation past the heating surface. The pump withdraws liquor from the flash chamber and forces it through the heat exchanger back to the flash chamber. Circulation is maintained regardless of the evaporation rate. The liquid velocity past the heating surface is limited only by the pumping power needed or available and by accelerated corrosion and erosion at higher velocities.

- Tube velocities normally range from a minimum of about 1.2 m/s (4 ft/s) in salt evaporators with copper or brass tubes and liquid containing 5% or more solids up to about 3 m/s (10 ft/s) in caustic evaporators having nickel tubes and liquid containing only a small amount of solids. Tube velocities can be as high as 4.5–6 m/s (15–20 ft/s).
- Forced circulation evaporator is widely used for a variety of applications, especially for handling fouling, viscous, and crystallizing liquids. Well suited as crystallizing evaporators for saline solutions.
- Natural circulation is not enough to maintain velocities suitable for good heat transfer and in crystallizing applications, it is essential to ensure suspension of the solids all the time in the system.
- Highest heat transfer coefficients are obtained in forced circulation evaporators when the liquid is allowed to boil in the tubes.
- This type of forced circulation evaporator is not well suited to salting solutions because boiling in the tubes increases the chances of salt deposit on the walls and the sudden flashing at the tube exits promotes excessive nucleation and production of fine crystals.

- What are the advantages, disadvantages, and applications of forced circulation evaporators?
  - *Advantages*:
    - High heat-transfer coefficients.
    - Positive circulation.
    - Relative freedom from salting, scaling, and fouling.
  - *Disadvantages*:
    - High cost.
    - Power required for circulating pump.
    - Relatively high holdup or residence time.
  - *Applications*:
    - Crystalline product.
    - Corrosive solutions.
    - Viscous solutions.
- What are the frequently encountered difficulties with forced circulation evaporators?
  - Plugging of tube inlets by salt deposits detached from walls of equipment.
  - Poor circulation due to higher than expected head losses.
  - Salting due to boiling in tubes.
  - Corrosion–erosion.
- What is the range of tube velocities in forced circulation evaporators?
  - Forced circulation tube velocities are generally in the 1–6 m/s (3–20 ft/s) range.
- What is the effect of foaming on the operation of an evaporator?
  - Foams consist of two-phase mixtures with tiny liquid droplets entrained in tiny vapor bubbles. Heat transfer is more difficult through foams than through homogeneous liquids.
  - Foaming results from the presence of colloids or of surface tension depressants and finely divided solids in the evaporating liquid. Foaming increases entrainment and require large residence times in entrainment separators for disengagement.
- How are foaming liquids handled in an evaporator?
  - Use of antifoam agents, steam jets impinging on the foam surface, removal of product at the surface layer where foaming agents are likely to concentrate and operation at very low liquid levels so that hot surfaces can break the foam are some of the methods effective in combating foams.
  - Impingement of the foamy liquid at high velocity against a baffle tends to break the foam mechanically. This is one of the reasons for the effectiveness of long tube vertical, forced circulation, and agitated film evaporators with foaming liquids.
  - Operating at lower temperatures and/or higher dissolved solids concentrations may also reduce foaming tendencies.
- What is an agitated film evaporator? What are its advantages and applications?
  - Agitated film evaporators use a heating surface consisting of one large diameter tube that may either be straight or tapered, horizontal or vertical.
  - Solution to be concentrated is made to flow as a film rather than as bulk liquid. Liquid is spread on the tube wall by a rotating assembly of blades that either maintain a close clearance from the wall or actually scrape the liquid film on the wall. The liquid film is subjected to mechanical agitation to induce high levels of turbulence and surface renewal on the heat transfer surfaces.
  - These evaporators overcome problems of product degradation due to long residence times, fouling and plugging of tubes/heat transfer surfaces, poor heat transfer and high pressure drops while handling high viscosity liquids involved in conventional evaporators.
  - Agitated film evaporators are used for concentrating heat-sensitive, viscous, and fouling or high boiling liquids.
  - Residence times of only a few seconds permit concentration of heat-sensitive materials at temperatures and temperature differences higher than in other types.

- Very high temperature differences can be used with high-temperature heating fluids such as Dowtherm or other high-temperature media. This permits achieving reasonable capacities in spite of the relatively low heat transfer coefficients and the small surface that can be provided in a single tube (about $20\,m^2$).
- High agitation, of the order of 12 m/s (40 ft/s) rotor tip speed and power intensities of $2\text{–}20\,kW/m^2$ permit handling extremely viscous materials. Fluids with viscosities up to 50,000 cP can be handled by these evaporators. High feed to product ratios can be handled without recirculation.
- A variety of evaporation processes can be handled by these evaporators, which include evaporation and chemical reaction, two-phase flow (immiscible fluids, slurries, suspensions), cocurrent evaporation, multiple effect evaporation, and others.
- The expensive construction limits application to the most difficult materials.
- The structural need for wall thicknesses of 6–13 mm, is a major reason for the relatively low heat transfer coefficients when evaporating low viscosity materials.

• What is the difference between an agitated film and wiped film evaporators?

- In agitated film evaporator, blades of the agitator do not touch the cylinder wall. Close clearance between the blades and the wall ensure a thin turbulent film.
- In wiped film evaporator, liquid is fed onto the inside wall of a cylinder with the heating medium outside wiper blades scrape the cylinder walls, preventing buildup of fouling and maintaining a thin liquid film.
- Wiped film evaporators are used for very viscous liquids (up to 1,000,000 cP) or fouling liquids. These evaporators have low surface areas and use high ΔT. These are used as product finishing evaporators.

• Explain why circulating evaporators are not well suited for concentrating heat-sensitive liquids.

- Circulating evaporators involve large residence times for the liquids, which deteriorates quality of the heat-sensitive liquids.

• Name some heat-sensitive liquids, which are concentrated in agitated film evaporators.

- Food products such as fruit juices, jellies, syrups, milk products, pharmaceuticals, and the like.

• What is thermocompression evaporation?

- Thermocompression involves compression of the vapor from an evaporator and using the compressed vapor as heating medium in the same or another evaporator. Compression increases temperature of the vapor so that ΔT for heat transfer will be available.
- The compression may be accomplished by mechanical means (mechanical recompression evaporators) or by a steam jet (thermocompression evaporators). In order to keep the compressor cost and power requirements within reasonable levels, the evaporator must work at small temperature differences, about 5.5–11°C. This requires a large evaporator heating surface, which makes vapor compression evaporator more expensive in first cost than a multiple effect evaporator.
- Substantial savings in operating cost are realized when electrical or mechanical power is available at a low cost relative to low-pressure steam, when only high-pressure steam is available to operate the evaporator, or when the cost of providing cooling water or other heat sink for a multiple effect evaporator is high.
- Centrifugal compressors are generally used, being the cheapest for the intermediate capacity ranges that are normally encountered.
- Care must be exercised to keep entrainment at a minimum, since the vapor becomes superheated on compression and any liquid present will evaporate, leaving the dissolved solids behind.
- A mechanical recompression evaporator usually requires more energy than is available from the compressed vapor. Some of this additional energy can be obtained by preheating the feed with the condensate and, if possible, with the product.
- While theoretical compressor power requirements are reduced slightly by using lower evaporating temperatures, the volume of vapor to be compressed, and hence compressor size and cost increase so rapidly that low-temperature operation is more expensive than high-temperature operation.
- Steam jet thermocompression is advantageous when steam is available at a pressure appreciably higher than can be used in the evaporator. The steam jet then serves as a reducing valve while doing some useful work.
- Multiple steam jets are used to overcome effect of low efficiencies obtainable, when wide variations in evaporation rate are expected.
- Because of the low first cost and the ability to handle large volumes of vapor, steam jet thermo-compression evaporators are used to increase the economy of evaporators that must operate at low temperatures.

• What is a mechanical vapor recompression evaporator? Illustrate with a diagram.

- The basic principle of a mechanical vapor recompression evaporator involves removing the vapor produced from the evaporation process, compressing

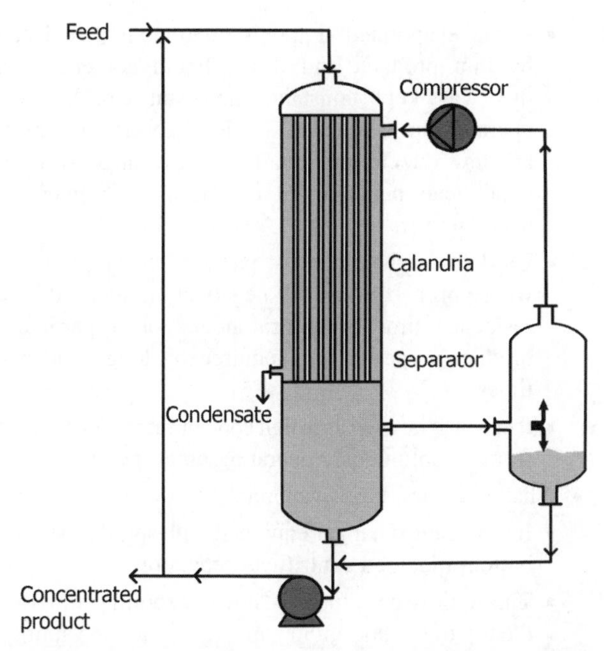

FIGURE 11.22 Mechanical vapor recompression evaporation system.

it in a mechanical compressor and using the resulting high-pressure vapor (steam) having higher saturation temperature, as heating medium for the evaporator as illustrated in Figure 11.22.

- The energy inputs are for the compression process which is very small compared to the advantage of utilizing the latent heat. Motor-compressor efficiency, heat and frictional losses for flow make its efficiency less than 100%.
- It is essential to maintain a constant suction pressure to the compressor to prevent it from surging. This can be done by operating the system using some makeup steam, which can be controlled through an automatic valve by the compressor suction pressure.
- Mechanical recompression can be applied to any evaporator design—natural or forced circulation or calandria type.
- A mechanical recompression unit normally does not require a condenser.
- A mechanical recompression evaporator evaporating water from a solution having a significant boiling point rise has to be designed for a higher compression ratio.
- Has good reliability and simpler control system than those for multiple effect evaporators.
- These systems are also used for distillation processes.
- Operating problems include handling moving parts and use of the compressor at elevated temperatures.

• What is *flash* or suppressed boiling evaporator?

- In a flash evaporator, boiling takes place inside the tubes, which is suppressed in the tubes due to hydrostatic head and superheated liquid is flashed into a separator operating at reduced pressure.
- Back pressure, either by a restriction in the piping or by elevation, prevents liquid from boiling as temperature rises.
- Liquid flashes into the separator.
- Plate exchangers can be designed as suppressed boiling evaporators.

• Where are flash evaporators used?

- Flash evaporators are used as multistage evaporators in desalination of seawater.

• What is a horizontal spray film evaporator? What are its advantages and disadvantages?

- This is essentially a horizontal, falling film evaporator in which the liquid is distributed by recirculation through a spray system. Sprayed liquid falls by gravity from tube to tube.
- *Advantages*:
  - Noncondensables are more easily vented.
  - Distribution is easily accomplished.
  - Precise leveling is not required.
  - Vapor separation is easily accomplished.
  - Reliable operation under scaling conditions.
  - Easily cleaned chemically.
- *Disadvantages*:
  - Limited operating viscosity range.
  - Crystals may adhere to tubes.
  - Cannot be used for sanitary construction.
  - More floor space is required.
  - More expensive for expensive alloy construction.
  - Limited application for once-through evaporation.
- Horizontal spray film units are suitable for multiple effects and for vapor compression.

• How does a plate evaporator operate? What are its particular features and fields of application?

- *Operation*:
  - Product and heating media are transferred in counter flow through their relevant passages.
  - Defined plate distances in conjunction with special plate shapes generate strong turbulence, resulting in optimum heat transfer.
  - Intensive heat transfer causes the product to boil while the vapor formed drives the residual liquid, as a rising film, into the vapor duct of the plate package.
  - Residual liquid and vapors are separated in the downstream centrifugal separator.

  - The wide inlet duct and the upward movement ensure optimum distribution over the total cross section of the heat exchanger.
- *Particular Features*:
  - Use of different heating media: Due to plate geometries, the system can be heated both with hot water as well as with steam.
  - High product quality: Due to uniform evaporation during single-pass operation.
  - Little space required: Due to compact design, short connecting lines, and small overall height of maximum 3–4 m.
  - Flexible evaporation rates: By adding or removing plates.
  - Ease of maintenance and cleaning: As plate packages can be easily opened.
- *Fields of Application*:
  - For low to medium evaporation rates.
  - For liquids containing only small amounts of undissolved solids and with no tendency to fouling.
  - For heat-sensitive products, for highly viscous products or extreme evaporation conditions, a product circulation design is chosen.

• What is a fluidized bed evaporator?
  - It consists of a vertical fluidized bed heat exchanger (on the tube side solid particles such as glass or ceramic beads, or stainless steel wire particles are entrained in the liquid). Other components are a flash vessel/separator and circulation pump.
  - The upward movement of the liquid entrains the solid particles, which provide a scouring/cleaning action. Together with the liquid they are transferred through the calandria tubes.
  - At the head of the calandria, the particles are separated from the liquid and are recycled to the calandria inlet chamber.
  - The superheated liquid is flashed to boiling temperature in the downstream separator and is partially evaporated.
  - Used for products that have high fouling tendencies, where fouling cannot be sufficiently prevented or retarded in standard, forced circulation evaporators and for liquids of low to medium, viscosity.

• What is a stirred tank evaporator? What are its applications?
  - It is a stirred vessel with an external jacket.
  - The liquid is supplied to the vessel in batches, is caused to boil while being continuously stirred and is evaporated to the required final concentration.
  - If the evaporated liquid is continuously replaced by thin product, and if the liquid content is in this way kept constant, the plant can be also operated in semibatch mode. Vaporization rates are low due to the small heat exchange surface, which can be enlarged by means of immersion heating coils.
  - Used for highly viscous, pasty or pulpy products, whose properties are not negatively influenced by a residence time of several hours, or if particular product properties are required by long residence times.
  - It can also be used as a high concentrator downstream from a continuously operating preevaporator.

• What is a spiral tube evaporator?
  - It is a heat exchanger equipped with spiral tubes and bottom mounted centrifugal separator.
  - The liquid to be evaporated flows as a boiling film from the top to the bottom in parallel flow to the vapor.
  - The expanding vapors produce a shear, or pushing effect on the liquid film.
  - The curvature of the path of flow induces turbulence considerably improving heat transfer, especially in the case of high viscosity liquids.
  - Used for high concentrations and viscosities, for example, for the concentration of gelatin.

• What is submerged combustion?
  - A fuel–air mixture is ignited in a cylindrical chamber, which is submerged in a liquid or slurry. Positive pressure evacuates the chamber of the liquid, thereby allowing adequate liquid-free volume for complete combustion before the hot exhaust gases contact the liquid.
  - When the gases reach the end of the chamber, they are divided into tiny bubbles that are in direct contact with the liquid. Initially there will be sensible heat transfer to the liquid and when the liquid gets saturated and further heat transfer involves latent heat, vaporizing the liquid.
  - Temperature approach between the gases and the liquid at the surface can reach to within about 5°C.

• What is a submerged combustion evaporator?
  - The submerged combustion evaporator makes use of hot flue from combustion of a fuel bubbling through the liquid as the heat transfer medium.
  - It consists of a tank to hold the liquid, a burner, and gas distributor that can be lowered into the liquid, and a combustion control system.
  - Since there are no heating surfaces on which scale can deposit, it is well suited for severely scaling liquids.

- The ease of constructing the tank and burner of special alloys or nonmetallic materials makes practical the handling of highly corrosive solutions.
- Since the vapor is mixed with large quantities of noncondensable gases, it is impossible to reuse the heat in this vapor, and installations are usually limited to areas of low fuel cost.
- It will involve high entrainment losses.
- These evaporators cannot be used in a crystallizing evaporator when control of crystal size is important. Submerged combustion evaporators generally are smaller than other types.
- Thermal efficiencies are high, ranging from 90% to 99% of the net heat input.
- Sometimes the heat in the vapor can be used as a preheater or preevaporator before the submerged combustion unit.
- Water vapor is produced in the combustion of fuel. To prevent its condensation into the liquid, the gas and liquor temperatures must exceed 60°C. Above that temperature the water derived from combustion is not sufficient to saturate the dry combustion gases, and water will be evaporated from the liquor to achieve saturation.
- Bubbling combustion gases through the liquid results in a boiling point depression. Thus under submerged combustion conditions, water boils at a temperature appreciably below its atmospheric boiling point.
- The submerged combustion boiling point depends upon the amount of noncondensable gases bubbled through the liquid.
- The depression in boiling point is not limited to water only but is also noticed with aqueous salt and acid solutions. The lowered boiling point can be advantageous in some applications and detrimental in others.

• What is the principal advantage and applications of submerged combustion evaporation systems?
  - Direct contact and so high heat transfer rates (no metal wall and scale resistances).
  - Useful for concentrating corrosive and scaling liquids, for example, acids, pickle liquors, etc.
  - It operates at near atmospheric pressure. No high pressures are involved, which gives better safety.
  - One application involves removal of ice from aircraft using ethylene or propylene glycol solutions and subsequent evaporation to concentrate the glycol solutions to above 50% for reuse.
  - In another application, submerged combustion is used to heat water to recover crude oil from tar sands. The water is reheated and recirculated to reduce environmental pollution.
  - Applications are limited to locations where fuel costs are low.
  - Not suitable when product contamination is detrimental to its quality.

11.3.1 Multiple Effect Evaporators

• How does a multiple effect evaporator work? What is its principal advantage?
  - In a multiple effect evaporator, steam from an outside source is condensed in the heating element of the first effect. If the feed to the effect is at a temperature near the boiling point in the first effect, 1 kg of steam will evaporate almost 1 kg of water. The first effect operates at a boiling temperature high enough so that the evaporated water can serve as the heating medium of the second effect. Here almost another kilogram of water is evaporated, and this may go to a condenser if the evaporator is a double effect or may be used as the heating medium of the third effect.
  - This method may be repeated for any number of effects. Large evaporators having six and seven effects are common in the pulp and paper industry, and evaporators having as many as 17 effects have been built.
  - Multiple effect evaporation is the principal means in use for economizing on energy consumption.
  - As a first approximation, the steam economy of a multiple effect evaporator will increase in proportion to the number of effects and usually will be somewhat less numerically than the number of effects.
  - The steam economy for an N-effect evaporator is nearly $0.8N$ kg evaporation/kg steam from mains used.
  - The total heat transfer surface will increase substantially in proportion to the number of effects in the evaporator.
  - For a double effect evaporator about half of the heat transfer is at a higher temperature level, where heat transfer coefficients are generally higher. On the other hand, operating at lower temperature differences reduces the heat transfer coefficient for many types of evaporator.
  - If the material has an appreciable boiling point elevation, this will also lower the available ΔT.
  - The increased steam economy of a multiple effect evaporator is at the expense of evaporator first costs.
  - For multiple effect evaporators, heat transfer relationships can be written, with two evaporators in

series,

$$\text{if } q_1 = q_2, U_1 A_1 \Delta T_1 = U_2 A_2 \Delta T_2. \quad (11.7)$$

$$\text{if } A_1 \text{ and } A_2 \text{ are equal}, U_2/U_1 = \Delta T_1/\Delta T_2. \quad (11.8)$$

(This can be extended to more than two effects.)

- From design, fabrication, and layout considerations, it would be simpler to use identical evaporators for the individual effects.

• What are the flow arrangements for the liquid being concentrated in a multiple effect evaporator?

- In *forward feed* operation, feed is introduced in the first effect and passed from effect to effect parallel to the steam flow. Product is withdrawn from the last effect.
  - ➢ This method is advantageous when the feed is hot or when the concentrated product is likely to be damaged or gives rise to scaling at high temperature. Steam temperatures are highest, giving higher ΔT in the first effect.
  - ➢ Liquid is transferred by pressure difference alone, thus eliminating the need for intermediate liquid pumps.
  - ➢ When the feed is cold, forward feed gives low steam economy an appreciable part of the input steam, which is needed to heat the feed to the boiling point, with no evaporation taking place.
  - ➢ If forward feed is necessary for a cold feed, steam economy can be improved by preheating the feed in stages with vapor from intermediate effects of the evaporator.
- Figure 11.23 illustrates a double effect evaporator with forward feed operation.
- In *backward feed* operation, the feed enters the last (coldest) effect, the discharge from this effect becomes the feed to the next-to-the-last effect, and so on until product is discharged from the first effect. A pump is necessary between each effect to move liquid from low pressure to high pressure operation as one moves from last effect toward first effect.
  - ➢ This method is advantageous when the feed is cold, since much less liquid must be heated to the higher temperature existing in the early effects.
  - ➢ It is also used when the product is viscous that high temperatures are needed to keep the viscosity low enough to give reasonable heat transfer coefficients.
  - ➢ When product viscosity is high but a hot product is *not* needed, the liquid from the first effect is sometimes flashed to a lower temperature in one or more stages and the flash vapor added to the vapor from one or more later effects of the evaporator.
  - ➢ Its main disadvantage is that the liquid is to be pumped between the effects as pressures progressively increase as the liquid moves from effect to effect from the last to the first one.
- Figure 11.24 illustrates backward feed operation for a double effect evaporator.

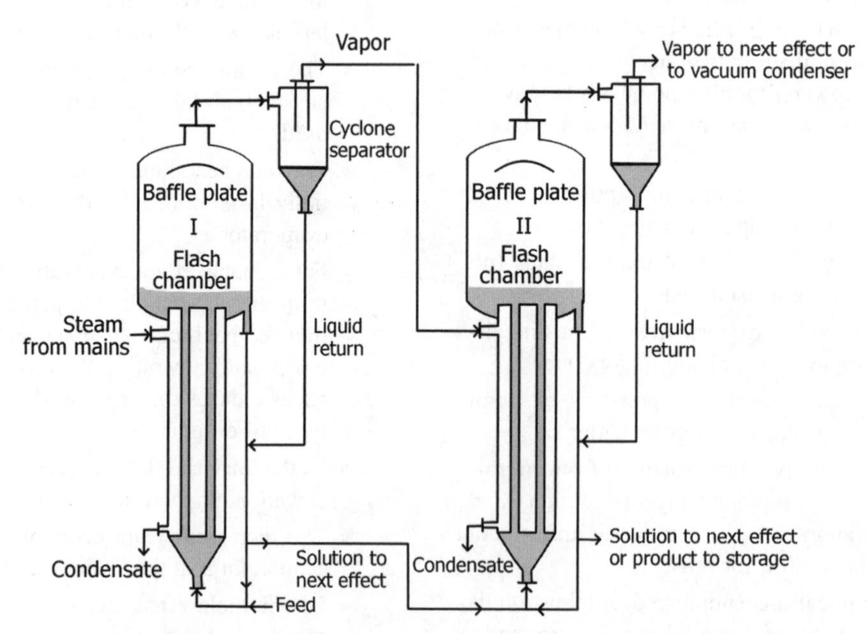

FIGURE 11.23 Double effect evaporator with forward feed operation.

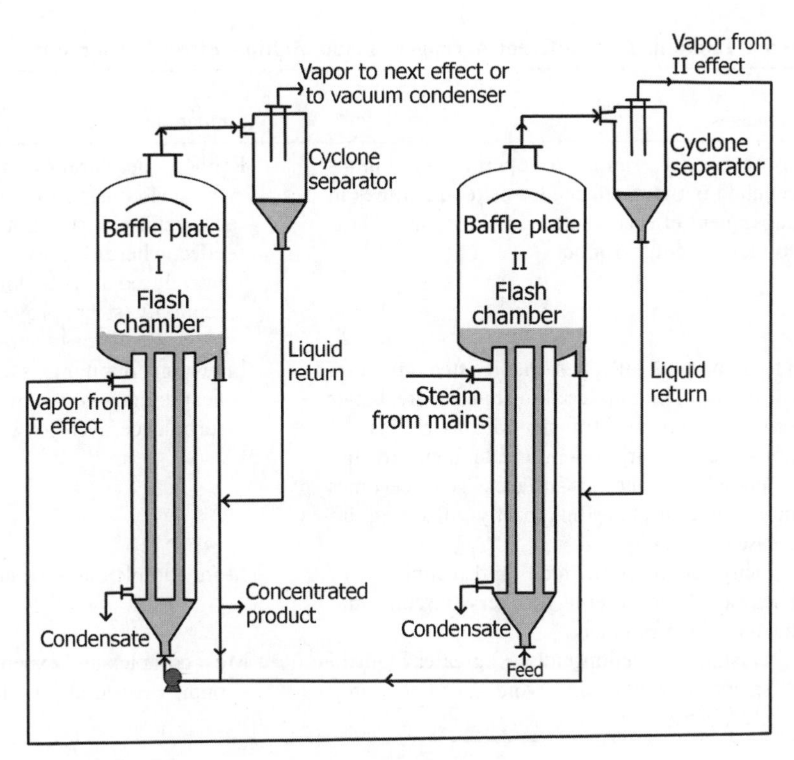

FIGURE 11.24 Backward feed operation for a double effect evaporator.

- *Mixed feed* operation is used only for special applications, for example, when liquid at an intermediate concentration and a certain temperature is desired for additional processing. In mixed feed operation, the feed enters at an intermediate effect, flows in forward feed through the later effects, and is then pumped back to the earlier effects for further concentration. Operation in the earlier effects can be either backward feed or forward feed. This eliminates some of the pumps needed in backward feed and permits final evaporation at the highest temperature. *Parallel feed* involves the introduction of the feed and withdrawal of product at *each effect* of the evaporator. There is no transfer of liquid from effect to effect.
  - It is used when the feed is nearly saturated and the product is a solid. An example is the evaporation of brine to make common salt. The product is withdrawn as a slurry. In such a case, parallel feed is desirable because the feed washes impurities from the salt leaving the body.
  - It is also used for crystallizing solutions.
  - An alternative for concentration of such solutions is to use the normal forward/backward feed systems up to certain concentration and then using separate evaporator involving crystallization.
- "Pressure in the last effect in a multiple effect evaporator is limited by cooling water temperature in the condenser." *Tue/False*?
  - *True*.
- "A steam jet may be used to remove noncondensables from an evaporator." *True/False*?
  - *True*.
- "A typical steam economy for a four effect evaporator without vapor recompression is 4.3." *True/False*?
  - *False*.
- Summarize the advantages and limitations of different types of feed arrangements for multiple effect evaporators.
  - Table 11.5 gives the advantages and limitations of different arrangements for multiple effect evaporators.
- "The overall temperature difference in a multiple effect evaporator will increase if the pressure in the last effect increases, while steam pressure in the first effect remains unchanged." *True/False*?
  - *False*.
- How is vapor recompression principle applied to multiple effect evaporation? What are its advantages and disadvantages?
  - The vapor leaving an *effect* of the multiple effect evaporation system is compressed to raise its saturation temperature and admitted as heating medium in the *next effect*. This will increase ΔT between the effect into which the compressed vapor is admitted and the boiling liquid in that effect, increasing heat transfer rates.

TABLE 11.5 Advantages and Limitations of Different Arrangements in Multiple Effect Evaporators

Arrangement of Effects	Advantages	Limitations
Forward feed	Least expensive, simple to operate, no pumps required between effects, lower temperatures in subsequent effects resulting in less risk of heat to more viscous product	Reduced heat transfer rate as solution becomes more viscous, rate of evaporation decreases with each effect, best quality steam is used in the first effect where evaporation is easiest. Feed must be introduced at its boiling point to prevent loss of economy (steam provides sensible heat, reducing available vapor to subsequent effects)
Backward feed	No feed pump initially, best quality steam used on the most difficult material to concentrate, better economy and heat transfer rate as effects are not subject to variation in feed temperature and solution meets hotter surfaces as it becomes more concentrated thus partly offsetting increase in viscosity	Interstage pumping is necessary, higher risk of heat damage to viscous liquids due to decreased turbulence over hot surfaces, risk of fouling
Mixed feed	Simplicity of forward feed and economy of backward feed, useful for very viscous and heat-sensitive products	More complex and expensive
Parallel feed	For crystal production, allows greater control over crystallization and avoids need to pump crystal slurries	Most complex and expensive of the arrangements, pumps required for each effect

  - Its disadvantages are that (i) the compressor is to operate at high temperatures with attending maintenance problems and (ii) any entrained liquid in the vapor can have damaging effects on the compressor as well as introduce corrosion problems to the shell and tubes of the evaporator(s) into which such vapors are admitted.
- "High pressure steam is used in thermal vapor-recompression (TVR) to recompress part of the exit vapors." *True/False*?
  - *True.*
- "Evaporators with thermal vapor recompression have a better steam economy than evaporators with mechanical vapor recompression." *True/False*?
  - *False.*
- What is staging in an evaporation system?
  - Often in multiple effect evaporators the concentration of the liquid being evaporated changes drastically from effect to effect, especially in the latter effects. In such cases, one or two *stages* can be used in later effects.
  - Staging is the operation of an effect by maintaining two or more sections in which liquids at different concentrations are all being evaporated at the same pressure.
  - The liquid from one stage is fed to the next stage. The heating medium is the same for all stages in a single effect, usually the vapor from the previous effect.
  - Staging can substantially reduce the cost of an evaporator system. The cost is reduced because the wide steps in concentration from effect to effect permit the stages to operate at intermediate concentrations which result in both better heat transfer rates and higher temperature differences. Staging can reduce the total heat transfer surface by as much as 25% when high boiling point rises occur.

11.3.2 Evaporator Performance

- Define (i) capacity and (ii) economy of an evaporator.
  - Capacity is quantity evaporated per unit time.
  - Economy is quantity vaporized per unit quantity of steam consumed.
- "The single largest cost in an evaporation system is energy cost." *True/False*?
  - *True.*
- What are the ways to increase energy economy in an evaporation system?
  - Energy efficiency of an evaporator can be improved by using multiple effects, heat pumps, by heat exchange and by using energy efficient condensers.
- Why vacuum is normally used in evaporators?
  - Heat-sensitive liquids require low boiling temperatures to prevent decomposition.
  - Operation under vacuum increases ΔT between steam and boiling liquid, thereby increasing heat flux.

- If the material is not heat-sensitive and high ΔT is provided by increasing temperature of heating medium, there is no advantage in maintaining vacuum. How is vacuum produced in evaporator systems?
  - By the use of ejectors/ejector–barometric condenser combinations.
- What is the disadvantage of using vacuum compared to pressure in an evaporation system?
  - Vacuum lowers boiling point thereby increasing viscosity.
  - Increased viscosities lower heat transfer coefficients.
  - Additional energy is required to produce and maintain vacuum.
  - Pressure operation, compared to vacuum operation, involves high boiling heat transfer rates, mainly because of lowering viscosities. Problems and costs associated with vacuum producing equipment can be eliminated.
  - Mechanical compression systems generally are more economical if operating pressures are above atmospheric. Operation at higher pressures (higher temperatures) in many cases is acceptable if adequate attention is given to residence time.
- What are the causes and effects of insufficient vacuum in a multiple effect evaporator system?
  - *Causes*:
    - Excessive air leakage into the system.
    - Inadequate condensation of the final effect vapor.
    - Any vapor that is not condensed must be removed by the vacuum system.
    - Condenser performance is limited by (i) fouling, (ii) low cooling water flow, and (iii) cooling water is warmer than the design temperature.
    - Condenser should be designed to provide subcooling of noncondensable steam to lower vapor pressure and amount of water vapor sent to vacuum system.
    - If temperature of line for noncondensables increases, this might indicate inadequate condensation.
  - *Effects*:
    - Boiling point in the last effect is set by the amount of vacuum that can be attained. If vacuum is low, boiling point in the last effect is high and total ΔT, that is, first effect condensing temperature versus final effect boiling temperature, is limited. This, in turn limits the amount of heat that can be transferred and evaporation capacity.
- What are the essential steps involved in multiple effect evaporator calculations?
  - In evaporator calculations, three relations must be satisfied: *material balance*, *required heat transfer rates*, and *heat balance*.
  - These relations must be applied to each effect in a multiple effect system as well as to the total system.
  - The equations describing evaporator systems can be solved algebraically but the process is tedious and time consuming. Trial and error procedures are generally used with the following steps:
    1. Assume values for boiling temperatures in all effects.
    2. From enthalpy balances determine rates of steam and liquid flows from effect to effect.
    3. Calculate rates of heat transfer and required heat transfer surface using the previously established temperature distribution.
    4. If the heating surface distribution is not as desired, establish a new temperature distribution and repeat steps 2 and 3 until desired results are achieved.
- What are the causes of poor heat transfer in an evaporator?
  - Fouling of heat transfer surfaces.
  - Poor venting and consequent accumulation of noncondensables.
  - Poor product distribution involving localized over concentration.
  - Dry spots on heat transfer surface causes increased fouling.
  - Condensing/boiling ΔPs or product BPR is higher than expected.
- What is boiling point rise and how it influences evaporator performance?
  - Pure water boils at a temperature corresponding to its pressure, but the presence of dissolved solids depresses vapor pressure in proportion to the solids concentration. As the liquid is concentrated in an evaporator, the temperature will rise even if the pressure is maintained at a constant level. This temperature rise is called *boiling point rise* (BPR) or *boiling point elevation* (BPE).
  - As evaporation proceeds, the liquor remaining in the evaporator becomes more concentrated and its boiling point will rise. The extent of the boiling point rise depends upon the nature of the material being evaporated and upon the concentration changes that are produced.
  - As the concentrations rise, the viscosity of the liquor also rises. The increase in the viscosity of the liquor affects the heat transfer and it often

imposes a limit on the extent of evaporation that is practicable.

- There is no straight forward method of predicting the extent of the boiling point rise in the concentrated solutions that are met in some evaporators in practical situations.
- In multiple effect evaporators, where the effects are fed in series, the boiling points will rise from one effect to the next as the concentrations rise.
- Effect of boiling point rise is to reduce ΔT between the steam and the liquid being concentrated. Temperature of the vapor rising from the evaporator and passing into the steam chest of the next effect will be saturation temperature at the prevailing pressure. Difference between this temperature and the solution boiling temperature will thus reduce, affecting heat transfer rates.

• What is Duhring rule?

- If a plot of BP of solution versus BP of water gives a straight line, it is said to be following Duhring rule.
- Duhring rule states that the ratio of the temperatures at which two solutions (one of which can be pure water) exert the same vapor pressure is constant.
- Thus, if one takes the vapor pressure–temperature relation of a reference liquid (usually water) and if one knows two points on the vapor pressure–temperature curve of the solution that is being evaporated, the boiling points of the solution to be evaporated at various pressures can be read off from the *Duhring plot*.

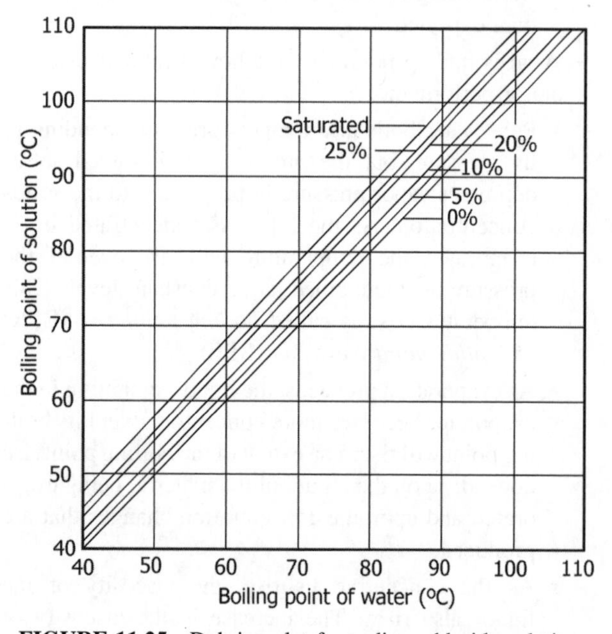

FIGURE 11.25 Duhring plot for sodium chloride solutions.

- Duhring plot gives the boiling point of solutions of different concentrations at different pressures, through interpolation along a line of constant composition (Figure 11.25).

• Give an equation for the estimation of boiling point rise and state its applicability.

$$\text{BPR} = 104.9\,N_2^{1.14} \tag{11.9}$$

where N_2 is the mole fraction of salts in solution. Correction to other pressures, when heats of solution are small, can be based on a constant ratio of vapor pressure of the solution to that of water at the same temperature.

- This correlation is claimed to be applicable for a number of inorganic salt solutions at high concentrations at atmospheric pressure boiling point.

• What is the effect of boiling point rise (BPR) or boiling point elevation (BPE) on the performance, evaporation rate/capacity of an evaporator?

- Boiling point elevation (BPE) reduces ΔT between the heating medium and the liquid being evaporated, decreasing evaporation rates and hence capacity of the evaporator.
- Boiling point elevation (BPE) as a result of having dissolved solids must be accounted for in the differences between the solution temperature and the temperature of the saturated vapor.
- Differences between solution and saturated vapor temperatures are generally between 1.5 and 6°C.

• Is there any influence of BPR on the economic number of effects in an evaporation system?

- Yes.
- Rules of thumb are as follows:
  - ➢ *Appreciable BPR (greater than 4°C)*: Number of effects in series (forward feed) are limited to 4–6.
  - ➢ *Small BPR*: More effects in series are typically more economical, depending on the cost of steam. Number of effects can be 8–10.

• What is the effect of hydrostatic head on evaporation rates in an evaporator?

- Hydrostatic head suppresses evaporation rates. The effect is equivalent to increase of boiling temperature with pressure. This will reduce ΔT for heat transfer.
- Boiling point elevation as a result of the hydrostatic head results in a temperature profile along the tube. The temperature of the liquid is increased in the sensible zone until it reaches the boiling point. From

that point on the temperature decreases as the boiling pressure decreases progressively up the tube.

- Lowering the liquid level in the evaporator reduces the boiling point elevation. However, it also reduces the liquid circulation rate there is less available pressure difference to drive the system.
- Generally, lowering the liquid level will result in less of the tube being required for the sensible zone and more of the tube available for the boiling zone. This results not only in higher heat transfer coefficients but also a higher driving force.
- Consequently, heat transfer is increased as liquid level is dropped to the optimum point. At the optimum point, the liquid entering the exchanger is entirely vaporized so that saturated vapor leaves the tube.
- A further drop in liquid level will result in superheating of the vapor, which is a very poor heat transfer mechanism.
- Therefore, at liquid levels below the optimum, heat transfer of the system is drastically reduced.
- Normally, evaporator exchangers are not designed to operate at the optimum liquid level. The design point for liquid level is usually at the top tube sheet because this gives the highest circulation rates.
- High circulation rates reduce chances of fouling.
- For systems where fouling does not occur, the optimum liquid level is utilized in the design.

• What are the sources of noncondensables in the steam used in multiple effect evaporators?

- Air leakage if preceding effect is under vacuum.
- Air entrained/dissolved in the feed.
- Gases liberated from decomposition reactions in the preceding evaporator.

• What is the effect of ΔP due to friction for flow of vapor through entrainment separators and piping on ΔT in a multiple effect evaporator?

- ΔT is reduced. Pressures of steam entering steam chest of next effect is reduced causing lower condensing temperatures with the consequence of reduced ΔT for the effect.

• What are the different types of liquid characteristics that require consideration in an evaporation process?

- Liquids not affected by high temperatures and heat-sensitive liquids.
- Evaporation accompanied by crystallization.
- Liquids with boiling temperatures same as those for water and liquids with BPR.

• "Effect of surface tension is generally to suppress boiling/vaporization." *True/False*?

- *True*. Surface tension holds vapor molecules escaping from the liquid.

• What is the main advantage of multiple effect evaporation system?

- Steam economy.
- The steam economy of an N-stage battery is approximately $0.8N$ kg evaporation/kg of outside steam.

• "Multiple effect evaporation is employed in dewatering waste sludges in sewage treatment plants." *True/False*?

- *True*. This method is finding increasing application in recent years due to environmental concerns in open drying.

• "The viscosity of a solution increases with increasing concentration of dissolved solids and decreasing temperature." Explain from a viscosity point of view when a forward feed arrangement is beneficial and when backward feed arrangement is beneficial.

- Viscosity of a solution generally increases with increase in concentration.
- In the forward feed arrangement, viscosity is lowest in the first effect and highest in the last effect. In the last effect, the temperature of the heating medium is the lowest as pressure decreases in progressive effects. Therefore, heat transfer rates will decrease progressively from the first to the last effect. Also BPR will be highest with the most concentrated solution, further reducing ΔT and hence heat transfer rates in the last effect.
- Therefore, where viscosity increases are considerable, backward feed arrangements are advantageous as the most concentrated solution with high BPR and high viscosity is boiled in the first effect where ΔT will be highest.
- It must be realized that backward feed arrangement involves use of pumps, which have to operate with high-temperature boiling liquids, requiring larger NPSH and maintenance problems.

• For what type of multiple effect evaporation system, parallel feed is preferred?

- For salt solutions/crystallizing systems.
- Alternatively, use of normal forward/backward feed systems up to certain concentration and then use of separate evaporator.

• "Parallel feed is used in the evaporation of saturated solutions such as brine to make salt." *True/False*?

- *True*.

• Under what circumstances backward feed is used in evaporators? Why?

- Cold feeds.
- High viscosity products.

  - High viscosity product leaves the first effect which is at the highest temperature, thereby reducing viscosity of product and increasing heat transfer coefficient.
- Explain why backward feed is advantageous for an evaporator system when the final product is highly viscous.
  - Reverse (backward) feed results in the more concentrated solution being heated with the hottest steam to minimize surface area.
- What is the disadvantage of backward feed?
  - The solution must be pumped from one stage to the next. The pumps are required to handle liquids at elevated temperatures with attending NPSH and maintenance problems.
- How does liquid hydrostatic head influence capacity of a multiple effect evaporator?
  - Liquid hydrostatic head increases boiling point of the liquid thereby reducing available ΔT in the evaporator.
- How would one increase interstage steam pressures in a multiple effect evaporation system?
  - Interstage steam pressures can be increased with ejectors (20–30% efficient) or mechanical compressors (70–75% efficient).
- Differentiate between mechanical and thermal vapor recompression evaporation systems.
  - In mechanical vapor recompression, a compressor is used to increase pressure of the vapor whereas in the thermal vapor recompression, steam jet ejectors are used for compressing vapors. The former involves more maintenance problems, as the mechanical compressors are to operate at elevated temperatures.
- What are the controlling factors for lowering heat transfer rates in evaporators?
  - Liquid circulation, that is, whether natural or forced circulation, and in case of natural circulation, whether long tube or short tube vertical evaporators are employed.
  - Whether the liquid evaporated is in bulk form or in film form and any agitation is employed.
  - Thermal properties of the liquids.
  - Liquid viscosities.
  - Desired concentration of the final product.
  - Boiling point rise of the solutions.
  - Hydrostatic head.
  - Vacuum available.
  - Pressure drop.
- What is chugging instability in evaporator tubes?
  - Chugging instability occurs when the liquid is heated well above its saturation temperature. When vapor formation eventually occurs, it does so very rapidly and may expel liquid in both directions.
  - The vapor formed can thus enter a region of subcooled liquid where it can collapse violently.
  - This situation can normally occur only at rather low heat fluxes.
- Give thumb rules for sizing nozzles in an evaporator.
  - Area of the vapor outlet nozzle should be approximately equal to the flow area of the tubes.
  - Diameter of the liquid inlet nozzle should be one-half that of the vapor outlet nozzle.
- What are the considerations involved in sizing downcomers in an evaporator?
  - Downcomers must be sized to be self-venting and to minimize liquid holdup on the top tube sheet.
  - In order to avoid entrainment of vapor in the downcomer, the superficial liquid velocity should not exceed 12 cm/s (0.4 ft/s), based on the expected liquid circulation rate.
  - Several downcomers are usually preferred in order to reduce the flow path liquid must take in order to reach the downcomer.
- How are baffles (deflectors) above tube sheet of an evaporator sized?
  - The baffle (deflector) diameter should be 0.15–0.3 m (6–12 in.) greater than the diameter of the heating element.
  - Height of the baffle above the tube sheet should be adequate to minimize pressure drop.
  - Flow velocities around the baffle should be approximately *half* that leaving the tubes.
  - Baffles must be adequately supported to avoid mechanical problems.

11.3.3 Auxiliary Equipment

- Give diagrams of typical entrainment separators used in an evaporator?
  - Figure 11.26 illustrates some commonly used entrainment separators in evaporation equipment.
- What is a barometric condenser? How does it work?
  - Barometric condenser is a direct contact type condenser in which vapor comes in contact with a spray of cooling water, most commonly, in countercurrent operation.
  - The condenser is located at a high point to facilitate discharge of water by gravity from the *vacuum* in the condenser.

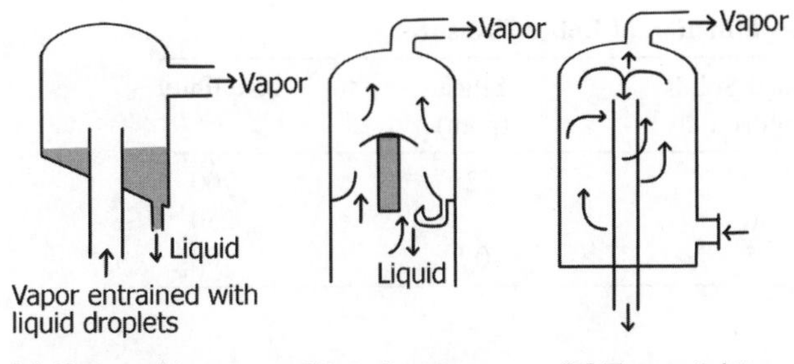

FIGURE 11.26 Different types of entrainment separators.

- Such condensers are inexpensive and are economical on water consumption.
- Barometric condenser works on the principle of a barometer with the barometric leg made up of a column of water. Figure 11.27 illustrates its working.
  - In a barometric leg water theoretically stands to a height of 10.33 m at sea level. Above this height in a sealed and evacuated space there will be absolute vacuum but in practice water exerts its vapor pressure and reduces the height of the barometric leg theoretically obtainable.
  - In a barometric condenser, water is at a much higher temperature and its vapor pressure is considerably higher. Moreover there will be flow, unlike in a static leg and also the condenser need

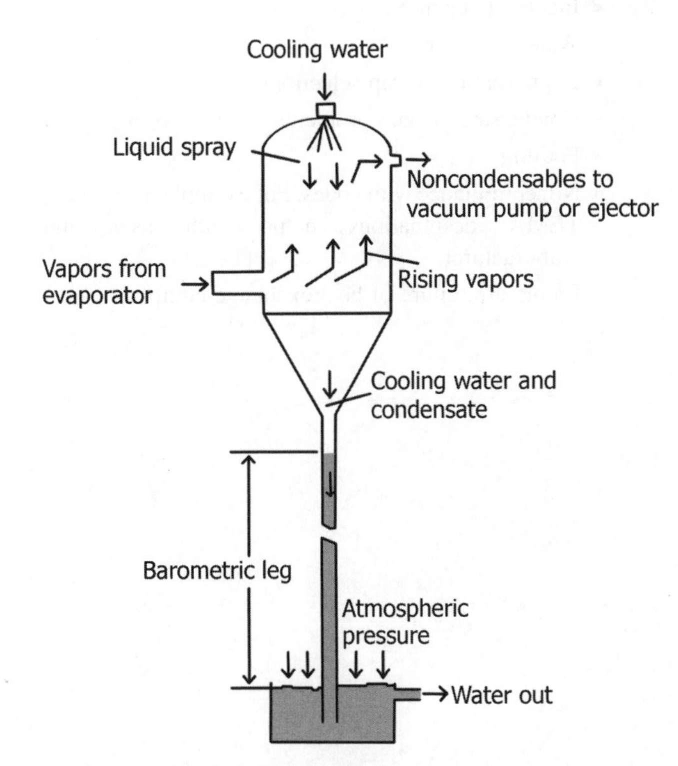

FIGURE 11.27 Barometric condenser.

not be working at sea level. Under these circumstances, height of water leg obtainable is much less and will be about 7–8 m instead of 10.33 m.

- Another type of direct contact condenser is the *jet condenser*, which makes use of high velocity jets of water both to condense the vapor and to force noncondensable gases out. This type of condenser is frequently placed below barometric height and requires a pump to remove the mixture of water and gases. Jet condensers usually require more water than the more common barometric type condensers.

- What are the variables that are to be considered in selecting an evaporation system, which gives the best balance between the capital and operating costs? Give guidelines.
  - The most important variables are evaporator type, number of effects in a multiple effect system, and the type of compression system.
  - Other variables include the following:
    - Initial steam pressure and its cost and/or availability.
    - Final pressure and its relation to cooling water temperature and wet and dry bulb air temperatures.
    - Final pressure and its relation to product quality.
    - Cost and availability of electrical power.
    - Materials of construction.
    - Maintenance and reliability.
    - Should more than one type of evaporator be used in the system.
    - Method of feeding and heat recovery.
    - Entrainment and separator type.
    - Evaporator accessory equipment.
    - Controls required.
  - Some guidelines are as follows:
    - The optimum number of effects increases as the steam costs increase or plants become larger.
    - Mechanical vapor compression looks attractive if electrical power is available at reasonable costs and boiling point rises are low.
    - Thermal compression looks attractive when electrical power is costly and boiling point rises are low.
    - When boiling point rises are large, combinations of vapor compression preconcentrators followed by multiple effect systems look attractive.
    - Larger plants enable heat recovery systems to be more easily justified.
    - Mechanical vacuum pumps are generally more energy efficient than steam jets.

TABLE 11.6 Boiler Feed Water Purity Requirements as Function of Boiler Pressure

Boiler Pressure (kPa gauge)	Total Solids (ppm)	Suspended Solids (ppm)	Silica (ppm)	Alkalinity
6.89–2,000	3,500	300	125	700
6,000–6,900	1,250	40	8	250
>14,000	500	5	0.5	100

- ➢ Frequently, it is necessary to consider other factors besides the evaporator design. Steam generation, electrical power, and cooling water distribution and supply costs may be sufficiently reduced to justify particular system designs.
- ➢ Sometimes evaporator efficiency can be sacrificed in order to utilize all of the heat available in a waste heat stream. Final operating pressure may be determined by the ability to recover the heat of condensation in some other part of the plant.

11.3.4 Utilities

- Name the common utilities made available in a process plant.
  - Steam: Boilers.
  - Cooling water: Cooling towers.
  - General water supply for process and sanitary requirements.
  - Electricity.
  - Instrument air.
  - Plant air.
  - Refrigeration system.
  - Fuel: Gas, liquid, and solid.
  - Flare system.
- What are the ranges of steam pressures in central power stations and process plants?
  - Central power stations: 12,000–24,000 kPa gauge (1,800–3,600 psig).
  - Process plants: 6,000–10,000 kPa gauge (900–1,500 psig).
- What are the positive and negative points in employing higher pressures in steam-generating units?
  - Energy savings due to higher efficiencies involved.
  - Lower operating costs.
  - Higher capital costs.
  - Higher capital and operating costs involved in water treatment.
  - Feed water treatment must produce purer water with increase in pressures.
  - For package boilers tighter limits are required.
  - Maximum limits for boiler water as given in Table 11.6.
- What are the common heat transfer issues involved when steam is used as heating medium?
  - Incorrect steam pressures.
  - Water hammer.
  - Improper steam trap selections.
  - Condensate backup problems.
  - Fouling.
  - Noncompliance with codes. For example, specifying TEMA designations helps both user and manufacturer.
  - Premature failure of heat exchange equipment.

12

REFRIGERATION, HEAT PUMPS, HEAT TRACING, COILED AND JACKETED VESSELS, STEAM TRAPS, AND IMMERSION HEATERS

12.1 Refrigeration 371
12.2 Heat Pumps 374
12.3 Heat Tracing 376
12.4 Coiled Vessels 378
12.5 Jacketed Vessels 379
12.6 Steam Traps 384
12.7 Immersion Heaters 393

12.1 REFRIGERATION

- What is a ton of refrigeration?
  - Heat equivalent to melting 1 ton (2000 lb or 907 kg) of ice in 24 h.
  - It is equal to 12,700 kJ/h (12,000 Btu/h).
- What is coefficient of performance (COP) of a heat pump?

$$\text{COP} = \text{ratio of energy delivered at higher temperature to energy input to compressor.} \quad (12.1)$$

  - It is the reciprocal of efficiency.
- "The higher the temperature difference between condenser and evaporator, the higher the COP." *True/False*?
  - *False.*
- What are the three most used refrigeration systems? Give the names of refrigerants used and the temperature ranges involved in these systems.
  - Table 12.1 lists refrigeration systems and refrigerants used.
- What are the temperature ranges involved for the use of (i) chilled brine and glycol solutions, (ii) ammonia, Freon, and butane, and (iii) ethane or propane as refrigerants?
  - Chilled brine and glycol solutions: −18 to −10°C.
  - Ammonia, Freon, and butane: −45 to −10°C.
  - Ethane or propane: −100 to −45°C.
- Why halocarbons as refrigerants are being phased out?
  - Halocarbons (CFCs) affect the ozone layer of the atmosphere causing an increase in cosmic radiation reaching the Earth.
- "Hydrofluorocarbons (HFCs) are replacing chlorofluorocarbons (CFCs) as refrigerants." *True/False*?
  - *True.*
- What is the principle of operation of vapor compression refrigeration? Illustrate.
  - Refrigerant vapor is compressed, increasing its temperature and pressure (points 1 and 2 in Figure 12.1), condensed with water or air (points 2 and 3), rejecting heat to the surrounding environment, and then expanded to a low pressure and correspondingly to low temperature through an expansion valve or an engine with power takeoff (points 3 and 4).
  - Because of the reduction in temperature and pressure, the refrigerant enters the evaporator as a saturated mixture. As the refrigerant passes through the evaporator, it absorbs heat energy from the environment that it is trying to cool.
  - The refrigerant exits the evaporator as a saturated vapor and returns to the compressor to begin the process all over again (points 4 and 1).
  - Due to fluid friction, heat transfer losses, and component inefficiency, the refrigeration cycle is unable to achieve complete thermodynamic saturation.
  - It operates on the principle of reverse Carnot cycle.

Fluid Mechanics, Heat Transfer, and Mass Transfer: Chemical Engineering Practice, By K. S. N. Raju

TABLE 12.1 Different Types of Refrigeration Systems and Refrigerants Used

System	Temperature Range (°C)	Refrigerant
Steam jet	2–20	Water
Absorption		
Water–lithium bromide	4–20	Aqueous lithium bromide solution
Ammonia	−40 to −1	Aqueous ammonia
Mechanical compression (reciprocating, centrifugal, rotary screw)	−130 to +4	Ammonia, halocarbons (phased out due to environmental concerns), propane, ethylene, propylene, and so on
Cryogenics	−850 to −130	Liquefied gases

- What are the different types of compressors used in refrigeration systems? Comment on their use in refrigeration systems.
  - *Centrifugal Compressors*: Three- or four-stage compressors are normally employed in refrigeration systems. The main drawback of centrifugal compressors is surging problems that necessitate use of recirculation that involves wasteful use of energy.
  - *Reciprocating Compressors*: Process temperatures generally dictate two-stage compression in reciprocating compressors. The first-stage cylinder is normally quite large as a result of the low suction pressure. As with centrifugal compressors, recirculation does result in wasted power. It is also possible to throttle the refrigerant suction pressure between the evaporator and the compressor in order to reduce cylinder capacity. However, suction pressure control can result in wasted horsepower and the possibility of below atmospheric suction pressure, which should be avoided.
  - *Screw Compressors*: Screws can operate over a wide range of suction and discharge pressures without system modifications. They have essentially no compression ratio limitations with ratios up to 10 being used, but operate more efficiently in the 2–7 ratio and are comparable in efficiency to reciprocating compressors within this range.
  - *Rotary Compressors*: There is a limited application of large rotary compressors. A rotary compressor serves the purpose of a high-volume booster compressor.
- "Compression of the refrigerant in the compressor takes place at constant enthalpy." *True/False*?
  - *False*.
- What are the commonly used evaporators (chillers) in refrigeration systems?
  - Kettle reboilers (most common) and plate-fin exchangers. Plate-fin exchangers offer significant savings for low-temperature applications where stainless steel is needed for shell and tube units. Significant pressure drop savings can be realized by using single or multiple units for chilling services.
- "Evaporation of the refrigerant in the evaporator takes place at constant pressure." *True/False*?
  - *True*.
- What is cascade refrigeration? Illustrate.
  - Figure 12.2 illustrates a cascade refrigeration cycle.
  - The cascade refrigeration system incorporates two or more refrigeration cycles in series (Figure 12.2) to acquire low temperatures, which cannot be achieved with a single refrigeration cycle.
  - Refrigerant enters the compressor as a saturated vapor and gets compressed, with increase in temperature and pressure (points 1 and 2).
  - From the compressor the refrigerant passes through the condenser, exchanging heat in the heat exchanger, which causes the refrigerant to cool and become a saturated liquid (points 2 and 3).

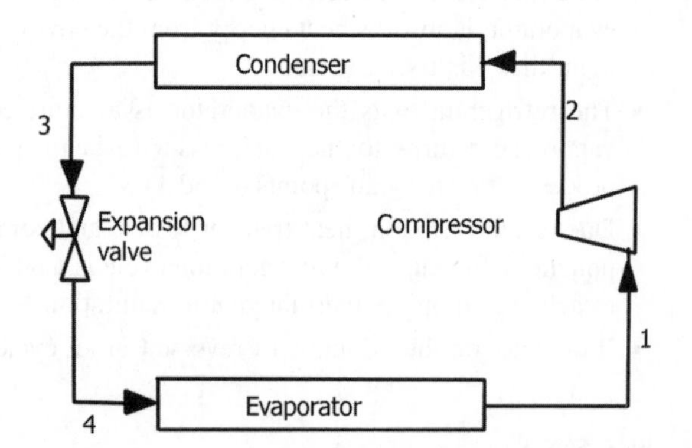

FIGURE 12.1 Typical vapor compression refrigeration cycle.

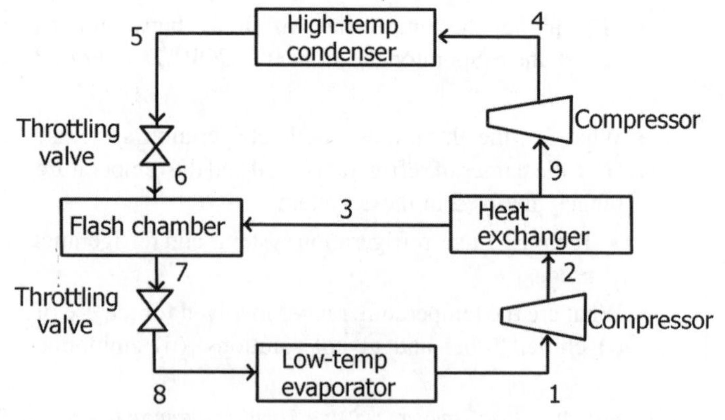

FIGURE 12.2 Cascade refrigeration cycle.

- The heat energy rejected by the first condenser is absorbed by the second system evaporator located in the heat exchanger (points 2 and 9), with the refrigerant passing through the second compressor, which increases the temperature of the refrigerant further (points 9 and 4).
- The refrigerant then passes through the second condenser giving off heat energy, which causes it to cool (points 4 and 5).
- It is then passed through the second throttling valve, causing expansion of the refrigerant with decrease in temperature (points 5 and 6). Both refrigerants from the first and second stages mix in the flash chamber creating a liquid–vapor mixture (points 6 and 7), which then passes through another throttling valve reducing its temperature still further (points 7 and 6).
- Finally, the refrigerant passes through the evaporator absorbing heat energy from the environment that it is trying to cool and exits the evaporator as a saturated vapor and returns to the compressor to begin the process all over again (points 8 and 1).
- Cascade refrigeration operates the same way as a regular refrigeration unit, except for the second stage of operation.
- It is used for industrial applications.

- When is cascade refrigeration involving two or three refrigerants used?
  - For below −62°C applications.
- What is the working principle in absorption refrigeration?
  - In absorption refrigeration, condensation is effected by absorption of vapor in a liquid at high pressure, and then cooling by expanding to a low pressure at which the solution becomes cold and flashed.
- How does absorption refrigeration, involving aqueous ammonia, work?
  - The principle of operation depends on the capacity of water to absorb large quantities of ammonia vapor, which can be released from the solution by the direct heat of the system, which thereby gets cooled.
  - The NH_3 vapors are drawn into the absorber by the affinity of NH_3 for aqueous ammonia solutions.
  - The heat of absorption plus latent heat released by condensation of NH_3 vapors is removed by cooling water circulating in tubes.
- "Tubing in ammonia refrigeration units is made of copper." True/False?
  - *False*. Ammonia attacks copper.
- "A disadvantage of ammonia as a refrigerant is its low latent heat of vaporization." *True/False*?
  - *False*.
- Explain the working of lithium bromide absorption refrigeration.
  - Capability of lithium bromide to absorb water vapor is used to evaporate and cool water in the system.
  - Process water for chilling is circulated from the process application where it is warmed from its normal low of 4.4°C to a maximum of 5.5–33°C.
  - Then the process water is circulated back to the refrigeration system evaporator coils for cooling back to system working temperatures of 4.4–7°C.
  - Similar in principle as an ammonia absorption refrigeration system.
- What is the principle involved in steam jet refrigeration?
  - The refrigerant, water, is evaporated at low pressures created by steam jets. Temperatures of about 13°C are normally attained.
  - Heat for evaporating water is taken from the system, thereby cooling it.
  - A barometric condenser is used to condense the steam.
  - The system is used to cool water.
    - Warm water enters the active compartment(s) and flashes due to reduced pressure maintained by steam jets.
    - The water boils (vaporizes) at this low pressure and the vapor is drawn into the booster jet (ejector) where it is compressed for condensation in the barometric condenser.
    - The cooled (chilled) water is pumped from the compartment with a pump capable of handling boiling water at this reduced pressure.
- What are the advantages of steam jet refrigeration?
  - No moving parts, except water pumps.
  - Water (nonhazardous) is the refrigerant involving low refrigerant costs.
  - System pressures are low.
  - System can be installed outdoors.
  - Steam condensate can be recovered in surface condensers.
  - Refrigeration tonnage can be varied with respect to amount and temperature level.
  - Simple for start-up and operation.
  - Barometric units can use dirty/wastewater.
  - Cost per ton of refrigeration is low.
- How is thermal efficiency related to ejector steam pressure?
  - Efficiency increases with increase in steam pressure. At 103 kPa: 30%.

At 690 kPa: 60%.

At 2756 kPa: 80%.

- Higher vacuum is produced at higher steam pressures.

- What is the advantage of using a number of stages in a steam jet water chiller? How many stages are normally employed in them?
  - A number of stages are employed to chill water to improve energy efficiency of the chiller.
  - Two to four chilling stages are normally used.
  - A typical steam jet ejector water chiller uses several *booster* ejectors, one or more per chilling stage.
    - A booster is an ejector that operates before a condenser. The booster suction pulls a vacuum, evaporating water and entraining dissolved noncondensables and air from leakages, and discharging them into the condenser. The motive steam and the water vapor are condensed by cooling water in a condenser, which must also operate under vacuum.
    - A two-stage steam jet system is commonly used to maintain at least the design vacuum in this condenser.
  - By evaporating some water under vacuum, the first chilling stage cools the remaining water to an intermediate temperature and each subsequent stage takes over until the final water temperature is reached.
- What are the common causes of failure of steam jet water chillers?
  - The most important reason is the failure of the vacuum system in the condenser. The following are some of the reasons for such failures.
    - *Air Leakage into the Chiller*: Often occurs through joints, flange connections, and eroded and corroded components and piping under vacuum. A hole of just few millimeters in diameter is sufficient to cause significant leakage.
    - *Pressure Drop and Blockages in the Line*: Pressure drop occurs due to blockages in the atmospheric discharge piping or improper assembly of joints during maintenance, for example, misplacement of gaskets in the vacuum lines.
    - *Insufficient Motive Steam Pressure to the Ejector*: Design steam pressure should be ensured and it should not be allowed to decrease below 10% of the design pressure of the ejector.
    - *High-Pressure Steam*: This could also cause problems, as substantially higher pressures above design pressures can cause choking at the throat of the nozzle of the ejector. Overpressures above 25% of design pressures could cause choking problems.
    - *Very High or Very Low Superheat in the Steam*: Superheat in the steam increases its specific volume. As the nozzle size is fixed, degree of superheat in the steam affects its performance, similar to effects involved with high-pressure steam.
    - *Excessive Moisture in the Steam*: Excessive moisture increases erosion problems and enlarged nozzle orifices alter the ejector performance.
    - *Cooling Water to the Barometric Condenser*: Hot water or water with lower flow rates can increase pressure in the condenser. High flow rates can flood the condenser.
- What are the effects of oil, from lubricated compressor bearings, contaminating refrigerants in the refrigerant evaporator in vapor compression cycle? How is lube oil contamination reduced?
  - In the refrigerant evaporator, lube oil accumulates due to its high boiling temperatures and forms thin viscous films on heat transfer surfaces of the evaporator. Even small amounts of oil (0.5–1% of the refrigerant charge) may be detrimental for the thermal capacity of the evaporator in one or more of the following ways:
    - Decreasing the evaporating heat transfer coefficients.
    - Increasing ΔP in the oil–refrigerant two-phase system.
    - Boiling point elevation of the refrigerant.
    - Preventing all the refrigerant from vaporizing.
    - Increasing foaming tendencies of the refrigerant.
    - Reducing LMTD.
  - Lube oil contamination is reduced by controlling the amount of compressor cylinder lubrication, using synthetic lubricants, providing a good compressor discharge vapor separator to eliminate free oil, and providing a good oil reclaiming system to remove oil accumulation.

12.2 HEAT PUMPS

- What is a heat pump?
  - A device for raising low-grade heat to a temperature at which the heat can be utilized.
  - It *pumps* the heat from a low-temperature source to a high-temperature sink, using a small amount of energy relative to the heat energy recovered.
- What are the basic functions of a heat pump?
  - Heat input from a waste heat source.
  - Increase of waste heat temperature.

- Delivery of the useful heat at the increased temperature.

• What are the three basic types of heat pumps?
  - Air-to-air heat pumps get their energy from the outside air.
  - Water loop heat pumps require access to a nearby well, pond, stream, or lake.
  - Ground source heat pumps take their heat from a circuit of pipes buried in the ground.

• What are the common types of industrial heat pumps?
  - *Closed cycle mechanical heat pump*, which uses mechanical compression of a working fluid, for example, a refrigerant, to achieve temperature increase. Drives include electrically operated systems, steam turbines, IC engines, and so on.
  - *Open cycle mechanical vapor compression (MVC) heat pumps*, which use a mechanical compressor to increase the pressure of waste vapor. Typically, these are used in evaporators, the working fluid being water vapor.
  - *Open cycle thermocompression heat pumps* use energy in high-pressure motive steam to increase the pressure of waste vapor using a jet ejector. Working fluid is steam. Used in evaporators.
  - *Closed cycle absorption heat pumps* use a two-component working fluid, the principle of operation being BPR and heat of absorption to achieve temperature rise. Key features of the absorption systems are that they can deliver a much higher temperature rise than the other systems.

• What are the applications of heat pumps?
  - Mechanical vapor compression (open cycle):
    - Separation of propane–propylene, butane–butylene, and ethane–ethylene mixtures.
    - Concentration of black liquor in pulp manufacture.
    - Concentration of different salt solutions.
    - Concentration of waste streams in waste treatment plants, cooling water blowdown in nuclear plants, desalination of seawater, radioactive wastes, electroplating wastes, solvent recovery from air, fermentation industries, sugar solutions, corn syrups, milk and whey, juices, and the like.
    - Compression of low-pressure steam.
  - Mechanical vapor compression (closed cycle):
    - Process water heating in pharmaceutical, paper, textiles, and other industries.
    - Space heating, heating of process solutions in food and soft drinks industries, and the like.
    - Dehumidification, drying of lumber, and the like.
  - Thermocompression (open cycle):
    - Flash steam recovery from steam stripping operations.
    - Concentration of dairy products, sugar solutions, and syrups.
  - Absorption (closed cycle):
    - Large-scale space heating.

• What is the thermodynamic principle on which heat pumps operate?
  - Carnot cycle.

• What is the difference between Carnot cycle and Rankine cycle? Illustrate.
  - The Rankine cycle operates on the principle involving degradation of high-grade thermal energy into lower grade thermal energy creating shaft work or power.
  - In contrast, mechanical heat pumps operate in the opposite manner, based on the Carnot cycle, converting lower temperature waste heat into useful, higher temperature heat, while consuming shaft work.

• Give a comparison between a steam turbine and a heat pump by means of an illustration.
  - Figure 12.3 illustrates the difference between a steam turbine and a heat pump.

• Illustrate the working of a heat pump system used in an evaporation process?
  - Figure 12.4 illustrates the working of a heat pump as applied to evaporation.
  - A heat pump uses a separate working fluid. Ammonia is the working fluid in this case.

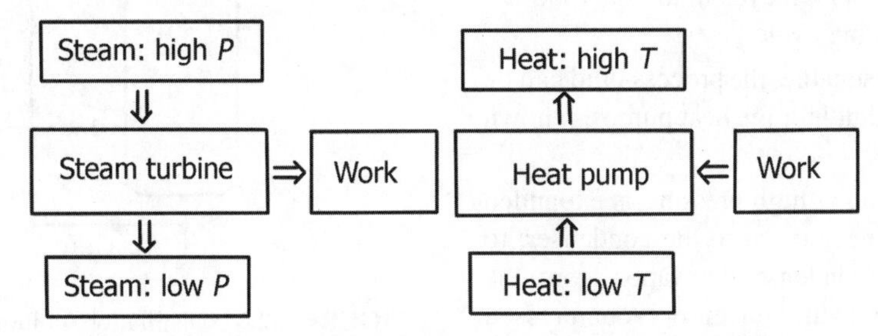

FIGURE 12.3 Comparison of a steam turbine and a heat pump.

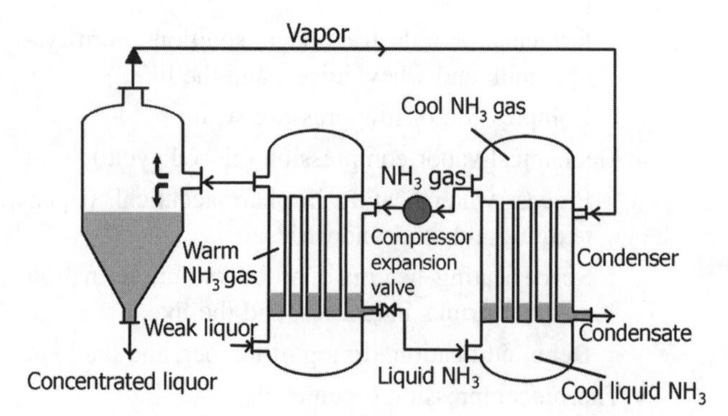

FIGURE 12.4 Heat pump as applied to evaporation.

- Ammonia gas (warm) vaporizes the feed liquor at a low temperature (e.g., at 288–313K) and ammonia gas condenses.
- Liquid ammonia then passes through an expansion valve and is cooled to much lower temperature.
- The cooled liquid ammonia then enters the condenser where it condenses the vapor leaving the separator and liquid ammonia vaporizes and leaves as a low-pressure gas that is compressed in a mechanical compressor and passed through the expansion valve to evaporate for the second cycle.
- The excess heat introduced by the compressor must be removed from ammonia by a cooler.
- The main advantage is that there is a great reduction of gas handled by the compressor.

• Illustrate the application of heat pump systems in distillation.
  - Figures 12.5 and 12.6 illustrate heat pump applications using (a) a refrigerant as the working fluid and (b) process fluid itself as the working fluid.
  - The working fluid (a commercial refrigerant) is fed to the reboiler as a vapor at high pressure and condenses, giving up heat to vaporize the process fluid. The liquid refrigerant from the reboiler is then expanded over a throttle valve and the resulting wet vapor is fed to the column condenser. In the condenser the wet refrigerant is dried, taking heat from the condensing process vapor. The refrigerant vapor is then compressed and recycled to the reboiler, completing the working cycle.
  - If the conditions are suitable, the process fluid can be used as the working fluid for the heat pump as shown in Figure 12.6.
  - The hot process liquid at high pressure is expanded over the throttle value and fed to the condenser, to provide cooling to condense the vapor from the column. The vapor from the condenser is compressed and returned to the base of the column.

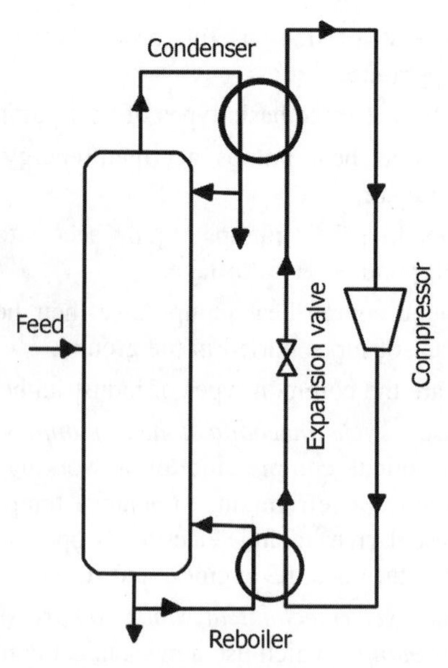

FIGURE 12.5 Distillation column with a separate refrigerant circuit.

  - In an alternative arrangement, the process vapor is taken from the top of the column, compressed, and fed to the reboiler to provide heating.

12.3 HEAT TRACING

• Under what circumstances heat tracing is necessary, over and above the insulation option?
  - Insulation alone cannot prevent freezing of lines that handle water or other aqueous solutions in a no flow condition if ambient temperature remains below

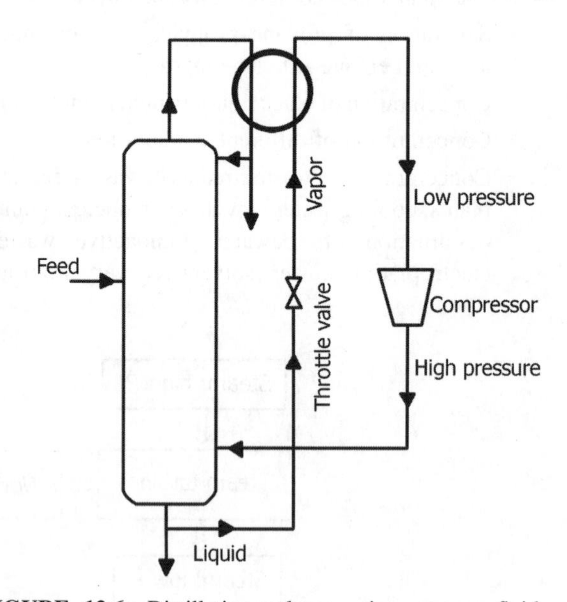

FIGURE 12.6 Distillation column using process fluid as a refrigerant.

freezing for an extended period of time. Insulation can only prolong the time required to freeze and can prevent freezing only if the water flow is maintained at a sufficient rate.

- Intermittent flow lines (and lines where it is not practical to drain or displace the process fluid on shutdown) should be traced if the pour point of the liquid is higher than the minimum ambient temperature that is likely to be experienced for a period of time, which will cause the fluid to solidify or become too viscous for pumping in an insulated line without tracing. The pumps can get highly overloaded or damaged in such cases under intermittent operation or during shutdowns.
- Instruments, safety and relief valves, analyzers, and other equipment under cold climate regions need *freeze protection* to ensure reliability and damage-free operation.

• What are the applications of heat tracing systems?

- Used to maintain process piping and flowing fluids above ambient temperatures. Examples are the following:
  - ➢ Preventing water piping from freezing.
  - ➢ Maintaining fuel oil pipes at temperatures above those at which viscosity of the fluids will not be high, creating flow problems.
  - ➢ Preventing condensation of liquids from a gas.
  - ➢ Preventing solidification of a liquid metal or a flowing liquid.

• What are the factors that need consideration in the selection of a heat tracing system?

- Where accurate temperature control is required, an electric tracing system is preferred.
- In some cases, steam is not available and installing a boiler only for tracing purposes is uneconomical.
- Providing steam supply from long distances is more expensive than providing electricity.
- Steam tracing gives high heat output.
- High reliability for heat supply with steam.
- Higher levels of safety with steam.
- Lower installation costs for electric tracing.
- Better temperature control with electric tracing.

• What are the types of heat tracing systems?

- *Fluid Tracing*: Steam tracing involves using a smaller pipe containing saturated steam as the *tracer*, parallel to the process fluid pipe. The two pipes are then insulated together with insulation and jacketed if necessary. Steam tracing is more labor intensive to install than electric heat tracing, but there are very few risks associated with it. The temperature of the tracer also cannot exceed the maximum saturation temperature of the steam, as it operates at specific steam pressures.
  - ➢ Other tracing fluids: ethylene glycol, hot oil, and other heat transfer fluids.
- *Electric Tracing*: An insulated electrical heating cable is spiraled around the process fluid pipe, after which the pipe and tracing are insulated with the appropriate type and thickness of insulation.
  - ➢ This method of heat tracing may be installed, maintained, modified, and reused with relative ease compared to steam tracing.
  - ➢ It is more expensive than steam and hot oil systems.
  - ➢ It poses several risks. The most important of these being the risk of electric spark, which may cause electric shock or ignite flammable substances resulting in fire and explosion. All points on a heat tracing line should be kept below 80% of self-ignition temperature of the most ignition-prone material in the area.
  - ➢ It may be necessary to provide insulation with vapor barriers and/or purge the system, which adds to the costs.
  - ➢ Ground fault protection is required.
  - ➢ If electric heat tracing is not carefully controlled, there is also the possibility that the cable could overheat and damage the pipe or insulation.

• What are the common types of electric heat tracing systems? What are their advantages and disadvantages?

- A *self-limiting* heat tracing system composed of conductors separated by a polymer that changes resistance as it heats up. It effectively cuts off at some point around 100–200°C, making it inherently safe.
  - ➢ Easy to install and modify and generally reusable.
  - ➢ Care must be exercised in sizing and selecting to ensure matching to the desired process temperature.
  - ➢ Adding extra tracing at a later stage has little effect on the installation, as the power of additional tape is often offset by a corresponding reduction in the power output of the already installed tape.
  - ➢ To ensure that the tape makes good thermal contact with the heated surface, closely spaced ties are generally required, increasing costs.
  - ➢ Start-up requires significantly more power than other alternatives.
  - ➢ Reaching temperatures above 100°C may be difficult.

- *Constant Wattage Type:*
  - Power remains constant regardless of operating temperature.
  - Used for temperature maintenance and not for protection against freezing.
  - Can reach much higher temperatures than self-limiting type.
  - Can exceed self-ignition temperatures of flammables in the area, making it not *inherently safe* and requiring over-temperature protection.
- *Resistance Type*: Typically consists of one or more resistance heating wires inside a fabric or polymer sheath.
  - Fabric type is suited for high temperatures (400–500°C) and elastomer type for lower temperatures (200–250°C).
  - Flexible, easy to wrap, and less expensive than other types.
  - Heater element wires normally exceed self-ignition temperatures of most flammables.
  - Requires a control system with appropriate over-temperature protection.
  - Good for laboratory use.
- *Thermocouple-Type Line Heaters*: Resistance wire is buried in cementitious insulation and placed inside a thin tubular outer sheath. Can reach very high temperatures.

- How is steam tracing applied?
  - There are two ways to apply steam tracing (Figure 12.7).
  - Bare steam tracing is the most popular choice as its installation and maintenance are relatively easy. It is well suited to lower temperature requirements. It is simply composed of a 12.7 or 19 mm (0.5 or 0.75 in.) copper tube carrying steam, attached to the process fluid pipe around and along process piping by straps, and both pipes are then insulated together.
  - The other available option is to make use of cemented steam tracing, in which heat conductive cement is placed around the steam tracer running parallel to the process fluid pipe. This increases the contact area available for heat transfer, between the tracer and the process fluid pipe. Figure 12.7 illustrates application of insulation for such systems.

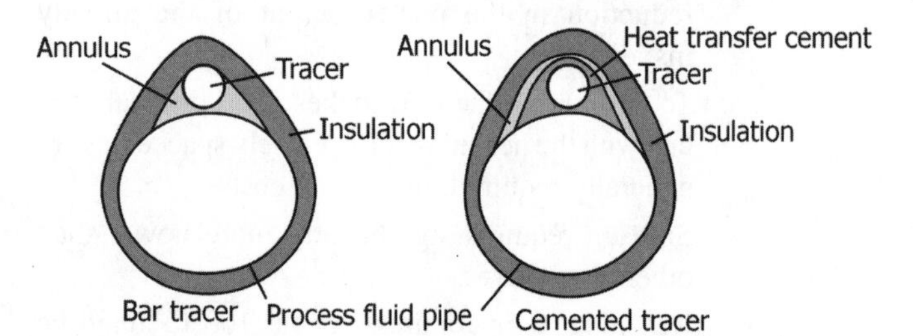

FIGURE 12.7 Heat tracer over a pipe carrying a fluid.

  - The annulus contains air that gets heated by the tracer and circulates inside, providing more heat transfer contact and reducing heat losses.
  - Inherently safe and rarely requires over-temperature protection, except for heat-sensitive process fluids.
  - Fire and explosion risks are minimal.
  - Easy to install and less expensive if steam is already available on-site.
  - Reliable and requires less maintenance, except for steam traps.
  - Steam pressure sets the temperature.
  - Wrapping around small piping and tubing and contact with process piping are difficult.
  - Condensation and ΔP become problems on long runs. Multiple short runs are better, though costs increase.
  - Regulators are necessary to set feed pressures.
  - Block valves for maintenance, check valves for preventing backflow, and strainers for rust and grit become necessary components.

- What are the plus and minus points of using hot oil heat tracing systems?
  - High-temperature heat transfer fluids, with no phase change, are used in the tracing systems.
  - Only sensible heat transfer is involved, requiring higher flow rates.
  - No condensation problems.
  - Systems are far larger than those for electric tracing.
  - Inherently safer than electric tracing.
  - More uniform temperatures are obtainable than electric tracing.
  - Higher installation and maintenance costs.
  - Fluids are sensitive to breakdown, giving rise to deposits.
  - Many heat transfer fluids for high-temperature applications are toxic or flammable.

12.4 COILED VESSELS

- How are tank contents heated?
  - By the use of heating coils through which steam or hot heat transfer fluid is circulated. The coils are installed at the bottom of the tank as shown in Figure 12.8.
  - Liquids that tend to solidify on cooling require heating by uniformly covering the bottom by a heat transfer surface such as a heating coil.

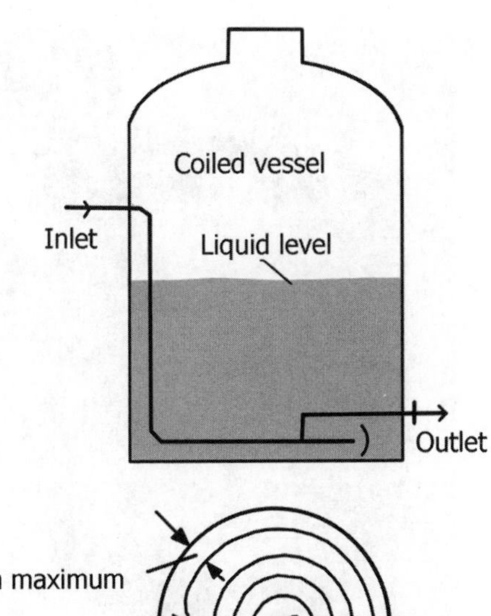

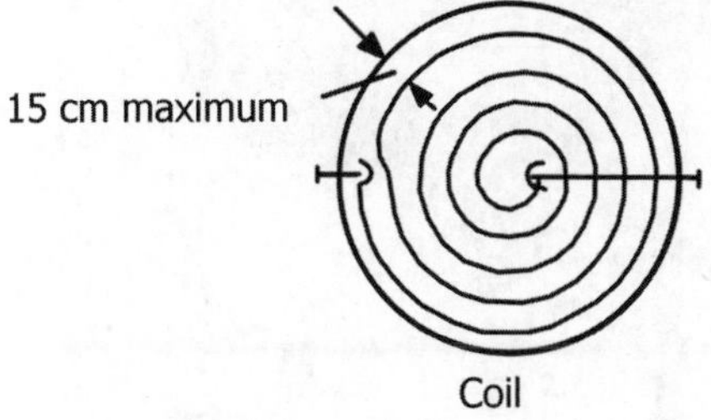

FIGURE 12.8 Heating coils.

- The coils should be fixed by clamping directly on the bottom or raised not more than 5–15 cm, depending on the difficulty of remelting the solids, in order to permit free movement of product within the vessel. The coil inlet should be above the liquid level. Coils may be sloped to facilitate drainage.

- Give an equation for the estimation of heat transfer coefficients in a helical coil.

$$h_h = h_s[1 + 3.5(D_i/D_c)], \quad (12.2)$$

where h_h is the heat transfer coefficient for the helical coil, h_s is the heat transfer coefficient for the straight tube, D_i is the inside tube diameter, and D_c is the diameter of the helix or coil.

- How is pressure drop due to friction estimated for helical coils? Give the equations.
  - The procedure used for straight tubes can be used to calculate the pressure drop in helical coils. For turbulent flow, a friction factor for curved flow is substituted for the friction factor for straight tubes. For laminar flow, the friction loss for a curved tube is expressed as an equivalent length of straight tube and the friction factor for straight tubes is used.
  - The Reynolds number required for turbulent flow is

$$N_{Rec} = 2100[1 + 12(D_i/D_c)^{1/2}], \quad (12.3)$$

where D_i is the inside diameter of the tube and D_c is the diameter of the helix or coil.

  - Friction factor for turbulent flow is obtained from the equation

$$f_c(D_i/D_c)^{1/2} = 0.0073 + 0.076[(D_iG/\mu)(D_i/D_c)^2]^{-1/4} \quad (12.4)$$

for values of $(D_iG/\mu)(D_i/D_c)^2$ between 0.034 and 300. f_c is the friction factor for curved flow.
  - For values of $(D_iG/\mu)(D_i/D_c)^2$ less than 0.034, the friction factor for curved flow is practically the same as that for straight tubes.
  - For laminar flow, the equivalent length L_e can be predicted as follows:
    - For $[(D_iG/\mu)(D_i/D_c)^2$ between 150 and 2000,

$$L_e/L = 0.23[(D_iG/\mu)(D_i/D_c)^2]^{0.4}. \quad (12.5)$$

    - For $[(D_iG/\mu)(D_i/D_c)^2$ between 10 and 150,

$$L_e/L = 0.63[(D_iG/\mu)(D_i/D_c)^2]^{0.2}. \quad (12.6)$$

    - For $[(D_iG/\mu)(D_i/D_c)^2$ less than 10,

$$L_e/L = 1, \quad (12.7)$$

where L is the straight length and L_e is the equivalent length of a curved tube in laminar flow.

12.5 JACKETED VESSELS

- What are the applications of jacketed vessels?
  - Jacketing is often used for vessels requiring frequent cleaning and for glass-lined vessels that are difficult to equip with internal coils. The jacket gives a better overall coefficient than external coils. The temperature and velocity of the heat transfer media can be accurately controlled. On the negative side, only a limited heat transfer area is available.
- What are the different types of jackets used in heat transfer equipment? What are their applications?
  - *Simple Conventional Jacket*: 5–7.5 cm wide annular space in which fluid flows at around 3 cm/s. The conventional jacket is of simple construction and is frequently used.
    - Conventional jackets require a greater shell thickness (compared to other types) along with expansion joints to eliminate stresses induced by the difference in thermal expansion when the jacket is not manufactured from the same material as that of the shell.
    - Natural convection predominates.
    - Nozzles, which set up a swirling motion in the jacket, are effective in improving heat transfer. It is most effective with a condensing vapor.

  - With services involving large volumes of water (used to maintain a high temperature difference), the conventional jacket usually offers the best solution.
  - The ability to vary the distance between the outer and inner vessel walls makes conventional jackets ideally suited to handle high-temperature heat transfer fluids such as Dowtherm vapors. Also, since Dowtherm vapor has a low enthalpy (1/10th that of steam), a large jacket space is needed for given heat flux.
- *Baffled Conventional Jacket*: Better turbulence than simple jacket.
  - Baffled jackets often utilize what is known as a spirally wound baffle. The baffle consists of a metal strip wound around the inner vessel wall from the utility inlet to the utility outlet of the jacket. The baffle directs the flow in a spiral path with a fluid velocity of 0.3–1.2 m/s (1–4 ft/s).
  - Conventional baffled jackets are usually applied with small vessels using high temperatures where the internal pressure is more than twice the jacket pressure.
- *Jacket with Agitation Nozzles*: Used to improve h for glass-lined vessels (glass lining reduces heat transfer) by providing localized turbulence (Figure 12.9).
- *Spirally Baffled Jacket*: Spirally wound ribbon strip runs as baffle inside jacket, giving increased local velocities/turbulence by directing flow in a spiral path.
  - Spirally baffled jackets are limited to a pressure of 690 kPa gauge (100 psig) because vessel wall thickness becomes large and the heat transfer is greatly reduced. In the case of an alloy steel reactor, a very expensive vessel can result.
  - A spiral-wound channel welded to the vessel wall is one type of jacketed vessel.
- *Dimple Jacket*: The dimple jacket is another type that offers structural advantages and is very economical for high jacket pressures. Depressions/dents in the outside wall of the jacket (Figures 12.10 and 12.11).
  - Dimple construction provides higher levels of turbulence in the jacket. The low volumetric capacity of the jacket produces a fast response to temperature changes.
  - The dimple jackets are generally limited to 2000 kPa gauge (300 psig) design pressure.
- *Half-Pipe Coil Jacket*: Continuous channel welded to vessel wall. Made by cutting pipe longitudinally into two halves and welding to vessel wall as a spiral. Spacing between the half-pipe sections is normally kept at 3/4 in. Spacing more than this increases heat losses unless adequately insulated. The half-pipe coil can use 1/4 in. thick carbon steel for jacketing but its economy compared to conventional jackets must be evaluated.
- Half-pipe coil jackets are generally made with either 180° or 120° central angles (D_{ci}) as illustrated in Figure 12.12.

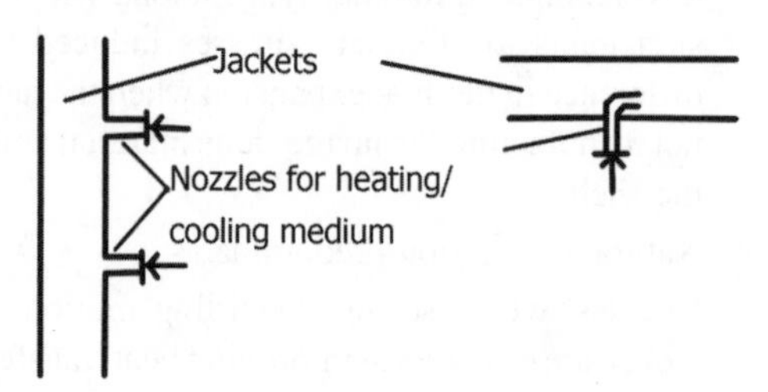

FIGURE 12.9 Jackets with nozzles to admit heat transfer fluids.

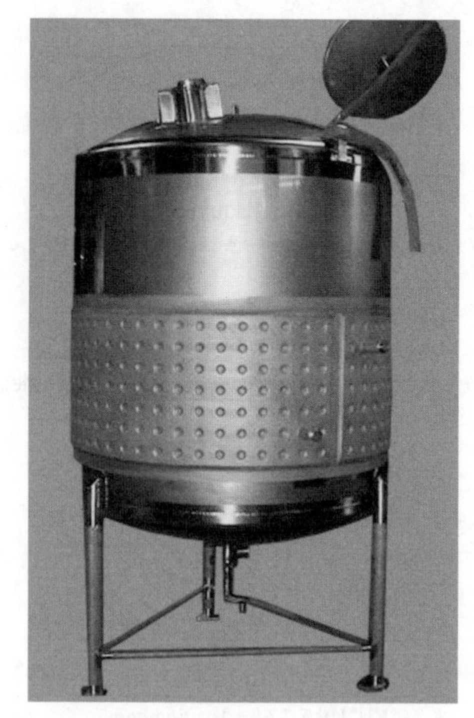

FIGURE 12.10 Dimple jacket vessel. (*Source*: www.reimec.co.za.)

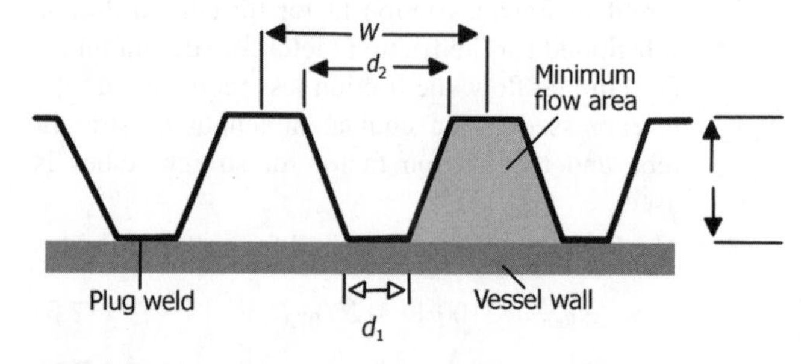

FIGURE 12.11 Dimple jacket cross section. (*Courtesy*: Santosh Singh (process.santosh@googlemail.com).)

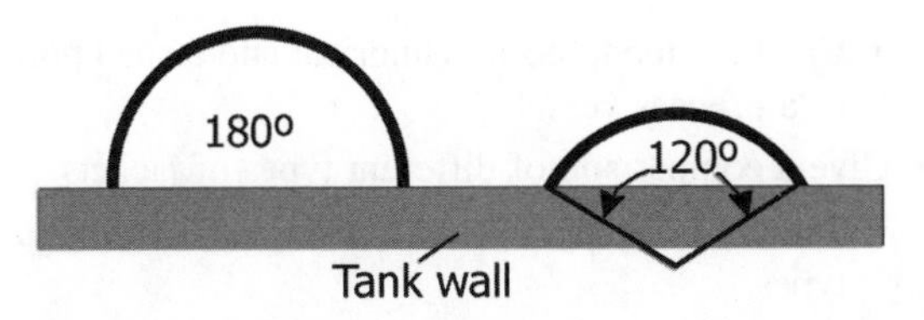

FIGURE 12.12 Half-pipe jacket angles. (*Courtesy*: Santosh Singh (process.santosh@googlemail.com).)

- For a 180° central angle,

$$\text{equivalent heat transfer diameter } D_e = P/4D_{ci}, \quad (12.8)$$

$$\text{cross-sectional area for flow } A_f = P/8D_{ci}^2. \quad (12.9)$$

- For a 120° central angle,

$$D_e = 0.708D_{ci}, \quad (12.10)$$

$$A_f = 0.154D_{ci}^2. \quad (12.11)$$

- The half-pipe jacket is used when high jacket pressures are required. The flow pattern of a liquid heat transfer fluid can be controlled and designed for effective heat transfer. Half-pipe coil jackets are well suited for use with high-pressure steam (up to design pressures of 5000 kPa gauge (750 psig)).
- Half-pipe coils are well suited for high-temperature applications where the heat transfer fluid is a liquid. Design temperatures can be up to 380°C. A carbon steel half-pipe jacket can be applied to a stainless steel vessel up to 150°C. Over 150°C, the jacket should also be stainless steel.
- There are no limitations on the number of inlet and outlet nozzles, so the jacket can be divided into multipass zones for maximum flexibility. The rigidity of the half-pipe coil design can also minimize the thickness of the inner vessel wall, which can be especially attractive when using alloy steels.
- Half-pipe coils provide high velocity and turbulence. The velocity can be closely controlled to achieve a good film coefficient.
- The good heat transfer rates, combined with the structural rigidity of the design, make half-pipe coils a good choice for a wide range of applications.
- A good design velocity for liquid utilities is 0.75–1.5 m/s (2.5–5 ft/s).
- For a half-pipe coil jacket, the higher heat flux rate may require multiple sections of jackets to avoid having condensate covering too much of the heat transfer area. For low-pressure steam services, convention jackets are a much more economical choice.
- Figures 12.13 and 12.14 illustrate half-pipe jackets and half-pipe coil dimensions, respectively.
- Hydraulic diameter for half-pipe coil:

$$D_H = 4(1/2)(\pi d_1^2/4)/[d_1 + (1/2)\pi d_1](12\text{ in./ft}) = 0.0509d_1. \quad (12.12)$$

- *Panel or Plate Coil Jacket*: Fabricated from two metal plates. For reactors, one plate is smooth and forms vessel wall. The outer plate is embossed to form a series of flow paths between the plates.

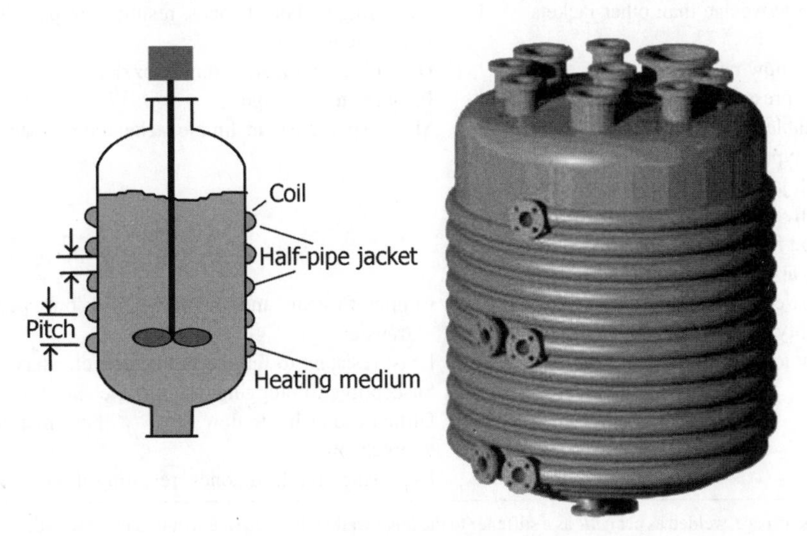

FIGURE 12.13 Jacketed vessel with a half-pipe jacket. (*Source*: www.reimec.co.za.)

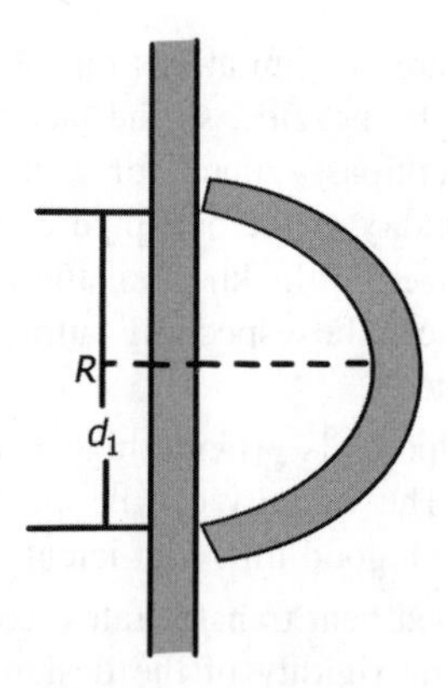

FIGURE 12.14 Half-pipe coil dimensions. (*Courtesy*: Santosh Singh (process.santosh@googlemail.com).)

- What are the advantages of dimple jackets?
  - Have far better heat transfer characteristics than conventional ones due to turbulence generated by dimples.
  - Heat transfer can be augmented by increasing jacket velocities through the use of jacket baffles.
  - For differential cooling or heating, jackets can be segmented into zones.
  - Relatively inexpensive to fabricate.
  - Relatively light gauge stainless steel can be used for pressures of 6–8 bar for the cooling medium, since the jacket is supported on a grid of plug welds.
  - Can be fitted to both cylindrical and dished portions of a process vessel.
- Give a comparison of different types of jackets.
  - Table 12.2 gives a comparison of different jacket types.
- Give suitable correlations for heat transfer coefficients for jacketed vessels.
  - Heat transfer coefficients for vessels with *conventional jackets* may be estimated using the correlation

$$(h_j D_e/k) = 1.02 N_{Re}^{0.45} N_{Pr}^{0.33} (D_e/L)^{0.4} (D_{jo}/D_{ji})^{0.8} N_{Gr}^{0.05}, \quad (12.13)$$

where h_j is the local heat transfer coefficient on the jacket side, D_e is the equivalent hydraulic diameter, L is the length of jacket passage, D_{jo} is the outer diameter of the jacket, D_{ji} is the inner diameter of the jacket, and N_{Gr} is the Graetz number.

➢ The equivalent diameter is defined as follows:

$$D_e = D_{jo} - D_{ji} \quad \text{for laminar flow,}$$
$$D_e = (D_{jo}^2 - D_{ji}^2)/D_{ji} \quad \text{for turbulent flow.} \quad (12.14)$$

- For *conventional jackets with baffles*, the following correlations can be used to calculate the heat transfer coefficient:

TABLE 12.2 Advantages and Disadvantages of Different Types of Jackets

Type of Jacket	Advantages	Disadvantages
Conventional (full enclosure)	High flow rates	Poor heat transfer coefficients due to low velocities
	Low pressure drop	Thick wall required to withstand jack pressure
	More coverage than other jackets	Bypassing and dead zones, resulting in poor heat transfer
		Highest cost
Half-pipe	High flow rates	Difficult to fabricate around nozzles
	Low pressure drop	Incomplete coverage
	Suitable for dirty fluids	More expensive than dimple jacket unless spacing is kept ≤25.4 mm (1 in.)
	No bypassing	
	High jacket pressure with no adverse effect on inner vessel wall thickness	
	Good fatigue resistance (cyclic service) if applied with full penetration welds	
Dimple jacket	Least expensive	Limited to steam and low flow liquids only performing maintenance heat transfer
	Ability to withstand high pressures	
	Easy to work around nozzle	Least resistant to fatigue failures (cyclic service)
		Susceptible to plugging, requiring clean fluid
		Difficult to estimate flow or ΔP without manufacturer's empirical procedures
		Bypassing and dead zones, resulting in poor heat transfer

Note: A spiral baffle in the jacket space, welded as per *code* as a stiffener to the inner wall, will reduce the required inner wall thickness and improve heat transfer due to increased fluid velocity.

$$h_jD_e/k = 0.027N_{Re}^{0.8}N_{Pr}^{0.33}(\mu/\mu_w)^{0.14}[1+3.5(D_e/D_c)] \quad (\text{for } N_{Re} > 10,000), \tag{12.15}$$

$$h_jD_e/k = 1.86(N_{Re}N_{Pr}D_c/D_e)^{0.33}(\mu/\mu_w)^{0.14} \quad (\text{for } N_{Re} < 2100). \tag{12.16}$$

- D_c is defined as the centerline diameter of the jacket passage and is calculated as

$$D_c = D_{ji} + [(D_{jo} - D_{ji})/2]. \tag{12.17}$$

- When calculating the heat transfer coefficients, an effective mass flow rate should be taken as 0.60 × feed mass flow rate, to account for the substantial bypassing that will be expected.
- D_e is defined as

$$D_e = 4 \times \text{jacket spacing}. \tag{12.18}$$

Flow cross-sectional area

$$= \text{baffle pitch} \times \text{jacket spacing}. \tag{12.19}$$

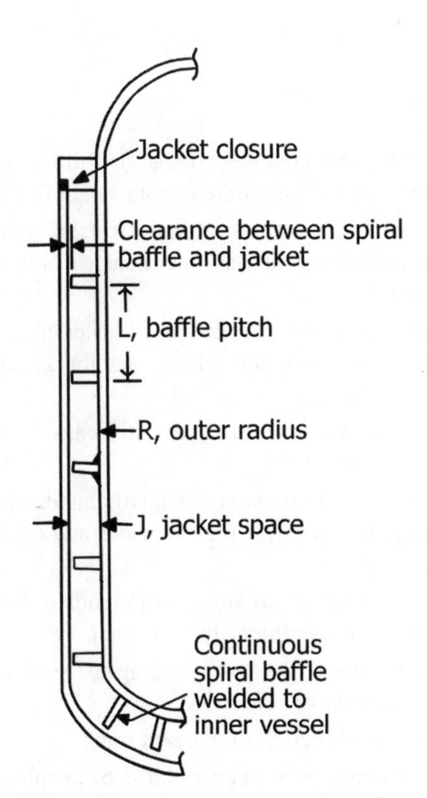

FIGURE 12.15 Conventional jacket with baffles. (*Courtesy*: Santosh Singh (process.santosh@googlemail.com).)

- Figure 12.15 illustrates a conventional jacket with baffles.
- Referring to Figure 12.15, the hydraulic radius is calculated as follows:

$$D_H = 4 \times \text{cross-sectional flow area/wetted perimeter} = 4L_{BP}J/(2L_{BP}+2J) = L_{BP}J/6(L_{BP}+J). \tag{12.20}$$

- *Heat transfer coefficients for dimple jackets* can be estimated by the following correlation:

$$h_iD_o/k = jN_{Re}N_{Pr}^{0.33} \quad \text{for } 1000 < N_{Re} < 50,000, \tag{12.21}$$

where

$$j = 0.0845(w/x)^{0.368}(A_{min}/A_{max})^{-0.383}N_{Re}^{-0.305}, \tag{12.22}$$

where w is the center-to-center distance between dimples and x is the center-to-center distance between dimples parallel to flow.
Note: w/x is equal to 1.0 for square spacing, which is common.

$$D_o = (d_1 + d_2)/2, \tag{12.23}$$

$$A_{min} = z(w - D_o), \tag{12.24}$$

$$A_{max} = zw. \tag{12.25}$$

- Give an equation for predicting heat transfer coefficients for agitated vessels.

$$hD_J/k = a(L2N\rho/\mu)^{2/3}(c_p\mu/k)^{1/3}(\mu/\mu_w)^{0.14}, \tag{12.26}$$

where D_J is the vessel diameter, L is the agitator diameter, and N is the speed of the agitator (rph).

- Values of a are given in Table 12.3.

TABLE 12.3 Values of *a* in Equation 12.26

Agitator Type	Surface	*a*
Turbine	Jacket	0.62
Turbine	Coil	1.5
Paddle	Jacket	0.36
Paddle	Coil	0.87
Propeller	Jacket	0.54
Propeller	Coil	0.83
Anchor	Jacket	0.46

- Give an equation for overall heat transfer coefficients for jacketed agitated vessels.
  - While calculating the overall heat transfer coefficient for a system, the vessel wall resistance and any jacket fouling must be taken into account:

$$1/U = 1/h_{\text{jacket}} + R_{\text{fouling}} + R_{\text{wall}} + 1/h_{\text{inside vessel}}. \tag{12.27}$$

- Give literature recommended values of overall heat transfer coefficients for agitated vessels.
  - Table 12.4 gives values of U for agitated vessels.
- Describe *constant flux* heat transfer control for variable geometry cooling/heating jacket on batch reactors.
  - Heat transfer surface is divided into multiple small elements, each of which could be switched on or off independently.
  - Each element is a small heat transfer pipe that wraps around the vessel and connects to a multiport control valve. There will be 20–60 such elements all connected separately to the valve.
  - As the valve position moves, it opens the elements in a cascade fashion to allow the flow of heat transfer fluid.
  - The pipes on the *coflux* jacket are evenly over the vessel surface, either clamped to vessel wall (on smaller vessels) or mounted on thermally conducting plates (on larger vessels).
  - Quantity of heat transfer fluid can be reduced as much as 90% without compromising flow capacity.
  - Dead spots can be eliminated in this type of design.
  - This type of system is claimed to be useful in calorimetry for measurement of heat.
  - Figure 12.16 illustrates constant flux heat transfer jacket.

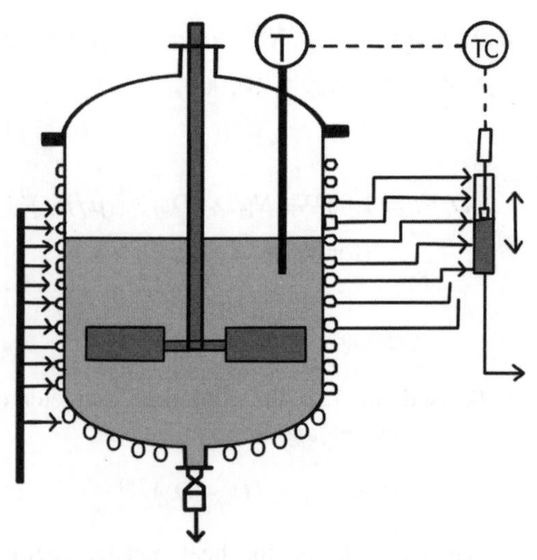

FIGURE 12.16 Constant flux heat transfer jacket.

TABLE 12.4 Values of Overall Heat Transfer Coefficients for Jacketed Vessels

Heat Transfer Fluid (Jacket Side)	Process Fluid	U (Including Fouling Effects) (W/(m² K)) Mild or No Agitation	Good Agitation
Steam	Water-like solution	710–1275	850–1560
Steam	Light oils	230–255	340–625
Steam	Medium lube oils	140–225	285–565
Steam	Fuel oil	60–170	340–455
Steam	Tar/asphalt	85–140	230–340
Steam	Molten sulfur	25–85	285–400
Steam	Molten paraffin	140–200	230–285
Steam	Air/gases	6–17	23–45
Steam	Molasses/corn syrup	85–170	340–455
Hot water	Water-like solution	400–565	625–910
Heat transfer oil	Tar/asphalt	55–115	170–285
Therminol	Tar/asphalt	70–115	170–285
Cooling services			
Water	Water-like solution	285–455	455–795
Water	Quench oil	40–55	85–140
Water	Medium lube oils	30–45	60–115
Water	Molasses/corn syrup	25–40	45–85
Water	Air/gases	6–17	23–45
Freon/ammonia	Water-like solution	115–200	225–340
Calcium/sodium brine	Water-like solution	285–425	455–710

12.6 STEAM TRAPS

- What are the adverse effects of failure to remove condensate, as it is formed, from steam lines?
  - *Loss of Efficiency in Steam Utilization*: Entrained water does not carry as much heat to a process as does steam.
  - *Damage to Nozzles*: Entrained water erodes nozzles and can adversely affect vacuum generation or atomization.
  - *Loss of Power*: Entrained water causes turbines to operate less efficiently.
  - *Increased Maintenance*: Water hammer can damage equipment such as turbine blades and control valve packing.
  - *Increased Risks to Safety of Personnel*: Water hammer can cause injury to personnel.
  - *Poor Process Control*: Flooding heat exchangers can lead to control problems.
- What is the function of a steam trap?
  - Steam traps prevent energy loss by keeping steam in a system and by maintaining steam pressure in the system.

- Steam traps function by removing condensate and venting off noncondensable gases such as air, carbon dioxide, and other noncondensable gases as they form within the steam piping and heat transfer equipment.
- These functions are achieved by the controlled opening and closing of an appropriately sized and controlled orifice.

• What are the effects of noncondensable gases present in steam on piping and heat transfer equipment?

- Noncondensable gases, being nonconductors, reduce heat transfer rates. In addition, dissolved carbon dioxide in condensate forms carbonic acid and causes corrosion of piping and equipment. The solubility of CO_2 in condensate increases if condensate is subcooled below saturation temperature of steam.

• What are the effects of malfunction and failure of steam traps on the system?

- A failed-open process trap dumps live steam to the condensate recovery system, with serious degradation in energy efficiency.
- The effect is even more critical when the condensate loop is not closed, since makeup water must be chemically pretreated and heated.
- A trap that has failed completely or is partially closed results in water hammer and wet steam, increased maintenance, longer start-up times, or total failure to heat transfer.
- As traps malfunction, boiler emissions load increases to make up for lost heat.
- A failed-open trap poses a risk of personnel exposure to live steam.
- A failed-closed trap allows a buildup of condensate to occur, with a risk of injury or damage caused by water hammer, necessitating downtime and expensive repairs.
- In cold climates, excess venting can result in icing and potential hazards to personnel.

• What are the different types of steam traps?

- There are three possible types of steam traps, namely, drip traps, tracer traps, and process traps.
- Drip traps help ensure that high-quality steam serves the processes through the effective draining of the mains. They also protect the distribution system by reducing the chance of water hammer.
- Tracer traps keep process fluids flowing properly in the transmission pipelines connecting the processes, around the process, and to and from storage tanks.
- Process traps ensure that the process is operating both effectively and efficiently by removing the condensate as it forms without the loss of live steam.
- Broadly, steam traps are divided into three basic types:
  - ➢ Mechanical traps.
  - ➢ Thermostatic traps.
  - ➢ Thermodynamic traps.
- *Mechanical traps* use density difference between steam and condensate to detect presence of condensate. These include the following:
  - ➢ Inverted bucket trap.
  - ➢ Float and thermostatic trap.
- *Thermostatic traps* operate on the principle that saturated process steam is hotter than either its condensate or steam mixed with noncondensable gas. When separated from steam, condensate cools to below steam temperature. A thermostatic trap opens its valve to discharge condensate when it detects its lower temperature. These include the following:
  - ➢ Liquid and wax expansion thermostatic trap.
  - ➢ Balanced pressure thermostatic trap.
  - ➢ Bimetallic trap.
- *Thermodynamic traps* use velocity and pressure of flash steam to operate the condensate discharge valve.

• How does an inverted bucket steam trap work? Explain by means of diagrams. What are its merits and demerits?

- The *inverted bucket steam trap*, shown in Figure 12.17, illustrates its method of operation. The inverted bucket is attached by a lever to a valve. It has a small air vent hole in the top of the bucket.
  - ➢ In (a), the bucket moves down, pulling the valve off its seat. Condensate flows through the bottom of the bucket, filling the body, and out through the outlet.
  - ➢ In (b), the flow of steam causes the bucket to become buoyant and then rises and closes the outlet.
  - ➢ In (c), the trap remains closed until the steam in the bucket has condensed or bubbled through the vent hole to the top of the trap body. It will then sink, pulling the main valve off its seat. Accumulated condensate is released and the cycle is repeated.
  - ➢ In (b), air reaching the trap at start-up will also give the bucket buoyancy and close the valve. The bucket vent hole is essential to allow air to escape into the top of the trap for eventual discharge through the main valve seat. The hole and the pressure difference are small making venting of air relatively slow. A certain amount of steam is also passed with air for the trap to operate once the air has cleared. A parallel air vent fitted outside the trap will reduce start-up times.

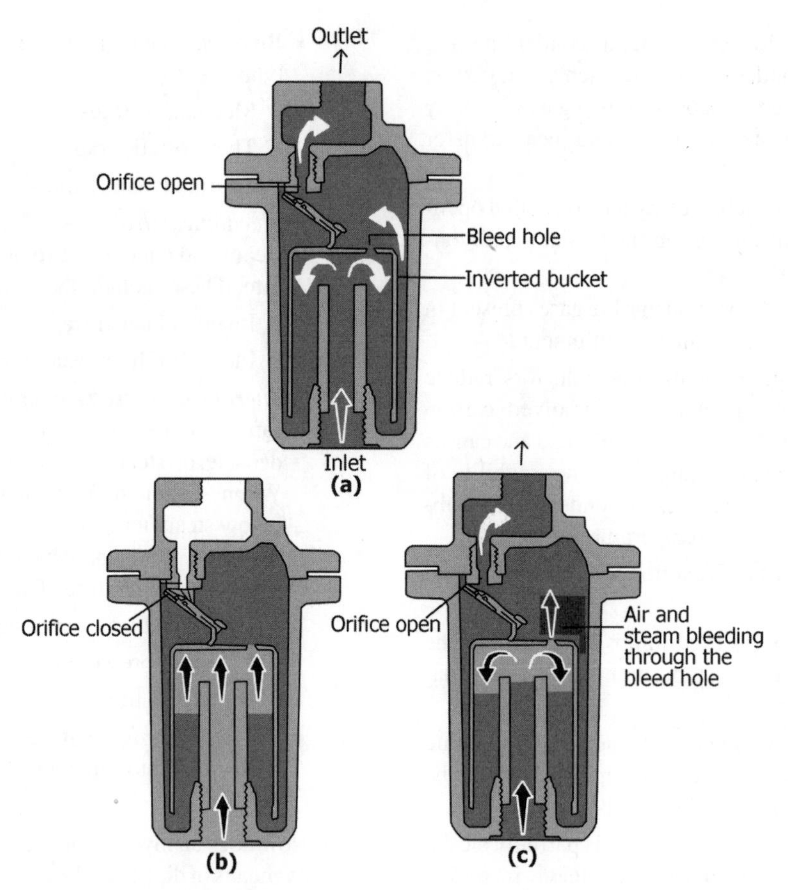

FIGURE 12.17 An inverted bucket steam trap. (*Courtesy*: Spirax Sarco.)

- *Merits*: It can be made to withstand high pressures. It has a good tolerance to water hammer conditions. Can be used on superheated steam lines with the addition of a check valve on the inlet. Higher temperature of superheated steam is likely to cause an inverted bucket trap to lose its water seal. Failure mode is usually open, so it is safer on those applications that require this feature, for example, turbine drains.
- *Demerits*: Condensate is discharged intermittently. This will be a problem if a pump is used for condensate removal. Discharge of air is very slow due to small size of the hole (has to be small as otherwise steam losses will be more). There should always be enough water in the trap body to act as a seal around the lip of the bucket. If the trap loses this water seal, steam can leak through the outlet valve. This can often happen on applications where there is a sudden drop in steam pressure, causing some of the condensate in the trap body to flash into steam. The bucket loses its buoyancy and sinks, allowing live steam to pass through the trap orifice. The inverted bucket trap is likely to suffer damage from freezing if installed in an exposed position with subzero ambient conditions. As with other types of mechanical traps, insulation can overcome this problem if conditions are not too severe. If ambient conditions well below zero are to be expected, a thermodynamic trap will be a better choice.

- Illustrate working of ball float and thermostatic traps.
  - Figure 12.18 illustrates the working of these traps.
  - The ball float-type trap operates by sensing the difference in density between steam and condensate. In the case of the trap shown in Figure 12.18a, condensate reaching the trap will cause the ball float to rise, lifting the valve off its seat and allowing the condensate to pass through. As can be seen, the valve is always flooded and neither steam nor air will pass through it. Venting of air is done manually by operating a cock at the top of the body. Modern traps use a thermostatic air vent, as shown in Figure 12.18b. This allows the initial air to pass while the trap is also handling condensate.
  - The automatic air vent uses a balanced pressure capsule element and is located in the steam space above the condensate level. After releasing the initial

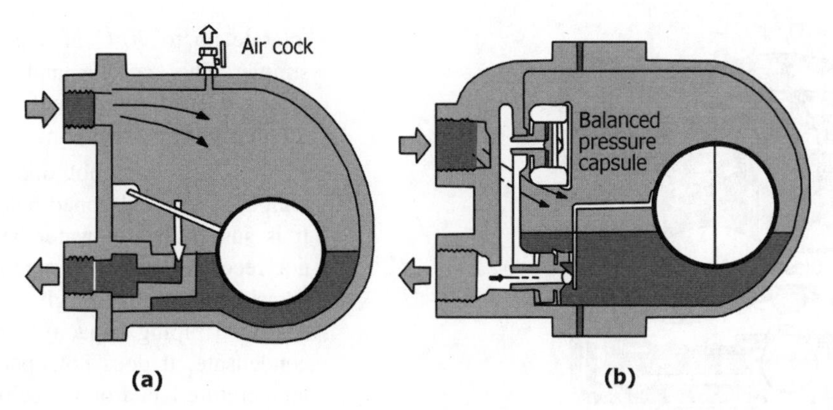

FIGURE 12.18 A ball float trap with (a) air cock and (b) thermostatic air vent. (*Courtesy*: Spirax Sarco.)

air, it remains closed until air or other noncondensable gases accumulate during normal running and cause it to open by reducing the temperature of the air/steam mixture.

- *Merits*: The thermostatic air vent significantly increases condensate capacity on cold start-up. It will discharge condensate as soon as it is formed, regardless of changes in steam pressure. This makes it the first choice for applications where the rate of heat transfer is high for the area of heating surface available. It is not affected by wide and sudden fluctuations of pressure or flow rate. It is unaffected by water hammer.
- *Demerits*: The float-type trap can be damaged by severe freezing and the body should be well insulated, and/or complemented with a small supplementary thermostatic drain trap, if it is to be fitted in an exposed position. As with all mechanical traps, different internals are required to allow operation over varying pressure ranges.

• What is a liquid expansion steam trap? Give a diagram and describe its merits and demerits.

- This is one of the simplest thermostatic traps. Figure 12.19 gives its constructional details. An oil-filled element expands when heated to close the valve against the seat.
- *Merits*: Liquid expansion traps can be adjusted to discharge at low temperatures. The adjustment allows the temperature of the trap discharge to be altered between 60 and 100°C, which makes it ideally suited to get rid of large quantities of air and cold condensate at start-up. The trap can be used as a start-up drain trap on low-pressure superheated steam mains where a long cooling leg is guaranteed to flood with cooler condensate. It is able to withstand vibration and water hammer conditions.
- *Demerits*: The flexible tubing of the element can be destroyed by corrosive condensate or superheat. Since the liquid expansion trap discharges condensate at low temperatures ($\leq$100°C), it should never be used on applications that demand immediate removal of condensate from the steam space. It must be well insulated in freezing atmospheres.

• How does a balanced pressure steam trap operate?

- The balanced pressure steam trap operates on the principle of liquid expansion due to an increase in temperature. The operating element is a bellows capsule, fixed at one end and partially filled with a special liquid and water mixture with a boiling point below that of water. (Figure 12.20). Integral to the bellows is a valve attached to its free end.

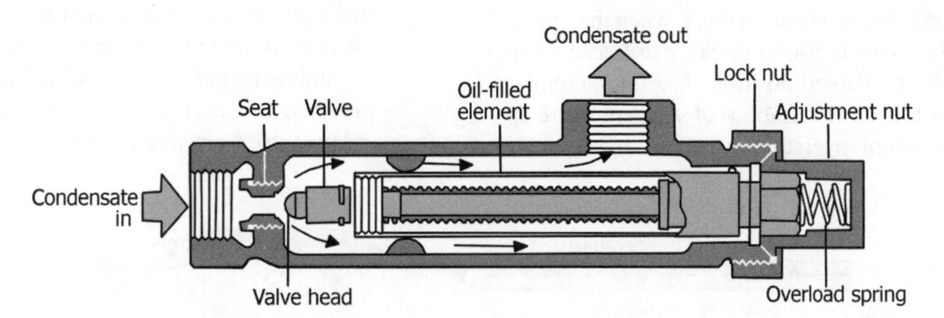

FIGURE 12.19 Liquid expansion steam trap. (*Courtesy*: Spirax Sarco.)

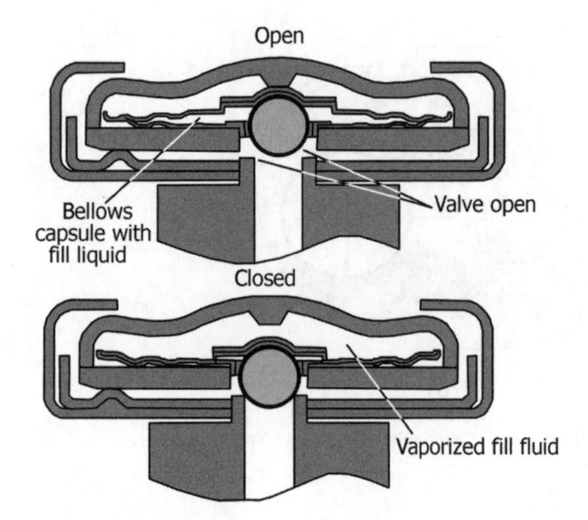

FIGURE 12.20 Balanced pressure steam trap. (*Courtesy*: Spirax Sarco.)

- Under ambient conditions, the bellows is contracted with the valve away from the seat. When steam, or condensate near its saturation point, comes in contact with the bellows, the liquid inside the bellows expands and drives the valve into the seat closing off steam flow.
- As the steam condenses, collects, and cools, the bellows will cool and contract, backing the valve away from the seat and allowing the accumulated condensate to pass. As the condensate passes through the trap and is replaced by steam, the bellows heats up again. As steam comes in contact with the bellows, the liquid inside the bellows gets heated up and expands the bellows resulting in closure of the valve, shutting off flow.
- The differential below steam temperature at which the trap operates is governed by the concentration of the liquid mixture in the capsule. The thin-walled element gives a rapid response to changes in pressure and temperature.
- *Merits*: Small, light, and has a large capacity for its size. The valve is fully open on start-up and remains in this position, allowing air and other noncondensable gases to be discharged freely and giving maximum condensate removal capacity, when the load is highest. This type of trap is unlikely to freeze when working in an exposed position. The trap automatically adjusts itself to variations of steam pressure up to its maximum operating pressure. It will also tolerate up to 70°C of superheat. Maintenance is simple. The capsule and valve seat are easily removed and replacements can be fitted in a few minutes without removing the trap from the line.
- *Demerits*: It does not open until the condensate temperature has dropped below steam temperature. It is susceptible to water hammer. It is generally not recommended for superheated steam service. Consideration must be given to providing sufficient upstream piping capacity for the accumulation of condensate. It does not open until the condensate temperature has dropped below steam temperature.

- What is a bimetallic steam trap? How does it operate? What are its merits and demerits?
  - Bimetallic steam traps are constructed using two strips of dissimilar metals welded together into one element. The element deflects when heated as illustrated in Figure 12.21.
  - There are two important points to consider regarding this simple element:
    - Operation of the steam trap takes place at a certain fixed temperature, which may not satisfy the requirements of a steam system possibly operating at varying pressures and temperatures.
    - Because the power exerted by a single bimetallic strip is small, a large mass would have be used that would be slow to react to temperature changes in the steam system.
  - To overcome the above constraints, a two-leaf design is developed to make the trap operate at different temperatures as shown in Figure 12.22.
  - A better design is the bimetallic disk-type steam trap as shown in Figure 12.23.
  - Bimetallic circular disks are used in which two sets of bimetallic laminates are joined at the perimeter with the metal layer of each bimetallic disk having the lower rate of expansion facing each other. Several sets of these joined disks may be stacked to increase the force applied when they expand.
  - Through the center of the stacked disks is a rod, which is attached to the uppermost disk. The rod runs through the sets of disks and then through a seated orifice. At the end of the rod is a valve. Under relaxed or ambient conditions, the disks are flat against one another. Under hot conditions, each set expands against itself causing the bellows to expand. As the

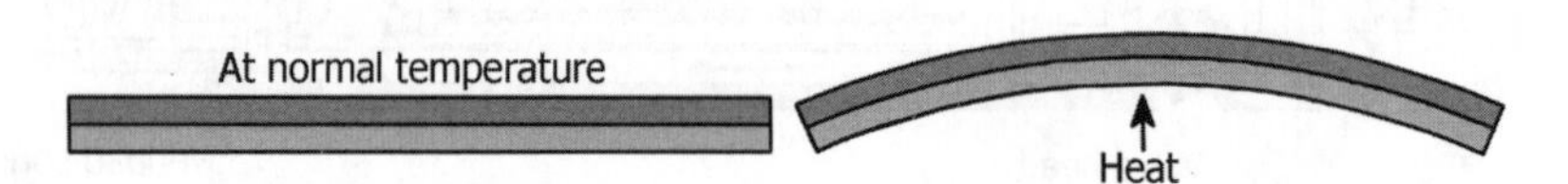

FIGURE 12.21 Bimetallic element made out of two laminated dissimilar metal strips. (*Courtesy*: Spirax Sarco.)

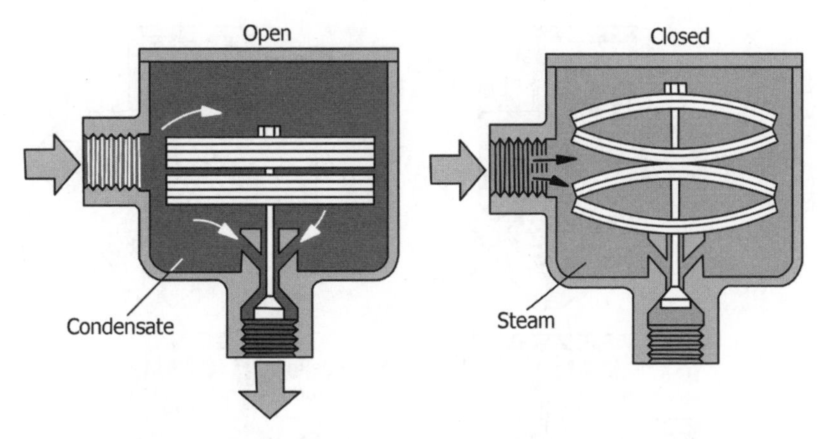

FIGURE 12.22 Operation of a bimetallic steam trap with a two-leaf element. (*Courtesy*: Spirax Sarco.)

bellows expands, it draws the valve into the seat of the orifice blocking off the flow of steam.

- *Merits*: These traps are generally compact but can have a large condensate capacity. The valve is wide open when the trap is cold, giving good air venting capability and maximum condensate discharge capacity under start-up conditions. Will not freeze when exposed to cold weather conditions. Able to withstand water hammer, corrosive condensates, and high steam pressures. Can work over a wide range of steam pressures without any need for a change in the size of the valve orifice. Maintenance is simple and the internals can be replaced without removing the trap body from the line.
- *Demerits*: As condensate is discharged below steam temperature, water logging of the steam space will occur unless the trap is fitted at the end of a long cooling leg, typically 1–3 m of unlagged pipe. Bimetallic steam traps are not suitable for fitting to process plants where immediate condensate removal is vital for maximum output to be achieved. This is particularly relevant in temperature-controlled plants. Some bimetallic steam traps are vulnerable to blockage from pipe dirt due to low internal flow velocities. If the bimetallic steam trap has to discharge against a significant backpressure, the condensate must cool to a lower temperature than is normally required before the valve will open. A 50% backpressure may cause up to a 50°C drop in discharge temperature. It may be necessary to increase the length of cooling leg to meet this condition. These traps do not respond quickly to changes in load or pressure because the element is slow to react.

- How does a thermodynamic steam trap operate? What are its features, merits, and demerits?
  - The thermodynamic trap operates by means of the dynamic effect of flash steam as it passes through the trap, as shown in Figure 12.24. The only moving part

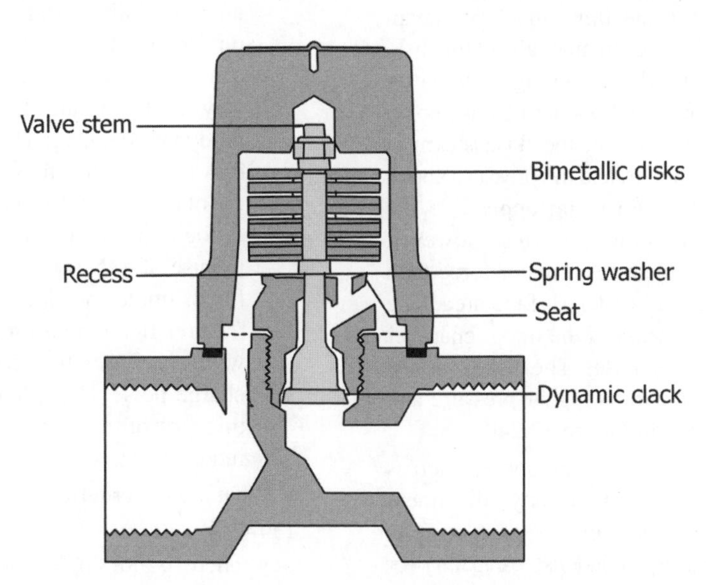

FIGURE 12.23 Multicross elements as used in the Spirax Sarco SM range of bimetallic steam traps. (*Courtesy*: Spirax Sarco.)

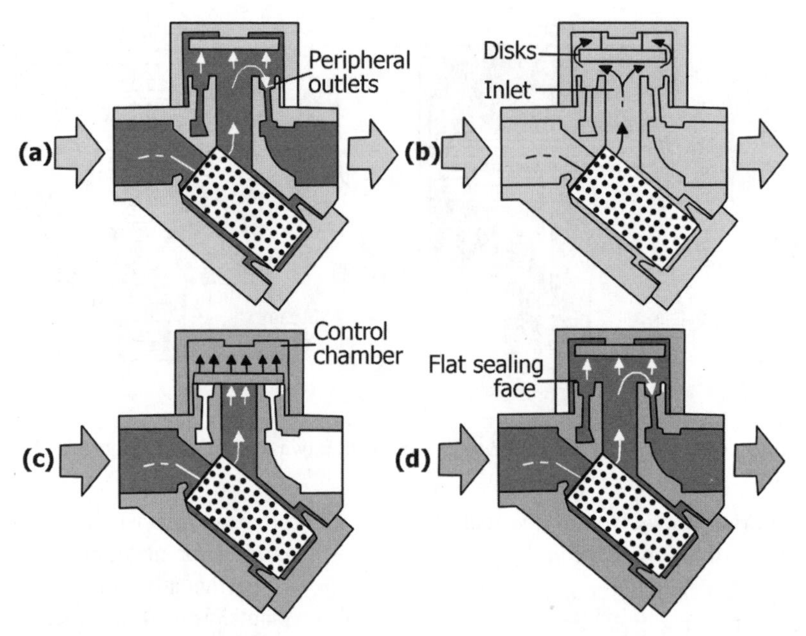

FIGURE 12.24 Operation of a thermodynamic steam trap. (*Courtesy*: Spirax Sarco.)

is the disk above the flat face inside the control chamber or cap. It is an extremely robust steam trap with a simple mode of operation.

- On start-up, incoming pressure raises the disk, and cool condensate plus air is immediately discharged from the inner ring, under the disk, and out through three peripheral outlets (only two are shown in Figure 12.24a.
- Hot condensate flowing through the inlet passage into the chamber under the disk drops in pressure and releases flash steam moving at high velocity. This high velocity creates a low-pressure area under the disk, moving it up toward its seat as shown in Figure 12.24b. At the same time, the flash steam pressure builds up inside the chamber above the disk, forcing it down against the incoming condensate until it seats on the inner and outer rings as shown in Figure 12.24c. At this point, the flash steam is trapped in the upper chamber, and the pressure above the disk equals the pressure being applied to the underside of the disk from the inner ring. However, the top of the disk is subject to a greater force than the underside, as it has a greater surface area. The pressure of the trapped steam in the upper chamber falls as the flash steam condenses. The disk is moved up by the condensate due to its higher pressure and the cycle repeats as seen in Figure 12.24d.
- *Merits*: Thermodynamic traps can operate across their entire working range without any adjustment or change of internals. They are compact, simple, lightweight, and have a large condensate capacity for their size. They can be used on high-pressure and superheated steam and are not affected by water hammer or vibration. The all stainless steel construction offers a high degree of resistance to corrosion. These traps are not damaged by freezing. As the disk is the only moving part, maintenance can be easily carried out without removing the trap from the line.
- *Demerits*: These traps will not work positively at very low differential pressures, as the velocity of flow across the underside of the disk is insufficient for lower pressure to occur. They are subjected to a minimum inlet pressure (typically 0.25 bar gauge) but can withstand a maximum backpressure of 80% of the inlet pressure. They can discharge a large amount of air on start-up if the inlet pressure builds up slowly. However, rapid pressure buildup will cause high-velocity air to shut the trap in the same way as steam, and it will "air bind." Modern thermodynamic traps can have an inbuilt anti-air-binding disk that prevents the build-up of air pressure on the top of the disk and allows air to escape. The discharge of the trap can be noisy and this factor may prohibit the use of a thermodynamic trap in some locations, for example, outside a hospital ward or operating theatre. If this is a problem, it can be easily fitted with a diffuser that considerably reduces the discharge noise. Care should be taken not to oversize this type of trap as this can increase cycle times and induce wear.

- What are the criteria involved in the selection of steam traps?
  - There is no single universally adopted steam trap technology for all applications. Several options are

available to accommodate all steam trap applications and performance levels.

- Equipment costs, installation and piping requirements, size and capacity of the traps, ease of maintenance and application or service in which the trap will be placed, and the control requirements are some considerations that go into the selection.

- What are the key design features for the selection of steam traps?
  - No steam losses under all operating conditions.
  - Rugged construction with low maintenance costs.
  - Fail-open feature for least disruption of process operation.
  - Self-draining design to prevent damage from freezing during plant shutdown.
  - Reduced valve seat wear.
  - Incorporation of features such as strainer, check valve, and air venting capability in the design.
  - Effective condensate discharge.
  - Fast plant start-up with continuous venting of air and CO_2.
- What are the important considerations involved in the selection of steam traps?
  - The considerations include venting of air during start-up, variations of system pressures and condensing loads, operating pressure and system load, continuous or intermittent operation of the system, usage of dry or wet return lines, and overall probability of water hammer.
    - *Venting of Air*: At start-up all steam piping, coils, drums, tracer lines, or steam spaces contain air. This air must be vented before steam can enter. Usually the steam trap must be capable of venting the air during this start-up period.
    - *Varying Loads and Pressures*: When a modulating steam regulator is used, such as on a heat exchanger, to maintain a constant temperature over a wide range of flow rates and varying inlet temperatures, the condensate load and differential pressure across the trap will change. When the condensate load varies, the steam trap must be capable of handling a wide range of conditions at constantly changing differential pressures across the trap.
    - *Differential Pressure Across the Trap*: To assure condensate drainage, there must be a positive differential pressure across the trap under all load conditions.
    - *Water Hammer*: When a trap drains high-temperature condensate into a wet return, flashing may occur. When the high-temperature condensate at saturation temperature discharges into a lower pressure area, this flashing causes steam pockets to occur in the piping, and when the latent heat in the steam pocket is released, the pocket implodes causing water hammer. Floats and bellows can be damaged by water hammer conditions. When traps drain into wet return lines, a check valve should be installed after the trap to prevent backflow. The check valve also reduces shock forces transmitted to the trap due to water hammer. Where possible, wet returns should be avoided.
    - *Process Equipment Design*: The design of the equipment being drained is an important element in the selection of the trap. Some equipment will permit the condensate to back up. When this occurs, the steam and condensate will mix and create water hammer ahead of the trap. A shell and tube heat exchanger has tube supports in the shell. If condensate backs up in the heat exchanger shell, steam flowing around the tube supports mixes into the condensate and causes steam pockets to occur in the condensate. When these steam pockets give up their latent heat, they implode and water hammer occurs often damaging the heat exchanger tube bundle. The trap selection for these types of conditions must completely drain condensate at saturation temperature under all load conditions.
    - Steam mains should be trapped to remove all condensate at saturation temperature. When condensate backs up in a steam main, steam flow through the condensate can cause water hammer. This is most likely to occur at expansion loops and near elbows in the steam main.
- Summarize steam trap selection criteria.
  - Table 12.5 gives briefly and broadly the steam trap selection criteria.
- What are the common practices in the installation/application of steam traps?
  - Using the traps for steam heaters and other heat transfer equipment involving use of steam.
  - Placing the traps at intervals along superheated steam lines to steam turbines/engines.
  - Installing the traps at the ground level, below the equipment, as close to the drain point as possible.
  - Providing strainers before the trap to prevent dirt entering the trap. During plant commissioning, the lines should be thoroughly flushed before fitting traps to prevent choking of the strainers.
  - The trap should be easily accessible for maintenance.
  - An isolation valve should be placed before the trap and another after the trap, the latter in case a condensate return header is used.

TABLE 12.5 Steam Trap Selection Criteria

			Mechanical		
Variable	Thermostatic	Thermodynamic	Float and Thermostatic	Inverted Bucket	Free Float
Response to load changes	Moderate	Slow	Fast	Moderate	Fast
Venting of noncondensables	High	Low	Medium to high	Low	High
Energy efficiency	High	Medium	Medium to high	Medium	Medium to high
Affected by ambient temperatures	No	Yes, unless insulated	No, but susceptible to freezing		
Relative cost	Low to medium	Low	Medium	Low	High
Capacity	Medium	Low	High	High	High
Size in relation to capacity	Small	Medium	Large	Large	Large
Service life	Moderate	Moderate	Moderate	Moderate	Long
Ease of maintenance	Very easy	Very easy	Moderate	Moderate	Moderate
Orientation limits	No	No	Yes	Yes	Yes
Applications	Drip legs[a] Process equipment[a] Steam tracing[a]	Drip legs[b] Not recommended Steam tracing[b]	Drip legs[b] Process equipment[a] Steam tracing[b]	Drip legs[c] Process equipment[b]	Drip legs[b] Process equipment[a]

[a] First choice.
[b] Second choice.
[c] Third choice.

- *Float-type* traps should be installed as close as possible below the equipment to be drained. It must be installed in horizontal position so that the float rises and falls vertically. The direction of flow should be as indicated on the float cover.
- *Inverted bucket trap* must be installed vertically with the inlet at the bottom and outlet at the top. At start-up, the inlet valve of the trap should be opened gradually allowing filling of the trap with condensate. Because the trap works on the principle of buoyancy, a minimum level of liquid must be maintained inside the trap. If superheated steam is involved, a minimum uninsulated pipe length must be ensured to prevent vaporization of the condensate seal in the trap.
- *Thermodynamic traps* can operate in any position, but horizontal position is preferred with the disk cap on the top. Under no circumstances the cap should be left loose after adjustment or maintenance, as a loosely fitted cap will leak steam from its joint with the body.
- *Thermostatic trap* should be installed at least 1.5 m of uninsulated pipe from the inlet header, as condensate must be cooled below the steam temperature for the trap to operate. The trap should be installed at least 1.5 m from the condensate outlet connection to avoid flash steam produced affecting the thermodynamic element. The trap should be in horizontal position and should not be installed in a vertical pipe run. The outlet side of the trap should have a continuous downward slope to avoid chances of freezing.

- Is there any need to use steam traps for lines carrying superheated steam feeding steam turbines?
  - Theoretically no, as there will not be any condensation as long as superheat is maintained in the steam. But during start-up conditions, condensate removal becomes a necessity to avoid steam losses.
  - Thermal insulating capability of materials and the insulation techniques have improved such that temperature drop along the steam lines is low.
    - For example, in an installation, superheater outlet pressure was 41 bar and temperature 380°C. Turbine, at the end of pipe length of 400 m, receives this steam at 40.4 bar and 365°C. Saturation temperature is about 250°C.
  - Steam of such a high superheat would not condense along the line, and under these circumstances, use of a steam trap becomes redundant.
- It is often observed that steam *escapes* from an otherwise perfectly functioning steam trap. Explain why.
  - Pressure of the steam, in steam lines and heat transfer equipment on which steam traps are used, is always higher than the atmospheric pressure.
  - In other words, the condensate inside a steam trap is at pressures above atmospheric pressure.
  - When the condensate is discharged to atmosphere, expansion takes place from high pressure to low pressure and part of the condensate vaporizes.
- What are the effects of oversizing of steam traps on their performance?

- Oversizing → rapid recycling → excessive wear and shortened trap life.
- Requires extra maintenance.
- Not desirable. It is the most common mistake.
- Misconception is that trap size is dictated by the diameter of the pipe.
- Trap should be sized according to the amount of condensate being drained and pressure condition.

• Why is it necessary to use superheated steam in steam turbines/engines?
  - The higher the temperature difference, the greater the output of the turbine/engine.
  - Process plants use steam at high pressure and high superheat (100–200°C superheat) to drive turbines, but extract it from these machines at a lower pressure for heating purposes.

• Summarize common problems of steam traps.
  - *Steam Leakage*: The valve seat of a steam trap might be damaged due to erosion and corrosion leading to leakage. Other reasons for leakage are oversizing of the trap, reduction of its load, and too low a pressure of the condensate.
  - *Improper Sizing*: An undersized trap will cause condensate to form a film on heat transfer surfaces and reduce its efficiency. Too much oversizing results in sluggish operation and generation of high back-pressure significantly reducing the life of the trap, besides being expensive.
  - *Dirt*: Scaling and corrosion products can erode trap valves, plugging the discharge valve or jamming it open. A strainer must be installed upstream to filter out the solids and maintained to prevent accumulation of solids.
  - *Noise*: Noise in a steam system is usually caused by lifting condensate in vertical return lines, water hammer, or a failed trap that leaks live steam to the condensate return line.
  - *Air Binding and Steam Locking*: When the trap is connected to the equipment by a long small diameter horizontal pipe, condensate backs up in the steam space by air binding or steam pockets and cannot flow to the trap. To prevent air binding, piping to the traps should be of large size and short. Another method is to put a vent valve at a high point in the system.
  - *Water Hammer*: Condensate at the bottom of a steam line can cause water hammer. High-velocity steam pushes this slug of condensate and creates water hammer when there is a change of direction, with condensate stopping or slowing down, giving up kinetic energy and thereby increasing pressure suddenly. This damages mechanism of float and balanced pressure thermodynamic traps.
  - *Freezing*: If the steam system is shut down with an appreciable amount of condensate in the trap, and the ambient temperature falls to subzero temperatures, freezing inside the trap occurs. Float and balanced pressure thermostatic traps are damaged by freezing (and expansion). Opening drain valves after shutdown is an option to prevent accumulation of the condensate in the trap.
  - *Loss of Prime*: This is a problem in inverted bucket traps, which start functioning only when some water is in the trap. Prime is lost when superheated steam enters the trap or there is sudden drop in pressure. A check valve in the trap inlet line prevents this problem.

12.7 IMMERSION HEATERS

• What are immersion heaters?
  - As the name implies, immersion heaters are directly immersed in the fluids being heated. These are electrical resistance heaters, with a protective metallic sheath, fabricated into different configurations to fit into a vessel, providing heat transfer surface to the fluid being heated.
  - These are compact and more efficient compared to heating or cooling coils through which steam or other heat transfer fluid or cooling water is circulated.
  - A wide range of fluids, including water, oils, viscous liquids, solvents, process solutions, molten metals, and gases, are heated with these heaters.
  - Range of designs includes size, kilowatt ratings, voltage, sheath materials, connections, and other accessories.

• What are the different types of immersion heaters?
  - The basic types include the screw plug, flange, pipe insert, and circulating.
  - These are generally available in either a round tubular or a flat tubular design.
  - The heaters are further classified as closed pressurized systems and nonpressurized, open tank systems.

• What are the applications of different types of immersion heaters?
  - Square-flange immersion heaters, consisting of round or flat design, are used in industrial water heaters and storage tanks with fuel oils, heat transfer fluids, caustic soda solutions, and solvents. These heaters are bolted directly to a mating companion flange that is welded to a tank wall or nozzle. This facilitates easy assembly and disassembly.

- Screw plug heaters are used for applications such as heating deionized process water, oils, antifreeze glycol solutions, liquid paraffin, and cleaning solution tanks, among others. These heaters are installed by inserting into threaded opening of a tank wall. These are typically used for pressures less than 6890 kPa (1000 psi) and maximum wattages of about 36 kW. For wattages up to 3000 kW or higher, flange heaters are used. These heaters can handle high-pressure applications up to 20,000 kPa (3000 psi) or more. Such heaters are typically used in pressurized and nonpressurized tanks handling superheated steam and compressed gases.
- Pipe insert or bayonet heaters, which are mounted inside a pressure-tight bayonet sealed pipe, are used for heating liquids in very large storage tanks with millions of liters capacity. The heater can be removed from the pipe without draining the tank.
- Circulating or in-line heaters come as complete units with an insulated tank with inlet and outlet piping and the liquid or gas gets heated as it flows through the tank.
- Booster heaters are a type of circulating heaters for low-wattage (3–6 kW or sometimes 12 kW) applications, for example, for engine start-up by circulating ethylene glycol in the engine.

• What is the basic information required for sizing an immersion heater?
  - Energy requirements for initial heating and heating during operational cycle.
  - Energy requirements for phase change, such as melting or vaporization, of materials, initial and operational cycles.
  - Energy requirements to account for heat losses.
  - Watt density, that is, total wattage divided by the active heater surface area, which is to be used in the design of heater elements—their size and length requirements.

13

COMPACT HEAT EXCHANGERS, REGENERATORS, AND RECUPERATORS

13.1 Compact Heat Exchangers 395
13.1.1 Air-Cooled Heat Exchangers 401
13.1.2 Plate Heat Exchangers 403
13.1.3 Plate-Fin Heat Exchangers 413
13.1.4 Spiral Heat Exchangers 413
13.1.5 Scraped Surface Heat Exchanger 415
13.1.6 Heat Pipes 416
13.1.7 Printed Circuit Heat Exchangers 420
13.1.8 Other Systems for Cooling Electronic Devices 421
13.2 Other Types of Heat Exchangers 422

13.1 COMPACT HEAT EXCHANGERS

- Classify compact and noncompact heat exchangers according to surface compactness.
  - Heat exchangers having surface area density >700 m^2/m^3 can be considered as compact heat exchangers.
  - Heat exchangers having surface area densities <700 m^2/m^3 can be treated as noncompact exchangers.
- What are the different techniques used to make a heat exchanger compact?
  - By the use of finned circular tubes.
  - By the use of fins between plates.
  - By the use of densely packed continuous or interrupted cylindrical flow passages of various shapes.
- What are the objectives of different heat transfer enhancement techniques?
  - A variety of heat transfer enhancement techniques have been developed for different applications over the years with the important objective of reducing heat transfer resistance in a heat exchanger by promoting higher convective heat transfer rate with or without surface area increases.
  - These results in reducing size of the heat exchanger, increasing heat duty of an existing heat exchanger, reducing ΔP and/or decreasing operating ΔT.
  - Decreasing ΔT is particularly useful in thermal processing of food, pharmaceutical, biochemical, and polymeric materials to avoid thermal degradation of the end products.
  - Heat exchange systems in electronic devices, spacecraft, and medical applications may primarily rely on enhanced heat transfer.
- Give a brief classification of heat transfer enhancement techniques.
  - Table 13.1 gives a classification of passive and active heat transfer enhancement techniques.

    Passive Techniques
  - *Treated surfaces* involve surface finish, coating techniques, and useful in boiling and condensation applications.
  - *Rough surfaces* are used to increase turbulence primarily for single-phase heat transfer.
  - *Extended surfaces* are finned surfaces to increase heat transfer area. Modified finned surfaces have dual role of increasing heat transfer area as well as increased turbulence.
  - *Displaced enhancement devices* are inserts primarily used in confined forced convection in tubes and ducts. They improve energy transport indirectly at the heat transfer surface by fluid displacement action over the surface.
  - *Swirl flow devices* produce and superimpose swirl or secondary recirculation on the axial flow in a channel.
  - *Coiled tubes* give good compactness. Curvature of the coils produces secondary flows that promote

Fluid Mechanics, Heat Transfer, and Mass Transfer: Chemical Engineering Practice, By K. S. N. Raju

TABLE 13.1 Classification of Passive and Active Heat Transfer Enhancement Techniques

Passive Techniques	Active Techniques
Treated surfaces	Mechanical aids
Rough surfaces	Surface vibration
Extended surfaces	Fluid vibration
Displaced enhancement devices	Electrostatic fields
Swirl flow devices	Injection
Coiled tubes	Suction
Surface tension devices	Jet impingement
Additives for liquids/gases	Combinations of active and passive techniques

higher heat transfer coefficients in single phase as well as boiling heat transfer.

- *Surface tension devices* consist of wicking or grooved surfaces, which direct and improve liquid flow to boiling surfaces and from condensing surfaces.
- *Additives for liquids* include solid particles, soluble trace additives, and gas bubbles in single-phase flows and trace additives to depress surface tension of the liquid for boiling systems.
- *Additives for gases* include liquid droplets or solid particles, which are introduced in single-phase gas flows in either dilute phase (gas–solid suspensions) or dense phase (fluidized beds).

Active Techniques

- *Mechanical aids* stir the fluid by mechanical means or by rotating the surface. Examples include rotating tube or scraped surface heat exchangers.
- *Surface vibration* has been applied at either low or high frequency in single-phase flows to obtain higher heat transfer coefficients.
- *Fluid vibration* or fluid pulsations with vibrations ranging from 1 Hz to ultrasound ($\approx$1 MHz) used in single-phase flows. Considered to be one of the most practical vibration enhancement techniques.
- *Electrostatic fields*, which could be electric or magnetic fields or combinations of the two, from DC or AC sources, can be applied in heat exchange systems involving dielectric fluids. Depending on the application, they can promote greater bulk fluid mixing and induce forced convection or electromagnetic pumping to enhance heat transfer.
- *Injection*, used only in single-phase flow, pertains to the method of injecting the same or a different fluid into the main bulk fluid either through a porous heat transfer interface or upstream of the heat transfer section.
- *Suction* involves either vapor removal through a porous heated surface in nucleate or film boiling or fluid withdrawal through a porous heated surface in single-phase flow.
- *Jet impingement* involves the direction of heating or cooling fluid perpendicularly or obliquely to the heat transfer surface. Single or multiple jets (in cluster or staged axially along the flow channel) may be used in both single phase and boiling applications.

- When can enhanced tubes be considered for use in place of plain tubes in a heat exchanger? What are the benefits of such an arrangement?
  - A rule of thumb says that enhancement may be considered when the thermal resistance of a fluid is *three times* that of the other fluid.
  - With increased experience with availability and use of tubes with enhancement on both sides, it would be beneficial to use such tubes, especially for new exchangers, which result in fewer tubes, smaller shell sizes, shorter tube lengths, and fewer tube passes with significant savings.
  - Sometimes, the enhanced tube unit can be half the size of a plain tube unit or two units in parallel can be replaced by one small enhanced tube unit. In such cases, cost savings can exceed 50% by using externally low-fin surface and internally microfinned surface or double-finned tubes.
- What are the common misconceptions with regard to application of heat transfer enhancement and how are these addressed?
  - The most common misconception is that although enhancement provides improved heat transfer at the expense of increased pressure drop. In reality the reverse is often true. Area and pressure drop reduction can often be achieved simultaneously.
  - The second misconception is that the enhancement devices that enhance flow are only worthwhile when applied to a stream having a controlling heat transfer coefficient, that is, having lower heat transfer coefficient.
  - The third is that fully turbulent flow is always the best option for heat transfer. Operating at lower tube velocity would appear to imply that exchangers employing heat transfer enhancement will have higher tube counts and consequently larger shell diameters than those using plain tubes. This may not necessarily be the case.
  - Mechanical constraints can play a significant role in the design of shell and tube heat exchangers. For example, restrictions are often be placed on the length of the tubes used. This can severely limit the pressure drop that can be used to promote heat transfer.

- In addition, duties involving close temperature approaches are often best handled in exchangers employing pure countercurrent flow. In the case of shell and tube heat exchangers, this means the use of single tube passes. However, if such duties also involve high heat loads, the designer can be faced with a serious problem. Once the maximum tube length has been reached, the surface area of the exchanger can only be increased by increasing the tube count. The result of increased tube count is reduced tube side velocity and reduced tube side heat transfer coefficient. If the reduction in tube side coefficient becomes large, or if the tube side Reynolds number approaches the transitional regime, the designer either has to artificially increase the *tube length* by increasing the number of shells or has to switch to a multiple tube pass design.
- The use of multiple tube passes can have several detrimental effects. First, it leads to a reduction in the number of tubes that can be accommodated in the shell. Second, for bundles having more than four tube passes, introducing pass partition lanes into the bundles results in increased bundle bypassing and a reduction in the shell side heat transfer coefficient. Third, it gives rise to wasted tube side pressure drop in the return headers. Finally, the use of multiple tube passes introduces the need to examine temperature cross considerations (i.e., where the cold stream outlet temperature is higher than the hot stream outlet temperature) and therefore reduction in the effective mean temperature difference and the need to split the duty between two or more exchangers operating in series.
- The net result of the problem is the decline in tube side coefficient as the tube count increases. Heat transfer enhancement can be used to overcome this problem.
- Devices that work very effectively in the laminar and transitional flow regimes can be used to maintain the heat transfer coefficient as the tube count increases, without incurring any pressure drop penalty.
- There are situations where the designer may deliberately choose a multishell design in order to make more effective use of available pressure drop. This happens when more pressure drop is available than can be used in a single shell and this additional pressure drop can be absorbed in a multishell design.
- The designer can then choose between a single-shell design or a series of shells of smaller diameter containing lower overall area. Again, the use of enhancement is an option that should be considered here. Plain tubes make poor use of pressure drop because friction in plain tubes is relatively low.
- Finally, one other consideration is also important in the context of effective use of pressure drop. There are limitations on the velocity that can be used. For example, a maximum allowable tube velocity of around 2–3 m/s (depending on the tube material) is often imposed for clean liquids. Velocities higher than this could possibly give rise to erosion problems.
- The more erosive the liquid, the lower the maximum velocity. Liquids containing suspended solids obviously require lower velocities.
- Some inserts (e.g., wire matrix inserts) operate with much lower shear than occurs in plain tubes. These allow fuller use of pressure drop without the danger of tube erosion.

• Name different types of heat exchangers that can be classed as compact heat exchangers. Give their characteristic features and applications.

- *Finned Tube (Tube-Fin) Heat Exchangers*: Finned surfaces are on only one side of the tubes. Have area densities of up to 3300 m^2/m^3. Used for gas–liquid applications. Tube side pressures of up to about 3000 kPa gauge and tube outside pressures of up to 100 kPa gauge. Can be designed for temperatures ranging from cryogenic to about 870°C. For moderate fouling outside tubes, plain uninterrupted longitudinal fins are used. These exchangers are predominant in cryogenic applications.
- *Finned Plate (Plate-Fin) Heat Exchangers*: High area densities of up to about 6000 m^2/m^3. Fins can be on both sides of plates. Commonly used for gas–gas heat exchangers. Generally designed for low-pressure applications (up to about 1000 kPa). Designed for temperatures ranging from cryogenic to about 800°C.
- *Plate Heat Exchangers*: Corrugated or embossed plates are packed to form channel passages for flow.
- *All-Welded Compact Heat Exchangers*: Consist of corrugated plates, welded together, eliminating the need for gaskets, making them suitable for processes involving high temperatures and aggressive fluids.
- *Regenerators*: Can have most compact surface area density compared to plate-fin or tube-fin surfaces, up to about 8800 m^2/m^3 for rotary-type regenerators and up to 16,000 m^2/m^3 surfaces for fixed-matrix regenerators. Used for gas to gas heat exchange involving low pressures, up to 615 kPa gauge for rotary type and even lower pressures for fixed-matrix regenerators. Have self-cleaning characteristics as hot and cold gases flow alternatively in the opposite directions through the same passage.

- What are the different techniques used to make a heat exchanger compact?
  - Use of fins between plates.
  - Use of finned circular tubes.
  - Use of densely packed continuous or interrupted cylindrical flow passages of various shapes.
- What are the applications of a finned tube heat exchanger?
  - Where heat transfer coefficients are very low and cannot be increased by heat transfer enhancement methods such as velocity/turbulence, for example, liquid to gas or gas to gas, as in an air-cooled exchanger. Viscous fluids such as lube oil, hydraulic oil, gear oil almost always necessitate use of finned tubes or extended surfaces to enhance the heat transfer coefficient on gas and oil side. An example of finned surfaces for room heating purposes involves oil heaters heating air.
  - Increasing surface by increasing number of tubes increases bulkiness of the exchanger and lowers tube side velocities, reducing heat transfer rates.
  - Fins not only increase heat transfer surface but also create turbulence around the tube as well as between the fins. This reduces the air side film resistance and increases air side heat transfer coefficient many fold.
  - Finned tubes increase heat transfer area by 2–2.5 times more than smooth tubes without increasing flow area and making the exchanger compact. Pressure drops are lower because of compactness of the exchanger.
- Will it be helpful to use finned tubes with fouling liquids?
  - No. High viscosity and fouling liquids create cleaning problems for finned surfaces. Fouling deposits on finned surfaces also create enhanced corrosion problems due to the imperfections introduced while making finned tubes.
- Why strips of metal (longitudinal rectangular fins) are welded to the outer casing of an electric motor?
  - Electric motor needs to be cooled, and air being poor electrical conductor; it is used as cooling medium by simply exposing the motor surfaces to ambient air.
  - Heat transfer coefficients being very low for air or other gases compared to liquids, surface area needs to be increased to compensate lower heat transfer coefficients.
  - Therefore metal strips are welded to the motor surface to increase heat transfer rates.
- What are the limitations in the use of finned tube surfaces for heat transfer applications?
  - Finned tube surfaces are difficult to clean compared to smooth surfaces, which poses a limitation on the use of fins on such surfaces.
  - Fluid must be clean (nonfouling) and relatively noncorrosive.
  - However, flue gases being hot and dry, cleaning of the surfaces can be done by blowing, as use of normal surfaces make the equipment too bulky on account of poor heat transfer coefficients for gases.
  - Operating temperatures and pressures are limited due to considerations of mechanical expansion, brazing, and other constructional features.
  - ΔP and hence fluid pumping power is more over finned surfaces.
  - For uniform flow distribution, header design for the exchanger is important.
- Are there any difficulties in fitting finned tubes into tube sheets of a shell and tube exchanger?
  - No. Finned tubes are available with plain ends so that these tubes are fitted in a similar way as plain tubes into the tube sheets. The plain ends of the tubes are a little longer than the tube sheet thickness so that they can be fitted by rolling or welding to the tube sheets.
- What is the function of car radiator? Is it a heat exchanger? If so, what type of heat exchanger it is?
  - Car radiator is a compact heat exchanger, its function being to cool the engine of the vehicle.
- "Fins are used on the outside surface of a heat exchanger tube when the heat transfer coefficient on the outside surface of the tube is higher than the heat transfer coefficient inside the tube." *True/False*?
  - *False*.
- Name different types of fins used on a finned tube heat exchanger.
  - Fins come in many shapes and sizes and can be broadly classified into fins of constant cross section, rectangular or pin (spike) fins, and fins of varying cross section, tapered fins.
  - *L-Footed Fins (Figure 13.1)*: Most commonly used fin type in air-cooled heat exchangers. The fin is

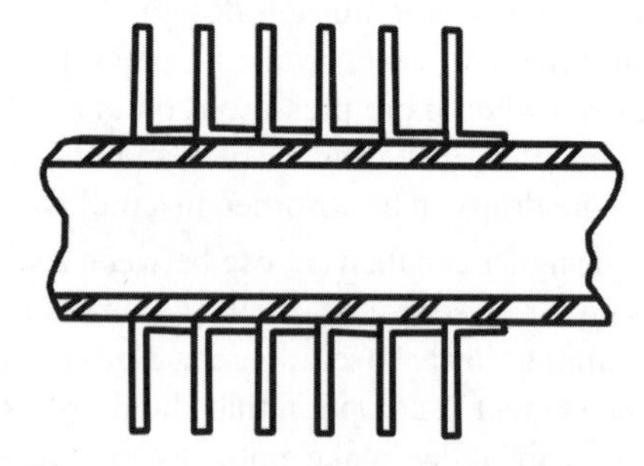

FIGURE 13.1 L-footed tension wound aluminum fin.

produced by wrapping an aluminum strip, that is, footed at the base around the tube by holding tension on the fin.

- The L-foot fin covers the tube more or less completely to protect the base tube against corrosive attack, but still leaves a potential corrosive site at the base of the fin adjacent to the preceding fin.
- This type of fin is used in services where the tube wall temperature does not exceed 180°C and air side corrosion is not extremely high. At the higher tube wall temperature, due to the difference in material between the tube and the fin, the fin will not maintain contact with the tube.
- *Embedded Fins (Figure 13.2)*: In high-temperature applications, an embedded process is employed to attach the fin to the tube wall. In this process, a groove is actually cut into the tube, the fin strip inserted, and the tube material then *plowed back* against the fin to bond it to the tube. Separation of the fin and tube due to corrosion or temperature differentials are not a factor with this fin type.
- In this type of construction, the tube is totally exposed to airside corrosion factors.
- Due to the groove cut into the tube, a thicker tube wall must be used to avoid over-pressuring the tube.
- The embedded fin is normally used for services greater than 180°C and less than 400°C.
- *Extruded Fins (Figure 13.3)*: Fins are extruded from the wall of an aluminum tube that is integrally bonded to the base tube for the full length.
- Since the tube is totally covered by the aluminum sleeve, the tube wall is protected from outside corrosion and the bond between the fin and the tube remains tight.
- For applications where atmospheric corrosion is critical, the extruded finned tube provides the best protection.

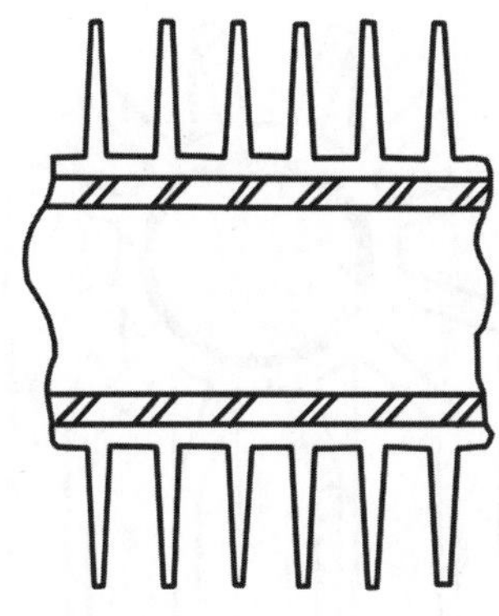

FIGURE 13.3 Extruded fin.

- The extruded finned tube is good for tube wall temperature up to 340°C. This is the most expensive finned tube to produce.
- *Double L-Footed Fins (Figure 13.4)*: This fin is similar to L-footed fin in that it is produced in much the same manner. In this process, a foot is formed on both sides of the upright portion of the fin, providing an overlapping of the fin. This provides a higher protection for the tube against atmospheric corrosion. This fin type is also referred to as an *overlapped* fin.
- Some configurations include the following:
  - *Extended Axial Fins*: See Figure 13.5.
  - The finned surface in Figure 13.6 is made by the use of solid fin formed by stretching a strip of metal around the tube.
  - The outer edge is stretched and the inner edge is compressed.
- A serrated fin (Figure 13.7) is formed by partially splitting the metal strip before winding it around the tube.

• What are low and high fin tubes?

- Fins are integral part of tube, for example, threaded tubes, tubes with groves, etc.

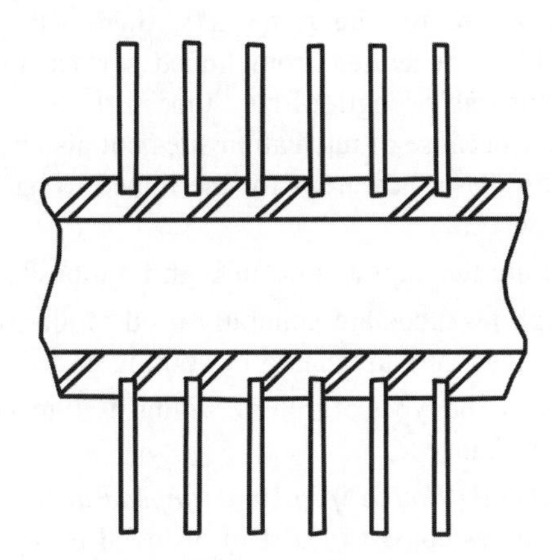

FIGURE 13.2 Embedded fin.

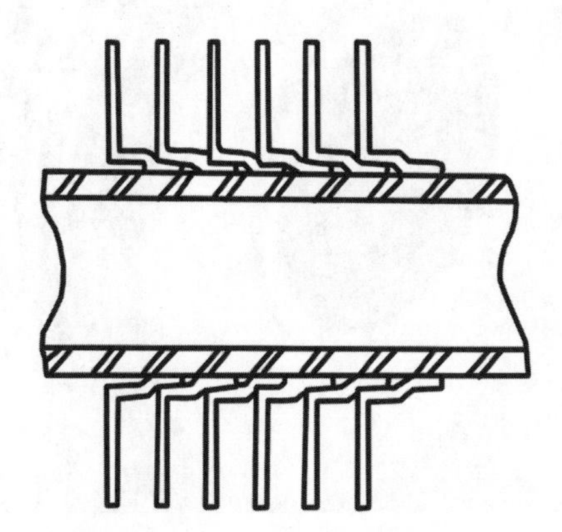

FIGURE 13.4 Double L-footed fin.

FIGURE 13.5 Extended axial finned tube.

- Usually there will be 6 or 8 fins/cm (16 or 19 fins/in.) of tube length.
- Have about 3.5 as the ratio of outside to inside surface.
- Adoptable to a conventional shell and tube heat exchanger and can be installed and handled in the same manner as plain tubes.
- The outside diameter of the fins is slightly less than the bare tube at the ends so that the tubes can be inserted through the holes in the tube sheet.

FIGURE 13.6 Continuous circular fins on a tube.

FIGURE 13.7 Serrated fins.

- High fin tubes have 2 or more per cm (5 or more fins per inch) with fins attached to tubes. The ratios of outside to inside surface areas can be up to 25.

- Under what circumstances use of low fin tubes are attractive?
  - When outside heat transfer coefficient is equal or less than 1/5th of the inside coefficient.
- What is the effect of replacing conventional tubes by low fin tubes on shell diameter requirement?
  - As number of tubes required decreases with the use of finned tubes, shell diameter gets reduced.
  - Usually shell diameter controls the exchanger cost.
- What are the advantages of using low fin tubes in a kettle reboiler?
  - In a kettle reboiler with low ΔT and in fouling media, low fin tube provides more nucleation sites per unit surface area than a bare tube. This is because finning operation involving alteration of tube surface with tools, tiny imperfections are introduced on finished finned surface.
  - Therefore for the same ΔTs, there will be more bubbles generated from finned surface than for a comparable length of bare tube surface. This is not only because of nucleation sites but also because of ratio of surface areas for the finned to bare tube is about 2.5.
- What are the applications of high fin tubes?
  - High fin tubes are generally used in air-cooled exchangers and in heating or cooling gases.
- What are the types of transverse fins used in a compact heat exchanger?
  - *Helically Wound Spiral or Crimped Fin Tubes:* These fin tubes consist of a strip of metal tension wound on tube.

  - *Wire Wound Fin Tubes:* Root soldered wire wound tubes consist of loops of wire helically wound over tube. The foot of each loop is root soldered to the base tube.
  - Transverse fins are also called radial fins.
- What are the common applications of transverse fins in a compact heat exchanger?
  - Transverse fins are used in cross flow and shell and tube configurations, in air-cooled exchangers.
    - High transverse fins are mainly used with low-pressure gases.
    - Low transverse fins are used for boiling and condensation of nonaqueous streams as well as for sensible heat transfer.
- Name some common applications for tubes with longitudinal fins.
  - Usually adapted to double pipe heat exchangers (which have limited surface areas) to increase area. Ratio of external to internal surface is about 10–15.
  - Also used in conventional tube bundles with special design considerations.
  - Other examples include cross-flow exchangers, kettle reboilers, chillers, and condensers.
- "Fins are more effective and beneficial when fin side heat transfer coefficient is lower than inside tube coefficient." *True/False*?
  - *True*.
- "Finned tube heat exchangers are suitable for heat transfer applications involving flue gases containing dust particles." *True/False*? Explain.
  - *False*. But these are used with effective blowing techniques to reduce bulkiness in the exchangers in flue gas ducts.
  - Dust particles will deposit on finned surfaces and increase fouling resistances. Also fouling deposits increase corrosion rates.
  - Cleaning of finned surfaces will be difficult.
  - However, finned surfaces are successful means of controlling temperature-driven fouling such as coking and scaling.
    - Fin spacing should be large enough to avoid entrainment of particulates in the fluid stream (minimum spacing = 5 mm).
- Define and explain what is meant by fin efficiency.
  - Fin efficiency is defined as the ratio of actual heat transfer from a fin to its ideal heat transfer, if the entire fin surface were at the same temperature as at the base.

$$\eta_{fin} = \text{Actual fin heat transfer rate}/\text{Heat transfer rate if the whole of the fin were at the root temperature} \quad (13.1)$$

  - Also defined as "*Ratio of mean temperature difference from fin surface to the fluid divided by temperature difference from fin to fluid at the base/root of the fin.*"
  - The additional area contributed by the fin is not as efficient for heat transfer as bare tube surface owing to resistance to conduction through the fin.
  - In order for a fin on a tube to conduct heat to or from a fluid, there must be a temperature gradient from the base of the fin on the tube to the tip or vice versa.
  - If temperature gradient is small, finned area will transfer almost as much heat per unit area as it would if the base and tip temperature were the same; that is, act like a bare tube and the efficiency of the surface would be high.
  - Fin efficiency is a function of heat transfer coefficient to the fin, thermal conductivity of the fin metal, height and thickness of the fin, and any contact resistance between the fin and the tube surface.
  - If heat transfer coefficient is high and the conduction ability of the fin is low, efficiency becomes poor.
- What is fin effectiveness?
  - The fin effectiveness, ε, is defined as the heat transfer from the fin compared with the bare surface transfer through the same base area.

13.1.1 Air-Cooled Heat Exchangers

- Illustrate, by means of diagrams, air-cooled heat exchangers.
  - Figure 13.8 illustrates the essential features of air-cooled heat exchangers.
- What are the size ranges for air-cooled finned surface heat exchangers?
  - Tube sizes: 19–25 mm (0.75–1 in.) OD.
  - Finned surface: 15–20 m^2/m^2 of bare tube surface.
- "Size of an air-cooled exchanger is much larger than a water-cooled heat exchanger for the same duty." Explain why?
  - Heat transfer coefficients for gases are much lower when compared to liquids, requiring large heat transfer areas for the same duty.
- What are the advantages of air-cooled heat exchangers?
  - Simple in construction even for high pressure and/or high temperatures.
  - No water problems, as associated with corrosion, algae, treating, fouling, etc.
  - Eliminates high cost of water including cost of treating water.
  - Thermal or chemical pollution of water resources is avoided.

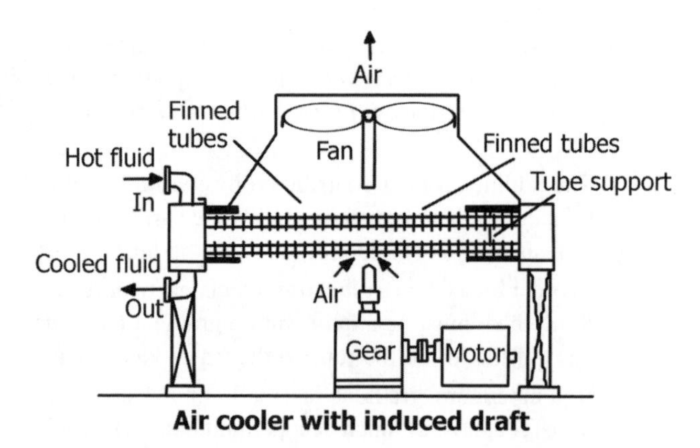

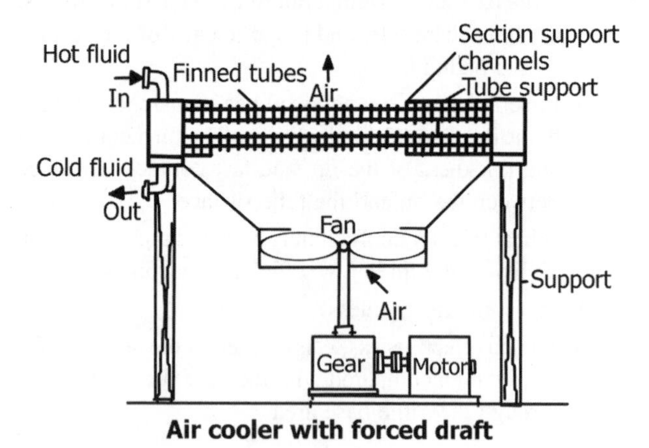

FIGURE 13.8 Air-cooled heat exchanger.

- Installation is simplified due to elimination of coolant water piping.
- Location of the air-cooled heat exchangers is independent of water supply location.
- Maintenance may be reduced due to elimination of water fouling characteristics, which require frequent cleaning of water-cooled heat exchangers.
- Air-cooled heat exchangers will continue to operate (but at reduced capacity) due to radiation and natural convection air circulation, in the event of power failure.
- Excellent for heat removal from high-temperature fluids (>90°C).
- Easy maintenance (≤1/3rd less than water cooling).
  - Fins can easily be cleaned by compressed air, brushes, etc., even while in operation.
- Lower operating costs.
- Floor space requirements are less than that for cooling towers.

• What are the limitations of air-cooled heat exchangers?

- High limitation on outlet fluid temperature.
- Generally most suitable for liquids or condensing vapors in tubes.
- Since air has relatively poor thermal transport properties compared to water, air-cooled heat exchangers require considerably more heat transfer surface area, which increases equipment costs (25–125% over water-cooled equipment for the same heat transfer duty) and space requirements.
- Temperature approach between the outlet process fluid temperature and the ambient air temperature is generally in the range of 10–5°C. Normally, water-cooled heat exchangers can be designed for closer approaches of 3–5°C.
- Outdoor operation in cold winter environments may require special consideration to prevent freezing of the tube side fluid or formation of ice on the outside surface.
- The movement of large volumes of air is accomplished by the rotation of large diameter fan blades rotating at high speeds. As a result, noise due to air turbulence and high fan tip speed is generated.
  - Rotating at high speeds, the fan blades must be balanced to ensure that centrifugal forces are not transmitted through the fan shaft to the drive or to the supporting structure. An unbalanced blade could result in severe vibration conditions.
- Limited use for gas cooling due to low inside heat transfer coefficient.
- If leaks occur to atmosphere (air side), air pollution, fire, or toxicity hazards are possible.

• Can an air-cooled exchanger be designed and operated without fans? Explain.

- Yes. Air-cooled exchangers without fans is an attractive option as it reduces energy costs and improves safety. However, operating an exchanger having fans but with fans turned off can have several disadvantages, the worst being tube blockage due to freezing and allowing the vapors to flow out due to the inability of the exchanger to condense them. Reliable performance of the exchangers without fans is important in the design and operation of such exchangers.

• Give a correlation for the estimation of the effective stack height above a forced draft air-cooled heat exchanger for still ambient conditions.

- The effective plume height may be considered as the height of the plume rising from an air-cooled exchanger in natural convection mode.
- As a plume source area increases, the pressure gradient caused by the upward movement of warm air is not easily neutralized by the entrainment of air from the surroundings. This pressure gradient drives cold air to replenish the vacuum created by the rising warm air.

- The area of the plume source is equivalent to the stack height and the extent of its effect is quantified by the equation

$$h_{op} = 10.1[L_B^{0.5}\,\beta^{0.8}\,\Delta T^{0.8}\eta^{0.3}\rho^{0.7}]/(\rho_a - \rho)g_n^{0.2} \quad (13.2)$$

  where h_{op} = effective plume stack height of a heated horizontal plate, m; L_B = horizontal plate breadth, m; β = coefficient of thermal expansion, $1/T_{ao}$; T_{ao} = air outlet temperature, K; η = dynamic viscosity, Pa s; ρ and ρ_a = warm air and ambient air densities, respectively, kg/m^3; g_n = gravitational constant, m/s^2.
- Equation 13.2 is applicable for natural convection. To apply it to forced convection air-cooled exchangers, the following correction is to be applied:

$$h_o = (c_{po}/c_{pop})h_{op} - h_b/2, \quad (13.3)$$

  where h_o = effective plume-stack height of the bundle, m; h_b = bundle depth, m; c_{po} and c_{pop} = specific heat capacities of air at the bundle exit and above horizontal flat plate respectively, kJ/kg K.

- What are the ranges of overall heat transfer coefficients obtainable in an air cooler?
  - 450–570 W/(m^2 (bare surface)°C).
- What are the power requirements for the fan input for finned tube air coolers?
  - 1.4–3.6 kW/(MJ h).
- What is the minimum temperature approach obtainable in an air-cooled heat exchanger?
  - 22°C.

13.1.2 Plate Heat Exchangers

- What is a plate heat exchanger (PHE)? Illustrate by a suitable diagram its constructional features.
  - A PHE consists of a series of pressed, corrugated metal plates fitted between a thick, carbon steel frame. Each plate flow channel is sealed with a gasket, weld, or an alternating combination of the two.
  - Figure 13.9 illustrates exploded view of a plate heat exchanger.
- How does a plate heat exchanger work?
  - Figure 13.10 illustrates the working of a plate heat exchanger.
  - When a pack of plates is pressed together, the holes at the corners form continuous manifolds, leading the fluids from the inlets into the plate pack, where they

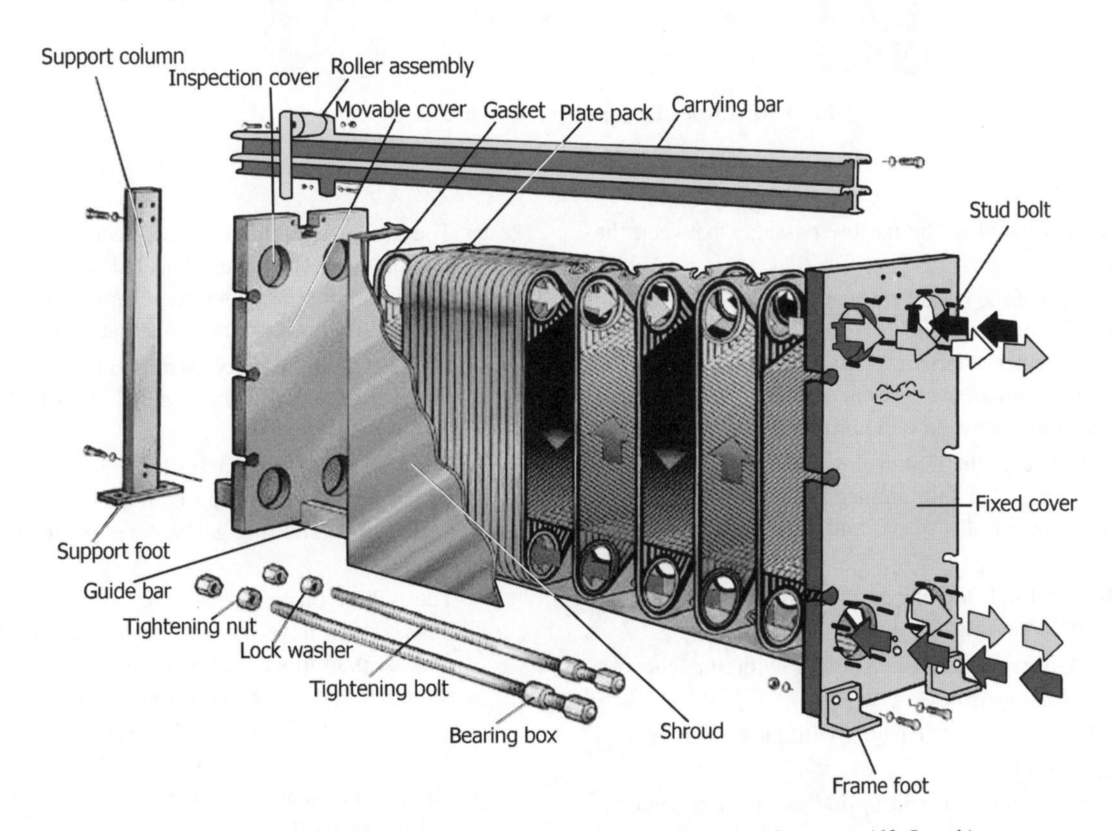

FIGURE 13.9 Construction of a typical plate heat exchanger. (*Courtesy*: Alfa Laval.)

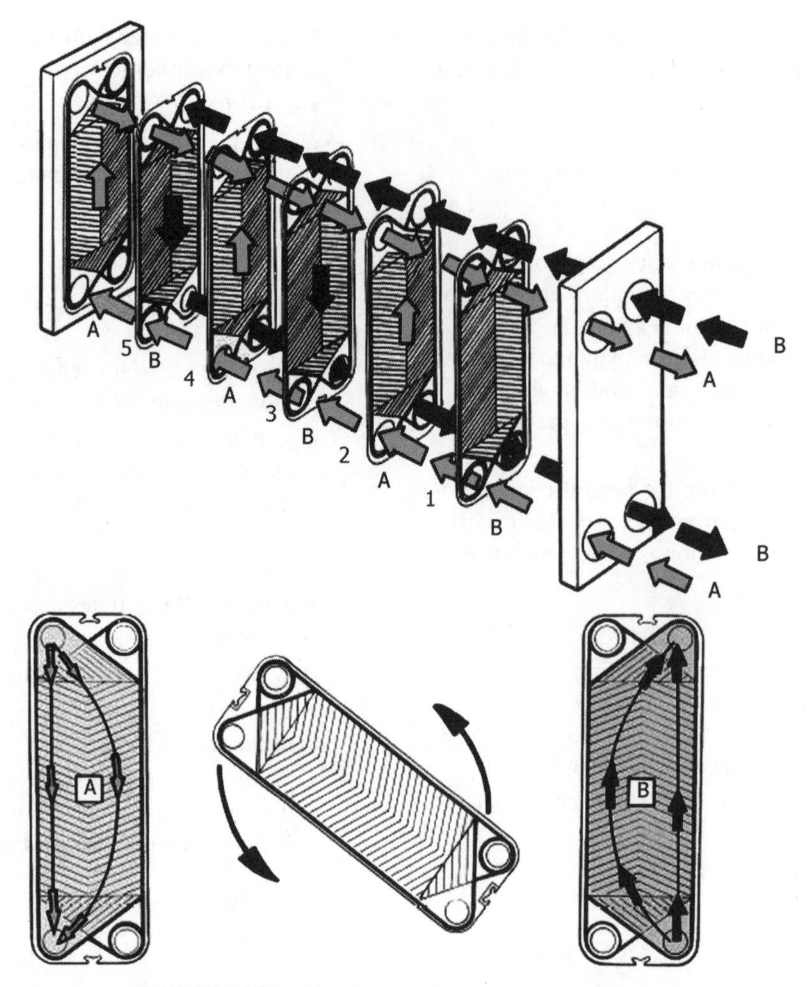

FIGURE 13.10 Flow patterns in a plate heat exchanger.

are distributed in the narrow passages between the plates.

- Because of the gasket arrangement on the plates and the placing of A and B plates alternately, the two liquids enter alternate passages, for example, the warm liquid between even number passages, and cold liquid between odd number passages.
- In most cases the liquids flow in opposite directions and finally, led into similar hole manifolds at the other end of the plates and discharged from the exchanger.
- The two fluids are effectively kept apart by the ring and field gaskets.
- An A-plate is a plate hanging with the chevron pointing downwards.
- A B-plate is a plate hanging with the chevron pointing upwards.
- If an A-plate is turned upside down, it becomes a B-plate.
- The gasket is molded in one piece. The material is normally an elastomer, selected to suit the actual combination of temperature, chemical environment, and possible other conditions that may be present.
- The one-piece gasket consists of one field gasket, two ring gaskets and links, shown in Figure 13.11 as 1, 2, and 3, respectively.
- The field gasket is by far the larger part containing the whole heat transfer area and the two corners connected to it. The ring gaskets seal off the remaining two corners.
- These three pieces are held together by a few short links, which have no sealing function at all. Their purpose is simply to tie the pieces together and to add some support in certain areas. On some plate heat exchangers, the gasket is held in place on the plate by means of a suitable cement or glue.
- The two fluids are effectively kept apart by the ring and field gaskets.

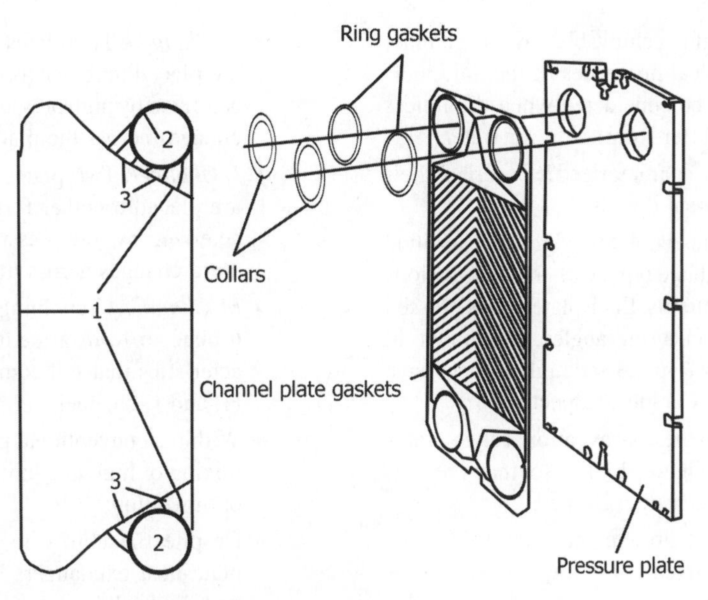

FIGURE 13.11 Gaskets for a plate exchanger.

- To prevent intermixing of the fluids in the corner areas (Figure 13.12) where field and ring gaskets are very close to each other, the link pieces have a number of slots which opens the area between the field and ring gaskets to atmosphere.
- Any leakage of media across either gasket will escape from the heat exchanger through the slots.
- It is important that these openings are kept clear. If they are not, there is a risk that in case a leak occurs in that region of the plate, there might be a local pressure build-up, which could allow one fluid to mix with the other.

* What is thermal length? What is its significance?
  - Thermal length is a dimensionless number that allows relating the performance characteristics of a channel geometry to those of a duty requirement.
  - The thermal length, θ, of a channel describes the ability of the channel to effect a temperature change based on the log mean temperature difference (LMTD).

$$\begin{aligned}\theta &= (\text{Temperature change})/\text{LMTD} \\ &= (T_{\text{in}}-T_{\text{out}})/\text{LMTD}.\end{aligned} \tag{13.4}$$

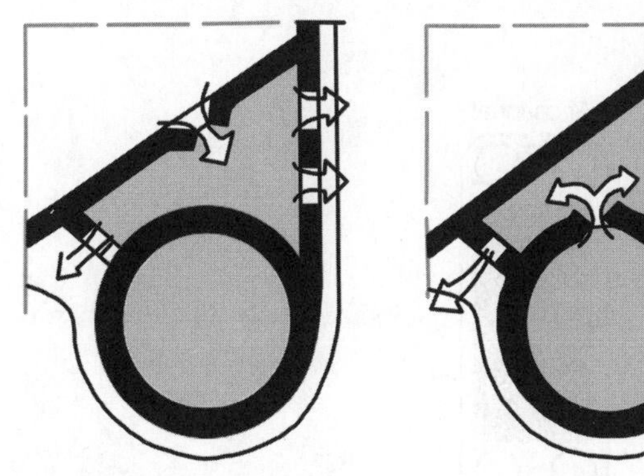

FIGURE 13.12 Ring and field gaskets to prevent intermixing of the fluids.

  - The thermal length of a channel is a function of the channel hydraulic diameter, plate length, and the angle of the corrugations, along with the physical properties of the process fluids and available pressure drop.
  - To properly design a PHE, the thermal length required by the duty must be matched with that achievable by the selected channel geometry.
  - For any chosen channel geometry, the thermal length required by the duty can:
    - Match the characteristics of the channel, thus the exchanger is optimally sized utilizing all the available pressure drop with no over dimensioning.
    - Exceed what is achievable by the channel at the allowable pressure drop, requiring that more plates be added and pressure drop reduced by lowering the velocity. Such a design is termed thermally controlled.
    - Be less than that achievable by the channel at the allowable pressure drop. This results in a greater temperature change across the plate than required, or over dimensioning. Such a design is termed pressure drop controlled.
  - To have the most economical and efficient exchanger it is critical to choose, for each fluid, a channel geometry that matches the thermal length requirement of each fluid.

- Since thermal length achievable by a channel depends on the physical properties of the fluid, correction factors must be considered when the fluid's physical properties differ from those for water.

- How are plate designs characterized? Describe their features and constructional details.
  - Conventional heat transfer plate designs are classified as chevron or herringbone type, with the corrugations forming a series of patterns. Each plate size is pressed with two different chevron angles, as shown in Figure 13.13, the low θ -plate and high θ -plate have acute and obtuse apex angles, respectively.
  - The gasket groove on these conventional style plates is recessed 100% (Figure 13.13) so that there is always a front and back to each plate.
  - By having the gasket groove recessed 100%, the plates can only be rotated about the *Z*-axis.
  - The channels are formed by alternately rotating adjacent plates 180° about their *Z*-axis so that the arrowheads of the chevron angles point in the opposite direction.
  - When two plates are adjacent to each other, the thermal and pressure drop characteristics of that channel depend strongly on the angle at which corrugations cross each other. With two different patterns, low and high θ, three distinctly different channels can be formed, each having their own hydrodynamic characteristics.
  - *H-Channel*: Two plates with obtuse angles and high θ are placed together forming a high θ channel, characterized by high pressure drop and high-temperature changes across the plate.
  - *L-Channel*: Two plates with acute angles and low θ are placed together forming a low θ-channel, characterized by low pressure drop and modest temperature changes across the plate.
  - *M-Channel*: Combining one high θ-plate and one low θ-plate to form a medium θ-channel, having characteristics that fall somewhere between those of an H- and L-channel.
  - Within a conventional plate pack, there can also be a mixing of high and low θ-channels for pressure drop optimization.
  - Despite the ability to mix channels, conventional plate heat exchangers have the major shortcoming that both fluids are subject to identical channel geometries since the channels are symmetrical. This symmetrical geometry is very effective when both fluids have the same thermal length requirement and pressure drop, but this is rarely the case as many applications involve unequal flow rates with varying thermal length requirements for the hot and cold fluids. When the duties are such, both fluids can never be totally optimized with symmetrical channels and the exchanger will not be the most economical.

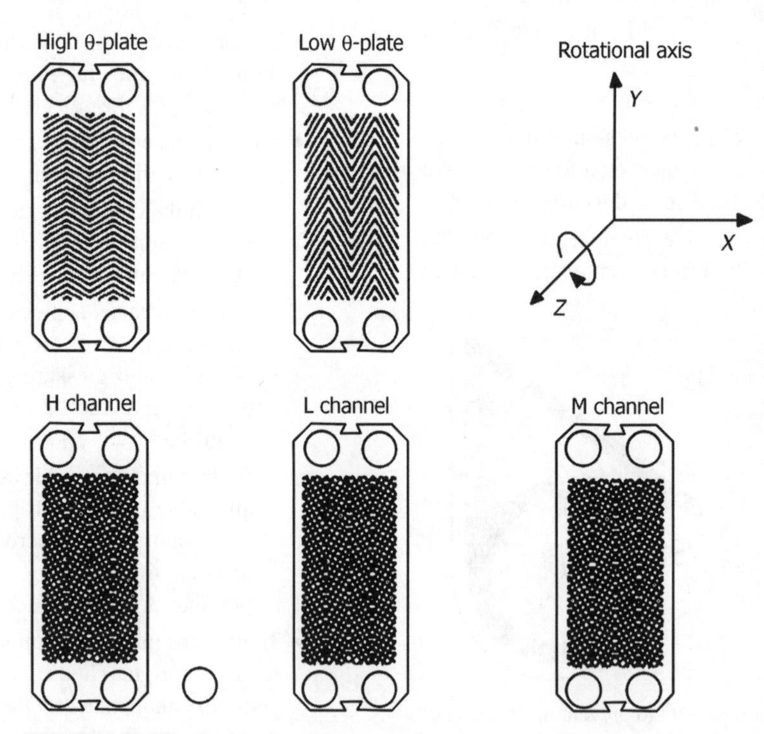

FIGURE 13.13 Conventional heat transfer plates and channel combinations.

- What is the advantage of asymmetrical channels?
  - With asymmetrical channels, the plate pack could be designed such that the hot-side channels would have a higher thermal length than those of the cold side. By doing this, both fluids would be individually optimized, making full use of both available pressure drops.
  - The asymmetrical plate has heat transfer section divided into four quadrants, with two different angles, B1 and B2, as shown in Figure 13.14.
- Show different channels formed in asymmetrical designs.
  - *HS Channel*: Two high-θ plates combined with arrowheads in the same direction, Figure 13.14(i).
  - *HD Channel*: Two high θ-plates combined with arrowheads in the opposite direction, Figure 13.14(ii).
  - *LS Channel*: Two low θ-plates combined with arrowheads in the same direction, Figure 13.14(iii).
  - *LD Channel*: Two low θ-plates combined with arrowheads in the opposite direction, Figure 13.14(iv).
  - *MS Channel*: Combining a high and low θ-plate with arrowheads in the same direction, Figure 13.14(v).
  - *MD Channel*: Combining a high and low θ-plate with arrowheads in the opposite direction, Figure 13.14(vi).
- What are the different possibilities for flow arrangements in a plate heat exchanger? Illustrate.
  - Several types of flow patterns can be achieved by appropriate gasket arrangements. These patterns can be divided into two basic types, namely, series and parallel.
  - In a series arrangement, a stream is continuous and changes direction after each vertical path.
  - In a parallel arrangement, the stream divides into a number of parallel flow channels and then recombines to flow through the exit in a single stream.
  - A system with parallel flow patterns in both streams is known as looped pattern.
  - Figure 13.15 illustrates series and parallel flows, looped patterns in U and Z arrangements and complex patterns.
  - In the U pattern, the flow distribution is less uniform than in the Z-arrangement.
  - The number of parallel passages is determined by the output of the exchanger and maximum allowable ΔP. A large number of parallel passages lower ΔP.
  - The number of series passages is determined by the plate characteristics and the heat transfer requirements. If a liquid being cooled gets into a viscous flow regime, the number of passages can be reduced to increase the velocity and hence heat transfer rates in the viscous region.
  - While a shell and tube exchanger can handle only two fluid streams, a single-plate heat exchanger can handle three or more process streams as illustrated in Figure 13.16 for a three-stream application, showing an equivalent shell and tube system for the same application.
- What is aspect ratio of a plate in a PHE and what is the normal recommended value for it?
  - Aspect ratio is the length to width ratio.
  - Recommended value is 2:1.

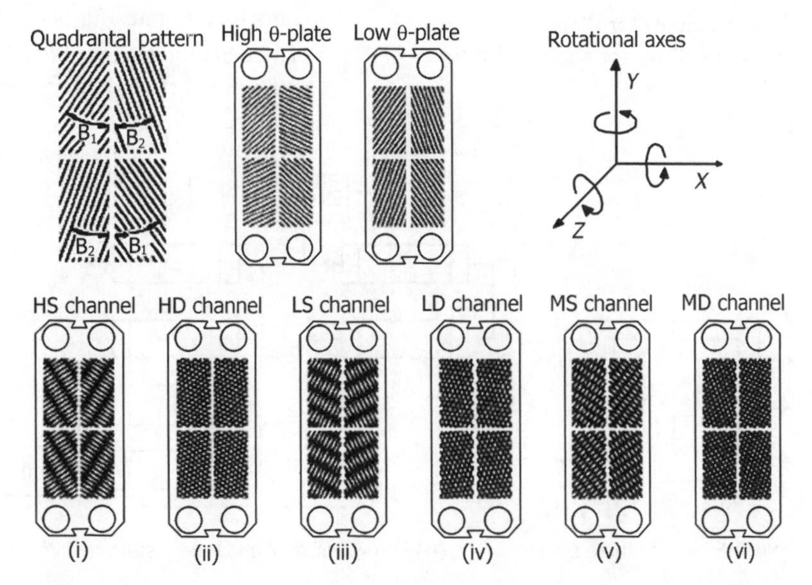

FIGURE 13.14 Asymmetric heat transfer plates and channel combinations.

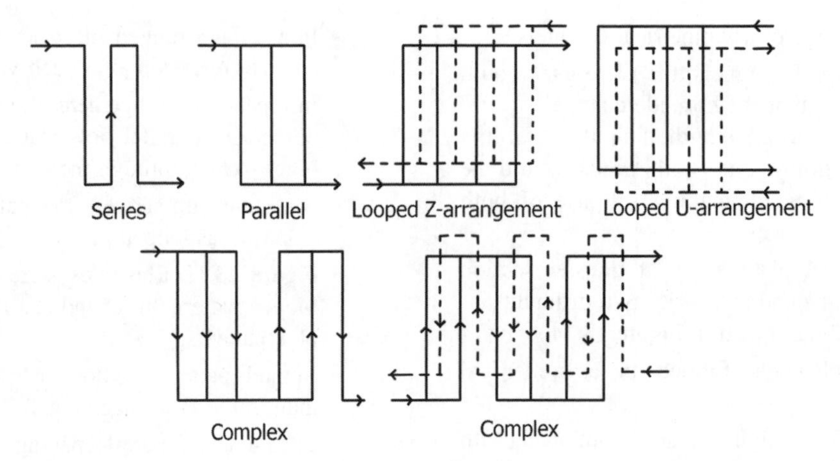

FIGURE 13.15 Flow patterns.

- Explain what is meant by *hard* and *soft* plates in a PHE.
  - Hard plates are long and narrow with high NTU and involve high ΔP and high heat transfer coefficients. Used for difficult duties, for example, regeneration applications, for which ΔT is small.
  - Soft plates are those where V-shaped angle is relatively small with low NTU and involve less ΔP and lower heat transfer coefficients.
- Define NTU for a plate heat exchanger.

$$\text{NTU} = |(t_o - t_i)|/\Delta t_{lm}. \tag{13.5}$$

where t_o and t_i are outlet and inlet temperatures, respectively.

- Compare heat transfer coefficients obtainable for PHEs with those obtainable in shell and tube exchangers for the same systems and flows.
  - Overall heat transfer coefficients for PHEs can be as high as three to four times those for shell and tube exchangers.
- When stainless steel is the material of construction, compare the costs for a shell and tube and plate heat exchanger for the same duty.
  - Plate heat exchanger is cheaper by 25–50% than shell and tube type (thin plates and consequently less material is required).
- What are the advantages of plate heat exchangers over shell and tube units?
  - Transition to turbulence at low N_{Re} (10–400). Well suited to low Reynolds number duties.
  - Good turbulence at these low Reynolds numbers by the corrugated surfaces inducing secondary flows.
  - Boundary layer separation and turbulence at relatively low Reynolds numbers lead to high heat transfer rates (very high heat transfer coefficients).
  - Low fouling rates further enhance heat transfer.
  - Thermal efficiencies of PHEs are three to four times those for shell and tube exchangers.

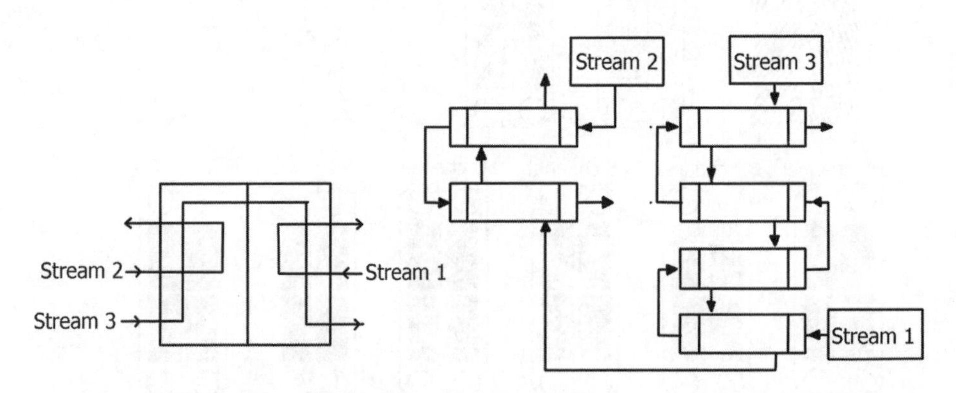

FIGURE 13.16 Single PHE handling three process streams.

- Shorter contact times with low liquid holdup make them well suited for applications involving heat-sensitive liquids.
- Can utilize up to about 82% of theoretical ΔT_{lm}, due to absence of cross flow, that is, truly countercurrent flow conditions prevail due to narrow passages. Shell and tube type utilize only 50% of ΔT_{lm}.
- Close temperature approaches (over 90% of heat can be recovered).
- Small number of passes due to high h results in low overall ΔP.
- The thin plates cut down metal wall resistance to minimum.
- Used with seawater (corrosion resistant plates—titanium and alloys) to cool demineralized water as plant cooling water.
- Easy accessibility and flexibility as these can be opened easily and heat transfer surface can be increased by adding more plates.
- It is easier to design plate heat exchangers for sanitary applications than shell and tube units. Plate designs allow use of polished plates and sanitary fluid ports, which are amenable for rapid disassembly to expose both process streams for mechanical cleaning, but it is customary to provide *clean-in-place* systems in such applications. Glueless, snap in or clip on gasketed plates can be cleaned or regasketed without removing the plate from the frame. However, for high fouling duties, where the plate pack must be opened frequently for cleaning, the glued gasket may reduce overall service costs.
- The flow channels between plates can be narrow with many contact points to increase thermal efficiency, or wide with few or no contact points to facilitate the passing of solids of fibers.
- As plate packs can be expanded, future anticipated process requirements can be incorporated more easily. Because fluid velocity decreases as plates are added in parallel, some loss in the overall heat transfer rate can be expected to occur, so future expansion needs should account for this factor.
- For applications involving cooling compressed gas, or for operating at high pressure or temperature, shell and tube technology is still preferable.
- Low heat losses to surroundings, because only narrow plate and gasket edges are exposed.
- Reduced sizes and requires less floor space. Give about four times the heat transfer rates per unit volume compared to shell and tube units.
- Small floor space requirements and less weight.
- Lower handling, transportation, and foundation costs.
- Thin plates in superior corrosion-resistant materials are comparable in costs to mild steel shell and tube exchangers.

• "Plate heat exchangers are used in coastal area plants for duties involving cooling process cooling water by seawater." What precautions would you suggest while pumping seawater from the sea intake to the heat exchanger circuit?

- Installing a screen on the suction side of the seawater pump to prevent large objects such as weeds, aquatic life, etc. entering the pump intake.
- Additional use of strainers or other suitable filters for the removal of finer solids.
  - In some plants use of strainers proved to be problematic, requiring frequent cleaning requirements for such strainers that were getting blocked by seaweed. This has necessitated replacement of strainers with more efficient and easily cleanable types of filters.
  - Large automatic back-flush strainers designed especially for seawater are used to overcome the problems of frequent blockages. Periodically, these strainers will reverse flow and blow-down debris to clear the strainer. This method has been used for many years with great success.

• "Approach temperatures in a shell and tube heat exchanger are much lower than the ones in a plate heat exchanger." *True/False*?

- *False*.
- Shell and tube type: 5–10°C.
- Plate type: As low as 1°C.

• Compare a plate heat exchanger with a shell and tube exchanger with respect to

- *Temperature Correction Factors*: Correction factors are higher for plate type as flow passages are narrower, making flow to be close to countercurrent. Temperature approaches are as close as 3°C.
- *Materials of Construction*: When expensive materials are required, plate exchangers are cheaper (because of smaller thicknesses required for plates compared to tubes).
- *High Viscosity Liquids*: Plate exchangers are better because of liquid flowing as thin films over heat transfer surfaces at increased turbulence created by embossed plates.
- *Fouling*: Low in-plate exchangers due to high turbulence in the corrugated flow passages and high wall shear stress exhibited by the flowing fluid. The higher the wall shear stress that the plate heat exchanger is designed for, the greater the resistance to *microfouling* due to fine particles, scaling,

crystallization, or similar surface mechanisms. Plate heat exchanger fouling factors can be as low as 1/10th that of shell and tube fouling factors for the same fluids.

- ➢ However, if suspended particles are going to be of such a size that they may not be able to pass through the corrugations of the plate channels, then *macrofouling* (plugging) of the heat exchanger is possible. A good rule of thumb is to keep particles below 2.5 mm, although special designs with *deep groove* or *wide gap* plates are available.
- ➢ Fibers can be particularly problematic in plate heat exchangers since the same plate corrugations that create turbulence and high heat transfer create contact points where fibers can get trapped, building up over relatively short periods of time and plugging flow paths, requiring that the heat exchanger be opened (perhaps too frequently) for cleaning.
- ➢ For such cases flow passages with wide gaps may be selected. These designs are advantageous for processes involving ethanol and bio-based processes with high fiber content. Increased gaps, however, lower turbulence. The best way is to use strainers or other filtration step before admitting such liquids into the exchanger.

- *Pressure*: Plate is not a good shape to resist pressure. Not suitable over 30 bar.
- *Pressure Drop*: Due to the short path through the heat exchanger, the pressure drop can be kept relatively low, although this depends on the number of passes and the phase of the fluid. For most liquid-to-liquid duties, a 70–100 kPa pressure drop is normal, while for a two-phase flow, the pressure drop can be as low as 2–5 kPa.
- *Gaskets*: Temperature limitations. Maximum 250°C.
- *Cleaning and Inspection*: Easy in-plate exchangers.
- *Space Requirements*: The units typically occupy only a fraction of the space needed for a shell and tube exchanger. Space savings are accompanied by savings on foundations and constructional steel work, and so on. The space needed for maintenance is also much smaller as no tube bundle access and pull out space is required.

• What are the applications of plate heat exchangers?

- *Viscous Liquids (Newtonian and non-Newtonian)*: Increased turbulence due to corrugated/embossed plates improves heat transfer coefficients (e.g., cooling of lube oils used in bearings of rolling mills and evaporation of high concentration liquids). Well suited for polymer solutions, pseudo-plastic, and thixotropic materials.
- *Heat-Sensitive Liquids*: Low residence times (low holdup), high h, uniform heat transfer and ease of cleaning, permit heat transfer applications to dairy (require cleaning once a day), food processing, brewing, and fine chemical industries.
- Same characteristics mentioned above with effective mixing make them suitable as chemical reactors.
- *Slurries*: Turbulence prevents solids deposition on heat transfer surfaces.
- *Corrosive Liquids*: Thin plates make it possible use of superior materials of construction economically for corrosive service. For example, cooling of chlorine gas saturated with water vapor from electrolysis units.
- Water-to-water closed circuit cooling systems, for example, seawater to desalinated/freshwater.

• What are the limitations of plate heat exchangers?

- Upper size is limited (due to fabrication problems) to size of presses available to stamp out plates from sheet metal up to around 1500 m^2.
- High ΔP due to zigzag and narrow passages. Due to high ΔP, capacities are limited as flows should be kept low.
- Temperatures are limited due to availability of gasket materials (usually not more than 250°C). Generally up to 160°C.
- Pressures are limited because of thin plates.
- Not preferred as air coolers/gas–gas/air–air/vacuum/condensing applications (due to excessive ΔP, limiting flow velocities, which limit capacities).
- Due to high turbulence, corrosion–erosion problems are high.
- Liquids with very high viscosities create problems of flow distribution/blockages.
- Flow velocities should not be less than 0.1 m/s, to prevent flow problems.

• "Plate heat exchangers are often used for interchanging duties due to their high efficiencies and ability to *cross* temperatures." *True/False*?

- *True*.

• What are the minimum recommended pressure drops across a plate heat exchanger to generate enough wall shear stress (frictional force of fluid against the heat transfer surface) to prevent fouling deposits?

- 50–70 kPa is the normally recommended ΔP. Higher values of ΔP may be necessary for some types of deposits and rougher surfaces.

- Does a PHE suitable for condensing vapors/boiling liquids?
  - Not normally. But used as flash evaporators in desalination applications requiring superior materials of construction to resist corrosion.
  - Heat transfer coefficients are already high for condensation or boiling and no additional advantage because of turbulence improving characteristics of plates.
  - Low flow areas limit these for liquid–liquid applications.
  - Temperature limitations imposed by gasket materials do not make these suitable for high temperatures (maximum temperatures: 180–250°C).
- Can plate heat exchangers be used as evaporators? What are their advantages over tubular exchangers?
  - Yes. The advantage is the high turbulence created (even at low Reynolds numbers) and short residence times in a plate exchanger, compared to shell and tube exchanger, makes it ideally suited for heat-sensitive and viscous liquids.
  - These are better alternatives to film evaporators with the added advantage of ease of cleaning and possibility of using superior materials of construction on cost effective basis.
  - These features make them well suited for evaporation processes involving food and pharmaceutical materials.
  - Added advantages include low headroom, smaller floor space, easy accessibility and flexibility, and low liquid holdup.
  - The vapors generated in the evaporation process are sent to a vapor–liquid separator as in a conventional evaporator and the liquid recirculated back to the exchanger.
- "A plate heat exchanger permits higher cooling water outlet temperature compared to a shell and tube heat exchanger, thereby reducing cooling water requirements and exchanger size for the same duty." *True/False*? Explain.
  - *True*. A plate exchanger involves more turbulence for the same flow rate increasing heat transfer coefficients and preventing fouling of its surfaces, permitting higher cooling water temperatures without giving rise to deposits.
  - The effects of higher turbulence and less fouling combined together keeps overall heat transfer coefficients very high thereby decreasing requirements of size and cooling water for the same duty.
  - With a shell and tube exchanger increasing cooling water rate increases heat transfer coefficients and hence decreases area requirements.
  - Increased water rates in a plate exchanger results in increased costs of the unit without additional benefits.
- What are the implications involved in adopting plate heat exchangers in place of shell and tube exchangers?
  - The individual exchangers are smaller and the spacing between the equipment can be reduced. Thus a smaller plot is needed for the plant, considering the large number of exchangers normally used in a process plant.
  - If the plant is to be housed in a building, the building can be smaller.
  - The amount of structural steel used to support the plant can be reduced, and because of the weight saving, the load on that structure is reduced. The weight advantage results in cheaper foundations.
  - Since the spacing between equipment is reduced, piping costs become less.
  - Adopting compact heat exchanger technologies can significantly reduce plant complexities by reducing the number of exchangers through improved thermal contacting and multistreaming.
  - This adds to savings associated with reduced size and weight and also has safety implications.
  - The simpler the plant structure, the easier it is for the plant operator to understand the plant.
  - In addition plant maintenance will be safer, easier, and more straight-forward.
- "Plate heat exchangers are not normally recommended for gas–gas applications." *True/False*?
  - *True*.
- Can a plate heat exchanger be used for high pressure/high-temperature applications?
  - No. Gaskets limit use for high temperatures.
  - Thin plates and gaskets do not permit high pressures.
- Is there any design of plate heat exchangers that make them suitable for high pressure and high-temperature applications?
  - Yes. Alfa Laval has introduced a new type of heat exchanger in recent times, which is called *Alfa Disc* plate and shell heat exchanger. This type of heat exchanger overcomes the limitation of the conventional gasketed plate heat exchanger for high pressure and high-temperature applications.
  - It has circular plates with corrugated patterns similar to patterns of plates of a conventional plate heat exchanger to promote high turbulence even at low Reynolds number flows.
  - Its all-welded circular plates and shell make it suitable for pressures up to 100 bar and temperatures in the range of −30 to +540°C.

- Its robust and leak-free construction makes it suitable for hazardous fluids and for cyclic operation.
- Approach temperatures can be as close as 1°C.

- Give reasons why fouling is much less in plate heat exchangers than in tubular units.
  - High turbulence maintains solids in suspension. High turbulence involves higher wall shear stress, which is responsible for lower fouling rates.
  - Velocity profiles across a plate are uniform with the absence of stagnant or low velocity zones.
  - Plate surfaces are smooth (alloys) and can be further electro-polished.
  - Deposits of corrosion products, which act as growth centers for fouling deposits, are absent/minimum due to low corrosion rates.
  - Moderately low metal wall temperatures in cooling services due to high heat transfer coefficient prevent crystallization growth of inverse solubility materials.
  - Deposits can be kept to minimum by frequent cleaning which is relatively easy because of smooth surfaces and ease of opening the unit.
- What is the advantage in using a plate heat exchanger for large-scale plant cooling loop using seawater to cool demineralized cooling water in process heat exchangers?
  - Plate heat exchangers are constructed with alloy steel plates which resists seawater corrosion at a comparatively lower costs compared to using shell and tube exchangers made of corrosion-resistant materials of construction.
  - Another aspect that favors plate exchangers for such an application is the relative ease of cleaning of deposits arising from seawater circulation.
- What are the commonly used gasket materials for a plate heat exchanger?
  - Styrene-butadiene rubber: up to 85°C.
  - Nitrile rubber: up to 140°C.
  - EPDM: up to 150°C.
  - Fluorocarbon rubbers: up to 180°C.
  - Compressed asbestos fiber (CAF): up to 260°C. Environmental considerations discourage use of asbestos products.
- What are the considerations involved in the selection of gaskets in a PHE?
  - Gasket life and resistance to operating temperatures and pressures are important considerations.
- "In a shell and tube heat exchanger there are two overall heat transfer coefficients based on inside and outside surface areas whereas in a plate heat exchanger there is only one value of U." Explain.
  - In a shell and tube heat exchanger, overall heat transfer coefficients are expressed based on the inside surfaces of the tubes or based on outside surfaces of the tubes. The surfaces are different from each other, the magnitudes of these differences are dependent on the tube thicknesses.
  - Thus, there are two overall heat transfer coefficients for a shell and tube heat exchanger.
  - In the case of a plate heat exchanger, surface areas on both sides of the plate are the same and therefore there is only one overall heat transfer coefficient.
- "Seawater is used as a cooling medium for demineralized cooling water circulated through process equipment in industry located in coastal areas." What type of heat exchanger would you recommend for the above purpose. Give reasons for your choice?
  - PHE is a preferred choice. Corrosion-resistant materials of construction, to withstand seawater corrosion, can be used in a PHE with nominal cost increases as the plates are thinner and material requirements are small compared to a shell and tube exchanger.
- While using seawater as a cooling medium what precautions would you suggest, while pumping seawater from sea to the heat exchanger?
  - Aquatic weeds, fish, etc. can enter pump suction lines and pieces may pass through the pump. When suction lines are blocked, pump may be damaged. Debris passing through the pump will foul heat transfer surfaces and reduce performance of the exchanger by increasing fouling resistances.
  - It would be necessary, to prevent entry of such material into suction lines, to provide primary screening and secondary filtration measures. Sometimes strainers are installed on suction side but this will work only if relatively debris-free water enters suction. Otherwise frequent cleaning of strainers can interrupt operation. Installation of special filters may be necessary for trouble-free operation.
- "Plate heat exchangers are cheaper, in stainless steel construction, than shell and tube heat exchangers." *True/False*?
  - *True*. About 25–50% cheaper.
- What is a welded plate heat exchanger? What are its applications?
  - In a welded-plate exchanger, the field gasket that normally contains the process fluid is replaced by a welded joint. The welded plates form a closed compartment.
  - In petrochemical and petroleum refinery applications, gaskets frequently cannot be used because aggressive media result in a short lifetime for the gaskets or because a potential risk of leakage is not

acceptable. In these cases, all-welded compact heat exchangers without gaskets should be considered.

- When plates are welded together at the periphery, leakage is prevented, making this design suitable for hazardous or aggressive fluids.
- Similar to gasketed designs, alternating flow channels are created to divert the flow of hot and cold fluids into adjacent channels. Aggressive fluids pass from one compartment to the next through an elastomer or Teflon ring gasket, while nonaggressive fluids are contained by standard elastomer gaskets.
- The use of welded joints can reduce total gasket area by 90% on the aggressive fluid side.
- Typical applications include exchangers handling vaporizing and condensing refrigerants, corrosive solvents, and amine solutions.

• What are the limitations of welded plate heat exchanger?
  - Some of the all-welded heat exchangers are sealed and cannot be opened for inspection and mechanical cleaning.
  - Because all-welded heat exchanger plates cannot be pressed in carbon.
  - Steel, plate packs are available only in stainless steel or higher-grade metals. The cost of an all-welded compact heat exchanger is higher than that of a gasketed plate heat exchanger.

13.1.3 Plate-Fin Heat Exchangers

• What is a plate-fin exchanger?
  - Plates are separated by corrugated sheets, which form fins.
  - These are made up of a block and called matrix exchangers.

• What are the applications of plate-fin exchangers? What is its principal disadvantage?
  - Used in cryogenic processes, for example, air separation units.
  - Used in offshore units where compactness and lighter weights are an advantage.
  - Not suitable for dirty and fouling fluids a cleaning is not possible.

• Compare the compactness of the following heat exchangers (i) plate and frame exchangers, (ii) plate-fin exchangers, and (iii) tube-fin exchangers.
  - Plate: 120–230 m^2/m^3.
  - Tube-fin: 330 m^2/m^3 (maximum).
  - Plate-fin: Up to 6000 m^2/m^3.

• Compare the compactness of a plate-fin exchanger with a shell and tube exchanger.
  - Plate-fin types have surface areas of about four times the heat transfer surface per unit area than shell and tube type.

• Compare a plate-fin heat exchanger with a tube-fin exchanger.
  - *Plate-Fin Type*:
    - ➢ High area densities up to about 6000 m^2/m^3.
    - ➢ Considerable amount of flexibility. Passage height on each side can be varied easily.
    - ➢ Different fins could be used for different applications.
    - ➢ Used for low pressure applications up to about 1000 kPa (10 bar).
    - ➢ Fins on each side can easily be arranged so that overall flow arrangement of the two fluids can result in cross flow, counter flow, or parallel flow.
    - ➢ Generally less expensive than tube-fin exchanger.
  - *Tube-Fin Type*:
    - ➢ For cases where extended surface is needed on one side (e.g., gas–liquid heat transfer) or
    - ➢ Operating pressure needs to be contained on one 'side.
    - ➢ Lower compactness than plate-fin type—maximum 3300 m^2/m^3 surface area density.
    - ➢ Tube side operating pressures can be much higher than plate-fin type—up to about 3000 kPa gauge or 30 bar.
    - ➢ Outside tube pressures up to 100 kPa.
    - ➢ Wide range of operating temperatures, cryogenic to about 870°C.
    - ➢ Generally more expensive than plate-fin type, since tubes are more expensive.

13.1.4 Spiral Heat Exchangers

• What is a spiral heat exchanger and what are its advantages?
  - A spiral heat exchanger can be considered as a plate heat exchanger in which plates are formed into a spiral.
  - The basic spiral element is constructed of two long metal sheets that are wound as spirals around a central core forming two concentric spiral channels. Normally these channels are alternately welded, ensuring that the hot and cold fluids cannot intermix.
  - The exchanger can be optimized for the process concerned by using different channel widths. Channel width is normally in the range 5–30 mm.

- Plate width along the exchanger axis may be 2 m, as can the exchanger diameter, giving heat transfer areas up to 600 m^2.

Advantages:

- Design of a spiral heat exchanger approaches the ideal in heat transfer equipment by obtaining identical flow characteristics for both media countercurrent to each other.
- With both fluids flowing spirally, countercurrent flow and long passage lengths enable close temperature approaches and precise temperature control.
- Spiral plates frequently can achieve heat recovery in a single unit, which would require several tubular exchangers in series.
- Even velocity and temperature distribution, with no dead and hot or cold spots.
- Good turbulence and ease of cleaning are its special attributes.
- More thermally efficient with higher heat transfer coefficients.
- It can be used for dirty fluids. Removal of one cover exposes the total surface area of one channel providing easy inspection cleaning and maintenance. Low fouling rates compared to shell and tube exchangers.
  - ➢ Any fouling that occurs, the spiral-plate can be effectively cleaned chemically because of the single flow path.
  - ➢ Because the spiral can be fabricated with identical flow passages for the two fluids, it is used for services in which the switching of fluids allows one fluid to remove the fouling deposited by the other.
- Single flow passages make it ideal for cooling or heating slurries or sludge.
  - ➢ Slurries can be handled in the spiral plate at velocities as low as 0.6 m/s (2 ft/s).
- Small holdup times and volumes.
- Copes with exit temperature overlap, or crossover, whereas shell and tube units require multiple shells in series to handle temperature crossover.
- Spiral plates avoid problems associated with differential thermal expansion in noncyclic services.
- It is compact, light weight, and involves low ΔP compared to shell and tube type.
- In axial flow, a large flow area affords a low pressure drop and can be of special advantage when condensing under vacuum or when used as a thermosiphon reboiler.
- For the same duty, a spiral heat exchanger heat transfer area would be 90 m^2 compared to 60 m^2 for a plate and frame design or 125 m^2 for a shell and tube design.
- In a *spiral tube bundle exchanger*, the spaces or gaps between the coils of the tube bundle becomes the shell side flow path when the bundle is placed in the shell. Tube side and shell side connections on the bottom or top of the assembly allow for different flow path configurations.
- The spiral shape of the flow for the tube side and shell side fluids creates centrifugal force and secondary circulating flow that enhances the heat transfer on both sides in a true counter flow arrangement.
- Well suited for heating or cooling viscous fluids because its length to diameter (L/D) ratio is lower than that of straight tubes or channels. Consequently, laminar flow heat transfer is much higher for spiral plates.
  - ➢ When heating or cooling a viscous fluid, the spiral should be oriented with the axis horizontal. With the axis vertical, the viscous fluid stratifies and the heat transfer is reduced as much as 50%.

• What are the limitations of a spiral heat exchanger?

- Repair in the field is difficult. A leak cannot be plugged as in a shell and tube exchanger. However, the possibility of leakage is much less in the spiral plate because it is fabricated from plate that is generally much thicker than tube walls and stresses associated with thermal expansion are virtually eliminated. However, in case of repairs, removing the covers exposes most of the welding of the spiral assembly. Repairs on the inner parts of the plates are complicated.
- The spiral-plate exchanger is sometimes precluded from service in which thermal cycling is frequent. When used in cycling services, the mechanical design must be altered to provide for the higher stresses associated with cyclic services.
  - ➢ Full-faced gaskets of compressed asbestos are not generally acceptable for cyclic services because the growth of the spiral plates cuts the gasket, which results in excessive fluid bypassing and, in some cases, erosion of the cover. Use of asbestos-based gaskets is discouraged due to hazardous nature of asbestos.
  - ➢ Metal to metal seals are generally necessary when frequent thermal cycling is expected.
- The spiral-plate exchanger usually should not be used when a hard deposit forms during operation, because the spacer studs prevent such deposits from being easily removed by mechanical cleaning. When spacer studs can be omitted, as for some pressures, this limitation is not present.

- Illustrate the working of a typical spiral heat exchanger.
  - In *Type 1 Alfa Laval design*, fluids are in full countercurrent flow. The hot fluid enters at the center of the unit and flows from the inside outward to the periphery, where it exits through a connection welded to the shell. The cold fluid enters through a peripheral connection and flows toward the center and exits through a center connection. Figure 13.17 illustrates the Type 1 Alfa Laval spiral heat exchanger.
- Explain how a spiral heat exchanger is well suited for fouling service.
  - As the fluid flows through a *single* continuous flow channel, if fouling starts to build up anywhere in the channel, the local cross-sectional area for the flow narrows down at that position. As a consequence, the local velocity will be higher than in the rest of the channel.
  - The shear rate between the liquid and the fouling deposit will increase in proportion to the square of the velocity and causes a scrubbing effect that helps to remove the fouling, which can be termed as *self-cleaning effect*.
  - In a multiple channel heat exchanger like a shell and tube type, partial plugging of one tube will lead to higher flow in the other tubes. The resulting smaller flow through the partially plugged tube may not be sufficient to scrub the tube to remove the deposit.
- What are the applications of a spiral heat exchanger?
  - As it is an ideal exchanger for fouling fluids due to the large channel widths and constant change in flow direction with shearing, used in the food industry (sauces, slush, and slurry) as well as in brewing and wine making.
  - Have many applications in the chemical industry including $TiCl_4$ cooling, PVC slurry duties, oleum processing, and heat recovery from many industrial effluents.
  - They also provide temperature control of sewage sludge digesters plus other public and industrial waste plants.
  - Due to their true countercurrent flow paths that permit the best possible overlap of exit temperatures, they can maximize the heat recovery on large scale cogeneration projects although they may be more expensive than plate designs.
  - One particular application where a spiral heat exchanger is used successfully is cooling of the bottom product containing up to 1% catalyst fines in fluid catalytic cracking units. Other refinery applications include interchangers of vis-breaker feeds/products, which contain fine coke and asphaltenes.
  - Other heat transfer applications include oil extraction from tar sands, coke oven plants, and polymer processing plants.
  - They can be mounted directly onto the head of distillation columns, acting in a condensing or reflux role. Specific advantages are ease of installation, low pressure drop and large flow cross section. Consequently, there are many condensing applications in all process industries particularly for condensing under vacuum.

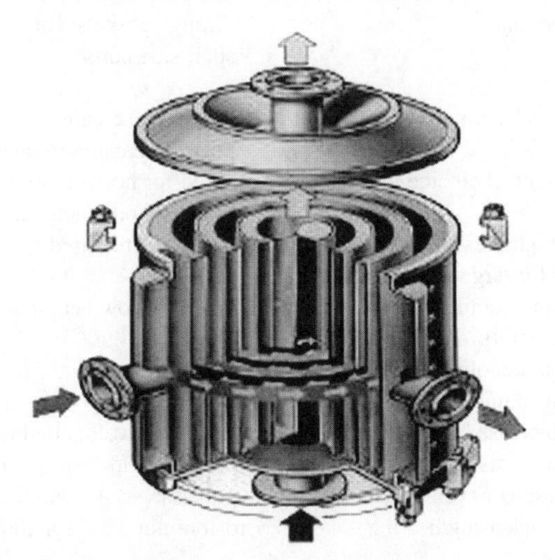

FIGURE 13.17 Spiral heat exchanger. (*Courtesy*: Alfa Laval.)

- What are the operating limits for spiral heat exchangers?
  - Typically, the maximum design temperature is 400°C set by the limits of the gasket material. Special designs without gaskets can operate with temperatures up to 850°C.
  - Maximum design pressure is usually 1500 kPa, with pressures up to 3000 kPa attainable with special designs.
- "Spiral heat exchangers are often used to slurry interchanger and other services containing solids." *True/False*?
  - *True.*

13.1.5 Scraped Surface Heat Exchanger

- What is a scraped surface heat exchanger? Illustrate by a suitable diagram.
  - Figure 13.18 illustrates a typical scraped surface heat exchanger.
- What are the applications of scrapped surface heat exchangers?
  - *Heat-Sensitive Products*: Delicate products that are adversely affected by prolonged exposure to heat are effectively processed in scraped surface heat

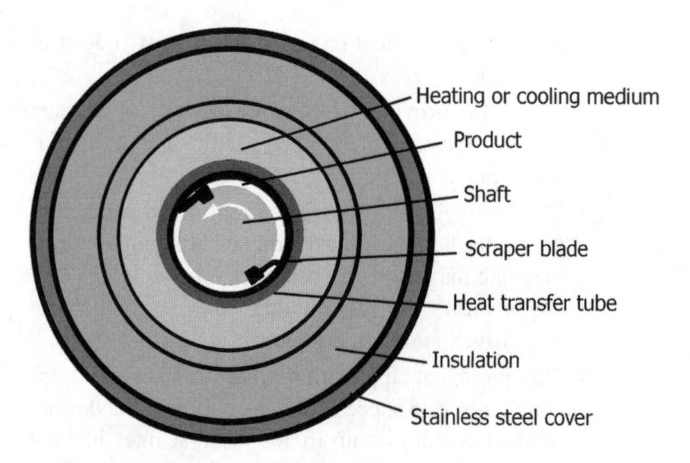

FIGURE 13.18 Scraped surface heat exchanger.

exchangers. The scraper blades prevent product from remaining on the heat transfer surface by continuously removing and renewing the film. Because only a small amount of product is exposed to heat for just a short time, charring is minimized or eliminated. These are employed by food and meat processing industries.

- *Viscous Products*: These exchangers process viscous products far more efficiently than conventional plate or tubular heat exchangers. Product *film* is continually scraped from the heat transfer wall to induce high heat transfer rates. Constant agitation causes turbulent flow and more consistent heating or cooling. Pressure drop is effectively controlled by the product annulus area. Agitation eliminates stagnant areas and product buildup. Cleaning is easier.
- *Particulate-Laden Products*: Products with particulates that tend to plug conventional heat exchangers are handled easily.
- *Crystallized Products*: Products that crystallize are handled efficiently. As product crystallizes on the heat transfer wall, the scraper blades remove it and keep the surface clean.
- For heat transfer in solvent extraction units, for example, paraffin wax plants, leaching plants, etc.

- Summarize selection guidelines for heat exchangers.
  - Table 13.2 gives selection guidelines for different types of heat exchangers.

13.1.6 Heat Pipes

- What is a heat pipe? Explain its principle of operation by means of a diagram.
  - Figure 13.19 illustrates the principle of heat pipes and their constructional features.
  - Heat pipe is a passive device with no moving parts.

TABLE 13.2 Types, Features, Applications, and Limitations of Heat Exchangers

Type	Feature(s)	Best-Suited Applications	Limitations
Fixed tube sheet	Both tube sheets fixed to shell	Condensers, liquid–liquid, gas–gas, gas–liquid, heating/cooling, horizontal/vertical, reboilers	For limited ΔT (<95°C)
Floating head, removable/nonremovable tube bundles	One tube sheet floats in shell; back cover can be removed for access to tubes	High ΔT (>95°C); easy cleanability; usually horizontal	Internal leakages possible through gaskets; corrosion to shell side parts
U-bundle	One tube sheet; removable bundle	High ΔT; easy cleaning on both tube and shell side	Bends must be carefully made; possible erosion in bends
Kettle	Shell enlarged for vapor disengagement	Boiling liquid on shell side used as reboilers	Limited for horizontal installation; large diameter shell
Double pipe	Annulus for shell side flow; modular assembly; fins over tubes	Suitable for high pressures; can be assembled for larger area needs	Large space required
Coiled vessels	Coils submerged in liquid in a vessel	Used for heating/cooling/condensing in stirred reactors	Limited for low heat loads
Air cooled	No shell is required; finned surfaces	Condensing and cooling	Low h
PHEs	Compact consisting of corrugated/embossed gasketed thin plate assemblies; compact	Low Reynolds number applications requiring increased turbulence; easy cleaning; viscous, corrosive liquids and slurries can be handled	For limited T and P applications; not well-suited for boiling/condensation not good for gas–gas applications
Spiral	Compact; concentric passages; high turbulence; no bypassing	Cross flow; condensing/heating	Erosion; not good for slurries

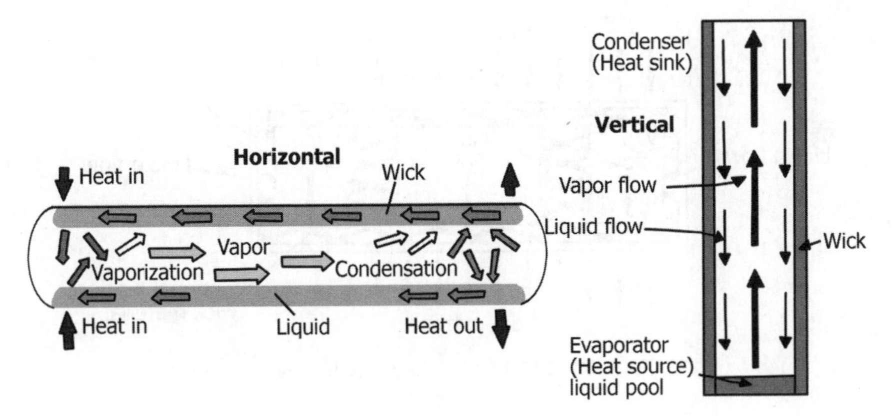

FIGURE 13.19 Heat pipes.

- Heat pipe is a sealed chamber in which a working fluid vaporizes taking heat from a heat source and condenses in a heat sink.
- It involves extremely high heat transfer rates, as it involves vaporization and condensation processes that are known to involve very high heat transfer coefficients compared to those involving sensible heat transfer processes.
- Key features of a heat pipe include the following:
  - Condenser to vaporizer surface area ratios are very high, as high as 25 or more.
  - Working fluids can be low temperature fluids such as ammonia, water, etc. for applications involving heat removal from electronic equipment.
  - In high heat flux or high-temperature systems, liquid metals such as cesium, potassium, or sodium are used as heat transfer fluids for applications involving nuclear or metallurgical systems.

• What are the advantages of heat pipes?
  - Very high heat transfer rates. For the same size, heat transfer capacity is up to 90 times more than a copper conductor.
  - Temperatures in condenser and vaporizer can remain nearly constant while vaporizer may have variable heat fluxes.
  - Condenser and vaporizer can have different areas to fit into different geometrical shapes and areas.
  - High heat flux inputs can be dissipated with low heat flux outputs using only natural or forced convection.
  - A constant condenser heat flux can be maintained while the vaporizer operates with variable heat fluxes.

• What are the applications of heat pipes?
  - *Electronic Cooling*: Small high performance components cause high heat fluxes and high heat dissipation demands. Used in cooling transistors and high-density semiconductors.
  - *Aerospace*: Cool satellite solar array and space shuttle leading edge during reentry.
  - *Heat Exchangers*: Power industries use heat pipe heat exchangers as air heaters on boilers.
  - *Other Applications*: Production tools, medicine, and human body temperature control, engines and automotive control.

• What are the types of heat pipes?
  - *Thermosiphon*: Gravity assisted wickless heat pipe. Gravity is used to force the condensate back into the vaporizer. Condenser must be above vaporizer in a gravity field.
  - *Leading Edge*: Placed in the leading edge of hypersonic vehicles to cool high heat fluxes near the wing leading edge.
  - *Rotating and Revolving*: Condensate is returned to the vaporizer through centrifugal force. No capillary wicks are required. Used to cool turbine components and armatures for electric motors.
  - *Cryogenic*: Low temperature heat pipe. Used to cool optical instruments in space.
  - *Flat Plate*: Rectangular in shape. Much like traditional cylindrical heat pipes. Used to cool and flatten temperatures of semiconductor or transistor packages assembled in arrays on top of the heat pipe.
  - Figure 13.20 illustrates a flat plate heat pipe.
  - *Micro-Heat Pipes*: Small heat pipes that are noncircular and use angled corners as liquid arteries. Used in cooling semiconductors, laser diodes, photovoltaic cells, medical devices, etc. Figure 13.21 illustrates a typical micro-heat pipe.
  - *Variable Conductance Heat Pipe (Figure 13.22)*: Permits variable heat fluxes into the vaporizer while the vaporizer temperature remains constant by pushing a noncondensable gas into the condenser when heat fluxes are low and moving the gases out of the condenser when heat fluxes are high, thereby increas-

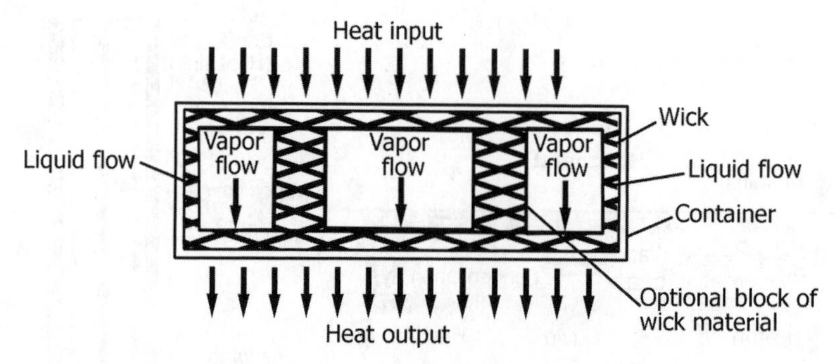

FIGURE 13.20 Flat plate type heat pipe.

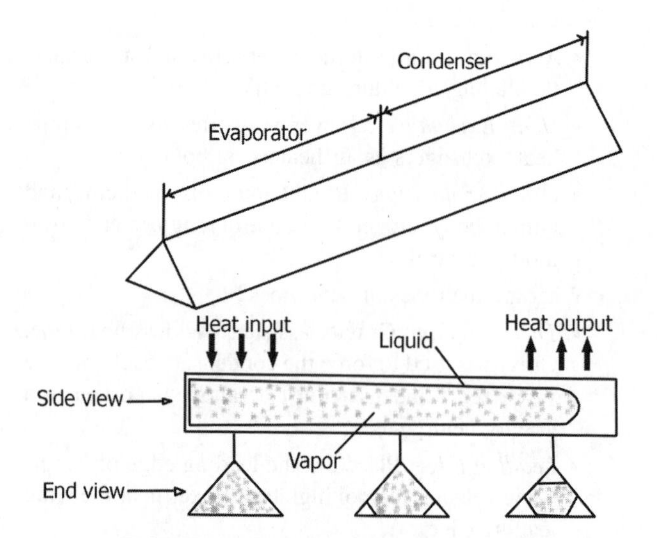

FIGURE 13.21 Micro-heat pipe operation.

ing condenser surface area. These heat pipes come in various forms such as excess liquid or gas loaded form. These are used in electronics cooling.

- *Capillary Pumped Loop Heat Pipe (Figure 13.23)*: For systems where heat fluxes are very high or the heat from the heat source needs to be moved far away. In the looped heat pipe, vapor travels around in a loop and condenses and returns to the vaporizer. Used in electronics cooling.

- What are the limitations of heat pipes?
  - *Capillary Limit*: This occurs when the capillary pressure is too low to provide enough liquid to the vaporizer from the condenser leading "dry out" in the vaporizer. This affects performance of the heat pipe.
    - The capillary pressure must be more than or equal to the sum of ΔPs due to inertial, viscous, and hydrostatic forces as well as pressure gradients.
    - If the above condition is not satisfied, the working fluid is not supplied to the vaporizer rapidly enough to compensate for the liquid loss through vaporization, leading to dry out conditions.
  - *Boiling Limit*: Occurs when the radial heat flux into the heat pipe causes the liquid in the wick boil and vaporizes causing *dry out*.
  - *Entrainment Limit*: At high vapor velocities, droplets of liquid are torn from the wick and get carried over with the vapor, again leading to *dry out*.
  - *Sonic Limit*: Occurs when the vapor velocity reaches sonic velocity at the vaporizer and any increase in ΔP will not speed up the flow, for example, choked flow in convergent–divergent nozzle. This usually occurs at start-up of the heat pipe.
  - *Viscous Limit*: At low temperatures, the vapor pressure difference between the condenser and vaporizer

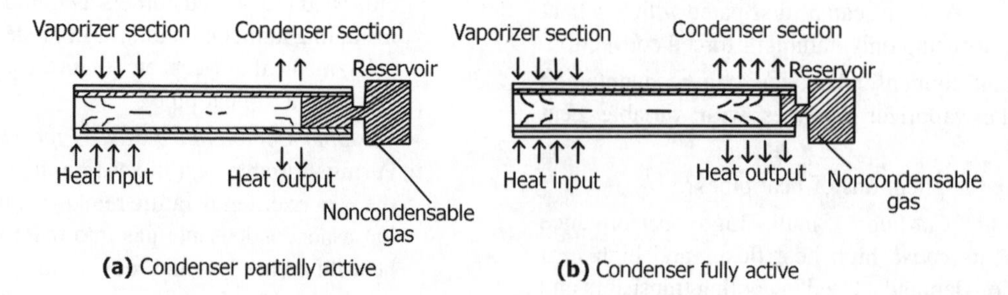

FIGURE 13.22 Variable conductance heat pipe.

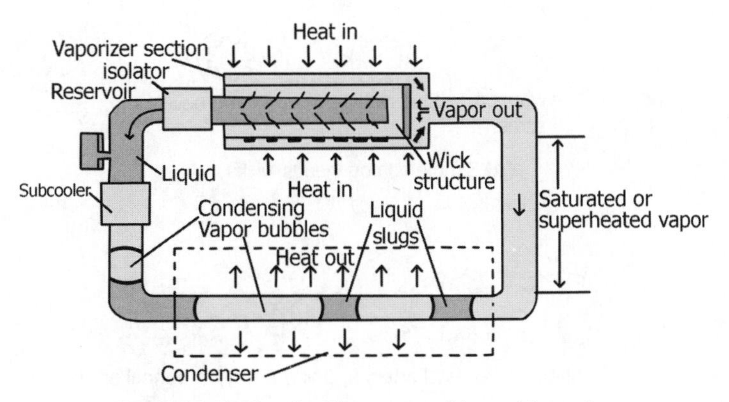

FIGURE 13.23 Capillary pumped looped heat pipe.

may not be enough to overcome viscous forces, resulting stoppage of vapor to the condenser.

- In practical situations, boiling and capillary limits are the most important limitations.
- The ranges of different limits are illustrated qualitatively in Figure 13.24.

- What are the types of wicks used in heat pipes and what are their desirable characteristics?
  - There are two types of wicks, namely, homogeneous and composite.
    - *Homogeneous*: These are made from one type of material or machining technique. Tend to have either high capillary pressure and low permeability or the other way round.
    - *Composite*: These are made from a combination of several materials with different types of porosities and/or configurations. Tend to have a higher capillary limit than the homogeneous type, but are expensive.

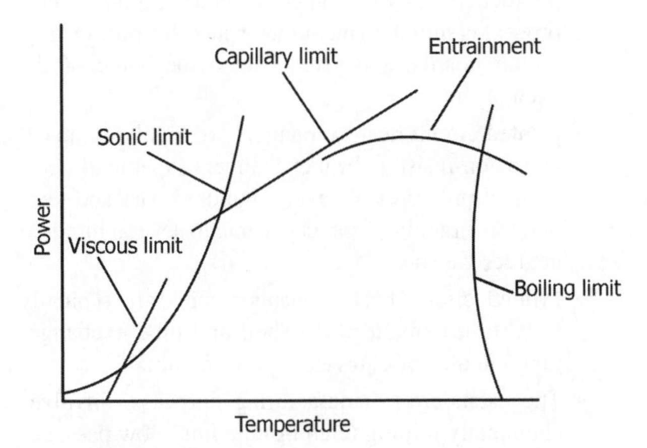

FIGURE 13.24 Heat pipe performance curves.

- What are the properties of heat pipe wicks that influence wick design?
  - *Pressure Drop*: Small capillary pore radius results in high pressure drop.
  - *Permeability*: Large pore radius results in low liquid pressure drops and low flow resistance. A balanced approach in wick design should be made between low liquid pressure drop and large capillary pressure. Composite wicks tend to find a compromise between these two variables.
  - *Thermal Conductivity*: High thermal conductivity results in small temperature difference for high heat fluxes.
- Illustrate with diagrams different types of wicks used in heat pipes?
  - There are several types of wick structures: screen, grooves, felt, and sintered powder. Figure 13.25 illustrates some of the structures.
  - Sintered powder metal wicks offer several advantages over other wick structures. One advantage of a heat pipe with a sintered powder wick is that it can work in any orientation, including against gravity (i.e., the heat source above the cooling source). The power transport capacity of the heat pipe will typically decrease as the angle of operation against gravity increases. Since groove and screen mesh wicks have very limited capillary force capability, they typically cannot overcome significant gravitational forces, and dry out generally occurs.
- Name some working fluids for heat pipes and their operating temperature ranges.
  - Table 13.3 gives some heat pipe working fluids along with their operating ranges.
  - Helium, nitrogen, argon, and neon are used in cryogenic applications.
  - Water and ammonia are the commonly used fluids for electronic cooling applications. Other fluids for

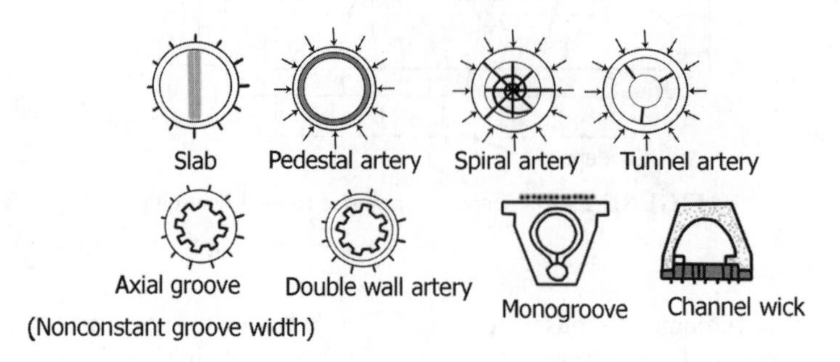

FIGURE 13.25 Typical heat pipe wick configurations and structures.

TABLE 13.3 Some Working Fluids for Heat Pipes

Fluid	Operating Range (°C)
Helium	−271 to −269
Nitrogen	−203 to −160
Ammonia	−60 to 100
Acetone	0–120
Methanol	10–130
Ethanol	0–130
Water	30–200
Toluene	50–200
Mercury	250–650
Sodium	600–1200
Lithium	1000–1800
Silver	1800–2300

room temperature applications include methanol, ethanol, and acetone.

- Mercury, sulfur, and organics such as naphthalene and biphenyl are typically used for medium temperature applications in the range of 270–430°C.
- Liquid metals such as sodium, potassium, and silver are used above 430°C.

• Illustrate by a suitable diagram application of heat pipes in electronic cooling.

- In single component cooling, vaporizer of the heat pipe may be attached to an individual heat source, such as a power transistor, thyristor, or a chip.
- The condenser is attached to a heat sink to dissipate the heat by natural or forced convection.
- Figure 13.26 illustrates typical heat pipe working as a heat sink for power transistors.

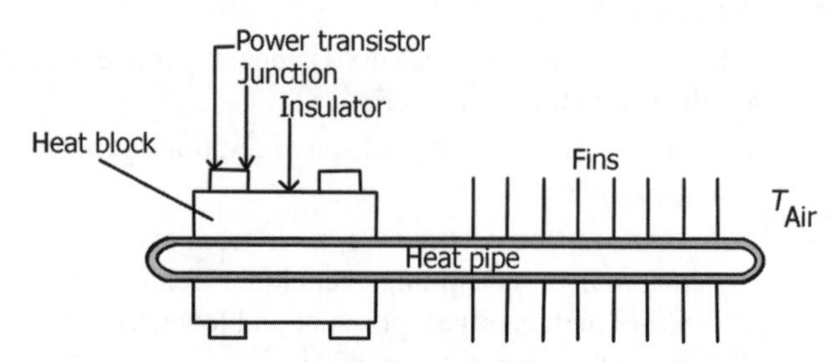

FIGURE 13.26 Heat pipe heat sink for power transistors.

13.1.7 Printed Circuit Heat Exchangers

• What is a printed circuit heat exchanger? What are its salient features?

- Printed circuit heat exchangers are constructed from flat alloy plates with fluid flow passages photochemically machined (etched) into them. This process is similar to manufacturing electronic printed circuit boards, and gives rise to the name of the exchangers.
- Printed circuit heat exchangers are highly compact, corrosion-resistant heat exchangers capable of operating at pressures of several hundred bars and temperatures ranging from cryogenic to several hundred degrees Celsius.
- Printed circuit heat exchanger cores are typically 5–10 times smaller than shell and tube exchanger tube bundles of equivalent performance.
- The standard manufacturing process involves chemically milling (etching) the fluid flow passages into the plates. This allows enormous flexibility in

thermal/hydraulic design, as complex new plate patterns require only minimal retooling costs.

- This plate/channel forming technique can produce a wide range of flow path sizes, the channels varying typically from 0.5 to 2.0 mm in depth.
- Stacks of etched plates, carrying flow passage designs tailored for each fluid, are diffusion bonded together to form a compact, strong, all-metal heat exchanger core.
- No gaskets or brazing materials are required for the assembly. Diffusion bonding allows the plates to be joined so that the bond acquires the same strength as the parent metal.
- The thermal capacity of the exchanger is built to the required level by welding together diffusion-bonded blocks to form the complete heat exchanger core.
- Fluid headers and nozzles are welded to the cores, in order to direct the fluids to the appropriate sets of passages.
- Printed circuit heat exchangers are all welded so there is no braze material employed in construction, and no gaskets are required. Hence, the potential for leakage and fluid compatibility difficulties are reduced and the high level of constructional integrity renders the designs exceptionally well suited to critical high-pressure applications, such as gas compression cooling exchangers on offshore platforms.
- Fluids handled may be liquid, gas or two phase, multistream and multipass configurations can be assembled and flow arrangements can be truly countercurrent, cocurrent or cross flow, or a combination of these, at any required pressure drop.
- Materials of construction include stainless steel (SS 300 series) and titanium as standard, with nickel and nickel alloys also being commonly used.

- What are the applications of a printed circuit heat exchanger?
  - Printed circuit heat exchangers extend the benefits of compact heat exchangers into applications where pressure, temperature, or corrosion prevents the use of conventional plate exchangers.
  - They are used in the following four major areas.
    - *Fuels Processing*
      - Gas processing, for example, compressor cooling, liquids recovery.
      - Dehydration.
      - Synthetic fuels production, for example, methanol.
      - Reactor feed/effluent exchange.
    - *Chemical Processing*
      - Acids, for example, nitric, phosphoric.
      - Alkalis, for example, caustic soda, caustic potash.
      - Fertilizers, for example, ammonia, urea.
      - Petrochemicals, for example, ethylene, ethylene oxide, propylene.
      - Pharmaceuticals.
      - Plastics, for example, formaldehyde, phenol.
    - *Power and Energy*
      - Feed water heating.
      - Geothermal generation.
      - Chemical heat pumps.
    - *Refrigeration*
      - Chillers and condensers.
      - Cascade condensers.
      - Absorption cycles.

13.1.8 Other Systems for Cooling Electronic Devices

- What are passive and active cooling systems that are used for cooling electronic devices?
  - *Passive cooling* devices use conduction, radiation, and natural convection heat transfer for heat removal from electronic devices. In these systems, advantage is taken to use all modes of heat transport. Typically, large heat sinks with wide fin-to-fin spacing are used for cooling. Television sets, set top boxes, pole or strand mounted telecommunications boxes are some of the typical examples in this category.
  - Design of these systems attempt to utilize conduction and radiation as the primary modes of heat transfer to maximize the thermal transport and to induce higher levels of natural convection.
  - In applications where convection is limited, for example, space shuttle cooling, radiant heat transfer becomes the sole mode of transport of heat from the source to the sink, sink being space, in the case of the shuttle.
  - These devices make use of a heat spreader (a conductive plate to spread the heat) and/or heat sinks specifically designed for such conditions. In space limiting applications, for example, laptops, heat pipes are often used to efficiently transport the heat from the device to a location where a larger space is available.
  - *Active cooling*, where the fluid motion is assisted by an external source, a fan in a forced air-cooled system, or pump and fan of an immersion or refrigeration cooled system. This class of solutions encompasses an array of cooling techniques that are diverse and extensive.

  - In the fan systems presently used in electronic equipment, flow in to the fan is found to be nonuniform, which is an inadequacy in modern electronic systems. Also noise generated by the fan is another problem.
- Extended surfaces with or without fans are by far the most commonly used as heat sinks in the electronics industry.
- In addition to the above, successful designs make use of fluid flow management, involving flow distribution in the region of interest and attempting to change the layout of the electronic devices to remove flow stagnation points.
- *Hybrid Systems*: A combination of liquid and air cooling for high power dissipation electronics, while minimizing contact resistance throughout the system. Avionics industry is an example of such systems, which produced cooling systems capable of removing high heat fluxes.
- *Air Jet Impingement*: The concept of using a concentrated jet for localized high heat flux cooling is similar to that used for metal quenching. Jet impingement offers not only the ability to remove high heat fluxes but also the ability to target hot spots or uneven heating.
  - One drawback is that a high pressure head is required which gets converted to high kinetic energy of the jet. The high speeds can give rise to noise problems.
- *Microchannels*: Microchannels are based on the very simple heat transfer concept that the heat transfer coefficient for laminar flow is *inversely proportional* to the hydraulic diameter. This means that the smaller the channel is, the higher will be the ability to draw heat from the source. Microchannels typically have sizes in the 5–100 μm range leading to a heat transfer coefficient as high as 80,000 W/m^2K. They are typically etched on the die surface in the shape of rectangular grooves.
  - Pressure drop and flow uniformity across the channels are the minus points. One solution for that is called *stacking*—instead of having a single layer of microchannels on top of the heat source, there can be two or more stacks.
  - Flow nonuniformity across the microchannels results in nonuniform cooling, which can affect the performance and reliability.
- *Spray Cooling*: A dielectric liquid is sprayed directly on the die resulting evaporating cooling. The vapor is condensed and recycled. Heat fluxes in excess of 100 W/cm^2 are obtainable.
- *Direct Contact Cooling*: The working liquid comes into direct contact with the chip.

- What is a thermo-electric heat exchanger?
  - Thermo-electric heat exchanger consists of two thermo-electric modules that use the Peltier effect to pump heat from one ceramic face to the other ceramic face when a DC current is applied. Conceptually, it is similar to a refrigerator that uses a refrigerant to move heat out into the environment, but the thermo-electric cooler (TEC) uses electrons instead of refrigerant.
  - When the current is applied, one side of the TEC module (or heat pump) is cold, and the other is hot. Reversing the polarity causes heat to flow in the opposite direction. Varying the current allows for tight temperature controls.
  - Fluid passing through the heat exchanger is cooled by the TEC Peltier effect and exits at nearly ambient air temperature.
  - Thermo-electric coolers are used to cool electronic components with applications to computer CPUs.
  - One area that has benefited greatly from TEC is optical devices where maintaining a laser temperature at a set level is a must for proper device operation.
  - Poor device efficiency, 30–50% and reliability are the two limiting factors in the use of TECs in the industry.

13.2 OTHER TYPES OF HEAT EXCHANGERS

- What is a storage type/regenerative type heat exchanger?
  - In this, heat transfer surface (generally of cellular in structure) comes into contact *alternately* with hot and cold fluids. Thus, same passage is used alternately at prescribed time intervals. The process is transient; that is, the temperature of the surface (and of the fluids themselves) varies with time during the heating and cooling of the common surface.
  - It involves two or more separate heat storage sections for uninterrupted operation.
  - Used to recover heat from flue gases (preheats air used for combustion).
  - Flue gases and combustion air take turns flowing through each regenerator, alternately heating the storage medium and then recovering heat from it.
  - In *rotating-type regenerators* heat is transferred between adjacent sources and sinks by alternate heating and cooling of the regenerator matrix as it passes between them. Rotational speed is low, typically 10 rpm. The rotating regenerator, sometimes called the heat wheel, can also transfer latent heat, but this is of more value in air conditioning applications.

   - *Advantages* include high efficiency and availability in a wide range of sizes. Can handle significant operating temperature ranges.
   - *Limitations* include possibility for cross-contamination, tolerance for only small pressure differences between source and sink and carry-over of moisture if condensation occurs.

- What are recuperators?
  - *Recuperators* are gas-to-gas heat exchangers in which heat transfer occurs between two fluid streams at different temperature levels in a space, that is, separated by a thin solid wall (a parting sheet or tube wall). Heat is transferred by convection from the hotter fluid to the wall surface and by convection from the wall surface to the cooler fluid. The recuperator is a surface heat exchanger.
  - Recuperators are placed on the furnace stack. Internal tubes or plates transfer heat from the outgoing flue gases to the incoming combustion air while keeping the two streams from mixing.
  - Tubular recuperators are used where size is not important but access for cleaning is essential. Glass tubes may be used for low temperature process heat recovery to facilitate cleaning.
    - *Advantages* include simple construction and ease of cleaning. Available in a wide range of materials.
    - *Limitations* include bulky sizes and lower efficiencies compared to other gas–gas systems.

14

RADIANT HEAT TRANSFER AND FIRED HEATERS

14.1	Radiant Heat Transfer	425
	14.1.1 Fired Heaters	427
	14.1.2 Design of Fired Heaters	434
	14.1.3 Operational Issues	438
	14.1.4 Furnace Tubes	442
	14.1.5 Burners	443
	14.1.6 NO_x Control	446
	14.1.7 Flares	448
	14.1.8 Incinerators	450
	14.1.9 Rotary Kilns	451
	14.1.10 Miscellaneous Topics	452

14.1 RADIANT HEAT TRANSFER

- What is the mechanism of radiation?
  - Energy is emitted by matter in the form of electromagnetic waves (or photons) as a result of changes in electronic configurations of atoms or molecules.
  - Unlike conduction and convection, energy transfer by radiation does not require presence of an intervening medium. It is the fastest mechanism for heat transfer, traveling at the speed of light.
  - Thermal radiation is due to temperature of the emitting body, unlike X-rays, γ-rays, radio waves, microwaves, TV waves, and so on, which are not related to temperature.
- Does radiation require any medium for it to pass through?
  - No.
- What is *solar constant*? What is its value?
  - Solar constant is the rate at which solar energy is incident on a surface on a plane normal to the rays of the Sun at the outer edge of the atmosphere when the Earth is at its mean distance from the Sun.
  - Its value is 1353 W/m^2.
- What are the temperatures above which radiant heat transfer becomes predominant?
  - Over 300°C.
- What are the three components involved when radiant energy is incident on a solid surface?
  - Absorption, transmission, and reflection.
  - The sum of absorptance α, reflectance ρ, and transmittance τ is unity, that is,

$$\alpha + \rho + \tau = 1. \tag{14.1}$$

- What is *albedo*?
  - Albedo is the ratio of scatter coefficient to the sum of scatter and absorption coefficients, that is, percent of incident radiation on a surface that is scattered.
- Compare values of albedo for ice and steel.
  - Ice reflects radiation to a great extent compared to steel.
  - The value of albedo for ice is therefore much higher than that for steel.
- "Radiation has much higher sensitivity to temperature than conduction and convection." *True/False*?
  - *True.*

$$Q \propto T^4. \tag{14.2}$$

- Define emissive power of a surface.
  - *Emissive power*, or *flux density*, is defined as (energy)/(time)(surface area), due to emission from it through a hemisphere.
- What are the characteristic properties of a blackbody?
  - Absorbs all radiation incident on its surface and the quantity and intensity of the radiation it emits are completely determined by its temperature.
  - A blackbody emits the maximum amount of heat at its absolute temperature.
- "A blackbody absorbs all the visible light but may reflect other wavelengths of the incident radiation." *True/False*?
  - *False.*

Fluid Mechanics, Heat Transfer, and Mass Transfer: Chemical Engineering Practice, By K. S. N. Raju

- What is emissivity?
  - The emissivity (ε) of a medium is the fraction of energy a body emits (E) at a given temperature compared to the amount it could emit (E_b) if it were a blackbody:

$$\varepsilon = E/E_b. \tag{14.3}$$

  - Real objects do not radiate as much heat as a perfect blackbody. They radiate less heat than a blackbody and are called gray bodies.
  - Emissivity is simply a factor by which the blackbody heat transfer is multiplied to take into account that the blackbody is the ideal case. Emissivity is a dimensionless number and has a maximum value of 1.0.
  - For a blackbody,

$$E = E_b \quad \text{and} \quad \varepsilon = 1. \tag{14.4}$$

  - Real bodies are typically not perfect blackbodies and do not absorb all the incident energy received on their surfaces or emit the maximum amount of energy possible.
  - The radiant heat absorbed by real surfaces is a function of the absorptivity of the surface, and the total radiant energy emitted by a body can be calculated from

$$Q = E = \varepsilon\sigma A T^4, \tag{14.5}$$

  where σ is the Stefan–Boltzmann constant (=5.669 $\times$ 10^{-8} W/(m^2 K^4)), A is the surface area of the body, and T is the absolute temperature.
  - The above equation, called the Stefan–Boltzmann equation, assumes that the emissivity is a constant value. The absorptivity and emissivity of some surfaces are a function of the temperature and the wavelength of radiation.
  - For radiant heat flux between two surfaces at absolute temperatures of T_1 and T_2,

$$Q/A = \sigma F(T_1^4 - T_2^4)/[1/\varepsilon_1 + 1/\varepsilon_2 - 1], \tag{14.6}$$

  where Q is the rate of heat transfer, σ is the Stefan–Boltzmann constant, F is the geometric view factor, and ε_1 and ε_2 are the emissivities of the surfaces, for example, combustion gases and wall, respectively.
  - The geometric view factor is the fraction of the surface area that is exposed to and absorbs radiant heat. It is based on the relative geometry, position, and shape of the two surfaces. The equation used to calculate F depends on the specific geometry. In using Equation 14.6, it is recommended that F should not be less than 0.67.
- Name some materials whose emissivities are (i) high and (ii) low.
  - *High* black paint (0.98), red brick (0.93–0.96), asbestos (0.93–0.97), human skin (0.95), water (0.96), marble (0.93–0.95), and soot (0.95).
  - *Low* aluminum foil (0.07), polished silver (0.02), polished copper (0.02), and gold (0.03).
- "Emissivity for aluminum is higher than that for steel." *True/False*?
  - *False.*
- Give a graphical representation of emissivity ranges for different materials.
  - Figure 14.1 represents emissivity ranges of different materials.
- "Rough surfaces have lower emissivity than polished surfaces." *True/False*?
  - *False.*
- Give normal values of emissivities of refractive materials used in furnace construction, indicating their temperature dependence.
  - Table 14.1 gives normal emissivity values of important refractories.
- What is absorptivity?
  - Fraction of radiation incident on a surface that is absorbed.
- What is Kirchoff's law?
  - Kirchoff's law states that emissivity ε and absorptivity α are equal at the same temperature and wavelength λ, that is,

$$\varepsilon(\lambda, T) = \alpha(\lambda, T). \tag{14.7}$$

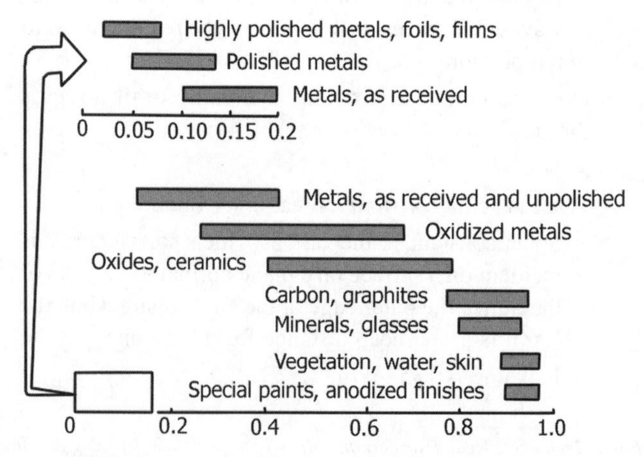

FIGURE 14.1 Emissivity ranges of different materials.

TABLE 14.1 Values of Emissivities of Some Materials as a Function of Temperature

	Temperature (C)						
Material	93	200	430	870	1090	1320	1340
Fireclay brick	0.90	0.90	0.90	0.81	0.76	0.72	0.68
Silica brick	0.90	–	–	0.82–0.65	0.78–0.60	0.74–0.57	0.67–0.52
Chrome-magnesite brick	–	–	–	0.87	0.82	0.75	0.67
Chrome brick	0.90	–	–	0.97	0.98	–	–
High-alumina brick	0.90	0.85	0.79	0.50	0.44	–	–
Mullite brick	–	–	–	0.53	0.53	0.62	0.63
Silicon carbide brick	–	–	–	0.92	0.89	0.87	0.86

Note: Values of temperatures are converted figures from °F to °C.

- Under equilibrium conditions, a surface absorbs and emits the same amount of radiation. This assumes that the surface condition remains the same.
- If the absorptivity and emissivity are independent of wavelength, the surface is said to be a *gray body*. In engineering calculations, most surfaces can be treated as gray bodies.
- In many practical situations, dependence of ε and α on temperature and wavelength is ignored and α and ε are taken to be equal.

- What is a gray body?
  - A body whose radiative properties are independent of wavelength.
  - For a gray body, absorptivity and emissivity are related by Kirchoff's law,

$$\alpha = \varepsilon. \quad (14.8)$$

- "Surfaces with a high emissivity to absorptivity ratio are suitable for collecting solar energy." *True/False*?
  - *False.*
- "A leaf is seen as green because it emits radiation at the green wavelength." *True/False*?
  - *False.*
- "A surface is seen as white because it reflects all the wavelengths of visible light." *True/False*?
  - *True.*
- What is the value of the Stefan–Boltzmann constant for blackbody radiation?

$$\sigma = 5.669 \times 10^{-8}\,\mathrm{W/(m^2\,K^4)} \text{ or } 0.173 \times 10^{-8}\,\mathrm{Btu/(ft^2\,h\,^{\circ}R^4)}. \quad (14.9)$$

- What is a Lambert surface?
  - A Lambert surface is the surface that emits or reflects radiation with an intensity independent of the angle.
- What is a *view factor*? What does it represent?
  - A view factor or exchange factor is the fraction of radiation leaving surface A_1 (radiating surface) in all directions that is intercepted by surface A_2 (receiving surface or sink).
  - It gives an indication of the amount of radiation from the source that is actually absorbed by the receiving surface.
- Aluminum paint is applied on storage tanks containing volatile liquid. Why sometimes this is done?
  - Aluminum paint has high reflective power for radiation from the Sun and reduces heat absorption by the tank. The result is that storage tanks remain cool, reducing vaporization tendencies of volatile liquids.
- In flat plate solar collectors, black chrome is plated on receiving surfaces. What is its advantage?
  - Black chrome captures more heat than other materials and therefore it is used in solar collectors.
- Is it true that radiant heat transfer is important in cryogenic insulation? If yes, why?
  - Yes, because of vacuum existing between particles, conduction and convection become insignificant and radiation is the predominant mode of heat transfer.
- What is quenching? Give examples of quenching processes.
  - Quenching involves bringing down temperatures of very hot bodies suddenly by water drenching. The water absorbs the heat through sensible and latent heat transfer processes.
  - Examples include quenching of red hot steel and other metal blocks, putting off fires, and the like.

14.1.1 Fired Heaters

- What is a fired heater? What are the types of fired heaters used in petroleum and petrochemical plants?
  - A fired heater is basically a tube still furnace, which is also called a direct fired heater.

- Vertical cylindrical type for small capacities and rectangular box type for large capacities.

- What are the major sections of a fired heater?
  - Radiant and convection sections.
- What are the important parts of a fired heater?
  - Radiant section, shield section (bridge wall), convection section, flue stack, air preheater, and burners.
    - *The radiant section* of a typical fired heater very often has two zones. The first is a lower firing zone that corresponds to a section of the heater wherein the fuel–air mixture leaving the burners is burned nearly to completion and the combustion products are simultaneously cooled by the surrounding heat transfer surfaces. The length of this section is very nearly equal to the flame length. In the second zone, located above the first, the combustion products are further cooled prior to entering the convection section.
    - *The convection* section typically preheats process fluids before they enter the radiant section. It consists of a refractory-lined enclosure that has a rectangular cross section. Inside the enclosure are multiple rows of closely spaced horizontal tubes. These tubes form channels through which combustion products leaving the radiant section pass at relatively high velocity. Heat is transferred, principally by convection, from the combustion products to the heating surfaces and process fluids. Combustion products typically leave the convection section at a reduced temperature (lower exhaust temperatures correspond to higher overall thermal efficiencies).
- What is the function of air preheater in a fired heater? Where is it located?
  - The air preheat system is used to preheat the combustion air going to the burners. Since it cools the flue gas further, while removing heat, it improves the efficiency of the heater. Using an air preheat system will frequently result in overall efficiencies above 90%. The air preheaters are recuperative tubular or regenerative types.
- What are the normal capacities of fired heaters?
  - 10,000–500,00 MJ/h.
- How are tubes arranged in a fired heater?
  - Tubes are arranged backed by the heater walls inside the radiant section. Both horizontal and vertical arrangements are practiced in fired heaters depending on their size and geometry.
  - Small size heaters are cylindrical in shape and tubes are arranged in vertical position backed by refractory walls. The tubes may be along the refractory wall, as in a circular pattern, or they may be exposed to the radiating flame from both sides, as in a cross or octagonal pattern.
  - Figure 14.2 shows these arrangements.
  - Large box-type furnaces are generally rectangular in shape in which tube arrangements are either vertical or horizontal, depending on the furnace volume and geometry.
  - Convection section tubes are in rows, horizontally arranged with triangular pitch for good heat transfer.
- What are the common shapes and tube arrangements in a fired heater?
  - Cylindrical shells with vertical tubes and cabin or box types with horizontal tubes.
  - Tubes are mounted approximately one tube diameter from the refractory walls. In the gap between tubes and refractory walls, there will be downward flow of gases due to cooling effects, contributing to convective heat transfer.
  - Usual center-to-center spacing is twice the outside tube diameter.
    - Wider spacing may be employed to lower the ratio of maximum flux at the front of the tube to the average flux. (Heat flux in a fired heater is defined as the total heat load of the heater divided by the total heat transfer surface area of the heater.)

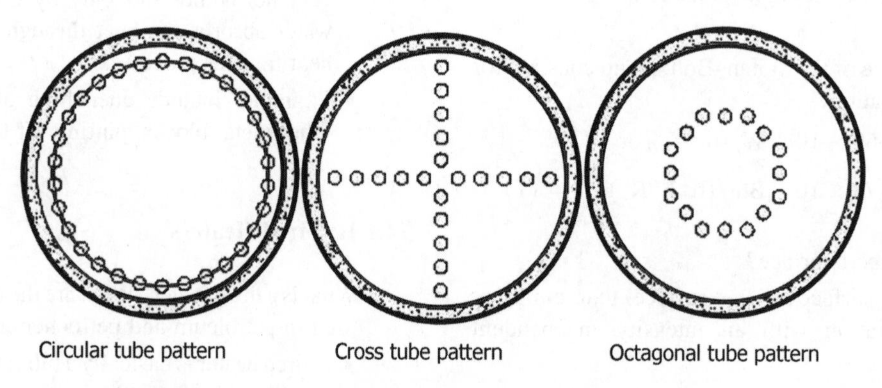

FIGURE 14.2 Tube arrangements in small cylindrical fired heaters.

➢ For single rows of tubes, some values of these ratios are given below.

Center-to-center spacing/tube diameter	1	1.5	2	2.5	3
Maximum flux/average flux	3.1	2.2	1.8	1.5	1.2

- In the convection section, tubes are typically arranged as horizontal tube banks.
- In some designs, a single horizontal tube bank is arranged in the center of the furnace with burners along each sidewall.

- How many rows of tubes are normally provided in a fired heater? If the number of rows is increased what happens to the effectiveness of the number of tube rows?
  - Maximum two rows. As the number of tube rows is increased, the rows behind the first row come under the *shadow* of the first row and effective surface receiving direct radiation from the flames by the second and third rows progressively decreases, making investment on the additional tube surfaces highly uneconomical.
- What are the considerations involved in deciding about the number of tube rows in a fired heater?
  - A second row of tubes on triangular spacing contributes only about 25% of the heat transfer of the front row.
  - Accordingly, new heaters employ only the more economical one-row construction.
  - Second rows are sometimes justifiable on revamp of existing equipment to marginally increase the duty.
- What is the approximate percent of total heat absorbed in the radiant section of a fired heater?
  - About 75–80%.
- Will there be any convective heat transfer in the radiant section of the fired heater?
  - Yes.
- If yes for the above question, what percentage of the heat transfer in the radiant section is due to convection?
  - About 10–20%.
- Explain how convection currents are induced in the radiant section of a fired heater and their effects?
  - Convection currents are induced by the rising flue gases inside the heater. The upward velocity of these gases past the tube surfaces transfers heat by convection to the tubes. Absorption of heat by the tubes cools the rising gases near the tubes, inducing a downward component of the velocity whose magnitude, especially on the backside of the tubes in the space between the cooler refractory walls and tubes, will be considerable.
  - Consequent to this cooling effect and downward velocity of the gases in the space between the refractory walls and backside of tube walls, heat transfer by convection is increased, partially compensating lower radiant heat transfer rates to these back surfaces of tubes on account of the absence of direct radiation.
  - Thus, back of these tube surfaces receives heat, in addition to reradiation from the refractory walls, by the increased convection rates.
- "The common types of heaters have cylindrical shells with vertical tubes and cabin or box types with horizontal tubes." *True/False*?
  - *True*.
- What are the performance objectives of a fired heater?
  - The performance objectives of fired heaters include maximizing heat delivery to the fluid in the tubes with varying fuel quality, minimization of fuel consumption, minimization of heater structural wear and stack emissions, and maximization of safety levels.
  - Requirement for efficiency increases is receiving greater attention as this factor involves escalating energy costs that are over 60% of operational costs of a major process plant. Furnace and heater fuel is the largest component of this cost.
- Give diagrams for typical vertical cylindrical and rectangular box-type heaters and describe their salient features.
  - A typical vertical cylindrical heater consists of tubes arranged in a vertical position backed by refractory cylindrical wall and centrally located multiple hearth burners firing vertically upward inside the tube circle (Figure 14.3). Natural draft is normally employed.
  - Combustion air flow is regulated by positioning the stack damper. Fuel to the burners is regulated by exit feed temperature.
  - The L/D ratio for the heater is normally in the range of 2–3, with 2.5 being commonly used. The overall height is about 20 m. Average radiant section heat flux is about 31.6 kW/m^2 (10,000 Btu/(h ft^2)). Maximum tube heat flux is about 174 kW/m^2 (55,000 Btu/(h ft^2)).
  - The radiant tubes are located along the walls in the radiant section of the heater and receive radiant heat directly from the burners or target wall. Horizontal tubes are located in box designs that are employed for large furnaces.
  - The radiant zone with its refractory lining is the most expensive part of the heater and most of the heat is transferred to the process fluid in the tubes there. This is also called the *firebox*.

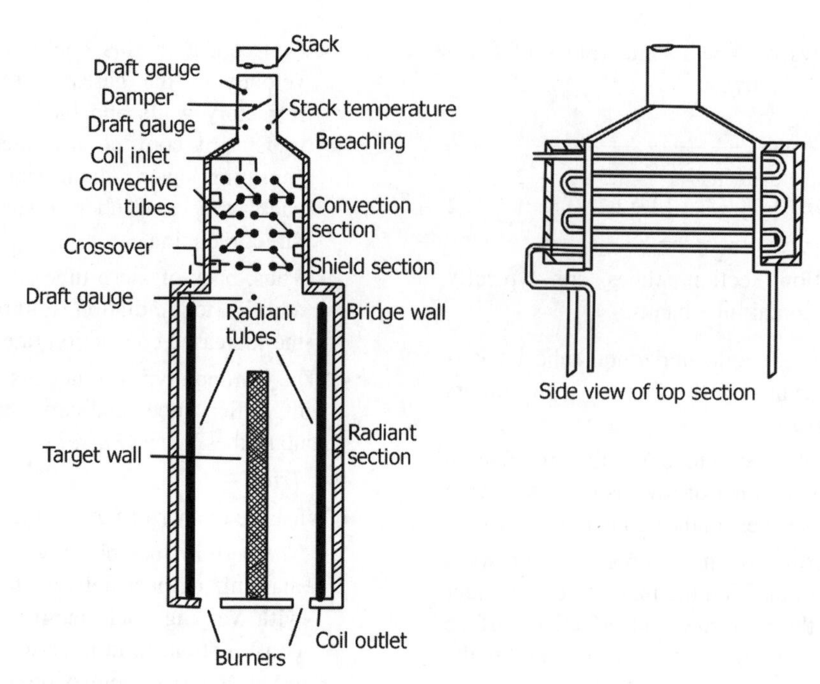

FIGURE 14.3 Fired heater with vertical radiant tubes and side view of top section.

- The feed enters the coil inlet in the convection section where it is preheated before transferring to the radiant tubes. The convection section removes heat from the flue gas to preheat the contents of the tubes and significantly reduces the temperature of the flue gas leaving the stack. If draft is high, heat transfer by convection will be high.
- *Shield section* is located just below the convection section, containing rows of tubing that *shield* the convection tubes from the direct radiant heat. Just below the shield tubes are two important monitoring points. The first is the *bridge wall* temperature, which is the temperature of the flue gas after the radiant heat is removed including the radiant heat to the shield tubes above, but before the convective heat to the shield tubes. The other is the draft measurement at this point, since for most heater designs, if it is negative at this point, it is negative throughout the heater.
- The first row of tubes in the convection section (shield tubes), above the radiant section, receives heat by convection due to increased flue gas velocities (as a result of decreased flow cross section) and at the same time receives heat by radiation as these tubes are *seen* by the burner flames. Unless increased heat absorption is provided, tubes might melt.
- Because of this reason, the first row of tubes is sometimes used for steam generation or to superheat steam or for admitting cold process fluid, to avoid excessive tube heat fluxes that can lead to tube failures. Passing steam will have a cooling effect on these tubes.
- Convection section tubes are generally finned type, to make convection section compact, because of low heat transfer coefficients involved for gases. Normally segmented, solid, or stud types of fins are used.
- The flue gases can be cooled to very low temperatures, but caution must be used to avoid cooling below the dew point of the flue gas as this could cause corrosion of the tubes or surface.
- Shield tubes are *never* finned type.
- Tube temperatures are monitored in both convection and radiant sections.
- Measurements of flue gas temperature, bridge wall temperature, draft, oxygen content of flue gases, and ppm of combustibles are normally made just below the shield section.
- Bridge wall temperature is the temperature of the flue gas leaving the radiant section and before it enters the convection section. Measurement of the draft at this point is also very important since this indicates how well the heater is set up.
- The transition from the convection section to the stack is called *breeching*.
- Flue gas exit temperature is important for determining the levels to which stack gases rise above the tip of the stack from atmospheric dispersion point of view. Measurements of stack emissions are important for compliance to environmental laws.

- Combustion air flow is regulated by positioning the stack damper. Fuel to the burners is regulated by exit feed temperature and firing rate is determined based on the design and operating parameters.
- A typical horizontal tube cabin heater consists of a rectangular firebox with refractory-backed horizontal tubes on the sidewalls, sometimes on the roof. One or more rows of burners are provided on the hearth centerline and fired vertically upward between the tube rows.
- The height of the heater is about 6 m and the ratio of length to width is in the range of 2.5–3.5 m.
- Gain is marginal by extending the ratio beyond 2.0.
- Excessive fluxes may damage the metal or result in skin temperatures that are harmful to the process fluid.
- A second row of tubes on triangular spacing contributes only about 25% of the heat transfer of the front row. Accordingly, new heaters employ only the more economical single-row construction.
- Second rows sometimes are justifiable on revamp of existing equipment to marginally increase the duty.
- Figure 14.4 illustrates a medium-capacity cabin heater, with a rectangular firebox, tubes being arranged horizontally.
- Figure 14.5 shows a large cabin heater with three separate radiant zones. A variety of tube arrangements are possible that are located along refractory walls. Primary air is supplied premixed with the fuel while secondary air is admitted from the bottom of the furnace.

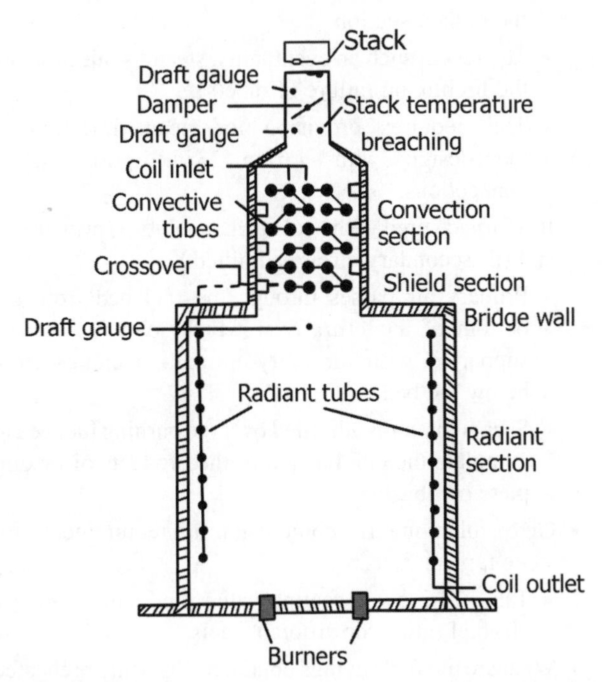

FIGURE 14.4 A cabin heater with horizontal tubes and a rectangular firebox.

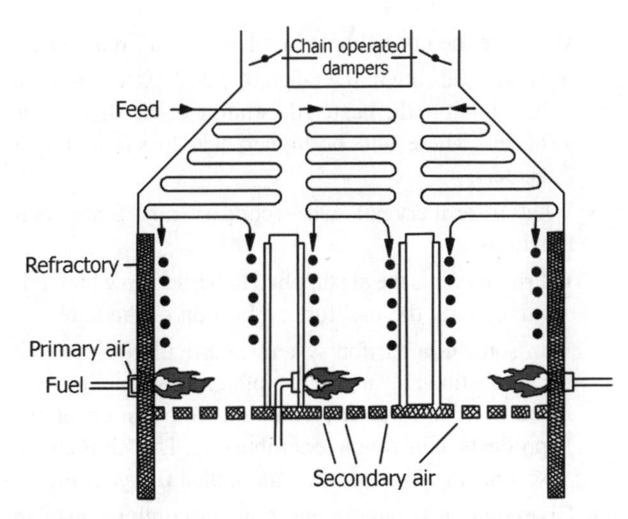

FIGURE 14.5 Large box-type cabin heater showing three separate radiant sections.

- What are the applications of fired heaters and what are the normally used fuels to fire them?
  - Vertical cylindrical and horizontal cabin heaters find application in refinery services, gas plants, petrochemicals, chemicals and synthetics, olefins, and ammonia and fertilizer plants. Examples in refinery service are as follows:
    - Crude distillation.
    - Preheating residue from atmospheric distillation of crude oils before admitting into a vacuum distillation column in a refinery.
    - Visbreaking furnace involving mild thermal cracking of residues at atmospheric pressure and 400–500°C to lower viscosity.
    - Furnace for catalytic reforming of naphtha.
    - Hydrocracking and hydrodesulfurization.
  - Heater fuels include light ends (e.g., refinery gas) from the crude units and reformers as well as waste gases blended with natural gas. Residual fuels such as tar, pitch, and Bunker C (heavy oil) are also used.
- What are the capacities that can be accommodated in heaters with single radiant chambers?
  - 3–60 MW (10–200 MBtu/h). (*Conversion*: 1 Btu/h = 0.292 W.)
  - For higher capacity duties, three or four radiant chambers with a common convection section are feasible.
- What is radiant efficiency of a fired heater?
  - Radiant efficiency of a fired heater is the ratio of heat absorbed in the radiant section to the net heat released.

- What are the normal heat loss levels in a fired heater?
  - Heat losses are in the range of 1.5–2.5% of the total heat input to the heater, depending on the size of the heater. There will be higher heat losses in larger heaters.
- What is primary air and secondary air? State their functions.
  - Primary air is the air supplied to burners in which it is mixed with the fuel for combustion of the latter.
  - In some combustion systems, particularly in which combustion may not be complete, air is supplied over the burners to ensure complete combustion of the products of incomplete combustion. This air is called secondary air and sometimes called *overfire* air.
- Give equations representing complete combustion of an organic.

$$C_cH_hO_oS_sN_n + (c + h/4 - o/2 + s)O_2 \rightarrow cCO_2 + (h/2)H_2O + sSO_2 + (n/2)N_2 \quad (14.10)$$

where c is the number of carbon atoms, h is the number of hydrogen atoms, o is the number of oxygen atoms, s is the number of sulfur atoms, and n is the number of nitrogen atoms in a molecule of the organic.

  - For each mole of oxygen required, 4.77 moles of air are required. For example, for complete combustion of methane, 9.54 moles of air are needed.
  - The theoretical air requirement, A_o, is represented by the equation

$$A_o = 4.77(c + 0.25h - 0.5o + s). \quad (14.11)$$

  - Percent excess air is calculated by

$$X = 100 \times (A_i - A_o)/A_o. \quad (14.12)$$

where X is percent excess air and A_i is moles of inlet air per mole of fuel.

- What are the minimum excess air requirements to ensure complete combustion for various types of fuels used in a furnace?
  - Gaseous fuels: 10%. Normally used: 15–25%.
  - Liquid fuels: 15–20%. Normally used: 30%.
  - Radiant panel burners: 2–5%.
- What are the consequences of using less excess air than the recommended minimum values?
  - There will be incomplete combustion resulting in the formation of products such as aldehydes, ketones, and carbon monoxide that give rise to pollution problems.
  - In addition to pollution, such gases, being at elevated temperatures, can give hazardous conditions of fire and explosion when they come in contact with atmospheric oxygen that might leak into the convection section of the furnace and reignite the hot unburnt fuel.
  - The flue gas temperature will increase from about 420 to 1093°C (7002000°F), which weakens the tubes leading to failure.
  - Another factor that reduces tube life is increased high-temperature corrosion.
  - Failed tubes result in spillage of their contents into the radiant section (firebox). Under these circumstances, the excess air used is not enough to cause complete combustion, with the appearance of black smoke at the exit of the stack.
  - Due to lack of enough oxygen, radiant section temperatures will not increase, making the flue gases too rich, that is, above *upper flammability limit* (UFL), preventing fire and explosion. (UFL is the fuel–air mixture composition above which combustion is not supported as the mixture becomes too rich in fuel to support combustion.)
  - If, however, leaking tubes are blocked and fuel supply is cut off, hydrocarbon content in the flue gas gradually decreases, that is, air/fuel ratio increases. This results in changeover from rich mixture passing through below UFL into flammability/explosive range. As refractory walls are still hot enough to initiate ignition, there is a possibility of explosion in the radiant section.
  - To prevent such an eventuality, steam is injected into the firebox on failure of tube(s).
  - This requires provision for steam injection in the design with 7.6 cm (3 in.) purge steam connections.
- In a furnace fired with coal at what points (i) primary air and (ii) secondary air are admitted?
  - Primary air passes through the fuel bed from the bottom of the *grate* over which solid fuel bed is supported with air entry through openings from below the bed.
  - Secondary air is admitted over the burning fuel bed to complete the combustion of the products of incomplete combustion.
- Give stoichiometric combustion air requirements for some typical fuels.
  - Table 14.2 gives heating values and stoichiometric air–fuel ratios for different fuels.
- What are the fuel savings obtainable by using preheated combustion air in a fired furnace?

TABLE 14.2 Heating Values and Stoichiometric Combustion Air Requirements for Different Fuels

Fuel	LHV (kcal/kg)	Combustion Air (kg/kg fuel)
Methane	12,000	17.2
Propane	11,000	15.2
Light fuel oil	9800	14.0
Heavy fuel oil	9700	13.8
Anthracite	6900	4.5

LHV: lower heating value.

- Fuel savings are functions of type of fuel, burner design, excess air used, temperature of the stack gases, and temperature of preheated air.
- Table 14.3 illustrates the savings obtainable using natural gas as fuel with 10% excess air.
- It can be seen that the savings are substantial when combustion air is preheated with the energy extracted from flue gases by installing recuperators in the furnace stack. These savings in terms of currency value encourage incorporating recuperators, where this is not done, particularly in view of the ever-escalating fuel costs.

- What are the gases in a furnace atmosphere that contribute to significant levels of radiant heat transfer?
  - Triatomic gases have high emissivities.
  - In furnace atmospheres, CO_2 and H_2O are present in significant amounts and radiant heat transfer from these gases will have to be taken into account in the design and operation of furnaces.
  - As the other triatomic gas, SO_2, is present only in small amounts, its effect is not significant.
  - The radiation from triatomic gases falls mainly in the region of infrared wavelengths.

TABLE 14.3 Fuel Savings as a Function of Preheated Air Temperature

Furnace Exit Temperature (°C)	Temperature of Preheated Air (°C) 316	427	538	630	742	854
	Percent Savings					
538	13	18	–	–	–	–
630	14	19	23	–	–	–
742	15	20	24	28	–	–
854	17	22	26	30	34	–
965	18	24	28	33	37	40
1076	20	26	31	35	39	43
1169	23	29	34	39	43	47
1262	26	32	38	43	47	51

Temperature values given are converted figures from °F to °C. Fuel: natural gas with 10% excess air.

- What are the factors on which emissivity contributions of CO_2 and H_2O vapor depend?
  - Gas temperature, partial pressures of CO_2 and H_2O, *thickness* of the radiating gas volume, quantified by what is known as *mean beam length* (which depends on shape and size of the furnace), and to a small extent on pressure.
- How does partial pressures of CO_2 and H_2O influence radiant heat transfer rates?
  - Increased partial pressures of the radiating gases, CO_2 and H_2O, increase radiant heat transfer rates.
- "The magnitude of radiation from CO_2 and H_2O vapors at furnace temperatures is higher than convective heat transfer at these temperatures." *True/False*?
  - *True.*
- Apart from radiation from flames and gases, is there any other item or items that contribute significantly to overall *radiation* in a fired heater?
  - Reradiation from refractory walls.
  - Radiation from incandescent soot, coke, and ash particles. These are formed due to decomposition–polymerization reactions of hydrocarbon fuels, especially heavy fuel oils that have strong radiation ability and can make the furnace atmosphere very bright. For coal-fired furnaces, soot formation is relatively insignificant as coal contains very small amounts of decomposable hydrocarbons. But coke particles are present and can emit strong radiation. Coke particles emit radiation close to that of a blackbody. Ash particles result in burning out of coke particles. The radiation intensity of ash particles is lower at higher temperatures and increases as the flue gas gets cooled.
- In any furnace is there any possibility of reradiation from refractory walls to tubes?
  - Yes. Refractory walls reradiate significant amounts of heat. This reradiation especially falls on the back-side of the tube surfaces.
- Comment on the operation of fired heaters with respect to the refractory walls and deposits on tube and refractory surfaces.
  - The emissivity of the refractory used in a furnace is important in determining the surface radiation (re-radiation) heat transfer between the walls, the load, and the flame. The heat transfer in many industrial combustion processes is dominated by radiation from the hot refractory walls.
  - It was shown that fuel consumption decreases as furnace wall emissivity increases, although there is no effect when the furnace atmosphere is gray.

Transient furnace operation or poor wall insulation reduces the beneficial effects of high wall emissivity.

- Normal emittances of common, commercially available refractories, including dense insulating firebrick and porous ceramic fiber, range from 0.3 to 0.7 at 1300 K.
- The emissivity of some high emittance coatings could extend the range of refractory emittances from 0.3 to 0.9 at 1300 K.
- In some industrial combustion processes, the surface absorptivity can change over time, which affects the performance of the system. This is particularly true of coal-fired processes, where the ash may be deposited on tube surfaces.
- The heat transfer from the combustion products to the tubes is primarily by radiation, with a lesser amount by convection heating. The heat must conduct through the ash deposits and the tube wall, before heating the fluid inside the tubes primarily by convection. As the ash deposit melts to form a slag, the absorptance increases dramatically to values approaching 0.9.
- The thermal conductivity of the deposit is highly dependent on its physical state, especially its porosity. It is interesting to note that the heat flux through the deposits decreases during the initial phases of its growth and then actually increases before reaching a steady-state value when the deposits have reached maturity.

* What is the effect of the presence of oxides of iron and chromium on the emissivities of refractory materials?
  - Emissivity of refractory materials increases with the presence of the above oxides.
* What are the revamping practices to improve performance of fired heaters?
  - There are several revamping schemes that can improve fired heater performance. The major ones are installing a convection section in an all-radiant heater, increasing heat transfer area of the convection section, converting a natural draft heater to a forced draft one, and adding air preheating or steam generation equipment.
  - *Installing Convection Section*: Heaters with a heat duty of up to 3 million kcal/h were built as radiant heaters. They generally have a thermal efficiency in the range of 55–65% and flue gas temperatures generally over 700°C. Adding a convection section in such heaters could recover additional heat and reduce their flue gas temperatures to within 50–100°C of their inlet feed temperatures. This could increase their thermal efficiencies to 80% or over. If the revamped heater is operated at the same heat duty, the radiant heat flux is reduced or additional heat duty is obtained in the convection section.
    - Such revamping requires checking foundations and structures for increased loads, additional space requirements if the convection section is installed externally, requirement of an induced draft fan or increasing stack height for additional natural draft to overcome the decreased draft, and increasing pump discharge pressures to overcome the additional ΔP through the tubes.
  - *Increasing Convection Section Heat Transfer Area*: This can be done by adding two more tube rows in the convection section and/or replacing plain tubes by finned tubes. This can increase surface areas two to five times, based on the choice of the type of finned surfaces. Increased areas will increase pressure drops as well as structural loads. These aspects should be taken care of by the above-mentioned solutions.
  - *Using Forced Draft Burners in Place of Natural Draft Burners*: These will improve turbulence and reduce excess air requirements.
  - *Preheating Combustion Air*: Air preheating is economically attractive, if flue gas temperature is above 350°C. It must be noted that use of air preheaters can increase corrosion of heat transfer surfaces and NO_x levels in flue gases.
  - *Steam Generation*: Waste heat can be utilized by installing waste heat boilers. The flue gas can be cooled to within 50–100°C of the inlet boiler feed water temperature, generating medium- or low-pressure steam.

14.1.2 Design of Fired Heaters

* Write the Stefan–Boltzmann equation for radiant heat transfer between two real surfaces at temperatures T_r and T_s (radiating and receiving surfaces, respectively).

$$Q_r = \sigma AF(T_r^4 - T_s^4), \tag{14.13}$$

where A is the area of one of the surfaces and F is an *exchange factor* that depends on the relative area and arrangement of the surfaces and on the emissivity and absorptivity of each of the surfaces.

  - Either the heat radiating or heat absorbing surface may be used as the basis for determining Q_r. However, the value of the exchange factor F depends on which surface is used.
  - The exchange factor F is the product of F_a and F_e, where F_a is the *shape factor*, which depends on the spatial arrangement of the two objects, and F_e is the emissivity factor, which depends on the emissivities of both objects.

- The product of F_a and F_e can be represented by the symbol F. F_a, F_e, and F are dimensionless quantities. F is sometimes called *exchange factor* or *radiation configuration factor*, which takes into account the emissivities of both bodies and their relative geometries.
- In a furnace, the heat absorbing surface is well defined and important in the design and operation of furnaces. Consequently, the heat absorbing surface, usually called *cold* surface, is used as the basis for estimating radiant heat transfer using the Stefan–Boltzmann equation.

• What is *equivalent cold plane area*? Explain.

- The heat absorbing surface in a furnace consists of a number of parallel tubes backed by a refractory wall. The tubes absorb part of the radiation from the flames and furnace gases and the remaining part passes through to the refractory walls on the back of the tubes and is reradiated back. Part of the reradiation from the refractory walls is absorbed by the tubes and the remaining passes through. It is complicated to quantify this situation for calculation purposes in which an area term is to be used.
- This situation is handled by replacing the tube bank by an *equivalent plane surface*, A_{cp}, which is designated as *cold plane area*:

$$A_{cp} = (\text{number of tubes})(\text{exposed length of the tube})(\text{center-to-center distance}). \quad (14.14)$$

- As the tube bank will not absorb all the heat falling on it, A_{cp} is to be corrected by an efficiency factor, which is the absorptivity α, by defining an *equivalent cold plane area*, αA_{cp}. It is the area of an ideal black plane having the same absorbing capacity as the actual tube bank.
- α is estimated from published charts for different tube arrangements. A typical chart is illustrated in Figure 14.6.
- In estimating the total equivalent cold plane area in a furnace, the applicable value of α is applied only to the refractory-backed tubes. The cold plane area of the tubes not backed by the refractory wall, that is, the cold plane area of the shield tubes at the entrance to the convection section, is to be taken at full value as α for such tubes may be taken as 1.0.

• How is the exchange factor F estimated?

- In order to estimate F, the emissivity of the flue gas must be estimated first. The components in the flue gas that significantly contribute to radiant emission in a furnace are CO_2 and H_2O vapor. This contribution depends on the partial pressures of each of these two

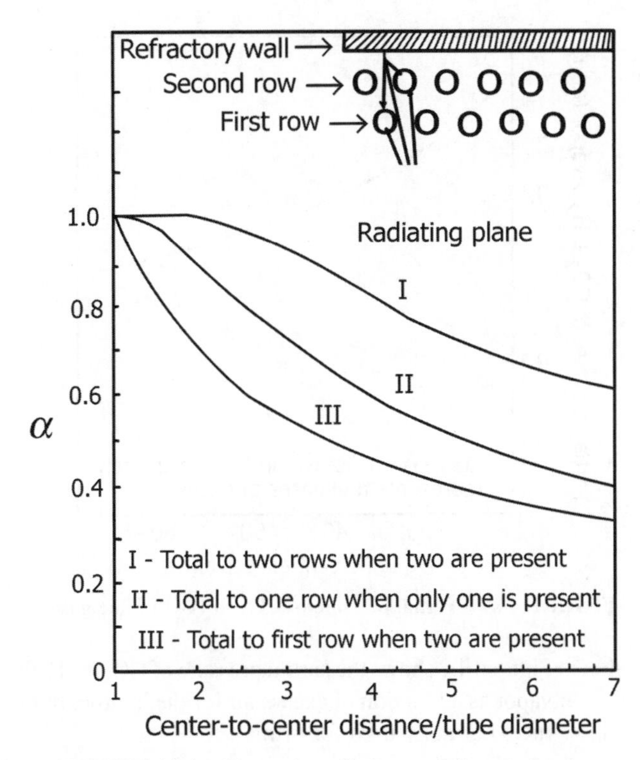

FIGURE 14.6 Absorption efficiency of the tube banks.

components, furnace dimensions, and temperatures of the gases and the absorbing surface.

- According to Hottel, the composition and dimensional effects can be accounted by a single term, the partial pressure of CO_2 plus H_2O vapor, and a term called *mean beam length* that accounts for the dimensional effects of the furnace.
- Relatively accurate and simple heat transfer calculations can be carried out if an isothermal, absorbing/emitting, but not scattering, medium is contained in an isothermal, black-walled enclosure. Though these conditions are very restrictive, they are met to some degree by conditions inside furnaces. For such cases, the local heat flux on a point of the surface may be calculated from

$$q = [1 - \alpha L_m]\varepsilon_{bw} - \varepsilon L_m \varepsilon_{bg}, \quad (14.15)$$

where ε_{bw} and ε_{bg} are blackbody emissive powers for the walls and medium (gas and/or particulates), respectively, and $\alpha(L_m)$ and $\varepsilon(L_m)$ are the total absorptivity and emissivity of the medium for a path length L_m through the medium.

- The length L_m, known as the *average mean beam length*, is defined as a directional average of the thickness of the medium as seen from the point on the surface.

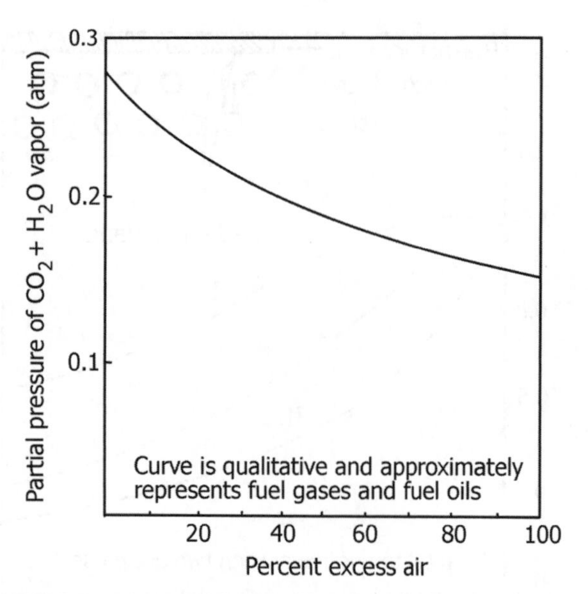

FIGURE 14.7 Partial pressure of $CO_2 + H_2O$ in flue gases.

- Figure 14.7 shows the partial pressure of $CO_2 + H_2O$ vapor as a function of excess air for the hydrocarbon fuels normally used for firing.
- Values of mean beam length L for various furnace shapes are given in Table 14.4.
- Mean beam length for different geometries, not listed above, may be estimated from the following relationship:

$$L \approx 0.9L_0 \approx 3.6V/A, \tag{14.16}$$

where L_0 is known as the geometric mean beam length, which is the mean beam length for the optical thin limit, and L is the spectral average of the mean beam length. V is the volume of the participating medium and A is its entire bounding surface area.

TABLE 14.4 Mean Beam Length for Different Furnace Shapes/Dimensions

Dimension Ratio	Mean Beam Length, L
Rectangular furnaces	
Length:width:height	
(in any order)	
1:1:1 to 1:1:3 and	2/3 of (furnace volume)$^{1/3}$
1:2:1 to 1:2:4	
1:1:4 to 1:1:∞	1 × (smallest dimension)
1:2:5 to 1:2:∞	1.3 × (smallest dimension)
1:3:3 to 1:∞:∞	1.8 × (smallest dimension)
Cylindrical furnaces	
Diameter:height	
1:1	2/3 × diameter
1:2 to 1:∞	1 × diameter

- Emissivity is correlated as a function of the product of partial pressure of $CO_2 + H_2O$ vapor and mean beam length, that is, pL, neglecting the effect of variations in tube wall temperature, which is marginal.
- Gas emissivities are estimated from the curves illustrated in Figure 14.8, which give the relationship between pL and gas emissivity, with gas temperature as a parameter.
- Using the emissivity estimated above, the exchange factor F is obtained from Figure 14.9, which gives F as a function of gas emissivity with $A_w/\alpha A_{cp}$ as a parameter. Here A_w is the exposed refractory area, which is defined as the area that would be exposed if the tube bank were replaced by the equivalent cold plane. That is, it is the total envelope area of the firebox, less the equivalent cold plane area of all the tubes.
- Figure 14.9 shows that the tubes do not completely absorb all the radiant energy that is incident on them. The curves are based on a tube surface absorptivity of 0.9, which is a commonly accepted value for oxidized metal surfaces.

- Give the final form of the equation for the estimation of radiant heat transfer in the firebox (radiant section) of a fired heater.

$$Q_r = \sigma\alpha A_{cp}F(T_r^4 - T_s^4), \tag{14.17}$$

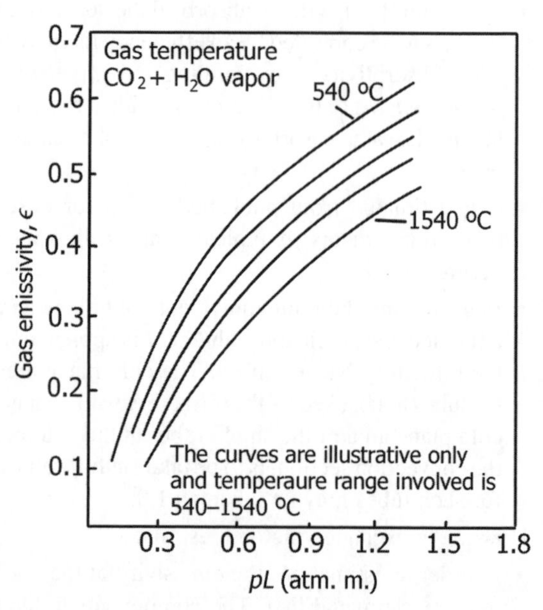

FIGURE 14.8 Gas emissivity as a function of gas temperature. (Lobo WE, Evans JE. Heat transfer in radiant section of petroleum heaters. *Transactions of the American Institute of Chemical Engineers* 1939;35:743.)

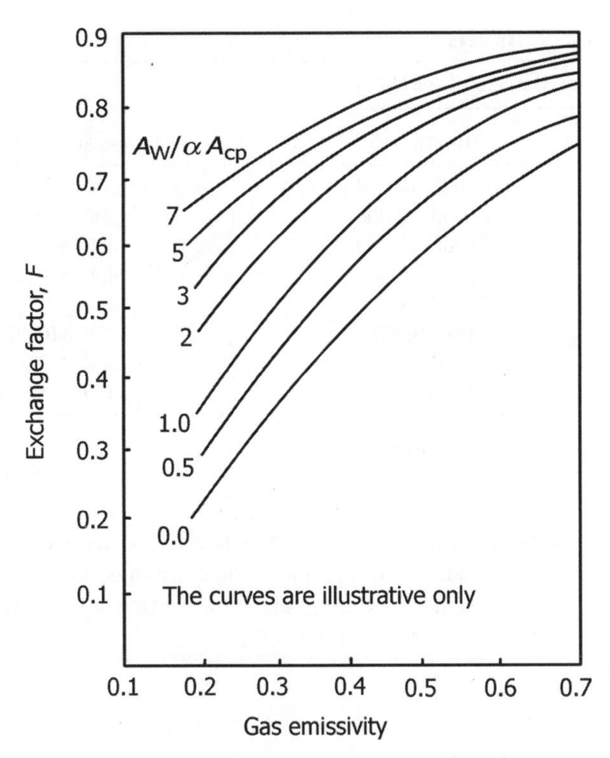

FIGURE 14.9 Overall radiant exchange factor F. (Lobo WE, Evans JE. Heat transfer in radiant section of petroleum heaters. *Transactions of the American Institute of Chemical Engineers* 1939;35:743.)

where T_r and T_s are the absolute temperatures of the radiating plane and the tube surface, respectively.

- How is convective heat transfer in the radiant section of a fired heater estimated?
  - The equation for the estimation of convective heat transfer in the radiant section can be written as

$$Q_{rc} = h_{rc}A_{rt}(T_g - T_t), \quad (14.18)$$

where Q_{rc} is the convective heat transfer rate, h_{rc} is the convective heat transfer coefficient, A_{rt} is the total effective heat transfer surface of the tubes, and T_g and T_t are the temperatures of the gas and tubes, respectively, in the radiant section of the furnace.

  - Normally for tube still heaters, h_{rc} is about 11.35 W/(m^2 °C) (2 Btu/(h ft^2 °F)) and A_{rt} can be approximated to about $2\alpha A_{cp}$. With these assumptions, the equation for convective heat transfer in the radiant section becomes

$$Q_{rc} = 23\alpha A_{cp}(T_g - T_t). \quad (14.19)$$

- Write the equation for total heat transfer rate in the radiant section.

$$Q_R = Q_r + Q_{rc} = \sigma\alpha A_{cp}F(T_r^4 - T_s^4) + h_{rc}A_{rt}(T_g - T_t). \quad (14.20)$$

- What are the design parameters used for a fired heater?
  - Radiant section heat flux: 37.6 kW/m^2 (12000 Btu/(h ft^2)).
  - Convection section heat flux: 12.5 kW/m^2 (4000 Btu/(h ft^2)).
  - Cold oil tube velocity: 1.8 m/s.
  - Thermal efficiency: approximately 70–90% based on lower heating value of the fuel.
  - Flue gas temperature above feed inlet: 140–195°C.
- How is the volume of a box-type heater fixed?
  - As a rough guide, about 1.22 m^3/m^2 (4 ft^3/ft^2) of radiant transfer surface, but the ultimate criterion is sufficient space to avoid flame impingement on the tubes.
- Give typical average heat fluxes and temperatures employed in process heaters.
  - Table 14.5 gives average heat fluxes and temperatures employed in commercial process heaters.
- What are the recommended allowable radiant heat fluxes for personal exposure?
  - Table 14.6 gives maximum permissible radiant heat fluxes for personal exposure.
- What are the reasons for limiting maximum heat fluxes in the radiant section of a fired heater?
  - While higher heat fluxes result in less expensive heaters, maximum heat fluxes are limited due to the following reasons:
    - High tube skin temperatures are generated. In many applications, it is the fluid film temperature, not the bulk fluid temperature, that limits the duty of the heater.
    - Process fluid might be sensitive to high temperatures.
    - Metal wall loses its strength at higher temperatures.
    - Corrosion rates will increase with increased temperatures.
  - Steam boilers permit much higher heat fluxes compared to oil heaters. For example, heat fluxes of the order of 130,000 Btu/(h ft^2) are used in boilers whereas the maximum heat fluxes are limited to about 20,000 Btu/(h ft^2) in hydrocarbon service.
- What is meant by radiant efficiency of a fired heater?
  - The ratio of heat absorbed in the radiant section to net heat released.
- What is the range of efficiencies obtainable in a fired heater?

TABLE 14.5 Average Heat Fluxes and Temperatures Employed in Process Heaters

Service	Average Radiant Heat Flux (Based on Tube O.D.)		
	(kW/m^2)	($Btu/(h\,ft^2)$)	Temperature (°C)
Crude heaters	31.55–44.16	10,000–14,000	200–370
Reboilers	31.55–37.86	10,000–12,000	200–300
Catalytic cracker feed heaters	31.55–34.7	10,000–11,000	480–590
Catalytic reformer feed heaters	23.66–37.86	7500–12,000	430–540
Lube vacuum heaters	23.66–26.81	7500–8500	450
Visbreakers	28.39 -31.55	9000–10,000	370–510
Hydrocracker/treaters	31.55	10,000	370–450
Steam superheaters	28.39–41	9000–13,000	370–815
Ethylene and propylene heaters	31.55–47.32	10,000–15,000	700–900
Convection section heat fluxes	12.62	About 4000	

  - About 70–75%. In modern furnaces, higher thermal efficiencies of the order of 80–90% are achieved, thereby conserving the ever-increasing fuel costs.
- What are the tube sizes employed in fired heaters?
  - 10–25 cm (4–10 in.) O.D. and 6–12 mm (1/4 to 1/2 in.) thick.
- How are furnaces lined?
  - Furnaces are lined with lightweight refractory brick 12.7–20 cm (5–8 in.) thick. These linings are topped by a 2.54 cm (1 in.) layer of insulating brick followed by a metal wall. While lining the furnaces with the above layers, consideration is to be given for structural integrity and the availability of standard manufactured sizes of the bricks.
  - Modern furnaces are lined with lightweight ceramic tiles or ceramic fiber blankets rather than the customary massive refractory bricks. A sandwich construction of this material in two densities is sometimes practiced.
  - In general, 7.6 cm (3 in.) of ceramic fiber blanket could do a better job than 15 cm (6 in.) of refractory and weighed much less.
  - The floor of a furnace uses castable refractory or brick, or both because it is more durable and stands better during maintenance.

TABLE 14.6 Maximum Permissible Radiant Heat Fluxes for Personal Exposure

Flux		Range of Exposure Times	Average Exposure Time
(kW/m^2)	($Btu/(h\,ft^2)$)		
1.9	600	Infinite	Infinite
2.84	900	30 min to 1 h	45 min
4.416	1400	2–5 min	3.5 min
6.3	12,000	15 s (escape only)	15 s

  - Because the ceramic fiber is porous, protective ceramic coating of ceramic furnace linings improves protection of the linings and improves the uniformity of reradiation from the walls.

14.1.3 Operational Issues

- What are the different means available for increasing boiler furnace efficiencies?
  - Feed water preheating using economizers.
  - Automatic blowdown control. Automatic blowdown controls can be installed that sense and respond to boiler water conductivity and pH.
  - Reduction of boiler steam pressure.
    - Reduction of boiler steam pressure is an effective means of reducing fuel consumption by as much as 1–2%. Lower steam pressure gives a lower saturated steam temperature and without stack heat recovery a similar reduction in the temperature of the flue gas is obtained. Steam is generated at pressures normally dictated by the highest pressure and temperature requirements for a particular process.
  - Combustion air preheating.
    - Combustion air preheating is an alternative to feed water heating. In order to improve thermal efficiency by 1%, the combustion air temperature must be raised by 20°C.
  - Minimization of incomplete combustion of fuel.
    - Incomplete combustion can arise from a shortage of air, surplus of fuel, or poor distribution of fuel. It can be identified from the color of the smoke. A quite frequent cause of incomplete combustion is the poor mixing of fuel and air at the burner.
    - The root cause can be the following:

1. For oil firing, it can be due to improper viscosity, worn tips, carbonization on tips, and deterioration of diffusers or spinner plates.
2. For coal firing, nonuniform fuel size could be one of the reasons for incomplete combustion. In chain grate stokers, large lumps will not burn out completely, while small pieces and fines may block the air passage, thus causing poor air distribution. In sprinkler stokers, stoker grate condition, fuel distributors, wind box air regulation, and overfire systems can affect carbon loss. Increase in the fines in pulverized coal also increases carbon loss.

- Excess air control.
  - ➢ The optimum excess air level varies with furnace design, type of burner, fuel, and process variables. It can be determined by conducting tests with different air–fuel ratios.
  - ➢ Controlling excess air to an optimum level always results in reduction in flue gas losses; for every 1% reduction in excess air, there is approximately 0.6% rise in efficiency.
  - ➢ Various methods are available to control the excess air:

1. Portable oxygen analyzers and draft gauges can be used to make periodic readings to guide the operator to manually adjust the flow of air for optimum operation. Excess air reduction up to 20% is feasible.
2. The most common method is the continuous oxygen analyzer with a local readout mounted draft gauge, by which the operator can adjust air flow.
3. The same continuous oxygen analyzer can have a remote-controlled pneumatic damper positioner, by which the operator in the control room can control a number of firing systems simultaneously.

- Flue gas temperature control.
  - ➢ The stack exit temperature should be as low as possible. However, it should not be so low that water vapor in the exit gases condenses on the stack walls. This is important in fuels containing significant sulfur as low temperature can lead to sulfur dew point corrosion. Stack temperatures greater than 200°C indicate potential for recovery of waste heat.
  - ➢ Typically, the flue gases leaving a modern three-pass shell boiler are at temperatures of 200–300°C. Thus, there is a potential to recover heat from these gases.
- Minimization of radiation and convection heat losses.
- Reduction of scaling and soot losses.
  - ➢ In oil- and coal-fired boilers, soot buildup on tubes acts as an insulator against heat transfer. Any such deposits should be removed on a regular basis. An estimated 1% efficiency loss occurs with every 22°C increase in stack temperature. Therefore, stack temperature should be checked and recorded regularly as an indicator of soot deposits. It is also estimated that 3 mm of soot can cause an increase in fuel consumption by 2.5% due to increased flue gas temperatures. Periodic off-line cleaning of radiant furnace surfaces, boiler tube banks, economizers, and air heaters may be necessary to remove deposits.
- Variable speed control.
  - ➢ Generally, combustion air control is affected by throttling dampers fitted at forced and induced draft fans. Though dampers are simple means of control, they lack accuracy, giving poor control characteristics at the top and bottom of the operating range. In general, if the load characteristic of the boiler is variable, the possibility of replacing the dampers by a variable speed control should be evaluated.
- Controlling boiler loading.
  - ➢ The maximum efficiency of the boiler does not occur at full load, but at about two-thirds of the full load. In general, efficiency of the boiler reduces significantly below 25% of the rated load and operation of boilers below this level should be avoided as far as possible.

- What are the causes of hot spots on furnace tubes?
  - *Flame Impingement*: Due to poorly designed and/or operated burners.
  - *Poor Radiant Section Heat Distribution*: High localized radiant heat flux more than 54,000 kcal/m^2 (>20,000 Btu/(h ft^2)), normally used heat flux being 32,600 kcal/m^2 (12,000 Btu/(h ft^2)). Causes are sagging tubes, too small a radiant section, and inadequate tube spacing. Remedy is use of more small burners instead of too few large burners to give more uniform distribution of the heat flux.
  - *Tube Side Fouling*: Due to deposits of coke, salts, and corrosion products, which form an insulating layer that restricts heat flow through the tube wall. Delayed coking service in refinery fired heaters experiences the most severe operating conditions in petroleum refining application.
    - ➢ To prevent/reduce fouling, tube velocities should be 6 m/s (20 ft/s) for two-phase flow and 3 m/s (8 ft/s) for single-phase liquid flow.

 - ➢ Since excessive internal tube fouling occurs above certain temperatures, designers often try to keep temperatures down by overspecifying the heat transfer surface area.
 - ➢ When heat transfer duty increases for a given amount of heat transfer surface area (as measured by the resulting increase in heat flux), an increase in peak tube metal temperatures and more rapid coke deposition would result.
 - *Dry Point Deposits:* Complete vaporization in sections of tubes results in single-phase vapor flows with no wetting of tube internal walls, which enhances deposit rates with eventual tube failure.

- How are fouling deposits removed from the heater tubes?
  - Periodic removal helps avoid tube damage while ensuring optimum heat transfer and system performance. When the tube wall reaches its design temperature, the heater must be shut down and decoked.
  - Methods of removal of deposits include use of steam–air decoking or controlled burning, or mechanical cleaning by using rotary cutting tools, among others.
  - The time interval between shutdowns is critical in the optimization of tube life, maintenance and nonproductive periods associated with such shutdowns, and energy savings through improvements in heat transfer.
- What are the reasons for overpressure development leading to fires in process heaters?
  - Delayed ignition during lighting, fuel leakage into the firebox, or insufficient firebox purging.
  - Failure to establish reliable pilot flames before opening main fuel supply.
  - Rapid readmission of air to correct insufficient air supply.
  - Flashback into waste gas supply manifold to incinerator.
  - Tube rupture due to thermal shock, overfiring, corrosion/erosion, or high temperature due to flame impingement or internal tube fouling.
  - Closure of flue gas damper or trip of induced draft fan.
- At what temperatures high-temperature creep results in furnace tubes and what is its consequence?
  - High-temperature creep results at temperatures above 700–760°C (1300–1400°F).
  - High-temperature creep results in the bulging of the tubes and eventual failure as a result of the internal pressure of the flowing fluid.
- What are the general problems in the operation of process furnaces?
  - Flame impingement of tubes.
  - Hazy combustion chamber due to insufficient air.
  - Poor atomization of liquid fuels.
  - Poor quality of steam used for fuel atomization.
  - Fouling of convection section tubes.
  - Ash deposits on tube surfaces.
  - High-temperature corrosion at 600°C or above, if flue gases contain vanadium, sulfur, and/or sodium. Effects of sodium and vanadium are reduced by crude desalting in petroleum processing.
  - Presence of salts and/or water in process fluids.
  - Incorrect draft condition at radiant section outlet.
  - Inversion problem at stack top at low loads.
  - Imbalanced flow distribution inside tube passes.
  - Low process fluid velocities in tubes.
- What is the principal cause for failure of furnace tubes?
  - Overheating of tubes. Overheating of the tubes above 700–750°C results in *high-temperature creep*, leading to thinning of the walls and consequent failure.
- How does the life of a tube in a furnace get affected with wall temperature variations?
  - Increased tube wall temperature results in decreased metal strength reducing its life. This strength reduction effect is compounded by temperature fluctuations due to cycling effects on the tube material.
  - Tube wall temperatures increase due to reduced tube side velocities and due to coke formation in refinery furnaces and other fouling deposits inside tubes. Outer side of the tubes might be covered by soot due to burner malfunctions and nature of the fuel used, which also increases tube wall temperatures. For example, when heavy fuel oils are used for firing, metal compounds such as vanadium, nickel, sodium, and iron, present in such fuels, result in deposits of ash.
  - A rule of thumb is that, for every 10°C increase in tube wall temperature, tube life reduces to half.
- In a furnace tube if the flow is stopped, for example, due to pump failure or plugging of an orifice hole in the orifice meter, what will be the consequences?
  - Tubes will soften and melt due to increase in tube wall temperatures in the absence of heat absorbing medium, that is, fluid inside the tubes. The cost of refurbishing and lost downtime will be extremely high.
  - In modern ceramic tiled furnaces, heat storage capacity of the refractory walls is very low, and consequently, reradiation effects are minimum. Therefore, acceleration of tube overheating is reduced, giving some time to cut off fuel to burners and restore flow through tubes, thus avoiding tube burnout.

- List out the common consequences of furnace tube overheating.
  - Flame impingement from improperly adjusted burners can overheat localized tube areas.
  - Coke deposition in petroleum tube still furnaces due to cracking and secondary reactions.
    - Coke provides insulation to the tube wall and retards heat flow from tube surface, which receives heat by radiation from flames to flowing fluid inside tubes.
    - Similar effects result from salt and corrosion product deposits.
    - The insulating effect will be to further increase tube wall temperatures.
  - Poor radiant heat distribution resulting in hot spots. This could be due to sagging of the tubes, small firebox volume for the heat released, inadequate tube spacing, or other reasons such as choosing few large burners in place of many small burners.
  - Accelerated corrosion.
  - Decreased flow rates of fluids flowing inside tubes, reducing heat absorption rates.
    - Reduced flows will result in increased tube wall temperatures as heat received by radiation will be more than heat absorbed by the fluid by conduction–convection processes.
  - Reduced excess air for combustion of fuel in burners increases flame temperatures and radiation rates and hence raises tube wall temperatures.
  - Rise in tube wall temperatures will result in decreased tube strength and hence tube life, accelerated rate of coke deposits, and softening and failure of tubes.
  - Heat absorption rates by the fluid flowing inside the tubes are not uniform and depend on the changeover from liquid phase to two-phase vapor–liquid conditions, resulting in wide variation in heat absorption rates and consequent temperature variations along the tubes.
  - There have been instances of large-scale tube failures, particularly due to flow stoppages or reduction of flows, with enormous losses in terms of replacement costs and costs due to loss of production.
- "Forced circulation is used in fired heaters." *True/False*? Comment.
  - *True.* Tube failure depends on flow rate through tubes. Reduced flows affect tube temperatures and hence their life. Forced circulation ensures adequate velocities through the tubes. Also, the higher ΔP through furnace tubes demands use of forced circulation using a pump.
- What are the methods used for controlling flame temperatures in a furnace?
  - *Water or Steam Injection*: Water or steam injection into the flame zone can be an effective means of reducing flame temperatures, thereby reducing thermal NO_x. Steam injection rates up to 30% by weight have been used.
  - *Flue Gas Recirculation*: Flue gas equivalent to 10–20% of combustion air is recirculated back to the furnace to reduce flame temperatures. The quenching effect from the CO_2-rich flue gas lowers the O_2 concentration in the combustion zone, as well as the adiabatic flame temperature.
  - Figure 14.10 illustrates a convective heater with flue gas recirculation.
  - A disadvantage is that flue gas has to be handled at high temperatures involving blower problems at these temperatures.
  - Also, since flue gas is not a clean gas, fouling might increase, increasing tube wall temperatures.
  - In addition to requiring large ducts, fans, and dampers, additional controls will be required.
  - The additional gas flow through the firebox and flues must be taken into account in the furnace design.
  - *Selective catalytic reduction* (SCR) is a post-flame treatment process to convert the NO_x *formed* to N_2.
  - A reducing agent is added to the combustion gas stream to take oxygen away from NO. For large heaters, ammonia is used as the reducing agent. The desired reaction is

$$6NO + 4NH_3 = 5N_2 + 6H_2O \qquad (14.21)$$

  - As there is always some oxygen present, reactions such as the following take place:

$$4NO + 4NH_3 + O_2 = 4N_2 + 6H_2O \qquad (14.22)$$

$$2NO_2 + 4NH_3 + O_2 = 3N_2 + 6H_2O \qquad (14.23)$$

  - Catalysts include zeolites, titanium, and vanadium.
  - The best temperature range for SCR catalyst activity and selectivity is from 300 to 500°C. Ammonia is injected downstream from the combustion zone and upstream of the catalyst.
  - In selective noncatalytic reduction (SNCR), ammonia or urea is injected into the flue gas to act as a reducing agent and the NO_x in the flue gases is converted back to N_2.
  - As the activity of the catalyst decreases over a period of time, the amount of excess ammonia required for the desired conversion increases.

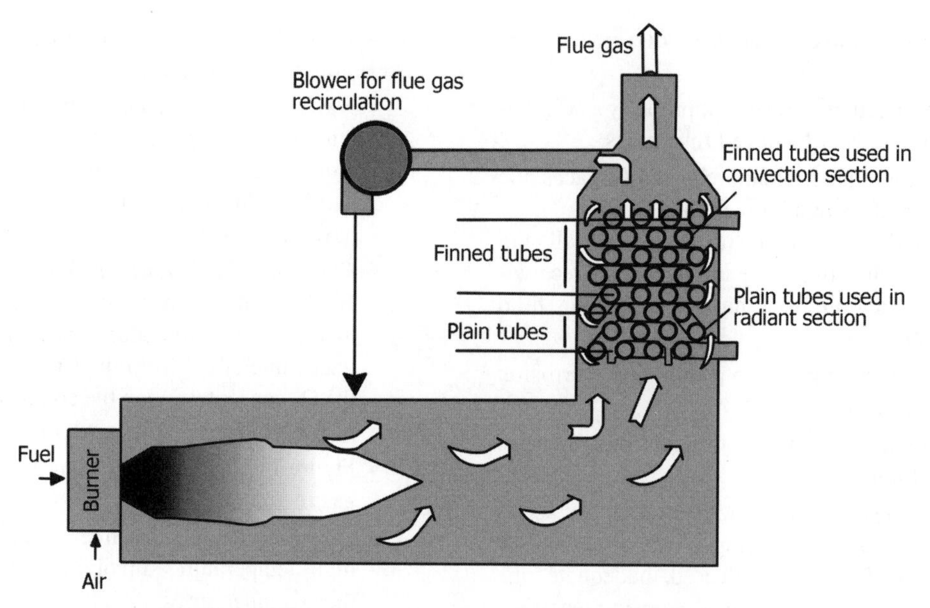

FIGURE 14.10 Convective heater with flue gas recirculation.

- *Increasing Excess Air*: Though this reduces flame temperatures, CO formation and high-temperature oxidative corrosion of the furnace tubes reduce tube life, besides reducing furnace efficiency.

14.1.4 Furnace Tubes

- What are the materials that are commonly used for furnace tubes?
  - High-chromium alloy steels (not less than 5% Cr).
  - 5Cr–0.5Mo, 7Cr–0.5Mo, and 9Cr–Mo steel tubes are commonly used. Modern furnaces use tubes with a chrome content in the range of 11–13%. Tubes with a chrome content of 11–13% can normally withstand tube *skin* (external tube surface) temperatures of 700–740°C. Low chrome content tubes of about 3% may be limited to 650°C.
  - Sometimes nickel and molybdenum content is increased for added temperature and corrosion resistance. A 0.5% silicon content enhances the oxidation resistance.
- Give reasons why finned surface tubes are frequently used in the convection section of a furnace. What are their minus points in such a practice?
  - Heat transfer coefficients for gases are very low and convective heat transfer in the convection section involves flue gases transferring heat to the tubes through which process fluid flows. In addition, if part of the convection section is used for superheating steam, gas-to-gas heat transfer both outside and inside the tubes is involved, with extremely low heat transfer rates.
  - Use of finned tubes increases heat transfer area per unit volume of the convection section. Also turbulence in the flowing flue gases will increase. Net result is to increase heat transfer *capacity* of the convection section.
  - Finned tubes increase pressure drop in the flue gas flow, decreasing the draft, with increase in the pressure of the hot gases.
  - Replacing bare tubes with finned tubes in a revamp of the tube still heater requires evaluation of the draft changes.
  - The other minus points include increased investment on the finned tubes over plain tubes and increased fouling of the tube surfaces with soot particles, which are relatively difficult to clean. Normally soot blowers are installed to clean soot deposits.
  - The finned tubes for convection duty are normally made out of low-chrome alloy steels that are suitable for the relatively lower convection section temperatures (≤700°C). These tubes are less expensive than the high-chrome alloy steels used in the radiant section.
- What is dimpled tube technology? What are its benefits?
  - The use of dimpled tube technology in convective sections of commercial high-temperature heaters has the potential to substantially improve energy efficiency and cost effectiveness.
  - In dimpled tube technology, spherical cavities are arranged on the outer surface of the convective section tube.
  - Each dimple works as a *vortex generator* that intensifies the rate of convective heat transfer

between the dimpled surface and the gas flowing over the surface.

- The use of dimpled tubes is claimed to increase the heat transfer coefficient by over 30% compared to bare tubes and has the potential to decrease fouling deposition rate compared to extended surface tubes.
- Applying dimpled tubes also improves heater thermal efficiency by as much as 5% without any significant increase in system draft loss and decreases CO, NO_x, and unburned hydrocarbon emissions.
- Figure 14.11 illustrates a dimpled tube.
- Dimpled tube technology will not increase the pressure drop across the bank of convective section tubes.
- Due to the vortex structures developed on the enhanced surface, a reduction in the fouling rate on that surface is expected.
- Computational fluid dynamics (CFD) modeling has demonstrated the existence of stable three-dimensional vortices in the dimples, as well as the internal structure of the vortex inside a single cavity.

• What are the conditions that increase internal corrosion of furnace tubes?
  - Sulfur is present in fluids (particularly in crude and heavy oils).
  - If sulfur is in the form of mercaptans, which are relatively unstable, the result is increased corrosion at the high temperatures involved in furnaces.
  - Coke formed on internal side of tubes contains high percent of sulfur, which initiates corrosion attack in overheated tube areas.

• What is carburization (with respect to refinery furnaces)?
  - Carburization is the mechanism in which carbon atoms diffuse through the metallic matrix of a metal. Carbon atoms squeeze through openings between interstitial sites to diffuse as crystals. The diffusion process is highly dependent on temperature.
  - At high temperatures, coke/carbon inside the furnace tubes results in carbon absorption into the metal.
  - Carbon absorption increases volume of metal and coefficient of expansion:
    - ➢ Results in strong internal stresses giving rise to premature tube failure.
    - ➢ Reduces creep strength and ductility of the metal (aging).
    - ➢ Fissures are initiated next to the carburized layer and propagate first to the outer wall and then to the inner wall.
  - When hydrocarbons are present with high percent CO, a strong carburizing reaction is likely at elevated temperatures.

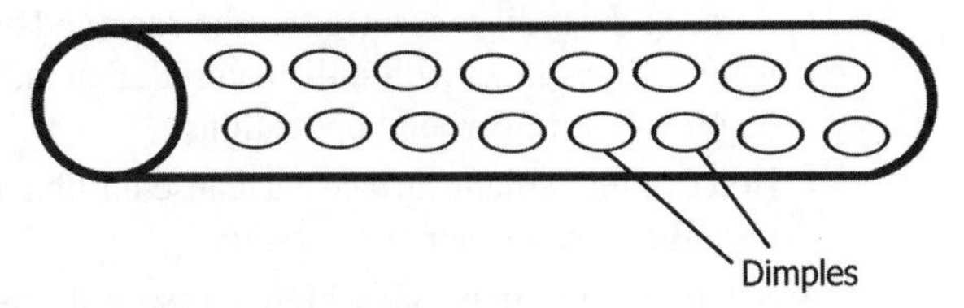

FIGURE 14.11 Dimpled tube.

14.1.5 Burners

• What are the characteristic features of a premix burner? Illustrate.
  - Figure 14.12 gives details of a premix gas burner. The pressure of the fuel gas supply is important since low gas pressure degrades performance. The primary air flow should be maximized without lifting the flame off the burner. Most of the air (as primary air) is delivered to the burner along with the fuel. Secondary air is introduced with appropriate controls, based on measurements and controls indicated in the previous paragraphs.
  - Increased or decreased levels of secondary air supply lead to poor combustion. A minimum excess air level is required for complete combustion but too much excess air reduces flame temperature and decreases efficiency.
  - When excess air falls below the recommended optimum value, incomplete combustion results, leading to the formation of large amounts of CO and H_2, which will reduce flame temperatures and can mislead operators to increase fuel supply, making matters worse. The condition may not be noticed because leakage in the convection section can hide that insufficient air is getting to the burner. Thus, hazardous conditions are introduced.
  - If combustion of CO and H_2 takes place in the convection section, convective tubes can get damaged.
  - Any changes in fuel type or quality can lead to variations in excess air requirements.

• What are radiant panel burners?
  - Radiant panel burners operate by burning a fossil fuel, which heats a solid surface that radiates infrared energy to a load. These burners are used in a number of lower temperature heating and drying applications.
  - In radiant panel burners, also called surface combustion burners, the incandescent walls are located 0.6–1 m (2–3 ft) from the tubes.
  - The furnace side of the panel may reach 1200°C whereas the outer side remains at 50°C because of continual cooling by the air–gas mixture.

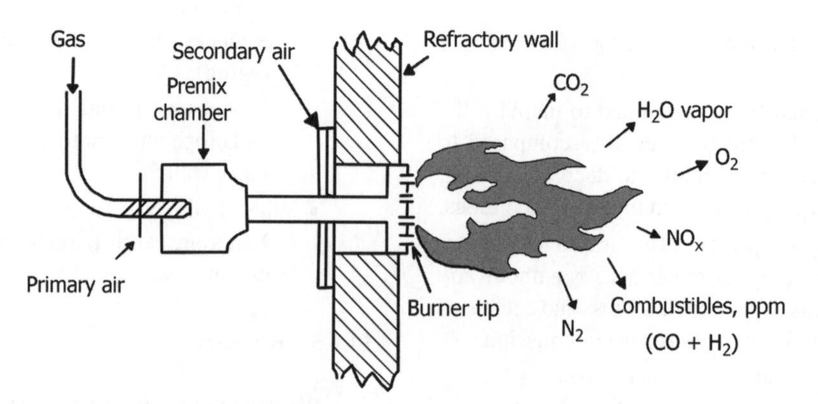

FIGURE 14.12 Premix gas burner.

- Radiant panel burners require only 2–5% excess air compared to 10–20% for conventional burners.
- Heaters equipped with radiant panels cost more but provide better control of temperatures of reactions such as pyrolysis of hydrocarbons to, for example, ethylene.
- Both gas-fired and electric radiant heaters are commonly used; however, gas-fired radiant burners have lower operating costs due to the difference between the cost of electricity and natural gas in most locations.

• Give examples of applications of radiant panel burners.
- Drying paper and cardboard in a paper mill.
- Paper finishing.
- Drying wood.
- Porcelain frit drying.
- Curing ceiling tiles.
- Teflon curing.
- Drying and curing coatings on paper and metals.
- Curing ink on paper and powder coat paints.
- Baking in mass food preparation.
- Setting dyes in textile and carpet production, sometimes referred to as predrying.
- Curing lens assemblies in automotive headlamp assembly manufacturing.
- Plastics curing.

• What are the advantages of radiant panel burners?
- Radiant burners are designed to produce a uniform surface temperature heat source for heating and melting a variety of materials. The uniform surface temperature produces more homogeneous heating of the materials, which normally improves the product quality compared to conventional burners that may produce hot spots. Other advantages of these burners may include the following:
  - High thermal efficiencies.
  - Low pollutant emissions.
  - Directional heating.
  - Very fast response time to load changes.
  - Very fast heating compared to convective heating.
  - Burner shape can be tailored to the shape of the heat load to optimize heat transfer.
  - Ability to segment a burner to produce a nonuniform heat output profile, which may be useful in certain types of heating and drying applications.
  - Certain types of radiant burners have very rapid heating and cooling times.
  - No open flames that could ignite certain types of materials (e.g., paper or textiles).
  - More control over the heating process because of the known and measurable surface temperature of the radiant surfaces compared to open flames, where the flame temperature is very difficult to measure.
  - Burners are very modular and can be configured in a wide variety of geometries to accommodate the process heating requirements.

• What are the limitations of radiant refractory panel burners?
- Relatively low temperature limit for the radiant surface due to the limits on the refractory material.
- Fuel and oxidizer must be clean to avoid plugging the porous radiant surface, which essentially precludes the use of fuels such as coal or heavy fuel oil.
- Some of the radiant surfaces can be damaged by water or by contact with solid materials that may be prevalent in certain applications.
- Holes in the radiant surface can cause flashback because these burners use premix.
- Some designs may have high pressure drops, which means that more energy is needed for the

blower for the combustion air flow through the ceramic burner material.

- Due to the limits in radiant surface temperatures, the firing rate density is usually limited.
- Some types of radiant burners using hard ceramic surfaces may have high heat capacitances that could ignite certain load materials upon a sudden line stoppage.

• What are the different types of radiant panel burners?

- Perforated ceramic or wire mesh radiant burners:
  - *Perforated ceramic burners* may consist of a pressed ceramic plate that may include pre-punched holes, where the flames heat the surface directly. The surface can be textured to further enhance the radiant efficiency of the burner. New developments in ceramic foams are being applied to this type of burner. These foams are often less expensive to make than perforated ceramics. They provide a higher surface area for radiation and a more uniform heating surface, compared to perforated ceramics. Many shapes are possible with the ceramic foams and the pore size is adjustable.
  - *Wire mesh burners* are made from high-temperature-resistant metals, such as stainless steels or Inconel. The open area in the mesh serves as the port area for the burner. However, due to the high thermal conductivity of metals, several layers of mesh are often required to prevent flashback. The thermal conductivity between the layers is much less than that through the mesh itself because of the contact resistance between the layers. An important problem with wire mesh radiant burners is the lower temperature limits compared to ceramic burners, due to the temperature limits of the metals.
- Flame impingement radiant burners:
  - In this type of radiant burner, the flame impinges on a hard ceramic surface, which then radiates to the load. The heat transfer from the flame to the tile is mostly by convection due to the direct flame impingement. The heat transfer from the tile to the load is purely by radiation. The hot exhaust gases from the burner heat up the surrounding wall by convection, but at a much lower rate compared to the burner tile.
  - One advantage of this type of burner is that there is no metal matrix with lower temperature limits compared to ceramic that has high temperature limits. Therefore, this type of burner can often be used in higher temperature applications.
  - There is also no matrix of porous ceramic fiber refractory that could get plugged up.
  - This type of burner can also fire a liquid fuel as opposed to many other radiant burners that use only gas.
  - A disadvantage is that the heat flux is not as uniform as other types of radiant burners. Another problem is that the burner tile has the typical problem of thermal cycling, which can cause the tile to disintegrate.
- Porous refractory radiant burners:
  - In this type of burner, the surface is made of a porous ceramic fiber. A relatively new type is made from a woven ceramic fiber mesh, similar to the wire mesh radiant burners except that ceramic fiber is used instead of metal.
  - The predominant shape used in porous refractory burners is a flat panel.
- Radiant wall burners:
  - This type of burner is commonly used in process heaters where they heat a refractory wall that radiates heat to tubes parallel to the wall.
  - These burners are similar to flame impingement radiant burners except that the flame in a radiant wall burner is directed along the wall and not at the wall or burner tile as in the case of impingement burners.
  - The objective of a radiant wall burner is to distribute heat as evenly as possible over a fairly wide area.
  - The impingement burner primarily heats its burner tile, which then radiates to the load.
- Radiant tube burners:
  - In some heating processes, it is not desirable to have the products of combustion come in contact with the load. One example is in certain types of heat treating applications where the exhaust gases from a combustion process could contaminate the surface of the parts being heated. In those cases, an indirect method of heating is needed.
  - Electric heaters are sometimes used but the energy costs are considerably higher than the cost of fossil fuel-fired heaters.
  - The radiant tube burner is useful for indirect heating.
  - The objective of the radiant tube burner is to efficiently transfer heat from the combustion gases to the radiant tube and then to efficiently radiate that energy to the load.
  - Typical tubes are constructed from high-temperature metal alloys or ceramics. Metal alloys can

be expensive and typically do not have as high a continuous operating temperature as ceramics.

- What are the sources of noise in burners?
  - Low-frequency noise sources:
    - Combustion noise.
    - Fan noise.
  - High-frequency noise sources:
    - Gas jet noise.
    - Piping and valve noise.

14.1.6 NO_x Control

- What is NO_x?
  - All fossil fuel burning processes produce NO_x. The principal oxides formed are nitric oxide (NO), which represents 90–95% of the NO_x formed, and nitrogen dioxide (NO_2), which represents most of the remaining nitrogen oxides.
- What are the primary sources of NO_x in a fired heater and the reactions involved?
  - Thermal NO_x and fuel NO_x.
  - The highest percentage of NO_x formed is the result of the high-temperature fixation reaction of atmospheric nitrogen and oxygen in the primary combustion zone.
  - Thermal NO_x is formed by the reaction of N_2 with oxygen in the flame front of a burner. High flame temperatures contribute to thermal NO_x formation.
    - Thermal NO_x formation is a route where high flame temperatures cause nitrogen molecules from the combustion air to break apart and combine with oxygen to form nitric oxide. The sequence is complicated, but the following two steps represent the essential features:

$$\begin{array}{c} O + N_2 \leftrightarrow NO + N \\ N + O_2 \leftrightarrow NO + O \\ \hline N_2 + O_2 \leftrightarrow 2NO \end{array} \tag{14.24}$$

    - Combustion is a free radical process whereby the reactions include intermediate species. In the above sequence, atomic oxygen attacks a nitrogen molecule to yield nitric oxide and a nitrogen radical. In the second step, the nitrogen radical attacks diatomic oxygen to produce another nitric oxide molecule and replenish the oxygen atom. The sequences sum to the overall reaction of one nitrogen and one oxygen molecule producing two nitric oxide molecules.
  - Fuel NO_x is formed by nitrogen atoms that are chemically bound in the fuel. Liquid fuels almost always contain some nitrogen, and as a rule, liquid fuels make more NO_x than burning gases. Generally, single nitrogen atoms are the most reactive. Amines and ammonia in the fuel contribute greatly to fuel NO_x formation.
- What is the effect of temperature on NO_x formation?
  - As flame temperature increases, NO_x levels increase almost linearly.
- Summarize the factors that affect NO_x emission rates.
  - NO_x emissions increase with increase in the following:
    - Furnace design and heat release rates. As furnace size and heat release rates increase, NO_x emissions increase. This results from a lower furnace surface-to-volume ratio, which leads to a higher furnace temperature and less rapid terminal quenching of the combustion process. Higher temperatures produce higher NO_x emissions.
    - Preheating of combustion air.
    - Fuel nitrogen content.
    - Turbulence in combustion zone.
    - Distance between burners and number of burners per unit. Interaction between closely spaced burners, especially in the center of a multiple burner installation, increases flame temperature at these locations.
    - The tighter spacing lowers the ability to radiate to cooling surfaces, and there is a greater tendency toward increased NO_x emissions. Burners operating under highly turbulent and intense burner flame conditions produce more NO_x. The more the bulk mixing of fuel and air in the primary combustion zone, the more the turbulence created. Flame color is an index of flame turbulence. Yellow hazy flames have low turbulence, whereas blue flames with good definition are considered highly turbulent.
  - NO_x emissions decrease with increase in the following:
    - Fuel moisture content.
    - Quenching rate in water-cooled furnaces.
- What type of firing in a furnace helps decrease NO_x formation? Give reasons.
  - Tangentially fired units generate least amount of NO_x because they operate with low levels of excess air and because bulk of the burning of the fuel takes place in a large volume of the furnace.

- A large amount of internal recirculation of bulk gas, coupled with slower mixing of fuel and air, provides a combustion system that is inherently low in NO_x formation for all types of fuels.

- What are the available technologies for post-combustion NO_x reduction?
  - SNCR, SCR, and catalytic oxidation/scrubbing.
  - SCR has typically been used for very tight NO_x control, SNCR for dirty and high-temperature services, and catalytic scrubbing for units that are already using scrubbers for a different purpose.
  - SNCR units can reduce NO_x levels by approximately 40–60%.
  - When used in combination with other technologies, SNCR can produce low NO_x concentrations. SNCR uses the reducing capability of ammonia to reduce NO_x to nitrogen. If urea is the reduction agent, it decomposes to ammonia, which reacts with the NO_x. The reaction is

$$CO(NH_2)_2 + 2NO + \tfrac{1}{2}O_2 \rightarrow 2N_2 + CO_2 + 2H_2O \tag{14.25}$$

  - The requirements for good conversion include good mixing, adequate residence time, and no impingement of the injected chemical against tubes. A more critical variable is the operating temperature of the system.
  - SNCR tends to work over a narrow temperature range, the optimum being approximately 930–1050°C. The process can be used at as low as 870°C and as high as 1200°C, but with decreased reaction rates.
  - The equations for the reaction of ammonia with NO_x are as follows:

$$4NO + 4NH_3 + O_2 \rightarrow 4N_2 + 6H_2O \; (NH_3 : NO = 1:1) \tag{14.26}$$

$$2NO + 4NH_3 + 2O_2 \rightarrow 3N_2 + 6H_2O \; (NH_3 : NO = 2:1) \tag{14.27}$$

  - For temperatures between 930 and 1050°C, the first equation dominates. The higher the temperature, the greater will be the contribution of the second equation.
  - Advantages of SNCR are that it requires a lower capital investment than SCR, can be used in dirty and fouling services (particulates and/or high sulfur), and can be combined with other technologies to achieve greater NO_x reduction.
  - SNCR may also be economical in conjunction with SCR. The SNCR can be used to reduce the required size of the SCR by reducing the inlet NO_x concentration to the SCR.
  - SCR is based on the following reactions between ammonia and NO_x at high temperatures:

$$6NO + 4NH_3 \rightarrow 5N_2 + 6H_2O \tag{14.28}$$

$$2NO + 4NH_3 + 2O_2 \rightarrow 3N_2 + 6H_2O \tag{14.29}$$

$$6NO_2 + 8NH_3 \rightarrow 7N_2 + 12H_2O \tag{14.30}$$

  - This is basically the same reaction that occurs in SNCR. The differences are that the catalyst allows the reaction to proceed at lower temperatures and that a greater NO_x reduction can be achieved.
  - There are three basic types of SCR technology:
    - Low-temperature catalysts, which have the advantages of low pressure drop and lower temperature operation, but are susceptible to sulfur and particulates.
    - Medium-temperature catalysts.
    - High-temperature catalysts, which tend to be zeolites.

- What are the methods used for NO_x control?
  - NO_x is formed by conversion of nitrogen in combustion air at high flame temperatures.
  - NO_x is controlled by reducing (i) peak flame temperatures, (ii) gas residence time, and/or (iii) oxygen concentration in flame zone.
    (i) Peak temperatures are reduced by the following:
      - Using a fuel-rich primary fuel zone.
      - Increasing the rate of flame cooling.
      - Decreasing the adiabatic flame temperature by dilution.

    (ii) Gas residence time is reduced in the hottest part of the flame zone by the following:
      - Changing the shape of the flame zone and using steps indicated under (i).

    (iii) O_2 content in the primary flame zone is reduced in the flame zone by the following:
      - Decreasing the overall excess air rates.
      - Controlled mixing of fuel and air.
      - Using a fuel-rich primary flame zone.
      - Flue gas recirculation (FGR). Flue gas recirculation requires high rates of flue gas recirculation to reduce NO_x to acceptable levels with attending significantly high fan capacities

and such high rates of recirculation causing operational instabilities in the burners.

  - Steam injection.

- Discuss flue gas recirculation as a method for NO_x reduction.
  - Flue gas recirculation has been extensively used in *boilers* to reduce NO_x. Recirculating the flue gas reduces the oxygen concentration and thus reduces flame temperatures. Reported flame temperatures are over 1900°C with no recirculation, and 1600°C at 20% FGR. To achieve very low NO_x in a boiler, FGR is required.
  - FGR can be economical in *fired heaters* with air preheat systems, if induced draft and forced draft fans are already available, so that new fans are not required.
  - Only a small number of fired heaters are equipped with FGR. This is partially because fired heaters tend to have lower firebox temperatures, so there is less need for FGR. In addition, many fired heaters do not have forced draft fans and it can be more expensive to install FGR in natural draft heaters.
  - Recirculation rates in the range of 15–20% have been used, with NO_x reductions in the 40–55% range. Recirculation of 40% is the maximum that is possible stoichiometrically.
  - In modern systems, recirculation close to 40% is practiced to meet the environmental regulations, with reductions of NO_x up to 70%. These systems require very tight control systems to maintain flame stability.
- What are low NO_x burners (LNBs) and ultralow NO_x burners (ULNBs)?
  - Low NO_x burners burn the fuel in two or more stages (*staged combustion*) rather than in a single stage.
    - In the first stage, a certain percent of the fuel is burned with substoichiometric air and the resulting heat is transferred to the tubes and the refractory walls. Temperature levels involved in this stage are high.
    - In the second stage, excess air is added to complete the combustion process by converting CO and residual hydrocarbons.
    - Because of two-stage combustion involved, maximum flame temperature is reduced.
    - Air or fuel staging is practiced. In *air staging*, only a portion of the air flows across the fuel injection zone and this forms a fuel-rich primary combustion zone where the fuel is only partially burned. The remainder of the air is injected downstream to complete the combustion. In *fuel staging*, only a portion of the fuel is injected into the primary combustion zone. The rest of the fuel required is introduced downstream.
    - The basis for the design is to develop a stratified flame structure with specific sections of the flame operating fuel-rich and other sections operating fuel-lean. The burner design thus provides for the internal staging of the flame to achieve reductions in NO_x emissions while maintaining a stable flame.
    - The fuel-rich and fuel-lean zones both cause combustion at lower peak temperatures than a uniform fuel–air mixture, resulting in lower thermal NO_x formation. The combustion products from these two zones then combine to complete the combustion process and result in the complete oxidation of the fuel. By creating a fuel-rich zone in the front part of the flame, one can also reduce the conversion of fuel-bound nitrogen to NO_x and thereby lower fuel NO_x formation.
    - To increase the flexibility of combustion staging and flame shaping capabilities, some low NO_x burner designs are equipped with a tertiary air zone installed at the outer periphery of the burner throat. The tertiary air, which can be given swirling motion to increase turbulence, is mixed in the furnace with the bulk furnace gas to achieve complete fuel combustion. This provides for the complete combustion of the fuel in the post-combustion zone where NO_x formation is inhibited by lower combustion temperature and reduced O_2 concentration.
  - Low NO_x burners reduce formation of NO_x and give more uniform heat release rates than conventional burners. These will reduce hot spots on the tubes with reduced coking problems in the tubes. Conventional burners involve complete mixing of fuel and air before combustion and give rise to higher flame temperatures promoting formation of NO_x, which gives rise to pollution problems.
  - In contrast to LNBs, ULNBs reduce NO_x by inducing the internal circulation of fuel gas within the heater.
  - Most ULNBs use internal flue gas recirculation to dilute and cool the flame. These burners require tight oxygen control to prevent/minimize air leaks into the heater. Tight controls are required.

14.1.7 Flares

- What is the function of a flare system?
  - In the safe, satisfactory operation of a process plant, the flare system is the single most important element

for operational or emergency relief of flammable substances in the liquid or gaseous phases.

- Flaring converts flammable, toxic, or corrosive gases and vapors into less hazardous and objectionable compounds by combustion.
- A typical flaring system handles the discharge of all relief vapors, gases, or liquids inside a designated unit or number of units. A flare system generally consists of the following major components and subsystems:
  - Collection piping within a unit, including pressure reliefs and vents.
  - A flare line to the site.
  - A knockout drum to recover liquid hydrocarbon from the gas stream.
  - A liquid seal to provide positive header pressure without surging and to protect against flashbacks.
  - A flare stack with flare tip.
  - An assist system to maintain smokeless burning.
  - A fuel gas system for pilots together with ignition source.
  - Controls and instrumentation.

- What are the two types of flares used in industry?
  - *Elevated Flares*: A single tall flare stack with pilot ignition source.
  - *Ground Flares*: These have usually several short flare stacks that may be mounted in the open or with protective hat barriers or may be placed in a large shallow pit.
- Name the three types of elevated flares and state their specific applications.
  - *Smoking*: Used with relatively clean burning gases, for example, NH_3, H_2S, and hydrocarbons, having high hydrogen/carbon contents ($\geq 25\%$).
  - *Smokeless*: Due to steam injection to produce a clear exhaust. Organics having lower H/C ratios ($<25\%$).
  - *Endothermic Flares*: Heat added by auxiliary means. Gases having heat content $<900\,\text{kcal/m}^3$ ($100\,\text{Btu/ft}^3$).
- "Flares cannot be used if a gas has a fuel value less than $1350\,\text{kcal/m}^3$ ($150\,\text{Btu/sft}^3$)". *True/False*?
  - *True*.
- What are the parameters (safety, environmental, and operational) that must be considered while designing a flare system?
  - Human activity in the area.
  - Residential areas outside the plant.
  - Commercial areas.
  - Roads.
  - Other elevated structures.
  - Anticipated solar radiation, and velocity and direction of wind.
  - Temperature inversions that affect atmospheric dispersion of released gases.
  - Capacity for future expansion.
  - Hazardous material storage areas.
  - Prevention of liquid carryover by providing knockout drums.
  - Gas rate and its compositions.
  - Anticipated frequency of flaring.
  - Flashback and air ingress protection.
    - When flashback occurs, flue gases burn inside the flare stack, gas collection header, or plant piping. To prevent this and minimize the risk of explosion, flares must be equipped with flashback protection. One such approach involves using liquid seals in the flare system that keep the flame front from propagating back into the plant piping.
    - Similarly, liquid seals are often included at the base of an elevated stack to help maintain a positive pressure in the downstream flare header, to ensure that any leaks in the flare header will be discharged into the atmosphere (rather than back into the plant), and to prevent air leakage into the header.
    - Air leakage into a flare header can be extremely dangerous, as it may result in a potentially hazardous gas–air mixture, with the possibility of an explosion in the stack or header.
    - If a liquid seal is not used at the base of the stack, the flare stack should be purged continuously with an oxygen-free gas, such as nitrogen.
    - Mechanical seals located at or below the flare tip can also be used to prevent flashback in flare systems. These seals significantly reduce the amount of continuous purge gas required to prevent air infiltration into the flare stack.
  - Thermal radiation.
    - In flare design, the effect on personnel and equipment of heat radiating from the combustion of large quantities of flammables must be evaluated. For safe operation, the flare height should be designed with consideration given to potential radiation levels around the flare.
    - The potential radiation levels should not exceed allowable radiation levels on the ground.
  - Smokeless operation.
    - The most common method to achieve smokeless combustion is to inject steam into the flame. The specific requirements for steam-assisted flow

operation depend on the molecular weight, heating value, carbon–hydrogen ratio of the gas, and flare tip design.

    - If steam is introduced at pressures below 70 kPa gauge (10 psig), the desired turbulence or air entrainment will not achieve smokeless operation due to insufficient momentum. On the other hand, too much steam will cool the flame and possibly put it out.
    - Where steam is not readily available, one must look for other means of smokeless flaring. Approaches include air blowers, high-velocity multitips that use the kinetic energy of a waste stream for smokeless flaring, and water sprays. Water sprays operate like steam but do not offer the same performance as steam because they have lower velocities and lower air mixing momentum levels. Lower velocity is less efficient in drawing outside air into a flare flame, while lower momentum reduces flame resistance to wind shear. Even a slight wind can tilt the flame out of the water spray zone, resulting in smoking.
    - Forced air flares are designed to provide a thorough mixing of air, from a forced air fan, with the waste gas as the two streams leave the flare tip. Since combustion takes place in a controlled zone immediately above the flare tip (and not within the tip or stack), these flares provide a long service life.
  - Flameout conditions.
    - To ensure safe flare operation during periods when the flame may have gone out, ground level concentrations for flammables should not be within the flammable or explosive limits of the gas.
  - Governmental regulations affecting height, noise, allowable toxic concentrations, and other important factors.
- What are the characteristics of enclosed ground flares?
  - Can burn both gaseous and liquid wastes by means of several burners.
  - State-of-the-art enclosed flares emit no smoke, heat, noise, or visible flame. Although they are more expensive than traditional designs, the advantages of enclosed ground flares have led many major companies to consider these designs for all but the most remote installations. For refineries, storage terminals, and chemical plants surrounded by congested urban areas, enclosed flares combine high efficiency for smokeless combustion with unsurpassed safety. Enclosed flares can also be very quiet, with maximum noise levels of about 60 dBA compared to about 75 dBA for an elevated flare measured at a distance of 30 m.
  - The enclosed flare encloses thermal radiation completely. Heat is dissipated by natural draft convection through the enclosure. A soft ceramic blanket lining absorbs noise to acceptable levels. Additional acoustical treatment of the air intake can be carried out to further reduce noise levels.
  - Consists of a refractory chamber or an open pit.
  - Design is more complex.
  - High capital cost.

14.1.8 Incinerators

- What is incineration?
  - Incineration involves burning solid, semisolid, liquid, or gaseous wastes. It is an efficient means of reducing waste volume. The solid, noncombustible residue of incineration (ash) is inert, sanitary, and odorless.
  - Emissions from an incinerator include CO_2, particulates, oxides of sulfur (SO_x), oxides of nitrogen (NO_x), chlorides, CO, water vapor, and VOCs.
- What are the different types of incinerators?
  - *Municipal Incinerators*: Small incinerators have capacities below 50 tons/day and larger ones have capacities over 50 tons/day. Municipal incinerators fall into three categories, namely, rectangular, circular vertical, and rotary kiln types.
    - *Rectangular Units*: These are either refractory-lined or water-cooled built as multiple cell units with a combustion chamber into which primary air is admitted below a grate on which solid waste burns, followed by a mixing chamber into which secondary air is introduced for completion of combustion. Ash is removed from pits in the bottom of all the chambers.
    - *Vertical Circular Units*: Waste is usually fed into the top of the refractory-lined chamber. The grate consists of a rotating cone in the center surrounded by a stationary section with a dumping section around it. Arms attached to the rotating cone agitate the waste and move the ash to the outside. Primary air is admitted below the grate and secondary air is admitted into the upper section of the chamber.
    - *Rotary Kiln Incinerators*: Rotary kiln incinerators are used to complete the combustion of the waste that has been dried and partially burned in a rectangular chamber. The waste is mixed with combustion air by the rotating action of the kiln.

Combustion is completed in the mixing chamber following the kiln where secondary air is added. Ash is discharged at the end of the kiln.

- *Industrial Incinerators*: There are a variety of incinerators in this category that include single chamber, multichamber, controlled air, and fluid bed types. Controlled air units have high efficiencies with low particulate emission rates. Fluid bed units are well suited for incineration of sludges.

- What is a regenerative thermal oxidizer (RTO)?
  - A regenerative thermal oxidizer burns very lean waste gases without using much fuel. An RTO is always more fuel efficient than any other type of oxidizer.
  - Operating temperature is about 850°C but the hot flue gas passes through a heat exchange module before reaching the stack. Waste gas enters at around 20°C and flue gas leaves at 90°C, contrary to normal incinerators in which flue gas leaves at least around 250°C, with considerable heat losses.
  - The module is an insulated box with heat exchange media, usually ceramic packing. Structured ceramic packing is more compact than random packing. At least two modules are used; while one absorbs heat from the flue gas, the other releases heat into the waste gas (Figure 14.13). When a box is saturated with heat, it is taken off-line, with waste gas passing through it backward until the box is cooled. Once cooled, it is returned to handle hot flue gas.
  - Two boxes are needed so that the flue gas always has a path to the exhaust stack. Specialized valves set on a timer switch each box from heating to cooling about every 5 min.
  - In waste gas incinerators, oxygen in air reacts with waste hydrocarbons at high temperature to produce a clean flue gas. A good incinerator operates at very high efficiencies close to 100% with very low fuel usage and zero emission of carbon monoxide and nitrogen oxides.
  - For good efficiencies, furnace temperature must be high, residence time of the flue gas in the furnace must be sufficiently long, and 2–3% O_2 must remain in the flue gas leaving the stack. For most incinerators, furnace temperatures are in the range of 750–850°C. Higher temperatures require more expensive refractory to avoid any damage. Furnace residence time typically is 0.5–1 s. Hydrocarbons that are hard to burn, such as pesticides, may require more time.

14.1.9 Rotary Kilns

- What are the major applications/uses of rotary kilns?
  - Cement making, roasting of ores, manufacture of titanium dioxide, calcination of bauxite to Al_2O_3, petroleum coking, preparation of nodules from fines of phosphate rock, production of Plaster of Paris from gypsum, and so on.
- What is the normal speed of rotation of a cement kiln?
  - 0.22 m/s.
- "Time of passage of a solid material being dried/processed in a rotary kiln depends on kiln length *L*, rpm *N*, slope of the kiln *S*, and inside diameter *D* of the kiln." How is the time affected when each of the variables is increased?

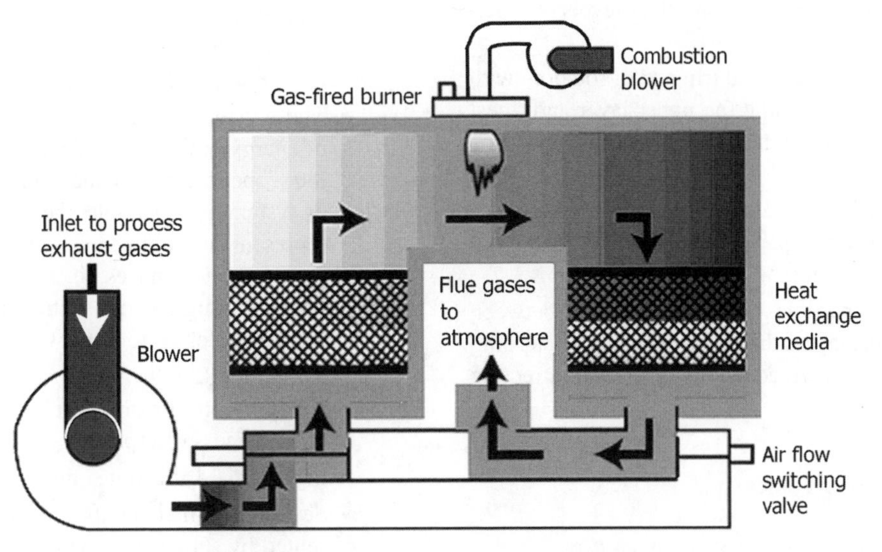

FIGURE 14.13 Regenerative two-bed oxidizer.

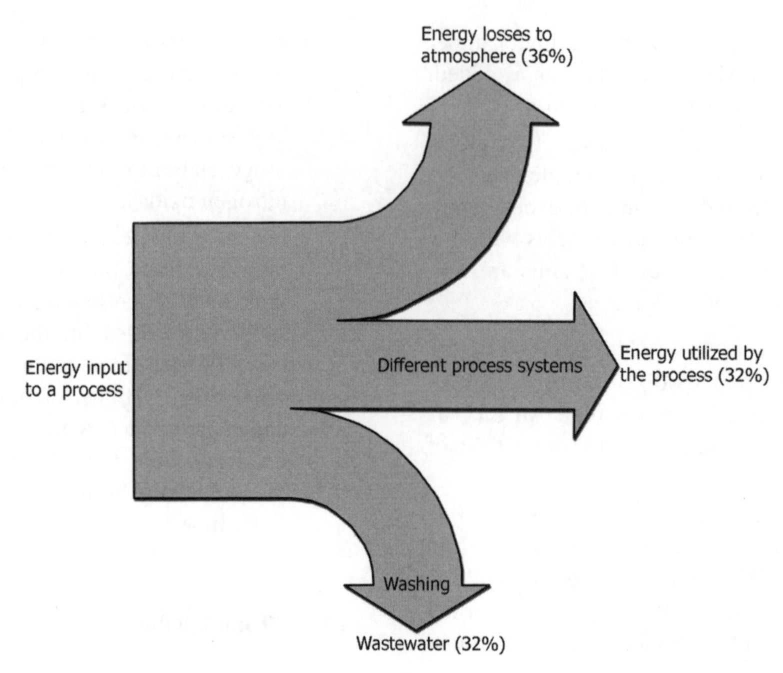

FIGURE 14.14 Sankey diagram.

- With L, time increases.
- With N, time decreases.
- With S, time decreases.
- With D, time decreases.

- What are the ways by which thermal efficiency of rotary kilns can be increased?
  - Increasing kiln length.
  - Use of chains inside kilns.
  - Use of heat recovery equipment on flue gases (e.g., waste heat boilers) and also on solids leaving the kiln (e.g., cooling with inlet air used for combustion).
- How does particle size variation affect levels of calcination in a rotary kiln?
  - Size segregation occurs during kiln rotation with coarser particles forming the upper layer and finest particles remaining at the bottom, in contact with the hot refractory lining.
  - Coarser particles are calcined by direct contact with hot gases and finest particles by direct contact with hot refractory lining.
  - Intermediate particles are sandwiched between coarse and fine layers that act as insulators, retarding heat transfer to intermediate layer, and thus affecting calcination levels.

14.1.10 Miscellaneous Topics

- What are the applications of Sager cones?
  - Sager cones are triangular ceramic cones of different compositions that are used for temperature monitoring and control in batch furnaces and kilns used for firing refractories and porcelain objects.
  - The cones are placed in these furnaces and kilns at strategic locations and their tips are kept under observation during firing. Melting of the tip indicates particular temperature levels attained. The observation gives indication of firing adequacy for different products during operation of the kiln/furnace.
- What is a Sankey diagram? Illustrate its utility.
  - Sankey diagrams are a specific type of flow diagrams, in which the width of the arrows is shown proportionate to the flow quantity.
  - These diagrams are used for pictorial representation of energy or material balances in order to determine the scope for performance improvement of a process or system. For example, these are frequently used to represent fuel efficiencies in combustion processes involved in furnaces, in order to identify energy losses through various streams and surfaces and device strategies to improve upon the processes for better energy utilization.
  - Using colors for different types of flows enhances effectiveness of representation of such flows.
  - Figure 14.14 illustrates the principle.
  - Several detailed inputs and outputs can be represented by showing branches on the diagram.

SECTION III

MASS TRANSFER

15

MASS TRANSFER BASICS

15.1	Introduction	455
15.2	Diffusion	455
15.3	Mass Transfer Coefficients	460
15.4	Dimensionless Numbers and Correlations in Mass Transfer	463
15.5	Analogies	464
15.6	Theories of Mass Transfer	465
15.7	Mass Transfer with Chemical Reaction	468
15.8	Drops, Bubbles, Sprays, Mists, Aerosols, and Foams	469

15.1 INTRODUCTION

- What is the importance of mass transfer?
  - Field of mass transfer involves some of the most typical chemical engineering problems.
    - ➢ Design and operation of chemical process equipment involves preparation of reactants, carrying out chemical reactions, and separation of the resultant products.
    - ➢ Ability to carry out the above operations largely rests on the proficiency of principles of mass transfer, which is largely the domain of chemical engineers.
    - ➢ Mass transfer is meant to be the tendency of a component in a mixture to travel from a region of higher concentration to that of lower concentration.
- Name the important mass transfer processes.
  - *Absorption:* Used for separation of acidic impurities from mixed gas streams, for example, CO_2, H_2S, SO_2, other organics such as carbonyl sulfide (COS) and mercaptans, NH_3, HCN, and NO_x, VOC control in environmental cleanup, odor control, wastewater stripping, and so on.
  - *Distillation:* Differential distillation, equilibrium flash vaporization, steam distillation, batch and continuous fractionation, binary and multicomponent distillation, azeotropic and extractive distillation, and so on.
  - *Liquid–Liquid and Solid–Liquid Extraction:* Pharmaceuticals and foods, extraction and refining of fats and oils, dewaxing and aromatic extraction from lube oils, leaching of ores, and so on.
  - Humidification, dehumidification, and water cooling.
  - Drying.
  - Adsorption, chromatography, and parametric pumping.
  - Ion exchange.
  - *Membrane Separation Processes:* Based on molecular size differences:
    - ➢ Gas–gas, gas–liquid, liquid–liquid.
    - ➢ Liquid membranes, for example, for H_2S removal.
  - Reverse osmosis, ultra- and microfiltration, pervaporation, bubble and foam fractionation, and so on.

15.2 DIFFUSION

- What is the difference between diffusion and mass transfer?
  - *Diffusion:* The phenomenon of diffusion is a result of the motion of molecules in a fluid medium. If one observes the motion of molecules in a fluid medium from the viewpoint of the molecular scale, the molecules are seen to be moving randomly in various directions and at different velocities. Physical stimulus caused by molecular types, sizes, and shapes, intermolecular forces, temperature effects, and other similar reasons cause the movement of a component through a mixture.

Fluid Mechanics, Heat Transfer, and Mass Transfer: Chemical Engineering Practice, By K. S. N. Raju

- *Mass Transfer:* Movement of a component from one phase to another due to concentration difference between the phases, for example, concentration gradient. Mass transfer encompasses diffusion and convection currents set up due to flows, laminar or turbulent, temperature differences, and concentration differences, which are complementary to one another, across phase boundaries.

• "Mass diffusivity has the same units as thermal diffusivity and kinematic viscosity." *True/False*?
  - *True.*

• What are the different types of diffusion? Explain.
  - *Molecular Diffusion:*
    - Gases: As a result of random motion of the molecules.
    - Liquids: Diffusivities are much less compared to gases, about 10,000 times less than those for gases. For example, crystals in an unsaturated solution dissolve with subsequent diffusion away from solid–liquid interface.
    - Solids: Mass diffusivities of atoms and molecules in solids are inversely proportional to the molecular weight of the diffusing species. In contrast with the thermal conductivity, the diffusivity is usually a very strong function of temperature. Diffusivities are thousands of times less than those for liquids.
    - Diffusion can occur in two or three dimensions. The most common is 3D diffusion, or migration of solute atoms in the bulk of a solid. 2D diffusion occurs on the surfaces of solids or along internal surfaces that separate the grains of polycrystalline solids. This is termed *grain boundary diffusion.* Estimation and experimentation is complex and involve very poor accuracies.
  - *Eddy Diffusion:* Caused by bulk mixing in a fluid, involving bulk flow and molecular diffusion, for example, dissipation of smoke from a stack. Turbulence causes mixing and transfer of smoke to the surrounding atmosphere.
  - *Thermal Diffusion:*
    - If two regions in a mixture are maintained at different temperatures so that there is heat flux, concentration gradient is set up.
    - In a binary mixture, one kind of molecule tends to travel toward hot region and other kind toward cold region; this is called *Soret effect.*
    - *Soret effect* has negligible influence on mass transfer but is useful in the separation of certain mixtures.
    - Tendency of generation of a temperature gradient in conjunction with mass transfer arising from a concentration gradient is called *Dufour effect.*
    - *Thermal diffusion in solids* refers to the transport of atoms driven by a temperature gradient.
    - Unlike ordinary diffusion in a concentration gradient, thermal diffusion is not a random walk process. The affected atoms flow along the temperature gradient, either up or down. Thermal diffusion affects impurity atoms in a host crystal, the important example being hydrogen in Zircaloy.
    - Thermal diffusion is determined by *heat of transport* property of the species in the solid.
  - *Pressure Diffusion:*
    - Occurs when there is a pressure gradient in a fluid mixture.
    - For example, in a deep closed well or a closed tube rotated around an axis perpendicular to axis of the tube—lighter components tend to travel toward region of low pressure, that is, top of well or end of tube near axis of rotation, for example, ultracentrifuge.
  - *Forced Diffusion:*
    - Caused by action of an external force other than gravity on the molecules.
    - External force must be such that it acts to a different degree on different species, for example, ions in an electrolyte in an electric field.
    - Ionic diffusion.
    - There are significant interactions between molecules/ions.
  - Diffusion in porous solids.
  - *Pores:* (a) Macropores and (b) micropores, for example, in catalyst pellets.
    - By far the most important and least understood phenomena.
    - Wide range of applications, which include solid catalyzed reactions, adsorption, drying, leaching, packed beds, fluidized beds, filtration, heat transfer, metallurgical processes, and the like.
    - Diffusion of moisture through cellular solids like wood can lead to deterioration, requiring treatment with preservatives. Swelling and shrinkage add to complexities to the diffusion process.
    - Effective diffusivity in a solid is less than diffusivity in a free fluid because
      - Tortuous nature of the path increases distance that a molecule must travel to advance a given distance in a solid.
      - Free cross-sectional area is restricted.

- For many catalyst pellets, $D_{effective}$ for a gaseous component is approximately equal to 1/10th of diffusivity in a free gas.

➢ *Knudsen Diffusion:* Occurs when size of pores approaches the mean free path of the gaseous molecules. A typical molecule under Knudsen diffusion collides predominantly with walls of pores rather than with other molecules.

➢ Knudsen diffusion coefficient is given by the expression

$$D_{KA} = v'_A d'/3, \qquad (15.1)$$

where v'_A is the mean molecular speed and d' is the average pore diameter.

$$v'_A = [8RT/(\pi M_A)]^{1/2}. \qquad (15.2)$$

- *Polymer Films and Melts:* Important in membrane separations, lining of tanks, pigment distribution in polymers, electrolysis processes, and so on.
  - ➢ More orderly than porous solids.
  - ➢ Diffusion depends on type of polymer, that is, crystalline or amorphous. Diffusion is mainly through amorphous regions.
- *Crystalline Solids:* There are several mechanisms involved. One complication is anisotropy, that is, diffusivities are different in different directions.
  - ➢ Mechanisms include direct exchange of lattice positions; migration through interstitial sites, to unoccupied sites, along grain boundaries; displacement atoms, and so on.
  - ➢ In silica and glass (crystalline), H_2 diffuses as a molecule.
  - ➢ Diffusion in glass becomes important for applications involving high vacuum.
- Metals, for example, diffusion of carbon atoms in iron when heated in a bed of coke, *case hardening* of steel, *doping* of semiconductors, *oxidation* of metals, solid-state *formation of compounds* from individual components. *Sintering* is the process by which an object made from powders becomes dense and strong.
- Diffusion of light gases through metals involve as a first step dissolution of the gas in the metal, for example, hydrogen dissolves and dissociates to an atom reacting to form hydrides.
- Hydrogen is known to diffuse through metal walls of a pressure vessel, thermocouple sheaths, and so on, and leak out.

- "The driving force for mass transfer by molecular diffusion is the difference in chemical potential." *True/False*?
  - *True*.
- What are the effects of water absorption in polymers?
  - Water in liquid or vapor form is absorbed into the polymer.
  - Water molecules fill the voids that are formed between polymer chains and then induce relaxation, or swelling of the polymer.
  - The presence of water in a polymer film has been shown to affect a variety of physical properties including tensile strength, hardness, net dielectric constant, and glass transition temperature. The swelling of polymer films causes strain and stress, and can induce cracking, degradation, and other signs of physical aging.
  - Plasticization, the softening and increase of flexibility of a polymer material, can occur.
- Why is it important to study water absorption in polymers?
  - The study of water sorption in polymers is important for many applications.
  - Polymers are used as coatings on metal surfaces to protect the metals from corrosion when subjected to ambient humidity. A hydrophilic coating may be needed to enhance adhesion or wettability of the metal surface. A hydrophobic coating can be used to prevent permeation of water.
  - Water is the most important factor in governing microbial spoilage in foods. Changes in moisture content is very important in food preservation. Therefore, polymers providing excellent barriers for water are needed for packaging.
  - Another area that can gain from knowledge of water uptake in polymers is sensors. Humidity sensors are used in quality control in industries such as food, paper, and electronics.
  - Drug delivery systems, which use polymers, is another major area related to water and polymers.
  - Water management is a critical problem and design issue for polymer electrolyte fuel cells.
  - In the cosmetics industry, hairsprays with polymers as active ingredients must have stiffness that stands up to humidity.
  - Understanding water reaction of photoresists for design of microelectronics is an important area.
- "Permeability refers to the diffusion of a gas in a solid and is used extensively in calculating mass transfer in packaging materials." *True/False*?
  - *True*.

- "Permeability is equal to the product of the diffusion coefficient and the solubility of the gas in the solid." *True/False*?
  - *True.*
- "Permeability decreases as the temperature increases." *True/False*?
  - *False.*
- "Polyethylene is a good water vapor barrier and serves as an adhesive to the next layer." *True/False*?
  - *True.*
- "Aluminum foil is a good gas barrier." *True/False*?
  - *True.*
- Explain what is meant by *vacancy mechanism* for diffusion in metals.
  - Unlike the migration of self-interstitials or interstitial impurities, the diffusion of substitutional impurity atoms in nearly all metals occurs by an atomic process called the *vacancy mechanism.*
  - A solute species entirely surrounded by sites occupied by host atoms cannot exchange places with one of them because the energy required is too high. Only if one of its nearest neighbors is vacant can the solute atom move into it, and thus proceed with the diffusion process.
- What are the differences and commonalities in diffusion in ionic crystals and diffusion in metals and elements?
  - *Differences:*
    (a) Ionic solids contain at least two components and the two are oppositely charged.
    (b) Cations and anions generally exhibit very large differences in intrinsic diffusivities, occasionally as large as seven orders of magnitude.
    (c) The type of defect that predominates controls the magnitudes of the diffusivities.
    (d) The effect of substitutional doping of ionic solids with cations of different valence from the host cation has a profound effect on the diffusivities, often of both ions.
    (e) The requirement of local electrical neutrality affects the movement of the ions.
    (f) Interstitial impurity diffusion is not as important in ionic solids as it is in metals and other elements.
  - *Commonalities:*
    (a) The diffusion coefficients obey the Einstein equation.
    (b) The mechanisms commonly found in metals and elements (vacancy and interstitial) also predominate in ionic solids.
    (c) The various types of diffusion coefficients discussed in the preceding section (intrinsic, tracer, self, mutual) are also present in ionic solids.
    (d) The relation between point defect properties (concentration and mobility) and the ion diffusivity is the same for ionic solids as it is for metals and other elements. However, motion of the ions is restricted to the sublattice of the same charge. All the four mechanisms are not operative in a particular ionic solid. The mechanism that dominates for the cations may be different from that controlling anion motion.
- What is the difference between molecular and eddy diffusion?
  - *Molecular Diffusion:* Transfer of different species of molecules takes place within a gas/vapor or liquid due to random velocities of molecules.
  - *Eddy Diffusion:* Transfer of different species of molecules takes place due to circulating or eddy currents present in a turbulent fluid.
- State Fick's law for diffusion.
  - When the concentration gradient drives J, the flux is given by *Fick's First Law*,

$$J = -D\partial c/\partial x, \tag{15.3}$$

    where J is the flux, $\partial c/\partial x$ is concentration gradient and x is the distance in the direction of transfer.
  - This equation follows the universal observation that matter diffuses from regions of high concentration to regions of low concentration, hence the minus sign.
  - D, called diffusion coefficient, is a function of temperature and *concentration only*, but *not of the concentration gradient*. The units of D are length squared per unit time, usually cm^2/s provided that J, c, and x are in consistent units.
  - Fick's Second Law is given by the equation,

$$\partial c/\partial t = \partial/\partial x[D\partial c/\partial x] + Q, \tag{15.4}$$

    where t is the time, x is the distance, and Q is the source term of the diffusing species, atoms (or moles) per unit volume.
  - This equation is also called the *diffusion equation*, by analogy to its heat transport counterpart, the heat conduction equation.
- "Fick's second law is used in unsteady-state mass transfer problems." *True/False*?
  - *True.*

- "The required time for a certain change in concentration due to diffusion is proportional to the square of the thickness of the body." *True/False*?
  - *True.*
- "Diameter is used as the characteristic dimension of a sphere for the solution of unsteady-state mass transfer problems." *True/False*?
  - *False.*
- State and explain Einstein equation for diffusion in solids.
  - One of the most important equations in diffusion theory is an equation attributed to Einstein. This equation connects the macroscopic property D, the diffusion coefficient, with the microscopic properties w and λ, the jump frequency and the jump distance, respectively.
  - The frequency with which the diffusing atom jumps to any available adjacent site is called the *total jump frequency* and is denoted by Γ. In Figure 15.1, only jumps to the left or the right are shown, making $\Gamma = 2w$, and the equation for diffusivity becomes

$$D = \frac{1}{2}\lambda^2\Gamma. \qquad (15.5)$$

  - Einstein equation for three dimensional case is

$$D = \frac{1}{6}\lambda^2\Gamma. \qquad (15.6)$$

  - The path followed by the diffusing species through the lattice from one equilibrium site to another, the diffusion path, will always be the one of least resistance, specifically the route that demands the lowest average energy increase to effect the change in position, or the *jump*.
  - The diffusing species spends most of the time in equilibrium sites and only occasionally moves to an adjacent site. This movement is called a *jump*, and the length of the jump is denoted by λ.

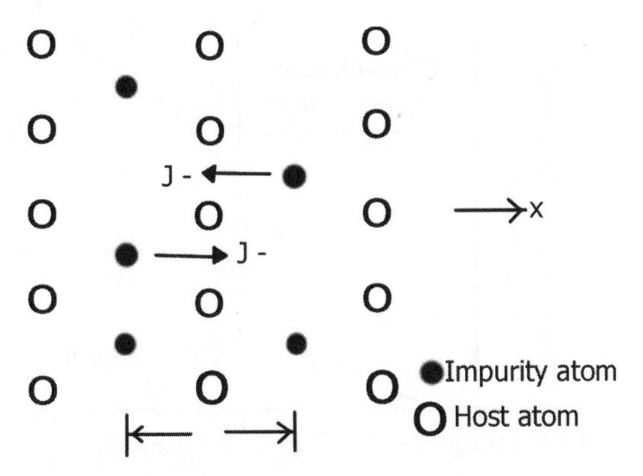

FIGURE 15.1 One-dimensional diffusion.

  - The rate at which impurity atoms jump from the left-hand plane to the right-hand plane is J_+ and in the opposite direction, the flux is J_-.
  - The quantity w is a one-way jump frequency, or the frequency with which an impurity atom jumps from an equilibrium site to a particular adjacent equilibrium site.
- What is equimolar counterdiffusion?
  - For diffusion of every molecule of a particular species in one direction, there will be diffusion of another molecule in the opposite direction.
  - For example, in distillation, for every molecule vaporized from the liquid phase, there will be one molecule condensing from the vapor phase, with the net result that the number molecules (not necessarily the same species) in each phase remains constant. This is possible if both phases are at their saturation temperatures and heats of vaporization, λ_v, are equal for different species of molecules in the system consisting of both the phases and the process is truly adiabatic, that is, no heat losses or heat gain are involved.
- "There is bulk flow in equimolar counterdiffusion." *True/False*?
  - *False.*
- Give typical equations for the estimation of molecular diffusivities for gases and liquids.
  - Gas phase diffusion coefficients in air can be estimated by the equation (Fuller's equation):

$$D_{BA} = 10^{-3}T^{1.75}\sqrt{M_r}/[P(V_A^{1/3} + V_B^{1/3}]^2, \qquad (15.7)$$

where $M_r = (M_A + M_B)/M_A M_B$. The subscripts B and A denote the solute and gas phase (or gas mixture), respectively, T is the temperature (K), M is the molecular weight, P is the pressure (atmospheres), and V_A and V_B are the molar volumes (cm^3/mol) of air and the gas in question, respectively.

  - Theoretical equation for binary mixtures, derived by Hirschfelder and coworkers from kinetic theory of gases is

$$D_{AB} = 1.858 \times 10^{-7}T^{3/2}[1/M_A + 1/M_B]^{1/2}/ [(P/101325)\sigma_{AB}^2\Omega(T_D^*)]\mathrm{m}^2/\mathrm{s}. \qquad (15.8)$$

$$\sigma_{AB} = \sigma_A + \sigma_B. \qquad (15.9)$$

σ_A and σ_B are the collision diameters of components A and B in Angstrom units, and $\Omega(T_D^*)$ [ ] is the collision integral.

$$T_D^* = kT/\varepsilon_{AB}. \quad (15.10)$$

$$k/\varepsilon_{AB} = \sqrt{(k/\varepsilon_A)(k/\varepsilon_B)}. \quad (15.11)$$

- ε_A/k and ε_B/k in units of K are the characteristic energies of components A and B, respectively, divided by Boltzmann constant k.
- There are other equations available in literature.
- Diffusion coefficients in water are given typically by Wilke and Chang equation:

$$D_{BW} = 7.4 \times 10^{-8}(\Phi_W M_W)^{1/2}T/\eta_W V_B^{0.6}\ \text{cm}^2/\text{s}, \quad (15.12)$$

where M_W is the molecular weight of water (g/mol), T is the temperature (K), η_W is the viscosity of water (cP), V_B is the molar volume of solute B at its normal boiling temperature (cm^3/mol), and Φ_W is the solvent association factor that is 2.6 for water.

- Diffusion coefficients for chemicals in liquids can be estimated, for example, from the correlation of Hayduk et al.:.

$$D_{BW} = 1.25 \times 10^{-8}(V_B^{-0.19} - 0.292)T^{1.52}\eta_W\varepsilon^{*}\ \text{cm}^2/\text{s}, \quad (15.13)$$

where V_B is the molar volume (cm^3/mol), η_W is the viscosity of water (cP) and

$$\varepsilon^* = (9.59/VB) - 1.12. \quad (15.14)$$

15.3 MASS TRANSFER COEFFICIENTS

- What is mass flux?
  - The rate of mass transfer of component i is usually expressed as the mass of component i passing through unit area of the interface per unit time, which is referred to as the *mass flux* of component i, N_i, kg/(m^2 s). N_i can be expressed in terms of *molar flux*, kmol/(m^2 s).

$$\text{Mass flux} = (\text{diffusional flux}) + (\text{convective mass flux}). \quad (15.15)$$

  - The fact that the mass flux is always accompanied by convective mass flux is a phenomenon characteristic to mass transfer and has no parallels in momentum or heat transfer.
- What is the significance of mass transfer coefficients?
  - Mass transfer coefficients generally represent rates that are much greater than those that occur by diffusion alone, as a result of convection or turbulence at the *interface* where mass transfer occurs.
  - There are several principles that relate mass transfer coefficient to diffusivity and other fluid properties and to the intensity of motion and geometry.
  - The rate equation can be expressed as

$$\text{Rate of mass transfer} = k(\text{interfacial area})(\text{concentration difference}). \quad (15.16)$$

k is a proportionality factor, called mass transfer coefficient. This is a simple way to define mass transfer coefficient.

  - Mass transfer coefficients involve transfer of species across phase boundaries (interface between the phases).
  - Figure 15.2 shows a schematic representation of the concentration distribution near an interface. Although the variation in the concentration near the interface is very sharp, it becomes more gradual in the region slightly further from the interface and the concentration slowly approaches that in the bulk fluid. Moreover, the concentration at the interface is in equilibrium between the phases.
  - In real problems, however, there is no direct means of evaluating concentration gradients at the interface, except in very exceptional cases.
  - Overall mass transfer coefficients eliminate the need of finding interfacial concentrations, which are difficult to measure.
  - Overall coefficients involve bulk phase concentrations that are easy to measure.
  - In contrast to heat transfer problems, various definitions of concentration are used in a case-by-case way, which leads to different definitions of mass transfer coefficients.

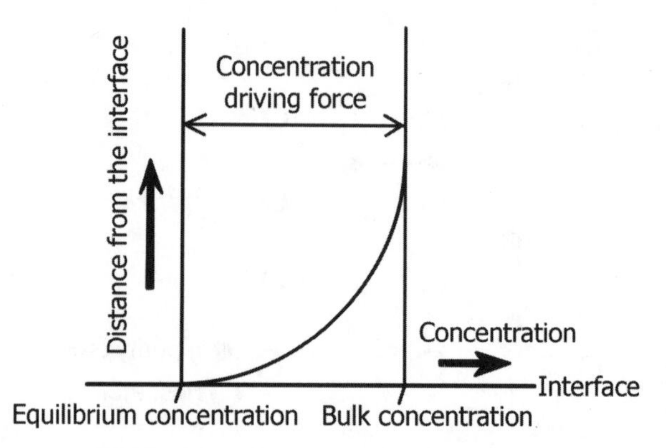

FIGURE 15.2 Concentration driving force.

  - These are expressed as mass fractions, mole fractions, mole ratios, partial pressures, absolute humidities, and so on.
  - Table 15.1 summarizes different definitions of mass transfer coefficients.
- One problem that remains complex is to measure interfacial area, particularly in packed beds.
  - Interfacial area depends on several factors including type and size of packing, way it is packed, wetting characteristics, and so on.
  - In order to overcome this problem *volumetric mass transfer coefficients*, $k_{G}a$ or $k_{L}a$, are introduced, which are products of mass transfer coefficient and a characteristic area, defined as area/unit volume of the packed bed.
  - The interfacial area, a, is the same for the gas film and for the liquid film.
  - The interfacial area, a, should not be confused with the geometric surface area of the column packing.
  - Interfacial area is not directly related to the wetted area.
  - For example, the increase in overall mass transfer coefficient for ceramic Intalox saddles compared to ceramic Raschig rings could, at least in part, be attributed to increased surface area per unit volume.
  - However, metal Pall rings show a substantial increase in mass transfer coefficient compared to metal Raschig rings, though both packings have the same surface area for the same size packing.
  - Another factor that adds to the complexity of determining interfacial area is that modern column packing shapes tend to be permeable, so that the gas and liquid can flow through the pores of the packing element as well as around its external shape. These packings are due in part to increased interstitial transfer within the internal pores.
- To complete defining mass transfer coefficients, these are expressed as coefficients based on *gas film controlled* or *liquid film controlled*, depending on which of the two films offer more resistance for mass transfer.

• What is meant by *gas film controlled* and *liquid film controlled* mass transfer processes?

- *Gas film controlled* processes are those mass transfer processes in which the resistance to mass transfer will be predominantly in the gas film, liquid film offering negligible resistance for mass transfer.
- *Liquid film controlled* processes are those mass transfer processes in which the resistance to mass transfer will be predominantly in the liquid film, gas film offering very little for mass transfer.
- In systems that are gas film controlled, the solute either will be highly soluble in the liquid phase or will react rapidly with a component in the liquid phase.
- In liquid film controlled systems the solute either has a low solubility in the liquid phase or reacts with a component in the liquid phase at a slow rate. In some cases, depending on the value of m, one may approximate the overall coefficient by neglecting the smaller resistance to mass transfer.
- In reality this is an extremely simplifying assumption because the percentage of gas and liquid film control varies with the concentration of solute as well as the physical properties that change the solubility of the solute in the liquid phase.

TABLE 15.1 Definitions of Mass Transfer Coefficients

Mass Transfer Coefficient	Units	Definition	Driving Force	Phase
k_y	kmol/(m^2 s)	$N^*_A = k_y(y_s - y_\infty)$	Δy	Gas phase
k_G	kmol/(m^2 s kPa)	$N^*_A N^*{}_A = k_G(p_s - p_\infty)$	Δp	Gas phase
k_Y	kmol/(m^2 s)	$N^*_A N^*{}_A = k_Y(Y_s - Y_\infty)$	ΔY	Gas phase
k	m/s	$N_A = \rho_{Gk}(m_{Gs} - m_{G\infty})$	Δm_G	Gas phase
k_H	kg/(m^2 s)	$N_A = k_H(H_s - H_\infty)$	ΔH	Gas phase
k_L	m/s	$N^*_A N^*{}_A = k_L(c_s - c_\infty)$	Δc	Liquid phase
k_x	kmol/(m^2 s)	$N^*_A N^*{}_A = k_x(x_s - x_\infty)$	Δx	Liquid phase
k_X	kmol/(m^2 s)	$N^*_A N^*{}_A = k_X(X_s - X_\infty)$	ΔX	Liquid phase
k	m/s	$N_A = \rho_{Gk}(m_{Ls} - m_{L\infty})$	Δm_L	Liquid phase

c = molar density, mol/m^3; H = absolute humidity; N_A = mass flux, kg/m^2 s; $N^*_A = N_A/M_A$ = molar flux, kmol/m^2 s; p = partial pressure, kPa; x, y = mole fraction; $X = x/(1-x)$; $Y = y/(1-y)$; m = mass fraction.

- "In liquid film controlled systems the value of K_Ga primarily is a function of the liquid flow rate." *True/False*?
  - *True.*
- "In gas film controlled absorption systems the value of K_Ga is a function of both the liquid and the gas flow rates." *True/False*?
  - *True.*
- "In gas film controlled systems the value of K_Ga is much greater than in liquid film controlled systems." Comment.
  - This is because the liquid film resistance reduces the value of the overall coefficient in liquid film controlled systems, while it has only a marginal effect on this value in gas film controlled systems.
  - For example, in absorption operations carried out at a constant liquid to gas ratio, the number of moles of solute to be absorbed will increase in direct proportion to the gas rate at a fixed gas composition. However, the K_Ga value for a liquid film controlled system will increase only by about the 0.37 power of the flow rates at a constant G/L ratio.
  - Give examples of gas film controlled mass transfer processes (highly soluble gases in water).
  - Absorption of NH_3 in water or aqueous ammonia solution or 10% H_2SO_4.
    - Because of the extremely high solubility of NH_3 gas in water, there have been incidents involving collapse of ammonia storage tanks, when water inadvertently entered the tanks containing ammonia vapors, as such tanks are not designed to withstand vacuum conditions. Cleaning such tanks with water spraying should precede suitable precautionary measures to prevent such failures.
  - Absorption of water vapor in strong acids.
  - Absorption of SO_3 in 98% H_2SO_4.
  - Absorption of HCl in water or weak HCl solution.
    - Because of practically no liquid-phase resistance in HCl absorption in water and high heat of absorption, design of HCl absorbers becomes more of a heat transfer problem than a mass transfer problem.
  - Absorption of Cl_2 in 5% NaOH solution.
  - Absorption of SO_2 in alkaline/ammonia solutions.
  - Absorption of H_2S in weak caustic soda solutions.
  - Evaporation and condensation of liquids.
- Give examples of liquid film controlled mass transfer processes.
  - Absorption of CO_2, H_2, O_2, Cl_2, and N_2 in water.
    - Oxygen absorption in water being a slow process due to high liquid-phase resistance, *turbulent conditions* need to be created to purify water and wastewaters through enhanced oxygen absorption. In natural waters, turbulent flow of rivers and streams in hilly areas ensures oxygenation and oxidation of organic loads in such waters in contrast to stagnant waters like in lakes and ponds. High organic loads can lead to eutrophication of lakes. Eutrophication refers to oxygen deficiency in waters, which affect aquatic life.
    - Another way to increase absorption rates in such cases is changing the solvent or using reversible chemical reactions.
  - Absorption of CO_2 in weak alkali solution.
- Give examples of mass transfer processes in which both gas and liquid films are controlling.
  - Absorption of SO_2 in water.
  - Absorption of acetone in water.
  - Absorption of nitrogen oxide in strong H_2SO_4.
  - Table 15.2 gives a typical industrially important list of the gas film or liquid film controlled processes.
- "In gas film controlled systems the value of K_Ga is much greater than in liquid film controlled systems." Why?
  - This is because the liquid film resistance reduces the value of the overall coefficient in liquid film controlled systems, while it has only a small effect on this value in gas film controlled systems.
- "For a liquid film controlled system, the effect of gas rate on the overall K_Ga value is small." *True/False*?
  - *True.*
- State Henry's law. For what type of systems it is applicable?

TABLE 15.2 Examples of Gas or Liquid Film Controlled Processes

Solute	Solvent Liquid	Controlling Phase
Oxygen	Water	Liquid
Chlorine	Water	Liquid
Carbon dioxide	Water	Liquid
Carbon dioxide	4% NaOH	Liquid
Carbon dioxide	12% MEA	Liquid
Water vapor	Water	Gas
Water vapor	93% H_2SO_4	Gas
Ammonia	Water	Gas
Ammonia	10% H_2SO_4	Gas
Sulfur dioxide	4% NaOH	Gas
Hydrogen chloride	Water	Gas
Chlorine	5% NaOH	Gas
Sulfur trioxide	98% H_2SO_4	Gas

$$P_A = H_{CA}. \quad (15.17)$$

- H is a dimensionless Henry's law constant.
- Applicable for dilute solutions.

- What are overall mass transfer coefficients? How are they related to individual film coefficients?

$$N_A^* = k_G(y_\infty - y_s) = k_L(x_s - x_\infty)$$
$$= K_G(y_\infty - y^*) = K_L(x^* - x_\infty). \quad (15.18)$$

k_L and k_G are the liquid and gas-phase mass transfer coefficients, K_L and K_G are the liquid and gas phase overall mass transfer coefficients, N_A^* is the molar flux, x_s is the equilibrium mole fraction of the liquid at the interface, x_∞ is the mole fraction of the bulk liquid, x^* is the mole fraction of the liquid in equilibrium with the bulk gas, y_s is the equilibrium mole fraction of the gas at the interface, y_∞ is the mole fraction of the bulk gas, and y^* is the mole fraction of the gas in equilibrium with the bulk liquid.

- The following equations relate the overall mass transfer coefficients to the individual mass transfer coefficients, assuming linear equilibrium relationship:

$$1/K_G = 1/k_G + m/k_L. \quad (15.19)$$

$$1/K_L = 1/mk_G + 1/k_L. \quad (15.20)$$

K_G and K_L are overall mass transfer coefficients based on gas phase and liquid phase, respectively; k_G and k_L are individual mass transfer coefficients based on gas and liquid phases, respectively; m is the slope of the equilibrium line. For systems following Henry's law, m is replaced with Henry's law constant, H. The overall and individual mass transfer coefficients for gas and liquid phases can be replaced by the respective volumetric coefficients, K_Ga, K_La, k_Ga, and k_La in the above equations that are written as

$$1/K_Ga = 1/k_Ga + m/k_La. \quad (15.21)$$

and

$$1/K_La = 1/mk_Ga + 1/k_La. \quad (15.22)$$

- Limiting cases are the following:
  - For highly soluble gas ($H \ll 1$),

$$K_G \approx k_G. \quad (15.23)$$

  - For sparingly soluble gas ($H \gg 1$),

$$K_L \approx k_L. \quad (15.24)$$

- What is the effect of liquid flow rate on K_Ga values for liquid film controlled absorption processes?
  - Liquid flow rate has a significant effect on the overall K_Ga value. A doubling of the liquid rate typically will increase the overall K_Ga value by 23%.
- "H_2 has high diffusivity in water at low concentrations." *True/False*?
  - *True.*
- In a wetted wall column experiment for determining mass transfer coefficient, what is the principal source of error?
  - Liquid film flow at $N_{Re} > 25$, surface waves will be caused to flowing liquid film and so surface area of the exposed liquid film will not be the same as the inner surface of the test section of the column (will be more).
  - Dirty surface of the inner wall of the column impairs wettability of the surface.

$$N_{Re} = 4\Gamma/\mu, \quad (15.25)$$

where Γ is the liquid mass flow rate per unit width of the surface.

- In a wetted wall column experiment for determining mass transfer coefficients, what are the sources of error?
  - Liquid film flow at NRe > 25(NRe $= 4\Gamma/\mu$, where Γ is liquid mass flow rate per unit width of the surface), surface waves will be caused in the flowing liquid film and so surface area of the exposed liquid film will be more than the inner surface section of the column.
  - Dirty surface of the inner wall of the column impairs wettability of the surface.

15.4 DIMENSIONLESS NUMBERS AND CORRELATIONS IN MASS TRANSFER

- What are the important dimensionless numbers that are used in correlating mass transfer coefficients?
  - Mass transfer coefficients are correlated by means of dimensionless numbers:
    - *Sherwood Number, kl/D:* Mass transfer velocity/diffusion velocity. The Sherwood number represents a dimensionless mass transfer from a fluid flow through a specified boundary surface. The nondimensionalizing quantities are the geometrical length scale and the relevant fluid species diffusion coefficient. Hence Sherwood number is a function of fluid and geometry properties. The Sherwood number can be used to determine species transfer by means of correlations with Reynolds number and Schmidt number. Note

that the Sherwood number can be defined locally or as an average overall value for a given surface.

- *Stanton Number, k/v^0:* Mass transfer velocity/ flow velocity.
- *Schmidt Number, N_{Sc}, v/D or $\mu/\rho D$:* Diffusivity of momentum/diffusivity of mass. Is a function only of fluid properties. Schmidt number is typically just less than 1 for gases and is much greater than 1 for most other cases of interest. Schmidt number is the mass transfer analog to Prandtl number. An important interpretation of Schmidt number is as the relative thickness of the velocity and concentration boundary layers. $N_{Pr}=1$ gives boundary layers of equal thickness, $N_{Pr}>1$ gives a thinner velocity boundary layer as momentum transfer is more rapid than mass transfer. Used in combined heat and mass transfer studies. Determines ratio of fluid to mass transfer boundary layer thickness.
- *Lewis Number, α/D:* Diffusivity of energy, that is, thermal diffusivity/diffusivity of mass, that is, thermal diffusivity/diffusion coefficient. It is used to characterize fluid flows where there is simultaneous heat and mass transfer by convection. α is thermal diffusivity and D is mass diffusivity. It can also be expressed as $N_{Le}=N_{Sc}/N_{Pr}$. Lewis number is a function only of fluid properties (which themselves may be a function of system pressure and/or temperature). Lewis number indicates the relative magnitude of thermal and species diffusivities and as such is a ratio of Schmidt number and Prandtl number.
- *Graedz Number, Gz, $(D_i/L)N_{Re}\cdot N_{Sc}$.*
- *Biot Number, kL/D_{solid}:* Ratio of diffusive to convective (or reactive) mass transfer resistance. Determines uniformity of concentration in the solid.

- Give an equation showing the ratio of thickness of concentration boundary layer to that of hydrodynamic boundary layer is approximately equal to $N_{Sc}^{-1/3}$.

$$\delta_c/\delta \approx (\mu/\rho D)^{-1/3} = N_{Sc}^{-1/3}. \quad (15.26)$$

δ_c and δ are the thicknesses of mass and momentum transfer boundary layers, respectively.

- Figure 15.3 illustrates the relative positions of laminar hydrodynamic and concentration boundary layers for a flat plate.

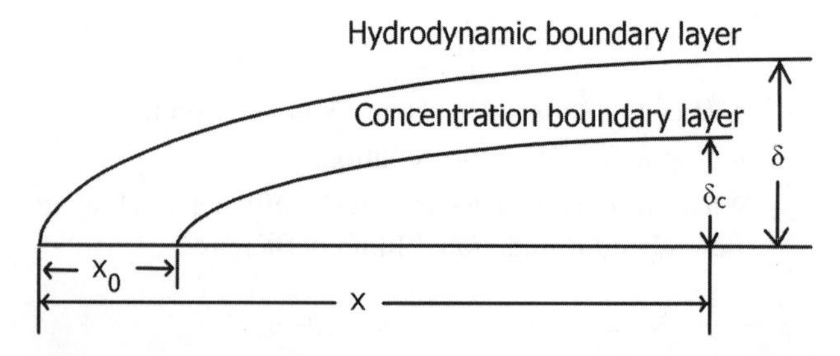

FIGURE 15.3 Laminar hydrodynamic and concentration boundary layers for a flat plate.

- What are the orders of magnitude of Prandtl and Schmidt numbers for common gases and common liquids?
  - Common gases:

$$N_{Pr} \approx N_{Sc} \approx 1. \quad (15.27)$$

  - Common liquids (except liquid metals):
    - $10 < N_{Pr} < 10^2$.
    - $400 < N_{Sc} < 10^4$.

- Give a correlation relating Sherwood number to Reynolds and Schmidt numbers.

$$N_{Sh} = kL/D = A(N_{Re})^n(N_{Sc})^{1/3}. \quad (15.28)$$

15.5 ANALOGIES

- Briefly explain the utility and basis of analogies among momentum, heat, and mass transfer.
  - Analogies among momentum, heat and mass transfer provide a method of generating, for example, mass transfer data from friction factors and heat transfer data, which are easily available compared to mass transfer data.
  - Analogous expressions for momentum, heat, and mass transfer that form the basis for the analogies:

$$\frac{\tau}{\rho} = -(v+\varepsilon_M)\frac{du}{dz}. \quad (15.29)$$

$$\frac{q}{C_P\rho} = -(\alpha+\varepsilon_H)\frac{dT}{dz}. \quad (15.30)$$

$$N_A = -(D_{AB}+\varepsilon_D)\frac{dC_A}{dz}. \quad (15.31)$$

where ε_M, ε_H, ε_D are momentum, heat, and eddy diffusivities; v is momentum diffusivity (kinematic viscosity), μ/ρ; α is thermal diffusivity, $k/\rho C_P$.

- As a first approximation, the three eddy diffusivities may be assumed equal.

- The analogous expressions suggest that there is a possibility of estimating heat and mass transfer rates in a turbulent flow, which, although approximate, is practically useful.
- The similarity of the transport equations relating to momentum, heat, and mass transfer suggests that the friction factors and the heat or diffusional fluxes at the wall are mutually interrelated, and so if any one of the above quantities is known by some means, then the other two quantities may be estimated by assuming mutual relationships. The problem then reduces to finding mutual relationships between the friction factors and the heat or diffusional fluxes or that between the heat and diffusional fluxes. This indirect method of predicting transport phenomena in turbulent flow is known as *analogy*.

• What are the important analogies in momentum, heat, and mass transfer?
  - *Reynolds Analogy:* In the turbulent core in the region near the wall, the contribution of the turbulent transport mechanism far outweighs that of the molecular mechanism. Reynolds analogy assumes the following equation:

$$\varepsilon_{\mathrm{M}} \approx \varepsilon_{\mathrm{H}} \approx \varepsilon_{\mathrm{D}}. \quad (15.32)$$

  - The relationships in Reynolds analogy are

$$\begin{aligned} f/2 &= N_{\mathrm{Nu}}/N_{\mathrm{Re}}N_{\mathrm{Pr}} \cong N_{\mathrm{StH}} \\ f/2 &= N_{\mathrm{Sh}}(1-m_{\mathrm{s}})/N_{\mathrm{Re}}N_{\mathrm{Sc}}. \end{aligned} \quad (15.33)$$

  - These two equations are referred to as the *Reynolds analogy*. It is only valid for the special case of $N_{\mathrm{Pr}} = N_{\mathrm{Sc}} = 1$.
  - *Chilton–Colburn Analogy:* Chilton and Colburn proposed a well-known correlation on the basis of wide ranges of heat and mass transfer data.
  - According to this analogy,

$$f/2 = J_{\mathrm{H}} = J_{\mathrm{D}}, \quad (15.34)$$

  where J_{H} and J_{D} are *J*-factors for heat and mass transfer, respectively, which are defined by the following equations:

$$J_{\mathrm{H}} = N_{\mathrm{Nu}}/N_{\mathrm{Re}}N_{\mathrm{Pr}}^{1/3}. \quad (15.35)$$

$$J_{\mathrm{D}} = N_{\mathrm{Sh}}(1-m_{\mathrm{s}})/N_{\mathrm{Re}}N_{\mathrm{Sh}}^{1/3}. \quad (15.36)$$

  - These equations are usually referred as *Chilton–Colburn analogy*.
  - *von Ka'rman Analogy:* The Reynolds analogy and the Chilton–Colburn analogy are based on the transport mechanisms for the two extreme cases, the turbulent core and the laminar sublayer. von Ka'rman proposed the following equations by considering the contribution of the transport mechanisms in all three of the regions, with the assumption that the velocity, temperature, and concentration distributions are similar in each region.
  - For heat transfer:

$$\begin{aligned} S_{\mathrm{tH}} = (f/2)/[1+5\sqrt{(f/2)}[N_{\mathrm{Pr}}-1 \\ +\ln\{1+5/6(N_{\mathrm{Pr}}-1)\}], \end{aligned} \quad (15.37)$$

$$\begin{aligned} S_{\mathrm{tM}}(1-m_{\mathrm{s}}) = (f/2)/[1+5\sqrt{(f/2)}[N_{\mathrm{Sc}}-1 \\ +\ln\{1+5/6(N_{\mathrm{Sc}}-1)\}]. \end{aligned} \quad (15.38)$$

  - The von Ka'rman analogy is not valid for liquid-phase mass transfer, where Schmidt numbers are very large.

15.6 THEORIES OF MASS TRANSFER

• Name and explain important theories of mass transfer.
  - Film theory.
  - Figures 15.4 and 15.5 illustrate the two-film theory.
  - In the bulk of the phases, turbulence is assumed that makes all composition variations negligible, bringing uniformity in the compositions.
  - The phase boundary is conceived of as a sharp transition between the two phases.
  - The phase boundary is considered to be rigid that causes damping of the turbulent eddies in the bulk near the interface.
  - Film theory assumes presence of a fictitious stagnant laminar fluid film for each phase next to the phase boundary and resistances for mass transfer are within the fictitious films and concentrations remain constant within the bulk phases.
    - ➢ According to two-film theory, mass transfer coefficient is proportional to diffusivity, that is,

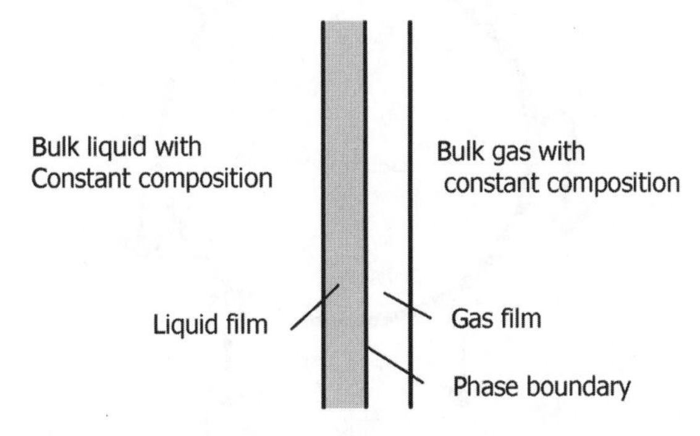

FIGURE 15.4 Simplified diagram illustrating two-film theory.

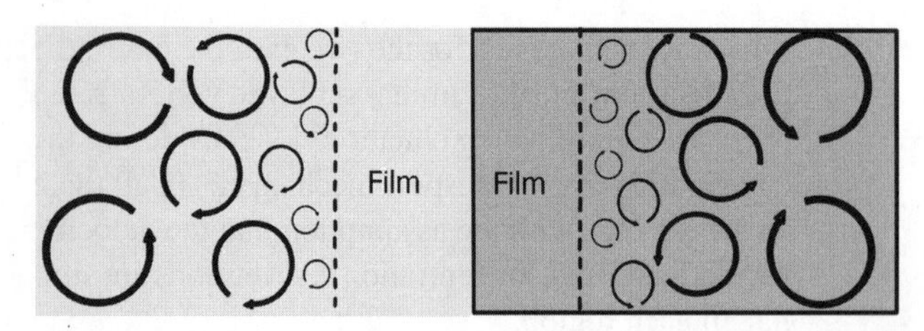

FIGURE 15.5 Schematic representation of the situation at the interface.

$$k_c \propto D_{AB}. \quad (15.39)$$

- In actual cases,

$$k_c \propto D_{AB}^{2/3}. \quad (15.40)$$

- In general,

$$k_c \propto D_{AB}^{n}, \quad (15.41)$$

where values of n are ranging from 0.5 to 2/3.

- For laminar flow conditions, film theory might represent, to an approximation, close prediction of mass transfer coefficients.
- For turbulent conditions, film theory gives results far from realistic.

- Higbie's penetration theory.
  - Time of exposure to mass transfer is short, so concentration gradient of film theory, characteristic of steady state would not have time to develop.
  - For example, rising gas bubble in a liquid (Figure 15.6): Particle of liquid, initially at top of bubble, is in contact with gas for time θ required for the bubble to rise a distance equal to diameter of the bubble.

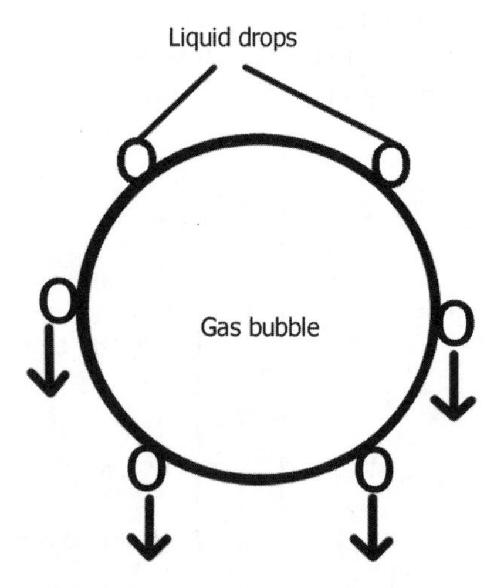

FIGURE 15.6 Rising gas bubble in liquid.

  - Time of exposure of eddies is assumed to be constant, that is, θ is constant.
- Denckwerts surface renewal theory.
  - Assumption of constant exposure time for fluid elements by penetration theory is not realistic and eddies are exposed for varying lengths of time in a random way (random distribution of periods of contact).
  - Mass transfer between the gas phase and the fluid elements takes place at the interface.
  - As rate of mass transfer depends upon exposure time, average rate for unit surface area must be determined by summation of individual values.

$$K_{\text{Lav}} = \sqrt{(D_{AB}S)} \quad (15.42)$$

and rate of mass transfer,

$$N_{\text{Aav}} = (C_{\text{Ai}} - C_{\text{Ao}})\sqrt{(D_{AB}S)}. \quad (15.43)$$

  - S is the surface renewal factor/fractional rate of replacement of eddies at the surface, θ^{-1}.
  - Eddy is swept to the surface (e.g., gas bubble) and undergoes unsteady-state diffusion and then swept away to the interface and so on (Figure 15.7).
- Film–penetration theory.
  - Principles of both film and penetration theories are incorporated.
  - Whole of resistance to mass transfer is assumed to be lying within a laminar film at the interface as in two-film theory. But mass transfer is considered as an unsteady-state process.
  - It is assumed that fresh surface is formed at intervals from the fluid that is brought from bulk

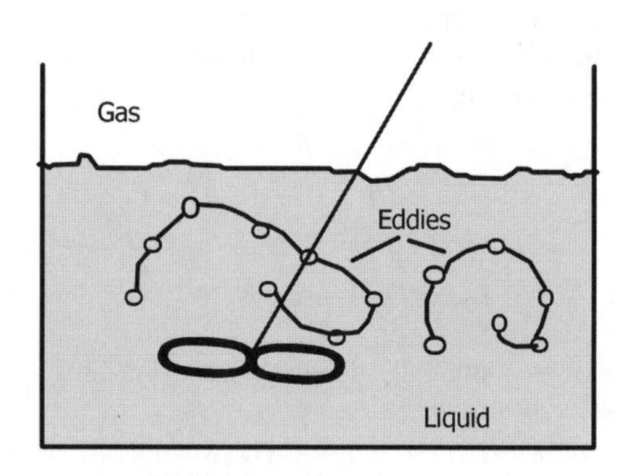

FIGURE 15.7 Gas–liquid contacting.

of the fluid to the interface by the action of eddy currents.

- Mass transfer takes place as in penetration theory, except resistance is confined to the finite film.
- Dobbins gives the following equation that represents combination of the film and penetration theories:
- $K_{Lav} = [(\sqrt{D_{AB}S})]\coth[\sqrt{(Sz_b^2/D_{AB})}], \quad (15.44)$

where z_b is the depth of penetration.

- Define enhancement factor as proposed by Denckwerts for gas–liquid mass transfer with chemical reaction.
  - Enhancement factor, I.

$$I = N_{A\ \text{with reaction}}/N_{A\ \text{without reaction}} = \{\sqrt{[D_{AB}(r+S)]}/\sqrt{(D_{AB}S)}[1+r/S]^{1/2}. \quad (15.45)$$

  - r represents reaction rate.
- What is Marangoni effect? Explain.
  - The phenomenon that liquid flows along a gas–liquid or a liquid–liquid interface from areas having low surface tension to areas having higher surface tension is named the Marangoni effect.
  - *Solutal Marangoni convection* is flow caused by surface tension gradients originating from concentration gradients, while *thermocapillarity* is flow caused by surface tension gradients originating from temperature gradients.
  - The Marangoni convection increases the mass transfer coefficient with respect to pure diffusive mass transfer. Furthermore, the Marangoni effect can also influence the shape and size of the mass transfer interfacial area.
  - Enhancement of Marangoni convection is largest for systems in which the liquid-phase mass transfer resistance is approximately equal to the gas-phase mass transfer resistance, that is, the Biot number is close to 1.
  - Large interfacial gradients may result in rapid motion of liquid at the interface, creating interfacial turbulence, called Marangoni effect. For example, Ether–water system involve large interfacial gradients.
- What is Marangoni number?
  - *Marangoni number* (Mg) is a dimensionless number.
  - Marangoni number may be regarded as proportional to (thermal) surface tension forces divided by viscous forces. It is, for example, applicable to bubble and foam research and propellant behavior calculations in spacecraft tanks.

$$\text{Mg} = -(d\sigma/dT)(1/\eta\alpha)L\Delta T. \quad (15.46)$$

$$\Sigma = \text{surface tension, N/m}. \quad (15.47)$$

L is the characteristic length, m, α is the thermal diffusivity, m^2/s, η is the dynamic viscosity, kg/(s m), and ΔT is the temperature difference, °C.

- What is Rayleigh number and Rayleigh convection?
  - Rayleigh number is defined as the product of the Grashof number, which describes the relationship between buoyancy and viscosity within a fluid, and the Prandtl number, which describes the relationship between momentum diffusivity and thermal diffusivity.
  - For free convection near a vertical wall, this number is

$$Ra_x = Gr \cdot \text{Pr} = (g\beta/\upsilon\alpha)(T_s - T_\infty)x^3, \quad (15.48)$$

where Ra_x is the Rayleigh number, Gr_x is the Grashof number, Pr is the Prandtl number, g is the acceleration due to gravity, x is the characteristic length (in this case, the distance from the leading edge), T_s is the surface temperature (temperature of the wall), T_∞ is the Quiescent temperature (fluid temperature far from the surface of the object), ν is the kinematic viscosity, α is the thermal diffusivity, and β is the thermal expansion coefficient.

  - Buoyancy, or Rayleigh convection, can occur at the same time as Marangoni convection.
  - The same concentration and temperature gradients that are responsible for surface tension gradients, also create density gradients. The intensity of the density driven convection is characterized by the Rayleigh number.
  - Usually, the Rayleigh effect dominates the flow in liquid layers with dimensions larger than 1 cm. The Marangoni effect usually dominates when the characteristic dimension is smaller than 1 mm.
- What are the causes for interfacial resistance for mass transfer?
  - Interfacial turbulence is attributed to hydrodynamic instability caused by fluctuations in interfacial tension associated with mass transfer across the interface.
  - Surface active agents concentrated at interface may
    - partially block interface for solute transfer;
    - make interfacial liquid layers more rigid;
    - interact with solute.
  - Net effect is reduction of mass transfer rates. For example, addition of hexadecanol to open ponds of water substantially reduce evaporation rates.

15.7 MASS TRANSFER WITH CHEMICAL REACTION

- Name some important areas of application of mass transfer with chemical reaction.
  - *Gas–Liquid Systems:*
    - Gas absorption and stripping for the removal gaseous components from a mixture of gases. Examples include CO, CO_2, H_2S, Cl_2, SO_2, SO_3, HCl, and so on.
    - Manufacture of products such as H_2SO_4, HNO_3, nitrates, phosphates, and so on.
    - Carrying out liquid-phase reactions such as oxidation, halogenation, polymerization, sulfonation, hydrogenation, and so on.
    - Water pollution control processes such as aerobic fermentation of sludges, biological waste treatment, and so on.
  - *Liquid–Liquid Systems:* Chemical reactions are used in liquid–liquid systems to achieve the following objectives:
    - Nitration and sulfonation of aromatics, alkylation and hydrolysis of esters, oxidation of cyclohexane, extraction of metals, and so on, in which chemical reaction is may be part of a process.
    - Removal of acidic solutes from hydrocarbon systems in which chemical reactions are introduced deliberately.
    - Yields and rates of formation of many single-phase reactions are increased by the controlled addition of an immiscible extractive phase to extract a product from the reactive liquid-phase *extractive reactions.*
  - *Gas–Liquid–Solid Systems:*
    - The solid may be reactive as in the cases of absorption of CO_2 in limestone and thermal coal liquefaction.
    - The solid functions as a catalyst, for example, when gases like hydrogen, water, ammonia, or oxygen are involved resulting in the hydrogenation, hydration, amination, or oxygenation.
- What are the advantages and attributes to be incorporated in equations for mass transfer with chemical reaction?
  - The main objective of coupling mass transfer and chemical reaction is to enhance overall mass transfer rates. Chemical reactions enhance the mass transfer rate, because they take out the diffusing solute in the region of the interface producing a steeper concentration gradient. This has very important implications for the design of gas absorbers and many mass transfer devices.
  - The general equation of mass transfer when accompanied by a chemical reaction is an unsteady-state mass transport equation, that incorporates not only diffusion but also convective mass transport and chemical reaction contributions.
    - The general equation can be written as

$$\partial C_i/\partial t + v\nabla C_i = D_i\nabla^2 C_i + R_i, \qquad (15.49)$$

      where v is the fluid velocity vector, $v\nabla C_i$ is the convective mass transport contribution, $D_i\nabla^2 C_i$ is the molecular diffusion contribution, R_i is the chemical reaction contribution, and D_i is the diffusion coefficient of i in the liquid.
    - For steady state,

$$D_i\nabla^2 C_i = 0. \qquad (15.50)$$

    - If the medium is stationary,

$$v = 0. \qquad (15.51)$$

    - The mechanism of steady-state diffusion with reaction distributed homogenously throughout the material is often adopted in the analysis of cases of mass transfer and chemical reaction.
    - There are, however, some reactions for which the distinction between homogenous and heterogeneous systems is not sharp as in the cases of
      - enzyme–substrate reactions
      - very rapid chemical reactions
  - In multiphase systems, chemical reactions affect the mass transfer rate in two distinct ways:
    - At low reaction rate, it serves to change the bulk concentration of the transferring solute, thus increasing the driving force.
    - On the other hand for reasonably fast reactions the concentration gradient near the interface is affected leading to an *enhancement* of the mass transfer rate.
  - It is convenient to represent the effect of the chemical reaction on mass transfer in terms of an *enhancement factor*.
  - The enhancement factor can be defined as the ratio of the average rates of mass transfer in presence of reaction to the average rate without reaction and is given by the equation:

$$\phi = k_L / k_L^0, \quad (15.52)$$

where k_L^0 and k_L are the mass transfer coefficients for physical mass transfer and for mass transfer accompanied by a chemical reaction, respectively.

- Relative rate of reaction and diffusion can be expressed by the dimensionless ratio,

$$\Phi = t_D / t_R, \quad (15.53)$$

where t_D is the measure of the time available for the molecular diffusion phenomena to take place before mixing of the liquid phase makes the concentration uniform. It, thus, decreases as the mixing or turbulence of the liquid phase is increased. The reaction time, t_R is the measure of the time required by the chemical reaction to proceed to the extent of changing by a significant amount the concentration of the limiting reactant.

- Three cases arise based on the magnitude of Φ:
  - $\Phi \ll 1$: The reaction is too slow to have any significant influence on the diffusion phenomena and essentially no rate enhancement will take place. The enhancement factor $\phi \approx 1$. This situation is referred to as the *slow reaction regime*. In this regime the chemical reaction only has the effect of keeping the concentration of solute low (i.e., a larger driving force). For this type of reactions, bubble columns and agitated reactors are recommended.
  - $\Phi \to \infty$: The limit of an infinitely fast reaction, *instantaneous reaction regime* is attained when all the resistance to mass transfer due to chemical reaction has been eliminated. At this point the enhancement factor,ϕ, for the instantaneous reaction will be very large. Values of ϕ are typically of the order 10^2–10^3. For this type of reactions, packed and plate columns or Venturi scrubbers are recommended.
  - $\Phi \gg 1$: The reaction is fast enough to result in a significant rate enhancement, yet it is not fast so as to be instantaneous. This intermediate region is referred to as the *fast reaction regime*. Agitated tanks, ejectors and plate columns are recommended for this type of reactions.
- For mass transfer involving chemical reactions, in addition to the above considerations, there can be considerable heat effects, interfacial turbulence, interfacial resistance, bulk flow effects, effects of products of reaction are to be taken into consideration.

15.8 DROPS, BUBBLES, SPRAYS, MISTS, AEROSOLS, AND FOAMS

- What is the importance of drops, bubbles, sprays, mists, and foams in mass transfer processes?
  - Mass transfer involves transfer of species (components) from one phase to the other through phase boundaries, that is, interface. The larger the interfacial area, the more will be the chances for the migration of the species through the interface.
  - Large interfacial areas are created by breaking up the phases into small droplets and bubbles.
  - A variety of ways are involved in the breakup of the phases, which include spraying, precipitation and condensation, agitation, emulsification, foam formation, and so on.
  - Though increasing interfacial area increases mass transfer, there are some opposing factors reducing overall transfer that include retarding phase separations once the phases are brought into intimate contact. Examples include coalescence, settling, entrainment separation, control of mist formation, emulsification and demulsification, foam destruction, and so on.
- Give examples of processes involving bubbly flows.
  - Oxygen supply to biological systems.
    - Aeration of lakes.
    - Bioreactors and fermenters.
    - Bioremediation plants.
  - Wet scrubbers.
    - Fly ash and particulate pollutant removal.
  - Bubble columns and airlift reactors.
    - Hydrogenations and carbonylations.
    - Fischer–Tropsch synthesis.
- Illustrate the relationships among droplet size, surface area, and droplet count.
  - Reducing droplet size by half produces eight times increase in number of droplets and twice of total surface area. Table 15.3 gives relationships among drop size, surface area, and drop count.
- Define (i) aerosol, (ii) mist, and (iii) spray.
  - *Aerosol:*
    - Suspended particulate system, either solid or liquid, which is slow to settle by gravity.
    - Particles range from submicron size to 10–20 μm.
  - *Mist:* Fine suspended liquid dispersions usually resulting from condensation and ranging upward in droplet size from around 0.1 μm.
  - *Spray:* Entrained liquid droplets. Droplet sizes are relatively large ranging from 5000 μm and less.

TABLE 15.3 Relationships Among Drop Size, Drop Surface Area, and Drop Count

Volume and Surface Area for Different Droplet Diameters

Droplet Diameter (μm)	Surface Area of a Single Droplet (mm^2)	Volume of a Single Droplet (mm^3)	No. of Droplets per Liter	Total Surface Area per Liter per m^2
2,000	12.6	4.19	239,000	3
1,000	3.14	0.524	1,910,000	6
500	0.785	0.0655	15,300,000	12
250	0.196	0.00819	122,000,000	24
125	0.0491	0.00102	977,000,000	48
60	0.0113	0.000113	8,840,000,000	100
30	0.00283	0.00283	70,700,000,000	200
15	0.000707	0.000707	565,000,000,000	400

Source: Welander P, Vincent TL. *Chemical Engineering Progress*, June 2001, p. 76.

- What are the size ranges for the following particulates?
  - Viruses: 0.003–0.08 μm.
  - Bacteria: 0.3–30 μm.
  - Large molecules: <0.01 μm.
  - Fumes: Up to 1 μm.
  - Smog: Up to 1 μm.
  - Fog and Clouds: 1–100 μm.
  - Mists: Up to 10 μm.
  - Sprays: 10 to >1000 μm.
  - Rain: 100–5000 μm.
  - Dusts: 1 to >1000 μm.
- What is the main difference between a spray and mist?
  - When liquid droplets form in gas streams from gas–liquid contacting equipment, a spray or mist is obtained.
  - If the droplets range in size from 10 to over 1000 μm, it is called a spray (1 μm = 0.001 mm).
  - If the droplets range in size from 10 to well below 1 μm, it is called a mist.
  - Droplets in the size range of 1–100 μm are called clouds/fog.
  - Droplets in the size range of 0.001–0.1 μm are called fumes.
- How are sprays and mists formed?
  - Sprays are formed through mechanical means, for example, by the use of nozzles.
  - Mists are generally formed by shock cooling of condensable vapor.
  - Acid mists are formed by gas-phase reaction of water vapor and SO_3 and subsequent shock cooling of H_2SO_4 vapor.
- What are the mechanisms involved in the formation of sprays and mists?
  - The mechanisms for formation of sprays and mists are quite different.
  - Sprays are formed by one or more of the following primary mechanisms:
    - Breakup of liquid jets in gas streams.
    - Breakup of *roll waves* caused by gas flow over liquid films.
    - Bursting of gas or vapor bubbles at gas–liquid interfaces.
  - A secondary mechanism that affects the size distribution of sprays is the subsequent breakup of the liquid droplets formed by the primary mechanisms and by collision with the solid surfaces in the gas–liquid contacting equipment.
- What is the difference between a bubble and foam?
  - Bubble is a globule of gas or vapor surrounded by a mass of liquid or a film of liquid.
  - Foam is a group of bubbles separated from one another by thin liquid films. It is dispersion of a gas in a liquid or solid.
  - Colloidal system containing large volumes of dispersed gas in a relatively small volume of liquid is an example of a foam.
- What is the difference between a foam and froth?
  - In a froth, gas bubbles are round in shape whereas in a foam, they are polyhedral in shape.
  - Froths are temporary dispersion of small gas bubbles in a liquid of no foaming ability. Froths collapse as spherical gas bubbles, move through the liquid and coalesce.
  - Foams are generated via froths. Liquid film between two bubbles thins to a lamella rather than rupturing as the bubbles approach.
- Give examples of processes involving formation of sprays/mists in condensation systems.
  - Condensers, condensate receivers, and so on.
  - Condensation occurs when two vapors react to form a liquid product.

  - For example, unabsorbed SO_3 in acid plants reacts with water vapor to form liquid H_2SO_4 mist.
  - H_2SO_4 in vapor form gives rise to very fine acid mist on cooling.
- Heat developed in compressors may vaporize some lubricating oil and on cooling gives rise to fine oil mist.
- Cooling of superheated gas fractions in distillation operations produce mists of extremely fine sizes (>1 μm).

- How are foams destroyed?
  - *Mechanical Methods*: Static or rotating breaker bars or slowly revolving paddles. → Foams are deformed causing rupture of lamella walls.
  - *Pressure and Acoustic Vibrations*: Pressure changes in a vessel containing foam, stress lamella walls by contracting and expanding gas in foam bubbles. Acoustic vibrations have similar effect.
  - *Thermal Methods*:
    - Heating greatly reduces surface viscosity of the liquid film and changes form from slow draining to fast draining foam.
    - Heating causes expansion of walls of foam leading to deformation and collapse.
    - Cooling to freezing temperatures destroys foams.
  - *Electrical Methods*:
    - Most foams have electrical double layers of charged ions.
    - Can be broken by electrical fields.
  - *Chemical Methods*:
    - Cause desorption of stabilizing agents from interface.
    - Chemical changes in adsorption areas is brought about by the use of chemicals like silicone oils, alcohols, fatty acid soaps, phosphates, and so on.

- What are the characteristic features of drops and bubbles?
  - The motion and mass transfer of bubbles and drops are closely related to the shapes of their surfaces, and the latter are quite sensitive to the conditions of flow as well as to the physical properties of the systems. In this regard, they are quite different from those of solid particles.
  - For small bubbles or droplets the surface tension is dominant at the interface, and as a result they tend to be spherical in shape. For intermediate sized bubbles and drops, however, the shapes are deformed and sometimes becomes unstable.
  - Figure 15.8 shows schematic pictures of bubbles in the liquid phase. Small bubbles of diameter <1 mm

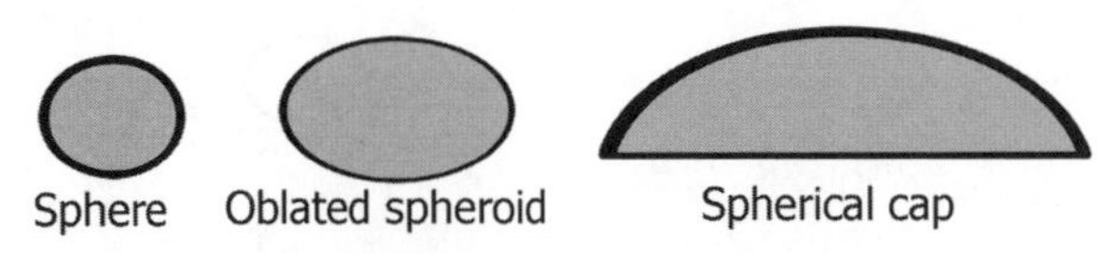

FIGURE 15.8 Shapes of bubbles.

are usually spherical in shape, whereas bubbles of intermediate size tend to deform from spherical to oblate spheroid, with the minor axis of the ellipse as the axis of rotation, as the bubble size increases. Larger bubbles are even more deformed to mushroom-like shapes, and their shapes are quite unstable and may even be oscillating.
  - In an ascending motion, small bubbles move straight upward, medium-sized bubbles ascend in a spiral motion, while large bubbles tend to oscillate in an irregular manner during their ascent.
  - Drops in the liquid phase of diameter <1 mm are spherical, whereas medium sized drops of several millimeters in diameter tend to deform from spherical to oblate spheroid.
  - Drops in the gas phase are rather stable in comparison with those in the liquid phase and are nearly spherical in shape if their sizes are less than a few millimeters in diameter.

- Why there is deformation in drops and bubbles?
  - Due to differences in pressures acting on various parts of the surface.
- Why drops tend to flatten on their forward faces?
  - Because impact and hydrostatic pressures will be greater on forward face than on rear face.
- Why there is elongation on the rear side of a drop?
  - Due to viscous drag.
- Why small drops tend to be spherical?
  - Deformation is opposed by surface tension forces, leading to retention of spherical shapes for small drops.
- What are the various stages of bubble collapse and droplet formation?
  - Bubble coming to rest at liquid surface forms a hemispherical dome, its internal pressure producing a depression of the interface (Figure 15.9a).
  - Liquid drains from the dome until the upper part is so weakened that the internal pressure causes formation of a second cap (Figure 15.9b).
  - The second cap subsequently disintegrates giving rise to droplets of few microns in size. These droplets are carried away by the rush of gas issuing from the perforated dome (Figure 15.9c).
  - The result is a system of standing waves and to leave a well-defined crater in the interface (Figure 15.9d).

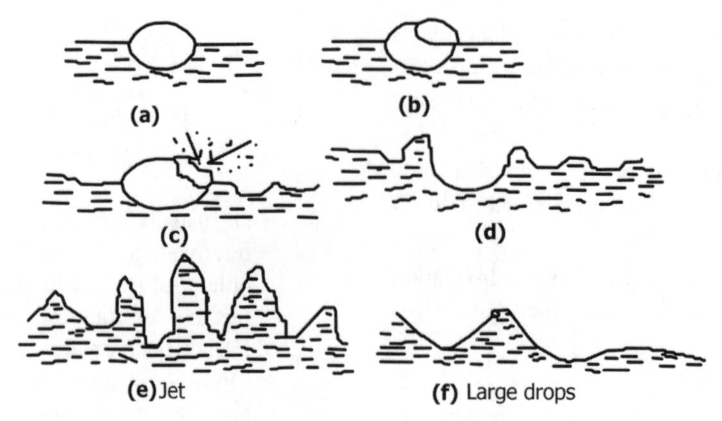

FIGURE 15.9 Bubble collapse and droplet formation phenomena.

- As the crater fills in, momentum of the inflowing liquid produces a jet, which rises at a high velocity and sometimes detaches one or more large drops from its apex (Figure 15.9e and f).
- Figure 15.9 illustrates the above-described phenomena.
- These large drops are responsible for the main entrainment losses.

• What are the characteristics of bubbles and drops with respect to *dispersed phase* and *continuous phase* mass transfer?

- Bubbles and drops are referred to as the *dispersed phase*, while the fluid containing the *dispersed phase* is referred to as the *continuous phase*.
- Since heat and mass transfer in the dispersed phase and those in the continuous phase are affected by quite different factors and mass transfer resistances of bubbles and drops exist on both sides of the interface, one must be careful in deciding on which side the resistance constitutes the rate-controlling process.
  - ➢ For example, the evaporation of drops in the gas phase is continuous phase mass transfer resistance controlled, whereas absorption by raindrops is dispersed phase controlled.
  - ➢ Absorption by bubbles in the liquid phase is usually continuous phase controlled.

• What is Brownian motion?

- Molecules of fluid bombard particles in a random manner.
- For very small particles, net resultant force acting at any instant may be large enough to cause a change in its direction of motion.

• What are the common applications of spray nozzles? Explain.

- *Humidification/Dehumidification:* By atomizing a liquid as fine droplets to ensure complete evaporation. If temperature of the liquid is below dew point of the surrounding gas, dehumidification takes place.
- *Evaporative Cooling:* Spraying a liquid into a hot gas uses sensible heat of the gas, warming up the droplets and evaporation from their surface. Evaporation cools the droplets further by taking heat of vaporization from them.
  - ➢ Sensible heat transfer is very low compared to the latent heat transfer and hence calling the cooling process evaporative cooling.
  - ➢ Droplet size must be small enough to cause complete evaporation of the liquid before the droplet strikes a solid wall.
  - ➢ Examples include spray drying, cooling of hot gases from a rotary kiln upstream of a bag house, cooling water in a cooling tower or spray pond, and so on.
- *Spray Drying:* Dividing liquid solution into fine droplets into a hot gas/air. Evaporation takes place leaving dried particulates. For example, spray drying of milk to produce milk powder, producing soluble coffee powder, and so on.
- *Chemical Injection:* Liquid is atomized and vaporized and reacted with another gas. For example, spray injection of aqueous ammonia into a combustion gas to control NO_x emissions.
- *Two-Phase Reaction:* Gas–liquid reactions without evaporating the liquid. For example, using limestone solution in acid gas scrubbing.
- *Gas Separation:* Liquid injected into a gas stream with the purpose of absorbing a solute gas component. For example, removal of ammonia gas by water sprays.
- *Spray Painting:* By spraying a paint onto a surface to be painted produces a smooth finish to the painted surface compared to brush painting.

- How are droplets formed in nozzle spraying?
  - Interaction between air resistance and surface tension breaks the sprays into individual droplets.
  - Large droplets break up into smaller ones if surrounding air resistance overcomes surface tension.
- What is the effect of surrounding gas pressure on droplet size in nozzle spraying?
  - If spraying is carried out into a vessel in which gas is under compression, smaller droplets result due to the increased density and resistance of the gas.
  - Spraying into vacuum results in larger droplets.
- What is the effect of spray velocity on droplet size?
  - Higher velocities produce smaller drops.
  - Injecting compressed gas into the liquid stream or injecting liquid into a high velocity gas stream produces very fine droplets.
- What are the differences with respect to sprays produced by pressure type and two fluid- or gas-atomized nozzles?
  - *Pressure Type:* Relatively coarser sprays ($100\,\mu m > 20\,\mu m$).
  - *Two Fluid Type:* With the available pneumatic energy, liquid breakup gives rise to finer droplets ($<30\,\mu m$).
    - For $<10\,\mu m$, high gas pressures and high gas to liquid mass transfer rates are required.
    - High power requirements.
    - Uneconomical for high capacities.
- Give examples of processes involving formation of sprays/mists in condensation systems.
  - Condensers and condensate receivers.
  - Condensation that occurs when two vapors react to form a liquid product. Example is unabsorbed SO_3 in acid plants reacts with water vapor to form acid mist. H_2SO_4 in vapor form gives rise to very fine acid mist on cooling.
  - Heat developed in compressors may vaporize some lubricating oil, giving rise to fine oil mist on cooling.
  - Cooling of supersaturated gas fractions in distillation operations give rise to mists of extremely fine sizes of less than $1\,\mu m$.
- What are the droplet/particle size ranges for which the following equipment are useful?
  - Fiber bed mist eliminators: Size ranges 0.01– 100 μm.
  - Electrostatic precipitators: 40 μm.
  - Gravity settling chambers: >80 μm.
  - Centrifugal separators: >0.1–1000 μm.
  - Wet scrubbers: 0.1–100 μm.
  - Impingement separators: >8 μm.
- What are microbubbles? What are the prospects of their advantages and applications?
  - Microbubbles are very small bubbles with sizes of the order of 20 μm, in contrast to conventional fine bubbles with size ranges of 1–3 mm.
  - The microbubbles, because of their much higher surface area per unit volume, can achieve 50-fold greater mass transfer rates, taking up to 18% less energy to produce, according to a recent claim.
    - They might result in enhancement of heat and mass transfer in many gas/liquid systems where bubbles are used, keeping materials in suspension or separating them by flotation processes.
    - Wastewater purification is claimed to be one of the promising applications.
    - Fluid oscillation techniques, involving a series of pulsations at frequencies of 1–100 Hz that force gas through microporous diffusers, are used in generating microbubbles.

16

MASS TRANSFER EQUIPMENT

16.1 Mass Transfer Equipment—General Aspects 475
- 16.1.1 Tray Columns 477
  - 16.1.1.1 Tray Efficiencies 483
  - 16.1.1.2 Entrainment 485
  - 16.1.1.3 Flooding 488
  - 16.1.1.4 Diameter for Tray Columns 491
- 16.1.2 Packed Columns 493
  - 16.1.2.1 Packed Column Internals 496
    - 16.1.2.1.1 Packings 496
    - 16.1.2.1.2 Structured Packings 498
    - 16.1.2.1.3 Other Internals 502
  - 16.1.2.2 Packed Versus Tray Columns 509
  - 16.1.2.3 Packed Column Diameter 511
  - 16.1.2.4 Operation and Installation Issues 513
- 16.1.3 Reflux Drums and Vapor/Gas–Liquid Separators 515

16.1 MASS TRANSFER EQUIPMENT—GENERAL ASPECTS

- What are the basic attributes of mass transfer equipment?
  - Large interfacial area/unit volume is obtained by breakup of liquid into bubbles, drops, sprays, films, foams, and so on.
  - High degree of turbulence is obtained by intense mixing and high flow velocities.
  - Low ΔP is obtained by proper selection of internals.
  - Low liquid holdup.
  - Ease of separation of the phases.
  - Ease of maintenance.
  - Cost effective.
- Name different types of mass transfer equipment.
  - Wetted wall columns.
  - *Spray and Bubble Columns:* Gas–liquid contact.
    - Spray columns employ several levels of spray nozzles that should be placed strategically so that the entire cross-sectional area is covered creating dense spray zones through which the gas must pass. Hollow cone nozzles are typically employed because the droplets produced by them are small enough to provide good interfacial area and large enough to prevent entrainment.
    - Spray columns are used for applications where pressure drop is critical, such as flue gas scrubbing.
    - They are also useful for slurries that might plug packing or trays.
    - Generally, they are not suitable for applications requiring several stages of contact or a close approach to equilibrium.
  - Figure 16.1 illustrates a typical spray column.
  - Bubble columns are used more frequently as reactors than as absorbers. When used as a reactor, the gas that is bubbled through the continuous liquid phase is normally one of the reactants that must be absorbed before it can react. Advantages of bubble columns include low maintenance due to the absence of moving parts, high heat transfer rates, ability to handle slurries without plugging, or erosion and long residence time for the liquid, which permits slow reactions to proceed.
  - One disadvantage compared to tray and packed columns is backmixing.
  - *Venturi Scrubbers:* For absorption involving dirty fluids.
    - Venturi scrubbers entrain large volumes of gas. Scrubbing liquid is sprayed through a nozzle and creates a draft that draws the gas into the moving stream. Different types of nozzles are used to spray the liquids. Hollow cone nozzles are usually employed for the purpose, because of the droplet sizes obtainable and resistance to blockages.
  - *Tray and Packed Columns:* Most common.

Fluid Mechanics, Heat Transfer, and Mass Transfer: Chemical Engineering Practice, By K. S. N. Raju

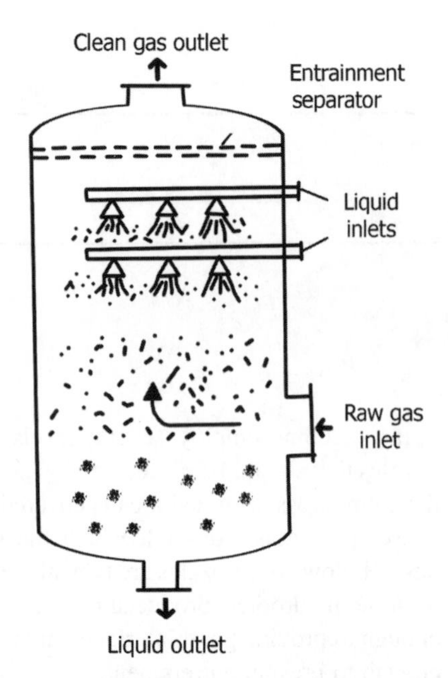

FIGURE 16.1 Schematic diagram for a spray column.

  - ➢ For absorption in packed columns, large interfacial area is provided by the use of full cone spray nozzles, which provide good spray coverage and distribute the liquid evenly through the packing. For difficultly soluble gases redistributors over multiple packed beds may be necessary.
- *Mixer–Settler Units:* Liquid–liquid extraction.
- *Centrifugal Equipment:* Liquid–liquid extraction involving phases with low ΔP.
- Cooling towers, rotary dryers, spray dryers, tray and tunnel dryers, and other types.
- *Special Types:* Solid–liquid extractors, adsorbers, membrane units, ion exchangers, and so on.

• What is a Venturi scrubber? How does it work? What are its applications?
  - A Venturi scrubber employs a gradually converging and then diverging section of a conduit to clean incoming gaseous streams.
  - Liquid is either introduced to the Venturi upstream of the throat of the conduit or injected directly into the throat where it is atomized by the gaseous stream. Once the liquid is atomized, it collects particles from the gas and discharges from the Venturi.
  - Venturi scrubbers are generally used for controlling particulate matter and sulfur dioxide.
  - They are designed for applications requiring high removal efficiencies of submicron particles, between 0.5 and 5.0 μm in diameter.
  - The high ΔP through these systems results in high energy use.
  - They are not generally used for the control of volatile organic compound (VOC) emissions in dilute concentration.
  - Venturi scrubbers are able to achieve a high degree of liquid–gas mixing, but have the disadvantage of a relatively short contact time that generally leads to poor absorption efficiency. However, for gases with high solubilities and proper selection of the scrubbing liquid, the Venturi can be a good choice.
  - Figure 16.2 illustrates essential details of a Venturi scrubbers.

• What are the plus and minus points in the use of Venturi scrubbers?
  - Low initial cost.
  - No blockages.
  - Very compact in size.

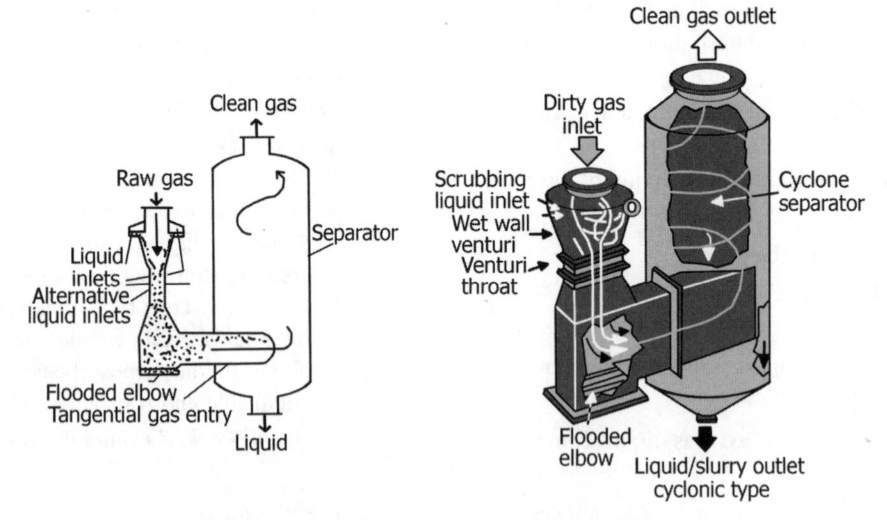

FIGURE 16.2 Details of Venturi scrubbers.

- Can remove both gases and particulates.
- High-energy requirements:
  - Energy is required primarily to operate fans to accelerate the gas.
  - Energy requirements are substantial as these operate at 40–50 cm (15–20 in.) water ΔP.
  - Typically 448–522 kW (600–700 hp), 1360–1415 sm^3/min (48,000–50,000 sf^3/min), and 46–51 cm (18–20 in.) pressure differentials are required.
- Short contact times between gas and liquid making it not attractive for difficultly soluble gases that require large contact times.

• What are the common problems involved in the use of spray nozzles in gas–liquid contact equipment?
  - *Erosion:* Gradual wear causes the nozzle orifice and internal flow passages to enlarge or become distorted. This results in increased flows and decreased pressure, creating irregular flow patterns and increased droplet sizes.
  - *Corrosion:* The effects are similar to erosion, with possible additional damage to the outside surface of the nozzle.
  - *Clogging and Caking:* Contaminant particles can cause blockages inside the nozzle orifices, restricting flows and disturbing uniformity of the sprays.
  - *High Temperatures:* Certain liquids must be sprayed at high temperatures and high temperature environments. In such cases, the nozzles can soften and distort unless special temperature-resistant materials of construction are used. This is of particular issue of concern in spray drying applications where temperatures in the dryer can be as high as 250°C.
  - *Improper Assembly/Reassembly or Damage:* Some nozzles require careful assembly or reassembly after cleaning so that internal gaskets, *o*-rings and valves are properly aligned as improper assembly can cause leakages and inefficient performance.

16.1.1 Tray Columns

• What is turndown ratio?
  - Ratio of minimum allowable throughput to operating throughput.

• Name the important internals of a tray column.
  - Internals of a tray column: Trays, downcomers, and weirs.

• What are the factors that influence selection of distillation column internals?
  - Maximum and minimum temperatures.
  - Fouling characteristics of the process fluids.
  - Reactive and corrosive characteristics of the fluids.
  - Sensitivity to contamination.
  - Hydrodynamic requirements.

• Name different types of trays and compare their characteristics.
  - *Trays:*
    - Bubble cap, sieve, valve, jet, v-grid or extruded valve caps, flexi, ripple, turbogrid, and other types.
    - Commonly used trays are sieve, valve, jet, v-grid, or extruded valve caps and bubble cap types.
    - Sieve and valve trays are comparable in capacity, efficiency, entrainment characteristics, and ΔP.
    - For valve trays turndown is much better than that for sieve trays. Turndown for valve trays is in the range of 4:1–5:1.
    - Valve trays are slightly more expensive than sieve type. Better energy efficiency. Justifies the low cost difference between sieve and valve trays. Most common in use.
    - Bubble cap type have best turndown characteristics and most suitable for extremely low liquid rates. Have lower capacity and efficiency, higher entrainment, and higher ΔP. Most expensive. Most troublesome to maintain.
    - Sieve trays are best suited for most services. Suited when fouling or corrosion are expected and if turndown capability is not of importance. Turndown ratio is poor (2:1).
  - Figure 16.3 illustrates a typical bubble cap.
    - Sieve trays are easy to maintain and least expensive. Have poor turndown. Hole diameters range from less than a millimeter to about 25 mm.

• What are the types of valve trays? Illustrate.
  - Valve trays can be fixed or floating types (Figure 16.4).
  - Flow area is varied in a floating type valve depending on flow variations during operation, thus providing

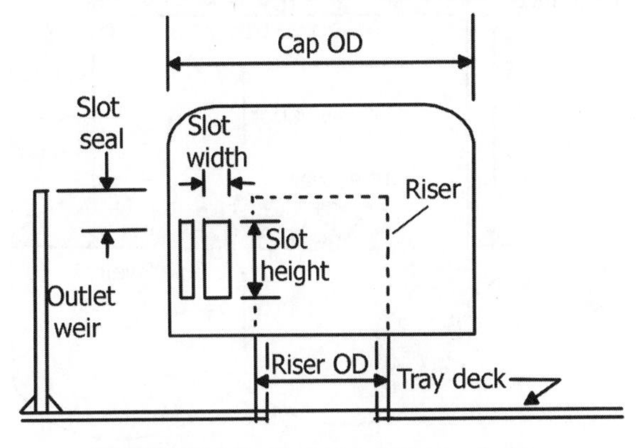

FIGURE 16.3 Typical bubble cap design.

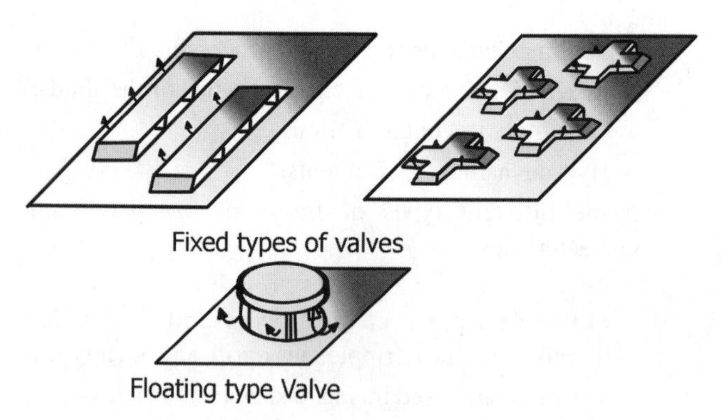

FIGURE 16.4 Types of valves used on valve trays.

flexibility for flow changes. More expensive than fixed type.

- What is net area (free area) in a tray column?
  - Total column cross-sectional area *minus* area at top of the downcomer (smallest area available for vapor flow in the intertray spacing).
  - Figure 16.5 illustrates a tray indicating different parameters such as free area, downcomer height and width, weir details, and tray spacing.
- State rules of thumb used in the selection and design of distillation columns.
  - The three most common types of trays are valve, sieve, and bubble cap.
  - Bubble cap trays are typically used only when a liquid level must be maintained at low turndown ratio. They can be designed for lower pressure drop than either sieve or valve tray.

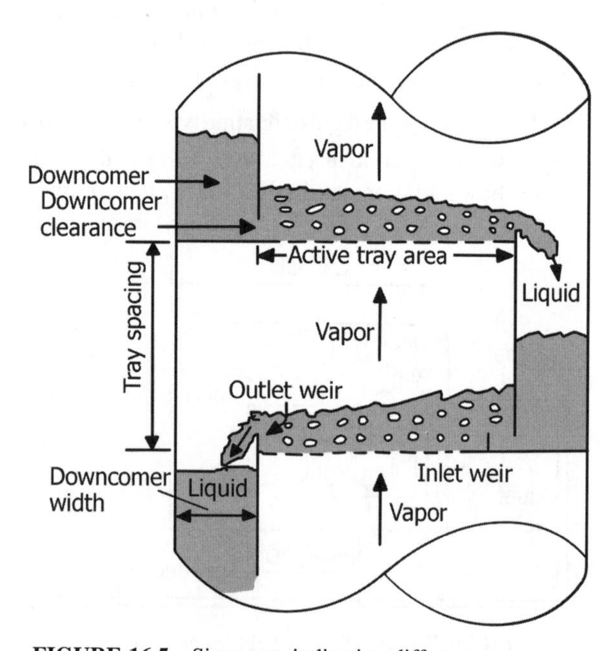

FIGURE 16.5 Sieve tray indicating different parameters.

  - Sieve tray holes are 0.6–0.7 cm (0.25–0.50 in.) diameter with the total hole area being about 10% of the total active tray area. Keeping fraction of open area around 5% is sometimes recommended. Smaller hole areas are preferred provided fouling is not a problem.
  - Valve trays usually are slightly expensive than sieve trays.
  - Valve trays typically have 3.8 cm (1.5 in.) diameter holes each provided with a lifting cap.
    - 130–150 caps/m^2 (12–14 caps/ft^2) of tray is a good benchmark.
    - Trays with small fixed valves can be used with success.
    - Such designs have a relatively large open area, but a smaller perforation size compared to larger fixed or floating valves.
    - Peak tray efficiencies occur at linear vapor velocities of 0.6 m/s (2 ft/s) at moderate pressures or 1.8 m/s (6 ft/s) under vacuum conditions.
    - Tray efficiencies for aqueous solutions are usually in the range of 60–90% while gas absorption and stripping typically have efficiencies closer to 10–20%.
    - The most common weir heights are 5 cm (2 and 3 in.) and the weir length is typically 75% of the tray diameter.
    - Maximum liquid rates on weirs are about 1.2 m^3/(min m) (8 gpm/in.) of weir length.
    - Multipass arrangements are used when liquid rates are higher (>1.2 m^3/(min m) of weir length).
  - For columns that are at least 0.9 m (3 ft) in diameter, 1.2 m (4 ft) should be added to the top for vapor release and 1.8 m (6 ft) should be added to the bottom to account for the liquid level and reboiler return.
  - Column heights should be limited to 53 m (175 ft) due to wind load and foundation considerations.
  - The L/D ratio of a column should not be more than 30 and preferably below 20.
  - Reflux pumps should be at least 25% overdesigned.
  - Peak efficiency of trays is at values of the vapor factor F_s $(= u\sqrt{\rho})$.
  - u is in the range of 0.3–0.36 m/s (1–1.2 ft/s) and ρ is in kg/m^3.
    - F_s establishes the diameter of the column.
    - Roughly, linear velocities are 0.6 m/s (2 ft/s) at moderate pressures and 1.8 m/s (6 ft/s) in vacuum.
  - A typical pressure drop per tray is 7.6 cm (3 in.) of water or 0.007 bar (0.1 psi).
  - For trays to function reasonably close to their best efficiency point, the dry tray pressure drop must be roughly equal to the hydraulic tray pressure drop.

- What are the considerations involved in recommending tray spacing for different operating pressures?
  - Spacing involves optimization of column height and diameter.
  - With larger tray spacing, the column height increases, while the required diameter decreases.
  - Height/diameter ratios more than 25–30 are generally not recommended.
  - The purpose of tray spacing is to remove entrainment of liquid droplets from rising vapors (For high-pressure columns it is easy to remove entrainment compared to vacuum columns.), provide for accessibility (Providing for man holes, nozzles, etc.), avoiding/reducing flooding, foam carryover, and so on.
  - Tray spacings are in the range of 450–900 mm.
  - Recommended tray spacings for the following applications (Table 16.1).
- "Tray spacing at a distillation column feed point should be more than normal spacing if vapor–liquid mixture is admitted as feed." Comment.
  - Amount of entrainment and velocities and turbulence involved will be more than normal, requiring larger spacing for disengagement of vapor bubbles from liquid portion and de-entrainment of droplets from rising vapors.
  - Feed nozzle is located half way between the two trays by increasing spacing to one spacing + diameter of the feed nozzle.
- Illustrate parallel flow and single pass trays in a distillation column.
  - Tray designs can be divided into cross-flow and parallel-flow types.
  - Figures 16.6 illustrate flow arrangements on trays.
  - Cross-flow trays are much more common in commercial practice than counterflow trays because of their higher separation efficiency, wider operating range and lower costs.
  - Parallel-flow trays, if properly designed, can provide higher efficiency that can be higher by 10% or more.
  - As the column diameter increases, the ratio of weir length to throughput decreases. Moreover, as the column diameter increases, the liquid load increases faster than the vapor load and the columns and trays begin having difficulty handling the flow of liquid, which restricts column capacity.
  - For larger diameter columns, multipass trays, that split the load, are often used for better liquid distribution. This allows half the liquid to go one way and half to move the other, which reduces the liquid load.
  - If the load is still too large, the trays can be split again into a four-pass tray, but it can become problematic because in one or two passes, there is good level of symmetry, but in four or more passes, symmetry is difficult to obtain.
  - The lack of symmetry causes maldistribution, which occurs when the liquid to vapor ratio on one pass significantly differs from that of another and causes the separation process to suffer and efficiency to be reduced, sometimes by a large magnitude.
  - Dual flow trays are ripple trays, with round slots and turbogrid trays with rectangular slots, are not very common. These have no downcomers.
- What are the causes for vapor cross-flow channeling on a distillation tray?
  - Vapor cross-flow channeling is mainly attributed to the liquid hydraulic gradients that often occur on

TABLE 16.1 Recommended Tray Spacings

Service	Recommended Spacing (mm)	Comments
Based on column pressure		
Atmospheric column	600 (24)	–
High-pressure column	450 (18)	–
Vacuum column	760 (30)	–
Based on column diameter		
$D > 3000$ mm	>600 (>24)	Larger spacing required due to support beams restricting maintenance access
$D = 1200–3000$ mm	600 (24)	Spacing sufficient for maintenance access
$D = 750–1200$ mm	450 (18)	Easy access to column wall through manways
Fouling and corrosive service	>600 (>24)	Requirement of frequent maintenance
Highly foaming systems or operation involving spray regime	≥450 (≥18)	Preferably ≥600 mm to avoid premature flooding
Column operation in spray regime	≥450 (≥18)	Preferably ≥600 mm to avoid excessive entrainment
Column operating in froth regime	<450 (<18)	Lower spacing restricts allowable vapor velocity, promoting operation in froth regime

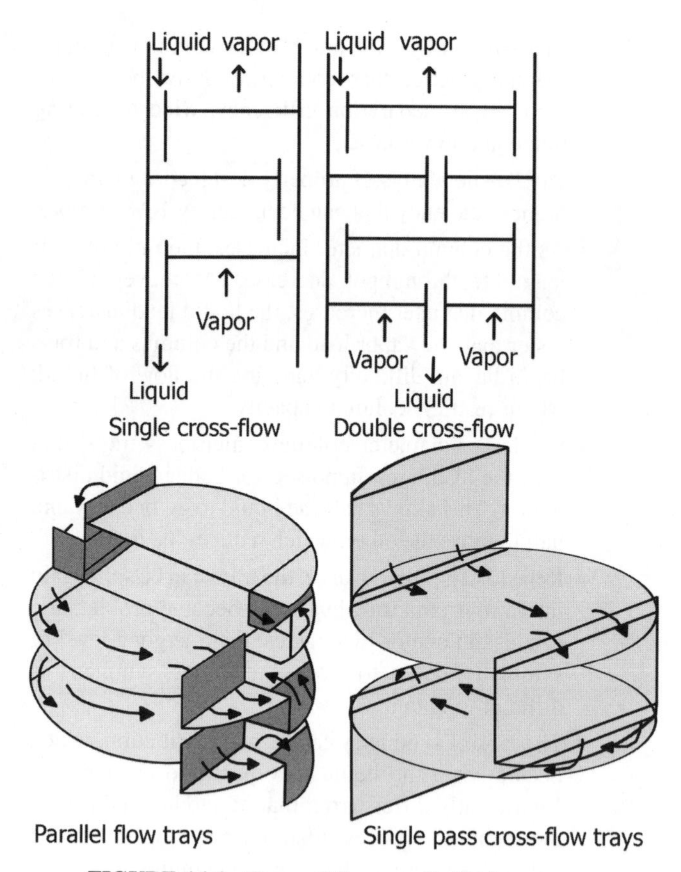

FIGURE 16.6 Types of flows on distillation trays.

trays having high liquid rates, long flow path lengths that are mainly the result of large tray diameters, short outlet weirs, elevated tray decks like with bubble cap trays, low dry pressure drops and excessive deck resistance. In response to this gradient, vapor flows preferentially through the tray outlet area.

- Clear liquid flows across an appreciable portion of the tray without any vapor contact and thus do not participate in the mass transfer process.
- Liquid weeping occurs near the tray inlet.

- What are the ways by which high liquid gradients minimized?
  - By the use of increased number of flow passes, high outlet weirs, use of sieve or low-rise valve decks and reduced deck open areas. Bubble cap trays are particularly prone to excessive gradients since the caps provide appreciable resistance to liquid flow.
- What are the criteria in the selection of number of tray passes?
  - Number of tray passes should be selected such that the liquid loads do not exceed 70–90 $m^3/(h\ m)$ of weir length.
  - After selecting the number of passes, the column diameter is adjusted in order to provide a minimum path length of 400 mm.
    - For column diameters less than 1200 mm, single pass trays should be selected.
    - For a column diameter of 1200–2100 mm, a maximum of two tray passes should be selected.
    - For a column diameter of 2100–3000 mm, a maximum of three tray passes should be selected. Three pass trays are rarely used because they are asymmetric and difficult to balance.
    - For a column diameter above 3000 mm, a maximum of four tray passes should be selected.
- What is the function of a downcomer?
  - The primary function of a downcomer is to convey liquid from a tray to the one immediately below.
    - It should provide sufficient residence time for disengagement of the entrapped vapors in the liquid being conveyed to the tray below.
- What are the recommended minimum residence times for liquid in a downcomer?
  - Table 16.2 gives recommended minimum residence times for liquid in a downcomer.
- "Highly foaming liquids require larger residence times in a downcomer." *True/False*?
  - *True*. Any significant amount of vapor traveling out from the downcomer reduces tray capacity and efficiency.
- What are the shapes of cross sections used for downcomers?
  - Circular, segmental, or rectangular shapes are used in the downcomer designs.
- What are the criteria adopted in the selection of liquid velocities in a downcomer?
  - Liquids with minimal foaming tendencies: 0.12–0.21 m/s.

TABLE 16.2 Recommended Minimum Residence Times for Liquid in a Downcomer

Nature of Liquid	Minimum Residence Time (s)
Low foaming liquids (low molecular weight hydrocarbons, alcohols)	3
Medium foaming liquids (medium mo lecular hydrocarbons)	4
High molecular weight hydrocarbons (mineral oil absorbers)	5
Very high molecular weight hydrocarbons (amines and glycols)	7

- Medium foaming liquids: 0.09–0.18 m/s.
- Highly foaming liquids: 0.06–0.09 m/s.

- What should be the downcomer area in relation to column cross-sectional area?
  - Not less than 5–8% of column cross-sectional area. Usually it is 5–30% of the column cross-sectional area, depending on the liquid load.
- What should be the downcomer width in relation to column cross section?
  - Not be less than 10% of the column diameter.
- How is downcomer clearance specified?
  - Downcomer clearance should be not less than the outlet weir height as otherwise, vapor will flow up the downcomer rather than through the tray deck above.
  - The downcomer clearance should be selected such that the liquid velocity under the downcomer does not exceed 0.45–0.50 m/s.
- What are the design guidelines with respect to seal pans?
  - The clearance between the seal pan floor and the bottom downcomer should exceed the clearance normally used under the tray downcomers. It should be at least 50 mm.
  - The distance that the downcomer extends downward within the seal pan should be about the same as the clearance between downcomer bottom and pan floor.
- What are the design guidelines with respect to weirs?
  - Weir heights in the froth regime are restricted to 50–80 mm.
  - Weir heights for columns operating in the spray regime should be 20–25 mm.
  - Weir loadings should be within the range of 15–70 $m^3/(h\,m)$ of weir length.
- What is a picket weir? When is it employed?
  - Figure 16.7 illustrates schematic of a picket weir.
  - Used when liquid flows are small.
- What are the recommended arrangements for the nozzles with respect to the bottom section of a distillation column?
  - Inlets for the bottom feed and reboiler return lines should be at least 300 mm above the high liquid level.
  - The bottom feed and reboiler return should not impinge on the bottom seal pan, seal pan overflow, or the bottom downcomer. The tops of both pipes should be at least 400–450 mm below the bottom tray.

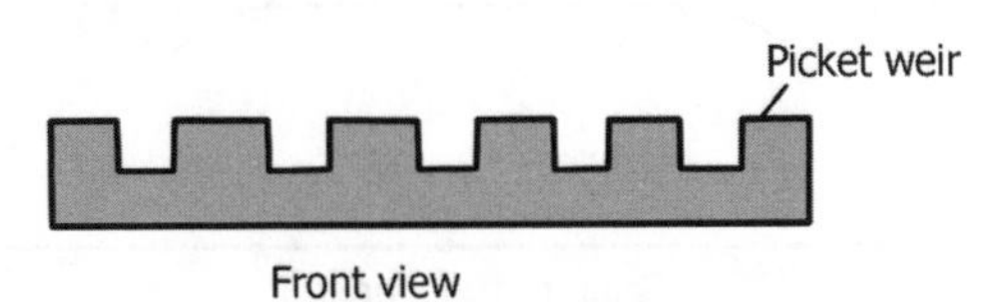

FIGURE 16.7 Picket weir.

- "Sieve trays are usually cheaper than valve trays." *True/False*?
  - *False*.
- What are the parameters involved in the evaluation of tray performance?
  - Tray ΔP depends on slot opening; static and dynamic slot seals; liquid height over the weir; and liquid gradient across the tray.
  - Downcomer conditions depend on liquid height; liquid residence time; and liquid throw over the weir into the downcomer.
  - Vapor distribution.
  - Entrainment.
  - Tray efficiency.
- What are the requisites for the satisfactory performance of a bubble cap tray?
  - Vapor flowing through all the caps.
  - Acceptable bubbling action from efficiency point of view: Neither too low nor excessive.
  - To keep entrainment within design limits.
  - Prevention of dumping (back flow) of the liquid down the risers.
  - No undesirable vapor jetting (coning) around the caps. Reduces effective interfacial area.
- Explain *liquid throw*, *coning*, *blowing*, *weeping*, *dumping*, *raining*, and *priming* in the operation of a distillation column.
  - *Liquid Throw:* Horizontal distance traveled by the liquid after flowing over weir.
  - *Coning:* A condition where the rising vapor pushes liquid back from the top of the hole and passes upward with poor liquid contact.
  - *Blowing:* A condition where the rising vapor punches holes through the liquid layer on a tray and usually carries large drops or slugs to the tray above.
  - *Weeping:*
    - Defined as vapor rate when weeping becomes first noticeable.
    - At low vapor rates (but somewhat higher than the rates leading to dumping) liquid flows partly through the holes and partly over the weir.
    - As vapor rate is decreased below weep point, weep rate increases with loss of efficiency. A rule of thumb states that 20% weeping leads to a 10% loss in efficiency.

> Larger liquid rates and higher weir heights increase liquid heads and hence weep rates.
> When the dry pressure drop falls below 12 mm H_2O, weeping becomes a problem.
> Main factor affecting is fractional hole area.
- Larger fractional hole area decreases ΔP for vapor flow thereby increasing weeping tendency.

- *Dumping:*
  - As vapor rate is lowered below weep point, fraction of liquid weeping increases until all liquid fed to the tray weeps through the holes.
  - At low vapor rates all the liquid falls through *some* holes (with no flow over the weir) and vapor rising through the *remaining* holes.
- *Raining:* Similar to dumping except liquid fall is through all the holes due to vapor rates lower than those for dumping.
- *Priming:* When velocity of gas or vapor bubbles is increased liquid entrainment will increase, carrying liquid droplets to the tray above. In the extreme, more froth/foam formation occurs that may build up occupying the entire spacing between the trays carrying considerable quantity of liquid to the tray above. This process is called *priming*.

- What are the causes for *blowing* to occur in the operation of a distillation tray?
  - High vapor–liquid ratio.
  - Low total liquid rate compared to the tray size.
- What is meant by spray *regime* in the operation of a distillation tray? How does it occur?
  - As vapor–liquid ratio increases, a stage is reached where the tray has liquid droplets suspended in vapor. Instead of liquid being continuous phase on the tray, vapor will be the continuous phase. This is called *spray regime*.
- What are the important causes of poor tray efficiencies?
  - Entrainment, weeping, flooding and dumping.
  - Tray efficiency depends upon column throughput.
  - Figure 16.8 gives generalized performance diagram for cross-flow trays.
  - Figure 16.9 illustrates the regions of stable operation indicating regions of drop in efficiency as a result of the causes indicated above for such a drop.
  - On either side of the stable region, performance drops off.
  - The drop off at low rates results first from weeping and then more significantly from dumping.
  - At high rates, heavy entrainment decreases efficiency and then performance dramatically drops when flooding occurs.

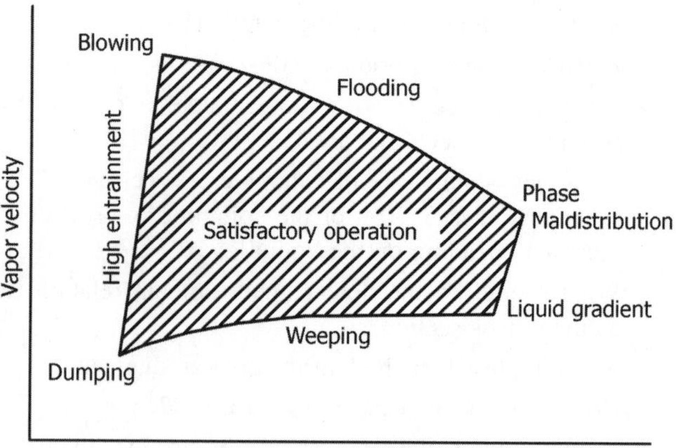

FIGURE 16.8 Generalized performance diagram for cross-flow trays.

- Explain the phenomena of weeping and dumping in the operation of a distillation tray.
  - During weeping, a minor fraction of liquid flows to the tray below through the tray perforations rather than through the downcomer.
    - This downward flowing liquid typically has been exposed to rising vapor and therefore, weeping only leads to a small reduction in overall tray efficiency.
  - In contrast, during dumping, a substantial portion of liquid flowing down the column passes through a region of the perforated tray deck.
    - Often, most of this liquid has not been exposed to the rising vapor and therefore, performance degrades significantly.
  - Weeping and dumping differ in their underlying mechanisms:
    - For sieve trays with smaller perforations, weeping is transient, resulting in spurts of liquid leaving a nonbubbling perforation.
    - The spurting occurs when there is a local and instantaneous downward pressure imbalance over the perforation.

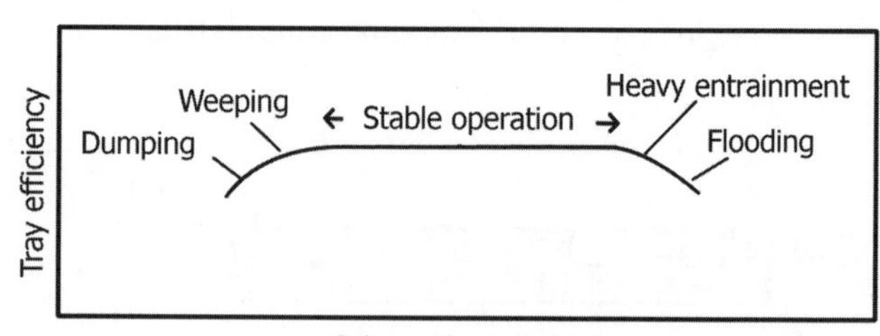

FIGURE 16.9 Tray performance versus throughput.

- ➢ The cause of this imbalance can be associated with the bubbling frequency or, because the flow on a large scale tray is very complex, the imbalance can be from local density and height variations of the froth waves traveling on the tray.
- ▪ The result is weeping regions tend to move around on the tray deck.
- ▪ Dumping is much more extreme and occurs, because, at the intended tray throughput, there is insufficient vapor ΔP to retain entering liquid on the tray deck.
  - ➢ Thus, significant quantities of liquid flow through a portion of the tray that has little or no vapor flow. Flow over the outlet weir can be zero. Normally minor phenomena like the hydraulic gradient or tray inlet liquid flow maldistribution, can have a significant impact on the minimum tray ΔP required to prevent dumping.
- What are the recommended limits for vapor and/or liquid loading for the flexible operation of a tray?
  - ▪ Not less than 50% and not more than 120% of vapor load.
  - ▪ Not less than 15% and not more than 130% of liquid load.
- What are the acceptable pressure drops for a bubble cap tray?
  - ▪ For columns operating under higher or normal pressures: 5–10 cm (2 4 in.) H_2O.
  - ▪ For columns operating under vacuum ($\leq$500 mmHg): 2–4 mmHg.
- What are the conditions that could lead to pulsing on a bubble cap tray?
  - ▪ Too low and unsteady vapor rate.
  - ▪ Low slot openings <12.7 mm (<0.5 in.).
  - ▪ Low liquid dynamic seal.

16.1.1.1 Tray Efficiencies

- Give the important correlations for overall column efficiencies.
  - ▪ O'Connell correlation is based on data collected from actual columns. It is based on bubble cap trays and is conservative for sieve and valve trays. It correlates the overall efficiency of the column with the product of the feed viscosity and the relative volatility of the *key component* in the mixture. These properties should be determined at the arithmetic mean of the column top and bottom temperatures.
  - ▪ Figure 16.10 and Equation 16.1 represent O'Connell correlation:

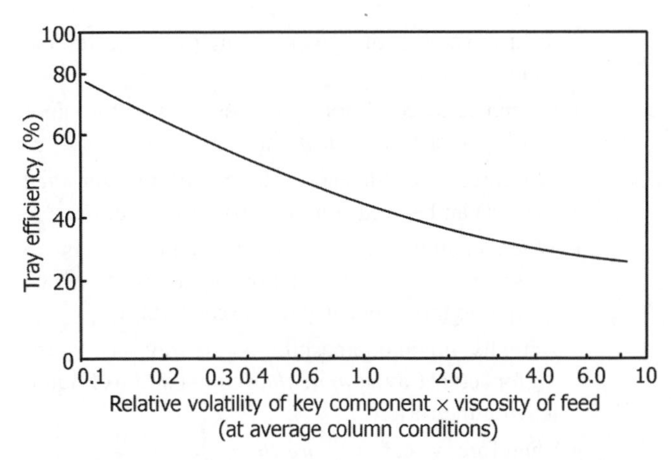

FIGURE 16.10 O'Connell correlation for the estimation of overall column efficiency.

$$\eta = 0.492(\alpha_{\text{key}}\mu_{\text{f}})^{-0.245}. \quad (16.1)$$

- ▪ Equation 16.2 based on O'Connell correlation and proposed by Eduljee, is

$$\eta = 51 - 32.5\log(\alpha_{\text{key}}\mu), \quad (16.2)$$

where α is the average relative volatility of light key and μ is the average liquid viscosity, mNs/m^2.

- ▪ O'Connell correlation is useful for preliminary estimates of overall column efficiencies. For hydrocarbon systems, O'Connell correlation gives overall efficiencies in the range of 50–85%.
- ▪ Drickamer–Bradford correlation for overall efficiencies is given by Equation 16.3.

$$\eta = 0.17 - 0.616\log_{10}\Sigma[x_{\text{F}}(\mu_{\text{L}}/\mu_{\text{W}})], \quad (16.3)$$

where μ_{L} is viscosity of the liquid at average column temperature and μ_{W} is viscosity of water at 293 K.

- ▪ Efficiency increases as viscosity and relative volatility decreases. Liquid viscosity is an important parameter as decreased viscosities increase diffusivities and turbulent mass transfer with thinner liquid and vapor films, giving better liquid-phase mass transfer coefficients.
- Define Murphree tray efficiency and explain where and why Murphree tray efficiency can sometimes be more than 100%.
  - ▪ Equation 16.4 defines Murphree tray efficiency.

$$E_{\text{MV}} = \text{actual separation divided by theoretical separation} = (y_n - y_{n+1})/(y_{n*} - y_{n+1}). \quad (16.4)$$

- ▪ y_n and y_{n+1} are compositions of vapor *leaving* and *entering* the nth tray, respectively, and are uniform as diffusion and mixing processes are fast for gases and vapors.

- y_{n^*} and x_n are equilibrium compositions for the vapor and liquid streams *leaving* nth tray.
- Murphree tray efficiencies can be different for *each tray* as can be seen from the defining equation.
- When column diameter is large and there is no backmixing in the liquid flowing across the tray, Murphree tray efficiency can exceed 100%.
- Actually, liquid composition on the tray, say, x_n^a (or $x_{n'}^a$ for actual tray) is *more than* composition of liquid leaving the tray, x_n.
- Therefore, y_n can be *more than* y_{n^*}.
- That is, numerator in the Murphree efficiency equation can be more than the denominator. Therefore, E_{MV} can be *more than* 100%.
- *Note*: All compositions are expressed in terms of mole fractions of more volatile component.
- Figure 16.11 illustrates the compositions of flows for actual and theoretical trays.
- Equation 16.5 gives overall efficiency, η_0, from Murphree efficiency, E_M, vapor and liquid flows, V and L, and slope of the equilibrium line, m:

$$\eta_0 = \log[1 + E_M(mV/L-1)]/\log(mV/L). \quad (16.5)$$

The equation is useful for dilute mixtures where equilibrium *curve* can be treated as straight line.

- Effect of entrainment can be accounted by Colburn Equation 16.6:

$$\eta_a = E_{MV}/[1 + E_{MV}\{e/(1-e)\}], \quad (16.6)$$

where η_a and e is fractional entrainment = entrained liquid/gross liquid flow.

- Van Winkle gave the following correlation (Equation 16.7) for Murphree efficiencies for bubble cap and sieve trays in terms of dimensionless groups:

$$E_{MV} = 0.07 D_g^{0.14} N_{Sc}^{0.25} N_{Re}^{0.08}, \quad (16.7)$$

where D_g is the surface tension number, $\sigma_L/\mu_L u_V$. u_V is the superficial vapor velocity, σ_L is the liquid surface tension, μ_L is the liquid viscosity, N_{Sc} is the liquid Schmidt number, $\mu_L/\rho_L D_{LK}$, D_{LK} is the liquid diffusivity for light key, N_{Re} is the $(h_w u_V \rho_V/\mu_L)(F_A)$, h_w is the weir height, ρ_V is the vapor density, F_A is the area of holes or risers/total column cross-sectional area.

- What is the effect of vapor velocities on tray efficiencies? Illustrate.
  - Peak tray efficiencies usually occur at linear vapor velocities of 0.6 m/s (2 ft/s) at moderate pressures or 1.8 m/s (6 ft/s) under vacuum conditions.
- What are the ranges of tray efficiencies for distillation and gas absorption?
  - Distillation of light hydrocarbons and aqueous solutions: 60–90%.
  - Gas absorption and stripping: 10–20%.
- What is the reason for very low tray efficiencies for gas–liquid contacting compared to vapor–liquid contacting?
  - The resistance offered by a gas film (noncondensing phase) for mass transfer is very high compared to that offered by a vapor film (condensable phase), giving rise to poor efficiencies in the former case.
- What are the causes of tray inefficiencies?
  - Uneven liquid flows.
  - Uneven vapor flows.
  - Out of level trays, that is, trays are not truly horizontal due to column out of plumb, due to tray support rings might not be level, sagging of trays due to pressure surges, large unsupported sections of trays, sloppy installation, and so on.
  - Loss of downcomer seal (explained earlier).
  - Liquid must flow over the weir. If all the liquid is weeping through the tray deck, there will be no overflow of liquid on the weir.

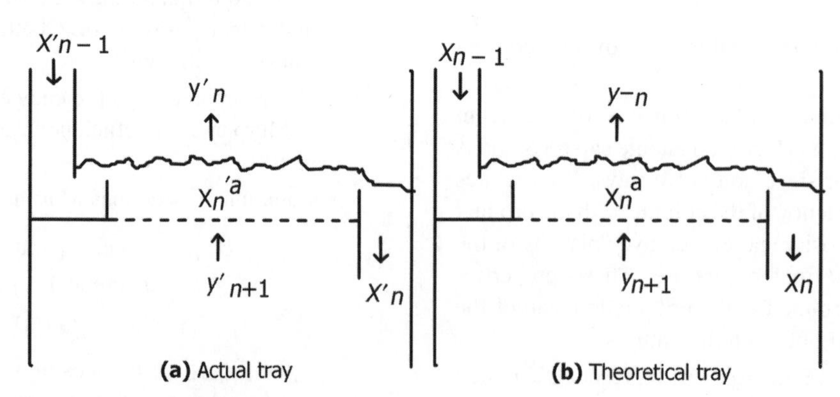

FIGURE 16.11 Murphree tray efficiencies illustrated.

- What is the effect of increased reflux ratio on tray efficiency?
  - Higher reflux ratio increases reboiler duty (assuming reboiler is on automatic temperature control).
  - Increased reboiler duty results in increased vapor flows and increased dry tray ΔP that might stop tray leakage/weeping.
  - Increased reflux ratio increases energy costs that are escalating due to increasing fuel prices.
  - Best way is to achieve good efficiency with minimum vapor flows by designing and installing tray decks and outlet weirs as level as possible.
  - Damaged tray decks should not be reused unless they are brought to proper levelness.
- What is the effect of percent flooding condition on tray efficiency?
  - In the absence of hydraulic limitations such as weeping, entrainment or channeling, percent flood has little effect on tray efficiency.
- What are the geometric variables that affect tray efficiencies?
  - *Flow Path Length:* Increasing the number of tray passes from 1 to 2 or 2 to 4 can reduce tray efficiency by as much as 10–20% at low pressures and (0.3–1.7 bar) and by 5–15% at higher pressures (11–28 bar).
  - Increasing fractional hole area on trays from 5% to 8% of the active area has a minor effect on efficiency, but a further increase to 14% reduces efficiency by 5–10%, as at larger hole areas there may be an increased sensitivity for leakage.
  - As hole diameters increase from 3 to 13 mm, efficiency increases by 5–10%, 13–25 mm, efficiency remains constant and above 25 mm, efficiency decreases.
  - At fractional hole areas of 8–9%, larger weir heights have minor effect on efficiencies and around 14%, such weirs may slightly enhance efficiencies.
  - Maldistribution in multipass trays may lead to further major reductions in tray efficiencies.
- What happens if top reflux is increased in the following example?
  - *Example:* Operation of propane–butane splitter in which pressure and bottoms temperature are kept constant, that is, percentage of propane in the bottom butane product is kept constant.
    - Top temperature increases.
    - Butane in overhead product decreases.
    - Bottoms temperature decreases.
    - Reboiler duty increases to restore column bottoms temperature to its set point.
    - Flow and velocity of vapor through the trays increase.
    - Spray height or entrainment between the trays increases.
    - When spray height from lower trays impacts the upper trays, the heavier, butane-rich liquid, contaminates the lighter liquid on the upper trays with heavier butane.
    - Further increase in reflux ratio, then act to increase rather than decrease butane content of overhead propane fraction.
- "Bubble cap trays may be operated over a far wider range of vapor flows and fractionate better in commercial service than valve/sieve trays." Why, then, are bubble cap trays rarely used in a modern distillation column?
  - Bubble cap trays are difficult to install because of their weight.
  - Have about 15% less capacity because, when vapor escapes from the slots, it is moving in a horizontal direction and must turn 90°. This change of direction promotes entrainment and causes jet flooding.
  - Involve higher ΔP and difficult to clean.
  - More expensive.

16.1.1.2 Entrainment

- What are the different sources of entrainment?
  - Boiling liquids.
  - Bubbling of gas through liquids.
  - Sprays from spray nozzles (pressure type/two-fluid atomizers).
  - Condensation:
    - Acid mist formation in condensers.
    - Vaporization of lube oils from compressor sources and subsequent cooling/condensation giving rise to oil mists.
  - Cooling of supersaturated gas fractions in distillation equipment giving rise to mists.
- What are the problems associated with entrainment?
  - Loss of efficiency.
  - Loss of valuable liquid.
  - Contamination of desirable overhead product. For example, salt may contaminate desalinated water. Heavy impurities/coloring matter in distillation overheads.
  - May cause downstream problems—undesirable chemical reactions, catalyst deactivation, corrosion, erosion, and fouling.
  - Water pollution.

- Air pollution (acid mist in a sulfuric acid plant).
- Odor emissions.
- Limiting throughput capacities.

- Give reasons why entrainment separators (mist eliminators) are employed in mass transfer equipment.
  - For improved emission control (reduction of air and water pollution).
  - Recovery of valuable products.
  - Improved product purities.
  - Protection of downstream equipment from
    - corrosive liquid droplets and mists carried to downstream equipment;
    - erosion by liquid droplets;
    - solids deposition;
    - fouling of valves and fittings.
- Name some process applications for which entrainment separation is necessary.
  - Separation of natural gas condensates from the gas at wellheads.
  - To prevent entrainment of heavy fractions from reboilers.
  - To separate acid mist in acid condensers.
  - To prevent liquid carryover in flash drums.
  - To remove water droplets from input steam to turbines.
  - To prevent carryover of radioactive materials in evaporators processing radioactive materials.
  - To prevent loss of product in vegetable oil deodorizers.
- What are the factors on which the amount and droplet size distribution of entrainment over a boiling/bubbling surface depend?
  - Height above the liquid surface.
  - Vapor velocity over the liquid surface.
  - Terminal velocity of the droplets.
  - Projection velocity of the droplets.
  - Relative amounts of large and small droplets.
- Explain how entrainment has a reducing effect on tray efficiency?
  - The separation obtained by transfer of a solute from vapor to liquid is reduced by remixing of part of the liquid back into the vapor in the form of entrainment, which gets carried to the next upper tray, for example, in a distillation column or to the relatively leaner liquid above in an absorber. Thus, part of the separation already obtained is lost through entrainment. This means that efficiency of separation decreases.
- What are the different mechanisms used in entrainment separation?
  - Centrifugation.
  - Sedimentation.
  - Inertial impaction.
  - Diffusion and Brownian motion: Motion is due to collisions with molecules of suspending fluid.
  - Interception.
  - Venturi contacting.
  - Electrostatic precipitation.
- Name different types of equipment/operations for which entrainment separators need to be employed.
  - Distillation equipment:
    - Blowing (fine fog/mist) due to high vapor velocities and low liquid loading.
    - Flooding.
    - Result in loss of efficiency.
    - Contamination of product.
  - Evaporators:
    - Result in loss of product, particularly high-value product.
    - Cause stream pollution.
    - Cause contamination of condensate.
- Name different types of entrainment separators/mist eliminators used in industrial equipment.
  - Settling chambers: For $>100\,\mu m$ droplets.
  - Baffle type mist eliminators for coarse sprays:
    - Provide number of direction changes.
    - Arranged in zigzag manner, may be vertical for horizontal gas flow or horizontal with vertical gas flow.
    - Louver type or blade type.
  - Rods and tubes: Vertical, either round or tear shape.
  - Vane type impingement separators.
  - Packed beds.
  - Centrifugal (cyclonic) separators.
  - Inertial separators: Inertial impaction through abrupt changes in direction is the main mechanism.
  - Diffusional separators: Capture droplets on a target surface by diffusion/Brownian movement (very fine droplets).
  - Wire mesh eliminators/demisters:
    - Knitted stainless steel wire mesh pads of about 10 cm (4 in.) thick.
    - Good for lower gas rates than baffle type.
    - For relatively clean gases.
  - Fiber bed eliminators:
    - Consist of fine fibers (generally glass fibers), packed to densities of 130–240 kg/m^3 and 5–10 cm thick.

    - ➢ Separation involving inertial impaction and Brownian diffusion.
    - ➢ For removal of fine mists formed by shock-cooling of condensable vapors, for example, acid mists, oil mists, and so on.
    - ➢ Give >99% collection efficiency for submicron particles.
    - ➢ Surface areas are 30–150 times higher than those for wire mesh eliminator.
    - ➢ Velocities are 4.5–12 m/min (15–40 ft/min).
    - ➢ $\Delta P = 12.5$–40 cm (5–15 in.) water.
    - ➢ Gas flows: For inertial impaction type.
    - ➢ Efficiency increases for Brownian movement type.
    - ➢ Geometries: Radial flow cylinder and axial flow panel.
  - ▪ Electrostatic precipitators: For acid mists.
  - ▪ Wet scrubbers:
    - ➢ Spray columns, baffled spray columns, packed beds, cyclones, venturis, and their combinations.
    - ➢ Not primarily for entrainment separation.
    - ➢ Dusts, mists, soluble gases like SO_2.

- What is a cyclonic separator? How does it work? What are its characteristics?
  - ▪ A cyclonic separator is a cylindrical tank with a tangential inlet or turning vanes. The tangential inlet or turning vanes impart a swirling motion to the droplet-laden gas stream. The droplets are thrown outward by centrifugal force to the walls of the cylinder. Here they coalesce and drop down the walls to a central location and are recycled to the mass transfer equipment.
  - ▪ These units are simple in construction, having no moving parts. Therefore, they have few plugging problems as long as continuous flow is maintained. Good separation of droplets 10–25 μm in diameter can be expected. The pressure drop across the separator is 10–15 cm (4–6 in.) of water for a 98% removal efficiency of droplets in the size range of 20–25 μm.
  - ▪ Separation of 5–10 μm size droplets is possible in a cyclonic separator.
- What is a rotary stream separator? Explain.
  - ▪ Rotary stream separators are similar to cyclone separators. While they use centrifugal forces for separation, but do not include tangential entry.
  - ▪ Internal vanes, baffles, and directional slots or combination of these are used to give a confined gas vortex that develops centrifugal force.
  - ▪ Variety of designs are available.
- What is the principal mechanism of an inertial entrainment separator?
  - ▪ Inertial impaction is the principal mechanism.
  - ▪ Gas stream is forced to make one or more abrupt changes in direction, thereby causing the liquid droplets, due to their inertia, to come into contact with a surface where the liquid forms a film.
  - ▪ When a number of droplets are added to the liquid film, weight of the liquid causes drainage of the liquid due to gravity.
- What is a vane type impingement separator? How does it work?
  - ▪ Vane impingement separator elements consist of parallel zigzag plates or vanes with included collection pockets.
  - ▪ As an entrainment laden gas moves through the separator, the liquid droplets are forced onto the vane surfaces by the inertial impaction mechanism.
  - ▪ The resultant liquid film flows to the collection pockets, where it drains by gravity.
- What are the factors that govern allowable gas velocities through wire mesh mist eliminators?
  - ▪ Gas and liquid density.
  - ▪ Surface tension.
  - ▪ Liquid viscosity.
  - ▪ Specific surface area of the wire mesh.
  - ▪ Liquid entrainment loading.
  - ▪ Suspended solids content.
- "Wire- or plastic-mesh pads are capable of removing smaller droplets than either cyclonic or blade separators." *True/False*?
  - ▪ *True*. However, they are also more susceptible to plugging.
- What is the correct answer for the following case? To prevent plugging, in general, wire mesh pads should be
  - (a) installed at a slant;
  - (b) sprayed from the bottom;
  - (c) sprayed from the top;
  - (d) sprayed from the top and bottom.
  - ▪ Answer: *d. Sprayed from the top and bottom.*
- What is the mechanism of capture of droplets in a fiber bed entrainment eliminator?
  - ▪ Brownian movement of particles.
- Give an equation for the estimation of entrainment in a tray column.

$$e = c(U/S)^{3.2}, \tag{16.8}$$

where e is the entrainment, kg/kg vapor, U is the superficial velocity, m/s, S is the effective tray spacing, m, and c is 0.73 for sieve trays and 0.21 for bubble cap trays.

- List five important characteristics of spray nozzles used in wet scrubbing systems.
  - Opening size, droplet size, spray pattern, operating mechanism, and power consumption.

16.1.1.3 Flooding

- What is flooding?
  - Flooding represents inoperable conditions in a column in which no separation is possible.
  - Mechanisms for flooding: Spray entrainment, froth entrainment, high vapor rates, downcomer backup flooding, downcomer choke flooding.
- What are the various mechanisms involved in flooding? Explain.
  - Spray entrainment flooding (most common mechanism):
    - At low liquid flows, trays operate in the spray regime where most of the liquid on the tray is in the form of drops.
    - As vapor velocity is raised, bulk of the drops are carried up (entrained) onto the tray above.
    - Liquid then accumulates on the tray above instead of flowing to tray below.
  - Froth entrainment flooding (encountered when tray spacing is small):
    - At higher liquid rates, dispersion on the tray will be in the form of froth.
    - As vapor velocity is increased, froth height increases.
    - When tray spacing is insufficient (<45 cm), upper froth surface approaches tray above and entrainment rapidly increases leading to liquid accumulation on the tray above.
    - With large spacing (>45 cm), froth has less chances to reach tray above.
    - As vapor velocity increases, a condition is reached where some of the froth converts into spray and flooding occurs by spray entrainment.
  - System flooding (also called vapor flood):
    - Ultimate vapor rate when exceeded, liquid will entrain or blow upward (irrespective of tray types).
    - Influence of liquid rate is secondary.
    - This limit is higher than other flooding mechanisms.
    - This type of flooding occurs, for example, in vacuum operations and low liquid to vapor ratios.
  - Downcomer backup flooding:
    - Aerated liquid backs up in the downcomer because of pressure drop and liquid height on the tray and friction losses in the downcomer.
    - All the above phenomena increase with increased liquid rates.
    - Tray ΔP also increases with increased vapor rate.
    - When backup of aerated liquid exceeds tray spacing, liquid accumulates on tray above, causing downcomer backup flooding.
    - Downcomer backup can be reduced by increasing tray spacing, decreasing outlet weir height, decreasing tray ΔP or increasing the downcomer clearance.
  - Downcomer choke flooding:
    - As liquid flow rises, so does velocity of aerated liquid down the downcomer.
    - When this velocity exceeds a certain limit, friction losses in the downcomer and its entrance become excessive and aerated liquid cannot be transported to tray below, causing liquid accumulation with the consequence of flooding on the trays above.
    - Downcomer choke flooding can be mitigated by increasing the downcomer size or by sloping the downcomer.
  - Downcomer flooding caused by excessive liquid rates is often called *liquid flood.*
    - Examples of this condition are observed in high pressure operations and high liquid to vapor ratios.
- What is the difference between *hydraulic flood* and *operational flood*? Explain.
  - There is a critical distinction between operational flood and hydraulic flood.
    - Hydraulic flood results when the downcomer, at a given column throughput, becomes fully loaded with liquid and entrained vapor and this mixture within the downcomer begins to impede flow over the outlet weir.
    - The added resistance increases tray liquid inventory and ΔP.
    - This, in turn, raises the two-phase mixture height in the downcomer area, further impeding flow over the outlet weir and boosting ΔP even more.
  - Finally, at hydraulic flood, all the liquid that enters the column no longer can leave the column. Liquid is accumulated above the flood point and ΔP increases rapidly.
  - In contrast, in operational flood, all the liquid entering the column section still leaves the column, even

though ΔP can be very high and efficiency is very poor or unstable.

- What is jet flooding?
  - Entrainment above froth levels on a tray is a function of (1) vapor velocity through the tray and (2) foam height on the tray.
  - High vapor velocities plus high foam levels result in the foam height hitting the underside of tray above. This causes mixing of the liquid from lower tray with liquid on the upper tray.
  - Result is lowering of efficiency or flooding.
  - When foam/froth/spray height from tray below hits tray above signals approaching of incipient flood point. This is called initiation of jet flooding, which is common cause for tray flooding. Jet flood is often confused with operational flood. Jetting often is used as a synonym for the spray like regime and an operational flood can occur for either spray like or froth like conditions.
- What are the circumstances that can lead to flooding in the downcomer of a distillation tray?
  - Downcomer cannot carry the necessary liquid flow either due to
    (1) too high a reflux ratio (liquid rates will be high, increasing ΔP);
    (2) too low the liquid density due to entrapped vapor bubbles.
- What are the factors that promote flooding in a tray column?
  - Low bubbling area [total column cross section − $\sum$(downcomer area + downcomer seal area + any other nonperforated areas on the tray)].
  - Low net area.
  - Low fractional hole area (<8% of bubbling area).
  - Low tray spacing.
  - High weirs and shorter weir lengths.
    - Reduce spray action and so slightly decrease tendency toward spray entrainment flooding.
    - Also increase froth height, liquid height on the tray, and tray ΔP.
    - Increase tendency toward entrainment flooding and downcomer backup flooding.
- What are the causes of flooding in a tray column?
  - Inadequately designed trays, short of vapor capacity.
  - Downcomer backup: Due to foaming liquid not draining quickly from the downcomer.
  - Downcomer clearance (distance between bottom edge of the downcomer and the tray below):
    - Due to loss of downcomer seal.
    - Proper downcomer seal should be neither too high nor too low.
    - If downcomer clearance is too small, liquid backs up in the downcomer and trays above flood.
    - If downcomer clearance is too large, vapor flows up the downcomer and trays above flood.
    - Height of outlet weir is below bottom edge of the downcomer.
    - Bottom edge of the downcomer is about 12.5 mm (1/2 in.) below top edge of outlet weir. This is a critical factor to prevent flooding.
    - Weir height should preferably be adjustable between 50 and 75 mm (2–3 in.).
  - Inlet weir height, $B >$ outlet weir height, A, as illustrated in Figure 16.12.
  - All trays above flood.
  - All trays below loose liquid levels and dry out.
  - Early indication of flooding: Loss of liquid level in column bottom/reboiler.
- What are the operational causes for premature column flooding?
  - Off-design feed composition or thermal condition.
  - Unexpected entrainment, for example, at a feed point or internal transition.
  - Internal damage or fouling, resulting in flow restriction or mass transfer efficiency.
  - Foaming problem leading to inaccurate level measurement and unexpected entrainment.
  - Level control difficulties: Maintaining high level in the column sump, kettle reboiler, or chimney tray.
  - Unstable control system.
  - Incorrect flow measurements of feed, product, heating medium, or reflux that might conceal true operating rate.
  - Presence of an immiscible liquid phase in the column, for example, water in oil.
- How would you detect flooding occurring inside a column?
  - Early indication of flooding is loss of liquid level in the column bottom.
  - Pressure drop in the column increases.

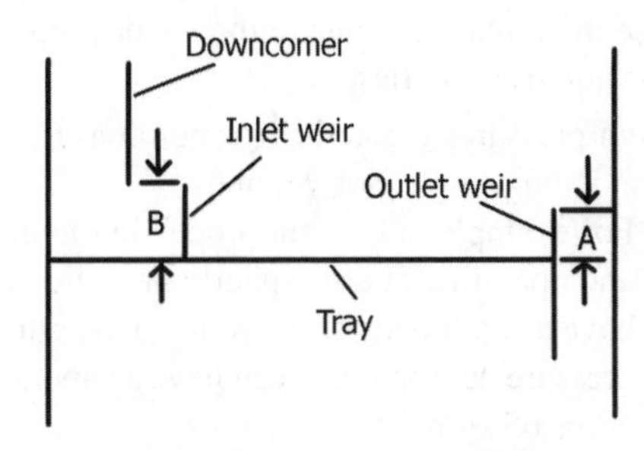

FIGURE 16.12 Weir height.

   - Vapor from top of column will carry increased entrainment and condenser liquid levels increase.
- How would you differentiate between flooding and weeping trays?
   - Measure ΔP.
      - If $\Delta P > 3$ times weir height, tray is flooding.
      - If $\Delta P <$ weir height, tray is under weeping.
- "When flooding starts on a tray, all trays above that tray will also flood but trays below will go dry." *True/False*?
   - *True.*
- What are the circumstances that can lead to flooding in the downcomer of a distillation tray?
   - Downcomer cannot carry the necessary liquid flow either due to
     (i) too high a reflux or
     (ii) liquid density is too low due to entrapped vapor bubbles.
- What are the important requisites for achieving intimate vapor–liquid contact on trays?
   - Vapor should bubble through perforations on the tray deck evenly.
   - Liquid should flow evenly across the tray deck.
   - Liquid gradients across the tray should be negligible.
- What is the important reason for uneven liquid flows on a tray?
   - Outlet weir being out of level.
   - Other reasons include uneven distribution of liquid across the tray for large diameter columns.
- What is the minimum weir loading for achieving reasonable crest height (height of liquid over the weir)?
   - Not Less than 3 l/(min cm) (2 gpm/in.) of weir length.
- What are the characteristics of high-pressure columns?
   - High-pressure distillation involves four characteristics that can influence tray design. They are
      - high liquid rates;
      - low density difference between vapor and liquid;
      - low surface tension;
      - low efficiency (above certain pressures).
- What are the reasons for high liquid rates in a high-pressure column?
   - The main reason is due to the vapor properties than the liquid properties.
   - High pressures create high vapor densities and corresponding low vapor volumes.
      - For example, a column processing hydrocarbons and operating at atmospheric pressure is likely to have a vapor density of around, 6.5 kg/m^3. A high pressure depropanizer can have a vapor density of 50 or 65 kg/m^3.
      - Since the column cross-sectional area is devoted to vapor handling, high-pressure columns will always have smaller diameters than lower pressure columns processing a similar mass rate.
      - Also, high-pressure distillation usually involves lower molecular weight components that have a lower liquid density. This can increase in the volumetric liquid flow rate by 40%.
      - The net result is that high-pressure distillation applications use a smaller diameter column while handling a much larger proportion of liquids.
- What are the consequences of the effects of density differences between vapor and liquid and surface tension on high-pressure column operation?
   - Vapor/liquid density difference:
      - High-pressure distillation applications have a significantly lower density difference between the vapor and the liquid phases.
      - As the density difference decreases, it becomes more difficult to separate the vapor and liquid phases because the vapor buoyancy in the liquid decreases.
      - This creates a need for significantly larger downcomers. The extreme case is critical pressure, where no separation can occur due to the lack of a density difference.
   - Surface tension:
      - High-pressure distillation applications typically have significantly lower surface tension.
      - High surface tension liquids, like water, form big droplets that are much easier to coalesce and less likely to be entrained to the tray above.
      - Low surface tension liquids form significantly smaller size droplets that are much more susceptible to entrainment to the tray above.
- What is the effect of column high pressures on tray efficiencies?
   - Above a certain pressure level, higher pressure systems have lower efficiencies.
   - With standard hydrocarbon separations, a debutanizer typically achieves the highest efficiency with commercial scale values of around 90%.
   - Separations of lighter hydrocarbons (depropanizers, deethanizers, and demethanizers) yield lower and lower efficiencies as the pressures increase.
   - This is generally due to increase in liquid viscosity and decrease in relative volatility, both of which directly lead to lower efficiency.
- What are the different reasons for providing sufficient time in liquid drawoff sump of a distillation column?

  - To disentrain vapor in the sump liquid. Vapor bubbles may otherwise move along with the liquid to downstream pump, creating cavitation in the pump and choke downstream piping.
  - To buffer downstream units from upstream and column upsets. This is important when sump is feeding sensitive units like furnaces.
  - To buffer the column from downstream upsets.
  - To give operator sufficient time to take corrective action if upsets occur (e.g., pump trips, loss or gain of level too fast).
  - To provide sufficient settling time if two liquid phases (e.g., hydrocarbons and water) are to be separated.
- What are the ways by which high liquid rates are handled in high-pressure columns?
  - Increasing the size of the downcomers.
  - Using multipass trays in which the liquid is split into multiple flow paths, reducing the weir loading significantly.
  - Using multiple downcomers. This type of tray is most commonly used in superfractionators (C_2 and C_3 splitters) where a very large number of trays are required in a high pressure, high liquid rate application.
- What are the mechanisms involved in downcomer flooding when liquid rates are high in high-pressure columns?
  - As the loading in the downcomer increases, the downcomer can flood from three main methods: choking, backup, and excessive vapor entrapment in liquid flowing into downcomer. Liquid entrapment impairs the ability of the downcomer in separating vapor and liquid prior to delivering the liquid to the tray below.
  - Downcomer choke occurs when the top area of the downcomer is inadequate to handle the high froth flow, preventing effective vapor disengagement of the vapor. This causes the liquid to backup onto the tray deck until the column floods.
  - Downcomer backup flooding occurs when the hydraulic head requirements to flow of the liquid through the downcomer exceed the height of the downcomer itself.
  - Downcomer backup is a function of the tray pressure drop, the head at the inlet side of the tray, and the frictional losses in the downcomer itself.
  - Downcomer vapor entrapment occurs when the froth in the downcomer does not completely breakup and vapor bubbles are allowed to travel fully through the downcomer and exit the bottoms. This is especially a problem in high-pressure columns, especially with shorter tray spacing.
- "Use of carbon steel trays, particularly valve trays, is often said to be a cause of flooding." Explain.
  - When valve caps are also made out of carbon steel, valves have a tendency to stick in a partially closed position. The results are as follows:
    - Increased ΔP_{vapor} through the valves:
      - Liquid level goes up in the downcomer draining the liquid.
      - Liquid then backs up onto the tray deck and promote jet flood due to entrainment.
    - Dirt, polymerized materials, gums, salts, and so on, cause a reduction in open area of tray deck, which also promote jet flooding.

16.1.1.4 Diameter for Tray Columns

- What are the considerations involved in determining column diameter?
  - Column diameter is estimated based on the operational constraints due to *flooding*. Knowledge of vapor and liquid loads throughout the column is needed but not number of trays to find the diameter. The number of actual trays and spacing are needed to get the column height.
  - The important factor that determines the column diameter is the vapor velocity that must be below that which would cause excessive liquid entrainment or a high pressure drop.
  - If the vapor/gas velocity is too high, liquid droplets are entrained, causing priming. Priming occurs when the vapor/gas velocity through the column is so high that it causes liquid on one tray to foam and then rise to the tray above.
  - Priming reduces column efficiency by inhibiting vapor/gas and liquid contact.
  - For the purpose of determining column diameter, priming in a tray column is analogous to the flooding point in a packed column. It determines the *minimum acceptable diameter*.
  - Diameter estimations are made for each section where the loading might be high. Top and bottom trays, above and below feed points, side draws, or heat addition or removal points should be considered for finding the diameter requirements for each section.
  - If estimated diameters differ significantly, two-diameter columns might be considered.
    - This possibility might be considered where the estimated diameter varies by 20% or more, but it

must be realized that it will be more expensive than a column with the same diameter.

  ➢ In petroleum distillations, especially atmospheric crude distillation and vacuum distillation of atmospheric column bottoms, columns with two diameters are common. These columns involve side-draw products.

- If diameter varies to a significant effect, sections with different diameters should be considered. This situation might arise for columns with multiple streams, for example, multiple feeds, side streams, and so on.
- Column design should take into consideration choice of actual reflux rate over the estimated minimum reflux rate and likely operational changes. An increase in the flows by about 20% might be wise for such a consideration.

• Write down Souders–Brown equation. For what purpose it is used?

$$V_f = C[(\rho_L - \rho_V)/\rho_V]^{1/2}(\sigma/20)^{0.2}, \tag{16.9}$$

where V_f is the maximum allowable vapor velocity, based on net area, A_N. A_N = active area + one downcomer area (shown in Figure 16.13). C is the capacity parameter, obtainable from the generalized correlation (given in the following plot), ρ_L and ρ_V are liquid and vapor densities, respectively, and σ is the surface tension of liquid.

- Figure 16.14 represents generalized correlation for obtaining maximum allowable vapor velocity.
- Flow parameter is obtained from vapor and liquid densities and given vapor rate through the column. Vapor rate = distillate rate + reflux rate.
- V_f, flooding velocity is calculated from Souders–Brown equation, once value of C is obtained from the graph.

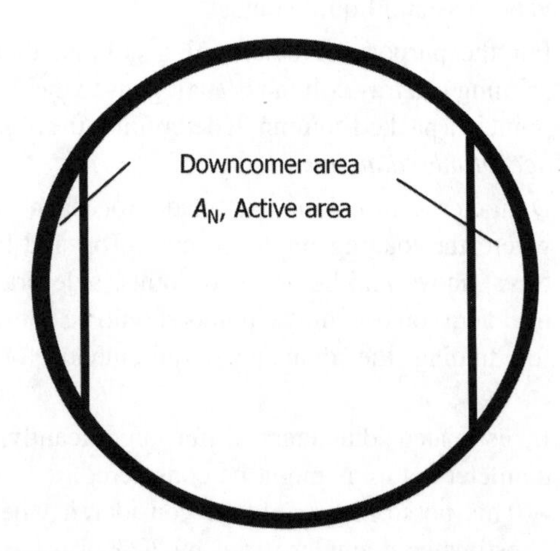

FIGURE 16.13 Downcomer and active areas illustrated.

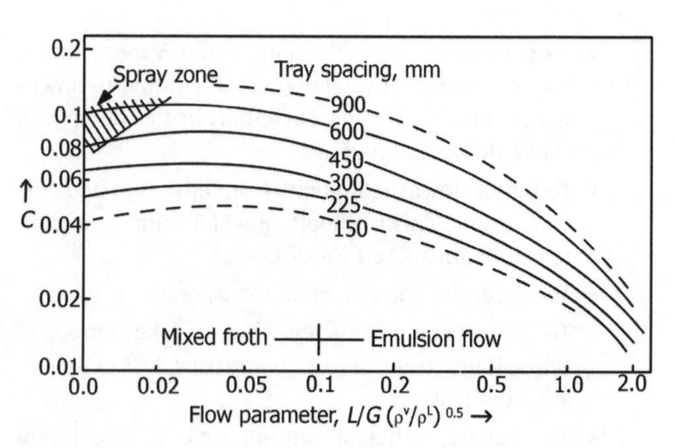

FIGURE 16.14 Flooding correlation for cross-flow trays (sieve, valve, and bubble cap trays).

- Using calculated V_f, column diameter for flooding conditions is found.
- Thumb rule values for C for various cases:
  ➢ Absorbers: 0.55.
  ➢ Petroleum distillation columns: 0.8–0.95.
  ➢ Stabilizers/strippers: 1.15.
- The correlation is used for the determination of column diameter, subject to the following restrictions:
  ➢ Low foaming to nonfoaming systems.
  ➢ Weir height less than 15% of tray spacing.
  ➢ Hole diameter 12.7 mm (0.5 in.) or less (sieve trays).
  ➢ Hole or riser area 10% or more of the active, or bubbling, area.
  ➢ Smaller hole areas tend to produce jetting because of the high hole velocities.
- Souders–Brown correlation considers entrainment as the controlling factor.
- It is a conservative equation for pressures 0.35–17 barg.
- V_f is to be multiplied by 1.05–1.15 to determine actual diameter to be used.
- Another equation, based on Souders–Brown correlation (Equation 16.10) can be used to estimate maximum allowable vapor velocity and hence the minimum diameter required:

$$V_{max} = (-0.171P_s^2 + 0.27P_s - 0.047)\,[(\rho_L - \rho_V)/\rho_V]^{1/2}, \tag{16.10}$$

where V_{max} is the maximum allowable vapor velocity, based on the total column cross-sectional area, m/s and P_s is the tray spacing, m (range 0.5–1.5).

- The column diameter is obtained from Equation 16.11,

$$D = \sqrt{(4W_V/\pi\rho_V V_{max})}, \tag{16.11}$$

where W_V is the maximum vapor rate, kg/s.
- This approximate estimate of the diameter should be revised when the detailed plate design is carried out.
- Another correlation for estimation of column diameter is given below.
- The smallest allowable diameter for a tray column is expressed by Equation 16.12:

$$d_t = \Psi(Q_G\sqrt{\rho_g})^{0.5}, \tag{16.12}$$

where Ψ is the empirical factor, $m^{0.25}h^{0.5}/kg^{0.25}$, ρ_g is the vapor/gas density, kg/m^3, and Q_G is the volumetric vapor/gas flow rate, m^3/h.
- The term Ψ is an empirical factor and is a function of both the tray spacing and the densities of the vapor/gas and liquid streams.
- Depending on operating conditions, trays are spaced with a minimum distance between them to allow the vapor/gas and liquid phases to separate before reaching the tray above.
- Consideration should be made in spacing to allow for easy maintenance and cleaning. Trays are normally spaced 45–70 cm (18–28 in.) apart depending upon the column operating pressure.
- Spacing for high-pressure columns should be on the lower side compared to atmospheric and vacuum columns. For vacuum columns spacing should be highest.

- What are the limiting values of column diameter while deciding on optimum diameter?
  - Diameter corresponding to flooding is the smallest.
  - Diameter based on minimum wetting rate is the largest.

16.1.2 Packed Columns

- Name different parts of a packed column.
  - Column, packing, support plate, liquid distributor/redistributor, hold-down grid, liquid collector, inlet and outlet nozzles, skirt and man ways.
  - Figure 16.15 illustrates a typical packed column.
- What are the desirable features of a packed column?
  - *For increased mass transfer efficiency*, wetted surface area per unit volume of the packed bed should be maximized.

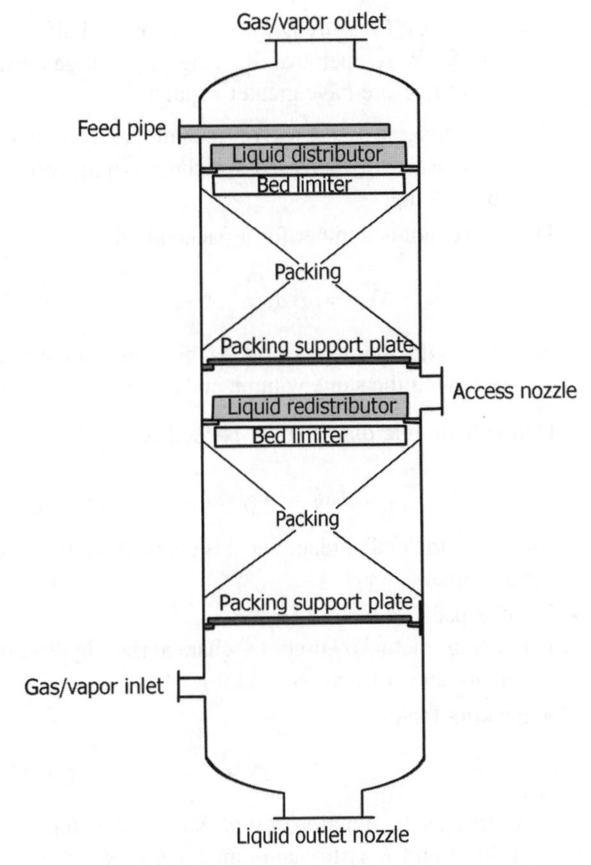

FIGURE 16.15 Packed column.

  - *For increased capacity*, average free cross-sectional area available for flow of vapors and liquid (area not blocked by the packing) should be maximized.
  - The above two objectives should be optimized while selecting packing.
  - For example, structured packing has about 50% more open area and two to three times more wetted area than Raschig rings. Due to this reason modern packed columns are equipped with structured packing and those in operation with conventional packing are revamping with structured packing.
  - The following objectives for packing lead to the goal of realizing these two overall objectives:
    - ➢ Maximize specific surface area.
    - ➢ Distribute surface area uniformly.
    - ➢ Maximize void space per unit column volume that minimizes resistance to vapor upflow thereby enhancing packing capacity.
      - For random packings capacity increases with packing size.
      - For structured packings capacity increases with space between adjacent layers.
      - Packing size that maximizes capacity also minimizes efficiency.

- *Minimize Pressure Drop:* For example, Pall rings are far more open than Raschig rings for gas flow and therefore have greater capacity.
- *Minimize Cost:* Cost of packing and supports increase with increased weight per unit volume of packing.

• Define Reynolds number for a packed bed.

$$N_{\text{Re}} = \rho_g v d_a / \mu_g (1-\varepsilon), \tag{16.13}$$

where d_a is the diameter of the packing equivalent to a sphere having the same volume and ε is void fraction.

• Define hydraulic diameter for packed beds.

$$d_h = 4\varepsilon / s, \tag{16.14}$$

where ε is the bed voidage and s is the packing specific surface area.

• What is packing factor? Define.

- Packing factor is used to characterize hydraulic performance of random packing.
- Packing factor,

$$F = Ks/\varepsilon^3, \tag{16.15}$$

where F is the function of type, size, and material of packing and K is the constant for a given class of dumped packing.

• "Packing factor is constant at low flows but is a function of flow rates at high flows." *True/False?*

- *True.*

• Explain what is meant by wall effect in a packed bed?

- In a packed bed liquid distributed on top of the column tends to flow toward the walls of the column as it descends through the bed. This effect is much more pronounced in small diameter columns.
- Once the liquid reaches the walls of the column, it flows down along the column walls without flowing toward the center of the packing, necessitating redistribution of the liquid into the central area of the column.
- This amounts to liquid and gas/vapor streams bypassing without contacting each other, which phenomena is known as *channeling*.

• What are the flow regimes in packed beds involving gas–liquid contact? Explain.

- Film flow regime, film spray regime, and spray regime.
  - *Film Flow Regime:* For structured packing made out of crimped wire mesh, wire gauge, and sheet metal, there will be thin liquid film on the packing surface.
  - *Film Spray Regime:* For random packing, a portion of the liquid flows as film on the packing and portion as liquid droplets falling from one piece of packing to the next.
  - *Spray Regime:* For low surface area random and large crimped structured packing, majority of liquid as droplets.

• Define wetting rate for a packed bed.

- Volumetric liquid rate per unit cross-sectional area divided by packing surface area per unit volume.

• What are the recommended minimum wetting rates for random packings?

- 0.35×10^3 to 1.4×10^3 m^3/(s m^2) cross section.

• What are the factors on which minimum wetting rate depends with respect to falling films?

- Depends on geometry and nature of vertical surface.
- Liquid surface tension.
- Mass transfer between the surrounding gas and liquid.
- For water at room temperature, minimum wetting rate varies from 0.03 to 0.3 kg/(m s).

• What are the differences between wetted area and effective surface area of packing in a packed column?

- Effective surface area, a_e, is the surface area of the interface between the gas/vapor and liquid phases per unit volume of the packing. It is the sum of the effective wetted surface area of the packing and the surface area of the jets and drops trickling in the free spaces. It is simply the effective surface area available for mass transfer processes.
- Wetted area of the packing, a_w, need not be same as effective surface area, a_e. The differences between them are as follows:
  - A part of the wetted area does not participate in the mass transfer process because of capillary forces at the contact points of packing elements. This area does not take part in mass transfer process and should not be considered for mass transfer.
  - The surface area of the drops and jets trickling in the free volume of the packing is not a part of the wetted area, but in some cases it is a significant part of the effective surface of the packing.
  - Both wetted and effective surface areas depend on the wettability of the material of the packing, the liquid properties, especially surface tension and on the liquid superficial velocity.

• What is angle of wettability? What is its influence on the effective surface area?

- The forces responsible for the holding of the liquid phase on the surface of the packing are the intermo-

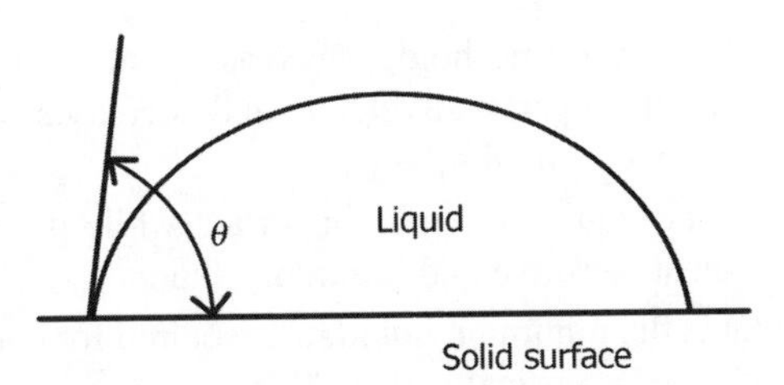

FIGURE 16.16 Angle of wettability.

lecular adhesion forces. Their effect can be measured through the angle of wettability, which is illustrated in Figure 16.16.

- The angle of wettability varies from 0° (fully wetted) to 90° (not wetted material).
- The difficulty in measuring this angle, dependent on the packing material, its preliminary treatment and gas/vapor and liquid properties, results in not using this parameter in correlations for calculating the effective packing area. As the angle is partially related to surface tension of the liquid, the later appears in the correlations. Decreased surface tension of the liquid increases wettability.
- As wettability is related to intermolecular forces, a nonpolar surface is better wetted by nonpolar liquids and vice versa. For example good wettability is achieved by treating a plastic packing with oxidants. This is one of the reasons why used plastic packing is wetted better than new packing.

• "Packed columns are not satisfactory when surface tension of the liquid is more than 10 dynes/cm." *True/False*? Comment.
  - *True.*
  - In packed columns, since liquid exists as a film, surface tension effects become more pronounced.
  - If pulsations are introduced, surface tension effects are minimized and liquids with surface tensions as high as 30–40 dynes/cm can be handled.

• "Packed columns almost always have lower pressure drop than comparable tray columns." *True/False*?
  - *True*. Three to five times lower. Pressure drops for trays are in the range of 10–12 cm/tray.

• Which of the following packings give lower ΔP? Raschig rings or Pal rings?
  - Pall rings.

• What are the generally recommended design pressure drops for various services of a packed column?
  - Design pressure drops are given in Table 16.3.

• Compare pressure drop characteristics of a packed column with a tray column.
  - Packed columns almost always have lower pressure drop than comparable tray columns.

TABLE 16.3 Recommended Design Pressure Drops for Packed Columns

	ΔP/Unit Packing Height	
Service	cm H_2O/m	H_2O/ft
Absorbers and regenerators		
Nonfoaming systems	2.1–3.3	0.25–0.4
Moderately foaming systems	0.8–2.1	0.15–0.25
Fume scrubbers		
Water scrubbing	3.3	0.4–0.6
Chemical solvent scrubbing	2.1–3.3	0.25–0.4
Atmospheric or pressure distillation	3.3–6.7	0.4–0.8
Vacuum distillation	0.8–3.3	0.15–0.4
Maximum for any system	8.33	1.0

  - Range of ΔP per tray for normal operation of a tray column is 3–8 mmHg.
  - For random packing in a packed column per HETP ΔP is 1–2 mmHg and for structured packing it is 0.01–0.8 mmHg per HETP.

• What is the most important consideration (other than contact area) in the selection of packing for a vacuum distillation column?
  - ΔP should be very low.

• "Packed columns should operate near (1) 90%, (2) 50%, or (3) 70% flooding conditions"? Choose the right answer.
  - 70% flooding condition.

• "A properly designed packed column can have 20–40% more capacity than a plate column." *True/False*?
  - *True.*

• What are the problems involved in the use of packed columns for distillation?
  - Poor liquid distribution and channeling problems.
  - Dry (unwetted) packing pockets.
  - Crumbling of packing due to thermal stresses or lack of strength.
  - Cleaning problems when dirty fluids are handled.

• Why conventional packed columns are not recommended for high capacities involving large diameters?
  - Uniform liquid distribution is critical in the performance of a packed column.
  - In large diameter columns, poor liquid distribution and channeling problems are critical and require properly designed liquid distributors/redistributors that are expensive leading to choice of tray columns.

• "A *conventional* packed column is normally recommended for capacities involving less than 1 m diameter rather than a tray column." *True/False*?

- *True*. This is because of flooding, channeling, and other maldistribution problems involved in conventional packed columns.
- Packed columns using structured, high-efficiency packings, with very low pressure drops, are being used for high capacities involving large diameter columns that are competitive to tray columns.

- Explain the differences between static and operating liquid holdup in a packed bed?
  - There are two different types of liquid holdup in a packed bed, namely, static and operating.
  - Static holdup represents that volume of liquid per unit volume of packing that remains in the bed after the gas and liquid flows stop and the bed has drained. The static holdup is dependent on the packing surface area, the roughness of the packing surface, and the contact angle between the packing surface and the liquid. The static holdup with no gas flow is a small value that can be assumed to be a constant independent of liquid flow rate.
  - In addition, capillary forces will hold liquid at junctions between individual packing elements.
  - Well-designed column packings normally do not trap stagnant pools of liquid within the packing element itself.
  - Operating holdup is that volume of liquid per unit volume of packing that drains out of the bed after the gas and liquid flows to the column stop.
  - Total liquid holdup is the sum of the static holdup and the operating holdup.
  - Operating holdup primarily is a function of the liquid flow rate.
  - Liquid holdup increases with increasing liquid viscosity. The following examples illustrate the variations of liquid holdup with viscosity.
    - If the liquid viscosity is increased from 1 to 2 cP, the holdup will increase by 10%. At 16 cP liquid viscosity, the holdup will be about 50% greater. If liquid viscosity is reduced to 0.45 cP, the holdup will be about 10% lower. At a liquid viscosity of only 0.15 cP, the holdup will be reduced by 20%.
- What is liquid holdup for a packed bed?
  - Volume of the liquid in the packed bed divided by the volume of the packing in the bed.
    - The liquid holdup consists of two components, namely, static liquid holdup when there is no flow (liquid held in the packing due to wetting on stoppage of flow) through the bed plus the dynamic holdup when there is liquid flow through the bed, that is, during operation of the packed column.
    - The dynamic holdup increases with increase in liquid superficial velocity and decreases with increased liquid density.
    - The liquid holdup is important while processing heat sensitive and hazardous fluids.
- What is the minimum liquid rate required for complete wetting of a vertical surface?
  - The minimum liquid rate required for complete wetting of a vertical surface, for example, vertical surface of a packing, is about 0.03–0.3 kg/(m^2 s) for water at room temperature. The minimum rate depends on the geometry and nature of the vertical surface, liquid surface tension, and mass transfer between surrounding gas and the liquid.

16.1.2.1 Packed Column Internals

16.1.2.1.1 Packings

- What are the various types of packings used in packed columns?
  - Raschig rings, Pal rings, partition rings, Berl saddles, Intalox saddles, Nutter rings and different types of structured packings.
  - Figure 16.17 illustrates some common packings.
- Name the first, second and third generation packings.
  - *First Generation:* Raschig rings and Berl saddles.
  - *Second Generation* Intalox saddles, super Intalox saddles, Pal rings, Hy-Pak, and so on.
  - *Third Generation*:
    - Intalox metal packing, cascade mini rings, Nutter rings, Fleximax, and so on.
    - Sulzer packing (corrugated sheet type and wire gauze type).
    - *Others:* Mellapak, Flexipac, Montz, and so on.
- What are the plus and minus points in Raschig rings as column packings?
  - Cheap.
  - Good structural integrity.
  - Considerable side thrust on the column.
  - Internal liquid channeling more.
  - Directs more liquid to walls.
  - Low efficiency.
- "Berl saddles are more efficient than Raschig rings." *True/False*?
  - *True.*
- Which of the packings, Raschig rings or Pal rings, give lower ΔP?
  - Pal rings.

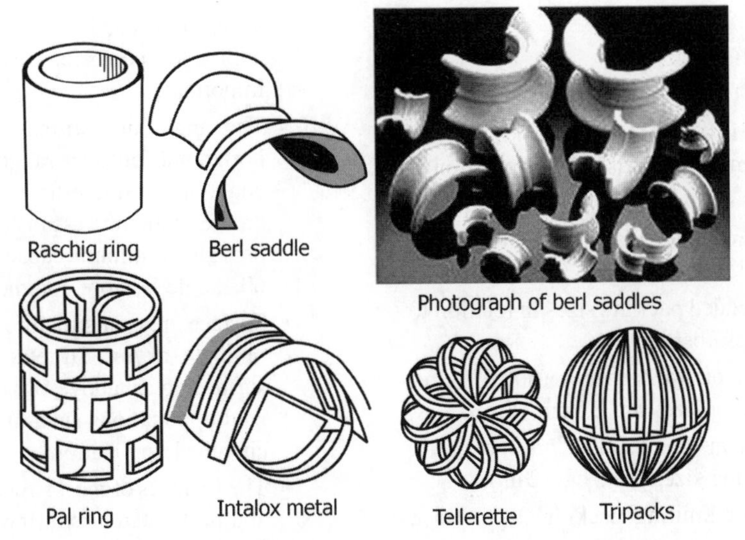

FIGURE 16.17 Some common types of random packings. (*Courtesy*: Koch Knight LLC for permission to use FLEXISADDLE^{TM}.)

- What are the general guidelines involved in the selection of packings for mass transfer equipment?
  - *Cost:* Generally plastic packing is less expensive than metal packing. Ceramic packing is the most expensive.
  - *Low* ΔP*:* Pressure drop is a function of the volume of void space in a tower when filled with packing. Larger packing size for a given bed size, gives lower ΔP.
  - *Corrosion Resistance:* Ceramic or porcelain packings are commonly used for corrosive atmospheres.
  - *Large Specific Area:* A large surface area per unit volume of packing, m^2/m^3 is desirable for mass transfer.
  - *Structural Strength:* Packing must be strong enough to withstand normal loads during installation, service, physical handling, and thermal fluctuations. Ceramic packing may crack under sudden temperature changes.
  - *Weight:* Heavier packing may require additional support materials or heavier tower construction. Plastics are much lighter than either ceramic or metal packings.
  - *Design Flexibility:* The efficiency of a column varies with the liquid and gas flow rates. Packing must be able to handle the process changes without substantially affecting efficiency.
  - *Arrangement:* Packing may be arranged in a column as random dumped packing or stacked packing. Randomly packed columns provide a higher surface area, m^2/m^3, but also cause a higher pressure drop than stacked packing. In addition to the lower pressure drop, stacked packing provides better liquid distribution over the entire surface of the packing. Installation costs are higher for stacked packing.
- What are the important geometric characteristics of packings?
  - *Size:*
    - Equivalent hydraulic diameter of the packing is given as $d_h = 4\varepsilon/a$ where ε is the void fraction and a is the specific surface area of the packing.
    - Another way of defining equivalent diameter is the diameter of a spherical packing having the same specific surface and void fraction as the real packing. It is given by equation,

$$d_e = 6(1-\varepsilon)/a. \tag{16.16}$$

  - *Specific Surface Area:* Packing surface area per unit bed volume.
  - Effective surface area of the packing, through which mass transfer takes place, is the wetted area of the packing including the area of the liquid drops and jets trickling through the free space of the packing per unit bed volume.
  - *Void Fraction:* Volume of the free space per unit volume of the packed bed.
- Is choice of packing size dependent on column diameter? If so, how do you relate packing size and column diameter?
  - Yes.
  - For random packing, size of packing should not be more than 1/8th of column internal diameter.
  - For different types of random packings, the maximum recommended sizes of packing are related to column diameter in the following manner:

- ➢ Raschig rings: 1:8 to 1:20.
- ➢ Berl saddles: 1:10.
- ➢ Intalox saddles: 1:8 to 1:10.

- What is the thumb rule in the selection of packing size in relation to column diameter for satisfactory operation of a packed column?
  - Ratio of column diameter to packing diameter should usually be at least 15.
- What are the recommended packing sizes in relation to gas flow rates in an absorber?
  - For gas flow rates of 14.2 m^3/min (500 ft^3/min), recommended packing sizes are 2.5 cm (1 in.).
  - For gas flows of 56.6 m^3/min (2000 ft^3/min) or more, recommended packing sizes are 5 cm (2 in.).
- What is the reason for limiting packing depths in a packed column?
  - Due to the possibility of deformation, plastic packing should be limited to an unsupported depth of 3–4 m (10–15 ft) while metallic packing can withstand 6–7.6 m (20–25 ft).
- What is the generally recommended percentage increase in packed height over the design value to account for uncertainties due to maldistribution and nonwetting of packing?
  - About 40% over design height.
- "Contact efficiency for a packed bed is lower for larger size packing." *True/False*?
  - *True*.
- "Replacing trays with packing allows greater throughput and separation in existing column shells." *True/False*?
  - *True*.
- "Packing is often retrofitted into existing tray columns to increase capacity or separation." *True/False?*
  - *True*.
- What are the ranges of HETP values obtainable with respect to pall rings?
  - 40–55 cm (1.3–1.8 ft) for 2.54 cm (1 in.) and 76–91 cm (2.5–3 ft) for 5 cm (2.0 in.).

16.1.2.1.2 Structured Packings

- What are structured packings?
  - Structured packings consist of mats of thin, corrugated, perforated metal sheets layered together in a vertical pattern. The metal sheets from which structured packings are made are generally thin with 0.2–0.2 mm in thicknesses. Since these are generally made from thin sheets, providing corrosion allowances may not be possible. In these cases careful choice of corrosion resistance material becomes important.
  - The angle of corrugations is typically 45° from horizontal, but can range from 30° to 60°. As the angle increases, efficiency and pressure drop decreases with increase in capacity. A 45° angle usually provides optimum capacity, efficiency and cost. A 60° angle is more common in gauge packings.
  - Surface areas of structured packings are in the range of 40–90 m^2/m^3. Structured packings with very high surface areas, above 500 m^2/m^3, are used in specialized applications such as in air separation and fine chemical applications.
  - The surfaces of most structured packings are textures and perforated. Texturing promotes easy wetting of the sheet surface. Perforations help liquid flow and pressure equalization between individual sheets. Sheets without surface texture and perforations give rise to lower efficiencies.
  - Wire gauze packings, fabricated from woven metal cloth are used for very high efficiency applications such as high vacuum service.
- Name suitable packings for low pressure drop and high-efficiency applications.
  - *Structured Packings:* For example, Sulzer, Flexi-Pak, and so on.
- Compare structured packings with random packings.
  - In general, structured packings achieve lower pressure drops and achieve better separation efficiencies than random packings.
  - Structured packings are more expensive than random packings and difficult to install.
  - Random packings are often preferred than structured packings or trays in corrosive services, especially where ceramics are required because of highly corrosive environments and high temperatures.
  - Smaller size random packings involve more specific surface than larger size packings but involve higher pressure drops and lower throughputs.
  - Structured packings with more sheets and smaller corrugations give rise to larger specific surfaces and higher efficiencies but involve higher pressure drops with lower throughputs.
  - Higher specific surface areas achieved are at a higher material costs. Chances of blockages are also higher.
- Name the important categories of structured packing.
  - *Knitted:* Wire mesh weavings/knittings.
  - *Nonknitted:* Sectionalized beds of corrugated or crimped sheets (usually somewhat thin) in contact with each other.

- Spiral-wound corrugated strips/ribbons coiling about a central axis:
  - Strips are made of knitted wire mesh or woven wire gauze.
  - Most suitable for small diameter columns.
- Grid type, open, heavy (usually metal) bar grid shapes stacked together.

• Give diagrams of typical structured packings.

- Figures 16.18–16.20 illustrate some structured packings.
- Efficiency increases with decrease in crimp angle.

• What are honeycomb packings? Illustrate.

- Honeycomb polymer packings (Figure 16.19) can be made from sintered plastics to obtain a packing to operate at very low liquid superficial velocities.

• What is flexeramic structured packing? What are its characteristics? Illustrate by means of a suitable picture.

- It is a ceramic structured packing manufactured by Koch Knight Company. Figure 16.21 gives a typical Koch Knight ceramic structured packing.
- These packings provide high capacity, higher efficiency, lower energy costs, and lower pressure drop than any random packing and most trays available.
- This packing, made out of ceramic material, is the best possible packing in acid applications.
- Ensure proper mixing.
- Its geometric construction reduces channeling of both the liquid and vapor and provides for more effective contact of liquid and gas.
- The corrugated sheets of ceramic are vertically oriented in the packed column, eliminating any horizontal surfaces that create resistance to fluid flow.
- Improve capacity by up to 50%, efficiency by up to 25%, and reduce ΔP by 60% or more over standard ceramic saddle packing.
- Good resistance to plugging.

• Show schematically, typical arrangement of expanded metal horizontal sheets of packing.

- In the arrangement, shown in Figure 16.22, liquid from sheet above drips onto the sheet opposingly arranged and breaks up, increasing turbulence for effective mass transfer, particularly for liquid film controlled systems.

• What are the advantages of structured packing over random packings or trays?

- Lower HETP (<0.5 m), that is, higher efficiency.

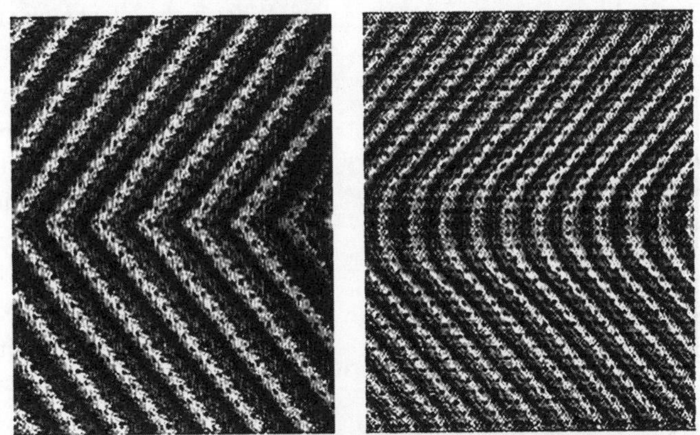

Sulzer Mellapak–smaller crimp angle Sulzer mellapak plus–larger crimp angle

Close-up picture of mellapack

FIGURE 16.18 Different structured packings (Mellapak). (*Courtesy*: Copyright © Sulzer Chemtech Ltd.)

FIGURE 16.19 Structured packing assembled to fit into a given column diameter. (*Courtesy*: Copyright © Sulzer Chemtech Ltd.)

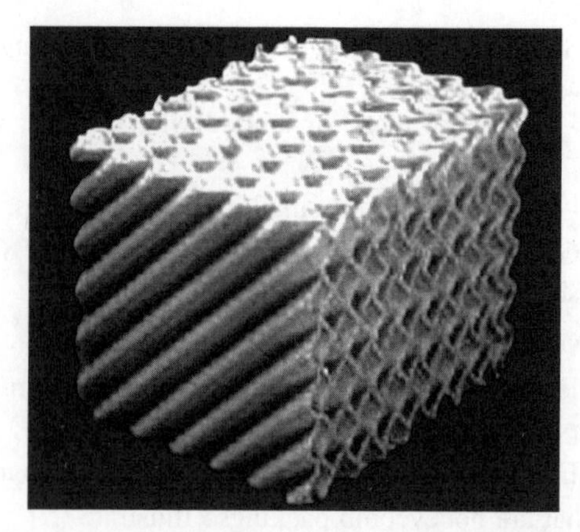

FIGURE 16.21 Ceramic structured packing. (*Courtesy*: Koch Knight LLC for permission to use FLEXERAMIC® TYPE 28 Packing.)

- Higher capacity.
- The wire gauze type structured packing is wetted by capillary action, so that the entire geometric surface area becomes available for mass transfer at *low liquid flow rates* compared to dumped packings.
- Lower ΔP ($\sim$100 Pa/m) and hence lower energy consumption. For example, the use of sheet metal structured packing will reduce ΔP per theoretical plate by a factor of 2 or more, compared to random packing. ΔP per theoretical tray for Intalox structured packings is only 40–65% of that for Intalox conventional metal packing of similar capacity.
  - These features make structured packing especially suited for use in vacuum distillation services where column size is controlled to a large degree by the ΔP per theoretical tray.
- Easier scale-up methods.
- Disadvantage is its high cost (proprietary in nature).

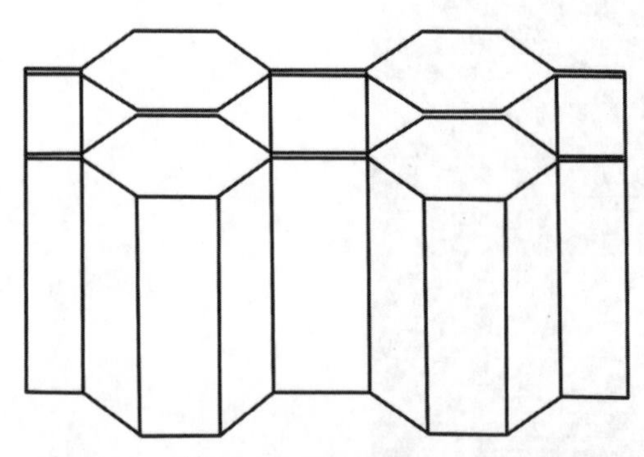

FIGURE 16.20 Honeycomb packings.

- What is the advantage of structured packing over conventional types for vacuum service?
  - Lower ΔP.
  - Permits lower pressures in the column and a reduction in vacuum requirements at the suction of vacuum-producing equipment.
  - Decreased bottoms temperatures with less degradation (cracking) of bottom product in vacuum distillation.
- What are the typical situations for which structured packings should be considered?
  - For debottlenecking to increase production capacities.
  - For energy savings through reduced ΔP or by allowing use of a heat pump and vapor recompression.

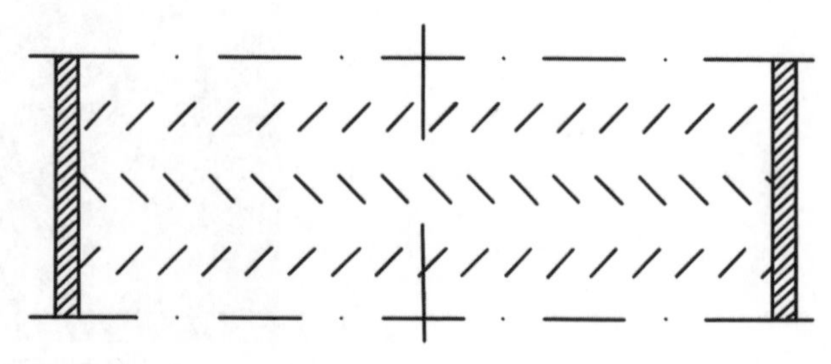

FIGURE 16.22 Typical arrangement of horizontal expanded metal sheets with opposing angles.

- To minimize inventory (holdup) in columns
  - when handling high-value products;
  - safety considerations when toxics/flammables are handled.
- Processing heat-sensitive materials preventing thermal degradation/polymerization.
- To increase product purity/yield in plants involving height or site limitations are involved.
- To cope with difficult separations requiring a large number of trays or large packing heights with conventional packings.
- Best suited for vacuum distillation (due to low ΔP and low liquid holdup).

• What are the mechanical constructional features of structured packing?
- Made from thin sheet metal with angled corrugation, textured, and perforated.
- Layers are segmented and rotated.
- Sheet metal typically 0.01–0.02 cm (0.004–0.008 in.) in thicknesses. Larger crimp packings may require more thickness.
- Essentially no corrosion allowance is provided.
- Material selection is critical.
- Gauze packings made from woven metal cloth are usually for very high efficiency applications.
- Typically textured and perforated.
  - Texturing promotes spreading of liquid on surface.
  - Perforation allows equalization of flows and pressures between sheets.
  - Lack of texture and/or perforation reduces efficiency.
- 45° angle is most common (Sulzer Y—Designation) and is usually the optimum angle for good efficiency, capacity, and cost.
- Second most common is 60° (Sulzer X—Designation). More often used in absorption and heat transfer applications where surface area is more important.
- Typical surface areas are 40–900 m^2/m^3. Lower surface area packings, which are often grid type, are used in heat transfer and absorption applications. Have areas in the range of 40–90 m^2/m^3.
- High surface area packings (>500 m^2/m^3) are used for air separation and fine chemicals applications requiring high efficiencies.

• Briefly state the applications of structured packing.
- High capacity.
- High efficiency.
- Low pressure drop.
- Proper distribution.
- Good for low pressure, high vapor velocities and low liquid rates.
- Ideal where large number of trays are required in distillation applications but not for super fractionators requiring high liquid rates and high pressures.

• What are the special applications of structured packing?
- For difficult separations requiring large number of conventional trays:
  - Isotope separation.
  - Separation of close boiling components.
  - For high-purity products.
- For high vacuum distillation (low ΔP).
- For column revamps (to increase capacity and reduce reflux ratio requirements).

• "Use of structured packings compared to conventional packings like Raschig rings can increase column capacity." Comment.
- Structured packings can be supported more easily. In other words, supporting grid made out of bars provides more spaces for flow.
- Bars can be placed less closely, increasing free cross section for flow, thereby increasing column capacity.

• "High performance structured packings can double the throughput (capacity) and triple mass transfer efficiency." Comment.
- Possible.
- Packing is relatively more expensive that may allow savings in auxiliary equipment with the result that overall investment may not increase.
- Instruments must be recalibrated when structured packing replaces trays in a column.

• What are the criteria for the selection of structured packing for distillation systems?
- Three criteria control the selection of structured packing: (1) liquid load, (2) vapor density, and (3) liquid to vapor density ratio. For distillation systems, the vapor density and the liquid to vapor density ratio are strongly linked.
- The lower the vapor density, the better structured packing performs compared to other devices. The higher the vapor density, the better trays or random packing perform compared to structured packing.
- For hydrocarbon systems the selection criteria for vapor density and liquid to vapor density ratio where structured packing are as follows:
  - Vapor density less than 24 kg/m^3.
  - Liquid to vapor density ratio greater than 18.
- Structured packings are typically used in lower liquid rates, below 50 $m^3/(m^2\,h)$, especially where minimization of column ΔP is important.

- When structured packings are used for liquid rates above 50 $m^3/(m^2 h)$, trays or random packings may be advantageous.

- What are the limitations in the use of structured packings?
  - Not good for high liquid rates. The definition of high liquid load varies greatly with packing geometry.
  - At high liquid rates 50–60 $m^3/(m^2 h)$ (>20–25 gpm/ft^2), efficiencies may decrease.
  - Liquid rates greater than about 53 $m^3/(m^2 h)$ (20 gpm/ft^2) for a 25 mm (1 in.) crimp structured packing should be used with caution.
  - High liquid loads make the liquid film on the structured packing thicker, reducing the area open for vapor flow. This increases vapor velocity. High vapor velocities make backmixing more likely.
  - Not good for high-pressure applications.
  - For hydrocarbon systems, structured packing should be restricted to applications with system pressures below 10 barg (165 psia) or lower. One recommendation restricts structured packing in hydrocarbon systems at pressures below 7 barg (115 psia).
  - For example, structured packing is a poor choice for most refinery gas and LNG gas plant columns.
  - An exception for high-pressure application is where the liquid density is independent of system pressure and remains high at all pressures is in glycol dehydration and amine sour gas scrubbing, for which structured packing are excellent.
  - For fractionation of high relative volatility liquids structured packings might prove to be more expensive than conventional packings.
  - High-viscosity liquids might give rise to plugging problems.
  - High surface tension may reduce performance.
- What are the process control problems that are likely to arise when a tray column is retrofitted with structured packing?
  - Column residence time comes down.
  - Response of instruments for any fluctuations of process variables such as temperature, pressure, stream compositions, and flow rates of liquid and vapor/gas will be *swift*.
  - Automatic control with precision instruments is a must for smooth operation.
- What is the usual range of pressure drops for structured packing?
  - 30–40 mm H_2O/m of packed height.
- How does liquid holdup compare for structured packing with conventional packing?
  - Much less than for conventional types.

16.1.2.1.3 Other Internals

- What are the common packed column internals other than packings?
  - Packing support plates.
  - Vapor distributors.
  - Bed limiters and hold-down plates.
  - Feed and reflux distributors.
  - Liquid redistributors.
  - Wall wipers.
  - Liquid collectors.
- What is the function of a support plate in a packed column?
  - The primary function of the packing support plate is to serve as a physical support for the column packing plus the weight of the liquid holdup, while allowing free passage of gas/vapor and liquid.
  - The support plate must pass both the downwardly flowing liquid phase as well as the upwardly flowing gas phase to the limit of the capacity of the tower packing itself.
    - ➢ It should be as open as possible. Net free cross-sectional area of support should be ≥65% of column area and more than free area of the packing itself. Ratio of open area/free area should be in the range of 70–100%.
    - ➢ If free flow area is restricted, the results will be higher ΔP, liquid build up on the plate and onset of flooding conditions.
  - ΔP should not exceed 0.75 cm (0.3 in.) H_2O for most applications.
  - It should prevent fall through of packings.
  - It should be noted that fouling causes increased blockages and higher ΔP and preventive and maintenance steps should be adequate.
  - In the design of the packing support plate, no allowance is made for the buoyancy due to the pressure drop through the packed bed nor for the support offered by the column walls.
  - In extremely corrosive services, ceramic support plates may be required.
  - In larger diameter columns (such as sulfuric acid plant columns), a series of ceramic grids is installed resting on brick arches or piers. Then a layer of cross-partition rings or grid blocks is stacked on these grid bars to support the dumped ceramic packing, as shown in Figure 16.23.
  - For structured packings, a simpler design will be adequate, because such packings are installed in discrete modules.

FIGURE 16.23 Stacked packing to support dumped packing.

- Support plates for such packings provide a horizontal contact surface designed to prevent distortion of the packing while possessing sufficient structural strength to support the weight of the packing and liquid holdup over the length of the span. Such support plates have a very high open area for vapor/gas and liquid passage and do not add any significant pressure drop.

• Name different types of support/grid plates.
  - Vapor–injection grids (Figures 16.24 and 16.25):
    - Separate paths for vapor and liquid flows (like in ripple trays).
    - This type of support plate permits openings with a greater flow area than cross section of the column.
    - Will not create condition of onset of flooding.
  - *Subway Gratings:* Light weight and made of corrugated expanded metal.
  - Load bearing capacity is limited 7200 kg/m^3 ($\approx$450 lb/ft^3).
  - Relatively low free area.
  - Lowest cost.
  - Suitable only for small columns.
  - Super support grid:
    - 95–97% free area.
    - Can handle relatively clean liquids.
    - Moderate to high vapor rates, in any diameter column.

FIGURE 16.24 Vapor injection support plate. (*Source*: Saint Gobain Norpro.)

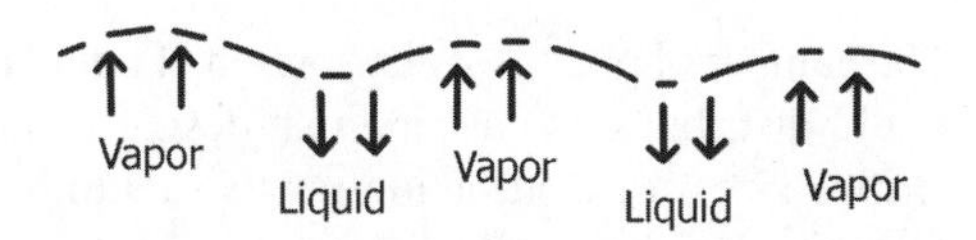

FIGURE 16.25 Schematic of vapor injection grid.

• What are the features of a vapor distributor?
  - Properly designed sieve or valve tray can work as vapor redistributor.
  - As shown in the figures for liquid distributors, chimneys are installed to cover the column cross section uniformly with chimney width of not more than 12.5 cm (6 in.).
  - For proper operation, a vapor distributor should have a reasonable pressure drop.
  - Distortion of valve caps during installation gives rise to poor distribution.
  - Usually only the packing support plate immediately above the vapor inlet needs to act as a vapor distributor. This support plate should be located at least one vapor inlet diameter plus 0.3 m above the centerline of the vapor inlet nozzle.
  - A vapor sparger could be used to produce uniform flow of vapor up the column. This approach to vapor distribution control frequently is used for high-pressure stripping steam, where sufficient pressure gradient is available to permit the use of multiple vapor orifices.

• Why is it necessary for liquid distributors in a packed column?
  - Effective packed column operation requires that the liquid and gas/vapor streams should be in intimate contact with each other throughout the entire column length and cross-sectional area.
  - Liquid introduced into the column at the top tends to flow down over a relatively small cross section of the column. The liquid flows in little streams down the column without wetting the entire packing area.
  - Liquid should be distributed over the entire cross section of the packing.
  - Once the liquid is distributed over the packing, it flows down through the packing, following the path of least resistance and tends to flow toward the column wall, where the void spaces are greater than in the center. The liquid flows straight down the column from the wall. This is called *channeling*.
  - Liquid redistributors send the liquid back over the entire surface of packing. It is recommended that redistributors be placed at intervals of no more than 3 m or 5 column diameters, whichever is smaller.

- The number of points for distribution of liquid from a liquid distributor is an important design parameter. Orifice sizes for distribution points should be carefully chosen. Too small orifices can be blocked due to fouling of the liquid.
- At high liquid rate, the cross-flow capability of a distributor and its predistribution system are important.
- Gravity fed distributors are dependent on liquid level to determine flow. Therefore, it is important to balance the liquid properly to provide uniform point-to-point flow.
- A distributor performs best at its design liquid flow rate. At some turndown rate, the point-to-point flow variation will fall outside of acceptable limits.

• What is the effect of liquid distributor design on the performance of a packed column?
  - Maintaining liquid to vapor ratio inside the column is critical in its operation. Any deviation from this ratio caused by poor or uneven liquid distribution affects separation. This lack of separation often is misinterpreted as poor packing efficiency, that is, a higher than expected HETP. The mass transfer capability of the packing might not have changed but poor liquid distributor design might have caused a change in L/V ratio and thus loss in separation efficiency.
  - The deviation from the desired L/V ratio in some areas of the column may result in equilibrium pinching, that is, a condition in which packing will not effect further separation, no matter what depth of packing is available.
  - Uniform liquid distribution is key to obtaining expected performance from a packed bed. The distributor should uniformly allocate the liquid for all anticipated flow rates, with an adequate number of liquid admission points for the size of the packing and sufficient open area for vapor passage.
  - The local variations, in terms of the L/V ratios, may cause compositional pinch leading to no change in composition along the column. Liquid maldistribution over a large section of the column cross-sectional area may lead to uneven liquid flow through the packing, concentrating near the wall.
  - When a liquid distributor is operated at rates below its designed turndown ratio, it will not provide uniform flow point to point and across the column cross section because of low liquid head above the orifices. As the flow uniformity diminishes separation efficiency deteriorates.

• What are the different types of liquid distributors?
  - Liquid is distributed over the packing material by the use of weirs, tubes, or spray nozzles.
  - Gravity type: V-Notched channel type that can handle liquids with solids and most widely used for ≥ 1 m (3 ft) diameter columns.
  - Trough and weir type (Figure 16.26).
  - The weir distributors must be installed level as otherwise distribution will be uneven.
  - Weir and notched trough distributors generally give poor liquid distribution. They also are susceptible to creating liquid entrainment, because they discharge liquid into the vapor riser area where vapor velocity typically is three times higher than the superficial column vapor velocity even when the packed bed is running at reasonable rates. That can lower separation efficiency by liquid backmixing.
  - However, this type of distributor may be a good choice if plugging and fouling are serious problems.
  - Pan or orifice type:
    - ➢ Similar to sieve tray in operation.
    - ➢ Preferred for small diameter columns.
    - ➢ Used for clean liquids.
    - ➢ Also used as liquid redistributors.
    - ➢ Pipe orifice headers:
      - High free area.
      - Can handle low liquid rates.
      - For any diameter.
      - For clean liquid surface.
  - A typical orifice type liquid distributor is illustrated in Figure 16.27.
    - ➢ Orifice plate distributor is the most common liquid distributor for general nonfouling applications.
    - ➢ Vapor flows through the large chimneys and liquid drains through the orifices.
    - ➢ Height of liquid on the distributor tray depends on densities of vapor and liquid and vapor velocity.

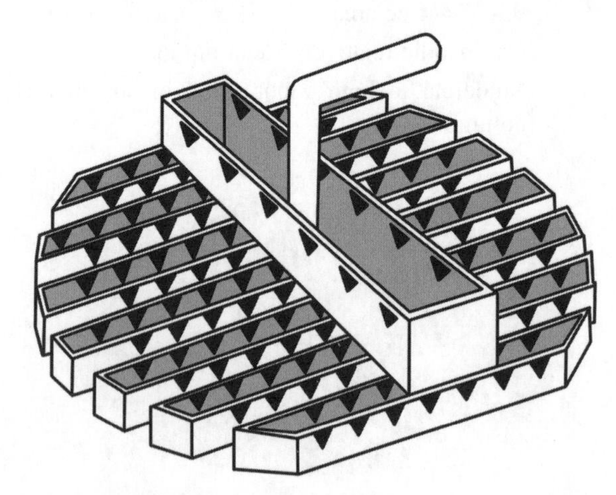

FIGURE 16.26 Trough and weir type distributor. (*Courtesy*: Kotch-Glitsch LP.)

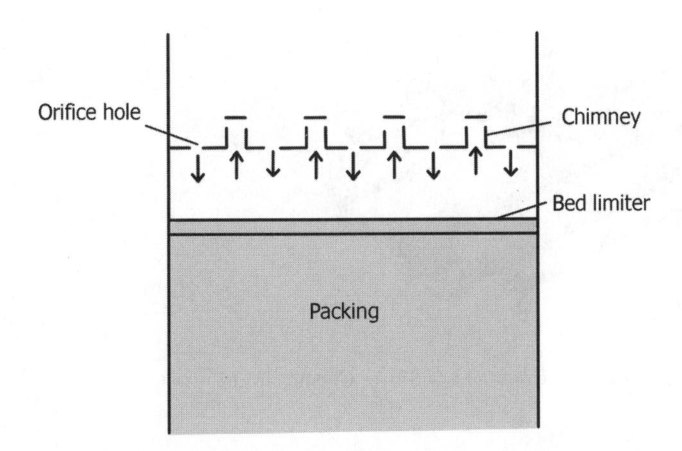

FIGURE 16.27 Orifice plate liquid distributor.

- ΔP of vapor through chimneys:

$$\Delta P_V = K_c(\rho_V/\rho_L)(v^2). \quad (16.17)$$

- ΔP_V is in inches of liquid.
- v is a vapor velocity, ft/s.
- K_c is a constant in the range of 0.4–0.6.

$$\Delta P_L = K_o V_l^2. \quad (16.18)$$

- ΔP_L is in inches of liquid.
- V_L is liquid velocity through the orifice holes in ft/s.
- Total height of liquid in the distributor = $\Delta P_V + \Delta P_L$.
- The distributor tray should be level as otherwise, maldistribution leading to channeling occurs.

- Perforated tube type is illustrated in Figure 16.28.

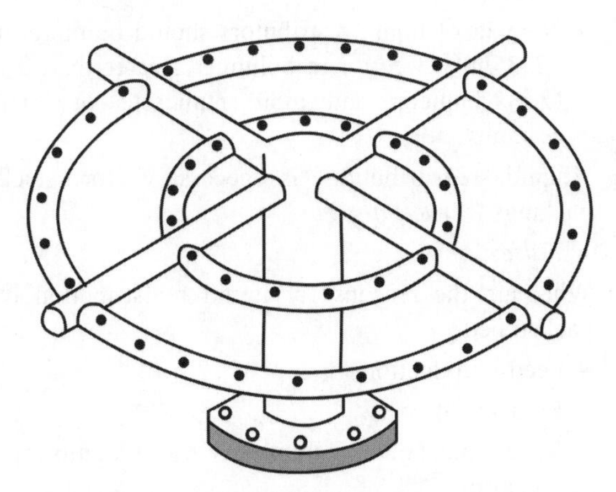

FIGURE 16.28 Perforated tube type distributor.

- The perforated tube provides good liquid distribution patterns, but the holes are subject to plugging if any particles or contaminants are in the liquid.
- The drilled tube is often buried within the packing. This prevents the liquid from being blown against the sidewalls of the column. Burying the tube also allows the packing above the tube to act as an entrainment separator for countercurrent flow columns.
- Packed columns, designed with spray nozzles to distribute liquid, operate better with a few large nozzles than with many small nozzles. Large nozzles are less susceptible to plugging.
- Small nozzles that produce a finer spray are not needed in a packed column.
- Figure 16.29 illustrates a typical spray nozzle type liquid distributor.
- The spray nozzle distributor poses entrainment problem because it breaks the liquid up into small droplets that can be easily carried back to the column.
- The advantages and disadvantages of weir, tube, and spray type liquid distributor are listed in Table 16.4.
- A multipan liquid distributor is illustrated in Figure 16.30.
  - Figure 16.30 is for illustrative purposes only. Depending on the diameter of the column, more number of pans incorporated one below the other with lateral tubes to cover the cross section uniformly.
  - More efficient than orifice type.
  - More expensive.
  - Difficult to design for small liquid flows.
  - Occupy more column height.
  - Difficult to install properly.
- Pressurized feed type:
  - Liquid feed is under pressure.
  - Minimizes splashing and mist generation.
- Flashing feed type:
  - Absorbs and controls destructive forces of incoming feed.
  - Permits complete disengagement of liquid and vapor.
  - Baffle type: For small diameter columns.
  - Gallery type: Good for foaming/frothing liquids.

- What are the parameters to be considered in the selection of liquid distributors?
  - Column diameter.
  - Specific liquid load, $m^3/(m^2\,h)$.
  - A variety of proprietary designs are available from different manufacturers, whose recommendations

FIGURE 16.29 Spray nozzle type liquid distributor. (*Courtesy*: Copyright © Sulzer Chemtech Ltd.)

TABLE 16.4 Advantages and Disadvantages of Different Types of Liquid Distributors

Distributor	Advantages	Disadvantages
Weir	Can handle dirty liquids with high solids content	Expensive
		Distribution is not as uniform as other types
	Can be easily inspected and maintained, if access is available	Weirs must be level
Tubes	Uniform liquid distribution	Easily plugged
	Can be buried below packing	Must use filter
	Generally least expensive	Difficult to know if holes are plugged when buried in packing
Spray nozzles	Uniform liquid distribution	Highest ΔP and operating costs
	Column need not be plumb	Easily plugged
	Can be easily inspected and maintained, if access is available	Must use filter

and prior experience, if any, will be the guiding points in the final selection for a particular application.

- What are the important criteria in the selection and design of a liquid distributor?
  - A well-designed liquid distributor should provide uniformity of flow, appropriate liquid admission point density across the column cross section, proper irrigation along the column wall area, sufficient open area for vapor flow, and entrainment prevention.

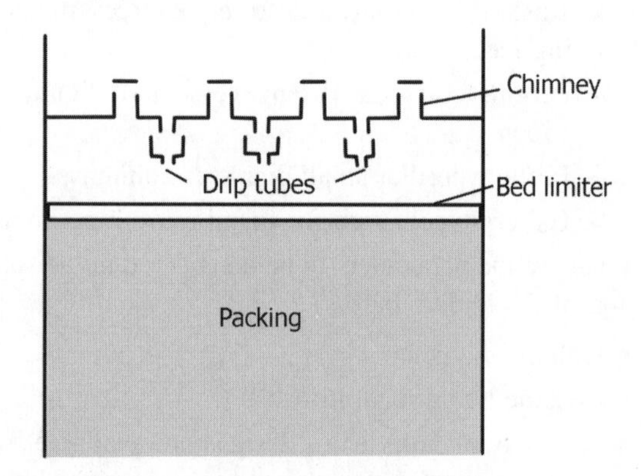

FIGURE 16.30 Multipan liquid distributor.

- What are the guidelines for placing liquid distributors in a packed column?
  - Liquid distributor should be placed every 5–10 column diameters (along the length) for pall rings and every 6.5 m (20 ft) for other types of random packings.
  - Number of liquid distributors should be more than 32–35/m^2 (3–5/ft^2) in columns greater than 0.9 m (3 ft) diameter and more numerous in smaller columns.
- "Liquid redistribution is necessary for stacked packings." *True/False*?
  - *False*.
- What are the reasons for liquid redistribution in a packed bed?
  - Feed introduction.
  - Product side draw.
  - Requirement of large number of transfer units for the separation.
  - Desire to cross-mix the liquid.

  - Liquid maldistribution.
  - Physical weight of the packed bed.
  - A conservative rule of thumb is to limit a single packed bed of plastic packing to not more than 10 transfer units.
  - In addition, the depth of a single bed should be not more than 15 times the column diameter.
- What is the function of a liquid collector?
  - For collection of the liquid from an upper bed for redistribution or liquid drawoff as side product.
- What are the different types of liquid collectors?
  - Riser type:
    - Similar to pan or orifice liquid distributor.
    - For atmospheric or higher pressure applications.
  - Chevron type:
    - Small ΔP.
    - Best suited for vacuum service.
    - Large free area (>90%).
- Why is it necessary to have packing restrainers/hold-down plates/bed limiters in a packed column?
  - To maintain bed top surface level. A nonlevel bed promotes vapor maldistribution and may cause liquid entrainment if the packing level approaches the distributor.
  - During normal operations, weight of packing will keep the packing in place.
  - At high vapor/gas rates or during an upset, such as a surge in pressure that accompanies, for example, a slug of liquid water entering a hot oil column due to maloperation or gas pockets or uneven loading, top layers of packing can be fluidized/breakup (particularly ceramic/carbon packings can break up), leading to blowing out or pieces filtering down and plugging the column.
  - Bed limiters are used to prevent expansion particularly with light-weight packings like plastic and metal packings, as well as to maintain the bed top surface level. If packing is lifted up and enter liquid distributor, performance of the later can be adversely affected.
  - Bed limiters are commonly used with metal or plastic column packings.
  - Bed limiters for structured packing are recommended only when there is the potential for packing displacement during upset or flooding conditions. If columns operate at a low pressure drop with a low percentage of flood and are not prone to sudden vapor surges, bed limiters are not needed.
  - In cases where upset conditions are not a concern, a bed limiter resting on the structured packing can be used as a support for a trough or pan distributor.
- How are bed limiters installed?
  - Normally, the upper surface of the packed bed is at least 12 cm below a liquid distributor or redistributor. The bed limiter or hold-down plate is located on top of the packed bed in this space.
  - It is important to provide such a space to permit gas disengagement from the packed bed. Such a space allows the vapor/gas to accelerate to the velocity necessary to pass through the distributor without exceeding the capacity of the packing.
  - Spot welding hold-down grid clamps, use of stainless steel tie rods between the support and hold-down grids, welding nuts are some other solutions to prevent upsets in packings.
  - Bed limiters are fabricated as a light-weight metal or plastic structure.
  - Because the bed limiter rests directly on the top of the packed bed, structurally it must be only sufficiently rigid to resist any upward forces acting on this packed bed.
  - In large diameter columns, the packed bed will not fluidize uniformly over the entire surface. Vapor surges fluidize random spots on the top of the bed, so that after return to normal operation, the bed top surface is quite irregular.
  - Thus, the liquid distribution can be affected adversely by such an occurrence. Usually a mesh backing is used to prevent passage of individual pieces of column packing.
- What is the difference between bed limiters and hold-down plates?
  - Hold-down plates are weighted plates used with ceramic or carbon column packings.
  - With these packings, it is especially important to prevent fluidization of the packed bed top surface.
  - These brittle materials can be fractured during an operating upset or at high vapor loadings, so that the resulting fragments migrate down into the packed bed where they can severely reduce column capacity.
  - The hold-down plate must rest freely on the top of the packed bed because beds of ceramic and carbon packings tend to settle during operation.
  - These plates usually act by their own weight to prevent bed expansion. The plates weigh 320–480 kg/m^3 (20 lb to 30 lb/ft^2).
  - The free space through these plates must be high enough not to restrict the capacity of the column packing.
- What is a wall wiper? What is its application?
  - In small diameter columns the column wall surface area is substantial when compared to the total packing surface area.

- As the column diameter becomes larger, the wall area diminishes in significance compared to the packing area.
  - ➢ Wall wipers are generally not needed if the column diameter is more than 10 times the packing diameter (size).
- If the liquid reaching the column wall continues to flow down the wall, it represents a bypassed stream that reduces the overall separation efficiency.
- In small columns with a high percentage of wall flow, wall wipers are installed frequently.
- Wall wipers fit tightly against the column wall to intercept all of the liquid flowing down the wall.
- The wall wiper is used to remove this liquid from the wall and place it into the packed bed where it will be adequately contracted with the rising vapor phase.
- Generally, wall wipers are required only in the lower portion of the stripping section.
- These devices are usually spaced apart by about two HETP.
- Because wall wipers are installed within the packed bed itself, they must be designed carefully to avoid severely reducing the column capacity.
- Figure 16.31 illustrates a typical design for a wall wiper.

FIGURE 16.31 Wall wiper liquid redistributor. (*Source*: Norton.)

- Why and where is it necessary to install liquid collectors in a column?
  - Sometimes it is necessary to intercept all of the liquid flowing down the column.
  - If the lower portion of the column is of a larger diameter than the upper portion, the liquid must be collected at the bottom of the smaller diameter section. It is then fed to a redistributor located at the top of the larger diameter section to irrigate uniformly the lower section.
  - If the lower portion of the column is of a smaller diameter than the upper portion, the liquid must be collected at the bottom of the larger diameter section. It is then fed to a redistributor at the top of the smaller diameter section to prevent excessive wall flow in the lower section.
  - In some operations, liquid is removed and recirculated back to the top of the same packed bed after cooling.
  - All such cases require liquid collection plates.
- What is the recommended range of heights of packed beds?
  - 6–9 m (20–30 ft).
- What is the recommended practice relating depth and type of packing with location of liquid distributors?
  - In general, liquid distributors should be placed every 5–10 column diameters for pall rings and every 6.5 m (20 ft) for other types of random packing.
  - *Raschig Rings:* Recommended depth is 3 column diameters and not more than 3.6 m (12 ft) over support plate. For multiple beds, each not exceeding 3–4.5 m (10–15 ft)/section.
  - *Saddles:* 5–10 column diameters. Not more than 4.5–6 m (15–20 ft) over support plate. For multiple beds, each not exceeding 3.5–6 m (12–20 ft)/section.
  - Depth of packing without intermediate supports is limited by its deformability. Metal construction is limited to depths of 6–7.6 m (20–25 ft) and plastic to 3–4.6 m (10–15 ft).
  - Intermediate supports and liquid redistributors are provided for deeper beds and at side stream withdrawal or feed points.
  - Liquid redistributors usually are needed every 2–3 column diameters for Raschig rings and every 5–10 diameters for pall rings, but at least every 6 m (20 ft).
- In the design of liquid redistributors, how are the streams (liquid outlets) distributed?
  - For redistribution, there should be 80–120 streams/m^2 of column area for columns larger than 1 m (3 ft) in diameter.
  - They should be even more numerous in smaller columns.
- What is the range of ΔP per HETP for (i) random packing and (ii) structured packing?
  - Random packing: 1–2 mmHg.
  - Structured packing: 0.01–0.8 mmHg.
- What are the recommended ranges of pressure drops for random packings in absorbers/strippers and distillation (atmospheric) columns?
  - Absorbers/strippers: 15–50 mm H_2O/m depth.
  - Atmospheric distillation columns: 40–80 mm H_2O/m depth.

- What are the advantages of small liquid holdup?
  - Sharper separation.
  - Small inventory of flammable/toxic liquids involves better safety.
  - Short liquid residence times are advantageous for heat-sensitive materials. Thermal degradation of fluids handled is minimized.
  - Lower ΔP_{gas}.
  - Shorter column drainage times on shutdown.
- What is the effect of packing size on liquid holdup?
  - Holdup decreases with increasing packing size.
- What is static liquid holdup in a packed column?
  - Represents liquid held in the packing after a period of drainage time, usually until constant weight of material is reached.
  - Independent of liquid and gas rates.
  - Total holdup = static holdup + operating holdup.

16.1.2.2 Packed Versus Tray Columns

- Compare packed and tray columns.

Packed Columns

- Packing is often retrofitted into existing tray column, to increase capacity or separation. Thus, same size of packed columns can handle more than tray columns.
- *Lower ΔP:* The interfacial area for packed columns is generated through liquid spreading on the packing surface, which is a low pressure drop phenomenon compared to the mechanism required to generate high mass transfer efficiency within a tray column.
  - Packing ΔP is much lower because the open area of the packing is much larger than that of trays and the liquid head on the trays incurs substantial ΔP, which is absent in packing. This is often the reason for specifying packed columns for vacuum service where reboiler pressures and temperatures are low.
  - Typically, random packing ΔP is about 1/3rd of that for trays. Typically, replacing 25 trays in a gas absorber by random packing reduces ΔP from about 20 kPa (3 psi) to about 7 kPa (1 psi). This ΔP can be reduced to about 3.5 kPa (0.5 psi) when structured packing is used, but this alone is not justified due to cost considerations involved in structured packing over random packing.
  - A packed column with structured packing may involve a pressure drop of 2.54 cm (1 in.) liquid, whereas a tray column may involve a pressure drop of 16 cm (6 in.) of liquid. This aspect is very important especially for vacuum columns.
- *Lower Liquid Holdup:* For example liquid holdup for tray columns is typically about 8–12% of column volume. For packed columns it is about 1–6% of column volume. For high-efficiency structured packings it can be much lower.
- *Foaming:* Good handling of foam forming liquids. The lower local velocities through packings suppress foam formation. The large open area of the larger random packing makes foam dispersal easy.
  - Foaming resistance of trays with well-designed downcomers is comparable to that of random packing, as foams are known to be bottlenecks to downcomers rather than trays.
  - Switching from trays to structured packing can aggravate foaming.
- Preferable for liquids with high foaming tendencies.
- Preferred for corrosives (ceramics can be used for column packing). Packed columns can be constructed with a fiberglass-reinforced polyester shell that is generally about half the cost of a carbon steel tray column.
- *Maldistribution:* Liquid and vapor/gas maldistribution is a common cause of failure in packed columns. Maldistribution problems are most severe in large diameter columns, low liquid flow rates and smaller packing sizes. Structured packing is generally more prone to maldistribution than random packing.
- Preferable to tray columns for small installations, corrosive service, foaming liquids, very high liquid to gas ratios, and low pressure drop applications.
- Packed columns offer greater flexibility because the packing can be changed with relative ease to modify column operating characteristics.
- Simpler and cheaper to construct.
- Other advantages of packed columns include requirement of shorter column heights for the same duty, mechanical simplicity, and ease of installation.

Tray Columns

- *Plugging:* Less susceptible to plugging. Trays can handle solids much better than packing. However, not all trays are resistant to fouling. Floats on moving valve trays tend to *stick* to deposits on tray deck. Fouling-resistant trays have larger sieve holes or larger fixed valves and these should be used when fouling and plugging are the primary considerations.
  - Both vapor/gas velocities on trays are much higher than those through packing, providing a sweeping action that keeps tray openings clear. Solids tend to accumulate in the voids of packing.
  - There are fewer locations on trays where solids can be deposited. Plugging of liquid distributors in packed columns is a common trouble spot. Distributors that resist plugging have larger hole

sizes. While such larger holes are readily acceptable for high liquid flow rates, they are a problem for low liquid flow rates.
 - ➢ Cleaning trays is much easier than random packing and cleaning of structured packing is next to impossible.
 - ➢ Preferred for fouling liquids or gases with particulates.
- ▪ Preferred for operations that involve difficult gases to absorb, requiring large number of transfer units, because of absence of channeling or that must handle large gas volumes. To achieve the same separation efficiency for difficult absorption processes, packed columns must have either deep packed beds or multiple beds. Packed columns can experience liquid channeling problems if the diameter or height of the column is too large. Redistribution trays must be installed in large diameter and tall packed columns to avoid channeling.
- ▪ Less of a problem with channeling.
- ▪ Better efficiency over wider range of liquid rates.
- ▪ Less sensitive to maldistribution.
- ▪ Temperature surges resulting in less damage.
- ▪ Preferred when internal cooling is required in the column. With appropriate tray design, cooling coils can be installed on individual trays, or alternatively, liquid can be removed from the column from one tray, cooled and returned to another tray.
- ▪ High turndown ratios. Packed columns cannot handle volume and temperature fluctuations as well as tray columns. Expansion or contraction due to temperature changes can crush or melt packing material.
- ▪ Tray columns are intrinsically cheaper than packed columns, because far less surface is needed for trays than for packings and trays require far lower cost internals than packings.
- ▪ Well suited for large installations, noncorrosive, and nonfoaming liquids and low to medium rate liquid flow rates.
- ▪ Robustness.
- ▪ Less weight.

Additional Comments

- ▪ When the service is nonfouling, either type of column can be designed to yield comparable theoretical tray counts per section height.
- ▪ Packing has the disadvantage of requiring redistributors if large tray counts are required.
- ▪ Fouling can pose problems for both trays and packing.
- ▪ Packed columns often can be designed with greater stable operating range than sieve trays.
 - ➢ Valve and bubble cap trays can have a stable operating range equal to or even greater than that of columns with random or normal corrugated structured packing.
 - ➢ Structured packing made of expensive woven materials can have a very broad operating range.
- ▪ The design of internals for a large stable operating range is much more difficult for packings than for trays.
- ▪ In general, trays are used in applications involving liquid flow rates of $30\,m^3/(m^2\,h)$ and above.
- ▪ For low-pressure distillation applications, liquid flow rates tend to be very low. This promotes spray like conditions with trays that, in turn, promote high entrainment and low tray efficiency.
 This can be mitigated with small perforations on trays, but packed columns are preferable, if liquid rates are sufficient for good wetting.

- Summarize disadvantages of packed columns.
 - ▪ At low liquid flows, liquid distribution will be difficult.
 - ▪ Proper vapor distribution is essential.
 - ▪ Possibilities of restrictions in vapor flow from grid support or hold-down plate.
 - ▪ Inspection of packing, especially inside packed sections will be very difficult. Packing can get deformed/damaged by maintenance personal who might trample them while installing.
- Compare a packed column with a tray column with respect to capacity to handle vapor and liquid loads.
 - ▪ A well-designed packed column will have 20–40% more capacity than a tray column.
 - ▪ Downcomers in tray columns typically occupy 30% of cross-sectional area of the column while the entire cross-sectional area of a packed column is available for vapor/gas–liquid contact.
- Compare tray and packed columns with respect to liquid holdup.
 - ▪ Liquid holdup for tray columns is about 8–12% of column volume.
 - ▪ For packed columns, liquid holdup is of the order of about 1–6% of column volume.
- "A properly designed packed column can have 20–40% more capacity than a tray column." *True/False?*
 - ▪ *True.*
- "Tray columns can operate efficiently over a wide range of liquid flow rates than can packed columns." *True/False?*
 - ▪ *True.*

- "Tray columns can operate efficiently over a wide range of liquid flow rates than packed columns." *True/False?*
  - *True.*
- "Packing ΔP is about 3–5 times *lower* than tray ΔP for the same HETP." *True/False*?
  - *True.*
- What is the range of ΔP per tray for normal operation of tray columns?
  - 0.4–1.0 kPa (3–8 mmHg).
- What are the recommended pressure drops across trays?
  - 10–13 cm (4 5 in.) H_2O.
- Give possible reasons for increase in ΔP in the operation of a packed column?
  - Increased liquid/vapor loads.
  - Compaction/deflection of plastic packings.
  - Breakage of ceramic packings.
  - Unsteady column operations.
  - Fouling of packings.
- What are the consequences that arise if packing is damaged during installation in a packed column?
  - Vapor and liquid flow paths get restricted resulting in increased ΔP.
  - In case of vacuum distillation, higher ΔP will cause rise in boiling temperatures, affecting quality of heat-sensitive materials.
  - Localized flooding may occur due to blockage of liquid and vapor flow paths. This can also occur if packing is fouled.
- How is localized flooding/damaged packing detected in a column operation?
  - By measuring ΔP across packed sections, if taps for such measurement are provided.
  - When $\Delta P \geq 10$ cm (4 in.) liquid, liquid drainage reduces/stops, leading to incipient flooding.
  - By using an isotope scanning technique.
- What are the possible reasons for fouling of packing?
  - Presence of contaminants as suspended solids.
  - Presence of oxygen may lead to bacterial growth (biofouling).
  - *Variation in Process Conditions:* Temperature and concentration effects can lead to precipitation of solids/crystals from solutions.
  - High temperatures (particularly in distillation) could lead to decomposition/cracking/reactions that might result in the formation of solids.
- How is fouling detected during a packed column operation?
  - ΔP increases.
  - Concentration changes in exit streams (due to decreased interfacial areas and hence mass transfer rates).
- How is prevention/minimization of cleaning requirements of packings accomplished?
  - Pretreatment/filtration of entering streams.
  - Acid or solvent flush/wash.
  - For water soluble systems, use of pH adjustment, use of pretreating with sequestering agents or biocides, ozone, and other steps.

16.1.2.3 Packed Column Diameter

- Describe a method for the estimation of a packed column diameter.
  - A common procedure for estimating flooding velocity (thus, setting a minimum column diameter) is to use a generalized flooding and pressure drop correlation.
  - One version of the flooding and pressure drop relationship for a packed column is in the *Sherwood correlation*, shown in Figure 16.32.
  - In Figure 16.32, L and G are mass flow rates (any consistent set of units may be used as long as the term is dimensionless), ρ_g is the density of the gas stream, and ρ_L is the density of the absorbing liquid.
  - G' is the mass velocity of gas per unit cross-sectional area of column, g/(s m^2), ρ_g is the density of the gas stream, kg/m^3, ρ_L is the density of the liquid, kg/m^3, g_c is the gravitational constant, 9.82 m/s^2, F is the packing factors for different types of packings (given in standard textbooks or manufacturer's literature), φ is the ratio of specific gravity of the liquid to that of water, and μ_L is the viscosity of liquid.

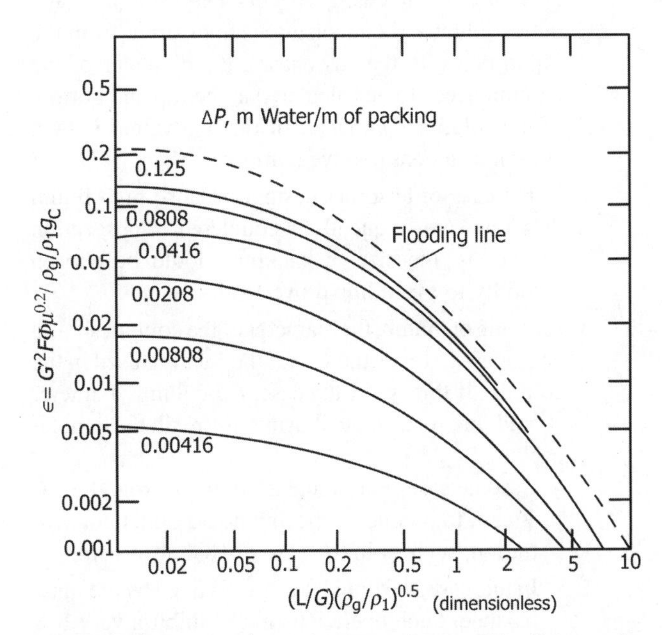

FIGURE 16.32 Generalized pressure drop and flooding correlation for packed columns.

- The x-axis (or abscissa) is a function of the physical properties of the vapor/gas and liquid streams. The y-axis (ordinate), is also a function of the gas and liquid properties as well as the packing material utilized.
- The graph is used to predict the conditions that will cause flooding.
- The procedure involves first calculation of the value of the abscissa from known flows and densities, which is then used to locate the point on the abscissa. The ordinate, ε, is read from the graph.
- Using this value, G' is estimated from the rearranged expression, which corresponds to flooding conditions:

$$G' = [\varepsilon \rho_g \rho_L g_c / F \varphi {\mu_L}^{0.2}]0.5. \qquad (16.19)$$

- G' at the operating conditions is estimated from $G'_{\text{operating}} = fG'_{\text{flooding}}$, where f is the percentage of flooding velocity, which is assumed to be in the range of 50–75% of the flooding velocity.
- Cross-sectional area is found from

$$A = G/G'_{\text{operating}}. \qquad (16.20)$$

- Column diameter is then found from

$$d_t = (4A/\pi)^{0.5} = 1.13A^{0.5}. \qquad (16.21)$$

- Since flooding is an unacceptable operating condition, this sets a minimum column diameter for a given set of gas and liquid conditions.
- Knowing minimum unacceptable diameter, a larger, *operating* diameter can be specified.
- If a substantial change occurs between inlet and outlet volumes (i.e., moisture is transferred from the liquid phase to the gas phase), the diameter of the column needs to be calculated at the top and bottom of the column. The larger of the two values is then chosen as a conservative number.
- In the case of absorber design, because high liquid flow rates are frequently encountered in absorption operations, the size of packing should be chosen carefully to avoid liquid overload.
- As a rule of thumb, the diameter of the column should be at least 15 times the size of the packing used in the column. If this is not the case, the column diameter should be recalculated using a smaller diameter packing.
  - In some cases, the solute is absorbed from the gas stream to produce a specific liquid outlet concentration, such as in the case of HCl absorbers.
  - In this case, the liquid flow rate is fixed by the mass balance. Such operations may exhibit a very low L/G ratio (less than 1.2). In these cases, it is advisable to use a larger size random packing or a structured packing in order to reduce column diameter and thus increase the liquid rate.
- The column cross-sectional area usually is specified by the pressure drop produced as a result of the design gas and liquid flow rates.
  - *Absorbers* normally are specified to give a pressure drop between 3.5 and 35 mm H_2O/m (0.1–10 in./ft) of packed depth.
  - In nonfoaming systems the design pressure drop is usually between 17 and 35 mm H_2O/m (0.25 and 0.40 in. H_2O/ft).
  - For systems that tend to foam moderately, the design pressure drop should be reduced to a maximum of 17 mm H_2O/m (0.25 in. H_2O/ft) at the point of highest loading. Such a design will avoid imparting energy from the gas stream to the liquid phase, which can promote additional foaming.
- Liquid rate used should completely wet the packing.
- For absorption involving significant heat effects, for example, acid gas absorption, higher liquid rates may have to be used to avoid rise in temperature that decreases absorption rates and may cause vaporization of the solvent, especially in the lower section of the absorber where amounts of solute absorbed will be high.
  - If solvent flow is to be kept low and avoid an equilibrium pinch in the column, the heated liquid phase must be removed from the column, cooled externally and then returned to the absorber.
- Another consideration for fixing diameter of the top section of the column is the gas velocity at the top of the column, which should be kept low to avoid entrainment of liquid, which represents loss of solvent and contamination of the outlet gas stream.
  - The main parameter affecting the size of a packed column is the gas velocity at which liquid droplets become entrained in the exiting gas stream.
  - As gas velocity is increased, a point will be reached where the liquid flowing down over the packing begins to be held in the void spaces between the packing.
  - This gas–liquid ratio is termed the *loading point*.
  - The pressure drop of the column begins to increase and the degree of mixing between the phases decreases.
  - A further increase in gas velocity will cause the liquid to completely fill the void spaces in the packing. The liquid forms a layer over the top of

the packing and no more liquid can flow down through the column.

- The pressure drop increases substantially, and mixing between the phases is minimal. This condition is referred to as flooding, and the gas velocity at which it occurs is the *flooding velocity*.
- Using an extremely large diameter column would eliminate this problem. However, as the diameter increases, velocities and hence turbulence decreases, reducing mass transfer rates. Besides, the cost of the column increases.
- Normal practice is to size a packed column diameter to operate at a certain percent of the flooding velocity. A typical operating range for the gas velocity through the columns is 50–75% of the flooding velocity. It is assumed that, by operating in this range, the gas velocity will also be below the loading point.

16.1.2.4 Operation and Installation Issues

- Summarize operational problems associated with packed columns (used as air strippers) and their causes.
  - Static ΔP increases:
    - Liquid flow rate to liquid distributor has increased and should be checked.
    - Packing in irrigated bed could be partially plugged due to solids deposition and may require cleaning.
    - Entrainment separator could be partially plugged and may require cleaning.
    - Packing support plate at bottom of packed section could be plugged, causing increased pressure drop, which will require cleaning.
    - Packing could be settling due to corrosion or solids deposition, again requiring cleaning or additional packing.
    - Air flow rate through column could have been increased by a change in damper setting, which may need readjustment.
  - ΔP decreases, slowly or rapidly:
    - Liquid flow rate to distributor has decreased and should be adjusted accordingly.
    - Air flow rate to scrubber has decreased due to a change in fan characteristics or due to a change in system damper settings.
    - Partial plugging of spray or liquid distributor, causing channeling through scrubber, could be occurring. Liquid distributor should be inspected to ensure that it is totally operable.
    - Packing support plate could have been damaged and fallen into bottom of the column, allowing packing to fall to bottom and produce a lower pressure drop. This should be checked.
  - Pressure or flow change in recycled liquid causing reduced liquid flow:
    - Plugged strainer or filter in recycle piping, plugged spray nozzles, or partially plugged piping, which may require cleaning.
    - Liquid level in sump could have decreased, causing pump cavitation.
    - Pump impeller could have been worn excessively.
    - Valve in either suction or discharge side of pump could have been inadvertently closed.
  - High liquid flow:
    - Break in the internal distributor piping.
    - Spray nozzle that has been inadvertently *uninstalled*.
    - Spray nozzle that may have come loose or eroded away, creating a low pressure drop.
    - Change in throttling valve setting on the discharge side of the pump, allowing larger liquid flow. Reset to the proper conditions.
  - Excessive liquid carryover:
    - Partially plugged entrainment separator, causing channeling and reentrainment of the collected liquid droplets.
    - Air flow rate to column could have increased above the design capability, causing reentrainment.
    - If a packed type entrainment separator was used, packing may not be level, causing channeling and reentrainment of liquid.
    - If a packed entrainment separator was used and a sudden surge of air/gas through the column occurred, this could have caused the packing to be carried out of column or to be blown aside, creating an open area *hole* through separator.
    - Velocity through column has decreased to a point that mass transfer does not effectively take place and poor performance is achieved.
  - Reading indicating low gas flow:
    - Packing in column may be plugged, causing a restriction to air/gas flow.
    - Liquid flow rate to column could have been increased inadvertently, again causing greater restriction and ΔP, creating lower air/gas flow rate.
    - Fan belts have worn or loosened, reducing air/gas flow to equipment.
    - Fan impeller could be partially corroded, reducing fan efficiency.
    - Ductwork to or from column could be partially plugged with solids and may need cleaning.

    - Damper in system has been inadvertently closed or setting changed.
    - Break or leak in duct could have occurred due to corrosion.
  - Increase in flow:
    - Sudden opening of damper in system.
    - Low liquid flow rate to column.
    - Packing has suddenly been damaged and has fallen to bottom of column.
  - Sudden decrease in column efficiency:
    - Liquid makeup rate to the column has been inadvertently shut off or throttled to a low level, decreasing column efficiency.
    - Set point on pH control may have to be adjusted to allow more chemical feed.
    - Problem may exist with chemical metering pump, control valve, or line plugging.
    - Liquid flow rate to column may be too low for effective removal.
- What are the reasons for overpressure developments in distillation units?
  - Migration of internals into lines resulting in blockages.
  - Blockage of packing/trays.
  - Liquid/vapor decomposition initiated by high temperature resulting from loss of vacuum.
  - Self-ignition of vapor caused by air leakage into equipment operating under vacuum leading to fires and explosions.
- What are the preinstallation checks made for the installation of column internals and the guidelines for correct installation?
  - Before installation checks should be made about the column internal diameter, especially around nozzles and manholes, tangent-to-tangent measurement, roundness, location and levelness of wall clips and support rings, and removal of scale and dirt.
  - After installation, checks should be made for roundness of the shell (large diameter shells can become elliptical), vertical condition of the column and presence of dirt and scales.
  - It should be noted that out of roundness can occur especially in the area of the manholes and nozzles. Out of roundness can occur during service and it should be checked and rectified before revamps.
  - In the case of packed columns, close attention should be paid for installation of support plate, packing, packing retainer, distributor and feed pipe.
  - It is important that the support beams fit the support ring correctly to avoid falling off the ring during vapor flow surges. If an upset is possible, it will be a good practice to clamp the support plate to the support ring.
  - In the case of structured packing, support grids are used that should be level as otherwise liquid distribution at the top of the packing gets affected.
  - Metal random packing normally is poured into a column from boxes or bags. The vertical distance a metal packing can be poured is about 6 m.
  - The exact distance depends upon the packing shape and the gauge of material from which it is made.
  - Packing should not be crushed into the support plate. A good practice is to lower the first packing onto the support until the plate is covered with 30 cm or more of packing.
  - Dirt and particulates should not be allowed into the column during installation because they may plug-up the liquid distributors after start-up.
  - Even filling of the column that should be free from void spaces in the bed or at the packing/column wall interface to ensure the expected apparent packing efficiency.
  - Providing a manhole or nozzle above the support plate of each packed bed in the column can ease packing removal.
  - Using vacuum to remove the packing out of the column is another option, particularly for large columns.
  - Metal structured packing comes in blocks that are normally sized to allow installation through a manhole.
  - It will be very helpful to lay out each layer of structured packing before installing it into the column to make sure the configuration is correctly understood and complied with. The layers are to be properly rotated. The orientation from one layer to the next is usually 90°.
  - Walking on the packing can distort/damage it.
  - The structural elements of the packing retainer should never interfere with the liquid leaving the liquid distributor and cause splashing.
  - Correctly designing and installing a liquid distributor is critical to obtaining the best performance from the packed column.
  - The performance of liquid distributors can be negatively affected by being installed with poorly sealed joints, out of level, with construction dirt left in the column and with poorly located feed pipe discharges.
  - Reflux or feed introduced onto a liquid distributor via a feed pipe must not splash, upset the liquid pool, or impinge directly onto the distributor orifices. Any of

these occurrences can increase the liquid flow variation out of the distributor and decrease column efficiency.

- Correct packing height should be ensured without allowing any gaps between the bundles.
- For tray columns, it is critical that the downcomer outlet clearance and tray weirs are set to the specified height and leveled.
- The valves on the valve tray deck must be free to move and the valve legs must be the correct length and spread properly so the valve is not blown out of the deck by the vapor.
- Sieve tray orifice size and deck thickness should be checked to confirm that they match the design values.

• What are the constructional issues involved with respect to piping and nozzles for feed lines and reboiler return lines for trouble-free operation of distillation columns?

- Feed lines:
  - *Two-Phase Feed Flow*: Tray sections and baffles that come in contact with entering liquid + vapor feed mixture can be subjected to abnormally high forces leading to structural damage. These sections and baffles, therefore, should be suitably strengthened.
  - The two-phase flows should be outside slug flow regime as slugging can result in hammer and column instability.
  - If vapor is present in the feed, tray spacing at the feed inlet should be increased by 0.5–1.0 m.
  - When liquid flow is involved, nozzle velocities should not be more than 1 m/s.
  - Feed pipe supports should be located near feed nozzle so that the pipe is not supported by the feed nozzle.
  - *Bottom feed and reboiler return inlets* should never be submerged below the liquid level, because the column of liquid above submerged inlet can vary in height and, under certain circumstances, slugs of liquid and vapor can be blown up the column and lift the trays off their supports or dislodge the packing support plates. Also this can cause excessive entrainment.
  - Tangential entry of bottom feed/reboiler return nozzles should be avoided as such nozzles impart swirl to the sump liquid, disturb liquid surface, and promote vortexing.
- Excessive velocities should be avoided by proper sizing of the reboiler return piping.
- The vapor disengaging space above the reboiler return line should be adequate to avoid entrainment carryover.
- Liquid overflowing from bottom seal pan should not be allowed to interfere with the rising vapors from the reboiler return.
- Figure 16.33 illustrates good and bad practices with such lines.

• What are the common reasons/mechanisms for the presence of vapor in a column liquid outlet line?

- Inadequate residence time for the vapor disengagement from the liquid.
- Frothing caused by impact of falling liquid on the liquid surface in the drawoff pan or sump.
- Flashing occurs when vapor pressure of the liquid exceeds static pressure in the outlet pipe. To avoid this condition sufficient liquid head should be available at the sump. Correct sizing of the outlet piping should be ensured.
- Vortexing due to intensification of swirling motion as the liquid converges toward an outlet. Vortexing promotes entrapment of vapor into the drawoff line. Installing vortex breakers avoids this problem. Vortex breakers introduce shear in the vicinity of the outlet, thus suppressing the swirl.

16.1.3 Reflux Drums and Vapor/Gas–Liquid Separators

• What are the principles involved in gas–liquid separation?

- There are three principles involved in gas/vapor–liquid physical separations, namely, momentum, gravity settling, and coalescing. Any separator may employ one or more of these principles, but the fluid phases must be *immiscible* and have different densities for separation to occur.

• What are the forces acting on a liquid droplet suspended in a gas stream? Illustrate.

- Liquid droplets will settle out of a gas phase if the gravitational force acting on the droplet is greater than the drag force of the gas flowing around the droplet, as illustrated in Figure 16.34.

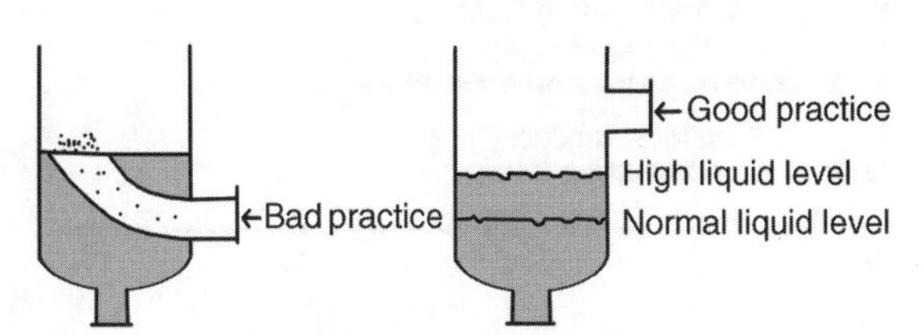

FIGURE 16.33 Practices of location of bottom feed or reboiler return lines.

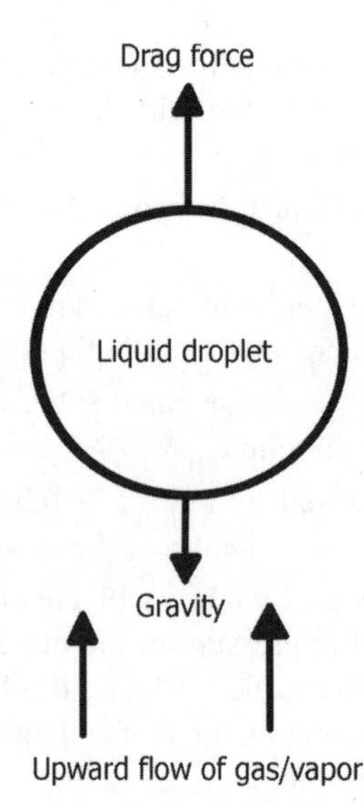

FIGURE 16.34 Forces acting on a liquid droplet suspended in a gas stream.

- Provide typical droplet size distribution from entrainment.
  - Figure 16.35 gives typical droplet size distribution from entrainment.
- What are the different types of vessels used in process industry?
  - Vessels are of two types, namely, those substantially without internals and those with internals. The main functions of the first kinds, called drums or tanks, are for intermediate storage or surge of a process stream for a limited or extended period or to provide a phase separation by settling.
  - The second type comprises the shells of equipment like heat exchangers, reactors, mixers, distillation columns, and other equipment. Their major dimensions are established by process requirements.
- What is the distinction between drums and tanks?
  - There is no hard and fast way to distinguish between drums and tanks, except the sizes involved.
  - Usually drums are cylindrical vessels with flat or curved ends, depending on the pressure and installed either horizontal or vertical.
  - In a continuous plant, drums have a holdup of a few minutes. They are located between major equipment or supply feed or accumulate product.
  - Surge drums between equipment absorb fluctuations in flows so that they are not transmitted freely along a chain, including those fluctuations that are characteristic of control instruments of normal sensitivity.
    - For example, reflux drums provide surge between a condenser and column and downstream equipment.
    - A drum upstream of a compressor will ensure that entrained liquid droplets do not enter the compressor, one upstream of a fired heater will protect the tubes from running dry and a drum following a reciprocating compressor will smooth out pressure surges.
  - Tanks are usually larger vessels of several hours holdup.
    - For example, the feed tank to a batch distillation may hold supply for a day and run-down tanks between equipment may provide several hours of holdup as protection of the main storage from possible off-specification product and opportunity for local repair and servicing without disrupting the entire process.
  - Storage tanks are generally installed in tank farms outside the process battery limits.
    - Time variations in the supply of raw materials and the demand for the products influence the sizes and numbers of storage tanks.
  - Liquid storage tanks are provided with a certain amount of vapor space, normally 15% below 2000 L capacities and 70% for higher capacities. Liquids with high vapor pressures, liquefied gases and gases at high pressure are stored in elongated horizontal vessels, less often in spherical ones.
- What are the major types of vapor/gas–liquid separators used in industry?
  - *Conventional:* Gravity settlers and demisters. Can be horizontal or vertical. Functions include acting as

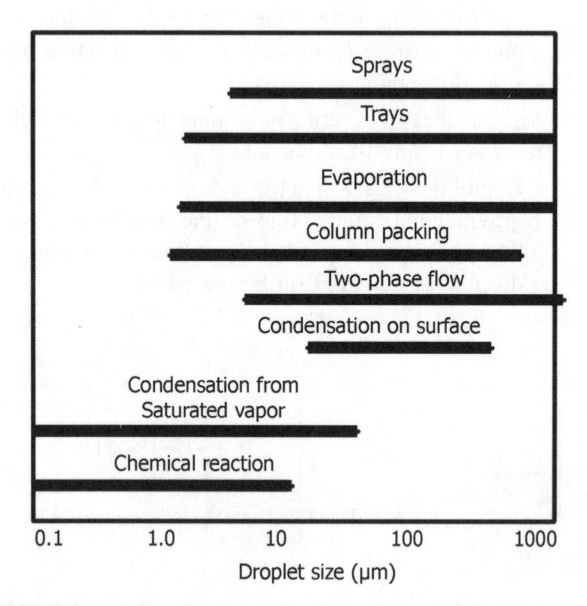

FIGURE 16.35 Typical droplet size distribution from entrainment.

flash tanks, oil-gas separators, slug catchers, and reflux and knockout drums.

- Centrifugal.
- Filter type.
- Flare knockout drums.
- Combinations involving gravity settling, impingement, and centrifugal separations.

• Under what conditions spherical separators are used?

- Spherical separators are used for high-pressure service where compact size is desired and liquid volumes are small.

• What is the normal liquid holdup/residence time in a reflux drum and feed drum in distillation applications?

- Minimum 5 min at half of the capacity of the reflux drum and 5–10 min for a feed drum.

• What are the recommended drum liquid residence times for different operating conditions in process units?

- Normally not less than 5 min storage. This is required to ensure reflux pump operation without NPSH problems and also to prevent vapor entrapment in the reflux liquid.
- Product to storage: 2–5 min.
- Product to another column: 10–12 min.
- Product to a heat exchanger: 5 min.
- Product to a fired heater: 10–30 min, half-full.
- Compressor feed liquid knockout drum: 10–20 min of liquid flow with a minimum volume of 10 min equivalent gas flow rate.

• What are the recommended residence times for liquid in the bottom section of a distillation column?

- 1–5 s based on the nature of controls used.

• What are the minimum recommended liquid holdup in separators from control point of view?

- Automatic control:
  - Product to storage: 4 min.
  - Feed to a furnace: 5 min.
  - Other applications: 4 min.
- Manual control: 20 min.
- For operator intervention during plant upsets between alarm and trip action: 5 min.

• What is the normally recommended ratio of vapor space to liquid space in a reflux drum?

- 50% (half-full).

• What is the normal L/D ratio for drums?

- 2.5–5.0 (3 is common), the smaller diameters at higher pressures and for liquid–liquid settling.

• What should be the minimum height of a vertical reflux drum above liquid level to avoid entrainment and for feed entry purposes?

- Not less than 1.2 m (4 ft).

• "Reflux drums are usually installed horizontal." *True/False*?

- *True*. Vertical drums increase entrainment losses while horizontal drums involve lower velocities for light vapors and released gases due to increased cross-sectional area of a horizontal drum of the same size as a vertical drum.

• What are the considerations involved in sizing reflux drums that receive water along with hydrocarbon condensate from the outlet of a distillation column condenser?

- When a small amount of a second liquid phase (for example, water in an immiscible organic) is present, it is collected in and drawn off a pot at the bottom of the drum.
- The diameter of the pot is sized on a linear velocity of 0.3 m/s (0.5 ft/s), is a minimum of 0.4 m (16 in.) diameter in drums of 1.2 2.4 m (4–8 ft) diameter and 0.6 m (24 in.) in larger sizes. The minimum vapor space above the high level is 20% of the drum diameter or 25.4 cm (10 in.).

• What are the considerations involved in designing flare system knockout drums? What is a split flow knockout drum used in a refinery flare system?

- Flammable liquid droplets greater than 150 microns should be stopped before reaching the tip of the flare.
- A knockout drum in a flare system is used to prevent the hazards associated with burning liquid droplets escaping from flare stack.
- Therefore, the drum must be of sufficient diameter to effect the desired liquid–vapor separation.
- During normal flaring conditions, which involve usually low flow rates, a simple settling chamber type knockout drum is effective in removing the liquid droplets. However, during emergency flaring conditions, drums of excessively large diameter are required to knockout liquid droplets.
- Split flow drums, where vapor enters in the middle of the drum and leaves it at both ends, have twice the capacity of single flow drums.
- Vertical drums usually require a larger diameter, about 1.4 times the diameter for single flow drums.
- The length to diameter ratio for a horizontal drum is between 3 and 4.

• "Gas–liquid separators are housed vertically." Comment.

- Not necessarily.
  - Vertical separators are usually selected when the gas–liquid ratio is high or total gas volumes are low.

  - Horizontal separators are most efficient when large volumes of liquid are involved. They are also generally preferred for three-phase separation applications.

- What are the normal considerations/practices involved in installing reflux drums or separators in vertical or horizontal positions?
  - Reflux drums are almost always horizontally mounted.
  - In cases involving total condensation of vapors and in the case of partial condensers with appreciably higher liquid rates compared to vapor rates, or where large volumes of total fluids and large amount of dissolved gas are present, horizontal installation is a preferred choice. Horizontal vessels can accommodate larger liquid slugs.
  - When vapor rates are high (partial condensers), vertical installation is preferred. This ensures sufficient liquid head for accurate liquid level measurements as well as smooth reflux pump operation.
  - Layout considerations like available headroom or floor space can influence the above choices.
  - In cases where there is a frequent fluctuation in inlet liquid flow or where revaporization or remixing of fluids in the vessel should be prevented, vertical separators should be preferred.
  - Vessels for the separation of two immiscible liquids usually are made horizontal and operate full, although some low-rate operations are handled conveniently in vertical vessels with an overflow weir for the lighter phase. With the L/D ratio of three or more, the travel distance of droplets to the separated phase is appreciably shorter in horizontal vessels.

- Show, by means of diagrams, the important dimensions of vertical and horizontal drums.
  - Figure 16.36 illustrates the important dimensions of vertical and horizontal knockout drums.
  - When calculated length to diameter ratio comes out to be less than 3, the length is increased arbitrarily to make the ratio 3 and when it comes out more than 5, a horizontal drum is preferably employed. The vapor space is made a minimum of 20% of the drum volume that corresponds to a minimum height of the vapor space of 25% of the diameter, but this should never be less than 0.3 m (12 in.).
  - When a relatively large amount of liquid must be held up in the drum, it may be advisable to increase the fraction of the cross section open to the vapor.

- Give diagrams for typical gas–liquid separators and describe their important features.
  - Regardless of shape, separation vessels usually contain four major sections plus the necessary controls. The inlet device is used to reduce the momentum of the inlet flow stream, perform an initial bulk separation of the gas and liquid phases and enhance gas flow distribution. There are a variety of inlet devices.
  - The gas gravity separation section is designed to utilize the force of gravity to separate entrained liquid droplets from the gas phase, preconditioning the gas for final flow through the mist eliminator (Demister). It consists of a portion of the vessel through which the gas moves at a relatively low velocity with little turbulence. In some horizontal designs, straightening vanes are used to reduce turbulence. The vanes also act as droplet coalescers, which reduces the horizontal length required for droplet removal from the gas stream.

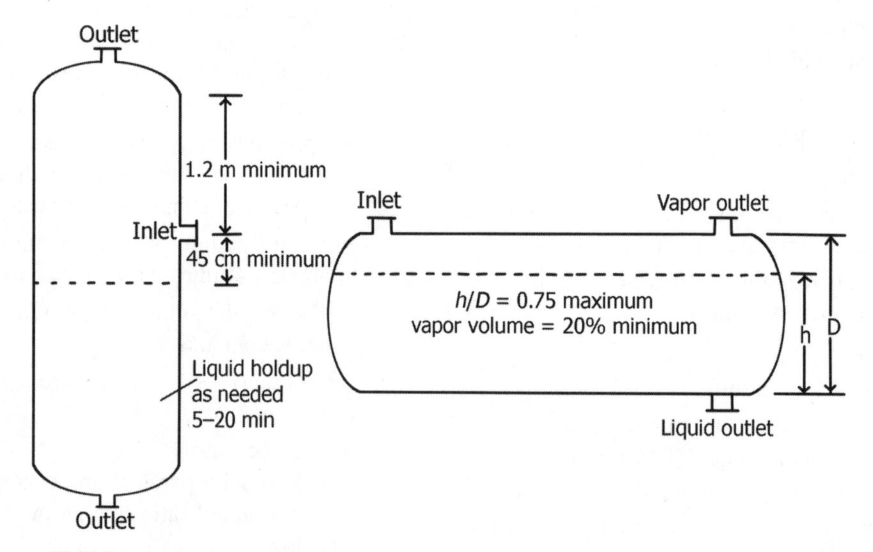

FIGURE 16.36 Important dimensions of vertical and horizontal knockout drums.

- The liquid gravity separation section acts as a receiver for all liquid removed from the gas/vapor in the inlet, upward movement of gas/vapor and demister sections.
- In two-phase separation applications, the liquid separation section provides residence time for disengagement of gas/vapor from the liquid. In three-phase separation applications this section also provides residence time to allow for separation of water droplets from a lighter hydrocarbon liquid phase and vice versa. Depending on the inlet flow characteristics, the liquid section should have a certain amount of surge volume, or slug catching capacity, in order to smooth out the flow passed on to downstream equipment or processes.
- Efficient gas/vapor removal may require a horizontal separator while emulsion separation may also require higher temperature, use of electrostatic fields and/or the addition of a demulsifier.
- Coalescing packs are sometimes used to promote hydrocarbon liquid–water separation, though they should not be used in applications that are prone to plugging, for example, wax, sand, and so on.
- The demister section utilizes a mist extractor that can consist of a knitted wire mesh pad, a series of vanes, or cyclone tubes. This section removes the very small droplets of liquid from the gas/vapor by impingement on a surface where they coalesce into larger droplets or liquid films, enabling separation from the gas/vapor phase.
- Figures 16.37 and 16.38 illustrate typical gas/vapor–liquid separator designs with demister pads.

- What are the advantages of using a horizontal separator as an amine flash vessel? Does it require large surge volume?
  - There is relatively large liquid surge volume required for longer retention time. This allows more complete release of the dissolved gas and, if necessary, surge volume for the circulating system.
  - There is more surface area per unit liquid volume to aid in more complete degassing.
  - The horizontal configuration handles a foaming liquid better than vertical.
  - The liquid level responds slowly to changes in liquid inventory, providing steady flow to downstream equipment.
- What are the effects of foam on entrainment separation?
  - A light frothy foam is more susceptible to entrainment than a clear liquid.
  - Once foam level rises above the feed inlet nozzle, vapor in the feed blows the foam up the drum, resulting massive carryover of liquid (or foam) into the vapor line.
- How are vapor–liquid separators sized? What are the issues involved in different applications? Illustrate.
  - Entrainment is one of the important issues in sizing and operation of vapor–liquid separators.
  - Figure 16.39 illustrates the common applications of vapor–liquid separators involving a single liquid phase, two liquid phases, and a vapor/gas phase, commonly found in a knockout drum from a petroleum distillation column, in which condensate consists of a water layer, hydrocarbon layer, and gas/vapor phase. There will be situations in which foams are generated in certain processes that are to be separated in the knockout drum.
  - For tendencies involving foaming, the following guidelines may be adopted:
    - For horizontal vessels, add the height of the foam above the high liquid level, to allow for the

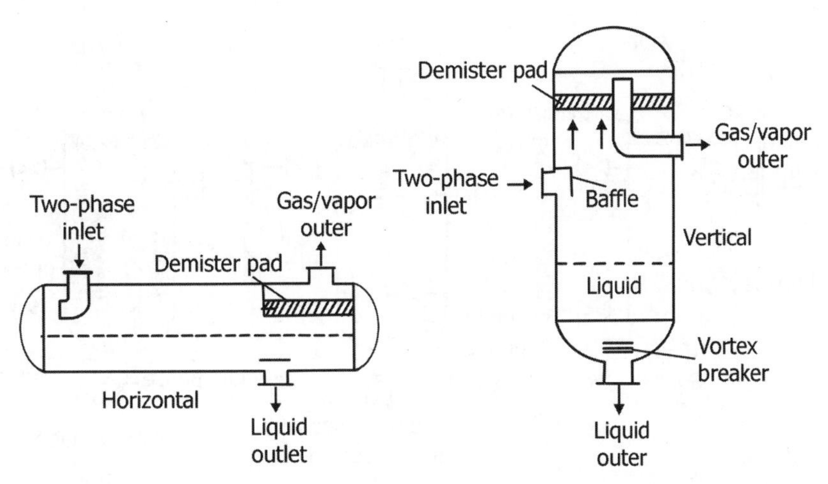

FIGURE 16.37 Typical gas/vapor–liquid separators. (*Courtesy*: Pace Engineering.)

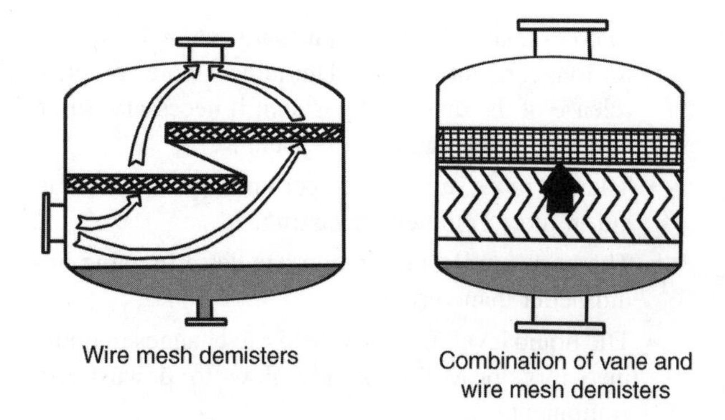

FIGURE 16.38 Typical gas/vapor–liquid separators. (*Courtesy*: Pace Engineering.)

reduced vapor space for vapor/liquid disengagement (typically 0.3 m).

➢ For vertical vessels, the downward liquid velocity may be limited to

$$v_L = 1.4 \times 10^{-4}[(\rho_L - \rho_g)/\eta]^{0.28}, \quad (16.22)$$

where η is the dynamic viscosity of the liquid, Pa s.

- As liquid circulating above the liquid outlet nozzle may result in vortex formation, the design shall include measures to avoid this. Allowing adequate liquid height above the outlet nozzle and installing a vortex breaker are two such measures.
- Normally vortex breakers are provided on outlet nozzles that serve as outlet to a pump.
- Liquid droplets escape as entrainment with the gas/vapor, which are usually captured by a demister pad in the top section of the knockout drum. As the droplets are not of uniform size, Stokes law, in its simple form, will not be adequate to size the drum.
- Therefore, sizing of the drum is usually done by using Souders–Brown equation that uses empirically determined coefficients (*K*-factors).
- A vertical drum, with a length to diameter ratio of about 3–4, and providing about 5 min of liquid inventory between the normal liquid level and the bottom of the vessel (with the normal liquid level being somewhat below the feed inlet).
- The maximum allowable vapor velocity is estimated using Souders–Brown equation:

$$V = K[(\rho_L - \rho_V)/\rho_V]^{0.5}, \quad (16.23)$$

where V is the maximum allowable vapor velocity, m/s and ρ_L, ρ_V are the liquid and vapor densities, respectively, kg/m^3.

K-values are dependent on the drum pressure and the recommended values, when a demister is used in the drum, are given in Table 16.5.

- What are the constructional details of wire mesh mist eliminators? How do they work?
  - Wire mesh demister pads are made by knitted wire, metal or plastic, into tightly packed layers that are then crimped and stacked to achieve the required pad thickness.
  - If removal of very small droplets, that is, less than 10 μm, is required, much finer fibers may be interwoven with the primary mesh to produce a coknit pad.
  - Mesh pads remove liquid droplets mainly by impingement of droplets onto the wires and/or coknit fibers followed by coalescence into droplets large enough to disengage from the bottom of the pad and drop through the rising gas flow into the liquid holding part of the separator.

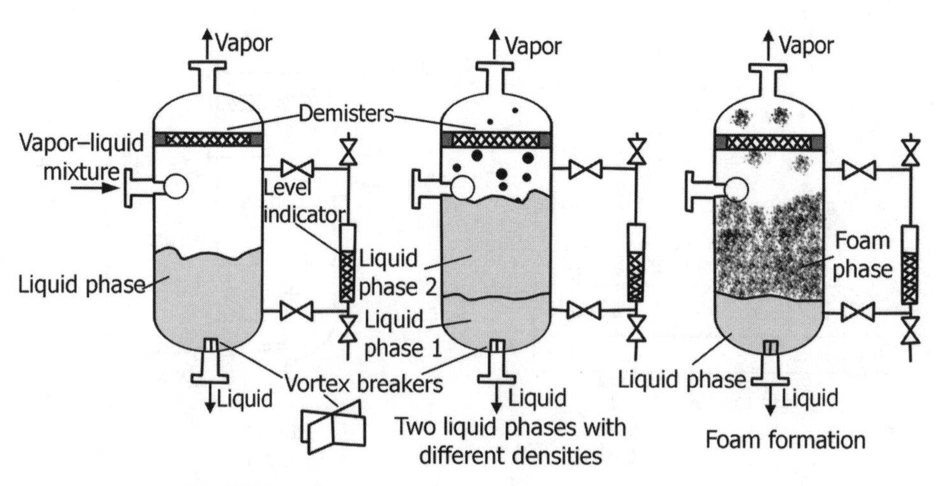

FIGURE 16.39 Vapor–liquid separator for different cases.

TABLE 16.5 Drum Pressure Versus *K*-Values

Gauge Pressure (bar)	*K*-Value (m/s)
0	0.107
7	0.107
21	0.101
42	0.092
63	0.083
105	0.065
Wet steam	0.076
Most vapors under vacuum	0.061

Notes: $K = 0.107$ at a gauge pressure of 7 bar. 0.003 is subtracted for every 7 bar above a gauge pressure of 7 bar. For glycol or amine solutions, above *K*-values are multiplied by 0.6–0.8. Half of the above *K*-values are recommended for approximate sizing the vertical drums *without the use of demisters*. *K* is multiplied by 0.7 for scrubbers used on compressor suction and expander inlet separators.

- Mesh pads are not recommended for dirty or fouling service as they tend to plug easily.
- These pads are normally installed horizontally with gas flow vertically upward through the pad. Performance is adversely affected if the pad is tilted more than 30° from the horizontal.
- Problems have been encountered where liquid flow through the pad to the sump is impaired due to dirt or sludge accumulation, causing a higher liquid level on one side. This gives rise to serious consequences of the pad being dislodged from its mounting brackets, making it useless or forcing parts of it into the outlet pipe.
- The top and bottom of the pad should be firmly secured so that it is not dislodged by high gas flows, such as when a pressure relief valve lifts or during an emergency blowdown situation.
- Most installations use a 15 cm (6 in.) thick pad with 144–192 kg/m^3 (9–12 lb/ft^3) bulk density. Minimum recommended pad thickness is 10 cm (4 in.).

• What are the efficiencies obtainable for wire mesh demisters?

- Generally droplet removal efficiencies for wire mesh demisters are in the range of 99–99.5% removal of 3–10 μm size droplets.
- Higher removal efficiencies are possible for denser, thicker mesh pads, and/or for smaller wire/coknit fiber diameter.

• What are vane or chevron mist eliminators? Illustrate.

- Vane or chevron type mist extractors (vane pack) use relatively closely spaced blades arranged to provide sinusoidal or zigzag gas flow paths. The changes in gas flow direction combined with the inertia of the entrained liquid droplets, cause impingement of the droplets onto the plate surface, followed by coalescence and drainage of the liquid to the liquid collection section of the separator.
- Figure 16.40 shows a typical vane type mist extractor. Vane packs may be installed in either horizontal or vertical orientations, though capacity is typically reduced significantly for vertical upflow applications.
- Vanes differ from wire mesh pads in that they typically do not drain the separated liquid back through the rising gas stream. The liquid can be routed into a downcomer that carries the liquid directly to the liquid holding section of the separator.
- Vane packs are better suited to dirty or fouling service as they are less likely to plug due to their relatively large flow passages.
- A number of different vane pack designs are available.

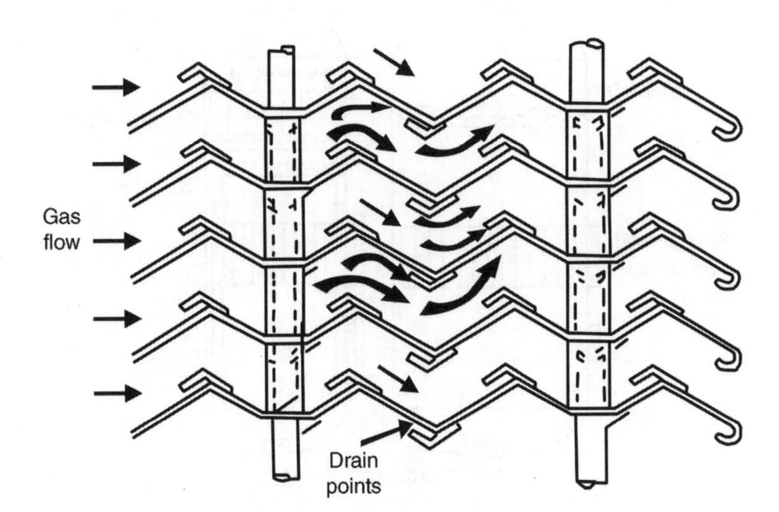

FIGURE 16.40 Cross section of vane element mist extractor showing corrugated plates with liquid drainage traps.

- Pack thicknesses are generally in the range of 15–30 cm (6–12 in.).
- Vanes are usually arranged in a zig-zag or sinusoidal pattern, with vane spacings of 2.54–3.8 cm (1–1.5 in.) typical.

- What are the droplet removal efficiencies of vane type of mist eliminators?
  - Removal efficiencies are about 99% for removing droplets greater than 10–40 μm. Higher removal efficiencies are obtainable for thicker packs, with closer vane spacings and more passes (bends).
- What are cyclonic gas–liquid separators? What are their characteristics?
  - There are several types of centrifugal separators that serve to separate entrained liquids and solids if present, from a gas stream.
  - For mist extraction applications, reverse flow, axial flow, and recycling axial flow cyclones are typically used in multicyclone *bundles*. Cyclonic mist extractors use centrifugal force to separate solids and liquid droplets from the gas phase based on density difference. Very high *G* forces are achieved that allows for efficient removal of small droplet sizes.
  - The main advantage of cyclonic mist extractors is that they provide good removal efficiency at very high gas capacity.
  - This generally allows for the smallest possible vessel diameter for a given gas flow.
  - Cyclonic mist extractors are often used in low liquid load gas scrubbing applications and for high-pressure gas–liquid separation.
- Describe a modern high-efficiency vapor/gas–liquid separator.
  - Figure 16.41 gives propriety designs of vapor/gas–liquid separators by Sulzer, incorporating special features:
    - Special feed inlet device for vapor distribution with bulk liquid removal.
    - Demisters that act as coalescers and separators, depending on the vapor/gas flow rate.
    - Shell swirl tube deck comprising multiple swirl tubes.
    - High liquid removal efficiency (>98–99%).
    - Very high turndown ratio (factor of 10).
    - Ability to handle slugs.
    - Maintains efficiency at high pressures.
- What are the considerations involved in sizing nozzles and piping for vapor/gas–liquid separators?
  - Normally vapor outlet nozzle should be sized to be equal to the diameter of the vapor outlet pipe, provided it satisfies the criteria: $\rho_g v_g^2 < 3750$, where ρ_g is gas/vapor density, kg/m^3 and v_g is vapor velocity, m/s.
  - Liquid outlet nozzle should be sized for a liquid velocity of 1 m/s and shall extend 100–150 mm above the bottom of the vessel.
  - There should be no valves, pipe expansions, or contractions within 10 pipe diameters of the inlet nozzle. If a valve in the feed line near to the separator cannot be avoided, it should preferably be of the gate or ball type fully open in normal operation.

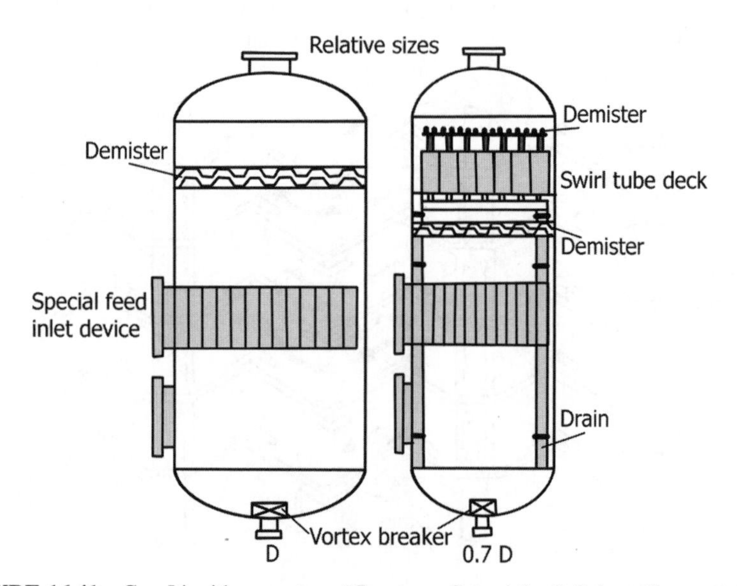

FIGURE 16.41 Gas–Liquid separators. (*Courtesy*: Copyright © Sulzer Chemtech Ltd.)

- There should be no bends within 10 pipe diameters of the inlet nozzle. except the following:
  - For knockout drums and demisters, a bend in the feed pipe is permitted if this is in a vertical plane through the axis of the feed nozzle.
  - For cyclones a bend in the feed pipe is permitted if this is in a horizontal plane and the curvature is in the same direction as the cyclone vortex.
- If desired, a pipe reducer may be used in the vapor line leading from the separator, but it should not be nearer to the top of the vessel than twice the outlet pipe diameter.
- High pressure drops that cause flashing and atomization should be avoided in the feed pipe. If a pressure reducing valve in the feed pipe cannot be avoided, it should be located as far upstream of the vessel as practicable.

- What are the reasons for failure of demisters?
  - Demister pad can be plugged with corrosion products, coke or other particles. The particulates stick to the fibers of the demister pad.
  - Moreover, because of the large surface area of the demister, a relatively low corrosion rate can produce a large amount of corrosion products.
  - When a demister plugs, it increases the pressure drop of the vapor and the demister will break.
  - Demister failure creates two problems:
    - Parts of the pad are blown into downstream equipment.
    - Failed demister promotes high localized velocities, creating more entrainment than without demister.
- Give some examples where vapor/gas–liquid separators are used.
  - Petroleum refinery operations.
  - Natural gas processing.
  - Petrochemical and chemical plants.
  - Refrigeration systems.
  - Air conditioning.
  - Compressor systems for air or other gases.
  - Gas pipelines.
  - Steam condensate flash drums.
- What is the function of a coalescer?
  - The function of a coalescer is to merge small droplets into formation of larger droplets (over 200 μm) so that the buoyancy forces are increased to facilitate easy and faster settling downstream of the coalescer element.
- What are the advantages of using high-efficiency coalescers, in new installations or revamps, over empty or baffled vessels as settling equipment?
  - Reduce size and cost of new liquid–liquid separators.
  - Improve product purity in existing installations.
    - Hazy distillates occur when dissolved water condenses from solution while the product cools in storage tanks. This can result in tying up product in inventory for days or weeks while the water settles to allow meeting haze specifications.
    - Proper coalescers permit almost immediate use of the product in blending and final product sale, reducing carrying costs.
  - Debottleneck existing reflux drums.
    - Reflux drums often become a bottleneck when distillation columns are retrofitted with higher capacity internals.
    - In most reflux drums, capacity can be at least doubled with the installation of the proper liquid–liquid separation media to prevent carryover of water condensate in the reflux or product.
  - Reduce downstream corrosion caused by corrosive liquid carryover.
  - Reduce losses of valuable chemicals.
  - Improve distillation column operation and reduce maintenance.
  - Reduce fugitive VOC emissions.
- Give typical arrangements of coalescers.
  - Coalescer vessels are normally arranged horizontally.
  - The coalescer packs may be placed across the axis of the vessel to achieve axial flow of the liquid through the media or, where a greater area is required, it may be placed along the central axis and the liquid sent in a cross-flow arrangement, as shown in Figure 16.42.
- What are the factors that influence design of coalescers in liquid–liquid contacting equipment?
  - Droplet size and size distribution is perhaps the most important factor.
  - The finer the droplets dispersed in an emulsion, the more stable it is, because the buoyancy force diminishes in magnitude as the diameter decreases.
- What are the stages involved in coalescence of droplets?
  - There are three stages involved in coalescence of droplets, namely, collection of individual droplets, combining of several small droplets into larger ones and rise/fall of the enlarged droplets by gravity.
  - *Droplet Capture:* The first stage of coalescing is to collect entrained droplets primarily either by intra-media Stokes settling or direct interception.
  - Elements that depend on intramedia Stokes settling confine the distance a droplet can rise or fall between parallel plates or crimps of packing sheets.

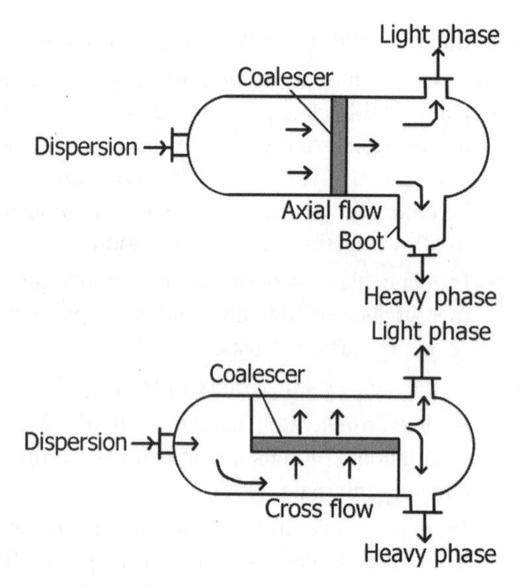

FIGURE 16.42 Coalescer plate pack orientations.

- Figure 16.43 shows plate pack and coknit mesh type coalescers.
- These coalescers make more efficient use of a vessel volume than a straight parallel plate coalescer since more specific surface area is packed per unit volume of the vessel.
- The plate packs can be arranged either in horizontal or vertical position. Vertical arrangement is better for fouling service as the plate pack retains less amount of solids. In vertical arrangement, the droplets have to travel a longer distance before capture compared to horizontal arrangement with some loss of efficiency. Pack surface areas are typically 125 and 250 m^2/m^3 and standard plate depths are 600 mm and standard plate angles are 60° and 45°.
- Figure 16.44 shows a typical three-phase horizontal coalescer.
- In simple gravity separators the traveling distance is equal to the entire height of the pool of liquid present in the separator.
- Meshes, knitted wire, and fibers and wire all depend primarily on direct interception where a multiplicity of fine wires or filaments collect fine droplets as they travel in the laminar flow streamlines around them.
- In general, they can capture smaller droplets than those that depend on enhanced Stokes settling.
- Finer coalescing media allow for the separation of finer or more stable emulsions.
- Table 16.6 gives a comparison of different media for droplet capture.
- As fine media will also capture or filter fine solid particulates from the process stream, unless the emulsion is very clean, an upstream duplex strainer or filter is needed to filter out the solids to protect a high-efficiency coalescer.
- *Droplet Coalescence:* The second step is to combine, aggregate, or coalesce captured droplets.
- Increasing the tendency for droplets to adhere to a medium, increases the probability that subsequent droplets will have the chance to strike and coalesce with those that already have been retained.

Plate pack coalescer

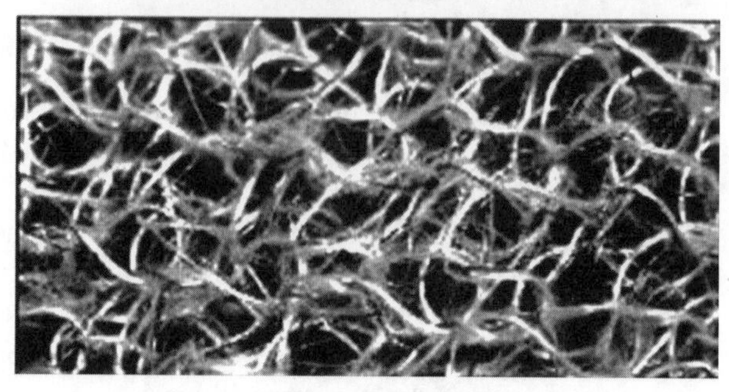
Coknit mesh coalescer

FIGURE 16.43 Typical coalescer designs.

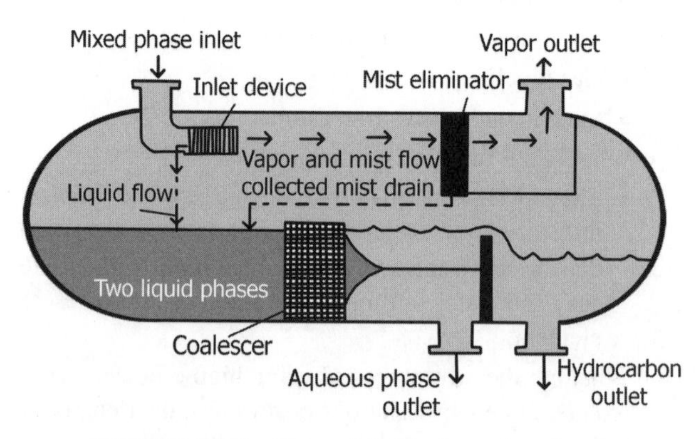

FIGURE 16.44 Three-phase horizontal coalescer.

TABLE 16.6 Comparison of Different Media for Droplet Capture

Media	Source	Maximum Droplet Diameter (μm)	Flow Range ($m^3/(h\,m^2)$)
Corrugated sheets	Separators with coarse emulsions and static mixers	40–1000	35–180
Wire mesh and wire wool	Overhead drums, extraction columns, distillation column feeds	20–300	20–110
Coknits of wire and polymer	Steam stripper bottoms, caustic wash drums, high ΔP mixing valves	10–200	20–110
Glass mat, coknits of wire, and fiberglass	Haze from cooling in bulk liquid phase, surfactants giving emulsions with very low interfacial tension	1–25	7.5–45

- Whether a coalescer medium is hydrophilic (like water) or oleophilic (like oil) depends on the solid/liquid interfacial tension between it and the dispersed phase.
- In general an organic dispersed phase *wets* organic (that is plastic or polymeric) media, as there is a relatively strong attraction between the two, while an aqueous dispersed phase preferably *wets* inorganic media, such as metals or glass. This helps in the coalescence step as the droplets adhere to the media longer.
- Also assisting coalescence is the density of media: Lower porosities yield more sites available for coalescing. In the case of yarns and wools, capillary forces are also important for retaining droplets. Judgments of the proper volume, and therefore residence time, in the coalescers are guided by experience and properties like interfacial tension, porosity, capillarity, density difference, and superficial velocity.
- Coalescers work better in laminar flow as residence time will be more and higher velocities will overcome surface tension forces and strip droplets out of the coalescer medium. For maximum performance Reynolds numbers should be around 500.
- The guidelines in the table are used for selecting the proper coalescer for a given source based on the droplet collection ability of the media and flow velocities.
- Stokes settling with coalesced droplets:
- The third step is the Stokes settling of the coalesced droplets downstream of the medium. The degree of separation primarily depends upon the geometry of the vessel and its ability to take advantage of the large coalesced droplets that were created through steps one and two as described above.

• What are the common process and equipment problems associated with an improperly sized separators in gas processing service? Give examples.

- An improperly sized separator is one of the leading causes of process and equipment problems.
- Inlet separation problems upstream of absorption systems (e.g., amine and glycol) can lead to foaming problems and upstream of adsorption systems (e.g., molecular sieve, activated alumina, and silica gel) can cause fouling, coking, and other damage to the bed.
- Equipment such as compressors and turbo expanders cannot tolerate liquid in the inlet gas steam, while pumps and control valves may have significant erosion and/or cavitation when vapors are present due to improper separation.
- Direct fired reboilers in amine and glycol service may experience tube failures due to hot spots caused by salt deposits resulting from water carryover into the feed gas.

• What are the important factors that should be considered in designing of separators?

- Three main factors should be considered in separator sizing:
  - *Vapor Capacity:* Will determine the cross-sectional area necessary for gravitational forces to remove the liquid from the vapor.
  - *Liquid Capacity:* Liquid capacity is typically set by determining the volume required to provide adequate residence time to *degas* the liquid or allow immiscible liquid phases to separate.
  - *Operability:* Operability issues include the ability of the separator to deal with solids if present, unsteady flow/liquid slugs, turndown, and so on.
- Also optimal design will usually result in an aspect ratio that satisfies these requirements in a vessel of reasonable cost.
- These factors often result in an iterative approach to the calculations.

• Give an example of a separator used without a demister in hydrocarbon processing industry? Give reasons for not using demisters and the procedure used in sizing such a separator.

- The most common application of a vapor–liquid separator that does not use a demister is a flare knockout drum.
- Demisters are rarely used in flare knockout drums because of the potential for plugging and the serious implications this would have for pressure relief.
- Typically a horizontal vessel, that utilizes gravity as the sole mechanism for separating the liquid and gas phases, is used.
- Gas and liquid enter through the inlet nozzle and are slowed to a velocity such that the liquid droplets can fall out of the gas phase. The dry gas passes into the outlet nozzle and the liquid is drained from the lower section of the vessel.
- To design a separator without a demister, the minimum size droplet to be removed must be set. Typically this size is in the range of 300–2000 μm.
- The length of the vessel required can then be calculated by assuming that the time for the gas flow from inlet to outlet is the same as the time for the liquid droplet of size d_p to fall from the top of the vessel to the liquid surface.

• What are the recommended residence times for (i) gas–liquid and liquid–liquid separators used in oil and gas industry?
  - Table 16.7 gives recommended residence times for gas–liquid and liquid–liquid separators.

TABLE 16.7 Recommended Residence Times for Gas–Liquid and Liquid–Liquid Separators

Application	Residence Time (min)
Gas–Liquid Separators	
Natural gas–condensate separators	2–4
Distillation column feed tank	10–15
Reflux drum	5–10
Distillation column sump (bottom section)	2
Amine flash tank	5–10
Refrigeration surge tank	5
Heating oil surge tank	5–10
Liquid–Liquid Separators	
Hydrocarbon–water separators	
Hydrocarbons above 35° API gravity	3–5
Hydrocarbons below 35° API gravity	
>37.8°C	5–10
26.7°C	10–20
15.6°C	20–30
Ethylene glycol–hydrocarbon cold separators	20–60
Amine–hydrocarbon separators	20–30
Hydrocarbon–water coalescers	
>37.8°C	5–10
26.7°C	10–20
15.6°C	20–30
Caustic soda–propane separator	30–45
Caustic soda–heavy gasoline	30–90

Courtesy: GPSA Engineering Data Book, 12th ed.

17

ABSORPTION, DISTILLATION, AND EXTRACTION

17.1 Absorbers and Strippers 527
17.1.1 Applications of Absorption/Stripping 530
17.1.2 Examples of Commercial Applications of Gas Absorption 532
17.1.3 Design of Absorbers 534
17.1.4 Transfer Unit Concepts in Design 539
17.1.5 Solvent Selection 542
17.2 Distillation 543
17.2.1 Phase Equilibria 544
17.2.2 Distillation Processes 548
17.2.3 Batch Fractionation 557
17.2.4 Multicomponent Distillation 559
17.2.5 Separation of Azeotropes and Close Boiling Mixtures 563
17.2.6 Reactive Distillation 570
17.2.7 High Pressure and Vacuum Distillation 571
17.2.8 Divided Wall and Petluk Distillation Columns 574
17.2.9 Operational Issues 579
17.2.10 Reboilers and Feed Heaters 583
17.2.11 Distillation Control 584
17.3 Liquid–Liquid Extraction 584
17.3.1 Liquid–Liquid Equilibria 587
17.3.2 Extraction Calculations 590
17.3.3 Extraction Solvents 592
17.3.4 Supercritical Solvents 595
17.3.5 Extraction Equipment 598
17.4 Solid–Liquid Extraction/Leaching 606

17.1 ABSORBERS AND STRIPPERS

- Define absorption and stripping.
  - Absorption is defined as the process of transfer of molecules of one substance directly to another substance. It may be either a physical or a chemical process. Physical absorption depends on the solubility of the substance absorbed and chemical absorption involves chemical reactions between the absorbed substance and the absorbing medium. It is a gas–liquid transfer process. It is used in the selective removal of a gaseous contaminant or product from a gas mixture. Removal is often effected by absorption in a liquid in which only the gas concerned is soluble.
  - It is sometimes also necessary to free a liquid from dissolved gases. This can be achieved by contact with an insoluble gas, in which case the process is known as *stripping* or *scrubbing*.
- What are the applications of absorption in industry?
  - Among major industrial uses are the absorption of SO_3 in oleum in the production of H_2SO_4 and of HCl and NO_2 in water in hydrochloric and nitric acid manufacturing.
  - Another major application is the purification of various process streams to prevent pollution, corrosion, catalyst poisoning, or condensation in subsequent treatment.
  - Examples of the applications include large scale removal of CO_2 from air or natural gas prior to liquefaction, absorption of sulfur compounds from sour natural gas, removal of acidic components from olefin plants, CO_2 removal from ammonia synthesis gas, absorption of formaldehyde in water to produce aqueous solutions, and drying of gases.
- What are the advantages and disadvantages of absorption systems using packed and tray columns in pollution control?
  - *Advantages*
    - Relatively low pressure drop.
    - Standardization in fiberglass-reinforced plastic (FRP) construction permitting operation in highly corrosive atmospheres.
    - Capable of achieving relatively high mass transfer efficiencies.

Fluid Mechanics, Heat Transfer, and Mass Transfer: Chemical Engineering Practice, By K. S. N. Raju

- ➢ Increasing the height and/or type of packing or number of trays capable of improving mass transfer without purchasing a new piece of equipment.
- ➢ Relatively low capital cost.
- ➢ Relatively small space requirements.
- ➢ Ability to collect particulates as well as gases.
- ➢ Collected substances may be recovered by distillation.

- *Disadvantages*
  - ➢ Possibility of creating water (or liquid) disposal problem.
  - ➢ Product collected wet.
  - ➢ Particulates deposition possibly causing plugging of the bed or trays.
  - ➢ When FRP construction is used, sensitive to temperature.
  - ➢ Relatively high maintenance costs.

- What are the different types of equipment used in absorption processes?
  - *Packed Columns*: Liquid stream is divided into thin films that flow through a continuous gas phase.
  - *Plate Columns*: Particularly well suited for large capacities involving clean, noncorrosive, nonfoaming liquids. They are also preferred when internal cooling is required in the column through the use of coils immersed in the liquid on the tray, or alternatively, liquid can be removed from the column at one tray, cooled, and returned to another tray.
  - *Venturi Scrubbers*: Liquid is atomized into gas stream.
  - *Spray Chambers*: Involve dispersion of the liquid as discrete droplets within a continuous gas phase.
  - *Agitated Vessels*: Particularly used as gas–liquid reactors. Gas is dispersed in the liquid.
  - *Spray Dryers*: The use of spray dryers for gas absorption is gaining favor for flue gas desulfurization, with the use of the heat available in the gas. The absorbent must contain a component that reacts with the absorbate to form a compound that can be dried to the solid form without decomposing.
- Name some types of wet scrubbers.
  - Spray columns.
  - Venturi scrubbers.
  - Centrifugal scrubbers.
- What are the applications of agitated vessels?
  - Agitated vessels are used mainly where absorption is accompanied by a slow chemical reaction between the dissolved gas and some constituent of the liquid and sufficient time has to be allowed for this reaction to proceed to the required extent.
  - Alternatively, they may be used where close control of the process is required in order to prevent the formation of undesirable by-products. Agitated vessels are suitable for batch operations. If used for a continuous process, countercurrent flow cannot be obtained within individual vessels and it becomes necessary to use a series of vessels arranged in multiple stages.
  - ΔP through agitated vessels is high compared to that for packed columns.
- What are the limitations of spray columns as absorption equipment?
  - Though spray columns are simple to operate and maintain and have relatively low energy requirements, they have the least effective mass transfer capability and are usually restricted to particulate removal and control of highly soluble gases such as ammonia and sulfur dioxide.
  - Effectiveness of spray columns depends on creation of fine liquid droplets to provide large interfacial area between the gas and liquid phase.
  - They also require higher solvent (water) recirculation rates and are inefficient at removing very small droplets that create entrainment problems. Coalescence of the droplets initially formed in the spraying process further reduces efficient operation of spray columns.
- Under what conditions a spray column is effective as a gas absorber?
  - When liquid to gas ratio is high.
- "A spray column cannot provide more than a single stage." *True/False*?
  - *True*. Due to poor contact/poor turbulence.
- What is the lowest size of particulates that are removed in a spray column?
  - 2–3 μm.
- Are there any gaseous components in flue gases that are to be removed along with particulates? If so, what type of equipment is suitable for the purpose?
  - SO_2 and NO_x.
  - A wet scrubber with a suitable solvent can remove both particulates and these gases.
  - However, suitability of such a scrubber also depends on the disposal of the wastes produced in the process.
  - A Venturi scrubber can be considered for the purpose.
    - ➢ Must be protected from freezing.
- What are the positive and negative points in the use of wet scrubbers?

Positive Points

- *Simultaneous Removal of Gases and Particulates*: A multistage wet scrubber that combines the removal of sulfur, hydrogen chloride, sulfuric acid mist and particulate matter in a single unit, for example, has reduced sulfur dioxide emissions by an average of 99.7% in its first large scale commercial installation on a coal/oil-fired boiler. Particulate emissions were claimed to have reduced to 12.5 mg/(N m^3).

- Temperature and moisture content of gas has no effect.
- Good performance over wide loading ranges.
- Reduced explosion hazards.
- Corrosive gases can be neutralized by proper choice of liquid.
- Space requirements are small compared to dry collectors.

Negative Points

- Capital and maintenance-intensive. High energy costs.
- Sludge disposal problems.
- Corrosion problems.
- Visible plumes.
- Reduction of plume buoyancy.

• What are the plus and minus points in the use of Venturi scrubbers?

Positive Points

- Low initial cost.
- No blockages.
- Very small particles along with gaseous components can be removed.

Negative Points

- High energy requirements:
  - Primarily to operate fans for movement of the gas. The fans operate at a ΔP of 38–50 cm (15–20 in.) water.
  - Typically fan power requirements of 450–520 kW (600 700 hp), moving gases at capacities of over 4000 sm^3/m (48,000–50,000 sft^3/m).
  - Short contact times between gas and liquid make Venturi absorbers unattractive for difficultly soluble gases that require large contact times.

• What type(s) of fans/blowers are used in gas scrubbers?

- Moderate gas flows are conveniently handled by centrifugal blowers. Such systems commonly are designed for a maximum pressure drop of 15 cm of water for the scrubber, including the duct work.
- Where large volumes of gas are handled, axial fans are normally used.
- Because of fan pressure limitations, the overall pressure drop through such a system usually is designed to be a maximum 10 cm of water.
- The blower or fan commonly is located following the fume scrubber. This location protects the fan from severe corrosion by chemical fumes or erosion by solid particulates.
- However, such a location means that the fan will be handling a gas stream that is saturated with water vapor. The fan discharge should be designed to prevent any condensed water being returned to this fan.

• Why is it necessary to install coolers for absorption of HCl gas in water?

- Absorption of acid gases is accompanied by evolution of heat of absorption, which is considerable. This leads to rise in temperature of the solution. Rise in temperature decreases solubility of the gas, and, instead of absorption, desorption might take place.
- At still higher temperatures, solvent (in this case water) starts vaporizing.
- Rate of absorption will be highest in the bottom section of the absorber, since the entering gas is richest with respect to HCl gas and consequently heat release rate and so temperature rise will be highest.
- Installing intercoolers, especially near the bottom section of the absorber, will keep the temperatures low, promoting absorption rates.
- Installation of intercoolers can be conveniently achieved if a tray absorber is used instead of a packed absorber.
- Cooling coils can be installed on the tray, submerged within the liquid, for which purpose provision of greater liquid depths in the tray design can be made.
- The increased costs of using trays with corrosion resistant material (stoneware packing in the case of a packed column) is more than offset requiring use of external coolers and additional internals like liquid distribution plates in a packed column.
- The materials of construction for the cooler–absorbers should be of high thermal conductivity and resistant to corrosion by HCl. Use of tantalum or graphite are common.

• What is reactive absorption?

- Reactive absorption involves absorption involving chemical reactions. Chemical reactions enhance absorption rates considerably, making reactive absorption an attractive process. Environmental considerations dictate use of reversible reactions for recovery and recycle of the solvent.
- Examples include sulfuric and nitric acid manufacture and H_2S removal from natural gas streams.
- Can handle gases with low concentrations of solutes with low solvent requirements.
- Liberation of heat of reaction (most reactions in reactive absorptions are exothermic in nature, which is a limitation, requiring use of intercoolers for heat removal.

- Solvent recovery is relatively difficult.
- Most reactive absorptions involve steady-state operations involving liquid-phase reactions, although some applications involve both liquid- and gas-phase reactions.
- Reactive absorption is a complex rate controlled process that occurs far from thermodynamic equilibrium. Therefore the equilibrium concept is insufficient to describe the process, requiring rate controlled kinetic models.
- Modeling and design of reactive absorption processes are usually based on the equilibrium stage model, which assumes that each gas stream leaving a tray or equivalent packed section (HETP) is in thermodynamic equilibrium with the corresponding liquid stream leaving. For reactive absorption, chemical reaction must also be taken into account.
- With very fast reactions, the reaction-separation process can be satisfactorily described assuming reaction equilibrium.
- A proper modeling approach is based on the nonreactive equilibrium stage model, which is extended by simultaneously considering the reaction equilibrium relationship and the tray/stage efficiency. If the reaction rate is slower than the mass transfer rate, the influence of reaction kinetics increases and becomes a dominant factor. This tendency is taken into account by integrating the reaction kinetics into the material and energy balances. This approach is widely used.
- In real reactive absorption processes, thermodynamic equilibrium is rarely reached. Therefore, parameters such as tray efficiencies or HETP values are introduced to adjust the equilibrium-based theoretical description to real column conditions. However, reactive absorption normally takes place in multicomponent systems, for which this simplified concept often fails.
- The mass transfer enhancement due to chemical reactions is often accounted through the use of the so-called enhancement factors, which are obtained from experimental data or from theoretical considerations with simplified model assumptions.

17.1.1 Applications of Absorption/Stripping

- What are the applications of absorption/stripping in pollution control?
  - Scrubbing of corrosive, obnoxious, or hazardous gases, vapors and particulates from a gas stream.
  - Packed scrubbers, which have low pressure drops and occupy a reasonably small floor space, are generally preferred equipment for such applications. They can handle high-temperature and high moisture content gas streams. Corrosion resistant construction is possible.

Particulate Removal

- Wet packed scrubbers are highly efficient for removing particles of 10 μm or larger. Removal efficiencies of 90–95% can be expected even on 6 μm size material.
- As the solids loading in the inlet gas stream increase, larger sizes of packings should be selected so as to avoid blockage with the recovered particulates. In addition, higher liquid rates should be employed in order to assure complete wetting of the packing surfaces, as well as to flush off deposited solids.

NO_x Removal

- A special design procedure is necessary for scrubbers intended to remove NO_x from gas streams.
- NO_x is usually a mixture of NO, NO_2 together with N_2O_4 and N_2O_5.
- NO is a relatively insoluble gas. It must be oxidized to NO_2 before it becomes soluble in water and allows effective absorption.
- N_2O_5 is readily soluble in water to produce a nitric acid solution.
- When NO_2 is dissolved in water, two-thirds of the moles react with water to form nitric acid. However, the other one-third of the moles are desorbed into the gas stream as NO:

$$3NO_2 + H_2O = 2HNO_3 + NO \tag{17.1}$$

- The by-product, NO, then must be oxidized to NO_2 for it to be reabsorbed.
- Because this oxidation reaction in the gas phase is time dependent, NO_2 removal efficiency usually is limited to about 80% in a single scrubber when water is the scrubbing liquid.
- When sodium hydroxide solutions are used for scrubbing NO_2, the resultant reaction produces equal moles of sodium nitrate and sodium nitrite:

$$2NO_2 + 2NaOH = NaNO_3 + NaNO_2 + H_2O \tag{17.2}$$

- NO_2 first must be absorbed by the water into the liquid film before it can react with the caustic.
- In such a liquid phase, two competing reactions take place, namely, NO_2 with water and NO_2 with NaOH.
- If the NaOH concentration in the liquid film is deficient, NO gas is released back into the gas phase in accordance with the first equation.

- The diffusion rate of NaOH into the liquid film is a function of the liquid-phase solution concentration. For this system, 3 N (12 wt%) NaOH seems to provide optimum absorption efficiency. Higher solution concentrations are less effective because the increase liquid viscosity retards the diffusion rate of the NaOH, as well as the products of reaction.
- Usually the gas stream entering the scrubber contains other oxides of nitrogen in addition to NO_2.
- Any residual NO in the gas stream leaving the scrubber tends to oxidize on mixing with the outside air.
- To meet increasingly stringent regulations, selective catalytic reduction of nitrogen oxides with ammonia is gaining favor where low NO_x levels are required in the discharge.
- Such systems give removal efficiencies of 90% or greater with a ΔP not exceeding 13 cm of water.

VOC Control

- The most common methods for controlling VOC releases are incineration, adsorption, condensation and absorption.
- Incineration is capable of handling even low concentrations of VOCs, but the energy requirements can be quite large.
- Carbon adsorption and absorption into a nonvolatile solvent normally are selected where the concentration of organic materials is at least 1000 ppm by volume because such processes permit recovery of these organics.
- Packed absorbers can remove over 95% of the entering organic vapors by contacting the gas stream with a low vapor pressure solvent to prevent its loss. It must be inert to the organic solutes being absorbed.
- Because many gas streams to be treated contain water vapor, the solvent must tolerate both water and the organic solutes without forming two liquid phases.

Odor Control

- Some contaminants in a gas stream are odorous materials that are considered a nuisance if vented into the atmosphere.
- Use of packed scrubbers is one method for controlling odorous emissions.
- A wet scrubber absorbs the odorous material into the liquid phase, but many of these substances have limited solubility in water. The effluent liquid may be a source of water pollution as the odorous substance in the effluent liquid is desorbed into the surrounding atmosphere.
- For effective odor control, normally a chemically reactive compound is added to the scrubbing liquid that converts the absorbed contaminant into a less objectionable substance.
- H_2S, for instance, can be absorbed by a solution of NaOH in which it is quite soluble. This is due to a chemical reaction with the absorbed solute that forms a stable sulfide.
- There are few inorganic odorous substances such as SO_2, HCl, NO_x, H_2S, HCN, O_3, Cl_2, and NH_3.
- Removing such contaminants by absorption has been practiced for some time, but it is becoming increasingly difficult to dispose of the spent scrubbing liquid.
- In large capacity systems, the reactive absorbent may be regenerated. Thus, H_2S absorbed into an alkaline solution can be oxidized to elemental sulfur, which can be recovered and the scrubbing liquid recycled.
- Removal of organic odors is a much more complex problem. Wet scrubbing using a combination of absorbent and oxidant has successfully reduced such odors. The four common oxidants used in the liquid phase are potassium permanganate, sodium hypochlorite, hydrogen peroxide, and chlorine dioxide.
- The operation of odor control systems requires careful control of the oxidant concentration as well as the solution pH.
- Degradation products also must be purged from the solution to prevent their buildup.

Wastewater Stripping

- Primary means of removal of volatile compounds from wastewater are air stripping and steam stripping.
- In air stripping removed contaminant should be recovered from air using another technique. Steam stripping requires more energy.
- Most of the organic pollutants can be steam stripped and less often air stripped. This is because the degree of separation increases with increasing temperature, making steam stripping more effective.
- Steam stripping allows for the removal of heavy soluble organics that other stripping techniques like air stripping will not remove.
- The overhead organic vapor from a steam stripper is condensed and recovered. The recovered hydrocarbons can either be treated in an incinerator or recycled back into the process.
- High VOC recovery (greater than 99%) can be expected in steam stripping. The purified water has very low contaminant concentrations and can be reused in plant operations.
- The materials removed by air/steam stripping are relatively hydrophobic with limited solubilities in

water. Such materials can also be removed using pervaporation that can handle smaller flow rate streams and is usually used as an additional technique to a common treatment system.

- The VOCs present in wastewater stream are usually in low concentrations. Typical organics present are benzene, toluene, xylenes (*o*, *m* and *p*), ethyl benzene, styrene, and chlorinated hydrocarbons.
- Packed or tray columns are used as strippers. Tray columns involve about 10 theoretical trays.
- Steam to water or air to water mole ratios employed are about 0.02.
- In addition to the VOCs found in water, ammonia is frequently present in wastewaters.
- Ammonia nitrogen exists in both the dissolved gas form (NH_3) and in true solution ($NH_4{}^+$). These two species are present in a dynamic equilibrium according to the equation:

$$NH_3 + H_2O = NH_4^+ + H_2O \qquad (17.3)$$

- This equilibrium is controlled by the solubility product that varies with temperature.
- Therefore, the relative concentrations of these two species depend on both the pH of the solution and the temperature. In general, at a temperature of 20°C and a pH of 7 or below, only ammonium ions are present. As the pH increases above 7, the chemical equilibrium is gradually shifted to the left in favor of the ammonia gas formation. At a pH of about 11.5–12, only the dissolved gas is present. In addition to converting all the ammonia to the dissolved gas phase, efficient ammonia stripping requires proper conditions to facilitate a rapid transfer of the dissolved gas from the liquid phase to the air.
- An ammonia stripping column is normally used for ammonia removal.
- The efficiency of an ammonia stripping operation depends primarily on five factors, namely, pH, temperature, rate of gas transfer, and air and liquid rates.

17.1.2 Examples of Commercial Applications of Gas Absorption

- What are the considerations and practices involved in natural gas dehydration processes?
  - Natural gas is dehydrated using a glycol in order to reduce the water content of the gas so as to prevent formation of solid hydrates.
  - Because of the cost of regeneration of the glycol, the flow of glycol usually is specified at 25–40 L of glycol/kg of water removed.
  - Because of the very low quantity of liquid compared to the gas flow, bubble cap trays have been widely employed in such absorbers.
  - When packed columns are used, such low liquid rates generally have discouraged the use of random packings in these absorbers.
  - Due to the high glycol viscosity, this system is liquid film controlled and tray efficiencies are of the order of 25%.
  - Structured sheet metal packings have replaced conventional random packings.
  - In modern plants, trays are replaced by structured packings, decreasing column height requirements by as much as 40%.
- What are the problems involved in gas drying processes and how these are solved?
  - Due to the high heat of vaporization of water, plus its high heat of solution in the dehydrating solvent, there can be a large heat load imparted to the liquid phase.
  - As the dehydrating solvent is diluted by the condensed water and as the liquid temperature increases, the vapor pressure of water above the liquid phase will increase.
  - Usually a high liquid circulation rate is used in each stage of drying to minimize both the temperature rise and the dilution of the liquid dehydrating agent.
  - When the amount of water vapor to be removed is large, a number of columns, each recycling cooled liquid, may be operated in series.
- What are the practices in SO_3 absorption in H_2SO_4 manufacture?
  - The hot gas stream from sulfur burners entering the SO_3 absorber usually contains 8–10 mol% SO_3, which must be almost completely absorbed so the gas leaving the column will contain only 30–40 mol ppm SO_3. The SO_3 is absorbed into 98 wt% sulfuric acid, in which it is highly soluble.
  - The temperature rise in the acid is due both to the heat of solution of SO_3 and the sensible heat removed by cooling the inlet gas stream.
  - This system is almost pure gas film controlled absorption.
  - The use of a design ΔP of 20–25 mm H_2O/m at the bottom of the absorber is common practice. This ΔP may double in 9–10 years of operation due to an accumulation of sulfation products in the packed bed. To minimize blockage of the packed bed, a 5 or 7.5 cm (2 or 3 in.) size ceramic packing normally is specified.
  - As a result of the use of modern column packings, as well as the development of improved packing support

systems and liquid distributor design, the superficial gas velocity within the column has been increased considerably in modern plants.

- What are the reactions involved in nitric acid manufacture?

Gas Phase

$$2NO + O_2 \rightarrow 2NO_2 \quad (17.4)$$

$$2NO_2 \leftrightarrow N_2O_4 \quad (17.5)$$

$$3NO_2 + H_2O \leftrightarrow 2HNO_3 + NO \quad (17.6)$$

$$NO + NO_2 \leftrightarrow N_2O_3 \quad (17.7)$$

$$NO + NO_2 + H_2O \leftrightarrow 2HNO_2 \quad (17.8)$$

Liquid Phase

$$N_2O_4 + H_2O \rightarrow HNO_2 + HNO_3 \quad (17.9)$$

$$3HNO_2 \rightarrow HNO_3 + H_2O + 2NO \quad (17.10)$$

$$N_2O_3 + H_2O \rightarrow 2HNO_2 \quad (17.11)$$

$$2NO_2 + H_2O \rightarrow HNO_2 + HNO_3 \quad (17.12)$$

- What are the special problems involved in the absorption of formaldehyde in water?
  - A fixed composition of the solution, usually at 50% concentration, is required for the formaldehyde product to be marketed. In such a case, the feed liquid rate to the absorber is set by a material balance.
  - Although, formaldehyde is highly soluble in water, it releases large amounts of heat when absorbed into water.
  - Because the vapor pressure of formaldehyde above its solutions increases with both concentration and temperature, it is necessary to withdraw the liquid, cool it and reintroduced into the column repeatedly. Multiple packed sections will be necessary in order to facilitate recirculating the solution through the absorber and external coolers.
  - This system is complicated by the tendency of formaldehyde solutions to produce a solid phase if the concentration is too high or the temperature too low. Thus, the 50 wt% formaldehyde solution must be discharged from the absorber at a temperature greater than 54°C.
- Give examples of commercial absorption processes involving chemical reactions.
  - Generally absorption processes involving reversible reactions are employed because of the pollution problems involved in disposal of effluents resulting from irreversible reactions.
  - Typical reversible reactions are involved in the absorption of H_2S into ethanol–amines, or the absorption of CO_2 into alkali carbonate solutions.
  - These reversible reactions permit the resultant solution to be regenerated so that the solute can be recovered in a concentrated form.
  - Some irreversible reactions are the absorption of NH_3 into dilute acids and the absorption of CO_2 into alkaline hydroxides. Regeneration of the solute is not possible. The purpose of such a reactant is to increase the solubility of the solute in the liquid phase and/or reduce the liquid film resistance to mass transfer.

CO_2 Absorption into NaOH Solution

- The absorption of CO_2 into a sodium hydroxide solution has been used to characterize the relative performance of column packings.
- The effect of the liquid-phase reaction is to reduce substantially the liquid film resistance.

CO_2 and H_2S Absorption in Amine Solutions

- One of the most widely used commercial absorption processes is the removal of CO_2 or H_2S from a gas stream by contacting it with a solution of monoethanolamine or diethanolamine.
- Both of these solutions are alkaline and combine chemically with 0.5 mol of acid gas/mol of amine.
- Although a chemical combination exists, still even at equilibrium, the acid gas exhibits a vapor pressure above the solution. Because this vapor pressure of the absorbed acid gas increases rapidly with temperature increase, it is possible to regenerate the rich amine solution by stripping the heated solvent.
- Due to the heat of solution and heat of reaction of the acid gas with the amine, the rich liquid effluent will be at a higher temperature than the lean liquid feed, unless cooled by the gas stream.
- If the gas stream to be treated contains condensable hydrocarbon vapors, the lean solvent temperature should be above the dew point temperature of these vapors in order to prevent condensation of an immiscible hydrocarbon liquid, which promotes foaming of the liquid phase in the absorber.
- Amines are considered to be moderately foaming systems.
- The K_Ga value decreases as the ratio of moles acid gas per mole MEA in the solvent increases. This effect on the K_Ga value is given by

$$K_G a \propto 1.375 - 2.5C, \quad (17.13)$$

where C is the number of moles of acid gas per mole of MEA.

- Increasing the concentration of MEA in the solvent normally would be expected to increase the $K_G a$ value because of increased diffusion of reactive amine into the liquid film.
- However, increasing the MEA concentration also increases the viscosity of the liquid phase, which reduces the rate of diffusion.
- Diethanolamine solutions find wide use in treating sulfur-bearing gas streams. This is because MEA forms a nonregenerable, stable chemical compound with carbonyl sulfide and carbon disulfide.
- $K_G a$ value for absorption into a DEA solvent is only 50–60% of that for an MEA solution of the same normality.

CO_2 and H_2S Absorption in Hot Potassium Carbonate Solutions

- Another very widely used commercial absorption process involves the removal of CO_2 or H_2S by contacting the gas stream with a hot potassium carbonate solution.
- Because the solvent is an aqueous inorganic salt solution, hydrocarbon solubility is very low. This process is widely used to purify high-pressure natural gas, as there is no loss of fuel gas by coabsorption with the acid gas.
- The reactions involved are

$$K_2CO_3 + CO_2 + H_2O = 2KHCO_3 \quad (17.14)$$

- When H_2S is absorbed, the reaction is

$$K_2CO_3 + H_2S = KHCO_3 + KHS \quad (17.15)$$

- A hot carbonate absorber is considered to be a slightly foaming system.
- The heat load on a hot carbonate system per unit quantity of CO_2 absorbed and regenerated is lower than for an amine solvent system.
- Generally, the hot carbonate process is much less corrosive to common metals than amine solvent systems.
- Potassium carbonate is a very stable compound that is not degraded at the usual high operating temperatures or by oxidizing agents.
- Hot carbonate systems, therefore, are used to remove CO_2 from recycle gas streams in organic oxidation reactions.

17.1.3 Design of Absorbers

- What is an absorption equilibrium diagram?
  - An absorption equilibrium diagram is a plot of the mole fraction of solute in the liquid phase, denoted as x, versus the mole fraction of solute in the gas phase, denoted as y.
- What is the effect of temperature on absorption equilibria? Illustrate.
  - At a constant mole fraction of solute in the gas, y, the mole fraction of solute that can be absorbed in the liquid, x, increases as the temperature decreases.
  - Figure 17.1 illustrates the effect of temperature on absorption equilibria for SO_2–water system at different temperatures.
  - Under certain conditions, Henry's law may also be used to express equilibrium solubility of gas–liquid systems.
    - Henry's law can be written as

$$y = H'x, \quad (17.16)$$

where y is the mole fraction of the solute gas in equilibrium with the liquid and H is the Henry's law *constant*, mole fraction in vapor per mole fraction in liquid, and x is the mole fraction of the solute in the equilibrium liquid.

    - Henry's law *constant* is not a true constant but has a significant nonlinear temperature dependence. Ignoring this temperature dependence can lead to serious inaccuracies in certain calculations.
    - H depends on the total pressure of the system.

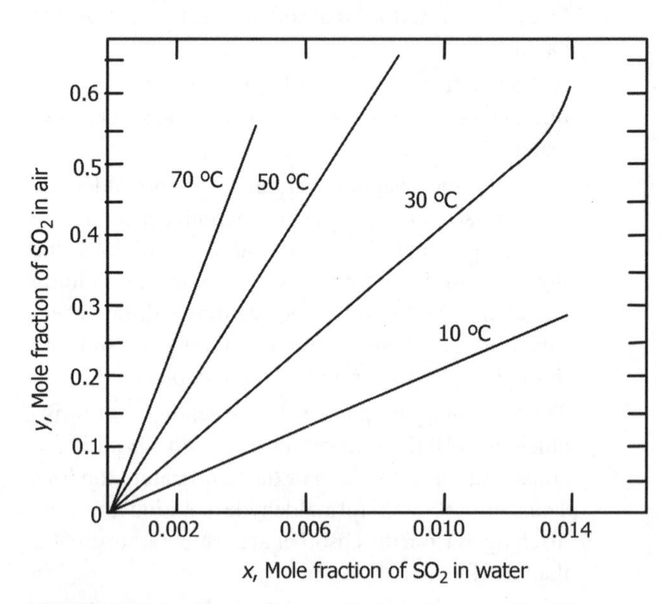

FIGURE 17.1 Absorption equilibrium diagrams for SO_2–water system at different temperatures.

- The thermodynamic definition for Henry's law *constant* is given by the equation

$$H_i = \lim(f_i/x_i),\ x_i \text{ tending to zero}, \qquad (17.17)$$

where f_i and x_i are the fugacity and mole fraction of the solute, respectively.

- Equation 17.17 can be applied at any temperature and pressure. The solvent may be a vapor, liquid, or supercritical fluid or even a solid. Most applications involve a liquid solvent. It is normally convenient to define H_i at a pressure equal to the vapor pressure of the solvent.
- Henry's law can be used to predict solubility *only* when the equilibrium line is straight.
- Equilibrium lines are usually straight when the solute concentrations are very dilute.
- Another constraint on using Henry's law is that it does not hold true for gases that react or dissociate upon dissolution. If this happens, the gas no longer exists as a simple molecule. For example, scrubbing HF or HCl gases with water causes both compounds to dissociate in solution. In these cases, the equilibrium lines are curved rather than straight.
- Data on systems that exhibit curved equilibrium lines must be obtained from experiments.

• How does Henry's law constant dependent on temperature?

- Henry's law constant typically increases with temperature at low temperatures, reaches a maximum and then decreases at higher temperatures. The temperature at which the maximum occurs depends on the specific solute–solvent pair. As a rule of thumb, the maximum tends to increase with increasing solute critical temperature for a given solvent and with increasing solvent critical temperature for a given solute.
- Use of a Henry's law constant obtained at 25°C at a different temperature can lead to serious design errors. Even a variation as small as 10°C can cause the Henry's law constant to change by a factor of two, which could have an impact on many designs.

• Describe briefly the graphical and analytical procedures involved in finding number of trays in the design of tray absorbers and strippers.

- Equilibrium diagram is plotted from experimental data corresponding to the temperature of absorption or from Henry's law.
- *Operating line* is drawn from material balance equation. This line defines operating conditions obtainable from material balance equation within the absorber.
- Equation for the operating line can be written as

$$Y_1 - Y_2 = (L_m/G_m)(X_1 - X_2). \qquad (17.18)$$

$$Y = \text{mole ratio of solute in the nonabsorbable (inert) gas} = y/(1-y). \qquad (17.19)$$

$$X = \text{mole ratio of solute in solute-free liquid} = x/(1-x). \qquad (17.20)$$

For dilute mixtures it can be approximated that

$$Y \approx y \quad \text{and} \quad X \approx x. \qquad (17.21)$$

L_m and G_m are molar flow rates of liquid and gas, respectively.

- Operating line with constant slope (straight line) is valid for dilute mixtures involving absorption of small amounts of solute. When concentrated mixtures are handled, involving considerable amounts of absorption, straight line operating line cannot be assumed, as flow rates of both gas and liquid vary inside the column. In such cases, curved operating line is constructed by dividing the column into small vertical sections, establishing material balances for each such section.
- The slope of the operating line is the liquid mass flow rate divided by the gas mass flow rate, which is the liquid to gas ratio, or L_m/G_m.
- When operating line touches the equilibrium line, driving force for mass transfer becomes zero, giving rise to inoperable conditions. For systems not involving heat effects (e.g., dilute mixtures and nonacidic gas absorption), the operating line touches equilibrium curve (pinch point) at the gas inlet conditions, that is, bottom of the absorber (Figure 17.2).
- Actual liquid rate, L_m, can be taken as 1.5 times $(L_m)_{min}$, as a thumb rule.
- Figure 17.2 represents the slopes of minimum and actual operating lines.
- In absorption involving acidic gases, significant amounts of heat are released, especially in the lower sections of the absorber, increasing the operating temperatures, particularly in the lower sections of the absorber, reducing the solubility of the gases, *lifting* the equilibrium curve as illustrated in the figure. In extreme cases, increased temperatures result in vaporization of the solvent, aggravating the effects, leading to *lifting* of the equilibrium curve and reducing the available driving force. Vaporization

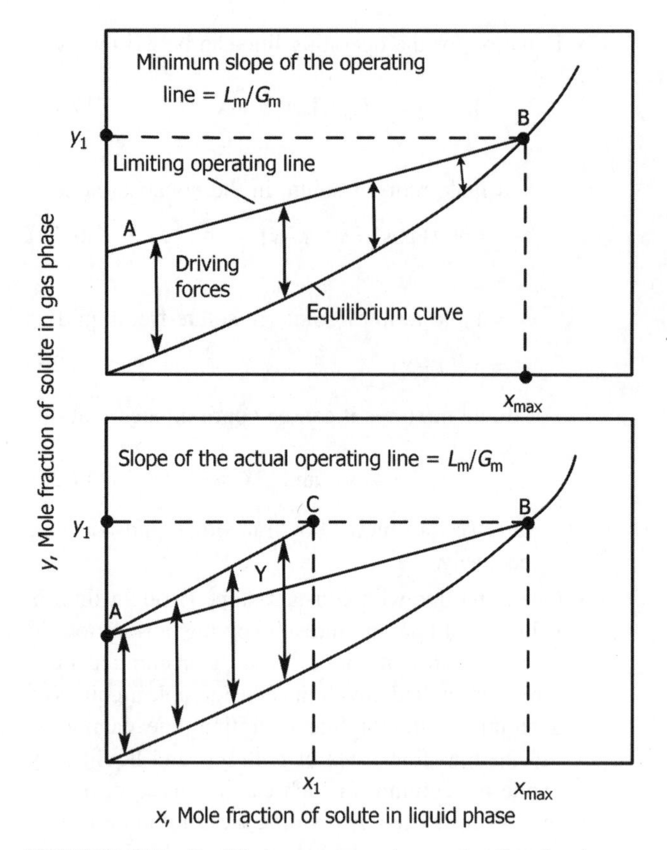

FIGURE 17.2 Equilibrium curve and operating line for absorption systems without heat effects.

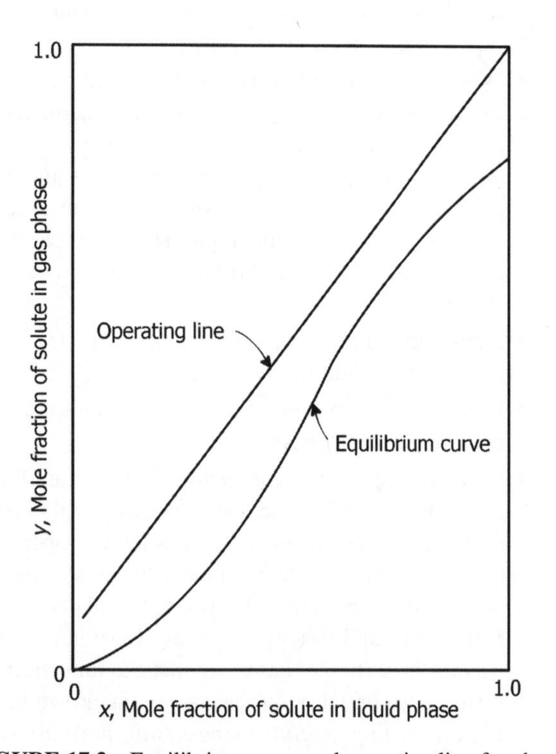

FIGURE 17.3 Equilibrium curve and operating line for absorption with heat effects.

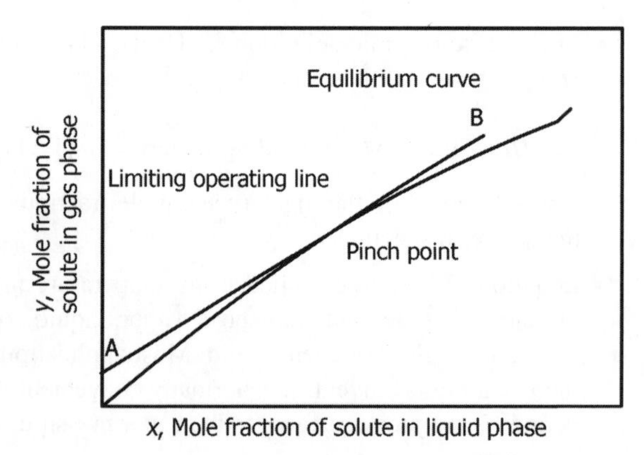

FIGURE 17.4 Limiting operating line for systems involving heat effects.

reduces the temperatures by taking latent heat of vaporization, making the equilibrium curve *fall back* as shown in Figure 17.3.

- In such situations, intercoolers should be installed in the lower sections of the absorber.
- This necessitates use of trays instead of packed sections so that cooling coils, which get submerged in the liquid, can be installed on the trays.

- Slope of the limiting operating line gives, minimum liquid rate, $(L_m)_{min}$ for a given gas flow capacity, G_m. Where acidic gas absorption involving considerable heat effects, pinch point can occur at a point inside the absorber as shown in Figure 17.4.
- Figure 17.5 illustrates the McCabe–Thiele graphical construction of number of theoretical trays for a

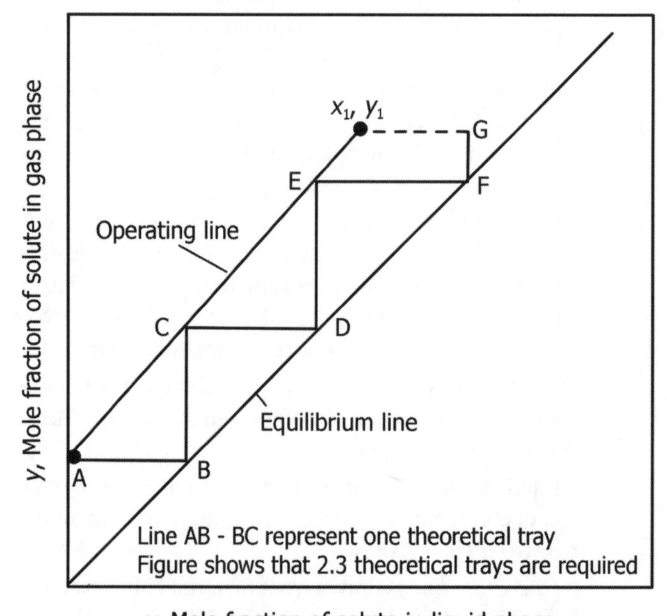

FIGURE 17.5 Graphical determination of number of trays for absorbers.

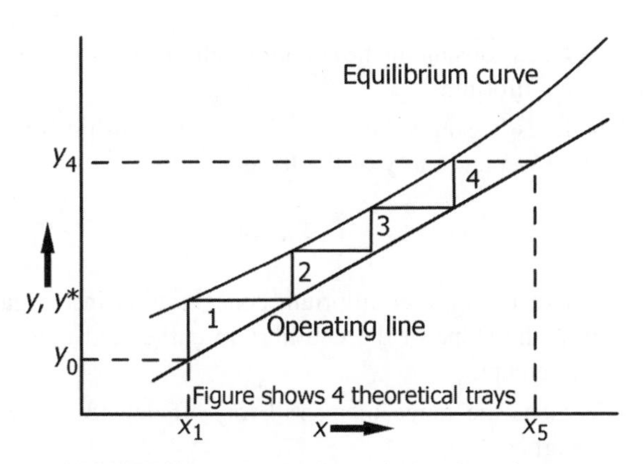

FIGURE 17.6 Tray column design for strippers.

given flow ratio of liquid to gas, L_m/G_m and a given separation (McCabe–Thiele method is discussed under distillation).

- For strippers, equilibrium curve will be above the operating line as shown in Figure 17.6.
- Analytical equation for number of theoretical trays for an absorber is given by

$$n_T = \ln[\{(Y_1 - mX_2)/(Y_2 - mX_2)\}\{1-(mG_m/L_m)\} + mG_m/L_m]/\ln\{L_m/mG_m\}. \quad (17.22)$$

- Number of actual trays is obtained by dividing n_T with tray efficiency.
- Here, n_T is number of theoretical trays.

• Define absorption and stripping factors.

- Absorption factor,

$$A = L/mG (= L/KV), \quad (17.23)$$

where K is the equilibrium ratio, that is, y/x and V is the vapor flow rate, for distillation.

- Stripping factor,

$$S = mG/L (\text{or } KV/L)\ (= 1/A). \quad (17.24)$$

• Write down Kremser–Brown–Souders equation for the determination of number of trays for an absorber. What is the important assumption involved in its application?

$$(Y_{n+1} - Y_1)/(Y_{n+1} - Y_0) = (A^{n+1} - A)/(A^{n+1} - 1). \quad (17.25)$$

- Solving the above equation gives n, number of theoretical trays.
- A convenient form of the equation for obtaining number of trays is

$$n = \ln\{[(Y_{n+1} - mX_0)/(Y_1 - mX_0)](1 - 1/A) + 1/A\}/\ln A. \quad (17.26)$$

- For the special case of $A = 1$ (or $S = 1$), the above equation reduces to

$$n = (X_0 - X_n)/(X_n - mY_{n+1}). \quad (17.27)$$

- These equations assume that the absorption factors (or stripping factors) remain constant from bottom to top of the column, that is, flows remain constant throughout the column.
- For a stripper, Kremser equation can be written as

$$(X_0 - X_n)/(X_0 - X^*_{n+1}) = (S^{n+1} - S)/(S^{n+1} - 1). \quad (17.28)$$

- These equations, for obtaining number of trays, are given in graphical form in a similar way as those for N_{OG} determination for packed columns (cf. plot given later).

• Write down Horton–Franklin absorption factor equations for analytical determination of number of theoretical trays for the case of variable flows, that is, involving concentrated mixtures.

- Absorption factor for each theoretical tray is defined as

$$A_n = L_{n+1}/K_n V_n. \quad (17.29)$$

$$(Y_0 - Y_n)/Y_0 = [(A_1A_2A_3 \ldots A_n + A_1A_2A_3 \ldots A_{n\text{-}2} + \cdots + A_1)/(A_1A_2A_3 \ldots A_n + A_1A_2A_3 \ldots A_{n\text{-}1} + A_1A_2A_3 \ldots A_{n\text{-}2} + A_1 + 1)] - (L_{n+1}X_{n+1}/V_0Y_0) [(A_1A_2A_3 \ldots A_{n\text{-}1} + A_1A_2A_3 \ldots A_{n\text{-}2} + A_1 + 1)/ (A_1A_2A_3 \ldots A_n + A_1A_2A_3 \ldots A_{n\text{-}1} + A_1A_2A_3 \ldots A_{n\text{-}2} + \cdots + A_1 + 1)]. \quad (17.30)$$

- When absorption factors for each tray are known, the above equation can be used to calculate the separation obtainable for a given number of trays.
- Edmister gives a short cut method described by the equations given below:

$$(Y_0 - Y_n)/Y_0 = [1 - (L_{n+1}X_{n+1})/A'V_0Y_0] [(A_c^{n+1} - A_c)/(A_c^{n+1} - 1)] \quad (17.31)$$

in which A_c and A' are defined by the following equations:

$$(A_c^{n+1}-A_c)/(A_c^{n+1}-1) = [(A_1A_2A_3\ldots A_n + A_1A_2A_3\ldots A_{n-2} + \cdots + A_1)/ (A_1A_2A_3\ldots A_n + A_1A_2A_3\ldots A_{n-1} + A_1A_2A_3\ldots A_{n-2} + A_1 + 1)] \quad (17.32)$$

and

$$(1/A')[(A_c^{n+1}-A_c)/(A_c^{n+1}-1)] = (A_1A_2A_3\ldots A_{n-1} + A_1A_2A_3\ldots A_{n-2} + A_1 + 1). \quad (17.33)$$

- Edmister approximated, for design purposes, *effective* absorption, and stripping factors A_e and S_e by the equations:

$$\text{Ae} = [\text{AB}(\text{AT}+1)+0.25]^{1/2}-0.5. \quad (17.34)$$

$$\text{Se} = [\text{ST}(\text{SB}+1)+0.25]^{1/2}-0.5. \quad (17.35)$$

  A_T, A_B and S_T, S_B are *absorption and stripping factors* corresponding to column top and bottom trays, respectively.
- For very approximate analyses of cases involving pure solvents and high solute recoveries as well as the assumptions implicit in the Kremser equation, the number of theoretical trays can be estimated by the use of the simple equation:

$$n = 6\log(Y_{n+1}/Y_1)-2 \quad \text{(Douglas equation)}. \quad (17.36)$$

- This is based on the use of the rule-of-thumb value of $A = 1.4$.
- For cases where the Murphree tray efficiency is known and relatively constant over the length of the column, the following correlation can be used to estimate the number of actual trays n_T as follows:

$$n_T = \ln[(y_{n+1}+\alpha)/(y_n+\alpha)]/\ln \beta_v \quad \text{(Nguyen equation)}, \quad (17.37)$$

  where n_T is total number of actual trays and β_v is a modified absorption factor, defined by the equation,

$$\beta_v = 1/[1+E_{MV}[(mG/L)-1] \quad (17.38)$$

  and

$$\alpha = y_1-[(L/mG)(mx_0+C]/[(L/mG)-1]. \quad (17.39)$$

  C is a constant in the equation defining the gas–liquid equilibrium.
- It is assumed that a linear relationship exists between x and y of the form

$$y^* = mx + C, \quad (17.40)$$

  where y^* is the equilibrium concentration in the gas, m is the slope of the equilibrium curve, and C is the y intercept.

• What is the range of overall tray efficiencies for gas absorbers?
  - 65–80%.

• What are the effects of operating pressure for a steam stripper?
  - The operating pressure of a steam stripper can influence the efficiency and reliability of the column. For example, lower operating pressures result better the removal efficiency (volatility).
  - Lower operating pressures will also result in lower operating temperatures.
  - Steam strippers operating under vacuum can be highly efficient, but equipment costs will increase due to requirement of vacuum pumps/ejectors, increased column diameters, utilities and maintenance.
  - Operation under vacuum allows the use of plastic internals, due to lower temperatures, which are better from corrosion point of view.
  - Typically, most steam strippers are designed at or near atmospheric operating pressures.
  - If the operating pressure of the column is increased, there are some unfavorable effects, such as
    - Increasing the pressure increases the solubility of the solute and increases the separation problems. More steam input to the column is required to achieve the same separation efficiency.
    - Increased pressures may also increase tendency for organic salt precipitation.
  - If the selected column pressure causes flashing of the feed liquid inside the column, this effect must be accounted for in the design of the upper section internals in order to avoid overloading and flooding near the top of the column.

• What are the expected operational and maintenance problems arising out of the improper design of steam strippers for wastewater treatment?
  - Poor knowledge about the feed stream composition.
  - Incorrect material selection for internals and piping. This could lead to stress corrosion cracking and other forms of material attack. The materials of construction must be chosen carefully due to the varying

nature of the composition of most wastewater streams.

- Loss of operational capacity due to foaming.
- Column fouling and plugging due to solid precipitation (e.g., salts).
- Reduction in column efficiency because of incorrect operational parameter design. Specifically, using the wrong concentration of organics to design the column could affect steam requirements in the column. If the column is not designed with enough steam capacity, removal of the organics to meet the design specifications.
- The characteristics of the feed stream must be determined before carrying out the design. This includes any possible composition variations that might occur.

17.1.4 Transfer Unit Concepts in Design

- Define H_{OG} and N_{OG} and state their significance and application to absorber design.
  - Frequently, it is convenient to calculate the packed height required by means of transfer units or equivalent theoretical plates. This is due to the fact that the height equivalent to a theoretical plate (HETP) for a particular system tends to be constant.
  - Further, the height of a transfer unit varies less with flow rates than do the mass transfer coefficients.
  - Measure of the efficiency of the absorption process in packed absorbers can therefore be expressed in terms of the overall height of gas-phase transfer units, H_{OG}.
  - The smaller the H_{OG}, the more efficient the absorption process will be. H_{OG} is defined as

$$H_{OG} = G_M/K_G aP. \tag{17.41}$$

  - The overall height of gas-phase transfer unit is related to individual gas and liquid-phase transfer units, m, slope of equilibrium line, and gas and liquid flow rates by the equation:

$$H_{OG} = H_G + (mG/L)H_L. \tag{17.42}$$

$K'_G a$ = Overall mass transfer coefficient, kmol/(m^3 s). H_G = Height of gas-phase transfer unit, m. H_L = Height of liquid-phase transfer unit, m. m = Slope of the equilibrium line. G, L = Gas and liquid mass velocities, respectively, kmol/(m^2 s).

  - The addition of resistances according to the two-film theory, in terms of mass transfer coefficients, cf. Chapter 15, also applies to the heights of transfer units:

$$H_{OG} = H_G + \lambda H_L, \tag{17.43}$$

where

$$H_G = G_M/k_G aP, \quad H_L = L_M/k_L a \tag{17.44}$$

and λ = Ratio of the slope of the equilibrium curve to the slope of the operating line

$$= mG_M/L_M. \tag{17.45}$$

G_M and L_M are gas and liquid molar flow rates on solute-free basis, which are constant throughout the column.

  - N_{OG} is defined as

$$N_{OG} = \int dY/(1-Y)(Y-Y^*). \tag{17.46}$$

  - For absorption involving low concentrations of solute, Henry's law can be used to describe equilibrium relationship and the equation for N_{OG} reduces to

$$N_{OG} = (Y_{Ao} - Y_{A1})/(Y_{Ao} - Y_A^*)\text{lm}. \tag{17.47}$$

  - N_{OG} is a function only of compositions and depends on the operating conditions, for example, G, L, T, P.
  - $(Y_{Ao} - Y_A^*)_{lm}$ is a log mean (average) driving force for absorption, which is given by

$$(Y_{Ao} - Y_A^*)_{lm} = (Y_1 - Y^*)/\ln[(Y_1 - Y^*)1/(Y_2 - Y^*)_2]. \tag{17.48}$$

  - In the above expressions mole ratios, rather than mole fractions, are used represented by capital Y.
  - For an absorption system in which the solute is highly soluble in the solvent, such that the equilibrium pressure of the solute over the absorbing liquid is very low, one can write

$$Y - Y^* = Y \text{ and } N_{OG} = \ln(Y_1/Y_2). \tag{17.49}$$

  - The above expression indicates that the number of overall mass transfer units, N_{OG}, is only controlled by the concentrations of the solute in the inlet and outlet gas streams.
  - Absorber height, Z, is obtained from

$$Z = H_{OG} \cdot N_{OG} = G'/K_G aP \int dY/(1-Y)(Y-Y^*). \tag{17.50}$$

  - The following Colburn plot (Figure 17.7) or the analytical form, Equation 17.41, can be used to estimate Z.

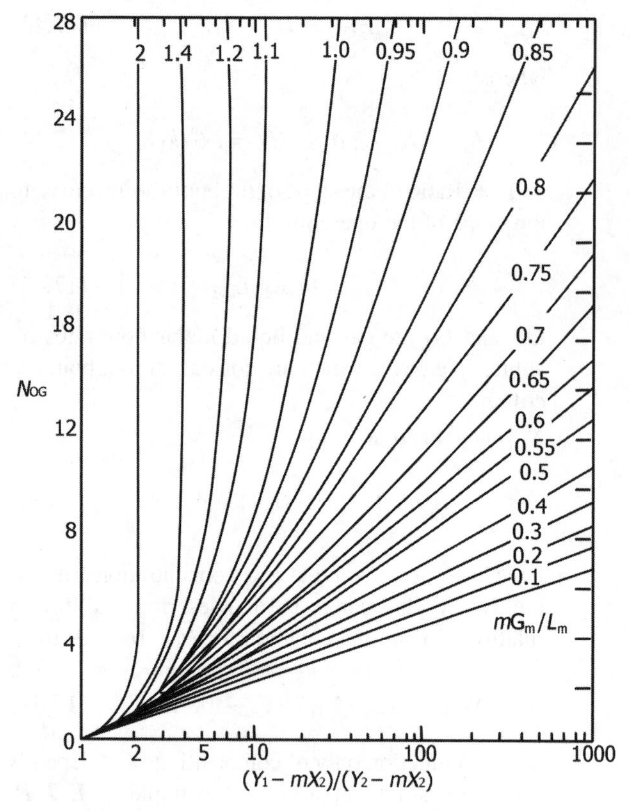

FIGURE 17.7 Colburn diagram for estimation of N_{OG}.

$$N_{OG} = \ln[\{(Y_1 - mX_2)/(Y_2 - mX_2)\} \{1 - (mG_m/L_m)\} + mG_m/L_m] / \{1 - (mG_m/L_m)\}. \quad (17.51)$$

- In terms of absorption factors,

$$N_{OG} = [A/(A-1)]\, ln[1 - \alpha/A/(1-\alpha)], \quad (17.52)$$

where A is absorption factor and $\alpha =$ Absorbed amount/theoretically maximum absorbable amount

$$= (Y_1 - Y_2)/(Y_1 - Y_2^*). \quad (17.53)$$

- In terms of stripping factors,

$$N_{OL} = [S/(S-1)]\ln[(1-\sigma/S)/(1-\sigma)], \quad (17.54)$$

where $\sigma =$ Amount stripped/theoretically maximum strippable amount

$$= (X_2 - X_1)/(X_2 - X_1^*). \quad (17.55)$$

- H_{OG} and N_{OG} are sometimes designated as HTU and NTU.
- In the transfer unit concept there is a clear separation between the number of transfer units, N_{OG}, which is based on driving forces and the height of a transfer unit, H_{OG}, which represents the rate of mass transfer.
- Thus, the specification for the amount of solute transferred from the gas phase determines the number of transfer units required and the height of a transfer unit depends mainly on the choice of column packing and flow rates.

- What is HETP?
  - HETP stands for height equivalent (of a packed bed) to a theoretical tray (plate).
  - The concept of HETP was introduced to enable comparison of efficiencies between tray and packed columns.
  - HETP is defined as the height of packing in packed column divided by number of theoretical trays for a given separation.
- How does HETP vary with size of packing? Illustrate by means of an example.
  - HETP decreases with decrease in packing size due to increased surface area.
  - For vapor–liquid contacting HETP for 2.5 cm (1 in.) Pall rings, is 0.4–0.56 m (1.3–1.8 ft).
  - For 5 cm (2 in.) Pall rings, HETP is 0.76–0.90 m (2.5–3.0 ft).
- Give the relationship between HETP and H_{OG} (or HTU).

$$\text{HETP} = H_{OG} \ln \lambda/(\lambda - 1) \quad \text{for } \lambda \neq 1, \quad (17.56)$$

where $\lambda = mG'/L'$.

$$\text{HETP} = H_{OG} \quad \text{for } \lambda = 1. \quad (17.57)$$

- Give an equation relating height of packing, Z, in a packed column, H_{OG}, N_{OG}, HETP, and number of theoretical trays, n_T, in a tray column.

$$Z = H_{OG} \cdot N_{OG} = (\text{HETP})/n_T. \quad (17.58)$$

- "HETP and H_{OG} approaches can be used to estimate packed bed heights." Which of the two approaches is generally preferred and why?
  - HETP approach is generally preferred due to its following advantages:
    - ➢ HETP approach is suitable for multicomponent systems while H_{OG} is difficult to use for these systems.
    - ➢ HETP approach can use tray-to-tray computer programs for multitray calculations.
    - ➢ H_{OG} concept is more complex and more difficult to use.

  - ➢ HETP approach enables easier comparison with tray columns.
  - The main advantage of H_{OG} concept is that it enables easier analysis in terms of mass transfer coefficients and therefore suitable for fundamental analysis and model development.
- What are the factors that influence HETP?
  - Packing type and size.
  - Generally packing efficiency increases (HETP decreases) when
    - ➢ Packing surface area per unit volume increases. Efficiency increases as packing size decreases (random packing) or as the channel size narrows (structured packing).
    - ➢ Packing surface is better distributed around packing element.
  - *Vapor and Liquid Loads*: For constant L/V operation below loading regime, generally
    - ➢ Vapor and liquid loads have little effect on HETP with random packings.
    - ➢ HETP increases with loadings with structured packings. The effect is more pronounced with wire mesh packing and much less with corrugated sheet packing. With larger crimp corrugated sheet structured packing, HETP is essentially independent of vapor and liquid loadings.
  - Distribution:
    - ➢ Maldistribution of vapor and liquid has profound effect on packing efficiency.
  - L/V ratio influences significantly HETP values.
  - Table 17.1 gives HETP values for different operations and different packing sizes.
- Give a simplified procedure for sizing a packed absorber.
  - *Step 1*: First the cross-sectional area, A, of the column in terms of the actual volumetric flow rate of the gas, q, and superficial velocity of the gas, v. Values of v are normally in the range of 1–2 m/s. Typically, 1.5 m/s may be selected:

$$A = q/v. \tag{17.59}$$

  - *Step 2*: Using A, column diameter, D, is calculated:

$$D = (4A/\pi)^{0.5}. \tag{17.60}$$

  - Packing sizes are selected based on D. As D increases increased packing sizes may be used. For $D = 0.9$ m (3 ft), a packing size of 25.4 mm (1 in.) is selected, for $D < 0.9$ m (<3 ft), a smaller packing may be used and $D > 0.9$ m, over 25.4 mm packing may be used.
  - *Step 3*: For estimating packed section height, Z, H_{OG}, and N_{OG} are required:

$$Z = H_{OG} \cdot N_{OG}. \tag{17.61}$$

In the event of equilibrium data not available, slope, m, of the equilibrium curve might be assumed to be approaching zero. This assumption is not unreasonable for most solvents that preferably absorb or react with the solute, involving low concentrations of the solute.

$$N_{OG} = \ln(y_1/y_2), \tag{17.62}$$

where y_1 and y_2 are specified inlet and outlet compositions, respectively. If the required removal efficiency, η, is specified in place of y_2, y_2 is obtained from solving equation:

$$\eta = (100 - y_2)/100. \tag{17.63}$$

  - *Step 4*: It is reasonable to use water or a solvent having physical and chemical properties as water as solvent for most scrubbing applications as a first choice. H_{OG} can be assigned values available in literature for water or aqueous systems.
  - *Step 5*: Equation 17.61 is used to find packing height of the absorber. The value obtained is multiplied by a factor ranging from 1.25 to 1.5 as a packing safety factor.
  - Height of the packed absorber is determined by adding a suitable height to the packed bed height to account for nozzles, space above and below the bed for distributors and support plates and access.
- In what ways design of a scrubber for emergency service differ from design for continuous service? Explain, giving the steps involved.
  - The design of a scrubber for emergency service is much more complicated than design for continuous process service.
  - Usually neither the composition of the emissions nor the emission rate are known.

TABLE 17.1 HETP Values for Preliminary Packed Column Design

Process	Packing Size	HETP
Distillation	25.4 mm (1 in.)	0.46 m (1.5 ft)
	38 mm (1.5 in.)	0.67 m (2.2 ft)
	50.8 mm (2 in.)	0.91 m (3.0 ft)
Vacuum distillation	25.4 mm (1 in.)	0.61 m (2 ft)
	38 mm (1.5 in.)	0.82 m (2.7 ft)
	50.8 mm (2 in.)	1.06 m (3.5 ft)
Absorption/stripping	All sizes	1.83 m (6 ft)

- During an emergency release, the emission rate is not continuous, and may initially be quite large.
- Scrubbers for emergency services operate once in a while, that is, only when emergency develops. Famous example is Bhopal tragedy in which emergency scrubbing and flare systems were not in operation when release occurred, killing and decapacitating thousands of people outside the battery limits of a pesticides plant, which released highly poisonous methyl isocyanate (MIC).
- The first step in designing an emergency scrubber for a particular application is to decide what process areas or potential events will be controlled by the enclosure and control system. This decision should be based on previous experience, if any and a detailed analysis of potential scenarios to assess their probability of their occurrence.
- The nature of the process or processes involved, whether there are temperature or pressure deviations that can occur due to runaway reactions or other causes must be understood with detail. Possible release scenarios must be determined.
- Next, the chemicals of concern must be identified and their mass release rates during a potential release must be quantified. The composition of a gaseous release may be quite different than the solid or liquid source from which it was derived due to chemical reaction(s) and reaction rates. Therefore, obtaining experimental data that represents the potential chemical composition of the release gas is critical.
- For example, an accelerating rate calorimeter can be used to study an exothermic polymerization in the laboratory.
- As a conservative approach, a constant rate equal to the maximum mass emission rate should be assumed for the estimated duration of the release. Estimating the duration of the release depends on the nature of the release and whether it is relatively short (15 min to 1 h) or long (several hours).
- Next, the permissible outlet concentrations of the chemical(s) involved must be set, using a safety factor over the published exposure limits such as *immediately dangerous to life and health (IDLH)* values or *short term exposure limits (STEL)*.
- Next step involves selecting the control system. Table 17.2 compares three common air pollution systems, namely, carbon adsorber, aqueous scrubber, and thermal oxidizer.

17.1.5 Solvent Selection

- What are the considerations involved in solvent selection for absorption processes?
  - Solubility of the solute is an important consideration as liquid circulation rates will be reduced.
  - Liquid viscosity influences mass transfer rates. High viscosity reduces mass transfer rates. In other words, the height of a liquid-phase transfer unit, H_{OL}, increases with increasing liquid viscosity.
  - When gas purification, rather than solute recovery, is required the solvent may be of low cost (such

TABLE 17.2 Comparison of Control Systems for Emergency Releases

Issues Involved	Carbon Adsorber	Aqueous Scrubber	Thermal Oxidizer
Operating temperature	Ambient	Ambient	760°C (430°C, if catalytic)
Delay in start-up	No delay	No delay	Heating required to get the desired temperature
Chemicals to be controlled	VOCs	Acids/bases	VOCs at high temperatures
		Scrubber may include neutralization, gases may be below dew point giving rise to corrosion problems	Can include organic particulates
Special issues	Ketones can give rise to fires	Nonaqueous organics will not be captured	Chloroorganics can form HCl and nitro compounds can form NO_x
Other operating issues	Fouling possible	Particulates captured	Particulates can poison catalysts
Waste treatment issues	Carbon can be regenerated	Wastewater treatment adds to costs	No wastes generated
Cost issues	Low capital costs	Medium capital costs	High operating costs
Overall feasibility	Suitable for VOCs. Not suitable for high particulate content or semivolatiles	Feasible for acids/bases particulates and some organics	Not suitable for emergency releases because of heating time requirements. High fuel costs if unit kept hot

as water) and, therefore, used on a once-through basis.

- On the other hand, an expensive solvent must be recycled through a regenerator. This is especially the case where solute recovery is the goal.
- Since the outlet gas stream will be saturated with solvent vapor, the cost of solvent losses through saturation of the outlet gases should be evaluated.
- The solvent selected should be chemically stable, noncorrosive, nontoxic, nonpolluting, and of low flammability.

17.2 DISTILLATION

- What are the essential differences between absorption and distillation?
  - In distillation, vapor is to be produced on each stage by partial vaporization of liquid at bubble point.
    - In absorption, liquid is well below its bubble point.
  - In distillation, counterdiffusion of molecules and in ideal system, equimolar counterdiffusion exists.
    - In absorption, gas molecules diffusing into liquid and negligible diffusion from liquid to gas.
  - In absorption, ratio of liquid to gas flow rate is considerably greater than for distillation.
  - Layout of trays is different for both cases.
  - With higher liquid rates in absorption, packed columns are more commonly used.
- What are the applications of distillation?
  - Purification/separation of liquid mixtures.
  - Separation of individual components/products involving groups of components with desired boiling ranges or to meet desired specifications.
  - Separations from naturally occurring mixtures.
  - Separations from mixtures resulting from chemical reactions/processes.
  - Separations from mixtures resulting from intermediate steps in a process, for example, solvent recovery from extraction or absorption processes.
- What are the situations under which distillation becomes unattractive as a separation process?

Mixtures with Low Relative Volatility or That Exhibit Azeotropic Behavior

- Some homogeneous liquid mixtures exhibit highly nonideal behavior that form constant boiling azeotropes. Separation of azeotropes cannot be carried out beyond the azeotropic composition using conventional distillation. The most common method used to deal with such problems is to add an entrainer to the distillation to alter the relative volatility of the key components in a favorable way and make the separation feasible (azeotropic distillation). If the separation is possible but extremely difficult because of low relative volatility between key components, <1.2, then an entrainer can also be used in these circumstances, in a similar way to that used for azeotropic systems (extractive distillation).
- In these situations azeotropic or extractive distillation, liquid–liquid extraction, membrane processes like pervaporation, crystallization, and adsorption may be used.

Separation of Materials with Low Boiling Components

- Low boiling materials are distilled at high pressure to increase their condensing temperature and to allow, if possible, the use of cooling water or air cooling in the column condenser. Very low boiling materials require refrigeration in the condenser in conjunction with high pressure. This significantly increases the cost of the separation since refrigeration is expensive.
- Absorption, adsorption, and membrane gas separations are the most commonly used alternatives to distillation for the separation of low boiling materials.

Separation of Heat-Sensitive and Reactive Materials

- High boiling materials are often heat sensitive and decompose if distilled at high temperature. Low boiling materials can also be heat sensitive, particularly when they are highly reactive. Such materials will normally be distilled under vacuum to reduce the boiling temperature.
- Giving rise to undesirable reactions at column temperatures.
  - Thermally unstable components in the feed can undergo reactions leading to significant product loss or formation of by-products difficult to separate.
  - Column fouling rates may be high due to production of solid precipitates or polymers.
  - Explosion hazards when feeds contain unstable materials causing free radical reactions.
  - Corrosion may necessitate use of rare materials of construction.
- Crystallization and liquid–liquid extraction can be used as alternatives to the separation of high boiling heat-sensitive materials.

Separation of Components with a Low Concentration

- If the product of interest is a high boiling component with low concentrations.
- If low concentrations of high boiling contaminants must be removed from a desired product, capital and energy costs will be high.

- Distillation is not well suited to the separation of products that form a low concentration in the feed mixture. Adsorption and absorption are both effective alternative means of separation in these cases.

Separation of Classes of Components

- If a class of components is to be separated (e.g., a mixture of aromatic components from a mixture of aliphatic components), then distillation can only separate according to boiling points, irrespective of the class of component. In a complex mixture where classes of components need to be separated, this might mean isolating many components unnecessarily. Liquid–liquid extraction and adsorption can be applied to the separation of classes of components.
- If the boiling range of one set of components overlaps the boiling range of another set of components from which it must be separated requires many distillation columns.

Separation of Mixtures of Condensable and Noncondensable Components

- If a vapor mixture contains both condensable and noncondensable components, then partial condensation followed by a simple phase separator often can give a good separation. This is essentially a single-stage distillation operation.

Other Situations

- Need for extreme temperatures and pressures. For temperatures more than 250°C or less than −40°C, energy costs increase immensely.
  - For pressures more than 5 MPa, column costs are high and for pressures less than 2 kPa column vacuum pump costs increase significantly.
- Low production rates, only a few tons/day or less. Does not scale down well, investment does not reduce as much as scale-up factor as production capacity is reduced. Availability of steam may be a problem, alternative energy sources such as electricity has a higher cost.
- In summary, distillation is not well suited for separating either low boiling or high boiling heat-sensitive materials. However, distillation might still be the best choice for these cases, since the basic advantages of distillation (potential for high capacities, any feed composition and high purity) still prevail.

17.2.1 Phase Equilibria

- State Raoult's law.
  - Raoult's law states that system pressure multiplied by vapor composition is equal to vapor pressure of the species multiplied by liquid composition and is represented by the equation,

$$Py_i = P_i^0 \cdot x_i, \tag{17.64}$$

where P is the system pressure, P_i^0 is the vapor pressure of the species and x_i, y_i are the equilibrium compositions in mole fractions of the species in liquid and vapor phases, respectively. Py_i is the partial pressure of the species.

  - Assumptions involved are that the vapor phase behaves as an ideal gas (low pressures) and liquid phase is an ideal solution (components are chemically similar).
  - When vapor phase is considered ideal but liquid phase cannot be considered ideal, nonideality of the liquid phase is described by the activity coefficient for the species and Raoult's law is modified as below:

$$Py_i = P_i^0 \cdot \gamma_i x_i. \tag{17.65}$$

  - Activity coefficients are obtained from experimental data, available for a large number of binary systems, or by activity coefficient correlations like those of Wilson, which can be extended to multicomponent systems with limited number of components. NRTL and UNIQUAC equations overcome the limitation of Wilson equation with respect to extension to partially miscible systems.
  - The UNIFAC model predicts liquid-phase activity coefficients for nonideal mixtures when no VLE data are available. This model uses a group contribution method with about 50 identified functional groups.
  - The liquid-phase activity coefficients are calculated from an equation by the use of molecular configuration. The parameters calculated are independent of temperature. This method is restricted to systems in which all components are condensable.
- Classify molecular interactions causing deviations from Raoult's law.
  - Table 17.3 gives a classification of molecular interactions, resulting deviations from Raoult's law.
- Classify molecules based on potential for forming hydrogen bonds.
  - Table 17.4 gives a classification of molecules based on their hydrogen bonding nature.
- Write Antoine equation. What for it is used?

$$\ln P^0 = A - B/(T + C). \tag{17.66}$$

$$\log_{10} P^0 = A - B/(T + C). \tag{17.67}$$

where P^0 is the vapor pressure, bar; T is the absolute temperature, K; A, B, C are Antoine constants.

TABLE 17.3 Deviations from Raoult's Law as Related to Molecular Interactions

Type of Deviation	Classes	Effect on Hydrogen Bonding
Always positive	III + IV	H-bonds formed only
Quasi-ideal; always positive or ideal	III + III III + V IV + IV IV + V V + V	No H-bonds involved
Usually positive, but some negative	I + I I + II I + III II + II II + III	H-bonds broken and formed
Always positive	I + IV II + IV	H-bonds broken and formed
Always positive	I + V II + V	H-bonds broken only

- Antoine equation is an empirical equation used for representing vapor pressure data generated from experimentation.
- When using either common form of Antoine equation, that is, Equation 17.52 or 17.53, one should keep in mind that the equation is dimensional, not dimensionless. The constants A, B, C apply when a specific set of units is used for temperature and vapor pressure and when a specific logarithm function, common log or natural log, is used. Therefore, care should be exercised in using tabulated values from different sources. Temperature range for which the values of the constants are applicable. Most references include this information with the tabulated constants and warn against extrapolation beyond the specified range.
- The applicable temperature range is not large and in most cases corresponds to a pressure interval of about 0.01–2 bar. This typically represents an upper temperature limit slightly above the normal boiling point and a lower limit approximately 50–150°C below the upper limit.
- Antoine equation should never be used outside the stated limits.
- The other variations of Antoine equation are

$$\ln P^0 = A - B/(T + C) + k_4 T + k_5 \ln T + k_6 T^{k7}, \qquad (17.68)$$

which reduces to Equation 17.52 when k_4, k_5, and k_6 are taken to be zero.

$$\ln P^0 = A - B/T. \qquad (17.69)$$

$$\ln P^0 = A - B/T + k_5 \ln T. \qquad (17.70)$$

- These equations, being empirical, are limited for use for moderate to high vapor pressures and temperatures approaching the critical point.
- There are several methods for prediction of vapor pressures using critical properties and a review and

TABLE 17.4 Classification of Molecules Based on Their Hydrogen Bonding Nature

Class	Description	Examples
I	Molecules capable of forming three-dimensional networks of strong H-bonds	Water, glycols, glycerol, amino alcohols, hydroxyl-amines, hydroxy-acids, polyphenols and amides Compounds such as nitro-methane and acetonitrile also form three dimensional hydrogen bond networks, but the bonds are much weaker than those involving OH and NH_2 groups. Therefore, these types of compounds are placed in Class II
II	Other molecules containing both active H-atoms and donor atoms (O, N, and F)	Alcohols, acids, phenols, primary and secondary amines, oximes, nitro and nitrile compounds with α-H-atoms, NH_3, hydrazine, HF, and HCN
III	Molecules containing donor atoms but no active H-atoms	Ethers, ketones, aldehydes, esters, tertiary amines (including pyridine type) and nitro and nitrile compounds without α-H-atoms
IV	Molecules containing active H-atoms but no donor atoms that have two or three Cl atoms on the same carbon atom as a hydrogen or one chlorine on the carbon atom and one or more Cl atoms on adjacent carbon atoms	$CHCl_3$, CH_2Cl_2, CH_3CHCl_2, $CH_2ClCHClCH_2Cl$, and $CH_2ClCHCl_2$
V	All other molecules having neither active H-atoms nor donor atoms	Hydrocarbons, CS_2, sulfides, mercaptans, and halohydrocarbons not in Class IV

recommendations are given in the book, *Properties of Gases, and Liquids,* by Poling, Reid and Sherwood.

- Define K-value? How are K-values estimated?
  - K is defined as the ratio of vapor-phase composition to the liquid-phase composition. $K_i = y_i/x_i$ and shows the volatility of an individual component, that is, the composition in the vapor phase relative to the composition in the liquid phase. It indicates the ease with which a component in a mixture undergoes phase change from liquid to vapor phase.
  - For the more volatile components in a mixture the K-values are greater than 1.0, whereas for the less volatile components they are less than 1.0.
  - In general K-values for all components in a mixture are function of the pressure, temperature, and composition of the vapor and liquid phases present.
  - For light hydrocarbons, the equilibrium ratio for any one component increases at higher temperatures and decreases at higher pressures.
  - In addition, the value of K is altered by the other components present in the system, as well as the concentration of the specified component in the liquid phase.
  - The value of the equilibrium ratio usually decreases as the molecular weight increases for a homologous series of compounds. This is because the vapor pressure at any fixed temperature decreases as molecular weight increases.
  - However, the variation of α with temperature is less than the variation of K with temperature, which adds to the convenience of calculative methods using the relative volatility.
  - The components making up the system plus temperature, pressure, composition, and degree of polarity affect the accuracy and applicability, and hence the selection of method for estimating K-values.
  - For many systems, for example, light hydrocarbons, K is a function of temperature and pressure and independent of composition.
  - K-values for such systems, K-values can be determined from De-Priester charts. In these charts, K-values for individual components are plotted on the ordinate as a function of temperature on the abscissa with pressure as a parameter. In each chart the pressure range is from 70 to 7000 kPa and the temperature range is from 4 to 250°C.
  - It has been observed that, if a hydrocarbon system of fixed overall composition was held at constant temperature and the pressure is increased, the K-values of all components converged toward a common value of unity (1.0) at some high pressure. This pressure is termed *convergence pressure* of the system and is used to correlate the effect of composition on K-values. Plotting this way permits generalized K-values to be presented in a moderate number of charts.
  - In some publications, K-values are plotted as a function of pressure on the abscissa with *convergence pressure* and temperature as parameters. In order to use these charts, convergence pressure should be determined first that involves a trial and error procedure. These plots are converted into equations for use with computers.
  - Using fugacity coefficients, which are obtainable from equations of state, K-values can be estimated by the equation,

$$y_i P \Phi_i^{\mathrm{V}} = x_i P \Phi_i^{\mathrm{L}} \quad \text{or} \quad K_i = \Phi_i^{\mathrm{L}} / \Phi_i^{\mathrm{V}}. \qquad (17.71)$$

  - Another equation is

$$y_i P \Phi_i^{\mathrm{V}} = x_i P_i^{\mathrm{sat}} \gamma_i \Phi_i^{\mathrm{sat}} \quad \text{or} \quad K_i = P_i^{\mathrm{sat}} \gamma_i \Phi_i^{\mathrm{sat}} / P \Phi_i^{\mathrm{V}}. \qquad (17.72)$$

  - Equations of state such as BWR, Soave–Redlich–Kwong and Peng–Robinson are used for estimating fugacity coefficients, Φ.
  - Several nomographs are available in literature for finding K-values. Notable examples are those developed by De-Priester, Hayden-Grayson, and Lenoir and their extensions.
  - These methods involve trial and error calculations.
  - If vapor phase is assumed to be ideal (as is the case for low pressures), the above equation reduces to

$$K_i = P_i^{\mathrm{sat}} \gamma_i / P. \qquad (17.73)$$

  - γ_i is liquid-phase activity coefficient that takes into consideration liquid phase nonideality.
  - If both phases are ideal, this equation reduces to Raoult's law.
- What is relative volatility?
  - For a two-component system, relative volatility compares the volatility of one component to the other.
  - It is defined as

$$\alpha = K_1/K_2 = y/x/(1-y)/(1-x)$$
$$= y(1-x)/x(1-y). \qquad (17.74)$$

  - A relative volatility of 1 indicates that both components are equally volatile and no separation takes place.
- "Relative volatility, α, is defined as $\alpha = K_i/K_j$." Under what circumstances, α becomes equal to P_i^0/P_j^0?
  - For ideal systems, that is, systems obeying Raoult's law,

$$\alpha_{ij} = P_i^0/P_i^0. \quad (17.75)$$

- For systems with liquid phase being nonideal,

$$K_i = P_i^0\gamma_i/P. \quad (17.76)$$

- For near-ideal systems, α can be estimated from a knowledge of boiling points and enthalpies of vaporization using the following empirical equation:

$$\alpha = \exp[0.25164(1/T_{b1} - 1/T_{b2})(\lambda_1 - \lambda_2) \quad \text{(Wagle)}, \quad (17.77)$$

where T_{b1} and T_{b2} and λ_1 and λ_2 are boiling points and enthalpies of vaporization of the components 1 and 2, respectively.

- How would you characterize negative and positive deviations from Raoult's law?
  - If the volatilities of all components are increased, the deviation from idealism is considered positive. If the volatilities of all components are reduced, the deviation from idealism is considered negative.

$$\text{Positive deviations: } \log\gamma > 0\,(\gamma > 1). \quad (17.78)$$

$$\text{Negative deviations: } \log\gamma < 0\,(\gamma < 1). \quad (17.79)$$

- State Henry's law and its applicability.
  - For species present as a very dilute solute in the liquid phase, Henry's law states that the partial pressure of the species in the vapor phase is directly proportional its liquid-phase mole fraction and is given by

$$y_i \cdot P = x_i \cdot H. \quad (17.80)$$

- Give commonly used equations of state for use in estimation of K-values.
  - Redlich–Kwong equation:

$$P = [RT/(v - b)] - a/(v^2 + bv), \quad (17.81)$$

where

$$b = 0.08664RT_c/P_c \quad (17.82)$$

and

$$a = 0.42748R^2T_c^{2.5}/P_cT^{0.5}. \quad (17.83)$$

  - Soave–Redlich–Kwong equation:

$$P = [RT/(v - b)] - a/(v^2 + bv), \quad (17.84)$$

where

$$b = 0.08664RT_c/P_c, \quad (17.85)$$

$$a = 0.42748R^2T_c^2[1 + f_\omega(1 - T_r^{0.5})]^2/P_c, \quad (17.86)$$

and

$$f_\omega = 0.48 + 1.574\omega - 0.176\omega^2 \quad (17.87)$$

where ω = acentric factor.

  - Peng–Robinson equation:

$$P = [RT/(v - b)] - a/(v^2 + 2bv - b^2), \quad (17.88)$$

where

$$b = 0.07780RT_c/P_c, \quad (17.89)$$

$$a = 0.45724R^2T_c^2[1 + f_\omega(1 - T_r^{0.5})]^2, \quad (17.90)$$

and

$$f_\omega = 0.37464 + 1.54226\omega - 0.26992\omega^2. \quad (17.91)$$

- Define bubble point and dew point for mixtures.
  - Temperature at which

$$\Sigma y_i = \Sigma K_i \cdot x_i = 1.0 \quad (17.92)$$

is called bubble point.

  - Temperature at which

$$\Sigma x_i = \Sigma y_i/K_i = 1.0 \quad (17.93)$$

is called dew point.

- How are bubble and dew points estimated?
  - Bubble point (liquid composition and pressure are known):
    - Find K-values from a suitable method at an assumed temperature.
    - Estimate $\Sigma y_i = \Sigma K_i \cdot x_i$. If estimated $\Sigma y_i > 1$, a *lower* value of temperature is assumed and the calculation repeated. If estimated $\Sigma y_i < 1$, a *higher* value of temperature is assumed and the calculation is repeated. This iterative process is repeated until $\Sigma y_i = 1$.
  - Dew point (vapor composition and pressure are known):
    - Find K-values from a suitable method at an assumed temperature.
    - Estimate $\Sigma x_i = \Sigma y_i/K_i$. If estimated $\Sigma x_i > 1$, a *higher* value of temperature is assumed and the calculation repeated. If estimated $\Sigma x_i < 1$, a *lower* value of temperature is assumed and the calculation is repeated. This iterative process is repeated until $\Sigma x_i = 1$.

- Newton–Raphson iterative procedure can be used to reduce number of iterations in the above calculations.

• What are *T–x–y* and *x–y* diagrams? Illustrate by means of examples.

- The *T–x–y* diagram represents the relationships between temperature and the corresponding vapor–liquid equilibrium relationships for a binary two-phase vapor–liquid system.
- When two phases are in equilibrium, VLE data relates the composition of a liquid phase to the composition of the vapor phase in equilibrium with the liquid phase. This type of diagram is called equilibrium diagram.
- Figures 17.8 and 17.9 illustrate such relationships.

• What is the effect of pressure on relative volatility? Illustrate.

- As system pressure increases, relative volatility decreases making separation by distillation difficult.
- Figure 17.10 illustrates the effect. As α increases, that is, pressure decreases, *x–y* diagram *bulges*, which makes separation easier.

17.2.2 Distillation Processes

• What are the different practices in distillation processes?

- Equilibrium flash vaporization.
- Carrier distillation:
  - Steam distillation for water–immiscible liquids, such as high boiling petroleum fractions, purification of high boiling chemical products that form immiscible phases with water, oils of vegetable origin, essential oils, and the like.

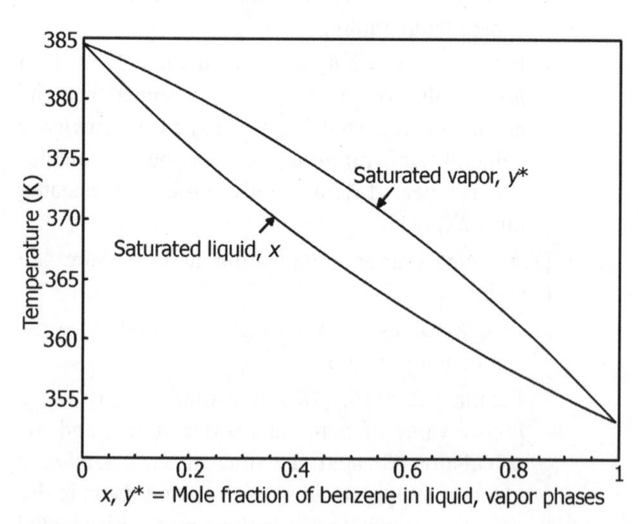

FIGURE 17.8 *T–x–y* diagram for benzene–toluene system at 1 atm.

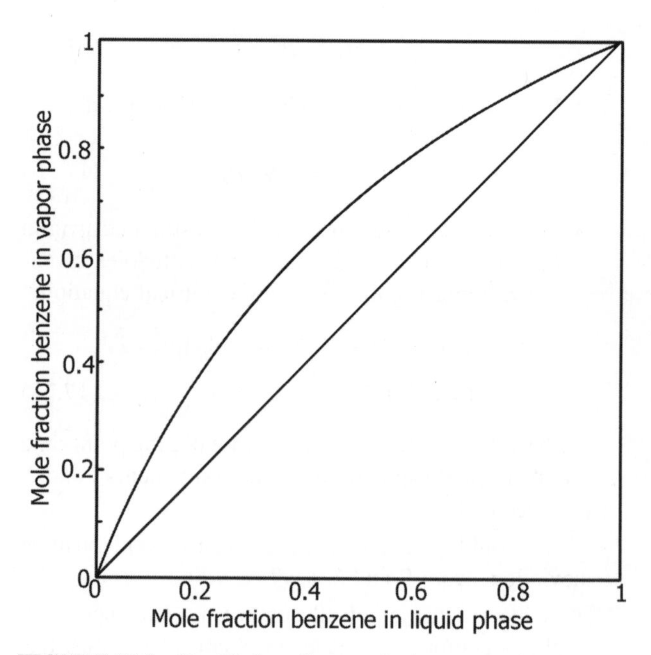

FIGURE 17.9 Equilibrium diagram for benzene–toluene system at 1 atm.

  - Use of carriers, other than steam, such as low boiling petroleum fractions such as propane for distillation of high boiling fractions.
- Differential distillation.
- Batch and continuous fractionation of binary or multicomponent mixtures.
- Steam stripping.
- Fractionations using open steam.
- Vacuum distillation.
- Molecular distillation.

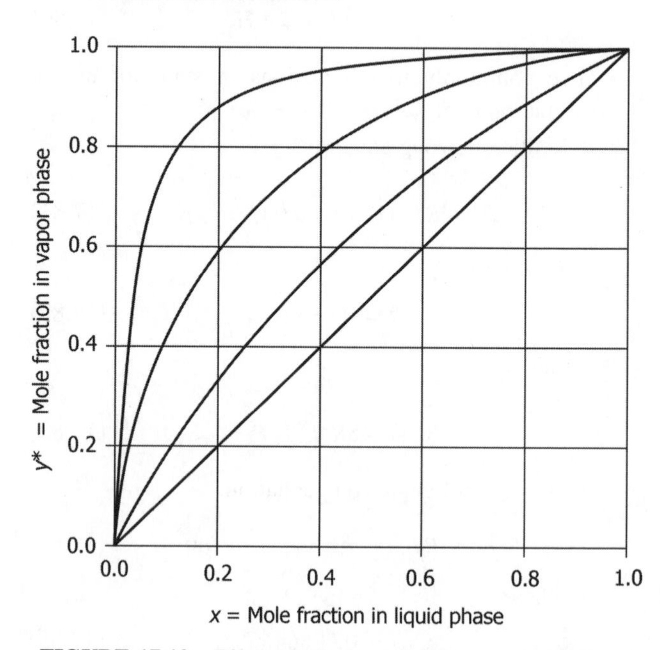

FIGURE 17.10 Effect of relative volatility on *x–y* diagrams.

- What are the advantages/applications of equilibrium flash vaporization?
  - In equilibrium flash vaporization (EFV), initial vapor generated is kept in contact with the liquid, which by virtue of its partial pressure effect, assists vaporization of the liquid.
  - Thus, at a given temperature, quantity of vapor produced will be more than that produced by fractionating type distillation.
  - In steam distillation processes, steam is used for providing partial pressure effect for enhanced vaporization.
  - The principle involved in EFV is used to advantage in distillation column feed heaters to provide vapors needed in the rectifying section of the column, thereby reducing heat load in reboilers that involve higher operating temperatures, as light components are not available for partial pressure effects.
- What is the difference between equilibrium flash vaporization and differential distillation?
  - In equilibrium flash vaporization, vapors generated continue to remain with the liquid providing their partial pressure effect for further vaporization of the remaining liquid.
  - In differential distillation, vapors generated are removed as they are formed and will not be available to assist vaporization of the higher boiling components in the remaining liquid.
  - Differential distillation can be considered as single-stage fractionation.
  - Equilibrium flash vaporization gives lower composition with respect to more volatile component, thus producing *impure* vapor. This is a nonfractionating type distillation.
- "ASTM distillation is a fractionating type distillation." *True/False*?
  - *False*. ASTM distillation is close to differential distillation, with very little fraction in the neck of the flask in which it is carried out providing some contact between vapor and liquid in the neck due to condensation as a result of heat losses.
  - ASTM distillations are rapid methods of laboratory distillations used to characterize petroleum fractions and involve no refluxing other than condensation on the neck of the flask due to heat losses.
- What is TBP distillation?
  - TBP stands for *true boiling point*. As component by component analysis of crude oils is not practically realizable, the composition of a crude oil sample is approximated by a *true boiling point curve* obtained on the sample in a high-efficiency batch distillation unit involving an equivalent of large number of trays, using high reflux ratios.
  - Normally TBP distillations are carried out on crude oils (not on petroleum fractions) as part of their evaluation. For petroleum products ASTM distillations are customarily carried out.
- What is an Older Shaw column?
  - A laboratory distillation column equivalent to a large number of trays. It is generally made of sieve trays with downcomers. By the use of ground glass joints, it is possible to alter its height, that is, number of trays.
  - These columns use more than 60 tray equivalent internals and high reflux ratios of over 5.
  - Column diameters range from 25 to 100 mm.
  - Adiabatic conditions are ensured with a silvered vacuum jacket.
  - These columns are used for bench scale simulation of distillations in order to evaluate the level of efficiencies expected upon scale-up. They provide a close approximation of point efficiencies (well mixed vapor and liquid) obtainable of larger trays.
  - TBP (true boiling point) distillations are carried out in this column to evaluate crude oils for yields of different petroleum products.
- Illustrate the difference between equilibrium flash vaporization and TBP distillation.
  - As stated earlier, equilibrium flash vaporization is essentially a nonfractionating type of distillation, practiced mainly to produce more vapors at a given temperature, thus supplementing energy requirements for distillation.
  - EFV can be run at higher pressures and vacuum, where as ASTM and TBP distillations are carried out at atmospheric pressure or under vacuum. EFV distillations are rarely carried out in laboratory distillations because of time and expense involved compared to the other types.
  - EFV initial boiling point is the bubble point of the liquid and final boiling point is its dew point.
  - Figure 17.11 illustrates the relative positions of EFV, ASTM, and TBP curves for a petroleum fraction.
- What is a spinning band column?
  - Spinning band distillation is a technique used to separate liquid mixtures that are similar in boiling points. When liquids with similar boiling points are distilled, the vapors are mixtures and not pure compounds. Fractionating columns help separate the mixture by allowing the mixed vapors to cool, condense, and vaporize again in accordance with Raoult's law. With each condensation–vaporization

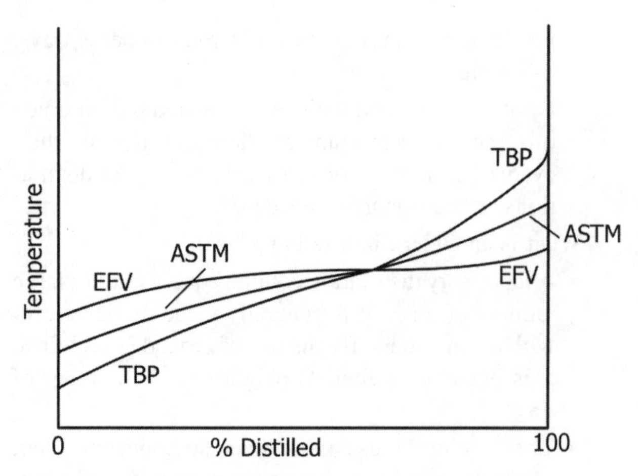

FIGURE 17.11 Relative positions of EFV, ASTM, and TBP curves on a plot of percent distilled versus temperature.

cycle, the vapors are enriched in a certain component. A larger surface area allows more cycles, improving separation.

- Spinning band distillation takes this concept one step further by using a spinning helical band made of an inert material such as metal or Teflon to push the rising vapors and descending condensate to the sides of the column, coming into close contact with each other. This speeds up equilibration and provides for a greater number of condensation–vaporization cycles.
- Spinning band distillation may sometimes be used to recycle waste solvents that contain different solvents and other chemical compounds.
- Spinning band columns can have up to 200 theoretical trays at atmospheric pressure. They can be stacked for even higher efficiencies. They are well suited for producing high-purity materials and making difficult separations.
  - *Low Liquid Holdup*: Because of their design spinning band distillation columns have very little liquid left behind in them after the distillation. The helical Teflon spinning band has a steep pitch, so liquids can easily drain into the boiling flask once the distillation is finished. Low holdup ensures that valuable sample is not lost in the distillation column. It also makes cleaning the equipment between distillations easy.
  - *Low Pressure Drop*: Spinning band distillation has very high free space. As a result, there is a very low pressure drop between the boiling flask and the top of the spinning band distillation column. Low pressure drop minimizes boiling temperatures and flood points.
- Applications include distillations for general purpose, solvent recycling, flavors and fragrances, essential oils, crude oil and products, and vacuum distillation.

- Write Rayleigh's equation. For what type of distillation it is applicable?

$$\ln(F/B) = \int dx/(y-x). \quad (17.94)$$

  - The integral in the above equation is evaluated by plotting x versus $1/(y-x)$ and finding the area under the curve between x_2 and x_1.
  - For constant α, the integral is found from the equation,

$$\ln(F/B) = [1/(\alpha-1)]\ln[x_f(1-x_b)/x_b(1-x_f)] + \ln[(1\text{-}x_b)/(1-x_f)]. \quad (17.95)$$

  - Applicable for differential/batch distillation.

- What are the advantages of batch distillation processes over continuous processes?
  - The main advantages of batch distillation over a continuous distillation lie in the use of a single column as opposed to multiple columns and its flexible operation.
  - For a multicomponent liquid mixture with n number of components, usually $n-1$ number of continuous columns will be necessary to separate all the components from the mixture.
    - For example, for a mixture with only four components and three distillation columns, there can be *five* alternative sequences of operations to separate all the components.
    - The number of alternative operations grows exponentially with the number of components in the mixture.
    - These alternative operations do not take into account the production of off-specification materials or provision for side streams.
  - On the other hand with conventional batch distillation, only one column is necessary and there is only one sequence of operation (with or without the production of off-specification materials) to separate all the components in a mixture. The only requirements are to divert the distillate products to different product tanks at specified times.
  - The continuous distillation columns are designed to operate for longer periods (typically 8000 h a year) and therefore each column (or a series of columns in case of a multicomponent mixture) is dedicated to the separation of a specific mixture.

- However, a single mixture (binary or multicomponent) can be separated into several products *(single separation duty)* and multiple mixtures (binary or multicomponent) can be processed, each producing a number of products *(multiple separation duties)* using only one conventional batch column.
- In pharmaceutical and food industries product tracking is very important in the face of strict quality control and batch wise production provides the *batch identity*.

- What is inverted batch distillation? What is its advantage?
  - Inverted batch distillation process combines the feed charge and the condenser reflux drum and operates in stripping mode with a small holdup reboiler.
  - This type of column operates exactly as the conventional batch column except that products are withdrawn from the bottom. Higher boiling products are withdrawn first, followed by the more volatile products.
  - This type of operation is supposed to eliminate the thermal decomposition problems of the high boiling products.
  - Figure 17.12 illustrates the process.
- What is constant level batch distillation? Give the relevant equation and state its advantage.
  - The batch is distilled to remove much of the original solvent (in solvent recovery operations).
  - In such operations involving solvents, switching of solvents is commonly used.
  - After removal of much of the original solvent, the replacement solvent (new solvent) is added and the batch is boiled down to remove the reminder of the original solvent.

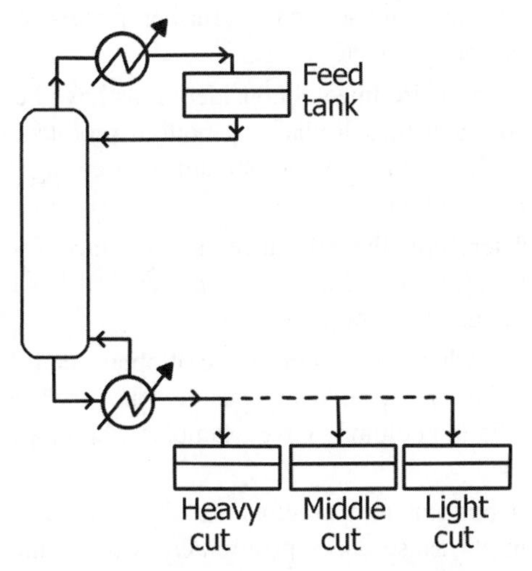

FIGURE 17.12 Inverted batch distillation.

  - In constant level operation, instead of adding replacement solvent prior to the start of the final distillation, the replacement solvent is continuously added to maintain a constant level during the batch distillation.
  - Equation for constant level batch distillation is

$$F/L = \mathrm{d}x/y \text{ (over the composition range } x_i \text{ to } x_F)$$

$$= 1/\alpha \ln[x_F/x_i] + [(\alpha - 1)/\alpha](x_F - x_i). \quad (17.96)$$

F is the liquid feed, moles; L is the liquid content of the still, moles.

  - Constant level process reduces loss of replacement solvent to the distillate (original solvent) by 50–70%.
  - Also constant level process will give a lower temperature profile than standard batch distillation involving exchange of solvents.
  - Constant level process results in less time–temperature exposure for the product.
  - It is easier to retrofit the constant level system (controls, etc.) to an existing facility.
- What are the situations in which steam distillation is commonly employed?
  - To separate relatively small amounts of a volatile impurity from large amounts of a liquid.
  - To separate appreciable quantities of higher boiling materials.
  - To recover high boiling materials from small amounts of relatively nonvolatile impurity with high boiling point. For example, purification of aniline.
  - To distill thermally unstable material or material that reacts with other compounds associated with it at high boiling point. For example, glycerin.
  - When material cannot be distilled by indirect heating even under low pressure because of high boiling point. For example, fatty acids.
  - Use of open steam in distillation columns to reduce bottoms temperatures for avoiding cracking in petroleum distillations. Steam when condensed along with top product: Two liquid layers that are separated by gravity.
- How is equilibrium flash vaporization carried out? Illustrate.
  - The liquid mixture is partially vaporized in a heating coil, while keeping the vapor formed in contact with the liquid and at the exit of the coil, the mixture is separated in a separating vessel.
  - Figure 17.13 illustrates the EFV process.
- Define an equilibrium tray?
  - If vapor and liquid streams *leaving* the tray are in equilibrium, then the tray is called equilibrium tray.

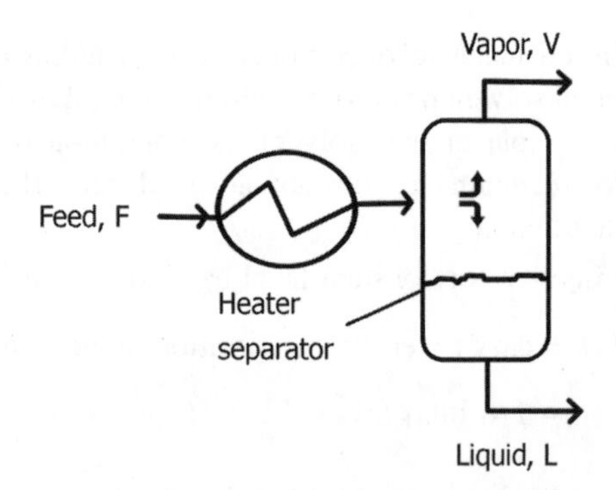

FIGURE 17.13 Equilibrium flash vaporization.

- "*Operating line* represents the relationship between vapor *entering* the tray to liquid *leaving* the tray." *True/False*?
  - *True.*
- What is reflux ratio?
  - Reflux ratio,

$$R = \text{flow returned to the column as reflux}/\text{flow of top product taken off} = L_R/D. \quad (17.97)$$

- Explain what is meant by minimum reflux.
  - As the reflux ratio is reduced, a *pinch point* will occur at which the mass transfer driving force becomes zero and infinite number of trays will be required for separating the mixture.
  - If reflux ratio is increased above minimum reflux ratio, the column will become operable, that is, separation becomes possible.
  - At or below R_m column becomes inoperable. Thus, R_m is the highest reflux ratio for inoperability of the column.
- Illustrate different parts of a simple distillation unit by means of a flow diagram.
  - Figure 17.14 illustrates the essential parts of a fractionating column and associated units for binary distillation, indicating *rectifying* and *stripping* sections.
- What is the basic functional difference between enriching/rectifying and stripping sections of a distillation column?
  - Function of the enriching/rectifying section is to *purify* the top product by preventing the higher boiling components from it by "condensing" them from the rising mixture and thereby moving them down the column.
  - In other words, rectifying section is to ensure the top product to meet its desired specifications by preventing higher boiling components reaching the top of the column.

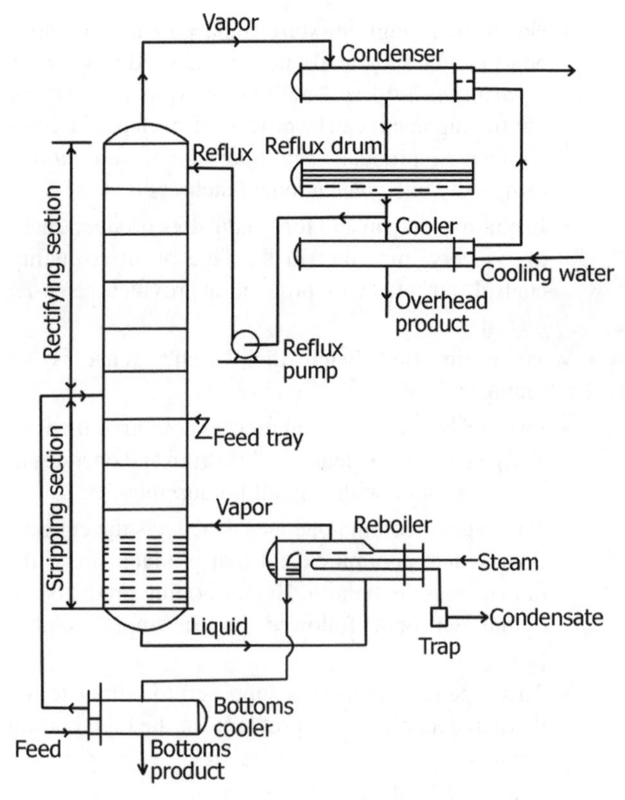

FIGURE 17.14 Operation of a simple distillation unit showing different parts.

  - The purpose of the stripping section is to *recover* light components by revaporizing and moving them up the column.
  - This nomenclature of rectifying and stripping sections was evolved in the past when the top product was considered to be the *valuable/desirable* product and the bottom product was considered to be a *waste* product. The present day concept is that there is nothing like a waste product and often valuable products are derived by further processing of the bottom product.
  - Also environmental considerations have become so stringent that discharging bottom products into the environment as wastes streams by treating them and reusing them.
  - Therefore, the relevance of continued use of the terms *rectifying* and *stripping* sections has lost their intended meanings.
- Under what circumstances use of open steam (in place of reboiler) is advantageous for the operation of a distillation column? Give examples where open steam is used.
  - If bottoms temperatures are high, use of a reboiler might cause decomposition/cracking of the bottom product. In such cases it is advisable to use open

steam that lowers boiling temperatures by its partial pressure effect.

- Open steam is used in petroleum atmospheric crude distillations or vacuum distillation of reduced crude so that bottoms temperatures are kept below about 150–200°C to avoid decomposition.
- Condensation of the top product results in two layer separation in the condensate drum.
- In alcohol distillations where bottom product is wastewater, use of open steam avoids use of a reboiler, reducing equipment costs.

• What is the effect of open steam in the operation of a distillation unit?

- To reduce column temperatures, through partial pressure effect of steam, avoiding decomposition of heat-sensitive materials being processed in the column.

• Explain why in a distillation column, temperatures are lowest at the top and highest at bottom.

- Lowest boiling components, because of ease of vaporization are collected at the top of the distillation column and highest boiling components, because of their high boiling nature, are collected at the bottom of the column. The top and bottom products leaving the column are at their saturation temperatures.
- In other words, due to the collection of low boiling components at the top of the column, bubble point temperature of the top product is the lowest and that of the bottom product is the highest.
- Due to the above reasons temperatures at the top of the column are the lowest and at the bottom are the highest.

• List out the factors required to be considered for the design of a distillation column for separating binary mixtures.

- Equilibrium relationship.
- Feed rate, composition, temperature, and pressure.
- Column pressure.
- Desired separation.
- R_m, R, N_m, and N.
- Tray or packed column/type of trays/type of packing.
- Column internals.
- Efficiency.
- Diameter of the column (capacity).
- Type of reboiler/condenser.

• What are the factors that affect operation of a distillation column with respect to changes in column operating pressure?

- As the operating pressure reduces
  - ➢ relative volatility increases;
  - ➢ tray leakage decreases;
  - ➢ entrainment increases.
- Increase in relative volatility and decrease in tray leakage help separation while increase in entrainment reduces separation.
- As column pressure is decreased, temperatures at top and bottom individually also decrease and this will not provide any information on changes in separation.
- But *temperature difference* between top and bottom will provide an indication on the separation. The larger the temperature difference, the better will be the separation.
- Figure 17.15 illustrates this relationship.

• What is optimum reflux ratio?

- As R increases, number of trays decreases.
- As R increases, V and L increase for a given capacity with the consequence that the column diameter increases.
- Also as R increases, larger condenser and reboiler are required.
- In addition operating cost involving steam, water, pumping, handling cooling water, condensates and steam, will increase.
- A point is reached where increase in diameter of the column is more rapid than decrease in number of trays.

$$\text{Cost} \propto (\text{number of trays}) \times (\text{cross-sectional area}). \tag{17.98}$$

- Optimum reflux ratio is the one that gives the lowest annual costs.
- As a thumb rule, optimum reflux ratio for many systems lies in the range of 1.2–1.5 times R_m.
- Figure 17.16 illustrates the various costs involved in finding optimum reflux ratio.

• What are the assumptions involved in McCabe–Thiele method? Under what circumstances these assumptions are justified?

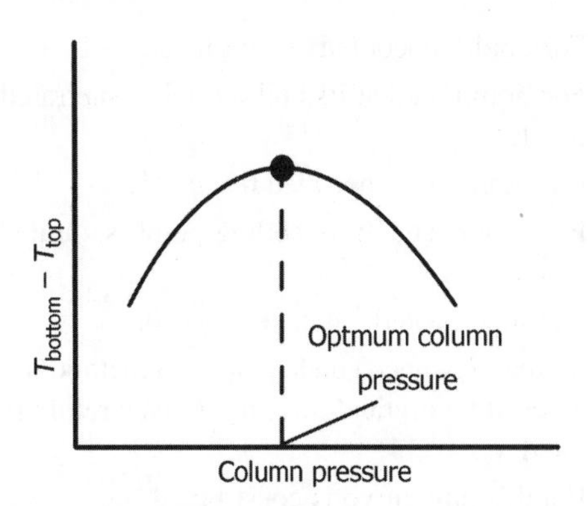

FIGURE 17.15 Optimum column pressure.

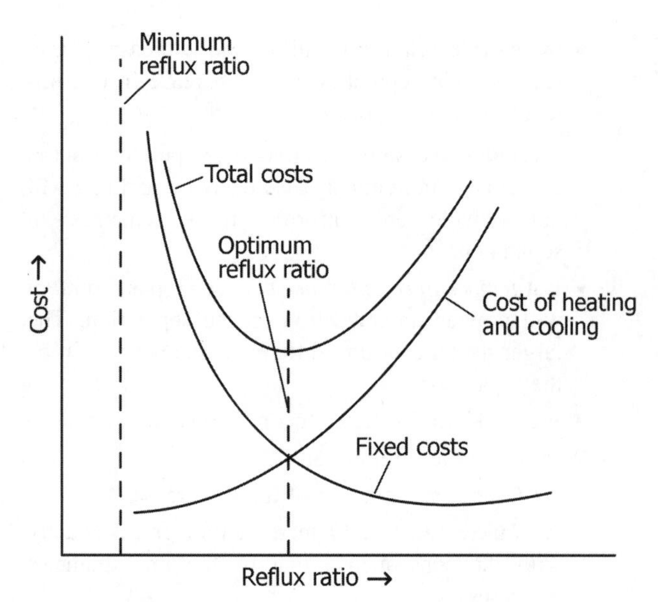

FIGURE 17.16 Optimum reflux ratio.

$$\lambda_1 = \lambda_2 = \text{constant throughout the column.} \quad (17.99)$$

- Sensible enthalpy changes, $(C_P \Delta T)$ and heats of mixing are negligible.
- Column is well insulated, that is, heat losses are negligible.
- Pressure in the column (top to bottom) is constant, that is, no ΔP.
- Above assumptions lead to the condition of constant molar flow rates in the rectifying section and stripping sections.

• Define q, which describes thermal condition of feed in a distillation process.

q = heat required to vaporize one mole of feed at entering conditions divided by molar latent heat of vaporization of feed
= moles liquid flow in the stripping section divided by moles feed. (17.100)

- For cold (subcooled) liquid feed $q > 1$.
- For liquid feed at its boiling point (saturated liquid) $q = 1$.
- For liquid + vapor feed $0 < q < 1$.
- For vapor feed at its boiling point (saturated vapor) $q = 0$.
- For superheated vapor feed $q < 0$.

• Describe McCabe–Thiele graphical method for finding number of theoretical trays for a given reflux ratio, x_D, x_B, and x_F.

- Equilibrium curve is constructed.
- Rectifying section operating line is drawn through the point x_D on the 45° line to intersect with the y-axis at $x_D/(R + 1)$.
  - ➢ For a column with a total condenser, the overhead vapor and distillate are of the same composition. The operating line for the rectifying section must pass through the distillate composition.
  - ➢ If a partial condenser is used, the reflux composition will be a liquid in equilibrium with the condenser outlet vapor composition. A partial condenser usually provides one additional theoretical tray.
- The q-line is then drawn that must pass through the point x_F on the 45° line with a slope $q/(q - 1)$.
- The operating line for the stripping section is drawn using the point of intersection of the enriching and q lines and the point x_B on the 45° line.
- Starting at the top on the 45° line and ending at the bottom on the 45°line, the equilibrium trays are stepped off between the operating lines and the equilibrium curve, switching from rectifying to stripping line as the feed composition is passed.
  - ➢ In a binary distillation, the feed should be introduced into the column at the equilibrium tray represented by the intersection of the two operating lines.
- Usually the reboiler is considered as one theoretical tray, so that the vapor to the bottom of the column is in equilibrium with the liquid bottoms product.
- The operating line for the stripping section must pass through the bottoms composition.
- Surprisingly, McCabe–Thiele method gives results comparable to those obtained from tray to tray calculations.
- Details of theoretical basis for McCabe–Thiele method and its extension to cases of different feed conditions, minimum reflux, total reflux, multiple feeds, side streams, and high-purity products are not considered here and reference may be made to text books.
- Figure 17.17 demonstrates McCabe–Thiele graphical procedure showing three theoretical trays in rectifying section and four theoretical trays in stripping section.

• What are the shortcomings of McCabe–Thiele method?

- When the feed is a subcooled liquid (or a superheated vapor), on the assumption that x_F, x_B, x_D, L, and D are all constants, the number of theoretical trays tends to a minimum according to the McCabe–Thiele method, which is not correct.
- When the feed is a subcooled liquid (when the temperature of the feed drops, the slope of the

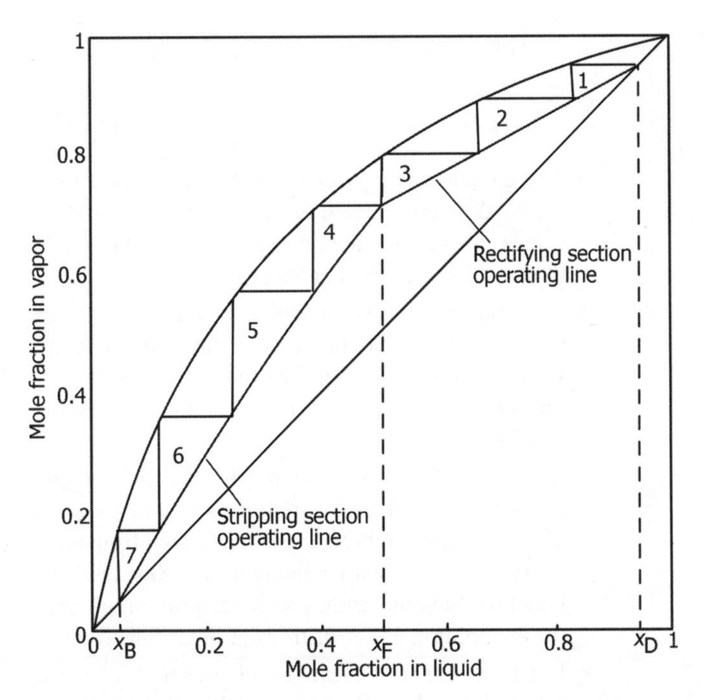

FIGURE 17.17 McCabe–Thiele construction for number of theoretical trays.

operating line will be close to the diagonal line of the x–y diagram). The number of theoretical trays tends to a minimum.

- Meanwhile, the number of theoretical trays for the enriching section tends to zero.
- Similarly, when the feed is superheated, the slope of the operating line will be very close to the diagonal line, the number of theoretical trays again tends to a minimum, and the number of theoretical trays for stripping section tends to zero.
- Since the separation process starts only when the feed is saturated, the more the thermal state of feed deviates from the saturated state, a larger number of theoretical trays will be needed. McCabe–Thiele graphical method does not indicate this fact.
- McCabe–Thiele method sometimes predicts negative minimum reflux ratio that is physically meaningless.
- According to McCabe–Thiele method, the limits of the minimum reflux ratio are $-\infty \le R_{\min} \le \infty$, which is also physically meaningless. These limits are derived when the feed is either a superheated vapor or a subcooled liquid (when the enriching section operating line coincides with the diagonal line on the x–y diagram).
- Some errors appear in the derivation of q-line, when the feed is a subcooled liquid (or a superheated vapor).
  - ➢ The equation of the feed line is derived by combining the equations of the two operating lines. McCabe–Thiele method uses the hypothesis that the intersection of the two operating lines always exists.
  - ➢ This is not the case for a subcooled liquid (or a superheated vapor) feed because the feed tray is not a theoretical tray in these cases.
  - ➢ There must be a heating (or cooling) section to bring the feed to its saturated state in order to start the separation process.
  - ➢ The trays of heating (or cooling) section are only heat exchange trays, so neither of these two operating lines can control this section.
  - ➢ Apparently the operating lines of the enriching and stripping sections will not meet together when the feed is superheated (or subcooled).
- Analytical methods are amenable for computer calculations than the classical graphical methods, one such method being that of smoker.
- R_m for a saturated liquid feed by McCabe–Thiele method can be obtained from the equation,

$$R_m = (y_D - y_f)/[(x_D - x_f) + (y_f - y_D)]. \quad (17.101)$$

- This equation is applicable when *pinch point* occurs at the feed point.
- Give Smoker equation for the estimation of number of theoretical trays for binary distillation, involving a composition change from x_n^* to x_0^*.

$$N = \ln[x_0^*(1-\beta x_n^*)/x_n^*(1-\beta x_0^*)/\ln(\alpha/mc^2), \quad (17.102)$$

where

$$\beta = mc(\alpha-1)/(\alpha-mc^2). \quad (17.103)$$

$$x^* = x-k \quad \text{for } x_0^* > k. \quad (17.104)$$

$$c = 1+(\alpha-1)k. \quad (17.105)$$

m is the slope of operating line.
For rectifying section,

$$x_0^* = x_D - k. \quad (17.106)$$

$$x_0^* = x_F - k \text{ and } m = R/(R+1). \quad (17.107)$$

$$\text{Intercept b of operating line} = R/(R+1). \quad (17.108)$$

For stripping section,

$$x_0^* = x_F - k \quad (17.109)$$

and

$$x_n^* = x_B - k. \quad (17.110)$$

$$m = [Rx_F + x_D - (R+1)x_B]/(R+1)(x_F - x_B). \quad (17.111)$$

$$b = (x_F - x_D)x_B/(R+1)(x_F - x_B). \quad (17.112)$$

If feed is not at bubble point,

$$x_F^* = [b + (x_F/(q-1))]/q/[(q-1)-m]. \quad (17.113)$$

- Explain the concepts of net flow and delta point in multistage separation processes.
  - Figure 17.18 illustrates the flow chart for multistage separations.
  - Consider material balances for a system including stages 1, 2, ..., j in the flow chart shown above, where stage j is some arbitrary stage number. One can write a total material balance, $n-1$ component material balances and an energy balance around this system:

$$L_0 + V_{j+1} = L_j + V_1. \quad (17.114)$$

$$L_0x_0 + V_{j+1}y_{ij+1} = L_jx_{ij} + V_1y_{i1}. \quad (17.115)$$

$$L_0HL_0 + V_{j+1}H_{vj+1} = L_jH_{Lj} + V_1H_{v1}. \quad (17.116)$$

  - These equations can be rearranged to give the equations of net flow between any two stages:

$$= L_j - V_{j+1} = L_0 - V_1d. \quad (17.117)$$

$$x_i = L_jx_{ij} - V_{j+1}\,y_{ij+1} = L_0x_{i0} - V_1y_{i1}. \quad (17.118)$$

$$H = L_jH_{Lj} - V_{j+1}H_{vj+1} = L_0H_{L0} - V_1H_{v1}. \quad (17.119)$$

  - Under steady-state conditions since flow rates are assumed to be constant, the net flow between any two stages is constant, independent of the position in the series.
  - It should be noted that the net flows may be positive or negative, depending on whether the L or V streams carry more material (or energy).
  - By combining the above net flow equations, a fictitious concentration for component i and fictitious enthalpy can now be defined.
  - These will be the definitions of the Δ-point, which will be used in graphical solutions to multistage processes:

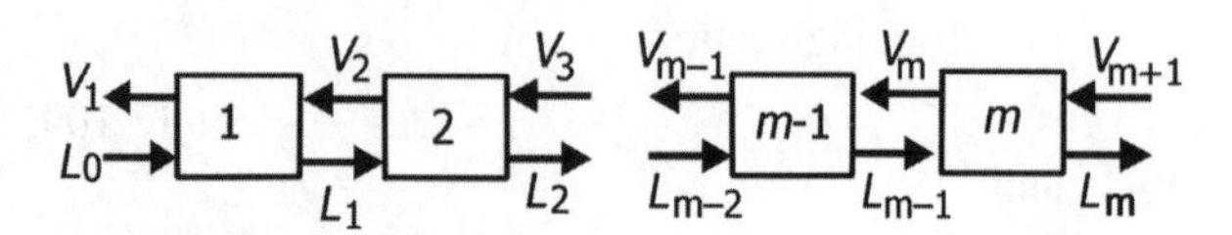

FIGURE 17.18 Flow chart for multistage separations.

$$x_i = (L_jx_{ij} - V_{j+1}y_{ij+1})/(L_j - V_{j+1}). \quad (17.120)$$

$$H = (L_jH_{Lj} - V_{j+1}H_{vj+1}). \quad (17.121)$$

  - These equations may be used on phase diagrams to locate Δ-points, which are constant for a series of stages, and provide the relationship between streams that are passing between stages.
  - For example, the composition of component i for the Δ-point ($x_{i\Delta}$) lies on a straight line through x_0 and y_1, and also on a straight line through x_m and y_{m+1}, the composition points at the other end of the series of stages.
  - Δ-points (relating streams passing between stages) and tie lines (relating equilibrium streams leaving any stage) are all that are needed to step from stage to stage on a phase diagram as illustrated for Ponchon–Savarit enthalpy–concentration method for obtaining number of trays in binary distillation.
  - If there are any significant heat losses for each stage, then the Δ-point will incrementally change position for each stage, making a graphical construction more difficult.
  - The same principles can be applied to other stage processes like multistage liquid–liquid extraction.
- Outline the steps involved in the determination of number of theoretical trays by Ponchon–Savarit enthalpy–concentration graphical method for a given reflux ratio.
  - Enthalpy–concentration diagram is constructed.
  - Compositions of feed, distillate, and bottoms and Δ-points, Δ_D (based on reflux ratio) and Δ_B, are located on the plot.
  - Tie lines are drawn from x, y data.
  - Series of operating lines are drawn from the top difference point Δ_D to connect the compositions of streams passing each other (e.g., x_n and y_{n-1},), the equilibrium values (e.g., x_n and y_n) being connected by tie lines. In this way, trays are stepped off from the top of the column down to the feed tray, at which point a shift is made to the bottom operating line. This latter line is based on a bottom difference point that is in line with the feed point, top difference point and with a composition equal to the bottoms composition.
  - The lower part of the column is covered by stepping off trays in a way similar to that in the upper part of the column and the final count of theoretical trays is then determined.
  - Figure 17.19 illustrates the construction for number of trays by Ponchon–Savarit method for a given reflux ratio.

- Ponchon–Savarit method requires enthalpy data at different compositions of vapor and liquid in addition to equilibrium data. It is useful when latent heats of vaporization of the components differ widely, invalidating McCabe–Thiele assumption of constant flow rates.
- The method can be used for the estimation of minimum number of trays, minimum reflux, side streams, and other similar cases as in the case of McCabe–Thiele method.
- Unlike McCabe–Thiele method, Ponchon–Savarit method cannot be extended for multicomponent systems.

• Write down expressions for R_m for binary mixtures.

$$(R_m x_F + q x_D)/(R_m + q - R_m x_F - q x_D) = \alpha[x_D(q-1) + x_F(R_m+1)]/[(R_m+1)(1-x_F) + (q-1)(1-x_D)]. \quad (17.122)$$

- The above equation is solved by trial and error for R_m.
- For case of $q = 1$ (feed all liquid at its bubble point), the equation reduces to

$$R_m = [1/(\alpha-1)]\{(\alpha x_D/x_F) - [\alpha(1-x_D)/(1-x_F)]\}. \quad (17.123)$$

- For the case of $q = 0$ (feed is all vapor at its dew point, $x_F = y_F$), the equation reduces to

$$R_m = [1/(\alpha-1)]\{(\alpha x_D/y_F) - [(1-x_D)/(1-y_F)]\}. \quad (17.124)$$

17.2.3 Batch Fractionation

• Illustrate how McCabe–Thiele method can be used for batch fractionation involving (a) constant R and (b) constant x_D.

Constant Reflux Ratio:

- At constant reflux, the distillate composition changes with time.
- In the example shown, a total of four equilibrium trays are provided.
- The slope of the operating line remains constant.
- As the distillate composition changes from x_{D1} to x_{D2}, the residue composition changes from x_{B1} to x_{B2}.
- As the purity of the accumulated product meets the product composition specification, distillation is stopped.
- Figure 17.20 illustrates McCabe–Thiele construction for constant reflux ratio.

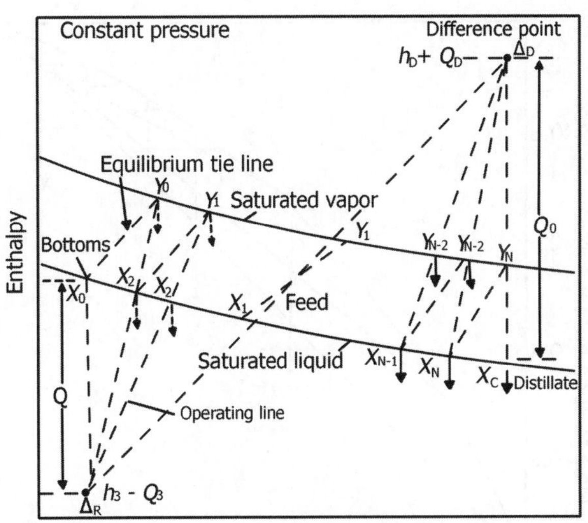

FIGURE 17.19 Enthalpy–concentration method for number of trays for binary distillation at a given reflux ratio

Constant Product Composition:

- The value of x_D is kept constant and the slope of the operating line is varied in accordance with the overall material balance.
- Figure 17.21 shows the conditions at the start of the distillation and time t.
- Four equilibrium trays are provided.
- Both modes, that is, with constant reflux ratio and constant product composition, usually are conducted with constant vaporization rate at an optimum value for the particular type of column construction.

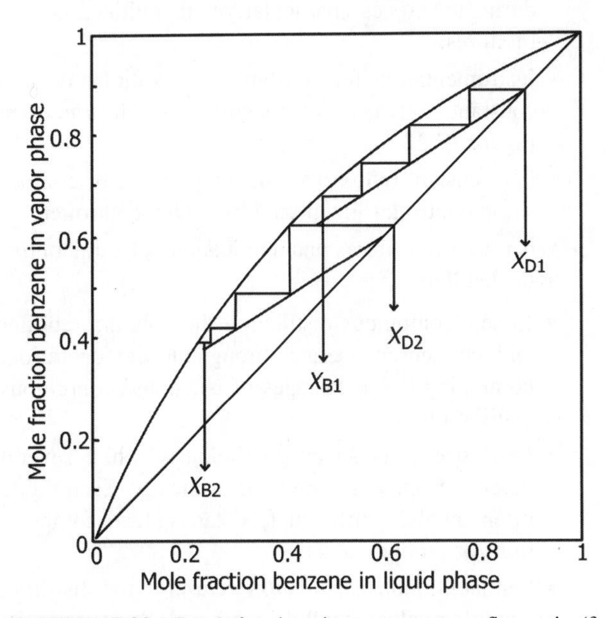

FIGURE 17.20 Batch fractionation at constant reflux ratio (for four theoretical trays).

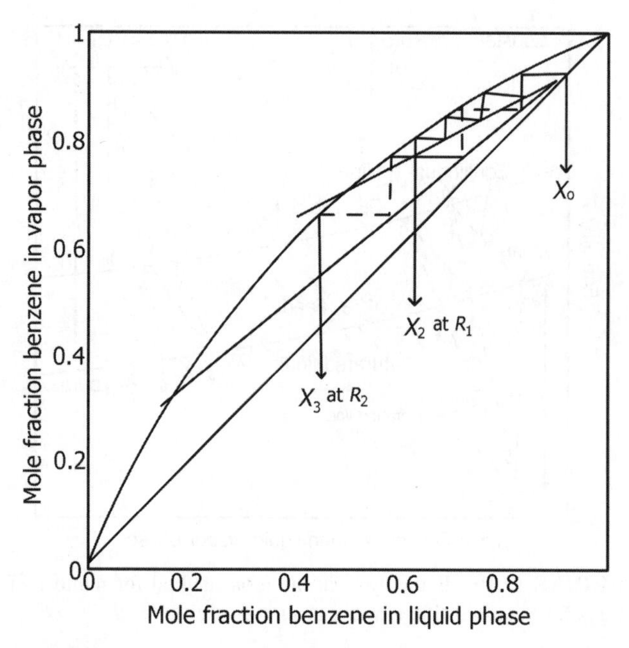

FIGURE 17.21 Batch fractionation at constant distillate composition (for four theoretical trays).

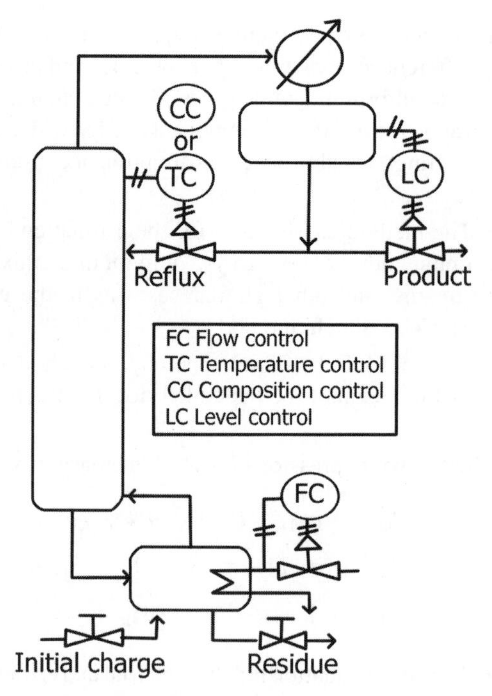

FIGURE 17.22 Instrumentation for constant vaporization rate and constant overhead composition in batch distillation.

- Small-scale distillations often are controlled manually, but can be controlled automatically also.
- Constant overhead composition can be assured by control of temperature or directly of composition at the top of the column.
- Constant reflux is assured by flow control on that stream.
- Sometimes there is an advantage in operating at several different reflux rates at different times during the process, particularly with multicomponent mixtures.
- Instrumentation for constant vaporization rate and constant overhead composition is illustrated in Figure 17.22.
- For constant reflux rate, the temperature or composition controller is replaced by a flow controller.

- What are the features and applications of semicontinuous distillation?
  - In semicontinuous distillation the column, reboiler and condenser operate throughout the operation, combining the advantages of batch and continuous distillation.
  - Feed stream is added continuously while several modes of operation are used cyclically. Each mode often involves different feed tanks, feed compositions or product tanks.
  - Semicontinuous distillation is useful for distilling new high value, small demand specialty chemicals, which are continuously coming into the markets.
  - A wide variety of mixtures, which include binaries, near-ideal ternaries, low boiling azeotropes and, with the use of an additional product tank, near-ideal quaternaries, can be separated.
- Illustrate batch, continuous and semicontinuous process schemes for the distillation of a near-ideal ternary mixture of equimolar *n*-hexane, *n*-heptane, and *n*-octane.
  - In batch process, initially the still is charged and *n*-hexane is collected as distillate.
  - Next, a mixture with high percentage of *n*-heptane is collected as distillate.
  - Finally, *n*-octane is collected as distillate.
  - Figure 17.23 illustrate batch, continuous, and semicontinuous process schemes.
  - Equimolar ternary mixture is charged into a near-empty feed tank.
  - Distillate, concentrated in *n*-hexane is removed continuously, although in decreasing quantities.
  - Bottoms, consisting of mostly *n*-octane, are also removed continuously in decreasing quantities.
  - After the tank, rich in *n*-heptane becomes sufficiently concentrated in *n*-heptane, it is nearly emptied and the cycle is repeated by charging equimolar mixture of the ternary to it.
- Compare batch, semicontinuous and continuous distillation processes.

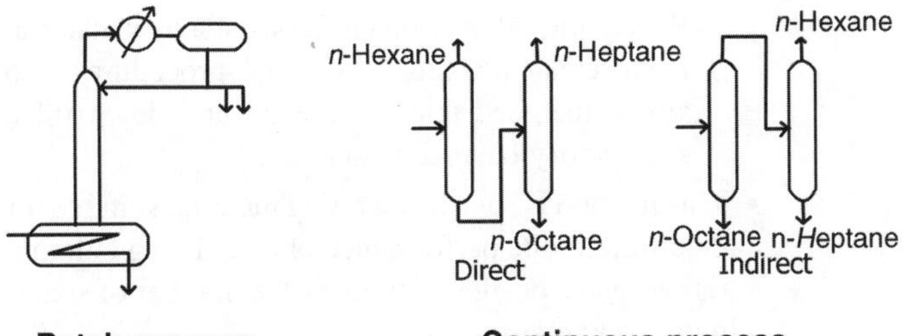

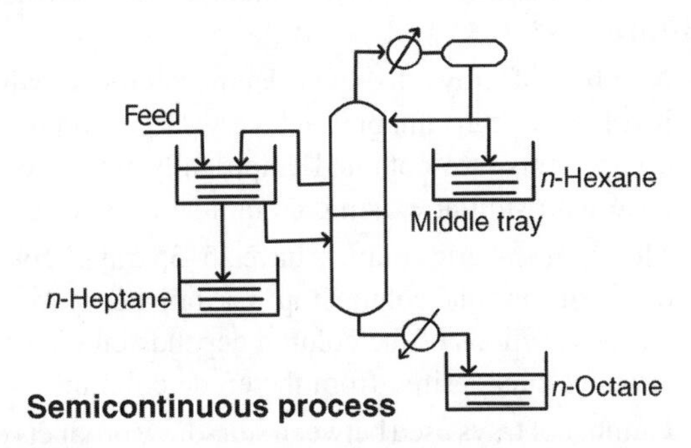

FIGURE 17.23 Batch, continuous, and semicontinuous processes for separation of *n*-hexane, *n*-heptane, and *n*-octane mixtures.

- Table 17.5 gives a comparison of batch, continuous, and semicontinuous processes.

17.2.4 Multicomponent Distillation

- What are meant by *key components* in distillation? What are the factors that are to be considered in the selection of key components?
  - For a multicomponent mixture to be split into two streams (distillate and bottoms), it is common to specify the separation in terms of two *key components* of the mixture. Separation between the key components serves as a good indication of the desired overall separation achieved.

TABLE 17.5 Comparison of Batch, Continuous, and Semicontinuous Distillation Processes

Variable	Batch	Semicontinuous	Continuous
Throughput	Low	Intermediate	High
Flexibility	Yes	Yes	No
Automatic control	Not common	Possible	Mostly
Heat integration	No	Possible	Yes
Single column for ternary separation	Yes	Yes	Not Often[a]
Investment	Lowest	Middle	Highest

[a] Ternary separation in a single column is possible with a side-draw stream over a narrow range of ternary compositions.

  - The *light key* will have a specified maximum limit in the bottoms product and the *heavy key* will have a specified maximum limit in the distillate product.
  - Normally, the keys are adjacent to each other in the ranking of the mixture components according to relative volatility but this is not always the case. All the components lighter then the light key go into the distillate and all the components heavier then the heavy key go into the bottoms product.
  - *Distributed components* may have volatilities intermediate to those of the keys and are present in both bottoms and top product. Components with negligible concentration ($<10^{-6}$) in one of the products are called *undistributed* components.
- Name the methods used to determine the number of trays required for multicomponent distillation.
  - One of the more widely used methods for estimating number of theoretical trays in multicomponent systems was developed by Lewis–Matheson. This method also involves the same McCabe–Thiele assumptions, but trial and error tray-to-tray calculations.
  - Method of Thiele–Geddes is a rating approach and can be used to find the distribution of components for a given number of theoretical trays and reflux ratio. The split of each component between overhead and bottoms can then be calculated. In other words, it *rates* a given column instead of designing it.
    - The method assumes a temperature profile throughout the column (number of trays being given) and, by starting at either end and working along the profile, arrives at a composition at the other end.
    - If the outcome of such an iteration is a split that does not check the material balance requirements, the profile is varied through successive iterations.
  - One advantage of McCabe–Thiele method is that it can be extended to multicomponent mixtures by reducing them into pseudobinaries. The method is described by Hengstebeck that involves the following steps:
    - Selection of key components and reducing the compositions of feed and products in terms of key components, as if the other components are not present.
    - Assumption of constant but different flows in rectifying and stripping sections as in the case of McCabe–Thiele method, which is arrived at based on the assumptions that the light key is absent in the stripping section and heavy key is absent in the rectifying section.

  - ➢ Construction of equilibrium diagram (using relative volatility between the key components) and operating lines and following McCabe–Thiele procedures for determining number of trays.
- ▪ Hengstebeck method is applied to light hydrocarbon mixture separations and claimed to be giving comparable results obtainable by more rigorous tray-to-tray calculation methods.

- Give Smith and Brinkley method for distribution of components in a multicomponent separation process.
  - ▪ For any component in the mixture, the fractional split of the component between feed and bottoms is given by the equation:

$$b/f = [(1-S_r^{N_r-N_s}) + R(1-S_r)]/[(1-S_r^{N_r-N_s}) + R(1-S_r) + G\cdot S_r^{N_r-N_s}(1-S_s^{N_s+1}), \quad (17.125)$$

where b/f is the fractional split of the component between the feed and the bottoms, N_r is the number of equilibrium stages above the feed, N_s is the number of equilibrium stages below the feed, S_r is the stripping factor, rectifying section $= K_iV/L$, S_s is the stripping factor, stripping section $= K'_iV'/L'$. V and L are the total molar vapor and liquid flow rates, and the superscript "′" denotes the stripping section.

  - ▪ G depends on the condition of the feed.
  - ▪ If the feed is mainly liquid,

$$G_i = (K'_iL/K_iL')[(1-S_r)/(1-S_s)]_I \quad (17.126)$$

and the feed stage is added to the stripping section.

  - ▪ If the feed is mainly vapor,

$$G_i = (L/L')[(1-S_r)/(1-S_s)]_I. \quad (17.127)$$

  - ▪ The procedure for using the Smith–Brinkley method is as follows:
    - ➢ Estimate the flow rates L, V, and L_0, V_0 from the specified component separations and reflux ratio.
    - ➢ Estimate the top and bottom temperatures by calculating the dew and bubble points for assumed top and bottom compositions.
    - ➢ Estimate the feed point temperature.
    - ➢ Estimate the average component K values in the stripping and rectifying sections.
    - ➢ Calculate the values of $S_{r,i}$ for the rectifying section and $S_{s,i}$ for the stripping section.
    - ➢ Calculate the fractional split of each component, and hence the top and bottom compositions.
    - ➢ Compare the calculated with the assumed values and check the overall column material balance.
    - ➢ Repeat the calculation until a satisfactory material balance is obtained. The usual procedure is to adjust the feed temperature up and down till a satisfactory balance is obtained.
  - ▪ This method is basically a rating method, suitable for determining the performance of an existing column, rather than a design method, as the number of stages must be known.

- How are number of trays obtained for petroleum distillations?
  - ▪ Number of trays for petroleum mixtures, which involve undeterminable and very large number of components, are obtained empirically from experience with similar existing columns in refineries.
  - ▪ Most atmospheric columns have 25–35 trays between the flash zone and column top. The number of trays in various sections of the column depends on the properties of cuts desired from the crude column.
  - ▪ Number of trays used between side-draw products of a crude distillation column is given in Table 17.6.
  - ▪ Kerosene and diesel product are withdrawn as side streams from the crude distillation column. They require adjustment of their flash points to meet the specifications by removing traces of lower boiling components that enter the side streams from the main crude column. This is done in side stream strippers having 4–6 trays, using superheated steam. The stripper overhead mixture with the steam is admitted into the main crude column.

- Write down Fenske equation for minimum number of trays (correspond to total reflux).

$$N_m = \ln[\{x_D/(1-x_D)\}\{(1-x_B)/x_B\}]/\ln \alpha_{av}. \quad (17.128)$$

$$\alpha_{av} = (\alpha_D\alpha_B)^{1/2}, \quad (17.129)$$

where α_D and α_B correspond to compositions x_D and x_B.

  - ▪ If there is wide variation of relative volatility between top and bottom, an additional value corresponding to x_F can be used in the average:

$$\alpha_{av} = (\alpha_D\alpha_F\alpha_B)^{1/3}. \quad (17.130)$$

TABLE 17.6 Number of Trays Used Between Side-Draw Products

Separation	Number of Trays
Naphtha–kerosene	8–9
Kerosene–light diesel	9–11
Light diesel–atmospheric residue	8–11
Flash zone–first draw tray	4–5
Steam stripper section	4–6

- For a multicomponent mixture the Fenske equation is given in terms of the fractional recoveries of component light key, lk, in the distillate, $(Fr_{lk})_D$ and component heavy key, hk, in the bottoms $(Fr_{hk})_B$:

$$N_m = \ln\{[(Fr_{lk})_D(Fr_{hk})_B]/[1-(Fr_{lk})_D][1-(Fr_{hk})_B]\}/\ln(\alpha_{lk-hk}), \quad (17.131)$$

where

$$(Fr_{lk})_D = D(x_{lk})/Fz_{lk}, \quad (Fr_{hk})_B = B(x_{hk})/Fz_{hk}. \quad (17.132)$$

- Give Underwood equations for minimum reflux ratio for binary systems.
  - Case I. Bubble point liquid feed:

$$R_m = [1/(\alpha-1)][(x_D/x_F)-\alpha(1-x_D)/(1-x_F)]. \quad (17.133)$$

  - Case II. Dew point vapor feed:

$$R_m = [1/(\alpha-1)][(\alpha x_D/x_F)-(1-x_D)/(1-x_F)]-1. \quad (17.134)$$

- Give the Underwood equations for the estimation of minimum reflux ratio for multicomponent distillation.
  - Knowing q of feed, relative volatilities of components with respect to heavy key, θ, the intermediate parameter is calculated using the equation:

$$1-q = \Sigma_i \cdot \alpha_i \cdot x_{Fi}/(\alpha_i-\theta). \quad (17.135)$$

  - From the relative volatilities of the components and value of θ obtained from the above equation, R_m is estimate using the equation:

$$R_m = \Sigma_i \cdot \alpha_i \cdot x_{Di}/(\alpha_i-\theta). \quad (17.136)$$

  - As the number of values of θ obtainable from the above equation are 1 + number of components between the keys, correct value of θ is that between values of α of light and heavy keys. If α is chosen to be that of light key with respect to heavy key, correct value of θ will be between α_{lk-hk} and 1.0.
  - Underwood equation is applicable for both binary and multicomponent separations.
  - Assumptions involved in Underwood equation are constant α, constant flows, and that the pinch occurs at the feed tray.
- Give Gilliland and Erbar–Maddox correlations for preliminary estimate of number of trays for a given separation.
  - These correlations require R, N_m, and R_m (Figures 17.24–17.26).

Gilliland Correlation

  - It is more convenient and easy to read when plotted on normal coordinates as given below.
  - Eduljee has given Gilliland correlation in terms of an equation:

$$(N-N_m)/(N+1) \approx 0.75-0.75[(R-R_m)/(R+1)]^{0.5668}. \quad (17.137)$$

Erbar–Maddox Correlation

- Give Kirkbride equation for the location of the feed tray.

$$N_r/N_s = [(x_{F,hk}/x_{F,lk})(x_{B,lk}/x_{D,hk})^2 B/D]^{0.206}, \quad (17.138)$$

where N_r and N_s are number of theoretical trays in rectifying and stripping sections, respectively. Usual notation for the rest of the variables.

- Give the steps involved for finding number of trays for a given separation and reflux ratio for a multicomponent mixture, using short cut methods.
  - Identify light and heavy key components.
  - Assume splits of the nonkey components and compositions of the distillate and bottoms products.
  - Calculate α_{lk-hk}.
  - Use Fenske equation to find N_m.
  - Calculate distribution of nonkey components.
  - Use Underwood method to find R_m.

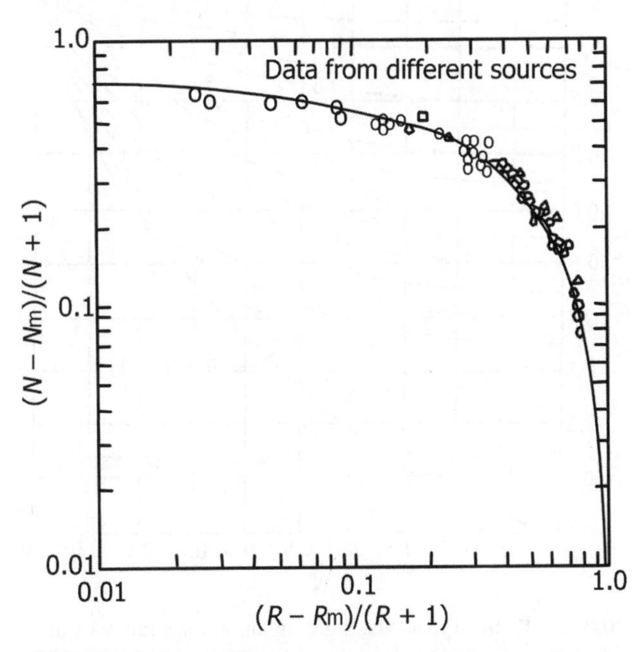

FIGURE 17.24 Gilliland correlation on log–log coordinates.

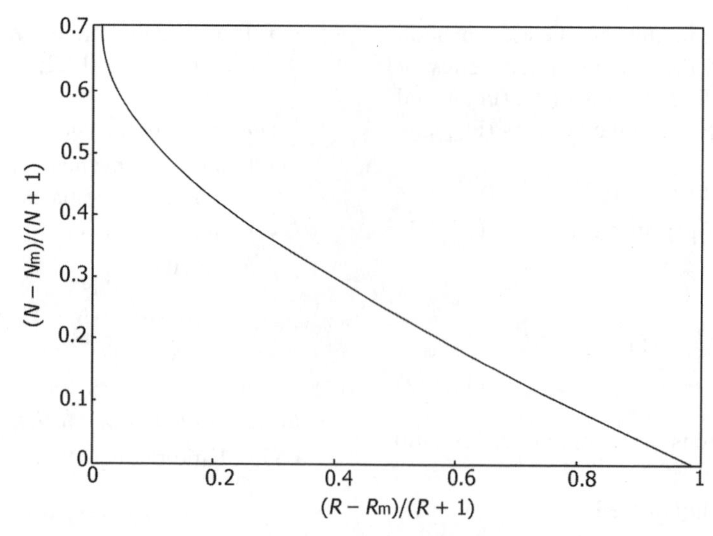

FIGURE 17.25 Gilliland correlation as a function of reflux ratio.

- Use Gilliland or Erbar–Maddox correlation to find actual number of ideal trays at the given operating reflux ratio.
- Use Kirkbride equation to locate the feed tray.

• State rules of thumb for the following:
  - Optimum reflux ratio: About 1.2–1.5 times of R_m.
  - Optimum number of trays: About twice N_m.
  - The optimum Kremser absorption factor is usually in the range of 1.25–2.00.

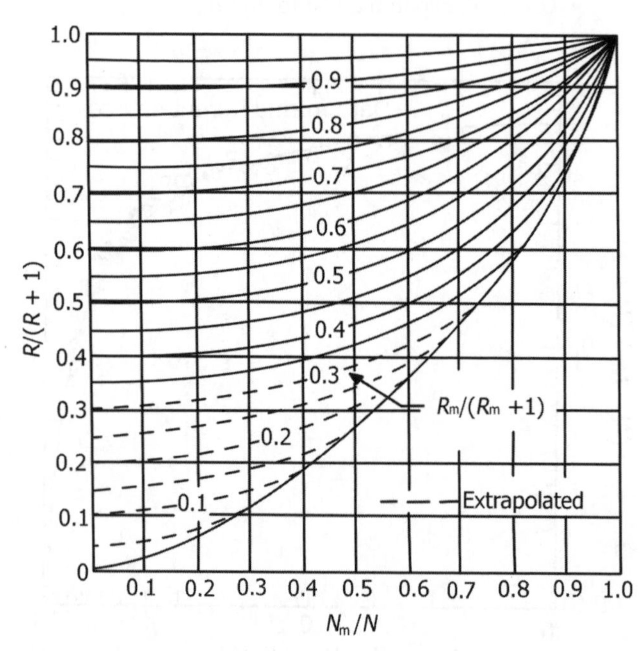

FIGURE 17.26 Erbar–Maddox correlation for number of theoretical trays.

  - Recommended excess number of trays over estimated ones: 10% more.
  - Peak tray efficiencies usually occur at linear vapor velocities of 0.6 m/s (2 ft/s) at moderate pressures, or 1.8 m/s (6 ft/s) under vacuum conditions.
  - Tray efficiencies for aqueous solutions are usually in the range of 60–90% while gas absorption and stripping typically have efficiencies closer to 10–20%.
  - Reflux pumps are made at least 25% oversize.
  - For columns that are at least 0.9 m (3 ft) in diameter, 1.2 m (4 ft) should be added to the top for vapor release and 1.8 m (6 ft) should be added to the bottom to account for the liquid level and reboiler return.
  - The length to diameter ratio of a column should not be more than 30 and preferably below 20.

• For distillations involving multicomponent systems and more than one column, how would you sequence the operations?
  - For sequencing columns:
    - ➢ Perform the easiest separation first (least trays and lowest reflux).
    - ➢ If neither relative volatility nor feed composition vary widely, take products off one at time as the overhead.
    - ➢ If the relative volatility of components do vary significantly, remove products in order of decreasing volatility.
    - ➢ If the concentrations of the feed vary significantly but the relative volatility do not, remove products in order of decreasing concentration in the feed.

17.2.5 Separation of Azeotropes and Close Boiling Mixtures

- What is an azeotrope?
  - An azeotrope is defined as a liquid mixture at a composition where changing the phase of the mixture from liquid to vapor or vice versa does not change the composition of the mixture, though the individual components involved are volatile in nature.
  - The components *travel* together into the vapor during vaporization and into the liquid during condensation at the given conditions of pressure and in the absence of any *contaminants* or *foreign material* in the mixture.
  - If at the equilibrium temperature the liquid mixture is homogeneous, the azeotrope is a *homoazeotrope*. If the vapor phase coexists with two liquid phases, it is a *heteroazeotrope*.
  - The azeotropes can be *minimum boiling* type, in which the boiling temperature is *less* than the boiling temperatures of either of the components, or *maximum boiling* type, in which boiling temperature of the azeotrope is *higher* than those of the individual components.
  - Systems that do not form azeotropes are called *zeotropic*.
- What causes formation of azeotropes?
  - Azeotropy is a characteristic of the highly nonideal phase equilibria of mixtures with strong molecular interactions.
  - Azeotropes are formed due to differences in intermolecular forces of attraction among the components in the mixture (hydrogen bonding and others). The particular deviation from ideality is determined by the balance between the physicochemical forces between identical and different components.
  - The tendency of a mixture to form an azeotrope depends on two factors, namely, the difference in the pure component boiling points and the degree of nonideality.
  - The closer the boiling points of the pure components and the less ideal the mixture is, the greater the likelihood of an azeotrope. A heuristic rule is that azeotropes occur infrequently between compounds whose boiling points differ by more than about 30°C.
    - An important exception to this rule is heteroazeotropic mixtures where the components may have a large difference in the boiling points of the pure components, and still exhibit strong nonideality and immiscibility regions. In heterogeneous mixtures, the different components may even repel each other. This is why only minimum boiling heteroazeotropes occur in nature.
  - The presence of some specific groups, particularly polar groups, containing oxygen, nitrogen, chlorine, and fluorine, often results in the formation of azeotropes.
- Give Ewell, Harrison, and Berg classification of possible entrainers (solvents) for azeotropic distillation.
  - Classified according to their hydrogen bonding characteristics (Table 17.7).
- Give examples of minimum boiling and maximum boiling azeotropes. Illustrate by means of boiling point diagrams.
  - Minimum boiling azeotropes: Ethanol–water (89.43 mol% ethanol), isopropyl ether–isopropanol, carbon disulfide–water.
  - These correspond to 101.3 kPa pressure.
  - Figures (17.27–17.29) are illustrative of minimum and maximum boiling azeotropes.
  - Maximum boiling azeotropes: Acetone–chloroform. Azeotropic composition is 0.64. Azeotropic temperature is 339.12 K (Figure 17.29).
- What is the difference between minimum boiling and maximum boiling azeotropes with respect to activity coefficients?
  - Minimum boiling azeotropes have activity coefficients greater than unity, while maximum boiling azeotropes have activity coefficients less than unity.
- Give an example of a heterogeneous azeotrope.
  - Very large positive deviations may lead to partial miscibility and formation of heterogeneous azeotropes.
  - 1-Butanol–water system is an example of a heterogeneous azeotrope (Figure 17.30).
  - Acetic acid–water–butyl acetate form heterogeneous ternary azeotrope.
- What are the ways by which azeotropic mixtures are separated?
  - *Azeotropic distillation*: involving addition of a third component, called entrainer.

TABLE 17.7 Ewell, Harrison, and Berg Classification for Entrainers

Strong	Weak
O → HO	N → HN
N → HO	O or N → $HCCl_3$
	HCCl-Cl
O → HN	$HCNO_2$
	HCCN

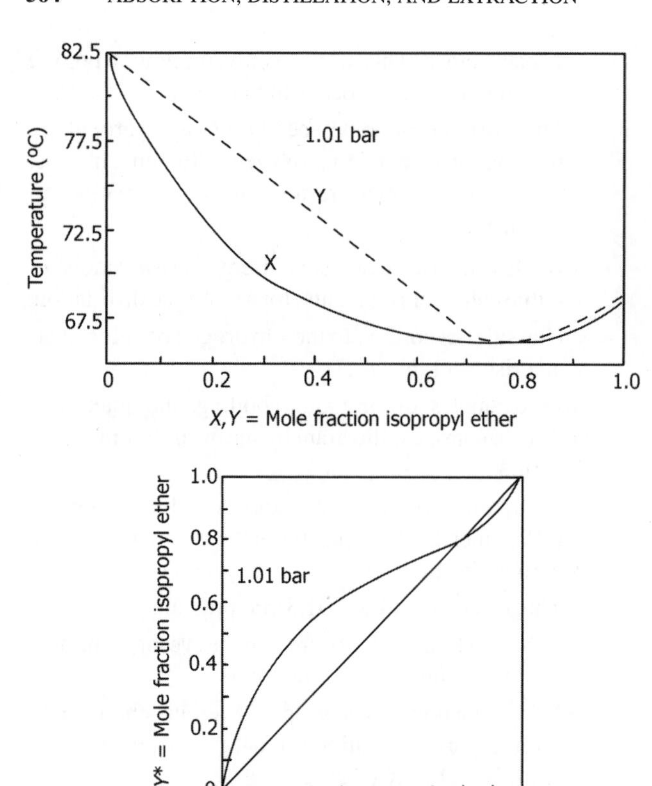

FIGURE 17.27 *T–x–y* and *x–y* diagrams for isopropyl ether–isopropyl alcohol system at 101.3 kPa pressure.

- *Pressure swing distillation*: based on the premise that azeotropic composition is shifted by altering distillation pressure.
- *Membrane processes*: especially involving pervaporation.
- *Adsorption techniques*: in which one of the components forming an azeotrope may be removed from the mixture by adsorption on an adsorbent.
- *Reactive distillation*: in which the entrainer reacts preferentially and reversibly with one of the original mixture components. The reaction product is distilled out from the nonreacting component and the reaction is reversed to recover the initial component. The distillation and reaction is usually carried out in one column (catalytic distillation).
- *Chemical Drying (Chemical Action and Distillation)*: The volatility of one of the original mixture components is reduced by chemical means. An example is dehydration by hydrate formation. Solid NaOH may be used as an entrainer to remove water from tetrahydrofuran (THF). The entrainer and water forms a 35–50% NaOH solution containing very little THF.
- *Distillation in the Presence of Salts*: The entrainer (salt) dissociates in the mixture and alters the relative volatilities sufficiently so that the separation becomes possible. A salt added to an azeotropic liquid mixture will reduce the vapor pressure of the component in which it is more soluble. Thus, extractive distillation can be applied using a salt solution as the entrainer. An example is the dehydration of ethanol using potassium acetate solution.

- Illustrate with the relevant diagrams, the operation of pressure swing distillation.
  - Pressure changes can have a large effect on the vapor–liquid equilibrium compositions of azeotropic mixtures and thereby affect the possibilities to separate the mixture by ordinary distillation.
  - By increasing or decreasing the operating pressure in individual columns one can move distillation boundaries in the composition space or even make azeotropes appear or disappear (or transform into heteroazeotropes).
  - For some mixtures, a simple change in pressure can result in a significant change in the azeotrope composition and enable a complete separation by pressure-swing distillation as illustrated in Figure 17.31.

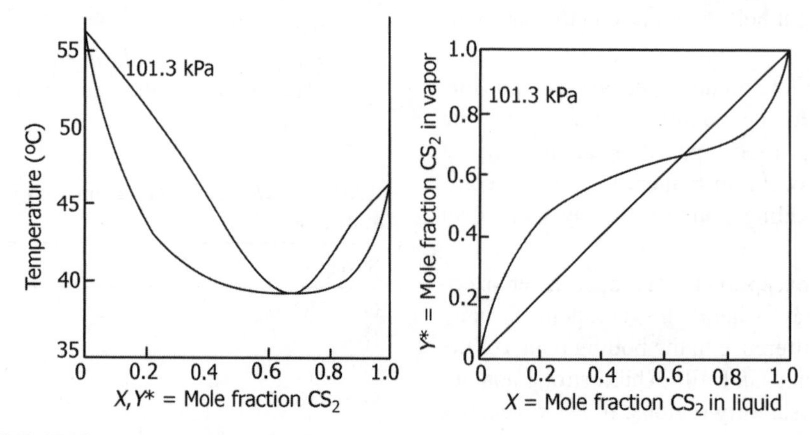

FIGURE 17.28 *T–x–y* and *x–y* diagrams for carbon disulfide–water system at 101.3 kPa pressure.

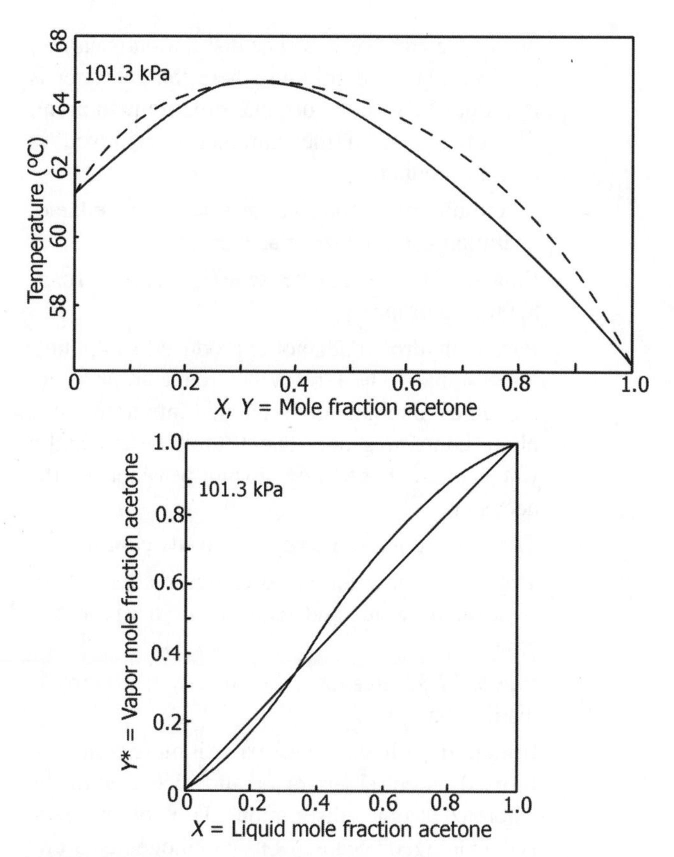

FIGURE 17.29 *T–x–y* and *x–y* diagrams for acetone–chloroform system at 101.3 kPa pressure.

- A binary homoazeotropic mixture is introduced as feed F_o to the low-pressure column.
- The bottom product from this P_1 column is relatively pure A, whereas the overhead is an azeotrope with x_{D1}. This azeotrope is fed to the high-pressure column, which produces relatively pure B_2 in the bottom and an azeotrope with composition x_{D2} in the overhead.
- This azeotrope is recycled into the feed of the low-pressure column.
- The smaller the change in azeotropic composition with pressure, the larger is the recycle in Figure 17.32.
- A change in azeotropic composition of at least 5% with a change in pressure is usually required.
- For example, tetrahydrofuran–water azeotrope may be separated by using two columns operated at P_1 = 101.3 kPa and P_2 = 810 kPa.
- Ethanol–water azeotrope is not considered to be sufficiently pressure sensitive for the pressure swing distillation process to be competitive.
- Although changing the vapor–liquid equilibrium (VLE) properties of an azeotropic mixture by purely physical means (pressure or temperature changes) is an attractive and definite possibility, it is generally not an economic option.

• Give some examples of azeotropes and close boiling mixtures that can be separated by pressure–swing distillation.

- Water–ethanol, methanol–benzene, methanol–methyl acetate, ethanol–benzene, ethanol–heptane, dimethyl amine–trimethyl amine, water–methyl ethyl ketone (MEK), water–propionic acid, water–tetrahydrofuran, benzene–hexane, CO_2–ethylene, and so on.

• What is azeotropic distillation? Illustrate.

- Physicochemical changes of the VLE behavior of an azeotropic mixture is brought about by the addition of an extraneous liquid component, called *entrainer*.
- There are three different conventional entrainer addition-based distillation methods depending on the properties and role of the entrainer and the process scheme.

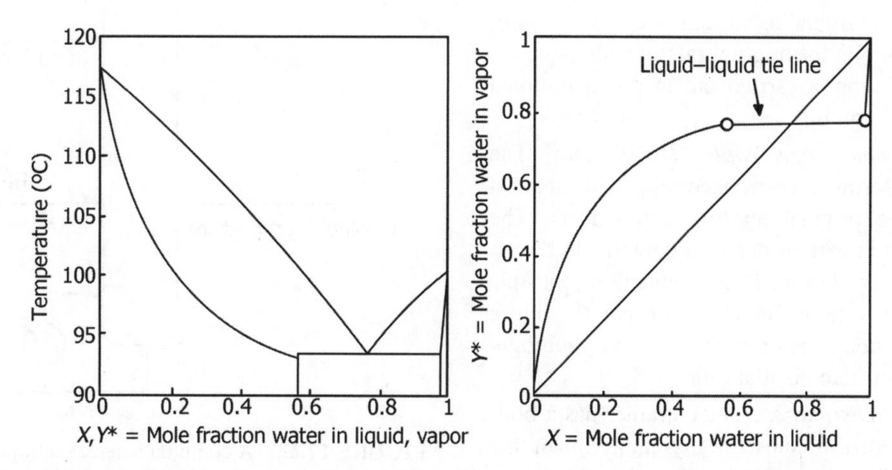

FIGURE 17.30 *T–x–y* and *x–y* diagrams for water–1-butanol system at 101.3 kPa. *Note*: The diagrams are based on NRTL equation.

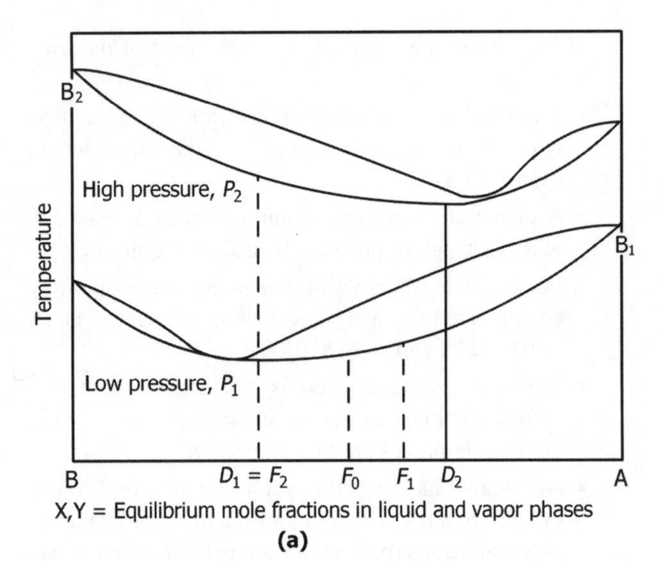

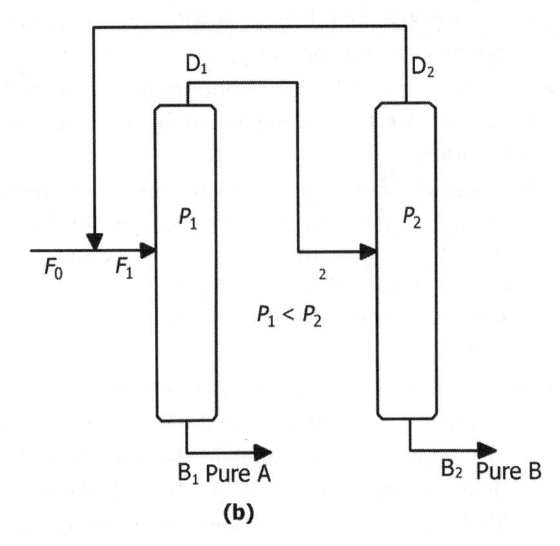

FIGURE 17.31 Pressure swing distillation for a minimum boiling binary azeotrope that is sensitive to changes to pressure.

- *Homogeneous Azeotropic Distillation*: The entrainer is completely miscible with the components of the original mixture. It may form homoazeotropes with the original mixture components. The distillation is carried out in a conventional single feed column.
- *Heterogeneous Azeotropic Distillation*: The entrainer forms a heteroazeotrope with at least one of the original mixture components. The distillation is carried out in a combined column and decanter system. Heterogeneous azeotropic distillation is more common for separating close boiling components or azeotropes than homogeneous azeotropic distillation.
- *Extractive Distillation*: The entrainer has a boiling point that is substantially higher than the original mixture components and is selective to one of the components. The distillation is carried out in a two-feed column where the entrainer is introduced above the original mixture feed point. The main part of the entrainer is removed as bottom product.

- An example of azeotropic distillation is ethanol dehydration using benzene as entrainer.
  - Ethanol (2)–water (1)–benzene (E) forms a ternary heteroazeotrope.
  - Pure (anhydrous) ethanol is produced as the bottom product. The ternary heteroazeotrope is removed as distillate and separated into an organic phase containing only one 1 mol% water and a water phase containing 36 mol% water in the decanter.
  - The organic phase is recycled to the column.
  - The water phase is sent to recovery columns for removal of water and recovery of benzene and ethanol.
  - Figure 17.32 illustrates the process of azeotropic distillation.
  - In modern systems, cyclohexane is more commonly used as entrainer for ethanol dehydration by heteroazeotropic distillation. Use of benzene is discouraged because of its carcinogenic nature.
  - Although the reflux usually is a single liquid phase, redistributors in the rectifying section may handle two liquid phases, making it necessary to use special designs for this service.
  - When two liquid phases are formed in a distillation column, foaming occurs on those trays with

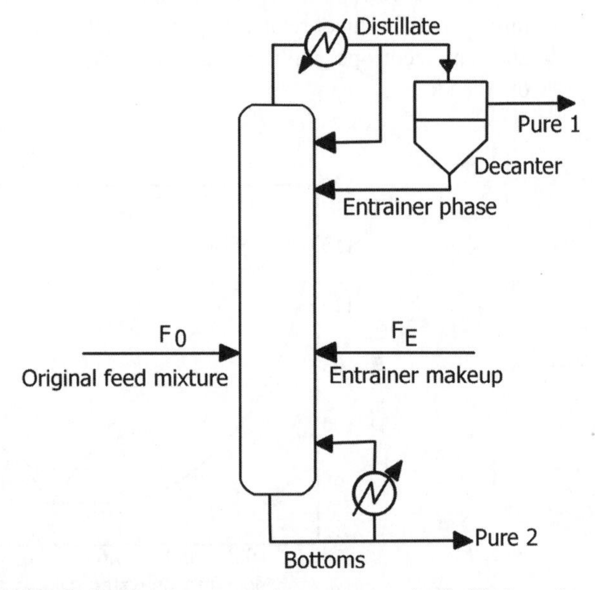

FIGURE 17.32 A common heteroazeotropic distillation scheme with distillate decanter.

a liquid composition near the two-phase transition.

- In azeotropic distillations, if the distillate forms a single liquid phase, the entrainer may be recovered by liquid extraction from the condensed overhead to permit its return to the azeotropic distillation column.
- The entrainer selected must be sufficiently volatile so that it can be stripped from the bottoms product to avoid its loss from the column and contamination of the bottoms liquid. At the same time the entrainer should not be *too* volatile, because it must be present in large concentrations in the liquid phase.

• What are the differences between azeotropic and extractive distillations?

- In azeotropic distillation, the entrainer must form a lower boiling azeotrope with at least one of the feed components than the azeotrope formed by the components in the feed to be separated.
  - Top product involves homogeneous or heterogeneous azeotropes.
  - Added solvent is volatile and goes as overhead product.
- In extractive distillation, no azeotrope formation occurs.
  - Solvent is relatively nonvolatile, added above the feed point of the column and goes with bottom product.
  - Above the solvent feed point is a wash bed to which some of the liquid distillate is returned as reflux. This arrangement prevents any solvent vapor being carried overhead from the column, which would represent a solvent loss from the system as well as contamination of the distillate product.
  - The solvent feed distributor may require special design if the mixture of reflux and solvent feed either foams or flashes.
  - Simpler than azeotropic distillation.
  - While normally the lower boiling feed components would be present in the distillate in greater concentrations than in the feed, the solvent may reverse the relative volatility relationship between the key components in the feed mixture.
  - Solvent selection is easier than that for azeotropic distillation.
  - High reflux may be harmful in extractive distillation because it weakens the extractive effect (decreases the entrainer concentration in the extractive column section).
  - Extractive distillation is usually preferred over azeotropic distillation, if both methods can be used to separate the feed components.
  - The bulk of the solvent in extractive distillation is not vaporized as compared to azeotropic distillation where the entrainer is recovered from the overhead vapor stream. The energy input necessary to effect separation usually is, therefore, lower for extractive distillation than for azeotropic distillation.
  - Also, an extractive distillation column can operate over a wider range of pressures than an azeotropic distillation column, because the azeotropic composition is a function of pressure.

• What are the considerations involved in the selection of a solvent for azeotropic distillation?

- Should preferably form low boiling heterogeneous azeotrope with one of the components.
- Preferably form azeotrope with the component having low composition to reduce heat loads.
- Should have sufficient volatility to make it readily separable from the rest and minimum amount lost with residue.
- Should be miscible with the components to be separated.
- Should boil within a limited range (0–30°C) of the components to be separated.
- Should have low λ_v to reduce energy requirements.
- Low viscosity to give better mass transfer and high efficiency.
- Low freezing point to facilitate handling and storage in cold weather.
- Should be cheap, readily available, chemically stable, nonreactive to the components being separated, noncorrosive and nontoxic.

• State the considerations involved in solvent selection for extractive distillation.

- High selectivity, that is, ability to alter α with the use of small amounts of solvent.
  - Selectivity is defined as

$$S_{12} = \gamma_1^\infty/\gamma_2^\infty. \qquad (17.139)$$

  - Separation factor is given by

$$\alpha_{12} = (\gamma_1^\infty/\gamma_2^\infty)(P_1^0/P_2^0). \qquad (17.140)$$

  - γ_1^∞ and γ_2^∞ are activity coefficients of components 1 and 2 at infinite dilution.
  - α_{12} is separation factor.
  - P_1^0 and P_2^0 are vapor pressures of components 1 and 2.
  - Activity coefficients at infinite dilution can be obtained from experimental data, databases, and

generated by the use of predictive methods such as group contribution methods like UNIFAC.

- High capacity, that is, ability to dissolve components in the mixture.
- Low volatility to prevent vaporization of solvent and ensuring its presence in appreciable concentrations in the liquid.
- The solvent should be miscible in all proportions with the feed components at distillation temperatures.
- The reduction in normal vapor pressure for the associated feed component varies with the concentration of solvent used.
- The effect of the solvent is dependent on intermolecular attractive forces, with hydrogen bonding having the most significance.
- Solvents usually are selected based on laboratory tests to determine their effect on the normal vapor pressure of a compound.
- Because the bottom column temperature increases due to the higher solvent boiling point, the reboiler duty should be evaluated in solvent selection.
- Because the process does not depend on the formation of an azeotrope and thus a greater choice of solvents is, in principle, possible.
- Easy separability.
- No azeotrope formation with the components.
- All other attributes listed under solvent selection for azeotropic distillation are applicable for extractive distillation solvents.

• Give examples of solvents used in extractive distillation processes.

- Table 17.8 gives examples of solvents used in different extractive distillation processes.

• For what type of mixtures extractive distillation is attractive?

- For separation of close boiling liquid mixtures.
- The very close boiling points produce low relative volatilities, which require a high reflux ratio and a large number of theoretical trays.
  - For example, the separation of butenes from butanes can require multiple columns with a total of 280 trays by conventional distillation.
  - The addition of a solvent to a mixture of C_4 hydrocarbons can greatly reduce the vapor pressures of the butenes while not affecting the vapor pressures of the butanes.
  - Thus, the distillate product from an extractive distillation employing such a solvent will be butanes and the bottoms will contain butenes.

TABLE 17.8 Examples of Solvents Used in Different Extractive Distillation Processes

Components to be Separated	Typical Solvents Used
Alcohols (ethanol, isopropanol, *tert*-butanol)–water	Ethylene glycol
Acetic acid–water	Tributyl amine
Acetone–methanol	Water, ethylene glycol
Methanol–methyl acetate	Water
Propylene–propane	Acetonitrile
C_4 hydrocarbons	Acetone, acetonitrile, dimethyl formamide, *N*-methyl-2-pyrrolidone, *N*-formylmorpholine
C_5 hydrocarbons	Dimethyl formamide
Aromatics–nonaromatics	Dimethyl formamide, *N*-methyl-2-pyrrolidone
Benzene–cyclohexane	Aniline
n-Heptane–toluene	Aniline, phenol
n-Heptane–methylcyclohexane	Aniline
Methylcyclohexane–toluene	Phenol
Isooctane–toluene	Phenol
Methyl ethyl ketone–water	Ethylene glycol monobutylether
Acetone–chloroform	Higher ketones and chloro compounds

• Illustrate, by means of flow diagrams, separation of a (i) two-component mixture and a (ii) multicomponent mixture by extractive distillation.

- Figures 17.33 and 17.34 illustrate separation schemes for binary and multicomponent mixtures by extractive distillation.

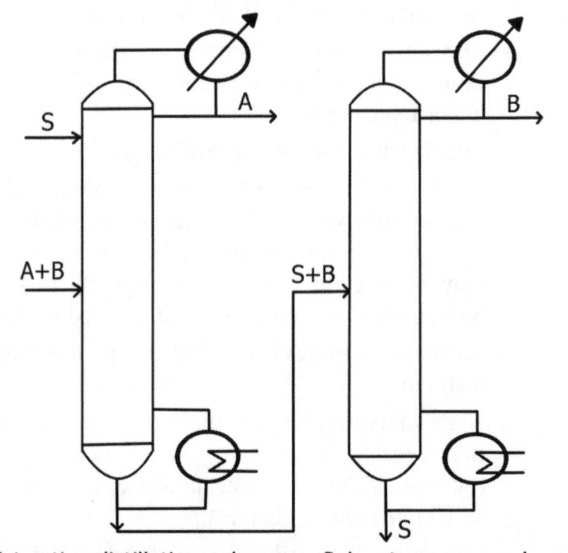

FIGURE 17.33 Two-column system for extractive distillation.

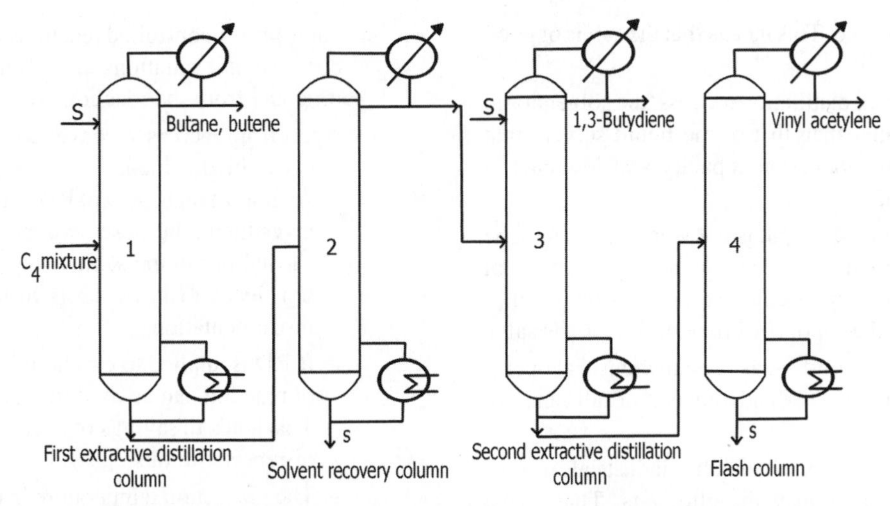

FIGURE 17.34 Extractive distillation process for separation of C_4 hydrocarbons.

- What are the differences, with respect to product purity and energy consumption, between extractive and azeotropic distillation processes?
  - Extractive distillation cannot obtain a very high purity of product, but consumes less energy while azeotropic distillation can obtain a very high purity of product, but consumes much more energy.
  - Higher energy consumption with respect to azeotropic distillation is due to vaporization of solvent in azeotropic distillation whereas the solvent in extractive distillation is nonvolatile and therefore no heat of vaporization is involved in the later.
  - Therefore, a separation method, involving the combination of extractive and azeotropic distillation processes eliminates the disadvantages of both extractive and azeotropic distillation processes and retains the advantages of both. This method has been used for the separation of 2-propanol and water while the high purity of 2-propanol over 99.8 wt% is required.
  - This process is designed according to the sequence, first, extractive distillation that ensures low energy consumption, and then azeotropic distillation that ensures high purity of 2-propanol as product.
  - The method is illustrated by Zhigang Lei, Biaohua Chen, and Zhongwei Ding in their book, *Special Distillation Processes*.
- Can a salt be used in extractive distillation processes in place of a liquid solvent? Explain and illustrate by means of an example.
  - In certain systems where solubility permits, it is feasible to use a salt dissolved into the liquid phase, rather than a liquid solvent, as the separating agent for extractive distillation.
  - Salt effect in vapor–liquid equilibrium refers to the ability of a salt that has been dissolved into a liquid phase consisting of two or more volatile components to alter the composition of the equilibrium vapor without itself being present in the vapor.
  - Influence of various salts on vapor–liquid equilibria for ethanol–water system is illustrated in Table 17.9.
  - It can be seen from Table 17.9, that the higher the valence of metal ion is, the more pronounced the salt effect is. The order of salt effect is $AlCl_3 > CaCl_2 > NaCl_2$ and $Al(NO_3)_3 > Cu(NO_3)_2 > KNO_3$.
- What are the advantages and disadvantages in using salts in extractive distillation processes?
  - In the systems where solubility considerations permit their use, solid salts as the separating agents have major advantages. The ions of a solid salt are typically capable of causing much large effects than the molecules of a liquid agent, both in the strength of attractive forces they can exert on feed component molecules and in the degree of

TABLE 17.9 Salt Effect on Vapor–Liquid Equilibria for Ethanol–Water System

Separating Agent	Relative Volatility
No solvent	1.01
Ethylene glycol	1.85
Calcium chloride (saturated solution)	3.13
Potassium acetate	4.05
Ethylene glycol + NaCl	2.31
Ethylene glycol + $CaCl_2$	2.56
Ethylene glycol + $SrCl_2$	2.60
Ethylene glycol + $AlCl_3$	4.15
Ethylene glycol + KNO_3	1.90
Ethylene glycol + $Cu(NO_3)_2$	2.35
Ethylene glycol + $Al(NO_3)_3$	2.87
Ethylene glycol + CH_3COOK	2.40
Ethylene glycol + K_2CO_3	2.60

selectivity exerted. This means that the salt is of good separation ability.

- In the extractive distillation process, the solvent ratio is much smaller than that of the liquid solvent that leads to high production capacity and low energy consumption.
- Moreover, since solid salt is not volatile, it cannot be entrained into the product. So as vapor does not contain solids there are no toxic effects involved in inhalation of the vapors by operators due to the salts.
- However, when solid salt is used in industrial operations, dissolution, reuse and transport of salt can pose handling problems.
- Fouling and erosion limit the industrial use of extractive distillation with solid salts. That is the reason for not using salts in extractive distillation in industrial scale widely.

17.2.6 Reactive Distillation

- What is a reactive separation process? What are the different types of reactive distillation processes?
  - Reactive separation processes such as reactive distillation, sorption-enhanced reaction absorption, reactive extraction, and reaction crystallization combine the essential tasks of reaction and separation in a single vessel.
  - The most important example of reactive separation processes is reactive distillation (RD), which is the combination of chemical reaction and distillation in a single column.
  - Reactive distillation processes can be divided into homogeneous and heterogeneous processes. The reactions involved are catalytic or noncatalytic. Catalytic distillation involves commonly use of a solid catalyst.
  - Reactive distillation is used with reversible liquid-phase reactions.
  - The equipment used for homogeneous reactive distillation processes consists of a trayed column, with weirs having larger heights than normal, so that the necessary liquid holdup and residence time needed for reaction is provided.
  - Heterogeneous reactive distillation uses a solid catalyst. The equipment consists primarily of catalyst-containing packing that allows for simultaneous reaction and mass transfer between vapor and liquid phases.
- What are the advantages of reactive distillation?
  Advantages
  - RD has many advantages over sequential processes. The most important advantage in the use of RD for equilibrium-controlled reactions is the elimination of conversion limitations by continuous removal of products from the reaction zone.
  - Apart from increased conversion, the following benefits can be obtained:
    - An important benefit of RD is a reduction in capital investment, because two process steps can be carried out in the same equipment. Such integration leads to lower costs in pumps, piping, and instrumentation.
    - If RD is applied to exothermic reactions, the heat of reaction can be used for vaporization of liquid. This leads to savings of energy costs by reduction of reboiler duties.
    - The maximum temperature in the reaction zone is limited to the boiling temperature of the reaction mixture so that the danger of hot spots in the catalyst is reduced and runaway reaction hazards are reduced significantly. A simple and reliable temperature control can be achieved.
    - Product selectivities can be improved due to fast removal of reactants or products from the reaction zone. Thus, the probability of consecutive reactions, which may occur in the sequential operation mode, is lowered.
  - Catalytic distillation (CD), carried out in fixed beds or as suspension of catalyst, has additional advantages in comparison with homogeneous reactive distillation:
    - An optimum configuration of the reaction and separation zones is permitted in a RD column whereas expensive recovery of liquid catalysts may be avoided.
    - If the reaction zone in the CD column can be placed above the feed point, poisoning of the catalyst (especially ion exchange resins) by metal ions can be avoided. This leads to longer catalyst life compared to conventional systems.
    - Many catalysts used in CD column are environmentally friendly such as supported molecular sieves, ion exchange resins, and so on.
- Illustrate, by means of an example, reactive extractive distillation.
  - Separation of acetic acid–water system is used as an example to illustrate reactive extractive distillation in which chemical reaction is involved.
    - Normally three methods are used in separation of acetic acid–water mixtures by distillation, namely, ordinary distillation, azeotropic distillation, and extractive distillation.
    - Energy consumption is large in ordinary distillation, requiring large number of trays for the

separation to give high-purity acetic acid required in many industrial processes.

➢ Though the number of trays required is reduced in azeotropic distillation, energy consumption remains large, with large amounts of azeotropic solvent to be vaporized.

➢ Extractive distillation is an attractive method for separating acetic acid–water mixtures.

- An alternative method uses reactive extractive distillation process with tributyl amine as solvent. The following reversible reaction takes place:

$$HAc + R_3N \leftrightarrow R_3NH^+ \cdot {}^-OOCCH_3 \qquad (17.141)$$

where HAc is acetic acid, R_3N is tributyl amine and $R_3NH^+ \cdot {}^-OOCCH_3$ is salt formed by the reaction, respectively.

- For the extractive distillation process, the forward reaction occurs in the extractive distillation column and the reverse reaction occurs in the solvent recovery column. The reversible chemical reaction represents a reaction between a weak acid (acetic acid) and a weak base (separating agent). Water is taken as overhead product in the extractive distillation column and acetic acid and tributyl amine are taken as bottom product. Solvent and acetic acid are separated by ordinary distillation in a second column.

• Give examples of reactive distillation processes.
  - Table 17.10 gives examples of reactive distillation processes.

17.2.7 High Pressure and Vacuum Distillation

• List the variables involved in the operation of a distillation unit.
  - Column pressure.
  - Reflux ratio.
  - Top and bottom temperatures.
  - Overhead and bottom product rates.
  - Overhead and bottom product compositions.
  - Reboiler and condenser duties.

• What are the considerations involved in the selection of distillation column pressure?
  - Column operating pressure is most often determined by the cooling medium used in the condenser.
  - Except while distilling heat-sensitive materials, the main consideration when selecting column pressure will be to ensure that the dew point of the distillate is above that which can be easily obtained with the plant cooling water. In other words, top product should be condensable with plant cooling water. Available cooling water temperatures are generally in the range of 20–40°C in tropical regions.
  - Use of refrigerated brine or other refrigerants to condense top product is normally expensive and should be considered as a special case to avoid using high pressures.
  - Higher column pressures should be considered only for difficultly condensable products, for example, gas fractionation that might require both higher pressures as well as use of refrigerant to condense top product.

TABLE 17.10 Examples of Reactive Distillation Processes

Reaction Type	Process	Comments
Esterification	Methyl acetate from methanol and acetic acid	Homo. or hetero.
	Ethyl acetate from ethanol and acetic acid	Hetero.
	Butyl acetate from butanol and acetic acid	Homo.
Transesterification	Ethyl acetate from ethanol and butyl acetate	Homo.
	Diethyl carbonate from ethanol and dimethyl carbonate	Hetero.
Hydrolysis	Acetic acid and methanol from methyl acetate and water	Hetero.
Etherification	Methyl tertiary butyl ether from isobutene and methanol	Hetero.
	Ethyl tertiary butyl ether from isobutene and ethanol	Hetero.
	Tertiary amyl methyl ether from *iso*-amylene and methanol	Hetero.
Condensation	Diacetone alcohol from acetone	Hetero.
	Bisphenol-A from phenol and acetone	
Hydration	Monoethylene glycol from ethylene oxide and water	Homo.
Nitration	4-Nitrochlorobenzene from chlorobenzene and nitric acid	Homo.
Alkylation	Ethylbenzene from ethylene and benzene	Hetero.
	Cumene from propylene and benzene	Hetero.
	Alkylation of benzene and 1-dodecene	Hetero.

Homo.: homogeneous and hetero.: heterogeneous.

- Product should not decompose at operating temperatures (high boiling/heat-sensitive liquids).
  - The maximum allowable reboiler temperature to avoid degradation of the process fluid should be the criteria for selecting column pressure while distilling heat-sensitive liquids.
  - Maximum allowable reboiler temperature may be around 185°C with steam as heat transfer medium at pressures of about 10 barg.
- Vacuum operation should be considered to reduce column temperatures for distilling heat-sensitive liquids.
- Column pressure has an influence on the equilibrium data used for the design of distillation columns.
  - High pressures reduce relative volatility and hence necessitate use of larger number of trays for a given separation.
  - Lower pressures increase relative volatility and hence increase separation obtainable for a given number of trays.
  - Errors can be introduced in the accuracy of equilibrium data used that might be obtained at a different pressure.
- Atmospheric distillations are considered to be those separations carried out at column pressures between 40 and 560 kPa (0.4 and 5.5 atm) absolute.
- Vacuum distillations are those operating at pressures below atmospheric, usually not greater than 40 kPa top column pressure.
- Pressure distillations normally use a top column pressure of 550 kPa or greater.

• How is distillation column pressure controlled?

- Columns that are directly linked to atmosphere through a breather vent over the condenser:
  - When pressure tends to fall, air is inhaled.
  - When vapor generation rate exceeds condensation rate, air is exhaled and some vapors are lost to the atmosphere.
  - This arrangement may not be permissible due to following factors:
    - Loss not tolerable from production point of view.
    - Vapor venting might be objectionable from environmental considerations.
    - Vapor venting or air inhalation might lead to fire and explosion.
    - Possible toxicity exposures due to vapor venting.
- Column pressure is more often controlled by manipulating condensation rate through either bypassing or flooding the condenser.

• What happens if column pressure gets too low?

- Reflux may not condense.
- Level in reflux drum falls.
- Reflux pump looses suction giving rise to cavitation.
- If pressure falls suddenly, liquid on the tray, being at bubble point, boils violently and surge in vapor flow may promote jet entrainment or flooding.
  - This saves reboiler duty and reduces energy costs. This reduces bottoms temperature that has the same effect of cutting down energy costs.
  - However, it must be understood that vacuum-producing equipment increases energy requirements and an optimization exercise is to be undertaken on overall energy costs.
- Net effect of lowering pressure gives better separation efficiency due to increased relative volatility; Tray deck leakage decreases; Entrainment increases with decreasing effect on efficiency.
- The operating column pressure that maximizes ΔT, ($T_{\text{bottoms}} - T_{\text{top}}$), across the column at a particular reflux ratio should be the criterion for the choice of the column pressure, other than available cooling water temperature levels.
- Any temperature control system, for optimizing column pressure, must operate on control of ΔT and *not* on bottom and top temperature readings.

• What happens if column pressure gets too high?

- The vapor flowing between trays, being at dew point, rapidly condenses if column pressure increases. This results in loss in vapor velocity through the tray deck holes leading to dumping.

• What are the applications of high-pressure distillation processes in industry?

- Separations of polymer grade propylene and ethylene from cracked gases at absolute pressures of 16 and 20 atm, respectively.
- Separations of very low boiling hydrocarbons like methane and ethane are carried out at pressures of around 70% of their criticals, whereas ethylene and ethane are carried out at 40 t0 55% of their criticals.
- All low boiling distillations must be carried out below their criticals in order to provide liquid reflux.

• What is retrograde phenomena?

- In normal distillation processes, as pressure is increased condensation is favored and vice versa.
- In super-critical distillations there are operating regions on the phase envelopes where increased pressure promotes vaporization instead of condensation. Similarly as pressure is decreased, instead of vaporization, condensation takes place. This

phenomena is called *retrograde condensation* and *retrograde vaporization*, respectively.

- In gas separations near critical regions, retrograde phenomena must be recognized and column pressures fixed accordingly as otherwise premature column flooding or blowing can occur.

• What are the objectives involved in preferring packed columns instead of tray columns in high-pressure distillation processes?

- To increase the capacity of downcomer-limited tray columns.
- To improve product purity or yield by accommodating a larger number of theoretical tray equivalents for a given packed height.
- To conserve energy by providing a larger number of theoretical tray equivalents in order to reduce the reflux ratio.
- To permit the use of vapor recompression due to the reduced pressure drop produced by column packings as compared to that developed by trays in the separation of close boiling materials.
- To reduce energy use by operation of columns in parallel, which because of the low pressure drop of column packings, permits the overhead vapor from the first column to supply reboiler heat to the second column.
- To lower the residence time for materials that degrade or polymerize.
- To reduce the inventory in the column of flammable or hazardous materials.
- To increase product recovery in batch distillations due to the lower liquid holdup in a packed column.

• What are the reasons for distillations carried out under vacuum?

- The relative volatility between components generally increases as the boiling temperature drops. This higher relative volatility improves the ease of separation, which lowers the number of theoretical stages needed for a given separation. If the number of theoretical trays is held constant, the reflux ratio required for the same separation can be reduced. In addition, if the number of theoretical trays and the reflux ratio are maintained constant, product purity will be increased.
- Lower distillation temperatures are desirable when processing heat-sensitive products. Lower bottoms temperatures retard undesirable reactions such as product decomposition, polymerization, or discoloration.
- Separations can be achieved for components with very low vapor pressures or compounds that degrade at temperatures near their atmospheric boiling point.
- Lower reboiler temperatures permit the use of less costly energy sources such as low-pressure steam or hot water.
- Usually the top column pressure is set by the selection of the vacuum-producing equipment.

• Why steam distillation is not preferred in place of vacuum distillation to achieve the goals listed in the above question?

- Steam required is costly because several moles of steam usually are needed for each mole of feed component vaporized.
- Presence of steam in the column usually increases the corrosion rate on commonly used materials.
- Steam condensate, after separation of the distillate product, usually requires waste treatment before it can be discharged to the environment.
- In special cases involving heat-sensitive products, the bottom column pressure may be fixed by the maximum allowable product temperature.
- Pressure drop considerations in the column become important in selection of column internals so that the bottoms temperature does not exceed that which can affect quality of the bottom product.
  - A well-designed sieve tray may involve a pressure drop as low as 0.4 kPa (3 mmHg) per theoretical tray.
  - Multiple columns operated in series are the alternative where more than 25 theoretical trays are required in the separation of heat-sensitive materials.
  - Another way is to use low ΔP structured packing in place of trays.
  - Structured packings can provide a large number of theoretical trays while using only one reboiler and one condenser, even if columns are operated in series.

• What are the applications of vacuum distillation in industry?

- Separation of azeotropes. Example is acetone–methanol system that form an azeotrope containing 80.0 mol% acetone when distilled at atmospheric pressure. However, if the pressure is lowered to 13.3 kPa absolute, these compounds can be separated by simple distillation, because no azeotrope is formed at this reduced pressure.
- Separation of high boiling liquid mixtures.
- Separation of ethyl benzene from styrene for which column top pressures are of the order of 6.65 kPa absolute.

- Separation of cyclohexanone from cyclohexanol in the production of caprolactam. Such a column normally operates at a top pressure of 6–9.3 kPa absolute and usually is equipped with 70–80 actual trays.
- Separation of dimethyl terephthalate (DMT) from the monomethyl terephthalate, with the top of the column operating at a pressure of 40 mmHg absolute.
- Separation of monoethylene glycol from higher glycols at column top pressures of 22 kPa absolute pressure.
- Refinery vacuum distillation columns for processing reduced crudes (atmospheric column bottoms) for lubricating oil production or preparation of reduced crudes for cracking processes. Absolute pressures used range from 4.7 to 10.6 kPa (35–80 mmHg).
- Fatty acid separations with column top pressures around 3.3–6.6 kPa (25–50 mmHg) absolute.

• What is molecular distillation?

- Distillation process that is carried out at such low pressures that the distance between the hot and condensing surfaces is less than the mean free path of the molecules.
- Each unit is a single stage, but several units in series are commonly employed.
- Molecular distillation is applied to thermally sensitive high molecular weight materials with molecular weights in the range of 250–1200. Contact times in commercial units may be as low as 0.001 s.
- In the centrifugal type still, the material that is charged to the bottom creeps up the heated, rotating conical surface as a thin film, vaporized, condensed, and discharged. The film thickness is of the order of 0.05–0.1 mm.
- Rotors are up to about 1.5 m diameter and rotate at 400–500 rpm. The rotating action permits handling much more viscous materials than possible in film evaporators.
- Evaporating surfaces are up to 4.5 m^2 per unit.
- Feed rates range from 200 to 700 L/h, and distillates range from 2 to 400 L/h, depending on the service.
- From three to seven stills in series are used for multiple redistillation of some products.
- A typical pumping train for a large still may comprise of a three-stage steam jet ejector, two oil boosters and a diffusion pump.
- Figure 17.35 illustrates a typical centrifugal molecular still.

17.2.8 Divided Wall and Petluk Distillation Columns

• What is a divided wall column (DVC)? What are its merits?

- A vertical wall is introduced in the central part of the column, creating a feed and drawoff section in this part of the column (Figure 17.36).
- The column may contain either trays or packing.
- The feed side of the two compartments acts as the prefractionator and the product side as the main column.
- The dividing wall, which is designed to be gas and liquid sealed, permits the low energy separation of the low and high boiling fractions in the feed section. The medium boiling fraction is concentrated in the drawoff part of the divided wall column. This arrangement saves a second column.

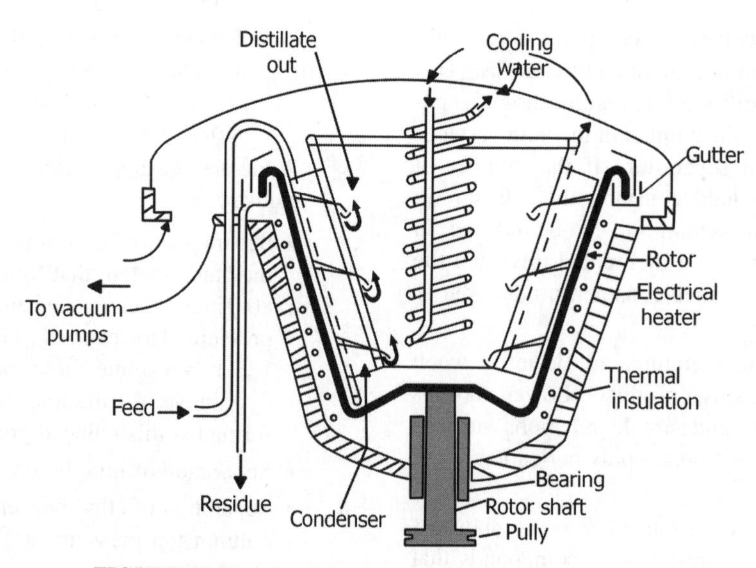

FIGURE 17.35 Centrifugal type molecular distillation still.

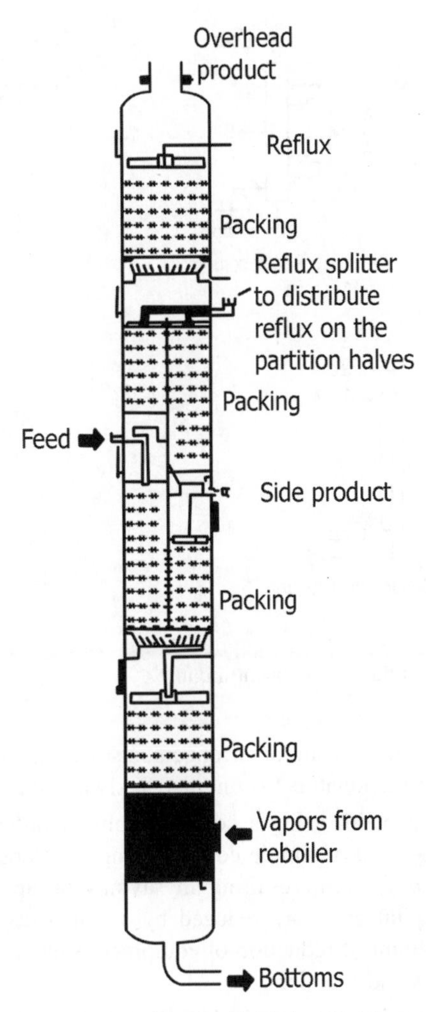

FIGURE 17.36 Details of a divided wall column (Montz).

- The column shell, internals, reboiler and condenser for a second column are not needed.
- Divided wall columns can be used wherever multicomponent mixtures have to be separated into pure fractions. They are particularly suited to obtain pure medium boiling fractions.
- The separation of a three-component mixture into its pure fractions in conventional systems requires a sequential system with at least two main columns with side columns. A single divided wall column can handle this task.
- The medium boiling fraction is concentrated in the drawoff part of the divided wall column.
- Control and maintenance work is significantly reduced.
- Divided wall columns are an alternative to multicolumn systems that can help to save investment and operating costs. Investment costs are cut by 20–30%, operating costs by around 25% compared to a traditional two-column system.
- The capital cost savings result from the reduction in the number of equipment (i.e., one column, reboiler, condenser, etc., instead of two of each).
- A DWC requires less plot area and, therefore, shorter piping and electrical runs, a smaller storm runoff system and other associated benefits.
- The flare loads are reduced because of the smaller heat input leading to a smaller flare system.

• Explain the concepts involved in Petluk and divided wall columns by comparing the separation arrangements involved in these systems with conventional distillation arrangements for three-component mixtures.

- Figure 17.37 illustrate simple and complex arrangements for separation of a three-component mixture into pure components along with Petlyuk and a divided wall column arrangements.
- Consider a mixture consisting components A, B and C where A is the lightest and C is the heaviest. Different schemes are illustrated as described below.
- Figure 17.37a illustrates direct column sequence of obtaining pure A, B and C using a two column arrangement. Figure 17.37b illustrates the working of a complex column, Figure 17.37c and d illustrates Petluk column arrangements, while Figure 17.37e shows a divided column arrangement (see next question for further details).
  - ➢ The first column separates the lightest product as the distillate and the rest is fed to the second column, which separates the two heavier components B and C. The column pressures can be optimized separately or individually to perform the separation of B and C.
  - ➢ The direct sequence is promising when the feed contains a high concentration of A and/or the separation between B and C relatively more difficult than that between A and B. This sequence is often employed for the separation of natural gas condensates into C_3, C_4 and C_5+.
  - ➢ For some mixtures in which B is the major component and the split between A and B is as easy as the split between B and C, this arrangement involves energy *inefficiencies*.
  - ➢ In the first column, concentration of B builds to a maximum at a tray near the bottom. On trays below this point, the amount of the heaviest component C continues to increase, diluting B, so that its concentration profile now decreases on each tray below toward the bottom of the column.
  - ➢ Figure 17.38 gives the concentration profiles.
  - ➢ Energy has been used to separate B to a maximum purity, but because it has not been removed at this

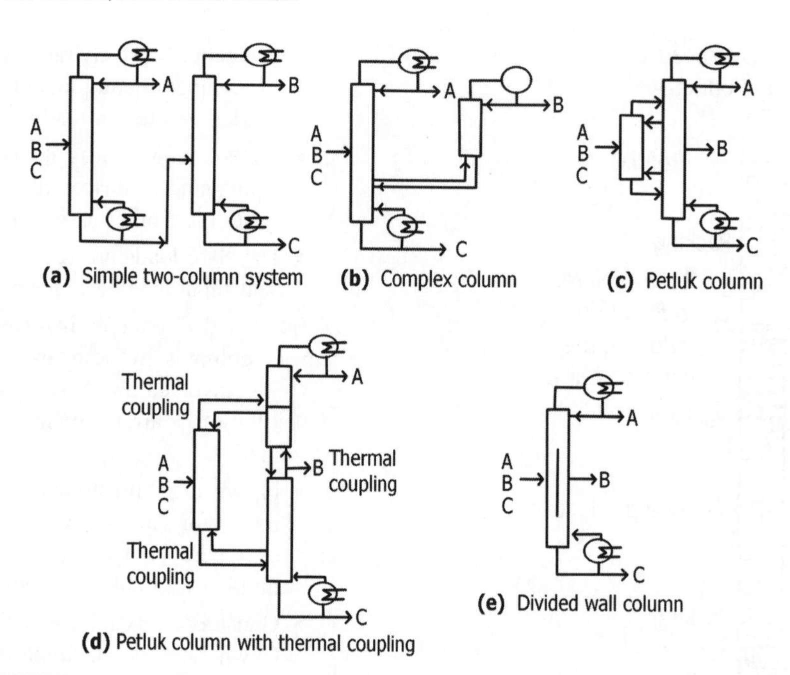

FIGURE 17.37 Different arrangements for separation of a three-component mixture.

point, it gets remixed and diluted to the level concentration at which it is removed with the bottoms. This remixing involves energy inefficiency.

- Employing complex column configurations, as shown in Figure 17.37b, can minimize these mixing losses by withdrawing liquid with maximum concentration of B before it can get remixed with component C, thus reducing energy consumption and decrease capital costs (number of auxiliary equipment like condensers and reboilers is reduced). Such columns promote greater interaction between vapor and liquid streams by introducing thermal coupling between different sections.

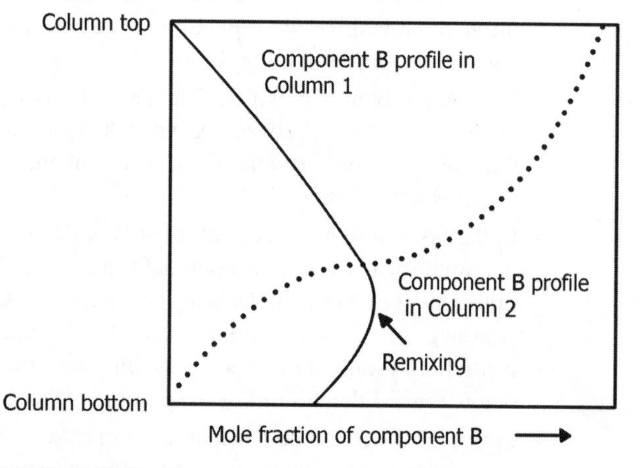

FIGURE 17.38 Remixing in the conventional direct distillation sequence.

- Complex column arrangements can also include prefractionators to minimize feed tray mixing losses.
- In general, using a complex column rather than a sequence of simple columns, improves energy efficiency, often resulting in savings of up to 30%. Capital costs are reduced by as much as 25% on account of reduction of equipment such as condensers and reboilers.
- Complex columns include Petluk and divided wall columns that accrue the above benefits.
- Figure 17.37c and d shows arrangements that eliminates the remixing problem. These prefractionator arrangements, which are known as Petluk columns, perform a sharp split between A and C in the first column, while allowing B to distribute between these two streams. All of A and some B are removed as overheads of the smaller prefractionator, while all of C and the remaining B are removed as bottoms.
- The upper portion of the second column, as shown in Figure17.37c, performs A and B separation, while the lower portion separates B and C. During design phase, fraction of B separated in the overhead of the prefractionator can be set to prevent the remixing. For a typical design, energy saving of 30–60% can be achieved. Petluk arrangement involves using vapor and liquid streams from the main column (second column) to provide vapor and liquid flows in the prefractionator. The only condenser and reboiler used in this arrangement is attached to the main column. Figure 17.37d is an alternative Petluk

arrangement involving thermal coupling, which minimizes separation losses and improves vapor–liquid interaction in all sections of both the columns.

- The Petluk column requires the least amount of stripping vapor or rectifying liquid among all of the options for a three-product system.
- Capital expenses are considerably reduced because Petluk arrangement involves fewer pieces of major equipment.
- The two-way vapor transfer between both columns imposes pressure constraints and poses control and operability challenges. These have been hurdles to the widespread use of Petluk designs.
- Figure 17.37e gives divided wall column (DVC) arrangement, which integrates the prefractionator into the main column, thus using a single column for the three-component separation, further reducing equipment costs. DVC is in essence is a Patluk column arranged in a single shell.
- The DVC introduces a vertical partition (wall) inside the shell to accommodate a prefractionator in the same structure.
- This column splits the stripping vapor (from main reboiler) and the rectifying liquid (from the main condenser) on the two sides of the wall to perform an effective separation. This allows substantial saving in the capital costs and separation efficiency at the same time.
- The only minor difference between the DVC and Petluk column is the heat transfer across the wall.

- What are the claimed benefits of a DVC?
  - Energy savings up to 30%. Capital cost savings up to 30%.
  - Higher purity middle component.
  - Smaller plot area requirements.
  - Reduced maintenance and operating costs due to fewer exchangers and pumps.
  - Less piping and services.
  - Lower flaring requirements.
  - The material is only reboiled once and its residence time in the high-temperature zones is minimized. This can be important if distilling heat-sensitive materials.
  - Standard distillation equipment can be used for the fabrication. Either packing or plates can be used, although packing is more commonly used in practice.
- What are the disadvantages of a divided wall column?
  - A single condenser and reboiler provide the entire reflux and reboiler duty for a DVC. The operating temperatures are the lowest for the condenser and highest for the reboiler. In a conventional system temperatures of the reboiler on the first column and condenser on the second column are at intermediate levels, facilitating the duties at intermediate levels. This may be advantageous for heat integration purposes or in cases of availability of less expensive intermediate duties.
  - The two columns in a conventional system might require significantly different operating pressures for reasons that may include restrictions in the operating temperatures, due to the available duties or restrictions on the bottoms temperature to prevent polymerization or decomposition of bottom product. In such situations a DVC is not suitable.
  - Where conventional sequence requires utilities at very different operating temperatures, DVCs are not favored.
    - If the condenser temperature is selected to fit the available utility, then the reboiler temperature is set by the pressure drop and the composition of the bottom stream. This may mean that a higher level of heating medium or a lower level of condensing medium is required to meet restrictions on the top or bottom temperature.
  - A DVC is likely to be of larger diameter and taller than the conventional columns.
  - Even though the arrangement might require less energy than a conventional arrangement, all of the heat must be supplied at the highest temperature and all of the heat rejected at the lowest temperature of the separation. This can be particularly important if the distillation is at low temperature using refrigeration for condensation. In such circumstances, minimizing the amount of condensation at the lowest temperatures can be very important. If differences in the temperature at which the heat is supplied or rejected are particularly important, then distributed distillation or a prefractionator might be better options as these offer the advantage of being able to supply and reject heat at different temperatures.
  - In general, DVCs are not suited to replace sequences of two simple columns that operate at different pressures.
  - If the two columns of a conventional arrangement require two different materials of construction, one being much more expensive than the other, then any savings in capital cost arising from using a DVC will be diminished. This is because the whole DVC will have to be fabricated from the more expensive material.
  - The hydraulic design of a DVC is such that the pressure must be balanced on either side of the wall.

This is usually achieved by designing with the same number of stages on each side of the dividing wall. This constrains the design.

- Alternatively, a different number of stages can be used on each side of the dividing wall, and different column internals with different pressure drop per stage used to balance the pressure drop.
- Also, if foaming is more likely to occur on one side of the wall than the other, this can lead to a hydraulic imbalance and the vapor split changing from design conditions.

- What are the fabrication aspects with respect to the shell and tray design of a DVC?
  - Adding a vertical partition to a conventional distillation column presents some challenges to fabricating the shell and trays. If the temperature gradient across the wall is too large, it may be necessary to install an insulated wall to prevent heat transfer that could affect the separation.
  - Additional manways may be needed in a DWC, so that the column is accessible on either side of the wall.
  - The wall within the DWC effectively changes the tray design, creating two distinct noncircular sections.
  - Also, if the wall is not exactly in the center of the column, the tray design becomes nonsymmetrical. A high performance tray is suited for this application because it has a 90° tray-to-tray orientation that makes designing easy within the noncircular section and asymmetrical sections.
  - The tray is designed for uniform flow distribution across the tray deck, which is important when the section is noncircular.
- Under what circumstances a DVC should be considered in place of a conventional two-column system?
  - When the middle boiling component is in excess in the feed.
  - When the desired purity of the middle boiling component is higher than that achievable by a simple side-draw column.
  - When the relative volatility distributions between the components is uniform.
- Under what circumstances a DVC is not an appropriate choice?
  - When there is a high difference in pressures between the columns in the conventional two-column sequence.
  - When the conventional sequence requires utilities, cooling water and refrigeration, at different temperatures for each column.
- What are the practical constraints in restricting sequencing options in distillation processes?
  - Process constraints often reduce the number of options that can be considered. Examples of constraints of this type are as follows:
    - Safety considerations might dictate that a particularly hazardous component be removed from the sequence as early as possible to minimize the inventory of that material.
    - Reactive and heat-sensitive components must be removed early to avoid problems of product degradation.
    - Corrosion problems often dictate that a particularly corrosive component be removed early to minimize the use of expensive materials of construction.
    - If thermal decomposition in the reboilers contaminates the product, then this dictates that finished products cannot be taken from the bottoms of columns.
    - Some compounds tend to polymerize when distilled unless chemicals are added to inhibit polymerization. These polymerization inhibitors tend to be nonvolatile, ending up in the column bottoms. If this is the case, it normally prevents finished products being taken from column bottoms.
    - There might be components in the feed to a distillation sequence that are difficult to condense. Total condensation of these components might require low temperature condensation using refrigeration and/or high operating pressures. Condensation using both refrigeration and operation at high pressure increases operating costs significantly.
    - Under these circumstances, the light components are normally removed from the top of the first column to minimize the use of refrigeration and high pressures in the sequence as a whole.
- Name and describe a typical process intensification equipment for distillation applications?
  - High gravity rotating contactor, known as Higee unit, involves compact design and can induce centrifugal forces of more than 1000 times that of gravity.
  - Higher gravity through centrifugal acceleration gives enhancements in mass transfer and throughput rates by one to two orders of magnitude and, consequently, drastically reduces column size for the same production objective.
  - A torus-shaped rotor spins at approximately 500 rpm. Vapor flows from the outside to the inside

of the rotor while liquid flows from inside to outside the rotor.

- This is also known as the rotating packed bed. Typically Higee reduces HETP to 5–12 cm from the 150 to 600 cm in a standard column.
- Packing can be PTFE coated aluminum sponge for corrosion resistance or other materials.
- In one Higee variant, the overhead condenser, reflux drum and the rectifying section are installed in one common vessel while the reboiler, column bottoms sump and striping section are combined in a second common vessel. Thus, a complete distillation column looks like two horizontal vessels with dished or hemispherical heads. This is claimed to provide a dramatic reduction in installed costs.
- These process intensification units are currently used in laboratory and research applications. These are often used to strip organics from wastewater. Future installations are expected to include various types of distillation applications and retrofits to existing columns.

• Summarize heuristics for the selection of the sequence for simple nonintegrated distillation columns.
  - Separations, where the relative volatility of the key components is close to unity or that exhibit azeotropic behavior, should be performed in the absence of nonkey components. In other words, the most difficult separation should be performed last.
  - Sequences that remove the lightest components alone one by one in column overheads should be favored. In other words, the direct sequence should be favored.
  - A component with the highest concentration in the feed should be removed first.
  - Splits in which the molar flow between top and bottom products in individual columns is as near equal as possible, should be favored.

17.2.9 Operational Issues

• What are the approaches for the optimum design of a distillation?
  - Optimum design procedure must be established as a first step for standard design and then the design must be tailored to optimize column operation for flexibility, low capital cost, energy efficiency, or handling difficult feeds.
  - For a standard design, column pressure must be established to obtain economical design of condensing heat exchanger. The cooling water or cooling air approach temperature usually sets the condensing temperature. If the pressure is set too low, a bulky and expensive exchanger will be required. If the pressure is higher, smaller condenser will be required but tray requirements will be more on account of lower relative volatilities between light and heavy keys at high pressures.
  - In other words, separation of lighter components requires higher pressures and separation of heavier components is accomplished at lower pressures.
  - The minimum number of trays and minimum reflux required can be estimated by Fenske and Underwood equations, respectively. Gilliland correlation can be used to determine number of trays versus reflux rates. To find the initial N_m, reflux rate may be kept 10% above the minimum reflux rate. Then several rigorous column simulations may be run to establish the appropriate number of trays. The minimum number of stages obtained by Underwood correlation may be used for the first run that may be increased by 5% for the second run.
  - Tray efficiencies may be determined by experimentation, use of different correlations or based on experience with existing columns or using recommendations from experienced engineers. Different sections of the column may operate at different efficiencies. Usually atmospheric columns have efficiencies in the range of 75–80% and vacuum columns 15–20%.
  - The optimum feed point location should be based on the ratio of light key to heavy key in the *liquid phase* of the feed.
  - To achieve operational flexibility, additional trays or structural packing sections may be installed. Provisions for increase in reflux may also be considered. Columns that are designed for feed variations are not optimized for any particular feed.
  - If product specifications are variable, the column must be designed for achieving most difficult separations. This must be justified as costs will be very high if large number of trays is to be used to achieve separations for high-purity products.
  - Reflux versus number of trays involves comparison of equipment costs and energy costs. High reflux ratios involve less number of tray requirements but increased column diameters, increased sizes of condensers and reboilers. This aspect must also be addressed to determine equipment costs.
  - Usually, condenser and reboiler are the most expensive equipment, followed by column shell and trays. Use of high-efficiency trays or packing and decreased tray spacing permit savings in column costs.
  - Other options include use of divided wall columns, heat pump systems, and pressure-staged columns.

- What are the recommended practices involved in the mechanical design of distillation columns?
  - For columns that are at least 0.9 m (3 ft) in diameter, 1.2 m (4 ft) should be added to the top for vapor release and 1.8 m (6 ft) should be added to the bottom to account for the liquid level and reboiler return.
  - Column heights should be limited to 53 m (175 ft) due to wind load and foundation considerations.
  - The length/diameter ratio of a column should be no more than 30 and preferably below 20.
- What are the disincentives involved when a product from a distillation unit is off specifications.
  - Low-grade product may be sold for less, provided a market exists for it.
  - Alternatives:
    - Reprocessing.
    - Using elsewhere in the plant, for example, as fuel.
  - Reprocessing:
    - Increases manufacturing costs by using additional energy.
    - Cuts into production capacity.
    - Can result in substantial composition and control disturbances on the unit.
- What are the ways by which wastes are generated in distillation columns? What are the remedies?
  - By allowing impurities in products that ultimately become wastes.
    - *Remedy*: Better separation. In some cases, normal product specifications (purities) must be exceeded.
  - By producing waste within the column itself. For example, high reboiler temperatures that cause cracking/polymerization.
    - *Remedy*: Maintain lower column temperatures by the use of open steam, vacuum and/or using low ΔP internals.
  - By having inadequate condensation resulting in vented/flared products.
    - *Remedy*: Improved condensation.
- What are the modifications in distillation operations that lead to waste minimization? Give their drawbacks.
  - Increased reflux ratios to increase separation. This is feasible if column capacity is adequate.
    - *Drawback*: Higher R involves higher flows (vapor and liquid): Higher ΔP leading to higher reboiler temperatures. Even if ΔP is kept low, bottoms temperature increases because of increased purity.
  - If the column is operating close to flooding, adding a new section, with increased number of trays/adding a packed section using a high-efficiency packing and using different diameter section increases separation and capacity.
  - Retraying/repacking the column either partly or all of the column with choice of high efficiency, lower ΔP trays/packing decreases reboiler temperatures.
  - Changing feed tray: Use multiple feed trays with appropriate valving. If feed condition is closer to top product (high concentration of light components), use higher level tray as feed tray and vice versa.
  - Insulating the column: Necessary to prevent heat losses.
  - Poor insulation:
    - Higher reboiler temperatures.
    - Column conditions fluctuate with weather.
    - Product specifications vary.
  - Better feed distribution, especially in packed columns to avoid channeling and maldistribution that affects efficiency.
  - Preheating feed:
    - Supplying heat through feed involve lower temperature levels than supplying through reboiler.
    - Reduces reboiler load.
  - If overhead product contains light impurities, high-purity product may be possible from a tray close to top tray. A bleed stream from overhead drum can be recycled back to purge light impurities.
    - Alternative to removal from a tray below top tray is to install a second column to remove small amounts of lights present in overhead product.
    - Light impurities if present in top product increase evaporation losses on storage:
      - Waste generation.
  - Increasing size of vapor line:
    - ΔP is critical in a vacuum/low-pressure column.
    - Large size vapor line reduces ΔP and decreases reboiler temperature.
  - For heat-sensitive liquids modifying reboiler design by replacing thermosiphon type (more contact time) by a falling film type.
  - *Using Spare Reboilers*: Shutdowns due to reboiler fouling can generate wastes, for example, material (holdup) in the column can become off-specification product. Economics of having spare reboiler may be evaluated.
  - *Reducing Reboiler Temperature*: Techniques such as using lower pressure steam, desuperheating steam, using an intermediate heat transfer fluid for reboiler, and so on.

- Lowering column pressure decreases reboiler temperatures.
  - This also lowers overhead temperatures causing condensation problems leading to loss of uncondensed vapors.
  - If an undersized condenser is used, retubing/replacing the condenser or adding a supplementary vent condenser might be considered to minimize losses.
  - Vent can be rerouted back to the process if process pressure is stable.
  - If refrigerated condenser is used, tube temperatures must be kept above 0°C to avoid moisture solidification.
- *Automatic Column Control*: Automated control systems respond to process fluctuations and product changes swiftly and smoothly minimizing waste production.
- Changing over to continuous process:
  - Batch processes involve more frequent start-ups and shutdowns that is a common source of waste and by-product formation.
  - Converting to continuous process mode and in a continuous process reducing shutdowns for maintenance/reducing faulty operations/troubleshooting lead to waste minimization.
  - May require equipment modification/proper selection, piping modification/process changes.
- Upgrading additives (stabilizers/inhibitors):
  - Some distillation processes use stabilizers that reduce formation of tars (from cracking/polymerization processes due to high bottoms temperatures).
  - Additives make the waste more viscous.
  - Viscous wastes trap the components supposed to be going with the product.

- What are the effects of weather on the operation of a distillation column?
  - Most distillation columns are open to the atmosphere. Although many of the columns are insulated, varying weather conditions can affect column operation.
  - Can limit production in summer.
  - Capable of condensing more vapor at night than during day.
  - When light components accumulate, condensation is impeded (with presence of noncondensables and light components, condensation heat transfer coefficients are reduced, reducing condenser capacity: solution is periodic venting of noncondensables).
  - The reboiler must be appropriately sized to ensure that enough vapor can be generated during cold and windy spells and that it can be turned down sufficiently during hot seasons.
  - The same guideline applies to condensers.
  - Other factors to consider include changing operating conditions and throughputs, brought about by changes in upstream conditions and changes in the demand for the products.
  - These factors, including the associated control system, should be considered at the design stages because once a column is built and installed, nothing much can be done to rectify the situation without incurring additional significant costs.

- What are the operating conditions in a distillation column that contribute to fouling?
  - Stagnant zones can contribute to fouling:
    - Mechanisms that can give rise to fouling include vaporization of volatile components, polymerization, condensation, sedimentation, and chemical reactions.
    - Stagnant zones increase the residence time of the process and lead to precipitation and buildup of polymer and coke. These zones can also promote unwanted chemical reactions because of high residence time and little or no movement of material.
    - Downcomers, weirs, and trays should be designed by providing a small slope and drain holes for liquid flow to prevent stagnant pools, especially during shutdown.
    - Sharp transitions and corners should be avoided in column internals for fouling service. Transitions and corners are the areas that seed growth of polymers. Packed column internals such as feed pipes and trough distributors can be areas for polymer and solids buildup.
    - If the fouling potential is in the vapor phase, the overhead vapor may be drawn from the side of the column instead of the column overhead to eliminate transition line length and an additional corner.
    - In hydrocarbon service where water is present, emulsion formation can create scope for fouling. If the pH of the water differs from the neutral ranges, emulsions with hydrocarbons can form.
    - Trace contaminants such as oxygen, nitrogen compounds, sulfur compounds, mercury, and chlorides have a significant effect on the performance of chemical process fouling and these need to be examined, if present. Contaminants with surfactant properties should be reviewed carefully.

  - ➢ Liquid distribution and packing residence time are important factors in selection of packing. Grid packing can handle fouling service, but have low efficiencies when compared to sieve trays, random and structured packings. Low ΔP, smooth surface and low residence time packings perform best in fouling service.
  - ➢ In fouling service, distributors are the ones where residence time is likely to be increased and fouling phenomena can occur. In high fouling services trough v-notch or other type of trough distributors are recommended over pan type distributors. Feed piping should also be reviewed.
  - ➢ Trays should be preferred over packing. The best trays to use in fouling services are sieve trays and dual flow trays. Moveable valve trays are less resistant to fouling because the valves are areas where fouling can seed and propagate.
  - ➢ A major problem is with downcomers as dead spots are likely in areas near the column wall, opposite to the downcomer outlet, near outlet weirs, and at the ends of the downcomer. Using modified downcomers that decrease the downcomer volume and using sloping downcomers helps to reduce the dead spots by keeping more of the downcomer volume fully agitated. For quick revamps of existing equipment an alternative is to install metal or ceramic shapes to modify the downcomer volume.
  - ➢ Influence of weir heights should be considered and use of inlet weirs might be avoided, if possible.
  - ➢ Dual flow trays are preferred for heavy fouling services. They have no downcomers, which are important sources of fouling. These trays are designed with enough open area on the tray decks to eliminate stagnation and promote backmixing. A disadvantage to dual flow trays is that their turn down ratios are limited. Ripple trays are examples in which vapor and liquid flows take place over the tray area. The continuous agitation of the liquid on the top side of the trays combined with continuous underside wetting/washing action makes this tray suitable for fouling services.
  - ➢ Selection of materials of construction is an important area to be considered in fouling service. Corrosion resistance is of prime concern as corrosion products are common foulants. Smooth finish of surfaces reduces fouling but these should not affect wetting rates of packed surfaces.
  - ➢ Use of additives is one way to combat fouling of the column internals. Additives chemically react with the fouling species to modify their fouling potential, or change the physical interaction between the foulant and the equipment surface or modify the deposit so that it can be removed by flowing fluids.

- What are the retrofit objectives in refinery distillations and how they are carried out?
  - ▪ Retrofit schemes in refineries mostly aim at reducing energy consumption and increasing capacities. Usually refineries aim to achieve these objectives by reusing the existing equipment efficiently rather than installing new columns and heat exchangers, which require substantial investments.
  - ▪ Some of the modifications involve changes to improve efficiencies including the installation of new internals with high efficiencies, for example, replacing trays with structured packing or replacing existing trays with more efficient ones, use of intermediate reboilers, installing pump-arounds at suitable locations in the column and adjusting cooling duties for the pump-arounds and installing preflash units or prefractionators before the crude distillation units.
  - ▪ Other retrofit schemes involve installing additional trays and reboilers to stripping columns, increasing column capacity by adding more pump-arounds, reducing operating pressures, and increasing the preflash overhead vapors.
  - ▪ Increasing energy efficiencies largely depends on heat exchanger network design. For example, the duty and ΔT of each pump-around affects the quantity of heat recovery and the heat exchanger areas and connections among the individual exchangers determine the actual amount of heat recovery achieved.
  - ▪ Use of pinch analysis has been a tool to identify modifications to the column and heat exchanger network. Different approaches are used in linking up the analyses made for the column and heat exchanger networks.
  - ▪ Overall, two levels of modifications are proposed, namely, relatively less expensive modifications and those requiring large investments.
    - ➢ Examples of inexpensive options include piping changes to avoid mixing of unlike streams together and adjusting stripping steam flow rates.
    - ➢ Capital intensive options include replacing internals and relocating feeds and side streams.
  - ▪ Finally, it is useful and effective to adopt a systems approach to retrofit issues, covering the column(s), heat exchanger network, pump-arounds and piping, controls and other components to achieve optimum

results toward the primary goal of achieving retrofit objectives.

- What is *component trapping* in distillation columns and what are its effects on the operation of the columns?
  - Feeds to a distillation column may contain components with boiling points between the key components. In some cases, column top temperature is too cold for these components to vaporize and leave the column along with top vapors and the bottom temperature is too hot to allow these components condense and leave the column along with bottom liquid.
  - Such components accumulate in the column, causing recycling, slugging and flooding.
  - If such intermediate component(s) is water or acidic, it may also cause accelerated corrosion and in refrigerated columns may produce hydrates.
  - A large difference between the top and bottom temperature, a large number of components and high tendencies to form azeotropes or two liquid phases are conducive for intermediate component accumulation.
  - The accumulation continues until the intermediate component concentrations in the overhead and bottoms allow removal of these components at the rate they enter or a flooding limitation is reached.
  - A typical symptom of unsteady-state accumulation is cyclic slugging, which tends to be self-correcting. The intermediate component builds in the column over a period of time, typically hours or days until the column floods or a slug rich in the accumulated component leaves either from the top or bottom.
  - Once a slug leaves, column operation returns to normal over relatively short period of time. The cycle then repeats itself.
  - Intermediate component accumulation may interfere with the control system. For example, a component trapped in the upper part of the column may warm up the control tray, with the controller increasing reflux, pushing the component down. As the component continues to accumulate, control tray warms up again increasing reflux. This phenomena continues until the column floods.
  - This phenomena can be compared with human stomach upsets leading to vomiting and/or purging!
  - There have been many examples of such problems in refinery light ends distillation units. Many times water accumulation occurred due to inadequate water removal facilities. In other instances, specific solutions for such problems have been used, including addition or bypassing of feed preheat systems, using internal water removal trays, foam control, and the like depending on the specific problems involved.

  An excellent review, with details of case studies, can be found in the article by Kister HZ. Component trapping in distillation towers: causes, symptoms and cures. *Chemical Engineering Progress*, August 2004.

17.2.10 Reboilers and Feed Heaters

- Give a thumb rule for rough estimate of reboiler duty.
  - A rough estimate of reboiler duty as a function of column diameter is given by

$$Q = 0.5D^2 \text{ for pressure distillation.} \quad (17.142)$$

$$Q = 0.3D^2 \text{ for atmospheric distillation.} \quad (17.143)$$

$$Q = 0.15D^2 \text{ for vacuum distillation.} \quad (17.144)$$

  where Q is in million Btu/h and D is the column diameter in ft.

- Briefly explain the impact of relative volatility on the operation of a thermosiphon reboiler.
  - High α leads to large temperature gradients that result in large variation in reboiler feed temperature due to compositional differences in the reboiler feed arising from different configurations.
  - High α mixtures: Tendencies for foaming of the liquids that in turn creates control operability, product quality, and stability problems.
- Illustrate how feed preheater duty can be optimized?
  - Feed preheater duty versus bottoms composition (with respect to more volatile component) is shown in Figure 17.39.
  - As feed preheater duty decreases, dumping is promoted.

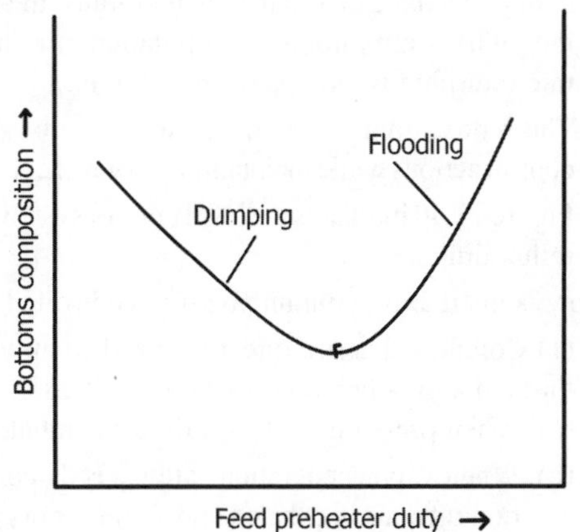

FIGURE 17.39 Feed preheater duty versus bottoms composition.

- As feed preheater duty increases, flooding is promoted.
- From the above figure, it can be noted that the optimum preheater duty is obtained at the change over from dumping to flooding condition.
- Increased feed preheat reduces requirement for reboiler heat duty:
  - ➢ Bottoms composition increases as vaporization of the more volatile component in the bottom product decreases.
  - ➢ Dumping increases because of decreased vapor rate in the stripping section.
  - ➢ Entrainment decreases resulting in increased efficiency.

17.2.11 Distillation Control

- What are the variables that may be considered for the control of a two-product distillation unit?
  - Flows of top and bottom products and reflux.
  - Heat input and heat removal.
- What are the controlled variables for the above?
  - Top and bottom product compositions.
  - Column pressure.
  - Reflux drum level.
  - Base liquid level.
- What is the effect of increasing reflux ratio, with the reboiler on automatic temperature control, on the column operation?
  - Tray and weir loadings and vapor velocity are increased, resulting increased tray ΔP and height of the liquid in the downcomer.
  - Net effect is to push the tray operation closer to incipient flood.
- How is reflux controlled in a distillation column?
  - The control strategy for a distillation column is to maintain a fixed distillate flow and adjust the level of the reflux drum through manipulation of reflux flow rate returning to the top of the column.
  - The reflux drum needs to be tuned for conservative control actions while maintaining the level constraints.
  - Figure 17.40 indicates the level control system for the reflux drum.
- How is distillation column pressure controlled?
  - (1) Columns that are directly linked to atmosphere through a breather vent over the condenser:
    - (a) When pressure tends to fall, air is inhaled.
    - (b) When vapor generation rate exceeds condensation rate, air is exhaled and some vapors are lost to the atmosphere.

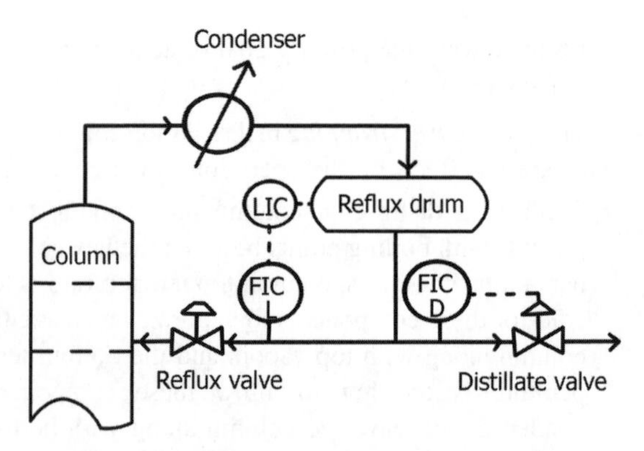

FIGURE 17.40 Level control for reflux drum.

This may not be permissible due to factors such as
- (i) Losses are not tolerable from production point of view.
- (ii) Vapor venting might be objectionable due to environmental considerations.
- (iii) Vapor venting or air inhalation might lead to hazardous conditions such as fire and explosion or toxic releases.

- (2) Column pressure is more often controlled by manipulating rate of condensation either bypassing or flooding the condenser, as illustrated in Figure 17.41.

17.3 LIQUID–LIQUID EXTRACTION

- What are the similarities and differences between extraction and distillation?
  - Both are mass transfer operations with concentration differences as driving forces and mass transfer rates are being influenced by factors such as mixing and separation, turbulence, diffusivities, interfacial areas between phases and properties like viscosity, molecular interactions, and the like.
  - Extraction is a process whereby a mixture of several substances in the liquid phase is at least partially separated upon addition of a liquid solvent in which the original substances have different solubilities.
  - When some of the original substances are solids, the process is called leaching. In a sense, the role of solvent in extraction is analogous to the role of enthalpy in distillation.
  - Differences in solubility, and hence of separability by extraction, are associated with differences in chemical structure, whereas differences in vapor pressure are the basis of separation by distillation.
  - Extraction often is effective at near-ambient temperatures, a valuable feature in the separation of thermally unstable natural mixtures or food and pharmaceutical substances.

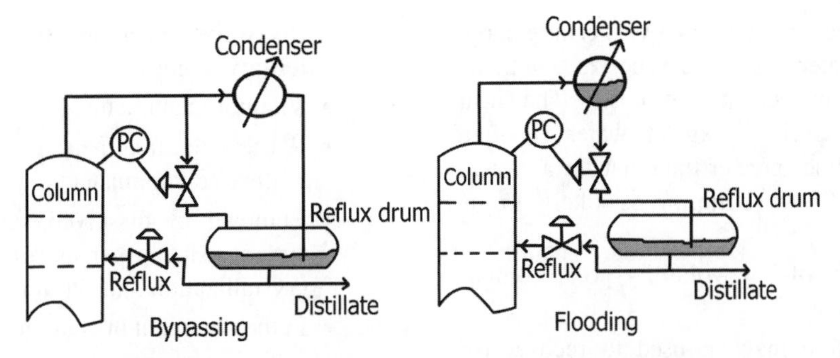

FIGURE 17.41 Bypassing and flooding the condenser illustrated.

- Table 17.11 summarizes the comparison between extraction and distillation.

- What are the situations where liquid–liquid extraction becomes attractive over distillation as a separation process?
  - Low relative volatility:
    - ➢ Capital and operating costs of a distillation process are highly dependent on the relative volatility of the mixture. A low volatility leads to large tray and reflux requirements. A large number of trays result in tall, expensive columns. High reflux ratios involve large heat exchangers and high energy costs. In general, liquid extraction may be considered when relative volatility is less than 1.2.
  - In certain cases, distillation option may involve addition of a solvent (extractive distillation) or an entrainer (azeotropic distillation) to enhance the relative volatility. Even in these cases, a liquid–liquid extraction process may offer advantages in terms of higher selectivity or lower solvent usage and lower energy consumption, depending upon the application.
  - Extraction may be preferred when the distillation option requires operation at pressures less than about 70 mbar (about 50 mmHg) and an unusually large diameter distillation column is required, or when most of the feed must be taken overhead to isolate a desired bottoms product.
  - Extraction may also be attractive when distillation requires use of high-pressure steam for the reboiler or refrigeration for overheads condensation, or when the desired product is temperature sensitive and extraction can provide a gentler separation process.
  - Liquid–liquid extraction may also be a useful option when the components of interest simply cannot be separated by using distillation methods. An example is the use of liquid–liquid extraction employing a steam strippable solvent to remove nonstrippable, low volatility contaminants from wastewater.
  - Removal/recovery of low volatility components from water:
    - ➢ Relative to distillation and stripping, liquid extraction can have real advantages for recovering heavy, essentially nonvolatile components from water.

TABLE 17.11 Comparison Between Extraction and Distillation

Extraction	Distillation
Extraction is an operation in which constituents of the liquid mixture are separated by using an immiscible liquid solvent	Constituents of the liquid mixture are separated by using thermal energy
Extraction utilizes the differences in solubilities of the components to effect separation	Relative volatility is used as a measure of degree of separation
A new immiscible liquid phase is created by addition of solvent to the original mixture	A new phase is created by addition of heat
Phases are hard to mix and harder to separate	Mixing and separation of phases is relatively easy and fast
Extraction does not give pure product and requires further processing	Gives nearly pure products
Offers more flexibility in choice of operating conditions	Less flexibility in choice of operating conditions
Requires mechanical energy for mixing and separation	Requires thermal energy
Does not require heating and cooling provisions	Requires provisions for heating and cooling
Often a secondary choice for separation of components of a liquid mixture	Usually the primary choice for separation of components of liquid mixture

  - Distillation becomes unattractive because large amounts of water has to be vaporized, with its high energy and equipment costs. The heat required to vaporize a kg of water is often three to four times greater than that for a typical hydrocarbon. Steam stripping is likely to be unattractive.
- Removal/recovery of low volatility polar components from organics:
  - Liquid extraction may be used to recover (or remove) salts and acids from an immiscible, organic liquid. This process is often called *washing*. Water with or without a complexing agent is used to recover the salts and acids. Caustic soda is often used as the solvent.
- Recovery of heat-sensitive components:
  - Distillation can cause problems with heat-sensitive materials. An example is recovering antibiotics from fermentation products. Extraction is a much gentler means of recovery.

• Name some applications of liquid–liquid extraction in petroleum and other industries.
- Extraction of lubricating oils for removal of aromatics.
- Separation of paraffins and aromatics. Recovery of BTX (benzene–toluene–xylenes) by using sulfolane, ethylene glycol, and so on, as solvents.
- Separation of *p*-xylene from mixed xylenes using HF, boron trifluoride, or other solvents. Distillation requires very large number of trays to achieve separation because of close boiling nature of these mixtures.
- Removal CO_2, H_2S, and other acidic contaminants from liquefied petroleum gases (LPGs) generated during operation of fluid catalytic crackers and cokers in petroleum refineries and from LNG. The acid gases are extracted from the liquefied hydrocarbons (primarily C_1 to C_3) by reversible reaction with amine solvents.
- Production of glacial acetic acid by separation from aqueous mixtures using ethyl or isopropyl acetate or methyl isobutyl ketone (MIBK) as solvents.
- Separation of fatty acids from oil using propane as solvent.
- Purification of caprolactum.
- Extraction of pharmaceuticals.
- Purification of penicillin.
- Extraction of tranquilizers.
- Recovery of antibiotics and other complex organics from fermentation broth by using a variety of oxygenated organic solvents such as acetates and ketones.
- Recovery of citric acid from fermentation broth with tertiary amines.
- Vitamin manufacture.
- Wastewater treatment involving removal of high boiling contaminants.
- Removal of dissolved salts from crude organic stream using water as solvent as an alternative to crystallization and filtration.
- In the treatment of water to remove trace amounts of organics, when the concentration of impurities in the feed is greater than about 20–50 ppm, liquid–liquid extraction may be more economical than adsorption of the impurities by using carbon beds, because the latter may require frequent and costly replacement of the adsorbent. At lower concentrations of impurities, adsorption may be the more economical option because the usable lifetime of the carbon bed is longer.
- Recovery of phenolic compounds from water by using methyl isobutyl ketone.

• Name some applications of extraction in nuclear and other metal industries.
- Extraction of uranium and plutonium.
- Extraction of zirconium, hafnium and niobium, using MIBK as solvent.
- Extraction of rare earths.
- Hydrometallurgy of copper.
- Extraction of cobalt, nickel, beryllium, and so on.

• What are the principal objectives of bringing two immiscible liquids in direct contact with each other?
- Separation of components in solution: Liquid–liquid extraction.
- Chemical reaction.
- Cooling/heating of a liquid by direct contact with another.
- Creating permanent emulsions.

• What are the advantages of liquid–liquid extraction?
- Large capacities are possible for applications in petroleum industry. Capacities of the order of 100,000 m^3/h or higher are possible.
- Separation of heat-sensitive materials at ambient or moderate temperatures.

• What are the factors affecting mass transfer rates in extraction?
- Phase composition: Governing diffusivity and interfacial turbulence.
- Temperature: By affecting diffusion.
- Degree and type of agitation: Governing film thickness and interfacial turbulence.
- Direction of mass transfer: Depending on the phase dispersed.

- Physical properties of the system: ρ, μ, interfacial tension, and so on.
  - Interfacial tension plays a predominant role in formation and dispersion of droplets. It is a measure of the energy or work required to increase the surface area of the liquid–liquid interface affecting the size of dispersed drops.
  - Systems having low mutual solubility exhibit high interfacial tension. Such a system tends to form relatively large dispersed drops and low interfacial area to minimize contact between the phases. Systems that have higher mutual solubility exhibit lower interfacial tension and more easily form small dispersed droplets.
- Interfacial area depends on
  - Phase composition: By affecting phase densities and interfacial tension, and so on.
  - Temperature: By affecting interfacial tension.
  - Degree and type of agitation: By creating a more intimate dispersion of the two phases.
  - Phase ratio.
  - Physical properties of the system: ρ, μ, interfacial tension, and so on.
- Concentration driving force, ΔC, depends on
  - bulk concentrations of the solute in both phases;
  - distribution coefficient: governing C_{Ai} and C_{Bi};
  - temperature: affecting distribution coefficient.

- "In a liquid–liquid extraction process, as droplet diameter is decreased, interfacial area increases, increasing mass transfer rates." Is there any disadvantage of using very fine droplet sizes (<0.1 mm)?
  - Leads to formation of stable emulsions that are difficult to separate by settling process.
  - A cloud/fog of very fine size droplets remains for long in the continuous phase.
- What is *extraction factor*?
  - Extraction factor, $\in$, is a process variable that characterizes the capacity of the extract phase to carry solute relative to the feed phase.
    - Its value largely determines the number of theoretical stages required to transfer solute from the feed to the extract.
  - The extraction factor is analogous to the stripping factor in distillation and is the ratio of the slope of the equilibrium line to the slope of the operating line in a McCabe–Thiele type of stage-wise graphical calculation.
    - For a standard extraction process with straight equilibrium and operating lines, $\in$ is constant and equal to the partition ratio for the solute of interest times the ratio of the solvent flow rate to the feed flow rate.
- Define distribution or partition coefficient.

$$K = y/x = (\text{weight fraction of solute in extract phase, y})/(\text{weight fraction of solute in raffinate phase, x}), \text{ at equilibrium.} \quad (17.145)$$

- What is selectivity?
  - Selectivity, β, of a solvent is the ratio of the two components in the extract–solvent phase divided by the ratio of the same components in the feed–solvent phase:

$$\beta = (x_A/x_B)_E/(x_A/x_B)_R. \quad (17.146)$$

- What is fractional extraction?
  - Fractional extraction refers to a process in which two or more solutes present in the feed are sharply separated from each other, one fraction leaving the extractor in the extract and the other in the raffinate.
- What are the undesirable effects of using a nonvolatile solvent for liquid–liquid extraction?
  - If the solvent is nonvolatile, it can cause accumulation of heavy impurities in an extraction loop that are surface active.
  - Even in trace concentrations, these impurities can have a devastating effect on extractor performance. They can reduce the coalescing rates of drops, and thus reduce column capacity.

17.3.1 Liquid–Liquid Equilibria

- "Liquid–liquid equilibrium data are represented on triangular diagrams." What are the characteristics of triangular diagrams?
  - Each of the three corners of the triangular diagram represents pure components A, B, and C.
  - The triangular diagrams can be equilateral (Figure 17.42) or right angled. Representation of equilibria on rectangular coordinates is also common.
  - Point M inside the triangle represents a mixture of A, B, and C. The perpendicular distance from M to the base BC represents the mass fraction of component A in the mixture, the distance from M to the base BC = mass fraction of C and the distance from M to the base AC = mass fraction of B.
- How are equilibria in multicomponent systems obtained?
  - Equilibria are representable in terms of activity coefficient correlations such as the UNIQUAC or NRTL.

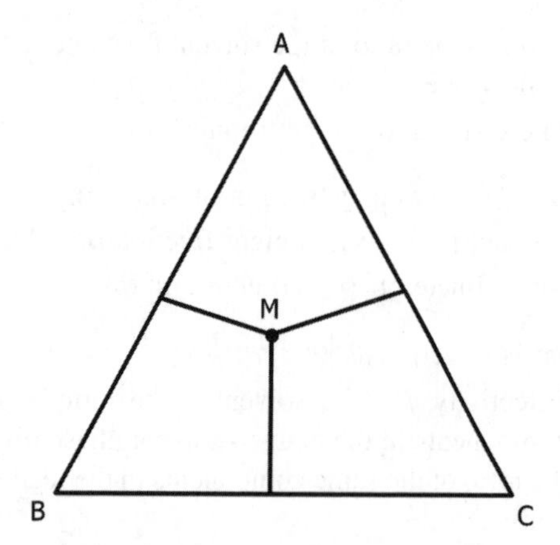

FIGURE 17.42 Equilateral triangular diagram.

- In theory, these correlations involve only parameters that are derivable from measurements on binary mixtures, but in practice the resulting accuracy may be poor and some multicomponent equilibrium measurements also should be used to find the parameters.

- What is a binodal curve?
  - Binodal curve is the ternary diagram commonly used to represent liquid–liquid equilibria involving three components, namely, solute, solvent, and rafinate component.
  - This curve only *looks* like a single continuous curve, but it consists of two branches, each of which represents a phase. They meet in what is called the *plait point*.
  - Once this continuous curve is plotted, one can no longer see which composition of the solvent phase is in equilibrium with which composition of the rafinate phase.
  - To overcome this minor difficulty, few *tie lines* are plotted, which connect equilibrium compositions of both phases.
  - These tie lines give an impression of how the components are distributed over the phases in equilibrium.
  - Ternary equilibrium diagrams can be represented on right-angled triangular plots or equilateral triangular plots or on rectangular coordinate plot on solvent-free basis. The rectangular coordinate plots are also known as Janecke diagrams.
  - To sum up, the limits of mutual solubilities are represented by the binodal curve and the compositions of phases in equilibrium by tie lines. The region within the dome is two phase and that outside is single phase.

- What is plait point? Explain.
  - Point on the phase envelope at which A-rich and B-rich (in a mixture of components A and B with solvent S) solubility curves merge.
  - The two phases are identical in composition at Plait Point.
  - Plait Point is not usually at the maximum concentration of B, the leftover component in the raffinate. Illustrated under the above question.

- What is spinodal curve?
  - There are areas within the demixing (separating) zone where a mixture must get over an energy barrier before it can demix. In the absence of external disturbances such a solution will remain mixed, and is called supersaturated or *metastable*. One can calculate the set of compositions that form the boundary between the metastable and the truly unstable compositions. This set of compositions, which form a curve within the demixing zone, is called the *spinodal curve*. It is shown in Figure 17.43.
  - Dotted curve represents the spinodal curve. Plait point is also shown.
  - At the plait point, the binodal and the spinodal curves always coincide.
  - Inside the spinodal curve, mixtures will always spontaneously demix, while mixtures in the zones between the binodal and the spinodal curves can be metastable. It is impossible to measure spinodal compositions.

- What are Type I and Type II liquid–liquid equilibria? Illustrate.
  - Liquid extraction exhibits at least two common forms on the triangular diagram. Figure 17.44 shows both

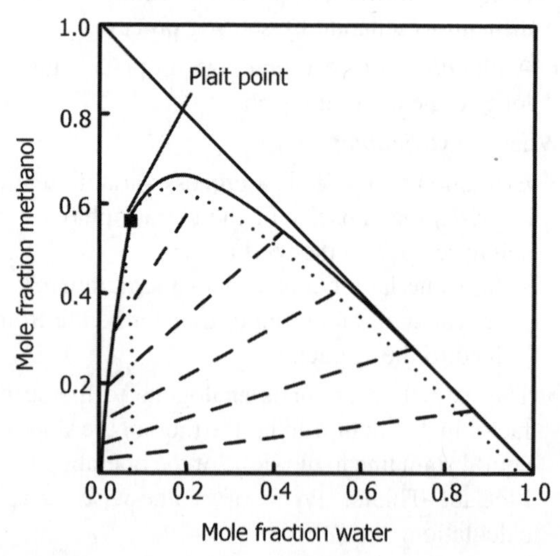

FIGURE 17.43 Binodal and spinodal curves illustrated. Solid curve represents binodal curve. Dotted curve represents the spinal curve. Five tie lines (dashed lines) for the system, water–methanol–benzene (estimated) and plait point are also shown.

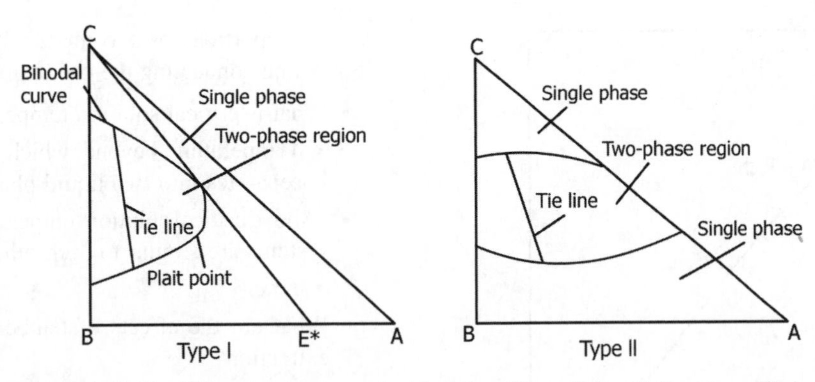

FIGURE 17.44 Triangular diagrams for Type I and Type II liquid–liquid equilibria.

Type I and Type II systems, the former being characterized by the *plait point*.

- In Type I systems, the tie lines linking equilibrium phases shrink to a point at the plait point, so that the two conjugate phases become identical.
  - The right triangular diagram is more convenient since it may be constructed readily to any scale, with enlargement of any particular region of interest.
- In Type II systems, there is no plait point.
- When reflux is used in a Type II system the feed is theoretically separable into pure A and B after solvent removal.
- This contrasts with Type I systems for which, even with the use of reflux, a feed mixture is theoretically separable only into pure B at one end of the unit and a *mixture* of A and B at the other end, after removal of solvent.
- Thus, the most concentrated extract obtainable is E^*

• Give examples for Type I and Type II liquid–liquid equilibria, illustrating by means of triangular diagrams for one system for each type.

- *Type I*:
  - 1-Hexene (A)–tetramethylene sulfone (B)–benzene (C) at 50°C.
  - Acetone (A)–water (B)–methyl isobutyl ketone (C).
  - Acetic acid (A)–water (B)–methyl isobutyl ketone (C).
  - Partial miscibility of solvent (MIBK) and diluent (H_2O).
  - Complete miscibility of solvent and component to be extracted (acetone).
  - Figure 17.45 represents Type I equilibria for 1-hexene (A)–tetramethylene sulfone (B)–benzene (C).
- *Type II*:
  - *n*-Hexane–aniline–methylcyclopentane.
  - Ethylbenzene–styrene–ethylene glycol.
  - Solvent (aniline) is only partially miscible with both the other components.
  - Figure 17.46 represents Type II equilibria for *n*-hexane–aniline–methylcyclopentane.
- Both equilateral and right triangular diagrams have the property that the compositions of mixtures of all

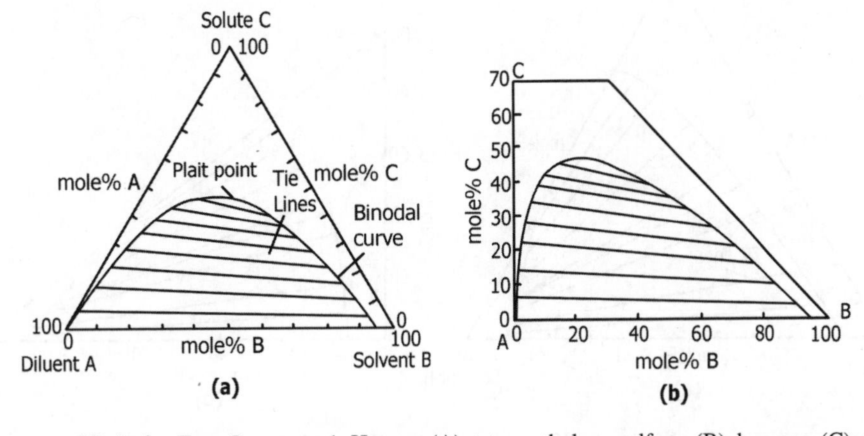

FIGURE 17.45 Ternary equilibria for Type I system. 1–Hexene (A)–tetramethylene sulfone (B)–benzene (C) at 50°C. (a) Equilateral triangular plot, (b) right angular triangular plot, and (c) rectangular coordinate plot (Janecke Diagram, solvent-free coordinates).

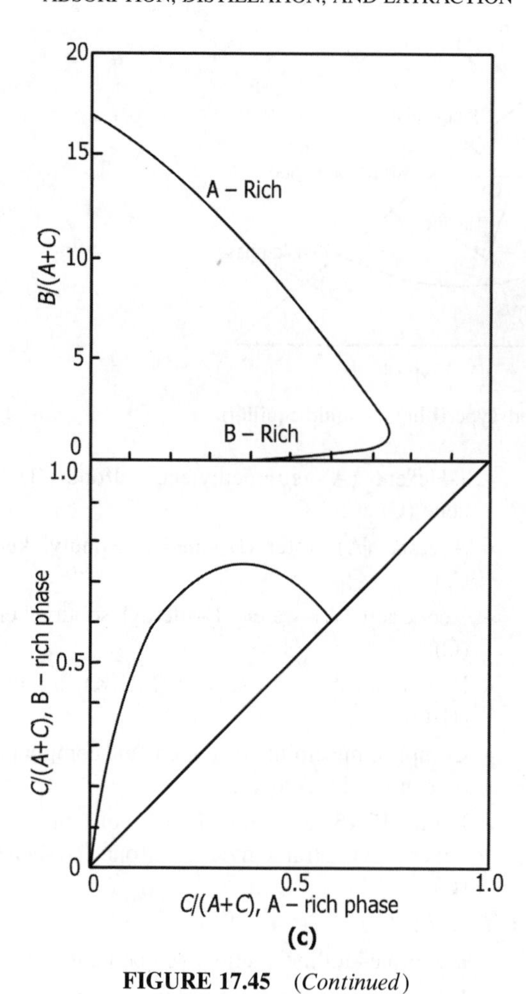

FIGURE 17.45 *(Continued)*

proportions of two mixtures appear on the straight line connecting the original mixtures.

- What is critical solution temperature?
  - Temperature beyond which the system no longer separates into two liquid phases.
- "Above critical solution temperature, Type I and Type II systems are similar to each other." *True/False*?
  - *True*.
- What are the effects of temperature on liquid–liquid extraction?
  - Value of the distribution coefficient is likely to be temperature sensitive.
  - The coefficient decreases with rise in temperature.
  - With rise in temperature, generally, concentration range for heterogeneous region (two-phase region) for the liquid–liquid system is narrowed down.
  - Increase in temperatures will increase rate of transfer.
  - Also, increase in temperature will enhance rate of coalescence/phase separation.
  - Temperature should not be higher than the critical solution temperature.

17.3.2 Extraction Calculations

- Illustrate the procedures for single-stage extraction calculations.
  - The process involves bringing the feed and solvent in intimate contact with each other in a mixer and separating them in a settler. Equilibrium conditions are assumed in the calculations.
  - Producing the mixture from feed F and solvent L:

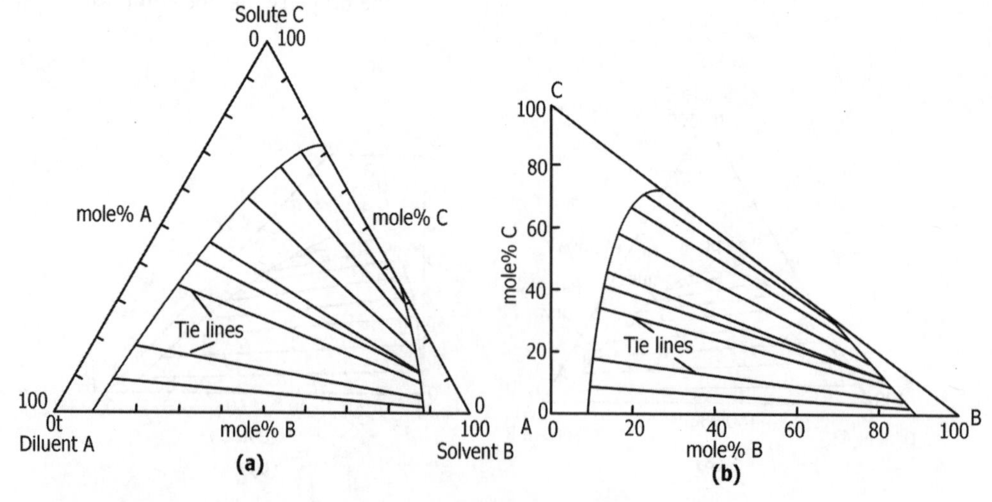

FIGURE 17.46 Ternary equilibria for Type II system. Hexane (A)–aniline (B)–methylcyclopentane (C) at 34.5°C. (a) Equilateral triangular plot, (b) right-angled triangular plot, and (c) rectangular coordinate plot (Janecke and solvent-free coordinates).

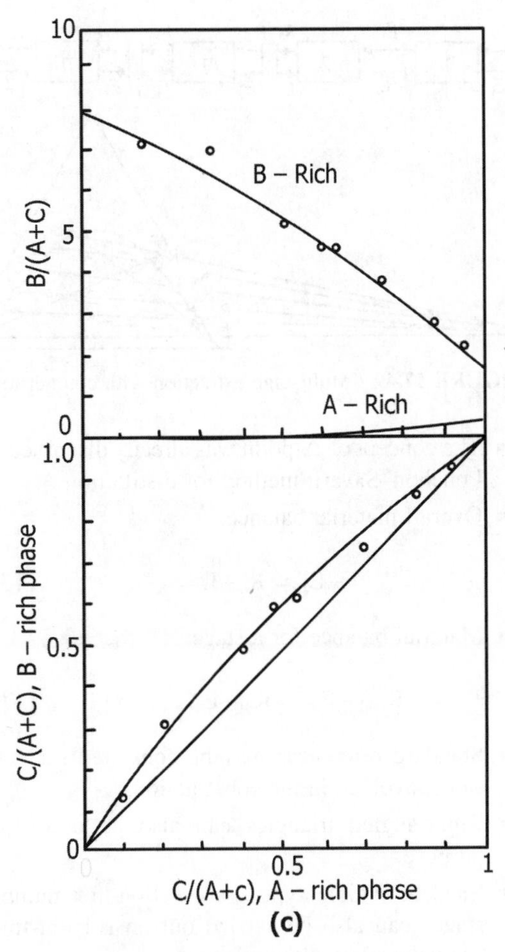

FIGURE 17.46 (*Continued*)

➢ Referring to Figure 17.47, the position of the mixing point M can be determined from the ratio of the distances

$$FM/ML = L/F. \quad (17.147)$$

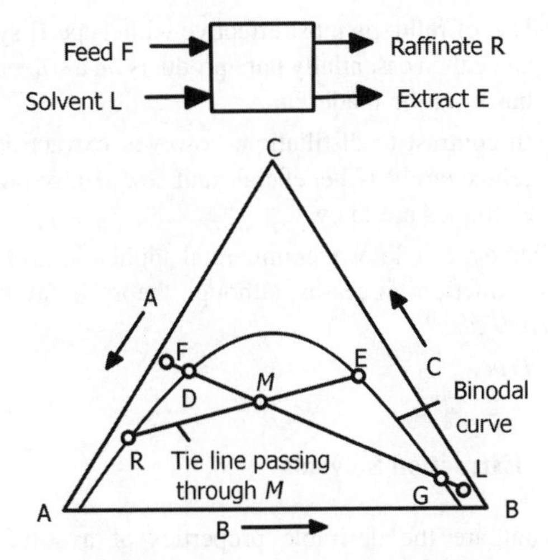

FIGURE 17.47 Single-stage extraction

➢ This can also be done from the material balance equations,

$$F + L = M, \quad (17.148)$$

$$Fx_{AF} + Lx_{AL} = Mx_{AM}, \quad (17.149)$$

and

$$x_{AM} = (Fx_{AF} + Lx_{AL})/(F + L). \quad (17.150)$$

- Separating the extract and raffinate phases:
  - ➢ The compositions of raffinate R and extract E can be determined by the tie line going through the mixing point M, using the relations,

$$RM/ME = E/R \text{ or from } E + R = M = F + L, \quad (17.151)$$

$$Ex_{AE} + Rx_{AR} = Mx_{AM} = Fx_{AF} + Lx_{AL}, \quad (17.152)$$

and

$$E = [(x_{AM} - x_{AR})/(x_{AE} - x_{AR})]M, \quad (17.153)$$

$$R = [(x_{AE} - x_{AM})/(x_{AE} - x_{AR})M. \quad (17.154)$$

- From the triangular diagram, Figure 17.47, it is obvious that a separation of the two phases is only possible if the mixing point M lies in the two-phase region. The crossing points of the line FL with the binodal curve are the extrema for M (minimum and maximum amount of solvent).
- M coinciding with D corresponds to requirement of minimum amount of solvent. If this amount of solvent is mixed with feed F, only raffinate, with composition D, is produced and there will be no extract.

$$M_{min} = F(x_{AF} - x_{AMmin})/(x_{AMmin} - x_{AL}). \quad (17.155)$$

$$FM_{min}/M_{min}L = M_{min}/F. \quad (17.156)$$

- M coinciding with G corresponds to maximum solvent requirement. Only extract and no rafinate is produced at this condition.

$$M_{max} = F(x_{AF} - x_{AMmax})/(x_{AMmax} - x_{AL}). \quad (17.157)$$

$$FM_{max}/M_{max}L = M_{max}/F. \quad (17.158)$$

- Illustrate the procedures for multistage cross-flow extraction calculations.
  - Feed goes through stages in series and to each stage, fresh solvent is added. This process is an extension of the single-stage extraction.
  - For multistage extraction with cross-flow, raffinate from each stage is contacted in the following stage with pure solvent. Extracts are withdrawn from each stage and sent to solvent recovery. The concentration of C in raffinate and extract decreases from stage to stage.
  - Figure 17.48 illustrates the graphical method.
  - If feed point F and solvent L are known, the first mixing point M_1 can be determined in the same way as for the single-stage extraction.
  - This mixing point separates into raffinate R_1 and extract E_1.
  - For the following stages the raffinate is the feed that is contacted with solvent L.
  - The total extract results from combining the extract of each of the stages.
- Outline the procedures for multistage counterflow extraction calculations.
  - Feed and solvent flow countercurrent to each other, entering at the opposite ends of the equipment.
  - While the raffinate is contacted with pure solvent, the extract is contacted with the feed.
  - Figure 17.49 illustrates the method.
  - The triangular diagram is constructed based on overall material balances and material balance for a stage.

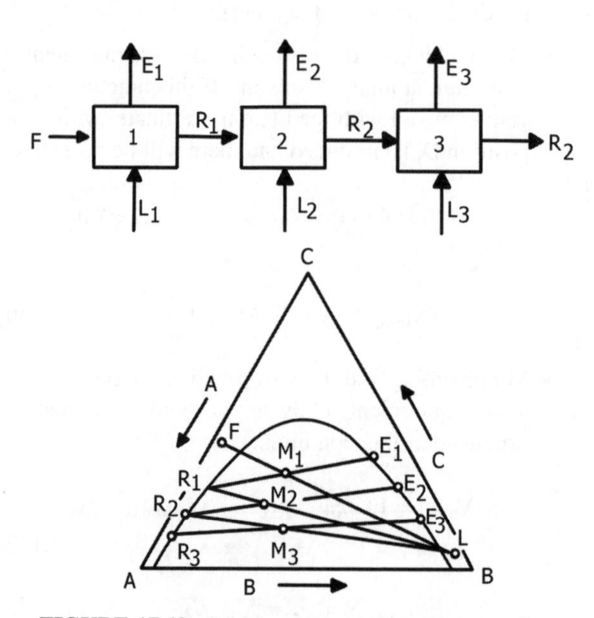

FIGURE 17.48 Multistage extraction with cross-flow.

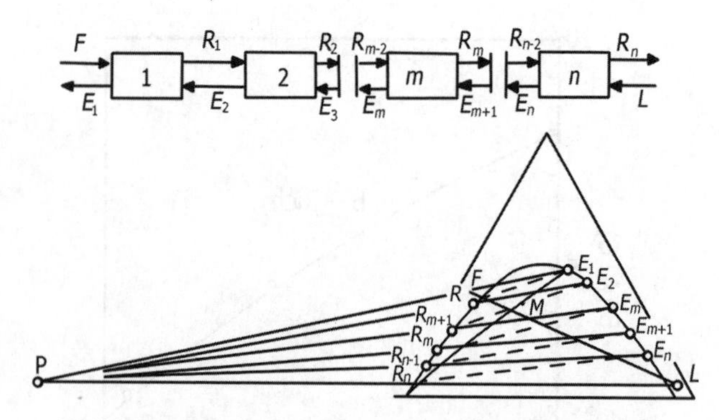

FIGURE 17.49 Multistage extraction with counterflow.

- The concept of Δ-point was already discussed under Ponchon–Savarit method for distillation.
- Overall material balance:

$$F - E_1 = R_n - L = . \quad (17.159)$$

- Material balance for a stage:

$$R_{m-1} - E_m = R_m - E_{m+1} = . \quad (17.160)$$

- Standard references may be consulted for detailed steps involved in the construction.
- Right-angled triangles can also be used for the purpose.
- McCabe–Thiele type construction for number of stages can also be carried out on solvent-free coordinates as an alternative.

- What is the advantage of using reflux in liquid–liquid extraction?
  - With ordinary countercurrent extraction, richest possible extract product is only in equilibrium with feed solution.
  - Use of reflux at extract end gives richer product like in distillation.
  - Use of reflux is most effective with Type II systems since then essentially pure products on a solvent-free basis can be made.
  - In contrast to distillation, however, extraction with reflux rarely is beneficial, and few if any practical examples are known.
- "There are no known commercial applications of reflux to extraction processes, although theory is favorable." *True/False*?
  - *True.*

17.3.3 Extraction Solvents

- What are the desirable properties of a solvent for liquid–liquid extraction?

- *Capacity*: Maximum concentration of solute the extract phase can hold before two liquid phases can no longer coexist or solute precipitates as a separate phase. If loading capacity is low, a high solvent to feed ratio may be needed even if the partition ratio is high.
  - Loading of solute/weight of solvent, that can be achieved in an extraction layer at plait point in a Type I system or at the solubility limit in a Type II system, should be high. This reduces requirement for solvent.
- *Distribution Coefficient*: Should be large, of the order of 10 or more, so that a low ratio of solvent to feed can be used.
  - Since high distribution coefficients generally allow for low solvent use, smaller and less expensive extraction equipment may be used and costs for solvent recovery and recycle are lower. In principle, distribution coefficients <1.0, may involve high solvent to feed ratios, with much higher costs.
  - A useful guide when selecting a solvent is that the solvent should be chemically similar to the solute.
  - A polar liquid like water is generally best suited for ionic and polar compounds. Nonpolar compounds like hexane are better for nonpolar compounds.
- *Selectivity or Separation Factor, β*: Selectivity of a solvent is the ratio of the two components in extraction solvent phase divided by the ratio of the same components in the feed phase.
  - The separation factor measures the tendency of one component to be extracted more readily than another. If the separation needs to separate one component from a feed relative to another, then it is necessary to have a separation factor greater than unity, and preferably as high as possible.
  - At $\beta = 1$, no separation is possible. At plait point, $\beta = 1$.
  - The more the deviation of β from 1, the more will be the ease of separation. Preference to only solute component for separation.
  - β varies with solute concentration. For dilute solutions, β will be the highest.
  - In certain applications, it is important not only to recover a desired solute from the feed, but also to separate it from other solutes present in the feed and thereby achieve a degree of solute purification.
  - When solvent blends are used in a commercial process, often it is because the blend provides higher selectivity and often at the expense of a somewhat lower distribution coefficient.
- *Recoverability*: Solvent must be easily and economically recoverable from both extract as well as raffinate for commercial success. As normally distillation is employed for solvent recovery, relative volatility, α, between solvent and nonsolvent should be sufficiently different from 1.0. Also latent heat of vaporization, λ, should be low to reduce heat loads. Other methods for solvent recovery, recycle and storage, include evaporation if the solute is highly volatile or crystallization if freezing points are high.
  - Solvent properties also should enable low cost methods for purging light and heavy impurities that may accumulate over a period of time, from the overall process.
  - One of the challenges often encountered in using a high boiling solvent involves accumulation of heavy impurities in the solvent phase and difficulty in removing them from the process.
  - Another consideration is the ease with which solvent residues can be reduced to low levels in final extract or raffinate products, particularly for food grade products and pharmaceuticals.
- *Mutual Solubility*: Low solubility of solvent in raffinate or feed nonsolute in extract involves low costs for solute recovery.
  - Low liquid–liquid mutual solubility between feed and solvent phases is desirable because it reduces the separation requirements for removing solvents from the extract and raffinate streams.
  - Low solubility of extraction solvent in the raffinate phase often results in high relative volatility for stripping the residual solvent in a raffinate stripper, allowing low cost desolventizing of the raffinate.
  - Low solubility of feed solvent in the extract phase reduces separation requirements for recovering solvent for recycle and producing a purified product solute.
  - In some cases, if the solubility of feed solvent in the extract is high, more than one distillation operation will be required to separate the extract phase.
  - If mutual solubility is nil (as for aliphatic hydrocarbons dissolved in water), the need for stripping or another treatment method may be avoided as long as efficient liquid–liquid phase separation can be accomplished without entrainment of solvent droplets into the raffinate. However, very low mutual solubility normally is achieved at the expense of a lower distribution coefficient for extracting the desired solute.
  - Mutual solubility also limits the solvent to feed ratios that can be used, since a point can be reached

where the solvent stream is so large it dissolves the entire feed stream, or the solvent stream is so small it is dissolved by the feed, and these can be real limitations for systems with high mutual solubility.

- *Density Difference*: Density difference between the two phases affects countercurrent flow rates that can be achieved in extraction equipment.
  - Affects coalescence rates, large density differences being beneficial for good coalescence rates.
  - At plait point, $\Delta\rho = 0$ and can invert phases at high concentrations.
  - Continuous contact equipment such as spray columns cannot cross phase inversions, that is, when $\Delta\rho$ passes through zero, continuous contact equipment cannot be specified.
  - Mixer–settler units can cross phase inversions and can be specified for such cases.
  - As a general rule, a difference in density between solvent and feed phases of the order of 0.1–0.3 g/ml is desirable. Too low a value results in poor or slow liquid–liquid phase separation and may require use of centrifugal extractor.
  - Too high a difference leads to difficult building of high dispersed droplet population density for good mass transfer, that is, it is difficult to mix the two phases together and maintain high holdup of the dispersed phase within the extractor. This also depends on the viscosity of the continuous phase.
- *Interfacial Tension*:
  - High interfacial tension can promote rapid coalescence. Require high mechanical agitation to produce small droplets (as dispersion is more difficult).
  - Interfacial tension usually decreases as solubility and solute concentration increases.
  - If surface tension < 1 mN/m, stable emulsions are produced.
  - $\sigma > 50$ mN/m, high amount of energy is required for dispersion and increased tendency for coalescence.
  - Preferred values for interfacial tension between the feed phase and the extraction solvent phase generally are in the range of 5–25 dyne/cm (1 dyne/cm is equivalent to 10^{-3} N/m).
  - Systems with lower values easily emulsify. Systems with higher values, droplets tend to coalesce easily, resulting in low interfacial area and poor mass transfer unless mechanical agitation is used.
  - Interfacial tension is zero at plait point.
- *Viscosity*: High viscosity leads to poorer mass transfer. Low viscosity is preferred since higher viscosity generally increases mass transfer resistance and difficulty in liquid–liquid phase separation.
  - Some times an extraction process is operated at an elevated temperature where viscosity is significantly lower for better mass transfer performance, even when this results in a lower distribution coefficient.
  - Low viscosity at ambient temperatures also facilitates transfer of solvent from storage to processing equipment.
- *Vapor Pressure*: High vapor pressure leads to higher solvent losses. High vapor pressure for an organic solvent will lead to the emission of volatile organic compounds (VOCs) from the process, potentially leading to environmental problems.
- *Freezing Point*: Should be low for ease of handling and storage.
  - Solvents that are liquids at all anticipated ambient temperatures are desirable since they avoid the need for freeze protection and/or thawing of frozen solvent prior to use.
  - Sometimes an *antifreeze* compound such as water or an aliphatic hydrocarbon can be added to the solvent, or the solvent is supplied as a mixture of related compounds instead of a single pure component, to suppress the freezing point.
- *Chemical Reactivity*: Chemically stable and inert to other components and to the common materials of construction.
- *Toxicity*: Solvents with low toxicity for inhalation and skin contact, having good warning properties, are desirable. Also toxicity affects aquatic life when discharged to streams and lakes.
  - A thorough review of the medical literature must be conducted to ascertain chronic toxicity issues.
  - Measures needed to avoid unsafe exposures must be incorporated into process designs and implemented in operating procedures.
- *Flammability*: Should be nonflammable.
- *Environmental Considerations*: The solvent must have physical or chemical properties that allow effective control of emissions from vents and other discharge streams.
  - Preferred properties include low aquatic toxicity and low potential for fugitive emissions from leaks or spills.
  - It also is desirable for a solvent to have low photoreactivity (to avoid smog formation) in the atmosphere and be biodegradable so it does not persist in the environment.

- ➢ Efficient technologies for capturing solvent vapors from vents and condensing them for recycle include activated carbon adsorption with steam regeneration and vacuum swing adsorption.
- ➢ Waste minimization and reduction of environmental impact are of necessity to comply with environmental regulations.

- *Multiple Uses*: It is desirable to use as the extraction solvent a material that can serve a number of purposes in the process plant. This avoids the cost of storing and handling multiple solvents.
  - ➢ It may be possible to use a single solvent for a number of different extraction processes used in the same facility, either in different equipment operated at the same time or by using the same equipment in a series of product applications.
  - ➢ In other cases, the solvent used for extraction may be one of the raw materials for a reaction carried out in the same facility, or a solvent used in another operation such as a crystallization.
- *Materials of Construction*: It is desirable for a solvent to allow the use of common, relatively inexpensive materials of construction at moderate temperatures and pressures.
- *Availability and Cost*: The solvent should be readily available at a reasonable cost. Considerations include the initial fill cost, the investment costs associated with maintaining a solvent inventory in the plant (particularly when expensive solvents are used), as well as the cost of makeup solvent.

- What is the potential in the use of ionic solvents in extraction processes? What are their characteristics and advantages?
  - Ionic solvents are finding their way as alternatives to conventional solvents for extraction processes. The reasons for the interest in ionic solvents include their existence as liquids over a wide range of temperatures and their possession of many attractive properties like low melting point, negligible vapor pressure, high heat capacity, high thermal conductivity, good thermal stability, and nonflammability. Hydrocarbon solvents pose risks of fire and because of their low thermal conductivities, invite ignition problems due to generation of static electricity during their handling.
  - Many areas of their use are emerging in petroleum, chemical, pharmaceutical, and other industries such as acid gas scrubbing, desulfurization of fuels, electroplating, reaction solvents, and others.
  - Unlike conventional solvents, the basic structural unit of which is a molecule, ionic liquids are salts consisting of ions in a dissociated form.
  - The base cation and anion affect the chemical and physical properties of a particular ionic liquid. The properties can be tailored by varying the length and branching of the alkyl groups on the ions.
  - Ionic liquids can perform some separations more easily than conventional solvents. Examples include recovery of aromatics from reformate, butadiene removal from a steam cracker C_4 stream, and separation of benzene from cyclohexane.
  - Many ionic liquids have higher selectivities and capacities for the solute than conventional solvents. Aromatics are more soluble in ionic liquids than aliphatics because of the electrostatic field around the aromatic molecule and its polarizability.

17.3.4 Supercritical Solvents

- What is a supercritical fluid?
  - Any fluid whose temperature is greater than its T_c and pressure is greater than its P_c is a supercritical fluid.
  - Figure 17.50 illustrates supercritical region on the phase diagram for a solvent. It is a plot of temperature versus pressure, showing the equilibrium coexistence of two phases along the *phase boundaries* separating different regions.
  - As one can see, the solid–liquid phase boundary is a plot of the freezing point at various pressures, the liquid–vapor boundary is a plot of vapor pressure of the liquid against temperature and the solid–vapor boundary is a plot of the sublimation vapor pressure against temperature.

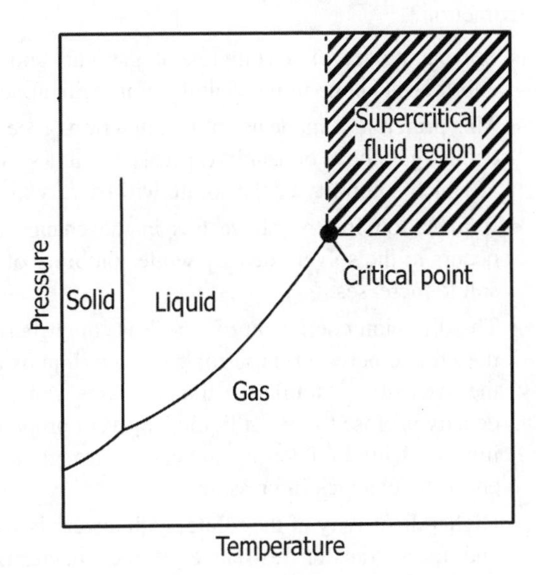

FIGURE 17.50 Pressure–temperature diagram for a pure compound.

- Solid, liquid, and gas phases are in equilibrium at the intersection of these three boundaries, known as the *triple point.*
- *Supercritical region* is shown in the figure. A fluid in this region is neither a liquid nor gas, but possesses some of the properties of each. At this stage, no matter how much pressure is applied, a fluid in this region will not condense and no matter how much the temperature is increased, it will not boil.
- At critical temperature, interface between the fluids disappear, at which condition the densities of liquid and vapor are equal. Above its critical values, liquid–vapor phase boundary of a compound does not exists and its fluid properties can be tailored by adjusting the pressure or temperature.
  - Although supercritical fluid has liquid-like density, it exhibits gas-like diffusivity, surface tension and viscosity. Its gas-like viscosity results in high mass transfer. Its low surface tension and viscosity lead to greater penetration into porous solids.
  - Because of its liquid-like density, the solvent strength of a supercritical fluid is comparable to that of a liquid.
  - At critical conditions, the molecular attraction in a supercritical fluid is counterbalanced by the kinetic energy. In this region, the fluid density and density-dependent properties are very sensitive to pressure and temperature changes.
  - The solvent power of a supercritical fluid is approximately proportional to its density. Thus, solvent power can be modified by varying its temperature and pressure.

- What is the principle involved in supercritical solvent extraction?
  - Solvent power of a compressed gas can undergo enormous change in the vicinity of its critical point.
  - This principle is made use of in extraction processes using a gas above or near its critical point as a solvent at which solubility of the solute will be very high.
  - Explanation for the above lies in the change that occurs to the solvent density while solubility of the solute increases.
  - The diffusion coefficient of a SCF is somewhere in the middle between those for gases and liquids and the viscosity is similar to that of gases while the density is close to that of liquids. Solvent properties improve with fluid density and can be strongly influenced by changes in pressure.
  - Higher diffusivity of the solute, with lower viscosity and mass transfer resistance of the supercritical solvent also contributes to the effectiveness of a supercritical gas as a solvent.
  - Supercritical fluids make ideal solvents because their density is only about 30% that of a normal fluid, a factor sufficient to provide for good solvent capability, but low enough for high diffusivity and rapid mass transfer.
  - They can be good solvents especially for highly fluorinated compounds. For example, fluorocarbons are insoluble in conventional solvents but are very soluble in supercritical carbon dioxide.

- What are the advantages of supercritical fluids as solvents?
  - Considerable solvent strength.
  - Exhibit better transport properties: Low viscosities and high diffusivities for the solute than liquid solvents.
  - Rapid diffusion of CO_2 through condensed phases, for example, polymers.
  - Density and solvent strength could be varied over a wide range by varying *T* and *P*. In other words, solvent strength is adjustable to tailor selectivities and yields.
  - Solvent recovery is fast and complete with no or minimum residue in the product.
    - With increasing scrutiny of solvent residues in pharmaceuticals, medical products and neutraceuticals, the use of supercritical fluids (SFCs) has become very attractive.
  - In addition to handling and disposal issues, organic solvents can pose a number of environmental concerns, such as atmospheric and land toxicity. Some organic solvents are under restriction due to their ozone layer depletion potential.
  - Regulatory and environmental pressures on VOC emissions make increased use of supercritical solvents in place of organic solvents have become attractive.
    - Supercritical fluid-based processes helped to eliminate the use of hexane and methylene chloride as solvents.
    - Supercritical carbon dioxide is an attractive alternative in place of traditional organic solvents. CO_2 is not considered a VOC. Although CO_2 is a greenhouse gas, if it is withdrawn from the environment, used in a process, and then returned to the environment, it does not contribute to the greenhouse effect.

- What are the issues involved in the use of supercritical solvents?
  - SCF technology requires sensitive process control, which is a challenge.
  - In addition, the phase transitions of the mixture of solutes and solvents have to be measured or predicted quite accurately.

TABLE 17.12 Critical Properties of Different Solvents used for Supercritical Extractions

Solvent	T_c (°C)	P_c (bar)
Carbon dioxide	31.1	73.8
Ethane	32.4	48.8
Ethylene	9.3	50.4
Propane	96.8	42.5
Propylene	91.9	46.2
Cyclohexane	287.5	43.4
Benzene	289.2	48.9
Toluene	318.6	41.1
Trichlorofluoromethane	198.1	44.1
Isopropanol	235.8	47.6
Ammonia	132.4	113.5
Water	374.4	221.2

- Generally the phase transitions in the critical region is rather complex and difficult to measure and predict.
- The processes involve use of high pressures and, with some solvents, high operating temperatures, increasing equipment costs.

• Name commonly used supercritical fluids and state their plus points.

- Table 17.12 gives some of the solvents along with their properties that find application in supercritical extractions.
- As Table 17.12 shows, the critical temperatures of gases and liquids can differ by hundreds of degrees. This difference suggests the use of *specific* supercritical fluids for *specific* applications. For example, because the critical temperatures of carbon dioxide, ethane, and ethylene are near ambient, they are attractive solvents for processing heat-sensitive flavors, pharmaceuticals, labile lipids, and reactive monomers. Substances that are less temperature sensitive, such as most industrial chemicals and polymers, are readily treated with C_3 and C_4 hydrocarbons with critical temperatures in the range 100–150°C.
- CO_2 and H_2O: CO_2 is extremely attractive for applications involving thermally sensitive materials or foods/pharmaceuticals. It is a good solvent for many nonpolar and a few polar, low molecular weight compounds.
- C_3 and C_4 hydrocarbons are generally better solvents for polymers than C_2 hydrocarbons.
- Still higher molecular weight hydrocarbons, such as cyclohexane and benzene, with high critical temperatures of 250–300°C, are used to process nonvolatile substances such as coal and high molecular weight petroleum fractions.

• What are the reasons for the higher application levels of CO_2 as a supercritical solvent?

- CO_2 has moderate P_c (which means less compression costs), high critical density, and T_c close to ambient temperature.
- High molecular diffusivity and low viscosity.
- Its solvent strength can be varied by adjusting its density.
- Ease of separation from extract (by lowering pressure). CO_2 leaves a lower amount of residue in products compared to conventional solvents.
- No solvent contamination of the product (advantageous for food and pharmaceutical products).
- Cheap and easily available in relatively pure form and in large quantities.
- Nonflammable.
- Nontoxic at low concentrations.

• Give examples of applications of supercritical solvents.

- Extraction of foods and pharmaceuticals.
- Removal of cholesterol from eggs and milk products.
- Purification of vitamin E.
- Extraction of fragrances, oils, and impurities from agricultural and food products.
- Extraction of flavors from hops in brewing industry.
- Removal of caffeine from coffee and tea.
- Removal of nicotine from tobacco.
- Recovery of mint oil from mint.
- Recovery of edible oils from oil-bearing seeds like soya beans, and so on.
- Extraction of different compounds from materials of biological origin and separation of biological fluids.
- Ethanol–water separation.
- Dissolving solids in supercritical solvents, precipitation of fine particles, nucleation, and recrystallization.
- Enhanced oil recovery and recovery of oil from oil shale.
- Crude deasphalting and lube dewaxing.
- Isomer separations, for example, separation of *ortho*-, *meta*-, and *para*-hydroxyl benzoic acids.
- Polymer/edible oil fractionation.
- Removal of monomers and solvent from polymers.
- Carrying out enzyme reactions in supercritical solvent media.
- Hydrogenation reactions in supercritical solvent media.
- Coal processing (reactive extraction and liquefaction).
- Carrying out high-temperature pyrolysis reactions in SCF solvent medium and removal of the products

from the high-temperature zone, thus avoiding further thermal decomposition.

- Pollution control: Clean up of contaminated soil and hazardous waste sites.
- Removal of contaminants from environment. For example, treatment of waste streams like those containing drilling fluids.
- Use as cleaning fluids (degreasing) for printed circuit boards, and so on.
- Drying of solvent cleaned surfaces, using supercritical CO_2.
  - The advantages over liquid solvents for parts drying are a supercritical fluid has no surface tension, so no bubble or meniscus is formed during the drying operation, CO_2 is more favorable than water due to its low heat of vaporization and subsequently lower energy costs and the component may not have to be subjected to high oven temperatures with subsequent part damage. One disadvantage, which must be considered with a pressurized CO_2 system is that when the pressure is released for parts removal, the residual CO_2 evaporates and cools the parts. This may lead to water condensation on the cold parts if they are not externally heated or an inert gas is used to purge the system.
- CO_2 is attractive for extraction of high-value products, such as flavors and fragrances, food supplements, nutraceuticals, speciality oils, herbal extracts, manufacture of medical devices, and so on.

- Give a simplified flow diagram for supercritical extraction process.
  - Figure 17.51 is a simplified flow diagram for supercritical solvent extraction process. It involves two vessels, the extractor operating under pressure and the separator, operating at low pressure.
  - Four major pieces of equipment are involved: an extraction vessel, a pressure reduction valve, a separator, and a compressor, apart from ancillary equipment like facilities like gas storage, piping, controls, and so on.
  - Addition of modifiers to a SCF (e.g., methanol to CO_2) changes polarity of CO_2 for increased selectivity for separation.

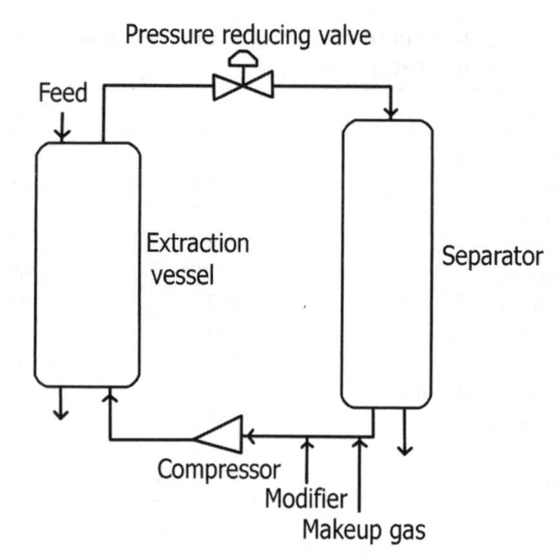

FIGURE 17.51 Simplified flow diagram for a supercritical solvent extraction process.

17.3.5 Extraction Equipment

- What are the different types of equipment used in liquid–liquid extraction?
  - There are varieties of solvent extraction equipment operating in the chemical, petrochemical, pharmaceutical, and metallurgical industries.
  - Unagitated columns like spray, packed, and perforated plate columns.
  - Mechanically agitated columns like pulsed columns, reciprocating columns, rotary columns like rotating disk column (RDC) and mixer–settlers.
  - Centrifugal extractors involve intense energy input and phase separation by centrifugal force leading to very compact size with residence times of the order of few seconds (5–10) as opposed to 3–5 min in liquid pulsed column and 5–10 min in mixer–settlers.
  - Centrifugal extractors are good for systems in which the density difference is less than 4%. This type of extractor should be used if the process requires many equilibrium stages.
- How is mixing accomplished in mixers?
  - By flat blade turbine impellers, with diameters of about one-third of vessel diameter, or circulating pumps.
  - The mixers are normally equipped with variable speed drives to allow the optimum speeds to be determined during commissioning and subsequent operation. Speeds of impellers range from 3 to 15 m/s.
  - Speeds control droplet sizes range 0.1–1 mm. Less than 0.1 mm size is not desirable even though mass transfer area increases but might lead to stable emulsions that are difficult to settle.
  - Sometimes flow mixers, consisting of a series of orifices or nozzles, are used to effect mixing. The two liquids are pumped cocurrently through them. They have limited application as mass transfer is poor and residence times are short.

- What are the advantages and disadvantages of mixer–settler extraction equipment?

  Advantages
  - Good contacting of the phases.
  - Can handle wide range of flow ratios.
  - Optimum droplet size can be obtained by adjusting agitator speed.
  - High efficiency, normally 80%.
  - Flexibility to add or remove stages. Can handle any number of stages.
  - Can tolerate suspended solids.
  - Reliable scale-up.
  - Easy start-up and shutdown.
  - Low headroom requirements.
  - Easy maintenance.
  - Low cost.

  Disadvantages
  - *Large Holdup*: This is particularly detrimental while handling valuable or hazardous liquids. Compared to mixers, settlers involve higher holdup.
  - High solvent inventory.
  - Backmixing possibilities in large diameter settlers.
  - Intense mixing can give rise to permanent emulsions.
  - High energy costs.
  - Interstage pumping may be necessary.
  - Large floor space requirements.
  - High equipment costs for multistage requirements.
  - Complex piping arrangements with increased chances of leakage and spillage problems.
- What are the desirable characteristics of mixer–settler units?
  - Must generate high interfacial area.
  - Must provide high mass transfer coefficients.
  - Low entrainment of air bubbles.
  - Low aspect ratio, that is, of flat geometry.
  - Must allow easy coalescence and phase separation.
  - Must allow easy collection of phases.
  - Volume fraction of the dispersed phase should not be more than 0.6–0.7 as above these values inversions might occur. That is, dispersed phase might become continuous and the other dispersed.
- What type of extraction equipment involves large holdup?
  - Mixer–settler.
- What are the assumptions involved in the design of settling units?
  - It is assumed that droplet diameters are 150 μm.
  - In open vessels, residence times of 30–60 min or superficial velocities of 0.15–0.46 m/min (0.5–1.5 ft/min) are provided.
- What are the factors that retard coalescence in the settler part of a mixer–settler unit?
  - Increase in viscosity of the continuous phase through decrease in temperatures.
  - Decrease in interfacial tension that causes a larger area of contact between the drop and the interface.
  - Decrease in density difference between the two phases.
- What are the extraction stage efficiencies assumed in the design of mixer–settler units?
  - About 80%.
- What are the maximum number of stages commonly used in mixer–settler extraction units?
  - 4–5.
- "Use of baffles in a cylindrical mixing vessel help prevent vortex formation." *True/False*?
  - *True*.
- How is the performance of settlers in a mixer–settler unit can be improved?
  - *Improved Feed Distribution*: By ensuring that the feed is distributed evenly across the full settler width, making all of the settler area effectively used. Addition of distribution vanes greatly enhances the ability to feed the settler over the full width.
  - *Elimination of Macroeddies Induced by the Feed and Discharge Arrangements*: The macroeddies that are inherent in the settler operation can be eliminated by either eliminating the cause of the eddy, or by dissipating the eddy flow pattern by use of baffles and other media. The use of random packed media in settler systems also assist with the elimination of the macroeddies. Use of smooth surfaces and elimination of sharp edges and corners helps reduce localized eddies.
  - *Improved Coalescence with Provision of Controlled Sections that Contain Deep Emulsion Bands*: The presence of unconstrained deep emulsion bands results in poor and highly erratic settler performance. Provision of picket fences to hold back a decreasing thickness of emulsion band is one way to improve coalescence.
  - *Improved Coalescence with In-Settler Equipment*: Further enhancement of settler operation can be achieved with coalescing systems. These include various physical forms such as reducing distance for free fall, trays, baffles, random packed media, extra picket fences, electrical means, and so on.

- *Use of Centrifugal Settlers*: Some times centrifugal force is used for improved settling.

- What are the approaches for design of settlers? Give an empirical equation for their sizing.
  - Adopting API procedures, used for oil–water separators, which involve an assumed droplet size of 150 μm.
  - Provision of sufficient residence time based on experimental studies. In open vessels residence times of 30–60 min or superficial velocities of 15–45 cm/min (0.5–1.5 ft/min) are provided. Longitudinal baffles can cut the residence time to 5–10 min. Coalescence with packing or wire mesh or electrical means cut these times substantially.
  - Determination of suitable flow rates to prevent spread of dispersion band thickness.
  - Settler diameter is estimated using the following empirical correlation:

$$D_S = 8.4(q_C + q_D)^{0.5}, \tag{17.161}$$

  where $q = q_C + q_D$, sum of flow rates of continuous and dispersed phases, respectively.
  - Settler height is determined by employing an L/D ratio of 4.0.

- Under what circumstances, recycle is practiced in a mixer–settler unit?
  - When the ratios of continuous and dispersed phases is far from unity.

- Give suitable alternative configurations for settlers.
  - Figures 17.52 and 17.53 illustrate typical configurations for settling chambers.
  - In the combination design, piping is simplified and leakage chances are minimized.

- Give a diagram of a two-stage mixer arrangement and explain its design.
  - The number of mixing compartments usually lies in the range of 1–3. One is suitable for small plants using extractants with rapid kinetics. Two or three compartments are more applicable to larger plants and longer mixing times.
  - In this particular design, the compartments are separate vessels joined by a launder. In other arrangements, especially for large operations, the mixing compartments may be formed by subdividing a single rectangular construction, which eliminates the connecting launder. In some cases, they have been made part of the settler. The first compartment is shown with a pump mix type of impeller, which provides head for interstage pumping of the solutions as well as mixing the two phases.

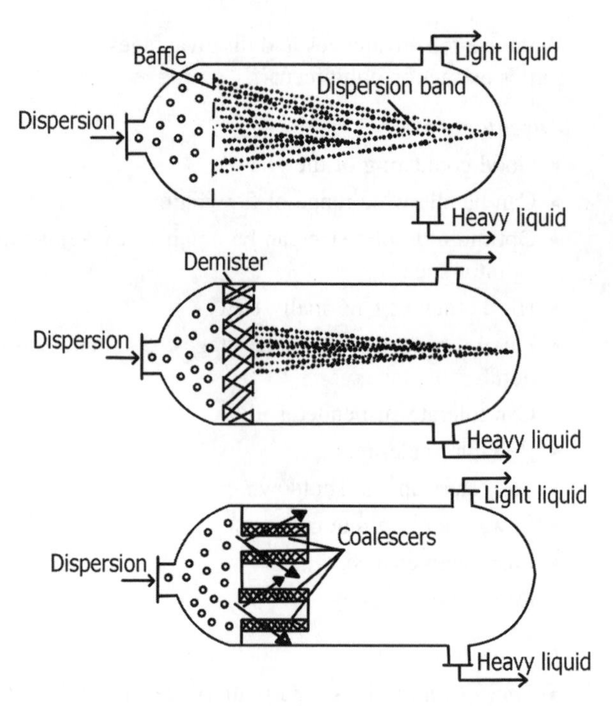

FIGURE 17.52 Settling chambers of different designs.

  - The second compartment is fitted with a mixing impeller designed for maximum mixing efficiency with minimum shear.
  - The first compartment is fitted with a false bottom to which the organic and aqueous feed and recycle pipes are connected. The aqueous and organic feeds must be kept separate by installing dividers in the false bottom, otherwise the organic flow will be hindered due to its lower density.
  - If the recycle line is designed for dual use (i.e., aqueous or organic), it must have its own separate compartment in the false bottom.
  - Vertical sidewall baffles are used to promote proper mixing patterns.
  - The shape of the settler is determined by specifying a width based on an organic phase velocity within the range of 3–6 cm/s.

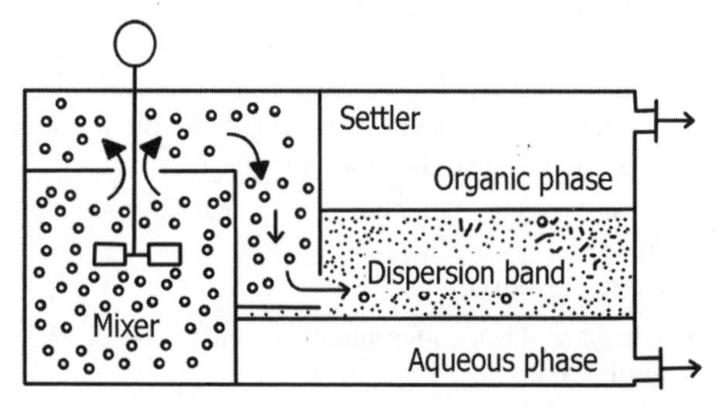

FIGURE 17.53 Combination of mixer–settler unit.

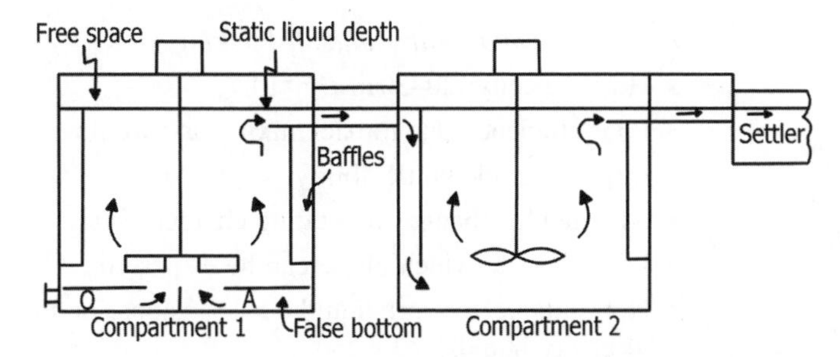

Note: False bottom is divided into organic (O) and aqueous (A) compartments. Orientation of the divider depends on piping layout and aqueous recycle.

(a) Mixer

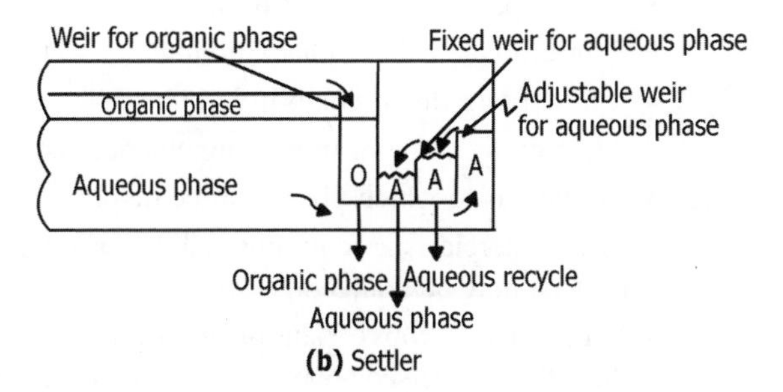

(b) Settler

FIGURE 17.54 (a) Two-compartment mixing system and (b) drop in weir box system.

- Figure 17.54 gives diagram for a two-compartment design.

- What are the hazards associated with settlers in a mixer–settler handling flammable materials?
  - One important factor that is not generally considered is the generation of static electricity in the phase separation process occurring in a settler. As droplets settle, they can generate static charge.
  - The presence of aqueous droplets in hydrocarbon systems is a source of static charge.
  - Use of demineralized water for washing/scrubbing operations should not be contemplated, unless its conductivity is increased with the addition of acid or electrolyte.
  - Activities that can generate static include the following:
    - Mixing of droplets.
    - Mixing or settling of solids.
    - Disruption of an interface.
    - Atomization.
    - Splashing.
    - Flow past a fixed boundary, for example, pipe or vessel wall.
  - Special attention is required in the following cases:
    - High-level entries into tanks (loaded and/or stripped organic tanks).
    - Free fall of hydrocarbons from inlets.
    - Excessive pipe velocities.
    - Entrainment of air into the mixers and subsequent separation in the settlers.
    - Entrainment from washing and scrubbing settlers.

- What are the desirable characteristics of dispersions for liquid–liquid extraction?
  - The objective in a liquid–liquid extraction process is to generate an unstable dispersion that provides reasonably high interfacial area for good mass transfer during extraction and yet is easily broken to allow rapid phase separation after extraction.
  - Given enough time, most dispersions will break on standing. Often this process occurs in two distinct periods. The first is a relatively short initial period during which an interface forms between two liquid layers, one or both of which remain cloudy or turbid.
  - This is followed by a longer period or secondary break during which the liquid layers become clarified.
  - During the primary break, the larger drops migrate to the interface where they accumulate and begin to coalesce. If the coalescence rate is relatively slow compared to the rate at which drops rise or fall to the interface, then a layer of coalescing drops or *dispersion band* will form at the interface.
  - The initial interface can form within a few minutes or less for drop sizes of the order of 100–1000 μm (0.1–1 mm), as, for example, in a water–toluene system.
  - When the drop size distribution in the feed dispersion is wide, smaller droplets remain suspended in one or both phases. Longer residence times are then required to break this secondary dispersion. In extreme cases, the secondary dispersion can take days or even longer to break.
  - Emulsions are broken by changing conditions to promote drop coalescence, either by disrupting the film formed at the interface between adjacent drops or by interfering with the electrical forces that stabilize the drops.
    - Physical techniques used to break emulsions include heating (including application of microwave radiation), freezing and thawing, adsorption of surface-active compounds, filtration of fine particles that stabilize films between drops, and application of an electric field.
    - Chemical techniques include adding a salt to alter the charges around drops, changing the pH of the system and adding a demulsifier compound (or

even another type of surfactant) to interact with and alter the surfactant layer.

- The stability of dispersion also may depend upon the surface properties of the container or equipment used to process the dispersion, since the walls of the vessel or the surfaces of any internal structures, may promote drop coalescence.
- In a liquid–liquid extractor or a liquid–liquid phase separator, the wetting of a solid surface by a liquid is a function of the interfacial tensions of both the liquid–solid and the liquid–liquid interfaces.
- In general, an aqueous liquid tends to wet a metal or ceramic surface better than an organic liquid, and an organic liquid tends to wet a polymer surface better than an aqueous liquid.
- In liquid–liquid extraction equipment, the internals generally should be preferentially wetted by the continuous phase, in order to maintain dispersed phase drops with a high holdup.
- If the dispersed phase preferentially wets the internals, then drops may coalescence on contact with these surfaces, resulting in loss of interfacial area for mass transfer. In an agitated extractor, this tendency may be mitigated to some extent by increasing the agitation intensity.

• What are the advantages and disadvantages of columns without agitation?
  - Advantages include low equipment and operating costs.
  - Disadvantages include poorer mixing and hence efficiency than mixer–settler units, high headroom requirements and difficult to scale-up.

• What are the advantages and disadvantages of columns with agitation?
  - Good points include efficient dispersion, suitable for multistage operations involving large number of stages and low investment costs. On the minus side, not suitable for high flow ratios and difficult to handle small density differences.

• What are the characteristic operational features of different types of extraction columns? Illustrate with diagrams.

Spray Columns

- High capacities because of their openness.
- Suitable when liquids contain suspended solids.
- Backmixing is severe.
- Commercially, spray columns are suitable for liquid–liquid processes in which rapid, irreversible chemical reactions occur, as in neutralization of waste acids.

Packed Columns with Random Packing

- High capacity: 20–30 $m^3/(m^2\,h)$.
- Poor efficiency due to backmixing and wetting.
- Limited turndown flexibility.
- Affected by changes in wetting characteristics.
- Limited as to which phase can be dispersed.
- Not suitable for dirt liquids, suspensions, or high viscosity liquids.
- Used where few stages are require.

Packed Columns with Structured Packing

- High capacity: 40–80 $m^3/(m^2\,h)$.
- Poor efficiency due to backmixing and wetting.
- Limited turn down flexibility.
- Affected by changes in wetting characteristics.
- Limited as to which phase can be dispersed.
- In commercial size columns, HETP of 0.60–1.5 m (2–5 ft) may be realized.
- Mass transfer drops off sharply with axial distance, so that the dispersed phase is redistributed every 0.6–2 m (5–7 ft).
- Packed columns with three or more beds are not uncommon and may be employed when 5–10 stages are involved.
- They are not satisfactory at interfacial tensions above 10 dynes/cm.
- Metal and ceramic packings tend to remain wetted with the continuous phase. Thermoplastics tend to be preferentially wetted with oil but they can be wetted by aqueous phase if immersed in it for several days.
- Intalox saddles and pall rings of 2.54–4 cm (1–1.5 in.) size are the most commonly used packings. Smaller sizes tend to be less effective since their voids are of the same order of magnitude as drop sizes.

Sieve Tray Columns

- Widely used.
- High capacity: 30–50 $m^3/(m^2\,h)$.
- Good efficiency due to minimum backmixing and droplets being produced at every tray.
- Multiple interfaces can be a problem.
- Limited turndown flexibility.
- Affected by changes in wetting characteristics. Liquid that wets the tray forms continuous phase and that does not wet the tray forms dispersed phase as it easily breaks away quickly from the tray to form bubbles.
- Limited as to which phase can be dispersed.
- Sieve tray columns work only at high density differences.

- Hole diameters are much smaller than for vapor–liquid contacting, being 3–8 mm, usually on triangular spacing of 2–3 hole diameters and occupy from 15% to 25% of the available tray area.
- Velocities through the holes are kept below about 25 cm/s to avoid formation of very small droplets.
- Both the reduced axial mixing because of the presence of the trays and the repeated dispersion tend to improve the efficiency over the other kinds of unagitated columns.
- Hole diameters are much smaller than for vapor–liquid contacting, being 3–8 mm, usually on triangular spacing of 2–3 diameters and occupy from 15% to 25% of the available tray area.
  Velocities through the holes are kept below about 24 cm/s (0.8 ft/s) to avoid formation of very small droplets. The head available for flow of the continuous phase is the tray spacing.

Pulsed Columns

- Rapid reciprocating pulse of short amplitude (5–25 mm) is hydraulically transmitted to the liquid phases, which results in improved mass transfer.
- In comparison with unagitated columns, which are limited to interfacial tensions below 10 dynes/cm, pulsed columns are not limited by interfacial tension up to 30–40 dynes/cm.
- Reasonable capacity: 20–30 $m^3/(m^2\,h)$.
- The action of the perforated trays is to disperse the heavy phase on the upstroke and the light phase on the downstroke.
- Best suited for nuclear applications due to lack of seal.
- Pulsating action can be accomplished without parts and bearings in contact with the process liquids.
- Suited for corrosive and hazardous applications as they can be constructed out of nonmetals.
- Limited stages due to backmixing.
- Pulsing is uniform across the cross section and accordingly the height needed to achieve a required extraction is substantially independent of the diameter as long as hydrodynamic similarity is preserved.
- In comparison with unagitated columns, which are limited to interfacial tensions below 10 dynes/cm, pulsed columns are not limited by interfacial tension up to 30–40 dynes/cm.
- Limited diameter/height due to the requirement of pulse energy.
- Figure 17.55 is a typical pulse column.

Karr Columns

- Highest capacity: 30–60 $m^3/(m^2\,h)$.

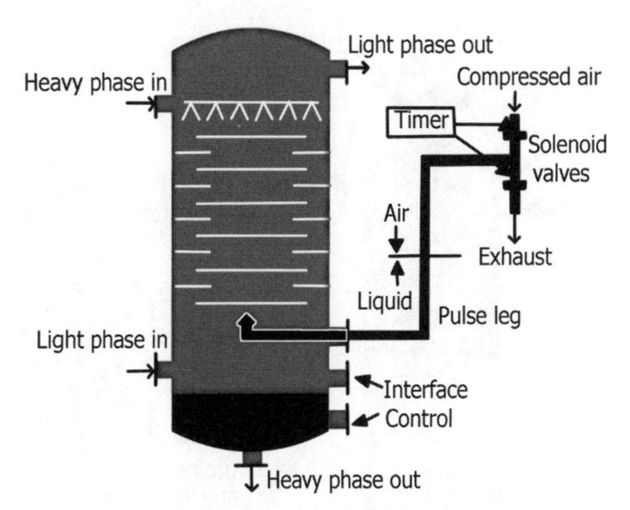

FIGURE 17.55 Pulse column.

- Good efficiency.
- Good turndown capability (25%).
- Uniform shear mixing.
- Best suited for systems that emulsify.
- Figure 17.56 illustrates Karr column.

Scheibel Columns

- Reasonable capacity: 15–25 $m^3/(m^2\,h)$.
- High efficiency due to internal baffling.
- Superior to packed columns.
- Good turndown capability (25%).

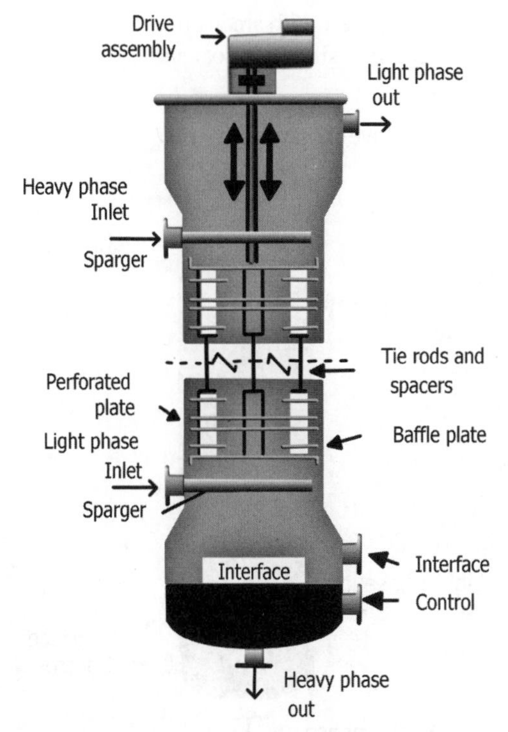

FIGURE 17.56 Karr extraction column.

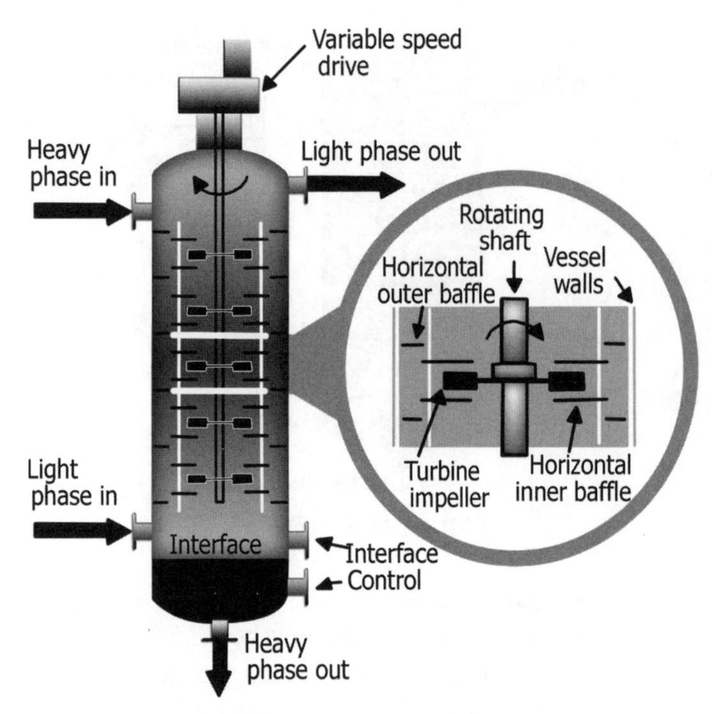

FIGURE 17.57 Scheibel column.

- Best suited when many stages are required.
- Not recommended for highly fouling systems or systems that tend to emulsify.
- Figure 17.57 illustrates Scheibel extraction column.

Rotating Disk Columns (RDCs)

- RDCs (Figure 17.58) are in effect mixer–settler units arranged in a single vertical shell.

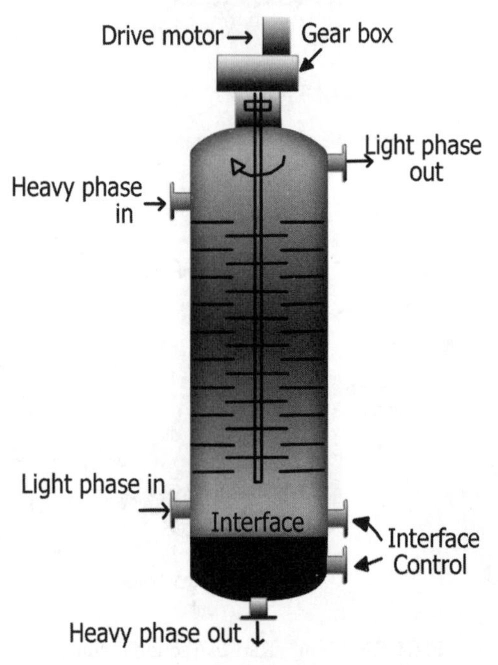

FIGURE 17.58 Rotating disk column.

- The impellers are flat disks and the mixing zones are separated by stationary baffles as shown in the diagram.
- Disks rotate at higher speeds than turbine impellers.
- Reasonable capacity: 20–30 $m^3/(m^2\,h)$.
- Limited efficiency due to axial backmixing.
- Reasonable turndown (40%)
- Suitable for viscous or fouling liquids.
- Sensitive to emulsions due to high shear mixing.
- Because of its geometrical simplicity and effectiveness, the RDC is one of the most widely used of agitated extractors.
- Not suitable when only a few stages are needed, in which case mixer–settlers will be satisfactory and cheaper, when large holdup and long residence times may be harmful to unstable substances, or for systems with low interfacial tensions and low density differences, because stable emulsions may be formed by the intense agitation.

• What are the advantages and disadvantages of centrifugal extractors?

Advantages

- Can handle systems with low density differences emulsifying tendencies, or require large solvent to feed ratios.
- Low holdup volume and short residence time.
- Small inventory of solvent.
- Small space requirements.

Disadvantages

- High initial and operating costs.
- Limited number of stages, although some units have as many as 20 stages.
- High maintenance costs.

• What are the special applications of centrifugal extractors?

- Centrifugal extractors are more attractive in fuel reprocessing step of fast breeder reactor (FBR) fuels.
- Due to higher activity of the aqueous solutions, the solvent tributyl phosphate (TBP) undergo degradation leading to loss of uranium and plutonium. Since the residence time in centrifugal extractors is short, these units are favored.

• Name a suitable extractor for handling low density difference liquids.

- Centrifugal extractor.

• In what type of extractor inventory of solvent is minimum?

- Centrifugal extractor.

• How is the diameter of an extraction column determined?

- The diameter of the column must be large enough to permit two phases to flow countercurrently through the column without flooding.
- Estimation of column diameter for liquid–liquid extraction is far more complex and uncertain than vapor–liquid contacting equipment due the larger number of important variables involved.
- Variables necessary for estimating extraction column diameter include the following:
  - Individual phase flow rates.
  - Density differences between the two phases.
  - Interfacial tension.
  - Direction of mass transfer.
  - Viscosity and density of continuous phase.
  - Geometry of internals.
- Column diameter may be best determined through scale-up of laboratory test runs.
- The necessary experimental data are obtained by the following:
  - Laboratory or pilot plant test unit should be operated with system components of interest.
  - Laboratory or pilot plant test unit, with a diameter of 2.5 cm (1 in.) or more, should be used.
  - Measurements of superficial velocities in each phase are made.
  - The sum of these velocities may be assumed to hold constant for larger scaled-up commercial units.
  - The superficial velocity data should be used to estimate the column diameter through the use of correlations based on the superficial velocity data.

- How is phase contacting achieved in a spray column?
  - Uses a sparger for light liquid and disperser for heavy liquid.
- "Even heights of 6–12 m (20–40 ft) for a spray column cannot provide performance equivalent to one stage of an extraction process." *True/False*?
  - *True.*
- "Spray columns cannot be depended on to function as more than a single stage even when they are as tall as 6–12 m (20–40 ft)." *True/False*?
  - *True.*
- Under what circumstances packed columns are advantageous in liquid–liquid extraction?
  - Suitable when 5–10 stages are sufficient.
  - Pal rings 2.5–3.8 cm (1–1.5 in.) size are best suited as packings.
- Using Pal rings as packing, what are the HETP values attainable in a packed column?
  - 1.5–3 m (5–10 ft).
- What are the dispersed phase loadings and distribution arrangements that are to be employed in a packed column?
  - Dispersed phase loadings should not exceed 10.2 m^3/(min m^2) (25 gal/(min ft^2)).
  - The dispersed phase must be redistributed every 1.5–2.1 m (5–7 ft).
- What are the maximum dispersed phase loadings for satisfactory operation of a packed extraction column?
  - 60 $m^3/(h\, m^2)$.
- "Packed columns are not satisfactory when surface tension is more than 10 dynes/cm." *True/False*?
  - *True.*
- What are the constructional and operational features involved in a sieve tray column employed for liquid–liquid extraction?
  - Sieve trays have holes of only 3–8 mm diameter.
  - Tray spacings are 15.2–60 cm (6–24 in.).
  - Velocities through the holes are kept below 0.24 m/s (0.8 ft/s) to avoid formation of small drops that are difficult to separate.
  - Tray efficiencies are in the range of 20–30%.
- What are the characteristics of pulsed sieve tray and packed columns?
  - Frequencies of operation are 90 cycles/min and amplitudes are 6–25 mm.
  - In large diameter columns, HETP values of about 1 m are possible.
  - Surface tensions as high as 30–40 dynes/cm have no adverse effect.
- What are the characteristics of reciprocating tray columns?
  - Desirable motion can be imparted to the liquids by reciprocating motion of the plates rather than by pulsing the entire liquid mass.
  - The holes of reciprocating plates are much larger than those of pulsed ones.
  - Can have holes of 1.5 cm (9/16 in.) diameter, 50–60% open area, stroke length of 1.9 cm (0.75 in.) and 100–150 strokes/min.
  - Tray spacing are in the range of 2.5–15 cm (1–6 in.). Normally 5 cm (2 in.).
  - HETPs are in the range of 50–65 cm for columns of about 75 cm.
  - Scale-up formula for HETP for different column diameters is

$$(\mathrm{HETP})_2/(\mathrm{HETP})_1 = (D_2/D_1)^{0.36}. \qquad (17.162)$$

  - Power requirements are much less than that for pulsed columns.
- What is the range of HETPs obtainable in a rotating disk column?
  - 0.1–0.5 m (0.33–1.64 ft). (HETP for packed columns are about 1.5–3 m and pulsed columns are about 1 m.)
- For difficult extractions involving large number of stages, recommend a suitable extractor.
  - Pulsating sieve tray column.
- What is the effect of using high velocities through the holes of a sieve tray column used for liquid–liquid extraction?
  - High hole velocities, above 25 cm/s (0.8 ft/s), are used, sizes of droplets formed will be small, which while increasing interfacial area for mass transfer, makes separation of the phases difficult.
- What is the principle and constructional and operational details of a Podbielniak extractor?
  - The Podbielniak contactor is a differential type, based on the use of centrifugal force differences between the phases.
  - It is constructed of several rotating perforated concentric cylinders on a horizontal shaft. Speeds are 30–85 rps.
  - Input and removal of the phases at each section are accomplished through radial tubes.
  - The flow is countercurrent with alternate mixing and separation occurring respectively at the perforations and between the bands.
  - Heavy liquid flows radially outward, displacing light liquid inwardly and both are led out through the shaft.
  - The position of the interface is controlled by the back pressure applied on the light phase outlet.
  - Residence time can be as short as 10 s.
  - Its basic cost is high but unlike other types, it requires fewer auxiliaries.
  - Used by pharmaceutical industry for the extraction of penicillin from natural broth.
- What are the problems involved in hindering extraction process in extraction equipment? How are they overcome?
  - Coalescence of droplets during extraction stage in the equipment.
  - In columns with no internals (empty columns with only the liquid phases), both phases seriously suffer from backmixing.
  - There are a few widely used solutions to these problems:
    - Rotating disk contactors (RDCs) and pulsed columns belong to the class of agitated columns, which are quite common.
    - In an RDC mixer disks mounted on a vertical shaft in the column keep the droplets small.
    - To reduce backmixing in an RDC, it is compartmentalized by rings, which reduce the cross section of the column between the mixer disks. So, each disk mixes its own compartment, and an RDC can be regarded as a series of nonideal mixers in series.
    - In a pulsed column, the liquid in the column is agitated by pulsing it.
    - There are also nonagitated columns, such as sieve tray columns, which are analogous to tray columns and packed columns. The internals of these columns keep the dispersed phase from coalescing in a passive way and simultaneously reduce the backmixing.
    - These types of columns are not so common, because their performance is even less predictable than that of agitated columns. This is mainly caused by the flow patterns in these columns.
- Summarize the general features of different extraction equipment.
  - Table 17.13 gives a summary of the general features of extraction equipment.

17.4 SOLID–LIQUID EXTRACTION/LEACHING

- What is leaching? What are the factors that influence leaching processes?
  - Removal of a soluble material from an insoluble solid phase with which it is associated, through solvent extraction. This is another name for solid–liquid extraction.
  - The process may be used either for the production of a concentrated solution of a valuable solid material, or in order to remove an insoluble solid, such as a pigment from a soluble material with which it is contaminated. The method used for the extraction is determined by the proportion of soluble constituent present, its distribution throughout the solid, the nature of the solid and the particle size.
  - If the solute is uniformly dispersed in the solid, the material near the surface will be dissolved first, leaving a porous structure in the solid residue. The solvent will then have to penetrate this outer layer before it can reach further solute, and the process will become progressively more difficult and the extraction rate will decrease.

TABLE 17.13 General Features of Liquid–Liquid Extraction Equipment

Extractor Type	General Features
Static extractors: spray columns, baffle columns	Simple construction with no internal moving parts; low to medium efficiency; best suited to systems with low to moderate interfacial tension; low capital, operating and operating costs, and high throughputs. Capacities of spray columns are high because of their openness and are suitable when liquids contain suspended solids. Backmixing is severe in spray columns
Mixer–settlers: mixing vessels with integral or external settling chambers	Good mixing;high efficiencies; long residence times; can handle high viscosity liquids and systems with low to high interfacial tensions; can handle wide solvent rates; good flexibility; high capacities; low headroom; number of stages can be added or removed offering good flexibility
Rotary agitated columns: RDCs, asymmetric RDCs, Scheibel column, Oldshue–Rushton column, Kuhni columns	Moderate to high efficiencies; many theoretical stages possible in a single column; suited to low to moderate viscosity liquids; can handle systems with moderate to high interfacial tension and with emulsifying tendencies; moderate HETPs; moderate capital costs and low operating and maintenance costs; moderate capacities
Reciprocating plate column: Karr column	Moderate to high efficiencies; can handle systems with low to moderate interfacial tension and with emulsifying tendencies; good versatility and flexibility; moderate capital costs and low operating costs; moderate capacities; simple construction; handles liquids with suspended solids; low HETPs
Pulsed columns: packed columns, sieve tray columns	Moderate to high efficiencies; no internal moving parts; can handle corrosive and toxic materials requiring hermitically sealed system; low HETPs; moderate capacities
Centrifugal extractors	Short residence time for unstable solutes; can handle systems with low density differences or tendencies for emulsification; minimal headroom and space requirements

- If the solute forms a very high percentage of the solid, the porous structure may break down almost immediately to give a fine deposit of insoluble residue, and access of solvent to the solute will not be impeded.
- The stages involved are (1) dissolution of the solute in the solvent, (2) diffusion of the solute through the solvent in the pores of the solid to the outside of the particle, and (3) transfer of the solute from the solution in contact with the particles to the bulk of the solution. Though the first process usually occurs so rapidly that it has a negligible effect on the overall rate. In some cases the soluble material is distributed in small isolated pockets in a material that is impermeable to the solvent, for example, gold dispersed in rock. In such cases the material is crushed so that all the soluble material is exposed to the solvent.
- If the solid has a cellular structure, the extraction rate will generally be comparatively low because the cell walls provide an additional resistance. In the extraction of sugar from beet, the cell walls perform the important function of impeding the extraction of undesirable constituents of relatively high molecular weight and the beet should therefore be prepared in long strips so that a relatively small proportion of the cells are ruptured. In the extraction of oil from seeds, the solute is itself liquid.
- Particle size, nature of the solvent used, temperature and agitation are the important factors that influence extraction rates.

- Name the ways by which leaching operations are carried out.
  - When solids form open permeable mass throughout leaching operation, solvent may be percolated through an unagitated bed of solids.
  - With impermeable solids/or materials that disintegrate during leaching, solids are dispersed in solvent and then separated.
- What is heap leaching? How is it carried out? Give examples of its applications.
  - Leaching is carried out by dumping crushed ores on sloping floors lined by an impervious polymeric membrane. Solvent is sprayed by means of drippers or sprayers distributed over the ore heap. The percolated solution flows from the bottom of the heap into a collection basin from where it is pumped to solvent recovery equipment where solvent is separated from the product liquid and recycled along with makeup solvent.
  - In the heap leaching of low-grade oxidized gold ores, a dilute alkaline solution of sodium cyanide is distributed over a heap of ore that typically has been crushed into about 25 mm size and agglomerated into lumps by a suitable additive to enhance percolation rates.
  - Heap leaching of very low-grade gold ores and many oxide copper ores is conducted on run of mine material. Heap leaching is the least expensive form of leaching.

- What is *in situ* leaching?
  - *In situ leaching*: involves pumping solvent through holes drilled and lined with perforated tubing into the mineral containing soil surrounding the deposit. The leaching solution is pumped down the injection wells and flows through the deposit and the solution is extracted from production wells, treated for solute recovery, reconstituted and reinjected.
  - *In situ leaching*: depends on the existing permeability of a subsurface deposit containing minerals or materials that are to be dissolved and extracted.
  - *In situ*: leaching is used for extraction of uranium, as well as for the removal of toxic or hazardous constituents from contaminated soil or groundwater.
- How are the solids prepared for extraction?
  - The capillary paths should be short so that only a short distance is involved in solvent diffusion into the solid. This is the reason why normally the raw material is crushed or chipped into small pieces.
  - For the percolation process the solid material must be modified in such a way that the solvent can easily flow through. For example oil seeds are pressed into thin flakes so that very short capillary paths are produced and further the cell walls, which include the oil material, are destroyed and a direct contact of solvent and extract becomes possible. Sugar beets are cut into thin wafers to help leaching process.
  - Pharmaceutical products from plant roots, leaves, and stems are predried that helps cell rupture and solute contact with solvent through ruptured cell walls.
  - On the other hand solids in very fine form (like fish meal) have very short capillary paths but the percolation rate is very low. Therefore, these fine powders are pelletized or flaked to get a granule with good percolation properties.
  - The percolation velocity has to be high enough to displace the extract solution that diffuses to the surface of the solid particles.
- Name the factors involved in the choice of a solvent for a leaching process.
  - High saturation limit and selectivity for the solute.
  - Capability to produce extracted material of quality unaffected by the solvent.
  - Ease and economy of recovery from extract.
  - Chemical stability.
  - Low viscosity, low vapor pressure, low toxicity and flammability, low density, low surface tension, and low cost.
- Give examples of leaching processes.
  - Leaching of minerals from ores, for example, copper minerals from ore using sulfuric acid or ammonical solutions, uranium ore slurries using acidic solutions, bauxite leaching, nickel and cobalt recovery, gold, and so on.
  - Vegetable oils from oil seeds.
  - Sugar from sugar beets.
  - Tannins from tree barks.
  - Pharmaceutical products from plant roots, and so on.
  - Extraction of tea/coffee from tea leaves/coffee seeds.
  - Washing of solids to free mother liquors.
- "In general use of higher temperatures helps leaching processes." Comment.
  - Better leaching due to higher solubility and lower viscosities.
  - Not good for heat-sensitive materials.
  - Undesirable materials are also leached out.
- What are the steps involved in leaching processes?
  - Dissolving soluble component in a solvent.
  - Separation of the solution (extract) from insoluble solid residue.
  - Washing the solid residue.
- What are the differences between using percolation and agitation in a leaching process?
  - Percolation is employed for coarse particles.
  - Agitation is employed for fine particles.
  - With coarse particles, leaching is slow and less thorough but requires less washing and gives more concentrated solution.
  - Leaching involving fine particles is more expensive (power required to grind the material) but more rapid and thorough. Dilute solutions result and solvent requirements are high.
- What types of equipment are used in batch solid–liquid extraction? Illustrate.
  - Figures 17.59 illustrates a vertical batch solid–liquid extraction vessel.
  - Figure 17.60 is a horizontal solid–liquid extractor. The false bottom construction for solution exit acts as a solids retainer while allowing the solution is exited. Multiple solution exits are shown for the liquid for large vessels.
- Illustrate operation of a multistage countercurrent solids extraction system.

  Figure 17.61 illustrates a multistage thickener units operating in series with the liquid flowing from stage to stage countercurrent to solids flow.
- What is Hildebrand extractor?

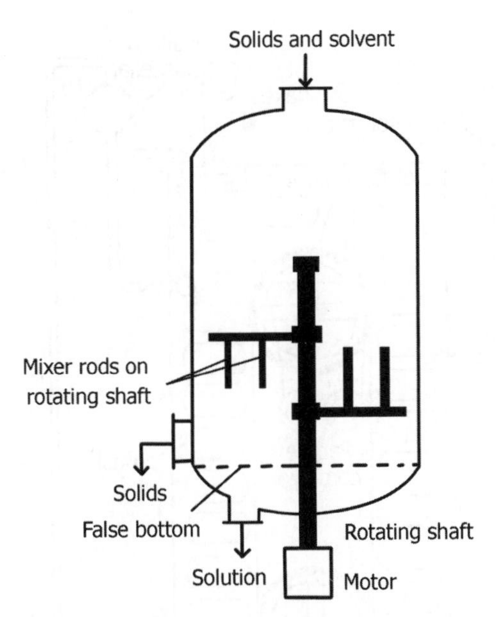

FIGURE 17.59 Batch solid–liquid extractor.

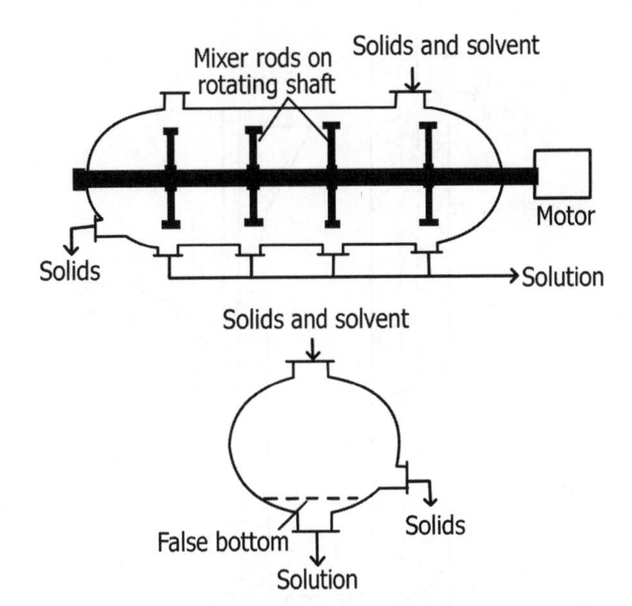

FIGURE 17.60 Batch solid–liquid extractor (horizontal).

- Figure 17.62 illustrates a screw type system called Hildebrand extractor with the flows of solids moving countercurrent to the solvent and extracted oil as shown.
- The helix surface is perforated so that solvent can pass through countercurrent to the solids.
- The screws are so designed to compact the solids during their passage through the unit.
- There are possibilities of some solvent loss and feed overflow, and successful operation is limited to light, permeable solids.
- Suitable for light permeable solids.

• What is a Bollman extractor?

- A bucket elevator unit with buckets perforated at bottom held on an endless moving belt used in solid–liquid extraction operations.
- Descending buckets are fed with solids (dry flakes) into the top bucket, moving downward, which are sprayed with the partially enriched solvent (*half miscella*) pumped from the bottom of the column and after the extraction, solids from the top bucket on the upward side, are dumped onto a conveyor as shown in Figure 17.63.
- The extractor is contained in a vapor-tight vessel and is used with seeds that do not disintegrate on extraction.
- Efficiency is not very high due to absence of agitation, although extract is a clear liquid. Thin flakes improve the diffusion of the solvent.
- Used to extract oils from oil seeds. Handles solids in the form of flakes, for example, soya beans.

• What is a rotocel extractor? Give a diagram.

- Rotocel extractor consists of walled compartments in the form of annular sectors with liquid-permeable floors revolving about a central axis.

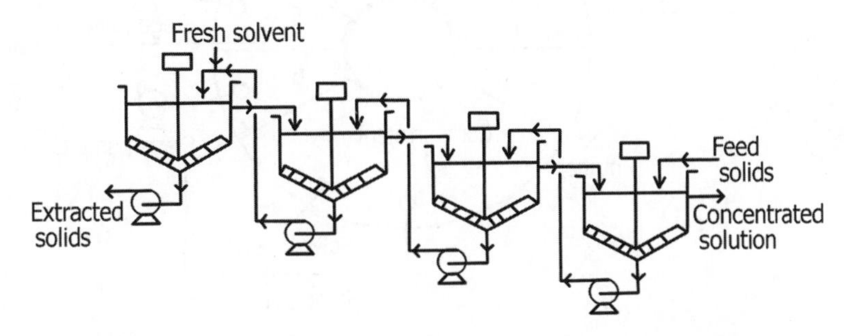

FIGURE 17.61 Thickener type countercurrent leaching equipment.

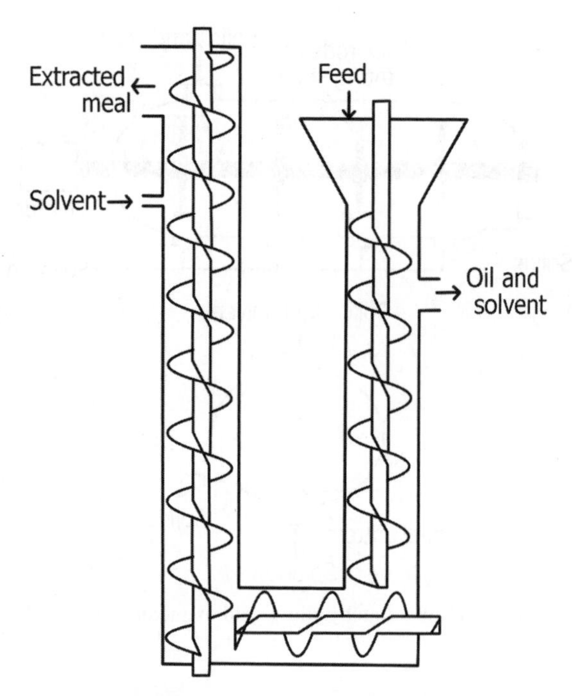

FIGURE 17.62 Hildebrand extractor.

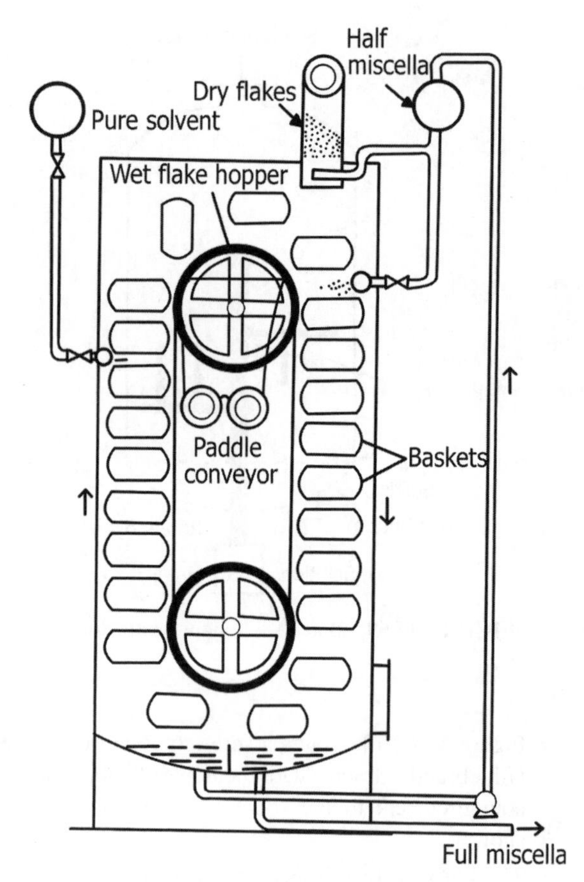

FIGURE 17.63 Bollman extractor.

- The wedge-shaped compartments successively pass a feed point, a number of solvent sprays, a drainage section, and a discharge station (where the floor opens to discharge the extracted solids). In the rotating shell there are about 18 such compartments.
- Countercurrent extraction is achieved by feeding fresh solvent only to the last compartment before dumping occurs and by washing the solids in each preceding compartment with the effluent from the succeeding one. The Rotocel extractor is simple and requires little headroom.
- Figure 17.64 illustrates the working of a rotocel extractor.

- What is a Pachuca tank?
  - Pachuka tank is an air-agitated vessel in which metal ores are leached.

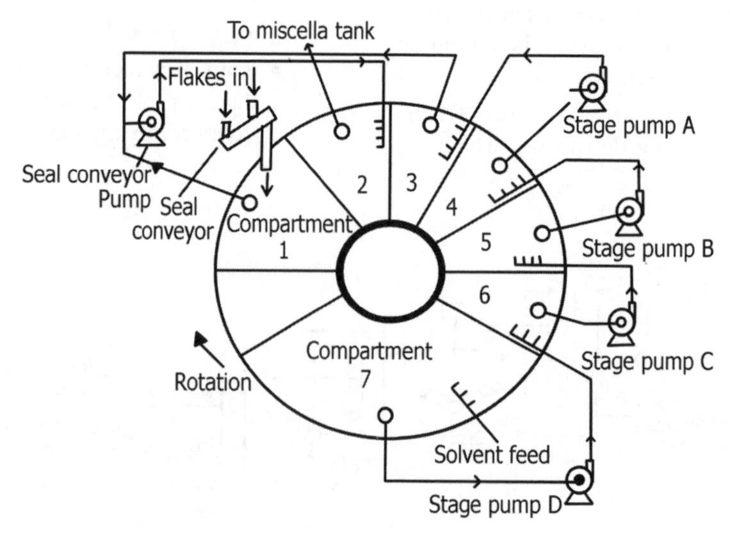

FIGURE 17.64 Rotocell extractor.

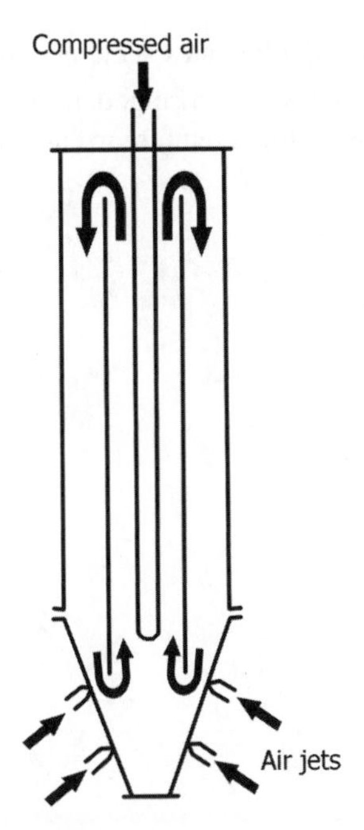

FIGURE 17.65 Pachuca tank.

- Typical tank is a vertical chamber with a conical bottom section, usually with a 60° included angle, 7 m in diameter and 14 m overall height.
- Figure 17.65 illustrates the operation of a Pachuca tank.

- What type of equipment is used for extraction of oils from oil seeds? Give a suitable diagram of an extractor used for extraction of oils from crushed seeds.
  - With seeds such as soya beans, ground nut and other seeds, containing about 15% or less of oil, solvent extraction is often used.
  - Light petroleum fractions are generally used as solvents. Trichloroethylene has been used where fire risks are serious with hydrocarbon solvents like hexane and acetone or ether where the material is very wet.
  - The extractor consists of a vertical cylindrical vessel divided into two sections by a slanting partition, as shown in Figure 17.66. The upper section is filled with the charge of seeds that is sprayed with fresh solvent via a distributor.
  - The solvent percolates through the bed of solids and drains into the lower compartment where, together with any water extracted from the seeds, it is continuously vaporized by means of a steam coil.
  - The vapors are passed to an external condenser, and the mixed liquid is passed to a separating box from which the solvent is continuously fed back to the plant and the water is run to waste recovery.

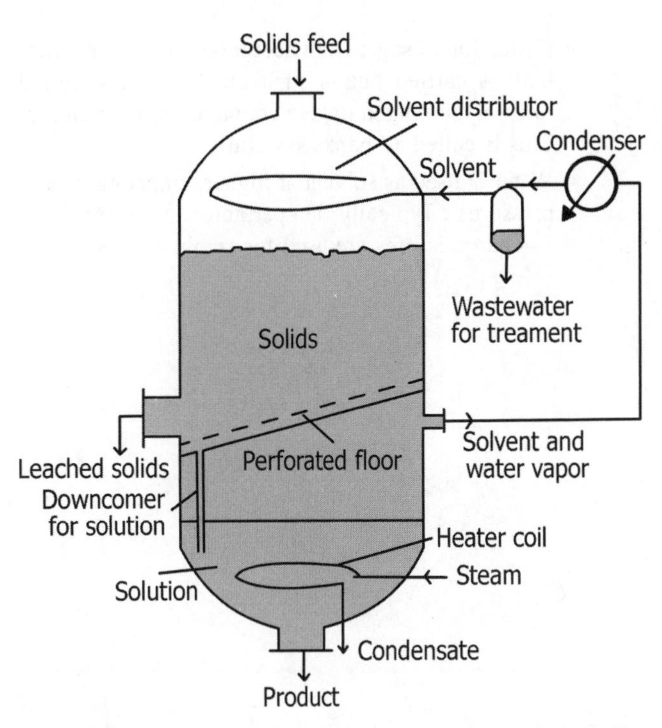

FIGURE 17.66 Percolator type extractor for extraction of oils from oil seeds.

- What are diffusers? Under what circumstances these are used for leaching?
  - These are closed vessels through which liquid is pumped through beds of solids.
  - These are used when ΔP is too high for gravity flow.

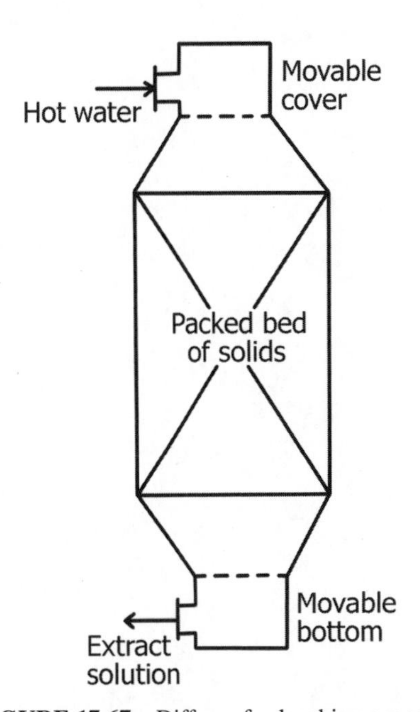

FIGURE 17.67 Diffuser for leaching process.

- Extraction of sugar from beet roots and tannins from bark is carried out in such closed vessels, called diffusers, arranged to give countercurrent operation. This is called a Shanks system.
- Water is used as solvent at higher temperatures and pressures. Typically temperatures of 120°C and 3–4 bar pressures are used for tannin extraction.
- Figure 17.67 illustrates such an extractor.
- These vessels are arranged in a battery of up to 16 vessels for countercurrent multiple contact operation.

18

CRYSTALLIZATION, AIR–WATER OPERATIONS, DRYING, ADSORPTION, MEMBRANE SEPARATIONS, AND OTHER MASS TRANSFER PROCESSES

18.1 Crystallization 613
18.1.1 Crystallization Equipment 621
18.1.2 Crystallization Types 627
18.2 Humidification and Water Cooling 630
18.2.1 Evaporative Cooling and Cooling Towers 636
18.3 Drying 644
18.3.1 Dryers 652
18.4 Adsorption 671
18.4.1 Adsorption Equilibria 680
18.4.2 Ion Exchange 682
18.5 Membrane Separations 684
18.5.1 Membranes 685
18.5.2 Membrane Modules 689
18.5.3 Membrane Separation Processes 693
18.5.3.1 Reverse Osmosis 695
18.5.3.2 Pervaporation 699
18.5.3.3 Other Membrane Separation Processes 701
18.6 Other Separation Processes 711

18.1 CRYSTALLIZATION

- What are the plus and minus points in crystallization processes for purification of materials?
  Plus Points
  - Pure product (solute) can be recovered in one separation stage. With care in design, product purity more than 99.0% can be attained in a single stage involving crystallization, separation, and washing.
  - A solid phase is formed that is subdivided into discrete particles. Generally, conditions are controlled so that the crystals have the desired physical form for direct packaging and sale.

  Minus Points
  - Purification of more than one component is not normally attainable in one stage.
  - The phase behavior of crystallizing systems prohibits full solute recovery in one stage. Thus, the use of additional equipment to remove solute completely from the remaining crystallizer solution is necessary.
- What are the different varieties of products for which crystallization processes are important?
  - Agrochemicals, catalysts, dyes/pigments, electronics, food/confectionery, health products, nanomaterials, nuclear fuel, personal products, and pharmaceuticals.
- What are crystals? What is crystallization?
  - A crystal may be defined as a solid composed of atoms arranged in an orderly, repetitive array. The interatomic distances in a crystal of any definite material are constant and characteristic of that material. Crystals are, in short, high-purity products with consistent shape and size, good appearance, high bulk density, and good handling characteristics.
  - Because the pattern or arrangement of the atoms is repeated in all directions, there are definite limitations on the shapes that crystals may assume. For each chemical compound, there are unique physical properties differentiating that material from others, so the formation of a crystalline material from its solution or mother liquor, is accompanied by unique growth and nucleation characteristics.
  - Crystals possess a well-defined three-dimensional order.
  - Crystallization is the process by which a chemical is separated from solution as a high purity, definitively shaped solid.

Fluid Mechanics, Heat Transfer, and Mass Transfer: Chemical Engineering Practice, By K. S. N. Raju

- While crystallization is a unit operation embracing well-known concepts of heat and mass transfer, it is nevertheless strongly influenced by the individual characteristics of each material handled. Therefore, each crystallization plant requires many unique features based upon well-established general principles.

- What are the properties of crystals that influence their quality?
  - Crystal properties include purity, appearance, agglomeration, caking tendency, specific surface area, surface characteristics, crystal size distri-bution, morphology/shape, polymorphic form, bulk density, bioavailability, flow properties, crystal strength, dissolution rate, and mother liquor and other impurities.
- What is a polymorphic system? What is its importance?
  - Polymorphic system is one that includes a chemical compound that can exist as two or more crystal structures with different internal unit cells, as measured by an X-ray diffraction pattern. This can result in different chemical and physical properties, which influence parameters like ease of formulation and bioavailability (e.g., solubility).
  - Polymorphism is displayed by both organic and inorganic compounds. For example, calcium carbonate has three forms of polymorphism, namely, aragonite, vaterite, and calcite. Calcite is the most stable of the three.
  - The identification and generation of polymorphic forms is an important source of patents and intellectual property rights for the agriculture and pharmaceutical industries.
  - Polymorphic form along with bioavailability, crystal size distribution, and dissolution characteristics are of considerable importance to pharmaceutical industry.
- What are the factors that affect crystal properties?
  - Solvent, classification, seeding, impurities, fines removal, temperature, method of generation of supersaturation, solids concentration, external equipment, crystallization kinetics, residence time, solution pH, hydrodynamics/agitation, feed inlet location to the crystallizer, and so on.
- What is supersaturation?
  - Supersaturation is defined as the excess dissolved solute in the mother liquor with reference to the solubility curve.
  - In order to drive the process of crystallization, the solution must be supersaturated with solute.
  - The solubility of the solution, related to crystal size, is defined by Kelvin equation:

$$\ln(c/c_s) = (4v_s\sigma_{s,L}/vRTD_p), \qquad (18.1)$$

where c is the mass solute concentration in the bulk supersaturated solution, c_s is the mass solute concentration in the solution at saturation, c/c_s is the supersaturation ratio, v_s is the molar volume of crystals, v is the number of ions/molecules of solute, $\sigma_{s,L}$ is the interfacial tension, D_p is the crystal diameter, R is the Gas constant, and T is the temperature.

  - Supersaturation can be quantified by the ratio of the mass solute concentration in the bulk solution to the concentration in the solution at the point of saturation.
  - Alternatively, it can be described by relative supersaturation, which is calculated by the following equation:

$$s = (c - c_s)/c_s, \qquad (18.2)$$

where s being relative supersaturation.

- What are the mechanisms involved in crystallization processes?
  - Crystallization from solution can be considered to be a two-step process.
    - The first step is the phase separation or birth of new crystals.
    - The second is the growth of these crystals to larger size.
    - These two processes are known as *nucleation* and *crystal growth*, respectively.
    - Analysis of industrial crystallization processes requires knowledge of both nucleation and crystal growth. These two processes together govern the crystal size distribution (CSD).
  - Mechanical design of a crystallizer has a significant influence on the nucleation rate due to *contact nucleation* (that which is caused by contact of crystals with each other and with the pump impeller, or propeller, when suspended in a supersaturated solution). This phenomenon yields varying rates of nucleation in scale-up and differences in the nucleation rates when the same equipment is used with different materials.
- What is nucleation? What are primary and secondary nucleation phenomena?
  - Nucleation results in the formation of new submicron size particles or nuclei by either primary or secondary mechanisms.
    - Primary mechanisms, those in which the product crystals do not participate, can be either homogeneous or heterogeneous. Homogeneous mechanism occurs in an ultraclean solution, whereas heterogeneous mechanism occurs in a solution containing foreign bodies such as dust and filter aids.

- First, molecules in the solution will associate into a microscopic cluster, which will either dissociate or continue to grow. When the cluster develops until it forms a lattice structure, it is then called an embryo.
- A stable crystalline nucleus is established when the crystal size exceeds D_p, given by Kelvin equation for the specific supersaturation ratio of the solution.
- Usually, the instantaneous formation of many nuclei can be observed out of the solution. The supersaturation driving force is created by a combination of high solute concentration and rapid cooling.
- In the salt example, cooling will be gradual so seeding is to be provided for the crystals to grow.
- Primary nucleation often occurs in precipitations with high levels of supersaturation, resulting in large numbers of small particles that compete for growth. If a disturbance/disruption of fine particles occurs during the initial crystallization event from an unseeded batch system, it may be difficult to grow an acceptable size distribution. The high surface area of the fines complicates the growth process as additional solute gets crystallized.
- On an industrial scale, a large supersaturation driving force is necessary to initiate primary nucleation.
- Secondary mechanisms are those in which the product crystals participate in the nucleation. These include contact, shear, fracture, and attrition processes.
- Secondary nucleation requires the presence of *parent* crystals that are involved in the initiation of further nucleation by breakage, attrition due to collisions, and removal of semiordered surface layers through fluid shear and collisions.
- The driving force for most industrial crystallizations is contact secondary nucleation, whereby the nuclei result from crystal to crystal, crystal to wall, or crystal to agitator contact.
- The contact can be between the solution and other crystals, a mixer blade, a pipe, a vessel wall, and so on. This phase of crystallization occurs at lower supersaturation (than primary nucleation), where crystal growth is optimal. The magnitude of secondary nucleation is significantly influenced by the level of supersaturation within the crystallizer, hydrodynamics, power input, shear forces, type of agitator, and concentration of solids.
- Secondary nucleation occurs in commercial crystallizers, where crystalline surfaces are present in order to produce large crystals.

- Nucleation occurs at the point that a crystal begins to form.
- As solids build up in a precipitator, the source of new nuclei is often primary nucleation at the feed point and secondary nucleation in the vessel.
- The relative rates of nucleation and growth are critical to crystallization kinetics, as they determine both crystal size and size distribution.
- Show on a diagram supersaturation, metastable, and unsaturated zones in crystallization process.

- Define metastable zone and state its importance.
  - Metastable zone is illustrated on Figure 18.1 The region between the solubility curve and the limit of stability (the dashed line) is called metastable zone.
  - It is the range of supersaturation in which growth of crystals occurs concurrently with secondary nucleation. This zone is system specific. Operation beyond this zone may result in undesirable primary nucleation. This zone has a supersaturation level where growth and secondary nucleation occur without primary nucleation.
  - An industrial crystallizer must operate below the solubility curve in order to produce a solid phase. However, since the objective is to avoid the creation of large number of small particles, the crystallizer

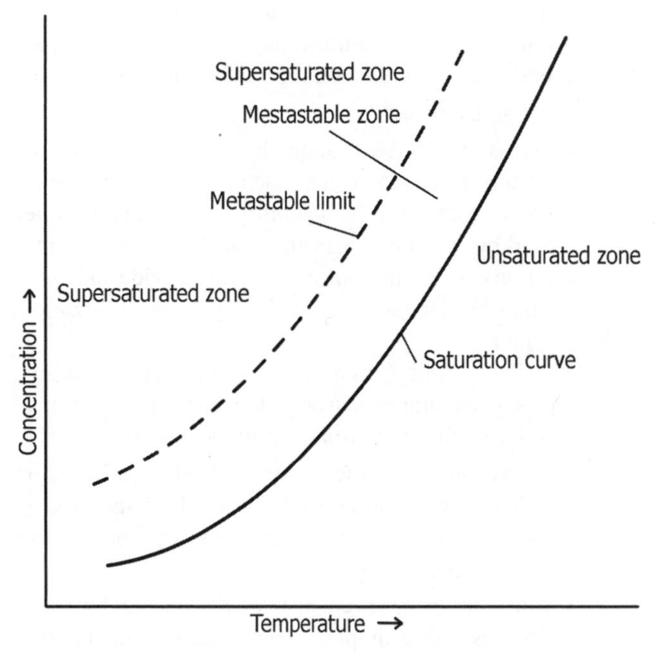

FIGURE 18.1 Depiction of supersaturation, metastable, and unsaturation zones in a crystallization process.

must not be operated in the unstable zone and its operation must be restricted to the metastable zone.
  - Seeding should occur within this zone in order to grow the crystals while minimizing the creation of additional nuclei.
- Give an equation for the rate of homogeneous nucleation.
  - Combining Kelvin equation with laws of chemical kinetics gives the rate of homogeneous nucleation, which is given below:

$$B_0 = A \exp[-16\pi v_s^2 \sigma_{s,L}^3 N_a / 3v^2 (RT)^3 \{\ln(c/c_s)\}^2], \quad (18.3)$$

    where B_0 is the rate of homogeneous primary nucleation, A is the frequency factor, and N_a is the Avogadros number. All other variables in the equation are as defined earlier.
- What are the common methods for obtaining supersaturation?
  - Decreasing solubility by cooling, which can be indirect or adiabatic evaporative. Example: crystallization of KCl from its aqueous solution by cooling.
  - Increasing concentration by removing a part of solvent, using, for example, evaporation. Example: crystallization of NaCl from its aqueous solution by evaporation.
  - *Drowning:* Drowning involves addition of an antisolvent (a liquid that, if added to a solution, reduces the solubility of the solvent) to the solution, which decreases the solubility of the solid and causes salting out. Example: crystallization of KNO_3 from its aqueous solution through salting out by isopropyl alcohol.
  - pH adjustment.
  - Increasing concentration by chemical reaction. Chemical reaction can be used to alter the dissolved solid to decrease its solubility in the solvent, thus working toward supersaturation. Example: precipitation of calcium sulfite from sulfur dioxide and lime/limestone found in wet flue gas desulfurization liquor.
    - For cooling and evaporative crystallization, supersaturation can be generated near a heat transfer surface and usually at moderate rates.
    - Drowning or reactive crystallization allows for localized, rapid crystallization where the mixing mechanism can exert significant influence on the product characteristics.
  - Where temperature changes are detrimental for some systems, for example, when dealing with protein-based drugs, solubility can be reduced by adjusting the pH to the isoelectric point, increasing ionic strength, adding nonionic polymers or adding a miscible nonsolvent.
- Explain the characteristics and differences between primary and secondary nucleation.
  - *Primary Nucleation:* Given a sufficient driving force, that is supersaturation in the case of solutions or supercooling in the case of a melt, a liquid to solid phase transformation commences with the initial formation of clusters, ordered collections of the crystallizing species. These clusters are precursors to the crystals eventually formed. In order to grow into a macroscopically detectable crystal, these nuclei have to reach a certain critical size. The critical radius is a measure of the critical nucleus size. Below the critical radius, redissolution of the nucleus is favorable and above the critical radius growth is more favorable. This nucleation mechanism is called *primary homogeneous nucleation*, which will occur if the system is free of impurities and mechanically undisturbed.
  - *Secondary Nucleation:* Under normal circumstances there are always impurities in the solution or melt, mechanical disturbances result from agitation of the solution, surface roughness of the equipment or vibrations from ancillary equipment. Under realistic conditions *heterogeneous nucleation* occurs at a much lower supersaturation and where impurities or rough vessel walls function as nuclei. The most frequently observed nucleation mechanism is called *secondary nucleation*. Secondary nucleation requires the presence of crystals of the material to be crystallized. Different mechanisms that lead to secondary nuclei are as follows:
    - *Initial Breeding:* Nuclei result from simply placing crystals into a supersaturated solution or supercooled melt via the washing off of dust particles from the surface of the crystals.
    - *Collision Breeding:* Nuclei result from fragments of existing crystals that are broken off due to mechanical impact on crystal faces due to crystal–crystal, crystal–wall, or crystal–stirrer (pump) collisions.
    - *Fluid Shear:* Nuclei result from clusters or outgrowths (sizes larger than the critical nucleus) being forced from the solid–liquid boundary layer due to shear forces resulting from liquid motion.
- How does secondary nucleation influence crystallizer performance and what are the measures to control and improve performance?
  - Secondary nucleation is the main source of new crystal nuclei without which a continuous crystallizer suffers lack of growing crystals.

- It is also the cause of production of fines, which reduce the quality of the product.
- Secondary nucleation rates can be controlled via diameter and tip speed of the impeller blade of the pump or agitator. Lower tip speeds and larger diameters result in lower secondary nucleation rates.
  - If control of tip speed and enlargement of impeller are not sufficient to reduce the amount of fines to below a value of the order of 10^{10} in terms of number per cm^3 (while still less than 1% in terms of mass percentage), other measures may have to be taken to control the product.
  - In general, not all nuclei will grow fast enough to become product size particles due to the fact that each crystal has its own, individual growth rate (this phenomenon is called growth rate dispersion).
  - In some cases small particles will agglomerate.
  - In all the other cases a fines trap has to be introduced to the crystallizer, either as an internal or an external design feature. A fines trap consists of an additional loop within the crystallizer in which the conditions are such, that all fines smaller than a given particle size passing through this loop are redissolved.
  - Nucleation is also a problem concerning the start-up of a crystallizer. Primary nucleation is difficult to control and unreliable, as it will not always occur at precisely the same supersaturation.
  - Primary heterogeneous nucleation depends on the number and the nature of the impurities and is therefore, within certain limits, a random event. As a consequence, reproducibility of product quality cannot be guaranteed and the performance of the crystallizer will vary. Product quality and purity are affected due to massive formation of small particles leading to agglomeration and caking tendency on storage.
  - If nucleation starts at too low a supersaturation, crystal growth may be slow and affect production rates. Crystallizer has to be of larger size in order to maintain required production.
  - In order to produce high-quality crystals in a reproducible manner, secondary nucleation by means of seeding is the preferred method of inducing the crystallization process. It is important to introduce the seed crystals at the same supersaturation.
  - Keeping the supersaturation constant subsequent to seeding normally results in high-quality crystals, provided the supersaturation selected coincides with the optimum growth rate for the system under consideration. The optimum growth rate is determined by laboratory tests.
  - In order to maintain constant growth rates requires good control of supersaturation and has to take into account the constantly increasing crystal surface area, which is ultimately responsible for the reduction of supersaturation.
  - Seed mass, seed size, seed surface quality (perfect or irregular), the seed state (dry or in suspension), and finally the point of introduction of the seeds into the crystallizer are important considerations in quality control.

- What are the normally kept c/c_s (concentration/saturated concentration) in crystallization processes?
  - During most crystallizations, c/c_s are kept near 1.02–1.05.
- What are the three types of mixing that occur during crystallization?
  - Mesomixing refers to turbulent dispersion of an incoming fresh feedstock plume by mixing it with the bulk liquid.
  - Macromixing refers to overall mixing performance during large-scale mixing and involves bulk fluid movement and blending. Macromixing reduces local differences in temperature, concentration, supersaturation, and suspended solids density.
  - Micromixing or turbulent mixing at the molecular level occurs locally in a small volume of the process, the materials coming in contact with each other a result of molecular diffusion.
- What is crystal habit and polymorphic state? Explain.
  - A crystalline particle is characterized by definite external and internal structures. Habit describes the external shape of a crystal, whereas polymorphic state refers to the definite arrangement of molecules inside the crystal lattice.
  - Use of different solvents and processing conditions may alter the habit of recrystallized particles, besides modifying the polymorphic state of the solid. Subtle changes in crystal habit at this stage can lead to significant variation in raw material characteristics.
  - Various indices such as particle orientation, flowability, packing, compaction, suspension stability, and dissolution can be altered even in the absence of significantly altered polymorphic state. These effects are a result of the physical effect of different crystal habits.
- What are the factors that restrict productivity and purity of crystals?
  - *Agglomeration:* Product size can be affected by agglomeration and fracture mechanisms. When

growing crystals collide, they may stick together and form new particles and when the collisions are inelastic, agglomerates form. The strength of the physical bonds thus form determines the stability of the agglomerates upon further collisions. For limiting crystal size, use of a surfactant can be effective to reduce stickiness. Agglomeration will be more extensive if there are more particles, which increase the probability of collisions or if the colliding particles are sticky. In general, agglomeration is associated with higher degree of supersaturation and even operation in the unstable zone, conditions under which large number of smaller size crystals are formed. To minimize agglomeration, supersaturation should be limited.

- *Liquid Inclusion in Individual Crystals and Agglomerates:* Liquid can get trapped inside growing crystals through several mechanisms. Higher growth rates are responsible for increased amounts of liquid entrapment. Thus, higher supersaturation could be problematic, particularly when large amounts of liquid impurities are present. Liquid can also get entrapped between colliding particles during agglomeration process. Again, the probability of this happening increases with increased supersaturation. In this regard, it is necessary to reduce the time for interaction and/or rapidly relieve the supersaturated condition.
- *Liquid Impurities Outside the Crystals:* When product collection is through filter cake, the amount of liquid outside the crystals, in the voids, is much larger than the amount within them. Since this quantity is, in general, inversely proportional to the square root of the mean crystal size, cakes containing small particles require larger quantities of wash liquid to achieve the same washing efficiency than cakes containing larger crystals.
- *Cavitation:* Though cavitation has been successfully used for homogenization, it has a destructive effect on the materials of construction and thus disrupting the process. Therefore more durable and expensive materials, such as polycrystalline diamond and stainless steel must be employed. It is also necessary to determine whether thermal or sonochemical reactions occur and if so their importance to the system under consideration.

- How are crystals of cubic and needle shapes produced?
  - Cubic shapes (highly desirable) of crystals are produced from a high solids content slurry and high power input from agitation.
  - Needles with high aspect ratio, that is, particle length to width ratio, are produced from low solids content slurry with low power input from agitation.
- What is the difference between magma and mother liquor?
  - The solid–liquid mixture obtained as a result of crystallization is referred to as *magma*.
  - The liquid or solution in which crystals are formed is called *mother liquor*.
- Name the two step theory of crystal growth.
  - In the first step, diffusion of solute molecules from the solution to the crystal–solution interface occurs.
  - In the second step, the kinetic step, a first-order reaction occurs at the surface of the crystal, during which solute molecules from the solution become incorporated into the crystal lattice structure, usually at dislocations in the crystal.
  - This theory is also known as *diffusion–reaction theory*.
  - The slurry velocity across the crystal surface determines which of the two mechanisms controls the process. High viscosity systems such as carbohydrates are often diffusion controlled.
  - Figure 18.2 illustrates diffusion-controlled regions and surface reaction resistance-controlled regions when plotted as growth rate as a function of solution mixing velocity.
- What are the factors that influence crystal growth rates?
  - Crystal growth rates depend not only on the temperature, pressure, and composition of the mother liquor but also on parameters such as supersaturation (under cooling), fluid flow conditions near the crystal surface, history of the crystals, the nature of the surfaces of crystals, and the presence or absence of additives (impurities) in the mother liquor. Crystal growth rates determine the retention time and therefore the size of the crystallizer, influencing investment costs.
  - Fast crystal growth encourages liquid inclusions in the crystals and often leads to rough surfaces.

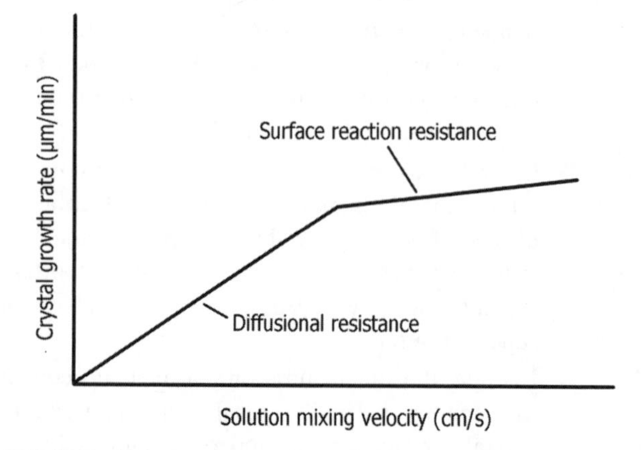

FIGURE 18.2 Crystal growth rate versus solution mixing velocity.

- How are crystal growth rates and crystal sizes controlled?
  - Crystal growth rates and crystal sizes are controlled by limiting the degree of supersaturation.
- What are the generally acceptable crystal growth rates?
  - A generally acceptable crystal growth rate is 0.10–0.80 mm/h.
- Why is it necessary to know eutectic points on a phase diagram for crystallization processes?
  - Determination of eutectic points is important since yield of a desired component is limited by the position of the eutectic.
- While larger crystals are generally desirable products in a crystallization process, give examples where small crystals are preferred products.
  - Nanotechnology and dyes and pigments are two areas where small crystals are the desired products.
  - Another area in which controlled production of uniformly small crystals is desirable, is the photographic materials industry. These small crystals and their controlled production is a typical domain of so called precipitation crystallization.
  - Crystals of table salt (sodium chloride), for example, must be small enough to fit through the openings of the salt cellar (generally smaller than 400 μm), yet large enough not to exhibit the typical properties of a powder that can lead to caking. Powders are generally not free-flowing since the attractive forces between the crystals are stronger than the gravitational force. The critical lower size limit where powder properties become noticeable is dependent on the material.
- "During crystallization by cooling/chilling, the temperature of the solution is kept 0.5–1.2°C below the saturation point at the given concentration." *True/False?*
  - *True.*
- What are the considerations involved in the selection of crystallization as a means of separation over distillation as the preferred method?
  - Solute is heat-sensitive and/or a high boiling material and decomposes at temperatures required to carry out distillation.
  - Low or no relative volatility exists between solute and contaminants and/or azeotropes formed between solute and contaminants.
  - Solute (product) is desired in particular form.
  - For example, if solute can be purified via distillation, then it must be solidified subsequently by flaking or prilling and crystallization may be a more convenient scheme to use in such cases.
  - *Comparative Economics Favor Crystallization:* If distillation requires high temperatures and energy usage, crystallization may offer economic incentives.
- What are the objectives involved in the design and operation of an industrial crystallizer?
  - Producing larger and more uniform crystals of the desired product.
  - Reducing the formation of agglomerates.
  - Reducing the amount of liquid impurities included in crystal agglomerates.
  - Reducing the liquid retained by the crystal cake after solid–liquid separation and washing.
- What are the mechanisms involved in the operation of an industrial crystallizer?
  - To generate supersaturation.
  - To relieve the supersaturation by providing the necessary active crystallization volume.
  - To control supersaturation by recirculation of mother liquor or preferably, magma (slurry) from the active crystallization volume.
  - Additionally to manage product crystal size, it is desirable to have mechanisms for dissolution of fines and for removal of classified product.
  - Table 18.1 illustrates the choice of the operating mechanism for the crystallizer.
- Name the instances in which solid crystals formed are not the commercial products of a crystallization process.
  - In some cases mother liquor might be the desired product. In some other instances, neither the mother liquor nor the crystals are the desired products. Examples include waste disposal processes in which crystallization is used to improve the disposal process. In such instances the solid crystals obtained after solid–liquid separation are disposed off with or without additional processing.
- Name some materials that are added to boiler feed water to prevent crystal growth.
  - Relatively large molecules, for example, glue, tannin, dextrin, or sodium hexametaphosphate.
  - Prevent formation of calcium carbonate crystals, reducing scale formation.
- Explain the phenomena of caking of crystals.
  - Tendency of caking is due to a small amount of dissolution taking place at the surface of crystals and subsequent reevaporation of solvent, leading to tight bonding of crystals.
  - Partial pressure of solution is *less than* P^0 for water and so some condensation occurs on crystal surfaces even though humidity of atmosphere is less than 100%.
  - Solution formed penetrates into pack of crystals by capillary action of small spaces between the crystals.

TABLE 18.1 Choice of Operating Mechanism for a Crystallizer

Mechanism	Should be Considered When	Should Not be Considered When
	Supersaturation Generation	
Cooling	The solubility of crystallizing component to moderately decreasing temperature. Very low crystallization temperatures are required	The solubility of the crystallizing component stays flat or increases with decreasing temperature
Evaporation	The solubility of the crystallizing component stays flat or increases with decreasing temperature	The solubility of the crystallizing component decreases steeply to moderately with decreasing temperature
Adiabatic evaporating cooling	The solubility of the crystallizing decreases steeply to moderately with decreasing temperature	The solubility of the crystallizing component stays flat or increases with decreasing temperature. The bubble point decreases very slowly with pressure
	Relieving the Supersaturation	
Mixed suspension	Magma recirculation is the mechanism of choice for control of supersaturation generation	Liquor recirculation is the mechanism of choice for control of supersaturation generation
Classified suspension	Liquor recirculation is the mechanism of choice for control of supersaturation generation	Magma recirculation is the mechanism for control of supersaturation generation
	Control of Supersaturation Generation	
Magma recirculation	The goal is to bring growing crystals into contact with liquid that becomes supersaturated. It is not possible or desired to maintain a classified suspension	Attrition and crystal breakage in the recirculation flow is to be avoided
Liquor recirculation	The goal is to have no crystals present when supersaturation is created and then to bring the stable supersaturated liquid into contact with the growing crystals. A classified suspension can be and is maintained in the active crystallization volume	Frequent changes and upsets in the operating conditions are likely to upset the classified suspension
	Particle Size Manipulation	
Fines dissolution and classified product removal	Additional control over particle distribution is desired. Crystals smaller than a certain size range and larger than a certain size range can be removed without affecting the active crystalli zation volume	These mechanisms are likely to interfere with the active crystallization volume

Source: Samant KD, O'Young L. *Chemical Engineering Progress*, October 2006, pp. 28–37.

   - Caking can occur if adsorbed moisture subsequently evaporates when percent humidity of atmosphere falls.
- How is caking prevented/reduced?
   - By forming large and uniform size crystals.
   - By adding small amounts of insoluble materials, for example, calcium phosphate, talc, water-repellant agents like stearic acid or a stearate.
- What are the considerations involved in selecting the operating mode of a crystallization process?
   - Lowering the temperature of the feed solution by direct or indirect cooling. If solute solubility is strongly temperature dependent, this is the preferred approach.
   - Adding heat to the system to remove solvent and thus solidify the solute. This technique is effective if solubility is insensitive to temperature.
   - Vacuum cooling the feed solution without external heating. If solubility is strongly dependent on temperature, this method is attractive.
   - *Combining Techniques:* Especially common is vacuum cooling supplemented by external heating for systems whose solubility has an intermediate dependence on temperature.
   - *Adding Nonsolvent:* This is a common technique for precipitating solute from solution and is useful as both a laboratory technique and as an industrial process for product recovery.
- What are the effects of impurities or additives on crystallization processes?
   - Impurities can lead to changes in the observed crystal shape. In terms of postcrystallization processing this can have a negative effect on filterability, drying, or on the handling of the dry solids.

- The effect of impurities can be of advantage to tailor the shape of the crystals. By exploiting the different interactions of different chemical species with the growing crystal faces, it is possible to manipulate the final macroscopic shape of the crystals by deliberately adding an impurity (additive) to the crystallization liquor.
- One strategy to influence the shape of the crystal is to find an additive that reduces the growth rate of faster growing faces while leaving the growth rates of the slower growing faces unchanged. This can be achieved by identifying additives that fit into the lattice on the fast growing surfaces (but not the others).
- The net effect of impurities is usually a retardation of crystal growth, likely due to adsorption of impurity on the surface of nucleus/crystal.

• What is the difference between evaporation and crystallization with respect to recovery of dissolved solids?
  - Complete recovery of dissolved solids is obtainable by evaporation.
  - Chilling leads to obtaining eutectic composition.
  - Melt crystallization also leads to eutectic compositions.

• What is the primary difference, with respect to size of crystals that are formed in crystallization process, between slow cooling and rapid cooling?
  - *Slow Cooling:* Large crystals are formed.
  - *Rapid Cooling:* Large crop of small crystals are formed.

• Name the important applications of industrial crystallization.
  - *p*-Xylene from mixed xylenes: Crystallization or adsorption processes are more common.
  - Naphthalene purification from coal tar: Melt crystallization.
  - Purification of acrylic acid above 98%: By melt crystallization.
  - Ultrapure bisphenol A: By melt crystallization.
  - Monochloroacetic acid: By melt crystallization.
  - Recovery of pickling bath effluents: By vacuum cooling crystallization.
  - Concentration of scrubber effluents from flue gas desulfurization units from thermal power plants: By evaporation crystallization.
  - Caffeine recovery from caffeine containing wastewaters resulting from supercritical caffeine extraction processes: By evaporation combined with surface cooling crystallization.

18.1.1 Crystallization Equipment

• Name different types of crystallization equipment.
  - Forced circulation (evaporative) crystallizer (Swenson): Most widely used crystallizer.
  - Oslo evaporative crystallizer.
  - Fluidized suspension crystallizer (Oslo).
  - Draft tube (DT) crystallizer.
  - Draft tube baffle (DTB) evaporation crystallizer.
  - Surface-cooled forced circulation crystallizer.
  - Direct contact refrigeration crystallizer.
  - Scraped-surface crystallizer.
  - Tank crystallizers (Static and agitated types).
  - Column crystallizers.

• What is the essential difference between evaporative crystallizers and cooling crystallizers?
  - *Evaporative Crystallizers* Near-isothermal conditions.
    - ➢ Supersaturation is achieved due to removal of solvent through evaporation.
  - *Cooling Crystallizers:* Supersaturation is achieved through lowering of temperature of solution:

    (i) Through sensible heat transfer.

    (ii) Through evaporative cooling.
  - Figure 18.3 illustrates the operation of cooling crystallizer.
  - *Operational Characteristics:* Feed enters at higher saturation temperature than is maintained in the crystallizer and is cooled in shell and tube or scraped-surface exchangers to remove sensible heat and heat of crystallization. Solids encrustation problems are generally confined to the cooling surfaces.

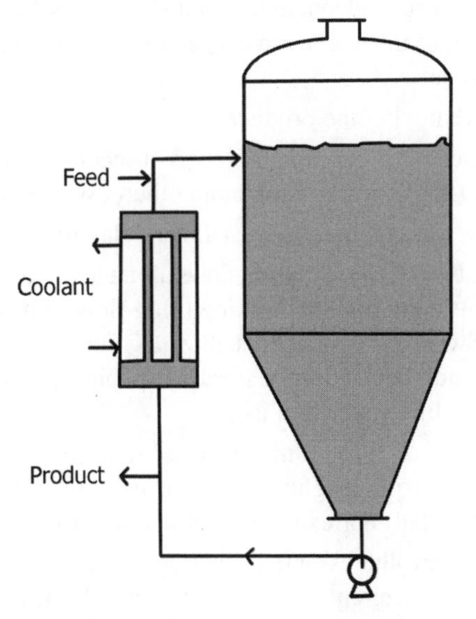

FIGURE 18.3 Cooling crystallizer.

- *Advantages:* Operation and control is simple. No vacuum equipment is necessary.
- *Disadvantages:* Care must be taken to prevent fouling of the cooling surfaces by maintaining low process and coolant temperature differences across cooling surfaces. If severe fouling is anticipated, scraped-surface heat exchangers may be necessary to assure reliable operation.

- What are the circumstances under which (a) evaporative crystallization and (b) cooling crystallization are used?
  - (a) Where solubility shows little variation with temperature. Examples include crystallization of NaCl, Na_2SO_4, $(NH_4)_2SO_4$, and $CaCl_2$.
  - (b) Where solubility variation is rather steep. Examples include crystallization of KCl, $NiSO_4{\cdot}6H_2O$, $CuSO_4{\cdot}5H_2O$, $AgNO_3$, and melamine.
- What is the advantage of using vacuum in crystallizers?
  - Vacuum is used in evaporative cooling type crystallization.
  - Under vacuum, flash evaporation of solution takes place, producing rapid cooling accompanied by a small increase in concentration.
  - Used in sugar industry where concentrated sugar solution is fed to an evaporator operating under vacuum.
- What are the functions of agitator in the design of crystallizers?
  - Keeping the crystals off the bottom of the crystallizer to maintain crystal growth.
  - Maintaining adequate mixing to prevent local excessive levels of supersaturation at feed points, cooling and evaporation surfaces, in order to avoid primary nucleation and encrustations.
  - Ensuring adequate heat transfer for cooling or heating so that the unit can operate within the metastable zone.
  - Removing the product.
  - Generating acceptable levels of secondary nucleation that will result in obtaining the desired CSD.
- How are continuous crystallizers classified?
  - *Linear Type:* Solution flows along a trough/pipe with little longitudinal mixing (plug flow). For example, Swenson–Walker crystallizer. Trough is divided into a number of sections, each capable of cooling with cooling water in a jacket.
  - *Stirred Type:* Uniform conditions are maintained by stirring. Highly viscous/heat sensitive liquids are handled. For example, Oslo crystallizer.
- What are the essential features and characteristics of a forced circulation crystallizer? Illustrate with a diagram.
  - Forced circulation is the most widely practiced method of crystallization and use evaporation or adiabatic cooling to generate supersaturation, provide a mixed suspension as the active volume for relieving the supersaturation, employ magma recirculation to control generation of supersaturation and can, in some cases, provide a mechanism for classified product removal. These units do not provide a mechanism for dissolution of fines.
  - Used for feeds where high rates of evaporation are required, where there are scaling components, or when crystallization must be achieved in solutions with inverted solubility.
  - This type of unit, also known as the *circulating magma crystallizer* or the *mixed suspension mixed product removal (MSMPR) crystallizer*, consists of a body sized for vapor release with a liquid level high enough to enclose the growing crystals. Suction from the lower portion of the body passes through a circulation pump and a heat exchanger and returns to the body through a tangent or vertical inlet. The heat exchanger is omitted when adiabatic cooling is sufficient to produce a yield of crystals.
  - The supersaturation is controlled so as to avoid spontaneous nucleation, by sufficient circulation capacity.
  - FC units operate with a high rate of circulation, which limits scaling, lending this unit to the evaporation of solutions with components that will scale, such as calcium sulfate. The FC unit uses a low shear, low head, and high volume axial flow pump that turns over the active volume (which includes the body, piping, and heat exchanger) about 1–4 times/min. Despite its low shear design, the circulating pump can still be a source of crystal breakage and secondary nucleation.
  - Figure 18.4 illustrates a typical forced circulation crystallizer of the Swenson design.
  - The unit consists of a closed vessel with a conical bottom. A combined stream consisting of fresh feed and recirculated slurry is pumped through the circulating pipe to a vertical or horizontal heat exchanger, where it is heated by condensing steam. This stream is introduced to the crystallizer below the liquid surface in the vapor body (where flashing occurs), mixes with the slurry near the point of feed entry and raises the local temperature enough to cause flashing at the liquid surface. Subsequent cooling at the surface results in supersaturation that is relieved either as crystal growth or the birth of new nuclei in the active volume. The vapors are removed by condensation.

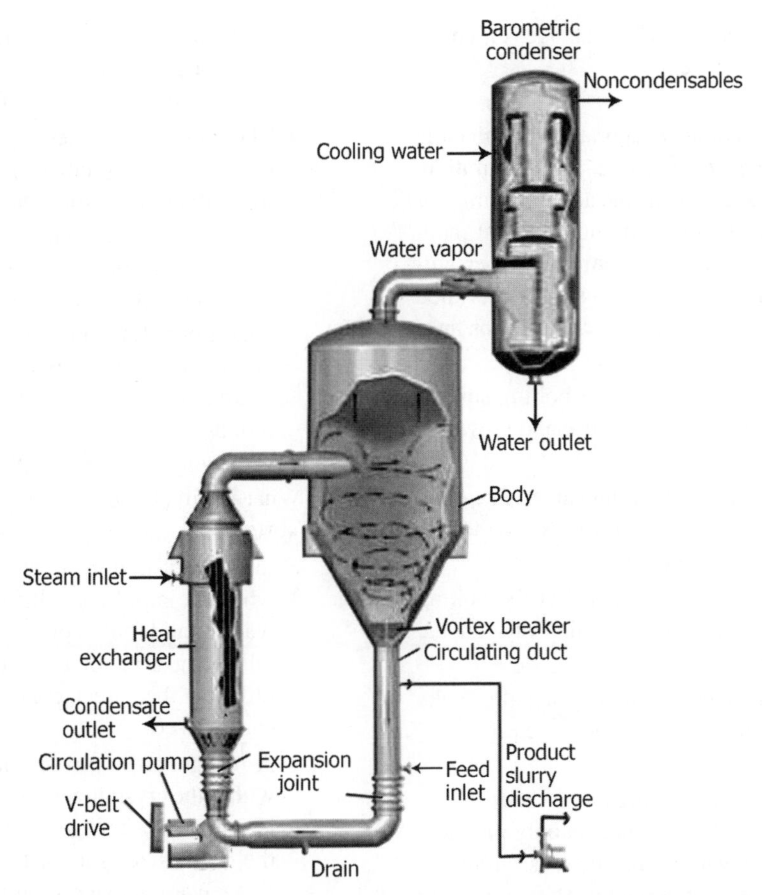

FIGURE 18.4 Forced circulation crystallizer. (*Courtesy*: Swenson Technology, Inc.)

- The product slurry goes to a thickener or separating device, such as a hydroclone, centrifuge, or filter for crystal separation.
- In some cases, classified product removal is achieved by the use of an elutriation leg, which is located at the bottom of the conical section of the vessel. Separation is achieved by upward flow of small amount of the mother liquor through the elutriation leg. This way crystals below a certain size are prevented from being withdrawn with the product slurry.
- The common use of FC crystallizer is as an evaporative crystallizer with materials having relatively flat or inverted solubility. It is also useful with compounds crystallized from solutions with scaling components.
- Most FC units operate under vacuum or at slight superatmospheric pressure.
- The heat exchanger is normally single or two pass type and is designed for relatively low temperature rises in the solution. This limits the supersaturation of scaling components when heating materials of inverted solubility. In most applications, the ΔT in the heat exchanger is limited to prevent bulk boiling within the tubes or vaporization at the tube wall.
- For applications where boiling point elevation is low, such as for sodium carbonate, it is possible to significantly reduce the cost of operation by vapor recompression.
- FC units typically range from diameters of 0.6–6 m, with some units approaching as large as 12 m in diameter.

• For what types of applications, forced circulation crystallizers are preferred?
  - Where high rates of evaporation are required.
  - Used where large crystal size is not a requirement.
  - Where there are scaling compounds.
  - Where crystallization is achieved in inverted solubility solutions.
  - Where the solution is of relatively high viscosity.
  - Examples of materials handled include sodium chloride, sodium sulfate, sodium carbonate monohydrate, citric acid, monosodium glutamate, urea and other similar crystalline materials.

- What are the basic principles and advantages of a draft tube baffle (DTB) crystallizer?

  Principles
  - Uses direct contact cooling, evaporation, or adiabatic evaporative cooling to generate supersaturation, provide a mixed suspension as the active volume for relieving the supersaturation, employ internal magma recirculation to control generation of supersaturation, normally provide a mechanism for classified product removal and provide a mechanism for fines dissolution when a baffle is present.
  - Growing crystals are brought to the boiling surface where supersaturation is most intense and growth is most rapid.
  - The baffle permits separation of unwanted fine crystals from the suspension of growing crystals, thereby effecting control of the product size.
  - Sufficient seed surface is maintained at the boiling surface to minimize harmful salt deposits on the equipment surfaces.
  - Low head loss in the internal circulation paths make large flows at low power requirements feasible.

  Advantages
  - Capable of producing large singular crystals.
    - Larger crystals can be obtained as only the clear solution, together with some fines, is circulated through the heat exchanger, while the crystals are circulated at low speed (60–125 rpm) in the equipment by means of a very large impeller pump (often exceeding 1 m in diameter) located within the draft tube. The low circulation rate results in less attrition of the crystals due to reduced mechanical impact with both pump components and other crystals.
  - Longer operating cycles.
  - Lower operating costs.
  - Minimum space requirements. Single support elevation.
  - Adoptable to most corrosion-resistant materials.
  - Can be easily instrument controlled.
  - Simplicity of operation, start-up and shutdown.
  - Produces a narrow crystal size distribution for easier drying and less caking.
  - Changes in capacity have minimal effect on product size.
- For what type of application a DTB crystallizer is preferred?
  - Where excess nucleation makes it difficult to achieve a crystal size in the range of 10–30 mesh.
  - This crystallizer, in both the adiabatic cooling and evaporative type, includes a baffle section surrounding a suspended magma of growing crystals has an annular settling zone, from which stream of mother liquor is removed containing excess fine crystals.
  - These fines can be destroyed by adding heat (as in an evaporating crystallizer) or by adding water or unsaturated feed solution.
  - The magma is suspended by means of a large, slow moving propeller circulator that fluidizes the suspension and maintains relatively uniform growth zone conditions. Examples of materials handled include ammonium sulfate, potassium chloride, diammonium phosphate, hypo, epsom salts, potassium sulfate, monosodium glutamate, borax, sodium carbonate decahydrate, trisodium phosphate, urea and so on.
- What is direct contact refrigeration crystallization?
  - When crystallization occurs at such a low temperature that it is impractical to use surface cooling or when the rapid crystallization of solids on the tube walls would foul a conventional surface-cooled crystallizer, a DTB or a forced circulation crystallizer utilizing the direct contact refrigeration technique can be used.
  - A refrigerant is mixed with the circulating magma within the crystallizer body where it absorbs heat and is vaporized. Refrigerant vapor leaves the surface of the crystallizer similar to water vapor in a conventional evaporative crystallizer. The refrigerant vapor is then compressed, condensed, and recirculated to the crystallizer to maintain continuous operation. Figure 18.5 illustrates direct contact refrigeration DTB crystallizer.

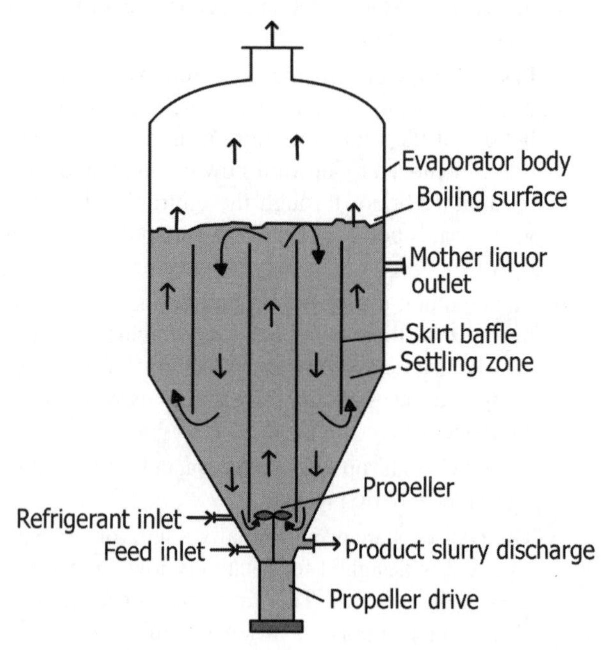

FIGURE 18.5 Direct contact refrigeration DTB crystallizer.

- Refrigerants must be relatively insoluble in the solutions processed and have the thermodynamic characteristics to minimize compressor power.
- Examples are the crystallization of caustic dehydrates with Freon or propane and that of *p*-xylene with liquid propane refrigerant.
- The annular baffle area functions as a settling zone through which a stream of magma liquid and fines are separated from the slurry in the active volume and removed via gravitational settling.
- A slow moving propeller (60–125 rpm) installed in a draft tube pumps the circulating slurry upward to the boiling surface, where the supersaturation created by concentration and cooling is relieved as crystal nucleation and growth.

• What are the salient features of surface-cooled crystallizers?
  - These crystallizers are of two types, namely, surface-cooled (SC) and surface-cooled baffled (SCB).
  - Surface cooling is used to generate supersaturation.
  - A mixed suspension is used as active volume for relieving supersaturation.
  - Magma recirculation is employed to control supersaturation generation.
  - No mechanism is provided to remove classified crystals.
  - In baffled type (not in unbaffled type), a mechanism is provided for fines dissolution.
  - Figure 18.6 illustrates a common SCB crystallizer. It consists of a shell and tube heat exchanger, a vessel with internal skirt baffle and a recirculation pump.
    - ➢ The baffle is positioned so that it acts as a partition between a settling zone and the active crystallization volume. The feed inlet is located on the recirculation pipe just prior to the recirculation pump.
    - ➢ The settling zone outside the baffle provides an outlet for the mother liquor.
    - ➢ The recirculating magma is returned to the crystallizer vessel through a central tube extending into the active crystallization volume.
    - ➢ The desired supersaturation is generated by cooling the mixed stream formed by the fresh feed and the recirculating slurry on the tube side of the heat exchanger. The tube side surface, being the coldest part of the process, is prone to solids buildup. Therefore, the exchanger is operated at a temperature difference between the tube and shell side, typically not exceeding 5–10°C.
    - ➢ The recirculating rates are sufficiently high to ensure that the supersaturation generated in the tubes is low.
    - ➢ Returning the recirculating slurry to a central tube, extending into the baffled volume, ensures thorough mixing inside the active crystallization volume.
    - ➢ Fines dissolution is done in a tank by drawing from the top of the settling zone created by the baffle as shown in the figure. This drawoff induces an upward flow into the settling zone. Fines below a certain size (whose settling velocity is less than the upward flow) from this zone are removed by the drawoff nozzle and sent to the dissolution tank and sent back to the crystallizer.

• What are characteristic features of an Oslo crystallizer? Illustrate.

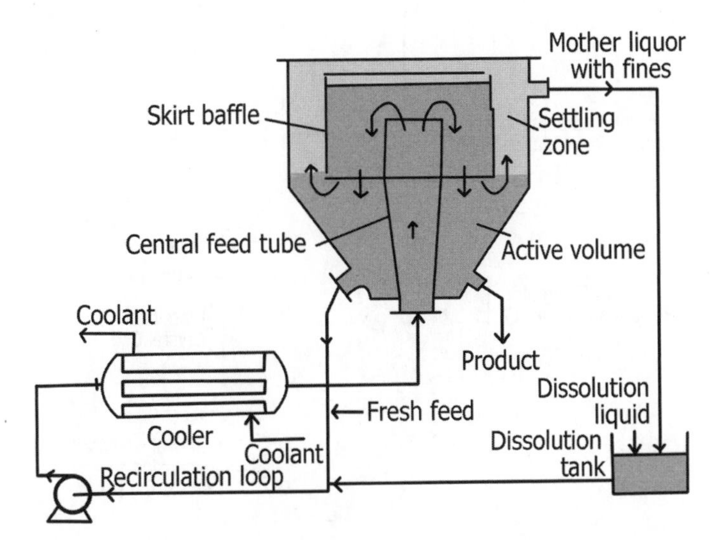

FIGURE 18.6 Surface-cooled baffled crystallizer using external heat exchanger surface to generate supersaturation by cooling.

- Oslo type crystallizer, also called classified suspension crystallizer, is used for the production of large, coarse crystals.
- Supersaturation is generated by surface cooling, evaporation, or adiabatic evaporative cooling.
- A classified suspension is provided as active volume for relieving supersaturation.
- Supersaturation generation is controlled by liquor recirculation. Liquor recirculation ensures that there will be no or minimal attrition and crystal breakage. The ability to maintain classified suspension is sensitive to changes in the recirculation rate.
- Built-in mechanisms are provided for fines dissolution and classified product removal.
- The basic design criteria are (a) desupersaturation of the mother liquor by contact with the largest crystals present in the crystallization chamber and (b) keeping most of the crystals in suspension without contact by a stirring device, thus enabling the production of large crystals of narrow size distribution.
- Figure 18.7 illustrates a typical design.
- The classifying crystallization chamber is the lower part of the unit. The upper part is the liquor–vapor separation area where supersaturation is developed by the removal of the solvent (water for most applications).
- The slightly supersaturated liquor flows down through a central pipe and the supersaturation is relieved by contact with the fluidized bed of crystals. Desupersaturation occurs progressively as the circulating mother liquor moves upward through the classifying bed before being collected in the top part of the chamber. Then it leaves via the circulating pipe and after addition of the fresh feed, passes through the heat exchanger where makeup heat is provided. It is then recycled to the upper part.
- The operating costs of the Oslo type crystallizer are much lower than with any other type when both large and coarse crystals are required.
- Since crystals are not in contact with any agitation device, the amount of fines to be destroyed is lower and so is the corresponding energy requirement.
- It is also used for reaction crystallization and separation crystallization when several chemical species are involved.
- Typical products are $(NH_4)_2SO_4$, Na_2SO_4, $AgNO_3$, hydrated monosodium glutamate and monoammonium phosphate (MAP).
- The crystallizer is best suited for use with compounds that have high settling rates, that is, greater than 20–40 mm/s. If the crystals have high settling velocities, the larger particles will settle out more quickly. Crystals with slow settling velocities mean large crystallizers and low crystal production rates.

• How does a fluidized suspension/Oslo crystallizer operate?
  - In the fluidized suspension/Oslo crystallizer supersaturation can be achieved through evaporating

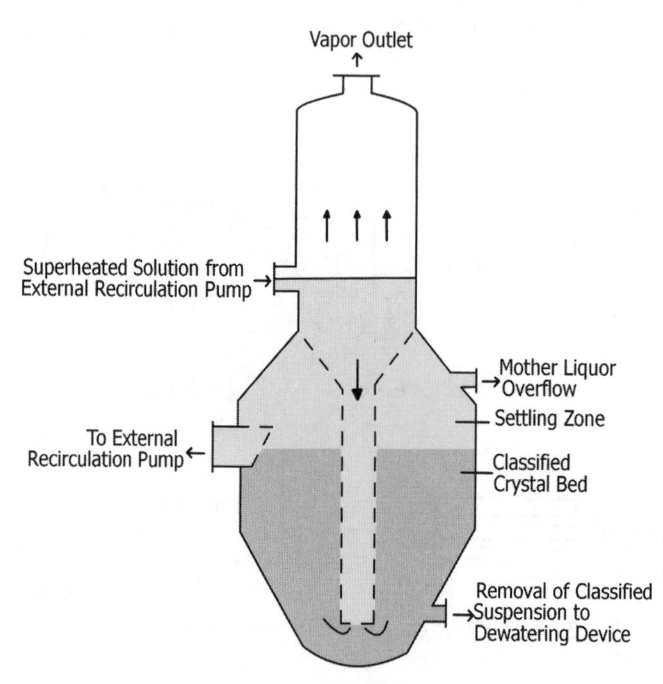

FIGURE 18.7 Oslo type crystallizer.

cooling, that is, adiabatically or by the evaporation and cooling of a heated stream leaving a heat exchanger within the recirculation loop and then relieved by passing the supersaturated liquid through a fluidized bed of crystals.

- The fluidized bed configuration, like the DTB, enables the production of larger crystal size distribution (CSD).

18.1.2 Crystallization Types

- What is the principle involved in melt crystallization? Illustrate the process with a flow sheet.
  - If an impure molten material is cooled to its freezing point and more heat is removed, then some of the material will solidify.
  - Impurities will concentrate in the remaining melt, the residue.
  - Purified product is recovered by separating the solid from the residue and remelting it.
  - The advantages of melt crystallization are in the relatively low energy demand of the freezing process and in the high selectivity of crystallization.
  - Melt crystallization can be carried out by a layer process, where the crystalline material to be separated forms on a cooled surface, or by crystallization from suspension.
  - In solid layer melt crystallization, the product is obtained by enforced encrustation. Heat is withdrawn from the melt through the crystalline layer by means of cooled walls within the crystallizer.
  - Figure 18.8 gives a flow sheet of a melt crystallization process.
  - In a wash column the crystals are flowing countercurrent to a stream of pure liquid, which replaces the impurities at the crystal surfaces. This liquid can be a saturated solution of the crystallizing component, or melt.
- What are the relative merits of layer and suspension crystallization processes?
  - In layer crystallization growth rates are about two orders of magnitude higher than in suspension crystallization.
  - In suspension crystallization high purities can be obtained in a single separation step due to a large surface area for mass transfer in comparison to layer crystallization processes. The surface area available for the same volume of crystals is approximately two to three orders of magnitude higher.
  - A further benefit of suspension crystallization is the applicability for continuous operation.
  - On the negative side, in suspension crystallization processes the next step of solid–liquid separation plays an important role on the final product purity. The melt, concentrated with impurities and high viscosity of the liquid phase make the separation more difficult. For this reason the solid–liquid separation in suspension melt crystallization processes is very often carried out in wash or crystallization columns.
  - Fouling of heat transfer surfaces is another problem arising in suspension crystallization. This requires use of scrapers or wipers as additions to the equipment.
  - Summary of comparison of layer and suspension crystallization processes is given in Table 18.2.

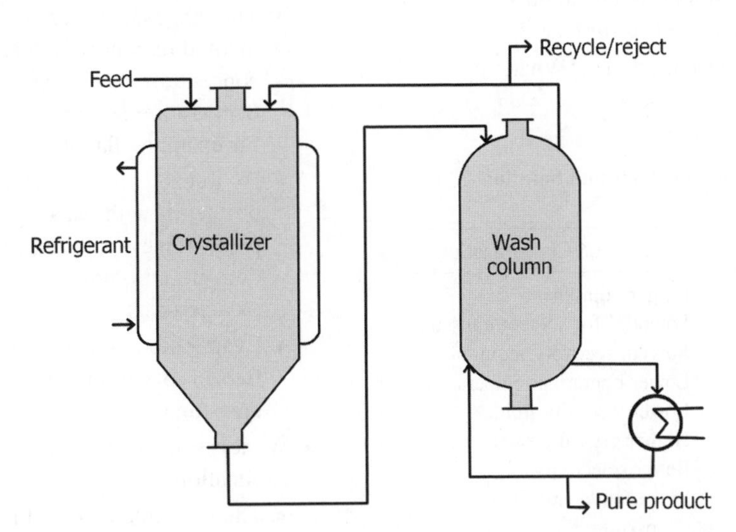

FIGURE 18.8 Generalized melt crystallization process with a wash column.

TABLE 18.2 Comparison of Layer and Suspension Crystallization Processes

Layer	Suspension
Batch	Continuous
Crystal fixed to the surface	Crystal free-flowing in suspension
Relatively simple components	More complex components
Simple scale-up	Scale-up requires specific know-how
Complex dynamic control	Simple steady-state control
Rapid growth requires sweating and multiple crystallization steps	Slow growth produces pure crystals in a single crystallization step
Simple gravity separation	Complex separation required due to massive crystal surface
Low efficiency leads to high utility and capital costs	High efficiency reduces utility and capital costs
Broad application base in small to medium capacity range	Limited references mainly for large capacity systems

- What are the differences between melt and solution crystallization?
  - Table 18.3 gives a summary of the differences between melt and solution crystallization processes.
- What is reactive crystallization?
  - Reactive crystallization, where a solid-phase crystalline material results from the reaction of two components, can often be performed more profitably in a crystallizer than in a separate reactor.
  - The DTB crystallizer is particularly suited for reactive crystallization. The reactants are mixed in the draft tube where a large volume of slurry is mixed continuously with the materials to minimize the driving force (supersaturation) created by the reaction.
  - Forced circulation type of crystallizer can also be used with reactants mixed in the circulating piping.
  - Heat of reaction is removed by vaporizing water or other solvents as in a conventional evaporative crystallizer.
  - An example of reactive crystallization is the production of ammonium sulfate from liquid or gaseous ammonia and concentrated sulfuric acid.
- What is freeze concentration? What are its applications?

TABLE 18.3 Differences Between Melt and Solution Crystallization

Melt Crystallization	Solution Crystallization
Compact equipment	Larger equipment
No solvent emissions	Potential for solvent emissions
No solvent recovery	Solvent recovery required
Higher operating temperatures	Lower operating temperatures
Higher viscosity fluid	Lower viscosity fluid
Moderate crystal growth rates	Higher crystal growth rates
Good selectivity	Better selectivity
Crystallization only by cooling	Evaporative crystallization possible

  - Freeze concentration is a crystallization process where water is crystallized out of an aqueous solution. Freeze concentration has been performed by both layer and suspension forms of crystallization, but suspension processes are more common.
  - Aqueous solutions form mainly eutectic systems, so ice can be crystallized in a very pure form. After the separation of the ice crystals from the mother liquor a more concentrated aqueous solution and pure water are obtained.
  - Freeze concentration is mainly used in food industry for concentrating of aqueous food stuffs and in chemical industry for purification of wastewater containing highly toxic components. The applications in food industry include concentration of coffee, juices, beer, milk, alcoholic beverage, and vinegar.
  - The applications for wastewater treatment are usually combined with an incinerator. The removal of water can cause significant savings to the investment and energy costs of the incinerator, due to the smaller amount and higher caloric value of the feed.
  - Though evaporation is the most commonly used method to concentrate aqueous solutions. By freeze concentration, it is possible to preserve volatile or temperature-sensitive components in the solution, for example, flavors and vitamins in foodstuffs.
  - By freeze concentration, it is also possible to avoid problems with vapor/gas handling and corrosion present in evaporation systems and achieve a reduction in emissions and transport, packaging and storage costs.
  - Evaporation is also more energy intensive, making freeze concentration attractive due to the ever increasing energy costs.
- What is sonocrystallization? What are its attributes and applications?
  - The principle involved in sonocrystallization is that high intensity sound initiates nucleation and helps to

control crystal size and habit to yield products that meet better required specifications.

- Application of ultrasound to crystallization improves quality of the product.
- The most important mechanism by which ultrasonic irradiation can influence crystallization is ultrasonic cavitation.
- Cavitation is particularly effective for inducing nucleation that helps reproducibility of the crystals.
- Crystal size distribution is effectively controlled and helps modify solid–liquid separation behavior, washing and product purity, product bulk density, and powder flow characteristics.
- Sonocrystallization eliminates the need to add seed crystals, which can be advantageous in contained, sterile operations involved in the manufacture of pharmaceuticals and fine chemicals.
- Controlled ultrasound at different stages of crystallization may be used to modify and tailor product properties to meet different requirements.

• What is ultrasonic cavitation?

- Effect of ultrasound on a fluid is to impose an oscillatory pressure on it. At low intensity, this pressure wave will induce motion and mixing within the fluid. This process is known as acoustic streaming.
- At higher intensities, the local pressure in the expansion phase of the cycle falls below the vapor pressure of the fluid, causing minute bubbles or cavities to grow.
- A further increase generates negative transient pressures within the fluid, enhancing bubble growth and producing new cavities on the fluid.
- These latter processes comprise the phenomenon of cavitation, which is the most important effect of ultrasound in chemical and crystallization systems.

• What are the common problems encountered during crystallization process (and their causes)?

- The problems are either crystal or particle related or equipment related.
- *Crystal or Particle-Related Problems:* These include (i) no or low crystal yield, (ii) poor solid–liquid separation characteristics, (iii) poor drying and fluidization characteristics, and (iv) poor end use properties of the product.
  - ➢ *Crystal Yield:* No or low crystal yield is encountered when a faulty measuring device such as temperature or concentration sensor is used or when there is a change in the feed composition or temperature. For a cooling crystallization process, a faulty temperature measurement, such as one that shows lower than actual temperature inside the crystallizer, lower yield results.
  - ➢ A change in the feed composition might be caused by changes in the upstream processes that include a decrease in solute concentration in the feed solution, a change in the solvent composition, including increased levels of impurities, that would increase solubility of the solute.
  - ➢ A low crystal yield may also result if an excessive number of particles are lost or ruined in the solid–liquid separation that follows the crystallization step. This can occur if too many fines are produced, during the crystallization step or formed due to disintegration during handling. In both cases, the particles are not effectively retained in the filtration process. Cycling of crystal size distribution (CSD) also can result poor yields.
  - ➢ *Solid–Liquid Separation Characteristics:* Poor solid–liquid separation results from low permeability of the filter or centrifuge cake. Low cake permeability might be due to too many small size crystals. Porosity tends to decrease with decrease in particle size and shape. Breakage, of needle shaped or plate shaped particles or large particles with poor strength, is relatively easy during handling. An increase in cake moisture might also be the cause of poor separation.
  - ➢ *Drying and Fluidization Characteristics:* Amount of cake moisture, particle surface area that is related to the size and shape of the particles influence drying rates and hence drying costs. If a fluid bed dryer is used, fluidization characteristics of the particles become an important parameter in the drying process. Minimum fluidization velocity, for instance, depends on particle size and shape.
  - ➢ *End Use Properties:* The desired properties that are associated with crystallized products include purity, bulk density, caking tendency, flowability, reactivity, dissolution rate, and safe handling characteristics. All of these properties can be controlled to some extent during crystallization. Particle size and shape determine the surface area, which influences impurities that are surface-laden. Particles are often coated by the mother liquor that might contain such impurities and washing is an effective means of removing them. The impurities might be mostly inside the particle, for instance in the lattice or as a negative crystal. The normal washing techniques might not be effective to remove such impurities. Transport, handling, and storage are influenced by properties like bulk density, flowability, and caking tendency.

Reactivity is influenced by particle size and shape and smaller particle sizes are preferred choice if increased reactivity and dissolution rate are desirable. From safety point of view, the tendency to generate dust clouds during handling should be avoided. Dry solids increase risks of static electricity generation and relative movement of particulates should be controlled to avoid dust explosions.

- *Equipment-Related Problems:* These problems can be divided into two types, namely, shutdown frequency (run life) and mechanical problems.
  - *Shutdown Frequency:* This depends primarily on fouling and encrustation rates and sometimes on mechanical reasons. Increased growth rate is a major reason for encrustation. Among other known factors affecting encrustation are solids concentration, localized flashing, and formation of cold spots (especially in cooling crystallizers). Particle shape is an important factor in determining particle velocity that influences encrustation rates. Fines are most likely to have velocities close to liquid velocities. Plate or needle-shaped particles behave as fines. Since fines have a greater likelihood of sticking to the inner surfaces of crystallizers, processes that produce excessive fines are more likely to have encrustation problems. Encrustation can be minimized by reducing formation of fines, creating localized turbulence by liquid sprays or ultrasonic vibrations.
  - *Mechanical Problems:* Various types of mechanical problems arise during operation of crystallization equipment that in turn affects shutdown frequency.

- What are the common problems, their probable causes and recommended remedies involved in the operation of crystallizers?
  - Table 18.4 summarizes the operational issues involved in the operation of crystallizers.

18.2 HUMIDIFICATION AND WATER COOLING

- Define absolute humidity, molar humidity, percent humidity, percent relative humidity, humid volume, saturated volume, humid heat, and dew point.
  - Absolute humidity = mass of water vapor/mass of dry air.
  - Molar humidity = moles of water vapor/moles of dry air.
  - Percent humidity = 100(humidity of air)/(humidity of saturated air).
  - Percent relative humidity = 100(partial pressure of water vapor in air)/(vapor pressure of water at the same temperature).
  - Humid volume = volume of unit volume of dry gas + its associated vapor at the prevailing temperature and pressure.
  - Saturated volume = volume of unit mass of dry air + its associated vapor.
  - Humid heat = heat required to raise temperature of unit mass of dry air + its associated vapor through 1°C at constant pressure.
  - Dew point:
    - Temperature at which vapor–air mixture is saturated when cooled at constant pressure out of contact with liquid.
    - Temperature at which first signs of fog formation on cooling occurs.
    - Below dew point condensation continues when air is further cooled.

- "Relative humidity of an air–water vapor mixture is independent of the temperature of the mixture." *True/False*?
  - *False.*

- "Humid heat of an air–water vapor mixture contains the sensible heat of dry air and the latent heat of the water vapor." *True/False*?
  - *True.*

- Define *adiabatic saturation temperature* (T_S) and *wet bulb temperature* (T_W) and explain the difference between them.
  - Adiabatic saturation temperature is the steady-state temperature of water/liquid being recirculated in a confined space through which air/gas is passed in intimate contact with the water/liquid. If contact between water/liquid droplets and air/gas is kept long enough, the leaving air/gas is saturated at the adiabatic saturation temperature.
  - Adiabatic saturation temperature is the *steady-state equilibrium temperature* attained when a *large amount of water/liquid* is contacted with the entering air/gas.
  - Figure 18.9 illustrates the principle involved in adiabatic saturation process.
  - Wet bulb temperature is the *steady-state nonequilibrium temperature* reached when *small amount of water/liquid* is contacted. Since liquid is in small amount, temperature and enthalpy of air/gas remain unchanged. For adiabatic saturation case, T and H change.
  - When a solid surface is covered with a liquid film that is kept in contact with air/gas, liquid evaporates

TABLE 18.4 Troubleshooting of Crystallizer Operation

Problem	Causes	Remedies
Deposits	*Local cooling due to lack of insulation* with increase in local supersaturation	All areas of the crystallizer and piping carrying saturated liquors, particularly protruding points such as reinforcing rings
	Low suspension density	Increase the crystals in suspension to maintain the design density
	Protruding gaskets, rough areas on process surface and growth	Remove protrusions and polish the rough areas
Crystal size too small	*Low suspension density*: decreases average retention time; time for crystal growth is insufficient	Increase crystals in suspension to maintain design density
	High circulation rate: creates increased voidage resulting in improper release of supersaturation	Maintain the circulation rate at design rate
	Solids in feed: introduces excess nuclei, than necessary for size control, into the crystallizer	Feed solution must be free of solids especially crystals
	Design feed rate exceeded: results in reduced average residence time	Within limits, this can be corrected by increasing the suspension density
	Excessive nucleation: caused by excessive turbulence, local cold spots, and subsurface boiling	Maintain level and pump or agitator speed at design point
	Difficulty in preventing excessive nucleation: results in high surface area to weight ratio, preventing proper growth	Remove fine salt (nuclei) from the system by dissolving or settling (*Fines removal system*)
Insufficient vacuum	*Obstructions in vacuum lines*: causes excessive ΔP	Remove obstructions like deposits due to solids carryover
	Insufficient cooling water or cooling water above design temperature: results in overloading the vacuum system	Cooling water at design flow rate and at/or below design temperature must be maintained
	Air leaks into system	Must be stopped to correct the problem
	Excessive back pressure on vacuum system: caused by an obstruction in the noncondensable discharge pipe or the discharge pipe sealed too deeply in the hot well	Obstruction must be removed or the depth of the seal in the hot well reduced
	Flooded intercondenser: usually caused by a blockage in the discharge line or by using an excess amount of cooling water	Remove blockage or reduce cooling water flow to design rate
	Low steam pressure to ejectors: *Caused due to* low line pressure, wet steam or blockage in the steam line	Corrective action to restore design steam pressure
	Low seal water flow: applies to mechanical vacuum pumps only. Reduces the subcooling of the noncondensables increasing load to the system	Seal water must be maintained at design flow
	Low rpm for vacuum pump: usually is caused by V-belt slippage or low voltage to the motor	Tighten V-belts or reduce load on electric circuit to motor
Instrument malfunction	*Air leaks*: cause erroneous readings	Seal air leaks
	Plugged purge line: causes erroneous readings	Should be given good flushing at least twice a shift
	Purge liquor boiling in purge line: due to higher vapor pressure of purge liquor than vapor pressure in crystallizer	Use purge liquor (usually water) at or below the maximum operating temperature of the crystallizer
	Improper adjustment	Proportioning band and reset should be adjusted to give smooth control. Damping must not be too high that sensitivity is lost
Foaming	*Foaming is not inherent to the solution*: it can usually be traced to air entering the circulating piping via the feed stream, leakage at the flanges, or by leakage through the pump packing	By eliminating the air leakage the problem is corrected
	Foaming is inherent to the solution	Use a suitable antifoaming agent Selection of antifoam must include the effects upon the crystal habit and growth rate as well as the amount required, availability and cost

TABLE 18.4 (*Continued*)

Problem	Causes	Remedies
Pump performance	*Loss of capacity*: usually caused by loose V-belts or line blockage	Check pump rpm and tighten V-belts if below design speed If pump speed is correct, check for piping blockage
	Packing leaks	Care must be taken to keep packing in good condition When repacked care must be taken to properly clean the housing before repacking
	Cavitation: caused by air entering into suction or insufficient NPSH	Check for air leaks or blockage in the pump suction piping Consult the pump curve for rpm and NPSH data
	Low solids content in product slurry: cause may be due to a restriction in pump slurry suction line	Flush the slurry line
	Slurry settling in line: usually caused by a heavy slurry or low slurry pump speed	
	Heavy slurry causing the problem	Dilute the slurry with mother liquor before pumping to the dewatering equipment
	Low slurry pump speed	Check pump rpm and tighten V-belts if necessary

taking latent heat of vaporization from the liquid film, lowering its temperature. The temperature driving force created between the liquid film and the surrounding air/gas, the later being higher, makes heat to transfer to the film. Vaporization process tends to lower the liquid temperature and heat transfer to the liquid from the surrounding air/gas tends to increase liquid temperature. In the ultimate, liquid temperature attains a constant value that is called wet bulb temperature.

- Figure 18.10 illustrates how wet bulb temperature can be obtained.

- What is psychometric ratio?

$$\text{Psychrometric ratio} = h/M_{\text{air}}k_y, \tag{18.4}$$

where h is the heat transfer coefficient, k_y is the mass transfer coefficient, and M_{air} is the molecular weight of air.

- For air–water vapor mixtures,

$$\text{psychrometric ratio} \approx 0.96\text{–}1.005, \tag{18.5}$$

which is close to humid heat, C_S.

- "Psychrometric ratio is a function of the heat and the mass transfer coefficients." *True/False*?
  - *True.*
- "Psychrometric ratio of an air–water vapor mixture is equal to the humid heat of that mixture." *True/False*?
  - *True.*
- "A wet material suspended in drying air, will be near the wet bulb temperature." *True/False*?
  - *True.*
- What is wet bulb depression?
  - The difference between dry bulb temperature and wet bulb temperature is called wet bulb depression.

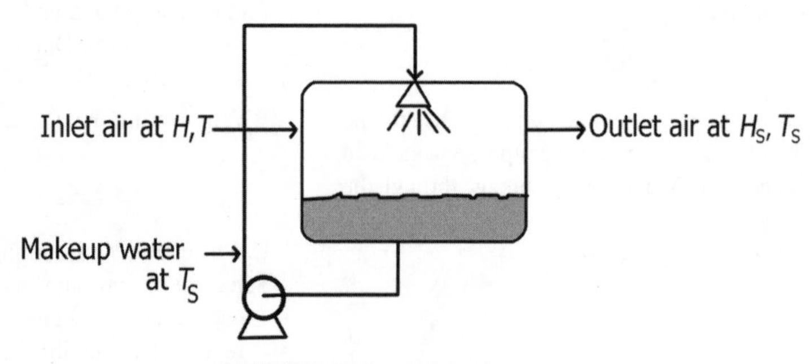

FIGURE 18.9 Adiabatic process.

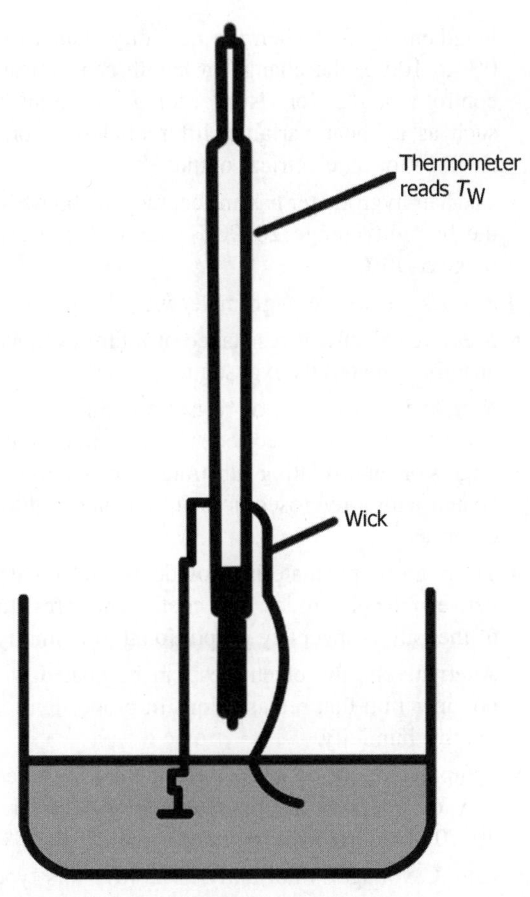

FIGURE 18.10 Illustration for obtaining wet bulb temperature.

- "The wet bulb temperature of an air–water vapor mixture is approximately equal to the adiabatic saturation temperature." *True/False*?
  - *True.*
- Illustrate, by means of a diagram, various curves/lines involved in a humidity chart?
  - Humidity chart shows the relationship between the temperature (abscissa) and absolute humidity (ordinate, in *g* moisture per kg dry air) of humid air at one atmosphere absolute pressure over a range of temperatures (Figure 18.11).
  - Lines representing percent humidity and adiabatic saturation are drawn according to the thermodynamic definitions of these terms.
  - Equations for the adiabatic saturation (as) and wet bulb (wb) temperature lines on the humidity chart are

$$(Y - Y_{as})/(T - T_{as}) = -C_s/\lambda_{as} = -(1.005 + 1.88Y)/l_{as}. \quad (18.6)$$

$$(Y - Y_{wb})/(T - T_{wb}) = -(h/M_{air}k_y)/\lambda_{wb}. \quad (18.7)$$

- "Adiabatic saturation lines in a humidity chart can also be used as wet bulb lines for air–water systems." *True/False*?
  - *True.* Adiabatic saturation and wet bulb temperatures are almost equal for the air–water system.
  - However, these temperatures are conceptually quite different. The adiabatic saturation temperature

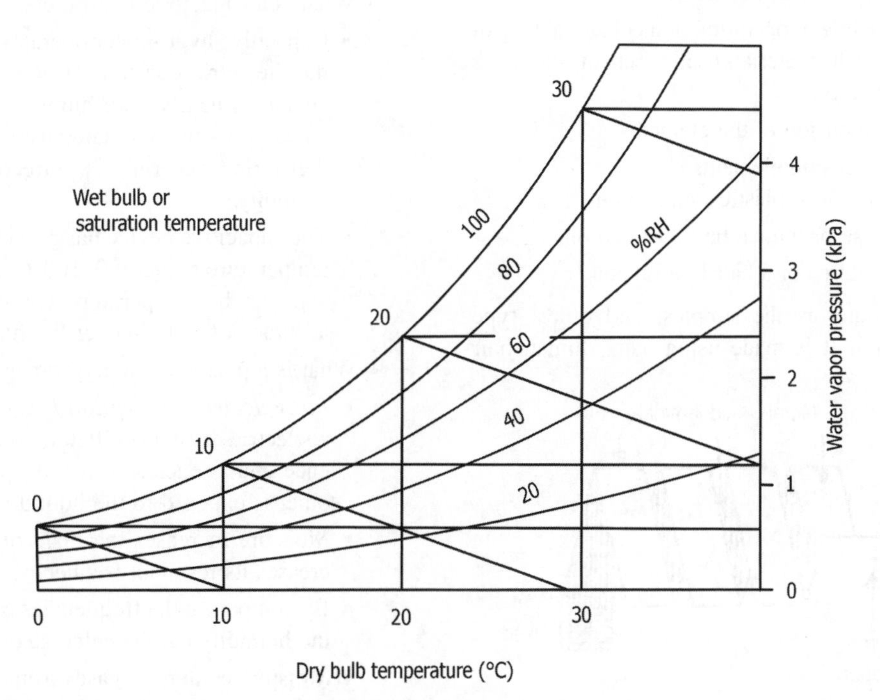

FIGURE 18.11 Humidity chart.

is a gas temperature and a thermodynamic entity while the wet bulb temperature is a heat and mass transfer rate based entity and refers to the temperature of the liquid phase.

- "Relative humidity of air can be determined if its dry bulb and wet bulb temperatures are known." *True/False*?
  - *True.*
- What are hygrometers?
  - Hygrometers are devices that indirectly measure humidity by sensing changes in physical or electrical properties in materials due to their moisture content. Materials such as hair, skin, membranes, and thin strips of wood change their length as they absorb water. The change in length is directly related to the humidity.
  - Such devices are used to measure relative humidity from 20% to 90%, with accuracies of about ±5%. Their operating temperature range is limited to less than 70°C.
  - Laminate hygrometer is made by attaching thin strips of wood to thin metal strips forming a laminate. The laminate is formed into a helix as shown in Figure 18.12.
  - As the humidity changes the helix flexes due to the change in the length of the wood. One end of the helix is anchored, the other is attached to a pointer (similar to a bimetallic strip used in temperature measurements).
  - The scale is graduated in percent humidity.
- What is the principle of operation of mechanical hygrometers? What are the materials used as sensing elements in such hygrometers?
  - Change in dimension of the element.
  - Examples of sensing elements:
    - Fine wires, fibers, plastics, and so on.
    - Camel, horse or human hairs, and so on.
- What are the features of a hair hygrometer?
  - Hair hygrometer is the simplest and oldest type of hygrometer. It is made using hair. Human hair lengthens by 3% when the humidity changes from 0% to 100%; the change in length can be used to control a pointer for visual readings or a transducer such as a linear variable differential transformers (LVDT) for an electrical output.
  - The hair hygrometer has an accuracy of about 5% for the humidity range 20–90% over the temperature range 5–40°C.

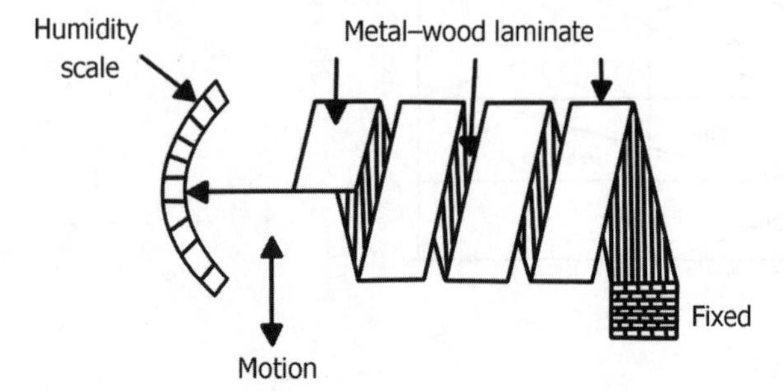

FIGURE 18.12 Hygrometer using metal–wood laminate.

- How does a resistive hygrometer work?
  - Measures electrical resistance of a film of moisture absorbing materials exposed to air.
  - *Resistive hygrometer* or resistive humidity sensors consist of two electrodes with interdigitated fingers on an insulating substrate. The electrodes are coated with a hygroscopic material such as lithium chloride.
  - The hygroscopic material provides a conductive path between the electrodes. The coefficient of resistance of the path is inversely proportional to humidity.
  - Alternatively, the electrodes can be coated with a polymer film that releases ions in proportion to the relative humidity.
  - Temperature correction can be applied for an accuracy of 2% over the operating temperature range 40–70°C and relative humidity from 2% to 98%.
  - An AC voltage is normally used with this type of device.
  - Variations of this device are the electrolytic and the resistance–capacitance hygrometer.
- What is a capacitive hygrometer?
  - Capacitive hygrometer operates on the principle that the dielectric constant of some thin polymer films changes linearly with humidity, so that the capacitance between two plates using the polymer as the dielectric material is directly proportional to humidity.
  - The capacitive device has good longevity, a working temperature range of 0–100°C, a fast response time, and can be temperature compensated to give an accuracy of ±0.5% over the full humidity range.
- What is a piezoelectric hygrometer?
  - *Piezoelectric or sorption hygrometers* use two piezoelectric crystal oscillators, one is used as a reference and is enclosed in a dry atmosphere, and the other is exposed to the humidity to be measured.
  - Moisture increases the mass of the crystal that decreases its resonant frequency.
  - By comparing the frequencies of the two oscillators, the humidity can be calculated.
  - Moisture content of gases from 1 to 25,000 ppm can be measured.

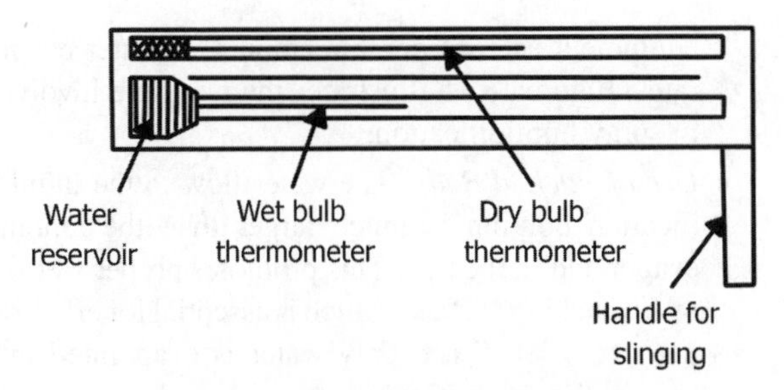

FIGURE 18.13 Sling hygrometer.

- What is the primary difference between a sling psychrometer and Assmann psychrometer?
  - Sling psychrometer, illustrated in Figure 18.13, is whirled to give the desired air velocity across the thermometer bulb.
  - In Assmann psychrometer, air is drawn past bulbs by a motor-driven fan.
- Under what conditions dehumidification takes place?
  - When a warm gas–vapor mixture is contacted with a cold liquid so that humidity of the bulk gas is more than that at the interface, dehumidification takes place.
- What are the methods by which dehumidification processes are carried out?
  - Cooling by the use of refrigeration coils.
    - Air to be dried passes through a cooling coil, which lowers the temperature of the air stream below its dew point. Water vapor condenses on the cooling coil surface and falls to the drain pan as liquid. In absolute terms air is dryer, but it also has a relative humidity close to 100%. If low relative humidity is needed in addition to a lower absolute amount of moisture, the air can be heated after it leaves the cooling coil.
    - Standard refrigeration equipment can produce dew points of about 4°C.
  - Figure 18.14 illustrates a typical vapor-compression cooling-based dehumidification process.
  - By the use of desiccants, for example, adsorption using silica gel, alumina, and so on, as adsorbents.
    - In a desiccant system, the process air stream passes through a desiccant medium, which adsorbs moisture from the air stream.
    - Desiccant dehumidifiers can produce dew points below −18°C.
    - The equipment uses differences in vapor pressure to remove moisture from air by chemical attraction. The driving force for the process is the difference in vapor pressures between the surface of the desiccant, which has a very low vapor pressure and the humid air.

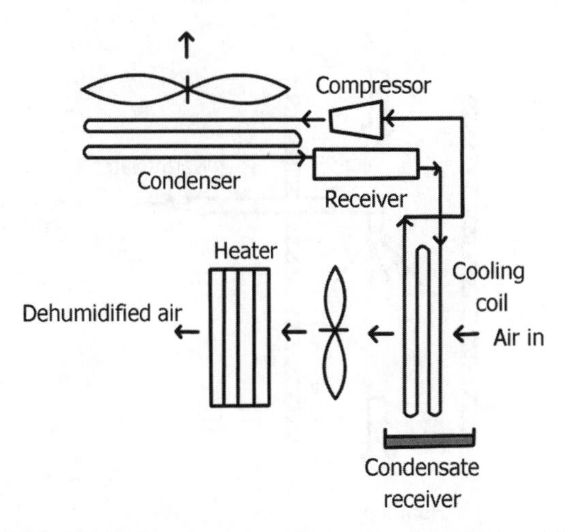

FIGURE 18.14 Vapor-compression cooling-based dehumidification process.

    - Eventually, the desiccant surface finally gets saturated with water vapor, requiring reactivation for reuse by drying.
- Compare applications of refrigeration-based and desiccant-based dehumidification processes.
  - Cooling-based dehumidification handles moisture at high dew points (>10°C) and desiccant-based dehumidification removes the moisture at low dew points.
  - Factors to consider include the following:
    - *Dew Point Control Level:* When the required moisture control level is relatively high cooling-based dehumidification is economical in terms of both operating and initial equipment costs.
    - Below 10°C, precautions need to be taken to avoid freezing the condensed water on the cooling coil.
    - Consequently, desiccants are more economical than cooling-based systems at lower dew points.
    - *Relative Humidity Sensitivity:* When a process needs a low moisture level but can tolerate a high relative humidity, cooling-based dehumidification without desiccants is less expensive.
    - In processes that require a low relative humidity, in addition to a low dew point, desiccant systems are used for humidity control, with supplementary cooling systems, to keep temperature within acceptable limits.
    - When a product is sensitive to relative humidity but not to temperature, a desiccant dehumidifier is used without a cooling unit to maintain a constant relative humidity.
    - *Temperature Tolerance:* If the application can tolerate a wide temperature range, then dehumidification alone may be enough. In most cases, both temperature and moisture must be maintained within set limits, so both cooling and desiccant

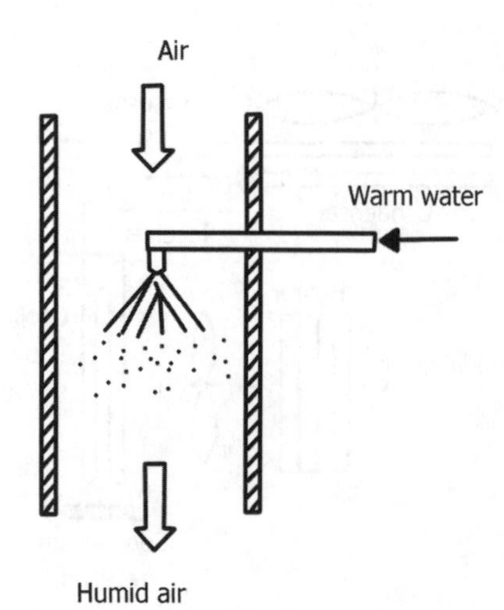

FIGURE 18.15 Spray humidification.

equipment are used in a combination to maintain control.

- What are the different methods used for humidification of compressed air/gases?
  - *Water Spray:* The main drawback of this method is that it puts extremely high requirements on water purity. Water that is not completely deionized contains small amount of salts, which deposit on equipment as considerable amount of water evaporates. Figure 18.15 illustrates the principle involved in spray humidification.
  - *Use of Packed Beds:* The water flow to the humidification column is much larger than the amount evaporated in the bed. This promotes proper wetting of the packing surface, which is essential for effective mass transfer. Since only water is evaporated into the air, the impurities, for example, salts, remain in the liquid water and exit the column with the blowdown from the water circuit. Figure 18.16 illustrates the use of packed bed humidifier.
  - *Use of Tubular Humidifiers:* Tubular humidifier is essentially a heat exchanger, with hot waste gases flowing on the shell side and gas to be humidified flows countercurrent to a falling film of water, just like in a wetted wall column. The water film is evaporated into the gas mainly due to heat transfer from the hot waste gases flowing on the shell side. Tubular humidifiers have a compact design and are considered to operate efficiently (Figure 18.17).

18.2.1 Evaporative Cooling and Cooling Towers

- What are the heat transfer processes involved in evaporative cooling?

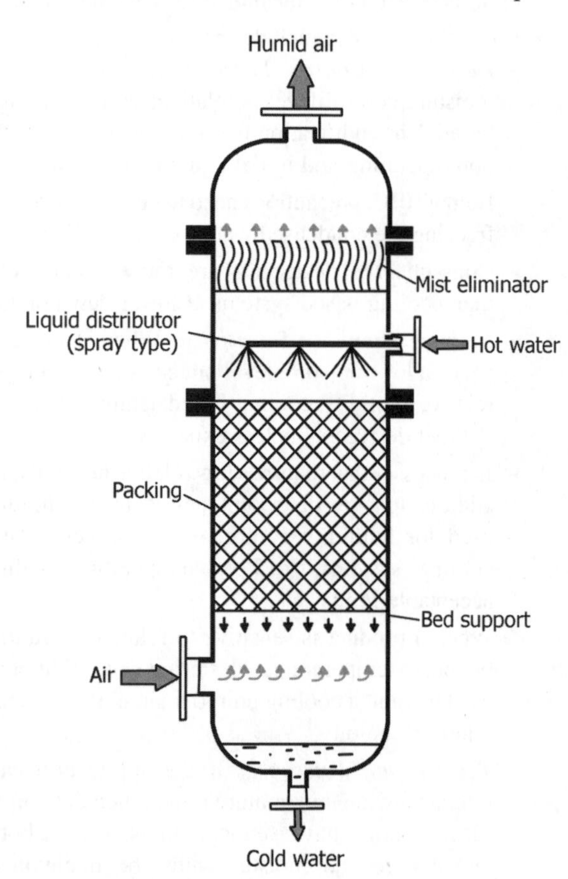

FIGURE 18.16 Packed bed humidifier.

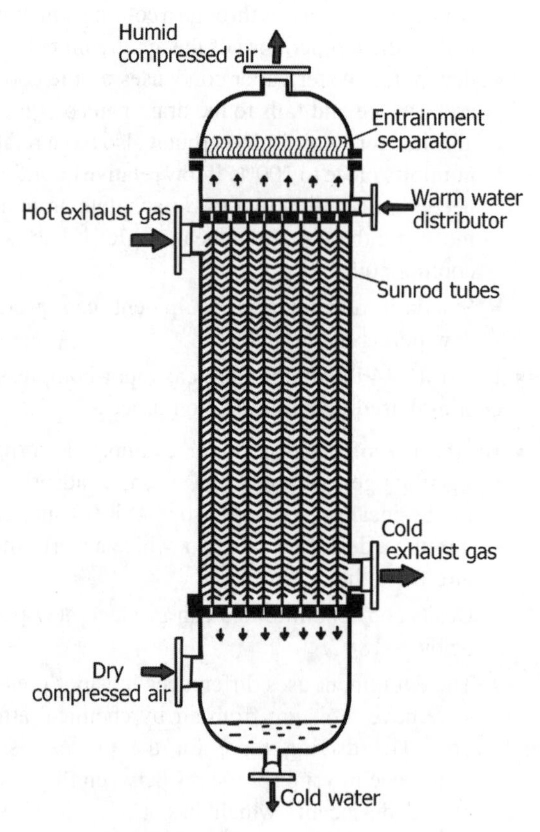

FIGURE 18.17 Tubular humidifier.

TABLE 18.5 Temperature Approach Versus Relative Tower Size

$T_{water} - T_{Wet\ bulb}$ (°C)	Relative Size
3	2.4
8	1.0
14	0.55

- Transfer of heat as latent heat of vaporization, which accounts to about 80% of total heat transferred.
- Sensible heat transfer due to temperature difference between air and hot liquid, which accounts to about 20% of the total heat transferred.

- "Water in contact with air under adiabatic conditions eventually cools to the wet bulb temperature." *True/False*?
  - *True.*
- What is *approach* in a cooling tower terminology?
  - Temperature difference between tower cold water outlet temperature and wet bulb temperature of air is called *approach.*
- "If *approach* is large, required tower size is large." *True/False*?
  - *False.*
- What approach is normally recommended for a cooling tower?
  - 5°C.
- What is *water temperature approach* for a cooling tower?
  - Difference between water temperature and wet bulb temperature.
- How does relative tower size depend on water temperature approach? Illustrate.
  - Relative tower size decreases with increase in temperature approach (Table 18.5)
- What is *range* for a cooling tower?
  - *Range* is the difference between inlet warm water and outlet cold water temperatures.
- What are the different types of equipment used in industry for water cooling?
  - Atmospheric draft cooling towers involve cross-flow of air. Depends on wind blowing horizontally across the tower.
  - Natural draft cooling towers involve positive movement of air through a tall tower, which makes use of density differences, that is, natural draft.
  - Figure 18.18 illustrates the differences between atmospheric and natural draft cooling towers.
  - Mechanical draft cooling towers:
    - Forced draft (fan is at lower end).
    - Induced draft (fan is at top).
    - Induced draft (cross-flow).
  - Spray chambers:
    - Used for adiabatic humidification-cooling operations with water recirculation.
    - Refrigeration coils for dehumidification.
  - Spray ponds:

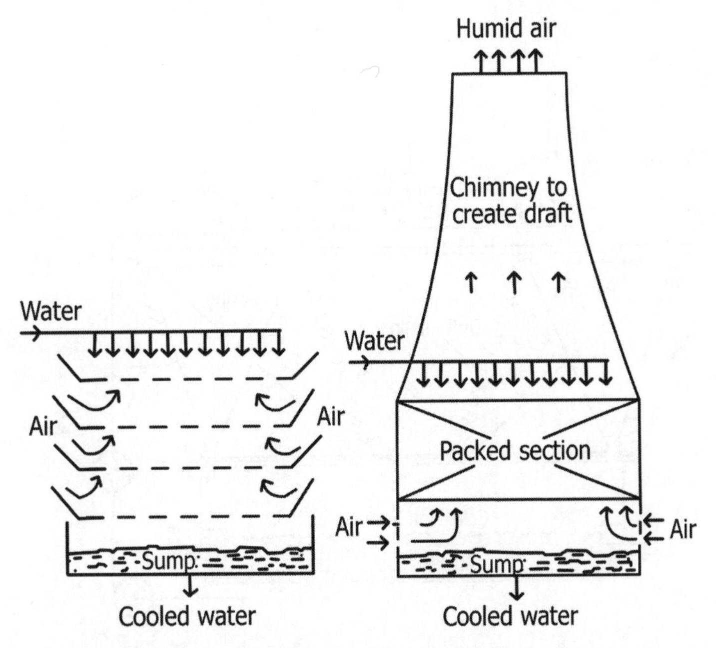

FIGURE 18.18 Atmospheric and natural draft cooling towers.

  - ➢ Used where close approach to wet bulb temperature is not required.
  - ➢ High drift losses.

- Illustrate the essential features of an induced draft cooling tower.
  - Figure 18.19 is a schematic diagram of an induced draft cooling tower.
  - Saturated air leaving the tower from its top forms a visible plume when the water vapor condenses upon contact with the outside cooler air.
- What is the function of baffles/louvers in a cooling tower?
  - Baffles minimize water loss as windage or drift when the air flow traps small droplets of cooling water.
- Can water be cooled to a temperature equal to wet bulb temperature?
  - No.
- "With industrial cooling towers, cooling to 90% of the ambient air saturation level is possible." *True/False*?
  - *True*.
- What approach to design wet bulb temperature is usually recommended for a cooling tower?
  - 5.5°C (10°F).
- What are the temperature ranges for the operation of a cooling tower?
  - Cooling tower water is received from the tower between 27 and 32°C and should be returned between 45 and 52°C depending on the size of the tower.
  - Seawater should be returned at temperatures not higher than 43°C to avoid scale formation.
- Water cooling has been normal practice in industry, but air cooling is acquiring importance. Give reasons for this shift.
  - Water shortages and costs involved in treatment and disposal of wastewaters originating from water cooling systems to meet pollution control requirements/regulations are making air cooling attractive.
  - Additional factors include elimination of piping to circulate cooling water, lower pressure drops for moving air across equipment than for moving water, absence of fouling which reduces heat transfer rates and requires frequent cleaning.
- What are the generally used (i) water and (ii) air circulation rates in a cooling tower?
  - Water circulation rates are generally 81–162 L/(min m^2) (2–4 gpm/ft^2).
  - Air circulation rates are 6350–8790 kg/(h m^2) (1300–1800 lb/(h)(ft^2).
- What are the common air velocities employed in a cooling tower?
  - Air velocities are usually 1.5–2.0 m/s (5–7 ft/s).
- What is an atmospheric cooling tower? What are its characteristics?
  - Atmospheric cooling tower depends on atmospheric wind to blow horizontally through the tower (require wind velocities of 7.24–10.46 km/h (4.5–6.5 mph).
  - Operate in cross-flow of wind contacting falling water.
  - Not capable of cooling water to temperatures much closer than 2°C of wet bulb temperature of entering air.

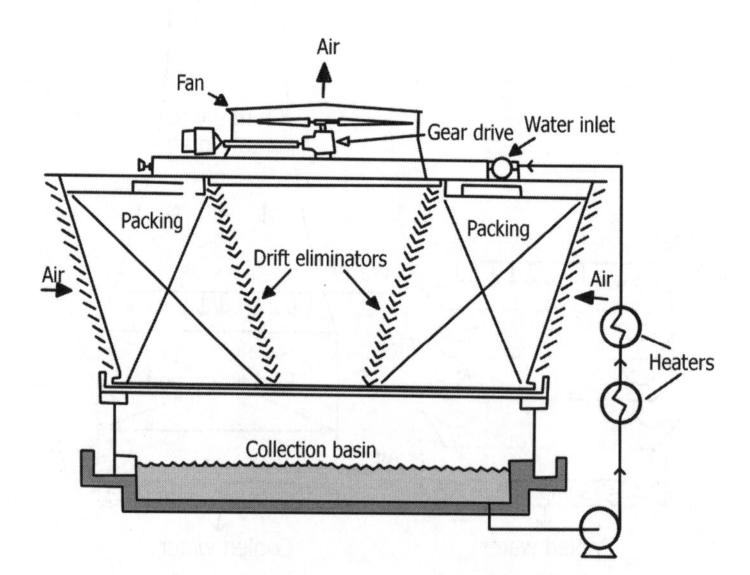

FIGURE 18.19 Induced draft cooling tower.

- Require no fan.
- Consume power only to pump water to top of the tower.
- Require large ground area.
- Require open areas around to permit unobstructed air flow.

- In a natural draft cooling tower, what causes air flow through the tower?
  - $\Delta\rho$ between cool inlet air and warm exit air, which is lighter than ambient air.
  - Called chimney effect.
- "Operation of a natural draft cooling tower depends on atmospheric temperature." *True/False*?
  - *True*.
- What is the effect of high humidity conditions on the draft obtainable in a natural draft cooling tower?
  - Draft will increase through the tower due to increase in available static pressure difference to promote air flow against internal resistances.
  - Natural draft cooling towers operate most efficiently in climates with high humidity. In climates with lower humidity, buoyancy effect may be enhanced in natural draft towers with the assistance of a fan.
- Why natural draft cooling towers are shaped hyperboloidically?
  - They have greater strength for a given thickness.
  - A tower of 75 m height has concrete walls of 13–15 cm thick.
- Why the cross section of a natural draft cooling tower is enlarged at the top?
  - To aid dispersion of exit humid air into the atmosphere (Venturi effect).
- What are the advantages and disadvantages of *forced draft and induced draft* employed in cooling towers?
  - Induced draft:

    Advantages
    - Better air distribution over cross section.
    - Low possibilities for short-circuiting of ejected air (outlet velocities are two and a half times inlet velocities, ~460 m/min).
    - Increased capacity in case of fan failure (compared to forced draft, stack effect is greater for induced draft).
  - Capable of cooling to within 1.1°C (2°F) of the wet bulb temperature.
  - A 2.8–5.5°C (5–10°F) approach is more common.

    Disadvantages
    - Higher power consumption due to fan located in hot air.
    - Bearings and other mechanical components are exposed to hot air giving rise to potential damage.
    - Poor accessibility of fan components for maintenance.
    - Above 180°C for fluids to be cooled, forced draft is used.
  - Forced draft:

    Advantages
    - Slightly lower power consumption (fan exposed to cold air). Power $\propto$ absolute temperature.
    - Better accessibility for maintenance.

    Disadvantages
    - Poor air distribution.
    - Increased possibility for hot air recirculation (low discharge velocity).
    - Low natural draft capability on fan failure (small stack effect).
    - Total exposure (fan) to Sun, rain, and so on.
- Which of the two, namely, induced or forced draft towers is common?
  - Countercurrent induced draft towers are the most common.
  - What is the temperature approach commonly attainable for induced draft towers?
  - A 2.8–5.5°C (5–10°F) approach is more common.
- What are the considerations involved in the design of cooling towers?
  - Cooling range, approach to wet bulb temperature, mass flow rate of water, web bulb temperature, air velocity through the tower, or individual cells of the tower and tower height.
- What factors should be compared when evaluating cooling tower bids?
  - Purchased cost, installed cost, fan energy consumption, pump energy consumption, water use, water treatment costs, expected maintenance costs, worker safety requirements, environmental safety, and expected service life.
- What are the factors that should be considered while selecting a cooling tower and its components?
  - Selection of a cooling tower and its components should include quality of the makeup water, fouling potential of the cooling water, heat load, site location limitations, past operating and reliability experience, and costs.
  - Considerations for internal selection should include natural versus forced or mechanical draft, film versus splash fill, and counterflow versus crossflow.

- Important components should include fans, fan shrouds, fan drive motors, packing, drift eliminators, air louvers, nozzles, and so on.
- Selection of materials of construction depends on tower size, water quality, fouling tendency.
  - Most cooling towers use stainless steel fan blades and drive shaft couplings. Small towers may use galvanized or fiberglass reinforced plastic or stainless components.
  - Use of plastic or plastic linings/coatings or what are known as *engineered plastic* for the shells of cooling towers improves life of the tower shells and reduces maintenance and operating costs.
- Additional selection criteria should include the following:
  - Cooling water circuits that might have the chances of process stream contamination should not use film packing due to the risk of fouling and packing failure.
  - Power plants that involve high heat loads, high water recirculation flow rates, and large cooling loads often use natural draft cooling towers with hyperbolic concrete towers.
  - Sites with nearby construction or chances of hot combustion exhaust gas flows should choose counterflow designs with special air intake systems.
  - Variable frequency fan drives involve high capital costs but provide greater operational flexibility and may be used where in towers involving more than two compartments.
  - Towers operating with sea or brackish waters must be approximately 5% more size than equivalent freshwater towers because sea and brackish waters have lower heat capacities than freshwaters.

- What are the undesirable phenomena and environmental problems in cooling tower operation?
  - Fog formation that results from mixing warm and saturated tower discharge air with cooler ambient air. Visible plumes that are of environmental nuisance, affecting visibility.
  - Visible steam plumes from cooling tower exhaust stacks.
  - Fine particulate (usually $<10\,\mu m$) air pollution, resulting from carryover of dried up total dissolved solids (TDS) in the drift loss.
  - Leakage of hydrocarbons and other process fluids into cooling water from aging heat exchangers, which subsequently
    - create air pollution problems by way of emission of volatile organic chemicals (VOCs) and hazardous air pollutants (HAPs), if leaked fluids are of low boiling type;
    - cause high BOD and COD levels in tower blowdown, if leaked fluids are high boiling type.
  - Leaching of wood treatment chemicals (when wood is used as packing for air–water contacting.
  - Free oil and grease leaked out from exchangers.
  - High discharge temperatures in blowdown.
  - Increase total dissolved solids (TDS).
  - Residual chlorine in the blowdown due to the non-oxidizing and oxidizing biocide for microbial control.
  - Increased pH due to treatment chemicals. In most cases, acid is added to the tower basin to reduce pH of blowdown.
  - Sludge buildup in the bottom of cooling tower basin and pump sumps from
    - suspended solids in the makeup raw water;
    - ambient air scrubbing;
    - algae;
    - piping corrosion products.
  - *Icing Problems:* Under certain conditions, a plume may give rise to fogging or icing problems in the surrounding air and on the drift eliminators.

- What are the problems involved in the even distribution of air and water in cooling tower operation?

  Cross-Flow Distribution
  - Cross-flow distribution systems are relatively simple and easy to clean. But in practice, these are not maintained properly.
  - As a result of poor maintenance, many cross-flow cooling towers have a significant number of nozzles plugged with mud, algae, wood, plastic, scales, sand, rust, and other debris. It is necessary to periodically inspect, clean the basin and plugged nozzles and replace the missing nozzles.
  - At least a minimum of 5–10% or more of the nozzles are plugged.
  - Plugged nozzles cause an increase in water level in the distribution basin, resulting in excessive drift losses. Also they cause an imbalance in the flows with deteriorated performance.
  - Instances are common that nozzles are washed out due to excessive water velocities and require proper fastening to prevent such situations.
  - Some nozzles loose their orifice plates, increasing their diameter thereby lowering water levels in the basin causing improper water distribution.
  - The distribution basins must all be maintained at the same depth and equal flow of water. Cross-flow

distribution valves must be kept operable in order to maintain equal flows between the basins.

Counterflow Distribution

- Counterflow distribution towers are more difficult to maintain compared to cross-flow towers.
- A rough rule of thumb is that a plugged nozzle costs the equivalent of twice the area covered. The net effect is channeling and imbalance in the distribution of air and water across the cross-sectional area. Air takes the path of least resistance aggravating the imbalance in distribution.
- It is essential that the nozzles be checked and cleaned at least at the beginning of hot season. All missing nozzles must be replaced.
- Another common problem is operation at reduced loads. To save energy, only sections or cells of the tower be used under such circumstances.

• What types of packing or fill are used in cooling towers?
  - Splash grids of wood or plastic construction are arranged so that the falling liquid must contact alternate rows of grids.
  - Splash type fill is designed to interrupt the progress of water as it falls through the cooling tower and to *splash* it into small droplets to maximize the efficiency of the cooling process. Splash fill gives low resistance to air flow and minimizes the potential for blocking in the tower.
  - All the grids in a row run in the same direction, although alternate rows could be transversely oriented. Such a fill commonly is to 7.5–9 m in depth.
  - Therefore, the grids are arranged to provide a high open area for air flow to minimize pressure drop.
  - Some newer plastic grids are contoured to encourage more streamlined air flow with reduced pressure drop.
  - Film type fill is designed to force the water to flow into a thin film over large, vertically oriented surfaces in the cooling tower. Film type fill will require a larger air side pressure loss and is much more susceptible to blockages due to the small areas available between the fill materials.
  - Because of their low pressure drop characteristics, structured plastic packings have been used extensively in modern cooling towers.
  - Generally, such packings are of designs especially developed for water cooling, rather than the types typically used in other mass transfer operations.
  - Such modern *fill* material for mechanical draft towers consists of vacuum formed plastic sheets installed as vertical modules to minimize pressure drop.
  - Arranging the sheets closely together provides a relatively large surface area for mass transfer.
  - Such a *fill* provides a ΔP per transfer unit of height similar to that of random packings, although the transfer unit height for such a *fill* may be *twice* that developed by the more efficient random tower packing.
  - While structured plastic cooling tower *fill* may be five times as efficient as wooden splash bars, uniform distribution of water is a prime requirement for maximum performance.
  - The circulating water must be treated to prevent buildup of bacterial growth or slime. Also, makeup water added to replace evaporated water must be treated to avoid solid deposits.

• What is the effect of altitude in cooling tower performance?
  - Air handling equipment, air cooled condensers, and the like are typically made to operate at higher speeds (or, with a steeper fan pitch) as altitude increases in order to maintain the same mass flow.
  - As an example, air at 1500 m height is approximately 17% less dense than at sea level and the fan speed increases by the same amount.
  - A cooling tower designed for operation at sea level will work well at 1500 m elevation without modification.
  - This is because air at reduced atmospheric pressure will accept increased amounts of water. The increased ability for the air to accept more water offsets the reduced air mass, resulting in a small net gain in capacity at altitude.

• Give reasons why a forced draft cooling tower involves less corrosion problems for its fan when compared to an induced draft cooling tower.
  - Handles only atmospheric air whereas an induced draft tower handles warm moist air that promotes corrosion.
  - Induced draft tower requires protection of the coated plastic or special metal blades and sealed motors and reduction gears.

• What are the normal types of losses, which are to be made up with fresh supplies of cooling water, in a circulating cooling water system?
  - Evaporation.
  - Drift.
  - Blowdown.

• What are the normal evaporation losses and drift losses in the operation of a cooling tower?
  - Evaporation losses are about 1% by mass of the circulation rate for every 5.5°C (10°F) of cooling.

- Drift losses are around 0.25% (0.1–0.3%) of the circulation rate.

- How are evaporation and drift losses estimated in the operation of a cooling tower?
  - Evaporation losses can be estimated using the following relationship:

$$\text{Evaporation losses in gpm} = \text{total water flow rate in gpm} \times \text{cooling range in °F} \times 0.0008. \quad (18.8)$$

$$\text{Drift losses in gpm} = \text{total water flow rate in gpm} \times 0.0002. \quad (18.9)$$

- Why is it necessary to have blowdown in a cooling system?
  - To maintain a predetermined water analysis with respect to chemicals and dissolved gases.
  - Buildup of solid or chemical concentration with continued evaporation leads to
    - fouling problems to exposed surfaces;
    - accumulation of sludge in the tower basin;
    - increased corrosion.

- What is percentage blowdown in the operation of a cooling tower? What is its significance? How is it estimated?

$$\text{Percent blowdown} = \frac{\text{ppm chlorides in the make up water}}{\text{ppm chlorides in the circulating water}} \times 100 \quad (18.10)$$

  - Blowdown is an important parameter as it is related to fouling and corrosion rates in heat exchangers and cooling towers in a plant.
  - Shows how much of make-up water is being wasted by going to the drain or leaving the tower as drift loss and how much is being usefully employed by being evaporated (evaporative cooling).
  - For example, if chlorides in makeup water are 50 ppm and in circulating water are 400 ppm, percent blowdown = (50 × 100)/400 = 12.5%, that is, of each 100 m^3 of makeup water, 12.5 m^3 are wasted and 86.5 m^3 are used.
  - Instead of chlorides, other constituents can be used in the calculations.
  - A rule of thumb determines the following maximum limits in the cooling tower water: 750 ppm chlorides, 1200 ppm sulfates, 1200 ppm calcium salts, 200 ppm sodium bicarbonate.
  - These limits determine which of these species are to be used in estimating blowdown rates. Normally chlorides are often the determining species. It should be noted that well waters contain significant levels of calcium salts.

- What is the normally recommended blowdown used in the operation of a cooling tower?
  - A blowdown of about 3% of the circulation rate is needed to prevent salt and chemical treatment buildup.

- What could be a possible cause for foaming in a cooling tower?
  - If a normally operating tower develops foaming, the reason could often be, leakages in a heat exchanger through which cooling water is flowing, resulting contamination of the cooling water by the fluids being cooled in the exchanger.
  - Even small leaks of some chemicals can cause such foaming problems.

- What are the considerations involved in cooling tower siting?
  - The tower orientation should match the direction of the prevailing winds in the area to facilitate the air flow into the tower.
  - Location of a new tower should not be within drift zone of an existing tower or in an area that will allow recirculation of the plume of an existing tower or other hot gas movement.
  - A cross-flow tower requires a large clearance around existing structures than a counterflow tower due to the inlet air flow requirements.
  - Hyperbolic natural draft towers are very large and require clearances similar to cross-flow towers.

- Give a typical cooling tower performance chart for illustrative purposes.
  - The curves (Figure 18.20) are between wet bulb temperature and cooling water temperature, with range as parameter. Separate curves for separate flow rates.

- What is water efficiency of a cooling tower?

$$\text{Ratio of actual cooling to theoretical cooling} = \frac{\text{hot water temperature} - \text{cold water temperature}}{\text{hot water temperature} - \text{wet bulb temperature}} \times 100 \quad (18.11)$$

- What is thermal efficiency of a cooling tower?
  - Percent water supplied to the tower that is evaporated is called thermal efficiency.

- What are the ways to increase thermal efficiency of a cooling tower?
  - Increasing amount of water evaporated by

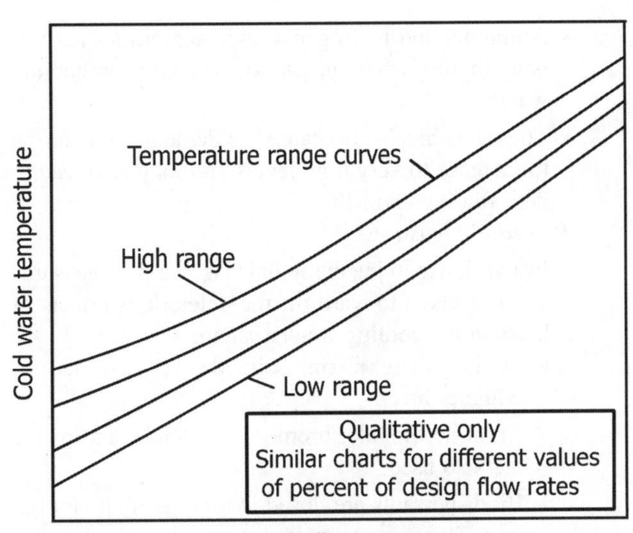

FIGURE 18.20 Cooling tower performance curves.

- > increasing temperature differential through the tower;
- > changing design to give better air–water contact.
- ▪ Decreasing amount of water lost as drift (not evaporated but lost as droplets):
  - > by decreasing drift losses;
  - > by decreasing leakages;
  - > by decreasing blowdown.

- What are the variables involved in the sizing of a cooling tower?
  - ▪ Cooling range $(T_{hot} - T_{cold})_{water}$.
  - ▪ Approach to wet bulb temperature ($T_{cold\ water} - T_{wet\ bulb}$).
  - ▪ Quantity of water to be cooled.
  - ▪ Wet bulb temperature.
  - ▪ Air velocity.
- What are the factors that influence the performance of a cooling pond?
  - ▪ Air temperature, relative humidity, wind speed, solar radiation, rain, heat transfer between Earth and the pond, changing temperature and humidity of air as it moves over the pond.
- What are the factors that help increase the performance of a spray pond?
  - ▪ Using a long narrow pond with its long axis at right angles to the prevailing air.
  - ▪ Decreasing water rate/unit of pond area.
  - ▪ Increasing height and fineness of drops in the spray.
  - ▪ Increasing nozzle height above the basin sides.
- How are drift losses reduced in a spray pond?
  - ▪ By providing louver fence, equal in height to maximum height of the spray.
- What is the cause for fog formation in gas quenching operations? How is it prevented?
  - ▪ If the exit water temperature is more than 10°C *cooler* than the inlet adiabatic gas temperature, fog formation may occur.
  - ▪ This occurs because the inlet gas is cooled at a faster rate by sensible heat transfer than it is dehumidified by mass transfer.
  - ▪ As a result, the gas is chilled below its dew point temperature. The resultant fog is of such small particle size that it is carried out of the gas quench tower along with the exit gas.
  - ▪ To prevent this, water flow to the quench tower should be reduced to raise the exit water temperature.
  - ▪ For example, for quenching flue gas from the combustion of natural gas at atmospheric pressure L/G ratio varies from 2 to 7.
- What are the effects of leaks of process fluids into cooling water systems in a process plant? What are the remedial measures?

  Effects of Leaks
  - ▪ *Biofouling:* Bacteria and algae stick to almost any surface in cooling water systems, particularly where water velocity is low.
  - ▪ The microorganisms produce a polysaccharide layer matrix, which is called slime or biofilm. This film further entraps inorganic matter, precipitates and corrosion products.
  - ▪ Numerous problems posed by the biofouling are given below:
    - > Loss of transfer and operational efficiency as these biofilms are four times more insulating than even calcium carbonate scales.
    - > Microbes produce localized concentrations of metabolites, such as corrosive gases and acids, which manifest in the form of pitting and grooving.
    - > Biofilms promote scale formation.
    - > Restriction of flow inside the cooler and condenser tubes.
    - > Typical chemical treatments can become ineffective when biofilms grow in volume, as the biocides cannot penetrate the impermeable structure.
    - > Biofilm promotes development of biocide-resistant strains due to sessile growth under and within deposits (sessile refers to microorganisms that are attached to the surface).

- ➢ Biofilms also harbor some harmful species that cause environmental and human safety-related concerns.
- *Scaling:* Organic acids and polymers produced by bacteria in biofilms combine with calcium and magnesium ions to form insoluble oxalates, acetates, and calcium–magnesium polymer complexes.
- These insoluble compound deposits are scales, which make biofilms even more impermeable. As a result, heat transfer efficiency drops significantly.
- *Microbiologically Induced Corrosion (MIC)*: This is another deteriorating effect of oil leaks in cooling water systems.
- Depending on the function of bacteria, they are grouped as aerobic or anaerobic. SRB is a typical example of anaerobic bacteria.
- Nitrifying bacteria, which produce nitric acids in the presence of ammonia, are aerobic.
- MIC results from various causes, including the following:
  - ➢ Cathodic depolarization of sulfur-reducing bacteria, such as *Desulfovibrio* and *Desulfurican*. Corrosion typically manifests in the form of localized pitting and grooving.
  - ➢ The production of corrosive metabolites, such as acids by *Thiobacillus* and *Thiooxidans* and other organic acids by various bacteria and fungi species.
  - ➢ Sometimes bacteria, such as *Gallionella* and *Clonothrix*, cause direct oxidation of metal, for example ferrous to ferric, and cause tubercles on metal surface. These are called iron oxidizing bacteria. Since areas under the tubercles are deficient in oxygen, they act as corrosion cells and result in deep internal grooving.
  - ➢ Some bacteria are acid-producing bacteria (APB) and thus corrode metals.
- MIC may be prevented by routine monitoring of TBC and SRB counts in the cooling water system.
- However, a common mistake is to measure the planktonic count (microbes present in the bulk water), which shows poor correlation with the sessile count on the metal surface.
- Sessile count monitoring, and identification of low velocity zones and fouling prone coolers and condensers are a must for formulating an effective water management program.
- It is important to select proper biocides to kill unwanted microorganisms and equally important to use biodispersants to disengage organisms from surfaces so that the biocides can act effectively.
- Ammonia and hydrogen sulfide sometimes accompany hydrocarbon gases to contaminate cooling water.
- These chemicals also cause the demand for chlorine to increase to very high levels and may also lead to both fouling and MIC.

Remedial Measures

- In addition to fixing the actual leak, parallel measures are also taken to ward off the deleterious effects of leaks in the cooling water system:
  - ➢ Dosing of oxidizing biocide is increased to a higher level.
  - ➢ Chlorine dioxide, bromo compounds, and ozone are also used.
  - ➢ Biodispersants are dosed at a higher than normal rate. They cause faster disengagement of organisms from the surface so that biocides act effectively. This enhances both planktonic and sessile efficacy.
  - ➢ Shock dosing of other nonoxidizing biocides, such as quaternary ammonium compounds, methyl bis-thiocyanate, and glutaraldehyde, is also done to kill the microbes.
  - ➢ Selection of particular biocides and the dosing rates is important because biocides possess different levels of efficacy against various microorganisms, and each cooling water system has its unique microbiological population.
  - ➢ An overflow and controlled blowdown of the cooling water sump eliminates oil, biomass and froth from the sump, which otherwise would circulate in the system and clog the coolers and condensers. Any blowdown, however, directly affects cost.

18.3 DRYING

- What is the difference between evaporation and drying?
  - Evaporation refers to removal of relatively large amounts of water as vapor at its boiling point, concentrating the solution. It is a heat transfer process.
  - In drying the water is usually removed as a vapor by air. It is simultaneous heat and mass transfer process.
- Why is it necessary to dry solid materials?
  - Need for easy handling of solids. Free-flowing solids are much easier to handle than wet solids.
  - Preservation and storage. Organic solids with moisture content deteriorate due to biological attack during storage. Dried foods can be stored for long periods of time.

- Reduction in cost of handling, packaging requirements, and transportation. Moisture in the solids increases their weight, apart from handling problems such as their bulk, stickiness, and so on.
- To avoid presence of moisture that may lead to corrosion.
- Achieving desired product quality. In many processes, improper drying may lead to irreversible damage to product quality.
  - The quality of a food product is judged by the amount of physical and biochemical degradation occurring during the drying process.
  - The drying time, temperature, and water activity (defined later) influence the final product quality. Low temperatures generally have a positive influence on the quality but require longer processing times.
  - Many dried foods are rehydrated before consumption. The structure, density, and particle size of the food plays an important role in reconstitution.
  - Drying or dehydration of foods is used as a preservation technique. Microorganisms that cause food spoilage and decay cannot grow and multiply in the absence of water. Also, many enzymes that cause chemical changes in food and other biological materials cannot function without water. When the water content is reduced below about 10 wt%, the microorganisms are not active. However, it is usually necessary to lower the moisture content below 5 wt% to preserve flavor and nutrition.
  - For vegetables, drying time is crucial to tenderness. The longer the drying time is, the less will be the flavor.
  - To retain the viability and the activity of biological materials such as blood plasma and fermentation products, the operation is carried out at very low temperatures, while more severe conditions can be applied to food stuffs and other materials.

- Why drying is considered to be an energy intensive process? What is the range of energy consumption involved in drying of materials in different industries?
  - Drying is one of the most energy intensive processes due to phase change and high latent heat of vaporization and the inherent inefficiency of using hot air as the (most common) drying medium.
  - Energy consumption for drying ranges from about 5% of chemical products to as high as 35% for cellulosic materials.
- What is (i) free moisture and (ii) bound moisture?
  - Free moisture:
    - Moisture content of a solid in excess of equilibrium moisture content (*unbound moisture*).
    - Exerts vapor pressure of pure water at the prevailing temperature.
  - Bound moisture:
    - Moisture retained by a solid in such a way that it exerts a vapor pressure *less than* vapor pressure of pure water at the same temperature.
    - Moisture contained inside cell walls of plant matter/structure or in loose chemical combination with cellulosic material. May be retained in capillaries and crevices throughout the solid or adsorbed on surfaces.
    - As a liquid solution of soluble components of solid or as solid solution in cell walls (Figure 18.21).
- What is equilibrium moisture content?
  - When a *hygroscopic* material is exposed to air at constant temperature and humidity for prolonged periods of time, the material attains a definite

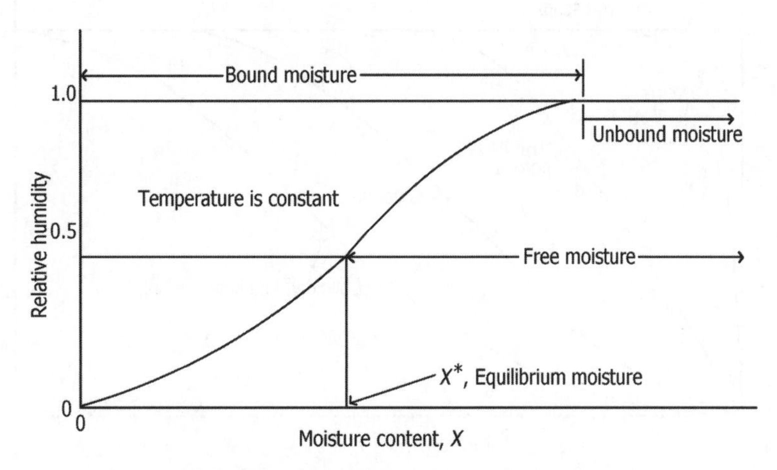

FIGURE 18.21 Types of moisture content.

moisture content that remains constant if temperature and humidity of air remains the same. Further exposure to air under the same conditions of humidity and temperature will not result any change in moisture content.

- In other words, it is the moisture content in a solid that is in equilibrium with the gas phase at a given partial pressure of vapor.
- Depends on temperature. With increase in temperature, equilibrium moisture content somewhat decreases, but for moderate changes in temperature, effect is not much.
- Equal to *zero* when there are no dissolved solids and no bound moisture (nonhygroscopic materials). Examples are inorganic salts like zinc oxide and other materials like glass wool and kaolin.
- Spongy or cellular materials like wood, leather, and tobacco have high equilibrium moisture content.
- At high relative humidities (60–80%), equilibrium moisture content increases rapidly with increase in relative humidity.
- At low relative humidities, equilibrium moisture content is *very high* for foods high in protein, starch, and high molecular weight polymers and *low* for foods with high percent solubles, crystalline salts, sugars, and so on.
- Depends on the direction from which equilibrium is approached, that is, whether a wet material is dried or a dry material adsorbs moisture. This is due to adsorption hysterisis.
- Figure 18.22 illustrates equilibrium moisture content for different types of solids.

- "Substances containing bound moisture are hygroscopic." *True/False*?
  - *True*.
- "For nonhygroscopic materials, equilibrium moisture content is essentially zero at all temperatures and humidities." *True/False*?
  - *True*.
- What type of materials have equilibrium moisture content?
  - Hygroscopic materials.
- Name some materials that have (i) low equilibrium moisture content and (ii) high equilibrium moisture content.
  - Low equilibrium moisture content: Glass wool, zinc oxide, kaolin, and so on.
  - High equilibrium moisture content: Spongy, cellular materials, for example, materials of vegetable/biological origin, like wood, leather, tobacco, and so on.
- When humidity of air is zero, what will be the value of equilibrium moisture content of a material?
  - Zero.
- For a given material, will the equilibrium moisture content normally be the same during drying process and during moisture absorption by a dry material?
  - Not necessarily. Many solids exhibit different equilibrium moisture contents depending on whether drying is taking place or moisture adsorption is taking place.
  - Because of irregular nature of pores, resistances depend on the direction of flow in the pores.
  - This phenomena is called *hysterisis*.
- "At high percent relative humidities (60–80%), equilibrium moisture content increases rapidly with increase in relative humidity." *True/False*?
  - *True*.
- What is water activity? Explain its significance.

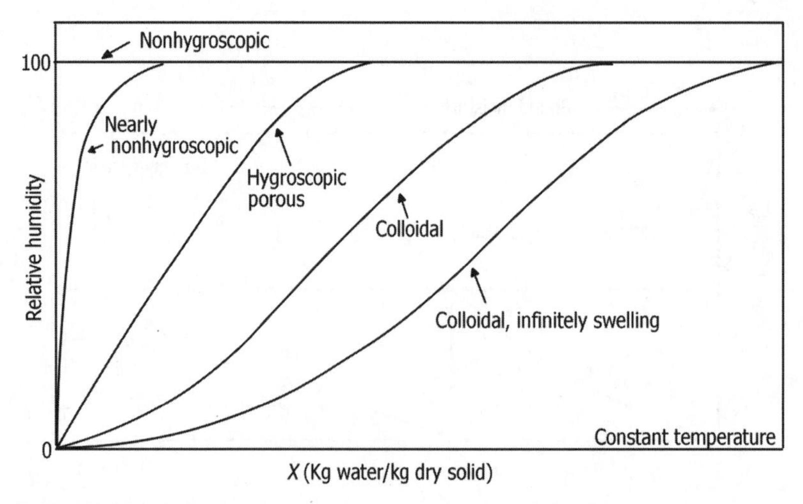

FIGURE 18.22 Equilibrium moisture content curves for different types of solids.

- In drying of some materials, which requires careful hygienic attention, for example, food, pharmaceuticals, and biological products, the availability of water for growth of microorganisms, germination of spores, and participation in several types of chemical reactions becomes an important issue.
- This availability, which depends on relative pressure, or *water activity*, a_w, which is defined as the ratio of the partial pressure, p, of water over the wet solid system to the equilibrium vapor pressure, p_w, of water at the same temperature. Thus, a_w, which is also equal to the relative humidity of the surrounding humid air, is defined as

$$a_w = p/p_w. \quad (18.12)$$

- Minimum measured values of water activity for microbial growth or spore germination for most bacteria are 0.91. The range for different materials is from about 0.6–0.98.
- If a_w is reduced below these values by dehydration or by adding water-binding agents like sugars, glycerol, or salt, microbial growth is inhibited.
- Such additives should not affect the flavor, taste, or other quality criteria.
- Since the amounts of soluble additives needed to depress a_w even by 0.1 is quite large, dehydration becomes particularly attractive for high moisture foods as a way to reduce a_w.
- Figure 18.23 illustrates water activity for different types of foods as function of moisture content.
- Low water activity retards or eliminates the growth of microorganisms, but results in higher lipid oxidation rates. Nonenzymatic browning reactions in food materials peak at intermediate water activities (0.6–0.7).
- Storage stability of a food product increases as the water activity decreases, and the products that have been dried at lower temperatures exhibit good storage stability.

- In a drying operation what are the driving forces for (a) heat transfer and (b) mass transfer?
  - (a) For heat transfer: (dry bulb temperature) – (wet bulb temperature).

$$-\mathrm{d}W/\mathrm{d}\theta = hA(T_D - T_W)/\lambda. \quad (18.13)$$

  - (b) For mass transfer: (pressure at wet bulb temperature) – (pressure at dew point.

$$-\mathrm{d}W/\mathrm{d}\theta = k'_g A_m(H_W - H_D). \quad (18.14)$$

- What is constant rate period?
  - When a solid surface is completely covered with water, rate of drying is fairly constant. This period is called constant rate period.
- What is falling rate period?
  - Begins when dry patches start appearing on the solid surface (i.e., the solid surface is no more *completely* covered with water).
  - Area for mass transfer progressively gets reduced (certain areas on the solid have moisture and certain areas are dry).

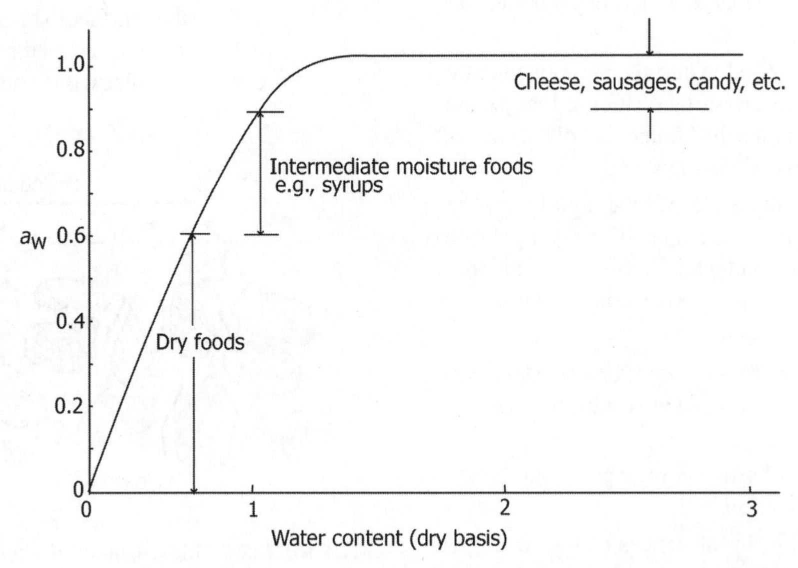

FIGURE 18.23 Water activity versus moisture content for different types of foods.

TABLE 18.6 Critical Moisture Content for Different Materials

Material	Critical Moisture Content ($kg\ H_2O/kg$ dry solid)
Salt crystals, rock salt, sand, wool	0.05–0.10
Brick clay, kaolin, crushed sand	0.10–0.20
Pigments, paper, soil, worsted wool fabric	0.20–0.40
Several foods, copper carbonate, sludges	0.40–0.80
Chrome leather, vegetables, fruits, gelatin, gels	>0.80

- What is critical moisture content? Give critical moisture content values for different solids.
  - Moisture content at the end of constant rate period is called *critical moisture content.*
  - Table 18.6 gives critical moisture content for different materials.
- What are the mechanisms involved in the transport of moisture during drying process?
  - *Liquid Diffusion:* Diffusion of moisture occurs when there is a concentration difference between the depths of the solid and the surface.
    - ➢ Takes place if the wet solid is at a temperature below the boiling point of the liquid and solids are close to homogeneity, from within to surface of solid due to concentration gradients setup during drying from the surface.
    - ➢ Transport of moisture usually occurs in relatively homogeneous nonporous solids where single-phase solutions are formed with the moisture, such as wood, leather, paper, starch, textiles, paste, clay, soap, gelatin, and glue.
    - ➢ In drying many food materials, the movement of moisture toward the surface (falling rate period) is mainly governed by molecular diffusion and therefore follows Fick's law.
  - *Vapor Diffusion:* Takes place if the liquid vaporizes within the material. For example, if heat is supplied to one surface of the solid while drying proceeds from another, moisture may vaporize beneath the surface and diffuse out as vapor.
  - *Knudsen Diffusion:* Takes place if drying takes place at very low temperatures and pressures, for example, in freeze drying.
  - *Surface Diffusion:* Diffusion of vapor from the surface into the bulk of air.
  - *Capillary Diffusion:* In substances with a large open pore structure and in beds of granular and porous solids, the unbound moisture flows from regions of low concentration to those of high concentration by capillary action. When granular and porous solids such as clays, sand, soil, plant pigments, and minerals are being dried, unbound or free moisture moves through the capillaries and voids of the solids by capillary action, not by diffusion. This mechanism, involving surface tension, is similar to the movement of oil in a lamp wick.
    - ➢ A porous solid contains interconnecting pores and channels of varying pore sizes. As water is evaporated, a meniscus of liquid water is formed across each pore in the depths of the solid.
    - ➢ This sets up capillary forces by the interfacial tension between the water and solid. These capillary forces provide the driving force for moving water through the pores to the surface.
    - ➢ Small pores develop greater forces than those developed by large pores.
    - ➢ Figure 18.24 illustrates movement of moisture through porous materials.
  - *Pressure Gradients:* By flow caused by pressure gradients.
  - *Hydrostatic pressure differences,* when internal vaporization rates exceed the rate of vapor transport through the solid to the surroundings.
  - *Effect of Shrinkage:* A factor often greatly affecting the drying rate is the shrinkage of the sold as moisture is removed.
    - ➢ When bound moisture is removed from rigid, porous or nonporous solids, they do not shrink appreciably, but colloidal and fibrous materials such as vegetables and other foodstuffs do undergo severe shrinkage.
    - ➢ When the surface shrinks against a constant volume core, it causes the material to warp, check, crack, or otherwise change its structure. The

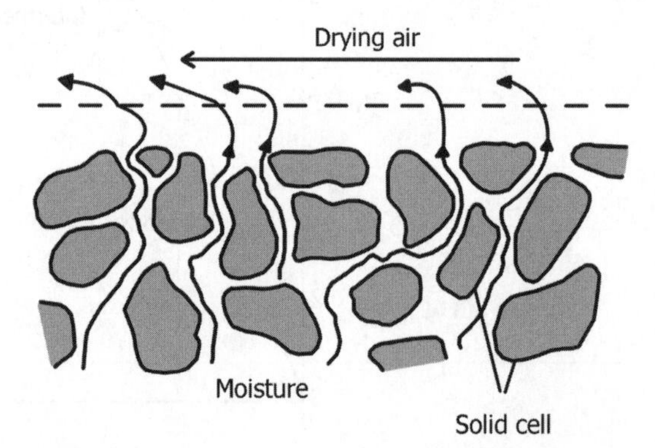

FIGURE 18.24 Movement of moisture during drying of porous materials.

reduced moisture content in the hardened outer layer increases the resistance to diffusion. This can happen in drying wood.

- ➢ The most serious effect is that a hard layer may be developed on the surface, which is impervious to the flow of liquid or vapor moisture and slows the drying rate. Examples are clay and soap.
- ➢ In many food stuffs, if drying occurs at too high a temperature, a layer of closely packed shrunken cells, which are sealed together, forms at the surface. This presents a barrier to moisture migration and is known as *case hardening*.

- *Gravity Flow:* By flow caused by gravity.
- Combinations of the above mechanisms.

- Under what circumstances moisture moves through a solid by vapor diffusion?
  - When temperature gradient is established by heating, which creates a vapor pressure gradient.
  - For example, n a slab of solid in which heating takes place at one surface and drying from the other.
- What is meant by *fiber saturation point* in solids?
  - Moisture content of cellular materials (e.g., wood) at which the cell walls are completely saturated while the cavities are liquid free.
  - May also be defined as equilibrium moisture content as humidity of the surrounding air approaches saturation.
  - Give examples of solid systems in which moisture and solid are mutually soluble.
  - Drying of soaps, glues, gelatins, and pastes.
- "If moisture moves to the surface through capillaries, the drying time is proportional to the square of the thickness of the solid." *True/False*?
  - *False.*
- "Free moisture content in a product is independent of the drying air temperature." *True/False*?
  - *False.*
- What are the variables involved in drying of different materials?
  - Product size and shape (microns to tens of centimeters thickness or depth).
  - Porosity may range from 0% to 99.9%.
  - Drying times range from 0.25 s (drying of tissue paper) to 5 months (for certain hardwood species).
  - Production capacities may range from 0.10 kg/h to 100 t/h.
  - Materials may be stationary or moving at as high a velocity as 2000 m/s (tissue paper).
  - Drying temperatures range from below the triple point to above the critical point of the liquid.
  - Operating pressure may range from fraction of a mbar to 25 atm.
  - Heat may be transferred continuously or intermittently by convection, conduction, radiation, or electromagnetic fields.
- What are the external variables involved in any drying process?
  - Wet and dry bulb temperatures, humidity, air velocity, state of subdivision of solid, agitation of solid, thickness of solid slab (no effect if moisture is surface moisture), method of supporting the solids, contact between hot surface and solid, surface finish of the solids, and so on.
  - Other factors that influence drying rates include the following:
    - ➢ Evaporation off a surface.
    - ➢ Surface may not dry uniformly and consequently effective surface may change with time.
    - ➢ Resistance to moisture diffusion and capillary flow of moisture may develop.
    - ➢ Shrinkage may occur on drying particularly near the surface, which hinders further movement of moisture outward.
    - ➢ Agglomerates of particles may disintegrate on partial drying.
- Draw a typical drying rate curve and briefly describe the different sections of the curve?
  - A plot of rate of drying versus free moisture content is known as drying rate curve (Figure 18.25).
  - At zero time the initial free moisture content is shown at point A.
    - ➢ In the beginning, the solid is usually at a colder temperature than its ultimate temperature, and the

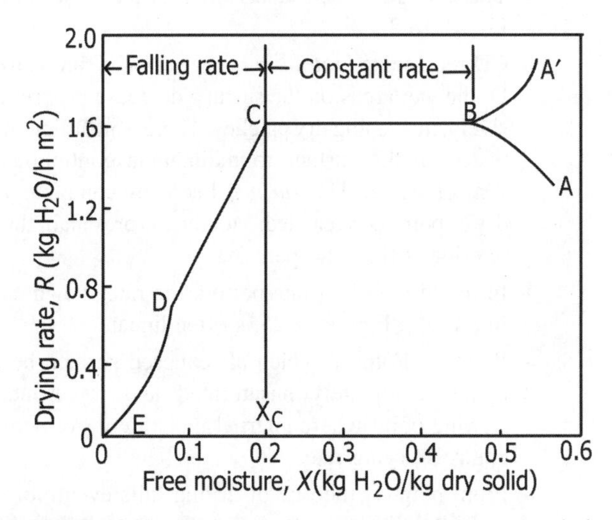

FIGURE 18.25 Typical drying rate curve for constant drying conditions.

evaporation rate will increase. Eventually at point B the surface temperature rises to its equilibrium value.

- ➢ Alternatively, if the solid is quite hot to start with, the rate may start at point A′. This initial unsteady-state adjustment period is usually quite short and it is often ignored in the analysis of times of drying.

- From point B to C in Figure 18.25, the line is straight, and hence the slope and rate are constant during this period. This constant rate of drying period is shown as line BC.
  - ➢ During the constant rate period, external heat/mass transfer rate is controlling. Water evaporates as if there is no solid present, and its rate of evaporation is not dependent on the solid. This continues until water from the interior is no longer freely available at the surface of the solid.
- Point C in Figure 18.25 is *critical moisture content*, X_C, which represents the point of transition from constant rate to falling rates. From point C, the drying rate starts to decrease.
- CE represents falling rate period during which internal heat/mass transfer rates control.
- The plane of evaporation slowly recedes from the surface. Heat for the evaporation is transferred through the solid to the zone of vaporization. Vaporized water moves *through* the solid into the air stream.
  - ➢ Although the amount of water removed in the falling rate period is relatively small, it can take considerably longer than in the constant rate period.
  - ➢ The falling rate period has two sections as seen in Figure 18.25.
  - ➢ CD represents *first falling rate period.* From C to D, the wet areas on the surface decrease progressively, increasing dry patches. There is insufficient water on the surface to maintain a continuous film of water. The *surface* becomes completely dry as point D is reached. Moisture is present in the interior of the solid pores.
  - ➢ In this first falling rate period, the rate shown as line CD in Figure 18.25 is often linear.
  - ➢ Point D: Point at which all exposed surface becomes completely unsaturated and represents starting point where internal moisture movement controls drying rate.
  - ➢ From point D the rate of drying falls even more rapidly, until it reaches point E, where the equilibrium moisture content is X^*.
  - ➢ When the *surface* is completely dry (point D), the evaporation process continues moving toward the center of the solid as shown by the curve DE, which represents the *second falling rate period* (internal moisture movement).
  - ➢ In some materials being dried, no sharp discontinuity occurs at point D, and the change from partially wetted to completely dry conditions at the surface is so gradual that no sharp change is detectable. CD may constitute all of the falling-rate period.
  - ➢ Nonhygroscopic materials have a single falling rate period whereas hygroscopic materials have two or more periods. In the first period, the plane of evaporation moves from the surface to inside the food, and water vapor diffuses through the dry solids to the drying air. The second period occurs when the partial pressure of water vapor is below the saturated vapor pressure and drying is by desorption.
- Many foods and agricultural products, however, do not display the constant rate period at all, since internal heat and mass transfer rates determine the rate at which water becomes available at the exposed evaporating surface.
- The following are normalized drying rate curves for some typical materials (Figure 18.26).

• "Critical moisture content in a product is independent of the initial drying rate." *True/False*?
  - *False.*

• "Equilibrium moisture content of a product in a dryer is independent of the air temperature in the dryer." *True/False*?

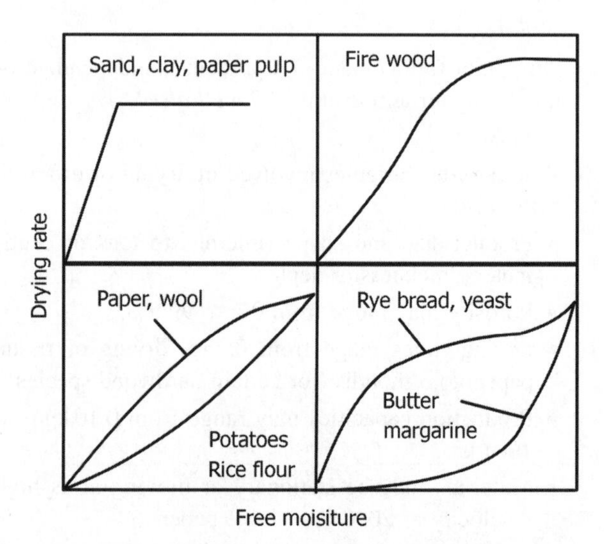

FIGURE 18.26 Examples of normalized drying rate curves for some typical materials.

- *False.*

- "The drying rate in a constant rate period is controlled by the internal resistance to heat and mass transfer." *True/False*?
  - *False.*
- "The drying rate in a falling rate period is controlled by the internal resistance to heat and mass transfer." *True/False*
  - *True.*
- "The product surface temperature during the constant rate period is always equal to the wet bulb temperature of drying air." *True/False*?
  - *False.*
- "A wet material, suspended in drying air, will be near the wet bulb temperature." *True/False*?
  - *True.*
- "Superheated steam is not a suitable drying medium for drying foodstuffs." *True/False*?
  - *False.*
- What are the ways in reducing the problems involved in drying processes?
  - The problems of shrinkage and case hardening during drying process can be minimized by reducing the drying rate, thereby flattening the moisture gradient into the solid. Since the drying behavior presents different characteristics in the two periods, namely, constant rate and falling rate, the design of the dryer should recognize these differences, that is, substances that exhibit predominantly a constant rate drying are subject to different design criteria than substances that exhibit a long falling rate period.
  - Since it is more expensive to remove moisture during the falling rate period than during the constant rate one, it is desirable to extend as long as possible the latter with respect to the former. Particle size reduction is a practical way to accomplish this because more drying area is created.
  - Sometimes to decrease the effects of shrinkage, it is desirable to dry with moist air. This decreases the rate of drying so that the effects of shrinkage on warping or hardening at the surface are greatly reduced.
  - As a result of change in physical structure of the solids during the drying process, for example, shrinkage and breakup, mechanisms might also change.
- Give equations for the estimation of drying times during the constant and falling rate periods.
  - Drying time is the difference in times corresponding to initial and final moisture contents from the drying rate curve.
  - Constant rate drying is controlled by heat or mass transfer rates:

$$N_c = h(T - T_s)/\lambda_s = k_y M_{air}(H_s - H), \quad (18.15)$$

where N_c is the rate of drying during constant rate period, T is the temperature, λ_s is the latent heat of vaporization of water, k_y is the mass transfer coefficient that remains constant under constant gas velocity and flow direction past the surface, and H_s and H are saturated humidity at liquid surface temperature T_s and prevailing humidity, respectively.

  - Rates from empirical data:

$$N = -(S_s/A)\mathrm{d}X/\mathrm{d}t, \quad (18.16)$$

where N is the drying rate, S_s is the mass of dry solid in a batch, X is the moisture content of the solid, t is the time, and A is the exposed surface area of the solid.

  - The above equation on rearrangement and integration gives time of drying under different conditions:

$$t = \int \mathrm{d}t = (S_s/A)\int \mathrm{d}X/N, \quad (18.17)$$

over the limits between X_1 and X_2.

  - Negative sign disappears as the integration is from X_2 to X_1.
  - *Constant Rate Period:* If drying takes place entirely within constant rate period, X_1 and $X_2 > X_c$ and $N = N_c$ and the equation reduces to

$$t = (S_s/AN_c)(X_1 - X_2) = S_s(X_1 - X_2)/Ak_y(H_s - H). \quad (18.18)$$

  - *Falling Rate Period:* If X_1 and X_2 are both less than X_c, drying takes place under conditions of changing N.
  - (a) General case: For any shape of falling rate curve, the integral can be obtained from graphical integration by plotting X versus $1/N$ between limits X_1 and X_2, using data from rate of drying curve.
  - (b) N is linear in X and can be represented by an equation for straight line:

$$N = mX + b, \quad (18.19)$$

where m is the slope of linear portion of the curve. The integral becomes

$$(S_s/mA)\ln[(mX_1 + b)/(mX_2 + b)] = (S_s/mA)\ln(N_1/N_2), \quad (18.20)$$

$$m = (N_1 - N_2)/(X_1 - X_2), \quad (18.21)$$

and

$$t = [S_s(X_1 - X_2)/A(N_1 - N_2)]\ln(N_1/N_2) = S_s(X_1 - X_2)/(AN_m), \quad (18.22)$$

where N_m is the log mean average of the rate N_1 at moisture content X_1 and N_2 at X_2.

- In many cases the entire falling rate curve can be taken as a straight line between points C and E shown on the drying rate curve.
- For this case,

$$N = m(X - X^*) = N_c(X - X^*)/(X_c - X^*) \quad (18.23)$$

and the equation for drying rate becomes

$$t = [S_s(X_c - X^*)/N_c A]\ln[(X_1 - X^*)/(X_2 - X^*)]. \quad (18.24)$$

- What properties of the materials being dried that are important in the selection of a dryer?
  - Physical characteristics when wet and when dry.
  - Uniformity in the final moisture content.
  - Decomposition characteristics.
  - Shrinkage.
  - Contamination.
  - Particle size.
  - Corrosiveness/abrasiveness.
  - Flammability/toxicity.
- What is stickiness?
  - Stickiness is a phenomenon that reflects the tendency of some materials that agglomerate and or adhere to contact surfaces.
  - Cohesion and adhesion phenomena contribute to stickiness.
- What are the factors that cause stickiness in materials?
  - High hygroscopicity, high solubility, low melting point materials, and materials with low glass transition temperature. If the product temperature is above its glass transition temperature, it will exhibit stickiness. At the glass transition temperature the amorphous material is converted to rubbery state (from its solid glassy state).
    - Highly hygroscopic amorphous materials can easily regain moisture when exposed to atmosphere.
- What are the materials that exhibit stickiness in dryers?
  - High organic acid and high fat foods and products with high sugar content like honey, fruit juices, molasses, sucrose, lactose, glucose, and fructose.
- Name glass transition-related problems in drying processes.
  - *Spray Drying:* Sticking onto the dryer wall, duct and cyclone, poor recovery of powder, agglomeration in the collection bag or container.
  - *Freeze Drying: Collapse* of structure while drying.
  - *Conventional Hot Air Solid Drying:* Poor fluidization, stick on the drying racks/shelves, soft product while drying but solid after cooling.
  - Stickiness causes buildup of solids on dryer surfaces. It can alter hydrodynamics of flow in dryers, leading, in extreme cases, to choking and blocking in the dryers.
  - Stickiness is not only undesirable from the point of view of deterioration and loss of product, but also hazardous as self-heating of the accumulated product on the walls of the dryer can result in explosions in dryers. There were several explosions that occurred in dryers on this account.
  - The design capacity of the dryers rapidly reduces on account of stickiness by affecting gas–solids contact and heat and mass transfer rates, residence time distribution and local solids holdup.
    - Observations pointed out that stickiness contributed to a fall of heat transfer coefficients by as much as 60% in a rotary dryer.
    - For example, drying sticky zinc sulfate from 17% to 8% moisture content in an industrial dryer was attributed to be the cause of difference in real residence time from calculated one by as much as 800%.
  - *Storage:* Clumping, agglomeration, caking, crystallization.
- State the methods by which sticky materials are handled in dryers.
  - Drying below the glass transition temperature (often not feasible).
  - Choosing mild drying temperature conditions.
  - Increasing the glass transition temperature of foods by adding high molecular weight materials (drying aids such as maltodextrins).
  - Cooling of the dryer walls and immediate cooling of the product by blowing cold air below its glass transition temperature.
  - By recycling part of the dried product to feed.
  - Using additives (drying aids) that modify material properties.

18.3.1 Dryers

- Give the key criteria used for the classification of dryers.
  - Table 18.7 gives the criteria for classification of dryers.
- What are the factors that are to be considered in the selection of a dryer?
  - The nature of the wet feed and the way it is introduced into the dryer.

TABLE 18.7 Criteria for Classification of Dryers

Criterion	Types
Mode of operation	Batch or continuous
Type of heat input	Convection, conduction, radiation and their combinations, electromagnetic fields
Material characteristics	Nature, size and shape of solids; wet, slurry, liquid
State of material in the dryer	Stationary, moving, agitated, dispersed
Relative movement between fluid and solids	Cocurrent, countercurrent, mixed flow
Method of conveyance of the solid	Belt, rotary, fluidized
Pressure	Atmospheric, vacuum
Drying medium	Air, superheated steam, flue gases
Temperature	Above boiling, below boiling, below freezing point
Stages	Single or multistage
Residence time and residence time distribution	Short (<1 min), medium (1–60 min), long (>1 h) Residence time distribution varies for moving particles
Energy and equipment costs	These vary from dryer to dryer and influence selection

- The way in which the moist material is supported.
- Feed condition: Solid, liquid, paste, crystals, powder.
- Feed concentration: Initial moisture/liquid content.
- Product specification: Dryness required physical form.
- Throughput.
- Heat sensitivity of the product.
- Nature of the material: Toxicity, flammability, dust explosion hazards.
- Nature of vapors produced: toxicity, flammability, environmentally objectionable for atmospheric release.
- In selecting a dryer for a particular application two steps are of primary importance:
  - A listing of the dryers that are capable of handling the material to be dried.
  - Eliminating the more costly alternatives on the basis of annual costs, capital charges + operating costs.
  - Once a group of possible dryers has been selected, the choice may be narrowed by deciding whether batch or continuous operation is to be employed and, in addition to restraints imposed by the nature of the material, whether heating by contact with a solid surface or directly by convection and radiation is preferred.
  - In general, continuous operation has the important advantage of ease of integration into the rest of the process coupled with a lower unit cost of drying.
  - As the rate of throughput of material becomes smaller, however, the capital cost becomes the major component in the total running costs and the relative cheapness of batch plant becomes more attractive.

- What are the drying times involved in different types of dryers?
  - Spray dryers: Few seconds.
  - Rotary dryers: Up to 1 h.
  - Tunnel shelf or belt dryers: Several hours or even several days.
  - Continuous tray and belt dryers for granular material or pellets of 3–15 mm size: 10–200 min.
  - Rotary cylindrical dryers: 5–90 min.
  - Drum dryers for pastes, milk-based sweets and slurries, or highly viscous liquids: 3–12 s.
  - Pneumatic dryers: 0.5–3 s for single pass and up to 60 s when recycling is used.
  - Fluid bed dryers: 1–2 min for continuous operation and 2–3 h for batch drying of some pharmaceutical products.
- What are contact dryers? What are their applications?
  - Contact or indirect dryers, where the heat is transferred to the material being dried through a heated surface. These dryers, also known as adiabatic dryers, use steam, hot water, glycol solutions, and commercially available heat transfer fluids for the supply of heat to the drying process.
  - Contact dryers include vacuum tray dryers, paddle dryers, spiral screw dryers, disk dryers, and contact kneader dryers.
  - These dryers handle a variety of materials that include low or high viscosity materials, flat and strip materials such as paper, textiles or cardboard, pasty or creamy materials, granular materials, polymers, sugar substitutes in food industry, crystallization of polymers in the plastics or rubber industries, and drying of block milk used in chocolate making,

pastry cooking, and dairy industries. Disk dryers are used for drying of sewage sludges.

- What are the advantages of contact dryers over direct dryers?
  - Avoid cross-contamination as heat transfer medium does not contact the material being dried.
  - Solvent recovery is easy because of the very small amounts of noncondensables present.
  - Extensive dust formation is generally avoided because of small volumes of vapors involved. But when high temperatures and low levels of vacuum are involved dust problems can arise with large volumes of vapors being evolved.
  - Vacuum operation is possible preventing product degradation.
  - The dried product, when granular or powdery, has a higher bulk density than the same material is dried in a spray dryer.
  - These dryers can be designed as pressure or shock resistant vessels.
  - Space requirements are usually small.
  - Closed design is possible, containing toxic and flammable hazards.
- What is meant by through circulation in a dryer?
  - Air penetrates and flows through interstices among the solids, circulating freely around the individual particles.
- What are the features of through circulation tray dryers?
  - Employ perforated or screen bottom tray construction.
  - Have baffles that force air through the solids bed.
  - Typical superficial velocities of air are around 50 m/min.
  - Drying rates are 1–10 kg/h m^2 tray area.
  - They are used primarily for small-scale manufacture of valuable materials.
  - Energy efficiencies are around 50%.
  - Tray or shelf dryers are commonly used for granular materials and for individual articles.
  - The material is placed on a series of trays that may be heated from below by steam coils and drying is carried out by the circulation of air over the material.
  - In some cases, the air is heated and then passed once through the oven, although, in the majority of dryers, some recirculation of air takes place, and the air is reheated before it is passed over each shelf.
  - Figure 18.27 illustrates working of a tray dryer.
- What are the operational problems involved in a tray dryer?
  - Nonuniform air flow.

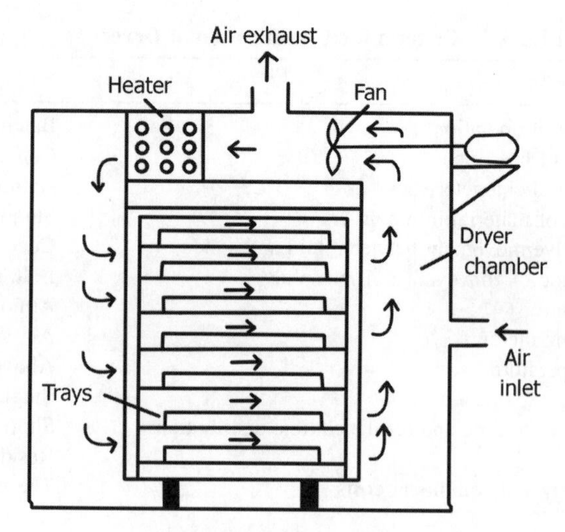

FIGURE 18.27 Tray dryer.

  - Difficulty in maintaining constant temperature.
  - Labor requirements in loading and unloading trays.
  - Blockages of vents that can result in pressure buildup and eventual explosions.
- What is a tunnel dryer?
  - Trays of wet material loaded on trucks may be moved slowly through a drying tunnel.
  - When a truck is dry, it is removed at one end of the tunnel, and a fresh one is introduced at the other end.
  - Fresh air inlets and humid air outlets are spaced along the length of the tunnel to suit the rate of evaporation over the drying curve. This mode of operation is suited particularly to long drying times, from 20 to 96 h for the materials.
- What are the drawbacks of tray dryers?
  - Charging, unloading, and cleaning are labor intensive and time consuming (typically 5–6 h for a large tray dryer).
- What type(s) of dryers are used for large capacity drying applications?
  - Direct and indirect rotary dryers.
  - Spray dryers.
  - Drum dryers.
  - Tunnel dryers.
  - Fluid bed dryers.
  - Pneumatic dryers.
- What type(s) of dryers are used for small-scale applications?
  - Vacuum tray dryers.
  - Agitated batch dryers.
  - Through circulation dryers.
  - Fluid bed dryers.
- What are the applications of vacuum shelf dryers?

- Drying of heat-sensitive/easily oxidizable materials.
- Pharmaceuticals of high value.
- Materials wet with toxic/flammable or valuable solvents.
- Recovery of solvents without coming into flammable range, due to low oxygen content under vacuum.
- Manufacture of semiconductor materials in controlled atmospheres and lower operating pressures.

• What are the characteristic features and applications of conveyor/belt dryers?

- Continuous conveyor/belt dryers are up to 20 m long and 3 m wide.
- Solids are dried on a mesh belt in beds 5–15 cm deep. The air flow is initially directed upward through the bed of the solids and then downward in later stages to prevent dried material from blowing out of the bed.
- Two- or three-stage dryers mix and repile the partly dried material into deeper beds (to 15–25 cm and then 250–900 cm in three-stage dryers). This improves uniformity of drying and saves floor space. Solids are dried to 10–15% moisture content and then finished in bin dryers. This equipment has good control over drying conditions and high production rates. It is used for large-scale drying of foods (e.g., up to 5.5 tons/h).
- Figure 18.28 illustrates a three-stage conveyor dryer.
- Dryers may have computer controlled independent drying zones and automatic loading and unloading to reduce labor costs.
- A second application of conveyor dryers is *foam mat drying* in which liquid foods are formed into stable foam by the addition of a stabilizer and aeration with nitrogen or air. The foam is spread on a perforated belt to a depth of 2–3 mm and dried rapidly in two stages by parallel and then countercurrent air flows.
- Foam mat drying is approximately three times faster than drying a similar thickness of liquid.
- The thin porous mat of dried food is then ground to a free-flowing powder that has good rehydration properties.
- Rapid drying and low product temperatures result in a high-quality product, but a large surface area is required for high production rates and capital costs are therefore high.

• What are the characteristics and operating parameters of drum dryers?

- Solutions, slurries, and pastes may be spread as thin films and dried on steam heated rotating drums.
- If a solution or slurry is run on to a slowly rotating steam heated drum, evaporation takes place and solids may be obtained in a dry form. This is the basic principle used in all drum dryers.
- Twin drums commonly rotate in opposite directions inward to nip the feed, but when lumps are present that could damage the drums, rotations are in the same direction.
- Drum dryers are used for highly viscous liquids, pastes, and slurries.
- Contact times are relatively short in the range of 3–12 s.
- Produce flakes 1–3 mm thick with evaporation rates of 15–30 kg/(m^2 h).
- The drum surface temperature is usually in the range 110–165°C.
- Diameters are generally 0.5–1.5 m (1.5–5 ft) and 0.2–3 m long.
- Rotation speeds are in the range of 2–10 rpm. Speeds depend on the consistency of the feeds. Thin liquids allow a high speed, thick pastes lower speeds.
- Maximum evaporation capacity is around 1350 kg/h (3000 lb/h) and evaporation rates are 15–30 kg/(m^2 h).

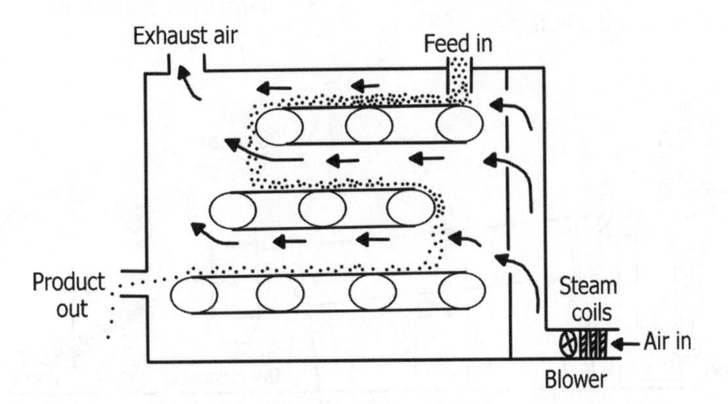

FIGURE 18.28 Three-stage conveyor dryer.

- Drum dryers have high drying rates and high-energy efficiencies and they are suitable for slurries in which the particles are too large for spray drying.

- What are the characteristics and operating parameters of a rotary cylindrical dryer? Illustrate with a diagram.
  - Rotary cylindrical dryers operate with air velocities of 1.5–3 m/s (5–10 ft/s), up to 10.5 m/s (35 ft/s) for drying coarse materials.
  - Residence times range from 5 to 90 min.
  - Solids holdup is about 7–8%.
  - For initial design purposes, an 85% free cross-sectional area is used.
  - Countercurrent design should give an exit gas temperature that is 10–20°C above the solids temperature.
  - Parallel flow should give solids temperatures of 100°C.
  - Rotation speeds of 4–5 rpm are common.
  - Diameters are around 1.2–3 m and lengths are 4–15 diameters.
  - The product of rpm and diameter (in meters), that is, rpm multiplied by diameter, should be 50–80.
  - Superficial gas velocities are 1.5–3 m/s, but lower values may be needed for fine products and rates up to 10 m/s may be allowable for coarse materials.
  - Flights attached to the shell lift up the material and shower it as a curtain through which the gas flows. The shape of flights is a compromise between effectiveness and ease of cleaning.
  - Dryers with jacketed shells or other kinds of heat transfer surfaces are employed when heating by direct contact with hot gases is not feasible because of contamination or excessive dusting.
  - Figure 18.29 illustrates a simplified diagram of a rotary dryer.

- How does an indirectly heated rotary dryer operate?
  - In indirect or nonadiabatic drying, the heat is transferred by conduction from a hot surface, first to the material surface and then into the bulk.
  - In one form of indirectly heated dryer hot gases pass through the innermost cylinder and then return through the annular space between the outer cylinder. This form of dryer can be arranged to give direct contact with the wet material during the return passage of the gases. Flights on the outer surface of the inner cylinder and the inner surface of the outer cylinder, assist in moving the material along the dryer.
  - This form of unit gives a better heat recovery than the single flow direct dryer, though it is more expensive.
  - In a simpler arrangement, a single shell is mounted inside a brickwork chamber, through which the hot gases are introduced.

- What are the advantages of indirect heat dryers over direct heat dryers?
  - It is not necessary to heat large volumes of air before drying commences and the thermal efficiency is therefore high. Typically heat consumption is 2000–3000 kJ/kg of water evaporated compared with 4000–10,000 kJ/kg of water evaporated for hot air dryers.
  - Drying may be carried out in the absence of oxygen to protect components of foods that are easily oxidized.

- What are the applications of a rotary dryer?
  - Suitable for continuous operation at large capacities for drying free-flowing granular materials that require drying times of 1 h or less.
  - Materials that tend to agglomerate because of wetness may be preconditioned by mixing with recycled dry product.
  - High thermal efficiency and low capital and labor costs.

- What is the purpose of *lifters* or *flights* in a rotary dryer?
  - To shower the material through open space of the dryer to provide more contact surface and turbulence in solids for better heat transfer.

- In a rotary dryer, what percentage of its total volume is normally occupied by solids?
  - 7–8%.

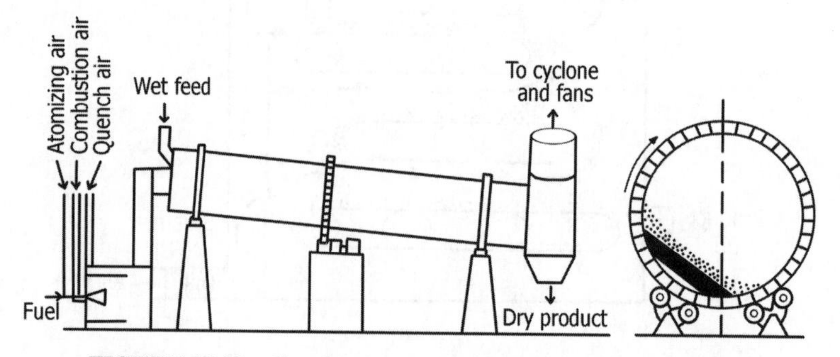

FIGURE 18.29 Simplified diagram of a direct heat rotary dryer.

- Some references suggest 10–15%.
- <10%: Insufficient to utilize lifters fully.
- >15%: Possibility of parts of solids surface unexposed to gases.

- Define time of passage of material in a rotary dryer?
  - Holdup/feed rate.
- What is the range of slopes of rotary dryers? Rotary kilns?
  - For rotary dryers: 0–8 cm/m.
  - For rotary kilns: 2–6 cm/m.
- Give a formula for power requirements for the rotation of a rotary dryer.
  - Power required to rotate the shell is given by

$$P = 0.45\, Wv_r + 0.12BDNf, \quad (18.25)$$

where P is in W, W is the weight in kg of the rotating parts, v_r is the peripheral speed of the carrying rollers (m/s), B is the holdup of solids (kg), D is the diameter of the shell (m), N is in rpm, and f is the number of flights along the periphery of the shell.

- For what type of applications vacuum rotary dryers are used?
  - When the temperature of the drying material must be kept as low as possible, for reasons of damage to product or change in its nature when exposed to high temperatures, vacuum drying is used.
  - Used for drying of large quantities of heat sensitive/granular/crystalline/fibrous solids.
  - When air at high temperatures causes oxidation to the product or leads to explosive conditions.
  - Rapid drying due to large ΔT. Shorter residence times.
  - When solvent recovery is required.
  - When materials must be dried to extremely low moisture levels.
- In a dryer operating under vacuum, what are the circumstances leading to deterioration of its performance?
  - Air leaks into the dryer is one of the important reasons for loss or decrease in available vacuum. This can increase drying temperature and loss of quality of the product, particularly for heat sensitive materials. Required drying times will increase to get the same level of drying in batch vacuum dryers.
  - Drying involving abrasive materials, product may contaminate seals that eventually leads to deterioration of the seal resulting in leakages through the seals. Improved designs of mechanical seal systems can overcome this type of problems.
- What is an agitated dryer? Where is it used?
  - In an agitated dryer, housing is stationary and solids movement is through the use of an internal mechanical agitator.
  - Used for obtaining free-flowing granular solids when discharged as product.
- What is the important advantage of an agitated dryer over a rotary dryer?
  - Large diameter seals are not required as housing is stationary with minimized leakage problems.
  - Shear forces in mixing process are useful in breaking up lumps and agglomerates.
- What is ball drying? Describe its operating features and plus and minus points.
  - In *ball drying*, a drying chamber is fitted with a slowly rotating screw and contains ceramic balls that are heated by hot air, blown into the chamber.
  - Particulate foods are dried mainly by conduction as a result of contact with the hot balls and are moved through the dryer by the screw, to be discharged at the base. The drying time is controlled by the speed of the screw and the temperature of the heated balls.
  - Suitable for small particles like vegetable pieces.
  - Advantages include low operating temperatures and rapid drying.
  - Continuous operation.
  - Good quality products.
  - Loss of product integrity.
  - Difficult to control.
- What is a fluid bed dryer? What are its advantages/disadvantages?
  - In a fluid bed dryer, drying air/gas is passed through the bed of solids at a velocity sufficient to keep the bed in a fluidized state.
    - Fluid bed dryers are operated at air velocities between the incipient and entrainment values. The larger and more dense the particles are, the higher the air velocity required to fluidize them.
    - Designing for an air velocity that is 1.7–2 times the minimum fluidization velocity is good practice. Air/gas velocities in the range 0.2–5.0 m/s are used.
    - Particles in the size range 20 μm to 10 mm in diameter can usually be fluidized. Work well with particles of few tenths of a mm and up to 4.0 mm in diameter. Fluid bed dryers may be used when the average particle diameter is $\leq$0.1 mm as other dryers may be too large to be feasible to handle small particles.
    - Inert gas may be used if there is the possibility of explosion of either the vapor or dust in the air.

- Turbulence promotes heat transfer and drying rates.
- Rapid and uniform heat transfer.
- Uniformity of temperature and composition throughout.
  - If the size, density, and moisture content of the feed particles are not uniform, the moisture content of the product may also not be uniform.
- Fluidized beds may be operated on a batch or continuous basis. Batch units are used for small-scale operations. Because of the mixing that occurs in such beds, uniform moisture contents are attainable.
- Short drying times.
  - Normally, drying times of 1–2 min are sufficient in continuous operation.
  - In batch applications involving some pharmaceutical products drying times of up to 2–3 h are used. (Slow drying to preserve quality of the product.)
- Free-flowing granular materials that require relatively short drying times are particularly suited to fluidized bed drying.
  - When longer drying times are necessary, multistaging, recirculation, or batch operation of fluidized beds still may have advantages over other modes.
- Particles must not be sticky or prone to mechanical damage.
- Good control of drying conditions.
- Low floor space requirements.
- High power requirements.
- Shallow beds are easier to maintain in stable fluidization and require a smaller load on the air blower.
- Use of multistage fluidized bed dryers is common. The partly dried product from the first stage is discharged onto the second stage, and so on. Up to six stages have been used by the food industry.
- Temperature of the air may be controlled at a different level in each stage. Such systems can result in savings in energy and better control over the quality of the product, as compared with a single stage unit.
- Fluidized beds may be mechanically vibrated. This enables them to handle particles with a wider size range than a standard bed. They can also accommodate sticky products and agglomerated particles better than a standard bed. Such fluidized beds are also known as *vibro-fluidizers*.

• What is a spouted bed dryer? Illustrate its working by a suitable diagram.

- Part of the heated air is introduced into the bottom of the bed in the form of a high velocity jet.
- A spout of fast moving particles is formed in the center of the bed. On reaching the top of the bed, the particles return slowly to the bottom of the bed in an annular channel surrounding the spout. Some of the heated air flows upward through the slow moving channel, countercurrent to the movement of the particles.
- High rates of evaporation are attained in the spout, while evaporative cooling keeps the particle temperature relatively low.
- Conditions in the spout are close to constant rate drying. Drying of the particles is completed in the annular channel.
- This type of dryer can handle larger particles than the conventional fluidized bed.
- In some spouted bed dryers the air is introduced tangentially into the base of the bed and a screw conveyor is located at its center to control the upward movement of the particles.
- Such dryers are suitable for drying relatively small particles. High rates of heat and mass transfer in the bed enable rapid drying of the small particles.
- Figure 18.30 illustrates the operation of a spouted bed dryer.

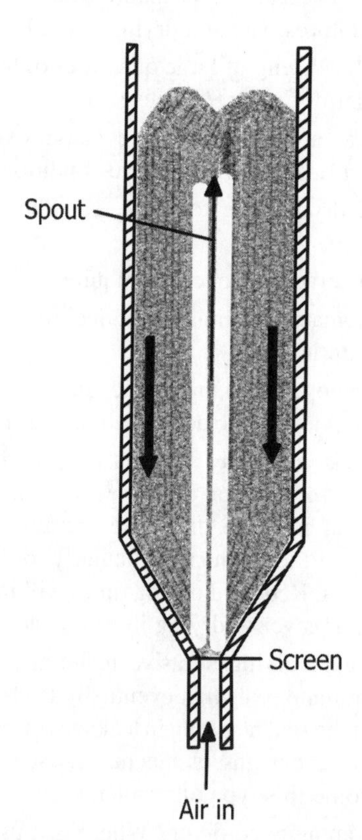

FIGURE 18.30 Principle of operation of a spouted bed dryer.

- What is a pneumatic or flash dryer? What are its features?
  - A flash dryer consists of an air heater, solids feeding device, vertical or inclined drying duct or tube, cyclone or other collector and an exhaust fan.
  - When air is the drying gas, a single induced draft fan can be used. Alternatively, both a supply air fan and an exhaust fan can be provided to obtain near ambient pressure at the feed entry point.
  - For heating the air, direct fired gas heaters or steam heat exchangers are commonly used, but indirect fired heaters can also be considered.
  - Product to be dried is dispersed into an upward flowing hot air and sometimes nitrogen stream. Flow of air and solids is cocurrent.
  - A cyclone or bag filter is used to separate dried particles from the outgoing gas. Cyclones are the least expensive means of product collection and will capture the bulk of the solid. However, they often fail to meet required emission limits, so bag filters are often used instead of or in addition to them. Where bag houses are unsuitable, the cyclone can be followed by a wet scrubber.
  - The duct or tube length should be of sufficient length to provide the required drying time.
  - In order to ensure proper dispersion of the wet solids into the gas, a venturi is sometimes incorporated to impart high gas velocity.
  - Provision for recycling some of the product generally is included. Backmixing of dry solids is practiced when dealing with a sticky and nonfriable feeds.
  - Free-flowing powders and granules may be dried while being conveyed in a high velocity air stream.
  - Single vertical dryers are used mainly for removing surface moisture.
  - These dryers are useful for moist, powdery, granular, and crystallized materials, including wet solids discharged from centrifuges, rotary filters, and filter presses. Particle size must be quite small, generally less than 500 μm, and the best feed is friable, and not sticky.
  - Due to the rapid drying process, flash dryers are not generally suitable for diffusion-controlled drying. When internal moisture is to be removed, longer drying times are required. Horizontal pneumatic dryers, or vertical dryers, consisting of a number of vertical columns in series, may be used for this purpose. The ducting may be in the form of a closed loop. The particles travel a number of times around the loop until they reach the desired moisture content. This type of dryer is known as a pneumatic ring dryer.

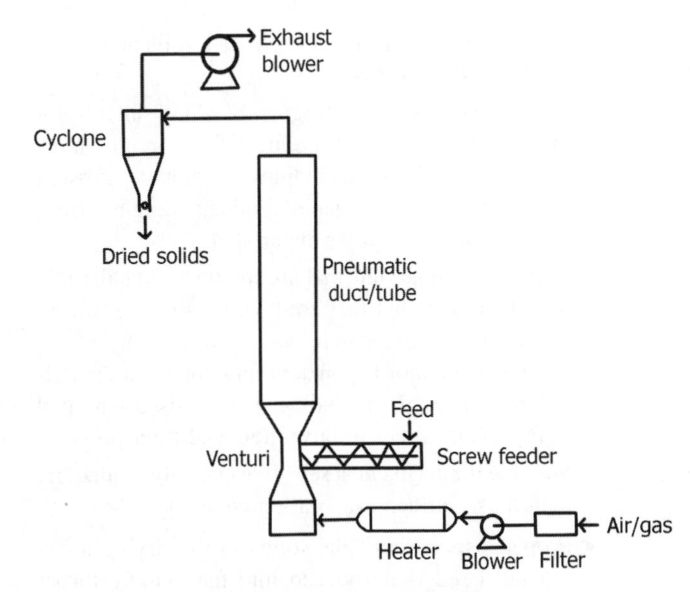

FIGURE 18.31 Pneumatic dryer.

  - Figure 18.31 illustrates a basic setup of a pneumatic dryer.
- What are the characteristics and advantages of pneumatic dryers?
  - Pneumatic dryers, also called flash dryers, are similar in operating principle to spray dryers.
  - Dryer sizes range from 0.2 to 0.3 m (0.6–1.0 ft) in diameter by 1–38 m (3.3–125 ft) in length.
  - High air velocities, in the range of 10–40 m/s, are used. The minimum upward velocity should be 2.5–3 m/s greater than the free fall velocity of the largest particles. Particles in the range of 1–2 mm correspond to an air velocity of 25 m/s.
  - Particle sizes handled must be small, 1–3 mm in diameter and in some cases up to 10 mm, generally less than 500 μm.
    - Appropriate for particles less than 1–3 mm in diameter and when moisture is mostly on the surface, particles up to 10 mm are handled.
  - Contact times are short, usually 0.5–3 s, for single pass drying, producing almost immediate *surface drying*.
    - With normal recycling, residence time can be near 1 min.
  - Most economical choice for drying solids that have been dewatered or inherently have low moisture content.
  - A single operation combines the necessary mixing and heat and mass transfer for drying a solid.
  - Best feed is friable and not sticky.
  - The dryer can operate closed cycle when drying solids from organic solvents and employ nitrogen

gas instead of air. The evaporated solvent is condensed and recovered.

- The simple flash dryer (Figure 18.31) is suitable for drying a wide range of products, for example, inorganic chemicals such as sodium bicarbonate, gypsum and alumina, and organic products ranging from starch to some polymer materials.
- Suitable for materials that are too fine to handle in a fluid bed dryer but that are heat sensitive requiring rapid drying (e.g., starch, flour, and resins).
  - Useful for moist, powdery, granular and crystallized materials, including wet solids discharged from centrifuges, rotary filters and filter presses.
- Suitable for drying heat sensitive or easily oxidizable materials. Surface evaporation cools the solids.
- Proper dispersion of the solids in the drying air is crucial. Feed is introduced into the venture throat where the high air velocity assists dispersion of the solids into the gas.
- Since attrition may be severe, fragile granules cannot be handled safely.
- Material inventory is low. This is advantageous while processing hazardous materials.
- Control is simple.
  - The low inventory in the dryer allows the control system to respond quickly to operational changes.
- Relatively simple and requires less floor space for installation.
- Lower capital costs but operating costs are higher.

• Name some materials that are typically dried in a pneumatic/flash dryer.

- Polymers, pigments, catalysts, zeolites, clay, proteins, cellulose, starches, animal feed, wood flour, and so on. Table 18.8 gives some materials dried in flash dryers.

• For what type of particulate drying applications, pneumatic dryers are *not* suitable?

TABLE 18.8 Some Materials Dried in Flash Dryers

Alumina	Pigments
Animal feed	PVC
Calcium carbonate	Polyethylene
Catalysts	Polypropylene
Kaolin	Polystyrene
Clay	Zeolites
Gypsum	Silica
Epsom salts	Sodium bicarbonate
Cellulose	Corn fibers
Proteins	Distiller's grain
Starches	Gluten
Synthetic resins	Wood flour

- Abrasive solids.
- Where size reduction of particulates is not desirable.
- Not suitable for sticky solids.
- Due to rapid drying process, not suitable for materials with internal pores for which drying processes are diffusion controlled.
  - Time required for such materials has to be sufficient for not only surface drying but also diffusion of fluids from interior of pores to the surface and then evaporate from the surface.

• What modifications are recommended for a pneumatic dryer when particulates are large or drying times required are more?

- A residence time chamber is recommended for incorporation between the flash/pneumatic tube and cyclone separator.
  - This decreases flow velocity and thereby increases contact time between the particles and hot exit gas.
  - This chamber might incorporate baffles.
  - This type of modification is sometimes used for drying of some polymers.
- A two-stage flash dryer is employed for increasing energy efficiency. Hot air/gas is introduced into the second stage and the outgoing air/gas is then introduced into the first stage.
- Sometimes outgoing air/gas is partially recycled. This method is used for drying cellulose pulp.
- Recycling dried product partially ensures better distribution of difficultly dispersible solids.

• What is a silo or bin dryer?

- It is a silo or bin fitted with a perforated base. The partly dried product is loaded into the silo or bin to up to 2 m deep. Dry, but relatively cool air, percolates up through the bed slowly, completing the drying of the product over an extended period, up to 36 h.
- A silo or bin for storage of solids, modified for slow diffusion-controlled drying to allow injection of a sweeping gas with or without heating can serve as a duel purpose equipment for dryer and storage. Such dryers are also known as gravity dryers.
- Such dryers have no mechanical agitators or other moving parts, eliminating the maintenance costs associated with such devices.
- They usually have longer residence times and suited for slow drying.

Spray Dryers

• For what types of applications spray drying is used?

- Spray drying is the most widely used industrial process involving particle formation and drying. It is highly suited for the continuous production of dry solids in either powder, granulate, or agglomerate form from liquid feed stocks as solutions, emulsions, and pumpable suspensions. Allows to control temperature and particle formation very accurately.

- Altering the process parameters in a spray dryer allows manufacture of complex powders that meet exact powder properties in terms of particle size and shape, particle size distribution, residual moisture content, bulk density, dispersibility, polymorphism, and flow properties in a very efficient manner.
- Spray drying is also applied in formulating products with unique properties. In the aroma industry, water-insoluble liquid aromas are encapsulated in a solid matrix of water-soluble carrier material and surface-active ingredients. After spray drying, the result is a powdery flavor with excellent shelf life and good redispersibility in water. The same is the case for oil soluble vitamin powders.
- Very fine powders, such as ceramics or hard metals, can be formulated into large compact particles of spherical shape with good flowability by the addition of binding agents. Being very uniform and with a consistent density, they can be used directly in pressing dies for forming ceramic products, cutting and mining tools, and other products.
- In dyestuff and pesticides industry, the nonsoluble active material can be formulated with binding and dispersing agents to produce a nondusting and water dispersible powder.
- Coating of suspended solids by spray drying the suspension is used for taste masking and controlled release of active materials in the pharmaceutical industry.
- The spray drying can also be applied for congealing, in which, a melted feedstock is atomized and turned into a free-flowing powder by cooling it in a stream of cold air or gas. Spray congealing is also applied for encapsulation.
- It finds use for several types of products from palm oil derivatives to special waxes, fats, glycerides, hydrates, and other inorganic or organic melts.
- If a potent or otherwise harmful chemical is suspended in a molten wax, it can be encapsulated, protecting the users from the bad effects.
- Many enzymes for the detergent industry are congealed this way.
- Spray drying is also applied for production of emulsion polyvinyl chloride and polyvinyl acetate, which are formulated by spray drying to produce high-quality powders.
- The spray drying process can be applied for carrying out chemical reactions. Dry absorption of SO_2 from flue gases from coal fired power plants and HCl and HF from waste incineration plants are some examples. The reaction takes place when the atomized liquid is suspended in the drying air/gas stream.
- Spray drying is used for bioactive products without destroying the bioactive elements. It is also applied in solid dosage pharmaceuticals to increase the bioavailability of the drug. Manufacture of active pharmaceutical ingredients (APIs), in an amorphous structure, with better bioavailability, can be made using spray drying process.

- What are the important characteristics of spray drying?
  - Short drying times that make a spray dryer suitable for drying heat sensitive materials. With the spray of liquid having a very large surface, heat transfer and mass transfer processes are very rapid. Higher temperatures can be used for these materials because of shorter residence times involved. Surface moisture is removed in about 5 s.
  - High porosity (low bulk density) and small rounded particles of the product: Desirable when material is to be dissolved (e.g., foods, detergents) or dispersed (e.g., pigments, inks).
  - Good control of product particle size, bulk density, and form.
  - Free-flowing product powders.
  - Can be used to produce powders and granules.
  - Spray drying plants can be designed for almost any capacity from very small quantities up to several tons per hour.
- What are the advantages and disadvantages of spray drying?

  Advantages
  - Properties of product are effectively controlled.
  - Heat sensitive foods, biological products, and pharmaceuticals can be dried at atmospheric pressure and low temperatures, employing inert atmospheres, if warranted.
  - Relatively simple equipment and can be used for high capacities in continuous operation.
  - Product comes in an anhydrous condition, simplifying corrosion problems and selection of materials of construction.
  - Produces relatively uniform spherical particles.

  Disadvantages
  - Not suitable for obtaining high bulk density products.
  - Generally not flexible, that is, a unit designed for obtaining fine particles may not be able to produce a coarser product or vice versa.
  - Feed must be pumpable.

  - Higher initial investment compared to other types of continuous dryers. Dust collection equipment adds to costs.
- What are the essential stages involved in spray drying?
  - Spray drying involves the following essential stages: Pumpable feed preparation, liquid atomization and hot air/gas contact, gas/droplet mixing and evaporation, particle shape formation, and drying and separation of the dried product from air/gas and discharge.
  - Drying rates involve both constant rate period and falling rate period. During constant rate period, drying is controlled at the surface of the liquid droplets, that is, heat transfer through the gas phase to the droplet surface and mass transfer of the water vapor from the droplet surface into the gas phase. During falling rate period, moisture removal is controlled by diffusion from interior of solid particles toward their surface.
  - Spray drying, being an expensive process requiring a very large chamber, if the entire drying, covering constant, and falling rate periods, is to be carried out in the spray dryer alone.
    - Multistage drying is often used to cut costs by adding one or more fluidized bed drying stages where the residence time is higher and the applied drying media temperatures are lower. The overall drying process is thus divided into a very rapid evaporation of surface moisture in the spray chamber part and an accurately controlled drying of the internal particle moisture in the fluidized bed. Integrated multistage involve direct mounting of the fluidized bed at the conical bottom of the spray dryer.
    - Removal of surface moisture in the first stage makes it nonsticky for handling in the second stage.
    - Quality of the dried product is further improved by classifying the powders in the fluidized bed and reintroducing the fines into the atomization zone. This way, powders with less dust, improved dispersibility, and a narrower particle size distribution can be produced.
- Define mean residence time of gas in a spray dryer.
  - Ratio of vessel volume/volumetric flow rate of gas.
- What is the normal residence time for completion of drying operation in a spray dryer?
  - 5–30 s.
- What are the flow arrangements for spray dryers with respect to dried particulates in spray dryers?

  Cocurrent
  - Drying air and particles move through the drying chamber in the same direction. Product temperatures on discharge from the dryer are lower than the exhaust air temperature and hence this is an ideal mode for drying heat-sensitive products. With the use of rotary atomizers, a high degree of turbulence is created leading to uniform temperatures throughout the drying chamber and producing fine particles.

  Countercurrent
  - Drying air and particles move through the drying chamber in opposite directions. This mode is suitable for products, which require a degree of heat exposure during drying. The temperature of the powder leaving the dryer is usually higher than the exhaust air temperatures.

  Mixed Flow
  - Particle movement through the drying chamber experiences both cocurrent and countercurrent phases.
  - This mode is suitable for heat stable products where coarse powder requirements necessitate the use of nozzle atomizers, spraying upward into an incoming air flow, or for heat-sensitive products where the atomizer sprays droplets downward toward an integrated fluid bed and the air inlet and outlet are located at the top of the drying chamber.
- "Parallel flow of air and solids is most common in spray dryers." *True/False*?
  - *True.* The atomized droplets are usually sprayed downward into a vertical chamber through which hot gases also pass downward.
- Give a diagram of the operation of a spray dryer.
  - Figure 18.32 gives a schematic line diagram of a spray dryer.

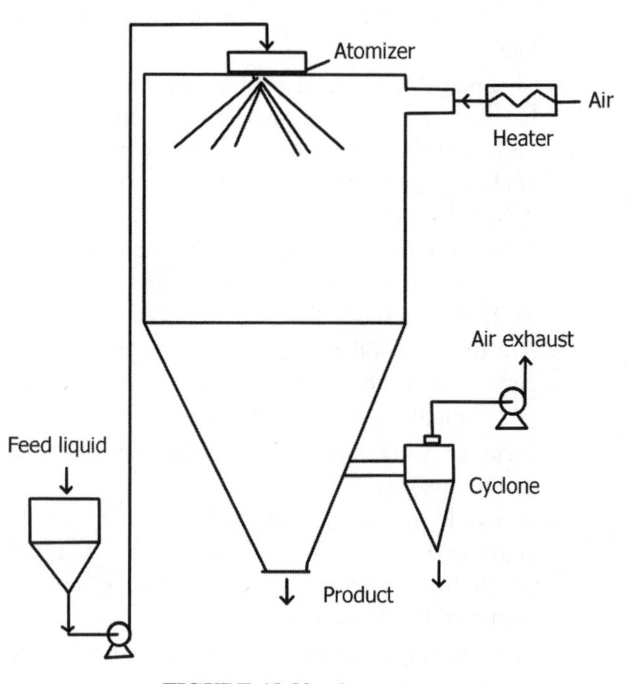

FIGURE 18.32 Spray dryer.

- What are the features that make a spray dryer suitable to obtain a high quality of product?
  - Drying is rapid.
  - Particles in the hot drying zone are wet and maintained near wet bulb temperatures through evaporation.
- What are the ways by which atomization of droplets is accomplished in a spray dryer?
  - *Centrifugal/Rotary Disk Atomizer:* Liquid is fed to the center of a rotating disk or bowl having a peripheral velocity of 90–200 m/s and 10,000–50,000 rpm. Droplets, 50–60 μm in diameter, are thrown from the edge to form a uniform spray. Fine particles are usually produced.
  - Atomization occurs at the wheel periphery and it is the peripheral speed normally in the range 100–300 m/s that controls particle size for any given product.
    - ➢ Small disks are used for low capacities (up to 10,000 rpm). The spray angle is about 180° and forms a broad cloud. Because of the horizontal trajectory these atomizers require larger chamber diameters. Usually the atomizers have radial vanes. The number and shape of the vanes influence product size distribution and capacity.
  - The rotary atomizer has the advantage that high feed rates can be accommodated by a single atomizer. It requires only a low-pressure feed system, which is resistant to abrasion or clogging, and the particle properties are not sensitive to feed rate.
  - *Spray nozzles* are simple with no moving parts and involve low energy consumption. Have tendency to blockages. Control during operation is not flexible. Nozzle atomizers are susceptible to blockage by particulates and abrasive solids gradually enlarge the nozzle tip (due to erosion) and increase the average droplet size.
    - ➢ *Pressure Nozzle Atomizer:* Operate on pressure energy to break up the droplets. Liquid is forced at a high pressure (2700–69,000 kPa/m^2) through a small aperture to form droplet sizes of 180–250 μm. Grooves on the inside of the nozzle cause the spray to form into a cone shape and therefore to use the full volume of the drying chamber.
    - ➢ *Two-Fluid Nozzle Atomizer:* Use compressed air for break up of droplets in addition to pressure energy. Compressed air creates turbulence that atomizes the liquid. The operating pressure is lower than the pressure nozzle, but a wider range of droplet sizes is produced. Used for small capacity units. Two-fluid design is not very common.
  - *Ultrasonic Nozzles:* Involve creation of high frequency sound by a sonic resonance cup placed in front of the nozzle. Considered to be promising for atomizing high viscosity long molecular chain structured materials and some non-Newtonian liquids, which form filaments instead of droplets.
    - ➢ Operate as a two-stage atomizer in which liquid is first atomized by a nozzle atomizer and then using ultrasonic energy induces further vibration.
- What are the mechanisms involved in the atomization process involving different atomizers in a spray dryer?
  - The droplet size from a given type of atomizer depends on the energy spent for breaking down the liquid into fragments, that is, increasing the overall surface of the liquid.
  - For most atomization systems, the liquid does not leave the atomizing head as a droplet, but as a fragment of a thin liquid film. The droplet formation takes place immediately after the liquid has left the atomizing head due to the surface tension of the liquid. The formation of a perfect droplet is therefore dependent on the rheological properties of the liquid and the interaction with the hot drying medium just outside the atomizing device.
- What are the sizes of openings and pressures employed in nozzle atomizers?
  - Nozzle openings are 0.3–4 mm.
  - Pressures employed are 20–275 bar.
- What are the differences with respect to sprays produced by pressure type and two-fluid or air/gas atomizing nozzles?
  - Pressure type nozzles produce relatively coarser droplets in the size range of 20–100 μm. Dryers utilizing nozzle atomization are generally of a smaller diameter but with increased cylindrical height. The increased particle trajectory achieved by the counter, cocurrent or fountain nozzle configuration in combination with the streamline air/gas flow allows production of coarser particles without wall deposits.
  - Two-fluid nozzles, with the available pneumatic energy, produce finer droplets of sizes less than 30 μm. For producing finer droplets of size ranges less than 10 μm, high gas pressures and high gas to liquid mass flow rates are required.
  - Two-fluid nozzles involve high power requirements and are uneconomical for high capacities. The energy for atomization is provided by the rapid expansion of air/gas that is mixed with the feed within the body of the nozzle (internally mixing) or at its tip (externally mixing).

- What is the essential functional difference between disk and nozzle atomizers in a spray dryer?
  - *Disk:* Lateral throw.
  - For disk operation, large diameter is required for the dryer body. L/D ratio of the dryer is 0.5–1.0. The larger the particle size desired in the final powder, the larger the diameter of the drying chamber.
  - *Nozzle:* Smaller diameter and greater depths for a given residence time. L/D ratio is 4–5 or more.
- What are the advantages of rotary disk atomizers?
  - The rotary disk atomizer is generally preferred due to its greater flexibility and ease of operation.
  - The advantages include the following:
    - Handling of high feed rates without the need for atomizer duplication, handling of abrasive feeds, no blockage problems, low-pressure feed system, and ease of droplet size control through wheel speed adjustment.
- What is the importance of the choice of atomizers?
  - The type of atomizer not only determines the energy requirements for formation of spray but also size, size distribution, trajectory and speed of the droplets produced, on which particle size of the product depends.
  - The body design of the dryer also is influenced on the choice of the atomizer.
- What are the ranges of droplet and particle sizes obtainable for different (a) atomizers used and (b) products obtained?
  - Table 18.9 gives droplet/particle sizes obtainable different atomizers and products dried.
- What is the generally obtainable ranges of surface areas by spraying 1 L of feed solution?
  - 20–600 m^2.
- What are the common causes of problems in a spray dryer?

TABLE 18.9 Droplet and Product Sizes (μm) Obtainable for Different Atomizers

Atomizers	
Disk atomizers	1–600
Pressure nozzles	10–800
Air atomizers	5–300
Products	
Milk	30–250
Coffee	80–400
Pigments	10–200
Ceramics	30–200
Pharmaceuticals	5–50
Chemicals	10–1000

  - *Clogging:* Centrifugal bowls or disks are less likely to become plugged apparently require less maintenance.
  - Caking and stickiness.
  - Temperature damage.
  - *Erosion and Wear:* Abrasive materials can cause problems with the atomizing devices.
  - Corrosion.
  - Improper assembly and reassembly.
  - Accidental damage.
- Why is it necessary to have a large diameter for a spray dryer?
  - Drops released by the atomizer must be completely dried before they strike the vessel walls to avoid build up of sticky scales on the walls that not only cause maintenance problems but also impairs product quality.
- "The droplet size in a spray dryer depends only on the type of atomizer used and is independent of the physical properties of the product." *True/False*?
  - *False.*
- How is separation of dried particles achieved in a spray drying process?
  - The separation of the particulates from the drying air can be achieved in one of two ways. The coarser material may be collected from the cone of the spray drying chamber with the finer particles recovered from the drying air/gas by a primary separation device like a cyclone or bag filter. The two powder streams may be subsequently mixed or segregated, often the coarser chamber fraction being recovered as product, whilst the fine material is recycled.
- What is the effect of different variables on the operation of a spray dryer?
  - Table 18.10 gives effect of different variables on the operation of a spray dryer.
- What are the special steps required to avoid contamination in spray drying of products for use in pharmaceutical, food processing and electronics industries to avoid contamination?
  - Incorporation of very high standards of filtration of the drying air/gas (HEPA filtration) immediately before its entry into the drying chamber.
  - Where practical/necessary the installation of the dryer itself or the product discharge in a controlled clean room environment.
  - The incorporation of effective *clean-in place* (CIP) systems on the main dryer components with easily dismantled ductwork.
  - Attention to surface finishes.

TABLE 18.10 Effect of Different Variables on the Operation of Spray Dryer

Variable Increased	Factors Increased	Factors Decreased
Chamber inlet temperature	Feed rate, product rate, particle size, product moisture, wall deposits	Bulk density
Chamber outlet temperature	Product thermal degradation	Feed rate, product rate, particle size, product moisture, wall deposits
Gas volume rate	Feed rate, product rate, particle size, product moisture, wall deposits	Residence time
Feed concentration	Product rate, bulk density, particle size	
Atomizer speed	Bulk density	Particle size, product moisture
Atomizer disk diameter	Coagulation, particle size, product moisture, wall deposits (*for unstable lattices*)	Wall deposits (*for stable lattices*)
Atomizer vane depth and number of vanes	Bulk density	Particle size, product moisture, wall deposits
Atomizer vane radial length		Particle size, wall deposits (*for unstable lattices*)
Feed surface tension	Bulk density	Particle size
Chamber inlet gas humidity	Product moisture, wall deposits	

- What are the safety issues involved in a spray drying process?
  - Due to the fine powders involved, hazards of dust inhalation and toxicity effects on plant personnel, fires, and dust explosions in spray dryers must be recognized.
  - *Material safety data sheets* (MSDS) on the powders must be made available to plant personnel and other staff with emphasis on training and retraining on the hazards involved and safety precautions to be followed. The data must include toxic effects, both inhalation and skin contact, particle size ranges, explosion concentrations, maximum dust explosion pressures and rate of pressure rises, minimum ignition energy and self and layer ignition temperatures of the dusts (the later arises with deposits accumulating within the equipment). First aid should be made available. Explosion suppression and firefighting systems should be provided and maintained.
- What is a toroidal dryer? What are its characteristics and applications?
  - The dryer works on a jet mill principle and contains no moving parts. Transport of solid material within the drying zone is accomplished entirely by high velocity air movement.
  - Heated process air is distributed through three manifold jets to the lower segment of the toroidal drying zone chamber. The air from one of the three jets is directed in such a way as to impinge upon the incoming wet feed material and propel this material into the drying zone, where particle size reduction and drying begins.
  - Additional jets in the drying zone convey the material into the toroid for additional drying, grinding, and classifying.
  - In a toroidal bed dryer heated air enters the drying chamber through blades or louvers, creating a fast moving, rotating bed of particles.
  - Process air and solids within the toroid move at a velocity of approximately 30 m/s.
  - The high velocity gas stream reduces the size of lumps or agglomerated feed material by impingement against the interior walls of the drying chamber and by collision with other particles. Wetter and heavier particles travel a path along the internal periphery of the dryer, whereas dryer and lighter particles are swept out with the gas stream and are removed from the drying zone. Heavy, wet particles stay in the dryer until they are broken up and dried.
  - The inlet temperature is usually controlled within the range of 260–760°C.
  - There is a sharp drop in the gas temperature within the dryer when the hot inlet gas stream meets the incoming wet solids.
  - Toroid dryers are developed for sludge drying in wastewater treatment plants.
  - The dryer exhaust temperature is usually controlled at a specific set point within the range of 90–150°C for sludge drying. The product temperature normally does not exceed 66°C.
  - The dried sludge particles exiting the toroid are sent to a cyclone where they are separated from the gas stream. A portion of the dried sludge is backmixed with the wet feed, and the remainder is transferred to

the product finishing section. There, the dried product may be extruded at a temperature of 60°C, cut into pellets, and bagged, if desired.

- What are kneader dryers? Describe their characteristics and features.
  - Kneader dryers are based on the operating principles of kneader mixers for solids and semisolids.
  - Kneaders dryers are positioned intermediate between screw type equipment, with relatively small volume and heat transfer areas and conventional paddle dryers with larger volumes, larger heat transfer area, but lack of a kneading effect and self-cleaning of heat exchange surfaces. Kneaders dryers combine the effective mixing and kneading action of screw type units.
  - The unit consists of a horizontal, cylindrical housing, and a concentric agitator shaft with disk elements perpendicular to the axis carrying peripheral mixing/kneading bars. Stationary hook shaped bars, set in the shell, interact with, and clean, the shaft and disk elements as they rotate.
  - The arrangement of the disk elements, the mixing/kneading bars and the shape of the static counter hooks impart a forward plug flow movement to the material.
  - These dryers have two parallel intermeshing agitator shafts rotating in a horizontal housing of a roughly *figure of eight* cross section.
  - The main agitator carries radially arranged disks welded on to the periphery of which are U-shaped kneading bars.
  - The second shaft is fitted with kneading disk elements that mesh with, and clean the main agitator's disks and bars.
  - Intermeshing of the two sets of elements generates an intensive mixing/kneading action and effective self-cleaning.
  - The arrangement of the internals is designed to provide a gradual forward conveyance of product, coupled with intensive lateral intermixing.
  - The shell housing, agitator shafts, and disk elements of kneader dryers can be heated or cooled, giving a very large heat exchange area in relation to volume.
  - The intensive mixing and kneading action, coupled with self-cleaning of the heating surfaces, combines to break up baked on crusts, agglomerates and lumps, ensuring a high rate of product surface renewal for both heat and vapor transfer.
  - Applications include drying of industrial sludges and residues.
  - Table 18.11 gives the characteristics and features of kneader dryers.
- What are the features that are incorporated/considered for heat dryers to reduce energy losses in drying?
  - Insulation of cabinets and ducting.
  - Recirculation of exhaust air through the drying chamber provided a high outlet temperature can be tolerated by the product and the reduction in evaporative capacity is acceptable.
  - Recovering heat from the exhaust air to heat incoming air using heat exchangers or prewarming the feed material.
  - Use of direct flame heating by natural gas and low nitrogen oxide burners to reduce product contamination by the products of combustion.
  - Two-stage drying (e.g., fluid bed drying followed by bin drying or spray drying followed by fluid bed drying).
  - Preconcentrating liquid to the highest solids content possible using multiple effect evaporation. Energy use per unit mass of water removed in evaporators can be several orders of magnitude less than that required for dehydration.

TABLE 18.11 Characteristics and Features of Kneader Dryers

Characteristics	Features
Intensive mixing/kneading action and high interface renewal rates	Enhanced heat and mass transfer; processes all product states/phases in a single unit; crushing of agglomerates
Large self-cleaning heating and cooling surfaces	Permit high rate of energy input and ensure precise temperature control; improved heat transfer
Large useful volume	High throughput and effective handling continuous processes with long residence times (0.2–3 h)
Minimal axial intermixing	Virtual plug flow ensuring narrow residence time distribution
Large cross-sectional area	Permits feeding and disengagement and flash evaporation of superheated feed solutions; low vapor speeds; minimal entrainment
Closed, contained construction	Allows vacuum processing and handling of toxic, explosive, or hazardous substances

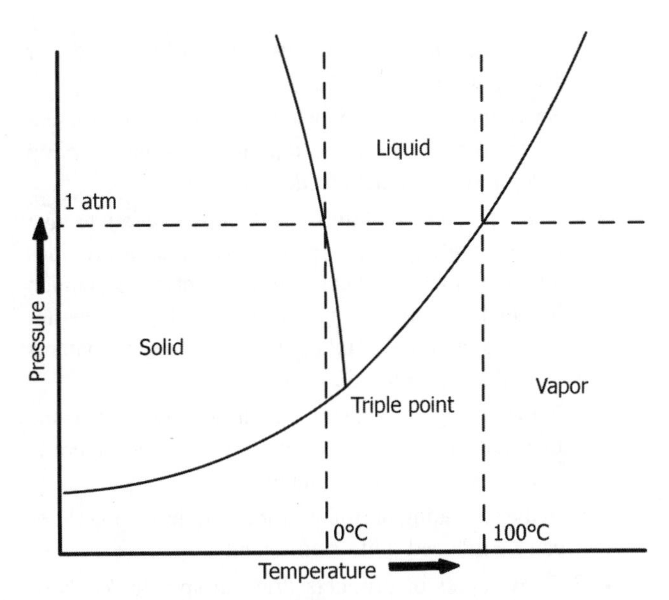

FIGURE 18.33 Phase diagram of water.

- Automatic control of air humidity by computer control.

- What is the principle of operation of a freeze dryer?
  - Removal of moisture from the solid is through sublimation, which takes place below triple point of water. Figure 18.33 illustrates phase diagram of water, showing its triple point.
- What is the advantage of freeze drying?
  - The basic idea of freeze drying is to completely remove water from some material, such as food, while leaving the basic structure, composition, and quality, such as flavor, of the material intact.
  - Freeze drying significantly reduces weight of the material and prolongs its shelf life.
- What are the applications of freeze dryers?
  - Pharmaceuticals (e.g., penicillin, blood plasma, vitamins, and vaccines), biological materials, food products like peeled cut pieces of apples, bananas, garlic, mushrooms, peas (perforated), mango pulp and pieces, chicken pieces, cauliflower and carrot, heat-sensitive materials, readily oxidizable materials, and astronaut foods.
  - For low capacity applications.
- Explain what is meant by primary and secondary processes in freeze drying.
  - The process of freeze drying involves three stages: (a) freezing the material, (b) subliming the ice (primary drying), and (c) removal of the small amount of water bound to the solids (secondary drying or desorption).
  - Sublimation phase in freeze drying is called primary drying. Ice will sublime when the water vapor pressure in the immediate surroundings is less than the vapor pressure of ice at the prevailing temperature. Ice will sublime when the water vapor pressure in the immediate surroundings is less than the vapor pressure of ice at the prevailing temperature.
  - At the end of the sublimation phase, that is, primary drying phase, practically all the ice disappears. The product temperature will start to increase, and tends to approach the control temperature of the shelf.
  - However, at this stage the product is not sufficiently dry for long-term storage. For most products, the residual moisture is in the region of 5–7%.
  - The product now enters the desorption phase, during which the last traces of water vapor are removed, along with traces of the *bound moisture* within the product structure. This phase is identified as *secondary drying*. The aim of this final phase is to reduce the product to the acceptable moisture levels needed for long-term storage, usually 3% to 1%.
  - Figure 18.34 gives the important components of a freeze dryer.
- What is osmotic drying?
  - When pieces of fresh fruits or vegetables are immersed in a sugar or salt solution, which has a higher osmotic pressure than the food, water passes from the food into the solution under the influence of the osmotic pressure gradient and the water activity of the food is lowered. This method of removing moisture from food is known as osmotic dehydration (drying).
- What is the principle involved in dielectric and microwave drying?
  - High frequency electricity generates heat internally and produces a high temperature within and on the surface of the material. Normally used for food heating and cooking.

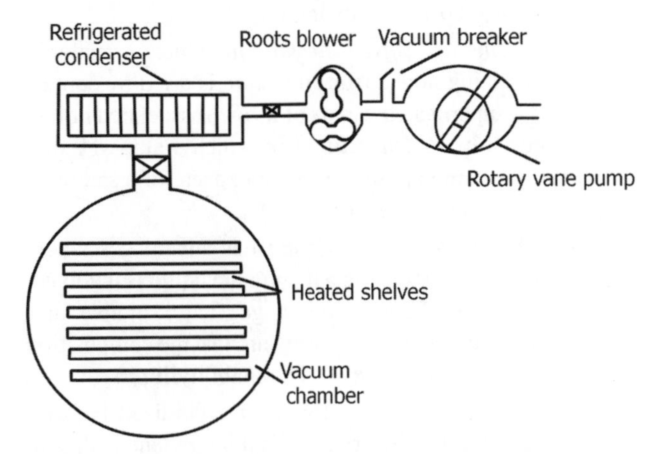

FIGURE 18.34 Main components of a batch freeze dryer.

- The molecular structure of water consists of a negatively charged oxygen atom, separated from positively charged hydrogen atoms and this forms an electric dipole.
- When a microwave or radio frequency electric field is applied to a food, dipoles in the water and in some ionic components such as salt, attempt to orient themselves to the field (in a similar way to a compass in a magnetic field). Since the rapidly oscillating electric field changes from positive to negative and back again several million times per second, the dipoles attempt to follow and these rapid reversals create frictional heat. The increase in temperature of water molecules heats surrounding components of the food by conduction and/or convection.
- Because of the widespread domestic use, some popular notions have arisen that microwaves *heat from the inside out*. What in fact occurs is that outer parts receive the same energy as inner parts, but the surface loses its heat faster to the surroundings by evaporative cooling.

- What are the advantages of a microwave dryer?
  - The radiation selectively heats moist/wet areas while leaving dry areas unaffected. This prevents damage to the material.
  - Drying times are short.
  - Uniform composition of final product.
  - Equipment is small, compact, clean-in operation and suited to automatic control.
  - There is no contamination of foods being dried by products of combustion or dust particles in drying air.
  - Microwaves and radio frequency energy overcome the barrier to heat transfer caused by the low thermal conductivity of foods. This prevents damage to the surface, improves moisture transfer during the later stages of drying and eliminates case hardening.
- What is the difference between superheated solvent drying and solvent dehydration?
  - *Superheated solvent drying* in which a material containing nonaqueous moisture is dried by contact with superheated vapors of its own associated liquid, and, *solvent dehydration* in which water-wet substances are exposed to an atmosphere of a saturated organic solvent vapor.
  - *Superheated solvent drying* has advantages where a material containing a flammable liquid is involved. Drying is effected with a gas with suitable air–moisture ratio in order to ensure that the composition is well below the lower flammability limit.
  - Superheated solvent drying in a fluidized bed has been used for the drying of polypropylene pellets to eliminate the need for water washing and for fractionating.
  - Solvent dehydration is used for kiln drying for seasoning timber where substantial reductions in drying times have been achieved.
  - The replacement of air by superheated steam to take up evaporating moisture is attractive as it provides a high temperature heat source that provides a much higher driving force for mass transfer since it does not become saturated at relatively low moisture contents as is the case with air.
  - In the drying of food stuffs, a further advantage is the fact that the steam is completely clean and there is much less oxidation damage.
  - In the seasoning of timber, for example, drying times can be reduced quite significantly.
- Give examples of products dried in specific kinds of drying equipment.
  - *Drum Dryers:* Skim milk, buttermilk, potatoes, cereals, dextrin, yeasts, instant oat meal, glue, sodium benzoate, sodium benzene sulfonate, polyacylamides, propionates, acetates, phosphates, chelates, calcium acetate-arsenate-carbonate-hydrate-phosphate, aluminum oxide, barium sulfate, *m*-disulfuric acid, caustic, sodium chloride, ferrous sulfate, lead arsenate, and so on.
    - ➢ Milk powder produced in drum dryers is coarser than that produced in spray dryers and useful in preparation of solid products like sweets, instead of using to get back milk through dissolution in water as and when required.
  - *Spray Dryers:* Skim milk (fine easily soluble particles), coconut milk, soluble coffee powder, foodstuffs, eggs, starch, yeast, proteins, flavors, pharmaceuticals, herbal extracts, biochemicals, enzymes, dyestuffs, fine/bulk chemicals, agrochemicals, tannins, animal blood, organic extract, malamine and urea formaldehyde resins, polyvinyl chloride, lignosulfonate wood waste, detergents, rubber chemicals, sulfonates, microspheres, catalysts, mineral concentrates, inorganic phosphates, pigments, inks, urea, salts, silica gel, ceramics, kaolin, sludges, effluents, and so on.
  - *Vacuum Drum Dryers:* Syrups, malted milk, skim milk, coffee, malt extract, glue, and so on.
  - *Vacuum Rotary Dryers:* Plastics, organic polymers, nylon chips, chemicals of all kinds, plastic fillers, plasticizers, organic thickeners, cellulose acetate, starch, sulfur flakes, and so on.
  - *Belt Conveyer Dryers:* Yeast, charcoal briquettes, synthetic rubber, catalysts, soap, glue, silica gel,

titanium dioxide, urea formaldehyde, clays, white lead, chrome yellow, metallic stearates, and so on.

- *Fluidized Bed Dryer:* Lactose base granules, pharmaceutical crystals, weed killer, coal, sand, lime stone, iron ore, PVC, asphalt, clay granules, granular desiccant, abrasive grit, salt, and so on.
- *Pneumatic/Flash Dryers:* Yeast, starch, whey, fruit pulp, synthetic casein, filter cake, sewage sludge, gypsum, copper sulfate, clay, chrome green, potassium sulfate, and so on.
- *Rotary Multitray Dryer:* Pectin, penicillin, fragile cereal products, caffeine, solvent-wet organic solids, waste sludge, boric acid, crystals melting near 40°C, prilled pitch, pyrophoric zinc powder, zinc oxide pellets, zinc sulfide, pulverized coal, calcium carbonate, calcium chloride flakes, inorganic fluorides, electronic grade phosphors, and so on.
- *Freeze Dryers:* Meat, sea food, vegetables, fruits, coffee, concentrated beverages, pharmaceuticals, biological products, veterinary medicines, blood plasma, and so on.
- *Dielectric Dryers:* Breakfast cereals, baked goods, furniture timber blanks, veneers, plyboard, plaster board, water-based foam plastic slabs, some textile products, and so on.
- *Infrared Drying:* Sheets of textiles, paper and films, surface finishes of paints and enamels, surface drying of bulky nonporous articles, and so on.

- What types of dryers are used for large capacity drying applications?
  - Rotary and spray dryers.
- Summarize data requirements for the selection of a dryer.
  - Capacity.
  - Initial moisture content.
  - Particle size distribution.
  - Drying rate curve.
  - Moisture isotherms.
  - Maximum allowable drying temperature.
  - Contamination by the drying gas.
  - Self-heating properties of the material.
  - Explosion characteristics of the material.
  - Toxic nature of the materials.
  - Corrosion aspects.
  - Material safety data sheets (MSDS) for the material.
  - Past experience.
  - Other physical and chemical data.
- What are the solids handling issues involved for a drying process to be successful?
  - For the drying process to be successful, a reliable transfer of thickened slurry, paste, or wet cake from the solid–liquid separation system to the dryer and reliable solids flow without interruption or buildup through the dryer is required. This step is probably the most difficult to execute properly during thermal drying operation.
  - Since the flow properties of bulk solids usually become worse with higher moisture content, problems often arise during the handling of the wet cake upstream of the dryer, feeding of the solids into the dryer and their flow through the dryer.
  - Feed hoppers, conveyers, and chutes are likely to accumulate solids, plug or exhibit erratic flow when handling wet solids.
  - Flow stoppages caused by *arching* and *rat holing* and loss of surge capacity due to solids build up on the walls, are very common in these systems. If flow from the surge bin stops, the entire process including drying comes to a stand still. Operator intervention is required in poking and prodding the solids flow continuously to the dryer becomes essential.
- How are liquids and gases dried?
  - Gases are dried by
    - *Compression:* The humidity of a gas may be reduced by compressing it, cooling it down to near its original temperature, and then draining off the water that has condensed.
    - During compression, the partial pressure of the water vapor increases and condensation occurs as soon as the saturation value is exceeded.
    - *Liquid Absorbents:* If the partial pressure of the water in the gas is greater than the equilibrium partial pressure at the surface of a liquid, condensation will take place as a result of contact between the gas and liquid. Thus, water vapor is frequently removed from a gas by bringing it into contact with concentrated sulfuric acid, phosphoric acid, or glycerol. Concentrated solutions of salts, such as calcium chloride, are also effective. Regeneration of the liquid is an essential part of the process, and this is usually effected by evaporation.
    - *Solid Adsorbents and Absorbents:* The use of silica gel or solid calcium chloride to remove water vapor from gases is a common operation in the laboratory. Moderately large units can be made, although the volume of packed space required is generally large because of the comparatively small transfer surface per unit volume. If the particle size is too small, the pressure drop

through the material becomes excessive. The solid desiccants are regenerated by heating.

- Gas is frequently dried by using a calcium chloride liquor containing about 0.56 kg calcium chloride/kg solution. The extent of recirculation of the liquor from the base of the tower is governed by heating effects, since the condensation of the water vapor gives rise to considerable heating. It is necessary to install heat exchangers to cool the liquor leaving the base of the tower.
- In the contact plant for the manufacture of sulfuric acid, sulfuric acid is itself used for drying the air for the oxidation of the sulfur.
- When drying hydrocarbons such as benzene, it is sometimes convenient to pass the material through a bed of solid caustic soda, although, if the quantity is appreciable, this method is expensive.
- The great advantage of materials such as silica gel and activated alumina is that they enable the gas to be almost completely dried. Thus, with silica gel, air may be dried down to a dew point of 203K.
- Small silica gel containers are frequently used to prevent moisture condensation in the low pressure lines of pneumatic control installations.

- Summarize applications of some common types of dryers.
  - Table 18.12 gives a summary of applications of different types of dryers.
- What are the causes of explosions in dryers?
  - Some dryers, including ductwork and associated equipment such as cyclones and dust collectors are prone to accumulation of deposits on dryer walls and duct work. Solids often accumulate on atomizers at the top of spray dryers where the temperatures can be highest.
  - The main cause of explosions in dryers is attributed to buildup of deposits on the heated surfaces of the dryer. When a layer of organic solids builds up, the material layer may undergo self-heating due to nondissipation of heat inside the solid deposit. When its temperature reaches *self-ignition temperature* (SIT) of the material, explosion can occur. The characteristics of materials deposited on walls or

TABLE 18.12 Applications of Different Types of Dryers

Dryer	Type of Applications	Type of Materials
Tray dryers	Batch or continuous, atmospheric or vacuum, small or medium scale	Hard or soft pastes, preformed hard granular sheets, free-flowing granular or crystalline so lids, oxidation or mechanically sensitive mate rials (under vacuum)
Through circulation	Batch or continuous, small or medium scale	Preformed, free-flowing granular or fibrous or special products, heat or mechanically sensitive
Agitated	Batch, small or medium scale	Liquid, slurry, paste, free-flowing granular or crystalline, dusts, toxic, or flammable materials
Fluid bed	Batch or continuous, small, medium, or large scale	Preformed hard and soft pastes, sludge, free-flowing granular or crystalline or fibrous materials, pharmaceuticals
Drum dryers	Continuous, medium (indirect) or large scale (direct), atmospheric or vacuum	Liquid, slurry, paste, syrup, sludge, granular, fibrous, sheet
Rotary dryers	Continuous, atmospheric or vacuum	Polymers and plastics, soda ash, food products, sludges
Spray dryers	Continuous, medium or large scale	Liquid, slurry, paste, oxidation-sensitive or special form products
Pneumatic dryers	Continuous	Soft paste, sludge, preformed paste, granular or fibrous, free-flowing granular or crystalline, heat-sensitive materials
Conveyor dryers	Continuous, medium or large scale	Clays, minerals, soaps, coal
Freeze dryers	Batch, small scale	Biological, pharmaceutical, and food products
Microwave and dielectric drying	Low temperature, batch or continuous, slow	High value and good quality products
Microwave augmented freeze drying	Low temperature, rapid	Good quality expensive products
Ball drying	Low temperature; rapid, continuous operation	Small particles, vegetable pieces Good quality products
Ultrasonic drying	Rapid	Low fat liquid foods
Explosive puff drying	Rapid	Good rehydration of food products

other surfaces may change during drying resulting change in SIT when the materials are exposed to high temperatures or other operating condition. There have been several instances of explosions in dryers. Smooth surfaces, elimination of potential points of solids accumulation can reduce solids deposits.

- Entrainment of fine solids inside the dryers can also lead to dust explosions inside dryers.
- To prevent such explosions, frequent monitoring and cleaning of the dryers is necessary.
- Many drying operations involve the vaporization of a flammable solvent from a combustible powder. This combination of a flammable vapor and combustible dust is called a *hybrid mixture*. Hybrid mixtures represent a greater explosion hazard than that presented by the combustible dust alone due to the following reasons:
  - The minimum energy required for a hybrid mixture to explode and minimum explosible concentration are usually lower than dust–air mixtures alone.
  - A hybrid mixture may cause more severe explosion than dust–air mixtures alone.
- Dryers and drying systems that can generate electrostatic charges must be properly bonded and grounded to drain off electrostatic charges and minimize the possibility of fires and explosions. Inerting is often needed to prevent such occurrences.
- Many dusts are heat sensitive and may decompose at high temperatures, resulting in an overpressure or fire.
- It is very important to determine if organic powders are thermally unstable and, if so, that they be tested for thermal stability to establish a safe operating temperature for the drying operation.

18.4 ADSORPTION

- Give definitions of terms used in adsorption processes.
  - *Adsorbent:* The surface of the solid material.
  - *Adsorbate:* The gas or vapor contacting the adsorbent.
  - *Adsorptive:* The gas or vapor in the bulk phase.
  - *Adsorption:* The process describing the accumulation of material on the adsorbent surface.
  - *Desorption:* The adsorbate in the condensed phase passes from the surface (of the adsorbent) to the fluid phase.
- What are the forces that are responsible for adsorbate components getting adsorbed on adsorbent surfaces?
  - In adsorption processes atoms, molecules, or ions in a gas or liquid phase diffuse to the surface of an adsorbent and bind to the surface as a result of London–van der Waals forces and/or nondispersion forces. The latter interactions are the result of polarization or nonreflected interactions.
- What are the differences between physical adsorption and chemisorption?
  - In physical adsorption, solute is held by the solid surface by weak (van der Waals) physical forces. Multilayer adsorption possible. Heat liberated is of the order of heat of condensation. Significant at relatively low temperatures. No electron transfer although polarization may occur. Always reversible. Adsorbed molecules maintain their identity.
  - In chemisorption (also called activated adsorption), forces are stronger and of the order of forces existing in chemical reactions.
  - *Monolayer Adsorption:* Heat liberated is of the order of heat of reaction (>2–3 times latent heat of adsorption). May involve dissociation of adsorbed species. Possible over a wide temperature range. Electron transfer leading to bond formation between adsorbate and adsorbent. Adsorbed molecules loose their identity. Can be irreversible.
- Give a summary of the differences of physical adsorption and chemisorption.
  - Table 18.13 gives a summary of the differences between physical adsorption and chemisorption.

TABLE 18.13 Differences Between Physical Adsorption and Chemisorption

Variable	Physical Adsorption	Chemisorption
Heat of adsorption	Small, same order as heat of vaporization (condensation)	Large, many times greater than the heat of vaporization (condensation)
Rate of adsorption	Controlled by resistance to mass transfer. Rapid rate at low temperatures	Controlled by resistance to surface reaction. Low rate at low temperatures
Specificity	Low, entire surface availability for physical adsorption	High, chemical adsorption limited to active sites on the surface
Surface coverage	Complete and extendable to multiple molecular layers	Incomplete and limited to a layer, one molecule thick
Activation energy	Low	High, corresponding to a chemical reaction
Quantity adsorbed per unit mass	High	Low

- What is coadsorption?
  - Several types of molecules can adsorb simultaneously on an adsorption surface, competing for active sites.
  - Highly polar molecules will displace less polar molecules, causing the less polar ones to emerge first.
  - For example, both water and CO_2 are present and then one can expect the less polar CO_2 to emerge first.
- What are the factors that influence adsorption and desorption processes?
  - The amount of adsorption depends on the chemical nature of the fluid and solid and is limited by the available surface and pore volume.
  - The rate of adsorption depends on the nature and amount of exposed surface of the adsorbent, its method of preparation/regeneration and rate of diffusion to the external and internal pore surfaces of the adsorbent. The internal surface constitutes the major part of the active surface.
  - Diffusion rates depend on temperature, pressure, and differences in concentration or partial pressures of the components.
  - The smaller the particle size, the greater will be the utilization of the internal surfaces, but also the greater will be the pressure drop for flow of bulk fluid through the adsorbent.
  - Individual components in multicomponent mixtures compete for the limited active surface of the adsorbent. Higher molecular weight members of homologous series adsorb preferentially on some adsorbents.
  - Repeated regeneration causes gradual deterioration of adsorbent and the capacity of regenerated adsorbent will generally be less than that of fresh adsorbent, because of permanent loss of adsorption sites to impurities (adsorption poisons) and structural changes of adsorbents.
  - Adsorption processes are sensitive to temperature changes and complicated by the substantial heats of adsorption involved during the adsorption process.
  - Desorption is facilitated by elevated temperatures and/or decreased pressure or washing techniques by suitable solvents.
- Define differential heat of adsorption.
  - Heat liberated at constant temperature when unit quantity of vapor is adsorbed on a *large quantity* (such an amount that adsorbate concentration is unchanged) of solid already containing adsorbate. This heat is directly measurable by calorimeter.
- Define integral heat of adsorption.
  - Integral heat of adsorption at any concentration, X, of adsorbate upon the solid is defined as the enthalpy of the adsorbate *plus* adsorbent combination *minus* sum of enthalpies of unit weight of pure solid adsorbent and sufficient pure adsorbed substance (before adsorption) to provide the required concentration, X, all at the same temperature.
- What are the applications of adsorption?
  - Drying of gases, refrigerants, solvents, transformer oils, and so on. Silica gel is the common adsorbent used in these applications and as desiccant in packaging.
  - Removal of organics, acetone, C_2H_4 and SO_2 from process off-gases, removal of CO_2 and sulfur compounds from natural gas and other gas streams.
  - Separation of N_2 and O_2 from air, H_2 from synthesis gas/hydrogenation processes, normal paraffins from isoparaffins and aromatics, *p*-xylene from C_8 aromatics, fructose–dextrose mixtures, ethanol from ethanol–water azeotrope and dehydration of aqueous azeotropes, and so on.
  - Carbons for odor control, recovery of solvent vapors, decolorizing syrups, sugars, and water purification (removal of phenols, halocarbons, pesticides, chlorine, etc.).
  - Active clays for treatment of vegetable oils, lube oils, removal of organic pigments, and so on.
  - Pollution control (removal of Hg, NO_X, and SO_X from gases).
  - Polymers (usually copolymers) and resins are used for water purification, separation of aliphatics from aromatics.
- What are the desirable characteristics of an adsorbent?
  - High selectivity to enable sharper separations.
  - High capacity to minimize amount of adsorbent needed.
  - Large internal surface area. The area should be accessible through pores large enough to admit the molecules to be adsorbed and also small enough to exclude molecules that it is desired not to adsorb.
  - Favorable kinetic and transport characteristics for rapid adsorption.
  - Chemical and thermal stability + extremely low solubility in the contacting fluid, to preserve the amount of adsorbent and its properties.
  - The adsorbent should not age rapidly, that is, lose its adsorptive capacity through continual recycling.
  - Hardness and mechanical strength to prevent crushing and erosion during handling.
  - Free-flowing tendency for ease of filling and emptying of vessels.

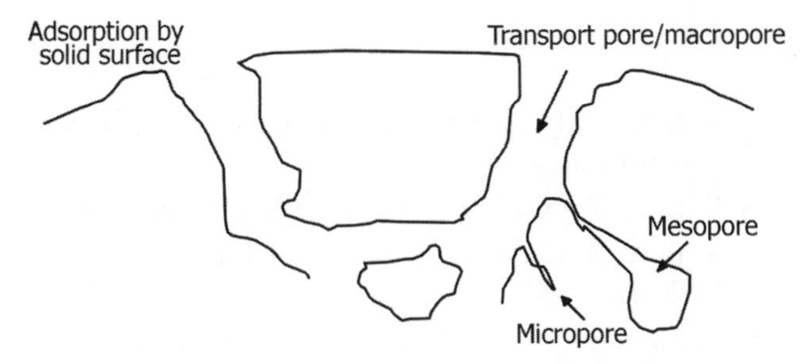

FIGURE 18.35 Types of pores on adsorbents.

- High resistance to fouling for long life.
- No tendency to promote undesirable chemical reactions.
- Capability for regeneration from trace amounts of high molecular weight impurities that are strongly adsorbed and difficult to desorb.
- Relatively low cost.

• What are the typical commercial characteristics of adsorbents?

- Shape: Granules, spheres, cylindrical pellets, flakes and/or powders.
- Size range: 50 μm to 1.2 cm.
- Specific surface areas: 300–1200 m^2/g.
- Particle porosity: 30–85 vol%.
- Average pore diameters: 10–200 Å.

• Illustrate different types of pores on adsorbents.

- Figure 18.35 illustrates different types of pores on adsorbents.
- *Transport Pore:* A pore that allows the transfer of the solute from the bulk phase to the interior of the adsorbent.
- *Macropore:* Pore of width $w \geq 50$ nm.
- *Mesopore:* Pore of width $50\,nm \leq w \leq 2$ nm.
- *Micropore:* Pore of width $w \leq 2$ nm.

• How are adsorbents classified? Give examples for each class.

- Adsorbents are classified into two categories, namely, hydrophobic and hydrophilic.
- *Hydrophobic (Nonpolar Surface):* For example, active carbons (surface areas are of the order of $10^6\,m^2/kg$), polymers, molecular sieves (600,000–700,000 m^2/kg). Coconut shells, fruit nuts, wood, coal, lignite, peat, petroleum residues, bones, and sometimes carbonized polymers (e.g., carbonized phenolic, acrylic, and viscose rayon filaments) are used to make active carbons.
  - ➢ Activated carbons in commercial use are mainly in two forms, namely, powder form and granular or palletized form. Powdered active carbons find applications in decolorization in food processing, for example, sugar refining, sodium glutamate manufacture, and wine making.
  - ➢ The main applications of granular active carbons in gas-phase processes are solvent recovery, air purification, gas purification, flue gas desulfurization, and bulk gas separation. Major applications in liquid-phase adsorption include decolorization in sugar refineries, removal of organics, odor and trace pollutants in drinking water treatment and advanced wastewater treatment.
  - ➢ Carbon molecular sieves are used in oxygen–nitrogen separation from air.
  - ➢ Active carbons used for gas adsorption have less pore volumes than liquid adsorption carbons, though both have same surface areas.
- *Hydrophilic (Polar Surface):* For example, silica gel (surface areas are of the order of 500,000 m^2/kg), activated alumina (surface areas are about 350,000 m^2/kg), and zeolites.
  - ➢ Silica gel is prepared by coagulation of a colloidal solution of silicic acid followed by controlled dehydration. Silica gels are of two types, Type A and Type B. Type A have pore sizes of 2–3 nm and Type B have larger pore sizes of about 7 nm. The main application is dehumidification of gases such as air and hydrocarbons. Type A is suitable for ordinary drying but Type B is more suitable for use at relative humidities higher than 50%.
  - ➢ Active alumina is prepared by dehydration of alumina trihydrate, $Al_2O_3{\cdot}3H_2O$ at controlled temperature conditions to a moisture content of about 6%. Specific surface area is in the range of 150 and 500 m^2/g with pore radius of 15–60 Å (1.5–6 nm). Porosity ranges from 0.4 to 0.76 (particle density of 1.8–0.8 g/cm^3). Used as a dehydrating agent and for removal of polar gases from hydrocarbon streams.

• What are molecular sieves? What are their applications?

- A molecular sieve is a material that can separate molecules based on size and shape. A subgroup of molecular sieves is zeolites.
- Molecular sieves are adsorbents composed of aluminosilicate crystalline microporous polymers (zeolites). They efficiently remove low concentrations of polar or polarizable contaminants such as H_2O, methanol, H_2S, CO_2, COS, mercaptans, sulfides, ammonia, aromatics, and mercury down to trace concentrations.
- These are available in the form of beads, granules and extrudates, including standard pellets.
- Surface areas of molecular sieves range from 600,000 to 700,000 m^2/kg.

- Zeolite films have large potential in many application areas such as sensors, catalysts, and membranes.

- What is a zeolite?
  - Zeolite is a crystalline aluminosilicate. Zeolites have the general stoichiometric unit cell formula: $M_{x/m}[(AlO_2)_x(SiO_2)_y]_z \cdot H_2O$, where M is the cation with valence m, z is the number of water molecules in each unit cell, and x and y are integers such that $y/x \geq 1$.
  - The cations balance the charge of the AlO_2 groups, each having a net charge of -1.
  - To activate the zeolite, raising the temperature and/or applying vacuum remove the water molecules. This process leaves the remaining atoms spatially intact in interconnected cage-like structures with six identical window apertures, each with a diameter of 0.38–1.0 nm (3.8–10 Å).
  - These apertures act as sieves, permitting small molecules to enter the crystal cage, but exclude large molecules.
  - Most commonly used zeolites are 3A, 4A, 5A, 10X, and 13X. These designations are mainly based on the aperture sizes, which are 0.29 nm for 3A, 0.38 nm for 4A type, 0.44 nm for 5A type, 0.8 for 10X type, and 0.84 nm for 13X type.
  - 3A type are used for drying of reactive gases. 4A are used for H_2O, CO_2 removal and air separation. 5A type are used for separation of air and linear paraffins. 10X and 13X are used for separation of air and removal of mercaptans.
  - Figure 18.36 shows the structures of Type A and Type X adsorbents.

- What are the applications of 13X type molecular sieves.
  - Gas drying.
  - H_2S and mercaptan removal from natural gas.

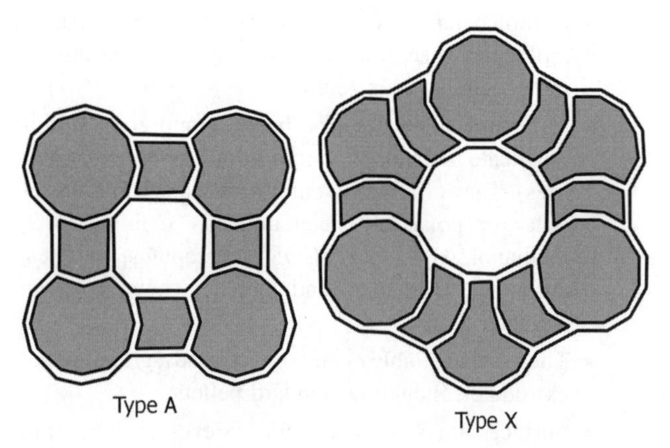

FIGURE 18.36 A and X type zeolites.

- What are the structures and applications of molecular sieves?
  - Table 18.14 gives the structures and applications of molecular sieves.

- What are the applications of Type X molecular sieves?
  - Table 18.15 gives applications of Type X molecular sieves.

- Name other types of adsorbents and give their important applications.
  - Bone char is similar to carbon adsorbents and contains calcium hydroxyl-apatite, which is an interesting adsorbent, which collects cations of heavy metals such as lead and cadmium.
  - Metal oxide adsorbents include oxides of magnesium, titanium, zirconium, and cerium.
  - Magnesium oxide (magnesia) is used for removing polar molecules such as color, acids, and sulfide compounds from gasoline and solvent for drycleaning purposes. Also it is effective in removing silica from water and magnesium trisilicate is used as a medicinal adsorbent. Magnesium hydroxide is a good adsorbent for phosphate removal in the advanced treatment of wastewaters.
  - Oxides of four valence metals, titanium, zirconium, and cerium, sometimes show selective adsorption characteristics in removing anions from water phases. Hydrous titanium oxide is known to be a selective adsorbent for recovering uranium in seawater, which is present in the form of carbonyl complex in concentrations as low as 3.2 ppb. Zirconium oxide in a monohydrated form is found to adsorb phosphate ion from wastewaters. Cerium oxide is effective in adsorption of fluoride ion in industrial wastewaters.

- What is a desiccant?
  - Desiccant is an adsorbent that shows primary selectivity for the removal of water.

- Name some commonly used solid desiccants.
  - Silica gel, alumina, and molecular sieves (remove water based on surface adsorption and capillary condensation). Related silicate adsorbents include magnesium silicate, calcium silicate, various clays, and diatomaceous earth.

- What are solid deliquescent desiccants?
  - Desiccants that remove water vapor by hydration reactions and dissolution, for example, $CaCl_2$, KOH.

- Summarize applications of common commercial adsorbents.
  - Table 18.16 gives applications of different commercially used adsorbents.

- Define dynamic or break loading.

TABLE 18.14 Structures and Applications of Different Molecular Sieves

Type	Cationic Form	Typical Structure	Applications
A	Na	$Na_{12}[(AlO_2)_{12}(SiO_2)_{12}]$	Desiccant; CO_2 removal from natural gas
	Ca	$Ca_5Na_2[(AlO_2)_{12}(SiO_2)_{12}]$	Separation of linear paraffins; air separation
	K	$K_{12}[(AlO_2)_{12}(SiO_2)_{12}]$	Drying of cracked gas containing C_2H_4, and so on
X	Na	$Na_{86}[(AlO_2)_{86}(SiO_2)_{106}]$	Pressure swing H_2 purification
	Ca	$Ca_{40}Na_6[(AlO_2)_{86}(SiO_2)_{106}]$	Removal of mercaptans from natural gas
	Sr·Ba	$Sr_{21}Ba_{22}[AlO_2)_{86}(SiO_2)_{106}]$	Xylenes separation
Y	Na	$Na_{56}[(AlO_2)_{56}(SiO_2)_{136}]$	Xylenes separation
	K	$K_{56}[(AlO_2)_{56}(SiO_2)_{136}]$	Xylenes separation
Mordenite	Ag	$Ag_8[(AlO_2)_8(SiO_2)_{40}]$	I and Kr removal from nuclear off-gases
	H	$H_8[(AlO_2)_8(SiO_2)_{40}]$	Same as above
Silicalite	–	$(SiO_2)_{96}$	Removal of organics from water

TABLE 18.15 Applications of Type X Molecular Sieves

Type	Applications
3A	Commercial dehydration of unsaturated hydrocarbon streams such as cracked gas, propylene, butadiene, and acetylene. It is also used for drying polar liquids such as methanol and ethanol
4A	Dehydration in gas or liquid system. Used as a static desiccant in household refrigeration systems; in packaging of drugs, electronic components, and perishable chemicals; as a water scavenger in paint and plastic systems. Also used commercially in drying saturated hydrocarbon streams
5A	Separates normal paraffins from branched chain and cyclic hydrocarbons through a selective adsorption process
10X	Aromatic hydrocarbon separation
13X	Used commercially for general gas drying, air plant feed purification, and liquid hydrocarbon and natural gas sweetening (H, S, and mercaptan removal)

TABLE 18.16 Applications of Some Commercially Used Adsorbents

Adsorbent	Applications
Silica gel	Drying of gases, refrigerants, organic solvents, transformer oils, desiccant in packaging and double glazing, dew point control of natural gas
Activated alumina	Drying of gases, organic solvents, transformer oils, removal of HCl from hydrogen, removal of fluorine and boron–fluorine compounds in alkylation processes
Carbons	Nitrogen from air, hydrogen from synthesis gas and hydrogenation processes, ethylene from methane and hydrogen, vinyl chloride monomer (VCM) from air, removal of odors from gases, recovery of solvent vapors, removal of SO_x and NO_x, purification of helium, cleanup of nuclear off-gases, decolorizing of syrups, sugars and molasses, water purification, including removal of phenol, halogenated compounds, pesticides, caprolactam, chlorine
Zeolites	Oxygen from air, drying of gases, removing water from azeotropes, sweetening sour gases and liquids, purification of hydrogen, separation of ammonia and hydrogen, recovery of carbon dioxide, separation of oxygen and argon, removal of acetylene, propane and butane from air, separation of xylenes and ethyl benzene, separation of normal paraffins from branched chain paraffins, separation of olefins and aromatics from paraffins, recovery of CO from methane and hydrogen, purification of nuclear off-gases, separation of cresols, drying of refrigerants and organic liquids, separation of solvent systems, purification of silanes, pollution control, including removal of Hg, NO_x, and SO_x from gases, recovery of fructose from corn syrup
Polymers and resins	Water purification, including removal of phenol, chlorophenols, ketones, alcohols, aromatics, aniline, indene, polynuclear aromatics, nitro- and chloraromatics, PCB, pesticides, antibiotics, detergents, emulsifiers, wetting agents, kraft mill effluents, dyestuffs, recovery and purification of steroids, amino acids and polypeptides, separation of fatty acids from water and toluene, separation of aromatics from aliphatics, separation of hydroquinone from monomers, recovery of proteins and enzymes, removal of colors from syrups, removal of organics from hydrogen peroxide
Clays (acid-treated and pillared)	Treatment of edible oils, removal of organic pigments, refining of mineral oils, removal of polychlorobiphenyl (PCB)

- It is the ratio of kg adsorbate per adsorption cycle per vessel divided by 100 kg of known adsorbent per vessel.

- What is pressure swing adsorption?
  - Adsorption is carried out at high pressures and desorption at low pressures by using two beds, one operating at pressures of several atmospheres and the other operating at atmospheric pressure as regenerator (by depressurization).
  - Pressure swing gas adsorption process is primarily used for the dehydration of air and for the separation of air into nitrogen and oxygen. Also used for hydrogen purification and separation of normal and isoparaffins.
  - The unit consists of two fixed bed absorbers for air drying as illustrated in Figure 18.37.
  - The above unit typically operates on a 10 min cycle, with 5 min for adsorption of water vapor from the air and 5 min for regeneration, which consists of repressurization, purging of the water vapor, and a 30 s repressurization. While one bed is adsorbing the other bed is regenerated.
  - During the 5 min adsorption period of the cycle, the capacity of the adsorbent for water must not be exceeded.

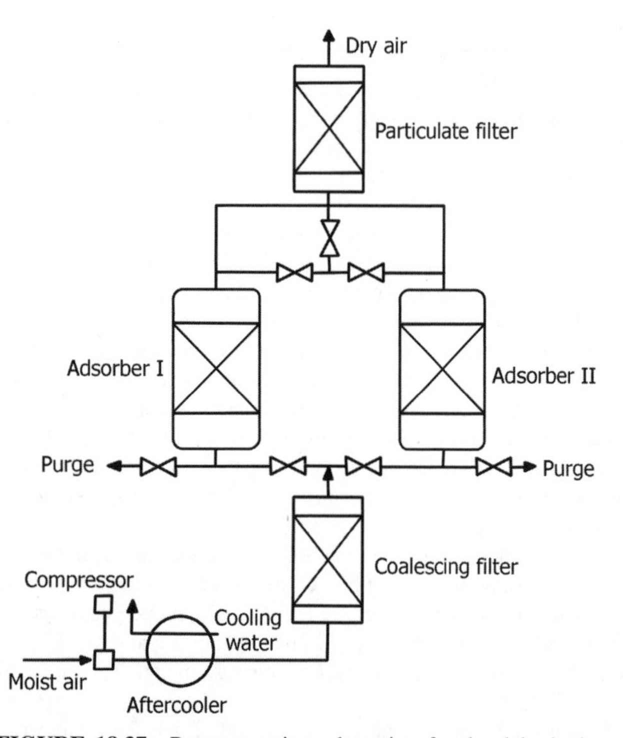

FIGURE 18.37 Pressure swing adsorption for the dehydration of air.

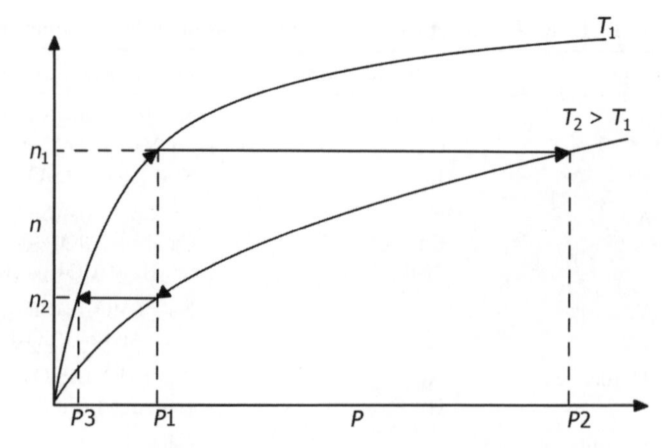

FIGURE 18.38 Temperature swing adsorption.

- What is thermal swing adsorption? Illustrate.
  - Figure 18.38 illustrates thermal swing adsorption process.
  - Steps involved:
    - *Step 1:* The feed fluid at p_1 and T_1 is passed through adsorbent. Equilibrium loading n_1 is reached.
    - *Step 2*: The temperature is raised to T_2 and the equilibrium partial pressure raises to p_2, causing desorption of the component from the adsorbent.
    - *Step 3:* Purge stream removes the component from the gas phase.
    - *Step 4:* Cooling returns back to T_1 line with some remaining loading n_2.
  - *Applications:* Drying, sweetening (H_2S, SO_2 removal from natural gas and hydrogen), NO_x removal, HCl removal from Cl_2, and the like.

- What are the advantages and disadvantages of adsorption systems in pollution control?

  Advantages
  - Possibility of product recovery.
  - Excellent control and response to process changes.
  - No chemical disposal problem when pollutant (product) recovered and returned to process.
  - Capability of systems for fully automatic, unattended operation.
  - Capability to remove gaseous or vapor contaminants from process streams to extremely low levels.

  Disadvantages
  - Product recovery possibly requiring an exotic, expensive distillation (or extraction) scheme.
  - Adsorbent progressively deteriorating in capacity as the number of cycles increases.
  - Adsorbent regeneration requiring a steam or vacuum source.
  - Relatively high capital cost.

- Prefiltering of gas stream possibly required to remove any particulates capable of plugging the adsorbent bed.
- Cooling of gas stream possibly required to get to the usual range of operation (<50°C).
- Relatively high steam requirements to desorb high molecular weight hydrocarbons.
- Spent adsorbent may be considered a hazardous waste.
- Some contaminants may undergo a violent exothermic reaction with the adsorbent.

• Discuss the criteria to be followed in design and operation of gas adsorbers using solid adsorbents.

- Allowance to be given for space and volume above and below the bed(s) for physical access for inspection, installation, and removal of internals such as adsorbent, bed supports, hold-down plates, distributors, and inert fillers like ceramic spheres.
- Use of several separated beds for varying the adsorbent, allowing for sampling, distributing weight and volume load, avoiding channeling, and so on.
- Depending on the diameter of the adsorber, a height of about 30–60% of internal height is generally used as nonadsorbent volume below as well as above the bed.
- With downward adsorption flow, a superficial velocity of 30 cm/s should be adequate for most applications. The adsorption velocities should be kept down below this value whenever possible as long as there is no risk of getting a preferential channeling effect by the process gas.
- A shock load imposed on an adsorption bed is a physical force, in the form of increased, spontaneous gas flow, such that the solid particle bed is made to move or *lift*. Sometimes these shocks can be of a magnitude that crushes or severely reduces some of the particles to dust.
- The source(s) of shock loads are as follows:
  - Process upsets upstream of the adsorption unit.
  - Quick, or abrupt opening of inlet or outlet valves when putting a bed into adsorption or regeneration service.
  - Sudden internal pressure changes within the bed, such as opening of pressure relief valve.
  - Sudden or rapid depressurization of a bed.
- The inherent cyclic nature of solid bed adsorption systems make them susceptible to required pressure changes.
- The potential danger of pressure shocks may be reduced by designing a pressurization/depressurization scheme whereby process gas is bled into the bed at a controlled rate (usually with a resistance orifice) prior to the bed being put back into adsorption service.
- Reducing the source of bed movement is one positive design step.
- Arresting or neutralizing the effects of any potential force that may cause bed movement is a desirable feature. This is done by fixing or *anchoring* the adsorbent bed. Mechanical design of the internals, in such a way that an external force will not be able to move or lift the bed, should be considered.
- One process technique is to use the direction of flow to arrest or neutralize such effects. Downward flow helps in this regard.
- As long as the bed support is strong enough, any incoming force will keep the bed pinned against the bottom support and movement should be nil.
- The use of *filler* inert material (such as ceramic spheres) is quite common in helping to support the bed in the bottom section and also retaining the bed in the upper section.
- A fixed bottom support is usually used and a strong, flexible, *floating* screen is employed between the top of the adsorbent bed and the inert spheres in the top section, to avoid voids within the adsorber, thereby reducing or eliminating internal space available for bed movement or migration.
- Use of spring-loaded screens in the top section to assure bed hold-down is another option.
- It should be noted that the internal material will expand and contract with alternative heating and cooling due to adsorption and regeneration. This is one of the reasons for use of floating screen at the top.
- Water can damage some adsorbents and should be removed prior to the adsorption unit with efficient separators.
- It should be ensured that the regeneration flow rate is always controlled and never *shocked* into the bed. Since regeneration is carried out in the upflow direction, the bed will have a tendency to *lift* if the regenerating flow is excessive, even for an instantaneous moment.

  Source: Cheresources Message Board, Mr. Art Montemayor.

• What basic information is required for the design of fixed bed adsorbers?

- Adsorbent properties like particle and bulk densities, void fractions, isotherms, or other equilibrium data and data on mass transfer kinetics involving intraparticle mass transfer resistance and fixed bed dynamics including fluid-to-particle transfer and pressure drop.

- How are fixed bed adsorbers designed? Give brief procedural steps involved.
  - *Step 1:* Selection of type of adsorbent and its particle size. The most common adsorbents are active carbon, active alumina, silica gel, and molecular sieves. Active carbon is mostly used in solvent recovery applications and the other adsorbents mentioned are mostly used in gas drying.
  - *Step 2:* Selection of cycle time that should normally be twice the regeneration time. Equal times may be selected for adsorption and regeneration.
  - *Step 3:* Velocity of the process stream through the bed may be taken to be in the range of 25–30 m/min.
  - *Step 4:* A suitable steam to solvent ratio for regeneration may be selected.
  - *Step 5:* Working capacity of the adsorbent may be obtained as a percentage of saturation capacity for the particular solvent, which must be obtained from experimental data or from operational data of similar plants.
  - *Step 6:* Mass of solvent adsorbed during one-half the cycle time as mass, $m = qc_i t_{ads}$, where q is the volumetric flow rate, c_i is the inlet solvent concentration and t_{ads} is the half-cycle time.
  - *Step 7:* Mass of adsorbent required may be estimated as m = (mass of solvent adsorbed during half the cycle time)/(working capacity).
  - *Step 8:* Volume of adsorbent required might be estimated from mass of adsorbent required from Step 7 divided by bulk density of adsorbent.
  - *Step 9:* Bed cross-sectional area may be obtained as volumetric flow rate divided by velocity through the bed.
  - *Step 10:* Bed height may be obtained as volume of adsorbent divided by cross-sectional area of the bed.
  - *Step 11:* Pressure drop may be estimated using a suitable correlation for packed beds.
  - *Step 12:* L/D ratio for the adsorber, including the space required for the nozzles, supports, access for maintenance, and space above the bed may be taken as 3:4, if velocities are kept below 10 m/min. Bed height should be kept below 10 m and diameter should be below 3 m for efficient operation.
  - *Step 13:* Mechanical design of the shell should be based on the consideration that it is completely filled with water.
  - *Step 14:* The adsorber may be installed in vertical position if actual volumetric flow rate is *less than* 70 m^3/min and installed in horizontal position if actual volumetric flow rate is *more than* 200 m^3/min.
  - The above-simplified procedure is for two units in parallel, one on adsorption cycle while the other is on regeneration cycle for adsorption of VOCs from air.
- What is chromatography?
  - Chromatography is a broad range of physical methods used to separate and or to analyze complex mixtures. The components to be separated are distributed between two phases: a *stationary phase* bed and a *mobile phase* that percolates through the stationary bed.
  - A mixture of different components enters a chromatographic column and the components move through the adsorption bed at different rates, due to repeated sorption–desorption cycles. These differential rates of migration as the mixture moves over adsorptive materials provide separation.
  - Repeated sorption/desorption acts that take place during the movement of the sample over the stationary bed determine the rates. The smaller the affinity a molecule has for the stationary phase, the shorter the time spent in a column.
  - It can separate complex mixtures with good precision. Chromatography can purify any soluble or volatile substance if the right adsorbent material, carrier fluid and operating conditions are employed.
  - Modern applications of chromatography employ a *column*. The column is where the actual separation takes place. It is usually a glass or metal tube of sufficient strength to withstand the pressures that may be applied across it.
  - The column contains the *stationary phase*. The *mobile phase* runs through the column and is adsorbed onto the stationary phase. The column can either be a *packed bed* or *open tubular* column.
  - A packed bed column is comprised of a stationary phase that is packed with a granular adsorbent. The stationary phase can be a liquid distributed over the surface of an inert porous solid through which the mobile phase passes through. The components to be separated have varying affinities for the stationary phase.
  - The carrier fluid and the sample to be separated flow through the column together. The carrier phase in gas chromatography is an inert gas. In liquid chromatography, it is a liquid of low viscosity that flows through the stationary phase bed. This bed may be an immiscible liquid coated onto a porous support, a thin film of liquid phase bonded to the surface of a sorbent, or a sorbent of controlled pore size.
  - Gas–liquid chromatography involves a bed of an inert solid support that is coated with a very viscous liquid that acts as an absorbent.

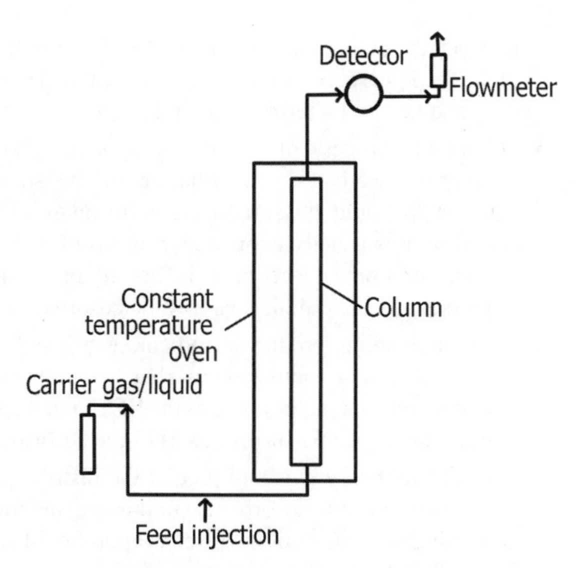

FIGURE 18.39 Chromatographic unit.

- Figure 18.39 illustrates the operation of a chromatographic unit.

• What are the different adsorbents used in chromatography?

- *Alumina:* Polar. Preferred for separation of weekly or moderately polar compounds. Polar compounds are retained more selectively and recovered from the column last. Basic adsorbent retains acidic components.
- *Silica gel:* Less polar; acidic.
- *Carbon:* Nonpolar. Highest attraction for larger nonpolar molecules.
- Adsorbent type sorbents are better suited for separation on the basis of chemical type (olefins, esters, acids, etc.)

• Classify analytical chromatographic systems.

- The following chart (Figure 18.40) gives a classification of chromatographic systems.

• What is partition chromatography?

- Partition chromatography is based on a thin liquid film formed on the surface of an inert solid support by a stationary liquid phase. Solute molecules equilibrate between the mobile phase and the stationary liquid.
- This method is suitable for separating individual members of homologous series.
- Partition chromatography uses a retained solvent, on the surface or within the grains or fibers of an inert solid supporting matrix as with paper chromatography, or takes advantage of some additional coulombic and/or hydrogen donor interaction with the solid support.

• What is HPLC?

- High performance liquid chromatography, or high pressure liquid chromatography, is a form of column chromatography used frequently in biochemistry and analytical chemistry to separate, identify, and quantify compounds. HPLC utilizes a column that holds chromatographic packing material (stationary phase), a pump that moves the mobile phase(s) through the column, and a detector that shows the retention times of the molecules. Retention time varies depending on the interactions between the

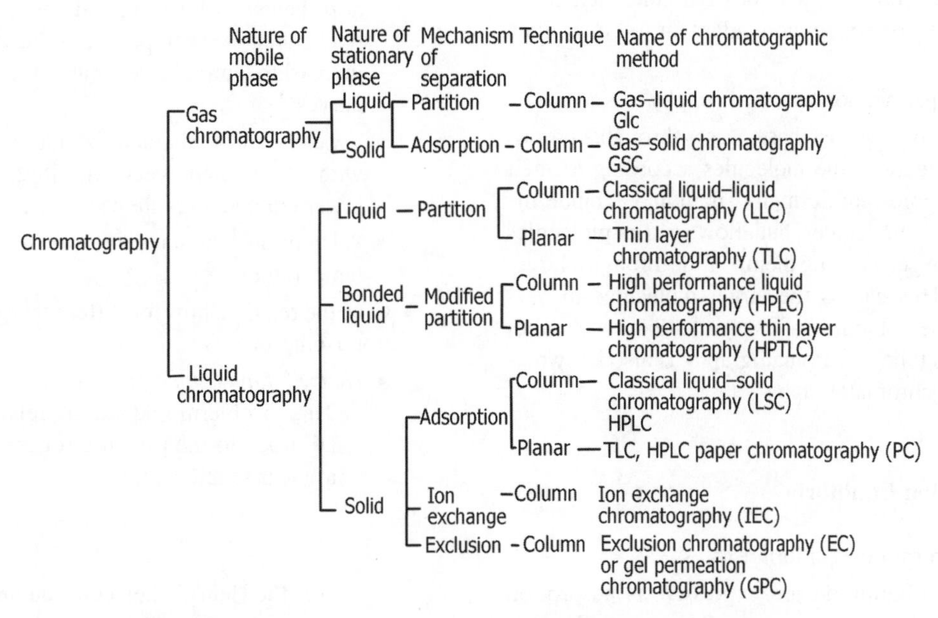

FIGURE 18.40 Classification of analytical chromatographic systems.

stationary phase, the molecules being analyzed and the solvent(s) used.

- How is thin layer chromatography (TLC) performed?
  - Thin layer chromatography is performed on a sheet of glass, plastic, or aluminum foil, which is coated with a thin layer of adsorbent material, usually silica gel, aluminum oxide, or cellulose. This layer of adsorbent is the stationary phase.
  - After the sample has been applied on the plate, a solvent or solvent mixture (mobile phase) is drawn up the plate via capillary action.
  - TLC plates are usually commercially available, with standard particle size ranges to improve reproducibility. They are prepared by mixing the adsorbent, such as silica gel, with a small amount of inert binder like calcium sulfate (gypsum) and water. This mixture is spread as a thick slurry on an unreactive carrier sheet, usually glass, thick aluminum foil, or plastic. The resultant plate is dried and activated by heating in an oven for 30 min at 110°C. The thickness of the adsorbent layer is typically around 0.1–0.25 mm for analytical purposes and around 1–2 mm for preparative TLC.
- What is ion exchange chromatography?
  - Ion exchange chromatography involves separation of ions and polar molecules based on the charge properties of the molecules. It can be used for almost any kind of charged molecule including large proteins, small nucleotides and amino acids. It is often used in protein purification, water analysis and quality control.
  - Types of ion exchangers used include polystyrene resins and cellulose and dextran ion exchangers (gels) and controlled pore glass or porous silica materials.
- What is gel permeation chromatography?
  - The liquid or gaseous phase passes through a porous gel that separates the molecules according to their size. The pores are normally small and exclude the larger solute molecules, but allow smaller molecules to enter the gel, causing them to flow through a larger volume. This causes the larger molecules to pass through the column at a faster rate than the smaller ones. This type of chromatography is also known as exclusion chromatography.

18.4.1 Adsorption Equilibria

- How are adsorption equilibria expressed?
  - Adsorption equilibria are expressed as adsorption isotherm that give the equilibrium relationship between the amount adsorbed and the concentration in the fluid phase. Equilibrium is temperature dependant and hence the term isotherm is used.
  - During the process of adsorption dynamic phase equilibrium leads to a distribution of the solute between the fluid phase and the solid phase. The equilibrium is usually expressed in terms of
    - concentration (if the fluid is liquid) or partial pressure (if the fluid is a gas) of the adsorbate and
    - solute loading on the adsorbent, expressed as mass, moles, or volume of adsorbate per unit mass or per unit area (determined as the BET monolayer equivalent specific surface area) of the adsorbent.
  - No suitable theory exists to predict the distribution of adsorbate and adsorbent similar to theories describing the distribution between vapor–liquid and liquid–liquid equilibria.
  - Experimental data are obtained for a particular solute, or mixture of solutes and/or solvent and the sample of the actual solid adsorbent material. If data are taken over a range of fluid concentrations at a constant temperature, a plot of solute loading on the adsorbent versus concentration or partial pressure leads to the definition of an adsorption isotherm.
- Classify gas adsorption isotherms.
  - Gas adsorption isotherms are classified by IUPAC into six types as illustrated in Figure 18.41.
  - Type I shows adsorption isotherms on microporous adsorbents for subcritical, near-critical, and supercritical conditions.
  - Types II and III are isotherms on macroporous adsorbents, with strong and weak affinities, respectively. For low temperatures these have steps but increasing temperatures transform them into smooth curves.
  - Types IV and V characterize mesoporous adsorbents with strong and weak affinities, respectively. At lower temperatures they show adsorption hysterisis.
  - Adsorption amount decreases with an increase in temperature.
- Give the relationships for different types of gas adsorption isotherms.
  - *Henry's Law (Linear Isotherm):*
    - This isotherm equation relates the amount adsorbed to the pressure (or relative pressure) as a linear relationship:

$$q = Hp \quad \text{or} \quad q = Hp/p^0. \qquad (18.26)$$

H is the Henry's law constant and p^0 is the vapor pressure.

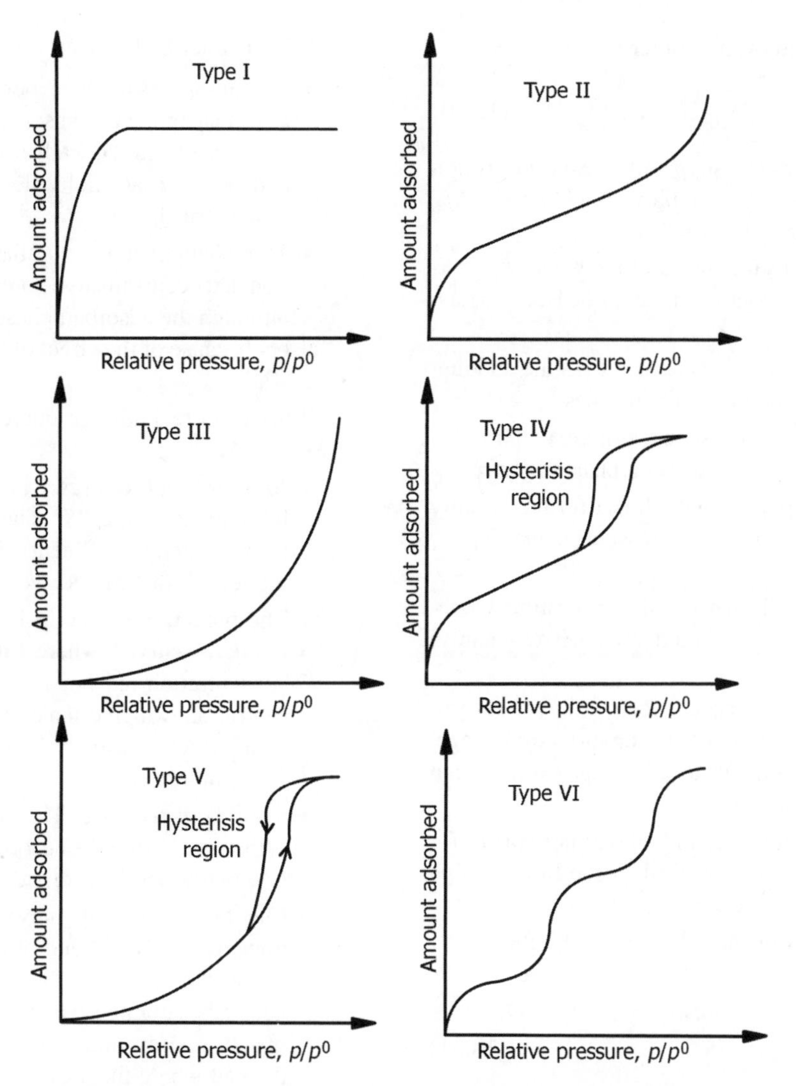

FIGURE 18.41 IUPAC classification of gas adsorption isotherms.

- *Freundlich (Strongly Favorable) Isotherm:*
  - This isotherm equation is empirical and nonlinear in pressure:

$$q = kp^{1/n}. \quad (18.27)$$

  - The constants k and n are temperature dependent constants attributed to the capacity (k) and the strength of adsorption ($1/n$). Generally, n lies in the range of 1–5. Of course, as n tends to 1, the isotherm tends to Henry's law. Experimentally, adsorption isotherm data are analyzed using a log–log plot as $\log q$ versus $\log p$. The slope of the graphical analysis gives $1/n$ and the intercept gives k. Although this analysis suggests a linear relationship, often the graph is curved.

$$\ln q = \ln k + 1/n \ln p. \quad (18.28)$$

  - Generally, k decreases with increasing temperature and n increases with increasing temperature, approaching unity at high temperatures.
  - Assumes exponential with p.
- *Langmuir (Favorable) Isotherm:*
  - The Langmuir isotherm is restricted to describing Type I adsorption isotherm shape.

$$q = q_{\mathrm{m}}Kp/(1 + Kp). \quad (18.29)$$

  - This expression may be rewritten as a linear expression to evaluate the statistical monolayer amount adsorbed, q_{m}, and the equilibrium constant for the adsorption and desorption processes, K.

- The linear form of the isotherm is given as

$$p/q = 1/Kq_m + p/q_m. \quad (18.30)$$

- Thus, a plot of p/q versus p will give the adsorption data as a straight line graph with a slope of $1/q_m$ and an intercept of $1/Kq_m$.
- At low pressures for the condition where $Kp \ll 1$, the Langmuir isotherm reduces to the Henry's Law expression.
- At high pressures, where $Kp \gg 1$, the amount adsorbed asymptotically approaches q_m.
- The isotherm may also be written in terms of relative pressure, where p is replaced by p/p^0.
- The isotherm plotted in the linear form is usually linear over the relative pressure range $0.05 \leq p/p^0 \leq 0.35$.
- In general, K will change with temperature whereas the value of q_m should remain (reasonably) constant.
- Langmuir fails to account for multilayer adsorption and dependency of ΔH_{ads} on surface coverage, that is, more difficult to adsorb gas in between adsorbed molecules.
- Langmuir postulate of monolayer is realistic for chemisorption and physical adsorption at high temperatures and low pressures.

- BET equation (Brunauer, Emmett, and Teller):

$$p/p^0/q(1-p/p^0) = (1/q_m c) + [(c-1)/q_m c]p/p^0. \quad (18.31)$$

- Assumptions involved in BET model: Uniform energies of adsorption on the surface.
- A number of layers of adsorbate molecules form at the surface and that the Langmuir equation applies to each layer.
- A given layer need not complete formation prior to initiation of subsequent layers; the equilibrium condition will therefore involve several types of surfaces.

- What are the mechanisms associated with adsorption of a solute on an adsorbent like alumina?
  - Adsorption of a solute on alumina is the sum of three different phenomena:
    - Chemisorption, forming the first layer at low partial pressures.
    - Physical adsorption, due to the formation of multiple layers by hydrogen bonding in the alumina pores.
    - Capillary condensation, where localized condensation takes place at temperatures above that of the dew point of the bulk fluid.

- What is micropore adsorption?
  - In micropores of size comparable to the size of adsorbate molecule, adsorption takes place by attractive force from the wall surrounding the micropores and the adsorbate molecules start to fill the micropore volumetrically.
  - This phenomenon is similar to capillary condensation that occurs in large pores at high partial pressure, although the adsorbed phase in micropores is different because of the effect of the force field of the pore wall.

- What is a breakthrough curve? What is mass transfer zone?
  - *Breakthrough Curves.* The breakthrough curve can be defined as the "*S*" shaped curve that typically results when the effluent adsorbate concentration is plotted against time or volume.
  - The breakthrough point is the point on the breakthrough curve where the effluent adsorbate concentration reaches its maximum allowable concentration, which often corresponds to the treatment goal, which is usually based on regulatory or risk-based numbers.
  - *Mass Transfer Zone:* The mass transfer zone (MTZ) is the area within the adsorbate bed where adsorbate is actually being adsorbed on the adsorbent.
  - The MTZ typically moves from the influent end toward the effluent end of the adsorbent bed during operation. That is, as the adsorbent near the influent becomes saturated (spent) with adsorbate, the zone of active adsorption moves toward the effluent end of the bed where the adsorbate is not yet saturated.
  - The MTZ is sometimes called the adsorption zone or critical bed depth. The MTZ is generally a band, between the spent adsorbent and the fresh adsorbent, where adsorbate is removed and the dissolved adsorbate concentration ranges from C_o to C_e.
  - Figure 18.42 illustrates an idealized breakthrough curve, indicating MTZ.

18.4.2 Ion Exchange

- What is ion exchange? Illustrate the reactions involved in demineralization of water by ion exchange.
  - Ion exchange is a chemical process that can be represented by a stoichiometric equation, for example, when ion A in solution replaces ion B in the solid phase:

$$\text{A(solution)} + \text{B(solid)} \leftrightarrow \text{A(solid)} + \text{B(solution)} \quad (18.32)$$

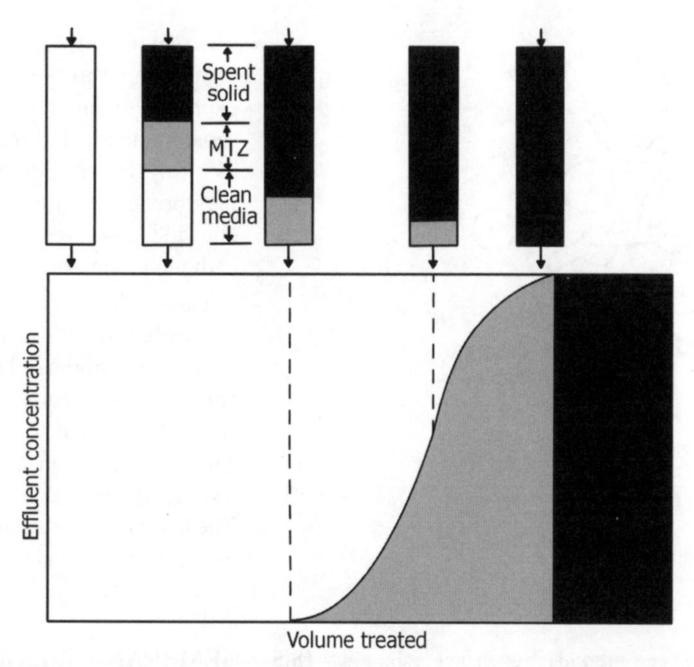

FIGURE 18.42 Adsorption column mass transfer zone and idealized breakthrough zone.

- Ion exchange processes function by replacing undesirable ions of a liquid with ions such as H^+ or OH^- from a solid material in which the ions are sufficiently mobile, usually some synthetic resin.
- The resin becomes exhausted on use and may be regenerated by contact with a small amount of solution with a high content of the desired ion.
- Resins can be tailored to have selective affinities for particular kinds of ions.
- Demineralization involves replacing all cations with hydrogen ions (H^+) and all anions with hydroxide ions (OH^-).
- Rates of ion exchange processes are affected by diffusional resistances of ions into and out of the solid particles as well as resistance to external surface diffusion.
- Ion exchange materials are like storage batteries; they must be recharged (regenerated) periodically to restore their exchange capacity.
- With proper design and operation, ion exchange processes are capable of removing selected ions almost completely (in some cases to a fraction of a ppm).
- Specific types of ions can be removed from water depending on the choice of exchange material and regenerant used.
- Table 18.17 lists different ion exchange processes.
- Figure 18.43 illustrates ion exchange process in solid ion resin.

• What are the characteristics of ion exchange resins?

- Ion exchange resins are polymers with cross-linking (connections between long carbon chains in a polymer). The resin has active groups in the form of electrically charged sites. At these sites, ions of opposite charge are attracted but may be replaced by other ions depending on their relative concentrations and affinities for the sites.
- Synthetic organic polymer resins based on styrene or acrylic acid type monomers are most widely used.
- Generally solid gels in spherical or granular form consisting of
  - a three-dimensional polymeric network,
  - ionic functional groups attached to the network,
  - counterions,
  - a solvent.

TABLE 18.17 Types of Ion Exchange Processes

Types of Minerals in Influent	Types of Exchanger	Minerals Converted to
$Ca(HCO_3)_2$	Na^+ exchanger	$NaHCO_3$
$CaSO_4$	Na^+ exchanger	Na_2SO_4
$Ca(HCO_3)_2$	H^+ exchanger	H_2CO_3
$CaSO_4$	H^+ Exchanger	H_2SO_4
$Ca(HCO_3)_2$	H^+ exchanger (weak acid)	H_2CO_3
Na_2SO_4	Cl^- exchanger	NaCl
$NaHCO_3$	Cl^- exchanger	NaCl
H_2CO_3	OH^- exchanger	H_2O
H_2SO_4	OH^- exchanger	H_2O

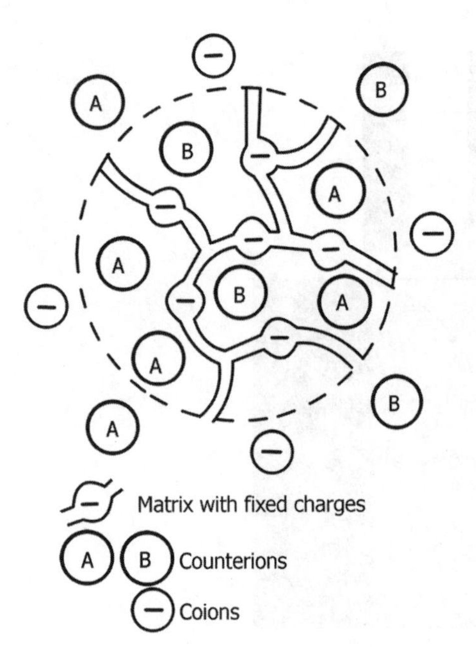

FIGURE 18.43 Ion exchange process in solid ion exchange resin.

- Strong acid, cation exchange resins, and strong base anion exchange resins can be produced.
- Two key factors determine the effectiveness of a given ion exchange resin: favorability of any given ion, and the number of active sites available for this exchange.
- To maximize the active sites, significant surface areas are generally desirable.
- The active sites are one of a few types of functional groups that can exchange ions with either plus or minus charge. Frequently, the resins are cast in the form of porous beads.
- Cross-linking, usually on the order of 0.5–15%, comes from adding divinyl benzene to the reaction mixture during production of the resin.
- The size of the particles also plays a role in the utility of the resin. Smaller particles usually are more effective because of increased surface area but cause large ΔP that increase energy costs.
- Temperature and pH also affect the effectiveness of ion exchange, since pH is inherently tied to the number of ions available for exchange, and temperature governs the kinetics of the process.

• What are the applications of ion exchange?
  - Water demineralization, decolorization of sugar solutions, recovery of uranium from acid leach solutions, recovery of gold, silver and platinum, recovery of antibiotics and vitamins from fermentation broth, pharmaceutical processing and formulations, artificial organs, solvent and reagent purification, and so on.
  - The major application for ion exchange membranes is still found in the chloralkali electrolysis. Cation exchange membranes, comprising fixed anions, separate anode and the cathode compartment in an electrolysis cell. Sodium ions can pass through the membrane and form sodium hydroxide, whereas chloride ions are repelled and cannot pollute the cathode compartment.
  - In electrodialysis, cations and anions can pass out of salt solutions through the respective oppositely charged membrane. Thus, a saline solution can be split into two streams, one with a lower and the other with a higher salt content than the original solution. The product of an electrodialysis process may be the desalinated stream, or the concentrated stream. The latter is, for example, important in the production of salt from seawater, as a preconcentration step.

18.5 MEMBRANE SEPARATIONS

• What are the important membrane-based separation processes?
  - Reverse osmosis.
  - Dialysis.
  - Electrodialysis.
  - Microfiltration.
  - Ultrafiltration.
  - Pervaporation.
  - Gas permeation.
  - Liquid membrane processes.

• What are the driving forces for mass transfer in membrane separation processes?
  - *Hydrostatic Pressure:* Separation is based on differences in hydrodynamic permeability of the membrane for different components, driven by hydrostatic pressure difference.
  - *Concentration Difference:* Concentration of different components in the membrane are different.
  - *Electrical Potential Differences:* Differences in charged particles give rise to different mobilities and concentrations in the membrane.

• What are the merits and demerits of membrane processes?
  - Merits:
    - ➢ Reduces the number of unit processes in treatment systems.
    - ➢ Potential for process automation and plant compactness.
    - ➢ Much smaller footprint than the conventional plants of the same capacity.

  - ➢ Easy scale-up, expansion, and retrofitting.
  - ➢ Less or no chemical use and provides highest quality water.
  - ➢ No formation of secondary chemical by-products.
  - ➢ Less sludge production.
  - ➢ Water reuse and recycling.
- ▪ Demerits:
  - ➢ Membrane fouling.
  - ➢ Low membrane lifetime.
  - ➢ Low selectivity.
  - ➢ High capital and operating costs.

- Define the terms *retentate* and *permeate* used in membrane separation processes.
  - ▪ *Retentate:* That part of the feed that does not pass through the membrane, that is, retained.
  - ▪ *Permeate:* The part of the feed that passes through the membrane.
- What is selectivity of a membrane?
  - ▪ For dilute aqueous mixtures of a solvent (water) and a solute (particles), the selectivity is expressed in terms of retention, R, toward the solute.

$$R = (C_F - C_P)/C_F. \qquad (18.33)$$

  - ▪ C_F and C_P are solute concentration in feed and permeate.
  - ▪ $R = 100\%$ (complete retention) of solute.
  - ▪ $R = 0\%$ (solute and solvent pass through the membrane).
- What are the desirable attributes of a membrane?
  - ▪ Good permeability.
  - ▪ High selectivity.
  - ▪ Chemical and mechanical compatibility with the processing environment.
  - ▪ Stability, freedom from fouling, and reasonable useful life.
  - ▪ Amenability to fabrication and packaging.
  - ▪ Ability to withstand large ΔP across the membrane.
- How is selectivity of a given membrane estimated?
  - ▪ The selectivity of membranes can be estimated by using the following dimensionless parameters:
  - ▪ Separation factor,

$$\alpha = Y_A/Y_B/Y_A/X_B = Y_A/(1-Y_A)/X_A/(1-X_A). \qquad (18.34)$$

  - ▪ Enrichment factor,

$$\beta = Y_A/X_A, \qquad (18.35)$$

where X_A is the weight fraction of preferentially permeating species in the feed phase, Y_A is the weight fraction of preferentially permeating species in the permeate phase, with

$$X_A + X_B = 1 \quad \text{and} \quad Y_A + Y_B = 1. \qquad (18.36)$$

  - ▪ Neither the separation factor nor the enrichment factor is constant. Both parameters are strong functions of feed composition.

18.5.1 Membranes

- What are the three primary requisites of a membrane?
  - ▪ The selectivity should be good, but the bipolar membrane layers are also permeable to salt coions, not only to the water splitting products in the respective layer.
  - ▪ The permeability should be unlimited, but the membrane layers form additional resistances to ion transport.
  - ▪ The long-term stability should be guaranteed, but the chemical stability, especially against concentrated alkaline solutions is too low at temperatures normally encountered in electrodialysis.
- How membranes are basically classified? What are the applications of nonporous membranes?
  - ▪ *Microporous* membranes discriminate according to size of particles or molecules.
  - ▪ *Nonporous* membranes discriminate according *to* chemical affinities between components and membrane materials.
    - ➢ Microporous membrane has a rigid, highly porous structure with randomly distributed, interconnected pores. However, these pores differ from those in a conventional filter by being extremely small, of the order of 0.01–10 μm in diameter. Any membrane, porous or nonporous, will act to some extent as a microporous medium. Even the membranes that are normally classified as *nonporous*, may have a number of minute pores whose diameter will be in the 5–10 Å range. However, such structures are considered nonporous.
    - ➢ All particles larger than the largest pores are completely rejected by the membrane. Particles smaller than the largest pores, but larger than the smallest pores are partially rejected, according to the pore size distribution of the membrane. Particles much smaller than the smallest pores will pass through the membrane. Thus, separation of solutes by microporous membranes is mainly a function of molecular size and pore size distribution. In

general, only molecules that differ considerably in size can be separated effectively by microporous membranes, for example, in ultrafiltration and microfiltration.

- ➢ *Nonporous, dense membranes* consist of a dense film through which permeants are transported by diffusion under the pressure, concentration, or electrical potential gradient as the driving force. Separation of different components of a solution is related directly to their relative transport rate within the membrane, based on their diffusivity and solubility in the membrane material.
- ➢ An important property of nonporous, dense membranes is that even permeants of similar size may be separated when their concentration in the membrane material (i.e., their solubility) differs significantly.
- ➢ Most gas separation, gas permeation, pervaporation, electrodialysis and reverse osmosis membranes use nonporous dense membranes to perform the separation.

- ▪ Membranes can be symmetrical or asymmetrical.
- ▪ Symmetrical membranes can be porous or dense, but the permeability of the membrane material does not change from point to point within the membrane. Dense symmetrical membranes are widely used in research and development and other laboratory work but are rarely used commercially, because the transmembrane flux is too low for practical separation processes.
- ▪ Asymmetric membranes, on the other hand, typically have a relatively dense, thin surface layer supported on an open, often microporous substrate. The surface layer generally performs the separation and is the principal barrier to flow through the membrane. The open support layer provides mechanical strength. In industrial applications, symmetrical membranes have been almost completely displaced by asymmetric membranes, which have much higher fluxes.
- ▪ Ceramic and metal membranes can be both symmetrical and asymmetric.
- ▪ Figure 18.44 illustrates symmetrical and asymmetrical membranes.
- ▪ In liquid membranes, the barrier is a liquid phase, often containing dissolved carriers that selectively react with specific permeants to enhance their transport rates through the membranes.
- ▪ In their simplest form, liquid membranes consist of an immobilized phase held in the pores of a microporous support membrane. In another form the liquid is contained as the skin of a globule in a liquid emulsion system.
  - ➢ A very large number of globules of emulsion can be formed to produce a very large membrane surface area for rapid mass transfer.
  - ➢ Figure 18.45 illustrates a liquid membrane.
- ▪ Figure 18.46 gives a classification of membranes.

- • What is a dense nonporous membrane?
  - ▪ A dense (nonporous) membrane is a membrane with no *detectable* pores.
- • Illustrate the structures of asymmetric composite membranes.
  - ▪ Figure 18.47 illustrates the structures of asymmetric composite membranes.
- • What are the applications of liquid membranes?
  - ▪ Tertiary treatment of wastewater involving contaminants like phenols, acids, ammonia, phosphates, chromates, mercury, cadmium, nitrates, and nitrites are examples in which liquid membranes have potential applications.
  - ▪ Liquid membranes possess a number of advantages over solvent extraction methods for the recovery of valuable metals from dilute aqueous solutions.
  - ▪ Recovery of nickel from electroplating solutions.
  - ▪ Recovery of zinc from wastewater in the viscose fiber industry.
  - ▪ Liquid membrane emulsions have potential applications as reagents in oil well problems like acid fracturing of rock and plugging of fractures.
  - ▪ Liquid membranes are used as heterogeneous catalysts or as reaction moderators in carrying out chemical reactions.

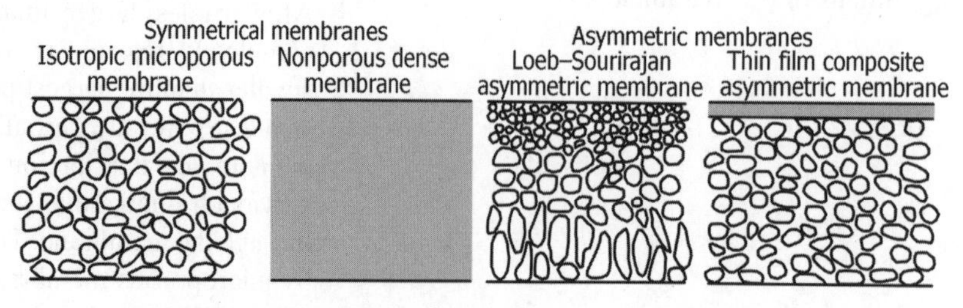

FIGURE 18.44 Symmetrical and asymmetrical membranes.

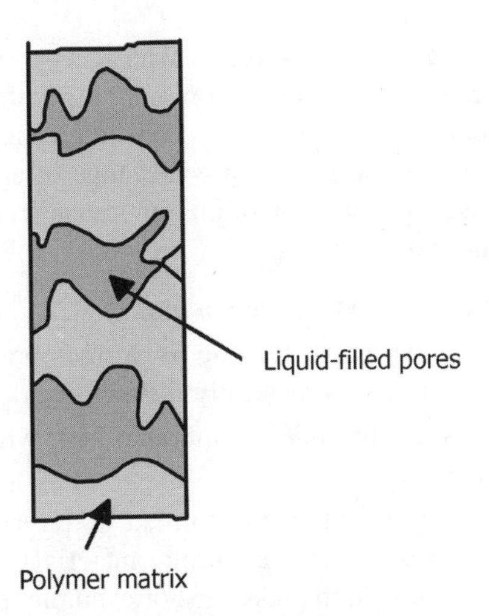

FIGURE 18.45 Liquid membrane.

- Use of liquid membranes is promising as bioreactors, biochemical and biomedical applications. For example, sensitive enzymes may be encapsulated to protect them from deactivating substances while maintaining free access to the substrate. Other examples include blood oxygenation, removal of toxins from blood, emergency treatment of drug overdose, slow release of enzymes and drugs, and development of artificial kidneys.

- How are membranes chosen?
  - The choice of the membrane strongly depends on the type of application. It is important which of the components should be separated from the mixture and whether this component is water or an organic liquid. Generally, the component with the smallest weight fraction in the mixture should preferentially be transported across the membrane.
  - Examples for permeation and vapor permeation processes are as follows:
    - ➢ Dehydration of organic liquids:
      - For the removal of water from water/organic liquid or vapor mixtures hydrophilic polymers are to be chosen.
      - Hydrophilicity is caused by groups present in the polymer chain that are able to interact with water molecules.
      - Examples of hydrophilic polymers are ionic polymers, polyvinyl alcohol (PVA), polyacrylonitrile (PAN), polyvinylpyrrolidone (PVPD).
    - ➢ Removal of organics from water or air streams:

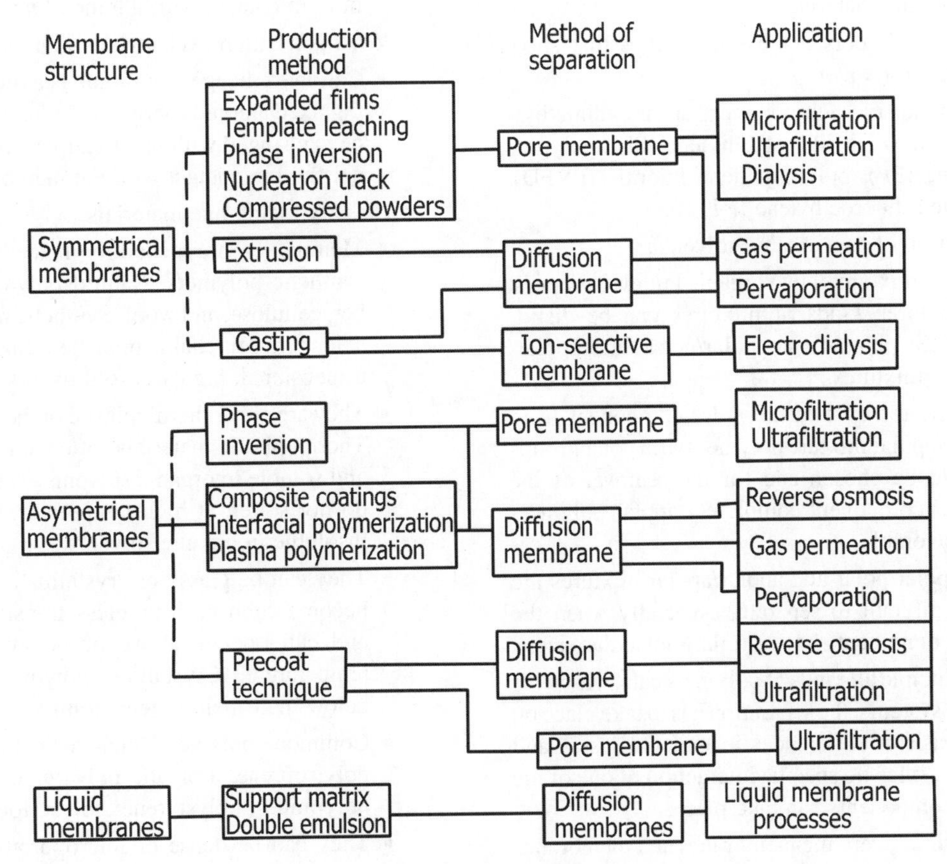

FIGURE 18.46 Classification of membranes.

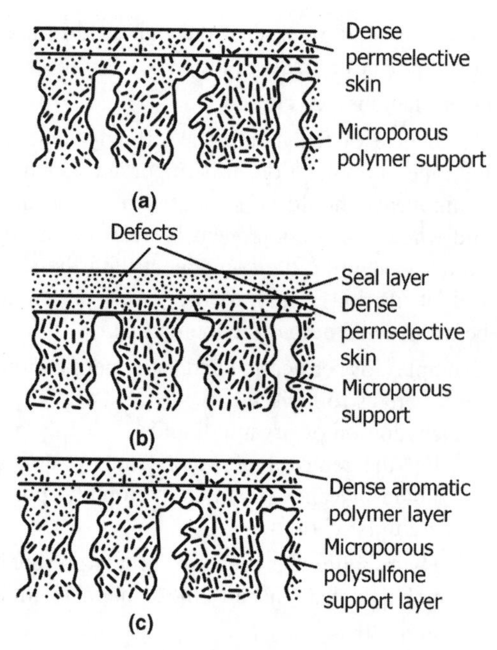

FIGURE 18.47 Asymmetric composite membrane.

- For the removal of an organic liquid from water/organic or organic/air mixture hydrophobic polymers are the most suitable polymers as membrane materials.
- These polymers possess no groups that show affinity for water.
- Examples of such polymers are polydimethylsiloxane (PDMS), polyethylene (PE), polypropylene (PP), polyvinylidene fluoride (PVFD), polytetrafluoroethylene (PTFE).

➢ Separation of two organic solvents:

- For the mixture of two organic liquids or vapors, again three kinds of mixtures can be distinguished: polar/apolar, polar/polar, and apolar/apolar mixtures.
- For the removal of the polar component from polar/apolar mixture polymers with polar groups should be chosen and for the removal of the apolar component completely apolar polymers are favorable.
- The polar/polar and apolar/apolar mixtures are very difficult to separate, especially when the two components have similar molecular sizes.
- In principle all kinds of polymers can be used for these systems, the separation has to take place on the basis of differences in molecular size and shape, since no specific interaction of one of the two components can take place.

▪ In recent times, ceramic membranes and membranes prepared from conducting polymers have also been used as the selective barriers in pervaporation. Ceramic membranes combine high thermal and chemical stability with very high performance. These can be used in a wide range of applications, including separation of mixtures at acid and alkaline conditions.

• What is a composite membrane?
  ▪ A composite membrane is a membrane having chemically or structurally distinct layers.
• What is a dynamic membrane and what are its characteristics?
  ▪ Dynamic membrane is formed by passing a dilute solution of membrane, forming material over a porous substrate at high pressure. For example, Polyacrylic acid and anhydrous zirconium oxide deposition on porous stainless steel or on ceramic tubes.
  ▪ Give higher yields, have high temperature capabilities and easy to replace if fouled up, by dissolving and depositing a new membrane without dismantling the equipment.
• What is an anion exchange membrane?
  ▪ Anion exchange membrane is a membrane containing fixed cationic charges and mobile anions that can be exchanged with other anions present in an external fluid in contact with the membrane.
• What is a cation exchange membrane?
  ▪ Cation exchange membrane is a membrane containing fixed anionic charges and mobile cations that can be exchanged with other cations present in an external fluid in contact with the membrane.
• Classify membrane materials.
  ▪ Membranes are generally made from natural or synthetic polymers. Natural polymers include rubber, cellulose, and wool. Synthetic membranes are of a long linear chain, branched chain or of a three dimensional, highly cross-linked structure.
  ▪ They are either thermoplastic or thermosetting types. Thermoplastic materials soften at high temperatures and soluble in organic solvents. Thermosetting type do not soften at high temperatures and are almost insoluble in organic solvents.
  ▪ They can be glassy or crystalline. Glassy polymers become rubbery near glass transition temperature and can operate above or below glass transition temperature. Crystalline polymers must operate below their melting temperature.
  ▪ Common polymers include cellulose triacetate, polyisoprene, aromatic polyamide, polycarbonates, polyimides, polystyrene, polysulfone, and Teflon.
  ▪ They can be dense or microporous or asymmetric composites.

- Ceramic- and zeolite-based membranes are multilayer composite structures formed by coating a thin selective ceramic or zeolite layer onto a microporous ceramic support. The membranes are made in tubular form. Extraordinarily high selectivities have been reported for these membranes and their ceramic nature allows operation at high temperatures, so fluxes are high. These are very expensive to make.

- Compare polymeric membranes and inorganic membranes as promising candidates for CO_2 capture.
  - Gas separation membranes can capture CO_2 by separating it from other flue gases.
  - There are several different mechanisms of separation including solution diffusion and molecular sieving. The solution diffusion mechanism involves the gas dissolving into the surface of the membrane and then diffusing through the membrane by mass transfer.
  - Molecular sieving involves physical separation of smaller molecules from larger ones by means of a very fine mesh.
  - Membranes are typically categorized by material type. Polymeric and inorganic are two common types.
  - Polymeric membranes transfer gases by the solution–diffusion mechanism. They are effective, inexpensive and can achieve a large ratio of membrane surface area to separation module volume (i.e., lower capital cost).
  - Hollow fiber polyimides are glassy polymers with a high affinity for carbon dioxide. The inclusion of ring structures along the polymer backbone increases the polymer strength, while the inclusion of fluoride groups as side chains adds to the CO_2 solubility.
  - Polymeric membranes are susceptible to degradation and their performance is low for certain gas flow characteristics.
  - Polymeric membranes have short useful life span. The membranes fail due to compaction, aging, or plasticization. This failure occurs by the chemical interaction of the carbon dioxide, or heavier hydrocarbons with the membrane material itself.
  - Inorganic membranes utilize many separation mechanisms, which allows for some level of optimization of the separation process.
  - In some areas inorganic membranes outperform polymeric membranes, but inorganic membranes are much more expensive and the ratio of membrane surface area to separation module volume is much worse than for polymeric membranes. These can be based on a zeolite matrix, carbon nanotubes, or other inorganic substrates.
  - The potential for inorganic membranes to be appropriate for CO_2 capture is considered more promising than for polymeric membranes.

- What is a bipolar membrane?
  - Bipolar membrane is a synthetic membrane containing two oppositely charged ion exchanging layers in contact with each other.
  - Unlike with membranes used for separation purposes, in a bipolar membrane, nothing should be transported from one side to the other.
  - The desired function is a reaction in the bipolar junction of the membrane where the anion and the cation permeable layers are in direct contact: water is split into hydroxide ions and protons by a disproportionation reaction.
  - The produced hydroxide ion and proton are separated by migration in the respective membrane layer out of the membrane.
  - Unlike water splitting at electrodes during electrolysis, no gases are formed as a side product to this reaction, nor are gases used up.
  - Electrodialysis with bipolar membranes (ED-BPM) can replace electrolysis with water splitting at the electrodes. ED-BPM can be used to produce acids and bases from a neutral salt. It is a membrane reactor process where a reaction and a separation occur in the same unit or even in the same membrane.

18.5.2 Membrane Modules

- What are membrane *modules* and what are their types and constructional features?
  - The membranes are packaged into *modules* designed to enable a large membrane area to be contained in a small volume. Membrane modules additionally offer protection to membranes in handling and operation.
  - *Stirred cell, flat sheet tangential flow* and *spiral wound* membrane modules use flat sheet membrane elements in their construction.
  - Stirred cell modules provide uniform conditions near the membrane surface and are used for small scale and research applications involving ultrafiltration and microfiltration.
  - Design of flat sheet tangential flow membranes is similar to that of plate and frame filter press and are easily assembled and disassembled for cleaning and membrane replacement. They have low packing density and can be used to filter suspended solids. Can handle viscous fluids. Suitable for ultrafiltration and microfiltration.

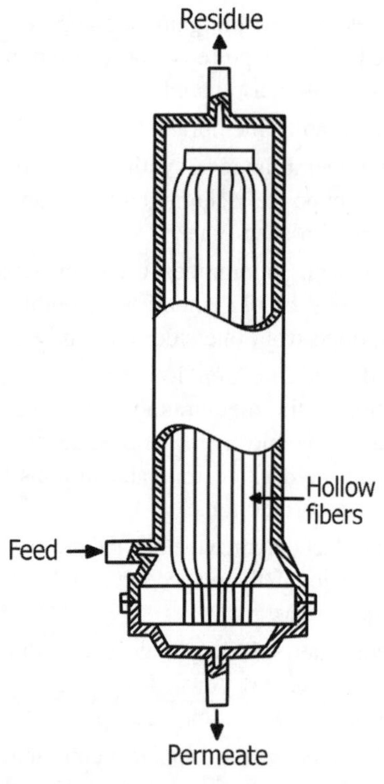

FIGURE 18.48 Hollow fiber module (vertical).

- *Tubular module* uses *several* tubular membrane elements arranged as in a shell and tube heat exchanger.
  - ➢ Feed enters on tube side, permeate passes through tube wall and collected on the shell side.
  - ➢ Retentate is collected on the other end of the tube.
  - ➢ Involves low fouling and cleaning is easy.
  - ➢ Can handle streams with suspended solids and involving viscous fluids.
  - ➢ Used for all types of pressure-driven processes.
  - ➢ Low packing density.
  - ➢ High capital and pumping costs.
- *Hollow fiber module* uses hollow fiber membrane elements.
- The two most successful configurations of modules are spiral-wound and hollow fiber types.
- Hollow fibers are made into bundles, several of which are contained within one housing, to form a single module or permeator. Similar in design to tubular membranes.
- Figures 18.48–18.50 show typical hollow fiber membrane modules.
  - ➢ A typical hollow fiber module contains 2.3 million individual fibers enclosed in a 20 cm diameter, 1.5 m long pressure vessel. They offer a very high membrane area at a relatively low cost.
  - ➢ One of the major operational problems involved with the hollow fiber membrane modules is fouling by suspended matter in the feed solution.
  - ➢ Advantages include very high packing density, that is, large surface area per unit volume, low pumping costs, two way filtration, that is, either *inside out or outside in* and ability to achieve high concentrations in the retentate.
  - ➢ Disadvantages are unsuitability for streams with suspended solids (fouling problems), fragility of the fibers, and high cost.
  - ➢ Two types of hollow fiber modules are used for gas separation, reverse osmosis, and ultrafiltration applications.
    - Shell side feed modules are generally used for high-pressure applications up to 7 MPa (1000 psig). Fouling on the feed side of the membrane can be a problem with this design and pretreatment of the feed stream to remove particulates is required.
    - Bore side feed modules are generally used for medium pressure feed reams up to 1 MPa (150 psig), where good flow control to minimize fouling and concentration polarization on the feed side of the membrane is desired.
- Spiral wound modules are the more widely used configuration, mainly because they are less susceptible to fouling. Cannot handle suspended solids.
  - ➢ In these modules, flat sheet membranes are made into long rolls, 100–150 cm wide.
  - ➢ Two to six spiral wound modules are placed in a single pressure vessel.

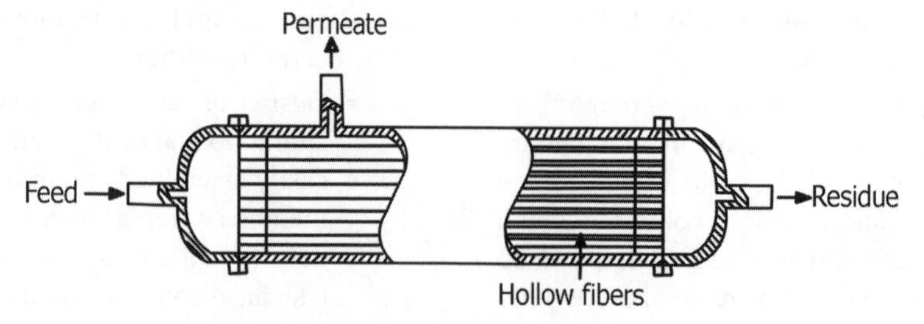

FIGURE 18.49 Hollow fiber module (horizontal).

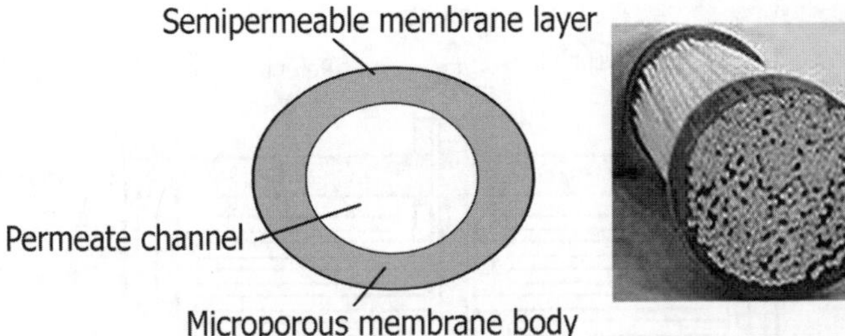

FIGURE 18.50 Hollow fiber membrane module.

  - Have high packing density and used in ultrafiltration and nanofiltration.
- Figure 18.51 illustrates a spiral wound membrane module.
- Plate and frame modules provide good flow control on both the permeate and feed side of the membrane, but the large number of spacer plates and seals lead to high module costs. The feed solution is directed across each plate in series. Permeate enters the membrane envelope and is collected through the central permeate collection channel.
- Figure 18.52 illustrates a plate and frame module.

- What is multichannel membrane? Illustrate with a diagram.
  - Multichannel honeycomb membranes are typically made out of ceramic supports over which ceramic membranes are coated.
  - A typical multichannel ceramic membrane, as illustrated in Figure 18.53, has multiple parallel passageways that run from a feed inlet end face to an outlet end face. The surfaces of the passageways are coated with perm selective membrane.

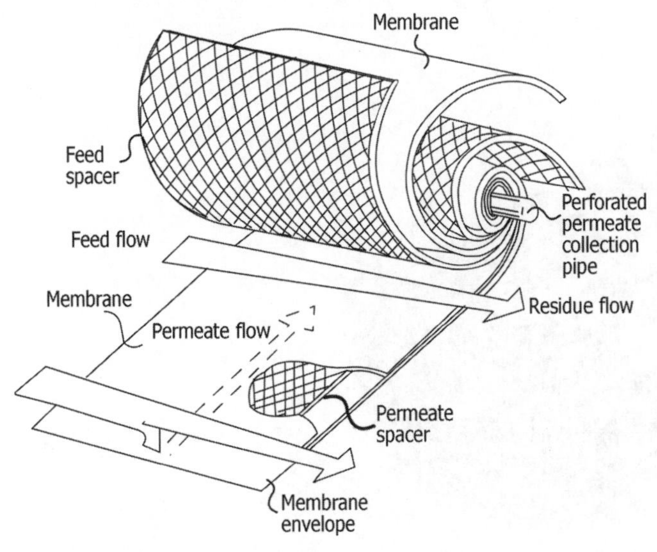

FIGURE 18.51 Schematic of a spiral wound membrane module.

  - A feed stream is introduced under pressure at the inlet end face, flows through the passageways over the membrane and is withdrawn at the downstream end face as retentate.
  - Material that passes through the membrane (permeate) flows into the porous monolith material.
  - Under an applied transmembrane pressure, the combined permeate from all the passageways flows through the porous monolith support to the periphery of the monolith and is removed at the monolith exterior surface.
  - Figure 18.54 is a cut section image of a typical membrane module.
  - There is a limitation to the use of monolith supports as described above. The long and tortuous path through which the permeate must flow to get to the outside of the monolith can lead to a large ΔP for permeate flow within the porous membrane support. The ΔP reduces the transmembrane pressure for the interior passageways. This limitation generally restricts the diameter of monoliths than can be used, at least for microfiltration and ultrafiltration, to approximately 50 mm.

- What are the materials that are commonly used in the manufacture of tubular membranes?
  - Ceramic, carbon, and a variety of thermoplastics are used in the manufacture of tubular membrane modules.
- What are the main criteria in choosing the membrane configuration?
  - Packing density and concentration polarization of each membrane configuration are important criteria for the selection of membrane configuration.
- What is the influence of packing density in the selection of membrane configuration?
  - Packing density is the membrane area per unit volume of membrane element. It is beneficial to pack as much membrane area into as small a volume as possible. The higher the packing density, the greater the membrane area enclosed in a device of a given

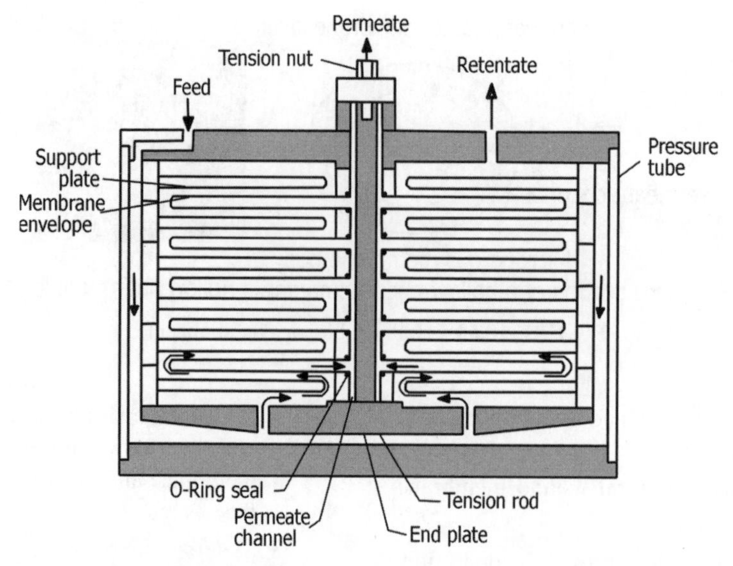

FIGURE 18.52 Plate and frame module.

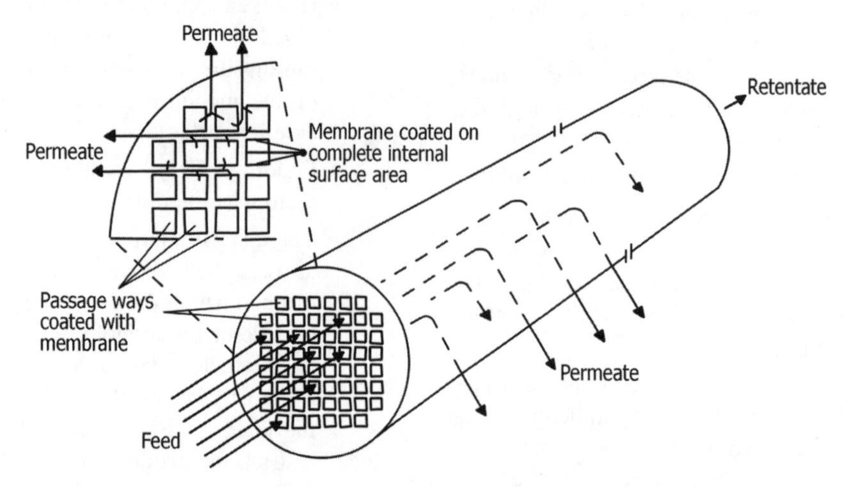

FIGURE 18.53 Schematic diagram of flows inside a multichannel membrane element operating in cross-flow mode.

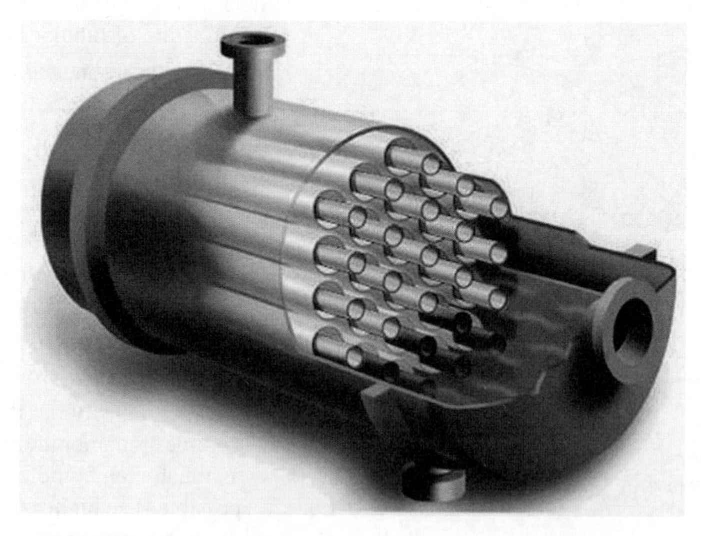

FIGURE 18.54 Cut section view of a typical membrane module.

TABLE 18.18 Comparison of Membrane Element Configurations

Membrane Configuration	Packing Density	Fouling Resistance
Tubular	Low	High
Spiral wound	Medium	Moderate
Capillary filter	Medium	High
Plate and frame	Low	High

volume, and, generally, the lower the cost of the membrane element.

- The disadvantage of membrane elements having high packing density is their greater tendencies for fouling.

- Compare different membrane configurations with respect to packing density and fouling resistance.
  - Table 18.18 gives a comparison of packing density and fouling characteristics for different membrane configurations.
  - Table 18.19 gives the characteristics of different types of membrane modules.

18.5.3 Membrane Separation Processes

- What are the applications of membrane systems?
  - Separation of gaseous mixtures by permeation through nonporous membranes is fairly common in specific industries. Applications include the following:
    - ➢ Hydrogen separation from a wide variety of slower permeating supercritical components such as CO, CH_4, and N_2.
    - ➢ Many ammonia-producing plants are equipped with a membrane system for the recovery and recirculation of hydrogen from purge gas.
    - ➢ Prepurification of natural gas by permeation of acid components like CO_2 or H_2S and dehydration through membranes is another application.
    - ➢ CO_2–CH_4 separation from gas recovery from enhanced oil recovery operations involving CO_2 injection into a production well.
    - ➢ Air is separated by membranes into an oxygen and a nitrogen enriched fraction. Especially the production of enriched nitrogen, with purities between 95% and 99% at small capacities has become a serious competitor for adsorption or cryogenic processes.
    - ➢ Pervaporation is another emerging membrane process using nonporous membranes.
    - ➢ Use of membranes in CO_2 capture from power plants is receiving increased attention as a means of emission control. *Postcombustion* and *precombustion* processes are being developed for CO_2 capture from power plants.
      - *Postcombustion* flue gas streams involve handling large volumes of flue gases, mainly due to the high volumes of N_2 in flue gases, operating at atmospheric pressures and CO_2 concentrations of 10–15 vol%.
      - Separating CO_2 from H_2 in the synthesis gas produced by integrated gasification combined cycle (IGCC) power plants in which higher pressures, smaller volume, CO_2 enriched *precombustion* synthesis gas streams. In these IGCC power plants, coal is first reacted with

TABLE 18.19 Different Types of Membrane Modules and their Characteristics

Hollow Fiber-Capillary	Plate and Frame	Spiral Wound
Very small diameter membranes (<1 mm)	Simple structure and easy membrane replacement	Formed from a plate and frame sheet wrapped around a center collection pipe
Large number of membranes in a module and self-supporting	Similar to filter press	Density is about 300–1000 m^2/m^3
Density about 600–1200 m^2/m^3 (for capillary membrane), up to 30,000 m^2/m^3 (hollow fiber)	Density about 100–400 m^2/m^3	Diameter can be up to 40 cm
Size is smaller than other module for given performance capacity	Membranes placed in a sandwich style with feed sides facing each other	Feed flows axial on cylindrical module and permeate flows into the central pipe
Process *inside out* permeate is collected outside of the membrane	Feed flows from its sides and permeate comes out from the top and bottom of the frame Membranes are held apart by a corrugated spacer	*Features*: • High pressure durability • Compactness • Low permeate ΔP and membrane contamination • Minimum concentration polarization

TABLE 18.20 Examples of Membrane Processes in Separation of Gas Mixtures

Gas Mixture Separation	Application Areas	Membranes
O_2/N_2	Oxygen enrichment, inert gas generation	Ethyl cellulose, silicone rubber
H_2/hydrocarbons	Refinery hydrogen recovery	Polyimide, polyethylene terephthalate, polyamide 6
H_2/N_2	Ammonia purge gas	–
H_2/CO	Synthesis gas ratio adjustment	–
H_2 purification	–	Palladium/silver alloys
CO_2/hydrocarbons	Acid gas treatment, land fill gas upgrading	–
CO_2 from air	–	Silicone rubber
H_2S/hydrocarbons	Sour gas treatment, H_2S from natural gas	Silicon rubber, polyvinylidene fluoride
H_2O/hydrocarbons	Natural gas dehydration	–
He/hydrocarbons	Helium separation from natural gas	–
He/N_2	Helium recovery	–
Hydrocarbons/air	Hydrocarbon recovery, air pollution control	–
H_2O/air	Air dehumidification	–
NH_3 from synthesis gas	–	Polyethylene terephthalate

O_2 or air in a gasifier and the resulting synthesis gas (mainly CO and H_2) is then fired in the power plant.

  - Adjustment of H_2/CO ratio in synthesis gas.
  - Recovery of helium from natural gas.

- Give examples of separation of gas mixtures by membrane processes and their application areas.
  - Table 18.20 gives examples of membrane processes used in separation of gas mixtures.
- What are the applications for metal membranes in gas separations?
  - Metal membranes are used for high temperature membrane reactor applications and for the preparation of pure hydrogen for fuel cells.
  - The metal membrane must be operated at high temperatures (>300°C) to obtain useful permeation rates and to prevent embrittlement and cracking of the metal by sorbed hydrogen.
  - Hydrogen permeable metal membranes are extraordinarily selective, being extremely permeable to hydrogen but essentially impermeable to all other gases. The gas transport mechanism is the key to this high selectivity.
- What are the applications of polymeric membrane systems in water treatment?
  - Drinking water treatment.
    - Treatment of municipal water, surface water, and well water for drinking.
  - Pretreatment and post treatment of raw water including seawater, surface water, well water, and sewage water.
  - Treatment of water for food and beverage industry.
  - Recycle use of condensation water.
  - Producing pyrogen-free water for pharmaceutical industry.
  - Deep treatments of wastewater for recycle.
  - Turbidity removal for wine.
  - Sterilization and turbidity removal for fruit wine, grape wine, and yellow wine.
  - Dialysis and blood filtration in medical treatment.
  - Membrane systems for wastewater treatment in an activated sludge process directly submerging hollow fiber membrane modules in an aeration tank.
- What is concentration polarization?
  - In membrane separation processes, a gas or liquid mixture contacts the feed side of the membrane and a permeate enriched in one of the components of the mixture is withdrawn from the downstream side of the membrane. Because the feed mixture components permeate at different rates, concentration gradients form in the fluids on both sides of the membrane. The phenomenon is called *concentration polarization.*
  - Concentration polarization, in simple terms, is the accumulation of rejected particles to an extent that transport to the membrane surface becomes limited. It reduces the permeability of the solvent and can lead to a limiting flux and a higher fouling tendency.
- Give an effective way to reduce concentration polarization in a membrane process.
  - One way to reduce concentration polarization is to increase turbulence at the membrane surface that reduces boundary layer thickness.
  - The most direct technique to promote mixing is to increase the fluid flow velocity past the membrane surface. Therefore, most membrane modules operate at relatively high feed fluid velocities.

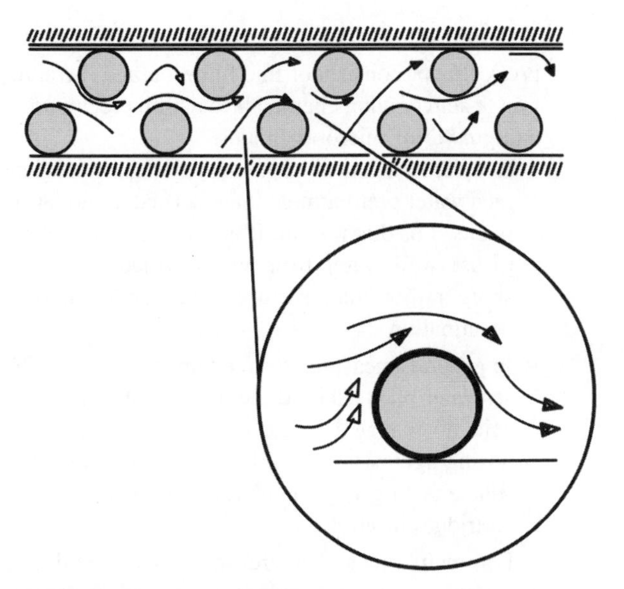

FIGURE 18.55 Flow disruption around spacer netting to promote turbulence.

- Membrane spacers are also widely used to promote turbulence by disrupting fluid flow in the module channels, as shown in Figure 18.55.
- Pulsing the feed fluid flow through the membrane module is another technique.
- However, the energy consumption of the pumps required and the pressure drops produced place a practical limit to the turbulence that can be obtained in a membrane module.

• What are the industrially established membrane processes? Compare pore sizes of the membranes used in these processes.
  - Reverse osmosis.
  - Microfiltration.
  - Ultrafiltration.
  - Dialysis.
  - Electrodialysis.
  - Reverse osmosis, ultrafiltration, microfiltration, and conventional filtration are related processes differing mainly in the average pore diameter of the membrane.
  - Reverse osmosis membranes are so dense that discrete pores do not exist and transport occurs via statistically distributed free volume areas.
  - The relative size of different solutes removed by each class of membrane process is illustrated in Figure 18.56.

18.5.3.1 Reverse Osmosis

• What is osmosis and reverse osmosis (RO)? Explain and illustrate.
  - If a pure solvent is placed on one side of a permeable membrane and a solution of the solvent + solute on the other side, the pure solvent will flow through the membrane and dilute the solution until the difference in heads of liquids on the two sides of the membrane is equal. This is called *osmotic pressure.*
  - Application of a pressure, more than osmotic pressure on the reverse side, reverses flow of the solvent. This process is called *reverse osmosis.*
  - Thus, reverse osmosis is a pressure-driven process that separates dissolved ions from water. Impurities are left behind (particle sizes retained: 0.0001–0.001 μm). For example, desalination, water purification, and so on.
  - Chemical potential of a mobile solvent molecule such as water is increased sufficiently to overcome *osmotic* factors opposing its separation from a solution.
  - The salt ions and other contaminants are rejected by the RO membrane, while the purified water is forced through the membrane by pressure. The permeate contains a very low concentration of dissolved solids. There is a positive correlation between the pressure applied to the membrane and the flow rate of the permeate and an inverse proportional relationship between the inlet feed concentration and permeate flow rate.

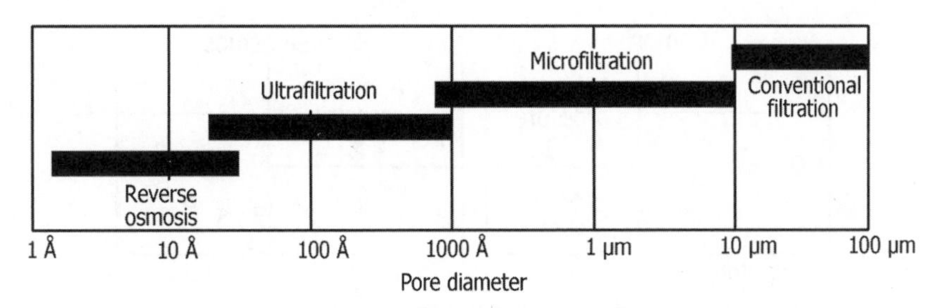

FIGURE 18.56 Range of pore diameters used in reverse osmosis, ultrafiltration, microfiltration, and conventional filtration. (*Note*: 1 Å = 10^{-4} μm.)

- To run a reverse osmosis process, applied operating pressure has to compensate *excessively* the osmotic pressure of the system that is a consequence of the different concentrations on both sides of the membrane.
- Operating pressures range from 1 to 10 MPa. High-pressure reverse osmosis shall be defined by an operating pressure of 10 MPa or higher.
- Figure 18.57 shows two compartments separated by a semipermeable RO membrane that allows water and some dissolved salts to pass through while retaining majority of dissolved solids. The concentration of dissolved solids in one comportment is higher than their concentration in the other.
- Osmosis is a natural process where water passes through the membrane from the low concentration compartment to high concentration comportment until the system has reached equilibrium until the concentration of dissolved solids is the same in both the compartments. At this equilibrium point, level of the solution is higher than its level in the other comportment. This difference corresponds to the osmotic pressure of the solution. The higher the concentration of the dissolved solids, the higher is the osmotic pressure.
- Reverse osmosis occurs when pressure exceeding osmotic pressure is applied to the original high concentration compartment, making the water move in the reverse direction, that is, from high concentration compartment to the low concentration compartment. As more water is removed, the concentration of dissolved solids will increase in the high concentration compartment, making the solution in the low concentration compartment more dilute.

• What are the basic components of an RO plant?
  - The basic components are pretreatment, high pressure pump, membrane assembly and post treatment.
  - Pretreatment is important because the membrane surface must be kept clean. To achieve this, suspended solids must be removed and the water must be pretreated so that salt precipitation and microbial growth does not occur on the membrane. Usually pretreatment consists of fine filtration and the addition of acid or other chemicals to inhibit precipitation and growth of microorganisms.
    - ➢ RO plants are much more sensitive to insufficient feed water pretreatment than thermal desalination plants. For this reason, it is accepted technology to use well water from wells drilled at the sea shore rather than surface water as in thermal desalination.
    - ➢ In general, measures for feed pretreatment of RO seawater plants should include pH adjustment by acid dosing against carbonate scaling, chlorine dosing against marine life in the plant, polyphosphate dosing against calcium sulfate scaling and cartridge filtration.
    - ➢ For membranes that are sensitive to oxidizing agents, active charcoal filtration and/or sodium bisulfite dosing is mandatory if chlorine has been used in the pretreatment.
    - ➢ A flushing and cleaning system is necessary.
  - The high pressure pump supplies the pressure needed to enable water to pass through the membrane and have the salts rejected. This pressure ranges from 15 to 25 bar for brackish water and from 54 to 80 bar for seawater.
  - Membrane assembly consists of a pressure vessel and a membrane that permits the feed water to be pressurized against the membrane. The membrane must be able to withstand the entire pressure drop across it.
  - Two types of membranes are most commonly used. These are cellulose acetate based and polyamide composites.
  - Normally polymers like polyamide and polysulfone composites are used for making RO membranes. Cellulose acetate membranes are the earliest membranes used and newer polymeric membranes are finding application with developments in polymer science.
  - Post treatment might consist of removal of gases and adjusting the pH.

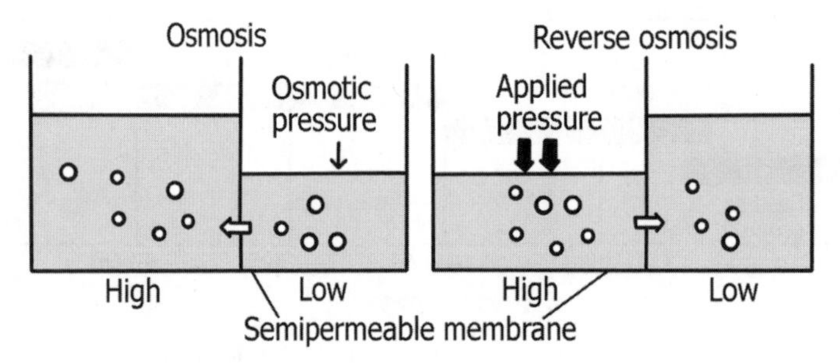

FIGURE 18.57 Osmosis and reverse osmosis.

- The nature of the spacer materials, especially the feed channel spacer material, is also an important aspect.

- What is salt rejection and percent recovery in reverse osmosis?
  - In reverse osmosis, the membrane selectivity is normally measured by a term called *salt rejection*, *R*, which is proportional to the fractional depletion of salt in the product water compared to the feed.
  - Salt rejection is defined as

$$S_r = 100 \times [1-(C_P/C_F)], \tag{18.37}$$

where S_r is the percent salt rejection and C_P and C_F are the salt concentrations in the product and feed waters.

  - Percent recovery = product water flow rate/feed water flow rate.
- What are the contaminants that generally foul or scale RO membranes?
  - Fouling contaminants include colloids, typically silicates, aluminosilicate, or iron silicate, microbes and organics that provide nutrients to sustain microbial populations, color that can foul membranes irreversibly, and metals, which precipitate when oxidized.
  - Aluminum fouling, which results from overfeeding alum to remove turbidity of feed waters is a very common metal fouling problem, especially for membranes operated on municipal water supplies originating from surface waters.
  - Membrane scaling involves precipitation of saturated salts on the membrane surfaces and on the feed water spacers.
- What are the effects of membrane fouling and scaling in RO processes?
  - Pressure drop increases, necessitating increase in pumping pressures.
  - Increased pumping pressures has a limit as higher pressures can damage membranes as well as fiberglass components of the system.
- What are the techniques available to pretreat waters before admitting them into a RO plant?
  - The treatment depends on the nature of contaminants in the feed water that might foul a RO membrane. A complete analysis of the water becomes a prerequisite for RO treatment.
  - For example, iron fouling might require oxidation followed by sand bed filtration, whereas biofouling might require oxidation followed by microfiltration.
  - A typical cleaning method consists of flushing the membrane modules by recirculating the cleaning solution at high speed through the module, followed by a soaking period, followed by a second flush, and so on.
  - The chemical cleaning agents commonly used are acids, alkalis, chelatants, detergents, formulated products, and sterilizers.
  - Acid cleaning agents such as hydrochloric, phosphoric, or citric acids effectively remove common scaling compounds. With cellulose acetate membranes, the pH of the solution should not go below 2.0 or else hydrolysis of the membrane will occur. Oxalic acid, typically a 0.2% solution, is particularly effective for removing iron deposits.
  - Acids such as citric acid are not very effective with calcium, magnesium, or barium sulfate scale. In this case a chelatant such as ethylenediaminetetraacetic acid (EDTA) may be used.
  - To remove bacteria, silt or precipitates from the membrane, alkalis combined with surfactant cleaners are often used. Laundry detergents containing enzyme additives are useful for removing biofoulants and some organic foulants.
  - Most large membrane module producers now distribute formulated products, which are a mixture of cleaning compounds. These products are designed for various common feed waters and often provide a better solution to membrane cleaning than devising a cleaning solution for a specific feed.
  - Sterilization of a membrane system is also required to control bacterial growth. For cellulose acetate membranes, chlorination of the feed water is sufficient to control bacteria.
  - Feed water to polyamide or interfacial composite membranes need not be sterile, because these membranes are usually fairly resistant to biological attack.
  - Periodic shock disinfection using formaldehyde, peroxide, or peracetic acid solutions as part of a regular cleaning schedule is usually enough to prevent biofouling.
  - Repeated cleaning gradually degrades reverse osmosis membranes.
  - Well-designed and maintained plants with good feed water pretreatment can usually expect membrane lifetimes of 3 years, and lifetimes of 5 years or more are not unusual.
  - As membranes approach the end of their useful life, the water flux will normally have dropped by at least 20%, and the salt rejection will have begun to fall. At this point the membrane may be treated with a dilute polymer solution. This surface treatment plugs microdefects and restores salt rejection. Typical

TABLE 18.21 Foulants and Pretreatment Techniques

Foulant	Pretreatment Technique
Silt, clay	Clarification, multimedia filtration
Bacteria	Chlorination, microfiltration
Organics	Clarification, ultrafiltration
Color	Clarification, nanofiltration

polymers are poly(vinyl alcohol)/vinyl acetate copolymers or poly(vinyl methyl ether).

- Table 18.21 lists some of the potential foulants and the pretreatment techniques.
- Chemical dosing is also required for scale control and for dechlorination, if the raw is chlorinated. Typical chemicals are as follows:
  - Sulfuric acid for pH control and calcium carbonate scale prevention.
  - Antiscalants/scale inhibitors for prevention of calcium carbonate and sulfate scale.
  - Sodium bisulfite for dechlorination and inhibition of fouling.

- How does degradation of RO membranes and membrane modules occur?
  - Degradation of RO membranes and membrane modules can take place due to oxidation and hydrolysis of membranes and delamination of the spiral-wound membrane module itself.
  - Oxidation and hydrolysis involve damage to the membrane polymer on a microscopic level. Cleaving of hydrogen bonds between polymer chains, as well as cleaving of polymer chains themselves are the mechanisms by which membranes can degrade.
  - Delamination of membrane modules can occur when membranes are under high stress due to increased pressure drops resulting from fouling and scaling of membranes. Temperature effects also have a deteriorating effect on membranes and their modules.
- What are the applications of reverse osmosis?
  - Table 18.22 gives applications of reverse osmosis.
- What are the considerations involved for the applications of high-pressure reverse osmosis? Give examples of applications favoring higher pressures.
  - If reverse osmosis is used as an alternative process in the field of evaporation, high concentrations are often required, for example, in wastewater treatment. The process must achieve very high water recovery rates (almost 100%), since all disposal methods for the concentrate are very cost intensive.
  - Another example is the concentration of fruit juices with reverse osmosis, which is a very careful and aroma preserving treatment, compared with evaporation. Because of the usually high content of sugar and acids in the juice, a high-pressure process is necessary to remove water from the juice.
  - But even in seawater desalination, the most important field of application for reverse osmosis membranes, a trend is observable to higher operating pressures. Pure water recovery should increase from 40% to 60%. This means that the osmotic pressure of the

TABLE 18.22 Applications of Reverse Osmosis

Industry	Applications
Desalination	Potable water from seawater, bore well water, and brackish water Municipal wastewater treatment
Ultrapure water	Semiconductor manufacturing Pharmaceuticals Medical uses
Utilities and power generation	Boiler feed water, cooling tower blowdown recycle
Domestic	Home RO
Chemical process industries	Process water purification and reuse Effluent disposal and water reuse Separation of water and organics Separation of organic liquid mixtures
Petroleum refining	Solvent separation from vacuum residues and lubricating oils
Metals and metal finishing	Mining effluent treatment Plating rinse water reuse Recovery of metals
Food processing	Dairy processing Concentration of sweeteners Concentration and purification of fruit juices, enzymes, fermentation liquors, and vegetable oils Concentration of wheat starch, citric acid, egg white, milk, coffee, syrups, natural extracts, and flavors Beverage processing Clarification of beer and wine Waste stream processing Recovery of proteins or other solids from distillation residues *Dealcoholization* to produce low alcohol beers, cider, and wines
Biotechnology/medical	Fermentation products recovery and purification
Textiles	Dyeing and finishing Chemical recovery Water reuse
Pulp and paper	Effluent disposal and water reuse
Analytical	Isolation, concentration, and identification of solutes and particles
Hazardous substance removal	Removal of environmental pollutants from surface and ground waters

retentate will increase from 4.5 to 7.0 MPa. Operating the system at higher pressure releases more freshwater out of a pressure vessel and this results in less energy and installation cost.

- Apart from seawater desalination plants (where the recovery with even 60% compared with other applications is not very high), high-pressure reverse osmosis units operate in combination with other reverse osmosis units at a lower pressure level. In a cascaded operation permeate of very high quality can be produced.
- Figure 18.58 shows the flow diagram of a possible wastewater treatment. The retentate of the 60 bar stage is further concentrated in a 200 bar stage, permeates of both stages are treated in a third stage to reach high quality.

• What are the design parameters that affect performance of membrane elements?
  - *Feed Water Composition:* Some RO systems experience fluctuation of feed water composition during operation. This may be due to seasonal fluctuation of feed water salinity or due to intermittent operation of a number of water sources of different salinity. As long as different feed water compositions will not require a change in the system recovery ratio, changing feed water composition will affect only the required feed pressure and permeate water salinity.
  - *Feed Temperature:* Change in feed water temperature results in the change in the rate of diffusion through the membrane.
  - *Feed Pressure:* RO systems equipped with spiral wound membrane elements are designed to operate at a constant flux rate (i.e., to produce a constant permeate flow). Over operating time, the feed pressure is adjusted to compensate for fluctuation of feed water temperature. Salinity and permeate flux decline due to fouling or compaction of the membrane. For the purpose of specifying the high pressure pump, it is usually assumed that specific flux of the membrane will decline by about 20% in 3 years. The pump has to be designed to provide feed pressure corresponding to the initial membrane performance and to compensate for expected flux decline.

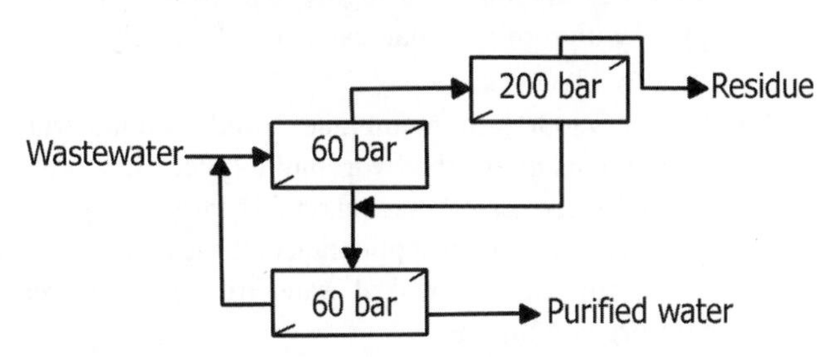

FIGURE 18.58 RO cascade to produce high-quality permeate.

  - *Membrane Compaction:* During operation of RO systems, membrane material is exposed to high pressure of the feed water. Exposure of membranes to high pressure may result in an increase in the density of membrane material (i.e., compaction), which decreases the rate of diffusion of water and dissolved constituents through the membrane. As a result of compaction, higher pressure has to be applied to maintain the design permeate flow. The effect of compaction is more significant in asymmetric cellulose membranes than in composite polyamide membranes. In seawater RO, where the feed pressure is much higher than in brackish applications, the compaction process will be more significant. Higher feed water temperature will also result in a higher compaction rate.
  - *Membrane Fouling:* Membrane fouling has a negative effect on membrane performance, and in extreme cases may result in irreversible membrane degradation.

18.5.3.2 Pervaporation

• What is pervaporation? What is vapor permeation?
  - Pervaporation is a membrane process in which a binary or multicomponent liquid mixture is separated by partial vaporization through a dense nonporous membrane. It differs from all other membrane processes because of the phase change of the permeate.
  - The minor component in a liquid mixture is permeated through the membrane preferentially, holding back the bulk fluid.
  - During pervaporation, the feed mixture is in direct contact with one side of the liophilic membrane whereas the permeate is removed in a vapor state from the opposite side.
  - The driving force for mass transfer of permeants from the feed side to the permeate side of the membrane is a chemical potential gradient, which is established by applying a difference in partial pressures of the permeants across the membrane.
  - The difference in partial pressures can be created either by reducing the total pressure on the permeate side of the membrane by using a vacuum pump system or by sweeping an inert gas on the permeate side of the membrane.
  - While pervaporation is used if the feed to the membrane is liquid, since the minor component or contaminant appears to evaporate through the membrane, vapor permeation involves a mixture of vapors or vapors and gases as feeds.

  - As in pervaporation, the permeate partial pressure is maintained by the use of a vacuum or an inert sweep gas.
  - There is no change of phase involved in its operation. Thus, compared to pervaporation, the addition of heat equivalent to the enthalpy of vaporization is not required in the membrane unit and there is no temperature drop along the membrane.
  - Operation in the vapor phase also eliminates the effect of the concentration polarization prevalent in liquid-phase separations, such as pervaporation.
  - Example of vapor permeation is hydrogen enrichment.
- Figure 18.59 illustrates the process.
- Pervaporation can be used for separation of azeotropic mixtures. The membrane itself acts as the separating agent, preferentially absorbing and diffusing one of the azeotrope forming components.
- A composite membrane is used that is selective for one of the azeotrope constituents. Used to purify chemicals, dehydration of solvents such as ethanol and isopropanol.
- It has promising application to enhance reactor performance, by integrating with the reaction step with purification of feeds or separating reaction products.
- Pervaporation is considered to be expensive both in terms of investment and processing costs (at high throughput). Because of its high selectivity, however, pervaporation is of interest in cases where conventional separation processes either fail or result in a high specific energy consumption or high investment costs.
- The most important category of separation tasks for which pervaporation is promising, is in separation of mixtures with a homoazeotrope or close boiling mixtures.

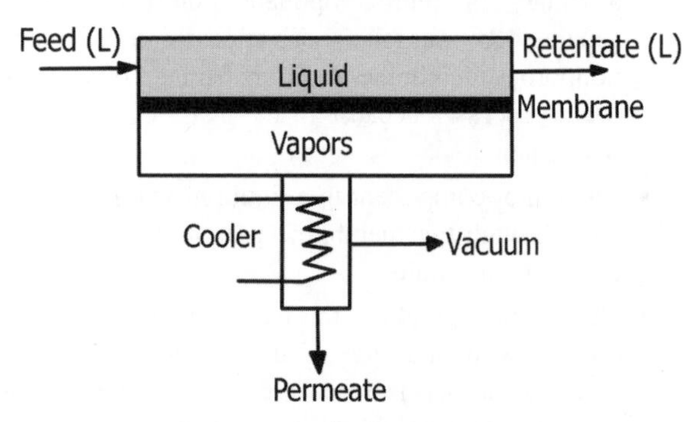

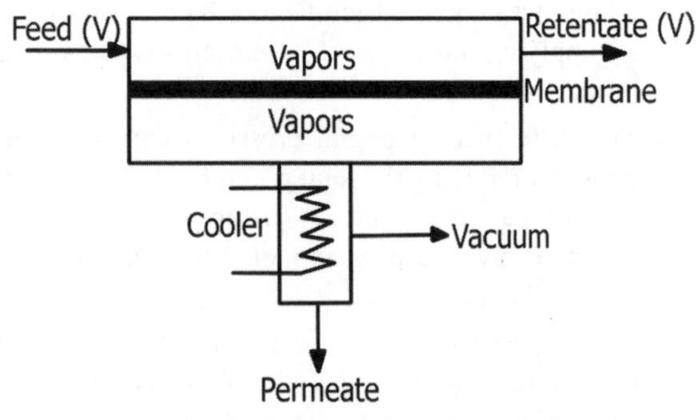

FIGURE 18.59 Pervaporation and vapor permeation processes.

- What is the difference between pervaporation, vapor permeation, and other membrane processes?
  - In pressure-driven membrane processes such as ultra-, micro- and nanofiltration, the bulk component is purified by passing it through a *porous membrane* that *holds back the minor component*. The membrane acts rather like a filter or strainer.
  - Reverse osmosis (RO) is similar, but uses *nonporous membranes*. The major component selectively permeates the membrane by preferential absorption, diffusion, and desorption. The solute or minor component is held back.
  - In pervaporation/vapor permeation processes, the reverse situation, that is, minor component preferentially permeates and *bulk fluid is held back* by the membrane.
  - Filtration and RO processes are used typically to purify water. In contrast, pervaporation/vapor permeation processes are commonly used to *remove water* (dehydrate) from organics. Pervaporation membranes are hydrophilic and therefore employed in dehydration of organics.
- What types of membranes are used in pervaporation processes?
  - Pervaporation membranes are usually of the composite type.
  - In a composite membrane the different fractions, like separation and mechanical stability, are attributed to different layers. Thus, composite membranes can combine very thin and highly selective separation layers with rigid, mechanically, and thermally stable backing layers.
  - The pervaporation membranes used in industrial applications are of the hydrophilic type. They preferentially permeate water but retain nearly all organic molecules. The main application of these membranes is therefore the removal of water from its mixtures with organic liquids.
- Name some applications of pervaporaion processes.

- Pervaporation and vapor permeation are used where distillation is difficult or expensive.
- Dehydrating ethanol–water and isopropanol–water azeotropes from less polar organics.
- Removal of methanol and ethanol from less polar organics. Methanol forms azeotropes with many materials, like esters and cannot be recovered from spent solvents by normal distillation processes. Pervaporation breaks such azeotropes.
- Continuous water removal from condensation reactions like esterification, acetylaton, and kelalization.
  - For example, esterification:

$$R'\text{-COOH} + \text{HO-R} = R'\text{COOR} + H_2O \quad (18.38)$$

- Batch condensation of ethanol and propanol.
- Continuous production of ethyl and propyl esters of low volatility acids.
- Continuous esterification with a heterogeneous catalyst.
- Transesterification using methyl esters.
- Continuous drying of reactor fed streams, replacing molecular sieves.
- Ceramic membranes (e.g., tubular ceramic membranes) allow extremely low water contents to be reached by operating the membrane at a higher temperature (as high as 200°C).
- The treatment of wastewater contaminated with organics.
- Pollution control applications. An example is removal of small quantities of VOCs from contaminated water.
- Recovery of valuable organic compounds from process side streams.
- Harvesting of organic substances from fermented broth.

18.5.3.3 Other Membrane Separation Processes

- Explain membrane-based seawater desalination process.
  - Involves anion exchange and cation exchange membranes.
  - Anion exchange membrane permits passage of anions such as Cl^-.
  - Cation exchange membrane permits passage of cations such as Na^+.
- Can the process mentioned in the above question, be called electrodialysis?
  - Yes.
- How is the equipment arranged?
  - The two types of membranes are alternately arranged in a large number of pairs with a set of electrodes at both ends. Compartments formed by this arrangement are filled with seawater and a DC current is passed between the end electrodes, Cl^- and Na^+ move toward anode and cathode, respectively.
- What is ultrafiltration? What are its applications?
  - *Ultrafiltration:* A type of reverse osmosis in which solute molecules are large. (Particle sizes retained by membranes: 0.001–0.02 μm.)
  - Employs membranes with smaller pore sizes than those used for microfiltration.
  - Most ultrafiltration processes operate in cross-flow mode.
  - The process, which uses pressures up to 1 MPa (145 psi), concentrates suspended solids and solutes with molecular weights greater than 1000.
  - Ultrafiltration is typically used for concentration of solutes by removal of solvent, purification of solvent by removal of solute and fractionation of solutes.
  - Salts, sugars, organic acids, and small peptides permeate the membranes, leaving behind proteins, fats, and polysaccharides.
  - An example of ultrafiltration membrane process is separation of a protein (macrosolute) from an aqueous saline solution. As the water and salts pass through the membrane, the protein is held back. The protein concentration increases and the salts, whose concentration relative to the solvent is unchanged, are depleted relative to the protein. The protein is, therefore, both concentrated and purified by the ultrafiltration.
  - Ultrafiltration membranes are used to separate colloidal materials from fluid mixtures, as well as macrosolutes from true solutions. Recovery of electrocoat paint is a prominent application.
  - Used to produce high-value products such as protein powder (a skim milk replacement and protein products, used as high value food ingredients, using whey from cheese and casein manufacture from milk.
  - Cheese production is one of the major areas for the application of ultrafiltration.
  - Clarification of fruit juices, for example, apple, orange, grape, and pear.
  - Concentration of textile sizing solutions.
  - Concentration of waste oily water into a stream of water suitable for a municipal sewer and an oily concentrate rich enough to support combustion, or for oil recovery.

- Concentration of gelatin from a dilute solution of hydrolyzed collagen.
- Removal of cells and cell debris from fermentation broth.
- Removal of virus from therapeutics.
- Color removal from Kraft black liquor in pulp making.
- Separation of oil–water emulsions from refinery wastes.
- Table 18.23 summarizes some applications of ultrafiltration.

• By means of an example explain how ultrafiltration is different from microfiltration and reverse osmosis.

- Using the same example of a protein solution, water and salt, reverse osmosis membrane will pass only the water, concentrating both salt and protein. The protein is concentrated, but not purified.
- The microfiltration membrane will pass water, salt, and protein. In this case, the protein will be neither concentrated nor purified, unless there is another larger component such as a bacterium is present.
- In ultrafiltration, solute molecules are large and it involves separation of macromolecules. Examples are proteins and polyvinyl alcohol.
- The distinction between microfiltration and ultrafiltration lies in the pore size of the membrane. Microfiltration membranes have larger pores and are used to separate particles in the 0.1–10 μm range, whereas ultrafiltration is generally considered to be limited to membranes with pore diameters from 10 to 1000 Å.

• What are the problems involved in ultrafiltration?

- Plugging of the flow channels by solids present in the feed is one of the problems. Apple juice is a good example of a solid-containing feed solution. The juice coming from a press may contain fairly high fiber content. As the juice is concentrated, the retained solids approach a level at which they do not flow.
- Since high juice yield is economically vital, the system is operated right up to the onset of plugging.
- Tubular membranes are resistant to fibers, dirt, and debris. Tubes are also very easy to clean, which made them a good choice for edible product applications.
- Fouling involves loss of throughput of a membrane because it has become chemically or physically changed by the process fluid.
- Fouled membranes are frequently restorable to their prior condition by cleaning.

• What are the characteristic features of ultrafiltration membranes?

- Ultrafiltration membranes are usually anisotropic structures.
- They have a finely porous surface layer or skin supported on a much more open microporous

TABLE 18.23 Applications of Ultrafiltration

Industry/Material	Applications
Pigments and dispersed dyes	Concentration/purification of organic pigment slurries; separation of solvents from pigment/resin in electropaints; concentration of pigments in printing effluents
Oil-in-water emulsion globules	Concentration of waste oils from metal working/textile scouring; concentration of lanolin/dirt from wool scouring
Metals/nonmetals/oxides/salts	Concentration of silver from photographic wastes; concentration of activated carbon slurries; concentration of inorganic sludges
Dirt, soils, and clays	Retention of particulates and colloids in turbid water supplies; concentration of fines in kaolin processing
Polymer lattices and dispersions	Concentration of emulsion polymers from reactors and washings
Synthetic water-soluble polymers	Concentration of PVA/CMC desize wastes
Polyphenolics	Concentration/purification of lignosulfonates
Microorganisms/plant/animal cellular materials	Retention of microbiological solids in activated sludge processing; concentration of viral/bacterial cell cultures; separation of fermentation products from broth; retention of cell debris in fruit juices; retention of cellular matter in brewery/distillery wastes
Proteins and polypeptides	Concentration/purification of enzymes; concentration of casein; concentration/purification of gluten/rein; concentration/purification of gelatin; concentration/purification of animal blood; retention of haze precursors in clear beverages, honey, and syrups; retention of antigens in antibiotics solutions; concentration/purification of vegetable protein extracts; concentration/purification of egg albumen; concentration/purification of fish protein extracts; retention of proteins in sugar diffusion juice
Polysaccharides and oligosaccharides	Concentration of starch effluents; concentration of pectin extracts
Dairy industry	Concentration of milk prior to manufacture of dairy products; concentration of whey to 20% solids or selectively to remove lactose and salts; concentration/purification of whey proteins; cheese manufacture

substrate. The finely porous surface layer performs the separation and the microporous substrate provides mechanical strength.

- What is polymer-enhanced ultrafiltration? Give a flow diagram illustrating its operation.
  - Polymer-enhanced ultrafiltration (PEUF) is based on the fact that it is easy to retain polymer–metal complexes in an aqueous solution according to their molecular structure, with the help of membrane filtration.
  - Using water-soluble polymeric reagents together with membrane filtration makes it possible to efficiently and selectively separate out inorganic ions.
  - Using water-soluble polymers that bind with specific ions makes it possible to easily filter out these complexes, provided an ultrafiltration membrane with the right pore size is chosen.
  - This method can be used to recover metals from dilute solutions.
  - Advantages include low energy requirements, fast reaction kinetics, aqueous-based processing, and high selectivity for separation if effective bonding agents are applied.
  - Polymer can be recovered and recycled through the complexation–decomplexation cycles. Figure 18.60 illustrates the process.
- What are the advantages and disadvantages of different membranes for application to reverse osmosis and ultrafiltration?
  - Table 18.24 summarizes the advantages and disadvantages of RO and UF membranes.
- What is microfiltration? What are its attributes?
  - Microfiltration involves simple screening mechanism. Used for separating colloidal and suspended micrometer-size particles.
  - Microfiltration is a pressure-driven cross-flow membrane-based separation process in which particles and dissolved macromolecules larger than 1 μm are rejected.
  - Microfilters are typically rated by pore size and, by convention, have pore diameters in the range 0.1–10 μm.
  - A microfiltration membrane is generally porous enough to pass molecules that are in true solution even if they are very large.
  - $\Delta P \approx 0.01$–0.5 MPa.
  - Low-pressure process.
  - Most effectively remove particles and microorganisms (bacteria).
  - High flux.
  - The liquid flows parallel to the membrane at high velocity and under pressure, thereby splitting the feed into two streams, one of which passes through the membrane. The continuous flow of liquid across the membrane performs a cleaning action, whereby fouling is reduced and the concentration on the surface is decreased to ease passage through the membrane.
  - Can be used for clarification of fermentation broth and biomass.
- What are the key characteristics necessary for microfiltration membranes?
  - Pore size uniformity, pore density, and the thinness of the active layer or the layer in which the pores are at their minimum diameter.
- What are the two types of membrane filters commonly used in microfiltration?
  - Screen and depth filters.
  - The screen filter pores are uniform and small and capture the retained particles on the membrane surface.
  - The depth filter pores are almost 5–10 times larger than the screen filter equivalent. A few large particles are captured on the surface of the membrane, but most are captured by adsorption in the membrane interior.
- What are the operating modes for microfiltration membranes and what are their relative merits/applications?

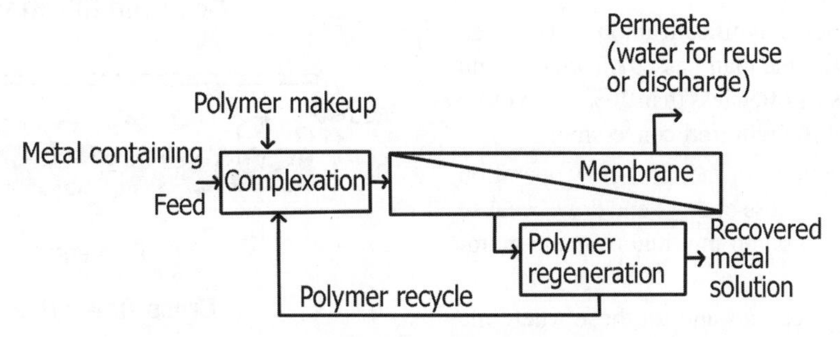

FIGURE 18.60 Polymer-enhanced ultrafiltration flow diagram.

TABLE 18.24 Advantages and Disadvantages of Different Types of Membranes for Applications to Reverse Osmosis and Ultrafiltration

Membrane Type	Advantages	Disadvantages
Cellulose acetate	High permeate flux Good salt rejection Easy to manufacture	Break down at high temperatures Sensitive to pH and can only operate between pH 3 and 6; cause cleaning and sanitation problems
Polymeric membranes	Polyamides have better pH resistance than cellulose acetate; polysulfones have greater temperature resistance, wider pH range, and better resistance to Cl_2 Easy to fabricate Wide range of pore sizes	Do not withstand high pressures and restricted to UF. Polyamides are more sensitive to Cl_2 than cellulose acetate
Composite or ceramic membranes	Inert; very wide range of operating temperatures and pH; resistance to Cl_2 and easy cleanability	Expensive

- Microfiltration membranes can be operated in two ways, as a straight through filter, known as dead-end filtration, or in cross-flow mode.
- In dead-end filtration, *all* of the feed solution is forced through the membrane by an applied pressure. Retained particles are collected *on* or *in* the membrane.
- Dead-end filtration requires only the energy necessary to force the fluid through the filter, which is low as compared to cross-flow filtration.
- In cross-flow microfiltration, the fluid to be filtered is pumped across the membrane parallel to its surface.
- Cross-flow microfiltration produces two streams, namely, a clear filtrate and a retentate containing most of the retained particles in the solution. By maintaining a high velocity across the membrane, the retained material is swept off the membrane surface. Thus, cross-flow is used when significant quantities of material will be retained by the membrane, resulting in plugging and fouling.
- An important difference in the operation of these two schemes is conversion per pass, or the amount of solution that passes through the membrane.
- In dead-end filtration, essentially all of the fluid entering the filter flows out as permeate, so the conversion is roughly 100%, occurring in the first pass.
- For a cross-flow filter, a significant amount of the feed passes *by* the membrane than passes *through* it, and conversion per pass is often less than 20%. Recycling is necessary to get the required conversion.
- The energy requirements of the cross-flow operation are much higher than those of dead-end flow, because energy is required to pump the fluid *over* the membrane *surface*.
- For high solids applications and for those where the solids would normally plug the filter when it is operating as a dead-end filter, cross-flow is the alternative.
- Figure 18.61 illustrates the operation of dead-end and cross-flow filtration systems.
- Membrane filters operating on feeds with medium loadings of solids (<0.5%) are generally operated in dead-end flow. Commonly, the surface of the membrane is protected by a guard filter made of packed glass or asbestos fibers, which entrains most of the larger solids before they reach the membrane.
  - Fluids with low solid loadings (<0.1%) are almost always filtered in dead-end flow, where the membrane acts as an absolute filter as fluid passes directly through it.

- What are the applications of microfiltration?
  - Microfiltration is one of the most widely used membrane process.

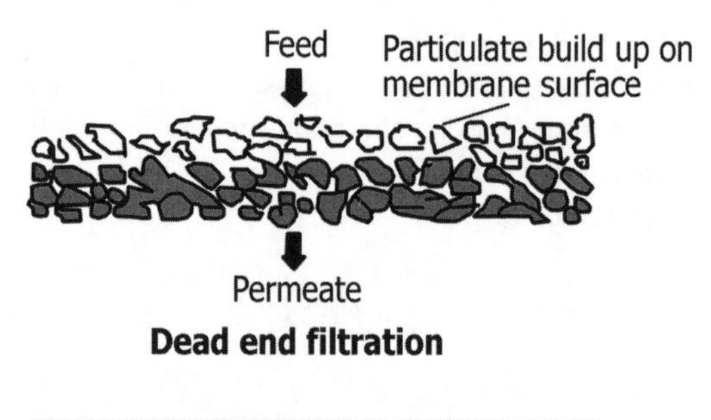

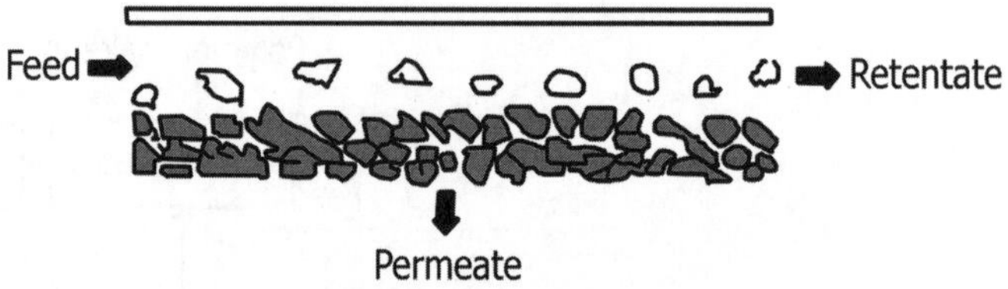

FIGURE 18.61 Dead-end filtration and cross-flow filtration.

- The heart of the microfiltration field is sterile filtration, using microfilters with pores so small that microorganisms cannot pass through them. They may be prepared with pores smaller than 0.3 μm, the diameter of the smallest bacterium, *Pseudomonas diminuta.*
- Microfiltration is used widely in the pharmaceutical industry to produce injectable drug solutions. Microfiltration removes particles but, more importantly, all viable bacteria, so a 0.22 μm-rated filter is usually used.
- Another major application for microfilters is in the electronics industry for the fabrication of semiconductors.
- Dirt particles represent potential short circuits in the semiconductor device. Microfiltration has been applied in filtering the gases and liquids used as reactants in making a chip.
- A particularly attractive application in semiconductor industry is final filtration of the water used to rinse semiconductors during fabrication.
- Most of this water is first treated by a reverse osmosis membrane. Since this is a much finer filter than a microfilter, this water contains only a small amount of dirt from the piping and equipment. Thus, microfilters used in this process have long lifetimes.
- Sterilization of wine and beer is another application for microfiltration.
- Drinking water supplies are filtered through microfilters.

- What is nanofiltration? What are its attributes?
  - Nanofiltration (NF) is a pressure-driven membrane-based separation process in which particles and dissolved lower molecular weight organic solutes and small inorganic ions smaller than about 2 nm are rejected.
  - NF removes molecules in the 0.001 μm range.
  - $\Delta P \approx 0.5$–6 MPa.
  - NF is essentially a lower pressure version of reverse osmosis.
  - NF performance characteristics between reverse osmosis and ultrafiltration.
  - Applications include water softening, removal of organic and coloring matter, desalting of organic reaction products.
- What are the considerations involved in the selection of RO and nanofiltration systems for water treatment?
  - *Membrane Selection:* Two types of membranes are most commonly used. These are cellulose acetate based and polyamide composites.
    - Membrane configurations typically include spiral wound and hollow fiber. Operational conditions and useful life vary depending on the type of membrane selected, quality of feed water, and process operating parameters.
    - Most current manufacturers only have the spiral wound option.
  - *Useful Life of the Membrane:* Membrane replacement and power consumption represent major components in the overall water production costs.
    - The relative contributions depend primarily on feed water salinity. In well-designed and operated RO systems, membranes have lasted 5 to over 10 years in suitable applications.
  - *Pretreatment Requirements:* Acceptable feed water characteristics are dependent on the type of membrane chosen and operational parameters of the system.
    - Without suitable pretreatment or acceptable feed water quality, the membrane may become fouled or scaled and consequently its useful life is shortened.
    - Pretreatment is usually needed for turbidity reduction, iron or manganese removal, stabilization of the water to prevent scale formation, microbial control, chlorine removal (for certain membrane types), and pH adjustment.
    - As a minimum pretreatment, cartridge filters are used for protection of the membranes against particulate matter.
  - *Treatment Efficiency:* Reverse osmosis is highly efficient in removing metallic salts and ions from raw water. Efficiencies do vary depending on the ion being removed and the membrane utilized.
    - For most commonly found ions, removal efficiencies will range from 85% to over 99%.
    - Organics removal is dependent on the molecular weight, shape, and charge of the organic molecule and the characteristics of the membrane utilized. Organic removal efficiencies may range from as high as 99% to less than 50%, depending on the membrane type and treatment objective.
  - *Bypass Water:* Reverse osmosis permeate will be virtually demineralized. If the raw water does not contain unacceptable contaminants, the design may provide for a portion of the raw water to bypass the unit and blend with RO permeate in order to maintain a stable water within the distribution system and improve process economics. Bypass/blend will reduce equipment size and power requirements.

- *Post treatment:* Post treatment typically includes degasification for carbon dioxide (if excessive) and hydrogen sulfide removal (if present), pH and hardness adjustment for corrosion control, and disinfection as a secondary pathogen control and for distribution system protection.
- *Desalting By-Product:* By-product water or the *concentrate* may range from 10% to 60% of the raw water pumped to the reverse osmosis unit.
  - For most brackish waters and ionic contaminant removal applications, the by-product is in the 10–25% range while for seawater it could be as high as 60%. The by-product volume should be evaluated in terms of availability of source water and cost of disposal.
  - Acceptable methods of by-product disposal typically include discharge to municipal sewer system or to waste treatment facilities, discharge to sea, or by deep well injection, depending on the by-product concentration and availability of the discharge option being considered.
- *Pilot Plant Study:* Prior to initiating the design of a reverse osmosis treatment facility, the reviewing agency should be contacted to determine if a pilot plant study will be required. In many cases, a pilot plant study will be necessary to determine the best membrane to use, type of pretreatment as well as post treatment, bypass ratio, amount of reject water, system recovery, process efficiency, and other design and operational parameters.
  Note: The material under the above question has been prepared as an educational tool by the American Membrane Technology Association (AMTA).

- What are the general transport mechanisms used to describe membrane separations of gases?
  - Knudsen diffusion, molecular sieving and solution diffusion.
    - The first type of separation is based on Knudsen diffusion and separation is achieved when the mean free paths of the molecules are large relative to the membrane pore radius. The separation factor from Knudsen diffusion is based on the inverse square root ratio of two molecular weights, assuming the gas mixture consists of only the two types of molecules. The process is limited to systems with large values for the molecular weight ratio, such as is found in H_2 separation. Due to their low selectivities, Knudsen diffusion membranes are not commercially attractive.
    - The molecular sieving mechanism describes the ideal condition for the separation of vapor compounds of different molecular sizes through a porous membrane. Smaller molecules have the highest diffusion rates. This process can happen only with sufficient driving force. In other words, the upstream partial pressure of the *faster* gas should be higher than the downstream partial pressure. The main limitation is that condensable gases cause fouling and alter the structure of the membrane. Therefore, it is only feasible commercially in robust systems, such as those that use ultramicroporous carbon or hollow fiberglass membranes.
    - Solution diffusion separation is based on both solubility and mobility factors. It is the most commonly used model in describing gas transport in nonporous membranes. Gas permeation can be seen as a three-stage process in the solution diffusion model:
      1. Adsorption and dissolution of gas at the polymer membrane interface.
      2. Diffusion of the gas in and through the bulk polymer.
      3. Desorption of gas into the external phase.
  - Figure 18.62 illustrates the three mechanisms.

- What are the applications of gas separation membrane systems?
  - Recovery of hydrogen from purge streams, for example, ammonia plants, retrofit chemical plants, and from hydroprocessors in refineries.
  - CO_2 recovery from wellhead gas in enhanced oil recovery and removal of H_2S and CO_2 from natural gas. Both H_2S and CO_2 permeate through membranes at a much higher rate than methane, enabling a high recovery of the acid gases without significant loss of pressure in the natural gas pipelines.
  - Dehydration of natural gas to bring it to pipeline specifications.
  - Separation of O_2 and N_2 in air.
  - Separation of methane from CO_2 formed during the decomposition of organic matter in land fills.
  - Hollow fiber membrane modules find to be attractive in gas separation processes.

- What is biofiltration?
  - Biofiltration is a general term applied to the removal and oxidation of organic gases (volatile organic compounds, or VOCs) from contaminated air by beds of compost or soil (biofilter media). Billions of indigenous microorganisms inherent within the biofilter media convert the organic compounds to carbon dioxide and water. These naturally occurring microorganisms consume the offending compounds in a safe, moist, oxygen-rich environment.

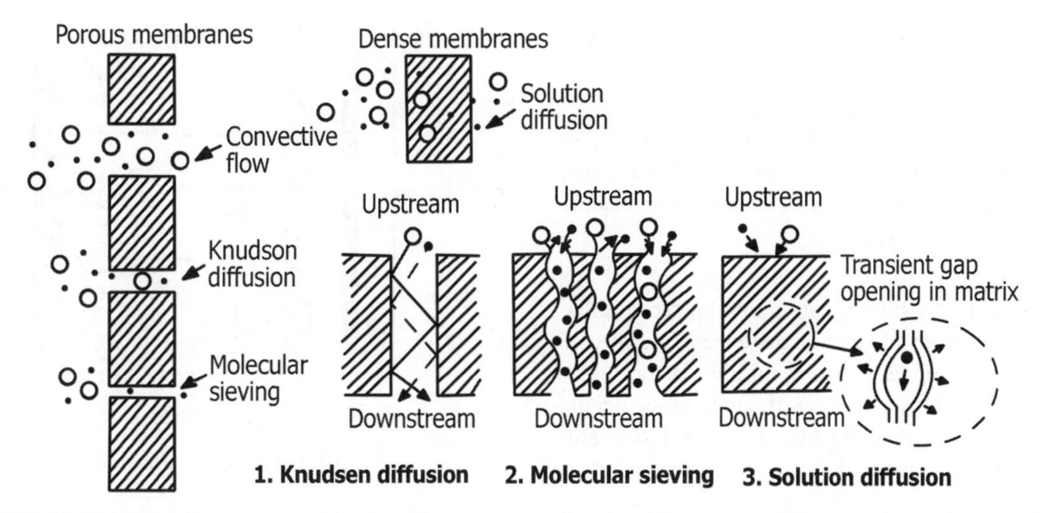

FIGURE 18.62 General transport mechanisms for gas permeation through porous and dense gas separation membranes.

- It is based on fundamental mechanisms of contaminant sorption and biodegradation.
- In biofiltration, off-gases containing biodegradable VOCs and other toxic or odorous compounds are passed through a biologically active bed of peat, soil, composted organic material such as wood or lawn waste, active carbon, ceramic or plastic packing, or other inert or semi-inert media. The media provides a surface for microorganism attachment and growth.
- Contaminant compounds diffuse from the gas phase to the liquid or solid phase in the media bed, transfer to the biofilm layer where microbial growth occurs and subsequently biodegraded.

- What are the mechanisms involved in the transfer of contaminants in the gas phase to the media in biofiltration?
  - Gas stream → adsorption on organic media → desorption/dissolution in aqueous phase → biodegradation.
  - Gas stream → direct adsorption in biofilm → biodegradation.
  - Gas stream → dissolution in aqueous phase → biodegradation.
  - Once the contaminants are adsorbed on the media, they are available to the microorganisms as a food source to support microbial life and growth.
- What are the positive points for biofiltration to be attractive for air pollution control over conventional technologies such as thermal oxidizers, scrubbers, and so on?
  - High removal efficiencies of over 90%.
  - Lower capital and operating costs in applications where air stream contains contaminants at relatively low concentrations and moderate to high flow rates.
  - Low energy requirements.
  - Produce low volume, low toxicity liquid wastes.
- What are the constraints in the use of biofiltration?
  - Does not achieve very high removal efficiencies, for example, over 99%.
  - Varying performance consistencies compared to technologies that do not depend on microorganisms.
  - Lack of application experience.
- What are the important characteristics of materials that are amenable to treatment by biofiltration?
  - Since biofiltration functions via contaminant sorption, dissolution, and biodegradation, contaminants that are amenable to treatment by biofiltration must have two characteristics.
  - *High Water Solubility:* This, coupled with low vapor pressure, results in a low Henry's Law constant and thus increases the rate at which compounds diffuse into the microbial film that develops on the media surface.
  - The classes of compounds that tend to exhibit moderate to high water solubility include inorganics, alcohols, aldehydes, ketones, and some simple aromatics; compounds that are highly oxygenated are generally removed more efficiently than simpler hydrocarbons. However, some biofilter designs have been developed for less water-soluble compounds such as petroleum hydrocarbons or chlorinated hydrocarbons.
  - *Ready Biodegradability:* Once a molecule is adsorbed on the organic material in filter media or in the biofilm layer, the contaminant must then be degraded. Otherwise, the filter bed concentration may increase to levels that are toxic to the microorganisms or detrimental to further mass transfer (sorption and dissolution). Either of these conditions

will result in decreased biofilter efficiency or even complete failure.

- More readily degradable organic components include those with lower molecular weights and those that are more water soluble and polar. Some inorganic compounds such as H_2S and NH_3 can also be oxidized biologically.

- What is dialysis?
  - Dialysis is a diffusion-based separation process that uses a semipermeable membrane to separate species by virtue of their different mobilities in the membrane. Solute transport across the membrane occurs by diffusion driven by the difference in solute chemical potential between the two membrane–solution interfaces.
  - It involves transfer of a solute through a membrane due to transmembrane solute concentration gradient rather than by pressure or electrical potential differences, across the thickness of a membrane.
- What is electrodialysis (ED)?
  - Electrodialysis is an electromembrane process in which ions are transported through ion exchange membranes from one solution to another under the influence of an electrical potential. The nature of the membranes and the driving force distinguish ED from the pressure-driven membrane processes such as gas permeation, reverse osmosis, and filtration. The electrical charges of ions allow them to be driven through solutions and water soaked membranes when a voltage is applied across these media. Since pressures are usually balanced across the membranes, the requirements for strength and support of the membranes and containment vessels are less demanding in electromembrane processes than in pressure-driven membrane processes. Instead, maintaining uniform distribution of solution flow and minimizing electrical resistance and current leakage become important considerations.
  - It is an electrolysis process for separating an aqueous electrolyte feed solution into a concentrate or brine and a diluate or desalted water under the influence of an electromotive force (voltage) applied across the ion-selective membrane.
  - *Electrodialysis* enhances the dialysis process with the aid of an electrical field and ion-selective membranes to separate ionic species from solution. It is used to separate an aqueous electrolyte solution into a concentrated and a dilute solution.
  - Cationic selective membranes allow only the positively charged ions such as copper, zinc, or nickel to pass through them, while conversely, anion-selective membranes allow only the passage of negatively charged ions such as chloride and sulfate, or cyanide complexes, and so on.

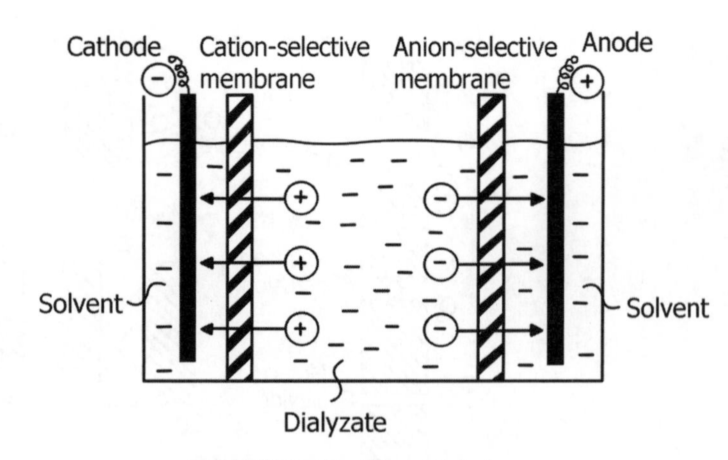

FIGURE 18.63 Electrodialysis.

  - Figure 18.63 illustrates the principle involved in electrodialysis.
  - The electrodes are chemically neutral. When a direct current charge is applied to the cell, the cations are attracted to the cathode (negatively charged) and anions are attracted to the anode (positively charged). Ions in the feed will pass through the appropriate membranes according to their charge, thus separating the ionic species.
- How is the equipment arranged in an electrodialysis process?
  - The membranes are thin sheets of plastic material, which have been subsequently impregnated to impart the appropriate ionic characteristic.
  - These are then, when arranged in parallel cells between two electrodes, positive and negative, along with specifically designed spacers and gaskets to separate the membranes into leak-tight cells, give the basic construction of an electrodialysis stack.
  - There are two general types of commercially available ion exchange membranes, namely, heterogeneous and homogeneous. Both types usually contain a reinforcing fabric to increase tensile strength and improve dimensional stability.
  - Heterogeneous membranes have two distinct polymer phases. Most commercially available membranes are of the homogeneous type with a continuous polymer phase containing ionic groups attached to the polymer chains.
  - An ED stack consists of alternating anion and cation exchange membranes with solution compartments between them. The solution compartments are bounded by perimeter gaskets that are pressed tightly between the membranes to confine the solutions within the compartments. The compartments usually

contain spacers that keep the membranes separated by a constant distance. The spacers also aid in distributing the solution velocity evenly throughout each compartment.

  - Each solution compartment has a means for introduction and removal of solutions.
  - The two types of membranes are alternately arranged in a large number of pairs with a set of electrodes at both ends.
  - Compartments formed by the above arrangement are filled with seawater and a DC current is passed between the end electrodes. Cl^- and Na^+ move toward anode and cathode, respectively.
  - ED differs from other desalting processes in the degree of desalting achieved in a single stage. In evaporation, reverse osmosis, and ion exchange the necessary degree of desalting can often be achieved with one pass through the device. In ED, the degree of desalting is usually limited to about 50% per pass, and some type of staging is needed for further desalting.

- What are the applications of electrodialysis?
  - One important application is desalination of brackish water.
  - A potential application area for ED is in the rough desalting of water that will be subjected to subsequent purification for use as boiler feed or rinse water in the electronics industry. Ion exchange has traditionally been used for preparing waters with low salinity, but the cost of regenerants and the magnitude of the waste disposal problem are proportional to the salinity of the feed water. The bulk of the dissolved solids can be removed more economically by ED or RO.
  - These two processes are competitive in cost, and both offer the advantage that they do not contribute additional water pollutants.
  - Since electrodialysis is suited only for the removal or concentration of ionic species, it is suited to recovery of metals from solution, recovery of ions from organic compounds, recovery of organic compounds from their salts, and so on.
  - Food processing provides many potential applications for ED because of its ability to separate electrolytes from nonelectrolytes. Whey treatment is an important application in food processing industry. Whey is the waste product from cheese making, and it contains useful quantities of proteins, lactose, and lactic acid. However, the high mineral content makes it unacceptable for human consumption and of marginal value as animal feed. Deashing by ED upgrades the whey so that subsequent processing can produce edible whey solids.
  - Metal finishing processes offer many applications for ED in pollution control and material recovery. The rinse streams from such processes pose particularly troublesome pollution problems. They are usually too dilute for direct metal recovery and too concentrated for disposal.
    - The used rinse water flows through the depleting compartments of the ED stack where the metal ions are transferred into the concentrate stream. The treated rinse water can then be reused in the process. The concentrate stream can be recirculated to build up its metal content to a level that is useful for further recovery or direct return to the plating bath.
  - Some of the most successful applications of ED have been on gold plating baths.
  - Other applications include silver plating, nickel plating, and chromium plating.
  - Electrodialysis plays an important role in the effluent treatment package of nickel and chromium plating units.

- Summarize size ranges of materials that can be separated by different membrane processes with examples of applications.
  - Table 18.25 gives summary of size ranges of materials separated by different membrane processes.

- What is facilitated transport?
  - Facilitated transport is a form of extraction carried out in a membrane.
  - It is different from most of the other membrane processes. To illustrate this point, for example, ultrafiltration is an alternative form of filtration, gas separations and pervaporation, depend on diffusion and solubility in thin polymer films.
  - In contrast, facilitated transport involves specific chemical reactions like those in extraction. An example is separation of metal ions with a carrier

TABLE 18.25 Summary of Size Ranges of Materials Separated by Different Membrane Processes

Process	Size Range	Examples
Reverse osmosis	<1 μm	Small particles, large colloids, microbial cells
Ultrafiltration	<0.1 μm to 5 nm	Emulsions, colloids, macromolecules, proteins
Nanofiltration	About 1 nm	Dissolved salts, organics
Electrodialysis	<5 nm	Dissolved salts
Dialysis	<5 nm	Treatment of renal failure

that selectively reacts with one ion and not with the others in the solution.

- Another type of facilitated transport process uses carriers that will selectively react with and transport one component of a gas mixture. One of the most widely studied gas separation facilitated transport processes is the separation of oxygen from air.
- Facilitated membrane diffusion can be much more selective than other forms of membrane transport, but the membranes used in the process are usually unstable. This instability is a tremendous disadvantage, and is the reason why this method is not commercially practiced.
- The high selectivity of facilitated transport comes from the chemical reactions between the diffusing solutes and the mobile carrier.
- Facilitated transport competes with gas absorption and liquid extraction, but not with evaporation. It uses pressure differences or chemical energy, not thermal energy. As a result, facilitated transport membranes will have little effect on the direct use of thermal energy.
- The advantage of facilitated transport membrane processes lies not in their energy consumption, but in their speed. Facilitated membrane processes are much faster than their conventional counterparts. Used in the form of membrane contactors, facilitated transport processes are about 80 times faster than absorption columns of equal volume and 600 times faster than extraction columns of equal volume.

• What are the areas for which facilitated transport is promising?
  - Separation and purification of metals, hydrocarbon separation, dehydration processes, and biochemical applications.

• What is coupled transport?
  - Coupled transport is a membrane process for concentrating ions and separating ions from aqueous solutions. The membrane used in this process consists of a water-insoluble liquid containing an ion complexing agent that is specific for the ion of interest. The desired ion is complexed at one interface of the membrane, forming a neutral-ion complex. The neutral-ion complex then diffuses across the membrane to the opposite interface, where the reaction is reversed by making appropriate changes in the external solution conditions.
  - The reformed complexing agent then diffuses back across the membrane, where it picks up more of the desired ion. Thus, the complexing agent acts as a shuttle to carry ions across the membrane.
  - Proper selection of complexing agent and conditions makes clean separations of metal ions possible.
  - Ions of interest can be concentrated against their concentration gradients.
  - Coupled transport membranes can thus be considered as chemical pumps. The energy for the pumping action derives from the flow of one species down its concentration gradient.
  - Coupled transport processes can be divided into two categories, depending on the type of reaction occurring between complexing agent and permeant. The first type is called countertransport.
  - The key feature of countertransport is that the fluxes of the two permeating ions move counter to each other across the membrane.
  - The second type of coupled transport is cotransport.
  - The key feature of cotransport is that the fluxes of the two permeating ions move in the same direction across the membrane.

• What are the applications of coupled transport?
  - *Separation and concentration of metals* from hydrometallurgical feeds or industrial effluent streams.
    - ➢ In general, the application of a coupled transport process to mining operations involves the installation of a very large plant, and mine operators are reluctant to risk this type of investment in as yet unproven processes. Thus, the commercial applications of coupled transport are likely to be smaller plants installed for pollution control applications.
  - *Recovery of copper* from *in situ* dump and heap leach streams. These streams are produced by the extractions of low-grade copper ores with dilute sulfuric acid.
  - *Cobalt and Nickel Recovery:* Cobalt and nickel are often found in complex ores such as laterites or deep sea nodules. The metals can be extracted from these ores by hydrometallurgy.
  - *Uranium Recovery:* Uranium streams are obtained by acid leaching of uranium ores or as by-product streams from the extraction of other metals.
  - Electroplating rinse waters.
  - *Copper Etchant Baths:* Renovation of circuit board etchant solutions that contain copper-coupled transport permits continuous on-site removal of copper from the etchant solutions and simultaneous regeneration of the etchant solution.

• What are the important medical applications of membrane processes?
  - Hemodialysis involving human kidneys.
  - Controlled drug delivery.
  - Blood oxygenators during surgery.

18.6 OTHER SEPARATION PROCESSES

- What is progressive freezing? How is it employed as a separation process?
  - Slow directional solidification of a melt is known as progressive freezing.
  - Slow solidification is obtained at the bottom or sides of a vessel/tube by indirect cooling.
  - The impurity is rejected into the liquid phase by the advancing solid interface.
  - By repeated solidifications and liquid rejections, a high-purity ingot can be obtained.
  - For example, aluminum is purified by continuous progressive freezing. Other examples are purification of naphthalene, *p*-dichlorobenzene, and so on.
- What is zone melting?
  - Zone melting or zone refining is a group of similar methods of purifying crystals, in which a narrow region of a crystal is molten, and this molten zone is passed through an ingot along the crystal from one end to the other by either a moving heater or by slowly drawing the material to be purified through a stationary heating zone.
  - The molten region melts impure solid at its forward edge and leaves a wake of purer material solidified behind it as it moves through the ingot. The impurities concentrate in the melt, and are moved to one end of the ingot.
  - A variety of heaters can be used for zone melting, with their most important characteristic being the ability to form short molten zones that move slowly and uniformly through the ingot. Induction coils, ring-wound resistance heaters, or gas flames are common methods.
  - Zone melting can be done as a batch process, or it can be done continuously, with fresh impure material being continually added at one end and purer material being removed from the other, with impure zone melt being removed at a rate based on the impurity of the feed stock.
  - Progressive freezing can be viewed as a special case of zone melting, which relies on the distribution of solute between liquid and solid phases to effect separation. One or more liquid zones are passed through the ingot.
  - Another related process is zone remelting, in which two solutes are distributed through a pure metal. This is important in the manufacture of semiconductors, where two solutes of opposite conductivity type are used.
- What are the applications of zone melting?
  - Purification of inorganic or organic materials, for example, semiconductors, solar cells intermetallic compounds, ionic salts, and oxides.
- What is fractional freezing?
  - Fractional freezing is a process used to separate two liquids with different melting points. It can be done by partial melting of a solid, for example in zone refining of silicon or metals, or by partial crystallization of a liquid.
  - Partial crystallization can also be achieved by adding a diluent solvent to the mixture and cooling and concentrating the mixture by evaporating the solvent.
  - Fractional freezing is generally used to produce ultrapure solids, or to concentrate heat-sensitive liquids.
- What is sublimation?
  - Sublimation of an element or compound is a transition from the solid to gas phase with no intermediate liquid stage.
  - It is an endothermic phase transition that occurs at temperatures and pressures below the triple point.
  - At normal pressures, most chemical compounds and elements possess three different states at different temperatures. In these cases, the transition from the solid to the gaseous state requires an intermediate liquid state. However, for some elements or substances at some pressures the material may pass directly from a solid into the gaseous state. This can occur if the atmospheric pressure exerted on the substance is too low to stop the molecules from escaping from the solid state.
- What is the difference between sublimation and quasisublimation?
  - *Sublimation:* Solid is vaporized and then condensed without going through the liquid phase.
  - *Quasisublimation:* Molten solid is vaporized and then condensed directly back to the solid.
- Define sublimation point/temperature.
  - Temperature at which vapor pressure of solid *equals* total pressure of gas phase in contact with it when the solid sublimes.
- Give examples of materials that can be sublimed.
  - Some substances, such as zinc and cadmium, will sublime at low pressures and thus may be a problem encountered in high vacuum applications.
  - Carbon dioxide is a common example of a chemical compound that sublimes at atmospheric pressure. A block of solid CO_2 (dry ice) at room temperature and at atmospheric pressure will change over into gas without becoming a liquid. Iodine is another example of a substance that produces fumes on gentle heating.

In contrast to CO_2, though, it is possible to obtain liquid iodine at atmospheric pressure by controlling the temperature at just above the melting point of iodine. Snow and other water ices also sublime, although more slowly, at below freezing temperatures.

- This phenomenon, used in freeze drying, allows wet cloth to be hung outdoors in freezing weather and retrieved later in a dry state. Naphthalene also sublimes easily. Arsenic can also sublime at high temperatures. Sublimation requires additional energy and is an endothermic change. The enthalpy of sublimation can be calculated as the enthalpy of fusion plus the enthalpy of vaporization.
- Other substances, such as ammonium chloride, appear to sublime because of chemical reactions. When heated, ammonium chloride decomposes into hydrogen chloride and ammonia in a reversible reaction:

$$NH_4Cl \rightarrow HCl + NH_3 \tag{18.39}$$

- Other examples include camphor, sulfur, uranium hexafluoride, zirconium tetrachloride.

• Give examples of commercial sublimation processes.
- Vapor deposition of metals.
- Freeze drying of foods.

• Classify separation processes involving bubbles and foams.
- Table 18.26 classifies separation processes involving bubbles and foams.

• What is foam separation? What is its mechanism?
- Foam separation is based on the adsorption of surfactants at the liquid/air interface and the association of various chemical species and particulates with these surfactants.
- Surfactant adsorption at the liquid/air interface takes place because interaction energy between the nonpolar hydrocarbon chains of the surfactant and the polar water molecules is less than the interaction energy between water molecules themselves and therefore the presence of the organic molecules in bulk water is energetically less favorable than their presence out of the bulk water at the interface.
  - As the size of the nonpolar chain increases, it becomes more and more energetically unfavorable for the chains to stay in the bulk water.
  - An increase in chain length therefore causes an increase in its adsorption at the liquid/air interface and therefore its percent removal by foaming.
  - If the number of polar groups or the number of double and triple bonds are increased, however, the adsorption and consequently the separation can be expected to be poorer.
- It is a technique for partially separating dissolved (or sometimes colloidal) material by adsorption surfaces of bubbles that rise through the solution to form foam that is then received overhead. Material to be removed is called *collagen*. Involves aeration at low flow rates in the absence of any agitation. It is a two-phase adsorptive bubble separation method. Surface-active material collects at bubble surfaces and leave in the foam product.
- Principle of separation is the tendency of surfactant molecules to accumulate at gas–liquid interface and rise with air bubbles. The basis for the separation by bubbles and foam (adsorptive bubble separation) is the difference in the surface activities of the various materials present in the solution or the suspension of interest.
- The material may be cellular or colloidal substances, crystals, minerals, ionic or molecular compounds,

TABLE 18.26 Classification and Principles of Major Bubble Separation Techniques

I. Nonfoaming adsorptive bubble separations
  A. Solvent sublation
  B. Bubble fractionation
II. Foam separations
  A. Foam fractionation (surface-active material removed at gas–solvent interface)
  B. Microgas dispersion (extremely small-sized bubbles)
  C. Froth flotation
    1. Ore flotation
    2. Precipitate flotation (formation *in situ*—the flotation of insoluble precipitates)
    3. Adsorbing colloid flotation (adsorption onto or coprecipitation with a carrier floc that is floated)
    4. Ion flotation (reaction with now surface-active material with surface-active collector surfactant to produce surface-active precipitate thatis then foamed)
    5. Molecular flotation (same principle as ion flotation)

precipitates, proteins, or bacteria, but it must be surface active at the air–liquid interface.

- These surface-active materials tend to attach preferentially to the air–liquid interfaces of the bubbles or foams. As the bubbles or foams rise through the column or pool of liquid, the attached material is removed.
- Created or added phase is gas and gas bubbles act as separating agents.
- In practice, foam separation consists of aeration at a low flow rate of the solution containing the species to be separated and a surfactant if the species are not naturally surface active, and separation of the adsorbed components by simply removing the foam mechanically and breaking it using various chemical, thermal, or mechanical methods.

- What is foam flotation?
  - Foam fractionation when carried out for microscopic size species that are naturally surface active is called foam flotation.
  - Examples of applications involve albumin, hemoglobin, algae, and methylcellulose.
- What is microflotation?
  - Flotation of colloidal size collagens with the aid of surfactants under mild agitation and aeration conditions is called microflotation. This technique has been used for removal of clays and other colloidal matter and microorganisms from wastewater.
- What are the applications of foam fractionation?
  - Used for removal and recovery of minerals, surfactants, separation of detergents from aqueous solutions, alkyl benzene sulfonates, amines, fatty acids, alcohols, surfactants from paper and pulp mill streams, enzymes, proteins, microorganisms, dyes, oils, different wastes, heavy metallic ions, and so on.
- What is bubble fractionation?
  - If an elongated vertical pool of liquid is bubbled, an appreciable vertical concentration gradient may form within it as a result of the carry up of surface-active material on the bubble surfaces.
  - Foaming is not required.
  - The gradient constitutes a partial separation, with stripping at the bottom of the pool where adsorption first occurs and enriching at the top where the bubbles deposit their carry up as they escape.
  - A concentration gradient is established between bottom and top.
  - This process is called bubble fractionation.
- What is solvent sublation?
  - A layer of some suitable immiscible liquid (usually nonaqueous) placed on the top of main liquid (usually aqueous), to trap adsorbed carry up by bubble fractionation.
- What is ion flotation?
  - If collagen is ionic, the technique is called ionic flotation.
  - Examples of applications include separation of cyanides, phosphates, silver, lead, mercury, strontium, and so on.
- What are the modes of operation of foam separation methods?
  - Column operation is either batch or continuous mode. Flow can be simple or countercurrent.
  - For countercurrent systems, there are three major modes of operation: stripping, enriching, and a combination:
  - In the stripping mode, influent liquid is introduced into the foam part of the way up the column and flows countercurrently to the rising foam. The section of the column above the point at which the influent is introduced permits foam drainage. In stripping columns, the objective is to reduce the effluent solute concentration to the lowest feasible level. Sometimes partial recycling of effluent is used to improve solute removals.
  - In enriching columns, some of the collapsed foamate is recycled back to the upper section of the column to increase the solute concentration in the foamate as much as possible.
  - Another option is the use of baffles. The use of baffles in continuous flow foam columns improves removal efficiency and increases maximum hydraulic loading rates by reducing channeling in the foam and other types of axial dispersion.
  - Figure 18.64 illustrates the modes of operation.
- What is froth flotation? What are its applications?
  - Froth flotation is a process for selectively separating hydrophobic materials from hydrophilic ones. This is used in several processing industries. Historically this was first used in the mining industry. Flotation can be performed in rectangular or cylindrical mechanically agitated cells or tanks, flotation columns, and other equipment.
  - Mechanical cells use a large mixer and diffuser mechanism at the bottom of the mixing tank to introduce air and provide mixing action. Flotation columns use air spargers to introduce air at the bottom of a tall column while introducing slurry above. The countercurrent motion of the slurry flowing down and the air flowing up provides mixing action.
  - Mechanical cells generally have a higher throughput rate, but produce material that is of lower quality,

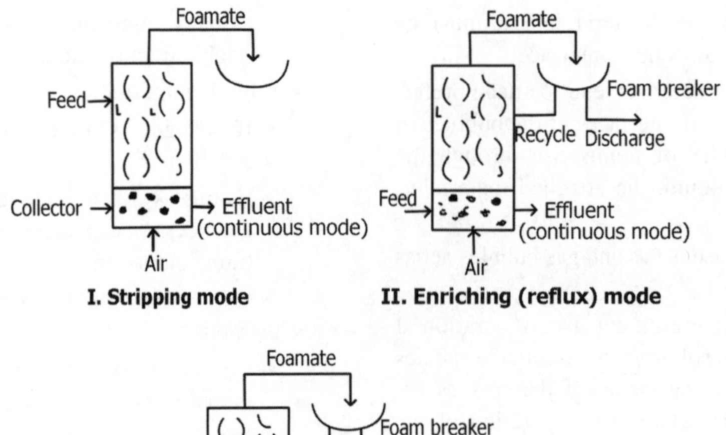

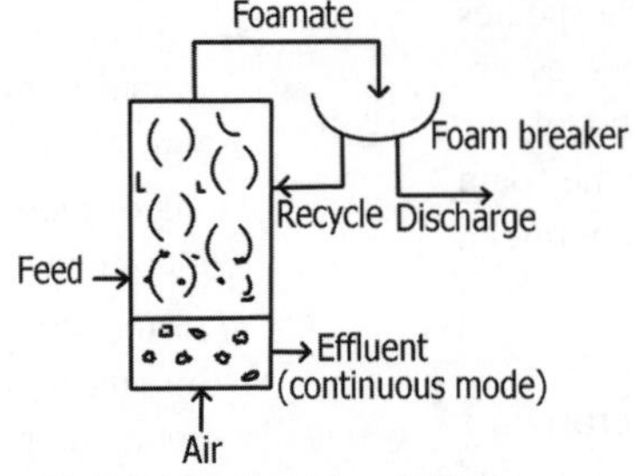

FIGURE 18.64 Illustration of the three operating modes for foam separating columns.

while flotation columns generally have a low-throughput rate but produce higher quality material. Figure 18.65 illustrates a typical froth flotation cell.

- Examples of froth flotation include separation of nonpolar minerals such as sulfur, coal, silica, calcium phosphate, potassium chloride, feldspar, precipitate floatation of ferric hydroxide, waste treatment, and so on.
- Table 18.27 gives some applications of floatation.

• What is parametric pumping?

- Parametric pumping is a dynamic separation technique, comprises alternating axial displacement of

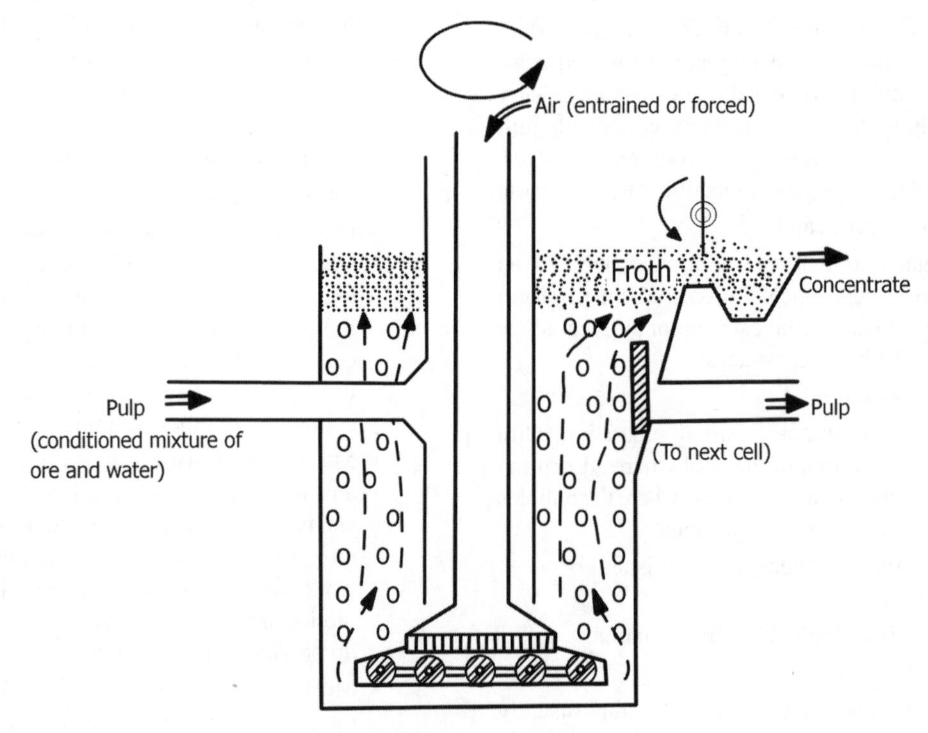

FIGURE 18.65 Froth flotation cell.

TABLE 18.27 Some Applications of Floatation

Environmental Applications Involving Solid–Liquid, Solid–Liquid–Liquid, or Liquid–Liquid Separations

- Effluents generated in the ore flotation plants, thickeners or gravimetric concentration of fines (cyclones, spirals, concentration tables)
- Treatment of organic compounds (solvent extraction plants), oils, fats, and dyes (agates)
- Treatment of effluent with heavy metals (Ag^{+}, Sn^{2+}, As^{3+}, Cr^{3+}/Cr^{6+}, Au^{2+}/Au^{4+}, Be^{2+}, Cd^{2+}, Co^{2+}, Ga^{2+}, Ge^{4+}, Hg^{2+}, Pb^{2+}, Mn^{2+}, Ni^{2+}, Cu^{2+}, Zn^{2+}, Sb^{3+}, Se^{2+}) and anions (CN, CrO_4, S^{2-} AsO_4, SO_4, PO_4, MoO_4, F^{-})
- Water recycle (filters): anions and calcium ions removal
- Treatment of acid mine drainage (AMD) and water reuse

Industrial Processes

- Proteins separation
- Removal of impurities in the sugar cane industry
- Separation of oils, fats, surfactants (soaps), odor removal, and solid wastes in the food industry
- Plastics recycle, pigments, dyes, and fibers
- Paper ink separation, rubber, resins, printer toner pigments
- Emulsified oil removal in the chemical and petrochemical industry
- Thickening of activated sludge
- Reuse (recycle) of industrial waters (PET, washing of vehicles, aeroplanes)

Others

- Removal–separation of microorganisms (algae, fungal, bacteria)
- Metals separation for analytical chemistry
- Treatment of soils: pesticides removal, oils, and radioactive elements
- Treatment of industrial waters in the corrosion control, removal of soaps, detergents
- Treatment of water for industrial and domestic use
- Treatment of sewage (removal of biological flocs, suspended solids)

a fluid mixture in a column of adsorptive particles upon which a synchronous cycling operating parameters such as temperature, pressure, pH, or electric field to which the transfer process is sensitive, is imposed. In other words, adsorption and desorption are induced by cyclic changes in such operating parameter.

- The process fluid is pumped through a particular kind of packed bed in one direction for sometime, then in the reverse direction. Each flow direction is at a different level of the operating parameter to which the transfer process is sensitive.
- Such a periodic and synchronized variation of the flow direction and the operating parameter is known as *parametric pumping*. The basic principle involves the dynamic coupling of unsteady-state mass transfer between two phases with cyclic flow of the mobile phase.
- Separation of heavy metals with variation of pH, treatment of phenolic wastewaters by thermal parametric pumping, electrochemical parametric pumping for desalination of seawater, and separation of proteins and enzymes through pH variation are some examples of promising applications of parametric pumping.

INDEX

Ablative material, 242
Absorber, 527
Absorption, 455, 527
  absorption and stripping factors, 537
  absorption efficiency, 538
  absorption equilibrium diagram, 534, 535
  absorption heat pumps, 375
  absorption of acid gases, 535
  applications of absorption/stripping, 530
  chemical absorption, 527
  design of absorbers, 534
  effective absorption and stripping factors, 538
  gas absorption, 532
  physical absorption, 527
    Edmister's short-cut method, 537, 538
    Horton–Franklin absorption factor equations, 537
    Kremser–Brown–Souders equation, 537
  reactive absorption, 529, 530
  venturi scrubber, 528
Absorptivity, 426, 427
Acentric factor, 547
ACGIH, 217
Acoustic velocity, 24
Activation energy, 671
Active alumina, 673
Active carbons, 673
Active tray area, 478
Activity coefficient, 544, 567
  NRTL correlation, 544
  UNIFAC model, 544
  UNIQUAC method, 544
  Wilson correlation, 544
Adiabatic cooling, 620
Adiabatic power, 157
Adiabatic process, 158
Adiabatic saturation temperature, 630, 633
Adsorption, 455, 671
  adsorbate, 671
  adsorbents, 672, 673, 675, 679
    metal oxide adsorbents, 674
  applications, 672
  breakthrough curves, 682
  chemisorption, 671, 682
  co-adsorption, 672
  equilibria, 680
  heat of adsorption, 671, 672
    differential heat of adsorption, 672
    integral heat of adsorption, 672
  hysterisis, 681
  isotherms, 680, 681
    BET equation, 682
    Freundlich Isotherm, 681
    Henry's law (linear) isotherm, 680
    IUPA classification, 681
    Langmuir isotherm, 681
    linear isotherms, 682
  micropore adsorption, 682
  monolayer adsorption, 671
  physical adsorption, 671, 682
  pressure swing adsorption, 676
  rate of adsorption, 671
  thermal swing adsorption, 676
  types of pores, 673
    macropore, 673
    mesopore, 673
    micropore, 673
    transport pore, 673

Fluid Mechanics, Heat Transfer, and Mass Transfer: Chemical Engineering Practice, By K. S. N. Raju

Aerators, 180
Aerosols, 469
Agglomerates, 618
Agglomeration, 617, 618
Agitation, 163, 164, 168. *See also* Mixers
  agitated vessel, 169, 528
  agitation/mixing intensity, 167
  agitation intensity number, 167
  bubble agitation 257
Air leakage, vacuum systems, 146, 151
Air preheater, 428
Albedo, 425
Alfa Laval, 403, 411, 415
  Alfa Laval disc plate, 411
Allowable pressure drop, 309
Allowable radiant heat flux, 437
Ammonia absorption refrigeration, 373
Analogies, 464
  Chilton–Colburn analogy, 465
  Colburn analogy, 247, 248
  Reynolds analogy, 248, 465
  Von Karman analogy, 465
Angle of wettability, 494, 495
Annular flow, 196, 198
Annulus, 24, 25
Antoine equation, 544–546
API, 600
API gravity, 4
Apparent fouling factor, 334
Apparent thermal conductivity, 234
Apparent viscosity, 5
Approach, 637
Aquifers, 200
Aspect ratio, 407
ASTM, 549
Atomization/Atomizers, 663
  centrifugal/rotary disc atomizer, 663
  nozzle atomizers, 663
    spray nozzles, 663
  pressure nozzle atomizer, 663
  rotary disc atomizers, 663, 664
  two-fluid nozzle atomizer, 663
  ultrasonic atomizers, 663
Attrition index, 208
Auxiliary equipment, 331, 368
Azeotrope, 563
  heterogeneous azeotrope, 563
  homogeneous azeotrope, 563, 565
  maximum boiling azeotropes, 563
  minimum boiling azetropes, 563

Backmixing, 599
Baffles, 169–172, 174, 287
  baffle cut, 289, 290
  baffle spacing, 291
  disc-and-doughnut baffles, 287, 288
  impingement baffle, 294
  orifice baffles, 288
  rod baffles, 288, 289
  segmental baffles, 287, 288
  side strip baffles, 292
Barometer, 10
  aneroid barograph, 12
  aneroid barometer, 11
  mercury barometer, 11
Barometric condenser, 272, 368, 369
Batch fractionation, 557
Batch mixing, 164
Bed expansion, 207
Bed height, 201
Bellows, 13
BEP, 111
Bernoulli's equation, 26
Bernoulli's principle, 25
Beta ratio, 62
Bingham plastic, 7, 165
Binodal curve, 588
Biodispersants, 644
Biofilms, 643, 644
Biofiltration, 706, 707
Biofouling, 298, 643
Black body, 425
Blasius equation, 36
Blending, 180
Blend time, 167
Blowdown, 268, 269, 642, 644
Blowers, 101, 152, 529
  centrifugal blowers, 152, 529
  roots blower, 143
Blowing, 481
Boiling, 251, 254
  bulk boiling, 257
  convective boiling, 259
  dispersed flow film boiling, 259
  film boiling, 254
  flow boiling, 257, 258
  inverted annular film boiling (IAFB), 258
  local boiling, 257
  natural convection boiling, 255
  nucleate boiling, 254, 255
  pool boiling, 255–257
  slug flow film boiling, 259
  transition boiling, 256
Booster ejector, 374
Boundary layer, 25
  concentration boundary layer, 464
  hydrodynamic boundary layer, 246, 247, 464
  thermal boundary layer, 246, 247
Bound moisture, 645
Bourdon tube, 12, 13
  helical bourdon tube, 12–14
Box type furnaces, 428
Break horse power (BHP), 110
Breakthrough curve, 682, 683
Breeching, 430
Bridge wall, 430
Brownian motion, 472, 486
Bubble columns, 475

Bubble flow, 195–197
Bubble point, 547
Bubbles, 469–471
  bubble collapse, 112, 471, 472
  shapes of bubbles, 471
Bubble separation, 712
  bubble fractionation, 712, 713
  solvent sublation, 712
Burke–Plummer equation, 200
Burn out point, 256
Brownian motion, 472
BWG, 38
BWR equation, 546

Caking, 619, 620
Calcite, 614
Capillary condensation, 682
Capsule device, 14
Carbon molecular sieves, 673
Carburization, 443
Carnot cycle, 375
Case hardening, 649
Catalytic cracking, 205–207
Cathodic depolarization, 644
Cations, 684
Cavitation, 56, 107, 112, 618, 629
  cavitation coefficient, 118
  cavitation corrosion, 115
  cavitation erosion, 114, 115
  cavitation index, 56
  discharge cavitation, 113, 114
  gas bubble cavitation, 113
  suction cavitation, 113, 114
  ultrasonic cavitation, 629
  vapor cavitation, 112
Center-to-center spacing, 428, 429
Centrifugal compressors, 155, 156, 158
  surge, 159, 161
Centrifugal fluidized bed, 203
Centrifugal scrubbers, 528
Ceramic fiber insulation, 238
Channel geometry, 405
Channeling, 494
Chemical cleaning, 304, 305
Chemical potential, 457, 695
Cheresources, 113, 261, 677
Chiller, 271, 372, 374
Choked flow, 33
Choking velocity, 210
Chromatographic unit, 679
Chromatography, 455, 678, 679
  exclusion chromatography, 679
  gas–liquid chromatography, 679
  gel permeation chromatography, 679, 680
  high performance liquid chromatography (HPLC), 679
  ion exchange chromatography, 679, 680
  liquid chromatography, 679
  liquid–liquid chromatography, 679
  liquid–solid chromatography, 679
  paper chromatography, 679
  partition chromatography, 679
  thin layer chromatography, 679, 680
Churchill equation, 36
Churn flow, 196
Cichelli and Bonilla, 257
Class C pipe, 39
Classification of chromatographic systems, 679
Classification of molecules, 545
Cleanliness factor, 336
Close boiling mixtures, 565, 568
CO_2 absorption into NaOH solution, 533
CO_2 and H_2S absorption in amine solutions, 533
CO_2 and H_2S absorption in hot potassium carbonate solutions, 534
Coalescence, 524, 599
  coalescer, 523–525
  coalescing plate packs, 524
Coanda effect, 27
Coating flows, 200
Coefficient of performance (COP), 371
Coiled vessels, 272, 378
Coils, heat transfer in, 379
Colbrook equation, 36
Colburn-type equation, 252, 253
Colburn J_H factor, 247
Colburn plot, 539, 540
Collision breeding, 616
Compact heat exchangers, 395
  air-cooled heat exchangers, 401, 402
  all-welded compact heat exchangers, 397
  asymmetrical plates and channels, 497
    H channel, 406
    HD channel, 407
    HS channel, 407
    L channel, 406
    LD channel, 407
    LS channel, 407
    M channel, 406
    MD channel, 407
    MS channel, 407
  finned plate (plate–fin) heat exchangers, 397
  finned tube (tube-fin) heat exchangers, 397, 398
  plate-fin heat exchangers, 397, 413
  plate heat exchangers, 271, 397, 403
    Alfa disc plate and shell heat exchanger, 408
    aspect ratio, 407
    flow patterns, 407, 408
    hard plates, 408
    soft plates, 408
    thermal length, 405, 406
  printed circuit heat exchangers, 420
  scraped heat exchanger, 271, 415, 416
  spiral heat exchanger, 271, 413–415
  welded plate heat exchanger, 412, 413
Compensated reference temperature systems, 228
Component trapping, 583
Compressibility factor, 5
Compressible fluids, 32
Compression fittings, 40

Compression ratio, 159
Compressors, 372
  axial flow, 155
  centrifugal, 372
  dynamic, 155
  jet, 154
  radial flow, 155
  reciprocating, 154, 372
  rotary, 372
  screw, 372
  sliding vane, 156
  surge, 159–161
  thermal, 156
Computational fluid dynamics, 443
Concentration gradient/driving force, 460
Concentration polarization, 694
Condensate backup, 332
Condensate depression, 333
Condensation, 251
  direct contact condensation, 251
  drop-wise condensation, 251
  effect of noncondensable gases, 334
  film condensation, 251
  freeze condensation, 251, 252
  homogeneous condensation, 251
Condensers, 331
  effect of sub-cooling, 332
  fog formation, 331, 332
  in parallel, 332
  in series, 332
Conduction heat transfer, 232
  composite wall, 234
  flat slab, 235, 236
  Fourier equation, 232
  hollow cylinder, 236
Conformal mapping, 236
Coning, 481
Conjugate heat transfer, 245
Continuity equation, 26
Control valve, 54, 55
Convection, 245
  forced, 245
  natural, 245
Convective heater, 442
Convective heat transfer, 245, 429
Convergence pressure, 546
Convergent–divergent nozzle, 33
Coolers, 271
Cooling electronic devices, 421
Cooling towers, 636
  atmospheric, 637, 638
  forced draft, 637, 639
  induced draft, 637–639
  mechanical draft, 637
  natural draft, 637, 638
    hyperbolic natural draft cooling tower, 642
  sizing of, 643
  spray chambers, 637
  spray ponds, 637
  thermal efficiency of, 642
  water efficiency of, 642
Cooling tower operation, 640
  air pollution, 640
  biofilms, 643
  biofouling, 643
  blowdown, 642
  drift losses, 642, 643
  evaporation losses, 641, 642
  foaming, 642, 643
  fog formation, 640
  icing, 640
  microbiologically induced corrosion (MIC), 644
  scaling, 544
  steam plumes, 640
Cooling tower performance chart/curves, 642, 643
Cooling tower selection, 639, 640
Cooling tower siting, 642
Counter-current flow and co-current flow, 311
Coupled transport, 710
Creeping flow, 28
Critical heat flux, 256, 267
Critical moisture content, 648, 650
Critical radius, 241
Critical solution temperature, 590
Critical temperatures, 596, 597
Cross flow heat exchanger, 312
Cross flow filtration, 704
Cryogenics, 372
Crystal growth, 615, 618, 619
  crystal growth rate, 618
Crystal habit, 617
Crystallization, 543, 613
  crystallization equipment, 621
  crystallization kinetics, 614
  crystallization types, 627
    evaporative crystallization, 622
    layer crystallization, 627, 628
    melt crystallization, 627, 628
    reactive crystallization, 628
    solution crystallization, 628
    sono-crystallization, 628, 629
    suspension crystallization, 627, 628
Crystallization equipment, 621
  circulating magma crystallizer, 622
  column crystallizers, 621
  continuous crystallizers, 622
  cooling crystallizers, 621
  direct contact refrigeration crystallizer, 621, 624
  draft tube baffle evaporation crystallizer, 621, 624
  draft tube crystallizer, 621
  evaporative crystallizers, 621
  fluidized suspension crystallizer (Oslo), 621, 626
  forced circulation crystallizer (Swenson), 621–623
  MSMPR crystallizer, 622
  Oslo evaporative crystallizer, 621, 625
  scraped surface crystallizer, 621
  surface cooled baffled, 625
  surface-cooled forced circulation crystallizer, 621, 625

tank crystallizers, 621
vacuum crystallizers, 622
Crystals, 613, 614
Crystal size, 619
Crystal size distribution (CSD), 614, 629
Crystal yield, 629
Cubic shapes, 618
Cyclonic gas-liquid separators, 522
Cylindrical shells, 428

D'Arcy equation, 35
D'Archy–Weisbach friction factor, 38
Dead end filtration, 704
Dead head speed, 131
Decolorization, 672
Dehumidification, 472, 635
cooling-based dehumidification, 635
desiccant-based dehumidification, 635
refrigeration-based dehumidification, 635
Dehydration, 668, 685
Δ-Point, 556, 557, 592
Demisters, 522–524, 526
Deposition velocity, 218
DePriester charts, 546
Desalination, 698, 699
Desiccant, 674
deliquescent desiccants, 674
desiccant dehumidifiers, 635
Design of adsorbers, 678
Desorption, 672
Dew point, 547, 630
Dialysis, 684, 695, 708
Diaphragms, 13
Differential heat of adsorption, 672
Differential shock, 30
Diffuser for leaching process, 611
Diffusion, 455–459
capillary diffusion, 648
coefficients, 460
diffusion in porous solids, 456
eddy diffusion, 458
equimolar counter diffusion, 459
forced diffusion, 456
Knudsen diffusion, 457, 648
liquid diffusion, 648
molecular, 456, 458
pressure diffusion, 456
surface diffusion, 648
thermal diffusion, 456
vapor diffusion, 648, 649
Diffusion coefficients, 460
Diffusional flux, 460
Diffusion membrane, 685
Diffusion-reaction theory, 618
Diffusivities, 456, 459
eddy diffusion, 456, 458
effective diffusivity, 456
mass diffusivity, 456
molecular diffusivity, 456
thermal diffusivity, 456
Dilatant, 6, 165
Dimensional analysis, 3
Buckingham's π-theorem, 3
Rayleigh method, 3
Dimensionless numbers, 3
agitation intensity number, 167
Archemedes number, 4
Biot number, 225, 464
bond number, 225
Cauchy number, 3
capillary number, 3
critical cavitation number, 3, 118
condensation number, 225
Deborah number, 4
drag coefficient, 4, 26
Euler number, 3
Fanning friction factor, 4
flow number, 4
Fourier number, 225
Froude number, 3, 165, 204,
Graetz number, 225, 246, 464
Grashof number, 225, 247
impeller blend number, 167
impeller force number, 167
impeller Reynolds number, 165
Knudsen number, 4
Lewis number, 464
Mach number, 4, 32
Marangoni number, 467
Nusselt number, 225
Peclet number, 225
power number, 165
pumping number, 166
Prandtl number, 225, 246
Rayleigh number, 225, 247, 467
Reynolds number, 3, 464
Schmidt number, 464
Sherwood number, 463, 464
Stanton number, 225, 248, 464
Strouhal number, 29
vapor condensation number, 225
Weber number, 3
Dimpled tube technology, 442
Dimple tube, 443
Discharge coefficient, 67
Dispersion, 600
Dispersion band, 600
Distillation, 376, 455, 542
ASTM distillation, 549
azeotropic distillation, 563, 565, 567
batch distillation/fractionation, 550, 557–559
constant composition, 558
constant reflux ratio, 557
heterogeneous azeotropic distillation, 566
homogeneous azeotropic distillation, 566
carrier distillation, 548
catalytic distillation, 570
constant level batch distillation, 551

Distillation (*Continued*)
continuous distillation/fractionation, 550, 559
differential distillation, 549
distillation processes, 548
enthalpy-concentration diagram, 557
equilibrium flash vaporization, 549, 552
extractive distillation, 566, 567, 569, 571
heteroeneous azeotropic distillation, 566
high pressure distillation, 571
inverted batch distillation, 551
molecular distillation, 574
multicomponent distillation, 559
operating pressure, 571
operational issues, 579
pressure swing distillation, 564
reactive distillation, 564, 570, 571
reactive extractive distillation, 570
semicontinuous distillation, 558, 559
steam distillation, 551, 573
TBP distillation, 549
vacuum distillation, 571, 573
Distillation control, 584
Distributed components, 559
Distribution/partition coefficient, 587, 593
liquid/vapor distributors, 503–506
Dittus–Boelter equation, 249
Divided wall column, 574–578
Dobbins equation, 467
Double pipe exchanger, 272, 273
Double suction, 105
Double tube sheet design, 281
Douglas equation, 538
Downcomer, 480
area, 481
clearance, 481
residence time, 480
width, 481
Draft tube, 177, 178,
Draft tube agitator, 177
Drag, 26
Drift losses, 641, 642
Drivers for pumps, 140
Droplet formation, 471, 472
Droplet flow, 196
Drops, 469, 471
deformation in drops, 471
Drowning, 616
Dryers, 653
agitated batch dryers, 654, 657
applications, 670
batch dryers, 653
classification, 652, 653
contact dryers, 653, 654
conveyor/belt dryers, 655, 668
three stage conveyor dryer, 655
dielectric dryers, 669
direct heat dryers, 656
drum dryers, 654–656, 668
vacuum drum dryers, 668
dryers with jacketed shells, 656
fluid bed dryers, 654, 657, 669
explosions, 670, 671
freeze dryer, 667, 669
indirect heat dryers, 656
kneader dryers, 666
microwave dryer, 668
pneumatic/flash dryers, 654, 658–660, 669
rotary dryers, 656, 657
multitray rotary dryers, 669
rotary cylindrical dryers, 656
vacuum rotary dryers, 668
selection, 652, 653
silo or bin dryer, 660
spouted bed dryer, 658
spray dryers, 528, 654, 660, 662–665, 668
three stage conveyor dryer, 655
through-circulation tray dryers, 654
toroidal dryer, 665
tray dryers, 654
vacuum tray dryers, 654
tunnel shelf dryers, 654
Drying, 455, 644, 645
ball drying, 657
chemical drying, 564
constant rate period, 647, 651
dielectric drying, 667
drying rate curve, 649
drying times, 653
explosive puff drying, 670
falling rate period, 647, 651
first falling rate period, 650
second falling rate period, 650
foam mat drying, 655
freeze drying, 712
gas drying, 669
infrared drying, 669
microwave drying, 667
osmotic drying, 667
superheated solvent drying, 668
spray drying, 472, 661, 662
Dufour effect, 456
Dumping, 482
Duplex, 41
Duplex tube, 282
Dusting segregation, 191
Dynamic compressors, 155
Dynamic or break loading, 674
Dynamic pressure, 10

Economy of evaporators, 364
Eddies, 21
Eddy viscosity, 22
Eduljee equation, 561
Effective stack/plume height, 403
Effect of altitude on cooling tower operation, 641
Effect of shrinkage, 649
Effect of temperature on absorption equilibria, 680

Einstein equation, 459
Ejectors, 145, 146, 147
  booster ejector, 149
  ejector compression ratio, 147
  multistage, 147
  single stage, 147
  steam jet ejectors, 148
  steam jet water chiller, 150
  three stage, 147
Electrical energy transfer, 233
Electrodialysis, 684, 695, 708, 709
Electronically controlled references, 228
Emergency scrubber, 541, 542
Emissivity, 426
  emissive power, 425
Emulsifiers, 179, 180
Enhancement factor, 467, 468
Enrichment factor, 685
Entrainers, 563, 564
Entrainment, 485. *See also* Tray columns
Entrainment separation, 519
  mechanisms, 486
    centrifugation, 486
    diffusion and Brownian motion, 486
    electrostatic precipitation, 486
    inertial impaction, 486
    interception, 486
    sedimentation, 486
    venturi contacting, 486
Entrainment separators, 369
  centrifugal separators, 486
  cyclone separator, 487
  demisters/wire mesh eliminators, 486
  diffusional separators, 486
  electrostatic precipitators, 487
  fiber bed mist eliminators, 486
  inertial separators, 486
  rotary stream separator, 487
  settling chambers, 486
  vane impingement separator, 486
  wet scrubbers, 487
Equations of state, 546
  Peng–Robinson, 546, 547
  Redlich–Kwong, 546, 547
  Soave, 546, 547
Equilateral triangular plot, 588
Equilibrium/*x*–*y* diagram, 548
Equilibrium moisture content, 645, 646, 650
Equilibrium tray, 551
Equivalent cold plane area, 435
Equivalent diameter, 24
Equivalent length, 37
Erbar–Maddox correlation, 561, 562
Ergun equation, 200
Eutectic points, 619
Eutrophication, 462
Evaporation, 347
Evaporation losses, 641, 642
Evaporative cooling, 472, 636
Evaporators, 347
  agitated/wiped film types, 348, 357, 358
  auxiliary equipment, 368, 369
  calendria type, 348, 350
  direct-heated type, 348
  falling film type, 348, 352, 353, 355
    Swenson falling film type, 354
  flash evaporator, 359
  fluidized bed type, 360
  forced circulation type, 356, 357
  long tube vertical (LTV), 348, 351, 353
  mechanical vapor recompression, 358, 359, 368
  multiple effect, 361, 367
    backward feed, 362–364, 368
    forward-feed, 362, 364
    mixed feed, 363, 364
    parallel feed, 364
  performance, 364
  plate type, 348
  rising film type, 348, 352
    Swenson rising film type, 353
  scraped surface, 348
  short tube, 348, 350
    spiral tube, 348, 360
  spray film type, 359
  submerged combustion evaporator, 360, 361
  stirred tank evaporator, 360
  thermal vapor recompression, 368
Evaporator performance, 364
  boiling point rise, 365, 366
  capacity, 364
  Duhring rule, 366
  economy, 364
  hydrostatic head, 366
  foaming, 357
Evaporator selection guide, 349
Ewell, Harrison, and Berg classification, 563
Excess air, 432
  percent excess air, 432
Exchange factors, 434
Explosions in dryers, 665, 670, 671
Expansion joints, 281
  shell expansion joint, 281
Extract, 591
Extraction calculations, 590
Extraction equipment, 598, 599
  batch solid-liquid extractor, 609
  Bollman extractor, 610
  centrifugal extractors, 604
  diffusers, 611
  Hildebrand extractor, 610
  horizontal solid–liquid extractor, 608, 609
  Karr extraction column, 603
  Kuhni column, 607
  mixer–settler units, 594, 599, 600
  Oldshue–Rushton column, 607
  Pachuka tank, 610, 611
  packed columns, 602, 605
  percolator type extractor, 611

Extraction equipment (*Continued*)
Podbelniak extractor, 606
pulse columns, 603, 605
reciprocating tray column, 605
rotary agitated columns
rotating disc column (RDC), 604, 606
Rotocel extractor, 610
Scheibel extraction column, 604
sieve tray columns, 602
spray columns, 602, 605
thickeners, 609
vertical solid- liquid extractor, 609
Extraction factor, 587
Extraction solvents, 592
Extruders, 183

Facilitated transport, 709, 710
Fanning equation, 35
Fans, 152, 529
axial flow type, 153, 529
centrifugal type, 153
lobe type, 154
Feed heaters, 583
Fenske equation, 560
FHP, 110
Fiber saturation point, 649
Fick's law, 458
Fick's second law, 458
Finned surfaces, 397
fin effectiveness, 401
fin efficiency, 401
finned tubes, 398
helically wound spiral or crimped fin tubes, 400
low and high finned tubes, 399, 400
wire wound fin tubes, 401
types of fins, 398
continuous circular fins, 400
double L-footed fins, 399
embedded fin, 399
extended axial fins, 399, 400
extruded fin, 399
L-footed fins, 398, 399
longitudinal fins, 401
serrated fins, 399, 400
transverse fins, 400, 401
Fired heaters, 427
bridge wall, 430
burners, 443
flame impingement radiant burners, 445
perforated ceramic burners, 445
porous refractory radiant burners, 445
premix gas burner, 443, 444
radiant panel burners, 443, 444
radiant tube burners, 445
radiant wall burners, 445
wire mesh burners, 445
cabin heater, 431
convection section, 428, 429
design, 434
flame impingement, 439
flares, 448
flue gas recirculation, 441
furnace tubes, 442
incinerators, 450
NO_x control, 446
operational issues, 438
radiant efficiency, 431
radiant section, 428, 429
rotary kilns, 451
shield section/tubes, 429
tube arrangements, 428
tube failure, 440
Fixed bed adsorbers, 678
Fixed tube sheet, 275
Flange joints, 40
Flange ratings, 40
Flare fittings, 40
Flashing flow, 197
Flat plate solar collectors, 427
Flooding, 488. *See also* Tray columns
downcomer backup flooding, 488, 490
downcomer choke flooding, 488
froth entrainment flooding, 488
hydraulic flood, 488
jet flooding, 489
operational flood, 488
spray entrainment flooding, 488
system flood/vapor flood, 488
Flow measurement, 59
Flow meters, 59, 60
anemometer, 96
hot wire anemometer, 96
heated thermocouple anemometer, 96
bubble flow meter, 97
Coriolis flow meter, 85–88
differential pressure flow meter, 60
sources of error, 67
Doppler flow meter, 82
dual-rotor turbine flow meter, 79
elbow flow meter, 70
electromagnetic flow meter, 83
electronic flow meters, 81
helical gear flow meter, 93
magnetic flow meter, 83, 84
mass flow meter, 85
thermal mass flow meter, 85
mechanical flow meters, 78
metering pumps, 90
nutating disc flow meter, 89
obstruction flow meters, 60
orifice meter, 60, 61
annular orifice, 64
concentric orifices, 64
conical edge plate, 65
eccentric orifices, 65
orifice coefficient, 66
orifice plate selection, 66
quadrant edge plate, 65

rangeability, 66
reliability/uncertainity/accuracy, 66
repeatability, 66
orifice taps, 62
corner taps, 62, 63
flange taps, 62, 63
pipe taps, 62, 63
radius taps, 62
vena contracta taps, 62
segmental orifice, 66
sharp edged orifice, 61
oscillating piston flow meter, 90, 91
oval gear lobe flow meter, 93
paddle wheel flow meter, 80
purge flow regulator, 75
positive displacement meters, 88
propeller flow meter, 80, 81
rotameter, 73, 74
different types of floats, 74
rotating vane positive displacement flow meter, 89, 90
shunt flow meter, 80
single piston flow meter, 91
target flow meters, 94
transit time flow meter, 81
turbine flow meters, 78
ultrasonic flow meters, 81
variable area flow meters, 73
V-cone flow meter, 71
V-element flow meter, 70
venturi meter, 66, 67
vena contracta, 61
venturi coefficient, 66
vortex shedding flow meters, 94, 95
Flow nozzle, 68
critical flow nozzle, 69
sonic nozzle, 69
Flow phenomena, 21
Flow shock, 31
Flow straighteners, 63
Fluid flow, 21
Bernouli's equation, 26
friction factors, 35
Fluidization, 200
aggregative fluidization, 203
boiling/bubbling fluidization, 202
centrifugal fluidized bed, 203
fluidization segregation, 191
incipient fluidization, 202
minimum fluidization velocity, 201
onset of fluidization, 201
particulate fluidization, 201, 203
recirculation fluidized bed, 203, 204
seal leg, 204, 206
slugging fluidization, 203
spouted bed, 202
tapered bed, 203
vibro-fluidization, 206
Fluidized bed catalytic cracking, 205, 206
Fluidized coal combustion, 205
Fluidized leaching, 205
Fluid–solid systems, 199
Fluid statics, 10
Foam flotation, 713
Foam fractionation/separation, 712
Foaming in a cooling tower, 642
Foams, 469, 470, 471
Forced convection, 245, 250
Form friction, 26
Fouling, 296
biofouling, 298
chemical reaction fouling, 298
combination mechanisms, 298
corrosion fouling, 298
idealized fouling curve, 300
particulate fouling, 297
precipitation or crystallization fouling, 297
sedimentation fouling, 298
solidification fouling, 298
Fouling factors, 261, 306, 307
apparent fouling factor, 334
Fractional freezing, 711
Free convection, 360
Free moisture, 645
Freeze concentration, 628
Friction, 35
Friction head, 106
Friction factor, 35, 247
Darcy–Fanning, 35
Friction losses, 38
Friction velocity, 23
Froth, 470
Froth flotation, 712–714
Fuel savings, 433
Fugacity coefficients, 546
Fuller's equation, 459
Furnace tubes, 442

Gas film controlled, 461, 462
Gaskets, 35, 43, 405
Gas permeation, 684
Gas–solids transport, 208
Gas/vapor–liquid flow, 195
Geometric view factor, 426
Geopolymers, 243
Gilliland correlation, 561, 562
GI pipe, 39
GPSA, 124, 127, 155, 160, 161, 279, 526
Granular active carbons, 673
Grenville equation, 167
Gray body, 427
Grooved tube holes, 280

Hagen–Poiseuille equation, 37
Hausen correlation, 249
Hayden–Grayson, 546
Hayduk correlation, 460
Heap leaching, 607
Heat capacity rate ration, 316

Heat conduction, 232, 233, 235
Heat conduction shape factors, 235, 236
Heat exchangers, 271
  cross flow heat exchanger, 271,
  double pipe heat exchanger, 272, 273
  finned tube/extended surface heat exchangers, 271
  floating head heat exchanger, 275, 276
  in parallel, 275
  in series, 275, 277
  selection guidelines, 416
  types 1–1, 1–2, 2–4, 273
  U-bundle heat exchanger, 272, 276, 278
Heat exchanger design margin, 321
Heat exchanger effectiveness, 323
Heater, 271
Heat Exchange Institute Standard Equation, 147
Heat flux, 234
Heat pipes, 416, 417
  capillary pumped loop type, 418, 419
  cryogenic, 417
  flat plate, 417, 418
  leading edge, 417
  microheat pipes, 417, 418
  rotating and revolving, 417
  thermosiphon, 417
  variable conductance heat pipe, 417, 418
  wicks, 419
    composite, 419, 420
    homogeneous, 419, 420
    sintered powder metal wicks, 419
Heat pumps, 374
  closed cycle absorption heat pump, 375
  closed cycle mechanical heat pump, 375
  open cycle mechanical vapor compression (MVC), 375
  open cycle thermocompression, 375
Heat tracing, 376, 377
  electrical tracing, 377
    constant wattage type, 378
    resistance type, 378
    self-limiting, 377
    thermocouple-type line heaters, 378
  fluid tracing, 377
  hot oil heat tracing systems, 378
  steam tracing, 378
Heat transfer coefficients, 246, 259
  ranges of overall heat transfer coefficients, 261
Heat transfer enhancement techniques, 395, 396
  active techniques, 395, 396
    electrostatic fields, 396
    fluid vibration, 396
    injection, 396
    jet impingement, 396
    mechanical aids, 396
    suction, 396
    surface vibration, 396
  passive techniques, 395, 396
    additives for liquids/gases, 395, 396
    coiled tubes, 395, 396
    displacement enhancement devices, 395, 396
    extended surfaces, 395, 396
    rough surfaces, 396
    surface tension devices, 396
    swirl flow devices, 395, 396
    treated surfaces, 395, 396
Heat transfer fluids., 262
  Dowcal, 264
  Dowfrost, 264
  Dowtherm, 264, 320
  flammability, 266
  fluorocarbons, 264
  fused salts, 264
  glycol-water mixtures, 264
  hydrocarbon oils, 263
  liquid metals, 263, 264
  refrigerants, 267
  silicones, 263
  steam, 265
  Syltherm, 264
  toxicity, 266
  trichloroethylene, 263
Heat transfer shape factors, 236
Heisler charts, 236
Helical coils, 379
Helixchanger, 293
Hemodialysis, 710
Hengstebeck method, 559, 560
Henry's law, 462, 463, 534, 547
Henry's law constant, 535
HETP, 540, 541, 605, 606
Higee (High gravity rotating contactor), 578
High pressure columns, 571
High temperature creep, 440
Hirschfelder, 459
H_{OG} and N_{OG}, 540
Hottel, 435
Hot spots, 439
Humidification, 455, 472, 630, 636
  spray humidification, 636
Humidifier, 636
  packed bed humidifier, 636
  tubular humidifier, 636
Humidity, 633
  absolute humidity, 630
  humid heat, 630
  humid volume, 630
  molar humidity, 630
  percent humidity, 630
  percent relative humidity, 630
  relative humidity, 634
Humidity chart, 633
Hydraulic diameter, 494
Hydraulic Institute, 102, 103
Hydraulic power, 110
Hydraulic shock, 30
Hydrogen bonding, 545
Hydrophilic adsorbents, 673
Hydrophobic adsorbents, 673

Hydrophilic, 673
Hydrophobic, 63
Hygrometers, 634
  Assmann psychrometer, 635
  capacitive hygrometer, 634
  hair hygrometer, 634
  laminate hygrometer, 634
  mechanical hygrometers, 634
  piezo-electric/sorption hygrometer, 634
  resistive hygrometer, 634
  sling psychrometer, 635
Hygroscopic material, 645, 646
Hysteresis, 646

Icing problems, 640
Immediately dangerous to life and health (IDLH), 542
Immersion heaters, 393
  circulating type, 303
  flange type, 393
  screw plug type, 393
  pipe insert type, 393
Impellers, 104, 172
  axial flow, 104, 110, 173
  closed, 104, 105
  double suction, 104, 105
  flat plate impeller, 175
  fluid foil impeller, 176
  gate impeller, 176
  hydrofoil impeller, 166
  leaf impeller, 176
  marine propeller, 174
  mixed flow, 104, 110
  open, 104, 105
  perforated propeller, 174
  pitched blade turbine, 175
  radial flow, 104, 109, 110, 173
  saw-toothed propeller, 174
  semiopen, 104, 105
  single suction, 104
  turbine impeller, 175
Incineration, 450
Incinerators, 450
Inertial forces, 3
Initial breeding, 616
In-line and staggered, 250
In situ leaching, 608
Instrument Society of America (ISA), 75
Insulation, 236–243
Intercoolers, 536
Interfacial area, 461, 475
Interfacial resistance, 467
Interfacial tension, 594
Interfacjal turbulence, 467
Internal design pressure (IDP), 39
International temperature scale, 226
Interstitial velocity, 200
Inverse/inverted solubility, 297
Ion exchange, 455, 682–684
  ion exchange materials, 683
  ion exchange processes, 683, 684
  ion exchange resins, 683
Ion flotation, 713
Ionic functional groups, 684
Ionic liquids, 594
Ionic solvents, 594
Ion selective membrane, 687
Ionization gauge, 15

Jacketed vessels, 272, 379
  baffled conventional jacket, 380
  constant flux heat transfer jacket, 384
  dimple jacket, 380, 382
  half-pipe coil jacket, 380
  jacket with agitation nozzles, 380
  panel or plate coil jacket, 381
  simple conventional jacket, 379
  spirally baffled jacket, 380
Janecke diagrams, 589, 590

Kelvin equation, 614, 616
Key components, 559
Kinematic viscosity, 5
Kinetic energy, 101
Kirchoff's law, 426
Kirkbride equation, 561, 562
Knock-out drums, 519
Koch–Glitch LP, 504
Koch Knight LLC, 497, 500
Kolmogorov length scale, 164
Kozney–Karman equation, 200
Kremser–Brown–Souders equation, 537
K-values, 545

Lagging fires, 238
Lambert surface, 427
Laminar flow, 22, 35
Leaking paths, 291, 292
Leidenfrost phenomena/point, 256, 258
Lenoir, 546
Lewis–Matheson method, 559
Lifters/flights, 657
Limitations of heat pipe, 418
  boiling limit, 418
  capillary limit, 418
  entrainment limit, 418
  sonic limit, 418
  viscous limit, 418
Limiting operating line, 536
Linde boiling surface, 254
Linear variable differential transformers (LVDT), 634
Liquid film controlled, 461, 462
Liquid holdup, 475, 509
Liquid inclusion, 618
Liquid level, 15
  capacitance, 17, 19
  continuous level detection, 15
  level sensors, 15

Liquid level (*Continued*)
liquid level measurement, 16
bubbler sensors, 18
differential pressure, 16
displacement, 16
electrical methods, 16
electromechanical devices, 16, 19
radar devices, 17, 19
FMCW radar, 17
guided wave radar, 18
pulse radar,
thermal methods, 17
ultrasonic methods, 17, 19
float, 15, 16
point level detection, 15
Liquid–liquid equilibria, 587
Liquid–liquid extraction, 455, 543, 584–586
Liquid membrane processes, 684
Liquid phase transfer unit, 539
Liquid throw, 481
Lithium bromide absorption refrigeration, 372, 373
Lobo and Evans, 436, 437
Loeb–Sourirajan asymmetric membrane, 686
Log mean temperature difference (LMTD), 311
LMTD correction factor, 315–317
Loss coefficients, 38
Loss of coolant accident (LOCA), 256
Louvers, 638
Lower heating value (LHV), 433
Low NO_x burners, 448

Macropores, 673
Magma, 618
Magneto-hydrodynamics, 27
Manometer, 12
Marangoni effect, 467
Marangoni convection, 467
Mass flux, 460
convective mass flux, 460
Mass transfer, 232, 455, 456,
Mass transfer coefficients, 460
overall mass transfer coefficients, 460
volumetric mass transfer coefficients, 461
Mass transfer equipment, 475
Mass transfer with chemical reaction, 468
Mass transfer zone (MTZ), 682
Mass velocity, 23
Material safety data sheets (MSDS), 665
McCabe–Thiele method, 537, 554, 555, 557, 559, 587, 592
McLeod gauge, 14
Mean beam length, 433, 435, 436
Mean temperature difference, 314
Mechanical heat pump, 375
Mechanical seals, 35, 43
Mechanical vapor compression (MVC), 375
Mechanisms in moisture transport, 648
capillary diffusion, 648
effect of shrinkage, 648
gravity flow, 649
hydrostatic pressure differences, 648
Knudsen diffusion, 648
liquid diffusion, 648
pressure gradients, 648
surface diffusion, 648
vapor diffusion, 648
Membrane compaction, 699
Membrane fouling, 699
Membrane modules, 689
bore side feed modules, 690
flat sheet tangential flow membranes, 689
hollow fiber module, 690, 691
plate-and-frame modules, 691, 692
spiral-wound modules, 689, 690
stirred cell modules, 689
tubular module, 690
Membranes, 685
anionic exchange membrane, 688, 701
asymmetric membranes, 686
asymmetric composite membrane, 686, 688
bipolar membrane, 689
cation exchange membrane, 688, 701
ceramic membranes, 688
classification, 687
composite membrane, 688
dynamic membrane, 688
inorganic membranes, 689
liquid membranes, 686
metal membranes, 694
microporous membranes, 685
multichannel membranes, 691
nonporous membranes, 685
dense nonporous membrane, 686
polymeric membranes, 689, 694
symmetrical membranes, 686
Membrane separation processes, 455, 564, 684, 693
Membrane spacers, 695
Metastable, 615
Metastable zone, 615
Microbiologically induced corrosion (MIC), 644
Microchannels, 422
Microfiltration, 684, 695, 703
Microflotation, 713
Micro pores, 673
Minimum acceptable diameter, 491
Minimum flow bypass, 124
Mists, 469, 470
Mist eliminators, 486, 521, 522
Misty flow, 197, 198, 341
Mixers, 598
anchor mixer, 181
baffled draft tube mixer, 170
baffled vessels, 169
draft tubes, 177
Banbury mixer, 181
change-can mixers, 181
convection mixers, 189
double arm kneading mixers, 181, 182
double cone mixer, 187

extruders, 181
fluidized bed mixers, 189
helical cone and anchor mixers, 183
helical ribbon mixers, 182
high shear mixers, 178
hopper mixers, 189
horizontal trough mixers, 190
in-line static mixers, 184
jet mixers, 172
kneaders, 181, 182
mixing rolls, 182
Muller mixer, 181
Nauta type mixer, 189
orifice plate mixers, 180
paddle mixers, 176
planetary mixers, 181
propeller mixers, 173, 174
ribbon blender, 187
rotor–stator mixer, 177
screw mixers, 181, 190
sigma blade mixers, 182
solids mixer selection, 190
solids static mixer,
static mixers, 183, 192
tee and injection mixers, 184
tumbler mixers, 173, 188
turbine mixers, 173
vortex mixer, 177
Z-blade mixers, 193
Mixer seals, 194
Mixing, 163, 598
acoustic/sonic mixing, 192
continuous mixing, 186
equipment, 169
flow-sensitive mixing, 167
macromixing, 164, 617
mesomixing, 164, 617
micromixing, 617
radial mixing, 617
solids mixing, 186, 191
Mixing intensity, 167
Mixing rolls, 181
Molecular interactions, 544, 545
Molecular sieves, 673, 674
Momentum transfer, 233
Moody chart/diagram, 35, 36
Mother liquor, 618, 619
Multicomponent distillation, 559
Multilayer flat slab, 235
Multilayer hollow cylinder, 235, 236
Multilayer hollow sphere, 236
Multistage compressors, 155, 156
Multistage extraction, 592
Mutual solubility, 593

Nanofiltration (NF), 705
Nanotechnology, 619
Natural convection, 245, 247
Needles, 618
Net flow, 556
Newtonian, 5
Newton–Raphson method, 548
Newton's equation, 4
Newton's law of cooling, 246
Nguyen equation, 538
Noise in burners, 446
Nonideal gas, 5
Non-Newtonian, 5, 165
Norton, 508
NO_x control, 446, 447
low NO_x burners, 448
Nozzles, 653, 664
NPIP, 107
$NPIP_A$, 108
$NPIP_R$, 108
NPSH, 107
$NPSH_A$, 107, 108
$NPSH_R$, 107, 108
$NPSH_{R_3}$, 108
NRTL, 544, 587
Nucleate boiling, 254
Nucleation, 614, 615, 617
homogeneous nucleation, 616
primary nucleation, 614–616
secondary nucleation, 614–617
collision breeding, 616
fluid shear, 616
initial breeding, 616
Number of theoretical trays, 555
Nusselt equation, 252, 253
NTU, 323, 408

Odor control, 531
Ohms law, 233
Oil damper, 31
Oldershaw column, 549
Open cycle, 375
Open steam, 552, 553
Operating holdup, 496
Operating line, 535
Optimum column pressure, 553
Optimum insulation thickness, 241
Optimum number of trays, 562
Optimum reflux ratio, 553, 554
O-ring, 43
Oslo, 626
Osmosis, 695
Osmotic pressure, 695
Overall heat transfer coefficients, 251, 261, 262
Overall mass transfer coefficients, 463

Pace Engineering, 519, 520
Packed absorbers, 541
Packed bed humidifier, 636
Packed beds, 199
wall effect, 494
wetting rate, 494
Packed columns, 493, 528

diameter, 511, 512
operation and installation issues, 513
packed column internals, 496
bed limiters, 502, 507
feed and reflux distributors, 502
hold-down grid, 507
liquid collectors, 502
liquid distributor, 503, 504
liquid redistributors, 502, 503, 506, 508
packing support plates, 502, 503
vapor distributors, 502, 503
wall wipers, 502, 507, 508
Packed vs. tray columns, 509
Packing density, 691, 693
Packing factor, 494
Packings 496
Berl saddles, 496–498
cascade mini rings, 496
Fleximax, 496
Flexipak, 496
Flexeramic, 499
honeycomb packings, 499, 500
Intalox saddles, 496, 498
Mellapak, 499
Montz, 496
Nutter rings, 496
Pal rings, 496
random packing, 498
Raschig rings, 496, 498
structured packings, 498
Parametric pumping, 455, 714
Particle attrition, 208
Pass partitions, 282
Peltier effect, 226
Percolation, 608
Permeability, 457, 458
Permeate, 685
Personal exposure, 438
Pervaporation, 684, 693, 699–701
Petluk distillation columns, 574, 576, 577
Petukhov–Kirillov equation, 249
Phase diagram, 667
Phase equilibria, 544
Pinch point, 535, 555
Pipe fittings, 38
Piping, 35
Pirani gauge, 14
Pitot tube, 72
annubar, 73
Pitot venturi flow element, 73
pitometer, 73
Plait point, 588, 594
Plastic piping, 41
Plug flow, 197
Pneumatic conveying, 208, 209
dense phase, 210
dilute phase, 210
Pneumatic conveyor, 213
Buhler Fluidstat system, 211
closed loop system, 212
Gatty system, 211
Molerus–Siebenhaar system, 211
negative (vacuum) system, 212, 213
positive pressure system, 212, 213
pressure negative (push-pull) combination system, 212
pulsed conveying system, 211
Takt–Schub system, 211
tracer air system, 211
Polymer films and melts, 457
Polymeric heat exchangers, 273
Polymorphic form, 614
Polymorphic system, 614
Polymorphism, 614
Polytropic efficiency, 158
Polytropic power, 158
Ponchon–Savarit enthalpy-concentration method, 556, 557, 592
Pores, 456
Porosity, 201
Potential flow, 23
Power consumption in mixing, 165
Power law fluids, 6, 165
Prandtl mixing length, 22
Precipitation, 297
Pressure drop, 308
Pressure gauge, 13
Pressure head, 106
Pressure–temperature phase diagram, 595
Primary air, 432
Priming, 482
Printed circuit heat exchangers, 420, 421
Process intensification, 578, 579
Progressive freezing, 711
Pseudo-plastic, 6, 165
Psychometric ratio, 632
Pulsations in flow, 129
Pumps, 101
acid egg, 140
air lift pump, 139
air operated double diaphragm pump, 136
axial flow, 130
bi-directional, 132
canned motor pump, 137, 138
centrifugal pumps, 101
affinity laws, 110
amperage, 122
diffuser casing, 106
double suction, 105
efficiency, 118
high flows, 123
impeller size, 122
inducers, 125
low flows, 123
minimum flow bypass, 124
parallel operation, 122–124
performance curve, 120, 121
recirculation, 120

series operation, 122–124
specific speed, 109, 110
volute casing, 106
system curve, 120, 21
dead head speed, 131
diaphragm pumps, 136
double-acting pumps, 128
duplex, 127, 128
dynamic head, 106
electromagnetic pump, 137
flow pulsations, 129
gear pumps, 131
external gear, 131, 132
internal gear, 131
rotary gear pumps, 131
jet pump, 146
kinetic pumps, 102
liquid ring pump, 131
lobe pumps, 135
magnetic drive pump, 137
mechanical efficiency, 129
metering/proportionating pumps, 134
multiplex, 127
multistage, 102, 103
performance curves, 120, 121, 124
peristaltic pumps, 140
piston pumps, 102, 103, 128
plunger pump, 128, 130
positive displacement pumps, 127
reciprocating pumps, 102, 103, 127, 128
regenerative pump, 140
rotary, 130
screw pump, 133, 144
scroll pump, 145
seal-less pumps, 137
self priming, 132
simplex, 127, 128
single-acting pumps, 128
single stage, 102, 103
sliding vane, 130
slip, 129
steam driven pumps, 140, 141
system curve, 120
triplex, 128
turbine pump, 138
vane pumps, 130
vertical pump, 138
volumetric efficiency, 111, 129

q, thermal condition of feed, 554
q-line/feed line, 554, 555
Quenching, 427

Radial flow, 104, 155
Radiant efficiency, 437
Radiant energy, 425
absorption, 425
reflection, 425
transmission, 425
Radiant heat flux, 426
Radiant heat transfer, 425
Radiation, 425
thermal radiation, 425
Rafinate, 591
Raining, 482
Rangeability, 74
Range for a cooling tower, 637
Rankine cycle, 375
Raoult's law, 544
Rayleigh convection, 467
Rayleigh's equation, 549
Reboilers, 336, 583
column internal reboiler, 340
configuration selection, 337, 338
flooded bundle, 336
forced circulation, 336
horizontal thermosiphon, 341
internal reboiler, 340, 345
kettle reboiler, 272, 339–341
natural circulation, 336
once through reboiler, 336
recirculating, 336
thermosiphon reboiler, 272, 341–343
horizontal thermosiphon, 336
vertical thermosiphon, 336, 341
Reboiler duty, 583
Reboiler selection guide, 347
Recirculation fluidized bed, 203
Recoverability, 593
Recuperators, 272, 423
Rectifying/enrichment section, 552
Reference temperatures, 226
compensated reference temperature systems, 228
electronically controlled references, 228
ice baths, 228
zone boxes, 228
Reflectance, 425
Reflux drums, 515
Reflux ratio, 485, 552
minimum reflux ratio, 552
optimum reflux ratio, 553, 554
Refractoriness, 242
Refractories, 242
Refractory walls, 433
Refrigerants, 371
Refrigeration, 371
absorption, 372, 373
cascade, 372, 373
mechanical compression, 372
steam jet, 372, 373
vapor compression, 371
Regenerative thermal oxidizer, 451
Regenerators, 272, 397
rotating type regenerators, 422
Reid vapor pressure, 107
Relative humidity sensitivity, 635

Relative volatility, 546
Removable tube bundle, 278
Reradiation, 433
Residence time, 526
Retentate, 685
Retrograde condensation, 573
Retrograde phenomena, 572
Retrograde vaporization, 573
Reverse Carnot cycle, 371
Reverse osmosis (RO), 455, 684, 695
Rheology, 5
Rheomalectiac fluids, 7
Rheopectic fluids, 7
Right angular triangular plot, 590
Riser, 477
Rodding, 304
Rotary kilns, 451
Rotated triangular pitch, 285
Rotating helical coil tube insert, 295
Rotating type regenerator, 422
Roughness factor, 35
Rubber ball cleaning, 305

SAE classification, 5, 40
Sager cones, 242, 452
Saint Gobain Norpro, 503
Saltation velocity, 210
Salt effect in vapor-liquid equilibria, 569
Salt rejection, 697
Sankey diagram, 452
Saturated boiling, 257
Saturated volume, 630
Scale-up, 169
Scale formation, 298
  conversion products, 299
  corrosion, 298
  electrochemical action, 299
  fluid oxidation, 298
  oxidation products, 298
  process deposits, 299
  pyrophoric iron sulfide scales, 304
  silicate scales, 299
Schedule number, 38
Screw joints, 40
Scrubbers, 527
  centrifugal scrubbers, 528
  venturi scrubbers, 476, 528, 529
  wet scrubbers, 528, 531
Scrubbing, 527
Seal leg, 206
Seal pans, 481
Seal strips, 292
Secondary air, 432
Sedimentation, 486
Seebeck effect, 226
Seeding, 616
Segregation, 190, 191
  dusting segregation, 191
  fluidization segregation, 191
  sifting segregation, 190
Seider–Tate, 35, 249
Selection of dryers, 652, 653
Semiconductor temperature sensors, 231
Selection of steam traps, 390–392
Selective catalytic reduction (SCR), 441
Selectivity, 567, 587, 593
Selectivity of a membrane, 685
Self-cleaning heat exchanger, 308
Self-priming pump, 109
Separation factor, 567, 593
Separation of azeotropes, 563
Sequencing columns, 562
Series flow arrangements, 407, 408
Sessile count monitoring, 644
Settling chambers, 486
Sewage sludges, 220
Shape factor, 434
Shear rate, 4
Shear stress, 4
Shear-thickening, 6
Shear-thinning, 6
Shell and tube heat exchangers, 271
  thermal design, 321
    Bell–Delaware method, 324
    ε-NTU method, 322
    Kern method, 321
Shell side vs. tube side, 309
Shell side feed modules, 690
Sherwood correlation, 511
Shield section, 430
Shield tubes, 430
Short term exposure limit (STEL), 542
Shrinkage, 648, 652
Shut-in pressure, 120
Shut-off head, 120
Sifting segregation, 189
Silica gel, 673
Simplex, 127, 128
Single mechanical seal, 194
Single stage extraction, 591
Skin friction, 26
Sliding bed, 219
Slip velocity, 198
Slope of equilibrium line, 539
Slug flow, 195, 196
Slugging, 203
Slurries, 217–219, 221
Slurry flow, 218–220
Smith and Brinkley method, 560
Smoker equation, 555
SO_3 absorption, 532
Solar collector, 427
Solar constant, 425
Solid–liquid extraction/leaching, 455, 606
  agitation, 608
  heap leaching. 607
  in situ leaching, 608
  percolation, 608

Solid-liquid (slurry) flow, 217
homogeneous flow, 218, 219
heterogeneous flow, 218, 219
intermediate regime, 218, 219
saltation regime, 218, 219
slurry pipelines, 220
brute force, 220
conventional, 220
Solubility product, 532
Solutal Marangoni convection, 467
Solvent-free coordinates, 592
Solvent selection, 542
Solvent sublation, 713
Sonic velocity, 32
Soot/ash particles, 319, 433
Spinning band column, 549, 550
Soret effect, 456
Souders–Brown equation, 492, 520
Sparger, 180
Specifications for tube and pipe, 38
Specification sheet, 311
Specific speed of pumps, 109
suction specific speed, 109
Specific supersaturation ratio
Sphering, 199
Spinodal curve, 588
Spirax Sarco, 386–390
Splash type fill, 639, 641
Splash grids, 641
Spouted bed, 202
Spray chambers, 528
Spray columns, 475, 476
Spray humidification, 636
Spray nozzles, 477
Spray ponds, 637
Sprays, 469, 470
Square pitch, 285, 286
Stack damper, 439
Stacked packing, 503
Stagnation point, 27
Stagnation pressure, 27
Static head, 106
Static holdup, 496
Static pressure, 10
Steam jet refrigeration, 373
Steam stripping, 531
Steam trap problems, 393
air binding, 393
dirt, 393
freezing, 393
improper sizing, 393
loss of prime, 393
noise, 393
steam leakage, 393
steam locking, 393
water hammer, 393
Steam traps, 384
balanced pressure trap, 387, 388
ball float trap, 386, 387, 392
bimetallic trap, 385, 388, 389
common problems, 393
drip traps, 365
inverted bucket trap, 385, 386, 392
liquid expansion trap, 387
mechanical traps, 385
process traps, 385
selection, 390, 391
thermodynamic traps, 385, 389, 390, 392
thermostatic traps, 385–387, 392
tracer traps, 385
Steam turbine, 375
Stefan–Boltzmann constant, 426
Stefan–Boltzmann law, 230, 434
STEL, 542
Stickiness, 652
Stoichiometric combustion air, 432
Storage/regenerative type heat exchanger, 422
Stokes flow, 24, 198
Strainer, 38, 39
Stratified flow, 197
Strippers, 537
Stripping, 527
Stripping section, 554
Sublimation, 711
quasi-sublimation, 711
sublimation point, 711
Suction energy, 119
Suction head, 106
Suction lift, 106
Suction specific speed, 109
Sulzer Chemtech, 499, 500, 506, 522
Supercritical distillation, 572
Supercritical extraction, 597, 598
Supercritical solvents, 595, 597
Superficial velocity, 23
Superheated-solvent drying, 668
Supersaturation, 614, 616, 619, 622, 626
Supersaturation ratio, 614
Supertherm, 237
Support plate, 503
Surface renewal factor, 466
Surge in compressors, 159
surge limit line, 159
Swenson Technology, Inc, 353, 354, 621, 623
Swirl tube deck, 522
SWG, 38

Taborak equation, 326, 327
Tapered bed, 203
TEMA, 273
TEMA designations, 274
TEMA fouling factors, 328
TEMA shell and tube heat exchangers, 274
Temperature approach, 637
water temperature approach, 637
Temperature correction factors, 315–317
Temperature gradient, 261

Temperature measurement, 226
  bimetallic thermometers, 230
    bimetallic strip thermometers, 230
  change-of-state temperature sensors, 230
  fluid expansion thermometers, 230
  infrared temperature sensor, 226
  pyrometers, 230, 231
  resistance temperature detectors (RTDs), 226
  semiconductor devices, 232
  thermistor, 226
  thermocouples, 226
  thermopile, 228
  vapor pressure thermometers, 231
Ternary equilibria, 588, 589, 590
Theories of mass transfer, 465
  Denckwerts surface renewal theory, 466
  film theory, 465
  film-penetration theory, 466
  Higbie's penetration theory, 466
Thermal conductivity, 232, 234
Thermal design of shell and tube heat exchangers, 321
  Bell–Delaware method, 324
  Kern method, 321
  ε-NTU method, 322
Thermal diffusivity, 234
Thermal insulation, 237
  ablative materials, 242
  cellular glass, 238
  ceramic fiber insulation panels, 238
  polyurethane foams, 238
  spray-on foam insulations, 241
Thermal length, 405
Thermal shock, 30
Thermo-compression heat pumps, 375
Thermocouple sheaths, 228
Thermoelectric heat exchanger, 422
Thermopile, 422
Thiele–Geddes method, 559
Thin film composite asymmetric membrane, 686
Thixotropic, 165
Thompson effect, 226
Tie-lines, 556, 557, 588–590
Time of passage, 657
Ton of refrigeration, 371
Total jump frequency, 459
Total dissolved solids (TDS), 268, 269
Total dynamic head (TDH), 106
Toxicity, 594
Transducer, 634
Transfer unit concepts, 539
Transition length, 24
Transmission, 425
Tray columns, 477
  diameter, 491
  entrainment, 485
  flooding, 488
  flow arrangements, 479
  net area, 478
  spacing, 479
Tray efficiency, 483, 562
  Colburn equation, 484
  Drickamer–Bradford correlation, 483
  Murphree tray efficiency, 483, 484, 538
  O'Connell correlation, 483
  overall efficiency, 484, 538
  Van Winkle correlation, 484
Tray passes, 480
Trays, 477
  bubble cap trays, 477
  cross flow trays, 479
  dual flow trays, 479
  jet trays, 477
  multipass trays, 479
  parallel flow trays, 479
  performance, 481, 482
  ripple trays, 477
  sieve trays, 477
  spray regime, 482
  turbo-grid trays
  valve trays, 477
  V-grid trays, 477
Tray spacing, 479
Transmittance, 425
Triangular diagrams, 587–589
Triangular notch, 99
Triangular pitch, 285, 290
Tri-atomic gases, 433
Triple point, 226
Tube arrangements, 428
Tube erosion, 282
Tube inserts, 294, 295
Tube pitch, 285–287
Tube side, 309, 310
Tube sizes, 284, 285
Tube supports, 282, 287
Tube-to-tube joints, 281
Tubular humidifier, 637
Turbulence, 21
Turbulent flow, 21, 23
Turn down ratio, 59, 60, 477
Twisted tapes, 294
Twisted tubes, 289
Two-film theory, 465
Two fluid nozzle, 662
Two phase flow, 195
T–x–y diagram, 548
Type I and Type II liquid–liquid equilibria, 589, 590
Types of centrifugal pump casings, 106
  circular casing, 106
  diffuser casing, 106
  volute casing, 106
Types of flow patterns, 195
  annular droplet flow, 196
  annular flow, 195, 196, 198, 199
  bubble flow, 195, 196, 199
  churn flow, 195, 196
  dispersed flow, 196, 199
  droplet flow, 196

flashing flow, 197
mist flow, 195, 196, 198
plug flow, 199
slug flow, 195, 197, 199
Stokes flow, 198
stratified flow, 197–199
wavy flow, 197–199
Types of heat pipes, 417
cryogenic, 417
flat plate, 417
leading edge, 417
rotating and revolving, 417
thermosiphon, 417
Types of pores on adsorbents, 673
macropore, 673
mesopore, 673
micropore, 673
transport pore, 673

U-bundle heat exchanger, 272, 276
Ultrafiltration, 684, 695, 701, 702
polymer-enhanced ultrafiltration, 703
Ultrasonic nozzles, 663
Underwood equations, 561
UNIFAC model, 544
UNIQUAC, 544
Unsaturation zone, 615
Unsteady state conduction, 233
Upper flammability limit (UFL), 432
Utilities, 370

Vacancy mechanism, 457
Vacuum pumps, 141
dry vacuum pump, 143, 144
liquid ring vacuum pump, 142
rotary claw vacuum pump, 143, 144
Vacuum producing equipment, 141
Valves, 35, 46, 47
ball valve, 48
block valve, 52
bleed valve, 52
butterfly valve, 50
check valves, 50
ball check valve, 51, 52
lift check valve, 51, 52
swing check valve, 51
diaphragm valve, 49
flexible valve, 53
flush bottom valve, 53
foot valve, 53
gate valve, 46
globe valve, 47, 48
needle valve, 48
pinch valve, 53
plug cock, 47
relief valve, 54
safety relief valve, 54
safety valve, 54
Vapor compression cooling, 635
Vapor compression refrigeration, 635
Vaporization, 257
Vapor-liquid exchange, 257
V-notch, 99
Vaporization, 254
Vapor/gas–liquid separators, 515
Vapor lock, 115, 116
Vapor permeation, 700
Vapor pressure thermometer, 231
Velocity gradient, 4
Velocity head, 106
Velocity profile, 23
Venturi effect, 26
Venturi scrubbers, 475, 476, 528
Vertical radiant tubes, 429
Vibrofluidization, 206
View factors for radiation, 427
Viscoelastic fluids, 7
Viscometers, 8
Brookfield viscometer, 8
Cambridge moving piston viscometer, 9
capillary viscometers, 10
cone and plate viscometers, 9
cup and bob viscometers, 9
Engler viscometer, 10
falling ball viscometers, 8
flow viscometers, 8
Oswald viscometer, 8
parallel plate viscometer, 8
Redwood viscometers, 8, 10
rheo-viscometers, 10
rising bubble viscometers, 8
rotational viscometers, 8
rotating disc viscometer, 8
Saybolt viscometers, 8, 10
Stormer viscmeter, 9
tube viscometers, 9
Viscosity, 4
Viscosity measurement, 8
Visible plumes, 640
VLE data, 548
VOC control, 531
Volumetric efficiency, 111
Von Kármán vortex street, 28
Vortex breaker, 522
Vortex shedding, 28

Wagle, 547
Wall drag, 27
Wall effect, 200
Wash column, 627
Waste water stripping, 531
Water activity, 646, 647
Water cooling, 630
Water hammer, 21, 30, 393
differential shock, 30
flow shock, 31
hydraulic shock, 30
thermal shock, 30

Water/steam injection, 449, 450
Water temperature approach, 637
Weeping, 481, 482
Weirs, 98, 99, 481
  picket weir, 481
Welded joints, 40
Weissenberg effect, 7
Wet bulb depression, 632
Wet bulb temperature, 630, 632, 633
Wet bulb temperature, 630, 633
Wet scrubbers, 528
Wettability, 494, 495
Wetted area, 494
Wetted wall column, 463
Wetting rate, 494
Wilke and Chang equation, 460
Wire mesh burners, 445
Wire mesh tube insert, 295

13X type molecular sieves, 674

Yield stress, 7

Zeolites, 674
  A and X type zeolites, 674
Zeotropic, 563
Zone boxes, 228
Zone melting, 711
Zwietering correlation, 193